sixth edition

Physical Science

Principles and Applications

sixth edition

Physical Science

Principles and Applications

Charles A. Payne ▪ William R. Falls ▪ Charles J. Whidden

Morehead State University · Morehead State University · Morehead State University

Wm. C. Brown Publishers

Book Team

Editor *Jeffrey L. Hahn*
Developmental Editor *Robert Fenchel*
Production Editor *Scott Sullivan/Dan Rapp*
Designer *Christopher E. Reese*
Art Editor *Margaret Rose Buhr*
Permissions Editor *Vicki Krug*
Visuals Processor *Jodi Wagner*

WCB **Wm. C. Brown Publishers**

President *G. Franklin Lewis*
Vice President, Publisher *George Wm. Bergquist*
Vice President, Operations and Production *Beverly Kolz*
National Sales Manager *Virginia S. Moffat*
Group Sales Manager *Vincent R. Di Blasi*
Vice President, Editor in Chief *Edward G. Jaffe*
Marketing Manager *John W. Calhoun*
Advertising Manager *Amy Schmitz*
Managing Editor, Production *Colleen A. Yonda*
Manager of Visuals and Design *Faye M. Schilling*
Production Editorial Manager *Julie A. Kennedy*
Production Editorial Manager *Ann Fuerste*
Publishing Services Manager *Karen J. Slaght*

WCB Group

President and Chief Executive Officer *Mark C. Falb*
Chairman of the Board *Wm. C. Brown*

Cover photo © Tracy Borland/FPG International

Copyeditor *Ann Mirels*

The credits section for this book begins on page 648, and is considered an extension of the copyright page.

Library of Congress Catalog Card Number: 91–73881

ISBN 0–697–09818–4 (paper)
0–697–13929–8 (cloth)

Printed in the United States of America by Wm. C. Brown Publishers, 2460 Kerper Boulevard, Dubuque, IA 52001

10 9 8 7 6 5 4 3 2

To Our Wives: *Essie C. Payne,*

Beatrice B. Falls,

Scharline Whidden

Contents

Preface

Science has deeply permeated our lives. We need only to consider our immediate surroundings to recognize the impact that current knowledge about the natural world has on the way we live—an impact that continues to become increasingly pervasive. Virtually each day there is a significant scientific breakthrough that will have far-reaching effects on humankind. The foundation of our society is built around the continuation of scientific investigation, the ongoing development of new scientific theories, and the constant application of scientific principles. A considerable portion of news media reports reminds us daily that we must attempt to stay abreast of scientific endeavors to be conscientious members of society. Colleges have realized this need and have included courses in the natural sciences as part of the general education requirements for all degree candidates. However, a significant number of college educators feel that many students are permitted to fulfill their science requirements by enrolling in classes that are too narrow in scope to encompass most of the major scientific fields. This textbook attends to all the major areas of the physical sciences and thus provides the student an opportunity to learn about those principal topics that constitute the study of our physical world.

The accomplishments of physical scientists span the spectrum from the world of subatomic particles to the farthest reaches of the cosmos. Yet these individuals must never cease their search for knowledge not currently available if humankind is to continue to flourish. A major portion of their endeavors must center around ways to meet our future energy requirements without further complicating our existing environmental problems. Surely one of their most important objectives should be the development of ways to reduce the rate by which our remaining natural resources are being consumed.

In their search for scientific knowledge, many students express considerable interest in the humanistic aspect of securing scientific information. For example, most students enjoy reading about the personal lives of such dedicated investigators as the Curies, Einstein, Galileo, Newton, and other individuals of their caliber. Within each chapter of the textbook boxed inserts enhance the student's understanding of science and point out, in many instances, how scientists strive to reach their respective goals.

As authors of this textbook, one of our major goals is to help students obtain an understanding of the scientific enterprise beyond their present level of knowledge. The fields of astronomy, chemistry, earth sciences, and physics are presented in an interrelated approach. The sixth edition is separated into five units, each introduced by a concise historical summary that should prove both interesting and informative to the student.

The sixth edition contains the most up-to-date discussions about scientific accomplishments and issues that publishing deadlines permit. The chapters on astronomy, earth science, and energy have been carefully revised through constant review of the latest scientific literature. This edition contains numerous new black-and-white photographs and original illustrations. Many new color photographs were furnished by EPA, JPL, NASA, NOAA, NWS, USGS, manufacturers, national museums, observatories, research institutes, scientists, and universities. Other photographs were taken by Carol Ann and Harold Clendenin of Clendenin Studios. Also included are several original drawings by artist Susan H. Huelsing.

Appendix 1 presents the International and English systems of measurement and discusses how the basic units were established and the manner in which various derived units are defined. Appendix 2 focuses on the mathematical concepts used throughout the textbook and should be reviewed before problem assignments are undertaken. Various segments of proper data manipulation, including graph analysis, are also presented in this appendix. Appendix 3 is a ready source of accepted symbols and abbreviations used throughout the textbook. This appendix was developed only after numerous authoritative sources were consulted. Appendix 4 contains the answers to selected end-of-chapter numerical problems, furnished to provide students with an opportunity to check their grasp of the various mathematical relationships. Appendix 5 consists of an alphabetized list of the elements and their symbols, atomic numbers, and accepted atomic weights.

We have attempted to write in a simple and direct manner so that the text will be in the student's "comfort zone" in terms of reading level and comprehension. In addition, we have emphasized the concepts of science throughout the textbook, at the same time providing all the factual data needed to explain the topics being discussed. We have, above all, attempted to capture the interest of the nonscience major for whom this text is primarily intended. The number of worked examples has been considerably increased; students will find many of them particularly relevant to their real-world encounters. Each chapter discussion concludes with a review summary as well as an equation summary, where applicable. In addition to the other features already mentioned, the sixth edition contains a most extensive glossary compared to other physical science textbooks. Student concerns regarding vocabulary deficiencies should be minimized through use of this glossary. The lengthy index has been carefully prepared in an attempt to assist the student in locating practically any scientific concept included in the textbook.

We would like to express our gratitude to various colleagues and students at Morehead State University for their assistance as we prepared the sixth edition, particularly Mary Whidden. We are also very grateful for the valuable comments received from professors at other institutions who have taken time to respond to the publisher's surveys. We are deeply indebted to the formal reviewers whose numerous constructive suggestions have been incorporated into this edition. They are Donald E. Rickard, Arkansas Tech University; Jerome J. Notkin, Hofstra University; Aaron W. Todd, Middle Tennessee State University; Robert J. Backes, Pittsburgh State University; Richard E. Baker, El Reno Junior College; J. H. Mehaffey, Francis Marion College; Steven L. Morris, Los Angeles Harbor College; Linda L. Payne, South Carolina State College; and Michael Rulison, Ogelthorpe University.

We are deeply indebted to our families for their assistance and encouragement during this revision. Also, we are most appreciative of the thorough editing service rendered by Ann Mirels. Finally, we are extremely grateful for the guidance and support given us by the textbook team at Wm. C. Brown that includes our editors, Bob Fenchel and Jeffrey L. Hahn.

sixth edition

Physical Science

Principles and Applications

PART ONE

Science, along with its applications, is as old as is the human race. Theoretical science, such as the formulation of rational thought about the observable, came later. Ultimately, the unobservable was theorized by the more intellectually active among the ancients. The requisites for their additional considerations of nature were curiosity and leisure time. Those who wondered about natural phenomena must have also communicated with others in order to have gained further insight.

There is ample proof that the ancients carefully observed their natural surroundings. The predictable movements of the heavenly bodies enabled the ancients to develop accurate calendars. Even the very early Britons, not generally considered to have been scientifically active, constructed Stonehenge in order to predict celestial events. There were other careful and intelligent observers, presumably among the early Greeks, who contrived the twelve "houses of the sun," the constellations that we know today as the signs of the zodiac.

Greek scholars of about the sixth century B.C. probably represent the first scientists, since they made the first attempt to establish coherence among the natural events they observed. They predicted celestial happenings and weather conditions with an acceptable degree of accuracy. The first organized school of mathematics was established by these ancient Greeks. Also, the progress they made in medicine during this period was significant, and this area of knowledge began to build upon a scientific, rather than a magical, foundation.

During such early times only a select few possessed scientific knowledge. It seemed reasonable to the ordinary citizen, who seldom ventured far from home, that Earth was flat and that various heavenly bodies moved across the observable portion of the universe as if attached to an inverted bowl. This vessel was supported above Earth, according to various myths, by an elephant, a turtle, the mountains, or Atlas. There are still some who maintain similar beliefs today, but organized support of such beliefs has declined since the advent of space ventures.

From the time the ancient Greeks conceptualized order in the universe to the present, only one period of history finds science held in low esteem. This interim, commonly known as "the Dark Ages," extended from the time of the fall of Rome, about A.D. 455, until the intellectual awakening that occurred in the West under Pope Sylvester II five centuries later. Among the most significant happenings in the fields of science during this period were the invention of the heavy-wheeled plow and the development of an efficient waterwheel along with an improved windmill. These devices were used to cultivate, irrigate, and process newly domesticated grains used as a major source of food.

The Dark Ages were followed by the Middle Ages, a period that lasted until the start of the Renaissance, which got underway in Italy during the fourteenth century. The Middle Ages is marked by the humanistic revival of classical influence, including the beginnings of modern science. The scientific revolution that occurred was brought on by a radical change in how one learned about the physical world. The new method was based on experimentation, direct observation, and mathematical analysis, rather than on the written word of the ancients, the foundation for medieval research.

The Renaissance lasted several centuries in various countries. This return to the classics brought about a decline in some areas of intellectual progress. However, during this period somewhat detailed plans were published, although not put to use, for powered flight, armored tanks, and submarines. The geocentric theory was, for the first time, strongly refuted. Alchemy was replaced with the more exact science of chemistry as the Renaissance continued and culminated toward the end of the seventeenth century.

The Foundations of Science

There was considerable scientific advancement in the eighteenth century. Halley accurately predicted the 1758 return of a comet to our portion of the solar system. For the first time, it was noted that stars were not fixed but freely moved about the heavens. The solid crust of Earth was surmised to be resting on a plastic-like layer. Ocean basins were studied and mountains were declared capable of rising from them. Organic chemistry became a new and exciting field of interest, although the execution of its leading researcher, Lavoisier, slowed its development as a major area of study. Various gases, such as carbon dioxide, ammonia, and oxygen, were separated from their sources. The thermometer was considerably improved during this century, and lightning was identified as a fascinating electrical display.

The humanistic trend established earlier continued into the nineteenth century with the appearance of romanticism. This movement centered around literature, but art, music, and philosophy were also popular among the scholars. However, a new social and economic attitude, brought on by the complexity of the times, was developing. The industrial revolution was initiated and gained momentum as steam became a viable source of power. Some scientists concentrated on the remaining mysteries of the subatomic world and rare elements. Others applied newly developed instruments and techniques to gain additional information about celestial objects. The understanding of electrical energy was greatly enhanced in this century, as one discovery shortly led to others. Electromagnetic radiation was investigated in depth and many of its related phenomena were expressed mathematically.

Scientific knowledge in the nineteenth century became quite diversified. No individual could adequately comprehend, much less stay cognizant of, all fields of study; hence, segregation of science into the various disciplines resulted. Laboratories to be shared by leading scholars were established at all major universities. Ironically, a considerable number of major contributions to the fields of science were made by individuals who held no academic position. Many countries came out of scientific isolation during the century and their scientists became active in the various learned societies that had been established. International efforts toward a specific objective became commonplace.

At the turn of the present century, society was beset with many problems, mostly of its own making, but had few valid solutions. With improved sanitation, better health and medical care, fewer infant deaths, and longer life spans, population increases threatened to upset the balance of nature. Because of the projected growth of society, food supplies were in danger of becoming insufficient. The rapid growth of cities and increased industrialization created problems of waste disposal, shortages of utilities, and deterioration of air, land, and water. Science and philosophy were emerging from a realistic, determinate, and ordered world to one that was more uncertain and unrealistic. This change in view of the universe was brought about by theorists who introduced the principle of uncertainty. According to their doctrine, the accurate measurement of an observable quantity inherently produces uncertainties in one's knowledge of the values of other observable phenomena. Scientists began to realize that the accuracy of their works was dependent on probabilities and that there was a need for better means of data collection and analysis. The rapid development of computer technology is the result of scientific efforts from a multitude of new and existing fields. With computer-assisted research, scientists can continue to arrive at a more likely molecular structure, nuclear reaction, or model of the universe. It is apparent that the foundations of science have become progressively stronger and will continue to do so as long as research is directed toward its development.

On Understanding Science

Chapter outline

Key terms/concepts

Internationally famous scientists conduct studies in all major scientific disciplines at the Brookhaven National Laboratory's research reactor facility.

cience and its techniques have changed human life more in the past 150 years than in all the rest of recorded history. The changes being wrought by science continue at an ever-increasing pace, and there is reason to believe that the rate of scientific breakthrough will escalate as we delve further into the unknown. Each time the answer to a question is found, new questions result, and a new impetus for seeking answers is born.

An important new factor has also clearly emerged in recent years. We ourselves are creating environmental changes to which we will have to adjust with unprecedented rapidity. Whether we can withstand the environmental changes that we have brought about, in addition to those caused by natural forces, remains to be seen.

Early Science

The earliest inhabitants of Earth saw the same natural phenomena that we see today. They watched the sun, moon, and other celestial bodies rise and set. They observed cloud formations, seemingly fixed stars, and changes brought about on Earth's surface by wind and water. They witnessed disease take their loved ones and fire destroy their surroundings. They were terrorized by phenomena such as avalanches, earthquakes, floods, and lightning displays.

Even later generations failed to understand the cause of the natural happenings that they observed, so they invented stories or myths to explain them. These early people were fearful of events not understood, but the tales helped them to be less afraid.

Often the early observers noted two things happening simultaneously and assumed that one event causes the other. At other times they observed that one event was preceded by another and considered the first to be a sign or omen of the other. A falling star signaled good luck; thick bark on a tree indicated that an extremely cold winter would follow.

Aristotle (384–322 B.C.) is credited as one of the first persons to study nature in a systematic fashion. Although he laid the grounds for many misconceptions, Aristotle also caused many people to question their surroundings. He classified everything known to him into combinations of four elements: air, earth, fire, and water. These elements determined the four properties of all substances in the universe: hot, cold, wet, and dry. Figure 1.1 depicts the relationships between the elements and their properties. As subsequent thinkers sought to know more about their natural surroundings, their discoveries established different classification schemes.

The greatest advances in early science came about during the sixteenth century, when people started to look for relationships rather than final causes and to test ideas through experimentation rather than simply offer "logical" explanations. Physicians of the mid-sixteenth century did not doubt the belief of the ancients that the human heart was composed of but three chambers. Aristotle taught that heavy objects fall faster than light objects (do rocks not fall faster than feathers?), and he believed matter to be continuous, not composed of little particles. These beliefs were accepted without question. To discount the words of Aristotle and other ancient philosophers meant severe punishment—sometimes even death. But observation began to gradually replace logical argument as the fundamental procedure of investigation.

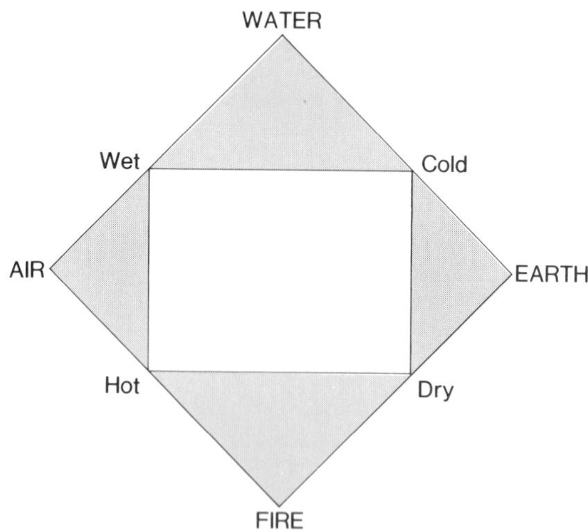

Figure 1.1 According to Aristotle and his followers, the four elements indicated determined the properties of all things in the universe.

The Search for Order

A large part of our intellectual activity involves building and refining models of the world and our position in it. For thousands of years various cultures have provided models that attempt to introduce order into the seemingly chaotic world. Often these models are simplified so that they are more readily understandable and can be considered in a more manageable way. Models are seldom totally accurate, but if they fit observable events and permit accurate predictions to be made about the concept involved, they are useful until further study modifies them.

Consider the early models of Earth and its place in the universe. Thales of Miletus, in the sixth century B.C., envisioned Earth to be a large flat disk, floating in water and composed of that basic substance from which all other things emerged. The ancient Hindus believed the heavens to be a vast sea of milk, the bottom of the universe. Earth rested on the backs of four elephants, each facing one of the four cardinal compass directions. The elephants stood on the back of a giant turtle that was swimming in the great white sea. A giant cobra was the keeper of the boundaries within which the other creatures were contained. Ancient Egyptians considered Earth to be a great coffin surrounded by vast fires. Darkness came as the coffin lid was closed. The stars were really views of the vast fires through tiny holes in the coffin lid. Later models of the universe will be presented in detail in Chapters 16–19 on astronomy.

Various models are used by scholars of most disciplines. Some models are designed after careful consideration of details; others

Aristotle

Aristotle is considered to be one of the greatest thinkers of all time. The ancient Greek philosopher was born in 384 B.C. at Stagira, on the northwest coast of the Aegean Sea. His father, a physician, served the king of Macedonia, and much of Aristotle's boyhood was spent at the court. At age 17 he went to Athens to study at the famous Academy directed by the philosopher Plato.

Aristotle became totally involved in Plato's pursuit of truth and goodness, and the young lad's talents were carefully developed at the Academy over the next twenty years. In later years he renounced some of Plato's theories and progressed far beyond him in breadth of knowledge. Aristotle became a teacher in a school on the coast of Asia Minor and spent two years studying marine biology on the island of Lesbos. In 342 B.C., Philip II invited Aristotle to return to the Macedonian court and teach his 13-year-old son, Alexander. This youth was to become known as Alexander the Great, conqueror of the world. Alexander became king at age 20 and gave his teacher a large sum of money to start a school in Athens. Aristotle taught brilliantly at the school, collected the first great library, and established a museum. He often strolled in the gardens of the school, discussing problems with his students. Athenians thus called his academy the Peripatetic (which means "to perform while moving") School. Aristotle directed his pupils in research in every known field of knowledge. His students dissected animals and studied the behavior of insects with the primitive instruments at their disposal. This science based on observation was new to the Greeks and frequently their conclusions were incorrect.

One of Aristotle's most important contributions was the development of classification schemes for the various branches of knowledge. He sorted them into logic, metaphysics, physics, psychology, poetics, and rhetoric, thus laying the foundation for many areas of knowledge studied today.

Anti-Macedonia feeling erupted in Athens in 323 B.C., and Athenians accused Aristotle of impiety. He fled so that the Athenians might not "twice sin against philosophy" (by killing him as they had Socrates). He escaped to Chaleis on the island of Eruboea, where a year later he died.

After his death, Aristotle's writings were scattered, stolen, and lost. In the early Middle Ages, the only works familiar to scholars of Europe were his recorded observations in logic. They became the basis for one of the three subjects of medieval teachings—grammar, logic, and rhetoric. Early in the thirteenth century, other books reached the West, some arriving from Constantinople, and others brought by the Arabs to Spain. Medieval scholars translated his writings into Latin.

Aristotle's best known writings that have been preserved are known by the following titles: *Constitution of Athens, De Anima* (on psychology), *History of Animals, Metaphysics, Nicomachean Ethics, Organon* (treatises on logic), *Poetics, Politics,* and *Rhetoric.*

Aristotle's views were treated with the highest esteem. They became the unquestionable authority and remained as such into the seventeenth century. Eventually, a number of Aristotle's conceptions were proven to be erroneous. However, his attempts to understand nature in a rational manner were not unreasonable within the framework of his experimental knowledge, and his influence brought about a dogmatic investigation of the natural world by later scholars ■

may be too hastily proposed. Some models are very limited in scope, whereas others encompass an entire way of thinking. For instance, psychology is founded on a systematic way of interpreting the mental or behavioral characteristics of an individual or a group. Philosophy is a systematic way to investigate the theories that underlie all knowledge and reality.

Biological science represents an organized way to study living things—their myriad forms, growth processes, evolutionary development, and death. This body of knowledge involves the study of the metabolic activities of growth, responsiveness, and reproduction present in all organisms. **Physical science** deals with inanimate matter as well as with energy and its interactions with material things. This field of study centers around our physical environment—the storehouse of knowledge we possess about it and the ways we attempt to increase our understanding of our surroundings. Together, biological science and physical science constitute the vast area of knowledge we know as **natural science.**

All these ways of thinking are artificially formulated by humans, not set up by nature. Without each way of thinking, no search for order can be complete, for each is dependent on the others. How can a psychologist understand the way the mind functions without understanding the biological processes present in the brain? How can a biologist understand the way the brain works without considering the chemistry of the brain cells or the manner in which electrical impulses are conducted through the nerve tissue of the brain? Further, how can experts in the various physical sciences understand the ways the material world functions without considering the way the human mind works? The circle of thinking, then, is complete if we consider the need for a valuable input from psychology. This input includes how we become cognizant of our natural world and the means by which we communicate this knowledge to others.

The Definition of Science

Science, as you may have surmised, represents the search for order in many different large areas of nature. Each area can be broken down further into more defined categories. For instance, biology

Science and Superstition

Many superstitions are derived from beliefs and practices that relate to a fear of the unknown. The ancients relied on numerous beliefs that they thought ruled their destinies, some of which were accepted as truths despite evidence to the contrary. As astrologers learned to predict with varying degrees of accuracy when an eclipse was to occur, certain cultures supposedly believed that a dragon was in the act of swallowing the sun or moon, and that by creating loud noises they could cause the monster to regurgitate the heavenly body. Noise was thought by later cultures to drive away other demons and spirits. Also, the unquestionable belief that Earth was the center of the universe caused people of Galileo's era to decline his invitation to view the moons of Jupiter through a telescope. The observations they could have made were later used to help dispel several superstitious beliefs.

The old superstitions that surround a black cat, a ladder, and a raised umbrella are only a few of those that continue to exist in various segments of our culture. Many people still trust in magic or chance to bring about or to prevent an occurrence they may or may not wish to happen.

Most superstitions exist because there are but few basic natural phenomena in which the actual cause is obvious. In fact, there often appear to be several possible causes for a single event. The ability to determine the correct one indicates the progress one has made toward scientific literacy ■

is often divided into the study of botany, zoology, and microbiology. These areas are further divided for advanced study. Physical science is readily divided into astronomy, chemistry, geology, meteorology, and physics, and each of these is subdivided into a multitude of defined areas.

Science is a way of thinking. If a person is asked to define the term, the response will reflect that individual's philosophy and experiences in the vast field of knowledge. Some students, according to the learning experiences they have had, may consider science to be a collection of facts, many apparently unrelated, that have had to be memorized. Other students may define science as a series of interesting experiments about nature. To the latter group, perhaps, science is basically what scientists do. Philosophers, teachers, technologists, physicians, and others approach science from a variety of standpoints, and thus may define it in terms of their perspectives.

An alternative way to explain the term is by means of an analogy. For instance, one might relate science to a familiar industry. Industrial processes undergo little change; industrial products are constantly changed. Many processes developed in the days of Henry Ford are still used to manufacture automobiles, although early automobile makers could not have conceived of the cars we now drive. Science, too, represents a series of processes through which the product known as knowledge is gained. Just as automobile models change yearly, our ideas about nature constantly undergo revision.

Science may also be thought of as a humanistic enterprise. It includes ongoing processes of pursuing explanations and understandings of the natural world, developing our storehouse of knowledge. Each time we gain new information we can correct ourselves and thus destroy faulty concepts. Many products developed by applying scientific techniques and principles are later found to be harmful in some way—through application of those same techniques and principles. Consider such products of scientific research as artificial sweeteners, aerosol spray propellants, and insecticides. Each product was developed through scientific research to meet a need, and each was modified as more research showed its harmful effects.

The term *science* itself is derived from the Latin *scientia* (from *scire,* "to learn" or "to know") and is the equivalent of the German term *Wissenschaft* (an "organized body of knowledge"). Many active scientists currently define **science** as *the systematized knowledge derived from observation, study, and experimentation carried on in order to determine the nature or principles of what is being studied.* In a broader sense, science includes such academic areas as history, philosophy, and other areas based on systematized knowledge, the accepted truths of which can be studied and learned. In a textbook about the study of the natural world, the term *science* refers to those general areas of nature that deal with the interactions of matter and energy.

Conceptual Schemes

Scientific facts originate from a collection of observations applied by an observer in a sincere attempt to develop tentative explanations for the order and behavior observed in nature. Tentative explanations are called **theories** and are subject to continual study, testing, revision, and rejection. These views of nature undergo constant changes as the results of further research are revealed. There are some theories that are not totally accepted by all scientists. However, most theories have been carefully examined by a host of independent investigators and are generally accepted. In fact, many theories presently considered valid reflect the thoughts of early scientists. Such theories are modified as the results of further research are publicized to make them more accurate and useful, yet still as simple as the accumulated data and observations permit.

Some theories are carefully considered, ordinarily over long spans of time, and are found to be consistently accurate, indicating they explicitly fit the observed. From these well-established theories, statements that describe how nature behaves under specific conditions may be derived and are known as **laws.** The theories involved are retained to correlate and interpret the laws, although the laws are ordinarily so fundamental they are seldom subjected to further scrutiny, and are simply accepted as truths. Realistically, laws are based upon experimental evidence which, because of the nature of measurement, can never be "absolutely proven." With these laws, which are occasionally modified in light of new evidence, science has developed the meaningful body of knowledge that is studied and built upon today.

Many of the seemingly diversified contributions made by researchers in the natural sciences are found to relate to the works of others who preceded them. For instance, the findings of some investigators may involve energy whereas the findings of others may center upon the understanding of living things. In fact, a portion of the related knowledge about a given subject can be arranged into a framework known as a **conceptual scheme.** The following sequence is a conceptual scheme summarizing our present views of matter and energy.

1. **Matter** is whatever occupies space and is perceptible to the senses in some manner. It is composed of so-called elementary particles such as electrons, protons, and neutrons. (Other elementary particles are presented in Table 15.2.) These particles are the building blocks of atoms. Under certain conditions, elementary particles can be transformed into energy, and conversely, energy can be converted into elementary particles. The properties of an elementary particle include such characteristics as its mass and electric charge, as well as its magnetic and rotational features.
2. Matter exists in units that can be classified by various degrees of complexity. For example, living things are composed of atoms that, in turn, form molecules. Aggregates of molecules, such as protoplasm, nuclei, and membrane, form cells that are components of tissues. Tissues fit into systems that form organisms, and organisms fall into natural groupings. Even the physical world can be organized into a hierarchy: first elementary particles; then atoms; followed by molecules and ions (parts of molecules or groups of atoms); aggregates of molecules and ions, such as grains of sand, automobiles, mountain ranges, rivers, and so forth; celestial bodies, including planets, moons, comets, and stars; solar systems; galaxies; galactic clusters; and lastly, presumably, the universe. A schematic representation of the hierarchy appears in Figure 1.2.
3. The behavior of all matter in the universe can be statistically determined, if we know all the variables involved. We can locate heavenly objects yet to be discovered as they interfere with the predicted paths of other objects, and we can predict the properties of elements yet to be synthesized in the laboratory, based on the consistency of nature.
4. Various units of matter interact with other units because of nuclear, electromagnetic, and gravitational forces. The binding of the particles within the nucleus of an atom is attributed to nuclear forces. The bonding between atoms and ions is electromagnetic in nature. The motions of planets, moons, and artificial satellites are produced by gravitational forces.
5. All interacting units of matter approach states of equilibrium in which the energy distribution is most random (see Fig. 1.3). The sum of available matter and energy in the universe remains constant. As matter interacts, a combination of heat energy, light energy, or electrical energy may result. As a chemical reaction takes place, changes occur in the composition of the interacting matter, yet the net gain or loss of matter and energy for the system is always zero. In addition, the direction of spontaneous changes in matter is always constant; heat always flows from the hotter object to the colder object. Complex matter may dissociate into simple building blocks, yet simple bits of matter do not necessarily form complex bits of matter. A cube of sugar dissolved in iced tea does not cause the sweetened tea, once evaporated, to yield a cube of sugar intact.
6. One of the forms of energy results from the motion of units of matter, living or nonliving. Such motion is responsible for the states of matter including solid, liquid, and gaseous, as well as for the heat and temperature of matter. The motion of particles in matter also accounts for the pressure exerted by matter in its gaseous state.
7. All matter exists in time and space. The units of matter constantly interact among and within other units of matter as time passes (see Fig. 1.4). The rates of interaction may be constant or variable. Celestial bodies are subject to changes in form, position, or composition as a result of the interaction of matter. New stars are the objects formed from stellar materials in outer space. The matter of which Earth is composed constantly undergoes transformation from one form to another. New rocks are formed from new molten deposits or from older rocks; thus, changes in composition occur. Water and sediment suspended in the rocks constantly undergo change in position, just as do Earth and the moon. In all cases, time is required for interactions to occur and motion to continue.

The Attitudes and Methods of Science

A report published some time ago by the National Education Association summarized the accepted belief that the spirit of science—the scientific attitude—is the longing to know and to understand, the questioning of all things, the searching for data and their meaning, the demanding for verification, the respecting of logic, the considering of premises, and the deliberating of consequences. These considerations often apply when investigators evaluate controversial issues, particularly those that lack scientific backing.

It is an established premise that science is based on the laws of nature. This vast accumulation of knowledge proceeds on the assumption that the universe is not erratic; rather, that the various phenomena observed under certain conditions in a specific sample

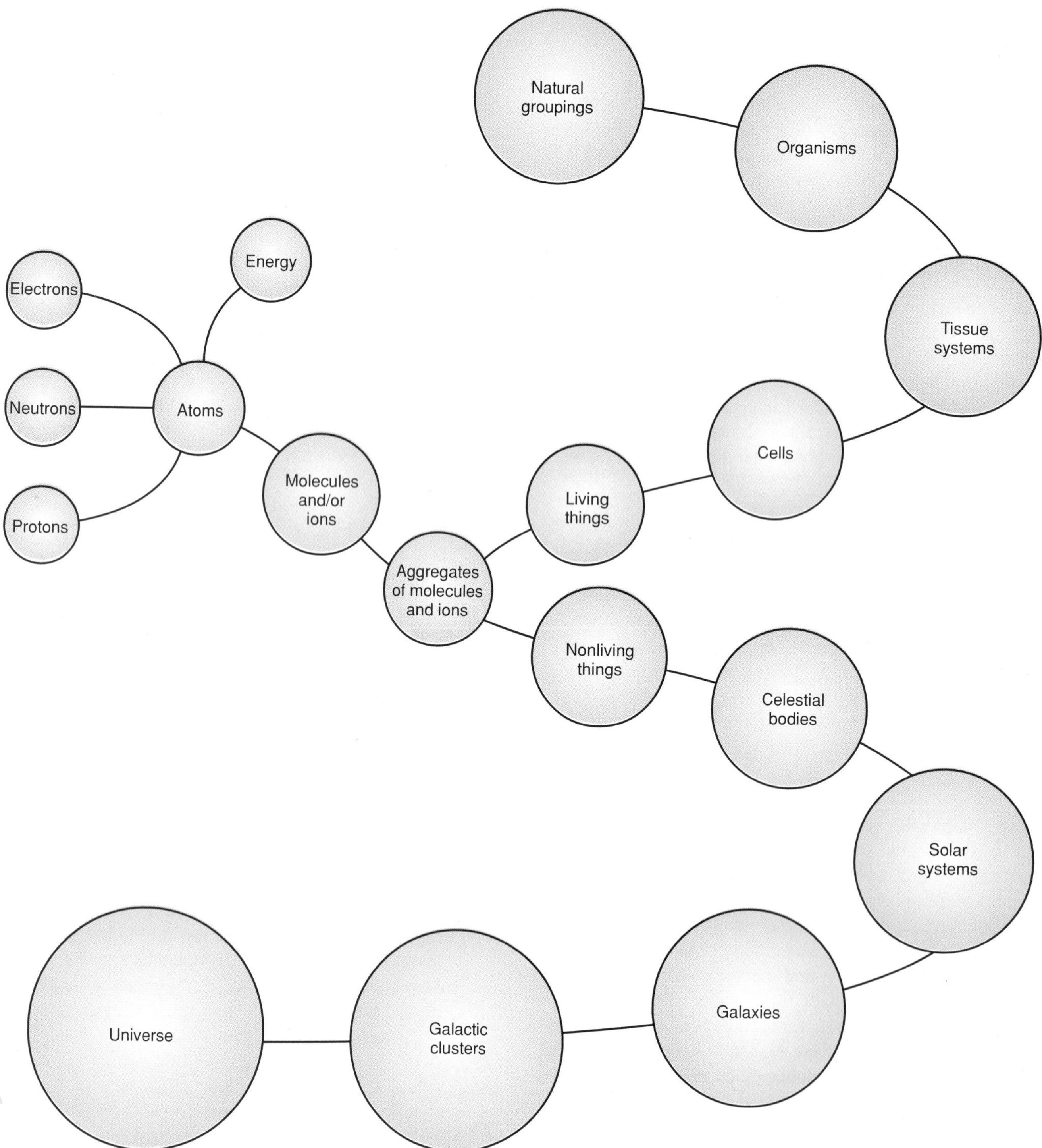

Figure 1.2 A schematic representation of how our views of matter fit into a conceptual scheme.

of matter occur under like conditions in other similar samples of matter. This consistency of nature permits us to gain insight as we probe the unknown.

The accumulated body of knowledge about nature is based on observations of samples of matter accessible to public investigation rather than on observations made from private research with samples of matter not available to others. The conclusions developed from scientific inquiry require support of evidence from other researchers. Few exceptions are made to this rigid condition, since the danger from actions based on invalid conclusions may affect all of society.

Scientific inquiry is piecemeal; no single problem is explored by all. The incredibly large number of specific problems under investigation justifies this statement. Neither the general public nor the scientific community would wish for all investigations to concentrate on one all-inclusive problem. Our knowledge must progress in all directions if we are to gain greater control over various aspects of our environment.

Measurement

The necessity for an acceptable system of weights and measures was realized long ago. The archeological records of the most ancient civilizations reveal well-developed concepts of weighing and measuring. The need to measure length apparently preceded the need to measure weight, volume, area, and other quantities.

The earliest organized measuring systems were developed by the ancient Babylonians, Egyptians, and Sumerians. Archeological discoveries definitely prove that by the time of the great Mesopotamian civilizations (about 3000 B.C.) and of the construction of Egyptian temples and pyramids (3000–1800 B.C.), systems of measurement were well developed.

As various societies throughout the civilized world developed, numerous systems of measurement were invented. Each system became more involved as various trades and occupations developed separate measurement systems for their specific use.

Once international trade increased, the countries involved realized the importance of establishing a system of measurement acceptable to each nation. Numerous systems were tried and discarded, mostly due to the lack of standardization.

Gabriel Moulton (1618–1694), a clergyman in Lyons, France, proposed a comprehensive decimal system of weights and measurements in 1670. Over a century later, the French Academy of Sciences undertook the task of choosing a suitable standard; hence, the ultimate development of the metric system. This system is presented in detail in Appendix I ■

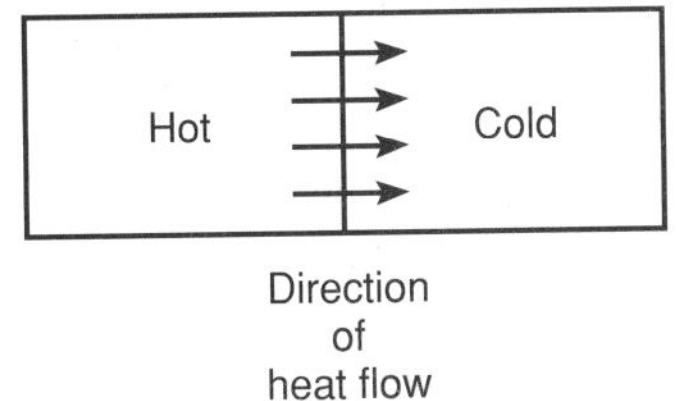

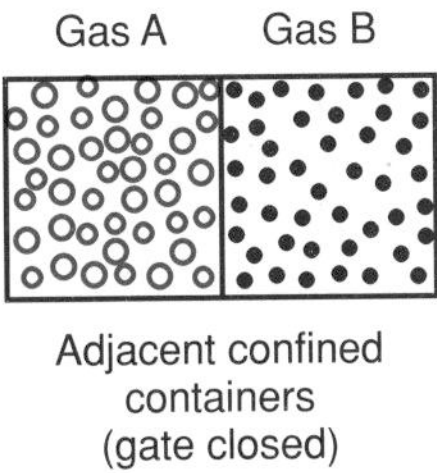

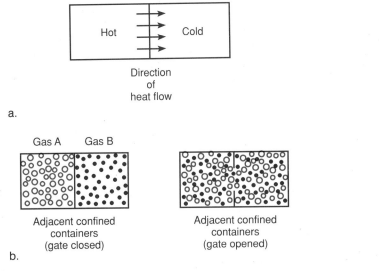

Figure 1.3 Two simple systems show how equilibrium is approached. (*a*) Two reservoirs in contact tend to approach the same temperature as heat (energy) flows from the hotter to the colder reservoir. (*b*) Gases approach equilibrium when permitted to mix.

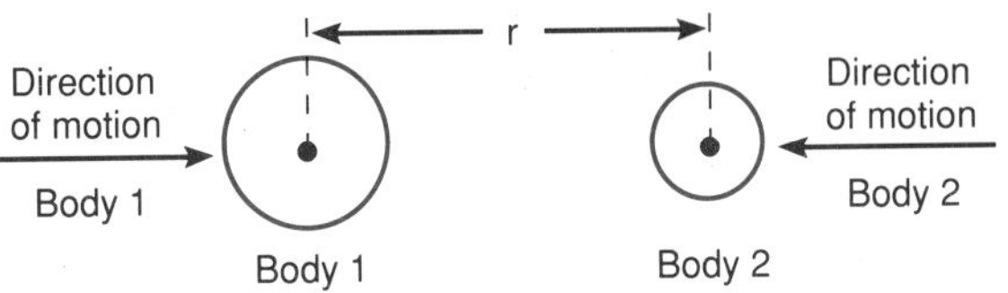

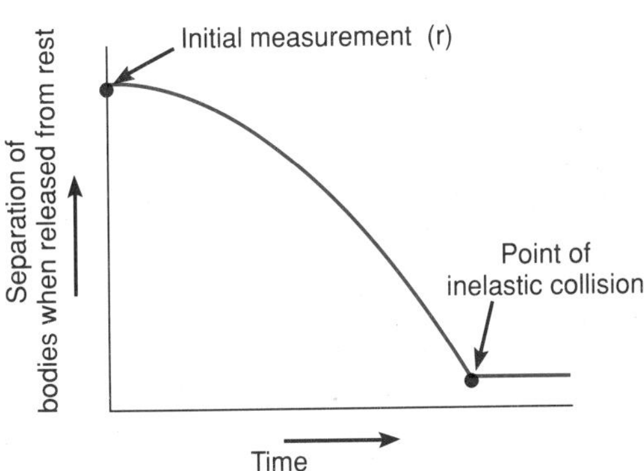

Figure 1.4 If the speed, direction of motion, and immediate distance between two unlike masses are known, a determination of their positions at any instant thereafter can be made.

Science can never be a finished enterprise. Each day many new discoveries are made, most of which lead to further needs for resolving new problems. Science, then, is an ongoing and cumulative enterprise, the pursuits of which lead to a better understanding of how things in the universe behave and interact.

Every branch of science is based on methods of obtaining quantitative measurements. Before new laws are formulated, rigorous and systematic checks of accumulated data must be made. Accuracy of measurement is necessary for duplication of conditions under which the problem was investigated. Measurement is extremely important to all phases of the scientific enterprise.

Society has developed the general attitude that scientists follow a rigid method by which they effectively gain and use information about a specific problem. Many textbooks include the system presumably set forth by Galileo in the seventeenth century by which scientific investigations are conducted. The steps in this **scientific method** are as follows:

1. Recognize that a problem exists and identify it.
2. Develop a hypothesis regarding the solution.
3. Predict a result based upon the hypothesis.
4. Devise and perform an experiment to test the prediction.
5. Develop the simplest theory that relates the result to previously existing knowledge.

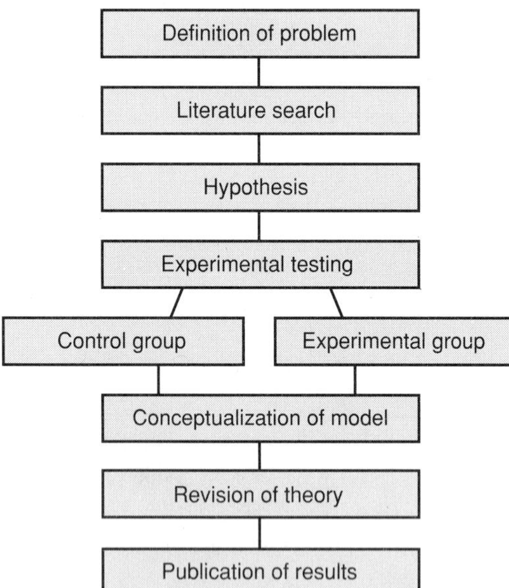

Figure 1.5 A representation of the scientific method as applicable to today's researchers. Most investigators typically start at the top of the flowchart and proceed downward.

In contrast, the following steps represent a more contemporary presentation of the scientific method (see Fig. 1.5):

1. The researcher determines whether a problem exists. This recognition usually results from the investigator's observation of the natural phenomenon. Questions involving the problem generally arise, so that it becomes more precisely defined.
2. A literature search usually follows to determine whether the problem has been previously attacked and to assess the efforts of prior researchers. The investigator may rely on computerized search programs or those conducted by the staffs of technical and university libraries. Other sources of information at the disposal of the investigator should be explored.
3. The investigator may conceptualize a possible answer or conclusion in a statement known as a **hypothesis.** The hypothesis reflects the beliefs of previous researchers and is based on all information available to them.
4. If the results of the literary search do not satisfy the curiosity of the investigator, controlled experimentation is then conducted to gain the additional information deemed necessary.
5. All the pertinent data derived from the various experiments conducted are analyzed. A model is conceptualized, based on the researcher's data analysis and supplemented by the information available from prior investigators. After all these steps are completed, a plausible theory is suggested.
6. More experiments are performed to test predictions. If the predictions of the theory are found to be inaccurate, the theory is revised accordingly. This additional experimentation may lead to the abandonment or modification of previously accepted theories. The very essence of the scientific method requires that every hypothesis and theory be tested.
7. The researcher reports—in scientific journals or in some other manner—the obtained results. These results must be presented in an honest and unbiased fashion; that is, all conclusions and judgments must be based on the actual observations made by the researcher, not on the predilections that often accompany prolonged, dedicated commitments.

Even today, the scientific method is not a magical prescription by which an investigator finds a solution to a problem. This method is often not followed by a researcher, at least in the step-by-step fashion ordinarily presented. Much of the progress in science is made by trial and error (guess and test) as well as by experimentation without the formulation of a hypothesis, using simply the systematic application of common sense. Consider that Thomas Edison's group of researchers reportedly performed over two thousand experiments, each based on the results of previous ones, before the materials to use in Edison's incandescent lamp were determined. Very often the solutions to problems are found by accident. Any number of discoveries are brought about by investigators who are working on another problem, related or unrelated. In concept, "discovery by accident" may be a misnomer. To illustrate this point, consider a statement attributed to the renowned chemist, Louis Pasteur: "Chance favors the prepared mind." The researcher is generally actively involved in the investigation of the natural world through experimentation, but by chance discovers some new scientific phenomenon that had not been previously considered. Trained scientists know to be on the lookout for unforeseen happenings. Among the list of discoveries made as researchers were attempting to solve other problems are the principles behind the vacuum tube, the laser, the transistor, and the X ray. Penicillin, too, was discovered by chance, as was the existence of radioactivity.

As emphasized earlier, scientists must be willing to accept the information they collect even though the results may not be those they would prefer. These researchers must strive to distinguish between what they see and what they hope to see. Intellectually honest people learn to change their minds about a possible solution to a problem. They seek not to defend their beliefs, but to search for more evidence that supports or rejects them in their search for truths. If the analysis of additional data does not follow the predicted path, they adjust their theories accordingly.

As in the past, our nation is endowed with many young scientists who will continue to assume bigger roles as advisors to those who must make decisions that affect our lives. According to several recent studies, these individuals express prime interest in many of the broad social and environmental issues that preoccupy the minds of the general public. These trained researchers generally are aware that science has its limitations and are interested in closing any existing gap between scientific endeavors and public concerns. Many are concerned about our future energy resources, the potential shortage of food, and predictions of overpopulation.

These young scientists desire to concentrate on medical advances, particularly in discovering cures and preventative measures for the various types of cancer and circulatory ailments. A large number in this group indicate a desire to become involved in controlling the arms race and in preventing nuclear war. Other promising young scientists are concerned with pollution, the environment, and related situations that affect the quality of life on Earth. A common agreement among the next generation's scientific leaders is that educating the general public about the nature and character of science is a high priority. Only in this way can the public take best advantage of the new knowledge that science and technology can provide. The scientific method and attitude furnish a scientist with the system to accomplish many things, but a humanitarian attitude must provide guidance in achievement.

The Differences between Science and Technology

Many individuals fail to distinguish between science and technology. Science offers the methods by which theoretical questions are answered; technology solves practical problems. Science seeks to discover truths about various observable natural phenomena and to establish theories that organize these truths into a conceptual scheme. **Technology,** on the other hand, is the application of scientific findings through the use of tools and manufacturing methods. It is the intent of technology to employ systematic means to provide objects for human sustenance and comfort.

Science and technology, in reality separate parts of the scientific community, must work closely together toward common goals. Both have assumed the task of creating the best of all possible worlds. Wise implementation of scientific and technological advances has already brought about a considerable alleviation of food and energy shortages. The concerns of science and technology have also led to the conclusion that greater care must be taken to prevent further damage to our environment.

Inherent feedback exists between scientific research and technology. As advancements are made in the one area, developments ensue in the other. For instance, as scientific knowledge concerning the properties of the electron was discovered, the electron microscope was developed. This device has revealed previously unknown characteristics of the atom. The same relationship prevails in numerous cases, including charged particles and particle accelerators, as well as electronics and integrated circuitry.

Science and technology collectively deserve much credit for developing many processes and products that enhance the quality of our existence. Society, however, must make decisions on many important matters that concern everyone. Guidance must be given to the scientific and technological communities regarding such issues as prolonging the lives of terminally ill patients, providing means of birth control, transplanting genes, developing germ warfare, releasing nuclear energy, and inventing new labor-saving devices. We all must very carefully consider these controversial items, along with a host of others. The need for all members of society to be well informed becomes more critical as we approach the twenty-first century.

The Physical Sciences

Natural science is divided into two major divisions: the biological sciences, which study living things, and the physical sciences, which study nonliving things. The physical sciences include astronomy, chemistry, geology, meteorology, and physics. These divisions are not always sharply defined, since all areas of study about the natural world are firmly interrelated and are only artificially separated for the convenience of researchers and students. A schematic representation of the natural sciences appears in Figure 1.6.

Astronomy is the area of science concerned with celestial bodies and their size, motion, structure, origin, fate, relationship, and composition. Astronomers have developed the meager understanding of the ancients into a concise, well-established body of knowledge. Of all scientific areas of the natural sciences, astronomy encompasses more unknowns than any other field. Astronomers are handicapped by not being able to gather information from hands-on activities available to most other scientists. Many of their growing number of discoveries about the universe have been accomplished by countless painstaking observations and carefully developed inferences since the invention of the telescope and other astronomical instruments. (It should be noted, however, that the accomplishments attributed to ancient observers before the invention of the telescope are actually amazing. Selected discoveries will be discussed in Chapter 16.)

Chemistry is the natural science that deals with the composition, properties, and conversions of matter. As matter interacts with other matter, transformations occur. These chemical and physical changes are studied by chemists, as are the conditions that might bring about further transformations. In addition, chemists create new substances for specific uses. The extraction of base metals such as copper, tin, lead, and iron from their ores is attributed to the ancient artisans commonly called alchemists. Undoubtedly, the discoveries of these processes came about by accident rather than by intensive chemical research. During the Middle Ages, some investigators concentrated on the impossible goals of extracting gold from the base metals and developing an elixir of life, a concoction expected to restore youth. Still others sought the universal solvent, a substance that would dissolve all other substances. Modern chemistry came into being in the latter part of the eighteenth century as chemical reactions began to be understood.

The natural science known as **geology** deals primarily with the science of Earth. Geology is perhaps the most varied of all of the natural sciences, for it encompasses the study of all aspects of our planet. This field of study became organized about 200 years ago and it includes the solid matter of Earth and other celestial bodies, along with the life recorded in the rocks. Geologists are interested in Earth's origin as well as its history and development over the 4.5 billion years of its existence. They determine the size and shape of Earth and study the many processes that alter its features. These scientists attempt to discover how life originated on Earth and what changes have occurred in living things. Geologists examine the soils, the rocks, and the ocean floors. They search Earth's rocky crust for fossils—traces or remains of plant and

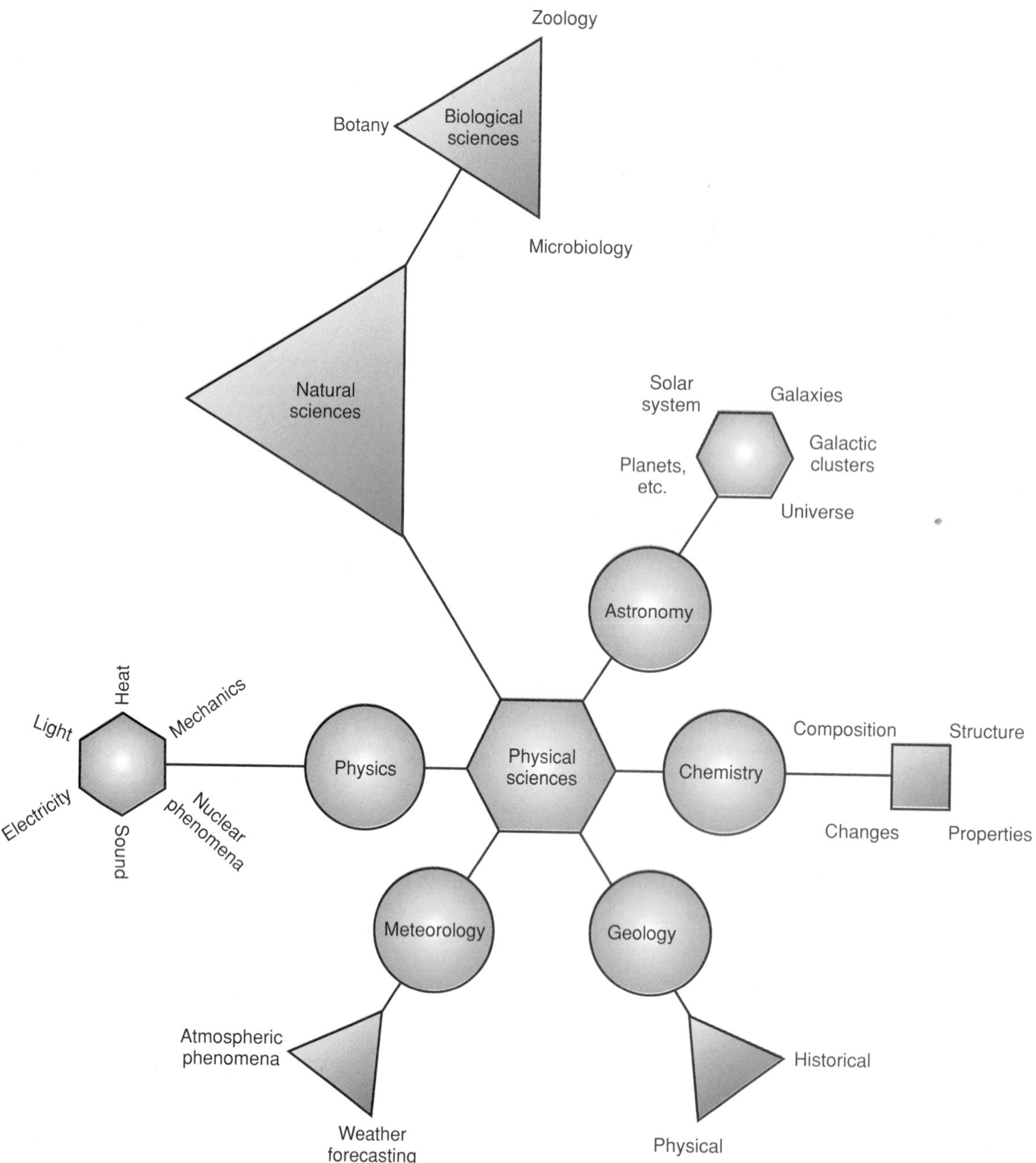

Figure 1.6 A schematic diagram relating the subdivision of the various areas of study about the natural world.

animal life long extinct on the planet. Geologists attempt to find the causes of earthquakes and volcanic eruptions and to ascertain how the mountains were formed. Many of today's geologists continue the search for Earth's natural resources and are involved, for safety reasons, in the construction of dams, highways, and building sites.

The atmosphere of Earth is more complicated than it appears. This body of air that surrounds Earth behaves in a consistent way describable by natural laws, and scientists have learned to predict its actions with considerable accuracy. The science of weather is called **meteorology,** from the Greek terms *meteōres* ("high in the air") and *logos* ("discourse"). Observers since ancient times have attempted to predict the weather. Various individuals learned to rely on signs such as dark sunsets and a halo about the moon to predict rain. Many weather proverbs have scientific validity; others are sheer superstitions. Meteorologists know that weather changes result from moving air masses. The directions in which the air masses move, along with moisture content and air temperatures, permit meteorologists to predict the weather conditions of a specific region for various intervals of time.

A solid comprehension of **physics** is essential to an understanding of the natural sciences. Its systematic development began some three centuries ago with the studies undertaken by Galileo, Kepler, and Newton as they sought to understand the mechanical motions of various visible objects. Physics continued to develop as dedicated researchers became interested in electrical and magnetic forces as well as in the nature of light of all sorts. This fundamental part of the natural sciences employs logic and mathematics on the one hand, and it leads to practical applications of science and technology on the other.

Physics is concerned with the basic laws of nature. The physical world in which we live appears to function according to basic fixed principles. Physics, then, holds the key position because it is concerned with the most fundamental aspects of matter and energy along with how they interact to make the physical universe function. Astronomy, geology, and meteorology rely on basic physics principles to interpret the meaning of observations and to make accurate predictions. Chemistry employs modern atomic theory, a branch of physics, to predict atomic structures and interactions between atoms. Although life may be inexplicable in terms of physical principles alone, living things are composed of matter and transform energy; thus, biology, too, must consider those applicable facets of physics.

Physics deals with phenomena that pertain to all classes of matter. One major division studies matter in motion. This area of study considers falling and projected objects, rotating bodies, and so forth. Another major division of physics investigates matter at rest, which may involve the designs of buildings, bridges, and highways.

The study of physics also includes the different kinds of energy that interact with matter, including electricity and magnetism, heat, light, and sound. From subdivisions of these major areas of study has grown modern physics, a branch of physics that includes electronics, nuclear physics, and other contemporary fields.

Science for Everyone

In a sense, we are all scientists because we have learned about the laws of nature through personal experiences from the time we were born. Obviously, countless others have made the same discoveries we have made. However, most individuals have not experienced the excitement of a truly original discovery.

Because of the complexities of many problems being researched today, most people, including many scientists with narrow specialties, lack the personal experience necessary to comprehend in depth many concepts that are being studied. However, once the analytical study of observations about a problem is completed, the researcher generally explains the results in the form of basic theories readily comprehended by others.

Many people have developed a distaste for science. Some fear science because they feel that it endangers our environment. Others blame science for the destruction they feel it brings about. These people fail to distinguish between science and its application. The role of mathematics in science adds to the dismay of many people, but scientists prefer the presentation of a natural law in mathematical terms to point out the precise relationships between concepts involved.

Regardless of individual feelings, however, the fact remains that our society has advanced to the point where scientific literacy is imperative. One of the most important goals of liberal education is to develop in our citizens an understanding of the natural world. This natural world is the world in which we live. How it came to be, how it works, and what is likely to happen to it should be of major concern to each of us. Study of the natural world can be fascinating and satisfying as well.

Summary

Science is an integral part of our lives. Through developments by scientists, the conditions in which we live are constantly undergoing changes—some beneficial, others harmful.

Science as a means of knowing about our natural world dates back to the civilized peoples of ancient times, particularly those of Early Greece and Rome. Their interpretations of natural phenomena are generally different from those of today, but without their early attempts at understanding nature, modern science would have had no foundation on which to build.

As we learn to interpret those things that we observe, we constantly change the conceptual models that represent our understanding. Our knowledge of the universe deepens through these revisions.

Science may be defined as the systematized knowledge derived from observation, study, and experimentation carried on to determine the nature or principles of what is being studied.

As scientists gather information, they attempt to establish order with their interpretations. This order often leads to a conceptual scheme that in turn fits into a specific pattern. This pattern may lead to a new area for study, and a greater understanding of our natural world results.

Science and technology are related, yet distinctly different. Science, through research, learns about the natural world and interprets its findings; technology basically develops practical applications of scientific knowledge and makes new products and processes available to society.

Science can be divided into many areas of study. Natural science is divided into the biological sciences and the physical sciences. The physical sciences include astronomy, chemistry, geology, meteorology, and physics. Physics is considered the physical science around which the other sciences are built.

We live in a technological world, one that will continue to advance as new scientific knowledge is obtained. In order to understand the role technology plays in our lives, we must understand the basic laws of nature. Our ability to offer assistance in directing technology is severely restricted without this comprehension of the natural world.

Questions and Problems

Early Science

1. When Galileo invited several scholars to view the surface of the moon for themselves through his telescope, they declined his invitation. In effect, they were admitting that they could not or would not believe their eyes. How do you think today's typical person would respond to such an invitation? Defend your answer.
2. Why did the ancients often depend on myths to explain natural phenomena?

The Search for Order

3. Even though various models about our natural surroundings may not be totally accurate, why are they still of some benefit?
4. Briefly discuss three of Aristotle's major contributions to science as presented in the chapter.

The Definition of Science

5. Science is both self-correcting and self-destroying. How would you interpret this assertion? Provide concrete examples to accompany your interpretation.
6. Science is considered both a process and a product. Can you provide an analogy relative to industry?

Conceptual Schemes

7. What is a theory? How does a theory differ from a law?
8. How do statistical data help us solve problems about our natural surroundings?

The Attitudes and Methods of Science

9. How does the consistency of nature affect our ability to gain greater knowledge about our surroundings?
10. Why do many researchers develop a scientific method as they attempt to solve a problem?

The Differences between Science and Technology

11. What areas, if any, do you think should be restricted in terms of research and intellectual inquiry? To what areas would you like to see research given higher priority?
12. Should we cease our continued investigations of the macroscopic and the microscopic worlds? Defend your answer.

The Physical Sciences

13. Define astronomy. How does this physical science differ from astrology?
14. How does the research of today's chemists differ from that of chemists of the Middle Ages?
15. Define the physical science known as physics.
16. How do meteorologists predict changes in our weather?

Science for Everyone

17. Do you believe UFOs exist? Defend your response.
18. According to the current news media, what science-oriented controversies are in the limelight?
19. Suppose scientists discover a totally effective chemical that would destroy all bacteria, germs, and insects. Would you recommend that this substance be universally used as soon as possible? If not, what restrictions would you place on its use?
20. Science has been likened to a Prometheus bringing gifts and also to a monster like Frankenstein's. Can you clarify each of these assertions?

2 Science before the Renaissance

Chapter outline

Key terms/concepts

The Parthenon as it stands atop the Acropolis. The structure was constructed during the fifth century B.C. and was used by temple priests as they translated the manuscripts of many civilizations into a common language and then maintained them for scholars to study.

It is impossible to tell precisely when an actual study of the natural world began, but one of the most important advances that humankind ever made was the control of fire, accomplished about a half million years ago. Another achievement of foremost importance was the invention of writing.

When *Homo erectus,* the ancient of ancients, learned to make and use fire, humans became the absolute masters over all other life forms on our planet. Humans were not only able to stand up against the greatest predators of the time with the first control of an energy source, but they were also able to pierce the darkness of night and to warm themselves when it was cold. They also discovered that many of their foods were tastier and easier to chew and digest after the chemical changes that result from cooking. Eventually, humans would use fire to bake clay into bricks and pottery, to secure various metals, and much later, to operate steam-powered machinery.

Figure 2.1 Arrowheads, scrapers, and blades that were fashioned by the ancients during the Stone Age. These points were made by striking one fragment against another until the desired shape was obtained. The two points in the upper right were believed to have been used by the Cro-Magnons; the balance were tools of the Neanderthals.

The Beginnings

Over a hundred thousand years ago, Neanderthals (*Homo neanderthalensis*) developed rudimentary knowledge of botany and zoology. It is assumed that they could identify various trees and plants and knew that the leaves and berries of certain botanical species could cause sickness and even death if eaten. They also must have observed that the internal structure of humans, saber-toothed tigers, and horses was reasonably similar. Realization of the close similarities of the internal organs of many animals that differed so widely in their external appearances must have provoked a few of the more intelligent to speculate on the kindredness of all animals.

The habitat of the Neanderthals apparently ranged throughout Europe, the Middle East, Africa, and Asia. One branch of these adventurous people was known as the *Mouseterians,* a society that advanced far beyond their predecessors. These people were cave dwellers and utilized tools of various sorts unknown to other tribes. The Mousterians fashioned scrapers, axes, and spear tips from skillfully chipped pieces of flint, bone, and wood as they stalked the various forms of animal life around them, including mammoth, bison, deer, and giant bear. A variety of Stone Age points is shown in Figure 2.1. Apparently, several households lived and hunted together, realizing the benefits of such collective endeavors. Such conclusions are reached by analyses of the enormous quantities of bones uncovered at various archaeological sites.

Another culture appeared perhaps 19,000 years ago, during the period in which glaciers retreated as climates warmed. The *Magdalenians,* named after the area in west central France where many artifacts such as spears, barbed fishing hooks, and harpoons attributed to them were found, apparently were closely related to the prehistoric race called *Cro-Magnons,* after the cave in France where considerable amounts of their remains were discovered. (The Cro-Magnons are thought to have thrived between 30,000 and 10,000 years ago.) These individuals refined their level of existence by developing needles and harpoons. They also learned to construct temporary shelters when they ventured away from their cave homes in search of food and other necessities, such as wood and water.

Although modern races had been in existence for many thousands of years, between 10,000 B.C. and 7000 B.C. many radical changes in life-styles occurred. As the climate became more temperate, adjustments in comfort needs resulted. The fertile areas of the Near East yielded wild grasses, such as wheat and barley. These nutritious plants were cultivated, and their seeds transported to other geographic areas. Many wild animals were domesticated and put to work assisting in transporting the harvests, as well as serving as a readily available source of food. A new age, then, was born as humankind finally settled in specific locations, bred and tamed the creatures they formerly hunted, planted crops regularly, and constructed more permanent homes. Now there was time for a period of cultural enlightenment to blossom. True craftsmen appeared, as evidenced by the woven baskets, the stone vessels, and the baked clay pottery discovered in various excavations of sites that date no later in time than 6000 B.C.

Also, the people of antiquity associated the celestial bodies with divine nature and worshiped them according to the degree of power the various gods or goddesses watching over them were assumed to possess. Over the centuries, worship of these deities increased, so that in ancient Greece, Egypt, and other centers of civilization, the practitioners of astrology systematically observed and studied the visible planets and stars.

In ancient times, the lands relatively close to the Mediterranean Sea, including the valley between the Tigris and the Euphrates rivers, had been explored by various adventurers. This particular region was settled by those people who had learned the values afforded by sound agricultural practices and had sought

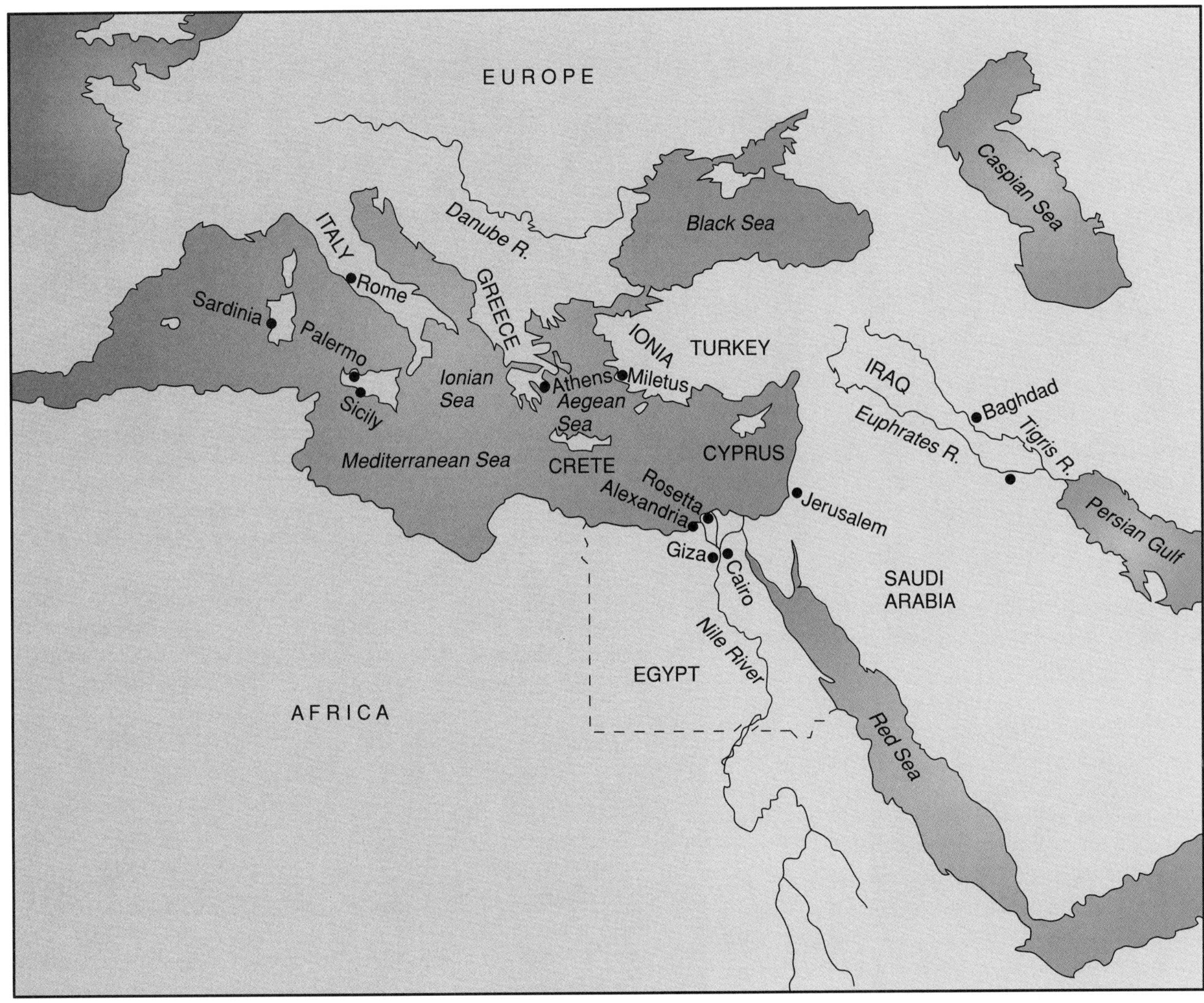

Figure 2.2 Ancient civilizations appeared to flourish along most all major bodies of water that surround the Mediterranean Sea. Ionia and Mesopotamia have been incorporated into other countries, yet most ancient cities remain.

more fertile lands to improve the living conditions for their families and those generations to follow. The ancient Greeks called the explored part of this crescent-shaped area "The Land Between the Rivers," or *Mesopotamia* (*mesos* meaning "between" and *potamos* meaning "river"). As noted in Figure 2.2, this region presently constitutes a large portion of modern day Iraq. The land slopes gently from Baghdad to the Persian Gulf. The rivers thus flow slowly, constantly depositing large quantities of fertile soil to replenish the soil that erodes away during flooding. However, the area had apparently been threatened by long periods of drought since it was first settled. The ancients had to develop means to irrigate their fields or face devastating famines. They constructed reservoirs and canals out of sun-dried bricks—projects that provided them with water for their fields and villages until they were destroyed by the Turks some centuries later. A scene considered typical for that period of time is depicted in Figure 2.3.

The early settlers in the Tigris-Euphrates valley also used similar bricks to build their homes, bridges, and protective walls. Even today, the ruins of some of their architectural achievements remain as monuments to the concentrated efforts of these civilizations.

The new cultures built strongholds that eventually became cities, the earliest of which presumably was in southern Mesopotamia. However, little remains of these settlements. Among the more famous was the city of Ur, founded in the fourth millennium B.C. This city supposedly was destroyed by an unusually devastating flooding of the Euphrates some several thousand years later. It was the site of the *ziggurats,* gigantic step-pyramids that supported the moon goddess Nanna, Ur's patron deity. Another ancient city, founded in the sixth millennium B.C., was called Lagash. This city became most famous under Gudea about 2125 B.C. The ruins of such cities clearly indicate that various cultures of the

Figure 2.3 Life in Mesopotamia was improved by the agricultural projects of the ancients, such as the irrigation projects depicted in this scene.

Mesopotamians were aware of basic architectural forms, such as the arch, the column, the dome, and the vault. Their comprehension of such structural designs illustrates the degree of advancement in mathematics and other natural sciences that they had made long before such feats were duplicated in other major civilizations.

The actual area between the two rivers was settled by the *Sumerians,* a culture that appeared prior to 3000 B.C. They undoubtedly constructed the cities Ur and Lagash, since this culture significantly advanced beyond that of the Semitic peoples who settled to the north. For instance, by 3000 B.C., the Sumerians had drained the marshes, irrigated the fields, and domesticated cattle and oxen, donkeys, sheep, and goats. They had also designed wheeled chariots, pulled by the various animals they had at their disposal. Later cultures used such chariots as weapons of war (note Fig. 2.4).

The Development of Writing

As the first great civilizations came into being, the need for means of communication other than oral and hand signals arose. Prior to the invention of writing, knowledge had to be passed on to others by word of mouth, a practice vulnerable to inaccuracies. It was vitally important that new generations know not only about the various gods of a particular civilization, but also about the lineage of great peoples, their differences, and their similarities. Furthermore, as the various cultures progressed, it became very necessary to have accurate and up-to-date information concerning such things as how much grain was needed to last the city-state through the winter, what residents had not completed their contribution of taxes for the year, how many men were available for construction projects, and which men could be conscripted for military service in the event of war. In addition, there would have been numerous reasons to accurately communicate with distant cities. For instance, trade routes were established between the various cities. A list of available items, raw materials, and the like had to be furnished to those who arranged the trades. Some description of the items offered for trade, along with an estimation of their worth, had to be established. With the advent of writing, scribes could also keep a day-by-day record of vital data in multiple copies as necessary.

Figure 2.4 A scene of battle, inscribed on a metal sheet by an ancient artist. The strength of invading armies was determined by the number of chariots used during the conflict.

As far back as 35,000 years ago, Cro-Magnons (*Homo sapiens*) learned to draw pictures that represented objects. Eventually, someone learned to make pictures that represented ideas, and finally, about 4000 B.C., some group of individuals developed a system of symbols that represented sounds. Writing apparently originated with the Sumerians about 6000 years ago, when they developed **cuneiform writing,** *wedge-shaped characters pressed or carved on clay tablets.* It is likely that the priests developed a system in order for them to inventory the surplus grain and other items that were stored in the temples for the future use of the people they served. Undoubtedly, their system was in the form of picture-symbols, such as symbols that represented a bag or a multiple of bags of wheat, ears of corn, numbers of various types of animals, and so forth. Evidence for the existence of such a style of communication is shown in the cuneiform writings in Figure 2.7.

It would be safe to assume that as time passed and needs developed, the systems became more precise as the scribes improved their confidence in communicating with each other. The earliest Egyptian hieroglyphs and the first forms of Chinese writing were picture-symbols, too, but the ingenuity of the Sumerians permitted sounds to be meaningfully associated with their system. Their rendition of picture-symbols also was adaptable to numerous languages and became the writing style used throughout the Middle East for centuries.

Apparently, the written communication of the priestly scribes of Mesopotamia and Egypt initially involved those disciplines that they had personally developed as they performed their duties. Much

Unraveling the Past

The excavations of the cities of Pompeii and Herculaneum, buried by the volcanic eruption of Mount Vesuvius in A.D. 79, began in 1738. This project marked the first of many that were to show how ancient civilizations could be studied and heralded the science of archaeology.

In 1799, during Napoleon's invasion of Egypt, a French officer stationed at a fort near Rosetta, a town on one of the many mouths of the Nile River, discovered a fine-grained black rock that had unusual characters carved on its surface. The rock, composed of basalt, came to be known as the *Rosetta stone.* A model of this invaluable relic from the ancient past is shown in Figure 2.5. It had three inscriptions that pointed out the capability of the ancient Greeks to decipher earlier writings of the Egyptians. The three writings, starting at the top of the stone, were found to be hieroglyphyic Egyptian priestly writings and pictures; hieratic (an abbreviated form of hieroglyphics, but Greek in origin); and demotic (characters representing popular script, which proved to be a simplified Egyptian version similar to the colloquial Greek writing of later times).

The Greek inscription proved to be a series of decrees associated with the reign of Ptolemy V around 196 B.C. The other two sets of carvings were correctly assumed to be the same message, but in different languages. Once the translations were completed, archaeologists at last had at their disposal a mechanism they could use to decipher many other carvings found at various excavation sites. By 1821, a French scholar had successfully transcribed the hieroglyphs and the demotic script into a more workable language. His efforts opened the way to decipher many other inscriptions found in the ruins of ancient Egypt.

Figure 2.5 A model of the Rosetta stone, a slab of basalt that played a primary role in the deciphering of ancient Egyptian inscriptions.

Later, a similar situation led to the successful transcription of the writings of the ancient Mesopotamians. Scholars found an inscription carved into the steep rocky side of a high cliff overlooking the village of Behistun in western Iran, apparently having had withstood the elements since about 520 B.C. The carvings were identified as the work of followers of the Persian emperor Darius I. The inscription told how the ruler had gained access to the throne and, to be certain that everyone could read the message, it had been translated into three languages—Persian, Sumerian, and Babylonian. The Sumerian and Babylonian writings were primarily pictographs, in use from perhaps before 3000 B.C. until the first century A.D. Examples of these cuneiform (wedge-shaped) scripts are presented in Figure 2.6. The translations of these messages required almost a decade to complete from their discovery in 1846. The dedicated efforts by those who transcribed the inscriptions permitted other scholars to unravel the history of the ancient civilizations that had lived between the Tigris and the Euphrates rivers.

Other expeditions into Egypt and Mesopotamia yielded other clay tablets from various ruins, and more information about the lives of the ancients was acquired. Only through such efforts do we know of such cultures as the Sumerians. Subsequent discoveries offered further proof that Homer's tales were not all legendary ■

Figure 2.6 Cuneiform writing, such as is shown on these tablets, supposedly originated with the Sumerians about 4000 B.C.

A Mystery of the Ancient World

Actual documentation concerning ancient civilizations is sparse. Even the events recorded in the Bible can be traced only to the reign of Saul, the first king of Israel, who is believed to have become ruler in 1025 B.C. However, strong legends were passed on by ancient Greek writers about the Trojan War of approximately 1200 B.C. and about a pre-Greek civilization on the island of Crete. According to historians, the people who were the earliest residents of the island of Crete (recall Fig. 2.2) represented a civilization whose achievements are considered outstanding even by scholars of the present. Ancient Egyptians called them Keftiu, but today they are known as the Minoans, after reference by Homer, the Greek poet, in his epic poems the *Iliad* and the *Odyssey* in the eighth century B.C., of their king, Minos. The culture of the Minoans apparently thrived from 3000 B.C. to 1400 B.C., and it represents the first authenticated civilization in Europe.

Adventurers who passed through mysterious Minoan Crete must have been enthralled with the beauty of the magnificent palaces, the villas, and the busy towns that dotted the mountainous countryside. Mighty fleets of sailing vessels set out to sea from the Minoan harbors, loaded with timber, pottery, and agricultural products. The ships visited Egypt, the Near East, and Greece, where they traded their cargoes for treasures of gold, ivory, and jewels.

The Minoans were excellent craftsmen and artists; they carved beautiful landscapes and scenes of dancing maidens on tiny stones, gems, and on the delicate pottery they made. They appeared to be a prosperous people, happy in their work and contented with their everyday lives. Then—silence. The Minoans, their origins unknown, their language alien to all others, and their fate obscure, faded away and were soon forgotten. Excavation of the ruins of their cities in the early twentieth century revealed evidence that perhaps a series of destructive earthquakes had struck (although the ruins showed they had survived and rebuilt after earlier earthquakes). Also, there were various portions of the uncovered Palace of Minos that indicated it might have been put to the torch. At this time one can only speculate as to what really led to the sudden demise of this entire civilization ■

Figure 2.7 Some artifacts indicate that the ancients maintained crude records of bartering that took place, perhaps between villages or people of various cultures.

of the early writings involved mathematics concerned with keeping accounts and surveying, astronomy for calendar-making and astrological prediction, and medicine for curing disease and driving away evil spirits. Not until later times were the chemical arts referenced, including instructions for the dyeing of various fabrics and other such practical applications of the knowledge they had obtained over many centuries of existence.

The Discovery of Metallurgy

Around 7000 years ago, in different places in the Near East, artisans discovered that heating a specific type of rock in a bed of glowing embers and coals caused the rock, an ore of copper, to yield this beautiful metal. Though copper is a very soft metal, they learned that it could be hardened by hammering its surface as it cooled. This hammering process caused crystallization to occur in the copper, a feature that significantly increased its hardness. Even so, because the copper tools and weapons were still soft in comparison to stoneware, implements made of the two substances were used in conjunction with each other. However, shortly after the technique to remove copper from its ore had been established, it was discovered that tin could be extracted from its ore by the same process. Although these ancient people did not understand the principles of chemistry involved in these transformations, they were really practicing this natural science. The knowledge known as **metallurgy** was considerably expanded around 3000 B.C., when some inventive people found that if a small amount of tin was added to molten copper, an *alloy,* or mixture of metals, called *bronze* was formed. This metallic substance had a hardness much superior to that of either copper or tin alone. Bronze was used to make all kinds of implements and treasured jewelry, but its greatest impact was in the fabrication of weapons. Shields and armor made of bronze withstood the blows of inferior copper weapons. The swords and axes had a much sharper edge than copper or stone weapons. Soldiers equipped with weapons other than bronze were no match for armies that carried the new metallic weapons and armor, as the Egyptians discovered when the bronze-wielding Hyksos invaders defeated them in 1750 B.C.

The next great advance in metallurgy came when smelters discovered how to extract iron from its ore. Iron is highly superior to bronze in strength, and about 1200 B.C. the Dorians, a barbaric Greek tribe, equipped with the new iron weapons, conquered the more organized but merely bronze-equipped armies of Greece. (Presumably without their knowledge, the Dorians' weapons were constructed of varying grades of steel, since some carbon in the smoke of the glowing charcoal used in the smelting process undoubtedly mixed with the iron to form a stronger, more suitable metal than iron alone.) By early in the so-called Iron Age, the basic equipment of most soldiers included arrows tipped with iron (or steel) points, sturdy daggers and swords, as well as sharp, penetrating, and massive axes. Protective armor, such as helmets, chest protectors, and hand-held shields, paralleled the development of weapons. The armored headgear of various designs permitted soldiers to identify friend or foe, a procedure followed somewhat even today. Those civilizations who failed to equip their warriors with iron weapons quickly perished as the great Bronze Age came to an end.

The **metalsmith** was held in varying degrees of esteem. On one hand, his appearance was one of low social order, since his clothes were often soiled and scorched, his hands were calloused, and his place of business was probably in some remote area of the village, simply due to how the operation was conducted. In contrast, the things he designed were considered useful, generally almost priceless, and often beautiful. He appeared to have the god-like ability to alter the very nature of matter. He could convert dull rocks into gleaming metal and had the skill to melt or solidify the metal as well as to make it pliable or rigid. The metals, then, remained both fascinating and mystical to most people. Philosophers, particularly the Greeks, theorized over what occurred as metals were heated and mixed together by the metalsmiths. The observations these scholars made led them to express their ideas about atomism and the elemental nature of matter.

Ancient Cultures on the American Continents

During a portion of the time the Egyptian and Babylonian cultures were dominant, other civilizations existed in parts of Central America and in Mexico. Although the origins of these peoples remain somewhat vague, perhaps at least some groups ventured across the land bridge that reportedly connected northeastern Siberia with the North American continent around 11,000 B.C. Most of them migrated southward, until they were free of the icecap that covered North America, in search of suitable lands on which to settle. Cultures became diversified, and residents of Mexico and Central America became among the best of architects and metal workers. The warm climate was conducive to the types of plants they cultivated: beans, chili peppers, and squash. Between 5000 and 3500 B.C. mutant forms of maize grew among wild forms of grain. Pottery that dates back to 2300 B.C. has been unearthed, but cultural advances apparently were slow. The natives of South America, presumed to be ancestors of the Incas, located along coastal plains and migrated onto the high plateaus of the Central Andes around 3000 B.C. Those who remained near the coasts learned to depend on the seas for their needs, along with the basins that provided rich supplies of clay, stone, copper, silver, and gold. Others who were tempted to venture to the higher elevations seemed content to lead a simple life, domesticating animals, such as the llama and alpaca, and cultivating vegetables, such as potatoes, in gardens that sloped as much as 60 degrees. There still stand the remains of great building projects that must have used a work force that numbered into the thousands, probably structures dedicated to their gods of the sun, of thunder, and of war.

Figure 2.8 Ruins from the Mayan civilization, probably constructed about 300 B.C. Note the steps that lead to the structure assumed to have been a temple.

The *Mayas,* who settled along the Yucatan Peninsula about 300 B.C., were among the best farmers of their time, cultivating corn and growing cotton. They are noted as a culture that settled down to one region, built a city (see Fig. 2.8), and assigned duties to members of their society. There is existing evidence that some members were water carriers, some corn growers, and others who cleared away the forests to make room for the settlements and agricultural areas. Those selected to be priests were among the powerful, and they worshipped gods of good, evil, economy, food, and so forth. The Mayas are noted for their accurate measurements of the solar year and the lunar month, as well as for the development of an accurate calendar. They also developed a number system that bears some similarity to the one we use today. Unfortunately their civilization underwent an unusually rapid rate of collapse, presumably due to numerous reasons, including intellectual exhaustion, climatic changes, disease, and even wars. Modern researchers, however, suspect the most serious difficulty that confronted them was a breakdown in their agricultural system. There is existing evidence that they burned away the foliage to gain more fertile soils until nothing remained but the grassy plains. Since they lacked the proper implements to cultivate the hardened regions with their grassy root systems, their agricultural practices failed. Of the estimated 2 million who survive today, it is noteworthy to mention that it is not unusual for their offspring to have a Mongolian spot, a blue patch of skin located near the base of the spine that disappears when the child is about 10 years old.

Early Science in India and the Orient

The beginning of civilization in India can be traced to the valley of Indus, dating prior to 3100 B.C. The early cities were situated in the area that is now Pakistan. There is evidence of early diversity among the cultures, but each group basically displayed considerable skill in planning for construction of their cities and overall economy. These cultures were exceptional in how they solved engineering problems, including the construction of drainage systems, public baths, paved streets, and so forth. In fact, scholars who study this region contend that various cultures in the specific area advanced beyond those of the same period in Mesopotamia. For instance, a Hindu lunar calendar has been discovered that apparently was developed in 3101 B.C., and there is considerable proof that decimals were in use about that time. Among the most famous people who lived during this period was *Buddha,* "the enlightened one," son of a ruler who lived along the Ganges River located in east-central India in 500 B.C. Buddha's influence led to the development of an early atomic theory that attempted to explain why things exist for but a short time. According to this theory, things are replaced by similar things, and thus objects are really composed of a series of similar objects that exist in a perpetual order. In other words, this atomic view explained how and why things change; for example, why children or plants grow. With the advent of Buddhism, monasteries and shrines were built from stones removed from rock quarries, such as those that form the caves on the island of Elephanta in Bombay Harbor. Interestingly enough, there is strong evidence to support the claim that our modern scheme of numerals was developed in India between 200 and 300 B.C., and thus it is not "Arabic" in origin, as originally thought.

Very little was known by Western cultures about the historical development of science in China until the 1960s. Perhaps this void of knowledge was largely due to the lack of scientific awareness in those scholars who first were involved in language translation. As a result, some authorities credit the Chinese with originating numerous scientific concepts and inventions, whereas others maintain that they made few, if any, notable contributions.

As studies continue, active scholars compare the culture of ancient China to that of Babylonia, particularly at about 1500 B.C. The earliest civilization appears to have settled along the Yellow River at Anyang. Here, specialized laborers worked with bronze, developed the *potter's wheel* for working with soft clay, cultivated and grew rice in the swampy regions, and bred silkworms to fabricate cloth garments and beautiful tapestries.

Ancient China was diversified with the formation of numerous *dynasties,* relatives who banded together under a common ruler. There apparently were many wars between these groups, and superior ones, such as the Shang dynasty of 1000 B.C., conquered and ruled large regions. Iron apparently was first used by the Ch'in dynasty around 513 B.C. This culture built waterworks for home use and irrigation and constructed the original Great Wall, about 2400 kilometers long, to protect most of China from the Mongolian invaders. The section of the wall depicted in Figure 2.9 is typical of the entire construction project. A relative of the

a.

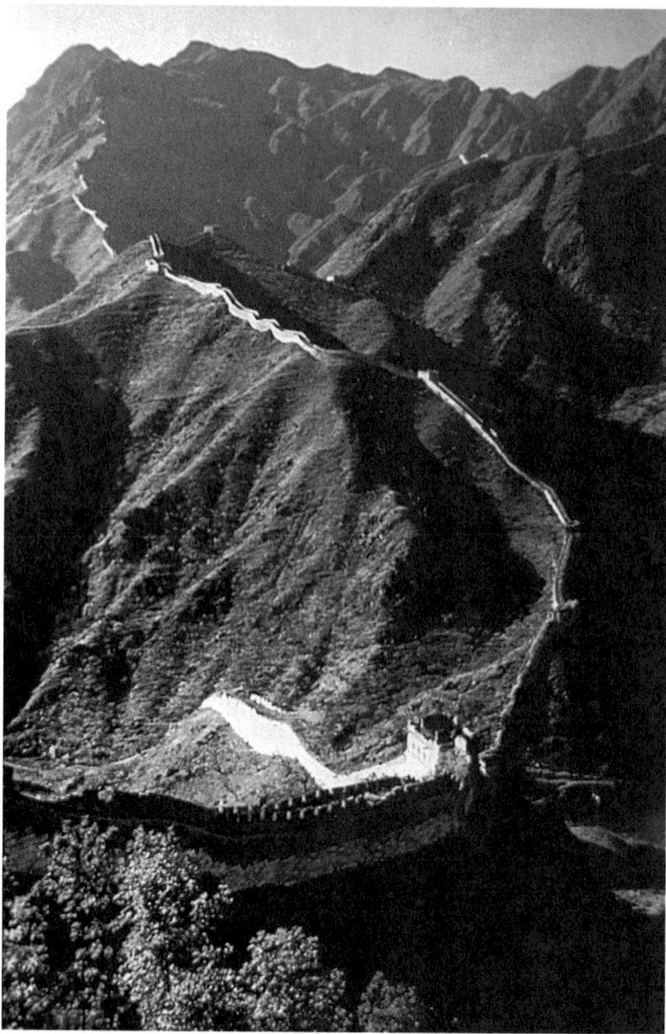

b.

Figure 2.9 (*a*) A section of the Great Wall of China, constructed during the rule of the Ch'in dynasty to protect all of China from invasion by the Mongolians. (*b*) The difficult task of constructing the Great Wall is apparent in this view of the desolate mountain peaks.

The Importance of the Circle

About 5000 years ago, the Egyptians discovered the *constant relationship between the circumference of a circle and its diameter.* The value of the constant is what we know today as π (**pi**), or 3.14. Prior to the determination of the constant, π, both the Egyptians and the Sumerians are assumed to have had access to the lever and the inclined plane. Undoubtedly, these simple machines played an essential role in the construction of such projects as the pyramids and the ziggurats.

At the same time the Egyptians were working with geometry, the Sumerians were beginning to use the wheel, a practical application of the circle. The invention of the potter's wheel allowed the Sumerians to shape their pottery symmetrically and to considerably decrease the length of time it took to produce pottery. They also were baking their pottery in ovens rather than over open fires. Fortunately for the Sumerians, the use of the wheel did not stop with pottery making; rather, some brilliant Sumerian attached the wheels to a boxlike structure to construct the first wheeled chariot. The Sumerian soldiers, clad in and armed with bronze and riding in wheeled chariots, were greatly feared and extremely successful on the battlefield ■

Ch'in dynasty conceived the Han dynasty (202 B.C.–A.D. 220), the culture that founded the Imperial University and then set up a bureau of scholars to rule the empire. The faculty first recorded their findings on bamboo slips, then on silk tapestry, and eventually on paper, invented about A.D. 105. (Specimens have been found that date back to about that time.) In addition to inventing paper, the Han dynasty is credited with other notable scientific discoveries. Their contributions include the study of magnetic materials (such as the compass); the method to cast iron; the development of simple machines, including the use of belts and pulleys; the procedure to construct a bellows to produce higher temperatures from glowing coals; the design of a water-driven mortar and pestle; and the construction of a vertical water wheel.

The various dynasties of the several centuries before the Christian era are also credited with developing the great philosophers, known as the legalists, the logicians, the Mohists, the Taoists, and the Confucians, according to the areas of concentration of study or the individual most prevalent among them.

Chinese *alchemy,* like that of many other civilizations, concentrated on discovering the pill of immortality (also known as the magic elixir), on developing the method to convert relatively inexpensive substances into gold, and on continuing the search for order among the various substances of which Earth was thought to be composed.

Figure 2.10 The Great Pyramid of Khufu was built from an estimated 2.3 million limestone blocks, each having an average mass of 2200 kg. According to the ancient Greek historian Herodotus, a workforce of 100,000 people built the structure. The pyramids apparently symbolized the primeval mount from which the world was formed; they all face the cardinal points of the compass.

The Soul of Science

If it were possible to observe ancient Egypt as it existed some 5000 years ago, there would be few, if any, among us who would not be fascinated with the accomplishments of individuals such as *Imhotep.* He is considered to be the founder of Egyptian medicine and served as personal physician to King Zoser. Later, Imhotep was accorded divine status as the patron god of medicine. Other great accomplishments are attributed to him, including designing and presumably directing at least a portion of the construction of the earliest step-type pyramid, located south of Cairo, to be the site of his king's final resting place. The achievements of this individual and his followers as they moved into place the gigantic blocks of stone transported from the rock quarries located so far away are still considered among the most marvelous tasks ever performed. **The Great Pyramid,** *the largest of the three pyramids located at Giza,* was built about a century later to hold the tomb of Khufu(Cheops) (see Fig. 2.10). Even with the availability of knowledge about some of the various simple machines, there can be only speculation as to how these great limestone blocks were moved from the quarry, much less lifted to their respective positions as high as a modern building of about 50 stories. One must strongly suspect that the grueling tasks of moving and lifting the stones were assigned to the same slaves who tamed and tended the domestic animals, built the various palaces and temples, made iron weapons and cooking utensils, and fashioned the gold jewelry worn by the royalty they served. These individuals undoubtedly even played the music and labored over the paintings enjoyed by the priests and nobles, those citizens who represented the higher level of society and accounted for perhaps a very small fraction of all residents of ancient Egypt. Life for the overwhelming majority of Egyptians was short and often plagued with disease, hunger, and brutality. Fear of natural catastrophes, such as floods, storms, and droughts, along with the constant danger from invading armies, was compounded by threats imposed by their ritual leaders who represented their many gods and goddesses.

Slowly, ever so slowly, small groups of individuals settled in the vast river valleys of India, China, and other fertile regions in Egypt and Persia. Life in these settlements became more secure, allowing for additional time to be spent in improving existing conditions. For these reasons, such areas became centers for the development of many useful inventions and valid scientific principles, although for the most part, the actual individuals responsible were never credited in the recorded literature of the time.

Another major portion of the soul of science was also spun slowly, particularly in the ancient world of *Ionian Greece,* located along the eastern shore of the Mediterranean Sea, and in Miletus, the ancient city built on the western coast of Turkey, now lying in irretrievable ruins. In such locations, scholars began to ask questions and to seek natural answers that excluded the previously unquestioned explanation that supernatural gods and demons were responsible for what was observed.

The **philosopher** (*a reflective thinker*) named *Thales of Miletus* (625–545 B.C.) is considered the founder of Greek science, mathematics, philosophy, and reportedly other academic areas. He obtained his wealth of knowledge and fame through his travels, including numerous trips to Mesopotamia and Egypt. *Anaximander,* a student of Thales, helped introduce the sciences of the ancient East to the Greeks. He is credited with making practical use of a sundial and with constructing one of the earliest relatively accurate maps of Earth. (Anaximander's thoughts about earthquakes, magnetism, and solar eclipses will be discussed later.) The speculative wisdom of this early Ionian natural scientist also was evidenced in his belief that all things in the universe are composed of a vast undefined substance in existence throughout eternity and from which will be formed more stars and planets. Another Greek philosopher, *Anaximenes,* who lived shortly after Thales, expressed his belief that air was the fundamental element of the universe. He contended that air, when compressed, became water, and further compression converted air into solid earth. Rarefied (less dense) air, according to his way of thinking, heated to become fire. This dedicated scholar of his natural surroundings felt he could distinguish between stars and the wandering pinpoints of light (the planets), and considered rainbows to be natural phenomena, not a beautiful display created by some contented goddess.

Xenophanes, late in the fifth century B.C., theorized that Earth changed in physical characteristics with time. He felt that mountains must have been covered by oceans at various periods in the past, thus explaining the life-forms he observed in the rock formations, and that the mountains rose higher than the seas as Earth underwent various physical changes, exposing the ancient seafloors. Still another Greek philosopher, *Pythagoras* (560–480 B.C.), is said to have traveled extensively throughout Egypt and numerous eastern countries. Perhaps a student of Anaximander (or even Thales), his travels helped him spread the philosophic views and scientific traditions of his teachers throughout Greece. He and numerous friends and fellow scholars pursued mathematics and the science of astronomy. The ancient Greek was also fascinated with how various sounds could be created, and he essentially proposed the laws that apply to stringed instruments. He developed a musical scale that is of considerable value even today. We know him best, however, for his mathematical contributions, specifically for the *Pythagorean theorem* that deals with right triangles. This proposition was included as the forty-seventh proposition of the first book of *Elements,* the treatise on geometry written in the third century B.C. by the Greek mathematician, *Euclid.*

The soul or the very basis for the natural sciences had been born. The daring first scientists had overcome the fear of the wrath of the gods and demons that traditionally had threatened their rights to question. They had fled to places where they found the freedom to study and to share their views with others. As one would expect, many conclusions the ancients reached were not totally valid, for, as with most generalizations, there are always exceptions and the need for further experimentation to develop accurate models of our natural world. Greek science did not conclude with the Ionians. In Athens, during the third and fourth centuries B.C., one could have discussed science and philosophy with such scholars as Aristotle, Plato, and Socrates. These individuals, however, became less interested in the physical theories contributed by other ancient Greeks and concentrated more on the moral, aesthetic, and metaphysical world. Experimental science, the best route to validate presumed truths, was apparently not of major interest to these Greek scholars who could have made significant contributions to the natural realm of the physical world about them. Could they have discouraged, rather than promoted, the more empirical science?

Summary

Perhaps we will never know precisely when the ancients fully became aware of their natural surroundings. The Neanderthals apparently extended the variety of plants, animals, and the like that could be safely eaten, compared to the limited diet of their ancestors. They are thought to have recognized the similarities and differences between the various forms of life they encountered. Apparently, they were brave, adventuresome individuals, for they explored and settled unfamiliar geographic areas.

The cultures in existence up to about 10,000 B.C. were characterized by a nomadic way of life and developed implements they chipped from stone. During the period from 10,000 to 7000 B.C., tools were significantly improved and the domestication of some wild animals was accomplished. Various plants were cultivated, and seeds from these plants were probably traded between certain cultures. There was a tendency for most of the civilizations to settle in certain locations where further development of specific skills became of prime importance.

Other cultures came into existence, such as those located between the Tigris and Euphrates rivers. Apparently, necessity led to the development of successful engineering feats, and life in some communities improved at a rapid pace. Still, there always remained the dangers from flooding, invasion, disease, and starvation.

Advancement in mathematics and the natural sciences varied from one civilization to another. The Sumerians seemed to have advanced in these categories more rapidly than other cultures. Their use of bronze implements undoubtedly contributed greatly to the progress they made. The invention of writing by the talented Sumerians about 6000 years ago, and the consequent recording of the numerous advancements they made, permitted later scholars and researchers to more fully comprehend what life was like in their communities.

There is little doubt that further excavations will yield even more information about the ancient civilizations. The copper-stone phase of cultural development occurred in the Near East about 5000 B.C., when copper was first obtained from its ore. The discovery of bronze, an alloy of tin and copper, probably changed civilization more than any previous single event, since better implements for agriculture, homes, and even war shortly became available. The extraction of iron from its ore around 1200 B.C. was an even greater accomplishment. Those individuals skilled in working with these metals were in high demand and undoubtedly were enticed to leave their homes for places where their services were better appreciated and where substantive rewards abounded.

Various civilizations eventually migrated onto the American continents, presumably across the land bridge that connected eastern Siberia with North America. These individuals may have been hunters in search of great herds of animals that would furnish the needed food for their communities. Most of these adventurers found their way into Mexico and Central America, where cultures such as the Incas and Mayas originated.

The ancient Indian culture apparently originated in the valley of Indus prior to 3200 B.C. However, numerous cultures seemed to have appeared in relatively short order. These groups constructed cities that demonstrated their considerable abilities in planning and in solving the various engineering problems they encountered. There is existing evidence that they advanced beyond the more refined culture of Mesopotamia. Among their outstanding accomplishments was the development of a system of numerals, essentially identical to the one we use today.

Our scientific knowledge about ancient China is quite limited, since the translation of the Chinese language into a form more familiar to science-oriented scholars was not initiated until the 1960s. Apparently, the Chinese culture appeared along the Yellow River in about 1500 B.C. The work force in the various communities was quite specialized. Excavations yield pottery of eloquent designs, beautiful silk tapestries, and evidence that cultivated rice was a staple food. Early China is known to have developed numerous dynasties, families, and followers under a common ruler. Remarkable accomplishments are credited to the Shang, Ch'in, and Han dynasties, among others.

Egyptian culture came into being about 3000 B.C. The early achievements in this region of various individuals make for a long and impressive list. The means by which they constructed the magnificent pyramids remain as mysteries, and all who view them marvel at the accomplishment of the ancients. Their many skills resulted in treasures of beautiful art of all sorts.

Perhaps the soul of science, though, centers around the many contributions of the ancient Greeks. The list of scholars who either were native to Greek soil or who migrated to Greek cities to study is a long one indeed. Ancient Greece was the center of culture, since scholars from all over studied at the libraries and relied on the great wisdom of the Greek philosophers when confronted with situations beyond their capabilities to comprehend.

Questions and Problems

The Beginnings

1. What were the two most outstanding achievements attributed to the earliest civilizations?
2. Approximately when were all modern races present on our planet? How did the climate essentially govern when they appeared?
3. What major reasons prompted the various civilizations to explore and settle unknown regions, such as the valley between the Tigris and Euphrates rivers?
4. How did the ancients overcome the effects of droughts and the dangers from excessive flooding of river basins?
5. Why were the Sumerians considered an advanced culture compared to other cultures that existed about the same time?

The Development of Writing

6. What is cuneiform writing? How was it different from other forms of written communication?
7. List five reasons the development of writing was so important to the ancients.

The Discovery of Metallurgy

8. Why were the cultures that developed bronze implements so superior to those that were content with their knowledge of stoneware?
9. How did the metalsmiths increase the hardness of a metal such as copper, then further improve its degree of hardness?
10. How did the discovery of a method to extract iron from its ore affect the armies of various civilizations?

Ancient Cultures on the American Continents

11. Many of those individuals who migrated onto the North American continent apparently moved continually southward into Mexico and Central America. What reasons might have brought about their further migration?
12. According to the information gathered during excavation of Mayan sites, what appeared to be the basic foods of the Mayans?
13. What evidence led investigators to conclude that the Mayans are direct descendants of the Mongolians?

Early Science in India and the Orient

14. Briefly discuss the atomic theory as attributed to Buddha and his followers.
15. Why should the "Arabic numerals" as we know them be credited to another culture?
16. What is meant by a dynasty? What effects did dynasties have on the rest of society?
17. What effects did the construction of the Great Wall of China have on the various civilizations?

The Soul of Science

18. What mechanical devices did the Egyptians apparently use in the construction of the pyramids?
19. Why were the pyramids constructed?
20. How did the Greek scholars Aristotle, Plato, and Socrates apparently perceive the role of experimental science?

PART TWO

The scientific realm of knowledge is subdivided in a somewhat arbitrary fashion. The field known as physics concerns itself with matter and energy at the most basic levels. The fundamental character of physics limits the number of variables involved in most problems, allowing them to be solved more or less exactly with mathematics. Thus, while physics is in one sense the simplest of the sciences, this very simplicity often requires the use of complex mathematics. It is no accident that some of the greatest physicists were also great mathematicians. They had to invent mathematical tools to solve the problems they encountered in physics. (The quantitative nature of physics is illustrated in this text by using only elementary algebra.)

The present field of physics began to assume its modern form about three centuries ago with the work of Galileo, Kepler, and Newton. Although some observations and principles of physics were handed down from earlier ages, it was in the seventeenth century that the laws describing the motions of bodies were first codified. The experiments of Galileo shaped the modern scientific method, demonstrating that observation and careful analysis were crucial in understanding natural phenomena. Kepler's meticulous work in ferreting out the laws of motion of the planets served as a stimulus to consider the underlying principles that controlled the motions of the celestial bodies. This setting provided the backdrop for the inspired work of Newton, which laid the foundations for the present-day area of physics known as mechanics.

Explorations continued in other areas, with attention being given to the areas of light, electricity, and magnetism. The nature of light had always intrigued those who studied it, with arguments centering on whether light was a stream of particles or a wave phenomenon. Experiments conducted around the beginning of the nineteenth century showed that light behaved as a wave, but the exact nature of the wave was unknown. There seemed to be little connection between light and the forces associated with electricity and magnetism until about the mid-1800s, when the labors of many researchers culminated in the work of Maxwell. His treatises unified the principles of electricity and magnetism and postulated that light is an electromagnetic wave. Later experimental confirmation reinforced the theoretical framework that had been built.

Around 1800, advances in discerning the nature of heat were also being made. The law relating the pressure and volume of gases was discovered in the seventeenth century, with the effects of temperature being included in the early nineteenth century. There was, however, no clear understanding of the nature of heat, which was believed to be some type of fluid. The observations of Thompson (Count von Rumford) led to the idea that heat was a form of energy, an idea that had a strong influence on the eventual publication by Helmholtz of the law of conservation of energy. The area of physics known as thermodynamics assumed its modern form when Carnot published the law describing conversion of heat into mechanical energy in the early part of the nineteenth century. The understanding that heat is statistical in nature, due to the combined effects of the motions of many tiny particles, was made later in the nineteenth century by Maxwell and Boltzman.

Near the end of the nineteenth century there was a belief among many investigators that all of the "fundamental" laws of physics had been discovered and all that remained was

Physics: Matter and Energy

to work out the consequences of the laws with sufficiently powerful mathematics. This recurrent theme in the history of science has been shown to be wrong over and over again.

Two great revolutions occurred in physics early in the twentieth century—one dealing with the very nature of space and time and the other with the fundamental structure of matter. These two upheavals led to the theory of relativity and the quantum theory, which are the bases of what physicists generally call "modern physics."

Concern with the propogation of electromagnetic waves, which includes light, had led to the idea that these waves must be transmitted through some type of medium. This "ether," as the medium was called, would have had to pervade all of space and have had very unusual properties of elasticity. Experiments done near the end of the nineteenth century yielded results that cast doubt on the existence of the ether. However, it was the theoretical work of Einstein, who apparently did not know of the experimental results about ether, that finally led most scientists to abandon the ether theory. His work showed that space and time are not independent of each other, as they seem to be in ordinary circumstances. The theory of special relativity showed that Newton's equations must be modified for bodies moving at speeds near that of light. Additionally, Einstein's theory of general relativity demonstrated that the force of gravitation could be considered to be a geometric curvature of space. Einstein's theories were consistent with the equations of electricity and magnetism.

Quantum theory, the theory that deals with the structure of matter, traces its beginnings to a search for an explanation of the manner in which electromagnetic radiation is emitted by a hot object. Early researchers had worked only with the "bulk" nature of matter; that is, they had not been concerned with its microscopic structure. The discovery of electrons and protons (the neutron was discovered somewhat later) soon led to a model of the atom based on quantum theory that not only could explain the salient features of electromagnetic radiation, including X rays, but also even explain the chemical characteristics of atoms and the nature of bonding between them. The predictions of quantum theory have led to numerous applications ranging from transistors to nuclear magnetic resonance imaging.

In attempting to understand the nature of matter, researchers have designed experiments that probe ever deeper into a search for its fundamental building blocks. Current research on fundamental particles has led to the discovery of a whole host of new particles. Theories of the way in which these particles fit into nature's scheme are beginning to emerge, but the fundamental research is still in progress.

One of the great lessons that has been learned from the physics of the twentieth century is that any theory, no matter how firmly it becomes entrenched within the scientific community, has limitations. New discoveries made with increasingly better tools admonish us to keep open minds and to be willing to modify our viewpoints when new evidence, or better interpretation of old evidence, warrants such change.

3 Newton's Laws of Motion

Chapter outline

Key terms/concepts

The changes in speed and direction of any object (an automobile, for example) may be calculated from a knowledge of the forces acting on the object and its mass.

The middle of the sixteenth century was a period of turmoil in the field of science. The practice of witchcraft was widespread, even in Italy—the center of the cultural world at that time. Many other superstitious beliefs and teachings were also popular. No one was free to speak against the teachings of Aristotle (384–322 B.C.), the Greek philosopher whose word was irrefutable, or against the Church, whose doctrine conflicted with certain scientific concepts.

Matter, according to Aristotle, could be maintained in motion only as long as the force that caused movement remained in direct contact with the matter. Should the force cease or lose contact, the object would stop abruptly. For example, when a stone or other projectile left a catapult, the medieval equivalent of a cannon, the propelled object was believed to be maintained in motion by the air that streamed in behind it and thus maintained a continuous and physical contact with the catapult. In addition, Aristotle contended that the weight of a falling object directly affected its rate of free fall. A 50-pound object, for example, fell decidedly faster than a 5-pound object. Both contentions were beyond reproach and were almost universally accepted well into the seventeenth century.

Sixteenth-century universities also taught other concepts that we now know to be incorrect. These included the premise that Earth stood still, that it was the center of the universe, and that the sun revolved about Earth, as did other heavenly bodies. These concepts were also the beliefs of the Church. The Church widely influenced much official thinking and would not tolerate interpretations of Holy Scripture that deviated from its own established dogma.

At the close of the sixteenth century, Galileo (1564–1642), a scholar of physics at the University of Pisa, dared question many of the teachings of Aristotle. Galileo was the first investigator to use a telescope to study the heavens. He discovered and studied the moons of Jupiter as they revolved about the planet and concluded that the moon also orbited Earth. As his studies continued, he expressed his belief that the heliocentric model of the universe proposed by Copernicus (1473–1543) was more nearly correct than the model presented by Aristotle. Galileo is also reputed to have taken his students to the Leaning Tower of Pisa, where they dropped various objects from different heights. As a result of this investigation, Galileo concluded that all objects, disregarding air friction, fall with the same rate of change of velocity.

Galileo tried many times to illustrate to his instructors the fallacies in Aristotle's teachings, but to his dismay he was rewarded only with abuse and the threat of poor marks in his classes. Galileo's writings also brought disfavor from the Church, since many concepts that he discussed were in defiance of the contemporary interpretations of Holy Scripture. Although he had become one of the outstanding men of science in Italy, his persistence could not be ignored by the Church, and eventually he was brought before the Inquisition. However, because of his age and fame, his punishment consisted of home confinement, where he was forced to recite the seven penitential psalms at least once a week for three years. It is said that at the end of each psalm Galileo added in a soft voice, "and still Earth moves."

Galileo's many experiments did much to disprove the misconceptions attributed to Aristotle and others. His standard practice of using experiments as an aid in discovering and verifying physical laws was one of his most valuable contributions. One of Galileo's major objectives was to develop an adequate description of motion, called *kinematics,* from which the laws of motion could be deduced. His contributions in the field of *mechanics,* the study of the motion of bodies under the influence of forces, along with those of Descartes (1596–1650) and Huygens (1629–1695), had a profound influence on the theory of motion developed by Isaac Newton (1642–1727).

The genius of Newton was made clear to the scientists of his time when his works were published in 1686 and 1687. In the three-volume series called *Philosophiae Naturalis Principia Mathematica,* or *Mathematical Principles of Natural Philosophy* (usually referred to by the shortened name, *Principia*), Newton showed that the motions of all bodies could be determined from only three laws of motion. The first two volumes were titled *On the Motion of Bodies* and contained elegant mathematical discussions of motion under various conditions. The third volume was called *The System of the World* and included the law of universal gravitation. This last volume then proceeded to demonstrate how the observed motions of planets, moons, and comets could be explained using the law of universal gravitation and the three laws of motion.

Newton made many other outstanding scientific contributions in addition to the laws of motion and the law of universal gravitation. He invented a form of calculus, the field of mathematics that deals with variations in functions and areas bounded by curves. (An alternate form was independently created by one of his contemporaries, Leibniz (1646–1716).) His experimental work in the area of optics established many of the principles in that field, although some of his theoretical work was later shown to be inadequate. In the opinion of many, the breadth of Newton's great intellectual achievements has never been equaled in the history of mankind. In a letter to Robert Hooke in 1676, Newton wrote, "If I have seen further than other men, it is because I have stood on the shoulders of giants." However, modesty cannot alter the fact that many others had the same shoulders upon which to stand.

Near the end of the nineteenth century, it became apparent to some physical scientists that a conflict existed between the laws of motion and certain laws of electricity and magnetism.

The Case against Galileo

In 1979, the Vatican decided to review its case against Galileo, whom it condemned for heresy in 1616, tried in 1633, and thereafter kept under house arrest until his death in 1642. Galileo was convicted for his announced beliefs in the Copernican theory of the solar system. He even cited scripture that he interpreted to mean Earth orbited the sun, but his inferences were strongly contested.

Galileo was declared a heretic for dogmatic rather than scientific reasons, and the arbitrary decision of his guilt revealed the lack of desire on the part of Inquisition members to pursue the truth of Earth's motion.

Religion undoubtedly searches for truth; science takes a literal rather than spiritual approach in its search. Religions and scientific bodies have disagreed on many occasions through the ages, such as in the case of Galileo. Today's religious leaders and scientists alike generally agree that when science and religion come to an impasse, it is beneficial to consider which group has jurisdiction before a theory is condemned by either.

The Vatican released all documents pertaining to Galileo's trial in 1984, and Pope John Paul II admitted that the Church might have erred in its condemnation of Galileo. The matter is still unresolved.

The publication of Albert Einstein's (1879–1955) theory of special relativity in 1905 resolved the conflict by showing that time was not an absolute quantity, but depended on the motion of the observer. Since time is an essential quantity in the description of motion, the theory of special relativity revealed that Newton's laws of motion were a special case of a more general theory. The more general theory *must* be applied to objects traveling near the speed of light, but for relatively slow-moving objects, Newton's laws give results in agreement with normal measurements. (At the fantastic speed of 18,600 miles per second, one-tenth the speed of light, Newton's laws are in error by about 0.5 percent. For comparison, the fastest spacecraft travel with speeds of about 7 or 8 miles per second.) Another generalization reached its culmination in the late 1920s in the form of quantum theory. Erwin Schrödinger (1887–1961) and Werner Heisenberg (1901–1976), working with others, presented two forms of the theory, which showed that Newton's laws of motion were inadequate for microscopic particles. Once again, Newton's laws of motion were shown to be a special case, applicable to macroscopic objects. These two modifications illustrate that *most scientific "laws" have a limited range of validity.* One of the goals of science is to determine the most general form of "laws."

Despite their limitations, Newton's laws of motion provide an adequate description of "ordinary" objects moving at "ordinary" speeds. Their general applicability has made them the basis for large segments of engineering. This chapter and Chapter 4 will discuss Newton's laws and illustrate some of their applications.

Newton's First Law of Motion and Inertia

Newton's **first law of motion** states: *Every body continues in its state of rest, or of uniform motion in a straight line, unless it is compelled to change that state by forces impressed upon it.*

The first law of motion deals with the concept known as inertia. **Inertia** is *the fundamental property of matter by which it tends to maintain its state of rest or of uniform motion (steady speed) in a straight line.* In other words, inertia is that attribute which enables a body to resist changes in motion. **Mass** is *the quantitative measure of inertia.* It is more difficult to change the motion of a body that has a large mass than it is to change the motion of a body having a small mass. For example, it is easier to change the motion of a baseball than that of an automobile, since the automobile has a greater mass; hence, a greater inertia. The numerical determination of mass is discussed later in connection with the second law of motion.

The condition for a body to maintain its state of uniform motion is that the forces acting upon it do not "compel" it to change this state. A **force** is *an action exerted upon a body that changes its state, either of rest, or of uniform motion in a straight line.* In other words, a force is a push or a pull. Every body in the universe has forces acting upon it, but because forces act in certain directions, the sum of the forces acting on a body can be zero. If *the sum of the forces acting on a body is zero,* the forces will not change the state of motion of the body and it will obey the first law of motion. Such a body is said to be in **equilibrium.** Consider the forces acting upon a rock lying on the ground (Fig. 3.1a). The force of gravity pulls downward on the rock, while at the same time the "contact" force of the material beneath the rock pushes upward on it. The sum of these two forces is zero; hence, the rock is in equilibrium. (This unpretentious example also illustrates how simplification serves in "modeling" a real situation. There may be other forces acting upon the rock that are not considered because they are assumed to be negligible.)

The airplane shown in Figure 3.1b may be flying at a speed of 600 miles per hour, but if the net force on it is zero, it is in equilibrium. The airplane will maintain a steady speed in a straight line. The second law of motion, discussed later, deals with the action of the unbalanced forces that had to act on the airplane to *change* its motion from rest, when it was on the ground, to its current state. Frictional forces, which generally oppose the motion of objects on Earth, usually make it necessary to continually apply a force to

Sir Isaac Newton: *Seventeenth-Century Genius*

Isaac Newton (1642–1727) was born prematurely on Christmas Day to Hannah Newton, two months after the death of his father. Isaac's mother remarried three years later and he was sent to live with his maternal grandmother. Newton received the Bachelor of Arts degree in 1665 from Trinity College, Cambridge, England, where his genius was recognized by all who knew him. He was appointed to the Faculty of Mathematics at the institution and soon committed to writing his mathematical methods of the calculus. In addition, he also discovered many characteristics of light, including the nature of color. Driven from Cambridge by the great plague, Newton went to his home at Woolsthorpe, where in 1666 the fall of an apple from a tree in his garden presumably suggested to him the law of universal gravitation. He returned to Cambridge and completed the Master of Arts degree in 1668.

In 1692 Newton suffered a nervous breakdown, brought about, according to some authorities, by a fire in his laboratory, caused when his dog overturned a candle. Much of his original manuscript of *Principia Mathematica* and 20 years of recorded observations about light were destroyed. Newton then turned his interests from his scientific works to the study of theology. He also served as Master of the Mint in London and was elected president of the Royal Society in 1703, a position he retained until his death.

Shortly before his death, Newton summarized his scientific achievements by modestly telling some unnamed companion, "I don't know what I may seem to the world, but, as to myself, I seem to have been only like a boy playing on the sea shore, and diverting myself in now and then finding a smoother pebble, or a prettier shell than ordinary, whilst the great ocean of truth lay all undiscovered before me." Newton was interred in Westminster Abbey, an honor now accorded only to royalty. ■

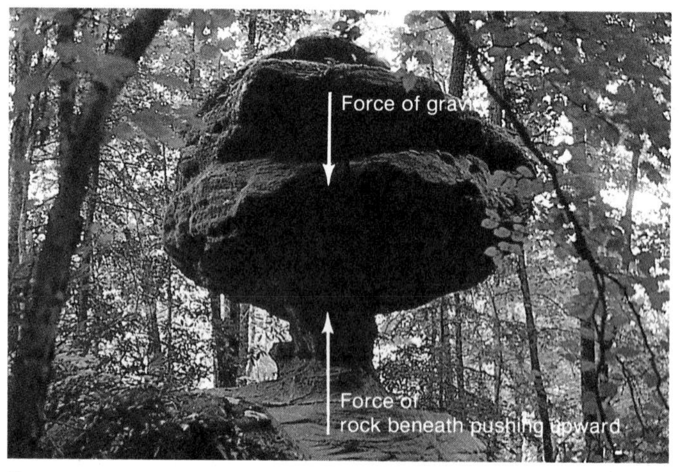

a.

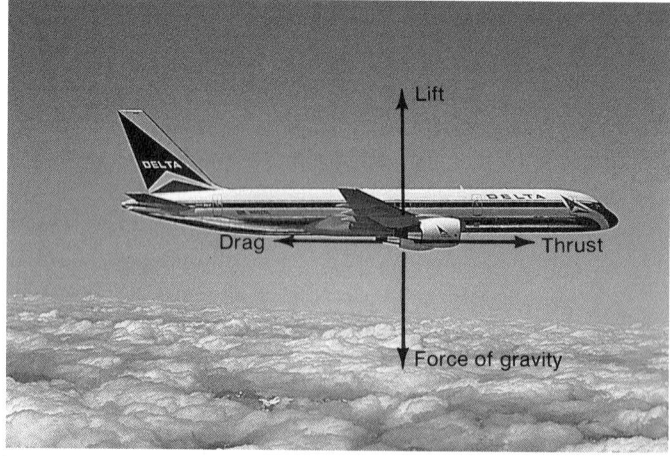

b.

Figure 3.1 Two objects in equilibrium. (*a*) The two major forces acting on the rock, the downward pull of gravity and the upward force of Earth, are equal in magnitude and opposite in direction. The net force on the rock is equal to zero. (*b*) An airplane flying at a constant speed in a straight line is also in equilibrium. The sum of all the forces is equal to zero.

an object to keep it moving. Galileo was the first to recognize that the applied force could be eliminated if there were no friction. Then, Newton refined the idea in his first law of motion.

Often, an object at rest is said to be in static equilibrium, while one in uniform motion in a straight line is said to be in dynamic equilibrium. The difference, however, lies only in the viewpoint of the observer. All motion must be measured *relative* to something. The usual reference for those of us on Earth is some point on the surface of Earth, itself. Earth is our "reference frame." *Any reference frame in which the first law of motion is obeyed* is said to be an **inertial reference frame.** Any reference frame moving at a constant velocity relative to an inertial reference frame is also, itself, an inertial reference frame. Let the reader consider his or her own experience when riding in an automobile. If the car moves at a steady speed in a straight line, the forces acting on a person seated in the car are exactly the same as if the car were sitting still. The moving car, in this case, is an inertial reference frame. If the car is moving and stops abruptly, the rider tends to maintain his or her speed because of inertia (Fig. 3.2a). This causes the rider to "pitch" forward *relative to the car.* Similarly, if the car makes a turn to the left, the rider tends to continue to move in a straight line because of inertia. The rider, attempting to follow the

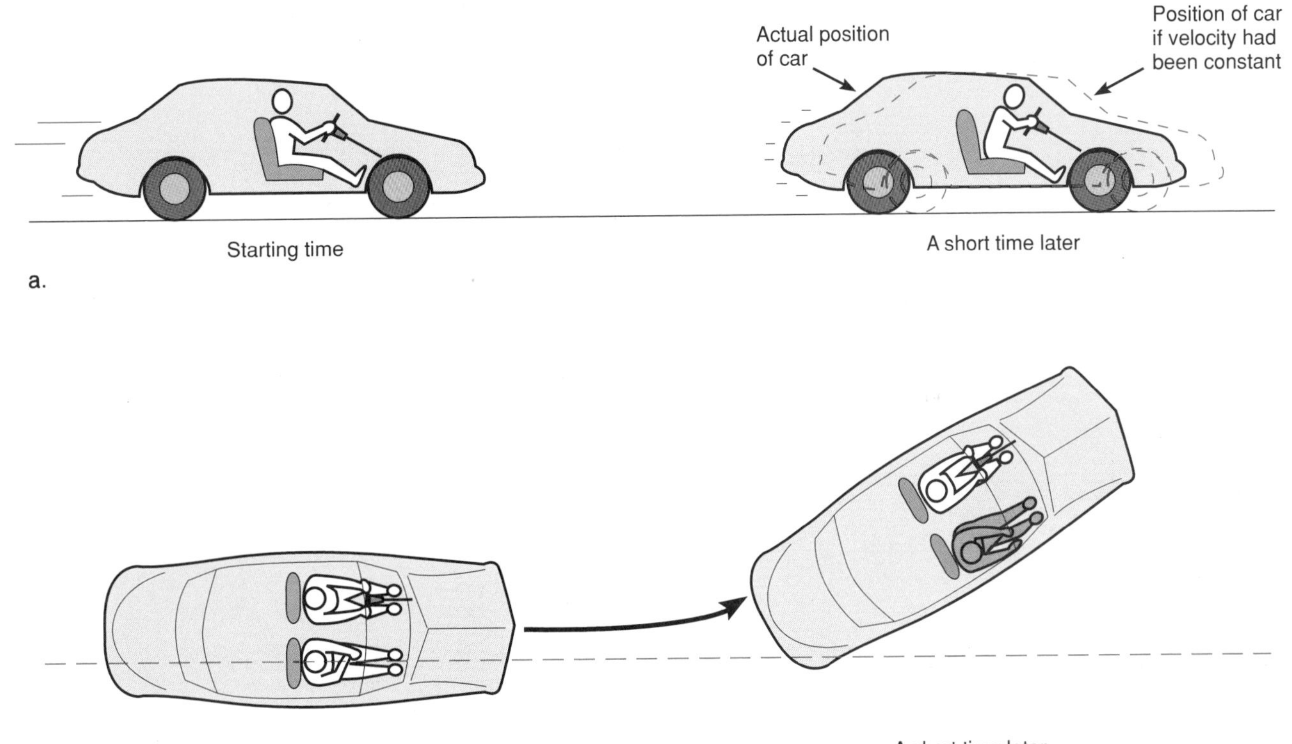

Figure 3.2 Because of their inertia, bodies tend to resist changes in speed or direction. (*a*) A passenger in a rapidly braking car tends to keep moving unless restrained by a seat belt or other part of the car. (*b*) When a car rounds a sharp curve, the passenger tends to move in a straight line. The passenger seems to be thrown "outward," but relative to Earth it can be seen that this is caused by the car moving to the left.

straight-line path, moves to the right *relative to the car* (Fig. 3.2b). In these last two cases, the car is *not* an inertial reference frame.

The surface of Earth is actually *not* an inertial reference frame. Earth is rotating, revolving about the sun, and so forth. For movements over small distances, these motions of Earth, itself, have no appreciable effect, but for movements covering large distances they can have a substantial effect. For example, atmospheric circulation patterns are influenced by the rotation of Earth.

Scalars and Vectors

A moment's reflection on the discussion of equilibrium brings to light the fact that to find the total force acting on a body it is not possible to merely add the numbers representing the forces; the directions of the forces must be considered as well. This leads to a classification of different types of quantities, depending on whether or not they have a direction.

Scalar quantities are *physical quantities that have only a magnitude (size), but do not have a direction.* Scalar quantities require only a single number (which may be positive or negative) and a unit to represent them. The unit is necessary to tell what quantity is being stated. Examples are time, distance, volume, energy, and mass (all of which will be discussed in this section). Thus, time is stated as 12 seconds, distance as 5.36 meters, etc.

Scalar quantities follow the ordinary laws of addition, subtraction, multiplication, and division. For instance, if something happened when a stopwatch read 12 seconds and something else happened when it read 43 seconds, the time elapsed between these two events would be 43 seconds minus 12 seconds, or 31 seconds. Also, if a piece of plywood is 2 meters wide by 3 meters long, its area is 6 square meters.

Vector quantities are *physical quantities that must be specified by both a magnitude (size) and direction.* If several forces act upon a body, the directions of the forces as well as their sizes are important; thus, force is a vector quantity. Some other examples of vector quantities that will be discussed in this chapter are displacement, velocity, acceleration, and momentum. Vector quantities, or *vectors,* are commonly represented graphically by arrows, the lengths of which are constructed to some suitable scale, and directions indicated by arrowheads on each vector. For example, if an object moves 5 meters eastward, an arrow 5 centimeters long drawn on a map pointing in an eastward direction might be used to represent the motion of the object (see Fig. 3.3).

Vectors can be added together to form a resultant vector, often simply referred to as a **resultant.** Each of the original vectors, then, is considered to be a component vector of the resultant. For example, if two forces act together and in the same direction, the combined effect, or resultant, is a force equal to the sum of the

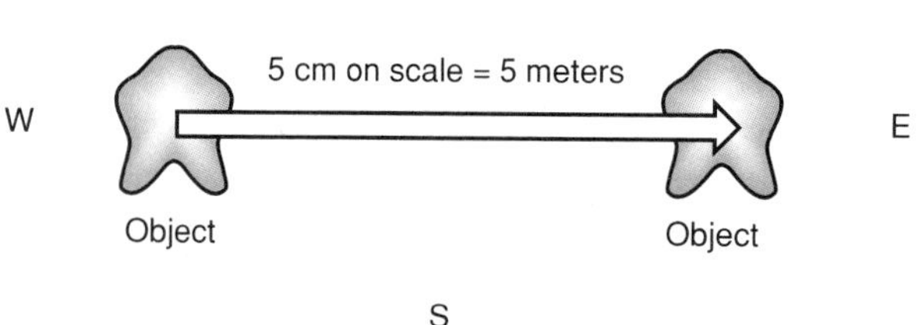

Figure 3.3 A vector is drawn to scale with its length representing the size of the vector, in this case the distance the object has moved. The arrowhead shows the direction of the vector.

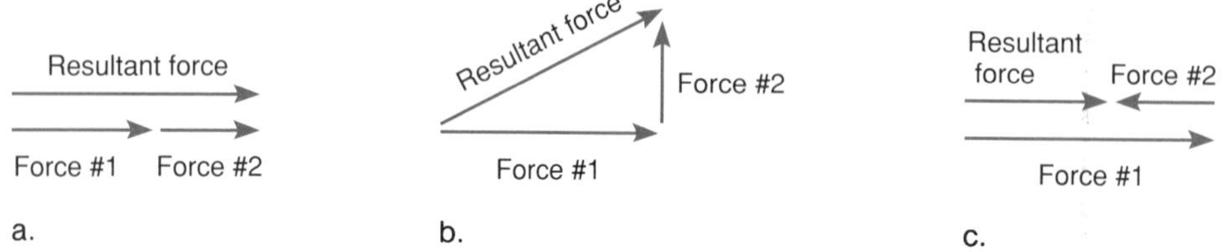

Figure 3.4 The resultant is the sum of two or more vectors. The directions of the vectors being summed must be taken into account as well as their magnitudes. In (*a*) the magnitude of the resultant is the sum of the magnitudes of the individual vectors, while in (*c*) it is the difference. In (*b*) the resultant is smaller than in (*a*) but larger than in (*c*).

two original forces. If the forces directly oppose each other, the resultant is a force equal to the difference between the forces and in the direction of the larger component. Vectors can be combined in various ways so that the resultant may be any value between their sum and difference (see Fig. 3.4).

If two forces act on the same body, but not along the same line, the resultant is determined by applying the rule of vector addition, in which the vectors are placed head-to-tail and the resultant is the vector drawn from the tail of the first vector to the head of the last vector. To add two vectors, a suitable scale is first chosen to represent the vectors; for example, 1 centimeter might represent 10 newtons. (The newton is a unit of force.) The vectors are then drawn with a ruler and a protractor to the proper size and direction, with the tail of the second vector touching the head of the first vector. The resultant (**R**) is the vector drawn from the tail of the first vector to the head of the second vector (see Fig. 3.5a). From a properly drawn diagram, both the magnitude and direction of the resultant can be determined using the ruler to determine the length, which from the scale chosen will give the magnitude, and the protractor to determine the direction.

An alternative method is to place the two vectors tail-to-tail and draw lines parallel to the two vectors to form a parallelogram with the vectors as two of the sides; then the resultant is the diagonal from the common point where the tails meet to the opposite corner (see Fig. 3.5b). It is easy to see the two methods are entirely equivalent.

Figure 3.6 shows a typical example of how two forces that act on the same body at a common point affect the motion of the body. The resultant indicates the direction and the useful effort gained from the two forces in pulling the boat into proper loading position on the trailer.

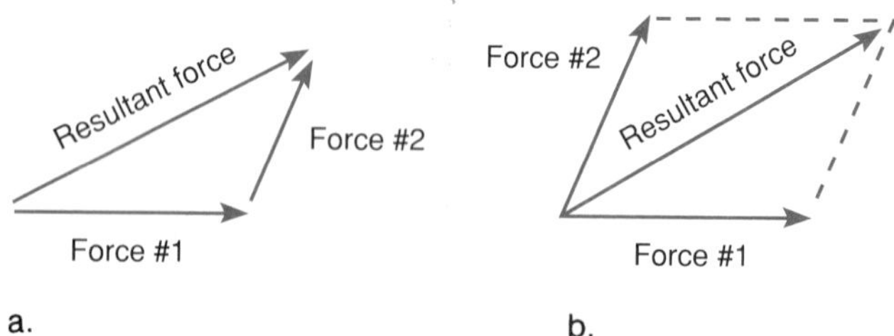

Figure 3.5 (*a*) Two vectors may be added by placing them head-to-tail and drawing the resultant from the tail of the first to the head of the last, or (*b*) by placing them tail-to-tail, forming a parallelogram with the vectors as two of the sides, and then drawing the resultant from the point where the tails intersect to the opposite corner of the parallelogram.

Consider a body that is in static equilibrium; that is, there is no acceleration, and hence no unbalanced force. In a tug-of-war, for instance, if two opposing teams are exerting equal forces on opposite ends of the rope, a condition of equilibrium exists. If team A exerts a force of 22,000 newtons eastward, team B must exert a force of 22,000 newtons westward for equilibrium to exist. The *tension* in the rope is 22,000 newtons rather than 44,000 newtons. In order to explain this apparent paradox, consider the reading on a scale if team A were to attach their end of the rope to a strong post. The other team, if it continued to exert a force of 22,000

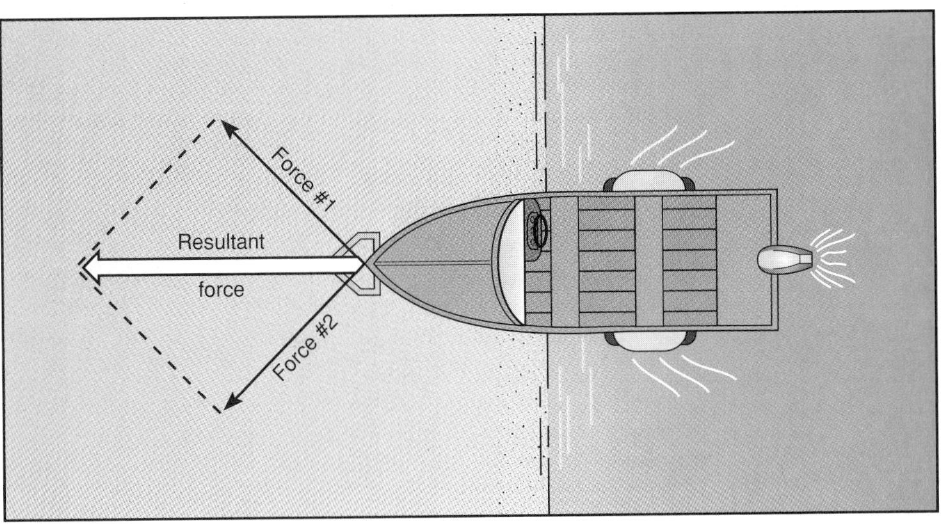

Figure 3.6 The combined effort of the two equal forces causes the boat to move straight and onto the trailer. The resultant force, then, is in the direction the boat moves.

a.

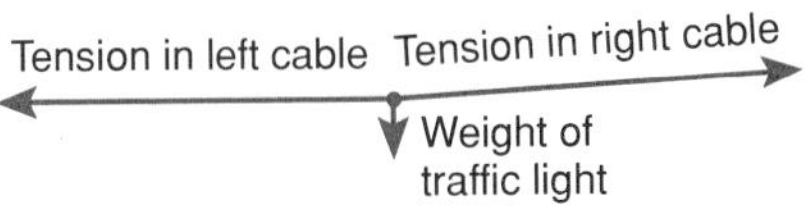

b.

Figure 3.7 (*a*) The traffic light exerts a downward force on the clamp attaching it to the supporting cable. (*b*) The tension in each of the supporting cables must be such that the total force on the clamp is zero. The tensions increase as the cables become more horizontal.

newtons, would maintain the condition of equilibrium and the scale would indicate a reading of 22,000 newtons. (Further discussion of opposing forces is continued in the next chapter.)

When three or more forces act on a body and it remains in equilibrium, the resultant of all the forces must be zero. Consider a traffic light suspended above an intersection, as shown in Figure 3.7a. The traffic light pulls downward on the clamp affixed to the supporting cable, so there are three forces acting on the clamp: the weight of the traffic light, the tension in the cable upward and to the left, and the tension in the cable upward and to the right. The upward pull of the cables must counterbalance the weight of the traffic light, and the pull to the left must be exactly balanced by the pull to the right.

Although a graphical determination of the tension in the cables is beyond the scope of this text, a sketch of the vector diagram is given in Figure 3.7b. The tension in the cables is much greater than would be the tension in a single wire that pulled directly upward on the object since only a portion of the tensions in the cables provides vertical support. The tension added to a telephone line or to an electric wire by the weight of ice that forms on it may be sufficient to break the wire.

Velocity, like force, is a vector quantity, and thus has both magnitude and direction. Typical examples of velocities that are additive, or capable of being combined, are illustrated in Figure 3.8, which shows how the speed of an airplane relative to the ground is determined. The airplane has a velocity of 500 miles per hour relative to the air. In each part of the figure, the air is moving in different directions relative to the ground at a velocity of 50 miles per hour. The resultant is the velocity of the airplane relative to the ground.

Entire books have been written on the subject of vectors. In this chapter, only vectors that lie along the same line, leading to straight-line motion, will be considered. Some of the interesting effects of simple curved-line motion will be considered in the next chapter.

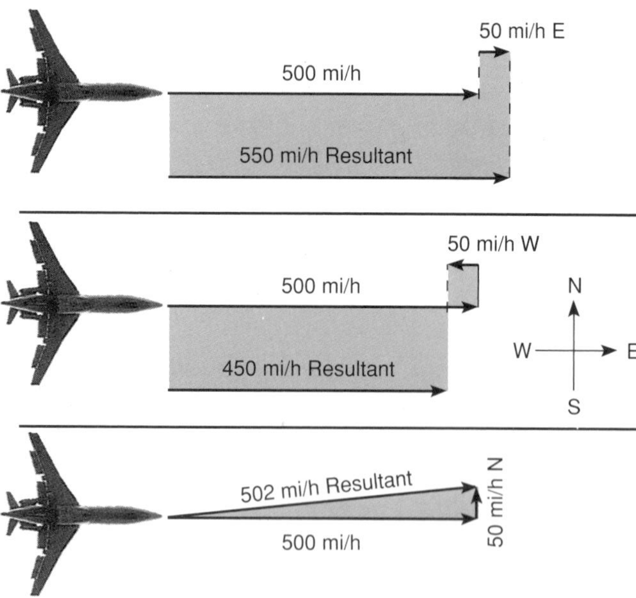

Figure 3.8 The plane has a speed relative to the air of 500 mi/h. The resultant gives the velocity of the plane relative to the ground when the wind velocity is 50 mi/h in the direction shown. The magnitude and direction of the resultant velocity depend not only on the speeds, but also on the relative directions of the velocities being added.

Straight-Line Motion

The study of how forces change the motion of bodies requires a brief look at how motion is described. General three-dimensional motion is beyond the scope of this text, but the major ideas incorporated in Newton's laws of motion can be illustrated with motion in one and two dimensions.

The position of a body relative to some reference point is called its **displacement** (Fig. 3.9a). Complete specification of position requires not only the distance from the reference point, but the direction as well; therefore, displacement is a vector quantity. Giving directions to the local theatre by telling someone to go four blocks in one direction, then to turn left and go three more blocks is in essence telling the person the displacement of the theatre relative to his or her current position. Although there are other methods, the displacement for a body confined to a flat surface may always be specified by giving two straight-line distances at right angles to each other, as shown in Figure 3.9b. This method allows one to simplify problems by considering two-dimensional motion as the sum of two one-dimensional motions. In addition, many bodies exhibit one-dimensional (straight-line) motion, such as a car moving on a straight road, or a ball thrown straight upward or dropped. As mentioned previously, only straight-line motion will be considered in this chapter.

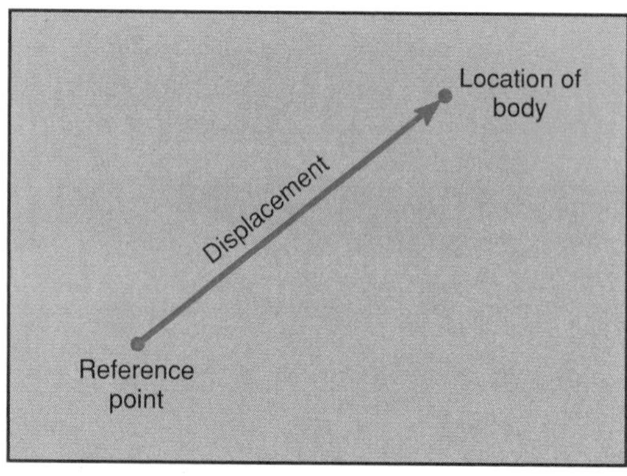

a.

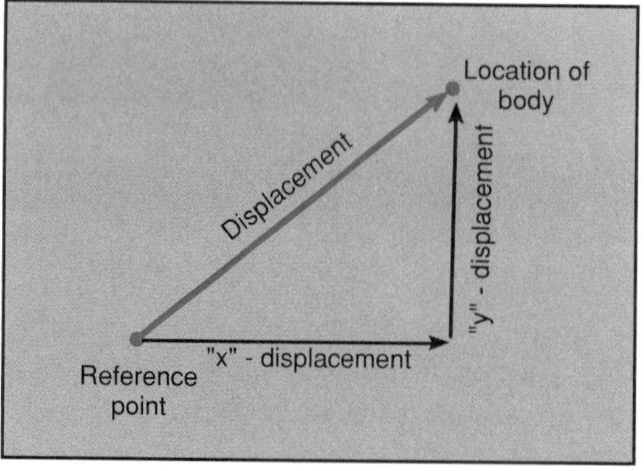

b.

Figure 3.9 (*a*) Displacement is a measure of the location of one point relative to another. Both the straight-line distance and the direction must be specified. (*b*) For flat surfaces, the displacement may be specified by stating the displacement along two mutually perpendicular directions. The advantage here is that direction can be indicated by the sign (+ or −) of the displacement.

a.

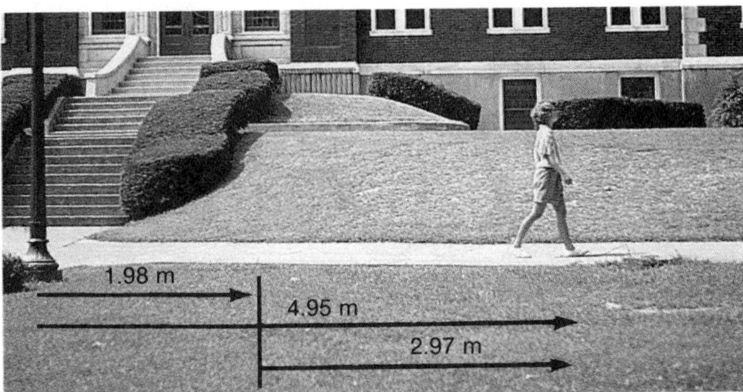

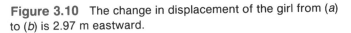

b.

Figure 3.10 The change in displacement of the girl from (*a*) to (*b*) is 2.97 m eastward.

The displacement of a body in one dimension is specified by giving the *distance and direction from the reference point.* The *distance* may be given in miles (mi), feet (ft), or some other unit, but in scientific applications the distance is expressed in meters (m) or some multiple of the meter, such as the centimeter (cm) or kilometer (km). Generally, this text will use metric units in calculations, but will often specify approximate English units for comparison purposes. The *direction* could be specified verbally, such as eastward or upward, but for one-dimensional motion it is more useful to use an algebraic sign to indicate direction. Thus, a plus (+) sign would indicate one direction and a minus (−) sign would indicate the other direction. For a ball thrown upward, (+) might represent upward and (−) downward. The algebraic symbol for displacement is $\mathbf{d}$, the boldface type indicating that the quantity is a vector and has a direction.

The advantage of using algebraic signs to indicate direction is that changes in displacement are easy to compute, and the sign of the change in displacement indicates its direction as well. That is, if an object moves in the positive (+) direction its change in displacement will be positive (+), and vice versa. Letting $\mathbf{d}_f$ represent the final displacement and $\mathbf{d}_i$ represent the initial displacement, the change in displacement is computed from the equation

$$\begin{aligned}\text{change in displacement} \\ &= \text{final displacement} - \text{initial displacement} \\ &= \mathbf{d}_f - \mathbf{d}_i.\end{aligned}$$

Example 3.1

A girl walks along a sidewalk. She moves from a position 1.98 meters east of a lamppost to a position 4.95 meters east of the post. What is her change in displacement relative to the lamppost? (see Fig. 3.10.)

Solution

Choosing eastward as the positive direction, the initial displacement ($\mathbf{d}_i$) of the girl is 1.98 m and her final displacement ($\mathbf{d}_f$) is + 4.95 m. Therefore,

$$\begin{aligned}\text{change in displacement} &= \mathbf{d}_f - \mathbf{d}_i \\ &= +4.95 \text{ m} - (+1.98 \text{ m}) \\ &= +2.97 \text{ m}.\end{aligned}$$

Therefore, the change in displacement of the girl is +2.97 m, and since eastward was chosen as (+), the girl is 2.97 m east of her starting point.

Example 3.2

A rock thrown upward is 5.90 meters above the ground at a certain time and 2.60 meters above the ground at a later time. What is the change in displacement of the rock? (See Fig. 3.11.)

Solution

Choosing upward as (+), $\mathbf{d}_i$ = +5.90 meters and $\mathbf{d}_f$ = +2.60 m. Then,

$$\begin{aligned}\text{change in displacement} &= +2.60 \text{ m} - (+5.90 \text{ m}) \\ &= -3.30 \text{ m}.\end{aligned}$$

Since the result is negative, the rock's change in displacement was therefore 3.30 m downward.

When an object is in motion, its **average speed** is *the total distance traveled divided by the time taken* to travel that distance. In equation form,

$$\text{average speed} = \frac{\text{total distance traveled}}{\text{time interval required}}$$

or

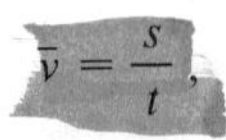

$$\bar{v} = \frac{s}{t},$$

AVERAGE SPEED

where $\bar{v}$ represents average speed, s represents total distance traveled, and t represents the time interval. Acceptable units of speed are miles per hour (mi/h), feet per second (ft/s), meters per second (m/s), and so forth. Note that *specification of speed does not involve direction* ($\bar{v}$ is *not* in bold print), and also that the distance traveled is the total distance, ignoring details such as turns or retracing the original path, and is not necessarily the same as the

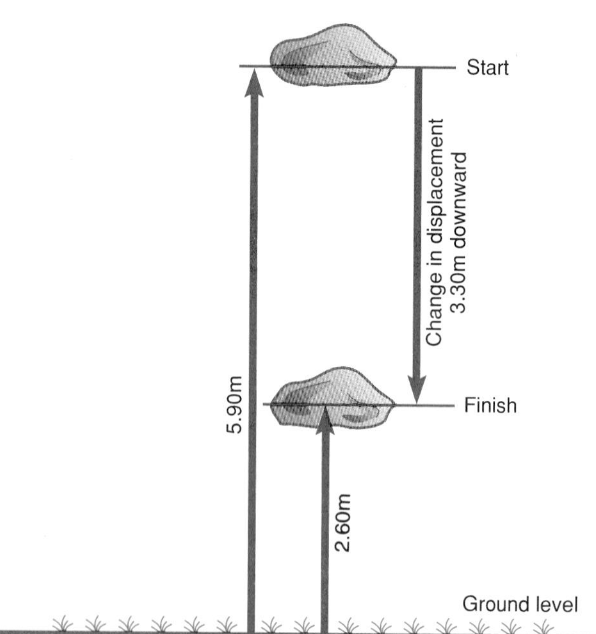

Figure 3.11 The change in displacement of a rock as it moves downward is negative if upward is chosen as the positive direction.

change in displacement. When a swimmer swims one "lap" in a pool, returning to the starting point, her change in displacement is zero but her total distance traveled is not.

Example 3.3

A swimmer swims from one end of a 25.0-meter-long pool to the other end and returns to his starting point in 27.0 seconds. What is his average speed? (See Fig. 3.12.)

Solution

The total distance traveled by the swimmer is 50.0 m, therefore $s = 50.0$ m. Then, with $t = 27.0$ s,

$$\bar{v} = \frac{50.0 \text{ m}}{27.0 \text{ s}} = 1.85 \text{ m/s}.$$

His average speed is 1.85 m/s.

Example 3.4

A driver planning a trip between Pittsburg and Indianapolis (Fig. 3.13) along Interstate 70 finds the total distance to be 363 miles. She thinks that she will be able to make the trip with an average speed of 50.0 miles per hour. At this average speed, what amount of time should she allow for the journey?

Figure 3.12 The average speed of a swimmer is computed by dividing the total distance traveled by the total time required to travel that distance. The details of the path taken do not enter into the calculation, so the swimmer may leave from and return to the same point.

Solution

The given quantities in this example are the total distance traveled, $s = 363$ mi, and the average speed, $\bar{v} = 50.0$ mi/h. The equation for average speed may be rearranged algebraically to give the time, *t*, in terms of the average speed, $\bar{v}$, and the total distance, *s*. This is accomplished by multiplying both sides of the equation for average speed by *t* and dividing both sides by $\bar{v}$, giving

$$t = \frac{s}{\bar{v}}.$$

Then, substituting the distance and average speed gives

$$t = \frac{363 \text{ mi}}{50.0 \text{ mi/h}}$$
$$t = 7.26 \text{ h}.$$

Therefore, the driver should allow 7.26 hours for the trip.

The speed at a particular time is called the **instantaneous speed.** It may be found by taking the average speed over a very short time interval. For example, the speed of a falling rock 3 seconds after it is dropped could be found by taking the distance traveled between 3.000 seconds and 3.001 seconds and dividing it by 0.001 second. The short distance traveled divided by the small elapsed time would be very nearly the speed at 3 seconds. (In principle, for exact results the time interval is made vanishingly small.) The speedometer of a car measures instantaneous speed.

The **velocity** of an object is a measure of *both its speed and its direction.* The **average velocity,** $\bar{v}$, is *the change in displacement divided by the time interval required for the change.* Therefore,

$$\bar{v} = \frac{\text{change in displacement}}{\text{elapsed time}}$$

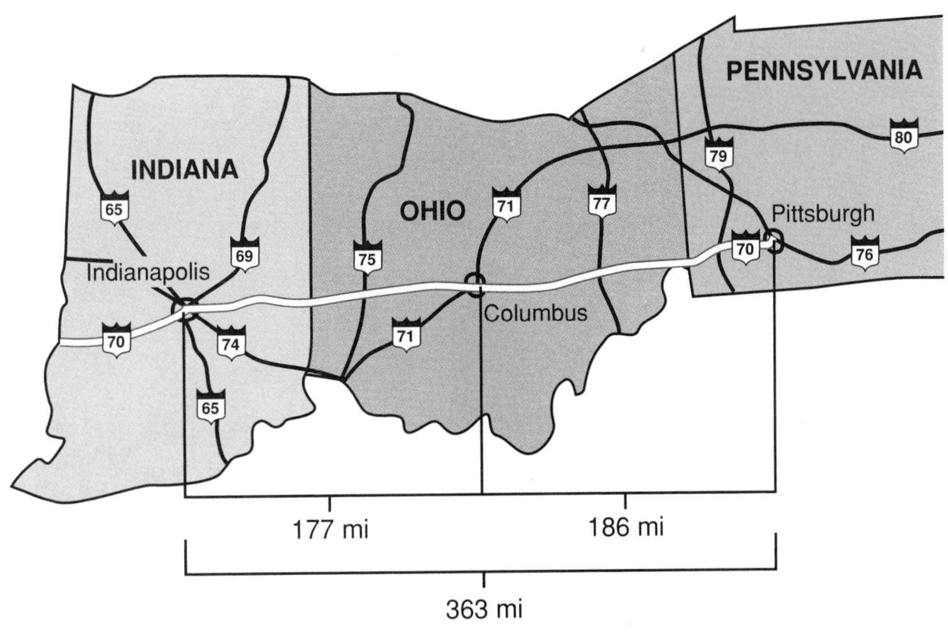

Figure 3.13 The total distance from Pittsburgh to Indianapolis along I-70 is 363 mi.

a.

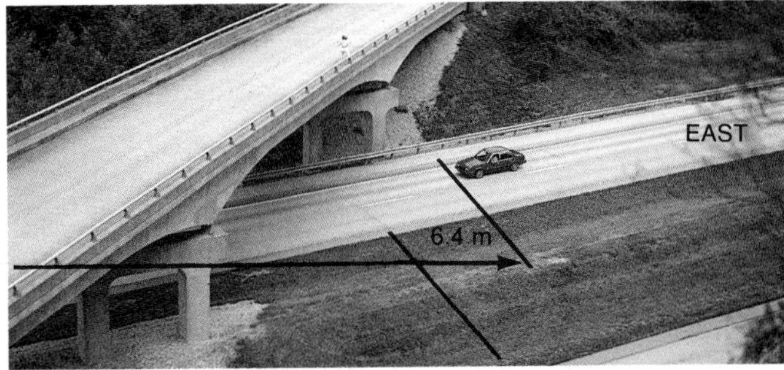

b.

Figure 3.14 The car is 6.4 m east of the reference point when the student starts timing (*b*) and is 61.2 m west of the reference point 2.39 s later (*a*). The average velocity of the car is computed in Example 3.5.

or algebraically,

$$\bar{v} = \frac{d_f - d_i}{t}. \qquad \text{AVERAGE VELOCITY}$$

Velocity is measured in the same units as speed, but for straight-line motion the sign, (+) or (−), indicates the direction. Note also that average velocity depends only on the position of starting and ending points of the motion and is independent of the details of the path between those points.

Example 3.5

A student with a stopwatch is standing on an overpass timing the motion of cars passing under the bridge (see Fig. 3.14). She observes that a car is 6.4 meters east of the bridge when she starts her stopwatch and is 61.2 meters west of the bridge when her stopwatch reads 2.39 seconds. What is the average velocity of the car during this interval?

Solution

The distances must be converted into displacements by first choosing a direction to be positive. Letting east be positive (+), $d_i = +6.4$ m, and $d_f = -61.2$ m. The elapsed time, is just the time on the stopwatch, 2.39 s, since the stopwatch started at 0. Substituting into the equation for average velocity gives

$$\begin{aligned}\bar{v} &= \frac{d_f - d_i}{t} \\ &= \frac{-61.2 \text{ m} - (+6.4 \text{ m})}{2.39 \text{ s}} \\ &= \frac{-67.6 \text{ m}}{2.39 \text{ s}} \\ &= -28.3 \text{ m/s}.\end{aligned}$$

Thus, the car has a velocity of 28.3 m/s (63.3 mi/h) westward, since east was chosen as positive.

The **instantaneous velocity** is *the velocity of a body at any instant of time.* It is computed in the same way as instantaneous speed, except that the change in displacement is taken over a very short interval of time instead of simply the distance traveled. The instantaneous velocity is indicated by using a subscript to signify the time at which the velocity has the given value. For example, the initial velocity of a body in motion is usually indicated by v_i and the final velocity by v_f.

Acceleration (*a*) is defined to be *the rate of change of velocity.* In equation form,

$$\text{acceleration} = \frac{\text{change in velocity}}{\text{elapsed time}},$$

$$\text{acceleration} = \frac{\text{final velocity} - \text{initial velocity}}{\text{elapsed time}},$$

or algebraically,

$$a = \frac{v_f - v_i}{t}. \qquad \text{ACCELERATION}$$

Technically, this equation gives average acceleration over the time interval, but if the acceleration is constant, the average acceleration and the instantaneous acceleration will be equal.

The units of acceleration are those of velocity divided by time. Acceptable units are (mi/h)/s, (m/s)/s, (km/h)/s, and so forth. Units that do not mix different time units, such as meters per second, per second, abbreviated (m/s)/s, are usually written with the time units combined in the denominator. Thus, meters per second per second is written meters per second squared (m/s^2).

Assuming constant acceleration, an auto that starts from rest ($v_i = 0$) and in 10.0 seconds attains a velocity of 25.0 meters per second ($v_f = +25.0$ m/s) increases its velocity during each second (25.0 m/s = 55.9 mi/h). If the increase is considered uniform; that is, if the velocity changes by equal amounts during equal time intervals, the change in velocity must amount to 2.50 meters per second for each of the 10.0 seconds, and the acceleration is 2.50 meters per second per second (see Fig. 3.15). If the acceleration remains constant, at the end of 15.0 s the auto reaches a velocity of 37.5 meters per second. If the brakes are then applied and the auto stops in 10.0 seconds, the acceleration will be −3.75 meters per second per second, since $v_f = 0$ and $v_i = +37.5$ mls, making $v_f - v_i = 0$ m/s − (+37.5 m/s) = −37.5 m/s for this part of the motion. The negative acceleration means that the acceleration is opposite the direction chosen as positive, which in this case is opposite the direction the auto was moving. Acceleration that is opposite in direction to the velocity slows the body down and is sometimes called deceleration. (It should be noted that a negative acceleration will cause the body to speed up if the velocity is also negative.)

A *change in direction of the velocity* also produces an acceleration. (Motion of bodies in circular paths is considered in the next chapter.)

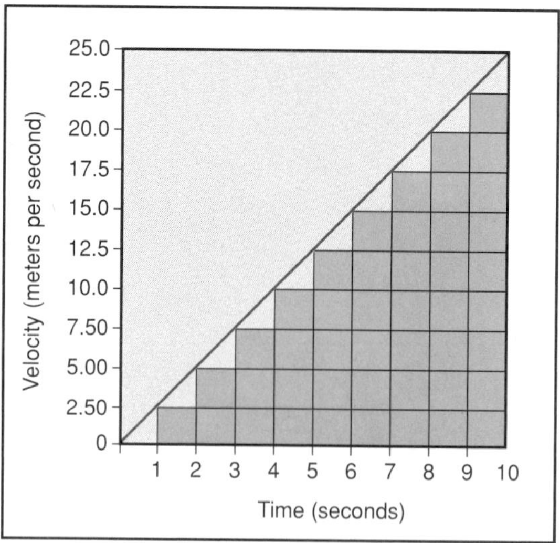

Figure 3.15 The time variation of the velocity of an auto undergoing uniform acceleration of 2.50 m/s/s is represented by the diagonal line.

The final velocity of an object with constant acceleration can be found from the definition of acceleration by multiplying both sides by *t* and adding v_i to both sides, giving

$$v_f = v_i + at. \qquad \text{FINAL VELOCITY}$$

This equation states that the final velocity is equal to the initial velocity plus the *change* in velocity; that is, ***at*** is the *change* in velocity.

The average velocity ($\bar{v}$) of a moving object, assuming constant acceleration, is equal to one-half the initial plus final velocities:

$$\bar{v} = \tfrac{1}{2}(v_i + v_f). \qquad \text{AVERAGE VELOCITY}$$

The final displacement of an object (d_f) is determined from the average velocity and time of travel. Rearranging the equation for the definition of average velocity on page 45 gives

$$d_f = d_i + \bar{v}t. \qquad \text{FINAL DISPLACEMENT}$$

Using algebraic manipulation, it can also be shown that

$$d_f = d_i + v_i t + \tfrac{1}{2}at^2. \qquad \text{FINAL DISPLACEMENT}$$

(More detailed forms of the equations for constant acceleration are given in Appendix 2.)

If an object accelerates and then assumes a constant velocity, finding the displacement requires several steps: The displacement at the end of the period of acceleration is found. The velocity at the end of the period of acceleration will be the constant velocity for the last part of the motion, allowing the displacement at the end of the interval of constant velocity to be found.

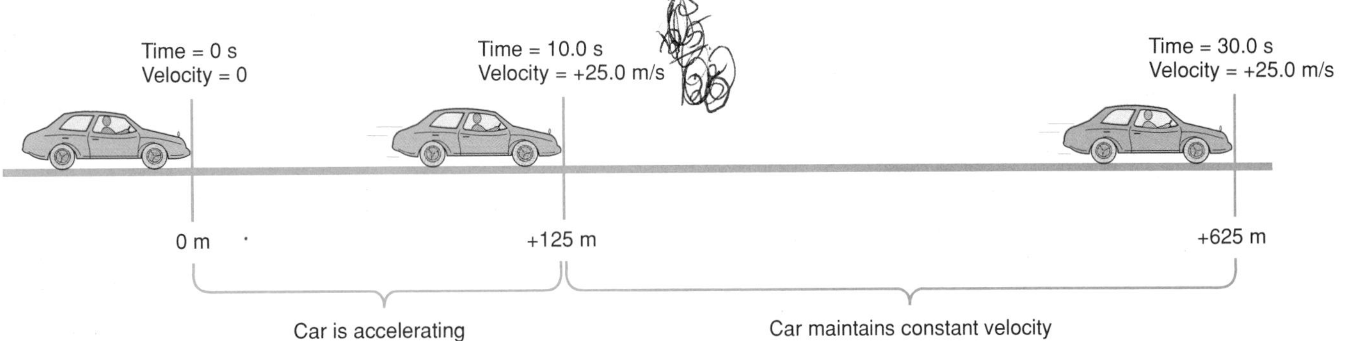

Figure 3.16 A car that is initially at rest accelerates for 10.0 s and then assumes a steady speed for the next 20.0 s. The diagram shows the displacement and velocity of the car at 10.0 s and 30.0 s after the car starts moving.

Example 3.6

An automobile on a straight road starts from rest and accelerates uniformly until it reaches a speed of 25.0 meters per second (55.9 mi/h) in 10.0 seconds. For the next 20.0 seconds the vehicle moves at a constant speed of 25.0 meters per second. What is the acceleration of the automobile during the first 10.0 seconds? (See Fig. 3.16.)

Solution

During the first 10.0 s, the initial velocity $v_i = 0$ m/s, and the final velocity $v_f = +25.0$ m/s (the positive direction is chosen to be that in which the car is moving). The acceleration is found from

$$\begin{aligned} a &= \frac{v_f - v_i}{t} \\ &= \frac{(+25.0\ \text{m/s}) - 0\ \text{m/s}}{10.0\ \text{s}} \\ &= +2.50\ \text{m/s}^2. \end{aligned}$$

Example 3.7

Find the displacement of the automobile in Example 3.6 at the end of the 30.0 seconds described.

Solution

This problem is broken into two parts: the first 10.0 s during which the car is accelerating and the last 20.0 s during which the velocity is constant and, hence, the acceleration is zero.

For the first 10.0 s, the initial velocity $v_i = 0$ m/s and the final velocity $v_f = +25.0$ m/s (again choosing positive to be the direction of travel). The average velocity is found from

$$\begin{aligned} \bar{v} &= \frac{1}{2}(v_i + v_f) \\ &= \frac{1}{2}[0\ \text{m/s} + (+25.0\ \text{m/s})] \\ &= +12.5\ \text{m/s}. \end{aligned}$$

The displacement for the first 10.0 s is then

$$\begin{aligned} d_f &= d_i + \bar{v}t \\ &= 0\ \text{m} + (+12.5\ \text{m/s})(10.0\ \text{s}) \\ &= +125\ \text{m/s}. \end{aligned}$$

Therefore, during the first 10.0 s the car moved 125 m in the positive direction.

During the last 20.0 s of the motion, the conditions in symbolic form are $d_i = +125$ m, $\bar{v} = v_i = v_f = +25.0$ m/s, and $t = 20$ s. The final displacement can be found from

$$\begin{aligned} d_f &= d_i + \bar{v}t \\ &= +125\ \text{m} + (+25.0\ \text{m/s})(20.0\ \text{s}) \\ &= +625\ \text{m}. \end{aligned}$$

Thus, the car has moved 625 m (2050 ft) from its starting point. The direction was not requested, but it is also shown. Figure 3.16 shows the displacement and velocity of the car at 0 s, 10.0 s, and 30.0 s.

Measurements and Computations: How Good Are They?

In evaluating the accuracy of computations, such as those done in the previous examples, a number of considerations should be taken into account. There is no problem with the mathematics, however such quantities as displacement, velocity, and time cannot be measured with perfect accuracy; that is, they are not exact numbers. Any real measurement has a limited degree of accuracy, which may be stated in several ways. For example, the time of 10.0 seconds in Example 3.6 implies that the time has been measured to the nearest tenth of a second. To indicate this, we could state the time as 10.0 ± 0.1 seconds. Normally, it is assumed that there might be some error in the last digit given, so that unless the time is written as shown, all that can be assumed is that the last digit has some uncertainty. A number such as 10.0 is said to have three significant digits. The number of **significant digits** is *the number of digits that are reliably known, including the first digit in which there is some uncertainty,* but not counting zeros, which are used to show the location of the decimal point. The numbers 2.16, 42.7, and 0.0118 all have three significant digits, whereas the numbers 6.8, 0.11, and 93 have two significant digits. What about a number such as 2050, indicating the number of feet the car has moved in Example 3.7? From the way the number is written, it is impossible to tell whether it has three or four significant digits, since the last zero may serve only to show the location of the decimal point. Such a problem may be remedied by writing numbers in scientific notation; thus, 2050 would be written 2.05×10^3 to show that the number has only three significant digits (see Appendix 2). The number 2050 feet was computed by converting 625 meters to feet. If the conversion is done very accurately, it is found that 625 meters is equal to 2050.5249 feet. Why not write all this out? The main reason is that it would imply an accuracy that is unjustified. If the distance of 625 meters is taken to mean a distance between 624 meters and 626 meters, the distance in feet would be between 2047 feet and 2054 feet. The number 2050 has three significant digits in it, just as 625 does, the last zero in 2050 serving only to locate the position of the decimal. In this text every attempt will be made to indicate the number of significant digits. Should there be any doubt, however, the reader may assume the numbers are accurate to no more than three significant digits. All answers to numerical problems should be given to three significant digits unless otherwise requested in the problem. Answers that include eight or ten digits are usually misleading ■

Example 3.8

A bullet leaves the end of the barrel of a rifle with a velocity of 800 meters per second, 0.00200 second after the rifle is fired. What is the acceleration of the bullet? Disregarding external forces in action on the projectile, how long after being fired will the bullet strike the target 300 meters away? (See Fig. 3.17.)

Figure 3.17 The bullet of a target rifle must have a tremendous acceleration to leave the end of the barrel with a velocity of 800 m/s, over twice the speed of sound.

Solution

The acceleration of the bullet is found from

$$\begin{aligned} a &= \frac{v_f - v_i}{t} \\ &= \frac{(+800\ \text{m/s}) - 0}{0.00200\ \text{s}} \\ &= +400{,}000\ (\text{or } 4.00 \times 10^5)\ \text{m/s}^2. \end{aligned}$$

At this velocity (letting $d_i = 0$) the time that the bullet takes to travel to the target is calculated from

$$\begin{aligned} d_f &= \bar{v}t \\ t &= \frac{d_f}{\bar{v}} \\ &= \frac{+300\ \text{m}}{+800\ \text{m/s}} \\ &= 0.375\ \text{s}. \end{aligned}$$

The total time the bullet takes to reach the target after being fired is found by adding the two times involved:

$$\begin{aligned} t &= t_1 + t_2 \\ &= 0.00200\ \text{s} + 0.375\ \text{s} \\ &= 0.377\ \text{s}. \end{aligned}$$

Forces: Countless Variety from Only Three

One of the difficulties in the application of Newton's second law of motion lies in determining the forces acting on a given body. An almost unlimited number of forces appear to be acting in even the simplest of problems. (One of the skills acquired in problem solving is learning to neglect all forces having little or no effect on the solution.)

It might come as somewhat of a surprise to most students to learn that there are only three known fundamental forces. They are the gravitational force, the electroweak force, and the strong nuclear force. (The electroweak force is a combination of the electromagnetic and the weak nuclear forces, formerly thought to be separate forces.)

Some manifestations of the gravitational force and the electroweak force are familiar to all of us. The gravitational force pulls all objects, including us, downward toward Earth. The part of the electroweak force called the electromagnetic force may cause us to be "shocked" when we touch an object after walking across a carpet on a crisp winter day. It also causes compass needles to point northward.

Other aspects of the electroweak force, called the weak force, are more obscure, such as the decay of neutrons into electrons, protons, and antineutrinos. The strong nuclear force is responsible for holding the nucleus of an atom together; this is vital to us, but not easily observed.

All of the other forces, such as friction, the elastic force of a spring, wind resistance, buoyancy, and so on, are simply combinations of these three fundamental forces (usually just the gravitational and electromagnetic forces). The reason for resorting to all these other forces is that it is not practical to begin each study at the level of fundamental forces. The complexity of the interactions among all of the particles making up a body is simply too great. Consider the simple problem of the force on the soles of your shoes when you stand on the floor. The electrical portion of the electromagnetic force between the electrons in the floor and those in the soles of your shoes is one of repulsion, but there are many billions of electrons. It is much simpler just to say there is a *contact force* between your shoes and the floor!

Some work seems to indicate that there may be another fundamental force. This force, if confirmed, would be much weaker than the gravitational force, which is the weakest of the three known fundamental forces ■

Newton's Second Law of Motion

Newton's **second law of motion** may be stated as follows: *The acceleration of a body is directly proportional to and in the same direction as the total force acting on the body and inversely proportional to the mass of the body.*

By definition, a force is an action exerted on a body that changes its state, either of rest or of uniform motion in a straight line. Newton's second law of motion states the numerical relationship between the force applied to a body and the acceleration experienced by the body.

Acceleration is always in the direction in which the force is applied. When a body, such as an automobile, moves in a straight line, an increase in velocity means that the force and acceleration are forward. If the automobile slows down, the force and acceleration are in opposition to the direction in which the automobile is moving.

According to Newton, **mass** is *the characteristic that gives a body its inertia,* in other words, its tendency to resist changes in motion. Newton also showed that the attractive force of gravity between two objects depends on the mass of the objects (Newton's law of universal gravitation is discussed in the next chapter). Thus, mass is used in two senses: When mass is used as *a measure of inertia* it is called **inertial mass;** when mass is used as *a factor to determine the force of gravity* it is called **gravitational mass.** It has been found experimentally that the inertial and gravitational mass of any particular object are the same to a high degree of accuracy, a relationship called the *principle of equivalence,* so that when using the term "mass" it is not necessary to always specify whether gravitational or inertial mass is meant. Since inertial mass is a measure of the resistance an object has to changes in its present state of motion, the greater an object's mass, the more difficult it is to impart motion to that object when it is stationary, or the more difficult it is to stop that object when it is moving.

Newton's second law in mathematical terms is written

$$\text{acceleration} = \frac{\text{force}}{\text{mass}},$$

or in symbols,

$$\boldsymbol{a} = \frac{\boldsymbol{F}}{m}. \qquad \text{ACCELERATION}$$

Mass has no direction, so the boldface lettering and the equality sign indicate that the direction of $\boldsymbol{a}$ and $\boldsymbol{F}$ are the same.

From Newton's second law it can be seen that if a force $\boldsymbol{F}$ is applied to an object of mass m, it will accelerate the object; and if the force is increased to twice its original value, the acceleration will be twice its original value. Conversely, note how mass is a measure of inertia. A force $\boldsymbol{F}$ applied to a body of mass m causes the mass to accelerate, while the same force applied to a different body having twice the mass will give the second body an acceleration that is half as great as that of the first body.

Newton's second law is usually written without the fraction in the rearranged form

$$\boldsymbol{F} = m\boldsymbol{a}. \qquad \text{FORCE}$$

If any two of the three quantities in Newton's second law of motion are known, the third can easily be found by rearranging the equation to solve for the unknown quantity.

The most widely used system of measurement (SI, discussed in more detail later in this section) uses the meter as the standard of length, the second as the standard of time, and the kilogram as the standard of mass. If these standards are used, the unit of force must, according to Newton's second law of motion, be in units of $kg \cdot m/s^2$. The **unit of force** known as the **newton** is defined as follows: *1 newton (N) is that unit of force that will impart an acceleration of 1 meter per second squared to a mass of 1 kilogram,* or in equation form,

$$1 \text{ N} \equiv 1 \text{ kg} \cdot \text{m/s}^2.$$

The **weight** of a body is *a measure of the attractive force of Earth or some other celestial body on an object due to gravity.* Thus, weight is a specialized term used for a particular force. In terms of weight, the expression of Newton's second law, $F = ma$, is usually written

$$w = mg. \qquad \text{WEIGHT}$$

Weight (w) is the force (F) with which Earth attracts an object of mass (m). The symbol g represents the acceleration due to gravity, which was found by Galileo to be the same for all freely falling objects, neglecting air resistance and the buoyancy of the air. The magnitude of the acceleration due to gravity has been measured with great precision; its accepted value is 9.80665 meters per second squared, as standardized by the International Committee on Weights and Measures. For *computational purposes, we will use the value 9.80 meters per second squared (980 cm/s²).* The acceleration due to gravity varies slightly with geographic location, so its actual value depends on position on Earth. For instance, the mean (average) acceleration due to gravity at Earth's equator has been calculated to be 9.78057 meters per second squared and the mean acceleration due to gravity at the poles has been found to be 9.83225 meters per second squared. (The variation occurs mostly because Earth is not quite spherical and is rotating.) Of course, the direction of the acceleration due to gravity is downward. (In English units, $g = 32.2$ feet per second squared.)

If the weight of an object is known, its mass can be calculated by rearranging the equation given for weight into the form

$$m = w/g. \qquad \text{MASS}$$

Since the mass of an object is constant, the relationship of w to g is such that if Earth's attraction for an object is less at one location than at another, the acceleration due to gravity is proportionally less. In experiments, the weight of an object is often measured by a scale that uses a tension spring. Since the tension the spring can exert is unaffected by location, the scale readings will vary slightly depending on the geographic location. Mass, on the other hand, is not affected by the location.

It should be noted that even though the preceding equation for mass is for a freely falling object, neither the mass nor the weight at a given location would be different if the object had other forces acting on it. Suppose, for example, a book is resting on a table. The table exerts an upward force on the book, but this upward force does not change the force (the weight) with which Earth attracts the book. The upward force is simply another force that must be considered when using Newton's second law to determine the acceleration of the book. In other words, the equation to determine the mass may still be used if the object is not in free fall. Experimental determination of mass may be accomplished by a beam balance, as illustrated in Figure 3.18.

Figure 3.18 Masses on each side of the scale are equal. A variation in weight may occur because of a variation in g at different locations, but the scale would remain in balance since the change would be proportional on both sides. Mass is independent of location; weight is not.

An excellent example of how mass and weight differ comes from one of our ventures into space. A 300-pound camera was taken by the astronauts to the moon. The camera was loaded prior to the launch by several strong technicians. However, when the astronauts reached the surface of the moon, the camera was removed from the landing vehicle and carried to the surface of the moon by a single astronaut, using an effort of only about 50 pounds. The camera was placed on a tripod, and the first pictures were taken. But when the astronaut wanted to aim the camera in a different direction, he had to apply the same force to turn the camera that he would have had to use on the surface of Earth. The force due to the gravity of the moon is about one-sixth that of the force due to the gravity of Earth; therefore, the weight of the camera was affected accordingly. However, the mass of the camera, a measure of its inertia, did not change. If the camera were to be taken to the planet Jupiter, the weight of the device would be about 780 pounds, since the force due to the gravity of Jupiter is about 2.6 times that of Earth. The effort required to turn the camera, however, would remain the same as on Earth or the moon. The mass of a body is a measure of the body's inertia—an intrinsic characteristic of the body, independent of the object's location.

Any object drifting through outer space has only a weak gravitational force acting on it; therefore, its weight may be considered negligible. We could assume, then, that the camera was weightless as it traveled in the space vehicle toward the moon. However, had an astronaut attempted to change the direction in which the camera pointed or to push or pull it out of the way, the same effort would have been required as for horizontal motion on the surfaces of Earth, Jupiter, or the moon.

Table 3.1

A comparison of various units of the three systems of measurement. BASE UNITS in each system are in UPPERCASE LETTERS, while derived units are in lowercase letters.

System: *Measurement:*	**SI**	**cgs**	**fps**
Length	METER (m)	CENTIMETER (cm)	FOOT (ft)
Mass	KILOGRAM (kg)	GRAM (g)	Slug (sl)
Time	SECOND (s)	SECOND (s)	SECOND (s)
Force	Newton (N)	Dyne (d)	POUND (lb)
Velocity	Meter per second (m/s)	Centimeter per second (cm/s)	Foot per second (ft/s)
Acceleration	Meter per second² (m/s²)	Centimeter per second² (cm/s²)	Foot per second² (ft/s²)

Convention plays an important role in our society. The example of the camera speaks of the weight of an object in pounds, yet earlier it was stated that force is usually expressed in newtons. A brief synopsis of the three major systems of units follows.

The International System of Units (Le Système International d'Unités in French), abbreviated SI in all languages, is the modern version of the "metric system." SI is based on seven *fundamental units,* three of which are the meter, the kilogram, and the second. (Some of the others are discussed in later chapters.) Most countries of the world have adopted SI as the legal standard, so that virtually all scientific work is carried out using SI. All other units in scientific and engineering work are based on the seven fundamental units (and some mathematical definitions), and are called *derived units.* For example, velocity is based on the meter and the second; force is based on the kilogram, the meter, and the second.

There are two other systems of units that are still in use today, even though their importance in scientific work has diminished. The centimeter-gram-second (*cgs*) system differs from SI in that it uses the centimeter as the standard of length and the gram as the standard of mass. The foot-pound-second system (*fps* or *"English" system*) uses the foot as the standard of length and, instead of having a standard for mass, uses the pound as the standard for force. The unit of mass in the *fps* system is a derived unit obtained from Newton's second law and is called the "slug." A summary of the fundamental and derived units of all three systems is given in Table 3.1, and all three are discussed in detail in Appendix 1. Also, conversion factors allowing one to convert units from one system to another can be found inside the back cover of this text.

Example 3.9

What is the weight, in newtons, of a 5.00-kilogram bag of sugar?

Solution

Applying Newton's second law (choosing downward as +):

$$F = ma \quad \text{or} \quad w = mg,$$

as presented earlier. Then

$$w = (5.00\text{ kg})(9.80\text{ m/s}^2) = 49.0\text{ N}.$$

So the bag of sugar weighs 49.0 N (about 11 lb) and, of course, the weight is downward.

Example 3.10

Determine the force, in newtons, a cable exerts on a crate, the mass of which is 100 kilograms, if the crate is held suspended at rest. (See Fig. 3.19.)

Solution

There are two forces acting on the crate: the weight of the crate pulling downward and the cable pulling upward. Since the mass of the crate is given as 100 kg, the weight of the crate can be calculated from the equation

$$w = mg = (100\text{ kg})(9.80\text{ m/s}^2) = 980\text{ N}.$$

Now the application of Newton's second law, with upward being chosen as positive (+), gives

$$F - 980\text{ N} = (100\text{ kg})(0) = 0,$$

because the acceleration of the crate is 0. Solving for F gives

$$F = 980\text{ N}.$$

Therefore, an upward force of 980 N is required to hold the crate suspended at rest.

Test

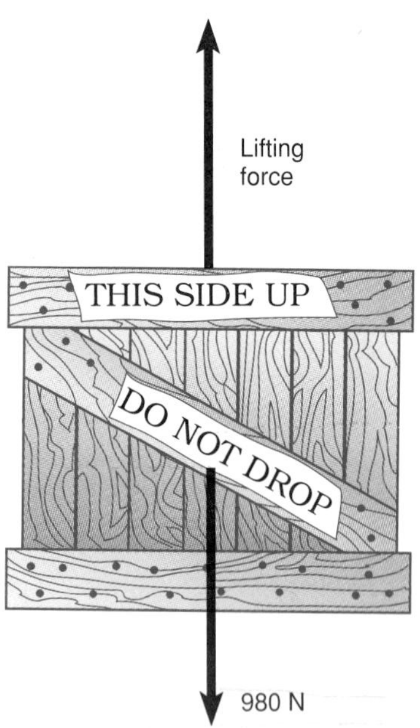

Figure 3.19 The lifting force required to hold a crate is equal and opposite to its weight.

Example 3.11

Determine the constant force in newtons a baseball bat must exert on a baseball of mass 0.143 kilograms to give it an acceleration of 9500 meters per second squared (see Fig. 3.20). Neglect the *weight* of the ball.

Solution

From Newton's second law,

$$\begin{aligned} F &= ma \\ &= (0.143\ \text{kg})(9500\ \text{m/s}^2) \\ &= 1360\ \text{kg}\cdot\text{m/s}^2 \\ &= 1360\ \text{N}. \end{aligned}$$

The bat must exert a force of 1360 N on the ball to give it the required acceleration. (The weight of the baseball is negligible, since it is only 1.40 N.)

Example 3.12

A small wagon that weighs 294 newtons is at rest on a level sidewalk. What force is required to give the wagon an acceleration of 2.00 meters per second squared?

(A force of 294 newtons would be required to lift the wagon; however, a lesser force, as the solution reveals, is needed to accelerate the wagon horizontally at the rate specified.)

Solution

In order to apply Newton's second law of motion, we must first calculate the mass of the wagon. From the previous discussion relating mass and weight,

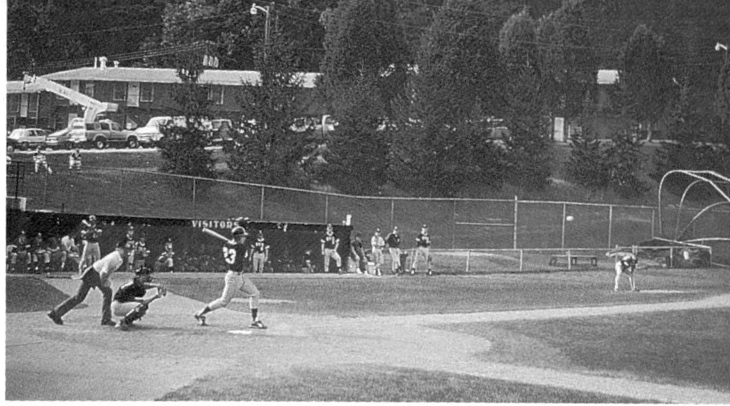

Figure 3.20 The bat must exert a relatively large force on the baseball to give it the required acceleration to reverse its direction.

$$\begin{aligned} m &= \frac{w}{g} \\ &= \frac{294\ \text{N}}{9.80\ \text{m/s}^2} \\ &= \frac{294\ \text{kg}\cdot\text{m/s}^2}{9.80\ \text{m/s}^2} \\ &= 30.0\ \text{kg}. \end{aligned}$$

Then (choosing the direction of the acceleration to be positive), the solution becomes

$$\begin{aligned} F &= ma \\ &= (30.0\ \text{kg})(+2.00\ \text{m/s}^2) \\ &= +60.0\ \text{kg}\cdot\text{m/s}^2 \\ &= +60.0\ \text{N}. \end{aligned}$$

The magnitude of the required force is 60.0 N and its direction is the same as that of the acceleration.

Example 3.13

In order to relate the laws of motion further, consider a fan cart (see Fig. 3.21), the mass of which, including a weight it carries, is 0.800 kilogram. As the fan is turned on, the cart moves across a table a distance of 3.00 meters in 5.00 seconds. Determine the following information about the fan cart in the described situation: (a) its initial velocity, (b) its average velocity, (c) its final velocity, (d) its acceleration, and (e) the force of the fan.

Solution

(a) The fan cart starts from rest, so its initial velocity is zero. That is, $v_i = 0$.

(b) The fan cart requires 5.00 s to travel 3.00 m; therefore, its average velocity is found from

$$\begin{aligned} \bar{v} &= \frac{d_f - d_i}{t} \\ &= \frac{+3.00\ \text{m}}{5.00\ \text{s}} \\ &= 0.600\ \text{m/s}. \end{aligned}$$

(c) The fan cart is assumed to have constant acceleration, so its final velocity is calculated by recognizing that if $v_i = 0$, then

$$\bar{v} = \tfrac{1}{2}(v_f + v_i) = \tfrac{1}{2}v_f,$$

and it follows that

$$v_f = 2\bar{v} = 2\,(+0.600 \text{ m/s}) = +1.20 \text{ m/s}.$$

(d) Then the acceleration can be found according to

$$a = \frac{v_f - v_i}{t} = \frac{(+1.20 \text{ m/s}) - 0}{5.00 \text{ s}} = +0.240 \text{ m/s}^2.$$

(e) The force of the fan can then be calculated from

$$F = ma = (0.800 \text{ kg})(+0.240 \text{ m/s}^2) = +0.192 \text{ kg}\cdot\text{m/s}^2 = +0.192 \text{ N}.$$

Figure 3.21 A fan cart is accelerated in a direction opposite to the direction in which the fan blows the air.

Newton's Third Law of Motion and Momentum

The **third law of motion** discussed in Newton's writings is applicable to all motion. The concept involves the realization that forces always occur in pairs. The third law may be stated as follows: *When one object exerts a force on another, the second object exerts an equal but opposite force on the first.* This relationship is often referred to as action-reaction. The illustrations in Figure 3.22 depict the concept from four slightly different points of view. In Figure 3.22a, the rocket moves upward because the exhaust gases are expelled downward; in 3.22b, the tennis ball is forced forward while in contact with the racket while the ball exerts a backward force on the racket. In 3.22c, the shot is forced out of the barrel by the expanding gases while the shotgun recoils; and in 3.22d, the automobile pushes downward while the compressed gases in the hydraulic lift push upward, maintaining a state of equilibrium. In all of these examples, note that the action-reaction pairs act on *different objects.* Two forces acting on the same object, even if they are equal and opposite, can never be an action-reaction pair. Simply stated, the action-reaction principle is the force of object 1 on object 2 = minus the force of object 2 on object 1.

When two objects interact, the relation of equal but opposite forces may be expressed mathematically by the equation $\boldsymbol{F_1} = -\boldsymbol{F_2}$, where $\boldsymbol{F_1}$ is acting on mass m_1 and $\boldsymbol{F_2}$ is acting on mass m_2. If the established equation of Newton's second law of motion, $\boldsymbol{F} = m\boldsymbol{a}$, is applied, the expression $\boldsymbol{F_1} = -\boldsymbol{F_2}$ can be written $m_1\boldsymbol{a_1} = -m_2\boldsymbol{a_2}$.

This last equation can be utilized to find the changes in velocities of the two objects, even when the forces, themselves, cannot

a.

b.

c.

d.

Figure 3.22 Various conditions that indicate action and reaction.

be determined. Recall that the acceleration $\boldsymbol{a}$ gives the change in velocity divided by the time; that is, $\boldsymbol{a} = (\boldsymbol{v}_f - \boldsymbol{v}_i)/t$. If the accelerations of each object are substituted into the equation $m_1\boldsymbol{a}_1 = -m_2\boldsymbol{a}_2$, it can be shown that

$$m_1v_{f1} + m_2v_{f2} = m_1v_{i1} + m_2v_{i2}, \qquad \text{CONSERVATION OF MOMENTUM}$$

where v_{f1} is the final velocity of mass 1, v_{i1} is the initial velocity of mass 1, etc. (See Appendix 2.) *The product of mass and velocity, mv,* is called **linear momentum.** (The angular momentum of rotating bodies is discussed in the next chapter.) The sum of the two momenta on the left side of the equation is the *total momentum after the interaction,* while the sum on the right side of the equation is the *total momentum before the interaction.*

This last equation is a statement of one of the most powerful laws in physics, the **law of conservation of momentum,** which says that *the total momentum of an isolated system before an interaction, usually a collision, is equal to the total momentum after the interaction.*

The momentum of a body is a vector quantity, having the same direction as the velocity, $\boldsymbol{v}$. The general application of the law of conservation of momentum is beyond the scope of this text, but straight-line (one-dimensional) interactions will be considered. In one-dimensional motion, one direction for velocity must be chosen as positive and the other as negative, just as in the previous work in one dimension.

Example 3.14

The bullet from a hunter's rifle has a mass of 0.300 kilogram. When the rifle is fired, the bullet leaves the rifle with a velocity of 370 meters per second. If the mass of the rifle is 5.00 kilograms, what is the velocity with which the rifle recoils?

Solution

This problem may be solved by using conservation of momentum. The momentum before the bullet is fired is zero (both masses have an initial velocity of zero), so the equation for conservation of momentum becomes

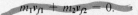

$$m_1v_{f1} + m_2v_{f2} = 0.$$

Letting object 1 be the bullet and object 2 the rifle (of course, they could be chosen the other way), then $m_1 = 0.0300$ kg, $m_2 = 5.00$ kg, $v_{f1} = +370$ m/s, and v_{f2} is unknown. Substituting into the equation gives

$$(0.0300 \text{ kg})(370 \text{ m/s}) + (5.00 \text{ kg})\,(v_{f2}) = 0,$$

and solving for v_{f2} gives

$$v_{f2} = -\frac{(0.0300 \text{ kg})(370 \text{ m/s})}{5.00 \text{ kg}} = -2.22 \text{ m/s}.$$

The minus sign indicates that the velocity of the rifle is in the negative direction. This observation is meaningful since the velocity of the bullet was taken to be +370 m/s (the + sign was indicated explicitly to point out that the direction is important), so the rifle "recoils." (It is the practice of an experienced shooter to brace the weapon tightly against the shoulder. This effectively increases the recoiling mass, thereby significantly lowering the recoil velocity.)

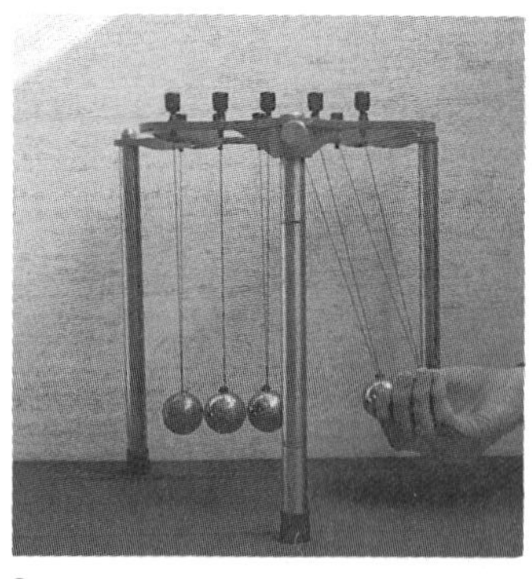
a.

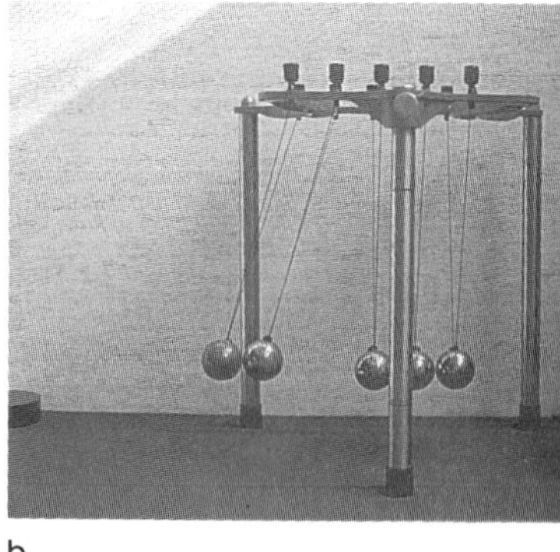
b.

Figure 3.23 (*a*) Two balls are pulled back, then released so as to strike three identical balls at rest. (*b*) Upon impact, the momentum of the two balls released is assumed by the two balls in front. Conservation of both momentum and energy require that two balls assume motion in the experiment shown.

The linear momentum of an individual body is changed when the net force acting on the body is not zero. The total change in momentum depends on the magnitude and direction of the force and the length of time during which the force acts (see Appendix 2 for the mathematical details).

The vectorial nature of momentum, combined with the conservation of momentum, is very useful in explaining the results of such events as explosions and collisions. For example, the momentum of the gases escaping from a rocket is in one direction, requiring that the rocket move in the other direction to conserve momentum. Even when objects undergo an inelastic collision, an interaction in which they stick together, their total momentum is The composite body they form assumes the total momentum the colliding bodies had before collision. The differences in mass and velocity, including direction, determine the velocity the composite body assumes.

The collision between two identical pool balls moving directly toward each other (in opposite directions) closely resembles an elastic collision—one in which the objects rebound from each other. If each ball were moving at the same speed before collision, they would maintain practically the same speed after collision, but each would move in a direction opposite to its original direction. The momentum of the system will be conserved. If a stationary pool ball is struck by a moving one that is cued so as to stop upon collision, the once-stationary ball assumes the momentum of the moving ball (ignoring all other forces that act on the two balls). Conservation of momentum is illustrated in a collision between balls in Figure 3.23.

Summary

The three laws of motion, conceived at least in part by Newton, are generally known as Newton's three laws of motion. Newton's first law of motion states that an object maintains its state of motion, either at rest or at constant speed in a straight line, unless acted upon by external forces that do not add to zero. Thus, bodies possess inertia, the property by which they tend to maintain their state of motion.

Newton's second law of motion describes quantitatively the effect forces have on the motion of objects. A force is a push or pull that affects the state of motion of an object. The force must be from some type of interaction with another object. Although there are only three known fundamental forces, it is convenient to think of two general classes of forces: direct "contact"-type forces, such as the push from a person, friction between surfaces, collisions with other objects, wind, water, and so forth; and "action at a distance" forces, such as gravity or the electrical forces between charged particles. A force is a vector quantity; that is, it has magnitude and direction. A force causes the object on which it acts to accelerate—to change its speed and/or direction. The acceleration is in the direction of the applied force and directly proportional to it. Objects also have a property called mass. Mass is the quantitative measure of inertia of an object, so that the greater the mass, the more difficult it is to change the state of motion of the object.

The weight of an object is the force of gravitational attraction of some astronomical body, such as Earth or the moon, on an object. Weight depends on the location of the body relative to the object causing the pull, but mass is independent of the location of the object; it is an actual property of the object. Objects dropped at the same location always fall with the same acceleration, neglecting air resistance. The accepted value of the acceleration of gravity near Earth's surface is about 9.81 m/s^2 (32.2 ft/s^2), allowing one to compute the weight of an object if its mass is known and vice versa.

Forces always occur in pairs according to Newton's third law. If body A exerts a force on body B, then body B must exert an equal and opposite force on body A. Such pairs of forces are often called action-reaction pairs. The third law can be used to find forces on objects exerted by other objects and leads to the important law of conservation of momentum. Momentum is the mass times velocity of an object, and since velocity is a vector quantity, momentum is also. The total momentum of interacting objects does not change if there are no external forces acting on the objects; in other words, if only action-reaction pairs are considered.

Newton's three laws of motion form the basis of most mechanical engineering, since they govern the motion of "ordinary" objects. They, as most other "laws," have limits. When considering atomic-sized particles, quantum theory often must be used, and when considering objects moving at speeds near the speed of light, relativity theory must be used.

Equation Summary

Displacement:

$$\text{change in displacement} = \boldsymbol{d_f} - \boldsymbol{d_i},$$

where the change in displacement (meters) is equal to the final displacement $\boldsymbol{d_f}$ (meters) minus the initial displacement $\boldsymbol{d_i}$ (meters). For one-dimensional motion, the directions of the displacements are indicated by their signs.

Average speed:

$$\bar{v} = \frac{s}{t},$$

where average speed $\bar{v}$ (meters per second) is equal to the total distance traveled s (meters) divided by the time t (seconds) required to travel the distance. Alternate forms of the equation are

$$s = \bar{v}t$$

and

$$t = \frac{s}{\bar{v}}.$$

Average velocity:

$$\bar{\boldsymbol{v}} = \frac{\boldsymbol{d_f} - \boldsymbol{d_i}}{t},$$

where the average velocity $\bar{\boldsymbol{v}}$ (meters per second) is equal to the change in displacement $\boldsymbol{d_f} - \boldsymbol{d_i}$ (meters) divided by the time t (seconds) required for the change in displacement to take place. For one-dimensional motion, direction is indicated by the sign of the velocity.

Acceleration:

$$\boldsymbol{a} = \frac{\boldsymbol{v_f} - \boldsymbol{v_i}}{t},$$

where the acceleration $\boldsymbol{a}$ (meters per second squared) equals the final velocity $\boldsymbol{v_f}$ (meters per second) minus the initial velocity $\boldsymbol{v_i}$ (meters per second) divided by the time t (seconds) required for the velocity to change. In one-dimensional motion, the direction of the acceleration is indicated by its sign.

For constant acceleration,

$$\boldsymbol{v_f} = \boldsymbol{v_i} + \boldsymbol{a}t,$$

$$\bar{\boldsymbol{v}} = \frac{1}{2}(\boldsymbol{v_i} + \boldsymbol{v_f}),$$

$$\boldsymbol{d_f} = \boldsymbol{d_i} + \bar{\boldsymbol{v}}t,$$

and

$$\boldsymbol{d_f} = \boldsymbol{d_i} + \boldsymbol{v_i}t + \frac{1}{2}\boldsymbol{a}t^2.$$

In these equations, $\boldsymbol{d_i}$ is the initial displacement (meters); $\boldsymbol{d_f}$ is the final displacement (meters), $\boldsymbol{v_i}$ is the initial velocity (meters per second); $\boldsymbol{v_f}$ is the final velocity (meters per second); $\bar{\boldsymbol{v}}$ is the average velocity (meters per second); $\boldsymbol{a}$ is the acceleration (meters per second squared); and t is the time (seconds). For one-dimensional motion, the direction of all quantities except t is indicated by the sign of the quantity.

Newton's second law of motion:

$$a = \frac{F}{m},$$

where the acceleration $\boldsymbol{a}$ (meters per second squared) is equal to the total force $\boldsymbol{F}$ (newtons) acting on an object divided by the mass m (kilograms) of the object. An alternate form of this equation is

$$F = ma.$$

Weight:

$$w = mg,$$

where the weight $\boldsymbol{w}$ (newtons) is equal to the mass m (kilograms) times the acceleration of gravity $\boldsymbol{g}$ (9.80 meters per second squared downward). The mass of the object can be found by dividing the magnitude of the weight by the magnitude of acceleration of gravity; thus,

$$m = \frac{w}{g},$$

Newton's third law of motion:

$$F_1 = -F_2,$$

where the force $\boldsymbol{F_1}$ (newtons) acting on object 1 exerted by object 2 is equal and opposite to the force $\boldsymbol{F_2}$ (newtons) acting on object 2 exerted by object 1.

Momentum:

$$\text{momentum} = mv,$$

where the momentum (kilogram-meter per second) of an object is equal to its mass m (kilograms) times its velocity $\boldsymbol{v}$ (meters per second). The direction of the momentum is the same as that of the velocity, and is indicated by the sign for one-dimensional motion.

Conservation of momentum:

$$m_1v_{f1} + m_2v_{f2} = m_1v_{i1} + m_2v_{i2},$$

where the left side of the equation is the sum of the momenta mv (kilogram-meter per second) before interaction (usually collision) and the right side is the sum of the momenta after the interaction. The direction of each of the momenta must be indicated by the sign for one-dimensional motion.

Questions and Problems

Newton's First Law of Motion and Inertia

1. Why is Newton's first law known as the "law of inertia"? Define inertia in your words.
2. Suppose a spacecraft is out in deep space, far from any objects. (a) Does it need to keep its rocket engine running to maintain a steady speed? Explain. (b) Does the rocket engine need to be fired if a change in direction of motion is desired? Explain.
3. (a) When the driver of a car slams on the brakes, why are the passengers "thrown" forward if they are not wearing seat belts? (b) Explain why passengers in a car seem to be "thrown" toward the right side of the car if the car makes a sharp turn to the left.
4. A book rests on a table. The downward force of gravity on the book (that is, its weight) is 5.25 N. What must be the direction of the force the table exerts on the book? What is the magnitude (size) of the force the table exerts on the book?

Scalars and Vectors

5. (a) State the meaning of the term *scalar* and give two examples of scalars. (b) State the meaning of the term *vector* and give two examples of vectors. (c) Using a drawing, show how two vectors that are in the same direction are added together. Show how two vectors that are in opposite directions are added together.
6. A student exits the door of her dormitory and walks along a straight sidewalk for 50.0 m when she hears a friend calling to her. If her friend is on the sidewalk 25.0 m from her, what is the greatest distance her friend could be from the door of the dormitory? What is the least distance her friend could be from the door of the dormitory?
7. A person in a parade is riding on a float that is moving at 10.0 mi/h eastward along the street. The person is throwing small pieces of bubble gum into the crowd at a speed of 5.00 mi/h. Relative to the street, what can be the maximum speed of a piece of gum? In what direction will a piece of gum be moving at the maximum speed? What is the minimum speed a piece of gum can have? In what direction will the gum be moving at the minimum speed?
8. A swimming course is laid out on a river so that the swimmers swim 1.00 km downstream and 1.00 km upstream to finish at the starting point. The water in the river is flowing at a speed of 0.500 m/s. If a swimmer can swim 1.20 m/s, (a) how long will it take him to swim 1.00 km down the river? (b) How long will it take him to swim 1.00 km back up the river? (c) What will be his average *speed* for the entire swim? (d) If the river was flowing fast enough, the swimmer would never be able to swim upstream to the starting point. How fast would the river have to be flowing for this to occur? Explain.

Straight-Line Motion

9. A car covers a distance of 150 km (93.2 mi) in a time of 6310 s (1¾ h). What is the average speed of the car? Is it possible with this information to determine the average velocity of the car? Is it possible to determine the speed at a particular time?
10. Running northward, a runner covers a distance 125 m in 27.2 s and continuing onward goes another 275 m in 87.3 s. (a) What is his average *velocity* during the first 27.2 s? (b) What is his average *velocity* during the next 87.3 s? (c) What is his average *velocity* over the entire time?

11. A girl rides a bicycle eastward along a straight road. Her distances from the starting point at different times are as follows:

Time (seconds)	*Distance (meters)*
0.00	0.00
1.00	2.00
2.00	5.00
3.00	9.00
4.00	14.0
5.00	19.0

(a) Find the girl's *change in displacement* for each second of the motion. (The change is the difference between the displacement at the beginning and end of the given second.) (b) Find her average velocity (not speed) during each second of the motion. (c) What is her average velocity over the entire 5.00 seconds? (d) What is her average acceleration between the first and second second? (e) What is her average acceleration between the fourth and fifth second?

12. A car starting from rest at a traffic light has a northward acceleration of 1.52 m/s². (a) What is the velocity of the car after 4.00 s? (b) What is the displacement of the car relative to its starting point after 4.00 s?

13. During an Independence Day celebration, a small skyrocket is launched. It starts from rest and after 1.65 s has an upward velocity of 97.6 m/s. (a) What is the average acceleration of the skyrocket during the first 1.65 s? (b) What was the displacement of the skyrocket 1.65 s after launch, assuming the acceleration to be constant?

Newton's Second Law of Motion

14. How are Newton's first and second laws of motion related?

15. The net force acting on a mass of 55.0 kg is 250 N toward the east. (a) What is the acceleration of the mass? (b) Assuming the force remains constant and the mass starts from rest, what is its velocity after 8.00 s?

16. Objects do not have to be "in contact" to exert certain types of forces on one another. What are two such types of forces?

17. A truck of mass 7500 kg is moving eastward along a level road when the driver slams on the brakes. The braking force gives the truck an acceleration of -2.36 m/s² (+ is taken as eastward). (a) Find the braking force acting on the truck (give magnitude and direction). (b) There are forces other than the braking force acting on the truck. What are these and why do they not affect the acceleration of the truck?

18. A 20.0 N weight is hanging from a wire in an elevator. While the elevator is accelerating upward, is the tension in the wire equal to 20.0 N, less than 20.0 N, or greater than 20.0 N? Explain.

19. (a) What is the mass of a person who weighs 725 N on Earth? (b) What is the mass of the same person on the moon? (c) If the acceleration of the moon is 1.66 m/s², what is the weight of this person on the moon?

Newton's Third Law of Motion and Momentum

20. If forces always come in equal and opposite pairs, why don't all forces cancel out so that nothing can ever accelerate?

21. When a person jumps from a table onto the floor according to Newton's third law of motion, the force of Earth on the person and the force of the person on Earth are equal and opposite. Why doesn't Earth accelerate up to meet the person? If it does, why aren't the acceleration of the person and Earth the same?

22. Sketch a rough vector diagram showing the equal and opposite force Newton's third law says exists for each force listed below. Show the object upon which the force acts, and label the vectors to indicate their magnitude and direction. (a) the weight of a freely falling 125-N rock; (b) the weight of a 125-N rock resting on the ground. (Hint: It's not the contact force of the ground on the rock); (c) the force with which a 125-N rock presses down on the surface of Earth; (d) the force of a car on a solid wall when the car crashes into it; (e) the force exerted by scales on a person who weighs 600 N standing on them if the person is not accelerating.

23. (a) Why does the nozzle of a garden hose push forcibly backward when the hose projects a swift stream of water forward? (b) Why does a carpenter find the chore of driving a nail into a loose board a difficult one?

24. A large truck with a mass of 40,000 kg has a velocity of 25.0 m/s eastward. At the same time, a car with a mass of 1500 kg has a westward velocity of 30 m/s. Let eastward be the positive (+) direction. (a) Multiply the mass of the truck by its velocity to find the momentum of the truck. Indicate the sign of the momentum to show its direction. (b) Repeat the procedure to find the momentum of the car. (c) Sum the momenta, including appropriate signs to indicate direction, to find the total momentum of the truck and car together. (d) If the car and truck collide head-on and "stick together," how much momentum is lost in the collision? (e) Find the velocity of the wreckage after the collision. (Since the car and truck "stick together," the mass of the wreckage is the sum of the two masses, and both car and truck must be moving with the same velocity. Add the masses together and multiply by v, the unknown velocity, then set that momentum equal to the total momentum found in part [c].)

25. (a) Why do scientists and engineers consider Newton's laws of motion to be so useful? (b) Why do students often find Newton's laws of motion hard to use?

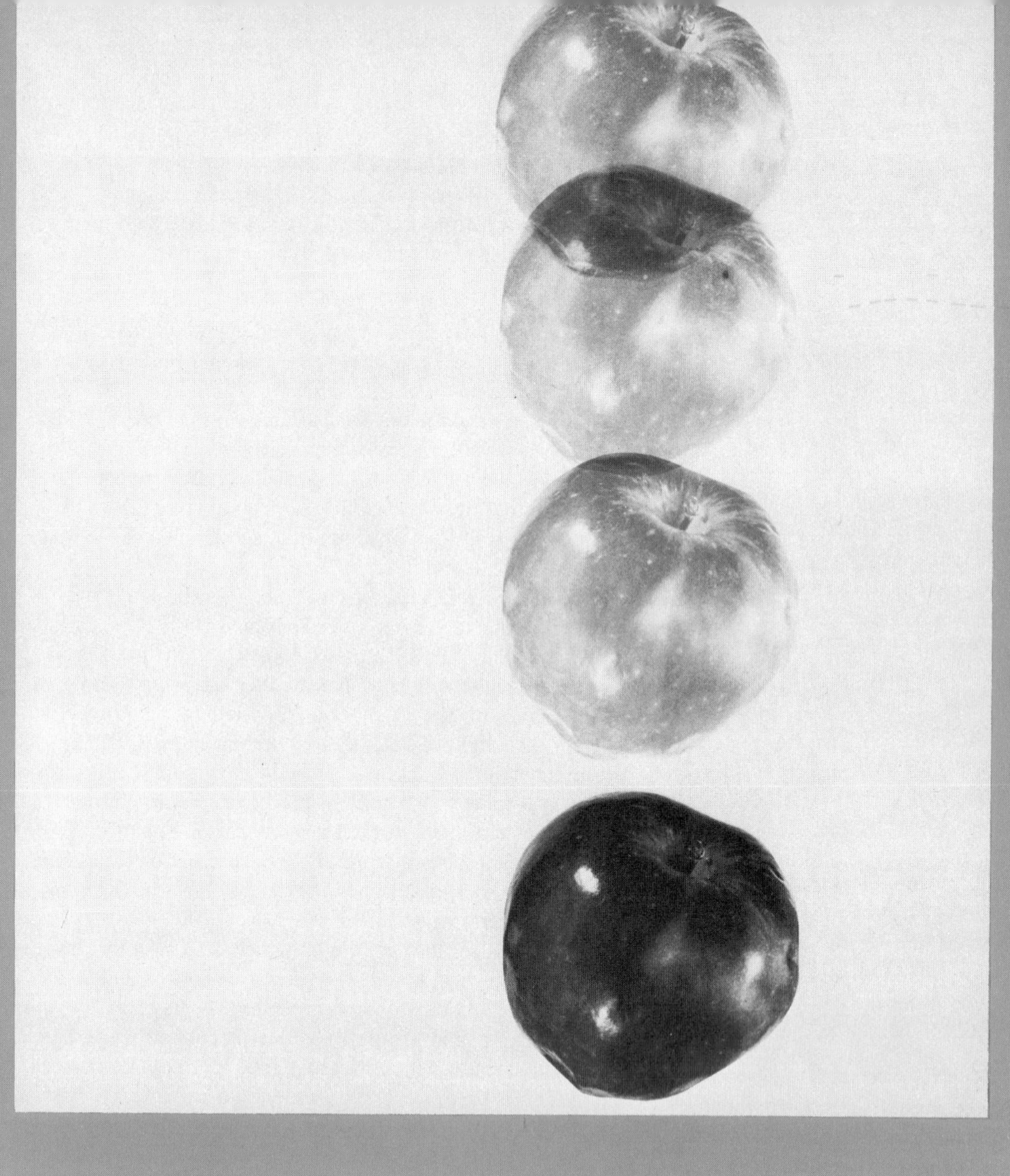

Universal Gravitation and Applications of Newton's Laws

Chapter outline

Key terms/concepts

Newton's observation of a falling apple and the orbiting moon inspired him to investigate the dependence of gravitational force on the distance from the center of Earth.

The application of Newton's three laws of motion enables scientists and engineers to theoretically work out the motions of objects and to study the forces acting on stationary objects. Uses extend from orbital mechanics (calculation of the paths of planets, spacecraft, stars, and even galaxies) to statics (studies of the forces and stresses on and in stationary objects). All of the natural sciences in some way use Newton's laws as do all of the disciplines of engineering.

The basic laws deal with the motions of the simplest of entities, the "particle." If the motion of an object does not depend on its physical size or orientation, or points of application of forces, but only on its mass, the object can usually be considered a particle. In everyday language, a particle is something small, but an astronomer studying the orbital motion of Earth around the sun may consider Earth to be a particle if the rotation of Earth is not important in the calculations. Investigation of the motions of particles under the influence of forces is an ongoing effort, even more than 300 years after the original publication of Newton's three laws. Exact solutions to problems that involve the motions of two particles can be computed; but when more than two particles are involved, the motions cannot usually be determined exactly. The study of many-particle systems may seem to be a hopeless endeavor, but many ingenious techniques have been developed to cope with the complexities. A system of particles such as the solar system, with the sun, nine planets, and many moons, can be dealt with quite effectively.

The advent of electronic computers has afforded scientists and engineers a powerful new tool for the study of motions. Since computers in a few seconds can perform calculations that may otherwise take a person months to complete, they can determine the motions of hundreds of particles that might make up a system. In addition, they can present the answers in a graphic or pictorial form that scientists or engineers can easily comprehend. Even though such computations made with computers are not "exact," they often can be made with sufficient accuracy to test some hypotheses, and they are routinely used to make engineering design decisions.

Of course, not all objects can be considered to be "particles" because rotation or physical size of objects is important in many applications. When this is the case, the motion of the object as a whole is usually easier to deal with than the motion of all the individual particles that make up the object. Indeed, a rotating wheel contains many billions of particles, but accounting for the motions of each one would be of little use when it is the motion of the wheel as a whole that is of significance. Newton, himself, recognized that extended objects (in other words, all of the particles treated collectively) needed to be considered so that the number of forces and equations could be kept to some manageable number. He, along with Leibniz, a German mathematician and a contemporary, created a new form of mathematics known as integral calculus to aid in such computations.

All of the ideas in the previous few paragraphs may be summarized in the following observations. When scientists and engineers want to apply Newton's laws, they must make up some "model" of the system they are studying. The model usually consists of several parts. First, a physical model is devised—some sort of "picture" representing the object under study. The second phase of the model involves determining interactions between the parts of the system as completely as possible. Third, a mathematical model is formulated from the second phase, using the interactions and Newton's laws (or equations derived using them as a basis), and this mathematical model is "solved." The last step is the interpretation of the results of the mathematical model. This last step is critical in evaluating the meaning of all the previous work. It enables scientists or engineers to design new experiments, to develop new machinery, or perhaps it forces them to reevaluate the assumptions made in one step of the modeling process.

In this chapter, the student should notice how the models are developed. Most are very simple models, since complete analysis of real systems is far beyond the limitations of this text. It should be remembered, however, that even the most sophisticated models use approximations to forces or neglect small effects. The old adage that "theoretically it should work this way, but actually it works some other way" usually is just a statement indicating that the model neglects some effect or effects that influence the results.

As a final word on models, it should be remembered that Newton's description of motion is, itself, a model, although it is so successful that it is dignified with the term "law."

Utilization of Newton's laws can generally take two forms: In one case the forces are all known, and the objective is to determine the motion of an object; in the second case it may be the motion that is known and the forces to be determined. An example of each case is involved in the flight of the Voyager spacecraft. To calculate the trajectory of the Voyager, the forces of gravity from the sun, planets, and moons near which it passes are utilized. On the other hand, it is possible to detect such small changes in the velocity of the Voyager that astronomers are actively using these motions to search for a possible tenth planet in the solar system. From its gravitational pull (if it exists), the location and size of the planet can be determined.

Newton's Law of Universal Gravitation

The story of Newton and the falling apple is well known. Almost every elementary school student has read the account of how Newton conceptualized gravity. Gravity, however, had been

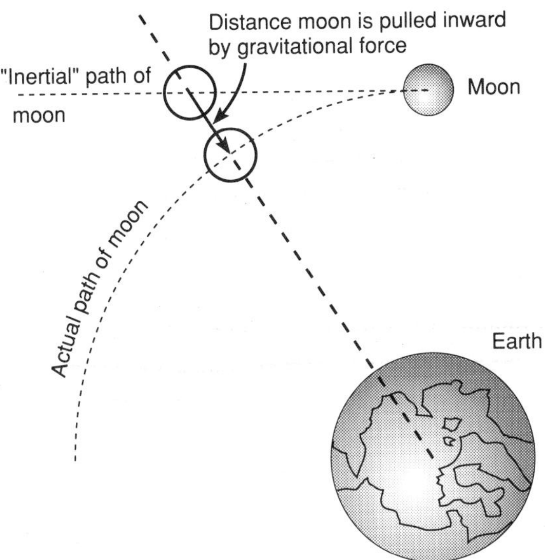

Figure 4.1 The moon's inertia tends to carry it in a straight line, but the gravitational pull of Earth causes it to be drawn inward.

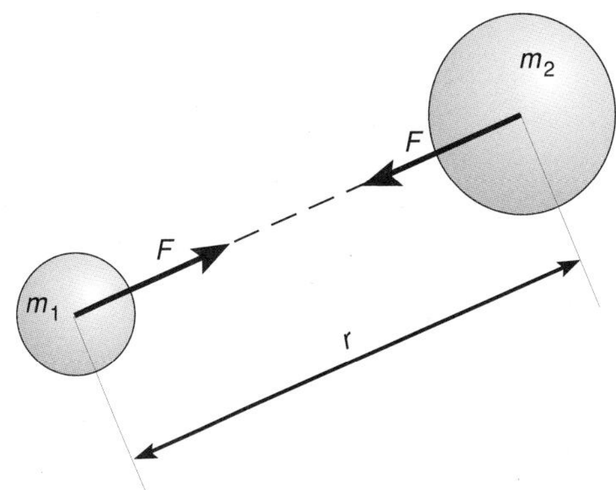

Figure 4.2 Every body in the universe attracts every other body. The force of attraction is directly proportional to the product of the masses of the bodies and inversely proportional to the square of the distance separating their centers (if the bodies are spherically symmetrical). The forces are equal and opposite, as required by Newton's third law of motion.

considered before Newton's time by such great scholars as Copernicus, Galileo, and Kepler. Newton, though, reasoned that even if the apple were to fall from a point some greater distance above the tree, it would still hit the ground. His thoughts continued until they encompassed the moon itself. From his considerations of the laws of motion that he had conceived a short time earlier, it was natural for Newton to reason that the moon falls toward Earth. The reason that it fails to reach Earth he attributed to his first law of motion. The inertia of the moon tends to make it move in a straight line, which would carry it further from Earth. The gravitational pull of Earth causes the moon to fall inward from that line, keeping it in its orbit (see Fig. 4.1). Because of the difficulty in proving that the distance between Earth and moon could be represented by the distance between their centers, Newton did not reveal his conclusions for some twenty years, but in the *Principia,* he stated his famous **law of universal gravitation:** *Every particle of matter (m_1) in the universe attracts every other particle of matter (m_2) with a force that is directly proportional to the product of their masses and inversely proportional to the square of the distance (r) between their centers* (Fig. 4.2). In mathematical terms, the proportion can be stated

$$F \propto \frac{m_1 m_2}{r^2}.$$

(The symbol $\propto$ means "proportional to.") As an equation,

$$F = G\frac{m_1 m_2}{r^2}. \qquad \text{LAW OF UNIVERSAL GRAVITATION}$$

The SI value of G, known as the universal gravitational constant, is 6.67×10^{-11} newton-meters squared per kilogram squared ($N \cdot m^2/kg^2$). The proportionality constant was determined experimentally by Henry Cavendish (1731–1810). By measuring the twist given to two masses attached to a light rigid rod suspended from a thin fiber when two large masses were placed near the small masses, he was able to determine the force of attraction. G could then be determined from the known masses, their separations, and the measured force (see Fig. 4.3). Scientists assume this value for G to be the same everywhere in the universe and to remain constant over time. All experimental evidence gathered indicates these assumptions are valid.

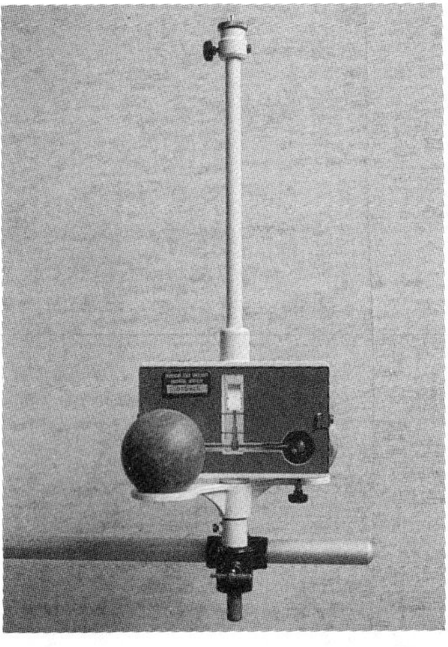

Figure 4.3 A modern version of the Cavendish experiment to determine ***g.*** The force of attraction between the large lead balls and the small lead balls in the enclosure is determined by measuring the twist of a fiber extending up the vertical tube when the large balls are rotated so that they are near the opposite small balls. A light beam reflected off the small mirror at the center is used to determine the amount of twist. From knowledge of the force, the masses, and their separations, ***g*** can be computed using Newton's law of universal gravitation.

There are many applications of Newton's law of universal gravitation, ranging from determining the mass of Earth and other celestial bodies to determining the motion of objects near these bodies and even the distribution of mass within such objects. Knowledge of this mass distribution permits the study of interiors of bodies, such as Earth and the moon. It is useful knowledge for prospecting for mineral deposits and developing various theories about celestial bodies and how they formed.

Falling Bodies

The decline of ancient Greek civilization marked the end of Aristotle's impact on science until the thirteenth century A.D. During the Middle Ages, however, Aristotle's writings—including those in astronomy, biology, literature, logic, philosophy, physics, and psychology—became part of the scholastic teachings of Christianity.

Although his numerous valuable contributions outweighed his misconceptions, many of Aristotle's observations and projections were not sound in theory. According to Aristotle, for example, an object that fell toward Earth reached a velocity commensurate with its innate characteristics very quickly. The factors that affected this maximum velocity included weight, the medium in which the object fell freely (objects fall faster in air than in water), and conceivably, even the color and temperature of the object.

The study of freely falling bodies was undertaken by several investigators, but was not undertaken in detail for several centuries after Aristotle's death. In the sixth century, John Philoponus of Alexandria concluded from his observations that the velocity of a freely falling object is determined by subtracting the resistance of the medium from the weight of the object instead of, as Aristotle had stated, dividing the weight by the resistance. But few people accepted Philoponus's findings because of Aristotle's great influence. Galileo, unlike Aristotle, emphasized the role of mathematics in terrestrial motion and explained with mathematical logic how falling objects and objects rolling down inclines actually performed. He was correct in his assumption that a falling object moves with a uniform change in velocity in any given time interval. The velocity, according to Galileo's way of reasoning, was directly related to the time during which an object falls freely.

Investigators who tested the reasoning of Galileo soon confirmed the existence of Earth's gravitational force on a body. This force causes an object to fall faster and faster in its path toward Earth, an aspect that is true for all objects regardless of mass. A freely falling body accelerates in its rate of fall about 9.80 meters per second squared.

Galileo's work was well known to Newton and was influential in Newton's development of the universal law of gravitation. The consistency of both ideas can be shown by considering a freely falling object with all forces other than gravity assumed to be negligible, the case for dense objects moving slowly enough that air resistance is not appreciable. By equating the force of gravity (the weight) on an object near Earth's surface with its mass times acceleration (Newton's second law),

$$\frac{Gm_E m}{r_E^2} = ma,$$

where m_E is the mass of Earth, m is the mass being accelerated, a is the acceleration, and r_E is the radius of Earth. This leads to

$$\frac{Gm_E}{r_E^2} = a = g.$$

Two points are noteworthy in the preceding derivation: First, the desired result that all accelerations of falling objects are the same since the three items on the left side of the equation are fixed quantities. The symbol g is used (as in the previous chapter) for this special value of acceleration, readily verified to be approximately 9.80 meters per second squared. Second, it is assumed that the *gravitational mass* and the *inertial mass* are equal. This assumption is why the mass of the object "cancels out" of both sides of the equation, making the mass of the falling object immaterial.

Since the acceleration of all falling objects is the same constant value, the equations developed in the previous chapter for objects moving with constant acceleration can be used for falling objects. To avoid confusion, the upward direction will be assumed positive (+) and the downward direction will be assumed negative (−). In this scheme, the constant acceleration equations may be written (with $d_f = d$ and $d_i = 0$):

$$v_f = v_i - gt,$$
$$\bar{v} = \tfrac{1}{2}(v_f + v_i),$$
$$d = \bar{v}t,$$
$$d = v_i t - \tfrac{1}{2}gt^2.$$

Positive values for velocities and displacements mean they are upward, and negative values mean they are downward. A few examples will illustrate the use of the equations.

Example 4.1

A steel ball was dropped from the top of the Empire State Building. The ball required 8.90 seconds to hit the ground. Determine (a) the velocity with which the ball hit the ground, (b) the average velocity of the ball, and (c) the distance the ball fell (the height of the building).

Solution

(a) The final velocity is found from

$$v_f = v_i - gt.$$

Because the ball was dropped, $v_i = 0$, so

$$v_f = 0 - (9.80 \text{ m/s}^2)(8.90 \text{ s}) = -87.2 \text{ m/s}.$$

The minus sign indicates that the ball was moving downward when it hit the ground.

(b) The average velocity is found from

$$\bar{v} = \tfrac{1}{2}(v_f + v_i)$$
$$= \tfrac{1}{2}(-87.2 \text{ m/s} + 0)$$
$$= -43.6 \text{ m/s}.$$

(c) The displacement of the ball can be found according to either of two solutions:

$$d = \bar{v}t$$
$$= (-43.6 \text{ m/s})(8.90 \text{ s})$$
$$= -388 \text{ m},$$

or

$$d = v_i t - \tfrac{1}{2}gt^2$$
$$= 0 - \tfrac{1}{2}(9.80 \text{ m/s}^2)(8.90 \text{ s})^2$$
$$= -388 \text{ m}.$$

The minus sign indicates that the ball ended its motion 388 m lower than it started. Since the ball only moved downward during its motion, the distance traveled in falling was 388 m.

Example 4.2

A rock is thrown with a velocity of 25.0 meters per second straight upward. (a) How long does the rock require to reach its maximum height? (b) How long does it take to return to Earth? (c) How far does the rock travel upward?

Solution

(a) The time required to reach maximum height is computed by noting that the velocity at the highest point is 0, so using

$$v_f = v_i - gt$$
$$0 = v_i - gt$$
$$-v_i = -gt$$
$$t = v_i/g$$

or

$$t = (25.0 \text{ m/s})/(9.80 \text{ m/s}^2)$$
$$= 2.55 \text{ s}.$$

(b) The rock required 2.55 s to reach its maximum height where its velocity is zero. It then becomes a freely falling object with an initial velocity of zero and is pulled back to Earth by gravity. As the rock falls, its downward velocity is a "mirror image" of its upward velocity. For example, the velocity 1.00 s before it reaches the top and 1.00 s after it reaches the top are the same, except the velocity is downward rather than upward. Because the magnitudes of the velocities are the same, the rock will have traveled the same distance in the opposite direction in the same length of time. The downward fall will take 2.55 s, just as did the upward flight. The total time is then 5.10 s. This result could have been obtained mathematically by noting that the displacement (distance from the starting point) is zero at the end of the motion as well as at the beginning, so

$$d = v_i t - \tfrac{1}{2}gt^2$$

becomes

$$0 = (25.0 \text{ m/s})t - \tfrac{1}{2}(9.80 \text{ m/s}^2)t^2.$$

Dividing by t, the equation becomes

$$0 = 25.0 \text{ m/s} - \tfrac{1}{2}(9.80 \text{ m/s}^2)t,$$

so that

$$t = 5.10 \text{ s}.$$

(c) If the initial velocity of the rock is 25.0 m/s and the final velocity is 0 (for the upward part of the motion), then the average velocity is 12.5 m/s. The distance the rock travels upward may be found from

$$d = \bar{v}t$$
$$= (12.5 \text{ m/s})(2.55 \text{ s})$$
$$= 31.9 \text{ m}$$

The rock, then, reaches a height of 31.9 m above Earth's surface before it starts the return journey home.

The motion of the rock is shown in Figure 4.4.

Projectile Motion

The acceleration due to gravity is only in a downward direction; therefore, it does not affect the horizontal motion an object might have. Conversely, the horizontal motion does not affect the vertical motion. A consequence of this is that if an object is projected horizontally, it will strike the ground (assuming it is level) at the same time as an object dropped from the same height. At the same time, inertia will have caused the object to maintain a constant horizontal velocity, and it will strike the ground some distance from the point of projection. A rifle bullet fired horizontally from 2 meters above the ground will hit the ground in about 0.6 seconds, as will a rifle bullet dropped vertically from the same height. The greater the velocity of the bullet fired from the rifle, the farther it will travel during the 0.6 second, and the farther from the rifle it will strike the ground. An illustration of this case would be similar to the concept presented in Figure 4.5.

A mathematical analysis of general projectile motion shows that if an object is given a certain initial speed, it will travel the greatest distance over horizontal land if the angle with which it is projected is 45 degrees above the horizontal, that is, halfway between the horizontal and vertical directions. If the angle is closer to the horizontal, gravity will pull the object to the ground in a short time and its range will be smaller than at 45 degrees. If the angle is closer to the vertical, it will remain in the air longer, but will not have enough velocity in the horizontal direction to go very far. The situation is shown in Figure 4.6 for three angles.

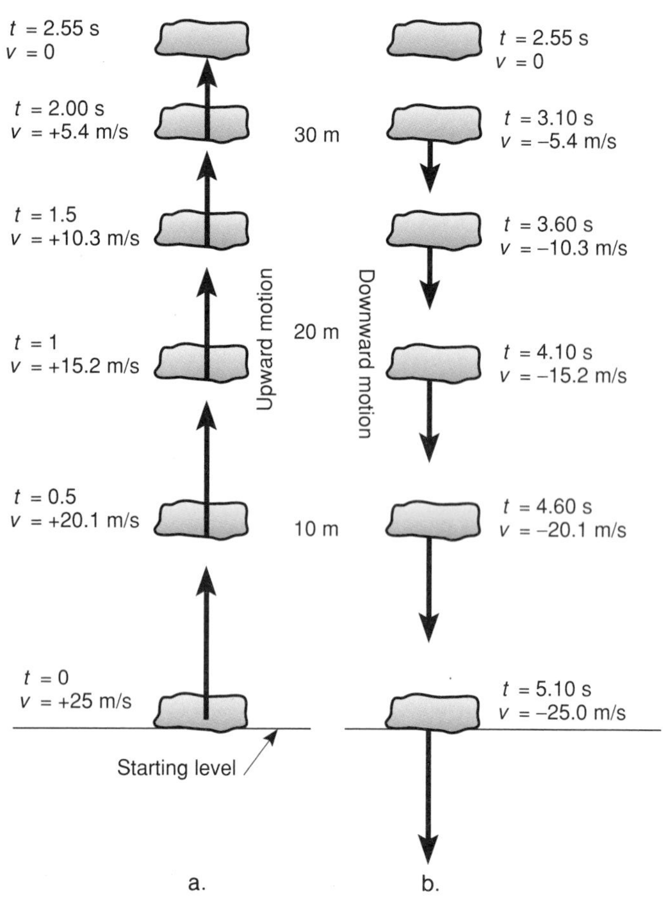

Figure 4.4 (*a*) A rock thrown upward with a velocity of 25 m/s reaches its maximum height of 31.9 m in a time of 2.55 s, slowing uniformly as it rises. (*b*) As the rock returns to its original level, the velocities are exactly reversed for each point.

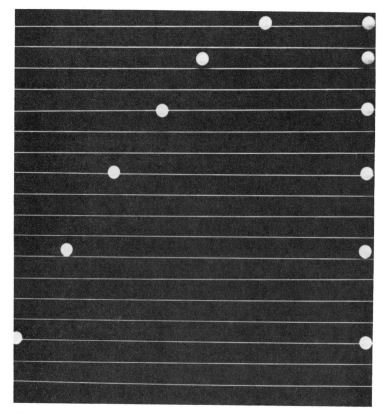

Figure 4.5 Two tennis balls in various conditions of free fall. One ball has velocity in a vertical direction only, the other in both horizontal and vertical directions. Note that both balls cross each time line (the parallel lines) at the same instant.

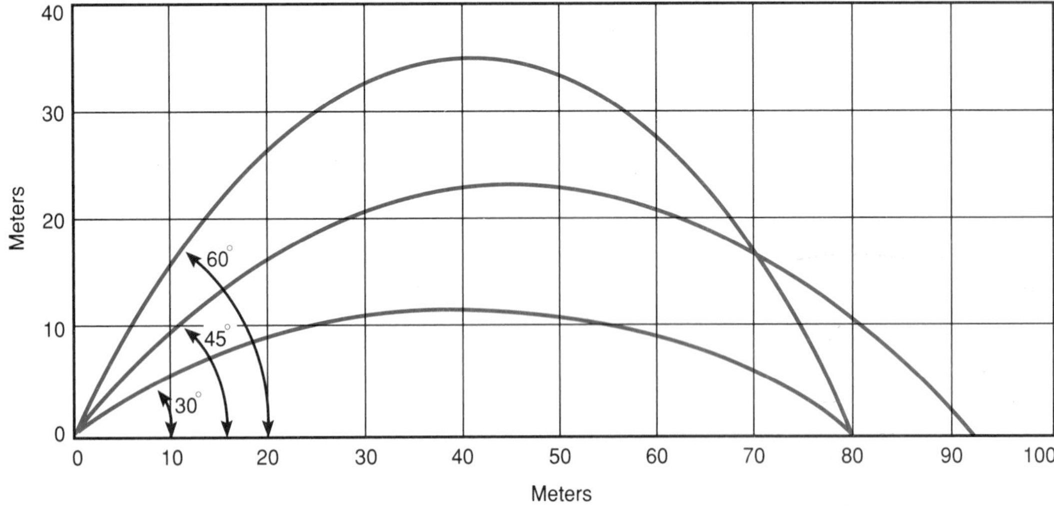

Figure 4.6 The horizontal distance traveled by a projectile given an initial velocity of 30 m/s for three different initial angles. Note that the maximum horizontal distance is achieved with an angle of 45°. (Trajectories shown neglect air resistance.)

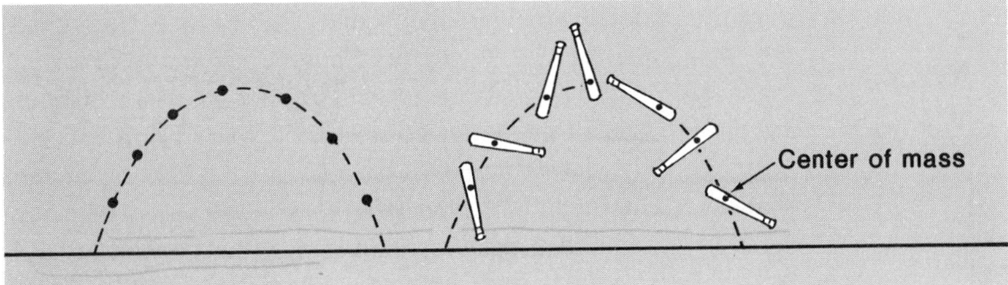

Figure 4.7 The center of mass of the ball and of the bat follow similar paths.

Rotational Motion

Up to this point, the discussion has centered around motion along a line, a phenomenon called **linear motion.** The same natural laws underlie linear motion as well as **rotational motion,** *the motion of a body that rotates about any axis inside or outside the body.* As an example of the latter case, Earth rotates daily about a central axis that passes through the geographic North and South poles. In addition, Earth rotates about an axis located in the sun, and the sun rotates about an axis near the center of the Milky Way galaxy. As a point of clarification, astronomers and others may prefer to apply the term "revolve" to orbital motion. In this context, Earth revolves about an axis located in the sun, and the sun revolves about an axis near the center of the Milky Way galaxy.

Have you observed closely the path that a rotating baseball follows as it is thrown into the air from one player to another? The path is classified as one of smooth parabolic trajectory. If a baseball bat were thrown in the same manner, it would assume a wobbly path. However, if one were to observe its center of mass, a point about which the bat could be brought to a level balance, its path would be similar to that of the baseball (see Fig. 4.7). The center of mass of a symmetrical body, such as a baseball or a cube, is at its geometric center. The center of mass of a hoop, would be at its geometric center, a point not on the hoop, itself. In a peculiarly shaped object such as a baseball bat, the center of mass is located toward the heavier end. An object also has a **center of gravity** where the force of gravity appears to act. This specific location is *the average position of weight.* If **g** does not vary over the volume of the object, the weight is proportional to the mass and the center of gravity and center of mass coincide.

The center of gravity of a race car is made low so the vehicle can negotiate turns at high speeds. An evenly loaded truck with trailer has a high center of gravity so it has a tendency to overturn quite easily. If the trailer were loaded with an equal weight of sheets of steel, it would be much more stable, assuming it thus would have a lower center of gravity. A crate, the weight of which evenly distributed, is more stable when it is resting on a side of large area than when placed on a smaller side. In addition, if two people are carrying an air conditioner, the person nearer the compressor, thus closer to the center of gravity of the unit, must exert a greater effort to lift the unit than does the other person. Finally, there is more weight on the tires nearer the engine of an automobile than on the other tires because the center of gravity of the vehicle is toward the end with the greater mass.

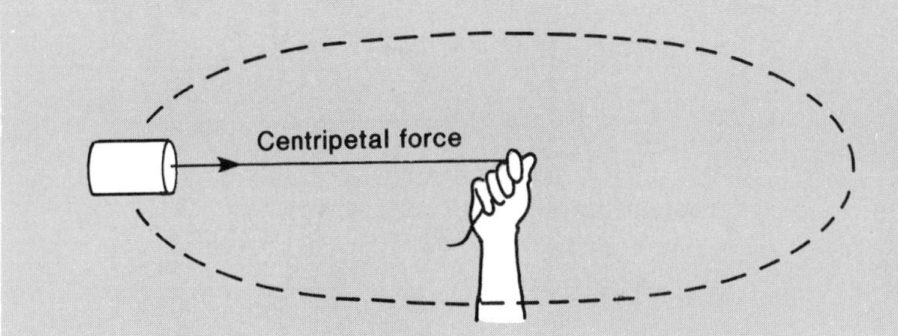

Figure 4.8 The only force that acts on the whirling can (ignoring gravity) is directed toward the center of circular motion and is called a centripetal force. No outward force acts on the can as it moves in a circular path.

For large bodies such as the moon, the farthest portion from Earth is in a region of weaker gravitation than the nearest part; thus, the center of gravity of the moon is not located at its center of mass.

The center of mass of the Earth-moon system is at a point about 1600 kilometers under Earth's surface, since the mass of Earth is much greater than that of the moon. The center of mass of the Earth-moon system follows a smooth path, making Earth wobble in its orbit around the sun. A similar wobble is present in the motion of the sun, since the center of mass of the solar system is located inside the sun. Wobbles have also been detected in other stars, an observation that convinces astronomers that these stars are binary star systems or, possibly, have planets.

Any force that causes an object to move in a circular path is called a **centripetal force.** Centripetal means "center-seeking"; therefore, a centripetal force acts toward the center of rotation. If the force ceases, the object can no longer maintain a circular path.

Consider a can attached to a string and caused to move in a circular path above your head, as illustrated in Figure 4.8. The centripetal force is transmitted from your hand to the can through the string. Similarly, the gravitational force that causes the moon to orbit Earth is transmitted through space and Earth. This gravitational force is a centripetal one, just as is the electrical force

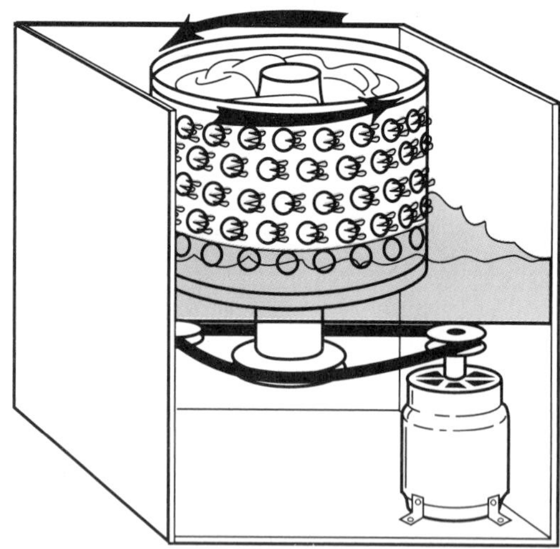

Figure 4.9 The clothes assume a circular path, but the water escapes through the holes in the tub.

that causes electrons to orbit the nucleus of an atom. Centripetal forces are always applied at right angles to the path of the body on which the force acts. The path of the body is a circular one as the result of the force.

An automatic clothes washer removes the water from the tub and clothes during the spin cycle. In this phase the tub rotates at a high rate and creates a centripetal force on the wet clothes. This causes them to assume a circular path against the inner wall of the tub. The water, however, because of inertia, tends to travel in a straight line, escaping through the holes in the tub to the drain. Therefore, by design, the tub does not exert the same force on the water as it does on the clothes (see Fig. 4.9).

As an automobile negotiates a curve, the friction between its tires and the pavement furnishes the centripetal force that permits the automobile to assume a circular path. If the linear momentum of the car overcomes the centripetal force provided by friction, the automobile fails to follow the curved path and may leave the road.

In various instances, an outward force *appears* to act on an object, a force also attributed to the circular motion of the object. This "fictitious" force is called centrifugal force, a term that means "center fleeing"; that is, away from the center of rotation. In the previous example of the can that is traveling in a circular path, one is tempted to state that a centrifugal force pulls outward on the can. However, if the string breaks, the can assumes a straight-line path because of its inertia, and it can be seen that there is no force pulling the can outward. Centrifugal forces *seem* to act when the observer is moving with a rotating reference frame. Such a reference frame is not an inertial reference frame as discussed in connection with Newton's first law of motion. When viewed from outside the rotating reference frame, it can be seen that *centrifugal forces are not real forces.*

Now imagine a large barrel attached to a merry-go-round. If you were to stand inside the barrel and the merry-go-round were to rotate rapidly, you would feel yourself forced to the far side of the barrel's path, as if pushed away from the center of the rotating merry-go-round by an outward force. From your frame of reference, rotating with the merry-go-round, the force seems real, but to a person who views your plight from outside the rotating system, the only real force in action on your body is the barrel's pushing against you—the centripetal force that holds you in a circular path.

A centripetal force, like other forces, causes a body to accelerate in the direction in which the force acts. Accelerations come about by changes in direction of velocity as well as by changes in speed. When a force acts in a direction perpendicular to the speed of an object, the object continually changes direction and moves in a curved path. If the force is always directed toward some point (in other words, if it is a centripetal force), the change in velocity can also be shown to be toward that point. *The rate of change of velocity toward the center of a circle* is called **centripetal acceleration.** If, in addition, the force is constant, the object will move in a circular path at a constant speed. Newton's second law of motion governs objects that travel in a circular path in the same manner that it governs objects that move in a straight line. Recall that acceleration is the change in velocity divided by the time required for the change. For an object moving in a circular path at a steady speed, the velocity changes direction but not magnitude. The acceleration is in the direction of the change in velocity which is *perpendicular to the velocity* (Fig. 4.10). The magnitude of the acceleration of an object moving in a circular path is calculated from the equation

$$a = \frac{v^2}{r} \qquad \text{CENTRIPETAL ACCELERATION}$$

where v^2 is the square of the magnitude of the velocity (the speed) and r is the radius of the path. (The derivation of this equation is beyond the scope of this text.) This is *not* a vector equation, because the acceleration is inward toward the center of the circle and the velocity is tangent to the circle. The application of Newton's second law then gives the centripetal force

$$F = ma = \frac{mv^2}{r}, \qquad \text{CENTRIPETAL FORCE}$$

where F is in the direction of the acceleration; that is, toward the center of the circle.

Example 4.3

A mass of 5.00 kilograms moves in a circular path having a radius of 2.00 meters with a speed of 10.0 meters per second. Calculate the magnitude of the centripetal force acting on the mass.

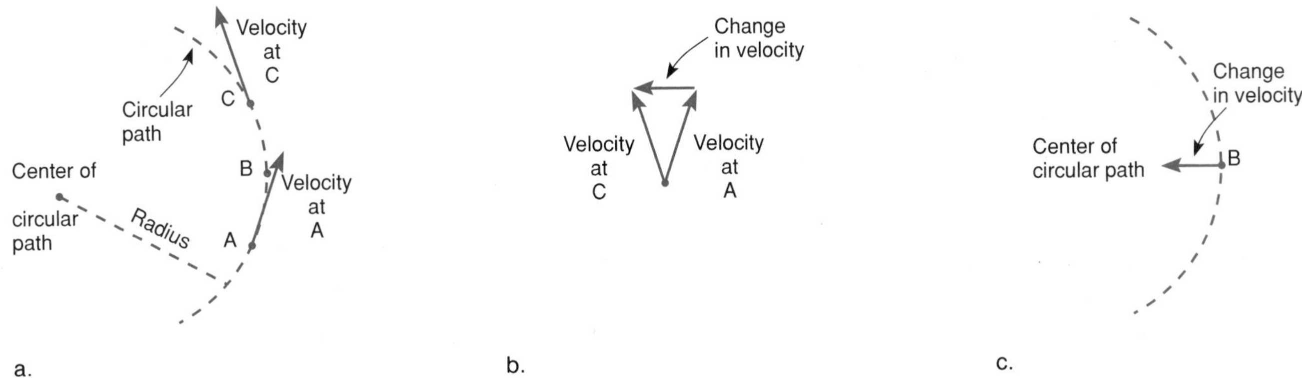

Figure 4.10 (*a*) As an object moves along a circular path from point A to point C at a steady speed, its velocity changes direction. Point B is the "average" location of the object during this motion. (*b*) It is the change in velocity that must be added to the velocity at point A to give the velocity at point C. (*c*) When the change in velocity is placed at the "average" location, it is seen to point inward toward the center of the circle. Thus, the centripetal acceleration, which is the change in velocity divided by the time, points inward toward the center.

Solution

Substitution may be made directly into the equation for centripetal force:

$$\text{force} = (\text{mass})(\text{velocity})^2 \div \text{radius}$$

$$F = \frac{mv^2}{r}$$

$$= \frac{(5.00 \text{ kg})(10.0 \text{ m/s})^2}{2.00 \text{ m}}$$

$$= \frac{(5.00 \text{ kg})(100 \text{ m}^2/\text{s}^2)}{2.00 \text{ m}}$$

$$= 250 \text{ kg}\cdot\text{m/s}^2$$

$$= 250 \text{ N}.$$

Thus, the centripetal force is 250 N inward toward the center of the circular path.

Example 4.4

Calculate the magnitude of the torque exerted to cause a door to swing open if a force of 10.0 newtons is applied 50.0 centimeters (0.500 m) from the hinges of the door and in a direction perpendicular to the door.

Solution

$$\text{Torque} = \text{force} \times \text{perpendicular distance from axis of rotation}$$

$$= Fd$$

$$= (10.0 \text{ N})(0.500 \text{ m})$$

$$= 5.00 \text{ N}\cdot\text{m}$$ (the unit of measurement of torque in SI units).

An object that spins about an axis continues its uniform rotation about the same axis unless acted upon by an external force that will overcome the rotation of the object. *The force that produces, tends to produce, stops, or tends to stop rotation* is said to exert a **torque** (τ). Torque consists of the product of the force and the perpendicular distance from the axis of rotation about which the force is applied. That is, torque equals force times distance from the axis of rotation. In order to open a door, two forces that exert a torque are necessary: a twist to turn the doorknob, and a push or a pull to move the door. The magnitude of a torque is

$$\tau = \text{force} \times \text{perpendicular distance to axis of rotation},$$ TORQUE

$$\tau = Fd.$$

The direction of a torque is usually specified by stating whether the torque tends to cause clockwise or counterclockwise rotation.

The reason bolts are more easily loosened by a wrench with a long handle than by one with a shorter handle is that the force applied at the end of the longer handle exerts greater torque than does the same force applied at the end of the shorter handle. In a like manner, a steering wheel of large diameter is easier to turn than a smaller steering wheel, if the same force is applied to each.

The tendency of a rotating body to continue rotating is called its *rotational inertia.* Extensions of Newton's laws of motion, beyond the scope of this text, show that rotational inertia is the principle that underlies the operation of the gyroscope, the flywheel, and other devices with rotating parts. Frictional forces usually bring rotating objects to rest eventually, just as they bring objects moving in linear paths with linear inertia to rest.

The rotational inertia of a body is dependent on the mass of the body and the way in which the mass is distributed with respect to the axis of rotation. The farther the major portion of the mass is located from the center of rotation, the greater the rotational

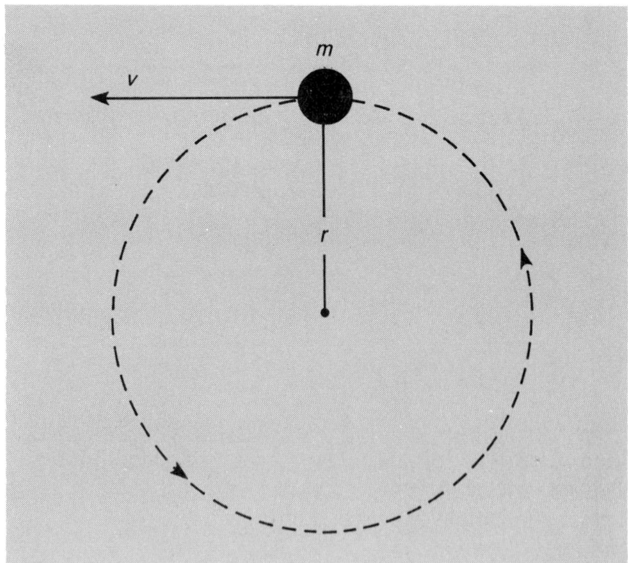

Figure 4.11 An object of mass m that revolves in a circular path, the radius of which is r and the velocity of which is $\mathbf{v}$, has an angular momentum of mvr.

inertia. Consider a bicycle wheel with heavy lead wire wrapped about its rim. Since a major portion of the mass of the wheel is concentrated along the rim, the wheel has a greater tendency to remain spinning than if its mass were concentrated closer to the axis of rotation. The greater the rotational inertia of an object, the harder it is to cause the object to accelerate in a rotational mode. To maintain their stability, steelworkers who walk across steel girders increase their rotational inertia by extending their arms. In a like manner, the greater the length of a pendulum, the greater its rotational inertia; thus, the slower it swings. Animals with long legs, such as horses and giraffes, run at slower gaits than do mice or dogs. A major reason that some athletes can run faster than others lies in the fact that the faster runners have the ability to reduce their rotational inertia by bending their legs more and can thus move them back and forth at higher rates than can slower athletes.

A hoop and a solid disk of equal diameter and mass have different amounts of rotational inertia. A hoop has its mass concentrated farther from the axis of rotation than does a disk; thus, it has a greater tendency to resist a change in the state of its rotation than does a solid disk or cylinder. A solid cylinder or disk, then, will travel faster down an incline than will a hoop.

A mass that moves in a straight line has linear momentum; a mass that follows a circular path has **angular momentum,** a measure of *the intensity of rotational motion.* The planets have angular momentum as they orbit the sun; so do an atom's orbiting electrons and objects we might whirl about our heads. Angular momentum is the product of the mass and velocity of an object, along with the radial distance between the mass and the axis about which it rotates (see Fig. 4.11). That is, at any instant, the magnitude of angular momentum = $m \times v \times r$.

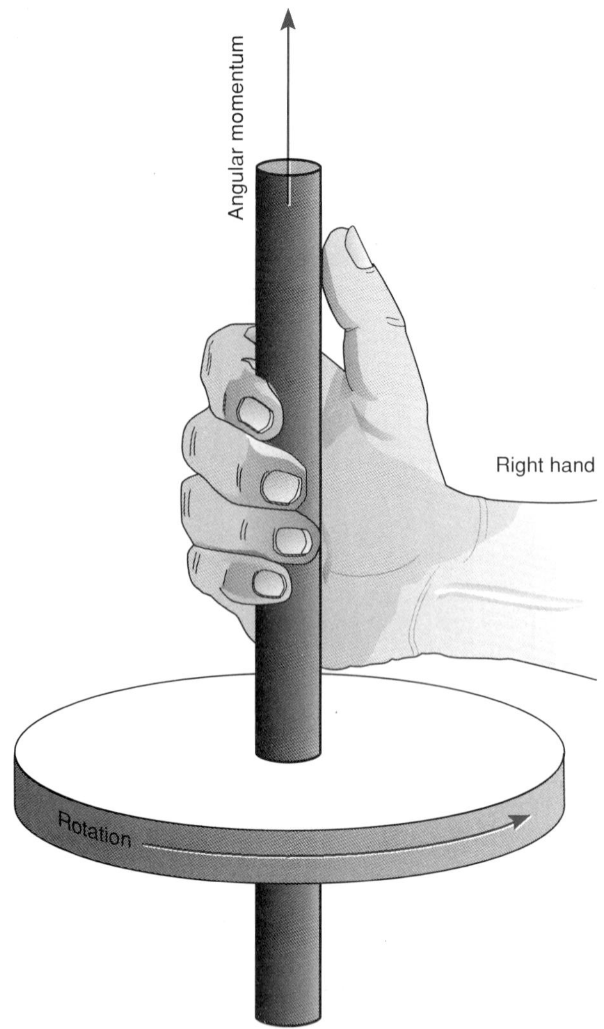

Figure 4.12 The direction of the angular momentum is found by wrapping the fingers of the right hand around the axis of rotation with the tips of the fingers pointing in the direction of rotation. The angular momentum lies along the axis in the direction the thumb is pointing.

Application of Newton's laws to rotating objects reveals that *an object or system of objects will maintain its angular momentum unless acted upon by a torque from some outside source,* a principle called the **conservation of angular momentum.**

Angular momentum also has a direction associated with it, just as does linear momentum. For a rotating object, such as a wheel, the direction is along the axis of rotation in the direction the thumb points when the right hand is wrapped around the axis with the fingers pointing in the direction of rotation (see Fig. 4.12). If a rotating object has a large angular momentum, this direction will be difficult to change, just as the direction of motion is difficult to change for an object with a large linear momentum. Thus, angular momentum is capable of providing stability to all systems that rotate.

In order for such a system to change direction, and thus angular momentum, a torque must be applied at some point other than at the axis of rotation. The *gyroscope,* a device that serves

Figure 4.13 The gyroscope. All applications of this instrument are governed by a special form of Newton's second law.

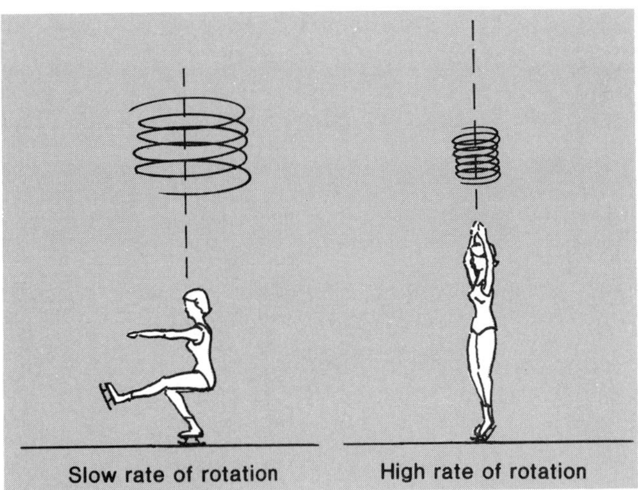

Figure 4.14 The rate of rotation of a skater can be varied by extending the arms and legs or bringing them closer to the axis of rotation.

as the central part of the automatic pilot on an airplane and the inertial guidance system of a rocket, relies on its constant angular momentum to maintain direction of the vehicle. The gyroscope shown in Figure 4.13 consists of a delicately balanced disk that is mounted in a system of gimbal rings so that the axis of rotation of the disk may point in any direction independent of the orientation of its support. As the disk is set into rotational motion with its axis pointing in a specific direction in space, the axis of the rotating disk will maintain its direction even though the base may be reoriented. The disk's axis of rotation is not altered by variations in a magnetic field, such as may affect a magnetic compass.

Rotation is also imparted to bodies other than gyroscopes to help maintain their stability. For example, a bullet fired from a rifle is caused to rotate by spiral grooves in the barrel of the rifle. The angular momentum of the bullet is quite high about an axis parallel to the direction of motion of the bullet, and the total torque caused by air resistance is not great enough to cause an appreciable change in the direction of rotation. The angular momentum impressed on the bullet by the rifle prevents the bullet from tumbling, and thus greatly improves its stability. The same effect can be seen in a properly thrown or punted football; in fact, a spinning ball used in any sport has greater stability than one that is not spinning.

Imagine a person sitting on a stool that can rotate. The person holds a large mass in the hand of each extended arm as an external force causes the person to rotate slowly. As the masses are brought toward the chest of the person, the rate of rotation (velocity) increases because the radius to the axis of rotation is decreased. Since the angular momentum remains constant, it is conserved and the product mvr remains the same. The mass of the system does not change; thus, the rotational velocity increases to balance the decrease in radius. The velocity of the mass, then, is inversely proportional to the radial distance, and mvr at a slow rate of rotation equals mvr at a faster rate. In a like manner, ice skaters often create a force that produces rotation while their arms are extended and their legs spread apart. Then as their legs are brought together and their arms are drawn toward their chests, their rate of rotation greatly increases (see Fig. 4.14).

Conversely, the rotation of Earth is slowed down very slightly because of the friction between the ocean floors and the water in the same manner as an automobile is slowed down by applying the brakes. As a result, Earth loses angular momentum to the moon, the other member of the system. In turn, the moon undergoes an increase in angular momentum and its orbital velocity increases; hence, the moon's orbital radius increases at a rate of 3 centimeters per year. Again, the law of conservation of angular momentum governs the action of the system.

Example 4.5

An object with a mass of 50.0 kilograms is caused to rotate about an axis of rotation, the radial distance of which is 2.00 meters from the mass. The speed of the rotating mass is 10.0 meters per second. Determine the magnitude of the angular momentum of the object.

Solution

$$\begin{aligned}\text{Angular momentum} &= \text{mass} \times \text{velocity} \times \text{radius}\\ &= mvr\\ &= (50.0\ \text{kg})(10.0\ \text{m/s})(2.00\ \text{m})\\ &= 1000\ \text{kg}\cdot\text{m}^2/\text{s}.\end{aligned}$$

Common Comparisons of Speed

The speed with which humans can run has increased throughout the years of modern sports as a result of better training techniques, including more beneficial diets for athletes. A competitive distance runner of today must be able to run a kilometer in less than 2.50 minutes. The average speed required to accomplish this feat can be determined from

$$\text{average speed} = \text{distance} \div \text{time}$$

$$\bar{v} = \frac{s}{t}$$

$$= \frac{1 \text{ km}}{(2.50 \text{ min})\left(\frac{1 \text{ h}}{60 \text{ min}}\right)}$$

$$= 24.0 \text{ km/h}.$$

Various track and field events still include a 1-mile race in which the athletes attempt to run this distance in less than 4.00 minutes. The average speed necessary to accomplish this feat would be

$$\bar{v} = \frac{s}{t}$$

$$= \frac{1 \text{ mi}}{(4.00 \text{ min})\left(\frac{1 \text{ h}}{60 \text{ min}}\right)}$$

$$= 15.0 \text{ mi/h}.$$

The athlete who competes in the 100-meter dash endeavors to run this distance in a time bettering the world record of 9.83 s. The average speed for this record would be computed from

$$\bar{v} = \frac{s}{t}$$

$$= \frac{100 \text{ m}}{9.83 \text{ s}}$$

$$= 10.2 \text{ m/s}.$$

For comparison, 10.2 m/s = 36.6 km/h = 22.8 mi/h.

How fast have we been able to travel with the aid of applied technology? An outstanding comparison may help us appreciate scientific accomplishments. The speed of a bullet from a common target rifle, the .22 caliber, is approximately 1200 feet per second, a speed equivalent to about 820 miles per hour (1300 km/h). Most of our sleek Air Force planes exceed this speed, as do some of our jet passenger planes. But consider the fantastic speeds at which our astronauts have traveled as they journeyed through space. In their conquest of outer space, the space pilots have guided their vehicles to speeds exceeding 25,000 miles per hour—over 30 times the speed of a target rifle! A comparison of the momenta of each of these objects shows even more difference. The momentum of the bullet is about 2 kilogram·meters per second, that of the airplane about 750,000 kilogram·meters per second, but the space shuttle has a momentum of approximately 6 million kilogram·meters per second.

The variables that affect the speed of an object have been discussed previously. The speed that an object of a given mass is capable of reaching is directly controlled by the magnitude of an external force in the direction of motion and the length of time the force is applied to the object. Falling objects naturally reach their velocities (speeds and directions) as a result of the force created by the attraction of Earth on them. Objects that slide down an inclined plane do so because of the same external force we know as gravity. Each object, sliding or falling freely, meets resistance in varying degrees. This reactive force assumes the form of friction, and is created by the surface contact of the object and the incline. In the case of objects in a state of free fall, friction is produced by the air. In both cases, frictional forces create a condition that limits the maximum velocity that Earth's gravitational attraction can impose on a given object. **Terminal velocity,** *the maximum velocity an object can attain under defined conditions,* varies with shape, size, mass, and the nature of the medium through which the object moves. This is the reason a feather takes so long to fall to Earth when it is dropped and is in free fall. An object such as a very small steel ball of equal mass would win the race with the feather consistently. The closer to Earth a falling object is, the more effect air resistance has on its velocity because of the increased density of the air that surrounds Earth.

Parachutists in free fall, just as all other freely falling objects, would plummet toward Earth with a continuous increase in velocity until terminal velocity was attained, a value for the human body of about 220 kilometers per hour (140 mi/h). With the aid of parachutes, however, they may slow their descent to 11 meters per second (25 mi/h) or less, depending upon body weight, the size of the parachute, ability to maneuver the body to offer greater air resistance, and wind currents that may be encountered.

Specifically, terminal velocity is reached when air resistance increases until it equals the gravitational force (the weight) acting on the falling object. At this point, acceleration reaches zero and the object falls at a constant velocity. A 900-newton (202-lb) parachutist would continue to accelerate until air resistance against the person and/or the parachute reached 900 newtons (202 lb). The purpose of the parachute, then, is to cause the person who is in a state of free fall to reach a lower terminal velocity; hence, return to the ground with a lesser impact. A parachutist of 600 newtons who jumped from the same altitude as a 900-newton person would reach a lower terminal velocity than the heavier person and, therefore, take longer to reach the ground.

Raindrops also reach a terminal velocity—a fact of extreme importance to all of us. Without the low terminal velocity the drops reach because of air resistance, imagine the danger to which we might be exposed in a sudden downpour. The raindrops would approach the velocity of the water in a powerful stream from a fire hose!

Automobiles and airplanes also reach a terminal velocity because of frictional forces. Streamlining of cars and planes means just what the name implies—choosing a design that will permit streamline flow of air at high velocities so that frictional forces that react against motion are minimized (see Fig. 4.15). Needless to say, objects set in motion by an external force have their actual paths and velocities affected by forces that resist their forward motion. An earlier discussion pointed out the effects of simultaneous forces in action on the same object. Both the speed and the direction of the object can be altered by air and wind resistance.

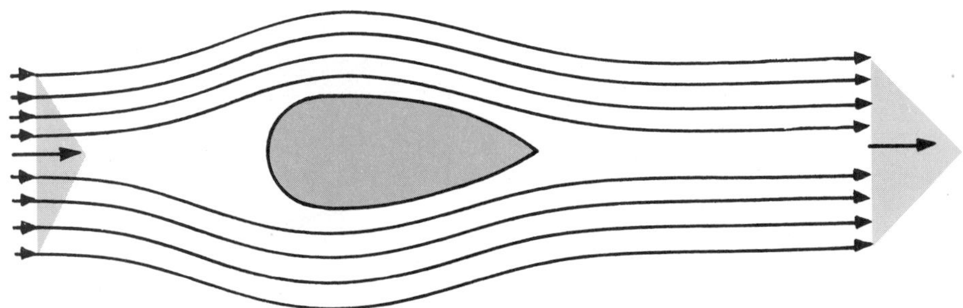

Figure 4.15 The fluid flow (air or water) around a body offers a minimal resistance if the flow is smooth and continuous.

Newton's Laws and Space Travel

One of the most exciting accomplishments of modern science occurred in 1957 as Soviet experimenters successfully projected *Sputnik,* an artificial satellite, into orbit around Earth. The public was amazed that objects could remain in space with no apparent external forces acting on them. That the moon orbits Earth and that the planets move about the sun were accepted facts, but so ingrained in our consciousness that they were never topics for popular conversation as was the flight of *Sputnik*. All objects obey the laws of motion formulated by Newton and are attracted to other objects according to Newton's law of universal gravitation.

The Law of universal gravitation explains why objects on the surface of Earth do not move off into space as a result of Earth's rotational and orbital motions. That is, the force with which Earth attracts an object is more than sufficient to cause the object to remain on Earth's surface. This force, of course, is called the weight of the object, as explained in the last chapter. Objects on Earth's surface also attract each other, but with significantly less force. For instance, a 50-kilogram lead ball and a 350-kilogram lead ball, the centers of which are 1 meter apart, attract each other with a force of 0.0000012 newton. The attractive force between two giant ships at sea is much greater because of greater masses, but even this force is of little significance when compared to the force with which Earth attracts each vessel.

If a suspended object were weighed by means of a spring scale on the moon, the scale reading would be only one-sixth of the reading on Earth. The variation in scale reading is caused by the differences in the value of **g,** the acceleration due to gravity. On the moon, the acceleration due to gravity is 1.6 meters per second squared; on the planet Mars, the value is 3.7 meters per second squared; and on the sun, the value is estimated to be about twenty-eight times the value of Earth's acceleration due to gravity, or approximately 270 meters per second squared.

The reasons for these differences can be easily understood by recalling that the weight of an object located on Earth's surface is given by $w = mg$. The magnitude of **g,** according to Newton's law of universal gravitation was shown earlier in the chapter to be given by

$$g = G\frac{m_E}{r_E^2}.$$

As applied to objects on Earth, m_E is the mass of Earth and r_E is the radius of Earth. *G,* of course, is the universal gravitational constant. This relationship is valid not only on Earth, but on every spherically symmetrical body, such as a moon, a planet, or a star, if the mass of Earth is replaced by the mass of the body, and the radius of Earth by the body's radius.

The variation in the weight of an object because of a difference in the value of the acceleration due to gravity brings about various interesting related effects. For instance, on the moon, a person could lift about six times as much mass as on Earth. The individual could also jump six times as high on the moon or throw a baseball six times as high with the same effort as would be expended on Earth.

The farther from the center of Earth or other celestial body an object is located, the less the attractional force due to gravity. However, the attractional forces acting on the object and on the celestial body are equal, but because of the object's insignificant mass compared to that of a celestial body, such as Earth, the object has a much larger acceleration than the larger body. For example, if the object is 6400 kilometers above the surface of Earth, or twice the distance from the center of Earth when it is located on the planet's surface, the attraction between Earth and the object is about one-fourth the attractive force when the object is on the surface of Earth. Other approximate values of the attractive force as determined by the application of the law of universal gravitation appear in Figure 4.16.

The astronaut shown in Figure 4.17 spent considerable time away from the space shuttle *Challenger* as the vehicle orbited Earth at an altitude of 360 kilometers in late 1984. He was able to maneuver about with the aid of a pack he wore on his back. The unit was propelled by nitrogen gas and was designed to be controlled by the astronaut as he wished to change speed or direction. The same force exerted on him by the device while in orbit about Earth was much more significant than the effect the force would have had if he were standing on the planet, itself, since he was in the "weightless" and frictionless environment of space.

Astronauts aboard the space shuttle *Discovery* used the technique to reclaim two satellites that had assumed errant orbits. Although the satellites weighed over 5400 newtons (1200 lb) each,

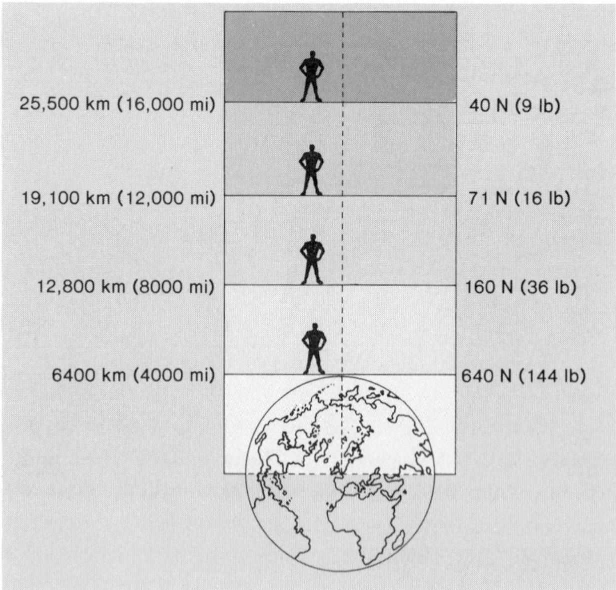

Figure 4.16 The weight of an astronaut decreases as the distance from the center of Earth increases.

Figure 4.17 Astronauts are able to move about outside the space shuttle through the application of Newton's laws of motion.

the astronauts had only to contend with the inertia attributed to their masses in the seemingly weightless Earth orbit in order to maneuver them into the shuttle's cargo bay.

When an object orbits Earth, it is falling freely in space. Consider the motion of a cannon shell projected along a horizontal path, as illustrated in Figure 4.18. Even though the projectile is given horizontal velocity, gravity pulls the shell toward the center of Earth, and so the shell accelerates toward Earth at 9.80 meters per second per second (disregarding friction). The curved path the projectile assumes, then, is due to the combined effect of the horizontal component of its velocity and the velocity it attains as it accelerates toward Earth's center. At a horizontal velocity of about 29,000 kilometers per hour (represented by V_5 in Fig. 4.18), the curvature of the projectile's path perfectly matches the curvature of Earth; therefore, the missile falls around Earth rather than to Earth's surface. If the shell were to meet no obstructions, including air resistance, it would become a satellite of Earth. The projectile would return to Earth only if an external force slowed it down. According to the governing natural laws, the velocity that an object must attain in order to orbit Earth or any other celestial body depends on the mass of the celestial body and the distance of the object's orbit from the celestial body.

It may be somewhat surprising to learn that an object deep in Earth's interior would weigh less than it would on Earth's surface. Newton correctly surmised that the pull of gravity of one hemisphere of Earth would counterbalance the pull of the opposite hemisphere; thus, an object at the center of gravity of Earth, or that of any relatively spherical celestial body, would be weightless.

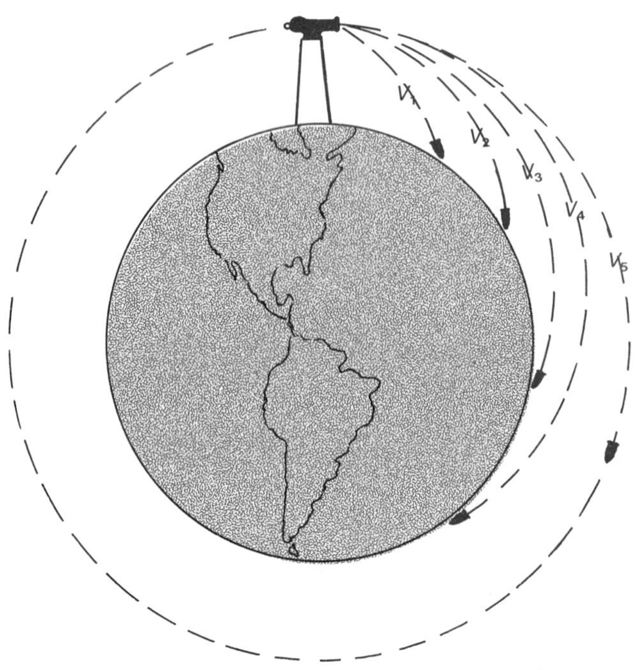

Figure 4.18 As the horizontal velocity of the projectile is increased, the range of the projectile is extended until gravity causes the projectile to "fall" around Earth.

Example 4.6

An apple drops from a tree. How far does it fall in 1.00 second? If the apple were to fall freely from a height of approximately 6400 kilometers above Earth's surface, how far would it fall in 1.00 second? How far does the moon "fall" about Earth in 1.00 second?

Solution

The distance the apple close to Earth's surface falls is the magnitude of the displacement (for straight-line motion), so

$$\begin{aligned} \text{distance} &= \tfrac{1}{2}(\text{acceleration})(\text{time})^2 \\ s &= \tfrac{1}{2}\, at^2 \\ &= \tfrac{1}{2}(9.80 \text{ m/s}^2)(1.00 \text{ s})^2 \\ &= 4.90 \text{ m.} \end{aligned}$$

According to the law of universal gravitation, at twice the distance from Earth's center, the force of attraction (***F***) of Earth on an object would decrease to ¼ of its value at the surface. According to the second law of motion, ***F*** = *m**a***. If ***F*** decreases to ¼, ***a*** must also decrease to ¼; hence, the apple would accelerate toward Earth at ¼ the rate that it would if it were near Earth's surface. (Recall that *m* is constant for an object at any location.) Therefore,

$$\begin{aligned} s &= \tfrac{1}{2}\, at^2 \\ &= \tfrac{1}{2}\left[\tfrac{1}{4}\left(\frac{9.80 \text{ m}}{\text{s}^2}\right)\right](1.00 \text{ s})^2 \\ &= 1.23 \text{ m.} \end{aligned}$$

And finally, the moon orbits Earth at about 380,000 kilometers above Earth's surface. This distance is 60 times the radius of Earth; therefore, the acceleration due to gravity would be $(1/60)^2$ the value on Earth's surface; thus,

$$\begin{aligned} s &= \tfrac{1}{2}\, at^2 \\ &= \tfrac{1}{2}\left[\left(\frac{1}{60}\right)^2\left(\frac{9.80 \text{ m}}{\text{s}^2}\right)\right](1.00 \text{ s})^2 \\ &= 0.00136 \text{ m.} \end{aligned}$$

Objects that orbit Earth appear to be weightless; however, the force of gravity of Earth on any object is always present. An orbiting space vehicle and objects within it still have weight, although all evidence of this property appears to have been lost. When we stand on a bathroom scale, we measure the apparent gravitational attraction between Earth and ourselves, a measurement called our weight. If we stand on a bathroom scale while riding an elevator, our weight appears to vary. As the elevator accelerates upward, our weight appears to increase, since the bathroom scale would push upward with a greater force than when the elevator was at rest. If the elevator were to accelerate downward, our weight would appear to decrease, since the force with which the bathroom scale pushes upward would decrease. If the elevator were to experience free fall, the scale would read zero; thus, according to the reading, we would appear to be weightless. If we recall that weight is a measure of the gravitational attraction of Earth on an object, we realize that we still have weight, for we fall downward with the same acceleration as the elevator does. From ***F*** = *m**a***, an unbalanced downward force must cause this acceleration.

Consider an astronaut in orbit about Earth or any other celestial body. This space traveler seems to be in a state of weightlessness, since he or she is not pushing against anything. A bathroom scale placed beneath the astronaut's feet would indicate zero, since the bathroom scale would be falling at the same rate as the astronaut. If the astronaut were to drop a steel ball, it would remain beside him or her for the same reason that no reading would appear on the bathroom scale. The astronaut, the scale, and the steel ball are all in a state of free fall. However, the objects involved are still under the influence of Earth's gravitational force; otherwise, all objects would move off along a straight line into outer space. The force of gravity pulls the objects inward; hence, gravitation is a centripetal force.

Numerous communication satellites have been launched into orbit around Earth to provide exceptionally improved transmission of telephone conversations and television programs. Most such satellites have been placed into *synchronous orbit;* that is, they have been launched so that they will constantly remain directly over a fixed point on Earth. To accomplish this feat, the period of revolution of the satellite must be precisely the same as Earth's period of rotation. Also, the orbit must follow an eastward path, as does Earth's rotation. Mathematical analysis shows that if a satellite is in an orbit at an altitude of about 43,000 kilometers and given a speed of approximately 11,000 kilometers per hour, the acceleration due to gravity and the centripetal acceleration as computed from v^2/r will be the same and the satellite will remain in orbit. It can be shown from the speed and circumference of the orbit that the time for one complete orbit will exactly match the time for Earth to make a complete rotation.

In order to propel a space vehicle into outer space *fast enough to overcome the effects of Earth's gravity,* the projectile must reach a velocity of about 41,000 kilometers per hour (7.1 mi/s), a value known as Earth's **escape velocity.** After a space vehicle has attained the escape velocity of Earth, its engines may be shut down and the vehicle will venture onward. It will gradually slow down due to Earth's gravitational pull, but this force is not sufficient to stop it. As the vehicle drifts further into space, perhaps passing near another celestial object, it will be accelerated by the gravitational pull of that object.

The maximum velocity produced by rocket engines, many times greater than the velocity of a rifle bullet, is directly determined by the combination of fuels used, the burning time, and the overall mass of the spaceship. For these reasons, the rocket contains a series of individual rocket engines such as aboard the giant vehicles shown in Figure 4.19. The first engine is designed to thrust the rocket upward at a velocity of about 3.2 kilometers per second. Then the second rocket engine ignites, and the spent stage of the rocket is ejected, obviously decreasing the overall mass of the rocket. The second engine increases the ship's velocity an additional 3.2 kilometers per second. This stage of the rocket is ejected, the third

Figure 4.19 Deliberate double exposure of rockets, actually 2.4 km apart, compares *Skylab 1* and *Skylab 2* space vehicles. *Skylab 1*, on right, is the modified *Saturn V* rocket that sent our first space station into orbit. *Skylab 2*, a *Saturn 1B* rocket, carried astronauts to rendezvous with the space station.

Figure 4.20 A modified *Saturn V* rocket is shown as it streaks upward to carry the *Skylab* space station into orbit. The rocket, the mass of which exceeded 90,000 kg (a weight of 100 T), projected the space station into its lofty orbit in about 10 min.

engine is fired, and so forth, until the escape velocity of about 11 kilometers per second is reached. The rocket, with its engines quieted, then travels through space toward its destination.

The principle of conservation of momentum governs the operation of a rocket just as it provides the means by which jet and propeller-driven airplanes function. As the rocket is poised prior to the launch, its momentum is zero, since its velocity is zero. As its engines are ignited, the momentum of the gas molecules ejected downward at any instant is equal to the momentum of the rocket in the opposite direction; that is, the direction in which the rocket begins to move. The total momentum of the system remains zero, since the momentum (a vector quantity) of the gases and the momentum of the rocket are equal in magnitude but opposite in direction.

A rocket does not operate by forcing the air backward as does a propeller-driven airplane. The expulsion of exhaust gases from the rear of the rocket provides the mechanism by which the vehicle is propelled through space. The operation of a rocket is similar in principle to the way that a balloon filled with air acts when the air is permitted to escape through the opening in the balloon. A rocket is well balanced; therefore, it does not assume a balloon's erratic path. The vertical lift of a rocket is provided by the escaping gases; an airplane is carried upward by the lift created by the difference in air pressure on its wings. An airplane changes direction by raising or lowering the ailerons (the moveable parts of the wings), thus changing the direction of the lift force. The rocket changes direction by expelling exhaust gases from its right or left side and slows down by ejecting gases in the direction opposite the desired path of the space vehicle.

Space stations, such as those launched by means of the modified Saturn V rocket, serve as a home where astronauts live and work as they seek to learn the effects of prolonged periods of apparent weightlessness. They also study Earth's environment and resources from their vantage point. The Saturn V version for this mission is depicted in Figure 4.20.

How feasible are manned space ventures—perhaps to the planet Mars and beyond? If a rocket were to travel through space at the escape velocity from Earth of 11 kilometers per second (25,000 mi/h) and to assume an optimum path toward Mars, the spaceship would require about 80 days to reach its target. Consider the amount of oxygen, food, and other supplies required for the trip. Also, the escape velocity from the gravitational pull of

a.

b.

Figure 4.21 (*a*) The space shuttle, along with its launching vehicle, about 18 stories tall, is shown with its three main engines, poised and ready for one of its many launches. (*b*) The space shuttle *Columbia*, along with its sister ships, has made numerous gentle landings in returning to Earth.

Mars is about 5 kilometers per second; thus, the kinetic energy required to escape from Mars's gravitational field would be about one-fourth the energy necessary to escape from Earth. All the fuel for the return journey would have to be stored in the vehicle, since there would be no auxiliary booster rockets available, as for Earth launches. How much additional fuel would be required to slow down the vehicle for a safe Earth landing? The problems, as one can readily ascertain, are tremendous, yet planning for such a feat continues.

The space shuttle *STS-1*, as shown in Figure 4.21, represents our first endeavor to place a manned space vehicle into orbit and to return it safely to Earth for additional flights. As various limitations are overcome and the causes of the space shuttle *Challenger* tragedy resolved and corrected, such vehicles may carry space explorers to the surface of Mars and beyond, then return to our planet with a wealth of information about our solar system.

Could visitors from other potentially existent solar systems reach Earth? The nearest star to our solar system is called Proxima Centauri and is 41 trillion kilometers (4.1×10^{13} km) away—a distance so great that its light requires about 4.3 years to reach us. Even if rockets were improved beyond all current expectations, a round trip to this star would require perhaps 20,000 years! Perhaps in the future, nuclear engines will propel humans at some (now) seemingly inconceivable velocity, but this could decrease the length of the trip to only about 100 years at best. Travel beyond our solar system, presumably, will never be accomplished by any technological advancements, hypothetical or otherwise, even remotely considered today.

From manned and unmanned flights, along with our knowledge of the observable celestial bodies, we can conclude that Newton's laws of motion apply throughout the universe. Each law governs the various aspects of space travel and will affect us as our attempts to explore outer space continue. Everywhere we look we see examples of the application of the laws, whether the application is our own doing or that of nature. Regardless of where we do go in space, we must apply Newton's laws of motion to leave Earth and rely on them to bring us back home.

A New Era in Space Exploration

The space shuttle *Columbia,* amid twin pillars of fire, a towering white cloud, and a deafening roar resulting from the ignition of the volatile propellants liquid hydrogen and liquid oxygen, was successfully launched on its maiden voyage into orbit about Earth on April 12, 1981. The vehicle, appearing quite cumbersome as it assumed the typical vertical launching position, was less than half the length of the Apollo/Saturn V rocket system used for space exploration and lunar missions, even with its boosters and its fuel tank attached (see Fig. 4.21). The spaceplane returned safely to Earth on April 15, gliding to a smooth, unpowered landing on a runway at Edwards Air Force Base in California's Mojave desert, essentially ready to return to a launching platform for future space flights. Only one-third as many launch personnel were required for the space shuttle as in the Saturn series, another factor that indicates what many see as the rapid progress made in our space program. The used spaceship was launched again on November 12, 1981. It completed most of its assigned mission and returned safely to Earth two days later.

The space shuttles can deploy and retrieve large satellites and service satellites already in orbit. The capability of the space shuttle to retrieve satellites was exploited in late 1989 when the Long Duration Exposure Facility was returned to Earth after remaining in orbit for over 5½ years. The satellite, about the size of a small bus, was used to test the performance of spacecraft materials, components, and systems that have been exposed to the space environment for a long time. The ability to launch large satellites was evidenced early in 1990 in launching the Hubble Space Telescope, a 43-foot-long 25,000-pound space observatory. Despite early troubles getting the space telescope to focus properly, it is hoped that eventually this instrument will be able to "see" planets, stars, and other objects in the universe about 10 times better than is now possible with the best optical telescopes on the ground. Now in the planning stage, a complete Earth Observing System deployed by the shuttle will allow us to understand global changes more fully, such as the ozone "hole" and heating of Earth, and how they are related to human activity ■

Summary

Newton's three laws of motion are extensively applied in all areas of science and engineering. Their general nature makes them useful in a wide variety of situations, but this generality also makes them difficult to use without thorough preparation on the part of the user. This qualification is one of the reasons it takes years to train engineers and scientists.

The three laws, in their basic form, apply to particles. A particle is an object, the rotational motion and physical size of which are not important for the motion under study. Often the forces on one or two particles can be computed and their motions can be determined. If more than two particles are interacting, their motions cannot be determined exactly, except in special circumstances. In these cases the power of computers is harnessed to supply approximate answers, but problems with many particles are still an active area of study and much work needs to be done to understand these complicated systems.

In addition to the three laws of motion, Newton stated one of the fundamental laws governing the force between masses—the law of universal gravitation. This law asserts that every object in the universe attracts every other object with a force directly proportional to the product of their masses and inversely proportional to the square of the distance between them. The law of universal gravitation governs the motions of falling objects, the flight of projectiles, the orbits of satellites and planets, and even the motions of stars and galaxies.

Particles moving in circular paths must have centripetal forces acting on them that accelerate them toward the center of the circle. These centripetal forces are important in the study of orbital motion of the particles, as well as rotational motion of extended objects. Even though an extended object can be considered to be a collection of particles governed by Newton's laws, it is more convenient, by applying mathematics, to extend the basic laws to cover the cases involved in rotation. This extension leads to a whole host of new concepts, such as angular momentum and torque. Torque tends to cause changes in the rotational motion of objects, just as force causes changes in linear motion. Angular momentum, because of its directional nature, is very useful in providing stability to rotating systems.

Speeds of ordinary objects are governed by several factors including mass, available force, resisting forces, and the time during which the forces are applied. Ordinary objects on Earth are subject to resistive forces that increase with speed; thus, they usually reach a terminal velocity after a period of time. Objects projected into space have little resistance from frictional forces once they leave Earth's atmosphere, and if given high enough velocity, can actually escape into deep space where the gravitational pull of Earth is negligible.

To compute motions of all objects, scientists and engineers develop models of the objects. The more detailed a model, the better it matches reality, but also the more complicated it is to apply. It is one of the tasks of the scientist or engineer to determine how detailed the model must be to produce the results sought, since decisions based on the model may be made concerning the feasibility of a scientific experiment or the practicality of an engineering design.

Equation Summary

Newton's law of universal gravitation:

$$F = G\frac{m_1 m_2}{r^2}\,.$$

where F is the force (newtons) with which the masses m_1 and m_2 (kilograms) attract each other if they are separated by a distance r (meters) and G is the universal gravitational constant, equal to 6.67×10^{-11} newton-meters squared per kilogram squared.

Freely falling objects:

All objects fall with a downward acceleration $\mathbf{g}$ of 9.80 meters per second squared when air resistance can be neglected. Using the upward direction as positive (+), the equations for a freely falling object are given by

$$v_f = v_i - gt,$$

$$\bar{v} = \frac{1}{2}(v_f + v_i),$$

$$d = \bar{v}t,$$

and

$$d = v_i\,t - \frac{1}{2}gt^2.$$

In these equations, v_f (meters per second) is the final velocity at time t (seconds) when the object started with an initial velocity v_i (meters per second); $\bar{v}$ (meters per second) is the average velocity; and $\mathbf{d}$ (meters) is the displacement (+ is upward), assuming the initial displacement is taken to be the starting point.

Centripetal acceleration and force:

An object moving in a circular path of radius r with a speed v has an acceleration toward the center of the circle called the centripetal acceleration, with magnitude given by

$$a = \frac{v^2}{r}\,,$$

where the magnitude of the centripetal acceleration a (meters per second squared) equals the square of the speed v (meters per second) divided by the radius r (meters) of the circle. There must be a force (called the centripetal force) toward the center of the circle of magnitude

$$F = m\frac{v^2}{r}\,.$$

The centripetal force F (newtons) is equal to the mass m (kilograms) times the centripetal acceleration as computed above.

Angular momentum:

An object having a mass m (kilograms) moving in a circular path of radius r (meters) with a speed v (meters per second) has angular momentum (kilogram-meters squared per second) given by

$$\text{angular momentum} = m \times v \times r.$$

Distance and velocity:

The average speed of an object $\bar{v}$ (meters per second) is the distance traveled s (meters) divided by the time t (seconds) required for the motion. This equation can be arranged in different ways:

$$\bar{v} = \frac{s}{t}\,,$$

$$s = \bar{v}t\,,$$

or

$$t = \frac{s}{\bar{v}}\,.$$

Questions and Problems

Introduction

1. (a) What is a particle? (b) What is an extended object? (c) In their simplest form, with what types of objects do Newton's laws deal?
2. Discuss the following statement: Newton's laws allow one to compute the motions of all objects (not considering quantum theory and relativity theory).
3. What is meant by a "model" of a physical system? What is the difference between the "physical" model and the "mathematical" model?

Newton's Law of Universal Gravitation

4. According to Newton's law of universal gravitation, the force of attraction between two objects decreases as the distance between them increases. (a) If the distance between two objects is doubled, how is the force affected? (b) If the distance is tripled, what is the change in force? (c) If two objects 1.00 meter apart are attracted by a force of 25.0 millionths of a newton (25.0×10^{-6} N), what is the force of attraction when the objects are separated by 5.00 meters?
5. Discuss the following statement made by a television news commentator during the first manned spaceflight to the moon: "The astronauts are now out of the gravitational pull of Earth and in the gravitational pull of the moon."
6. What is wrong with the statement: It can be deduced from Newton's law of universal gravitation that the sun attracts Earth but not the moon, since the moon is known to be in orbit around Earth.

Falling Bodies

7. Explain how Newton's law of universal gravitation validates Galileo's assertion that all objects falling near Earth's surface experience the same acceleration.
8. In general, would you expect the acceleration due to gravity to be larger or smaller at the top of a tall mountain than it is at sea level? Explain.

9. The equation on page 62 gives the acceleration due to gravity, g, in terms of the mass and radius of Earth. (a) If you were on a planet the same size as Earth, but this planet had twice the mass of Earth, what would be the value of g for this planet? (b) If the planet described in part (a) also had twice the radius of Earth, what would be the value of g? (Give your answer in meters per second squared.) (c) What would the mass of a 3.00-kg object be if it were placed on Earth and on the two planets described in parts (a) and (b)? (d) What would the weight of the same 3.00-kg object be on the three planets?
10. A rock is *dropped* from a bridge and takes 3 s to hit the water below. Neglecting air resistance, determine the velocity with which the rock will strike the water.
11. A student living in a multistory apartment building drops a coin out the window and using a stopwatch finds that it takes 1.70 s to hit the ground. (a) How far did the coin fall? (b) If it is 3.00 m between floors in the apartment building, what floor does the student live on?
12. A ball is thrown straight up with a velocity of 30.0 m/s. (a) How high will the ball go? (b) How long will it take to return to the ballplayer's glove?

Projectile Motion

13. The same rifle is used to fire two bullets. The first shot is at an angle of 60° above the horizontal and the second at an angle of 45° above the horizontal. (a) Which bullet will remain in the air longer? (b) Which bullet will strike the ground farther from the rifle? (Neglect air resistance.)

Rotational Motion

14. Is a rock being whirled around on the end of a string at a steady speed in equilibrium? Why or why not?
15. (a) What is a centripetal force? (b) When you are riding in the right front seat of a car that makes a sharp turn to the left, what forces act on you to make you follow the curved path of the car? (c) Is there a force that "pulls" you outward? Explain.
16. A spacecraft orbiting Earth at an altitude of 600 km makes the radius of the orbit about 7.00×10^6 m (its center being at Earth's center). To remain in a circular orbit at this altitude, the spacecraft must have a speed of 7570 m/s. Calculate the centripetal acceleration of this spacecraft. How does it compare with the value of 9.80 m/s^2? Is this what you would expect? Why?
17. An auto with a mass of 1800 kg is traveling at a steady speed of 25.0 m/s (about 56 mi/h) around a curve that has a radius of 75.0 m. (a) What is the acceleration of the automobile? (b) What is the centripetal force of the pavement on the tires? (c) Using the conversion factor 1.00 N = 0.223 lb, convert the answer in (b) to pounds.
18. (a) What does "center of gravity" mean? (b) Suppose the outlines shown below are objects cut from a piece of plywood. For each object, state whether the center of gravity is nearer the left end, the right end, or at the center.

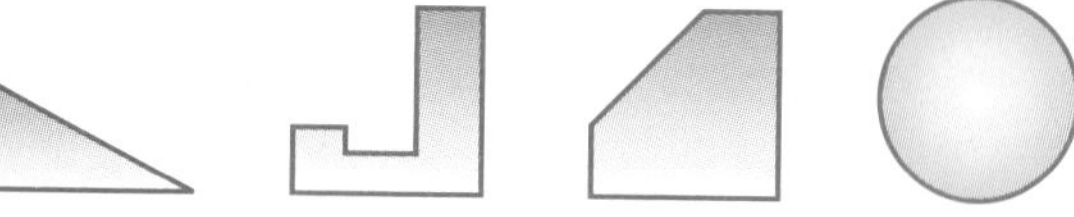

19. Why is it easier to open a door by pushing it farther from the edge with the hinges? Use the idea of torque to help explain your answer.
20. A mechanic trying to loosen a bolt on an automobile engine has two wrenches available, one with a handle 0.150 m long and the other with a handle 0.200 m long. If the greatest force the mechanic can exert is 300 N (about 70 lb), (a) what is the maximum torque he can exert on the bolt with each wrench? (Assume he can apply the force at the end of the handle.) (b) What should be the direction in which he applies the force relative to the handle? Why?
21. (a) What is angular momentum? (b) Why does Earth's axis remain in almost the same direction as Earth orbits the sun?
22. Why is it easier to stop a ball having a mass of 1.50 kg and a radius of 0.100 m from rotating than a bicycle wheel having the same mass but a radius of 0.330 m if they are both spinning at the same number of revolutions per minute?

Common Comparisons of Speed

23. What is "terminal velocity"? Why don't objects that fall from great heights continue to accelerate as they fall?
24. A sky diver weighing 820 N jumps from an airplane and reaches a terminal velocity of 52.2 m/s downward. What can be said about the magnitude and direction of the "drag" force on the sky diver when the terminal velocity has been attained?
25. The distance to Proxima Centauri, the nearest star other than the sun, is approximately 4.1×10^{13} km. How long would it take a spacecraft traveling 40,000 km/h to make a round trip from Earth to Proxima Centauri? Give your answer in hours and in years. Would you be willing to go on such a trip?

Newton's Laws and Space Travel

26. An astronaut is taken into space a distance equal to one Earth radius above the surface of Earth; that is, the astronaut is now two Earth radii from Earth's center. The astronaut is released *from rest*. Describe what would happen to the astronaut.

27. The weight of the astronaut in the previous question is found to be 600 N on the surface of Earth. Don't forget the weight is just the force of gravity. What is the force of gravity on her when in space? Give both the magnitude and direction of the force.

28. When the astronauts traveled to the moon, they left Earth with a velocity near the "escape velocity" of 11 km/s. As they moved away from Earth toward the moon, they "drifted" without any propulsion except for a few small rockets which they used to change their direction slightly. Describe how the gravity of Earth would affect the motion of their spacecraft as they moved away from Earth. Would you expect their average speed to be greater or less than 11 km/s? (See Problem 29.)

29. The astronauts took about 3 days to travel from Earth to the moon. The Earth-moon distance is about 400,000 km. What was their average speed in kilometers per second?

30. A satellite in "near-Earth orbit" takes about 90 min to orbit Earth once. What is the average speed of the satellite if the distance around its orbit is 41,000 km? How does this compare to the "escape velocity" of 11 km/s?

5 Work, Energy, and the States of Matter

Chapter outline

Key terms/concepts

Complex machinery makes the task of our labor force more productive and provides sources of chemical energy from our natural resources.

A progressive society such as ours becomes increasingly dependent upon products of its own creation. These products apply many principles of physics. In our modern homes, we are concerned with health, comfort, recreation, and work. Our heating and cooling systems, our radios and television sets, and our kitchen appliances with automatic controls have greatly contributed to the improvement of our quality of life. Many devices around the home do work for us. All the devices have two characteristics in common: they are all composed of matter, and they all require energy to do useful work. Of what value is a stove without heat energy, a lamp without light energy, or a sweeper without electrical energy? A mixing spoon is of little use without energy transmitted to it so work can be performed. We all are interested in conserving energy, increasing efficiency of our devices, and determining the usefulness of each device to us. In order to reach these objectives, a greater knowledge of the underlying principles by which devices in our everyday lives function is essential. This chapter is designed to develop an understanding of such concepts as efficiency, energy, and work.

Machines and Work

The many mechanical devices that help us perform tasks are known as **machines.** Machines of all sorts play important and integral parts in our lives. While a machine may be a complicated bit of ingenuity, each individual part is simple in its design. A machine performs one of three functions: (1) it lifts a heavy load through the application of a comparatively small force; (2) it moves objects very rapidly, while the applied force changes position slowly; or (3) it causes work to be done in a more convenient way than if the applied force were applied directly to the object to be moved.

The amount of work done is frequently expressed in terms of how long it takes to complete a given task or how tired one becomes in doing so. This idea of work does not apply to a machine, since it would not experience the same sensation of fatigue. A more technical means of defining work is thus necessary. More specifically, **work** is defined as *the product of the magnitude of the applied force and the parallel displacement (distance) through which the force acts.* In an equivalent fashion, work is determined by multiplying the displacement by the component of the force parallel to the displacement. That is,

$$\text{work} = \text{force} \times \text{distance},$$

or

$$W = Fs. \qquad \text{WORK}$$

If the distance an object moves when a force is applied is a value other than zero, work is done, since the object is moved. The applied force could assume various forms, such as mechanical, gravitational, or human. The units of work are obtained from the product of a unit of force and a unit of distance. Work is a scalar quantity, since its direction is not specified. The SI unit of work is defined as the *newton-meter* (N·m). (The *fps* system expresses work in foot-pounds, abbreviated ft·lb.) The derived unit, the **joule** (J), in honor of the English scientist, James Prescott Joule (1818–1889), is commonly used in preference to the newton-meter. For comparison, 1 joule is defined to be 1 newton-meter. (Further comparisons appear in Appendix 1.)

Example 5.1

Calculate: (a) the work done as a 100-pound force is applied to slide a crate 5.00 feet along a level surface, and (b) the amount of work required to lift a mass of 50.0 kilograms a distance of 5.00 meters.

Solution

(a) The amount of work is calculated from

$$\begin{aligned} \text{work} &= \text{force} \times \text{distance} \\ W &= Fs \\ &= (100\ \text{lb})(5.00\ \text{ft}) \\ &= 500\ \text{ft}\cdot\text{lb}. \end{aligned}$$

(b) Since work done against gravity is essentially the product of force and distance, the force required to lift the 50.0-kg mass (that is, the weight) must be determined. Using

$$\begin{aligned} \text{force} &= \text{mass} \times \text{acceleration} \\ F &= ma \text{ (and in these stated conditions, } F = w = mg) \\ &= (50.0\ \text{kg})\left(\frac{9.80\ \text{m}}{\text{s}^2}\right) \\ &= 490\ \text{kg}\cdot\text{m/s}^2 \\ &= 490\ \text{N}. \end{aligned}$$

Then the amount of work is found from

$$\begin{aligned} W &= Fs \\ &= (490\ \text{N})(5.00\ \text{m}) \\ &= 2450\ \text{N}\cdot\text{m} \\ &= 2450\ \text{joules (J)}. \end{aligned}$$

(The answer can be justified to three significant digits on the basis of the discussion about measurements in Chapter 3.)

In no case is a machine capable of doing an equal or a greater amount of work than the applied force is capable of doing on the machine. If more work could be done by the machine than was done on it, the machine would have the incredible ability to do work within itself. Any machine loses some of the energy applied to it because of friction between its moving parts. For this reason no machine can be totally efficient; that is, it cannot yield as output the amount of work that is applied to the machine as input. The efficiency may be calculated from the following relation:

$$\text{efficiency (\%)} = \frac{\text{output work}}{\text{input work}} \times 100 \qquad \text{EFFICIENCY}$$

Because of its increased number of moving parts, a complex machine generally has a lower efficiency than a simpler machine.

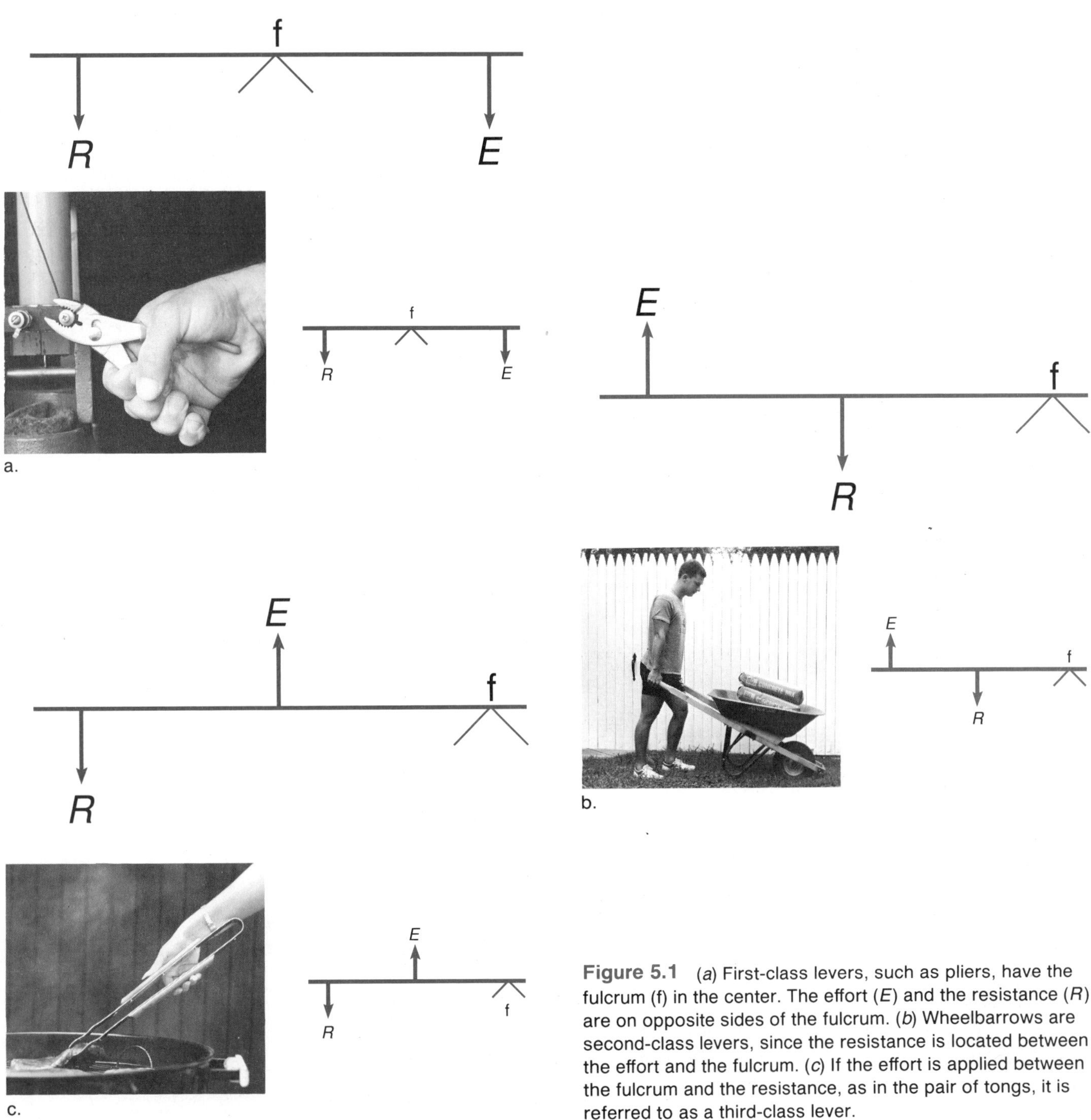

Figure 5.1 (*a*) First-class levers, such as pliers, have the fulcrum (f) in the center. The effort (*E*) and the resistance (*R*) are on opposite sides of the fulcrum. (*b*) Wheelbarrows are second-class levers, since the resistance is located between the effort and the fulcrum. (*c*) If the effort is applied between the fulcrum and the resistance, as in the pair of tongs, it is referred to as a third-class lever.

All machines are composed of varying numbers and arrangements of fundamental components called *simple machines.* The origin of each—the lever, the wheel and axle, the pulley, the wedge, the screw, and the inclined plane—is long-lost history. The most commonly used types of simple machines are the lever, the pulley and its systems, and the inclined plane. The others mentioned are actually special applications of these three simple machines. For example, the wheel and axle is a special application of a lever. The wedge and screw are, in effect, distinctive applications of an inclined plane.

In a basic form, the *lever* consists of a rod or pole that rests on an object at a point called a *fulcrum* (*f* in Fig. 5.1). The object to be lifted, commonly referred to as the *resistance* (*R*), may be placed in various positions with respect to the fulcrum and the

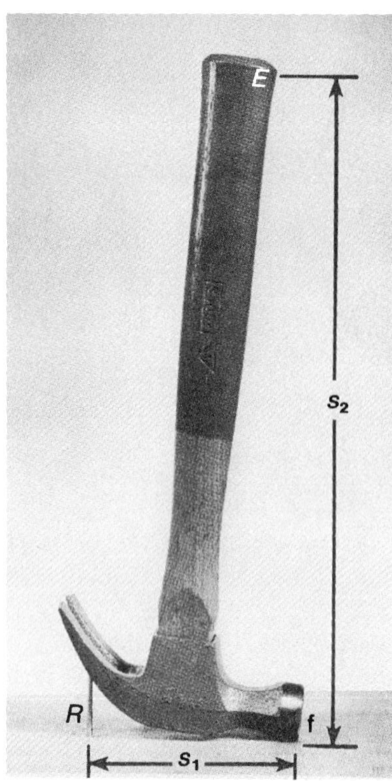

Figure 5.2 The hammer is often used as a lever. Note the type of lever illustrated.

applied *effort* (E), a force designed to counteract and move the resistance. The three possible arrangements are illustrated as stick diagrams in Figure 5.1, along with typical applications of each one.

Regardless of the arrangement of $E, f,$ and R, analysis follows a similar pattern.

The use of a hammer to remove a nail from a board is shown in Figure 5.2. The nail is held in place by the friction between it and the board and offers resistance at a point 7.50 centimeters (0.0750 m) (s_1) from the head of the hammer. The head, then, serves as the fulcrum with this application. The effort (E) is applied at the end of the handle (s_2) 30.0 centimeters (0.300 m) from the fulcrum. If the effort required to remove the nail is 100 newtons, the resistance offered by the nail is determined from resistance × distance from fulcrum = effort × distance from fulcrum; therefore,

LEVER PROBLEMS

$$Rs_1 = Es_2$$
$$R = \frac{Es_2}{s_1}$$
$$= \frac{(100 \text{ N})(0.300 \text{ m})}{0.0750 \text{ m}}$$
$$= 400 \text{ N}$$

A system of pulleys is closely related to the lever. With pulleys, the resistance can be moved slowly with less effort than would otherwise be required to move it, or the object can be moved very quickly with the exertion of more effort than that actually required to move the resistance.

A single fixed pulley (Fig. 5.3a) is used to change the direction the applied force moves the object. That is, the resistance (R) is moved upward as the effort (E) is applied downward. The individual who applies the effort can use his or her weight to lift the resistance. If the resistance is greater than the weight of the person applying the effort, then the single fixed pulley is useless. A single movable pulley (Fig. 5.3b) can be used to lift an object with one-half the effort required with the single fixed pulley. The single movable pulley then produces a **mechanical advantage** (*m.a.*) of 2. That is, *the ratio of the magnitudes of the output force (the resistance) to the input force (the effort)* is two to one. (Assume the mechanical advantage calculated or given in all cases to be accurate to three significant digits for problem solving.) Combinations of pulleys permit decidedly greater mechanical advantages, as displayed in Figure 5.3c. Pulley systems are quite common around the modern home. For example, these systems are component parts of radios, drapery rods, garage doors, and many other home appliances.

Example 5.2

The mechanical advantage of a pulley system such as that illustrated in Figure 5.3c is 4. If an object requires a force of 56.0 newtons to directly lift it, what effort theoretically is required to lift the weight with the system shown?

Solution

The effort required to lift the weight with the pulley system can be determined from

MECHANICAL ADVANTAGE

$$\text{mechanical advantage} = \frac{\text{resistance}}{\text{effort}}$$
$$m.a. = \frac{R}{E}$$
$$E = \frac{R}{m.a.}$$
$$= \frac{56.0 \text{ N}}{4}$$
$$= 14.0 \text{ N}.$$

As noted earlier, *a comparison of the output work to the input work* provides a means of determining the **efficiency** of a machine; therefore,

$$\text{efficiency } (\%) = \frac{\text{output work}}{\text{input work}} \times 100.$$

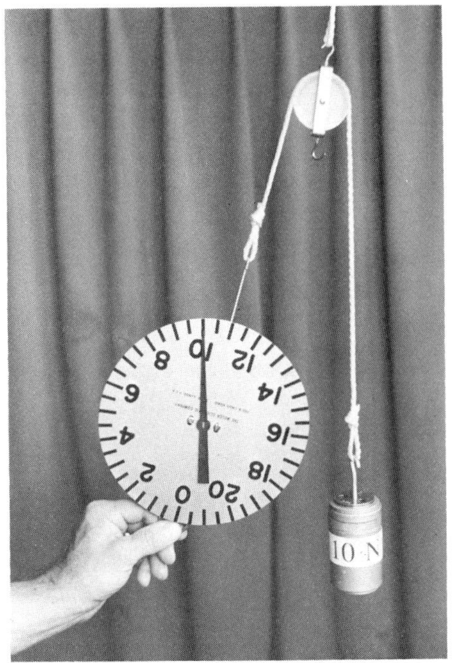

a.

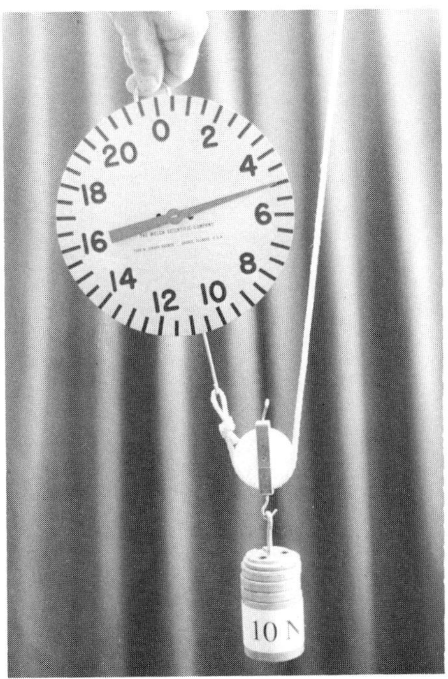

b.

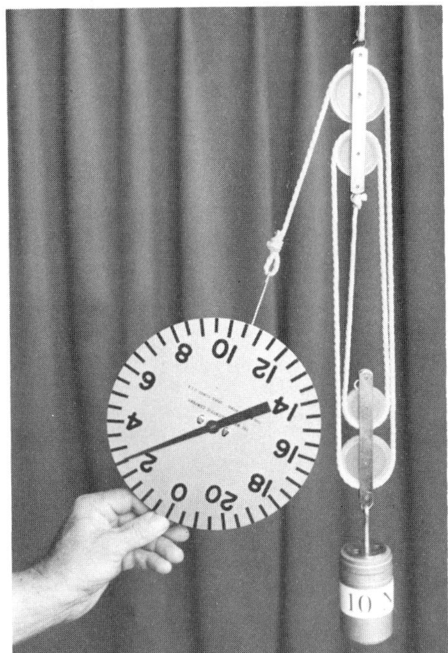

c.

Figure 5.3 (*a*) The single fixed pulley is used to lift objects by the application of a force in a downward direction. (*b*) The single movable pulley has a mechanical advantage of 2. (*c*) The block and tackle is composed of a series of fixed and movable pulleys, and is employed to lift heavy objects. Four strands of the small rope support the 10-N mass; thus, the mechanical advantage is 4.

If the work actually required to lift the 56.0-newton object a distance of 0.500 meter is found to be 7.50 newton-meters, the efficiency of the pulley system can be calculated by substitution:

$$\begin{aligned} \text{efficiency} &= \frac{(14.0\ \text{N})(0.500\ \text{m})}{7.50\ \text{N}\cdot\text{m}} \times 100 \\ &= 93.3\ \text{percent.} \end{aligned}$$

A third common type of simple machine is the inclined plane. In various forms it is applicable to many situations. Generally, it is a ramp that connects two levels, whether stairs between floors or a sloping driveway. In a circular form it makes its appearance as a screw or a bolt. The efficiency is high since frictional forces are minimal. The mechanical advantage is calculated by the comparison of the length of the inclined plane to its height (or, in more descriptive terms, "rise over run"). Thus,

$$m.a. = \frac{\text{length of plane}}{\text{height}} = \frac{l}{h}. \qquad \text{INCLINED PLANE}$$

The photographs in Figure 5.4 illustrate inclined planes. Typical applications around the home employ such tools as an ax, a letter opener, and a knife. As mentioned earlier, the screw is a special application of the inclined plane. In order to see more clearly that the screw is an inclined plane, imagine a piece of paper similar to the shape of the inclined plane illustrated in Figure 5.4b or 5.4c. If the piece of paper were rolled into a cylinder, starting with side *BC* and progressing toward angle *A*, the line *AB*, although spiral, would still be an incline. Other examples of circular inclined planes are the spiral staircase and the typical highway through mountainous terrain.

Example 5.3

A certain exit ramp on a major freeway is 200 meters long, and the upper end is 10.0 meters above the highway. Determine the theoretical effort required to move a truck with trailer, the mass of which is 20,000 kilograms, to the end of the ramp.

Solution

The mechanical advantage of the ramp, an inclined plane, is determined from

$$\begin{aligned} m.a. &= \frac{\text{length } (l)}{\text{height } (h)} \\ &= \frac{200\ \text{m}}{10.0\ \text{m}} \\ &= 20.0 \end{aligned}$$

To express the resistance, *R*, in newtons,

$$\begin{aligned} R = w = mg &= 20{,}000\ \text{kg} \times 9.80\ \text{m/s}^2 \\ &= 196{,}000\ \text{N}. \end{aligned}$$

Figure 5.4 Three applications of the inclined plane: (*a*) A series of stairs leading to upper floors; (*b*) a loading ramp; and (*c*) a treadmill.

Then, the theoretical effort can be determined from

$$\text{mechanical advantage} = \text{resistance} \div \text{effort}$$

$$m.a. = \frac{R}{E}$$

$$E = \frac{R}{m.a.}$$

$$= \frac{196{,}000 \text{ N}}{20.0}$$

$$= 9800 \text{ N}.$$

Another simple machine known as the wheel and axle (see Fig. 5.5) operates in principle as a continuous lever. In the ice-cream freezer, the rotating can opener, and the doorknob, the effort is applied in a circular path of large circumference as the resistance sweeps out a smaller circular path. The bicycle pedal, the automobile steering wheel, and the screwdriver are other examples of the wheel and axle.

In any machine, regardless of complexity, a frictional force acts to oppose the applied force, and thus opposes the motion brought about by the applied force. For an object to move at a

The Science of Mechanics

In the operation of simple machines, individually or in complex groups such as those used in construction, manufacturing, and transportation, certain general principles always apply. These principles underlie the science of *mechanics.*

The principles of mechanics apply throughout the universe—to celestial bodies as well as to the elementary particles of the atom. They explain how an airplane is able to fly and how an automobile moves. Mechanics involves the behavior of all kinds of matter under the influences of forces. The branch of mechanics that deals with bodies at rest is called *statics;* that which deals with bodies in motion is known as *dynamics.* Statics considers such objects as buildings or bridges, as well as the forces that act on them. These structures are essentially at rest, but the force of gravity on the steel beams, concrete, and other construction materials is always present. In addition, these structures must support the weight of the various objects contained within them, or which pass across them.

When various forces act on a body in such a manner as to provide an imbalance of forces, the resultant force produces a change in the motion of the body. The principle of dynamics explains how Earth moves through space and why a golf ball assumes a certain path, be it straight or curved. Dynamics is involved in the design of rockets, guided missiles, and submarines. Various applications of dynamics are quite complicated, but many of its principles are well demonstrated by simple machines ■

constant velocity, a force equal to the opposing force of friction must be present. No force other than that necessary to overcome frictional forces is necessary to maintain the motion of a body. The magnitude of the opposing force due to friction varies with type of surface and degree of smoothness.

The effect of friction on the amount of force required to maintain the motion of a block of wood as it is pulled along a rough surface and along a smooth surface is displayed in Figure 5.6. Note the position of the pointer on the spring balance in both cases. In Figure 5.6b the reading of the balance as the block is drawn over a heavily waxed surface is considerably less than the reading in Figure 5.6a, where the surface is unwaxed.

Frictional force is decidedly greater on the rougher of the surfaces shown in the illustrations. However, even extremely smooth surfaces have areas of irregularity one can readily see under a microscope—areas that act as obstructions to motion. Smooth surfaces have many small areas of contact, whereas rough surfaces have fewer but larger contact areas. Many investigations reveal that friction is not created primarily by surface irregularities but by the attractive forces between atoms or molecules at the point of contact between surfaces. Scientists have also noted that given the same degree of smoothness, surfaces of the same material produce greater friction between them than do surfaces of different materials. Mostly for this reason are machine bearings generally made of different metals than the shafts about which they rotate.

Friction occurs whenever two bodies move relative to each other. If a lubricant is placed between the two surfaces, the frictional relationship that results is quite complicated. The purpose of the lubricant, however, is to lower frictional forces so that less force is required to move one surface across the surface of the other. The lubricant tends to smooth out the irregularities in both surfaces by filling in the depressions inherent in both bodies. When one body slides over the other, *sliding friction* results. *Rolling friction,* generally a much smaller resistive force, results when one body rolls along the surface of another. *Fluid friction,* present in

Figure 5.5 An example of the wheel and axle. The theoretical mechanical advantage of this machine is determined from the comparison of the radius, diameter, or circumference of the wheel to a like measurement of the axle.

a.

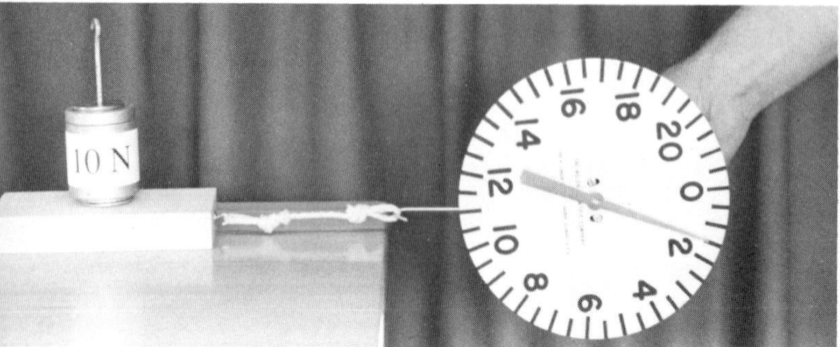

b.

Figure 5.6 (*a*) The effort required to pull the block and weight across the rough surface is substantial. (*b*) The block and weight are pulled across the smooth surface with considerably less effort than indicated in (*a*).

liquids and gases, is created as an object moves through the fluid and pushes it aside. Objects, such as meteors, that enter Earth's atmosphere at high velocities, are heated to the point of incandescence by fluid friction.

The force required to slide one body over the surface of another is dependent on two factors: (1) the weight of the moving body and (2) the type of surfaces involved. The magnitude of the force required to slide an object across a given horizontal surface compared with the force necessary to lift the object (its weight) is known as the *coefficient of sliding friction:*

$$\text{coefficient of sliding friction} = \frac{\text{force}}{\text{weight}}$$

SLIDING FRICTION

$$\mu = \frac{F}{w}.$$

Since friction between surfaces varies, the coefficient of friction assumes values commensurate with the surfaces involved. For instance, a dry wooden block that weighs 10 newtons requires an effort of 2.5 newtons to slide it across a horizontal wooden surface. Therefore, the coefficient of sliding friction (μ) for wood on wood is $F/w = 2.5 \text{ N}/10 \text{ N} = 0.25$. If both horizontal surfaces are steel, only 1.8 newtons of effort are required to slide the 10-newton steel object. The coefficients of sliding friction between various surfaces are presented in Table 5.1. A discussion of friction between nonhorizontal surfaces is beyond the scope of this text.

Table 5.1

Coefficients of sliding friction for smooth contact surfaces.

Materials	Sliding coefficients ($\mu = F/w$)*
Wood on wood, dry	0.25 to 0.50 (varies with type of wood)
Oak on oak	0.25
Pine on pine	0.35
Metals on oak	0.55
Rubber on oak	0.46
Metal on metal, dry	0.15 to 0.20 (varies with type of metal)
Steel on steel	0.18
Greased surfaces	0.05
Rubber tire on dry concrete	0.70 to 0.95 (varies with tread and composition)
Rubber tire on wet concrete	0.50 to 0.70 (varies with tread and composition)
Rubber tire on ice	0.20 to 0.40 (varies with tread and composition)
Iron on concrete	0.30
Glass on glass	0.40
Ski wax on dry snow	0.04
Ski wax on wet snow	0.14
Leather soles on ice	0.20
Cushion-crepe soles on ice	0.50

*(Note that coefficients of friction are approximate and may vary significantly under various surface conditions. Assume the coefficients to be accurate to three significant digits for problem solving.)

Example 5.4

What force is required to slide a 100-kilogram metal filing cabinet across an oak floor?

Solution

The weight of the filing cabinet may be calculated from

$$\begin{aligned} w &= mg \\ &= 100 \text{ kg} \times 9.80 \text{ m/s}^2 \\ &= 980 \text{ N}. \end{aligned}$$

Since the coefficient of friction = force ÷ weight, then

$$\begin{aligned} F &= \mu w \\ &= (0.55)(980 \text{ N}) \\ &= 539 \text{ N}. \end{aligned}$$

The coefficient of rolling friction is calculated and interpreted in a similar way. The coefficient of rolling friction for steel, as between a steel ball or cylinder and a steel plate, is about 0.002. In other words, a 10-kilogram steel ball requires a force of less than 0.02 newton to keep the ball rolling at a constant speed once inertia has been overcome. A selected list of rolling friction coefficients appears in Table 5.2.

When one compares the values of the coefficients of rolling friction to those of sliding friction, it becomes obvious why wheels, rather than runners, are used to move heavy machinery, and why lubricants composed of microspheres are superior to other types of lubricants in lowering friction between moving parts.

There are some cases, however, when we use the friction between moving parts to our benefit. The friction between the soles of our shoes and the sidewalk permits us to walk with ease. The friction between our automobile tires and the highway allows us to travel and to bring our vehicles to a stop. Without friction we could not build our homes using nails, since the nails are held in place primarily by friction between nails and the wood.

Friction is present in various degrees in all machines. If a lever acts on a very sharp, hard fulcrum, friction is almost negligible. But in such machines as screws and pulley systems, a great deal of friction is present—and this lowers the mechanical advantage from its theoretical value. Because of the presence of friction, the applied force must do work against friction as well as furnish the energy required to move the resistance. Thus, the useful energy output is less than ideal because part of the input energy is expended to overcome friction as well as gravity and inertia. The work done against friction is usually not recoverable. However, work expended in overcoming gravity or inertia is recoverable and assumes the form of potential energy, a concept discussed in the section that follows.

In general, as has been previously mentioned, the more moving parts a mechanical device has, the greater the amount of friction present and the lower the efficiency of the complex machine. Research teams representing the oil and automobile industries have strived to increase the efficiency of the automobile and its engine. New lubricating oils and smoother contact surfaces have been developed, the applications of which have had varying effects on improving efficiency. If greater success is met, one might foresee an engine that would propel an automobile over 100 miles per gallon of gasoline. Moving parts in machines would seldom have to be replaced. Many lifetimes have been dedicated to the task of developing a frictionless machine, but all attempts have resulted in some degree of failure. However, unwanted friction has been minimized through the application of ingenious ideas. The United States Patent Office still receives requests for patents on perpetual motion devices, contraptions in which, once inertia has been overcome, the moving parts should continue to move and to do work indefinitely. Many seemingly workable blueprints are on file with the Office, but no advance in technology has provided a totally efficient device. It would be safe to assume that such a device will never be developed, because friction between moving parts cannot be completely eliminated; thus, efficiency cannot equal 100 percent—the requirement for perpetual motion to be realized.

Table 5.2
Coefficients of rolling friction for selected contact surfaces.

Materials	Rolling coefficients*
Cast iron on steel rails	0.004
Rubber tires on concrete	0.030
Ball bearing in rolling contact	0.001 to 0.003
Roller bearing in rolling contact	0.002 to 0.007

*Assume coefficients accurate to three significant digits for problem solving.

Power and Energy

Work, as stated previously, is the product of the force exerted on an object and the distance through which the force moves. The same amount of work is done in moving a heavy weight up a flight of stairs whether the person who applies the force runs or walks. When one wishes to take into account the time in which the work is done, a time-rate unit called **power** is introduced. Power is defined as the work done per unit time:

$$\text{power} = \frac{\text{work}}{\text{length of time to do the work}}.$$

Making use of the proper symbols, this relationship is stated

$$p = \frac{W}{t}.$$

POWER

Power is generally expressed in terms of the SI unit, the *watt*. It is symbolized by W, and is equivalent to one joule of work done in one second. The *horsepower* (hp) is equal to 746 watts.

Example 5.5

Determine the power of an engine capable of lifting a mass of 500 kilograms to a height of 10.0 meters in 30.0 seconds.

Solution

Since power is work per unit time, one must first determine the force required to lift the object:

$$
\begin{aligned}
F &= w = mg \\
&= (500\ \text{kg})\left(\frac{9.80\ \text{m}}{\text{s}^2}\right) \\
&= 4900\ \text{kg}\cdot\text{m/s}^2 \\
&= 4900\ \text{N}.
\end{aligned}
$$

Then the amount of work done in lifting the object to the required height is found from

$$
\begin{aligned}
W &= Fs \\
&= (4900\ \text{N})(10.0\ \text{m}) \\
&= 49{,}000\ \text{N}\cdot\text{m} \\
&= 49{,}000\ (\text{or } 4.90 \times 10^4)\ \text{J}.
\end{aligned}
$$

Finally, the power is determined from

$$
\begin{aligned}
\text{power} &= \text{work} \div \text{time} \\
p &= \frac{W}{t} \\
&= \frac{49{,}000\ \text{J}}{30.0\ \text{s}} \\
&= 1630\ \text{J/s} \\
&= 1630\ \text{W}.
\end{aligned}
$$

Example 5.6

A motor can provide a force of 5000 newtons. How much work is done as it applies its maximum force on an object and lifts it 8.00 meters? If the work is performed in 10.0 seconds, how much power is generated and what is the equivalent in horsepower?

Solution

$$
\begin{aligned}
W &= Fs \\
&= (5000\ \text{N})(8.00\ \text{m}) \\
&= 40{,}000\ \text{N}\cdot\text{m} \\
&= 40{,}000\ \text{J}.
\end{aligned}
$$

To determine the power required,

$$
\begin{aligned}
p &= \frac{W}{t} \\
&= \frac{40{,}000\ \text{J}}{10.0\ \text{s}} \\
&= 4000\ \text{J/s} \\
&= 4000\ \text{W}.
\end{aligned}
$$

Since 1 horsepower equals 746 watts, the horsepower equivalent may be determined by the following computation:

$$4000\ \text{W}\left(\frac{1.00\ \text{hp}}{746\ \text{W}}\right) = 5.36\ \text{hp}.$$

The greater the power of an engine, the faster it can do work. An engine that will move a car to the top of a steep hill in about 30 seconds develops greater power than an engine that takes a car of comparable weight to the top of the same hill in a longer period of time.

The horsepower as a unit can be traced back to the days of James Watt and his steam engine. At that time the horse was used in the coal mines to perform tasks that the steam engine could perform. As Watt tried to sell his steam engine, he was asked by many mine operators how many horses he thought his engine could replace. Eventually, he realized that he needed to determine this equivalence. To accomplish his task, he used some of the horses that were on hand at one of the mines to find how long it would take each to raise a heavy weight out of a deep well. He eventually determined that an average workhorse could perform considerably more than 600 foot-pounds of work per second over a typical workday. In order to facilitate the sale of his engine, Watt allowed himself room for experimental error and rated each of his steam engines in terms of the power of a horse, but at 550 foot-pounds per second. The miners were elated with the efficiency and operation of the engine, and sales soared, particularly of engines in the 20- to 30-horsepower range. The horsepower unit is also commonly expressed with unit time in minutes, which becomes 33,000 foot-pounds per minute (60 s/min × 550 ft·lb/s). The unit is still generally used to rate small appliance motors, boat motors, and air conditioners.

Energy may be defined as *the ability to do work.* In order for a machine or any object to do work, energy must be provided to it. If something has energy, it can exert a force upon other objects, and thus it has the capability to do work. If a force is applied to an object and work is done, the object is set into motion or it experiences a change in its velocity. The object thus possesses energy and hence has the ability to do work. For instance, a cuckoo clock is wound by a force that lifts a suspended weight. As gravity pulls on the weight, the weight moves downward. The intricate gear system turns; thus, the clock hands move accordingly.

Mechanical energy appears in two forms: (1) as the energy stored by an object as a result of its height above the ground or as stored through some configuration, such as in a stretched rubber band or compressed spring; or (2) as the energy concerned with motion. (Of course, an object may possess both forms of energy simultaneously, as with falling objects.) *Energy stored in an object as a result of its position in a field of force or because of some configuration* is referred to as **potential energy.** A box that is lifted from its position on the floor and placed on a table has been given (gravitational) potential energy as a result of the relative heights. Work was performed to lift the box from its original position to the top of the table. Whenever work is done on such an object, the object gains potential energy. The potential energy gained is equal to the work done. That is, work $= Fs = mgh$, where h is the distance the box was lifted and g is the acceleration due to gravity. Thus, potential energy for such a situation can be calculated by

$$P.E. = Fs = mgh \qquad \text{GRAVITATIONAL POTENTIAL ENERGY}$$

(or weight × height, if the weight is known rather than the mass). If the box has 50 joules of work done on it, it gains 50 joules of potential energy.

Example 5.7

Calculate the potential energy of a vehicle, the mass of which is 2500 kilograms, and which is resting on a hydraulic lift 2.00 meters above the floor of a service station.

Solution

The potential energy is calculated from

$$\begin{aligned} P.E. &= mgh \\ &= (2500\ \text{kg})(9.80\ \text{m/s}^2)(2.00\ \text{m}) \\ &= 49{,}000\ \text{kg}\cdot\text{m}^2/\text{s}^2 \\ &= 49{,}000\ \text{N}\cdot\text{m} \\ &= 49{,}000\ \text{joules (J)}. \end{aligned}$$

The water at the top of Niagara Falls has potential energy because of its position, as does a skier poised on a mountain slope. Other objects that have potential energy because of stored energy include such things as a cocked pistol, a stretched spring or rubber band, and a poised hammer ready to strike a nail. Energy stored in an auto battery, although chemical in nature, is also potential energy. Other sources of chemical energy are gasoline, dynamite, coal, and food. All are capable of doing work as potential energy is released.

A moving object has energy as a result of its motion. For an object to move, a force must be applied to it, according to Newton's second law. Therefore, work must be done to move the object. Any moving object has the capability of doing work as it strikes another object and causes the second object to move. Although the second object may roll or slide to a stop without doing useful work, the ability remains as long as the object is in motion. *The ability a moving object has to do work* is called **kinetic energy.** Kinetic energy is equal to one-half the product of the object's mass and the square of the velocity at which the object moves; that is

$$K.E. = \tfrac{1}{2}mv^2. \qquad \text{KINETIC ENERGY}$$

The usefulness of kinetic energy comes from its relationship to work. By using Newton's second law, $F = ma$, and recalling the definition of work, $W = Fs$, where F is the force applied to an object and s is the distance through which the force acts in the direction of the force, it can be shown that *the total work done on an object equals its change in kinetic energy.* (This amount of work done on an object is against the object's inertia.) Starting with the second law, $F = ma$, and mutliplying both sides by s, $Fs = mas$. Recall, the definition of acceleration, a, is $a = (v_f - v_i)/t$, and it was noted that the distance traveled while an object is accelerating, s, is given by

$$s = \bar{v}t = \tfrac{1}{2}(v_f + v_i)t.$$

Substituting a and s into the preceding equation,

$$Fs = m\left(\frac{v_f - v_i}{t}\right)\tfrac{1}{2}(v_f + v_i)t.$$

Completing all the algebraic manipulations,

$$Fs = \tfrac{1}{2}mv_f^2 - \tfrac{1}{2}mv_i^2.$$

This series of substitutions point out that the work done on an object equals its change in kinetic energy. If the force is given in newtons and the distance in meters, both the work and kinetic energy would be expressed in joules.

Example 5.8

Determine the kinetic energy of a vehicle with a mass of 6000 kilograms that is moving at a constant velocity of 30.0 meters per second.

Solution

The kinetic energy is determined as follows:

$$\begin{aligned} K.E. &= \tfrac{1}{2}(\text{mass}) \times (\text{velocity})^2 \\ &= \tfrac{1}{2}mv^2 \\ &= \tfrac{1}{2}(6000\ \text{kg})\left(\frac{30.0\ \text{m}}{\text{s}}\right)^2 \\ &= 2{,}700{,}000\ \text{kg}\cdot\text{m}^2/\text{s}^2 \\ &= 2{,}700{,}000\ \text{N}\cdot\text{m} \\ &= 2{,}700{,}000\ \text{J}. \end{aligned}$$

Consider now the relationship between work, kinetic energy, and potential energy. The total work done on an object equals its change in kinetic energy, as just discussed. When an object is moved upward near the surface of Earth, the force of gravity, being in a downward direction, does a negative amount of work. Since there is no way to escape this work of gravity, it is separated from the work of all other forces and moved to the other side of the equation. Clearly, this work can be recovered by letting the object return to the lower level from which it started. It is simply this ability of gravity to do work that is known as the gravitational potential energy. With the work of gravity included on the side of the equation with energy, the equation points out that work equals the change in total energy, where *the total energy is the sum of kinetic energy and potential energy.* In equation form,

$$E = K.E. + P.E.$$

Scientists often refer to this combined energy as the total **mechanical energy.**

Energy is continually being converted from one form to another, yet the total energy of the universe is conserved. That is, if the sum of energy present in various forms throughout the universe could be determined from all possible perspectives, the same value would always result. Since the universe represents the largest system that can be visualized, the total energy must exist somewhere in some form; thus, it remains constant. These observations relate to one of the most important principles in the physical sciences, known as the **law of conservation of energy:** *The total energy is neither increased nor decreased by any process. Energy can be transformed from one form to another, and transferred from one body to another, but the total amount remains constant.*

Consider a child who is swinging at a playground. As the child swings back and forth, she or he alternately gains and loses both potential energy and kinetic energy. At the peak of the arc the swing makes, potential energy is at maximum, and for an instant, kinetic energy is zero. As the child starts downward from the top of the arc the swing makes, potential energy decreases and kinetic energy increases. At the point where the child is nearest the ground, kinetic energy is at maximum and potential energy is at minimum. At any point in the arc, the sum of kinetic energy and potential energy is a constant, so energy is conserved. It should be noted that a force other than gravity acts on the swing—the tension in the rope or chain—but this force does no work, since it always remains perpendicular to the motion of the child.

The States of Matter

Many scientists investigate the constantly growing list of substances present in our environment or synthesized in the laboratory. These substances are generally composed of those building blocks we know as *atoms.* Some scientists study the chemical properties of the substances, whereas other scientists study how the substances, representing matter of all sorts, differ in physical properties. Both groups concentrate on how matter and energy interact with each other. There is always the steadfast attempt to establish greater comprehension of how energy dissipates in matter, how energy brings about changes in matter, and the role energy plays as we learn about the characteristics of matter.

Matter ordinarily exists in one of three common states—*solids, liquids,* or *gases.* It apparently can exist beyond the accepted limitations of the three common states, as will be discussed in detail in Chapters 9 and 15. These unusual forms are known as plasma and liquid crystals.

The substance that we have most often observed existing at various times in all three ordinary states is water. It is composed of two atoms of hydrogen and one atom of oxygen that have chemically combined to form a *molecule.* With the exception of some specific gaseous elements, molecules are composed of one or more atoms of different elements. Disregarding the gases, in this form the combination is also known as a *compound.* Molecules simply refer to individual units of compounds.

With regard to water, we have seen this compound in its solid state (ice or snow) and in its liquid state, as we primarily make use of it. We have, however, never seen it in its true gaseous state. A gas, unless it has a definite color, is invisible, just as is the mixture of nitrogen and oxygen that primarily makes up our atmosphere. Some gases, such as chlorine (yellowish-green) and bromine (reddish-brown), are visible in concentrated amounts, because they have intrinsic color. But water in the gaseous state is colorless and, therefore, invisible. We have, however, made observations that prove to us the presence of gaseous water. Many times we have observed water droplets form on the outside of a cold glass. The appearance of liquid water droplets on a cold surface, seemingly from nowhere, is the result of gaseous water being cooled to the point where it changes from the gaseous state to the liquid state.

Why should there be such an apparent difference between the solid, liquid, and gaseous states of a substance? Solids are rigid substances and can hold a certain shape even though they are not in a container. This property would suggest that the atoms or molecules in a solid must be held together by certain attractive forces that produce the rigidity of a solid; that is, a solid's ability to hold a definite shape. This concept assumes that the atoms or molecules of a solid are restricted in their movements, since excessive movement of a solid's atoms or molecules would result in a continual change in a solid's shape, with a corresponding reduction in rigidity.

A liquid has no rigidity, nor can it hold its shape without being confined in a container. The attractive forces between atoms and molecules of a liquid, then, must be considerably less than the same forces in a solid. Liquids, like solids, are not very compressible, and therefore the atoms or molecules in a liquid, like those in a solid, must be quite close together so that an applied pressure cannot squeeze the atoms or molecules much closer together than they normally are. The atoms or molecules of liquids must be relatively free to move about within the body of the liquid, since a colored liquid and a colorless one mixed together will soon assume a single shade of color, which could be the case only if the atoms or molecules of the two liquids were able to mingle and mix freely.

Unlike solids and liquids, gases must consist of atoms and molecules that are very far apart, since gases are very easily compressible. Huge volumes of air can be forced into an extremely small place. This large compression could occur only if a given volume of a gas were mostly empty space. Unlike those of liquids and solids, the atoms or molecules of a gas are capable of moving about with very little restraint. If you open a bottle of perfume in one corner of a room, it does not take very long before a person on the other side of the room can smell the perfume. The molecules of the perfume in the gaseous state are obviously quite free to move and mingle with the air molecules in the room.

We have, then, developed the following picture of solids, liquids, and gases:

1. Solids consist of atoms or molecules held together in a rigid structure because of rather substantial attractive forces between the atoms or molecules. The atoms or molecules of a solid are close together and not easily compressible. Solids, then, maintain a fixed shape and a fixed volume.
2. Liquids are composed of atoms or molecules that are free to move about and mingle with each other, but attractive

Check Your Oil?

In order to gain the highest efficiency from a machine, such as an automobile, we use lubricants of all sorts to reduce the friction between moving parts. These substances may be gases, liquids, or solids, depending on the components involved.

One property of liquids we must consider is called *viscosity,* a measure of resistance to fluid flow. The greater the viscosity, the slower the fluid flows. Some fluids flow faster than others at a given temperature, and, in general, an increase in temperature lowers the viscosity of a fluid. In addition, the rate of flow of a liquid is less when the pressure on a liquid is increased. Thus, a comparison of the viscosity of one liquid to that of another is meaningful only if specific conditions are constant. Most viscosity measurements are made at specific temperatures and at standard atmospheric pressure.

One of the more popular instruments to measure the viscosity of oils is the Seybolt universal viscometer. This device heats a given volume of an oil (generally 0.06 liters) to a specific temperature, then determines the length of time, in seconds, it takes for the sample to empty into another vessel. This measurement of time is known as the Seybolt Universal Seconds (SUS). The SUS measurement, then, is a relative measure of the viscosity of the sample. The greater the viscosity of the oil, the higher the SUS number.

The Society of Automotive Engineers (SAE) has also established a means to standardize this characteristic of engine oils. The numbering system they adopted corresponds in general with the range of SUS units. Basically, then, the higher the SAE number, the greater the viscosity. Each number represents a range of viscosity at a certain temperature. In warm weather, an SAE number such as 30 should be used for proper engine performance. In the winter, a lower number, perhaps SAE 10W, is more favorable (the "W" indicates the proper number for temperatures below freezing). Oils, such as 10 W–20 W–40, commonly known as SAE 10 W–40, are classified as multigrade oils. Collective research efforts have developed various oil additives that extend the viscosity range of such lubricants, permitting us to use the same grade of oil for all seasons, excluding the extreme temperature changes experienced in some geographic areas ■

forces keep the atoms or molecules within the main body of the liquid; that is, they will not uniformly fill a closed container. Assuming sufficient volume, liquids conform to the shape of the confining vessel, hence, do not maintain a fixed shape.

3. Gases are made up of atoms or molecules that are completely free to move about at random and that will, therefore, uniformly fill a closed container. A gas is mostly empty space, the distance between atoms or molecules being very large compared to the size of the atoms or molecules themselves. The attractive forces between the atoms or molecules of a gas are very small. A gas, then, has neither a fixed shape nor a fixed volume. It simply expands to fill its container. Since liquids and gases do not maintain a fixed shape, they are free to flow; thus, either liquid or gas may be referred to as a *fluid.*

The Kinetic Molecular Theory of Gases

A body in motion has the capacity to do work; therefore, it possesses energy. Consequently, all moving bodies have kinetic energy because of their motion. The kinetic molecular theory of gases assumes the following:

1. Gases are composed of discrete particles (atoms and molecules) that travel in random directions.
2. The molecules of a gas travel at a wide variety of speeds; that is, different molecules possess different amounts of kinetic energy.
3. Colliding molecules conserve their kinetic energy.
4. The distance between molecules is very large in comparison with the size of the molecules themselves. Thus, the total volume of the molecules of a gas is negligible in comparison to the volume of the gas itself.

The preceding assumptions of the kinetic molecular theory of gases explain rather readily such observations as the diffusion of perfume throughout a large room, the escape of a gas from a pinpoint hole (such as of air from a tiny tire puncture), and the compression of a gas to a very small volume. Furthermore, gases are very light in weight because they represent mostly empty space. *The magnitude of the force exerted by a gas (or liquid) on a given area of its container* is known as **pressure.** (A discussion of the units for measuring pressure is forthcoming.) This pressure, as in balloons and bubbles, is due to the incessant bombardment of the container walls by the molecules of the gas.

Pressure and Its Applications

A pressure cooker makes cooking much quicker and easier than it once was. In this appliance, foods are cooked under increased pressure, so they are ready for serving long before similar foods cooked in conventional ways. Keep in mind that cooking is in effect a process by which kinetic energy is transferred from one body to another because of temperature differences. Note also that the temperature of a body is an indication of the average kinetic energy of its molecules. Water boils at a much higher temperature when heated under the pressure of the pressure cooker than when it is heated in an open utensil; thus, the food surrounded by the hot water is exposed to a higher temperature.

Table 5.3
The effect of pressure on the boiling point of water.

Existing conditions when measured	Pressure (atmospheres)	Barometric pressure (cm)	Boiling point (°C)
Upper reaches of Earth's atmosphere	0.0061	0.46	0
[altitude approximately 18 km (60,000 ft)]	0.068	5.17	39
[altitude approximately 3 km (10,000 ft)]	0.68	51.70	90
[sea level–1.01 × 10^5 N/m^2 (14.7 lb/in^2)]	1.00	76.00	100
Typical auto radiator system or pressure cooker	2.04	155.10	121

Observations show that the boiling point of water varies as a function of height above sea level because of differences in air pressure. At an altitude of 5800 meters, water boils at 81°C (178°F); at 4300 meters, it boils at 86°C; at 305 meters, 99°C; and at sea level, 100°C (212°F). On Pike's Peak, where the boiling point of water is 91°C, visitors find that making coffee and cooking eggs by boiling are difficult. At altitudes below sea level, the boiling point of water increases accordingly. When one considers the effect of altitude, the relationship between boiling point and effective air pressure becomes evident; that is, the boiling point of a liquid increases as the air pressure increases.

We often observe evaporation taking place well below the surface of liquids as we see bubbles of gas rising to the top of the liquid, where the gas escapes. The velocity of the molecules that form the gas is high enough to exert an outward pressure on the inside of the bubble greater than the combined pressure exerted by the liquid and the air above its surface. As the bubble rises, it continually increases in size, since the pressure exerted by a liquid is a function of depth. When the bubble reaches the surface, liquid pressure approaches zero, and the bubble bursts because the vapor pressure inside the bubble is significantly greater than the air pressure above the surface of the liquid. Without sufficient vapor pressure the bubbles will collapse, and no appreciable amount of the substance can escape. As the pressure on the surface of the liquid is increased, the molecules must move faster to exert a vapor pressure within the bubble greater than the increased pressure. Therefore, an increase in air pressure on the surface of a liquid raises the liquid's boiling point. The significance of the effect of air pressure on the boiling point of water is presented in Table 5.3.

The time required to cook foods in boiling water is much longer at high elevations than at low elevations, as many who seek outdoor recreation have discovered. The wise camper who anticipates spending some time in areas above normal elevations certainly includes a pressure cooker among the necessities. A typical pressure cooker, shown in Figure 5.7, blocks the escape of the steam produced by heat and increases the internal pressure to about 2×10^5 newtons per square meter (30 lb/in^2). (One should never force open a pressure cooker, since serious burns could result.) This procedure raises the temperature at which water boils to about

a.

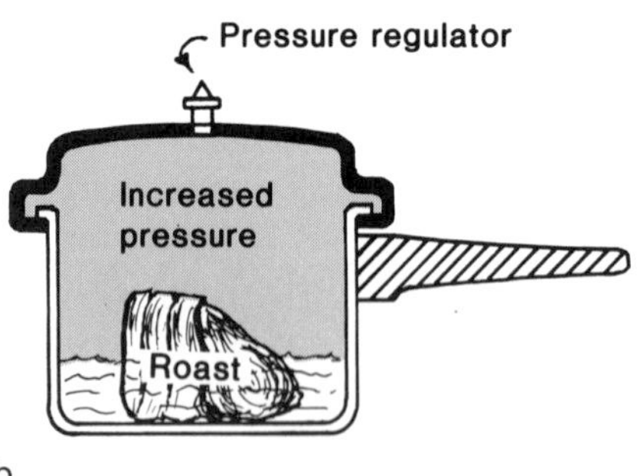

b.

Figure 5.7 The pressure cooker is a common cooking utensil in today's modern home. The components of the pressure cooker are standard in practically all models.

121°C. Meats that are typically tough when boiled under ordinary atmospheric pressure become more tender when cooked in water that boils under increased pressure at higher temperatures. The effect of increased pressure on the cooking times of several foods is presented in Table 5.4.

Pressure is ordinarily defined as the force exerted divided by the area over which the force is distributed:

$$\text{pressure} = \frac{\text{force}}{\text{area}}$$

$$P = \frac{F}{A}. \qquad \text{PRESSURE}$$

Table 5.4

The comparison of cooking times of common foods as affected by increased pressure.

Food	Cooking time	
	Regular	*Pressurized*
Green beans	50 min	15 min
Chicken	90 min	30 min
Ham	240 min	45 min
Potatoes	30 min	10 min
Pot roast	120 min	35 min

Example 5.9

If the internal pressure of a pressure cooker has reached 2.00×10^5 newtons per square meter, calculate the total force exerted on the inside of the lid. Assume the circular lid to have an inner radius of 0.150 meter.

Solution

The area of the inner surface of the lid is calculated from

$$\begin{aligned} A &= \pi r^2 \\ &= 3.14 \times 0.150 \text{ m} \times 0.150 \text{ m} \\ &= 0.0707 \text{ m}^2. \end{aligned}$$

Since $P = F/A$, the total force exerted on the interior of the lid would equal the product of the pressure and the total area, or

$$\begin{aligned} F &= P \times A \\ &= (2.00 \times 10^5 \text{ N/m}^2)(0.0707 \text{ m}^2) \\ &= 1.41 \times 10^4 \text{ N}. \end{aligned}$$

Example 5.10

Determine the pressure exerted on a floor by each roller of a refrigerator, the mass of which is 240 kilograms. Assume the mass to be evenly distributed among the four rollers, each of which has 3.20 square centimeters of its outer surface in contact with the floor.

Solution

First, one must determine the weight of the refrigerator. From $w = mg$, $w = (240 \text{ kg})(9.80 \text{ m/s}^2) = 2350$ N.

Since the weight is assumed to be evenly distributed, each roller must support

$$\frac{2350 \text{ N}}{4} = 588 \text{ N}.$$

The derived SI unit of pressure is called the *pascal.* As presented in Appendix 1, Table B, the pascal (Pa) is equivalent to one newton per meter squared. That is, $1 \text{ Pa} = 1 \text{ N/m}^2$. Therefore, it is necessary to express the area of the contact surface of the roller in meters squared. One meter squared is equal to 10,000 (10^4) centimeters squared, thus

$$\begin{aligned} 3.20 \text{ cm}^2 &= \frac{3.20 \text{ cm}^2}{10{,}000 \text{ cm}^2/\text{m}^2} \\ &= 0.000{,}320 \text{ m}^2. \end{aligned}$$

Then, using the equation for pressure,

$$\begin{aligned} P &= \frac{F}{A} \\ &= \frac{588 \text{ N}}{0.000{,}320 \text{ m}^2} \\ &= 1{,}800{,}000 \text{ N/m}^2 \\ &= 1{,}800{,}000 \text{ pascals (Pa)}. \end{aligned}$$

Earth is surrounded by a blanket of air that exerts tremendous pressure upon us. This pressure is due to air above the point of measure pressing down on the air below, compressing the air and thus increasing its density. In effect, air pressure is an indication of the weight of the air. At sea level each square inch of our bodies is exposed to about 14.7 pounds of air pressure, the weight of the column of air at this altitude. (This pressure is the equivalent of 1.01×10^5 newtons per square meter or 1.01×10^5 pascals. The value is also known as one *atmosphere.* [Note Table 5.3].) For instance, an object with a surface area of 1000 square inches is subject to a total pressure of 14,700 pounds on its outer surface. If this great pressure prevails on our bodies, what mysterious force keeps us from being crushed under the sheer weight of the air? The countering force of air pressure inside our bodies keeps us from being crushed to death and will continue to do so as long as we can keep the two forces approximately equal. When we gradually change altitudes as we ride in an airplane or automobile, we become aware of this difference. We can compensate for the difference by equalizing the pressure by yawning or swallowing. What, then, is meant by an automobile tire gauge that indicates tire pressure at 30.0 pounds per square inch? This means that the pressure inside the tire is 30.0 pounds per square inch greater than the outside pressure; that is, 30.0 + 14.7, or 44.7 pounds per square inch.

The pressure of our atmosphere permits the use of vacuum pumps, suction cups, and other such devices that rely on differences of air pressure to function. A straw used in a drink is a typical application of this phenomenon. The action of sipping on the straw reduces the pressure at the top end of the straw. The atmospheric pressure on the surface of the liquid forces the beverage up through the straw. The suction cup on a child's dart gun seemingly adheres to a smooth surface as the projectile is flattened out, creating an imbalance of pressure between the inner and outer surfaces. The human heart functions by means of a pressure differential. During contraction, the maximum pressure (*systolic pressure*) causes blood to leave the heart via the arteries. As the heart relaxes, the blood returns to the heart through the veins as pressure on the fluid decreases. This lower pressure is called the *diastolic pressure.* Blood pressure is indicated as the systolic over the diastolic. A reading of 120/80 in millimeters of mercury is considered normal. The measurement pertains to the height of a column of mercury in the manometer (pressure gauge) as the two pressures are determined.

Liquids as well as gases exert pressure on objects that they surround. Water, since it is many times heavier than air, exerts a much greater pressure than air. A container in the shape of a cube, one meter in all dimensions, would hold 1000 kilograms of water. The area of the bottom of the cube would be one square meter, and the weight of the water would exert a force of $m \times g = 1000$ kg $\times$ 9.80 m/s^2 = 9800 newtons. The pressure exerted by the column of water one meter deep would thus be 9800 newtons per square meter.

For every meter of depth an object is submerged, the pressure increases accordingly. A submarine that reaches a depth of 100 meters must be able to withstand a pressure of 9.80×10^5 newtons per square meter. *Trieste,* a deep-diving bathyscaphe or vessel with little or no capability to move horizontally, carried two U.S. Navy researchers to a record-setting depth of almost 12,000 meters (7.45 miles) off the coast of southwest Guam in 1960. The pressure exerted on the vessel, including the spherical observation chamber mounted underneath it, exceeded 1.16×10^8 newtons per square meter. This value corresponds to 16,900 pounds per square inch, since pressure exerted by water increases 0.433 pound per square inch for each foot (0.433 lb/in^2/ft) of depth. Compare this pressure to that determined in Example 5.10 as created by the weight of a refrigerator!

The pressure exerted by liquids or air is constant in all directions at a given depth. A submerged object that cannot equalize its internal pressure and the outside pressure is subject to collapse if the external pressure is great enough. The pressure of liquids is evident when an observer notes that the liquid in connected vessels, such as those illustrated in Figure 5.8, reaches a common height in each vessel. Liquid placed in the leftmost container (a) will flow through the connecting trough into b, c, d, and e until the height in each is identical. The pressures at the bottom of each container are equal and indicate that the shape of the container has no effect on pressure. The factors that affect pressure are the height of the column and the density of the liquid. The application of the observation that liquids seek their own level in connected vessels has led us to our method of transporting water from distant storage tanks or reservoirs to our homes. The force of water in the conducting pipes is dependent on the relation of the height of the reservoir to the height of the house. If the reservoir is not high enough to yield sufficient water pressure, the system must include a water tower (see Fig. 5.9) into which the water is pumped before it is made available to the consumer.

Pressure, as it relates to liquids and gases, is dependent on two variables. One of these is the height of the substance above the point at which the pressure is measured. For example, the atmospheric pressure is greater at the surface of Earth than it is at any point above it. The pressure that the air exerts is due to the weight of the air above the point of measure. Pressure, a measure

Figure 5.8 The shape of the container has no effect on the height of the column of liquid in interconnected vessels.

Figure 5.9 The water tower stores water well above ground level to provide the water pressure necessary for consumers in various parts of cities and rural areas.

of force per unit area, also results from the weight of a column of liquid directly above a given area, such as the base of each vessel in Figure 5.8. Since the column is cylindrical, the column's cross-sectional view is circular. The area of a circle is calculated from $A = \pi r^2$, where π denotes the ratio of the circumference of a circle to its diameter and r represents the radius, the distance from the center of a circle to its periphery.

The other variable that determines the pressure a substance exerts is **density,** *the mass (or weight) of one unit of volume of the substance.* (Density is commonly expressed as weight per unit volume in the English system of measurement rather than as mass per unit volume because the unit of mass in the English system, the slug, is not well known.) The density of water, for example, is ordinarily expressed as 1000 kilograms per cubic meter in SI units and as 1 gram per cubic centimeter in the *cgs* system of measurement, but as 62.4 pounds per cubic foot in the English system. Also, the density of air is noted as 0.001,29 gram per cubic centimeter (0.0817 lb/ft^3) and that of gold as 19.3 grams per cubic centimeter (1200 lb/ft^3). The densities of other substances are listed on the inside back cover of this text. The masses of the atoms composing a quantity of matter, along with the amount of space between the atoms, determine the density of a substance. Gases are lighter than liquids or solids because the distance between atoms is greater in gases than in the other states of matter. Solids are generally heavier than like volumes of liquids or gases because the space between atoms in solids is very small. The density of any substance can be found from the expression

$$\text{density} = \frac{\text{mass (or weight)}}{\text{volume}}, \qquad \text{DENSITY}$$

or

$$\rho^* = \frac{m(w)}{v}.$$

(*The Greek letter, rho (ρ) is currently used to designate density.)

(When the density of a substance is determined with respect to weight, rather than mass, per unit volume, the value obtained is often referred to as weight density. Units of weight density include newtons per cubic meter, dynes per cubic centimeter, and pounds per cubic foot.)

Example 5.11

Determine the density of a rectangular plate of metal alloy, the dimensions of which are 2.00 meters by 1.00 meter by 0.100 meter. The mass of the plate is 880 kilograms.

Solution

The volume of a rectangular solid is determined by the product of its dimensions:

$$\begin{aligned} \text{volume} &= \text{length} \times \text{width} \times \text{height} \\ v &= lwh \\ &= (2.00\ \text{m})(1.00\ \text{m})(0.100\ \text{m}) \\ &= 0.200\ \text{m}^3. \end{aligned}$$

Its density, then, is calculated from

$$\begin{aligned} \text{density} &= \text{mass} \div \text{volume} \\ \rho &= \frac{m}{v} \\ &= \frac{880\ \text{kg}}{0.200\ \text{m}^3} \\ &= 4400\ \text{kg/m}^3. \end{aligned}$$

Example 5.12

The volume of a certain sample of liquid is 125 milliliters (or 125 cm^3) and its mass is 98.8 grams. Determine the density of the liquid and identify the substance, using the chart on the inside back cover of the text. Finally, calculate the mass of 1.00 liter of the substance.

Solution

The density of the sample is found using

$$\begin{aligned} \rho &= \frac{m}{v} \\ &= \frac{98.8\ \text{g}}{125\ \text{mL}} \\ &= 0.790\ \text{g/mL}. \end{aligned}$$

Comparing 0.790 with the values in the chart, we determine that the substance is most likely alcohol.

Then, if 1.00 mL has a mass of 0.790 g and 1 L = 1000 mL, the mass of 1 L can be calculated in proper SI units:

$$\begin{aligned} \rho &= \frac{m}{v} \\ m &= \rho v \\ &= \left(\frac{0.790\ \text{g}}{\text{mL}}\right)(1000\ \text{mL})\left(\frac{1.00\ \text{kg}}{1000\ \text{g}}\right) \\ &= 0.790\ \text{kg} \end{aligned}$$

In order to relate density and the height of the column of a substance to the pressure exerted by the substance on the bottom of its container or on an object submerged in it, note the following:

$$\begin{aligned} \text{pressure} &= \frac{\text{force}}{\text{area}} = \frac{\text{weight}}{\text{area}} = \frac{\text{mass} \times g}{\text{area}} \\ \text{P} &= \frac{\text{density} \times \text{volume} \times g}{\text{area}} \\ &= \frac{\text{density} \times \text{area} \times \text{height} \times g}{\text{area}} \\ &= \text{density} \times \text{height} \times g \\ &= \rho h g. \end{aligned}$$

A Bad Day for the Goldsmiths

The Greek scholar and mathematician, Archimedes, in about 250 B.C., was reportedly challenged by Hieron, the king, to determine if a goldsmith who had been contracted to build a new crown was indeed worthy of the trust bestowed upon him. Although the crown weighed the same as the gold furnished the goldsmith, the king had reason to suspect that some less-expensive metal, perhaps silver, had been used to replace a portion of the gold.

Legend has it that Archimedes gave the problem his undivided attention and conceived of a way to test the goldsmith's integrity as he stepped into the bathtub he had unwittingly filled to its maximum capacity. The tub, of course, overflowed until the amount of water that spilled onto the floor equaled the volume of his body. The significance of such an observation was immediately recognized by Archimedes, and he hurried off to the king's quarters without regard for his lack of proper attire. *"Eureka!"* (I have found it), he reportedly shouted.

It is said that in the presence of the king and the goldsmith, Archimedes submerged an identical weight of gold as was to be in the crown into a vessel filled with water and measured the volume of water that was displaced. He repeated the process with the crown. The beautiful symbol of royalty displaced more water than did the gold, proving that less-dense metals such as silver had been substituted for some of the gold. The goldsmith was meted out his just reward for his intentions to deceive the king.

Apparently, this criminal act was among the earliest uncovered by scientific investigation. Today, we know the concept of comparing the weight of a given volume of a substance (density) to that of an equal volume of water (or air, in the cases of gases) as *specific gravity.* To apply this relationship, divide the density of gold, as listed on the inside back cover, by the density of water. Repeat the process for silver, and note the discrepancy ■

Example 5.13

Determine the pressure a column of mercury 0.760 meter high exerts on the bottom of its container. (See inside back cover.)

Solution

To express the answer in meaningful terms, the measurements must be given in SI units:

$$\text{pressure} = \text{density} \times \text{column height} \times g$$

$$\text{P} = \frac{13{,}600 \text{ kg}}{\text{m}^3} \times 0.760 \text{ m} \times 9.80 \text{ m/s}^2$$

$$= 100{,}000 \text{ N/m}^2$$

$$= 100{,}000\ (1.00 \times 10^5) \text{ pascals (Pa)}.$$

Note: The value, 0.760 m represents the pressure of the atmosphere under standard conditions at sea level. The atmosphere will support a column of mercury 0.760 m high if all the air above the column is removed, as in a mercury barometer.

Solids, as well as liquids and gases, exert pressure on the surfaces that support them. Many floors covered with tile or linoleum are pitted by chairs and tables that have legs that taper to a small end at the point of contact with the floor. These ends may have an area as small as 1 square inch or less. Imagine a 200-pound person sitting on a chair and leaning back until only two chair legs are in contact with the floor. In this position the contact surface area may be as little as 0.250 square inch, and assuming the weight of the person to be evenly distributed between the two points of contact, the pressure on the floor at each point of contact would be 400 pounds per square inch. This amount of pressure is often great enough to dent the surface permanently.

Baseball cleats, golf shoes, and women's sometimes-stylish spiked heels can have a similar effect. In another area, as we note the mounting commercial traffic on our highways, we realize the necessity of improving our road construction materials and developing wider tires that will reduce the pressure on our highway surfaces. These things need to be done, however, without creating an appreciable increase in friction, for this increase would lower the efficiency of commercial vehicles.

The principle that the smaller the area of contact, the greater the pressure that results, accounts for the reason that nails as well as stakes designed to be driven into the ground are sharpened to a point. The force that drives the nail or stake is multiplied by a factor related to the degree to which the object is sharpened. The snowshoe applies the principle in reverse as it increases the contact area of the wearer's foot and thus proportionally lowers the pressure exerted on the snow.

Pressure can also be controlled in the opposite direction. When objects or systems are subjected to extremely low air pressure, characteristically different properties are revealed. Prepared food maintained under reduced air pressure and the accompanying sterilized conditions keeps its taste and does not spoil, even though the food is not refrigerated.

The absence of air pressure creates a condition known as a *vacuum* and has generally been assumed to exist in outer space. The probability that a complete vacuum exists is extremely small, for even the most nearly void areas of outer space are estimated to contain 2 to 3 million atoms per cubic meter and to have a density on the order of 10^{-18} kilogram per cubic meter. Perhaps the most completely void space in nature exists within the atom itself.

Summary

Machines are designed to make tasks easier to perform. Machines transfer energy, transform energy, change direction of applied forces, and multiply force or speed. Regardless of complexity, all devices are composed of one or more levers, pulleys, or inclined planes. Both humans and their machines can do work. The amount of work done is dependent upon the amount of force exerted and the distance the force moves. Quantitatively, work is equal to the product of the force exerted and the distance that the force moves. Work is measured in joules (N·m), and occasionally in ergs or foot-pounds. The efficiency of a machine is a comparison of the output work to the input work. Efficiency is inversely related to the friction between the machine's moving parts.

Work done per unit time is called power. The common units of power are watts (joules per second) and horsepower. In order for work to be done, energy must be converted to a different form. Stored energy is referred to as potential energy. Potential energy is present in a battery, in gasoline, and in a hammer poised over a nail. Kinetic energy is the energy of motion. As an object loses potential energy, it gains kinetic energy proportionally. Since energy is always completely conserved, the total energy a body contains is a sum of its potential energy and its kinetic energy.

In a solid, the intermolecular forces that hold the molecules together are strong enough to bind the molecules into a three-dimensional rigid lattice. In the liquid state, the intermolecular forces are reduced to the point that the molecules of the liquid are free to move in relation to each other. In the gaseous state, the molecules have sufficient kinetic energy and are far enough apart that the intermolecular forces are extremely weak; consequently, the intermolecular forces do little to restrict movements of gaseous molecules.

Any object in motion, be it a ball or a molecule of a gas, has kinetic energy. The kinetic molecular theory of gases describes the motion of gaseous molecules in a logical sequence.

Increased pressure raises the boiling point of liquids. Conversely, the boiling point of a given liquid is lowered if the liquid is exposed to a decrease in pressure. In order to cook food more rapidly, pressure cookers (devices that increase the pressure to which the food is exposed) are used. Liquids and gases both exert pressure as a result of their respective weights. For example, the pressure on a body submerged in water is increased 0.433 pound per square inch per foot of depth. The pressure a fluid (gas or liquid) exerts is equal to the product of the weight density (weight per unit area) of the fluid and the depth at which the pressure is to be determined. Metals are generally the most dense substances and gases are typically the least.

Equation Summary

Work:

$$W = Fs.$$

The work W done by a force F depends on the distance s through which the force acts.

Efficiency:

$$\text{efficiency } (\%) = \frac{\text{output work}}{\text{input work}} \times 100.$$

The efficiency of a machine is determined by comparing the percentage of the work done with the device (output work) to the actual work (input work) furnished it.

Lever Problems:

$$Rs = Es.$$

The effort E required to overcome the resistance R provided by an object is directly related to the distance s both E and R are located from the fulcrum.

Coefficient of sliding friction:

$$\mu = \frac{F}{w}.$$

The coefficient of sliding friction μ is determined by comparing the force F applied to the weight w that is to be moved.

Mechanical advantage:

$$m.a. = \frac{R}{E}.$$

The mechanical advantage of a machine is found by comparing the resistance R to the effort E applied. When applicable, it can also be determined by comparing the distances s that R and E move. For an inclined plane, divide length l by height h.

Power:

$$p = \frac{W}{t}.$$

Power, generally expressed in watts W, is determined by comparing the work W done per unit time t.

Gravitational potential energy:

$$P.E. = mgh.$$

The potential energy $P.E.$ given to an object by raising it to a higher position is directly related to the object's weight w or mg and the height h it was moved.

Kinetic energy:

$$K.E. = \frac{1}{2} mv^2.$$

The determination of the work a moving object can accomplish ($K.E.$) is directly related to its mass m and the velocity v (or speed) with which it moves.

Total mechanical energy:

$$E = P.E. + K.E.$$

The total mechanical energy of a body is the sum of its potential energy $P.E.$ and its kinetic energy $K.E.$

Pressure:

$$P = \frac{F}{A}.$$

The pressure a substance or object can exert is found by comparing the force $\boldsymbol{F}$ it exerts to the area A over which the force is distributed.

Density:

$$\rho = \frac{m \text{ (or weight } w)}{v}.$$

The density of a substance if found by comparing its overall mass m (or weight w) to its volume v.

Questions and Problems

Machines and Work

1. What vertical force must be exerted on the handles of a wheelbarrow 1.10 m from the axle in order to lift a 490-N load of dirt, the center of gravity of which is concentrated at a point 0.500 m from the axle? (Neglect the weight of the wheelbarrow.)

2. A loaded truck with trailer weighs 334,000 N. The loading ramp is so constructed that the truck must move 100 m along the ramp to reach a height of 4.00 m above the horizontal surface. What effective force, neglecting friction, must the truck be able to provide in order to move slowly up the ramp?

3. (a) Briefly discuss three reasons why we use machines. (b) Do machines really save us work in the strictest sense? Defend your answer.

4. (a) What is a perpetual motion machine? (b) Can you suggest some models that might be investigated? (c) Is an artificial satellite or the moon a perpetually moving object as either orbits Earth?

5. Determine the braking force to bring a 5000-lb automobile to a skidding stop on a dry concrete highway.

6. A series of pulleys, called a block and tackle, has four ropes directly supporting a 200-kg crate. What actual effort must be applied to lift the crate if the efficiency of the block and tackle is 80 percent?

7. Theoretically, how much effort is required to move a 300-kg object up a smooth board 5.00 m long if one end of the board is placed in a windowsill 1.50 m above the ground?

Power and Energy

8. How much work is done in lifting a 500-kg mass to a height of 20.0 m above the ground? How much energy is stored in the object at this height?

9. Does a 10.0-kg object located 20.0 m above the ground have more potential energy than a 20.0-kg object resting 10.0 m above the ground? Support your answer.

10. A person climbs a flight of stairs to the second floor in 10.0 s and runs up the next flight to the third floor in 4.00 s. Compare (a) the force required, (b) the power involved, and (c) the work done in each case.

11. A mass of 5.00 kg is placed on a table 0.800 m high. What potential energy, in joules, is assumed by the object?

12. A baseball has a mass of 0.100 kg. If the ball is thrown toward the batter with a velocity of 40.0 m/s (about 90 mi/h), what is its kinetic energy?

The States of Matter

13. (a) Determine the mass of a gold brick, the dimensions of which are 15.0 cm × 8.00 cm × 3.00 cm. (b) Determine the mass of a similar brick made of aluminum. (See inside back cover.)

14. List five states in which matter has been found to exist.

The Kinetic Molecular Theory of Gases

15. Why is the total volume of the molecules in a gas considered negligible when compared to the volume of the gas itself?

16. Considering the weight of air, how can it exert such relatively high pressure on the interior of a balloon?

Pressure and Its Applications

17. (a) What effect does lowering the pressure on a liquid's surface have on its boiling point? (b) How would increasing the pressure affect the liquid's boiling point?

18. In order to recover an item dropped into a lake 70.0 ft deep, how much additional pressure must the diver's body be able to withstand at this depth?

19. (a) How is the pressure on the surface of water in a pressure cooker increased? (b) How does the increased pressure lower the necessary cooking time for foods?

20. Why do campers who spend a considerable part of their trip near the top of a high mountain find a pressure cooker of greater benefit there than at lower altitudes?

6 Heat and Thermodynamics

Chapter outline

Key terms/concepts

The molten ores observed in this scene are evidence of the tremendous energy requirements necessary as matter is processed to fulfill our needs.

cientists continually strive to find causes for the changes they observe. An investigator offers an explanation, or *theory,* that is usually scrutinized through further observation and experimentation by others. One such theory attempts to explain the behavior of atoms and molecules (two or more atoms chemically bound together). It points out how molecular motion can be affected and how the motion of the molecules influences the matter that contains them.

According to the theory, molecular motion is affected by heat. The nature of this source of energy baffled investigators until the rise of modern science. Such complex and difficult mysteries as those that involve the motions of the planets were solved long before a satisfactory explanation concerning the meaning of hot and cold was developed.

When an object is brought near a fire or something hotter than itself, heat is acquired by the object, since heat flows from hotter objects to colder ones. Even in the eighteenth century, heat was conceived as a fluid—a material substance that was invisible and weightless. The idea that heat was not fluidlike is attributed to a soldier of fortune named Benjamin Thompson, better known in history as Count von Rumford, as he was appointed by the Holy Roman Emperor, the Duke of Bavaria.

The development of the modern atomic theory was initiated by the writings of John Dalton, a British investigator and teacher who lived in the early nineteenth century. Dalton's concepts about the atom permitted many properties of matter to be understood. Eventually atoms were known to be held together by a chemical bond to form molecules. It is the motion of these atoms and molecules in a body that we now interpret as heat.

Temperature—Molecules in Motion

It has been noted that the molecules of a gas travel at many different speeds. Just how fast do molecules travel? Let's assume that we can see the molecules of oxygen gas in a container. What can be observed? We can see some molecules that have practically no motion at all, whereas other molecules are hurtling along very rapidly. The majority of the molecules have rates of motion between these extremes. If the oxygen gas is at 0°C, about 1 percent of the molecules will have rates of motion ranging up to about 400 kilometers per hour. At the other extreme, approximately 8 percent will travel at speeds that approach 2700 kilometers per hour. The balance of the molecules will have rates of motion between the two values. The average speed of all the oxygen molecules would be almost 1700 kilometers per hour.

How will the average speed of these oxygen molecules be affected if the temperature of the gas is increased? At 30°C the average speed of oxygen molecules is found to be approximately 1800 kilometers per hour, and at 100°C the average speed increases to about 2000 kilometers per hour. Obviously, the higher the "temperature" of a gas, the greater the average speed of the gas molecules. In other words, a thermometer used to measure the temperature of an object is actually a device to measure the average speed, or rate of motion of that object's molecules. More correctly, **temperature** is *the property that determines whether or not thermal energy is transferred between two objects in thermal contact with each other.* It is indicative of the average kinetic energy of the molecules of each object. Thermal energy is transferred from the object with a higher temperature to the one with a lower one.

How might this conceptual view of temperature apply to the degree of comfort a person experiences? Consider being in a confined space, such as in a single room at a motel. The molecules of air in the room constantly bombard the molecules of the skin, so that eventually an equilibrium between the motion of the molecules in the air and the motion of the molecules of the skin is established. (It should be pointed out here that although the molecules of a solid must move within a very restricted volume—we call this movement vibration—the molecules are nonetheless in constant motion.) If the temperature of the room is increased, then the average speed of the molecules of air in the room increases; the air molecules have greater kinetic energy at the higher temperature. When the faster moving molecules come in contact with the molecules of the skin, they impart extra kinetic energy to the molecules of the skin and cause the skin molecules to vibrate at a faster rate, or to move at a higher average speed. It is this increase in the average speed or rate of vibration of the skin molecules that we interpret physiologically as an increase in skin temperature.

Conversely, if the temperature of the room is decreased, the average kinetic energy of the air molecules becomes less than the average kinetic energy of the molecules of the skin. Now the molecules of the skin, on collision with air molecules, impart some of their extra kinetic energy to the molecules in the room. The exchange of kinetic energy results in a decrease in the rate of motion of the skin molecules, with the consequence that we physiologically detect a cooling of our skin.

There is an upper limit as to how hot any object can become and remain intact. At some point its atoms will be torn apart, destroying all or most of the physical characteristics of the object. Neither can an object become infinitely colder. As the temperature of an object is lowered, the average speed (therefore, the average kinetic energy) of the molecules decreases. If the decrease in temperature is continued, the kinetic energy and the rate of motion of the molecules are further decreased until eventually molecular motion is at a minimum. When the speed of the molecules reaches this point, kinetic energy is at a minimum. (According to the analysis of recently obtained data, there will always remain a slight amount of vibration among the molecules, and a trace of translatory motion will exist at the lowest temperature possible. Hence, the kinetic energy of the molecules of an object can never reach zero.) Since no additional energy can be removed, *the molecules are at the lowest level of energy they can possibly attain.* The temperature that creates this effect is called **absolute zero.** Scientists predict that even at this temperature a minimal amount of

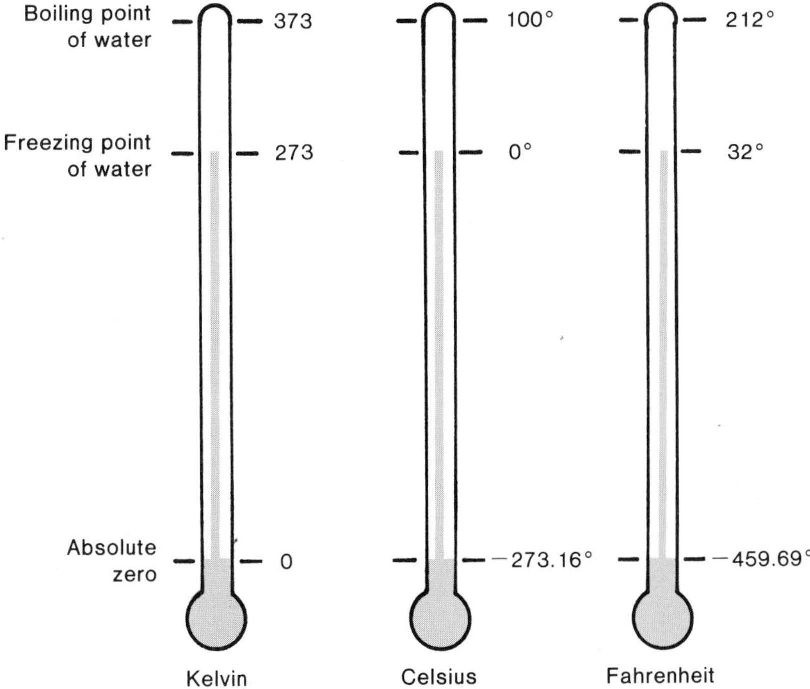

Figure 6.1 A comparison of the Kelvin (absolute), Celsius, and Fahrenheit temperature scales.

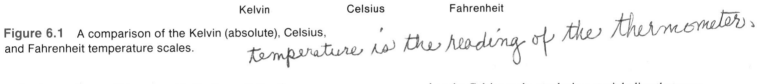

temperature is the reading of the thermometer.

molecular motion would be present in the form of vibrations among the atoms. These vibrations are referred to as zero-point vibrations and the energy associated with these vibrations is called the *zero-point energy.*

Scientists who study the effects of low temperatures on materials customarily refer to temperatures in kelvins. On this scale, absolute zero is designated 0 K. (See Appendix 1 for an explanation concerning the omission of the degree symbol on Kelvin temperature measurements.) There are no temperatures existing below zero on the Kelvin scale, so there are no minus numbers. The equivalent readings on the Celsius temperature scale and on the Fahrenheit temperature scale are −273.16°C and −459.69°F, respectively. Other comparisons are shown in Figure 6.1 and a range of temperatures is indicated in Figure 6.2.

The coldest temperature attained by scientists is reported to be within about 1 millionth of a Celsius degree of absolute zero, accomplished in 1990 at the University of Colorado. If this temperature could be maintained, the accuracy of atomic clocks could be improved by a factor of 100.

Various investigations have revealed that all gases, regardless of their densities, expand proportionally with the same increase in temperature. Although different gases have been studied and a wide range of volumes have been used, a graph of the effect of temperature change on volume, if pressure remains constant, yields a straight line for each gas. The fractional change in volume for any gas between the freezing point and the boiling point of water is the same, 100/273.16. A graph of volume versus temperature, as measured on the Celsius scale, results in a straight line that may intersect the vertical axis identified as 0°C at various points, yet, if extrapolated (theoretically extended beyond available data) to touch the horizontal axis, they all touch at the same point; namely, −273.16°C. This observation infers that the volume of the gas would be zero at this temperature. Consequently, if no volume remains, then this measure of average kinetic energy marks the lowest attainable temperature. The extrapolation of the gases to the temperature of absolute zero, as shown in Figure 6.3, is based on a hypothetical situation in which they retain their gaseous state over the entire range of temperatures presented. In reality, all known gases would have frozen and thus would be solids before absolute zero was reached.

The Calorie and Changes in State

Heat represents *the energy that flows when a body of higher temperature interacts with a body of lower temperature until thermal equilibrium is reached.* Heat, then, is manifest only when a body interacts with its immediate surroundings, so that the body's temperature changes toward that point where a change in state occurs. To distinguish heat from temperature, the latter concept represents whether or not thermal equilibrium exists between a specific object and its surroundings, whereas heat represents the total amount of energy transferred before thermal equilibrium is reached. The total energy that is transferred is associated with the random atomic and molecular motions of a substance. For

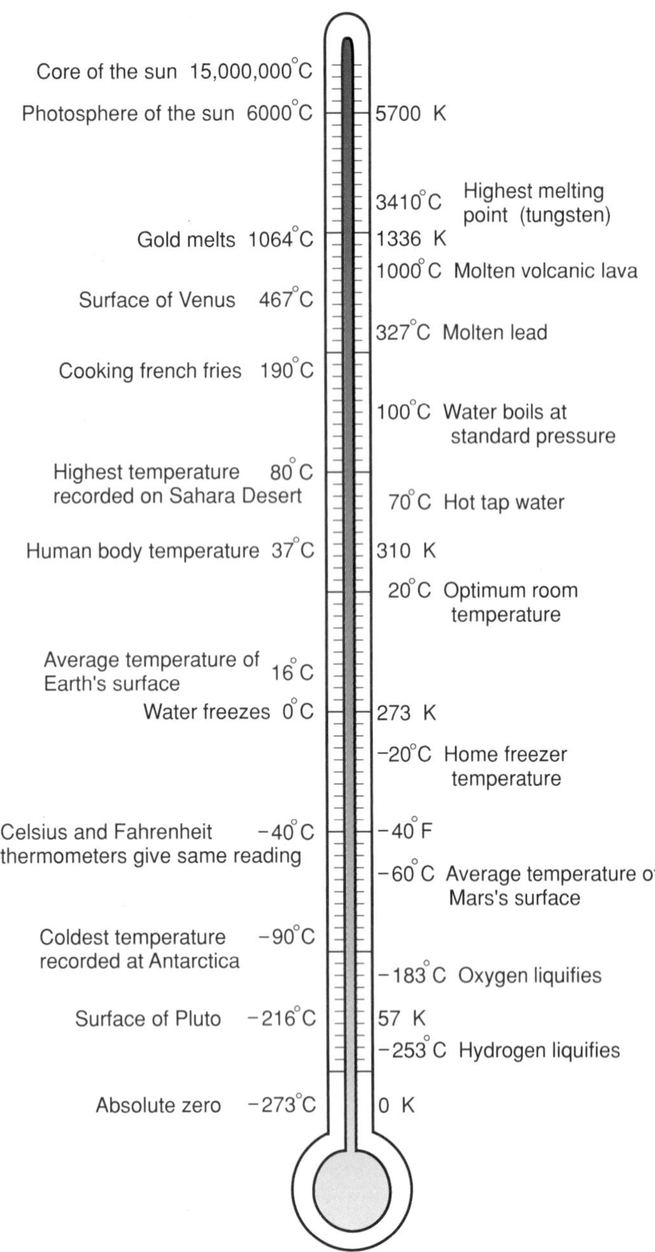

Figure 6.2 The temperatures in the universe range from about the coldest possible to temperatures that considerably exceed that of the sun's core, such as those in the interior of other stars.

example, the amount of energy that can flow from a sinkful of water at room temperature is considerably more than the amount of energy that can flow from a small pan of water at the boiling point. Of course, the individual molecules in boiling water have more kinetic energy per molecule than the individual molecules of water at room temperature, but there are so many more molecules present in the sinkful of water than in the small pan, that the total amount of energy that can transfer from the sinkful of water is greater.

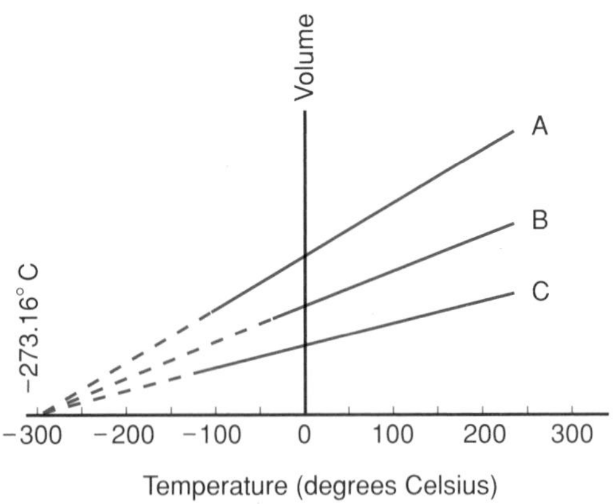

Figure 6.3 The volume of gases A, B, and C in this hypothetical situation would be zero at −273.16°C (0 K).

The amount of heat transferred to an object is measured in calories. A **calorie** is defined as *the amount of heat required to raise the temperature of one gram of water one Celsius degree.* For example, if the temperature of 1 gram of water is raised from 40° to 45°C, 5 calories must be supplied to the water. It also takes 5 calories to raise the temperature of 5 grams of water from 40° to 41°C. Conversely, it requires the release of 5 calories to lower the temperature of 1 gram of water from 40° to 35°C or of 5 grams of water from 50° to 49°C.

As a point of interest, the "calorie" that nutritionists talk about is actually a kilocalorie (1000 calories). Since we have become so conditioned in our society to counting calories, a person who is dieting might well go into cardiac arrest on learning that the peanut butter sandwich that he or she just consumed contained not 365 calories but 365,000! The kilocalorie is defined by international agreement as equal to 4.184×10^3 joules.

It has been stated that the application of heat to a substance increases the temperature of the substance. However, observing a thermometer submerged in boiling water will show that the continued application of heat to the boiling water does not increase the temperature of the water. Since the boiling water slowly decreases in volume, one must assume that the added energy is being used simply to convert the liquid water into gaseous water. In other words, the application of heat is necessary to bring about a change in the state of water from a liquid to a gas without a change in temperature. *The amount of heat (energy) required to bring about a change in state of a unit mass* (such as a gram) *of a substance from a liquid at its boiling point to a gas* is called the **heat of vaporization** of the substance.

Approximately 540 calories are required to vaporize 1 gram of liquid water at 100°C into 1 gram of gaseous water still at 100°C. Thus, the heat of vaporization of water is 540 calories per gram. Different substances require different quantities of energy to convert 1 gram of liquid at its boiling point to 1 gram of gas.

Discounting the Caloric Theory

The scientific relationships between the various forms of energy arose from the study of the generation of heat through friction and of the mechanical energy derived from the steam engine. Even during the seventeenth century, scientists such as Boyle, Hooke, and Newton considered heat to be the mechanical motion of minute particles in a body. The motion increased as the temperature of the body was increased. In the eighteenth century, however, heat came to be regarded as a weightless material substance, a fluid known as "caloric." The melting of a solid and the boiling of a liquid were thought to be the result of a chemical reaction between the material heat and the matter contained in the solid or liquid. Hot objects were thought to contain more of the fluid than colder objects; as hot and cold objects were brought into contact with each other, caloric flowed from the hot object to the colder one, warming one body and cooling the other.

According to the caloric theory, heat is caused by a release of the material heat when two bodies are rubbed together. The amount of caloric released depends on the number of times the surfaces are permitted to interact with each other, along with the duration of the interactions. However, Count von Rumford, a cannon borer by trade, noted in 1798 that the amount of heat produced and the amount of borings he obtained were essentially inversely proportional to each other. Blunt drill bits released more heat than sharp ones, he noted—an observation that contradicted the caloric theory. In fact, according to the caloric theory, sharp boring instruments should abrade the metal of the cannon more than dull ones and thus release a greater amount of caloric contained in the metal. Rumford observed that a dull drilling bit caused little or no abrasion, yet it generated sufficient heat to boil large quantities of water. Such an amount of heat was produced by friction, and Rumford concluded that heat was, in reality, a form of mechanical motion. He based his conclusion on the premise that anything that can be produced as long as the friction is continued cannot possibly be a material substance.

The mechanical theory of heat was further investigated by Sir Humphry Davy and later by Thomas Young in 1807. The energy theory gained little support and the material theory of heat, the idea of caloric, remained in vogue until about the middle of the nineteenth century ■

Since the temperature of boiling water does not change with the application of additional heat, where does the heat go? If the applied energy increased the kinetic energy of the water molecules, the temperature of the water would have to increase. Apparently, then, the energy is not being used to increase the kinetic energy of the water molecules. Instead, the energy is being used to overcome the intermolecular forces of attraction that hold water molecules together in the liquid state; hence, to increase the potential energy of the molecules. This *latent* (hidden) *heat* of 540 calories per gram is recovered when the gaseous water at 100°C condenses to liquid water at 100°C (in keeping with the law of conservation of energy). Steam heat in buildings utilizes this effect by transporting steam from the boiler and allowing it to condense in radiators, releasing its latent heat.

The considerations of heat application to a change in state of a liquid to a gas are also applicable to the change in state of a substance from a solid to a liquid. Thus, in the case of water, one must supply almost 80 calories to 1 gram of ice at 0°C in order to change its state from a solid at 0°C to a liquid at 0°C. Conversely, when 1 gram of water at 0° C is converted to 1 gram of ice at 0°C, 80 calories are liberated. *The amount of heat (energy), in calories, necessary to bring about this change in state of a substance from a solid to a liquid* is called the **heat of fusion** of a substance. Both the heat of fusion and the heat of vaporization for water are illustrated in the graph of heat supplied to a gram of water ice versus temperature and denoted as Figure 6.4. Note that the rate of temperature change for ice and steam is the same.

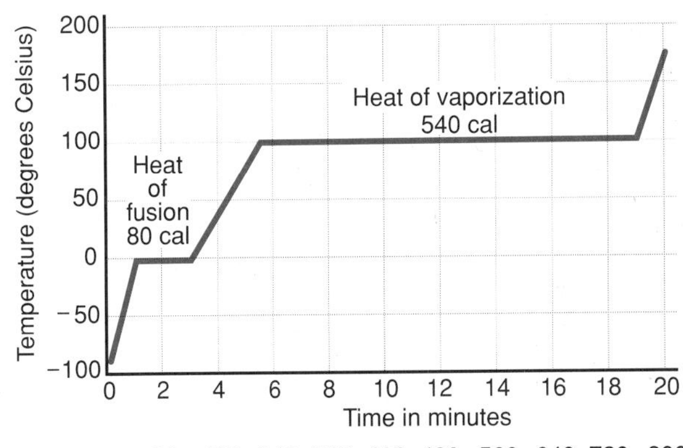

Figure 6.4 When heat is supplied to 1 gram of ice, the temperature of which is −80°C, at the rate of 40 calories per minute, note the length of time and the number of calories required to convert the gram of ice to steam.

Figure 6.5 The grinding wheel uses friction to sharpen and shape various tools.

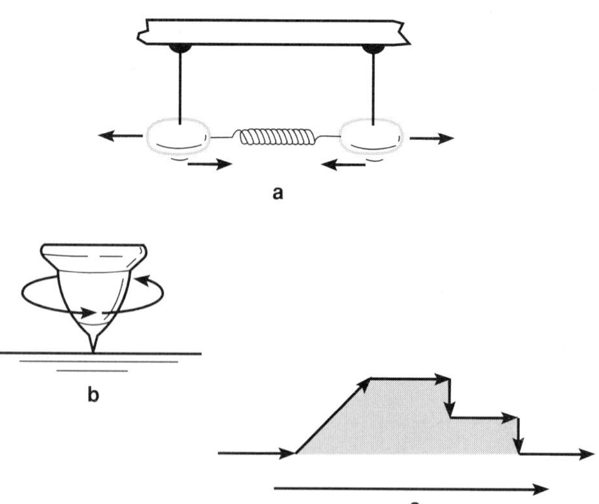

Figure 6.6 The three types of motion that an object or its molecules can possess: (*a*) vibratory, (*b*) rotary, and (*c*) translatory.

Example 6.1

Determine the heat required to convert (a) 50 grams of water at 100°C to steam at 100°C and (b) 50 grams of ice at 0° C to water at 0°C.

Solution

(a) The heat of vaporization for water is noted to be 540 cal/g; thus, the total heat required would be calculated as

$$\text{total heat} = (540\ \text{cal/g})(50\ \text{g}) = 27{,}000\ \text{cal.} \qquad \text{HEAT OF VAPORIZATION}$$

(b) The heat of fusion for ice is equal to about 80 cal/g; therefore, the total heat required would be

$$\text{total heat} = (80\ \text{cal/g})(50\ \text{g}) = 4000\ \text{cal.} \qquad \text{HEAT OF FUSION}$$

Heat, Friction, and Types of Motion

Even prehistoric people were aware of heat and some of its related phenomena. It was long ago observed that heat could be produced by friction. If two objects are rubbed briskly together, the amount of energy they contain increases. If a wire is bent back and forth, the wire gets warm in the area where it is being bent. Similarly, if a nail is driven into a board, the amount of energy that the nail contains increases, and if an object is exposed to the light (primarily infrared) energy from the sun, the object increases in temperature as the sun's energy is transferred to it.

In Figure 6.5 the sparks (fragments heated to the point of incandescence) provide evidence of the heat produced by friction between the grinding wheel and the tool. The same phenomenon is visible when a train is brought to an abrupt halt as the wheels are locked and thus slide along the tracks. Similarly, many observers have witnessed sleek jet planes as they approach a landing site with the front of their wings glowing a dull red produced by the friction between molecules of air and those of the wing. In order to prevent a space capsule from heating to the point of incandescence as it enters Earth's atmosphere, the vehicle is equipped with a heat shield.

The molecules of all substances are in a state of constant motion. This motion can be classified in three separate categories: vibratory, rotary, and translatory. *Vibratory motion* can be readily demonstrated by considering two objects attached to each other by means of a common spring. If the two masses are each supported by a string, attached to each other by a spring, and then pulled apart, as illustrated in Figure 6.6a, the two masses first move toward each other, and then in the opposite direction as the motion continues. This type of motion is generally known as vibration.

Rotary motion is illustrated in Figure 6.6b. This type of motion is evident everywhere in the universe, from the largest of stars to the smallest of atoms. Many objects and systems rotate; that is, each turns on its own axis.

The third type of motion is illustrated in Figure 6.6c. *Translatory motion* permits travel between streets, cities, and even bodies in or beyond our solar system. Generally, translatory motion involves movement along a straight line, but we obviously turn many corners as we walk and negotiate many curves in driving from city to city. Translatory motion is then qualified to mean that as a result of this motion, the ultimate direction of travel is a straight line. The same concept is prevalent in nature, even at the microscopic level, and is called *Brownian motion*, after Robert Brown (1773–1858). He is credited with being the first person to study the

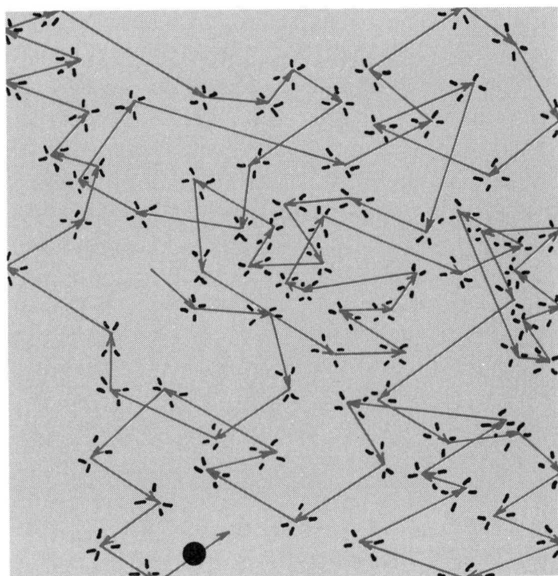

Figure 6.7 A molecule moves in random directions because of collisions with other molecules, including those that form the container.

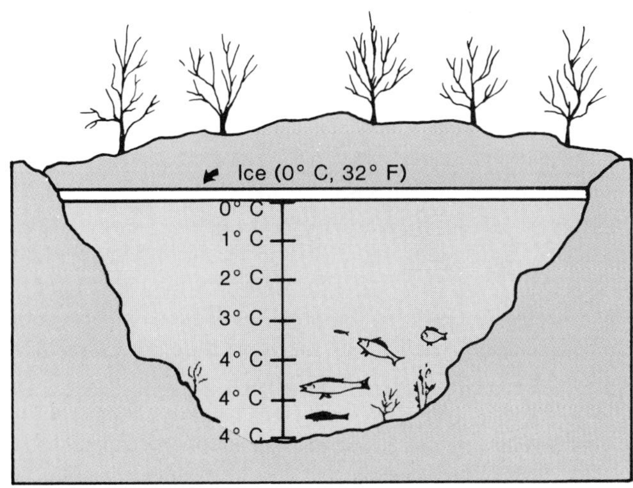

Figure 6.8 The stabilizing of water temperature above the freezing point in ponds, streams, and lakes of sufficient depth permits aquatic life to survive in cold climates.

erratic motion of pollen grains suspended in water. His investigations provided the first experimental evidence that molecules do exist. Figure 6.7 illustrates the typical random path that a molecule might assume as a result of numerous collisions with other molecules in a confined area, such as in an inflated balloon. The molecules present in the enclosed container gain kinetic energy as heat is transferred to that area from another body. Although much of the energy absorbed causes the molecules to vibrate more rapidly, as well as perhaps to increase their rate of rotation, part of the absorbed energy causes the molecules to increase in translatory motion. The ultimate result is an increase in the number of collisions between molecules and the walls of their container. This point helps explain why balloons, automobile tires, and the like undergo expansion as air molecules are added to them or as the temperature of air molecules inside them increases.

The Expansion of Liquids and Gases

Practically all liquids, because of the increased motions of their molecules, expand as they absorb heat. When they lose heat, the liquids contract as the molecules slow down because of loss of kinetic energy. Water is one of the exceptions, within a limited range of temperatures. When its temperature is above 4°C, water reacts in the same way that most other substances do. However, from about 4°C (39.2°F) to its freezing point, water slightly expands. As it changes to a solid, its volume increases significantly, a feature readily observed by the shape of a carton of milk after being placed in a freezer.

The expansion of water is a powerful force in nature. Much of the damage to our highways, streets, and sidewalks can be attributed to the freezing and expanding of water after it has saturated the ground beneath them.

The peculiar manner in which water reacts to changes in temperature has permitted aquatic life to thrive in areas where freezing temperatures are reached. Without the property of expanding as it cools within a specific temperature range, water would freeze solidly at all depths, and most life in lakes and rivers would end. Near the temperature at which water freezes, 0°C, the water that is coldest has a lower density than the rest of the water in the stream, so it rises to the surface and causes water to freeze at this point. The surface ice forms a protective and insulating cover that retards the rate at which the lower levels of water freeze. Generally speaking, the lower levels of water stabilize at approximately 4°C, as illustrated in Figure 6.8. The strange effect of temperature change on water produces another unusual situation. Equal volumes of water with temperatures of about 2°C and 6°C weigh approximately the same. Few substances—no common ones—react in the same manner.

The mass of any substance, in whatever state it exists, is not affected by expansion. The volume, however, is increased as thermal energy is absorbed, with the exception just noted, so the density of the substance is decreased. For this reason warmer air in a room is located near the ceiling, and colder air is found close to the floor. Hot-water tanks have the hottest water at the top of the tank, since hot water is of lower density than water of lower temperature.

The air in an automobile tire filled on a cool morning at a pressure of "30 pounds" may increase in pressure to 34 pounds per square inch as the temperature of the air inside the tire increases to about 25°C. The increase in pressure and temperature is caused by the exposure of the tire either to the sun and the warmed highway or to the heat produced by friction as the car is driven. A similar increase in pressure causes a cake to expand in volume as gaseous bubbles are formed during baking. The cake rises, and thus becomes light and porous.

The pressure, temperature, and volume of a confined gas are related to each other. In order to establish their relationships, imagine a tire pump with a faulty valve that will not permit the compressed air to escape into the tire as the pump handle is forced downward. If a sudden force were applied to the pump handle—a force sufficiently great to compress the confined air into a very small space in the cylinder—a vast rise in the temperature of the confined air would result. In fact, a piece of paper placed in the cylinder would be ignited as the greatly increased temperature of the confined air reached the kindling point of the paper.

Conversely, if the compressed gas in a welder's full tank suddenly escaped, the outer surfaces of the tank would instantly become very cold, and any substance in the path of the escaping gas would experience a great drop in temperature. From these examples it can be seen that the volume of a gas is inversely proportional to the pressure exerted on it and that the temperature of a gas is directly proportional to the pressure to which it is exposed.

As the previous discussion indicates, there are three variables that are related with regard to a confined gas: temperature (T), pressure (P), and volume (V). For instance, the product of the pressure exerted on a given volume of a confined gas is a constant, $PV = k$. *If the temperature of the gas remains constant, a change in pressure brings about a change in volume, so that* $P_1V_1 = P_2V_2$, where P_1V_1 represents the product of the initial pressure and volume, and P_2V_2 equals the product of the second pressure and volume readings.

Example 6.2

A soap bubble has a volume of 0.300 cubic meter when the atmospheric pressure on its surface is 1.00×10^5 newtons per square meter. What would be its volume if the pressure increased to 1.50×10^5 newtons per square meter?

Solution

From the equation

$$P_1V_1 = P_2V_2, \qquad \text{BOYLE'S LAW}$$

and solving for V_2,

$$V_2 = \frac{P_1V_1}{P_2}$$
$$= \frac{(1.00 \times 10^5 \text{ N/m}^2)(0.300 \text{ m}^3)}{(1.50 \times 10^5 \text{ N/m}^2)}$$
$$= 0.200 \text{ m}^3.$$

The relationship between the pressure and volume of a fixed amount of a gas maintained at a constant temperature is known as **Boyle's law,** so named after the seventeenth-century physicist who studied the relationship. The law applies to all confined gases, except when such gases are exposed to unusually high pressure. *If the pressure is held constant and the temperature is increased or decreased, then the volume is affected in proportion to the temperature change,* or $V \propto T$ when T is expressed in kelvins. To determine the value of T, find the sum of the temperature measured in degrees Celsius and use the constant converting number, 273. That is, K = °C + 273. For example, 0°C = 273 K (read as 273 kelvins); 20°C = 293 K; and −73°C = 200 K. (Note again, the absence of the degree symbol.) As an equation, $V_1/T_1 = V_2/T_2$.

Example 6.3

The temperature of a confined gas is found to be 20.0°C. If the volume of this gas is 2.25 cubic meters at this temperature, what would be its volume if the temperature increased to 35.0°C?

Solution

According to the equation

$$\frac{V_1}{T_1} = \frac{V_2}{T_2} \qquad \text{CHARLES' LAW}$$
$$V_2 = \frac{V_1T_2}{T_1}$$
$$= \frac{(2.25 \text{ m}^3)(35.0°\text{C} + 273)}{(20.0°\text{C} + 273)}$$
$$= 2.37 \text{ m}^3.$$

The aforementioned relationship is known as **Charles' law,** named after the Frenchman who, in 1787, noted how temperature affected the change in volume of a confined gas. This law was later used to determine the value of absolute zero, about −273°C. Boyle's law and Charles' law are combined to form the *general gas law:*

$$\frac{P_1V_1}{T_1} = \frac{P_2V_2}{T_2},$$

where T_1 and T_2 represent the initial and final absolute temperature, as measured on the Kelvin scale. The general gas law points out that the volume of a confined gas is inversely proportional to the pressure that acts on it and is directly proportional to the absolute temperature. (The letters denoting pressure, volume, and temperature are usually capitalized in gas laws.)

Example 6.4

A balloon is partially inflated with 50.0 cubic meters of helium gas at ground level where the air pressure is at 1 atmosphere, 1.01×10^5 newtons per square meter, and the temperature is 20.0°C. Determine the volume of gas in the balloon when the balloon rises to 15,000 meters, where the air pressure is 1.00×10^4 newtons per square meter and the temperature is −50.0°C.

Solution

From the expression

$$\frac{P_1V_1}{T_1} = \frac{P_2V_2}{T_2}, \qquad \text{GENERAL GAS LAW}$$

solve for V_2 by dividing both sides by P_2 and multiplying both sides by T_2. So, rearranging the equation to solve for the new volume,

$$\begin{aligned} V_2 &= \frac{P_1V_1T_2}{P_2T_1} \\ &= \frac{(1.01 \times 10^5\ \text{N/m}^2)(50.0\ \text{m}^3)(-50.0°\text{C} + 273)}{(1.00 \times 10^4\ \text{N/m}^2)(20.0°\text{C} + 273)} \\ &= 384\ \text{m}^3. \end{aligned}$$

How Solids Expand

Solids that expand when exposed to heat expand in all dimensions—length, width, and height. Since volume is always a function of these three dimensions, this expansion causes a considerable increase in volume. For this reason sidewalks, streets, and bridges are constructed in small sections that are placed in conjunction with the surrounding sections so as to allow room for expansion. If the material experiences a drastic temperature increase (due sometimes to air temperature but usually to a catastrophe such as fire or explosion), the rate of expansion becomes an agent of destruction. Many streets and bridges have been extensively damaged by unusual amounts of expansion. Of course, cold temperatures cause materials to contract, a phenomenon that could, if not compensated for by design engineers, produce vast destruction as the materials in bridges or railroad tracks pull apart and leave gaping spaces.

Substances change in size with changes in temperature. With few exceptions, substances increase in all dimensions with increasing temperature and correspondingly decrease with decreasing temperature. An increase in temperature brings about a greater vibration of the molecules; hence, a greater distance between the particles in a solid or a liquid. The result of increased distance between particles is an increase in all physical dimensions. Essentially, all solids expand at different rates and thus various metals, plastics, and other such substances cannot be used in conjunction with other substances if differences in expansion rate is of consequence. Rates of expansion have been measured for most substances and are known as *coefficients of linear expansion* and *coefficients of volumetric expansion.* These unique measures reveal the amount of expansion that occurs in a given length or volume of a substance as the temperature increases per unit degree.

To clarify the manner in which the coefficients of linear expansion and volumetric expansion are determined, we can examine the coefficient of linear expansion more closely. Almost all solids expand uniformly as they are heated. If we assign the length of a metal rod the symbol l and the temperature the symbol t, then the change in length, Δl, associated with the change in temperature, Δt, indicates the rate of expansion. (The symbol, Δ, when used in equations means "the change in.") More specifically, the coefficient of linear expansion equals the change in length divided by the product of the original length, l_i, and the change in temperature. To express the relation algebraically, the coefficient of linear expansion, α, becomes

$$\alpha = \frac{l_f - l_i}{l_i\,(t_f - t_i)} = \frac{\Delta l}{l_i \Delta t},$$

and

$$\Delta l = \alpha l_i \Delta t,$$

where l_i and t_i equal the original length and temperature, and l_f and t_f equal the length and temperature after the temperature change.

Generally, the temperature is expressed in Celsius degrees, and the unit of length and its corresponding increase (or decrease, if the temperature is decreased) are expressed in meters or any other convenient unit of linear measure. Some common substances and their coefficients of linear expansion (length, width, or height) are listed in Table 6.1. The unit increase is in proportion to the unit by which the original length is expressed; that is, if the coefficient of linear expansion of a substance is 0.000,017, a change in temperature of 1 Celsius degree would produce a change in original length from 1.000,000 to 1.000,017, regardless of the unit chosen to make the original linear measure. For example, an aluminum rod 5.0000 meters long at precisely 0°C would increase in length to 5.0049 meters at 40.0°C; that is, 5.0000 + [5.0000(40.0)(0.000,024,5)]. If the original length of the aluminum rod had been 5.0000 kilometers, its new length would be 5.0049 kilometers.

Table 6.1

The coefficient of linear expansion of some common substances (per Celsius degree).

Aluminum	0.000,024,5
Brass	0.000,018,5
Copper	0.000,016,6
Concrete	0.000,012,0
Glass, ordinary	0.000,009,0
Glass, Pyrex	0.000,003,2
Iron	0.000,011,5
Steel	0.000,012,0

Example 6.5

A steel bridge is 1.60 kilometers long. How much will the length of the bridge increase when the temperature increases from 10.0°C to 30.0°C?

Solution

The coefficient of linear expansion (α) for steel, according to Table 6.1, is noted to be 0.000,012,0/C°. The increase in length of the bridge can be determined from

$$\begin{aligned}\Delta l &= \alpha l_i \Delta t \\ &= (0.000{,}012{,}0/\text{C}°)(1.60\text{ km})(30.0°\text{C} - 10.0°\text{C}) \\ &= (0.000{,}012{,}0/\text{C}°)(1.60\text{ km})(20.0\text{ C}°) \\ &= 0.000{,}384\text{ km} \\ &= 0.384\text{ m.}\end{aligned}$$

Since the length of the bridge would increase by 0.384 m as a result of the increase in temperature, engineers obviously must allow for expansion (and contraction) in design of bridges and other structures.

There are many common examples of the effects of variation in expansion rates among materials. On glass jars, for example, we find that metal lids that are too tight to be removed readily can be easily loosened by pouring hot water over the lid. Often in making household repairs, problems involving the removal of a rusty nut and bolt are resolved by heating only the nut and then removing it by a conventional means. On the other hand, electric stoves and other devices that make use of a heating element can be ruined through being bent out of shape if the device is heated above the recommended temperature or if the element is jarred or otherwise disturbed while it is excessively hot.

Why, then, do construction workers heat rivets to glowing before they install them in steel beams? The reason, of course, is to pull the steel beams closely together as the rivets contract on cooling. Why does a cold glass crack when suddenly submerged in boiling water, or a hot light bulb shatter when touched by a drop or two of cold water? Obviously, expansion and contraction are involved. Similarly, the canning of foods meets with success only if the containers are sealed tightly while hot so that they contract and are fixed very firmly in place on cooling. The pressure inside the containers is lowered below normal atmospheric pressure as the lid becomes airtight. If a baby's prepared milk formula is placed in bottles while cold, the milk will sometimes overflow as the bottles are heated, since the milk expands at a greater rate than does the plastic or glass.

Air Conditioning and Thermodynamics

The conditioning of the air, in a broad sense, refers to our attempts to control our comfort, health, and efficiency. Air conditioning is installed in stores to increase sales by means of customer comfort. In offices and schools, air conditioning is used to increase the efficiency of the individuals who work there. Air conditioning is a unique application of two areas of physics known as *thermodynamics* and *aerodynamics*. From the former, we obtain basic data concerned with the thermal properties of such gas mixtures and vapors as would be found in an average environment of dry air and related water vapor. Aerodynamics deals with the motion of air and other gaseous fluids and with the forces that act on bodies when they move through such fluids or when such fluids move against or around the bodies.

Associated with every object in the universe is the physical quantity we know as energy. This quantity can move from body to body, often taking the form of heat. Thermal energy usually accompanies the transformation of energy from one form to another and is released or absorbed as chemical reactions occur. *The total amount of energy always remains the same; that is, energy can neither be created nor destroyed.* This expression obtained from the observed results of many experiments is often referred to as the **first law of thermodynamics.** (The statement is just another way to express the law of conservation of energy, as presented in Chapter 5.) If two bodies of unequal temperature are placed in contact with each other, the cooler body becomes warmer and the warmer body becomes cooler. However, there is no net loss or gain of energy as heat is transferred from one to the other.

Suppose we place several solids such as lead, iron, and aluminum, as well as a container of water, on a large block of ice. The mass of each substance is 1 kilogram and the temperature of each is 100°C. As the substances cool to the temperature of the ice, we note that the aluminum melts about one-fifth as much ice as the hot water, the iron melts about one-ninth as much ice as the hot water, and the lead melts about one-thirtieth as much ice as the hot water. We can conclude that aluminum requires one-fifth as much heat as an equal mass of water to undergo the same temperature change. This value (more precisely 0.22 cal/g · C°)

The R-Value

The greatest single cause of energy loss in most homes is inadequate insulation. A properly insulated house saves about 25 percent of the energy used to heat it in the winter and 10 percent of the energy used to cool it in the summer.

Insulation works by reducing the heat transfer that occurs from the heated living space in a home to unheated attics, garages, basements, and the outdoors. Heat can flow through the building envelope composed of the roof, walls, and floors, as well as through small openings around doors and window frames. Heat always flows from a warmer to a cooler area or object. Insulation provides a barrier that decreases the loss due to heat flow.

The effectiveness of various types of insulation is measured in terms of *R-values,* a rating scale that indicates the resistance to heat flow of a specific material. The higher the R-value, the more effective the insulation. This measurement is indicative of both the type and the thickness of the insulating material. An acceptable R-value depends on such factors as the climate, the location of the house, the type of heating used in the home, and the specific living space to be heated. Much of the heat loss in homes occurs through the ceilings; hence, the attic should be insulated so as to have the highest R-value ■

Table 6.2

The specific heat of selected substances, expressed in cal/g·C° or Btu/lb·F°.

Substance	Specific heat (*c*)	Substance	Specific heat (*c*)
Water	1.00	Asbestos	0.20
Ice	0.50	Ashes, fireplace	0.20
Steam	0.50	Gasoline	0.50
Aluminum	0.22	Sand	0.20
Iron	0.11	Cooking oil	0.40
Lead	0.031	Petroleum	0.50
Glass, crown	0.16	Seawater	0.94
Copper	0.092	Alcohol, absolute	0.58
Bakelite (plastic)	0.35	Nitrogen	0.25
Wood	0.58	Oxygen	0.22
Silver	0.056	Chlorine	0.11
Stainless steel	0.11	Humus (soil)	0.44

is called the **specific heat** of aluminum. (If mass is not specified, the relationship may be referred to as the *heat capacity.*) The specific heat of any substance is *the ratio of the heat that it gives off to the heat given off by an equal mass of water,* the temperature change being the same. By definition, the specific heat of water is 1 calorie per gram. Specific heat also indicates the amount of heat, measured in calories, required to raise or lower the temperature of 1 gram of a substance 1 C°. Thus, 1 gram of aluminum would increase in temperature 1 C° as it absorbed 0.22 calorie, 1 gram of iron would require 0.11 calorie, and 1 gram of lead would require 0.031 calorie. In Table 6.2, note the specific heat of substances used to cook foods and those used as handles on pots, pans, and other utensils.

To determine the amount of thermal energy required to produce a certain temperature change in a given substance, the following general formula is applied:

$$Q = mc\Delta t. \qquad \text{THERMAL ENERGY}$$

Q represents the amount of thermal energy required to produce a given change in temperature ($\Delta t = t_f - t_i$) in a mass, m, of a substance that has a specific heat, c. If m is expressed in grams and the change of temperature, Δt, in C°, Q will be determined in calories; if m (w) is in pounds, and Δt is in F°, Q will be calculated in **British thermal units** (Btu). The Btu is defined as *the amount of thermal energy required to raise the temperature of 1 pound of water 1 F°.* The concept of "pound-mass" on which the Btu is based has been discontinued for the most part, and the unit has not been redefined.

Example 6.6

Determine the amount of heat in calories required to raise the temperature of 5.00 kilograms of water from 20.0°C to 50.0°C.

Solution

The calorie is defined in terms of mass expressed in grams, so 5.00 kg = 5000 g, and for water $c = 1.0$ cal/g·C°. The amount of heat required is calculated by the product of mass, specific heat, and temperature change. That is,

$$\begin{aligned} Q &= mc\Delta t \\ &= (5000\text{g})(1.00\ \text{cal/g}\cdot\text{C}°)(50.0°\text{C} - 20.0°\text{C}) \\ &= (5000\ \text{g})(1.00\ \text{cal/g}\cdot\text{C}°)(30.0\ \text{C}°) \\ &= 150{,}000\ \text{cal}. \end{aligned}$$

Example 6.7

A cube of iron weighs 5.00 pounds. How many Btu's are required to raise the temperature of the iron from 45.0°F to 300°F? From Table 6.2, the specific heat of iron is found to be 0.11 Btu per pound-Fahrenheit degree.

Solution

$$\begin{aligned} Q &= mc\Delta t \\ &= (5.00\ \text{lb})(0.11\ \text{Btu/lb}\cdot\text{F}^\circ)(300^\circ\text{F} - 45.0^\circ\text{F}) \\ &= (5.00\ \text{lb})(0.11\ \text{Btu/lb}\cdot\text{F}^\circ)(255\ \text{F}^\circ) \\ &= 140\ \text{Btu}. \end{aligned}$$

As energy transformation occurs, some of the original energy is converted into thermal energy that is not available for further transformation. But where does the heat contained by a substance, such as hot coffee, go as the substance cools? The heat given off by the coffee is given to the atmosphere, which in turn radiates the thermal energy into the cold regions of outer space. As electrical energy is transformed into mechanical energy by an electric motor, there is a loss of useful energy in the form of heat. The amount of available useful energy decreases with each continued transformation until no useful energy remains. Wasted energy, then, cannot be regained; the available energy we have is continually assuming a lower, less useful form. These facts are described by the **second law of thermodynamics,** interpreted to point out that *each time energy is transformed, some of it is converted to thermal energy and its availability to do work is lost.* Further examination of the first and second laws of thermodynamics reveals that we cannot get more work out of a process than we put into it, with the second law explicitly describing why we should never expect to break even. Another way to express the second law is in terms of entropy. **Entropy** may be defined as *a measure of the disorder of a system.* If the system is highly disordered, then the inability of the system to do work is great, and the system is said to have a high degree of entropy. In contrast, a highly organized system, such as one that has a large reservoir of heat at a high temperature and other reservoirs at a very low temperature, is said to have a low degree of entropy. The highest degree of entropy exists in a system in which all reservoirs of heat energy have achieved the same temperature and no work can be done. In terms of entropy, the second law of thermodynamics can be stated: *The only possible processes that can occur in an isolated system are those during which the net entropy either increases or remains constant.* As one might surmise, the second law of thermodynamics, like other laws, can be correctly stated in numerous ways.

In air conditioning, more than simply the removal or addition of heat energy to a home is involved. Such variables as moisture content (humidity), gas composition (including carbon dioxide), and "freshness" should be controlled for optimum comfort. Most air conditioning units are primarily concerned with temperature and humidity control, and thus provide little adjustment of the other variables. Removing particles from the air is becoming common in the modern home and is accomplished by venting the flowing air through filters that are often electrostatically charged to attract dust and pollen grains. Humidifiers and dehumidifiers adjust moisture content according to the home temperature setting. The control of the *relative humidity,* the ratio of the amount of moisture in the air to the amount of moisture the air will hold at that specific temperature, is perhaps as important to our comfort as is the control of temperature. Relative humidity is expressed as a percentage; the combination of relative humidity and temperature determines how comfortable we feel under given conditions. Figure 6.9 shows a range of optimum conditions.

The basic requirement of comfort conditioning is the maintenance of an indoor environment that helps those persons in these given conditions develop a thermal balance between body energy production and body heat loss. This balance must be provided without subjecting the individual to unnecessary strain or excessive operation of body mechanisms, such as sweat glands or the muscles and nerves that induce shivering. The body is similar to the internal-combustion engine, since it receives energy from fuel (food) and dissipates it as work and heat energy. Heat energy is measured in kilocalories; a kilocalorie is the amount of heat required to change the temperature of 1 kilogram of water 1 C°. As was noted earlier in this chapter, the "calorie" used in rating the thermal energy of foods is actually a kilocalorie (kcal). The average adult requires about 100 kilocalories per hour to accomplish the internal work necessary to maintain life, such as breathing, pumping the blood, and releasing wasted thermal energy. This wasted energy is given to the surrounding air as the body tries to maintain equilibrium. (Have you noticed how a small crowded room eventually becomes unpleasantly warm?) If the rate of total energy loss is less than the production rate, the body temperature increases. The rise in temperature produces a maladjustment of the physiological mechanisms and creates a possible hazard to the individual's health. The average person may develop a fever of almost 2 C° in about an hour if prevented from losing heat energy by wearing excessive clothing or being confined to a warm area with insufficient ventilation. Digestion of a gram of protein or carbohydrate provides heat energy of about 4 kilocalories, whereas the digestion of a gram of fat releases almost 9 kilocalories. Interestingly enough, the complete combustion of a gram of gasoline yields about 12 kilocalories.

A Matter of Convention

The unit of energy called the calorie obtained its name from misconceptions about heat. Prior to the nineteenth century, heat was thought to be an invisible fluid known as caloric. But for this mistake in identity, heat would have been measured in such units as ergs or joules. (Actually there is an ongoing attempt in various scientific circles, in the interest of standardization, to express heat energy in joules.) The process of removing an established unit from the fields of science is a difficult one, since many units are derived from, or expressed in terms of, other units. Each attempt has generally met with much opposition from scientists and nonscientists alike. It appears that only upon unanimous recommendation of the members of various influential scientific organizations is a change likely to be universally accepted ■

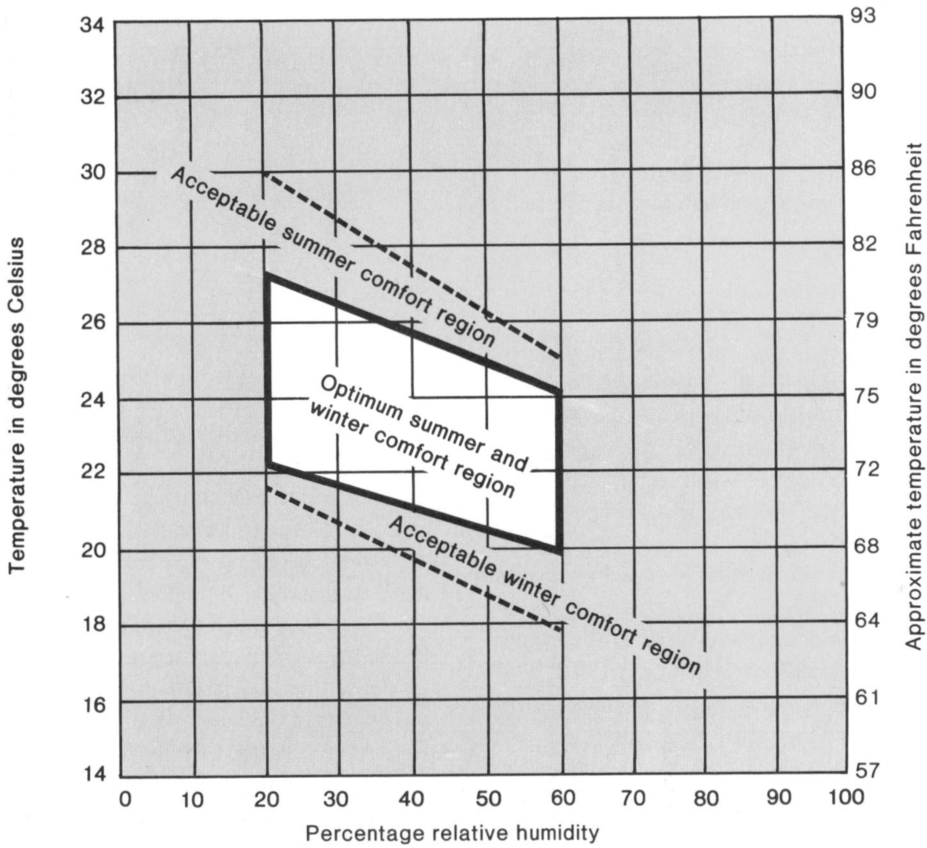

Figure 6.9 The range of optimum conditions for summer and winter comfort overlap, providing a range for constant conditioning control. The maximum and minimum acceptable comfort regions vary according to individual preference; hence, they are not absolute.

a.

b.

c.

Figure 6.10 Heat is transferred from one body to another by the three methods illustrated: (*a*) conduction, (*b*) radiation, and (*c*) convection.

Energy in Transit

Generally speaking, the *thermal energy* of the human body is dissipated at fixed rates through three primary methods: (1) **radiation**—*energy in the form of electromagnetic waves is transferred from exposed skin or clothing to the surrounding air* and amounts to about 45 kilocalories per hour at 21°C (normal room temperature); (2) **convection**—*energy is transferred from the body and clothing to the surrounding air by air currents* such as are produced by a fan in the room at the rate of 30 kilocalories per hour; and (3) **conduction**—*energy is transferred from molecule to molecule in direct contact* and is involved in the evaporation of about 45 grams of moisture hourly in the form of perspiration as well as in the measurable amount of moisture released by direct evaporation from living surfaces to respiratory air. (The energy transferred to the surrounding air by conduction amounts to approximately 25 kilocalories per hour.) These three methods by which energy is transferred from one object to another are illustrated in Figure 6.10.

The objectives generally taken into consideration for comfort air conditioning follow. First, humidity is controlled to prevent the sensation of dryness or clamminess of the skin. However, within the comfort region, variation will not compensate for major changes in temperature, as is pointed out in Figure 6.9. Second, temperature is controlled to provide an average of 21°C (70°F). The comfort equation, in which the sum of the air temperature and the immediate wall temperature is 60°C (140°F), explains why, for proper control, a temperature of 21°C cannot be fixed. As one would expect, the surface temperature of the walls varies directly, to a large extent, with the outside air temperature.

We require temperature control in our homes for comfort's sake. In some homes that use a floor furnace located near the center of the living area, the air over the furnace is heated so that it increases in kinetic energy and thus expands. This less dense air is forced upward by the colder, more dense air moving along the floor. As the warmer air reaches the ceiling, it spreads out toward the walls and loses kinetic energy by virtue of contact with colder air, so it returns to lower levels of the room. From this transfer, convection currents are created by unequal air temperature, and heat is transmitted throughout the home as a result.

The process of convection permits the ventilation of the home. As the air we breathe becomes increasingly warmer, it becomes less dense than the air from outdoors. Therefore, with proper arrangements of the windows, air in the rooms is easily changed. A window opened near the top of the room will permit the less dense air to be forced outward if a window is also opened near the floor. The fan installed over many kitchen cooking units exhausts the warm and less dense air that contains cooking odors without relying on convection currents initiated by warm and cold air.

Convection plays a major role in the heating of homes that have hot- and cold-air returns distributed throughout the homes. The hot-air vent is placed under a window if possible. This arrangement causes convection currents because the cold window glass comes in contact with the warm air of the room. Another use of convection is in fireplaces. The heated air in the chimney expands and, in conjunction with the smoke and gases, is forced upward by the cooler, denser air present in the room. A tall chimney is more efficient than a shorter one due to the greater difference in weight between the hot air and gases in the chimney and an equal volume of cold air outside it. The principle of convection is also central to such natural occurrences as land breezes, sea breezes, windstorms, tornadoes, and hurricanes.

Scientists have determined that control of the air is necessary for better efficiency, comfort, health, and possibly survival. They are also aware that we in the United States are pouring some 3.5×10^8 kilograms (a weight of 390,000 tons) of waste into the atmosphere every year. Unless the amount of waste products, gases, and particulate matter is drastically reduced, future humans may have to carry their own air supplies, as astronauts do. This would be virtually impossible, since an average adult requires some 16 kilograms of air a day, or about 6000 kilograms (a weight of 6.5 tons) of air per year, to survive!

Summary

The temperature of a body is in reality an expression of the average kinetic energy of its molecules. The higher the temperature of a gas, the greater is the velocity of its molecules. At absolute zero, about $-273°C$, all kinetic energy, except for that resulting from vibrations among the molecules and from a slight trace of translatory motion, has been removed. The energy that remains is called the zero-point energy.

Heat is the energy that flows from a body of higher temperature to one of lower temperature. The unit by which heat is measured is the calorie. The so-called food calorie is in reality the kilocalorie.

The heat of fusion of a solid represents the amount of heat (energy) necessary to overcome most of the intermolecular forces present in the solid, with the result that disruption of the crystal lattice occurs and the solid is converted into a liquid. The heat of vaporization of a liquid represents the amount of heat (energy) necessary to overcome much of the remaining intermolecular forces of the molecules of the liquid so that the molecules are able to move with sufficient freedom to pass into the gaseous state.

The molecules in a body are in constant motion. This motion is of three types: vibratory, rotary, and translatory. Molecular motion is mostly random and is caused by collisions between molecules.

Generally speaking, all solids and liquids contract as the kinetic energy of the molecules of the substance decreases. Water expands rather than contracts as its temperature is decreased below $4°C$. This unusual property of water permits aquatic life to survive in cold areas, since ice floats to the top and forms a generally effective barrier against total freezing of the water. All substances vary in their rates of expansion, the measure of which is known as the coefficient of linear (or volumetric) expansion.

The effect of changing the pressure on a confined gas is indicated by the relationship known as Boyle's law. The relationship of temperature to the volume of a confined gas is referred to as Charles' law. The general gas law describes how either pressure or volume of a confined gas is affected by a change in temperature.

Air conditioning is an application of thermodynamics and hydrodynamics. The variables in air conditioning include temperature, humidity, gas composition, and freshness. Air conditioning is designed to help an individual maintain a thermal balance between body energy production and body heat loss.

Energy is conserved, a statement attributed to the laws of thermodynamics. The second law of thermodynamics can be expressed in terms of entropy, a measure of the disorder of a system. If the system is highly disordered, then the inability of the system to do work is great; hence, the system has a high degree of entropy.

Energy in the form of heat is dissipated at fixed rates from our bodies by three primary methods: radiation, convection, and conduction. The three methods also represent the manners in which energy (heat) is transferred from one object to another.

Equation Summary

Heat of vaporization:

$$\text{total heat} = (540 \text{ cal/g})(m).$$

The heat of vaporization for water is 540 calories per gram. The total heat required to convert water in liquid form into steam (or the total heat that must be released to convert steam into liquid water) is directly related to the mass of water involved.

Heat of fusion:

$$\text{total heat} = (80 \text{ cal/g})(m).$$

The heat of fusion for ice is 80 calories per gram. The total heat required to convert ice into its liquid form is directly related to the mass of ice involved. (The total heat required to change water at the freezing point into ice is determined in the same manner.)

General gas law:

$$\frac{P_1V_1}{T_1} = \frac{P_2V_2}{T_2}.$$

The volume of a confined gas is inversely related to the pressure (P) and directly related to the temperature (T) to which it is exposed.

The effect of temperature change on length:

$$\Delta l = \alpha l_i \Delta t.$$

The change in length of a solid depends on the coefficient of linear expansion α for the material, its length at the original temperature l_i, and the magnitude of temperature change Δt.

Thermal energy:

$$Q = mc\Delta t.$$

The amount of energy required to bring about a given change in temperature of a substance is directly related to the mass m of the substance, its specific heat c, and the magnitude of temperature change Δt.

Questions and Problems

Temperature—Molecules in Motion

1. (a) What is the reading on the three major temperature scales at the accepted freezing point of water? (b) What are their readings at absolute zero?

2. Explain how perspiration lowers the temperature of the skin.

3. On a hot summer day a Celsius thermometer might read 35°. What would be the corresponding reading on a Kelvin thermometer?

The Calorie and Changes in State

4. Does increasing the temperature of an electric stove burner under a pot of boiling water cause submerged eggs to cook faster? Defend your answer.

5. How much heat is released when 5000 g of water at 0°C are allowed to freeze without a change in temperature?

6. How many calories of heat would be required to convert 50.0 g of ice at −30.0°C to steam at 150°C?

7. How many calories of heat would be required to raise the temperature of 500 g of ice at −100°C to its melting temperature, 0°C?

8. Describe how ice cubes placed into a soft drink cool the beverage.

9. How many calories of heat would be released when 200 g of water was cooled from 80.0°C to 20.0°C?

Heat, Friction, and Types of Motion

10. (a) What is heat? (b) How does it relate to temperature?

11. Briefly define each of the three types of motion and cite an example of each.

12. Why do microscopic particles, such as present in cigarette smoke, assume a random path?

The Expansion of Liquids and Gases

13. An inflated balloon expands as the air inside it is heated. Can one accurately conclude that if a volume of air is expanded it warms? Defend your answer.

14. At −23.0°C, the volume of a gas in a balloon is 0.0160 m^3. (a) What would be the volume of this gas at 27.0°C? (b) If the temperature were to remain at −23.0°C, what would be the volume of this gas if the pressure exerted on the balloon were to increase from 8.50×10^4 N/m^2 to 1.10×10^5 N/m^2?

How Solids Expand

15. A concrete highway is constructed of slabs of concrete separated by thin expansion joints. If the slabs are 12.0 m long, how much will the length of each slab increase if the ambient temperature rises from 5.00°C to 35.0°C? (See Table 6.1.)

16. The framework of many older tall buildings was assembled by construction workers who heated rivets to glowing as they were used to fasten steel girders to each other. Why were the rivets heated?

17. A glass rod and a copper rod are noted to be 2.30 m long when measured at room temperature, 20.0°C. How would their lengths compare if the rods were heated to 200°C? (See Table 6.1.)

Air Conditioning and Thermodynamics

18. For a system to achieve a low degree of entropy, what conditions must be met?

19. As you eat a piece of freshly baked apple pie, you may find the filling uncomfortably hot, but the crust considerably cooler. Why?

20. (a) Why does a silver spoon submerged in hot coffee readily approach the temperature of the beverage, whereas a stainless steel spoon submerged simultaneously may only slightly increase in temperature? (b) How does a plastic or a wooden spoon compare to a stainless steel spoon under the same conditions?

21. How many calories of heat would be required to raise the temperature of a 2.50-kg iron skillet from 20.0°C to 100°C? (See Table 6.2.)

22. A book and a metal desk top in a classroom have assumed the same temperature, yet the desk top feels much colder. Why?

23. How many calories of heat would be required to raise the temperature of 500 g of cooking oil at 20.0°C to 250°C? (See Table 6.2.)

24. How many calories of heat would be required to raise the temperature of a 300-g aluminum cooking utensil from 20.0°C to 200°C?

Thermal Energy in Transit

25. Briefly discuss the three means by which heat is transferred from one object to another.

26. Briefly explain the process by which most homes are heated.

95
twi - light, when the lights are low,
Still to us at twi - light comes love's old song,

7 The Science of Sound and Music

Chapter outline

Key terms/concepts

Musical instruments of various sorts, along with the renditions of those who play them, have pleased listeners and players alike since ancient times.

If a tree falls in a forest where no one can hear it, is there a sound? This question, which you no doubt have heard before, addresses the nature of sound. Those who consider sound a subjective measure typically respond in the negative, since they conclude that there is no sound without a listener. Sound to them is a sensation perceived by our sense of hearing—therefore, an *effect.* Another group responds affirmatively. This group considers sound an objective measure; hence, there is sound produced without a listener since there is a *cause.* The difference in opinion of the two factions results from their accepted definition of sound.

The scientist defines **sound** as *a form of energy.* Various characteristics of a sound determine if it is capable of being heard. But we have also learned to identify sounds as pleasing or displeasing to us. Music is an art in which intelligible combinations of sounds are structured to produce pleasant listening. But one who wishes to understand music must first understand some basic characteristics of sound. The information in this chapter will be helpful in understanding the scientific concepts that provide the basis for music.

The relationship between sound and music was made much clearer by the contributions made to the field of music by Herman Ludwig Ferdinand von Helmholtz, a German physiologist and physicist of the early nineteenth century. Helmholtz's book *Sensation of Tone* described such musical concepts as pitch, intensity, quality of musical notes, and general acoustics. His contributions to the understanding of sound have made it much easier for modern composers to achieve the musical effects they want.

It is clear that sound has its origin in the vibration of matter. The vibrating object disturbs the air in all directions, and the disturbance creates in the air concentric spherical shells of waves that center about the vibrating object (see Fig. 7.1). The particles in the disturbance vibrate back and forth about some equilibrium position, but do not move along with the wave. As these disturbances reach the ear, they cause the eardrum to vibrate, and the brain interprets these vibrations as sound. The musical sound created by a violin being played is caused by the vibrations of the strings as they are bowed. The sound of a saxophone results from the air being caused to vibrate in the hollow curved tube, a vibration brought about by the flexible reed in the mouthpiece of the instrument. If the speaker of a phonograph is touched while a record is playing, the components of the speaker can be felt to vibrate as the speaker transforms electrical impulses into sound. Sound is also produced by the striking of one object against another, such as a hammer against a gong.

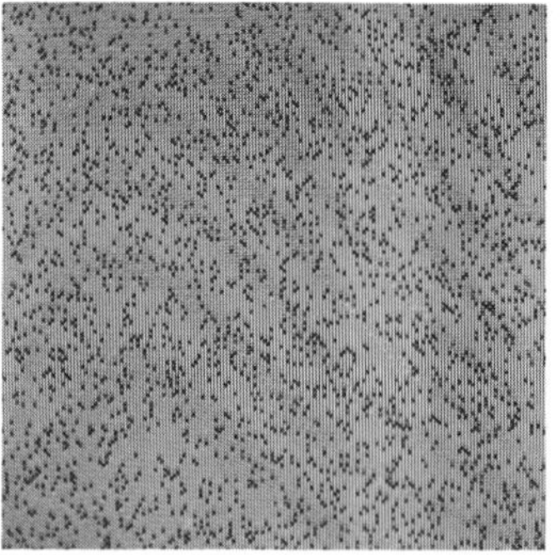

a.

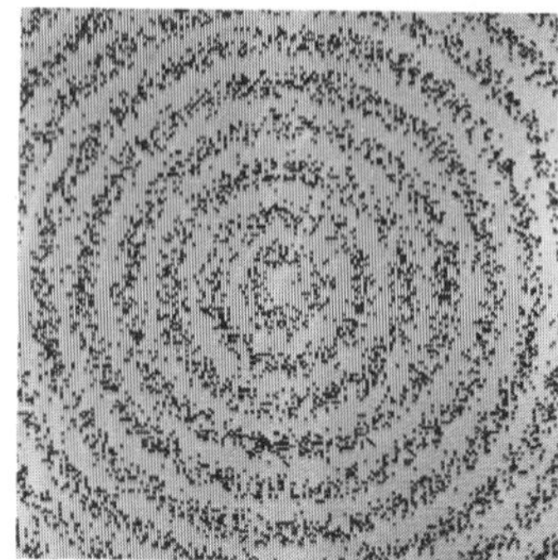

b.

Figure 7.1 (*a*) A representative view of air when no disturbance is present. (*b*) The spherical shells of disturbance formed in air about a vibrating object (figures somewhat idealized).

The Speed of Sound

Sound waves must have some medium through which they can be transmitted. Generally, this wave motion is produced in and transmitted through air. The disturbances produced in the air are illustrated in Figure 7.2. The air is disturbed by the vibrations produced by electrical impulses in the speaker. These disturbances abruptly push the air and thus compress it, and it exits from the front of the speaker in a conical pattern. Since the air is compressed, other air in the immediate vicinity decreases in density and becomes rarefied. The concentric circles of such disturbances move from the source at a speed commensurate with the speed of sound in the medium, generally air. If the disturbance of the air is created by an explosion, the sudden and forceful **compression** and resultant **rarefaction** of the air is transferred to objects in the path of the powerful force, and fragile objects such as window glass may vibrate to such a violent degree that they are shattered.

The speed of sound in air varies directly with the temperature. At 0°C the speed of sound is about 331 meters per second (1087 ft/s or 740 mi/h). As the temperature of the air rises, the speed

How We Hear

The human ear is divided into three areas: the outer ear, the middle ear, and the inner ear. The outer ear is composed of the *pinna,* the external part of the ear, and the *auditory canal* that carries sound waves to the thin sheet of tissue called the *eardrum.* The sound waves cause the eardrum to vibrate inward and outward in an alternating fashion. Opposite the eardrum are the three small bones of the middle ear—the *anvil, hammer,* and *stirrup*—so named because of the shapes they resemble. These bones serve to conduct the sound waves to the inner ear where the *cochlea* is located. This coiled tube, shaped somewhat like a snail shell, is about 2 centimeters long. It is in contact with the stirrup at its oval window. The cochlea is divided along its length by a membrane that contains hair-like projections and nerve fibers. Sound waves induce the projections to be stretched, creating nerve impulses that are transmitted by the auditory nerve to the brain. The brain interprets these impulses as sounds of varying frequencies, according to the location along the membrane of the excited nerve and the rate at which the impulses are transmitted ■

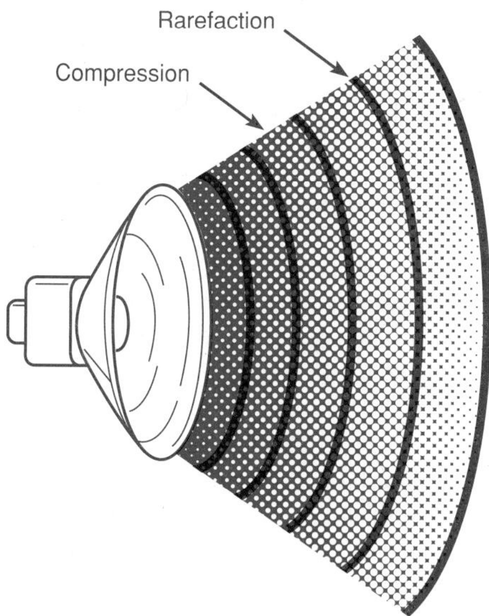

Figure 7.2 The disturbances created in the air by the type of speaker used in a radio or television set.

of sound increases; if the temperature of the air drops, the speed of sound decreases. On the Celsius scale, the speed of sound varies 0.610 meter per second per degree change. If the temperature is measured on the Fahrenheit scale, sound varies in speed about 1.1 feet per second per degree change. For practical purposes, the speed of sound is assumed to be 335 meters per second (1100 ft/s) unless otherwise specified. However, to clarify further how air temperature affects the speed of sound, note the examples that follow.

Example 7.1

The temperature of the air on a given summer day is 35.0°C. Compare the speed of sound in meters per second at this temperature with the speed of sound at an air temperature of −12.0°C.

Solution

The speed of sound in air at 0°C is 331 m/s. The speed of sound in air varies 0.610 m/s per C°. Thus, at 35.0°C ($T_{°C}$) the speed of sound in air is found from

$$v = 331 \text{ m/s} + \left[(T_{°C})\frac{0.610 \text{ m/s}}{\text{C}°}\right]$$

SPEED

$$= 331 \text{ m/s} + \left[(35.0°\text{C})\frac{0.610 \text{ m/s}}{\text{C}°}\right]$$
$$= 331 \text{ m/s} + 21.0 \text{ m/s}$$
$$= 352 \text{ m/s}.$$

At −12.0°C the speed of sound in air is determined in the same way:

$$v = 331 \text{ m/s} + \left[(T_{°C})\frac{0.610 \text{ m/s}}{\text{C}°}\right]$$
$$= 331 \text{ m/s} + \left[(-12.0°\text{C})\frac{0.610 \text{ m/s}}{\text{C}°}\right]$$
$$= 331 \text{ m/s} + (-7.32 \text{ m/s})$$
$$= 324 \text{ m/s}.$$

Light travels at about 300,000 kilometers per second (186,000 mi/s); therefore, light that originates on or near Earth reaches us almost instantly. Sound, however, travels only at about 0.32 kilometer (0.20 mi) per second. During a thunderstorm, if we count the number of seconds from the time we see lightning until we

Table 7.1

The speed of sound in various substances.

Substance	Speed (m/s)	(ft/s)
Gases:		
Air (at 0°C)	331	1087
Carbon dioxide	258	846
Hydrogen	1268	4160
Liquids:		
Alcohol	1186	3890
Water	1435	4708
Solids:		
Glass	5503	18,050
Iron	5128	16,820
Wood (oak)	3848	12,620

hear the thunder that accompanies it and divide the result by five, we can determine approximately how many miles away the storm is. For example, if 10 seconds lapse from the time we see the lightning until we hear the thunder, the storm associated with the lightning is approximately 2 miles (3 km) from us.

This ability to measure the speed of sound accurately has been applied in many ways. During World War I, for example, a technique was developed to determine the position of enemy guns by using the sound of the cannon in action. Sound ranging, as the technique was called, made use of microphones placed at strategic intervals over an area, with the position of each microphone oriented very carefully on a map. The exact time was recorded at each spot when the sound of the cannons reached it. Through the use of triangulation, a common trigonometric technique, the gun's position was accurately determined and immediately bombarded by the Allies.

The speed of sound is generally greater in solids and liquids than in gases because of the greater "stiffness" of substances in the solid or liquid state. For example, sound travels about 4.5 times as fast in water as it does in air. Some examples of the approximate speeds of sound in various mediums appear in Table 7.1. The whistle of an approaching train, for example, may be heard twice—first as the sound travels through the railroad track and again as it travels through the air.

Example 7.2

Determine the distance that a given sound would travel in 8.00 seconds if the medium were (a) air at 0°C, (b) water, (c) steel (iron).

Solution

According to Table 7.1, at 0°C, sound travels through air at 331 m/s. Its speeds through water and iron (steel) are 1435 m/s and 5128 m/s, respectively. (a) In the air the sound would travel the distance, *s*, according to

$$s = \bar{v}t = \left(\frac{331 \text{ m}}{\text{s}}\right)(8.00 \text{ s}) = 2650 \text{ m}.$$

(b) In water

$$s = \bar{v}t = \left(\frac{1435 \text{ m}}{\text{s}}\right)(8.00 \text{ s}) = 11{,}500 \text{ m}.$$

(c) The distance the sound travels in iron is also found using

$$s = \bar{v}t = \left(\frac{5128 \text{ m}}{\text{s}}\right)(8.00 \text{ s}) = 41{,}000 \text{ m}.$$

Kinds of Waves

If an investigator fastens both ends of a stretched spring and plucks the spring at right angles to its length, a disturbance is created in the form of a wave that travels along the length of the spring and back to the origin of the disturbance. Such a wave, in which *the particles vibrate at right angles to the path that the wave travels,* is known as a **transverse wave** (Fig. 7.3a). Some classic examples of such a wave disturbance, in which there is little or no actual forward movement of the medium, include a plucked string on a musical instrument and a garden hose as it is moved over an obstruction. The movement of the particles in the wave is up and down, while the crests and troughs of the wave move horizontally.

If several turns of the stretched spring are compressed and then suddenly released, the compressions move along the length of the spring in a **longitudinal wave** (Fig. 7.3b). Longitudinal waves are those in which *the direction of particle motion is parallel to the direction of wave propagation.* Close observation reveals rarefied areas (R) immediately preceding and following the compressed area (C) (see also Fig. 7.4). As in the transverse wave, little movement occurs in the medium along the length of the spring. Sound, of course, as it moves through a given medium, is an example of this longitudinal wave pattern.

Characteristics of Sound

Sound waves, like other waves, have three fundamental physical characteristics: wavelength, frequency, and amplitude. **Wavelength,** as illustrated in Figure 7.4, is *the distance between the centers of consecutive compressions* (or of consecutive rarefactions). **Frequency** is *a concept measured in terms of the number of times*

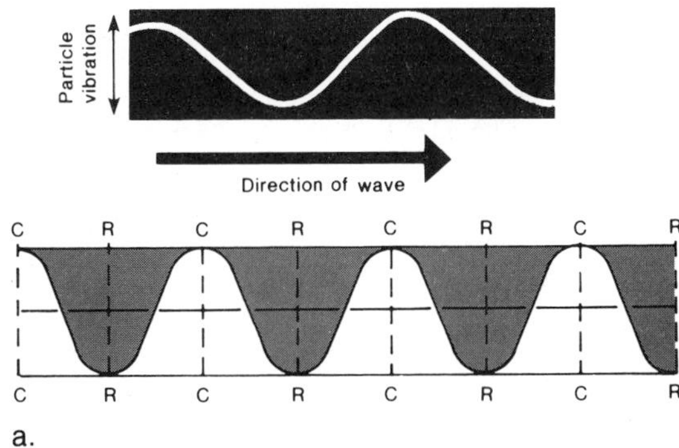

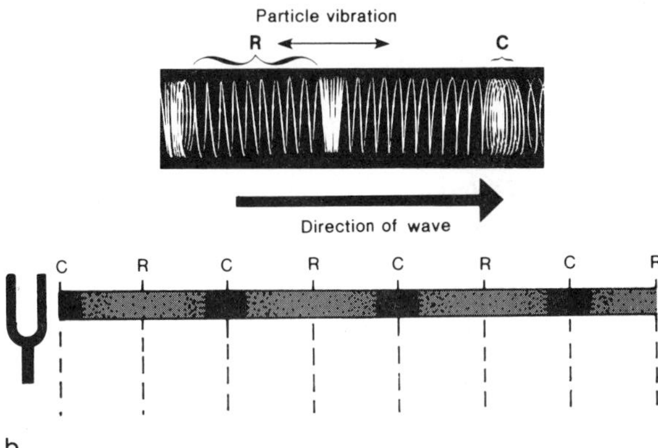

Figure 7.3 (*a*) A transverse wave and (*b*) a longitudinal wave.

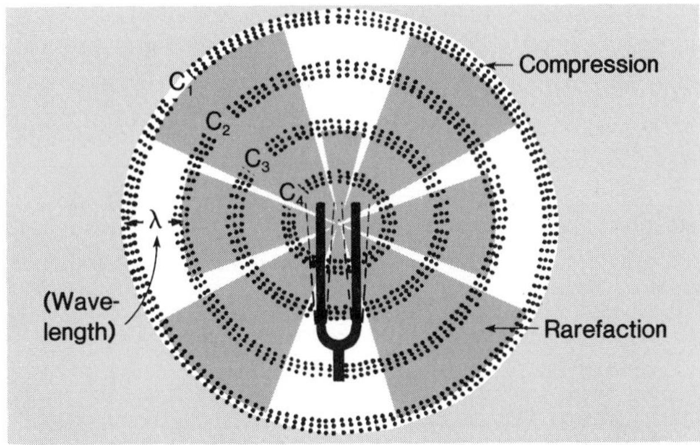

Figure 7.4 Representation of a series of longitudinal waves created by a tuning fork. The wavelength (λ) is a measure of the distance between C_1 and C_2, C_2 and C_3, and so forth.

that an object vibrates (or oscillates) in a given length of time, usually in seconds. As the particles continue to oscillate, or the object to vibrate, the time for one complete oscillation (or vibration) is called the period (T). The frequency and period are each other's reciprocal. That is, frequency and period are related by $T = 1/f$ and $f = 1/T$. Amplitude, as represented in Figure 7.5, is discussed in the next section of this chapter.

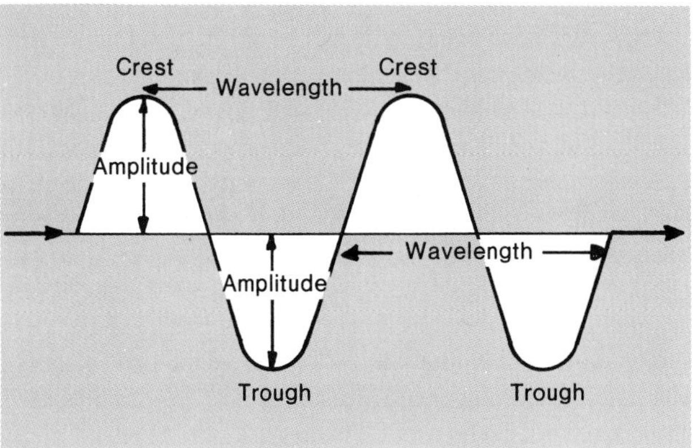

Figure 7.5 The quantitative characteristics of transverse waves.

Typical measurements of frequency (f) are recorded in vibrations per second (vps or vib/s), in cycles per second (cps or c/s), or in hertz (Hz). (Note how the frequency unit in the numerator is ignored in problem situations, such as in Example 7.3.) The terms are numerically equivalent, but the hertz, named for nineteenth-century physicist Heinrich Hertz, is rapidly replacing the other two terms in common usage and perhaps should be stressed.

Another useful relationship for waves may be developed from a basic equation provided in the discussion of motion in Chapter 3. Recall that average speed is the distance traveled divided by the time required to cover that distance:

$$\text{speed} = \frac{\text{distance}}{\text{time}},$$

or in symbolic form,

$$\bar{v} = \frac{s}{t}.$$

In the current discussion, suppose an object such as a speaker is caused to vibrate and, hence, emit sound waves. When the speaker moves in the direction the wave travels, it will compress the air. This region of compressed air will move away from the speaker (at the speed of sound) while the speaker begins to move in the opposite direction, an effect that creates a rarefied region of air. During one complete vibrational cycle, the original compressed region will have moved exactly one wavelength. So the speed at which the compression moves through the air must be related to the wavelength (λ) and the period of vibration (T) by

$$\begin{aligned} v &= \frac{\lambda}{T} \\ &= \left(\frac{1}{T}\right)\lambda \\ &= f\lambda. \end{aligned}$$

WAVE EQUATION

From this **wave equation** it can be noted, since the speed of the wave depends on the medium, that if the frequency f increases, the wavelength λ decreases, and vice versa. In other words, frequency and wavelength are inversely proportional to each other.

Several additional points should be noted. The wave equation applies to all types of waves, including water waves, waves in strings, radio waves, and light waves, as well as sound waves. It applies to both transverse and longitudinal waves. Finally, it should be pointed out that the velocity (or speed) in the wave equation is the rate with which the "disturbance" (in sound waves, the compression or rarefaction) travels, but not the speed of the individual particles in the wave. The speed of the particles depends on the frequency and amplitude of the wave, as well as on the time in the vibrational cycle at which the particle is observed.

Sound waves immediately above the range of human hearing (above about 20,000 hertz) are classified as *ultrasonic*. Sounds below the range of human hearing (below 16 hertz), are known as *infrasonic*. A third class, called *audiosonics*, includes those frequencies between 16 hertz and about 20,000 hertz. This range, although subjective (it varies with the listener), is known as "the range of hearing." The human ear appears to be most sensitive to frequencies of between 2000 hertz and 4000 hertz. This sensitivity diminishes rapidly for most individuals as frequencies increase or decrease (see Chapter 27). Considering the maximum range of hearing as discussed, corresponding wavelengths would fall between 20.7 meters and 0.166 meter (in air at 0°C).

Example 7.3

Calculate (a) the speed of a wave, the frequency of which is 60 hertz (vib/s or c/s) and the wavelength of which is 5.60 meters and (b) the wavelength of a wave, the speed of which is 343 meters per second and the frequency of which is 256 hertz.

Solution

Both solutions involve the same formula.
(a) In the first instance, we solve for v:

$$\begin{aligned} v &= f\lambda \\ &= (60.0 \text{ vib/s})\,(5.60 \text{ m}) \\ &= 336 \text{ m/s}. \end{aligned}$$

(b) In the second case, we solve for λ:

$$\begin{aligned} v &= f\lambda \\ \lambda &= \frac{v}{f} \\ &= \frac{343 \text{ m/s}}{256 \text{ vib/s}} \\ &= 1.34 \text{ m}. \end{aligned}$$

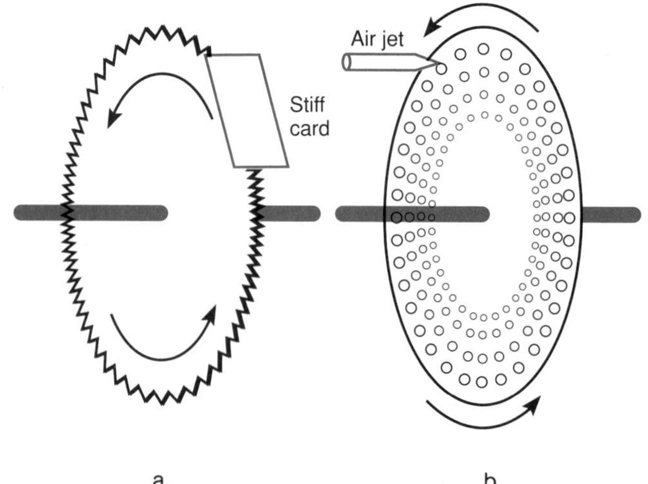

Figure 7.6 (*a*) A toothed disk, known as a Savart's disk (or wheel), produces a musical note when a stiff card is held against the teeth. As the wheel speeds up, the vibration frequency of the card increases and the note rises in pitch. (*b*) A constant blast of air is interrupted by a rotating disk that contains several rings of holes, an effect that produces a musical note. A variation in speed of rotation of the disk causes a change in the musical note produced, as in (*a*).

Subjective Measures of Sound

When a sound is produced, is it a musical sound or a noise? The distinction between the two has always been determined by culture and is somewhat determined by generation as well. If the sound is considered pleasant to the listener, it is a musical sound, but if it is displeasing, it is classified as noise. Therefore, one can assign either category to a given sound, depending on the culture and mood of the listener.

In a broad sense, the general attributes of a musical sound help discriminate between it and noise. These attributes are *pitch, loudness,* and *quality*. All are psychological terms that indicate sensations, or effects produced on the senses. One could also include *duration,* the length of time that the sound lasts. Each of the first three attributes has a counterpart that is physical in nature, and thus can be measured. These counterparts in order are *frequency, amplitude* or *intensity,* and *overtone structure*.

The pitch of a musical note refers to its position on a musical scale. It is determined by the frequency of the sound impulses produced by a musical instrument, such as a violin. The relationship of pitch to frequency can be readily demonstrated with a toothed disk rotating at various speeds and with a rotating siren disk (see Fig. 7.6). Factory whistles commonly use the principle of the rotating siren disk. Pitch, as it applies to music, is a subjective measurement; hence, it varies among individuals. The variation in each listener's degree of sensitivity creates the sensation of pitch. This

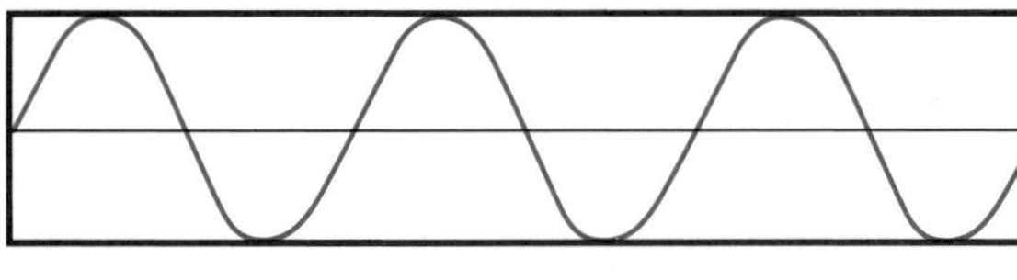

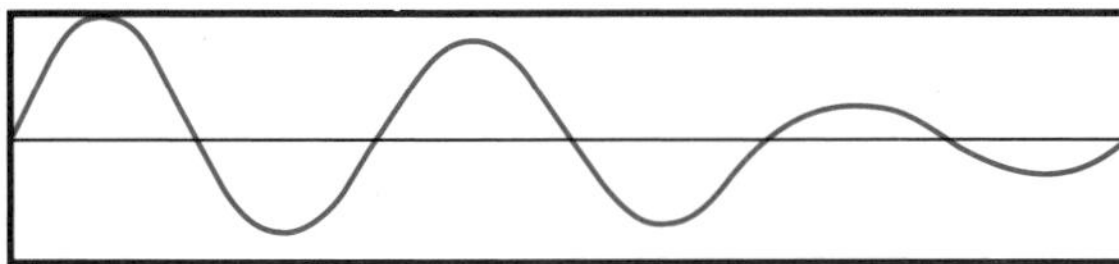

Figure 7.7 Representations of constant amplitude and decreasing amplitude.

sensitivity to pitch is innate, although it can be instilled in some people with proper and intense training. Frequency, on the other hand, is a physical measurement of the number of times the object that creates the sound vibrates in one second.

The sensation of pitch is related to frequency in that high-frequency sounds produce high pitch, and low-frequency sounds produce low pitch. Some musicians, however, contend that loudness affects the sensation of pitch to some degree. This group feels that as an object is caused to vibrate more vigorously but at a low constant rate, the tone produced not only becomes louder but also changes to a slightly lower pitch. Conversely, as an object is forced to vibrate at a relatively high rate, the trained listener apparently can note an increase in the pitch as the intensity of the sound is increased. Because of these observations and the result of various studies in which trained and untrained observers noted significant differences when asked to compare sounds of varying degrees of loudness, pitch and frequencies are not considered interchangeable terms.

In order to summarize the concept of pitch, let us review its relation to frequency. As has been pointed out, the greater the frequency of a vibrating object, the higher the pitch produced. In addition, wavelength has been found to be inversely proportional to frequency; that is, a shorter wavelength has a higher frequency and a longer wavelength a lower frequency, each quantity varying in an inverse manner because of the constant speed of the wave in any given medium.

Amplitude is the physical counterpart of the subjective characteristic known as loudness. Amplitude, as is indicated in Figures 7.5 and 7.7, is a measure of the deviation of the crest of the wave from an imaginary line that represents the surface of the medium when all disturbances cease, such as the surface of a body of water. Amplitude is also, and more commonly, represented as a deviation of an object, such as a violin string, from its position at rest. The greater the deviation of the string or the greater the disturbance on the body of water, the greater the amplitude of the wave produced; hence, the louder the sound it creates.

Table 7.2

Representative values of sound intensity levels (see also Fig. 27.12).

Source of sound	Sound intensity levels		Intensity
	in bels (B)	*in decibels* (dB)	*in watts/m²* (W/m²)
Threshold of hearing	0	0	10^{-12}
Rustle of leaves	1	10	10^{-11}
Air conditioner, window unit	4	40	10^{-8}
Vacuum sweeper on rug	6	60	10^{-6}
Automobile	7	70	10^{-5}
Pneumatic drill	8	80	10^{-4}
Airplane engine	11	110	10^{-1}
Threshold of pain	12	120	1
Jet airplane, at 30 m	13	130	10^{1}

Frequently, intense sounds do not appear to be very loud because of the poor transmission of the sound through a given medium or the great distance between the sound and the listener. Loudness is relative to the effect it has on the listener's ears and, therefore, is subjective. The intensity of a sound is dependent on the energy of the sound waves and is amplified by increasing the amplitude and the area of the object in vibration.

One can conclude, then, that loudness, expressed in a subjective manner, is a measure of sound power based on the effect that the sound has on the listener's senses. Intensity can be measured objectively as the amount of sound power produced. A convenient method of measure is to compare the power produced by one sound with that produced by another. The unit that is conventionally used is the bel(B), in honor of Alexander Graham Bell, inventor of the telephone. The human ear can detect a very wide range of intensities. For this reason scientists express the intensity level of sounds in a compressed manner through the use of logarithms, exponents that indicate the power to which a base number (commonly the number 10) is raised to produce a given level of intensity. For instance, the intensity level of a sound ten times that of another has an intensity-level difference of 1 bel; 100 (10×10, or 10^2) times as intense signifies a difference of 2 bels; 10,000 ($10 \times 10 \times 10 \times 10$, or 10^4) times as intense is equivalent to 4 bels, and so forth. For convenience the bel is divided into ten equal parts called decibels (dB), the unit of loudness most commonly used. The *decibel* is also described as a unit for evaluating the ratio of a power to a reference power, as the measurements relate to sound intensity level. Some comparative values can be found in Table 7.2. More values are listed in Figure 27.12.

The intensity level of a sound decreases as it moves from its source and dissipates throughout the conducting medium, generally air. The loss in intensity level of the sound is described by the inverse square law that applies to the dissipation of energies of all

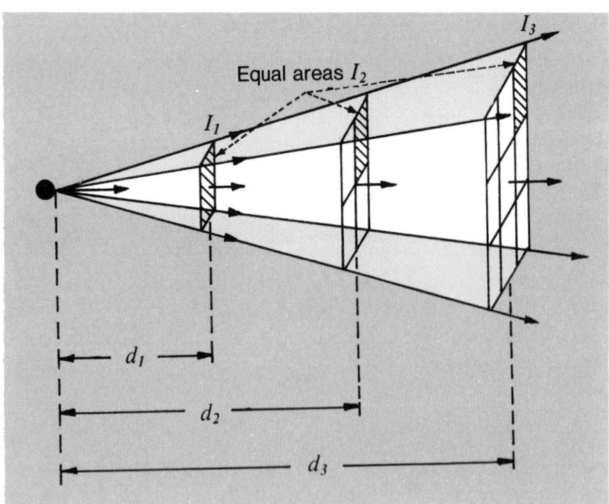

Figure 7.8 The inverse square law as it applies to the level of sound intensity. I is measured in W/m².

types. Sound intensity level, as an objective measure, is the amount of energy that the sound wave carries away from the source in one second through a specific area perpendicular to the source. The unit of sound intensity level generally used by scientists is the watt per square meter. The inverse square law as it applies to the intensity level of sound can be stated,

$$\frac{I_1}{I_2} = \frac{d_2^2}{d_1^2}, \qquad \text{SOUND INTENSITY LEVEL}$$

where I_1 and I_2 are intensity level measures of a sound in watts per square meter (W/m²) at distances in meters of d_1 and d_2, respectively. A representation of the inverse square law appears in Figure 7.8. A comparison of the energy of sound waves in watts per square meter is also included in Table 7.2.

Example 7.4

The level of intensity of a sound measured 10.0 meters from its source is determined to be 100 decibels (1.00×10^{-2} W/m²). Calculate the intensity level at a distance of 20.0 meters from its source.

Solution

The inverse square law as it applies to sound is valid only when the units of intensity level are expressed in terms of power per unit area:

$$\frac{I_1}{I_2} = \frac{d_2^2}{d_1^2}$$

$$I_2 = \frac{I_1 d_1^2}{d_2^2}$$

$$= \frac{(1.00 \times 10^{-2}\ \text{W/m}^2)(10.0\ \text{m})^2}{(20.0\ \text{m})^2}$$

$$= \frac{(1.00 \times 10^{-2}\ \text{W/m}^2)(100\ \text{m}^2)}{400\ \text{m}^2}$$

$$= \frac{1.00\ \text{W}}{400\ \text{m}^2}$$

$= 0.00250$ W/m². (For an estimate of sound intensity level, refer to Table 7.2.)

As has been mentioned, sound waves spread in all directions from the vibrating source. If one uses a megaphone, however, the sound is projected toward the direction in which the larger end of the megaphone is pointed. Doubling the distance in this case does not decrease the loudness to one-fourth the initial measurement for those observers in front of the person using the megaphone. The principle of the megaphone is also involved in the design of most musical instruments that are intended to be somewhat directional. One can observe the effectiveness of design by noting the decrease of loudness as a band performing at a football game pivots away to play for the fans in the opposite stands.

Quality, also known as overtone structure and as *timbre,* is the fundamental characteristic that requires the greatest amount of discussion, since it is the property that permits a listener to discriminate between sounds that are identical in frequency and intensity.

Two objects vibrating at the same rate may produce sounds that differ in the degree to which they are pleasing to the ear. Two instruments or two voices that produce the same pitch are distinguishable, even though the pitches are the same and each is equally loud. The way in which the two sounds differ is in quality. If a bow is applied near the middle of a violin string and thus causes the string to vibrate, the sound produced is known as the *fundamental tone* for that string. If the string is touched at the middle very lightly while it is vibrating, both the fundamental tone and the first *overtone* can be heard, because the string begins to vibrate in two equal segments. The quality of the new sound is richer and fuller than the original one because of the addition of the overtone to the fundamental. The fundamental tone can still be identified, since it is the component tone of lowest pitch in a complex tone. A well-played violin is bowed so as to produce the fundamental and numerous overtones. The quality of a sound is dependent on the number of overtones present, and also on how well each one can be distinguished.

The sonometer, an instrument used in the laboratory to study vibrating strings, is a hollow rectangular box over which several strings of varying size and composition are stretched. The sounds produced by the strings are varied by changing the tension in each string, using a fixed spring scale or slotted weights, and the length of the vibrating segment by use of a movable bridge. The instrument is illustrated in Figure 7.9.

Figure 7.9 The sonometer is an ideal device for studying the laws of strings since string diameter, length, and applied tension can be readily determined.

The meaning of the term **overtone** is easily demonstrated. It is possible to show that a string can be made to vibrate as a whole as well as in segments simultaneously (see Fig. 7.10). A string vibrating in two equal segments produces the first overtone, which is twice the frequency of the fundamental. If the string is made to vibrate in all three equal segments, the sound produced has three times the frequency of the fundamental. A C string used on a sonometer vibrates at 256 hertz. The first overtone of a C string is C′, which vibrates at 512 (or 2 × 256) hertz. The second overtone of C would be 3 × 256, or 768 hertz, which corresponds to the note known as G′. The third overtone of C has 1024 (or 4 × 256) hertz, and thus is the note C″. The fourth overtone of C is E″, at 1280 hertz.

A note that has a vibrating rate that is a whole-number (or integral) multiple of the fundamental frequency is known as an overtone, or a harmonic. In mathematical terms, the frequency (f) of each overtone or harmonic is expressed as $f = nf_1$, where

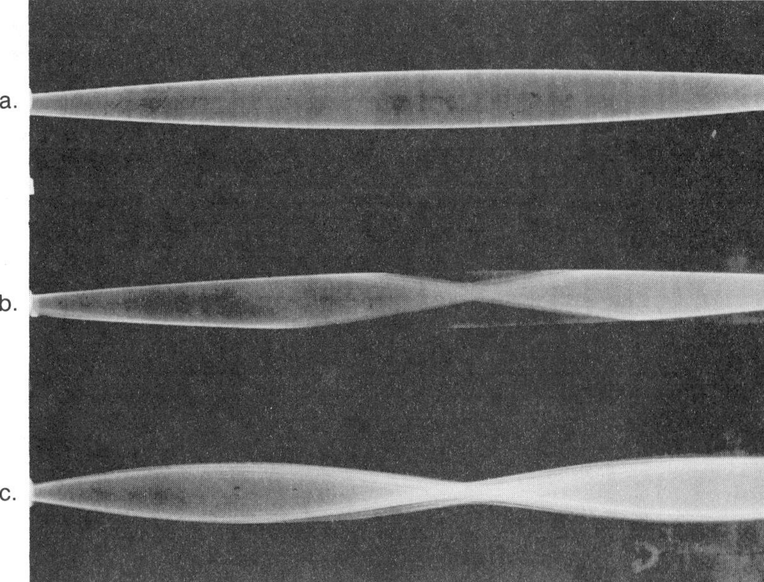

Figure 7.10 Standing waves in a vibrating string. (*a*) The string vibrating as a whole. (*b*) The string vibrating as a whole and in two equal segments. (*c*) The string vibrating in equal segments to produce harmonics.

f_1 is the fundamental frequency and n is the order of the whole-number multiple (see Fig. 7.10c).

The quality of a tone is subjective in that the fundamental frequency and its first overtone, illustrated in Figures 7.11a and 7.11b, combine and reach the listener and thereby cause his or her eardrum to vibrate accordingly (Fig. 7.11c). In Figure 7.12 some sound wave patterns appear as they are produced by a wave-analysis device called an *oscilloscope*. The instrument interprets variations in fluctuating electric current, converts them into wave forms, and projects them on a screen much like that of a television set.

The Laws of Vibrating Strings

There are several laws that describe the rate at which strings vibrate. If we investigate the various strings in a piano, we note that they vary in length, in diameter, in tension, and in composition (hence, density). Through the investigation of each variable, each factor can be shown to affect the sound produced by a given vibrating string. For instance, if all factors other than length remain constant, the vibrating rate of a string is found to be inversely proportional to its length. Expressed algebraically,

$$\frac{l_1}{l_2} = \frac{f_2}{f_1}. \qquad \text{STRING LENGTH}$$

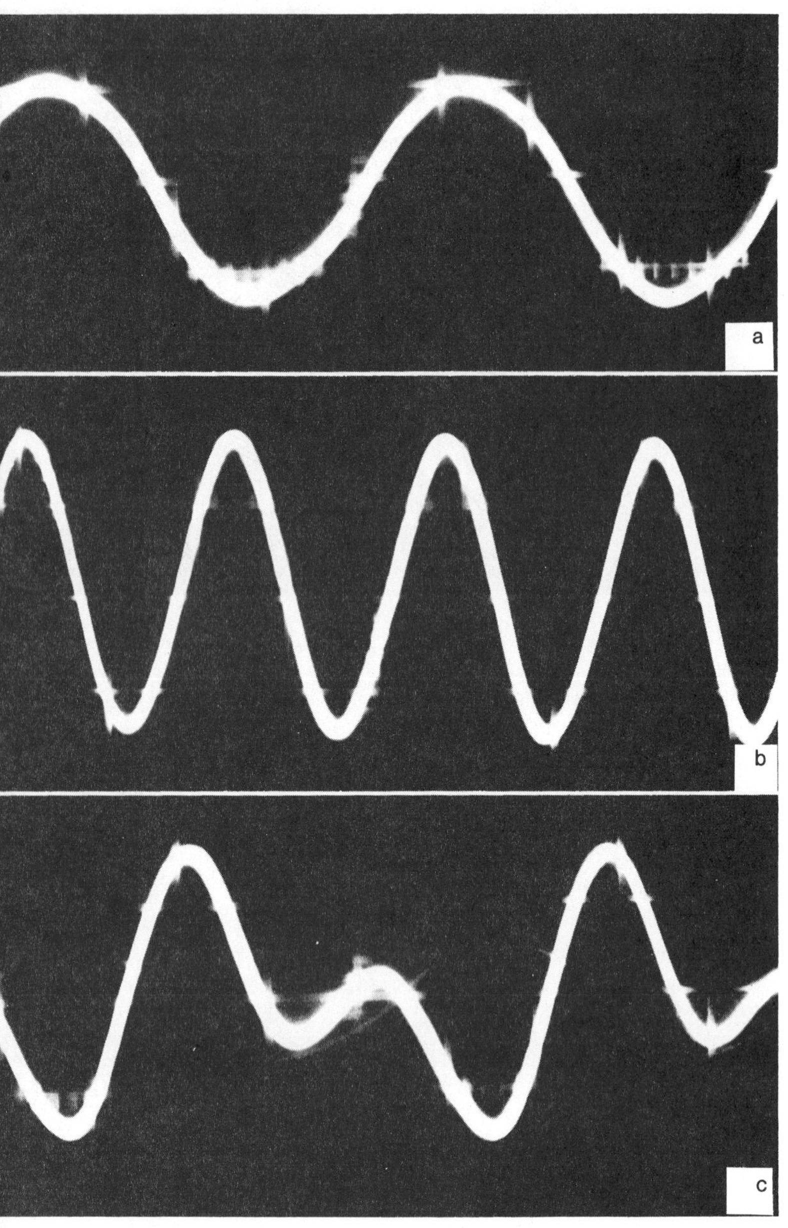

Figure 7.11 Oscilloscope patterns of (*a*) a fundamental tone, (*b*) the first overtone, and (*c*) the composite wave produced by combining the fundamental and first overtone.

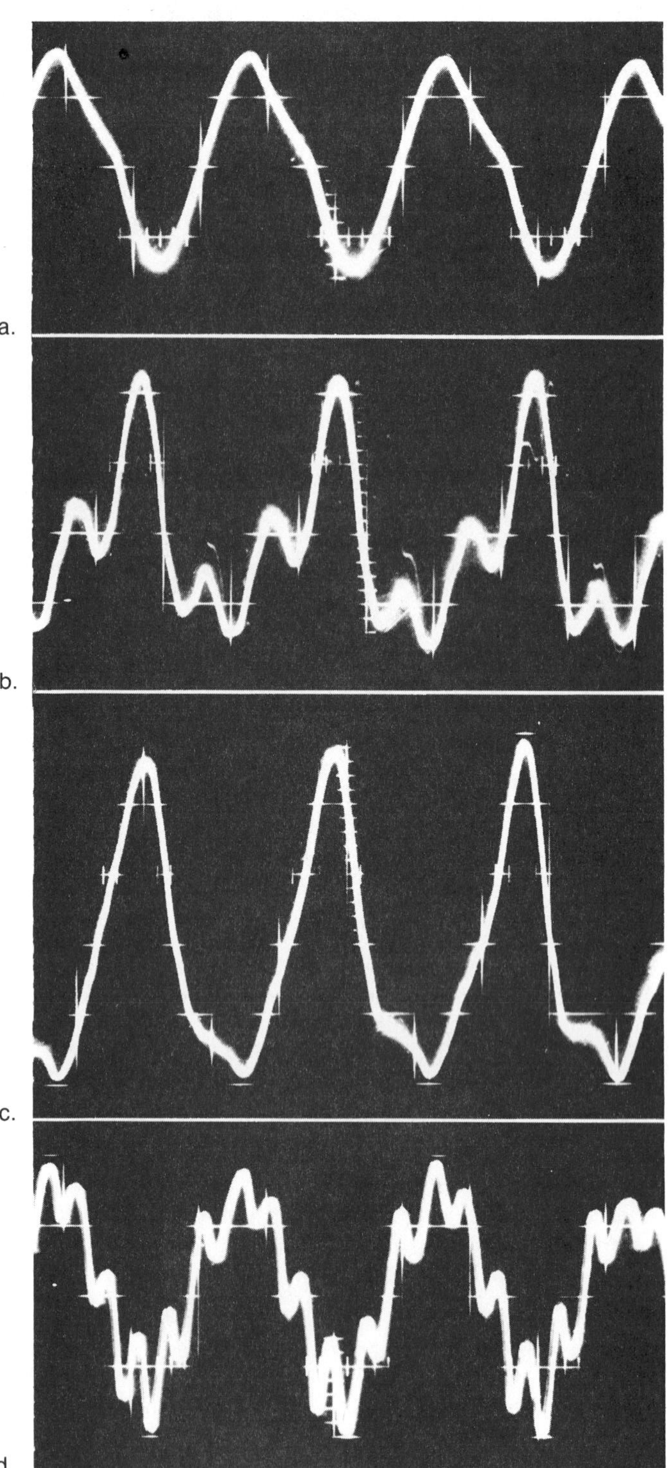

Figure 7.12 Oscilloscope patterns produced by various sounds. Illustrations (*a*), (*b*), (*c*), and (*d*) are of the same frequency but differ in the various overtones present. (*a*) Is a relatively pure note produced by an audio generator, (*b*) is a soprano voice, (*c*) is produced by a piano, and (*d*) is produced by a factory whistle.

Resonance and "Galloping Gertie"

When successive impulses are applied to an object, it may vibrate. If the impulses are applied to the vibrating object to coincide with its natural frequency, resonance occurs, and an increase in amplitude results.

In November 1940, just four months after its dedication, the Tacoma Narrows Bridge at Puget Sound, Washington, suffered almost total destruction. A constant wind of 67 kilometers per hour (42 mi/h) caused the flexible 854-meter (2800-ft) span to oscillate vertically at its natural frequency, and brought about a steady increase in the amplitude of vibration of the span. Shortly before its collapse, the bridge abruptly changed from undergoing a vertical oscillation to a twisting mode, visible in the photograph of Figure 7.13.

In 1990, a team of scientists, after spending over six years analyzing various mathematical models and other bridge structures that had collapsed prior to 1940, concluded that the twisting mode, rather than the vertical mode, actually had been responsible for the bridge's collapse. The twisting-type vibration appears to have been brought about by an alternate loosening and tightening of the stays that connected the roadbed to the main cable of the bridge. The failure of this structure, nicknamed Galloping Gertie, to withstand the relatively common high winds led engineers to change designs from thin, narrow roadbeds to wider, less flexible ones. The Tacoma Bridge was replaced by a four-lane rigid structure in 1950. Other suspension bridges, such as the Golden Gate, were also built to resist wind-induced motion. Scientists are still concerned, however, about the oscillations of suspension bridges caused by earthquakes, and design studies continue ■

Figure 7.13 A view of the twisting motion that brought about the 1940 collapse of the bridge at Tacoma Narrows in the state of Washington.

Example 7.5

If the D string on a violin is 72.0 centimeters long, what must be its effective length to produce the note C′, assuming all other variables such as tension remain unchanged? (See Fig. 7.19 for frequencies of notes.)

Solution

The note D is produced at 294 Hz and the note C′ at 523 Hz. Let l_1 represent the unknown length of the string and solve the proportion:

$$\frac{l_1}{l_2} = \frac{f_2}{f_1}$$

$$l_1 = \frac{f_2 l_2}{f_1}$$

$$= \frac{(293 \text{ Hz})(72.0 \text{ cm})}{(523 \text{ Hz})}$$

$$= 40.5 \text{ cm.}$$

Thus, the musician, in order to produce the note C′ on the D string of the violin, must touch the string to the fingerboard in such a way that the string is shortened by 31.5 (that is, 72.0 − 40.5) centimeters.

The strings on most stringed instruments vary in diameter. The larger diameter strings produce lower frequencies than the smaller diameter strings. A string twice the diameter of another vibrates at about one-half the rate of the smaller diameter one. If all other variables remain constant, the vibrating rate of a string is inversely proportional to its diameter.

A musician who wishes to tune a stringed instrument raises the frequency of the vibrating string by increasing the tension on it or by tightening the string; to lower the frequency, the string is loosened. The change in frequency resulting from a variation in tension is not a constant ratio. Assuming that all other factors remain constant, the vibrating rate of the string is found to be directly proportional to the square root of the tension on the string. That is,

$$\frac{\sqrt{T_1}}{\sqrt{T_2}} = \frac{f_1}{f_2}.$$

STRING TENSION

Example 7.6

A 25.0-newton force attached to one end of a string on a sonometer produces the note D (294 hertz) when the string is struck. In comparison, what is the total force required to produce the note D′ (588 hertz) from the same string?

Solution

If the vibrating rate is directly proportional to the square root of the tension, then letting T_2 represent the unknown tension,

$$\frac{\sqrt{T_1}}{\sqrt{T_2}} = \frac{f_1}{f_2}.$$

Solving for T_2,

$$\begin{aligned}\sqrt{T_2} &= \frac{\sqrt{T_1}f_1}{f_2}\\ &= \frac{(\sqrt{25.0\ \text{N}})(588\ \text{Hz})}{294\ \text{Hz}}\\ \sqrt{T_2} &= 10.0\sqrt{\text{N}}\\ T_2 &= 100\ \text{N}.\end{aligned}$$

Thus, a 100-N force applied to the string causes it to vibrate 588 times per second.

Finally, if one string is denser than another, the denser string will vibrate more slowly. Guitars, harps, and other instruments have some of their strings wrapped with copper wire for the purpose of increasing the string's overall density. Experimenters have determined that *the frequency of a vibrating string is inversely proportional to the square root of its density, if all other factors are constant;* that is,

$$f = \frac{1}{\sqrt{\rho}}.$$

So a string four times as dense as another will vibrate at one-half the rate of the less dense, if all other factors hold constant.

Resonance

Have you ever observed a small child attempting to swing a playmate who is much larger? The small child can transfer energy in small quantities to the larger child by pushing at the right moment, thus building up a vibration of great amplitude for the child who is swinging. This energy transfer by small increments is commonly referred to as **resonance:** *the building up of a large vibration by small impulses,* the frequency of which equals one of the natural frequencies of the resonating body. As to the large child in the swing, a large amplitude results only if the small impulses synchronize with the natural period of vibration of the swing. A small variation in synchronization results in a decrease in amplitude of vibration of the swing.

All objects have a natural frequency at which they vibrate most easily. If a sound wave of the same frequency as the natural frequency of the object comes in contact with the object, the object vibrates with a sympathetic vibration. If one depresses the loud pedal on a piano and hums, whistles, or otherwise creates a sound of the same frequency as the fundamental frequency of one of the piano strings, the string will be set in vibration by the sound and will continue to vibrate after the initial sound ceases.

Bridges have collapsed due to mechanical resonance built up by gusts of wind, by marching soldiers, by idling trucks, and by other objects vibrating at some constant rate. A certain rattle or vibration in the family automobile may occur at 44 kilometers per hour but not at 34 kilometers or 54 kilometers per hour or at most other speeds.

But how do two sounds perhaps produced by two identical tuning forks, and therefore having exactly the same frequency, affect each other? An experimenter will conclude that the two forks sounded simultaneously will reinforce the amplitude of each other in a form of constructive interference. Destructive interference can be created by causing the crest of one wave to coincide with the trough of the other, thus canceling each other in terms of amplitude. Waves set up by sounds of different frequencies can have a similar effect (see Fig. 7.14).

If the two tuning forks vibrate at slightly different frequencies, a third phenomenon known as *beats* results when the two forks are simultaneously struck. The new wave that results is sometimes reinforced and sometimes diminished, as displayed in Figure 7.15. The variation in loudness that results, varying between loud and soft, causes a throbbing sound instead of one reinforced sound, as when the two frequencies are identical. If a tuning fork that vibrates at 256 hertz is struck at the same time as a tuning fork of 288 hertz, 32 beats per second are produced, creating a sound displeasing to the ear. We obtain the number of beats produced per second by determining the difference between the frequencies of the two forks; that is, $f_1 - f_2 = \Delta f$, or 288 Hz − 256 Hz = 32 Hz.

Almost everyone is captivated by the pipe organ. The organ produces sound by directing a stream of air across the holes or openings in the pipes. The flute is closely related in principle to the organ, since sound is produced in both by a vibrating air column. Instruments such as the saxophone and the clarinet produce sound by a vibrating reed on the mouthpiece.

The pitch of the sound produced by a musical instrument composed of a pipe can be controlled by two factors: how long the pipe is and whether it is open or closed. A closed pipe, as shown in Figure 7.16a, produces a note that has a wavelength four times the length of the pipe itself. Thus, a closed pipe 2 meters long produces a note with a wavelength of 8 meters. An open pipe, as shown in Figure 7.16b, produces a note with a wavelength twice

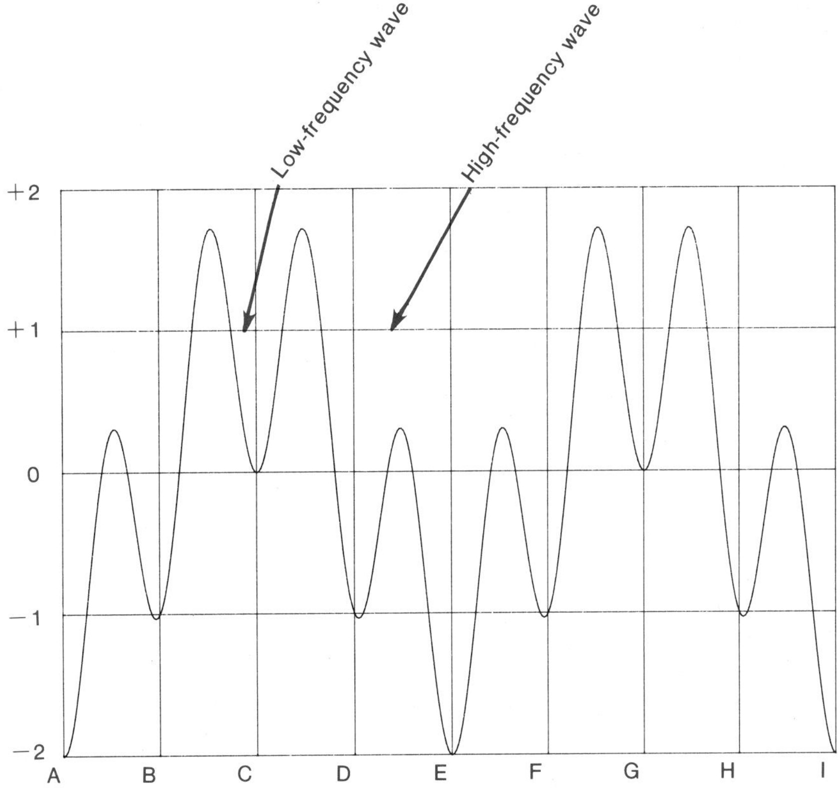

Figure 7.14 Waves of varying length can interact both constructively and destructively. The low-frequency wave of 200 Hz and the high-frequency wave of 800 Hz are exactly in phase at A, E, and I. They are exactly out of phase at C and G, thus canceling each other.

the length of the pipe; thus the 2-meter open pipe produces a note twice the frequency of the note produced by the closed pipe 2 meters long. Organ builders use both open and closed pipes because each type produces different harmonics, thus increasing the range of tone "colors" available to the organist.

Example 7.7

An organ has two pipes that are each 2.30 meters long; one pipe is open and the other is closed. Determine the frequency of the sound each pipe will produce when the temperature is 22.0°C.

Solution

Since the speed of sound in air at 0°C is 331 m/s, the speed of sound in air at the temperature given is determined from

$$\begin{aligned} v &= 331\ \text{m/s} + \left[(22.0°\text{C})\left(\frac{0.610\ \text{m/s}}{\text{C}°}\right)\right] \\ &= 331\ \text{m/s} + 13.0\ \text{m/s} \\ &= 344\ \text{m/s}. \end{aligned}$$

An open pipe produces a note with a wavelength twice the length of the pipe; therefore, the wavelength of the sound produced by the open pipe is determined from

$$\begin{aligned} \lambda &= (2)(2.30\ \text{m}) \\ &= 4.60\ \text{m} \end{aligned} \qquad \text{OPEN PIPE}$$

and the frequency from

$$\begin{aligned} f\lambda &= v \\ f &= \frac{v}{\lambda} \\ &= \frac{344\ \text{m/s}}{4.60\ \text{m}} \\ &= 74.8\ \text{Hz}. \end{aligned}$$

For a closed pipe the wavelength of the sound produced is four times the length of the pipe, so

$$\begin{aligned} \lambda &= (4)(2.30\ \text{m}) \\ &= 9.20\ \text{m}. \end{aligned} \qquad \text{CLOSED PIPE}$$

The frequency is then obtained from

$$\begin{aligned} v &= f\lambda \\ f &= \frac{v}{\lambda} \\ &= \frac{344\ \text{m/s}}{9.20\ \text{m}} \\ &= 37.4\ \text{Hz}. \end{aligned}$$

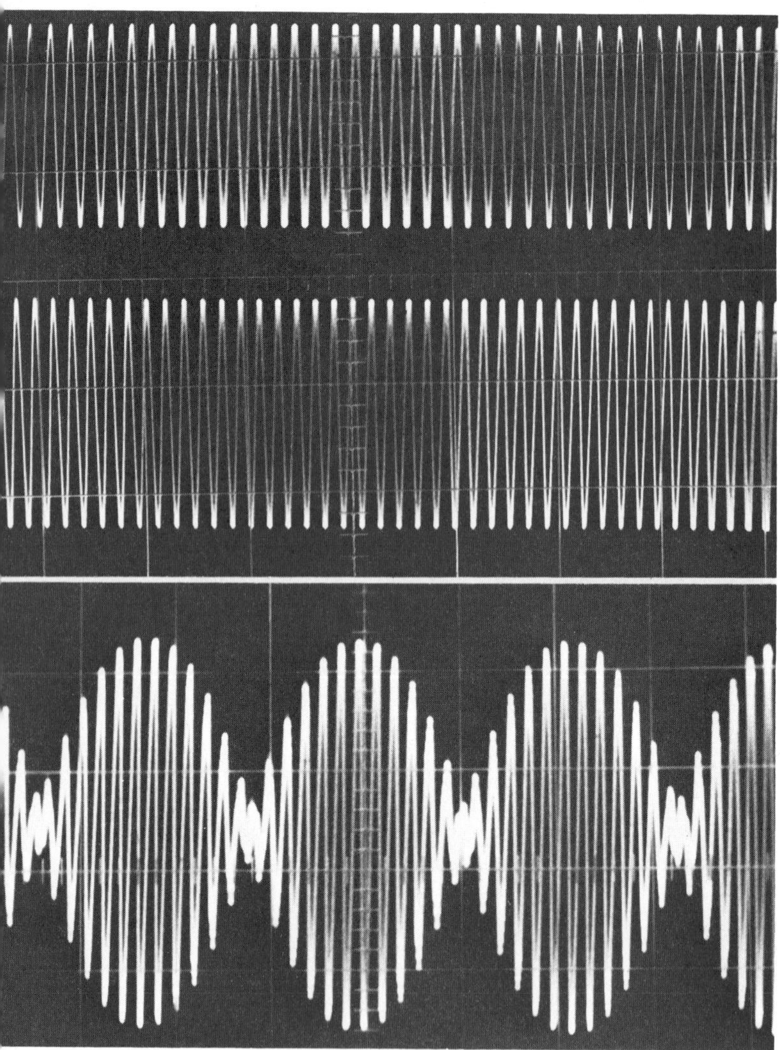

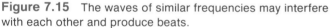

Figure 7.15 The waves of similar frequencies may interfere with each other and produce beats.

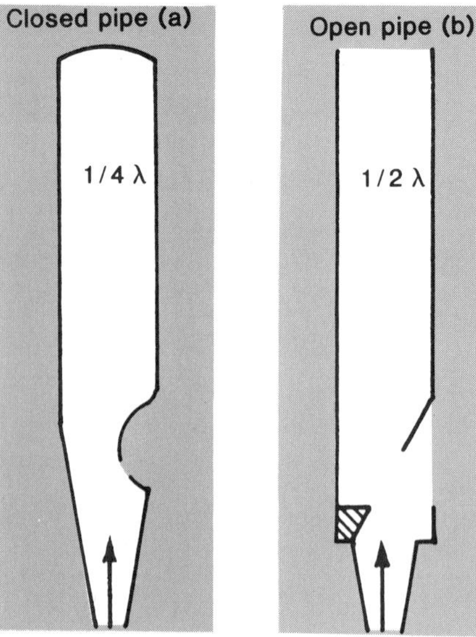

Figure 7.16 The two different kinds of pipe arrangements. Both types are necessary to permit musicians to have a full range of harmonics at their disposal.

The lips of the player serve as the vibrating object that produces the sound for brass instruments such as the trumpet, the French horn, the trombone, and the tuba. The frequency produced, however, is dependent on the length of the tube, regardless of the amount the tube is curved. In the trumpet, as well as the cornet, the holes or openings in the valves align with the curves of the tube in one route when all valves are up, and in different routes when the various valves are depressed. The effective length of the tube so constructed varies, depending on whether the first valve, the second, the third, or some combination is depressed. The different lengths produce different pitches (see Fig. 7.17).

In some other instruments, the vibrating part is a membrane stretched over a hollow cylinder, such as that of a drum or a tambourine. The sound is produced by striking the membrane and causing it to vibrate. The air column inside reinforces the loudness of the sound produced. Cymbals and bells, like drums, do not produce overtones that are harmonics of the fundamental note. The notes produced are dependent on the shape of the cymbal, bell, or membrane, as well as on the overall dimensions and type of material from which the instrument is constructed. The human voice also belongs in this category. Its sound is produced by means of vocal folds (cords) that vibrate when air is blown through them. When caused to vibrate, this pair of membranes stretched across the larynx (or Adam's apple) tightens, producing a high pitch. The sound produced when the vocal cords are relaxed is of lower pitch. The range between these two pitches varies with the individual, since the length, diameter, and tension of the vocal cords are different for everyone. On the average, the pitch of the male speaking voice is 150 hertz and that of the female 230 hertz. As children grow, their vocal folds increase in all dimensions; thus, the range of the voice increases. But where is the resonating air column that causes the sound produced to be much louder and of higher quality than that produced by the vocal cords alone? Of course, one makes use of the passages in the head, nose, throat, and mouth. The size and shape of each passage also determine the quality of the individual's voice. The various sounds produced when

Music from Foreign Countries

Through the ages, various cultures came to like the sound of certain tones that were played together or in sequence. The early musicians chose the notes according to their innate abilities; then performers learned to select and arrange sounds into various musical systems.

Fundamental to the music of all ages, tastes, and countries is the *octave,* the interval between two notes that have a frequency ratio of two to one. During the Middle Ages, musicians used an octave scale that consisted of seven tones derived from progressions of fifths and fourths and developed by the Pythagorean school of ancient Greece. The Pythagorean musical scale was modified in the sixteenth century and became known as the natural diatonic scale. Further modification led to the equally tempered scale generally in use today.

The musical scale with which we are familiar is by no means the only possible arrangement of musical intervals. Other scales have been developed, each most pleasing to the originator. The Chinese divide their musical scale into twelve intervals as we do, but mostly rely on the five tones that correspond to the black notes on a piano keyboard. Thus, this scale is known as a pentatonic scale. Hindu music divides the octave into twenty-two intervals, but the melodic structure is ordinarily based on just seven notes. The intervals between notes are the same as those intervals in the equally tempered scale, but Hindu musicians also use very small intervals referred to as grace notes. Snake charmers employ this unusual scale as they play their musical pipes to entertain their deadly pets ■

Figure 7.17 The popular cornet. Consider the length of tubing necessary to construct such an instrument.

an individual is engaged in ordinary conversation are generally all at the same frequency. The noticeable difference in the sounds results from the individual's changing the position of the tongue, lips, and cheeks. These changes create passages of various size through which the voice can resonate.

Musical Scales

Music scholars have developed several scales in which each note produced varies in some regular way from a standard. The most basic of the scales are the diatonic and the chromatic.

Two sounds are said to differ by one octave when their frequencies are in a ratio of 2:1. The piano has a range that includes practically all frequencies in the realm of musical sounds—more than seven octaves (see Fig. 7.18). The lowest C found on the piano has a frequency of 32.70 hertz. Seven octaves above this C, one would find a C with a frequency of 2^7 (or 128) times 32.70 hertz, or 418.6 hertz. (The lowest note produced by a piano is A, 27.50 hertz.) The ranges of other instruments, voices, and selected sound systems are also shown in Figure 7.18. It is common knowledge that there are eighty-eight keys on a piano, but how do these eighty-eight notes differ? How are they related? The answer comes from a brief study of the common musical scales.

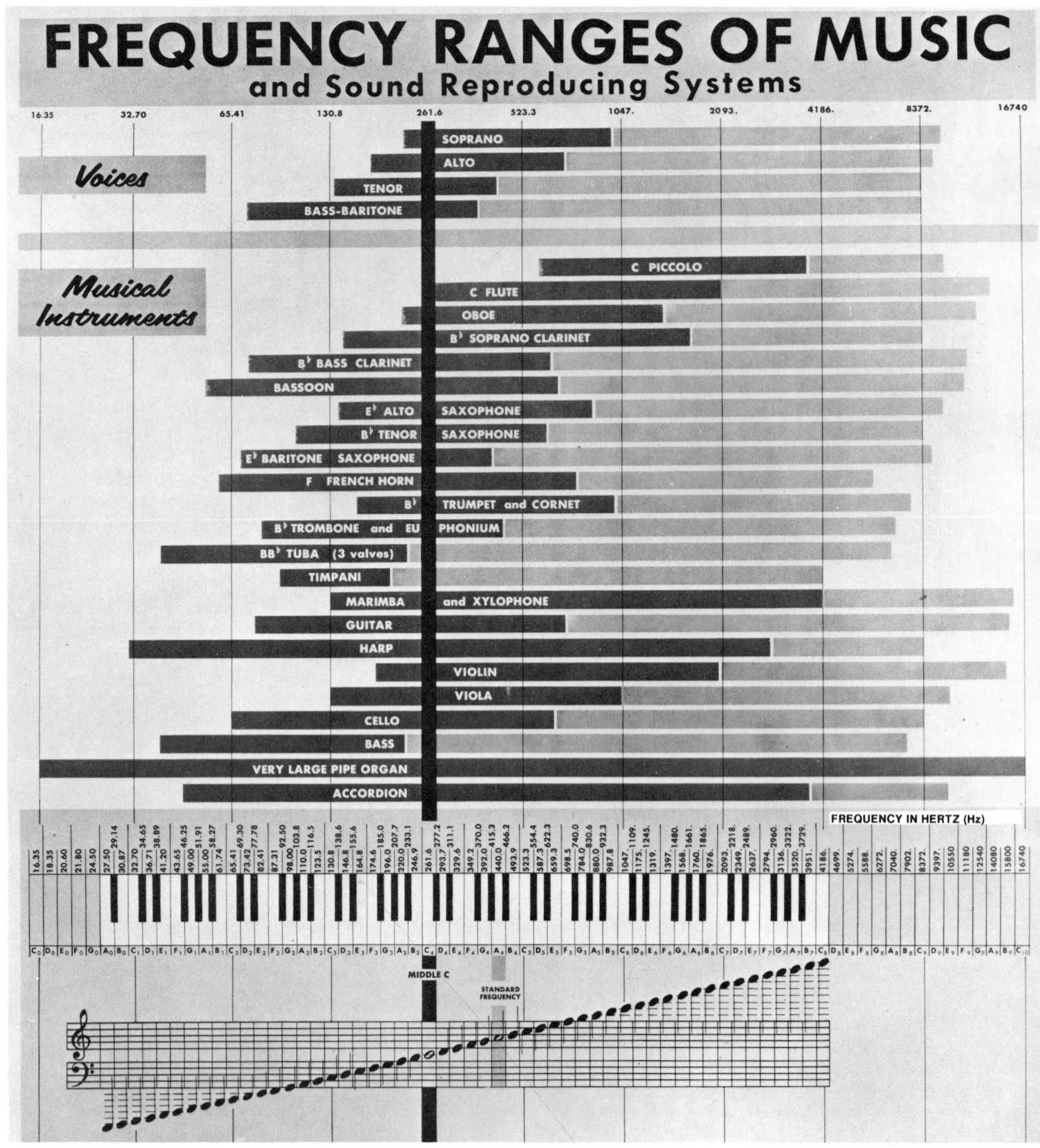

Figure 7.18 A comparison of the frequency ranges of voices and of various musical instruments. Note that the horizontal axis is not linear.

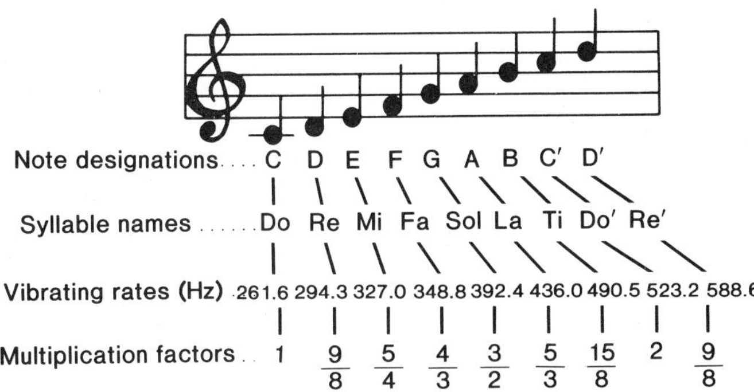

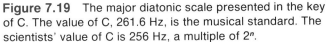

Figure 7.19 The major diatonic scale presented in the key of C. The value of C, 261.6 Hz, is the musical standard. The scientists' value of C is 256 Hz, a multiple of 2^n.

The *diatonic* scale is composed of the eight standard tones taught to us as students in elementary school. The notes and their respective syllable names are illustrated in Figure 7.19.

The major scale has its characteristic sound because the relationship between the frequencies of the right notes is based on mathematical ratios. These ratios, or multiplication factors, are also shown in Figure 7.19, as are the vibration rates of each frequency calculated from those ratios using 261.6 hertz (C) as the musical standard.

From the figure, one will note that D is approximately 294/262, or 9/8, times the frequency of C. The note E is 5/4 times the frequency of C, G is 3/2 times the frequency of C, and so on.

The notes with the very simplest ratios to C are particularly pleasing when sounded with C. Because it encompasses five diatonic notes—C, D, E, F, and G—the interval between C and G is called a fifth, as are any two notes in the ratio 3/2. The interval between C and E, with a ratio of 5/4, is known as a major third, and that between E and G, with a ratio of 6/5 (392/327), as a minor third. The chord C-E-G contains a fifth, a major third, and a minor third. The relative frequencies of the three notes are 4, 5, and 6. Any three notes, the vibrating rates of which are in this ratio, produce what is called a major chord when sounded together. Other major chords are G-B-D′ and F-A-C′.

If we try to build another scale based on the note A instead of C, using the same multiplication factors (compare Figs. 7.19 and 7.20), we encounter some significant problems. The notes E, B, and, of course, A and A′, have the same frequencies that they had in the scale of C, but the other notes will be found to differ by varying amounts. If a scale were similarly built using each of the scale notes as a standard, we would end up with seventy different frequencies for our eight basic notes. This exactness would not be very practical for musicians—or for piano tuners.

To solve this problem, musicians several centuries ago abandoned scales based purely on natural mathematical ratios and substituted a system called equal temperament, which smoothes out the differences from scale to scale by simply dividing each octave into twelve equal intervals. The notes added to the eight diatonic notes are called sharps and flats. The twelve notes played in succession make up the *chromatic* scale. The frequency of each tone differs from the previous one by the product factor of $\sqrt[12]{2}$, or approximately 1.06. This is the system used to tune the seven white keys and five black keys of every octave on the piano. These twelve tones are sufficient for practical purposes, because many of the differences between the seventy mathematically derived frequencies are so small as to be barely detectable by most individuals. So, while theoretically C-sharp and D-flat do not have identical frequencies, they are in fact assumed to be the same on today's pianos, and our ears are quite accustomed to this compromise. Players of stringed instruments, however, since they have control over the pitch of each note, do often play C-sharp and D-flat differently.

Example 7.8

A musician wishes to use the natural mathematical ratios to tune an instrument in the key of E, the frequency of which is 327. Determine the frequencies of the diatonic scale this instrument will then produce.

Solution

The key of E is developed by multiplying the specified frequency by the factors listed in Figure 7.19, so

$$
\begin{aligned}
E &= 327 \text{ Hz} \\
F &= 327 \text{ Hz} \times 9/8 &&= 368 \text{ Hz} \\
G &= 327 \text{ Hz} \times 5/4 &&= 409 \text{ Hz} \\
A &= 327 \text{ Hz} \times 4/3 &&= 436 \text{ Hz} \\
B &= 327 \text{ Hz} \times 3/2 &&= 491 \text{ Hz} \\
C' &= 327 \text{ Hz} \times 5/3 &&= 545 \text{ Hz} \\
D' &= 327 \text{ Hz} \times 15/8 &&= 613 \text{ Hz} \\
E' &= 327 \text{ Hz} \times 2 &&= 654 \text{ Hz.}
\end{aligned}
$$

SCALE FREQUENCIES

Other Characteristics of Sound

The noticeable change in the pitch of the sound made by racing cars as they rapidly approach, then speed by the TV camera and microphone en route to completing laps around the track is called the **Doppler effect.** The phenomenon, initially explained by the Austrian physicist, J. C. Doppler (1803–1853), is caused by a variation in the number of waves that reach the microphone in successive seconds. The same effect is witnessed by the fans who attend the event and hear the racers as they roar by the stands. In either case, as the racing car approaches, the pitch of the sound its engine makes is higher until the vehicle is at its nearest point to the observer. As the vehicle travels past, the pitch (frequency)

Note	A	B	C	D	E	F	G	A′
Multiplication factor	1	$\frac{9}{8}$	$\frac{5}{4}$	$\frac{4}{3}$	$\frac{3}{2}$	$\frac{5}{3}$	$\frac{15}{8}$	2
Key of A (Hz)	436	491	545	581	654	727	818	872

Figure 7.20 Note the variation in frequencies of some notes when A (436 Hz), rather than C, is used as the standard. (See Fig. 7.19.)

becomes lower (see Fig. 7.21). The higher the speed of the source of the sound, the more obvious the change in pitch. That is, the speed of the source of the sound must be added to or subtracted from the speed of sound, depending on whether the object's motion is toward or away from the observer. If the listener, as well as the source, is in motion, the overall effect is more evident if their directions are opposite and is less detectable if both are moving in the same direction.

Doppler also suspected that a similar frequency shift occurs in light, a supposition proved valid by later experimentation. As will be discussed in Chapter 18, the frequency shifts created by the Doppler effect of light (known also as *Doppler shifts*) reaching us from distant celestial objects can be used to gather vital information about the universe.

Another concept is responsible for causing the listener to lack proper appreciation of some musical performances. Much of a musical sound can be lost by poor acoustics. An orchestra that performs in a concert hall is sometimes hindered by *the reflection of the sound* it produces. This reflection is known as an **echo,** and can occur any time a surface in the hall is in excess of about 17 meters away from the source of sound. At approximately 335 meters per second, sound would return to its source in about 0.1 second, the length of time that must pass before the average listener can distinguish two separate sounds. One can stop the reflection of sound by installing curtains on the opposite walls and by placing acoustical tile on the ceiling. This tile has small holes that individually absorb some of the sound. A certain degree of reflection, however, is advantageous to an orchestra. If the total amount of sound is not absorbed, it will reflect from all surfaces, producing the effect of having the orchestra surrounding the audience; thus the sound will continue for a short time, a property known as *reverberation.* The best reverberation time for listening to an orchestra perform is about 1.5 seconds. When both echo and reverberation are adjusted for proper listening, the room is said to have good acoustics, another subjective measure.

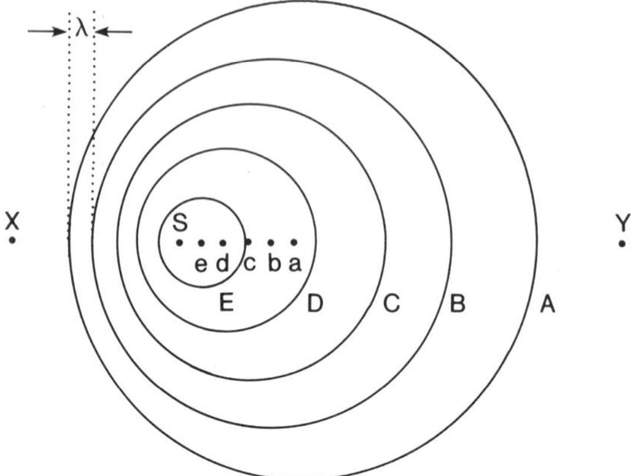

Figure 7.21 The Doppler effect. A rendition of the result of motion on a constant frequency emitted by the source, S, as it moves toward point X. Each crest of the wave, as identified in capital letters, corresponds to the respective point in lowercase where the wave originated. Wavelength decreases in the direction the source moves, and the frequency of the wave increases proportionally. Recall $v = f\lambda$. An observer near point X would note an increase in pitch; one near point Y, a decrease.

Summary

Sound travels through various mediums at different speeds. Generally speaking, the speed of sound increases with the density of the medium. The temperature of the medium also affects the speed of sound. For each Celsius degree the air temperature increases, the speed of sound in air increases 0.610 meter per second (2.00 ft/s). As the air temperature decreases, the speed of sound decreases accordingly.

There are two general types of waves—longitudinal and transverse. In longitudinal waves, the vibrating particles move back and forth parallel to the direction in which the wave is traveling. In transverse waves, the vibrations are at right angles to the direction in which the wave is traveling. Sound is an example of a longitudinal wave.

Frequency, wavelength, and amplitude are the fundamental characteristics of all waves. The speed of a wave is calculated from the product of its frequency and wavelength. Amplitude is an indication of loudness.

The general attributes of sound help classify a sound as noise or music. Since these inherent characteristics are subjective in nature, the given attributes of sound are pitch, loudness, and quality.

The rate at which a string vibrates is determined by its length, diameter, composition, density, and tension. The vibration rate of a string is inversely related to length, diameter, and density of the string. The vibration rate of a string increases with the square root of the tension applied to the string.

Resonance is the building of a large vibration by small impulses, the frequency of which is equal to one of the natural frequencies of the resonating body. If the impulses are slightly out of synchronization with one of these natural frequencies, a decrease in amplitude of the resonating body results.

The three most popular musical scales are the diatonic, the chromatic, and the tempered scales. Most common among the three scales is the diatonic scale. The tempered scale, however, is the accepted scale used in playing most instruments.

The Doppler effect, as it applies to sound, is the effect on pitch caused by the motion of the source relative to that of the listener. As the source moves toward the listener, the pitch is higher. As the source moves away, the pitch is lower. The Doppler effect can occur with the listener, the source, or both in motion. A sound reflected from a surface is called an echo. A series of echos of the same sound is known as reverberation.

Equation Summary

Speed of sound in air:

$$v = 331\ \text{m/s} + \left[(T_{°C})\frac{0.610\ \text{m/s}}{\text{C°}}\right].$$

The speed of sound v is directly related to the temperature of the air ($T_{°C}$). At temperatures below the freezing point of water, the second part of the equation is subtracted from 331 meters per second, the speed of sound at 0°C.

The wave equation:

$$v = f\lambda.$$

The speed v of a disturbance through a medium, such as sound through air, is directly related to the frequency f of the disturbance and to its wavelength λ.

Sound intensity:

$$\frac{I_1}{I_2} = \frac{d_2^2}{d_1^2}.$$

The inverse square applies to the intensity of sound where I_1 is determined at distance d_1 from the source. The intensity I_2 then can be calculated at distance d_2 from the source. Thus, the intensity of sound is inversely related to the square of its distance from the source.

String length:

$$\frac{l_1}{l_2} = \frac{f_2}{f_1}.$$

When the frequency f_1 of a given string of length l_1 is known, the frequency f_2 at any given length l_2 of the string can be calculated, all other factors held constant. As the equation indicates, the frequency f of a sound made by a vibrating string is inversely related to the length l of the string.

String tension:

$$\frac{\sqrt{T_1}}{\sqrt{T_2}} = \frac{f_1}{f_2}.$$

With all other factors held constant, the frequency f of a vibrating string is found to be directly related to the square root of its tension T, generally expressed in newtons (N).

Questions and Problems

The Speed of Sound

1. Marchers at a significant distance from a performing band are generally out of step with the marchers nearer the band. Why?
2. (a) What factors determine the speed with which a sound travels through a given medium? (b) Why does sound travel faster in warm air than in cold air?
3. If the speed of sound decreases 0.610 m/s per C° as the air gets colder, at what Celsius temperature would the speed of sound theoretically equal zero? According to the temperature established as absolute zero, is it possible for the speed of sound to reach zero?

4. A whistle sounded from a train one kilometer from a railroad crossing reached a pedestrian both through the iron track and through the wintry air. Approximately how long did the sound take to reach the person through each medium?

Kinds of Waves

5. How do transverse waves differ from longitudinal waves?
6. How does a longitudinal wave, such as sound, travel through a medium?

Characteristics of Sound

7. How many octaves of the note C can the normal human ear hear?
8. How does a change in frequency of a sound affect its wavelength, assuming the speed of sound remains constant?

Subjective Measures of Sound

9. (a) How does the pitch of a sound relate to its frequency? (b) How does the loudness of a sound relate to its intensity?
10. Why do notes from plucked strings on a guitar sound different from the same notes played on a banjo?
11. How many times greater is the intensity level of a sound of 80 dB than one of 40 dB? How many times greater is the intensity of a sound of 80 dB than one of 10 dB? (See Table 7.2.)
12. The intensity level of the sound produced by a moving truck with trailer 30.0 m away from a listener was measured to be 30 dB (1.00×10^{-9} W/m^2). Determine the intensity of the sound at 10.0 m from the vehicle. (See Table 7.2.)
13. How does the distance from the source affect the level of intensity of a sound?

The Laws of Vibrating Strings

14. (a) In tuning a violin, how can the pitch of a string be increased? (b) How can a violin or other stringed instrument produce so many notes with so few strings?
15. If the G string (392 Hz) on a musical instrument were 80.0 cm long, what must be its effective length to produce the note B, 491 Hz?

Resonance

16. A tuning fork has a frequency of 440 Hz. If another tuning fork is sounded at the same time, five beats per second are heard. What are the two possible frequencies of the second tuning fork?
17. What is the frequency of the tone produced by (a) an open organ pipe 2.00 m long and (b) the same pipe when the top of the pipe is closed? (Assume the speed of sound to be 340 m/s.)

Musical Scales

18. Calculate the wavelengths of the lowest note and the highest note on a piano if the air temperature is 20.0°C. (See Fig. 7.18.)
19. Calculate the musical scale that uses the natural mathematical ratios to tune an instrument in the key of G, 384 Hz. (See Fig. 7.20.)

Other Characteristics of Sound

20. (a) Why does the pitch of a given factory whistle vary on a windy day? (b) Why does the same whistle sound louder on days with high humidity than on days with low humidity?
21. If an echo is detected 6.00 s after the original sound is produced, how far from the source is the reflecting surface, assuming the speed of sound to be 340 m/s?

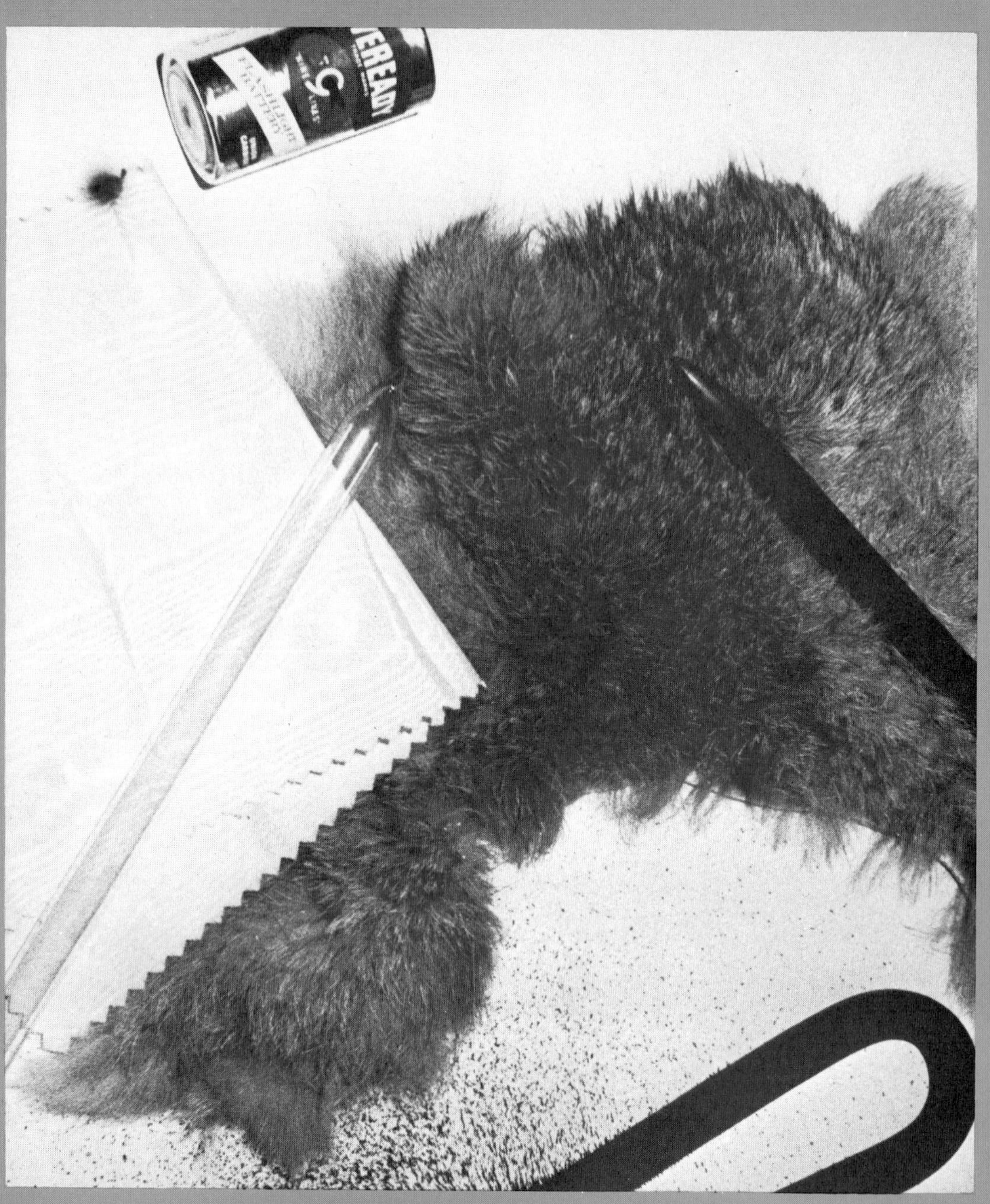

8 Basic Electricity and Magnetism

Chapter outline

Key terms/concepts

The fundamental concepts involved in electricity and magnetism have been under development for many centuries.

Some 2500 years ago the Greeks realized that a piece of highly polished material rubbed with cloth acquired peculiar properties. The substance, known as *amber,* was a yellowish or brownish fossil resin from which the Greeks made beads and necklaces. The amber would attract lightweight objects, such as animal hairs and small seeds, which would then stick to its surface. This seemingly insignificant discovery eventually led to our understanding of electricity, a term derived from the Greek *elektron,* which means "amber."

Little attention was paid to this mysterious attraction of one object for another until about 1600, when William Gilbert (1540–1603), Queen Elizabeth's physician, observed that a molten mixture of sulfur and sealing wax hardened with a glossy surface that, when rubbed with fur, attracted things to it. Gilbert, therefore, surmised that rubbed materials became filled with a light, airy fluid, which he named *electricity.* He could not explain the nature of the electrical fluid, but he noted that it appeared to flow from the cloth into the object being rubbed.

Further investigations by other experimenters revealed that some substances, such as glass and well-polished wood, retained the electric fluid placed on them, while others, such as copper and iron, were found to lose the electrical fluid instantly. The results of the combined investigation led to the classification of substances into two categories: those substances that stored the fluid, called *insulators,* and those substances that permitted electricity to flow through them, called *conductors.*

Experiments made by the French chemist Charles DuFay (1698–1739) showed that electricity must come in two forms instead of one. The American investigator, Benjamin Franklin (1706–1790), is generally considered responsible for naming the two types of electricity *positive* and *negative.*

Electric charge is *a property of some particles, including electrons and protons, that causes them to exert forces upon one another.* Thus, the two types of electricity have come to be known as **positive charge** and **negative charge.** When a glass rod is rubbed with silk, the glass rod is said to acquire a positive charge, while the silk acquires a negative charge.

Electricity at Rest

The early investigations of electricity mostly dealt with *charges that were at rest,* called **static electricity.** Later investigations of charges in motion, called *electric current,* showed many other new effects that will be studied later in this chapter.

A summary of the experiments done in the study of static electricity and how they are interpreted based on present-day understanding indicates:

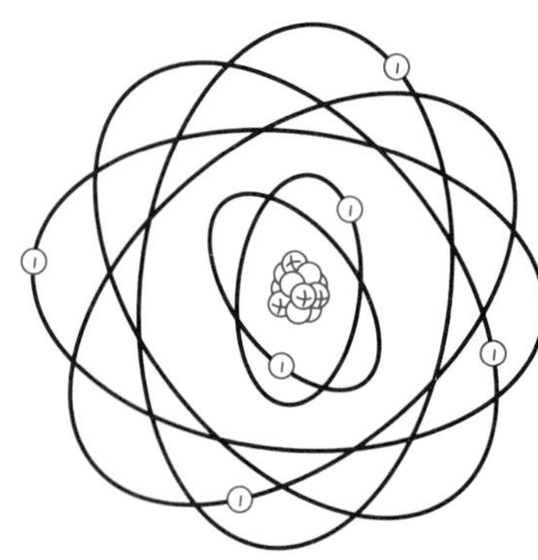

Figure 8.1 The planetary model of the atom. Negatively charged electrons "orbit" a positively charged nucleus made of protons (positive charge) and neutrons (no charge). Protons and neutrons have masses about 2000 times greater than electrons. Normally, atoms have the same number of electrons and protons and are neutral. The transfer of outer electrons, which are not as tightly bound as inner electrons, causes objects to become charged.

1. When objects are electrified by touching or rubbing the objects together, electric charge is transferred from one object to another. All matter is made of atoms. Every atom is composed of a *nucleus* that is positively charged, around which are distributed a number of *negatively charged* **electrons** (Fig. 8.1). The nucleus contains two types of particles: **protons,** which are *positively charged,* and **neutrons,** which have *no charge.* Protons and electrons have charges of exactly the same magnitude and exist in ordinary atoms in exactly the same numbers; therefore, normal materials, even though they contain enormous numbers of charges, are *neutral;* that is, they have no net charge.

   When an object is charged by rubbing or touching it with some other material, electrons are transferred from one object to the other (some objects having a greater affinity for electrons than others). The object to which the electrons are transferred becomes negatively charged, and the object from which the electrons have been removed now has an excess of positive charges and is said to be positively charged. No charge is created during the charging process. The generalization that *charge is neither created nor destroyed during ordinary processes is called the* **law of conservation of charge.**
2. When two objects are charged by rubbing them together, it is found that they tend to attract each other. This effect can be seen easily when dry hair is combed with a plastic comb on a crisp winter day. The hair is attracted to the comb, making combing difficult. On the other hand, if two similar objects, say two thin pieces of plastic, are charged by

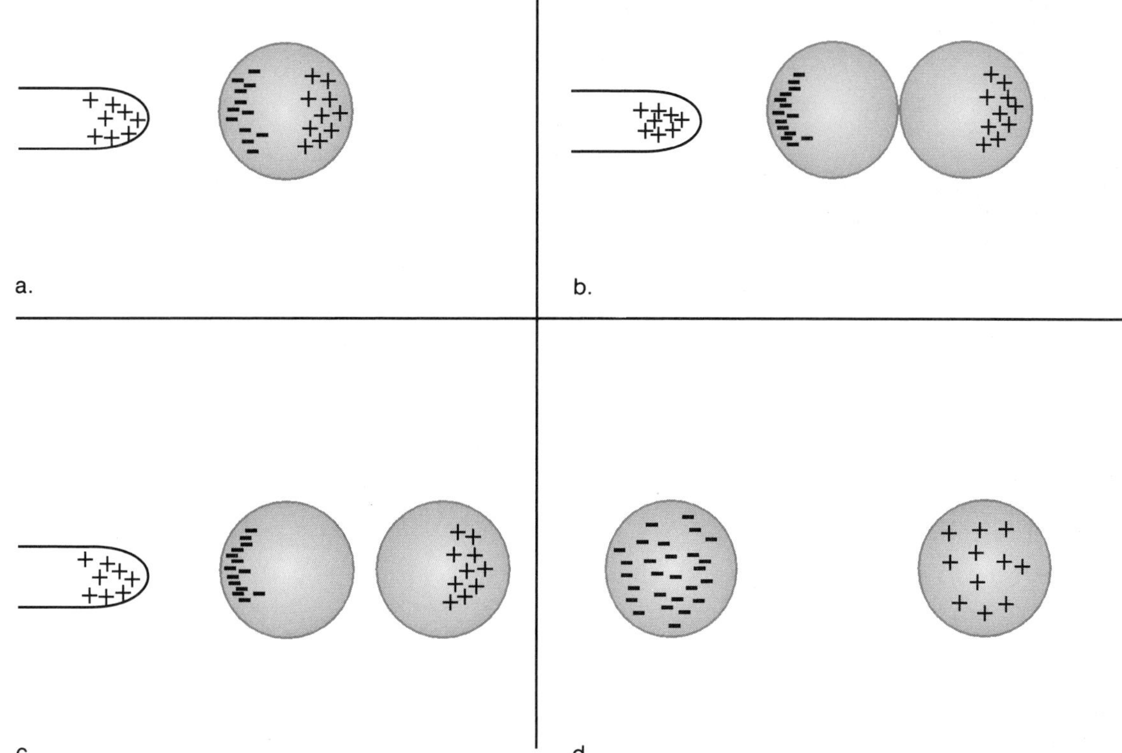

Figure 8.2 Charging by induction. When a positively charged object is brought near a conductor, freely moving electrons are attracted or repelled. A second conductor allows for the "trapping" of charges.

rubbing them with a piece of wool, the two objects will be found to repel each other. In the first example, the hair and comb have acquired unlike charges, whereas in the second example, both pieces of plastic have like charges. This experiment shows that *unlike charges attract each other,* and that *like charges repel each other.*

3. All materials are made of atoms, usually either of individual atoms of the same type (as in copper or iron) or of atoms of different types bonded together to form neutral units called molecules (as in water and sugar). In solid materials, the atoms or molecules are bound in relatively fixed positions with respect to each other. In some substances, because of the way the atoms and molecules are bound together, the electrons may move freely from one atom or molecule to another, in which case *charge may flow easily through the material.* Such a material is called a **conductor.** Most metals are conductors. If the electrons are bound tightly, the material is a nonconductor, or an **insulator,** and *charges will not flow freely in the material.* Most nonmetallic materials are insulators. Insulators, such as rubber or plastic, are used to prevent transfer of charges from one object to another.

Liquids also may be conductors or insulators. Most pure liquids are insulators, but many impure liquids contain salts, making them excellent conductors. An example is water. Pure water is a good insulator, but ordinary water is not usually pure, and only a slight amount of salt, such as from perspiration, is enough to make it an excellent conductor. Electrical properties of solutions will be discussed in more detail later.

4. Conductors may be easily charged by a process known as **induction.** Suppose a positively charged object is brought near a conductor, as shown in Figure 8.2a. The electrons in the conductor are free to move and will be attracted toward the object, making the part of the conductor near the object negatively charged and the part farther from the object positively charged, as indicated by the (−) and (+) signs. If the conductor is touched with another conductor in the region of the excess positive charge, negative charge from the second conductor will be attracted to the positive charge and will "neutralize" it (Fig. 8.2b). If the second conductor now is removed before the positively charged object is taken away, the negative charge is "trapped" on the original conductor (and an equal positive charge on the second conductor, if it is insulated) (see Figs. 8.2c and 8.2d).

a.

b.

c.

Figure 8.3 (*a*) This spectacular display of lightning is typical of those that accompany violent storms. (*b*) A cloud often dissipates its charge to the ground or nearby objects. (*c*) A few clouds arrange charges oppositely and discharge by interaction with close-by clouds.

A familiar example of charging by both friction and induction in nature can be found by examining charges on clouds and on objects beneath them during a thunderstorm. Friction from the wind and moving raindrops charges the clouds, while nearby objects, such as other clouds or objects on the ground, may be charged by induction (Fig. 8.3). When the charges become large, the electrical forces on the atoms in the air grow to the point that some of the electrons are literally ripped off the atoms. These electrons and the positive ions (**ions** are *atoms or molecules with a deficiency or excess of electrons*) speed through the air colliding with other atoms, knocking off more electrons and causing a tremendous discharge—a lightning bolt.

Objects on the ground can be protected somewhat from lightning bolts by discharging the induced charges with lightning rods. Figure 8.4 shows how charges tend to collect on the sharp point of a conductor. If the conductor is surrounded by air, air molecules near the sharp point become ionized and allow the charges to "leak off." Thus, lightning rods have sharp points that tend to dissipate the (usually positive) charges induced during thunderstorms. Lightning rods also allow for a safe path for the discharge should they be unable to dissipate the accumulated charge rapidly enough.

Coulomb's Law

In 1789, Charles de Coulomb (1736–1806), a French physicist, released the results of his research, a verification of other earlier reports about electrical charges. His research also showed how the force of attraction or repulsion between two charges could be determined.

It will be recalled from Chapter 4 that Newton's law of universal gravitation allows one to determine the gravitational attraction between objects. All objects possess mass, and the force with which two objects attract each other is proportional to the product of their gravitational masses and inversely proportional to the square of the distance separating their centers (if they have spherical symmetry). No effort was made to try to explain why masses attract each other. In fact, the nature of mass itself as used in the law of universal gravitation was not discussed. Such topics delve deeply into theoretical physics and are beyond the scope of this text. Mass is treated as a property of the objects from which the force of attraction can be determined.

Coulomb found that exactly the same form of force law holds for static charges. **Coulomb's law** states that *the force between two charges possessing spherical symmetry (or concentrated near two points) is directly proportional to the product of the charges and inversely proportional to the square of the distance between the charges* (Fig. 8.5). In equation form it reads

$$F = k\frac{q_1 q_2}{r^2}, \qquad \text{COULOMB'S LAW}$$

Benjamin Franklin

The contributions attributed to Benjamin Franklin (1706–1790) reveal a man of many talents. He was a printer, a diplomat, a scientist, an inventor, a philosopher, an educator, and a public servant. Several of his many accomplishments were, individually, enough to make him famous. He is credited with organizing the first lending library in America as well as with the inventions of many useful items, including the lightning rod. He attracted attention throughout the scientific world with his experiments in electricity alone.

Although Franklin had but two years of formal training, he became fond of books at the age of 10 and spent his leisure hours reading. While an apprentice in his half-brother's printing shop, the young Franklin studied arithmetic, navigation, and grammar. After becoming an expert printer, he moved to Pennsylvania in 1730 where he opened a successful printing shop. Franklin became Philadelphia's outstanding citizen and helped organize the street-paving, lighting, and fire-fighting units in that city. He closed the printing shop in 1748 and purchased a farm near Burlington, New Jersey, where he devoted his waking hours to science and public service. Franklin was an active inventor for most of his adult life. Among his most famous contributions was the Franklin stove. He declined to obtain a patent for it, reportedly so that the stoves could be made more cheaply and be available to all. In fact, he patented none of his many inventions.

Franklin realized that lightning was a discharge of electricity from the clouds. He tested this theory in 1752 with the assistance of his son. The two investigators went to a shed in a meadow during a thunderstorm, released a kite until it was high in the air, and permitted a charge of electricity to flow down the wet string that supported the kite. Franklin noticed that the unraveled threads of the kite string repelled each other. When he brought his knuckle near a key he had tied to the end of the string, he was "rewarded" by an electrical shock, proving that lightning was a spectacular display of electricity. (Flying a kite in an electrical storm is extremely dangerous.) His experiments confirmed the accuracy of earlier findings of French scientists. Franklin further pursued his investigation of lightning and invented the lightning rod to protect buildings from damage during electrical storms. He conducted other experiments in electricity and wrote a book on the subject. The principles he set forth in the publication partially formed the basis for modern electrical theory■

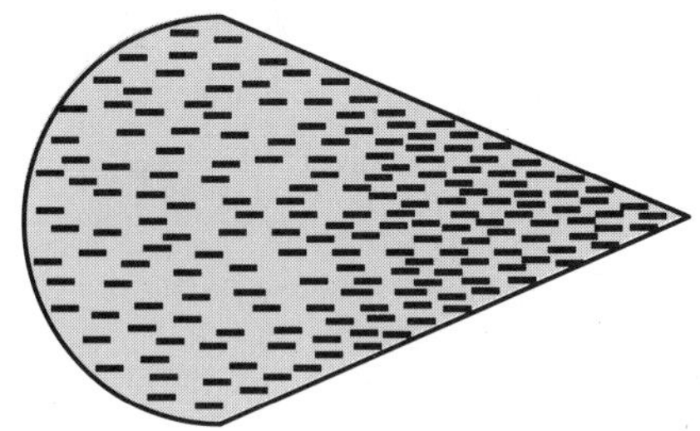

Figure 8.4 An electrical charge accumulates on curved surfaces in accordance with the extent of curvature.

where F is the force (often called the **electrostatic force**), q_1 and q_2 are the magnitudes of the two charges, r is the distance between the charges, and k is a constant of proportionality that must be determined from experiment. In SI units (see Appendix 1), the force is in newtons and the distance is in meters. The charge is measured in a unit called the **coulomb** (C). (The coulomb is not a base unit as is the kilogram. It is more practical to have a standard of electric current and to base the unit of electric charge on that standard, as will be seen later in this chapter's section on electric current.) The constant of proportionality, k, in Coulomb's law was determined experimentally to be $9.00 \times 10^9\ N \cdot m^2/C^2$. This means that if two 1.00-coulomb charges were placed 1.00 meter apart, the force between the charges would be 9.00×10^9 N, or 9.00 billion newtons!

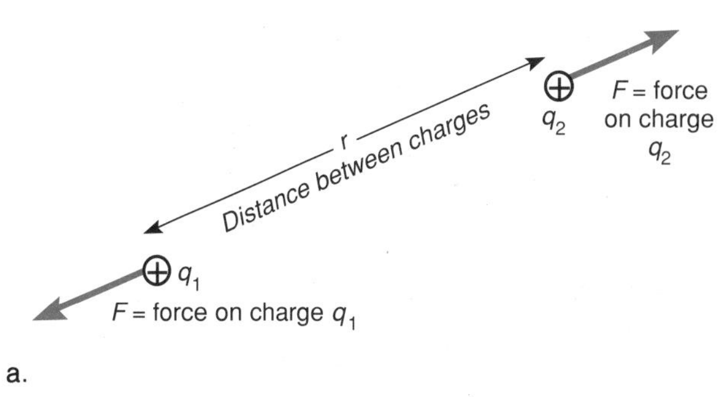

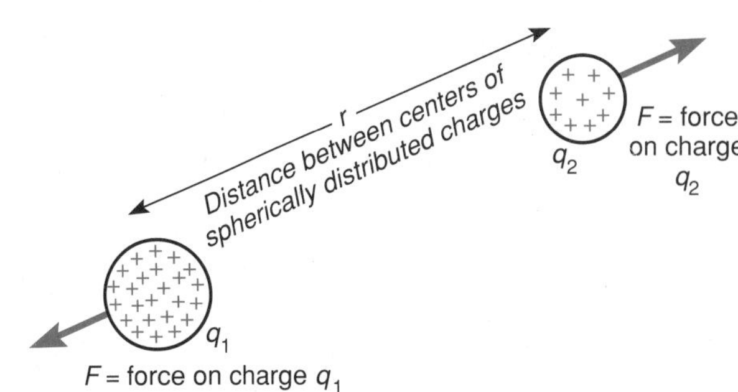

Figure 8.5 Geometry of charges in Coulomb's law. Charges may be "point charges," as in (*a*), or spherically distributed, as in (*b*), where the center-to-center distance is used. Charge distributions on spheres may be distorted by induction, as in Figure 8.3, thus requiring more complicated calculations.

If the two charges are both of the same sign (both positive or both negative), the charges repel each other and the force is termed *repulsive.* If the charges are of opposite sign (one positive and the other negative), the charges attract each other and the force is called *attractive.* The magnitude of the force on each charge is the same and the direction is opposite, in accordance with Newton's third law of motion.

No attempt will be made here to try to explain why charges attract or repel each other, just as no attempt was made to explain why masses attract each other. The attraction and repulsion will be treated as properties of charges.

Static charges as large as a coulomb are rarely achieved in practice, since the forces involved are so great. However, charges on the order of microcoulombs (μC, equal to 10^{-6} C) are often realized, as in Example 8.1.

Example 8.1

(a) What is the force between a charge of 5.00 microcoulombs and a charge of 3.00 microcoulombs separated by 0.200 meter? (b) What is the force between the same charges if the separation is increased to 0.400 meter?

Solution

(a) This is simply a matter of putting the proper quantities into Coulomb's law. Let $q_1 = 5.00\ \mu C = 5.00 \times 10^{-6}$ C, $q_2 = 3.00\ \mu C = 3.00 \times 10^{-6}$ C, $r = 0.200$ m, and, of course, $k = 9.00 \times 10^9\ N \cdot m^2/C^2$. Then,

$$F = \frac{(9.00 \times 10^9\ N \cdot m^2/C^2)\ (5.00 \times 10^{-6}\ C)\ (3.00 \times 10^{-6}\ C)}{(0.200\ m)^2}$$
$$= 3.38\ N.$$

(b) All quantities are the same except now $r = 0.400$ m, so

$$F = \frac{(9.00 \times 10^9\ N \cdot m^2/C^2)\ (5.00 \times 10^{-6}\ C)\ (3.00 \times 10^{-6}\ C)}{(0.400\ m)^2}$$
$$= 0.844\ N.$$

Note that in part (b) of the example, although the distance has doubled from part (a), the force is one-fourth as much, since the distance appears in the denominator squared. (Compare this with Figure 4.16, where the dependence of the gravitational force with distance is seen to follow the same type of inverse-square law.)

Coulomb's law and the concept of induction make it easy to see how a charged object, such as a comb, can pick up small objects that are neutral. When the comb is brought near a scrap of paper, for example, charges are induced on the paper even though the paper is an insulator. This happens because the "orbits" of the electrons are distorted by the nearby charge, causing *a slight separation of the average location of the charges,* called **polarization.**

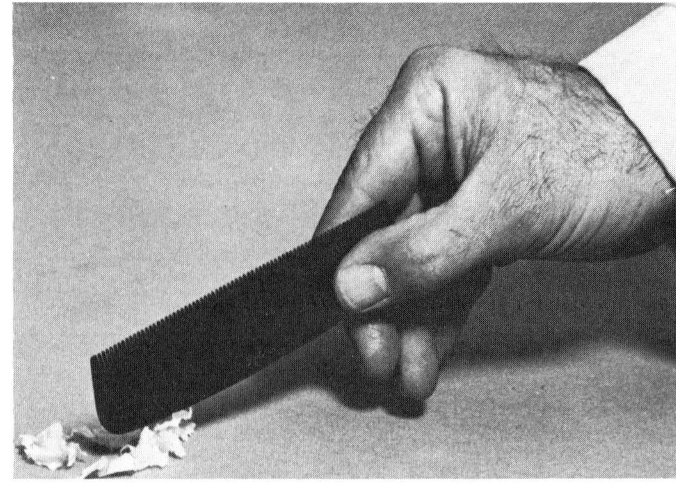

a.

Comb

Paper

b.

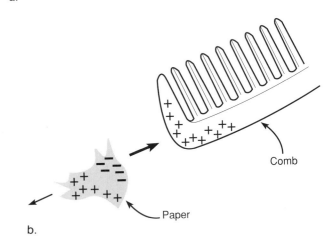

Figure 8.6 (*a*) Small bits of neutral paper are attracted to a charged comb. (*b*) Paper remains neutral, but differences in induced charges and distances cause the attractive force to be greater than the repulsive force.

The part of the paper nearest the comb acquires a charge opposite that of the comb, while the part of the paper farthest from the comb acquires the same type of charge as that of the comb. Since the force between charges decreases as the distance squared, there is a greater attraction between the closer charges than there is repulsion between the farther charges, giving a net force of attraction. The idea is illustrated in Figure 8.6.

The Charge on the Electron

When Coulomb's law was formulated, there was no knowledge of the composition of the "electrical fluid." Experimental work done by the American physicist Robert Andrews Millikan (1868–1953) at the University of Chicago from 1909 to 1913 showed that electric charge only comes in discrete units, and the charge is said to be "quantized." That is, there is a "natural" unit of electric charge.

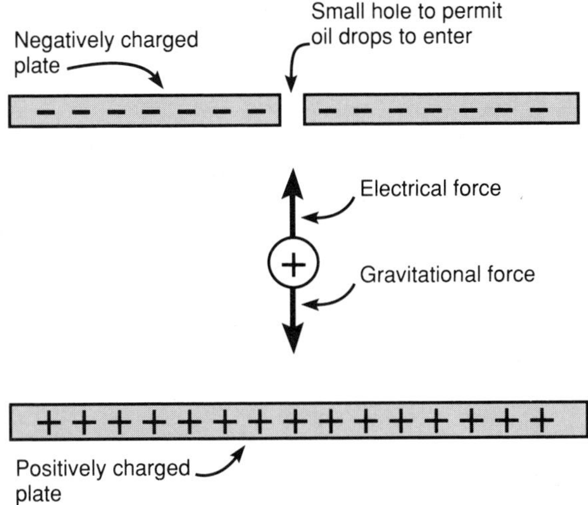

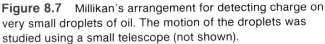

Figure 8.7 Millikan's arrangement for detecting charge on very small droplets of oil. The motion of the droplets was studied using a small telescope (not shown).

The charge on the electron and the proton is the smallest that has ever been observed.

Millikan's work was done by observing very small drops of oil that were sprayed between two flat charged conductors called "plates" (Fig. 8.7). When the drops were sprayed between the plates, some were charged by friction. If a drop had a positive charge, it would be repelled by the positive plate and attracted by the negative plate; negatively charged drops exhibited exactly the opposite behavior. The amount of charge and the weight of the drop determined the speed with which it moved. Some of the drops gained or lost charges while moving between the plates, but Millikan found that the gain or loss was always some multiple of the charge -1.60×10^{-19} coulombs. His conclusion was that the smallest unit of charge in nature, the charge on the electron, was equal to this value. Since the electron and proton have the same magnitude of charge, the charge on the proton must be $+1.60 \times 10^{-19}$ coulombs. The symbol e is used to represent the magnitude of the charge on the electron, that is, $e = 1.60 \times 10^{-19}$ coulombs.

The charge on electrons and protons is so small that it takes a tremendous number of them to give easily measured charge quantities. Electricity comes in discrete units, but the units are very small.

Example 8.2

The upper sphere on a Van de Graaff machine (Fig. 8.8) used in a classroom demonstration acquires a negative charge of 0.500×10^{-6} coulombs. How many excess electrons are on the sphere?

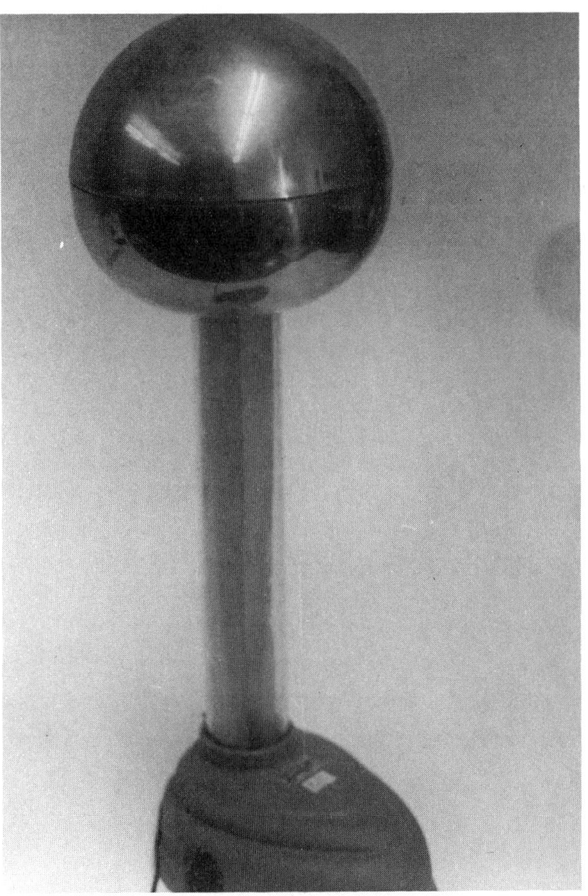

Figure 8.8 A Van de Graaff machine. A belt charged by friction is driven by a small motor and transports the electrical charge from the lower housing to the upper spherical conductor.

Solution

Since the charge on one electron is -1.60×10^{-19} C, the charge on N electrons (where N is the quantity to be determined) is N $(-1.60 \times 10^{-19}$ C). Then,

$$N(-1.60 \times 10^{-19}\text{ C}) = -0.500 \times 10^{-6}\text{ C}.$$

Solving for N gives

$$N = 3.13 \times 10^{12}.$$

Even though it takes an extremely large number of electrons to give 0.500 microcoulomb of charge, there is only one extra electron for about every 10 trillion atoms making up the sphere, so on a percentage basis, the excess number of electrons is very small.

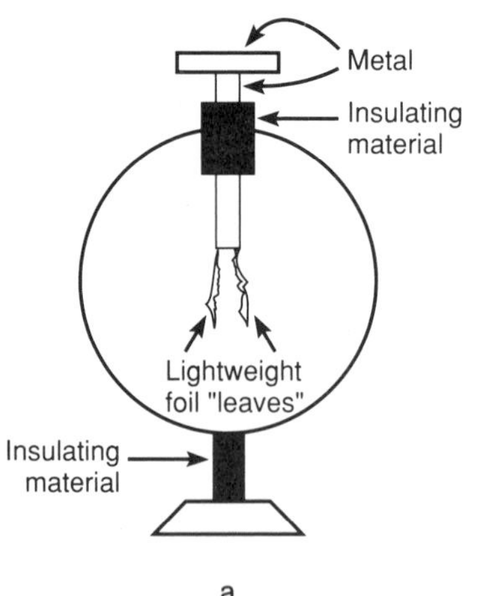

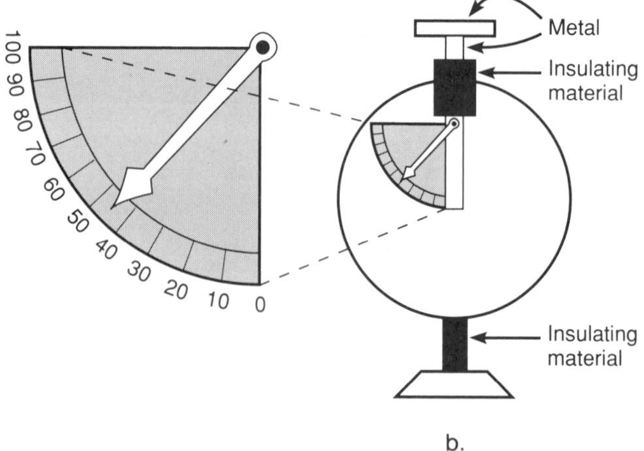

Figure 8.9 Electroscopes. The electroscope utilizes repulsive forces between charges (*a*) to force the lightweight foil leaves apart or (*b*) to push the lightweight pivoted arm outward. The arm in (*b*) may be calibrated to give an indication of the relative amount of charge.

Because like electric charges repel each other, a simple device called an electroscope can be used to determine whether an object is charged, and it may even be calibrated to show the amount of charge. Two electroscopes are diagrammed in Figure 8.9. Some of the charges from a charged object brought into contact with the piece of metal at the top will move along the conducting rod, causing the "leaves" of the electroscope in Figure 8.9a to repel each other. The lightweight loose arm of the electroscope in Figure 8.9b serves the same purpose as the leaves, but it can be calibrated in relative units. A "human electroscope" is shown in Figure 8.10.

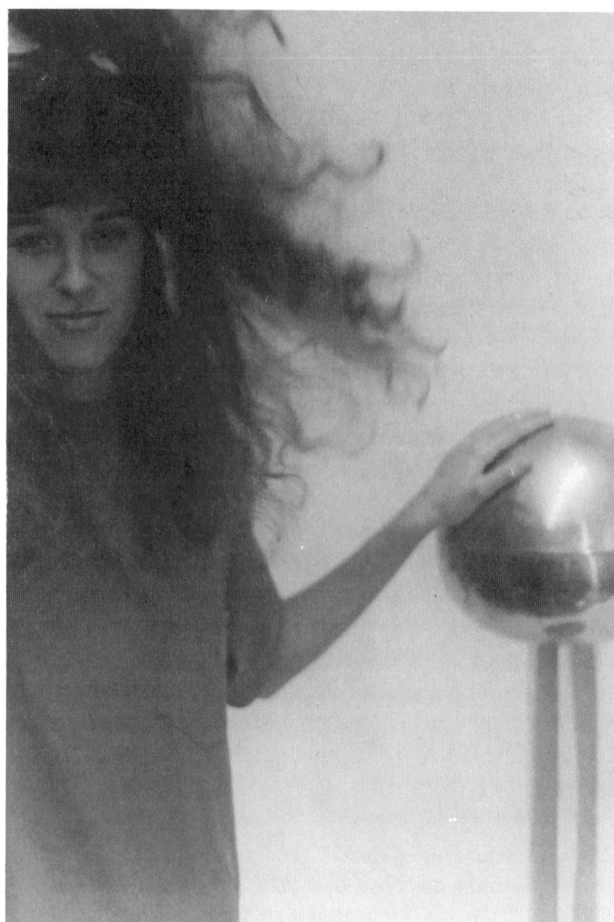

Figure 8.10 A human electroscope. The charges acquired by the person in touching the Van de Graaff machine cause the individual strands of hair to repel each other.

Electric Fields

Most situations that arise in the study of electricity do not have two charges located at points or situated in a spherically symmetrical manner. This makes it difficult to apply Coulomb's law when one needs to determine the force on a charge. Consider, for example, the charges on the plates in Millikan's oil-drop experiment shown in Figure 8.7. The oil drop is almost a point, meaning that the distance from the charge on the drop is essentially the distance to any point on the drop. However, the charges on the plates are more or less uniformly distributed, so that no single distance from the charge on the plates can be determined. Advanced mathematical techniques can be used to determine the force on the charged oil drop directly from Coulomb's law, but an indirect method, somewhat more abstract but widely applicable, is preferred.

A short digression will help explain the basic ideas of electric fields. Any place one travels on Earth, there is always a force pulling downward on objects. *Every position near Earth is under the influence of Earth,* or it could be said that a **gravitational field** surrounds Earth. Any object placed in this field is subject to the force of gravity. It is convenient to define the field in this case by taking

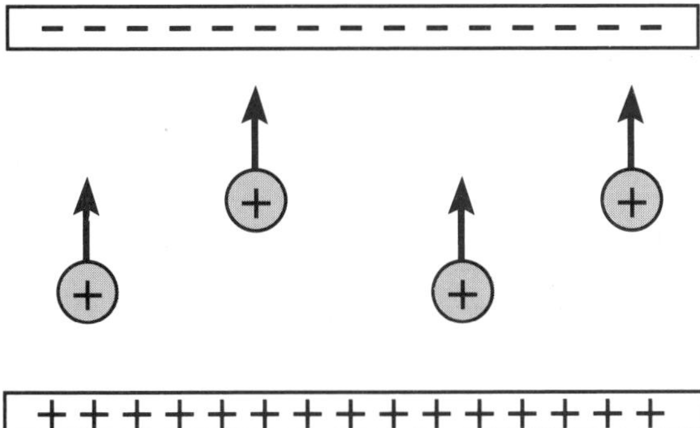

Figure 8.11 Charges between two charged plates. A charge placed between two conducting plates is repelled by like charges and attracted by opposite charges. The force on the charge between the plates is the same at any point between the plates if the charge is not too near the edges of the plates.

the force on an object and dividing by its mass. Since the force is mg in a downward direction, the field would be $mg/m = g$, in a downward direction. It will be recalled how useful it was to know g in Chapter 4. This usefulness stems in large part because g is independent of the mass of the object; that is, it will work for any mass. The units of g used in Chapter 4 were meters per second squared, but a glance at the previous idea of g as the gravitational field would show that g could also be thought of in terms of newtons per kilogram, the force per unit of mass.

In a similar way, *every point near electric charges is under the influence of those charges,* so that an **electric field** exists in the region near the charges. Any other charge brought into the region is influenced by the field. The electric field, E, produced at some point by a configuration of charges, is taken to be the force exerted on a positive charge placed at that point in the field divided by its charge; that is, $E = F/q$. The force is always proportional to the charge, so that the magnitude of electric field, E, is independent of the charge used in testing for the field, just as the magnitude of gravitational field, g, is independent of the mass of an object placed in it. The direction of the electric field is the same as the direction of the force on the positive charge placed in the field to determine its magnitude.

The force on a positive charge between two charged plates at several locations is shown in Figure 8.11. The force is constant and is directed from the positive toward the negative plate, which means that the electric field between the plates is also constant and directed from the positive toward the negative plate. A vector could be drawn at every point between the plates to represent the field, but the diagram would become very confusing. A more useful way to represent the field is shown in Figure 8.12. Lines are drawn with small arrowheads representing the direction of the field at many points. The lines allow one to estimate the direction of the field even at points between the lines. The strength of the field is shown by drawing the lines close together where the field is strong and far apart where the field is weak. Several examples of **electric field lines** are shown in Figure 8.13 for various charge configurations.

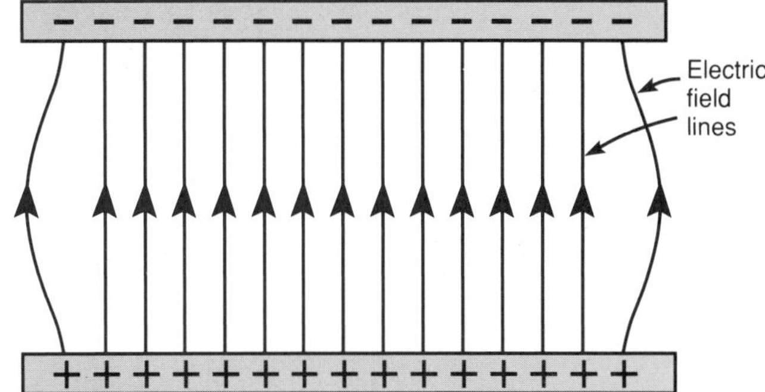

Figure 8.12 Electric field lines between charged plates. The electric field between the plates is constant, both in magnitude and direction, except for a slight "fringing" at the edges.

Computation of the field strength for the case of two flat plates, as in the oil-drop experiment, is beyond the scope of this text. Once found, the electrostatic force on the droplet is easily found by reversing the definition of the electric field and using $F = qE$.

Electric Potential

An electric charge situated in a region where there is an electric field has, by definition, a force acting on it. If the charge is moved in a direction opposite to that of the force, work must be done on the charge. It will be recalled from Chapter 5 that work, W, is equal to the magnitude of the force, F, times the distance in the direction of the force, d, that an object moves, or $W = Fd$, where work is expressed in joules if the force is in newtons and the distance is in meters. The work done on the charge in moving it between two points can be recovered if the charge is returned to the original point, regardless of the return path chosen (see Fig. 8.14). (This path independence is crucial later in working with electric circuits, where charges gain energy along one part of the path and lose energy in another part of the path.) It will also be recalled from Chapter 5 that potential energy is just recoverable work, so that the work done on the charge leads to a gain in potential energy of the charge.

When an electric field exists in a region of space, it is convenient to use a general method of determining the potential energy that can be applied to any charge, not just one specific charge. Thus, the definition: **electric potential** is *the electric potential energy per unit of charge.* In equation form,

$$\text{electric potential} = \frac{\text{potential energy}}{\text{charge}}.$$

In SI units, the potential energy is given in joules (J) (equal to the work done in moving the charge against the force); the charge in coulombs (C); and the electric potential in **volts** (V); that is 1 volt $\equiv$ 1 joule/1 coulomb (1 V = 1 J/1 C). The symbol for electric

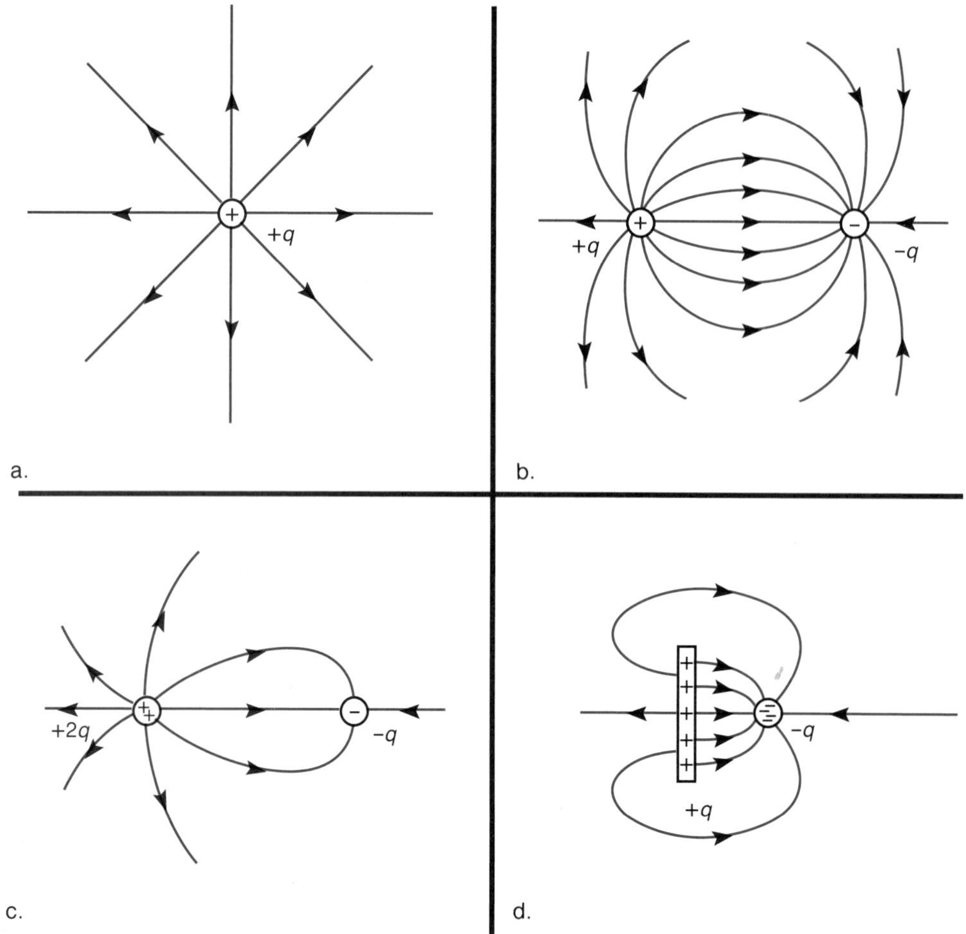

Figure 8.13 Electric field lines for various charge configurations. In general, electric field lines start on positive charges and end on negative charges (some of the charges could be very far away and not shown on the diagram). Larger charges have more field lines, indicating a stronger field near them.

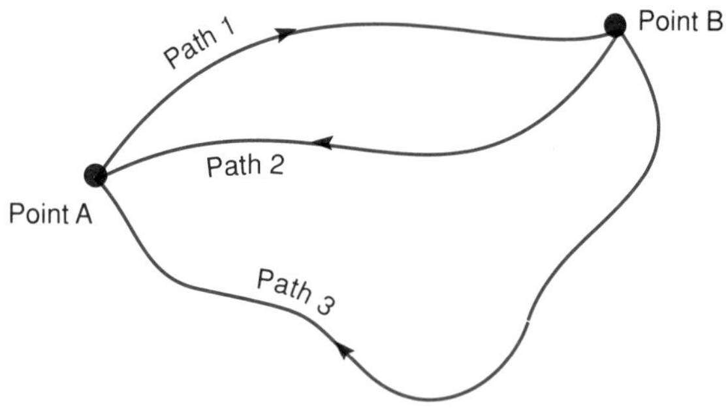

Figure 8.14 Work done in moving a charge between two points, A and B. If a charge is moved from A to B along path 1 and work is done in moving it against the electric field, all the work can be recovered if the charge moves along any path, such as paths 2 or 3, as it returns from B to A.

potential is V. (The student must be extremely careful not to confuse the symbol for potential, V, which is italicized, with the abbreviation for the unit of potential, the volt, V.)

The symbolic form of the equation for potential is

$$V = \frac{W}{q}, \qquad \text{ELECTRIC POTENTIAL}$$

where V is the electric potential, W is the work done in moving the charge against the field, and q is the charge.

Work is done only when a force acts through a distance, so that electric potential as previously defined is the difference in potential between the starting and ending points used in computing the work. It is often convenient to define a certain point to have a potential of 0 volt, then all other potentials are specified relative to this point. The point of zero potential is usually called *ground,* because Earth is a convenient reference point that is usually available and often a wire is connected directly from the point to Earth.

When a dry cell used in a flashlight is said to have a potential of 1.5 volts, it means that the potential of the positive terminal relative to the negative terminal is 1.5 volts. It might be convenient for some purposes to let the potential of the positive terminal be zero, which would make the potential of the negative terminal -1.5 volts. As far as operation of the cell in a circuit, only the difference in potential matters.

Example 8.3

A dry cell used in a flashlight has a potential difference of 1.50 volts between its terminals. If a lamp is connected between the terminals and 8.00 coulombs of charge flow through the lamp, how much energy is "lost" by the charge? (This energy would be changed into heat energy; thus, the filament in the bulb would get hot, giving off heat and light.)

Solution

The energy lost by the charge must equal the work done in giving the charge its potential energy, W. In the equation for electric potential $V = 1.50$ V and $q = 8.00$ C. Then

$$V = \frac{W}{q}$$

becomes

$$\begin{aligned} W &= Vq \\ &= (1.50 \text{ V})(8.00 \text{ C}) \\ &= 12.0 \text{ J}. \end{aligned}$$

So, 12.0 J of potential energy are converted into heat and light energy in the filament of the lamp.

Electric Charge in Motion

Whenever a potential difference exists between two points and there are charges that are free to move, the positive charges will tend to move from the regions of high potential to the regions of low potential, just as a rock falls toward Earth where the gravitational potential is lower. However, the forces on negative charges are always in the opposite direction from the forces on positive charges, so negative charges tend to move from regions of low potential to regions of high potential. As the charges move, their energy changes. This change in electrical energy is converted into other useful forms of energy by electrical devices. The fact that charges can be easily moved from place to place, along with their attendant energy, makes an understanding of the flow of charges indispensable in modern technology.

Electric Current

When charge flows from one place to another, not only is the difference in potential between the two places important, but the rate at which the charge flows is also important. **Electric current** is *the rate of flow of electric charge.* The electric current can be either a flow of positive or of negative charges or even a combination of positive and negative charges, but generally in electric currents it is the electrons that move.

In Figure 8.15a, suppose that two reservoirs of charge existed as shown: one with $+12$ coulombs of charge and the other with -8 coulombs of charge. Now, suppose that the two reservoirs are connected and that charge is transferred by some method from the reservoir on the left to the reservoir on the right, giving the resultant charges shown in Figure 8.15b. This result could have been achieved by either transferring $+3$ coulombs of charge from the reservoir on the left to the reservoir on the right (Fig. 8.15c) or by transferring -3 coulombs of charge from the reservoir on the right to the one on the left (Fig. 8.15d). Since the net transfer of charge is the same, it is immaterial which type of charge was transferred. The direction of conventional current is taken as that of the positive charges. However, in most conductors, it is actually negatively charged electrons that are flowing, so the direction of electron flow is opposite to the direction of the conventional current. When the direction of the current is given, one must determine whether it is conventional current or the direction of the electron flow that is being stated.

It should also be noted that current is the transfer of charge. Charge is not created nor destroyed by devices through which current passes, be they generators, batteries, motors, or any other ordinary electrical apparatus.

The quantitative definition of electric current, denoted by the symbol I, is given by

$$\text{electric current} = \frac{\text{charge transferred}}{\text{time for transfer}},$$

or, in symbolic form,

$$I = \frac{q}{t}. \qquad \text{ELECTRIC CURRENT}$$

The unit of current, the **ampere** (A), is defined as follows: 1 ampere $\equiv$ 1 coulomb/second, or 1 A = 1 C/s. The ampere is a fundamental SI unit just as are the meter, the kilogram, and the second. When the ampere is defined, the coulomb is automatically defined by the relationship $1\text{C} = 1\text{A} \cdot 1\text{s}$.

André-Marie Ampère: The Father of Electrodynamics

André-Marie Ampère (1775–1836) suffered many hardships as a youth. He had to overcome the shock of having witnessed the execution of his father during the French Revolution. He also had to fight extreme poverty in order to keep from being committed to debtors' prison. Through his hard work he settled his debts and attended college, earning degrees in mathematics and physics. He accepted the position of mathematics professor at the Polytechnic School of Paris and later was appointed professor of physics at the Collège de France. In 1820 he became aware that the Danish scientist Oersted had discovered the relationship between electricity and magnetism. Ampère immediately designed and conducted experiments to confirm this connection and outlined many aspects of the theory of electricity and magnetism that we use today. Five years later he formulated a complete theoretical explanation of how magnets interact with electric currents. Ampère became known as the father of modern electrodynamics, and the electrical unit of current was named for him ■

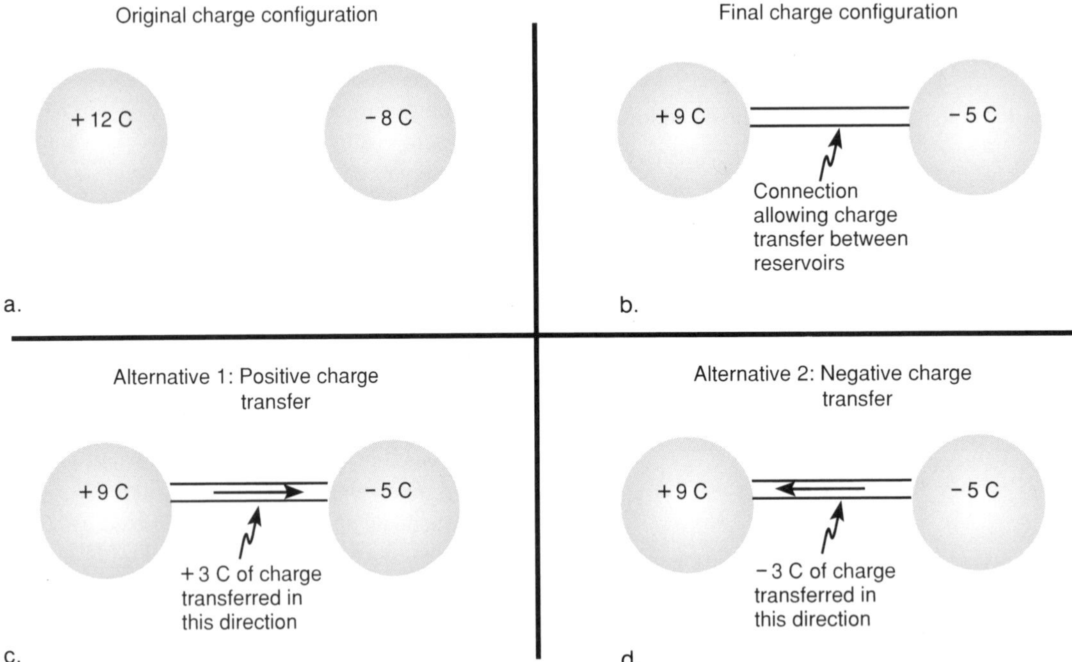

Figure 8.15 Charge transfer and the direction of current. When charge is transferred from one place to another, it may be done by transferring either positive or negative charges. Conventional current is the direction of positive charge transfer (*c*), but electron flow in metallic conductors is actually in the opposite direction (*d*). Note that the net charge in the two reservoirs is +4 C in all cases.

Example 8.4

A starter motor in a car requires a current of 60.0 amperes to operate. If it takes 3.00 seconds to start the car's engine, how much charge passes through the starter motor?

Solution

The symbols are used to represent the quantities given in the problem, $I = 60.0$ A, $t = 3.00$ s, and, in symbolic form, the question is asking for q. Thus,

$$I = \frac{q}{t}$$

becomes

$$\begin{aligned} q &= It \\ &= (60.0 \text{ A})\ (3.00 \text{ s}) \\ &= 180 \text{ A} \cdot \text{s} \\ &= 180 \text{ C}. \end{aligned}$$

So the charge passing through the starter motor is 180 C.

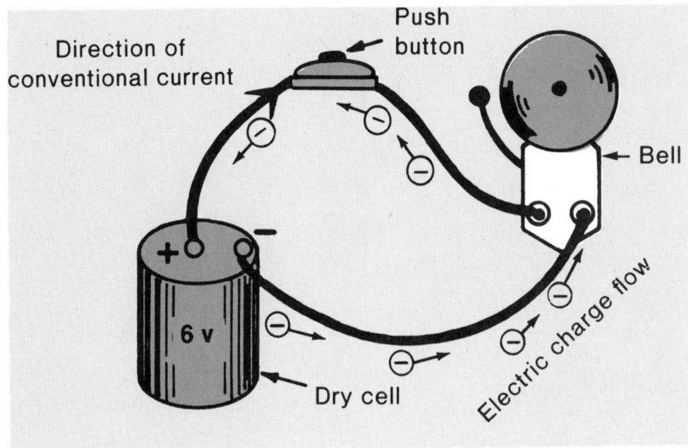

Figure 8.16 A complete electric circuit. A complete closed loop with a seat of electromotive force (emf) is necessary for a continuous current.

Electromotive Force

When two reservoirs of charge, such as those in Figure 8.15, are connected by a conductor, the charges are usually drained very rapidly—within a fraction of a second. To maintain a useful flow of charge requires some method of keeping the charges in motion and at the same time keeping them from collecting in large quantities, so that the buildup of charges will not prevent the flow of any additional charges. The second of these requirements is met by having *a closed electric circuit;* that is, a loop around which the charges may flow (Fig. 8.16).

To meet the first requirement, the charges are "pushed" around the closed circuit by a device that gives them energy, much as a water pump pushes water through a pipe by increasing its pressure. A **seat of electromotive force,** or *emf,* is *a device that can more or less continuously supply energy to charges.* In Figure 8.16, the dry cell is the seat of electromotive force. Recall that the voltage is a measure of energy per unit of charge, so an emf increases the voltage of the charges that pass through it. Emf's may take several forms in addition to batteries; for example, generators and solar cells. Seats of electromotive force change energy from some other form; in these cases, chemical, mechanical, and light, respectively, into electrical energy.

A detailed study of the ordinary dry cell (Fig. 8.17) as an emf will illustrate the salient features of these devices. The ordinary dry cell contains two terminals, one positive and the other negative, called *electrodes.* The positive electrode and the negative electrode are made of two dissimilar conductors; in the case of the ordinary dry cell, the positive electrode is a carbon rod and the negative electrode is a zinc can. The carbon rod is placed in the center of the zinc can, and the space between is filled with an *electrolyte,* a nonmetallic conductor in which charge is transported by the movement of *ions.* The electrolyte is a moist mix that actively dissolves the zinc, and in so doing releases chemical energy that causes the negative ions to move to the zinc can and the positive ions to move to the carbon rod. The carbon rod thus becomes positively charged and the zinc can negatively charged. The force that causes the positive ions to move toward the carbon rod must be a nonelectrostatic force because the electrostatic force would repel positive ions and prevent their collecting on the carbon rod. This nonelectrostatic force is the electromotive force. In a sense, it forces the ions to move "against their will." That is, the positive ions are forced to move to the positive electrode, overcoming the repulsion of the positive charges already there. Likewise, the negative ions are forced to move to the negative electrode, overcoming the repulsion of the negative charges on this electrode. Figure 8.18 shows how these forces act on the positive ions within the electrolyte.

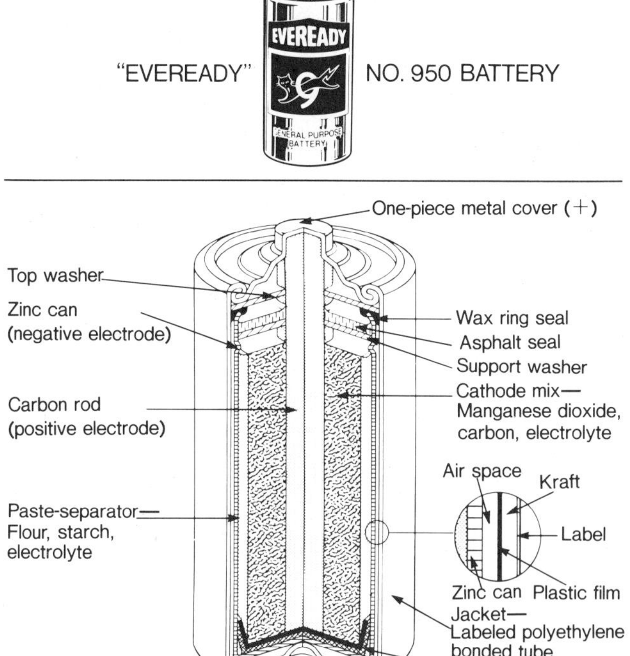

Figure 8.17 The illustration reveals the intricate components of a modern 1.5–V dry cell. Courtesy of Consumer Products Division, Union Carbide Corporation.

Alessandro Volta (1745–1827), an Italian experimenter, was the first to develop a cell that supplied a reasonable amount of current at useful voltages. Such cells are collectively referred to as *voltaic cells.*

Dry cells are limited by the amount of chemical energy that can be stored. Since the electrolyte dissolves the zinc can, even while the cell is not being used, they cannot be "recharged," and they also have a limited shelf life. (Recharging is a poor choice of terms because the cell itself does not store charge, rather it stores chemical energy.) Newer cells, such as alkaline cells and rechargeable nickel-cadmium cells (Fig. 8.19), overcome some of these limitations.

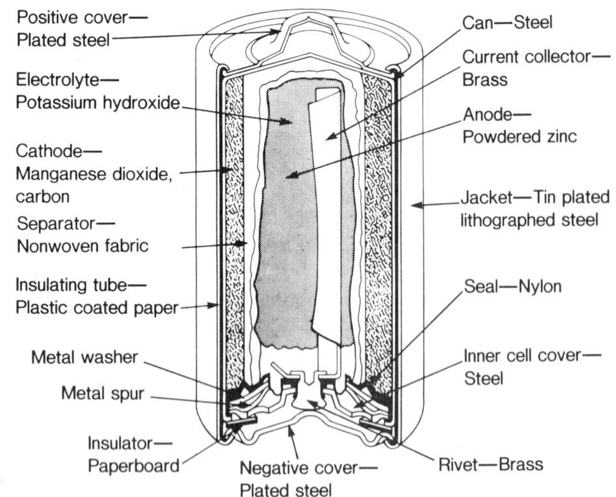

a.

SEALED NICKEL-CADMIUM RECHARGEABLE CYLINDRICAL TYPE CELL EXPANDED VIEW

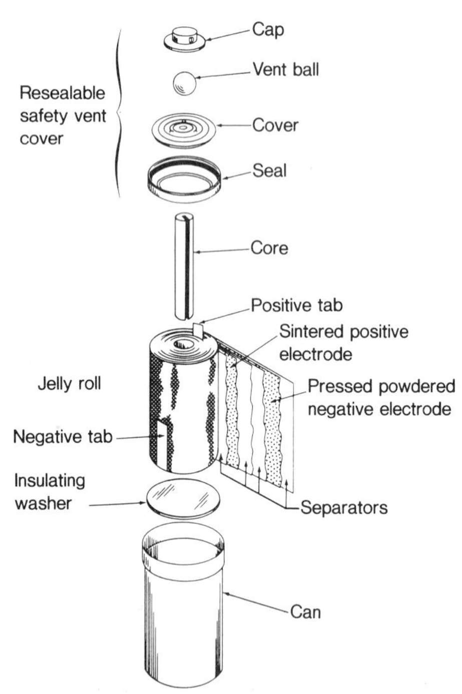

b.

Figure 8.19 (*a*) Alkaline dry cells that contain manganese oxide, silver oxide, or mercuric oxide as the cathode have an extensive shelf life and operate well at low temperatures. (*b*) The rechargeable nickel-cadmium dry cells are used in portable tools and appliances, calculators, tape recorders, toys, and a host of other devices. They, too, have good low-temperature performance. Courtesy of Consumer Products Division, Union Carbide Corporation.

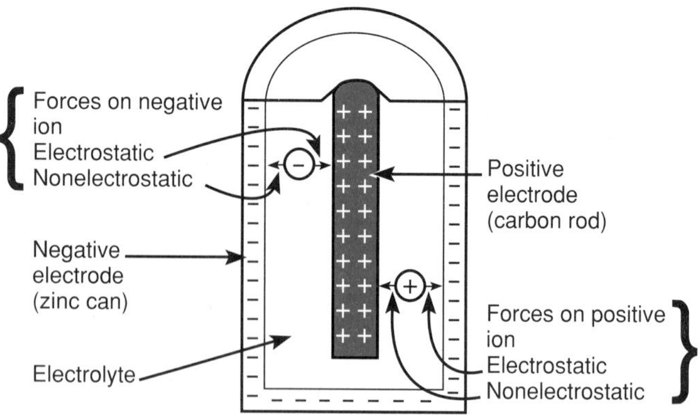

Figure 8.18 A schematic of electromotive force. The nonelectrostatic forces due to chemical reactions in the electrolyte force the negative ions toward the negative electrode and the positive ions toward the positive electrode. If a conductor is connected between the positive and negative electrodes, the electrostatic force is reduced because the positive and negative charges on the electrodes are reduced. The nonelectrostatic force remains the same, however.

The significant feature of seats of emf, then, is that they have a nonelectrostatic force acting on the charges, forcing them to increase their potential energy (move to a higher potential if they are positive and to a lower potential if they are negative). This increase in potential energy comes about because the nonelectrostatic force does work on the charges. When the work on a charge is divided by the charge, a voltage can be computed just as if the charge were moving in an ordinary electric field. The symbol $\mathcal{E}$ is used to indicate a voltage derived from a seat of emf, to distinguish it from V, the voltage associated with an ordinary electric field, even though both are measured in volts. It is customary to speak of the emf in volts; thus one says, "The battery has an emf of 1.5 volts."

The voltage $\mathcal{E}$ is defined by the following equation:

$$\mathcal{E} = \frac{\text{energy supplied to charge by emf,}}{\text{charge}},$$

or, in symbolic form,

$$\mathcal{E} = \frac{W}{q}, \qquad \text{ELECTROMOTIVE FORCE (emf)}$$

where W is now the energy supplied by the seat of electromotive force.

Example 8.5

If the battery used to operate a starter motor in a car has an emf of 12.0 volts and 180 coulombs of charge pass through the battery while the starter motor is cranking the car, how much energy is supplied by the battery?

Solution

Using the defining equation for emf with $\mathcal{E} = 12.0$ V and $q = 180$ C,

$$\mathcal{E} = \frac{W}{q}$$

becomes, on multiplying both sides by q,

$$W = \mathcal{E}q\ .$$

Upon substitution this gives

$$\begin{aligned} W &= (12.0\ \text{V})\ (180\ \text{C}) \\ &= 2160\ \text{V} \cdot \text{C} \\ &= 2160\ \text{J}. \end{aligned}$$

Thus, the energy supplied by the battery is found to be 2160 joules.

Power Supplied by an Electromotive Force

Power is the rate of doing work or supplying energy; that is, the work done or energy supplied divided by the time required (see Chapter 5). The work done on the charges by an emf, because of conservation of energy, must be supplied to the electrical circuit in some form. From algebra, if the numerator and denominator of a fraction are divided by the same quantity, the fraction is unchanged. Taking the definition of emf just given and dividing the top and bottom of the fraction on the right by the time, t, gives

$$\mathcal{E} = \frac{W/t}{q/t}\ .$$

The numerator of the fraction is the energy divided by the time, which, by definition, is power; symbolically, $P = W/t$. The denominator, by definition, is just the electric current, or $I = q/t$. The equation, using these definitions becomes

$$\mathcal{E} = \frac{P}{I}\ .$$

Multiplying both sides of the equation by I gives, in compact form, the power P supplied by an emf $\mathcal{E}$ when a current I is being provided to an external circuit:

$$P = \mathcal{E}I. \qquad \text{POWER SUPPLIED BY EMF}$$

Since electrical units are based on SI units, the unit of power is the watt, which we will discuss further in Chapter 9.

Example 8.6

Returning once again to the battery and starter motor of the two previous examples, find the power supplied by the battery to the starter motor. Restating the important facts, $\mathcal{E} = 12.0$ volts and $I = 60$ amperes, then

$$\begin{aligned} P &= \mathcal{E}I \\ &= (12.0\ \text{V})\ (60.0\ \text{A}) \\ &= 720\ \text{W}. \end{aligned}$$

Thus, the battery supplies 720 watts of power to the starter motor (actually, a small portion of the power would be dissipated in the connecting wires). If English units are desired, the conversion factor 746 W = 1 hp may be used, giving 0.965 hp.

Power consumption in electric circuits is discussed in more detail in Chapter 9.

Resistance and Ohm's Law

Voltage, current, and electrical resistance are related to one another. As pointed out by Georg Simon Ohm (1789–1854) in 1827, *the current in a given circuit is directly proportional to the impressed voltage and inversely proportional to a quantity called the resistance of the circuit.* That is,

$$\text{current} = \frac{\text{voltage}}{\text{resistance}}\ ,$$

a relationship that is now called **Ohm's law.** In terms of symbols,

$$I = \frac{V}{R}\ . \qquad \text{OHM'S LAW}$$

Alternative forms of Ohm's law are found by solving for V and R, giving

$$V = IR$$

and

$$R = \frac{V}{I}\ .$$

R, the **resistance,** is *a measure of how difficult it is to force charges to flow around the circuit.* If the voltage, V, is constant, a small value of the resistance, R, will permit a large current to flow through the circuit, as can be seen from the first form of Ohm's law. Conversely, a large resistance limits the current to a small value.

The unit of resistance, from the last form of Ohm's law previously given, must be expressed as volts per ampere. The unit of resistance is called the **ohm,** abbreviated by the uppercase Greek omega (Ω). The definition is 1 ohm ≡ 1 volt per ampere, or 1 Ω = 1 V/A. One ohm of electrical resistance is present between two points in a circuit when a voltage difference of 1 volt between the two points causes a current of 1 ampere to flow through the circuit between these points. The resistance in a typical lamp cord or household extension cord is less than 1 Ω, but the resistance of a 60-watt lamp is about 240 Ω.

The resistance of a given conductor is dependent on several factors:

1. Electrons are less tightly bound to some atoms than others, giving rise to variations in the ability of different substances to conduct electric current. For example, the resistance of iron wire is about seven times that of copper wire of the same length and diameter.
2. As the length of an electrical conductor is increased, the resistance of the conductor is increased proportionally.
3. The resistance of a conductor is inversely proportional to its cross-sectional area. A conductor with a cross-sectional area of 1 square centimeter has twice the resistance of a conductor with a cross-sectional area of 2 square centimeters, assuming that the two conductors have the same composition and equal length.
4. The resistances of most conductors increase as their temperatures increase. When it is glowing white hot, the filament in a light bulb has some ten times the resistance that it has when the light bulb has been off for some time. Other materials, such as electrolytes, carbon, and semiconductors, have a lower resistance when they are heated than they do at lower temperatures. One special class of materials, called *superconductors,* loses all resistance as the temperature is lowered past a certain critical value; that is, their resistance actually becomes zero. Until recently, the temperatures required were near zero K (absolute zero), but recent research has pushed the temperature much higher. (See the box titled "Superconductivity: Reaching for New Highs.")

Example 8.7

How much current is required to operate an electric iron that is plugged into a 120-volt electrical outlet if it has a resistance of 20.0 ohms?

Solution

This requires the simple application of Ohm's law:

$$I = \frac{V}{R} \qquad \text{CURRENT}$$
$$= \frac{120 \text{ V}}{20.0 \ \Omega}$$
$$= 6.00 \text{ A}.$$

Example 8.8

A 100-watt lamp plugged into a 120-volt outlet "draws" a current of 0.833 ampere. What is the resistance of this lamp when it is operating?

Solution

The resistance can be determined by using Ohm's law in the form $R = V/I$, where $V = 120$ V and $I = 0.833$ A. Then,

$$R = \frac{V}{I} \qquad \text{RESISTANCE}$$
$$= \frac{120 \text{ V}}{0.833 \text{ A}}$$
$$= 144 \ \Omega.$$

So, 144 Ω is the resistance of the lamp at operating temperature. As was noted previously, the resistance is lower when the filament is cool.

Example 8.9

An electronics engineer is building an amplifier for a stereo and needs to use a 100-ohm resistance at a certain place in the amplifier. He has a resistor (a device that has a certain amount of resistance) that will "burn out" if more than 0.0500 ampere of current pass through it. What is the greatest voltage that can be sustained across this resistor without "burning it out"?

Solution

Once again, Ohm's law is applicable, this time in the form $V = IR$. From this form, it can be seen that when the current is a maximum through a given resistance, the voltage is also a maximum. Then,

$$V = IR \qquad \text{ELECTRIC POTENTIAL}$$
$$= (0.0500 \text{ A})(100 \ \Omega)$$
$$= 5.00 \text{ V}.$$

Thus, the maximum voltage that can be applied to the resistor is 5.00 V.

Superconductivity: *Reaching for New Highs*

Magnetically levitated trains and superfast computers are two images immediately conjured up when superconductivity is mentioned. Recent breakthroughs in superconductivity have reinforced the public's ideas that we are on the threshold of these and many other technological marvels.

In 1911, Dutch physicist Kammerlingh Onnes discovered that when mercury is cooled below 4 K (452 degrees below zero Fahrenheit) all electrical resistance vanishes. He termed this effect *superconductivity.* (There are also other effects of superconductivity.) Onnes soon found other materials that exhibited superconductivity when cooled below temperatures near absolute zero, called their *critical temperatures.* In one key experiment, he found that a current started in a superconducting loop of lead did not measurably decrease over the period of a year. He envisioned the construction of large electromagnets with no loss of energy caused by the resistance of the wire, but he found that large currents destroyed the superconductivity.

In the decades that followed Onnes's discovery, other materials that exhibited superconductivity were discovered, some of which could carry large currents. By the early 1980s, superconducting magnets were in everyday use in specialized applications, but researchers had not been able to find any material for which the critical temperature was above 23 K (−418° F). Since cooling to these low temperatures is accomplished with liquid helium (boiling point 4.2 K), the required elaborate insulating schemes limit the general usefulness of superconductivity. (Liquid hydrogen could be used, but it is dangerous to handle.)

A breakthrough in raising the critical temperature was reported in 1986 by two researchers, Alex Müller and Georg Bednorz, at the Zurich laboratories of IBM. Their discovery of a compound that became superconducting at a temperature of 30 K (−406° F) soon led to the preparation of other compounds with critical temperatures of over 90 K (−298° F). These compounds can be cooled with liquid nitrogen, which may be held in ordinary "thermos" bottles. Recently, researchers have even discovered some compounds that show signs of superconductivity at room temperature.

This discovery of materials with high critical temperatures has created a flurry of activity in the field of superconductivity. In addition to the search for materials with ever-higher critical temperatures, research continues on methods to make the newly discovered materials more stable over long periods of time. Also, while these materials are easy to make, most are difficult to fabricate into useful devices. Another problem, the same one as found earlier by Onnes, is that they lose their superconductive properties when large currents are passed through them. In spite of present problems, these recent breakthroughs have left little doubt that many of the dreams of earlier researchers will soon become reality ■

How Magnetism Developed

Forces of attraction by objects that are not in contact are familiar to all of us. Gravity pulls objects toward Earth and, as mentioned previously, combing or brushing dry hair on a cool crisp day can be exasperating because the comb or brush attracts the dry hair. However, forces of repulsion with which we are familiar usually involve "contact" with some other object. Nonetheless, repulsion without contact was known even to the ancients. Historical records indicate that Thales, who lived in Greece during the sixth century B.C., was among the first to study the magnetic effects of a particular iron ore mined near the City of Magnesia in Thessaly. This ore became known as "the Magnesia rock." Materials that were attracted to the iron ore came to be known as magnets. Others knew the iron ore as *lodestone,* meaning "a stone that leads," because the Greeks, as well as the Chinese, noted that a suspended piece of the iron ore or iron that had been stroked with a lodestone always rotated to a given direction. Of course, the lodestone became the compass by which the ancients found their way to distant lands and back again. The early investigators considered this observation an isolated phenomenon, with no relation to other properties of magnetism that had been determined, such as attraction, repulsion, and induction.

At the close of the sixteenth century, William Gilbert published *De Magnete,* which discussed the characteristics of magnetism known at that time. Although many of its features have since been investigated in much greater detail, natural magnetism was relatively well understood during Gilbert's time.

Several important discoveries have been added to our understanding of magnetism in the last 150 years. Hans Christian Oersted discovered in 1820 that an electric current in the vicinity of a compass would cause the needle, in reality a small balanced magnet, to be deflected. This new observation led to the development by William Sturgeon of the **electromagnet,** a device that has been refined to the point where it can support automobiles and other heavy objects to be moved about and released upon disruption of the circuit. Another important discovery was the development of permanently magnetized materials. Recently, technologists have developed permanently magnetic materials, such as *alnico,* an alloy of aluminum, nickel, cobalt, and a small amount

Figure 8.20 Permanent magnets differ in size, shape, and composition.

of copper. Magnets composed of this alloy are capable of attracting and lifting objects many times their own weight. Other permanent magnets include those made of cobalt steel, chrome steel, or ceramic materials. Another common permanent magnet is composed of iron, nickel, and a larger portion of copper than is used in the alnico magnet. Various types of permanent magnets are illustrated in Figure 8.20.

Theories of Magnetism

The process of making a magnet from a piece of steel by stroking it with a magnet is known as *magnetic induction,* which was the only rapid way of making a magnet before the discovery of electromagnetism (see Fig. 8.21). If a piece of soft iron or low-carbon steel is magnetized through induction, the magnetism is found to be short-lived, whereas if hard steel is magnetized, the property is retained permanently. The magnetism can be removed, however, by hammering, dropping, or heating the magnet. The soft iron that becomes a temporary magnet is very important and will be discussed in detail in the section on Electromagnetism.

Figure 8.21 Magnetism may be induced in a piece of steel by stroking it in one direction with a permanent magnet.

Table 8.1

Relative permeabilities of some common materials.

Material	Relative permeability
Bismuth	0.999,833
Quartz	0.999,985
Water	0.999,991
Copper	0.999,995
Air (S.T.P.)	1.000,000,4
Oxygen (S.T.P.)	1.000,001,8
Aluminum	1.000,021,4
Liquid oxygen	1.003,46
Nickel	400.0
Iron	5000.0
Permalloy	100,000.0
Supermalloy	800,000.0

Just as materials vary in their ability to retain magnetic properties, they vary in *permeability,* the capacity to afford a path for transmission of magnetism through or around them. Iron and nickel, along with various alloys of these and other metals, have a high permeability; thus, the presence of these substances affects the portion of space near a magnetic body. *This space around a magnetic body* is the site of a **magnetic field.** Other substances—including plastics, wood, copper, and aluminum—have a low permeability. Generally the permeability of a substance determines its potential to develop and maintain a permanent magnetic field. The relative permeabilities of various materials are listed in Table 8.1.

The strength of the magnetic field that surrounds a magnetic body varies with the distance from the body and from point to point on the body itself. Michael Faraday, after prolonged investigations of magnetic fields, concluded that the strength and direction of the magnetic field could be represented by *magnetic*

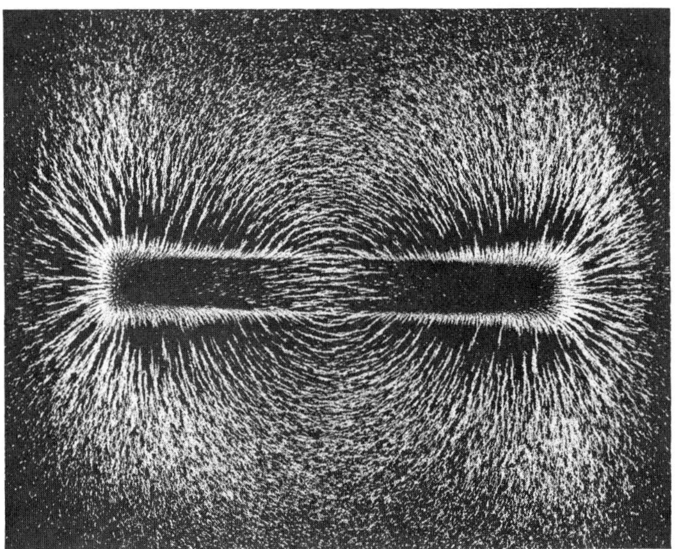

Figure 8.22 A representation of the magnetic field around a bar magnet.

lines of force, now called *magnetic field lines.* (These lines are parallel to the force on the north pole of a magnet, but the force on a charge moving through the field is perpendicular to the magnetic field lines.) The more lines of force at any location, the stronger the field (just as for lines of electric field). Faraday found that the lines of magnetic field follow definite paths through the space around a magnetic body, which upon closer examination were found to be actually continuous loops that passed through the bodies themselves. When precise measurements of the strength of a magnetic field around a magnet are made, it is found that the forces and fields follow an inverse square law similar to Coulomb's law for charges, but are complicated by the fact that magnets always have two poles.

If a bar magnet is placed beneath a sheet of paper and iron filings are sprinkled over the paper, as in Figure 8.22, the small fragments of iron will align themselves and form continuous loops that follow the magnetic field of flux. The number of loops that enter and leave near the ends, or *poles,* of the magnet display the variation in the potential of the magnet to attract or repel another magnet.

If two magnets are placed adjacent to each other under the paper, and iron filings are sprinkled on the paper, the results will vary, depending on the arrangement of the magnets. If the magnets are arranged so that two opposite poles, a north and a south, are adjacent to each other (Fig. 8.23a), an attractive force field exists. If like poles are placed adjacent to each other (Fig. 8.23b), the force field is disturbed by the repulsive forces present. These illustrations disclose why like poles of magnets repel and unlike poles attract, the latter evidenced by the smooth continuation of magnetic fields between the two magnets. An unusual arrangement of magnets is illustrated in Figure 8.23c, in which four magnets create both attractive and repulsive fields through the influence of one magnetic field on the other. If the pole at the top of the illustration is known to be a north pole, the identity of each pole

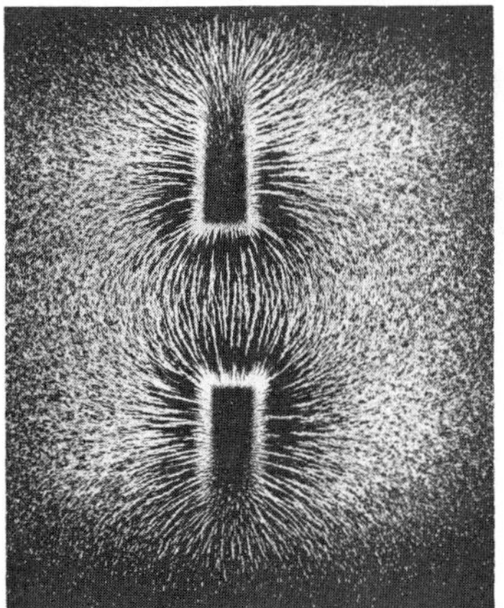

a.

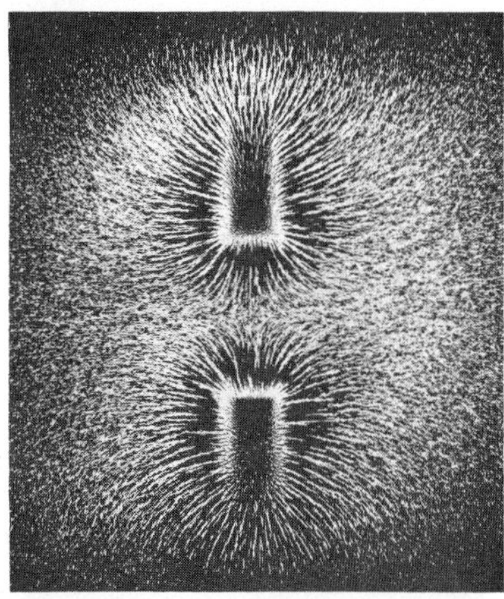

b.

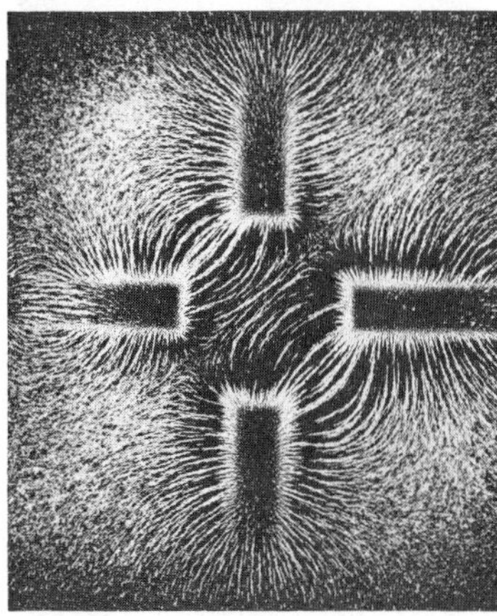

c.

Figure 8.23 (*a*) The magnetic field that interacts between opposite poles of two permanent magnets. (*b*) The magnetic field produced by two like poles of separate magnets. (*c*) The magnetic field produced by four poles from four separate magnets. The upper pole is considered a north pole. Can you identify the polarity at the ends of the other magnets?

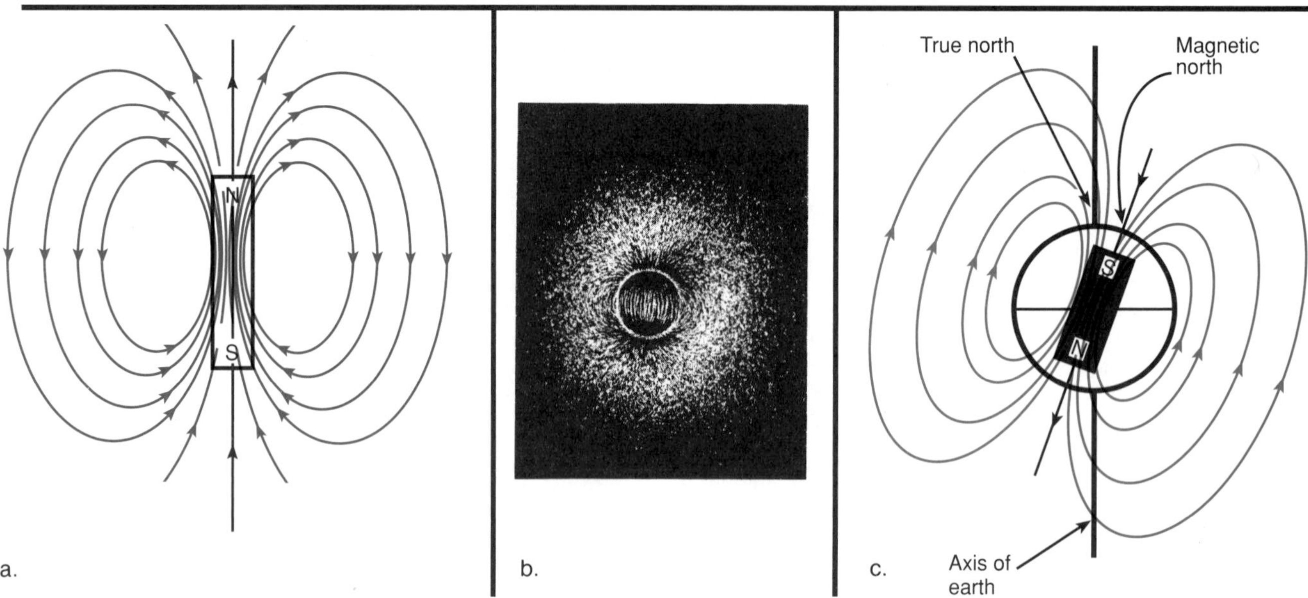

Figure 8.24 (*a*) A representation of the magnetic field of a bar magnet. (*b*) The magnetic field about a disk or sphere as shown by iron filings. (*c*) A representation of the magnetic field of Earth.

can be determined by its effect on the magnetic force field of the known pole.

Earth is considered a huge magnet. The way in which a compass aligns with Earth's magnetic field is further evidence of the continuity of the field about a magnet. The end of the compass that points to the north magnetic pole of Earth, regardless of the compass location on Earth's surface, is called the north-seeking pole. The end that points southward is thus the south-seeking pole. The names of the compass poles have been shortened to the *north pole* and the *south pole,* respectively. Since the north pole always points northward, the magnetic field of Earth or any magnet is continuous. Also, because opposite poles attract, the north pole of Earth is actually a south magnetic pole. The direction of the continuous magnetic field is conventionally considered to be from the north to the south pole on the exterior of the magnet, and from the south to the north pole within the magnet, as illustrated in Figure 8.24a. A spherical object, such as displayed in Figure 8.24b, has a magnetic field similar to that of the bar magnet. Earth, represented in Figure 8.24c, has a similar magnetic field. The field direction is the same as the external field of any magnet, and the internal field has a direction identical to that of both bar magnets and spherical magnets.

The position of the north magnetic pole of Earth shifts slightly from year to year, but the annual changes are quite small. The pole is considered to be located in the region of Hudson Bay, some 1100 miles from the geographic north pole. The difference between true north and the compass reading at a given position on the surface of Earth is called *magnetic declination.* The degree of declination is illustrated in Figure 8.25. The magnetic pole in the Southern Hemisphere is located south of Australia.

The reason that Earth and other celestial bodies have magnetic fields is still not definitely settled. Although Earth's magnetic field is like the magnetic field of a bar magnet, neither the construction nor the shape of the planet resembles that of a bar magnet. Many scientists theorize that there are electrical currents present in Earth's molten iron core created by the planet's rotation. These currents develop a dynamo effect similar to the one that produces a magnetic field in an electrical generator.

Jupiter has a rather intense magnetic field, an observation that reinforces this theory of how magnetic fields in such celestial bodies are produced. The currents in the light elements that form its interior are created by the rapid rotation of the planet. Further support for the theory comes from the fact that the moon rotates slowly and has a very weak magnetic field, while Venus, which turns on its axis more slowly than does the moon, has no detectable magnetic field (according to instruments aboard space probes of that planet).

Scientists have long wondered about the nature of magnetism. They have recently concluded that magnetically permeable materials have a crystal structure and that the crystals can be aligned in such a manner that *the individual magnetic properties become additive and cause the entire piece of material to become magnetized.* These crystal structures, or **magnetic domains,** are usually arranged in no definite pattern, but change orientation and relative size when they are immersed in a magnetic field, so that the north pole of one domain falls in line with the south pole of the adjacent domain. This alignment causes their individual magnetic field strength to become additive (see Fig. 8.26). When all domains in the material are so aligned, the material is said to have reached saturation and cannot become magnetically stronger. The fact that

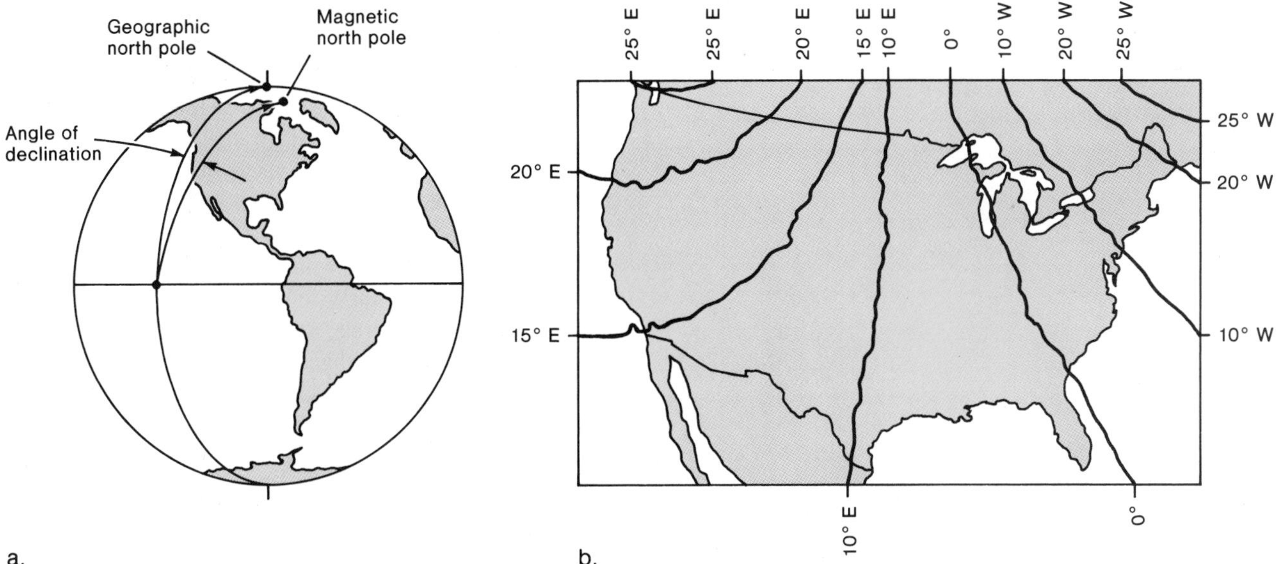

Figure 8.25 (*a*) A representation of Earth's magnetic declination. Note how the angle of declination varies with geographic location. (*b*) The degree of magnetic declination across the United States.

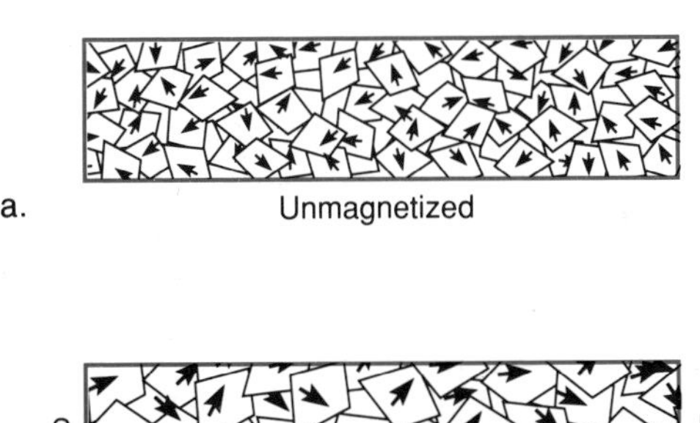

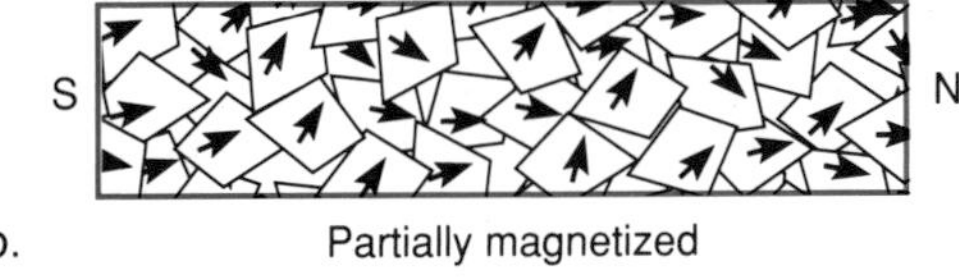

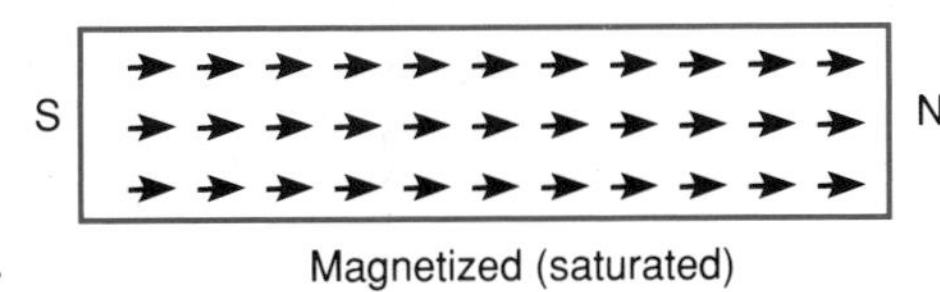

Figure 8.26 (*a*) Magnetic domains in an unmagnetized piece of iron are random in size and the fields point in random directions. (*b*) A partially magnetized sample of iron has more and larger domains pointing in one direction. (*c*) Saturated iron has all domains aligned. In (*d*), a magnet is broken to produce two magnets, both still having north and south poles.

some materials can exist with permanently aligned magnetic domains has made the permanent magnet extremely useful. Permanent magnets are found in electric motors, loudspeakers, telephones, timers, and in countless other devices.

The theory of domains is supported by the fact that a physical blow to a magnet will diminish the magnet's strength; such a shock jars the domains and knocks them out of alignment. In contrast, a steel bar may become magnetized if struck with a hammer while the bar is being held parallel to a magnetic field. The blows cause the domains to shift so as to line up with the magnetic field in a way much like magnetic induction. Further evidence in support of the *magnetic domain theory* is offered by the fact that a magnet becomes weaker when it has been heated. The critical temperature above which an ordinary magnet appears to lose its magnetic properties is called the *Curie temperature,* after the French scientist, Pierre Curie. Magnetic properties are also affected as the device is jostled or stored improperly with respect to Earth's magnetic field or that of a strong magnet nearby.

Electromagnetism

Hans Christian Oersted, a Danish physicist, discovered in 1819 that magnetic effects were produced by an electric current. His discovery led to the important principle that *magnetic fields are produced by moving charges.* (Even in a permanent magnet, the field is produced by charge motion on the atomic level.) An electric current is surrounded by a magnetic field. When the current is confined to a wire, the magnetic field lines form circular loops around the wire. The field is tangent to these circles at any point and its direction is found by using a right-hand rule: when the

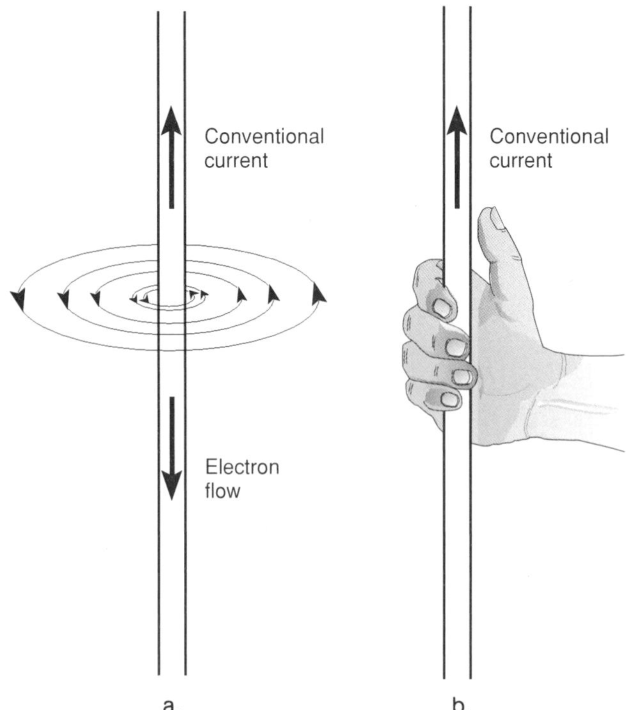

Figure 8.27 (*a*) Lines of magnetic field form circles around a straight wire carrying a current. (*b*) The direction of the field is that of the tips of the fingers when the right hand is wrapped around the wire with the thumb in the direction of conventional current.

fingers of the right hand are wrapped around the wire with the thumb in the direction of conventional current (opposite to the electron flow) the tips of the fingers point in the direction of the magnetic field (see Fig. 8.27).

The strength of the magnetic field surrounding a wire carrying a current depends on the current, the distance from the wire, and the magnetic nature of the materials (the magnetic permeability) near the wire. For a *straight* wire in air, the field strength is directly proportional to the current and inversely proportional to the distance from the wire. Thus, doubling the current would double the field strength, while doubling the distance from the wire would halve the field strength. Much stronger magnetic fields are generated if the wire is made into a coil, for then the currents in each loop of the coil generate magnetic fields that add together. Often, the coil is wound in a helical (spiral) pattern and is then called a *solenoid.* The strength of the magnetic field inside the solenoid is directly proportional to the current and the number of turns of the solenoid per meter of length.

The polarity of the magnetic field, the direction of north and south, is dependent on the direction of current flowing in the conductor (note the illustrations in Fig. 8.28). The *right-hand rule* can be used to determine the direction of the magnetic field that results. If the fingers of the right hand curl around the coil in the direction of conventional current (+ to −), the thumb will point to the north pole of the magnetic field produced.

The magnetic field created by an electric current can also be strengthened by the insertion of a soft-iron core into the loops that form the coil. This arrangement is the usual form of the *electromagnet.* The iron core offers an easy path for the magnetic field inside the coil and thus provides a minimum of magnetic resistance. *Reluctance,* as this property is called, is related to magnetism as resistance is related to electricity. The reluctance of a core of iron is much less than a core of air or one of cardboard.

When electromagnets were first developed, many potential uses were immediately realized. For the first time magnetism could be turned on or off at will. Electromagnets were soon developed that could lift objects many times their weight, move them to a desired location, and release them. Others were developed that could readily separate permeable materials from materials less permeable, such as fragments of iron from bits of aluminum or nickel. Technologists have developed many other applications for the electromagnet.

Alternating Current

We have seen how an electric current can produce a magnetic field or can create a magnet from a piece of nonmagnetized iron simply by sending a current through a wire wrapped around the iron. A magnet is also capable of producing an electric current in a coil of wire. In Figure 8.29 we see a bar magnet being inserted and withdrawn from inside a coil of wire. The ends of the coil of wire are attached to a galvanometer, a device capable of detecting an electric current. As the magnet is plunged through the coil, the needle deflects to the right of center, then returns to center (zero). As the magnet is removed, the needle deflects in the opposite direction, then returns to center. In order for a current to be maintained, the magnet must be in continuous motion. That is, the coil must be actively engaged in cutting through the magnetic field for current to be produced.

The fact that the needle is deflected in opposite directions leads the observer to deduce that the current must change directions. This change in current direction is the condition necessary to produce alternating current. **Alternating current** (**AC**) *flows forward and backward in a conductor at definite time intervals.* In contrast, the type of current produced by chemical action, discussed earlier in the chapter, *flows in one direction only* and is thus called **direct current** (**DC**). The alternating current that serves home and industry typically has an effective voltage of 120 volts and alternates in direction 120 times per second, or a total of 60 complete cycles per second (60 hertz). Alternating current provided to households in Europe is generally at an electric potential of 220

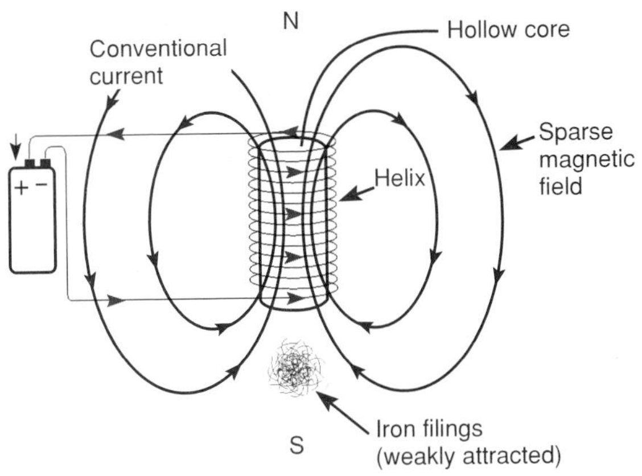

a.

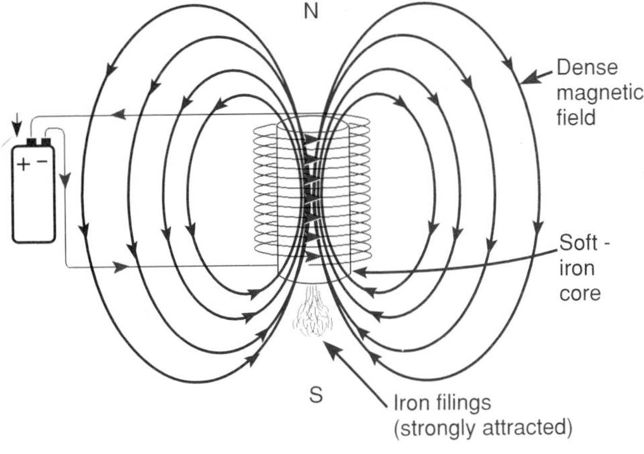

b.

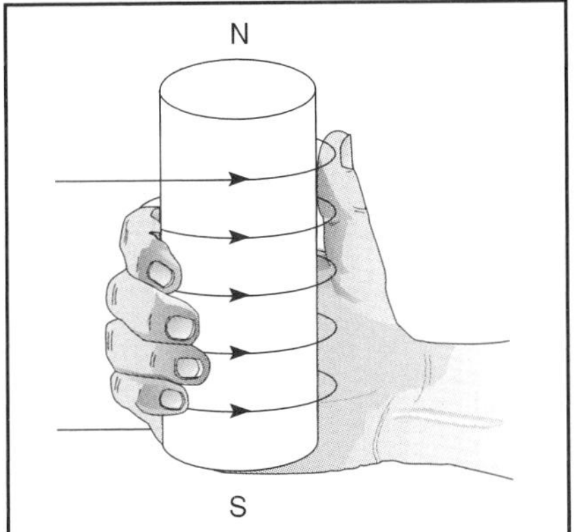

c.

Figure 8.28 (*a*) The magnetic field produced when a current flows through a solenoid is similar to the field of a bar magnet with a "north" and "south" pole near the ends of the solenoid. (*b*) The field is greatly strengthened if a soft-iron core is placed inside the windings. (*c*) The north pole of the solenoid is at the end to which the thumb points when the right hand is wrapped around the solenoid with the fingers pointing in the direction of conventional current.

a.

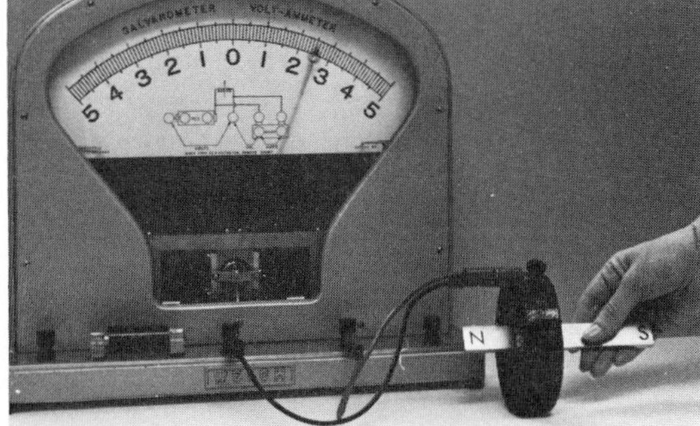

b.

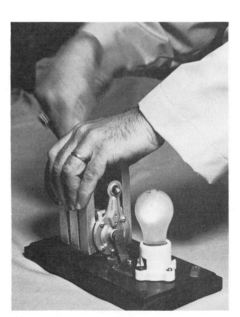

c.

Figure 8.29 (*a*) Electric current is produced by the insertion of a magnet into a coil of wire. (*b*) The direction of the electric current is reversed as the magnet is withdrawn from the coil. (*c*) The glowing light bulb offers further evidence that electric current results from the disturbance of a magnetic field by a coil of conducting wire.

volts, with a frequency of 50 hertz. Other foreign countries generate alternating current with frequencies between 35 and 85 hertz and voltages between 100 and 250. The tourist who travels worldwide finds electrical appliances designed to operate at 60 hertz and 120 volts are useless without proper adapting devices.

Alternating current for home and industrial use is generated by rotating a coil of wire 60 times per second in a region having a steady magnetic field. This rotation causes the wires in the coil to cut the magnetic field lines and to generate a voltage (an emf) which, in turn, produces a current if the coil is part of a complete circuit. The voltage and the current oscillate back and forth 60 times per second since the voltage generated during the second half of a rotation is in the opposite direction from that generated during the first half of the rotation. The voltage generated (and hence the current) depends on the *rate* at which the wires in the coil cut the magnetic field lines. Figure 8.30a shows an end view of a single loop of wire rotating in a magnetic field (the edges of the loop extend into the paper). When the end of the loop is aligned along the direction AE the edges of the loop are moving parallel to the field, and the rate at which field lines are being cut is zero. Conversely, when the coil is aligned along the direction of CG the edges of the coil are moving perpendicular to the field lines and are cutting them at a maximum rate. Thus the current at AE would be zero and at CG it would be a maximum, as shown in Figure 8.30b under A and C, respectively. By contrast, direct current reaches a steady value and remains there until something in the circuit changes (Fig. 8.30c).

A voltmeter specifically designed to measure direct current shows the actual value of the potential difference. The AC voltmeter, altogether a different instrument from the DC voltmeter, reflects the *effective value* of the AC potential difference. The effective value is a method of averaging AC potential difference so that it supplies the same power to a resistance as an equivalent DC potential. Using this averaging, it is found that the maximum AC potential, V_{max}, is $\sqrt{2}$ times the effective potential ($\sqrt{2} = 1.41$ to three significant figures). Then of course, the effective potential is $1/\sqrt{2}$ times the maximum, or $0.707V_{max}$, to three significant figures. As an example, standard 60-hertz, 120-volt household alternating current actually reaches a peak value of $\sqrt{2} \cdot 120\text{ V} = 170\text{ V}$ twice each complete cycle (once positive and once negative), or 120 times per second.

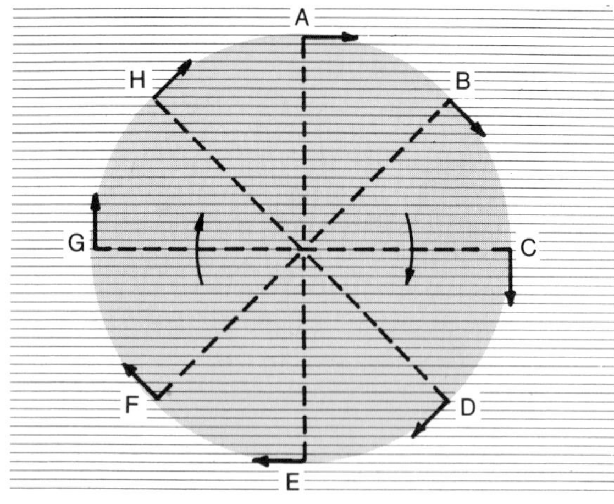

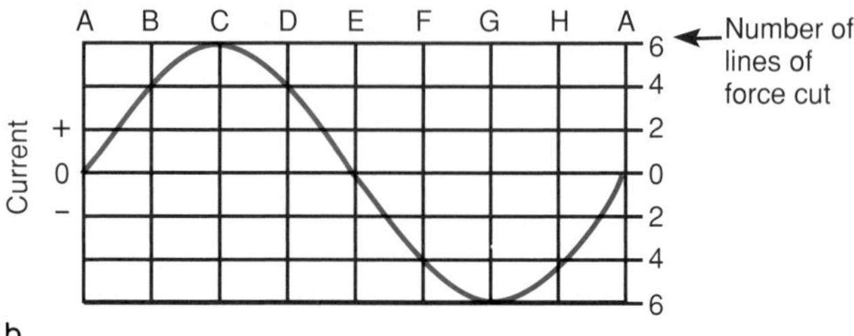

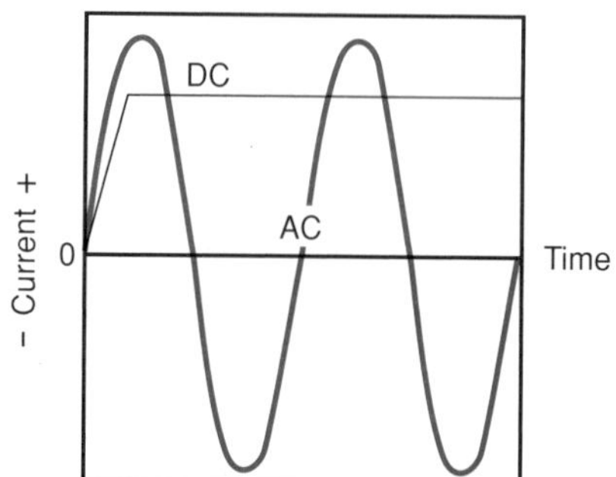

Figure 8.30 (*a*) The parallel lines represent the lines of magnetic field that are actively being cut by a conductor, the number of which can be noted to vary. The motion of the conductor, depicted by the arrows, at A, B, C, etc., represents the motion in a short time interval. (*b*) The amount of current varies with the number of lines actively cut. The capital letters relate to those in (*a*). (*c*) A comparison of direct current (DC) with alternating current (AC) in terms of variation in current intensity with time.

Example 8.10

The effective voltage of an AC source is 234 volts. What is the value of the maximum value of the source?

Solution

$$\begin{aligned} V_{max} &= \sqrt{2} \cdot 234\text{ V} \quad \text{MAXIMUM VOLTAGE} \\ &= 1.41 \cdot 234\text{ V} \\ &= 331\text{ V}. \end{aligned}$$

Commercially, alternating current is preferred over direct current for two primary reasons. First, at typical operating temperatures, production costs for alternating current are significantly less than those for direct current. Second, unlike direct current, AC voltages can be effectively increased by transformers. This feature is very significant, since as electric current is transmitted from the generating plant to the consumer, the voltage must be increased many times the generated value to minimize the amount of electrical energy converted to heat and hence wasted. These topics are discussed in more detail in Chapter 9.

Electricity in General

Electricity is perhaps our most widely used source of energy, yet its properties are not well understood and are often misconstrued by most of us. Even scientists and experts in various technological fields have many questions and interpretations concerning the exact nature of electricity. The following discussion presents a model of what occurs in a simple electrical circuit.

A sudden surge of energy is supplied by electrons to an automobile headlamp practically at the instant the switch is turned on to complete the electrical circuit. The energy necessary to heat the filament in the bulb to a temperature of white hot is conducted from the source of energy through wires made primarily of copper. Such electrical conductors are composed of atoms, the outermost orbiting electrons of which are loosely attracted to the individual atoms. These electrons from the tremendous number of atoms in the wire are somewhat free to roam about. They move at high velocities in the conducting wire, but in no definite direction; thus, there is no appreciable electric current in the conductor. A detectable flow of electrical charge would result only if the electrons were caused to move in the same direction by an attractive or repulsive electrical force. In other words, an electrical potential difference between the ends of the wire must be maintained, such as by a battery. The electric field that results moves through the electrical circuit as direction is given to the motion of the relatively free electrons, and they are forced toward the positive terminal of the battery. The directed path of the electrons is a short one, since they collide with the stationary atoms. The velocity of these electrons is very low, but the energy carried by the electric field they provide travels at relatively high velocity. As an example, long-distance communication is provided by the motion of electrical energy in a wire. During a telephone conversation with a person in Europe, a voice may be delayed in transmission long enough to permit the individual who is talking to pronounce the next syllable before the previous one is received by the listener. Precise measurements have been made indicating that electrical energy travels several million miles per hour, whereas light energy travels in excess of 600 million miles per hour. In addition, it is assumed that the electrons that furnish the electrical energy vibrate in many directions in a conductor, and move in a random fashion as they are accelerated toward the positive terminal of the source by the effect of an external source of electrical energy. The accumulated charge, or *electric field,* so produced—not the electrons—travels at a velocity of several million miles per hour.

Most users of electrical energy have sustained various degrees of electric shock. In order to receive an electric shock, two parts of your body must have an existing difference in electrical potential. If a difference exists, current can flow along a path in your body between these two regions. The damaging effects of electric shock are brought about by the current itself, rather than by the voltage to which one is exposed. One can survive a shock of millions of volts or can succumb to a shock of 50 volts or less. A current of 0.05 ampere is lethal in certain instances; victims seldom survive a current of 0.1 ampere if it flows through the heart. Often a person who is electrocuted has sweaty hands and is using a poorly "grounded" electric drill, a faulty electric hedge trimmer, or a similar device and is standing on wet grass or a damp concrete floor. Others suffer fatal electric shocks by simply touching an electrical appliance, light switch, or electrical outlet while they are bathing. Many unfortunate accidents occur when a person in a bathtub drops a hair dryer, radio, or similar appliance into the bathwater. Whether the appliance is turned on or off makes no difference; if it is plugged into an electrical outlet, a potential of 120 volts is applied to the impure water (a good conductor), and the result is the same.

According to Ohm's law, the current that flows through a circuit depends on the applied voltage and the resistance. The resistance of the human body can vary from 100 to over 500,000 ohms, depending on various circumstances. For instance, sweaty hands are good conductors of electricity because of the high content of salt in perspiration; they may offer a resistance of less than 800 ohms. The same hands when clean and dry may provide a resistance of 100,000 ohms. Even with moist skin, most individuals would not sense an electric shock received from an automobile battery. However, one should never touch any device capable of delivering a dangerous electric shock, such as the ignition system components of an idling automobile engine, with both hands simultaneously. The path for the current that this situation would provide includes the upper part of the body, and thus the heart.

Also, if a dangerous source such as a damaged conducting wire is handled, the sudden shock often causes the muscles in the hand to contract, and the victim cannot release the conductor before suffering a serious prolonged shock. If, for some reason, you must handle a potentially dangerous source, first touch the source with the back of your hand. The natural reflex action of the hand separates the hand from the conductor.

Summary

Electric charge is a property of certain particles that manifests itself as a force between the particles. There are two types of charge, called positive and negative. Electrons have negative charge and protons have positive charge, with the magnitudes of the charges being equal. A normal atom has equal numbers of electrons and protons and is neutral, while an ion is an atom (or a molecule) with an excess or deficiency of electrons. The electrons, because of their location in the outer parts of the atom, are the carriers that usually transport charge. Insulators offer great resistance to the flow of electric charge, whereas conductors offer an easy path through which the electric charge may travel.

Electricity at rest is called static electricity. Electric charge in the form of an excess or deficiency of electrons can be stored on insulators such as plastic or glass. Static electricity is produced by friction between two different insulating substances. One substance assumes a negative charge, the other a positive charge. Like electrical charges repel; opposite charges attract. An electroscope is a device to detect static electricity.

An electric charge is an accumulation or deficiency of electrons. Work done to move this charge against an electric field increases the electric potential energy of the charge. The electric potential energy per unit charge is known as electric potential, a value measured in volts. Electric potentials can be developed by friction between insulating substances, by chemical action, and by electric generators. The unit of charge is the coulomb, and the rate of charge flow, or current, is measured in amperes.

The quantity of charge that flows in a circuit depends on the electrical potential energy available and the opposition to flow the charge encounters, known as resistance. Several factors regulate the electrical resistance in a circuit; among these are the length, diameter, composition, and temperature of the conductor.

Current, voltage, and resistance are related according to Ohm's law; that is, $I = V/R$. The current, measured in amperes, is directly proportional to the impressed voltage (the voltage provided by an external source) and inversely proportional to the resistance in the circuit.

The ancient Greeks are given credit for discovering that naturally occurring iron ores attract bits of iron. Later, other investigators learned that iron bars stroked with lodestone attract small bits of iron. Iron alloys poured in the form of bars and magnetized were found to lift iron objects many times their own weight. William Gilbert published his findings about magnetism in the sixteenth century. The relationship of electricity and magnetism was first conceived in the early part of the eighteenth century.

Permanent magnets may be made from iron and selected iron alloys by magnetic induction. The strength of the magnet produced depends on the alignment of the magnetic domains within the magnetic substance. Permeability is a measure of how well the domains stay aligned. Regardless of the length or shape of the magnet, the aligned domains produce a north and a south pole near the magnet's ends. Earth is a gigantic magnet, the magnetic poles of which slightly deviate from the geographic poles. A measure of the deviation at a specific location is called the magnetic declination.

A conductor through which an electric current flows is surrounded by a magnetic field. If the conductor is wrapped around a soft-iron core, the magnetic field is enhanced by the core, and the looped conductor and the core form a temporary magnet. Essentially, the moment the current ceases to flow, the core loses its magnetic field. The electromagnet, as the device is called, is very useful in industrial and in home applications.

When a conductor cuts across lines of magnetic field, whether the conductor is moving or the field is changing, an emf is produced and a current may be generated. The emf (voltage) produced depends on the rate at which the magnetic field lines are cut. If the conductor is a coil of wire rotating in a steady magnetic field, the current will oscillate back and forth in the coil, giving what is called alternating current, the type of electricity typically used in the home.

Current electricity results from the propogation of an electric field in a conductor, not from the flow of electrons. The electric field moves through the conductor at a velocity that approaches the velocity of light. The danger from electric shock lies in the amount of current that flows, rather than in the electric potential, as is commonly thought. An electric current will exist in a conductor when there is a potential difference between two points in the conductor.

Equation Summary

Coulomb's law:

$$F = k\frac{q_1 q_2}{r^2},$$

where F is the force (newtons) of attraction or repulsion of charges q_1 and q_2 (coulombs) when they are separated by a distance r (meters) and k is a constant equal to $9.00 \times 10^9 \text{ N}\cdot\text{m}^2/\text{C}^2$.

Electric field:

$$E = \frac{F}{q},$$

where E is the electric field (newtons per coulomb) at a location where the force F (newtons) acts on the charge q (coulombs). This equation may be rearranged to find the force on the charge q if the electric field is known. Thus,

$$F = qE.$$

Electric potential:

$$V = \frac{W}{q},$$

where V is the electric potential difference (volts) between two points when it requires an amount of energy W (joules) to move the charge q (coulombs) from one point to the other. V is positive if the energy required is positive and the charge is positive. The energy difference of the charge when moved between the two points can be computed from

$$W = qV.$$

If W is positive, work must be done to move the charge between the two points, but if W is negative a positive charge will lose this amount of energy when moving between the two points (which can be utilized in operating electrical devices, for example).

Electric current:

$$I = \frac{q}{t},$$

where the current I (amperes) is equal to the charge q (coulombs) that flows in the time t (seconds). Rearranging to find the charge if the current and time are known gives

$$q = It .$$

Electromotive force:

$$\mathcal{E} = \frac{W}{q},$$

where the electromotive force $\mathcal{E}$ (volts) is equal to the energy W (joules) given to a charge q (coulombs) by a nonelectrostatic force.

Power and electromotive force:

$$P = \mathcal{E}I ,$$

where the power P (watts) supplied by an electromotive force $\mathcal{E}$ (volts) is equal to the electromotive force $\mathcal{E}$ times the current I (amperes) passing through the seat of electromotive force (a battery, for example).

Ohm's law:

Three forms of Ohm's law are

$$V = I \cdot R ,$$
$$I = \frac{V}{R},$$

and

$$R = \frac{V}{I},$$

where V (volts) is the potential difference across a resistance R (ohms) while the current through the resistance is I (amperes).

Alternating current:

$$V_{max} = \sqrt{2}\, V_{eff} = 1.41\, V_{eff}$$

and

$$V_{eff} = V_{max}/\sqrt{2} = 0.707\, V_{max} ,$$

where V_{eff} (volts) is the effective potential for an alternating voltage, while V_{max} (volts) is the maximum value of the voltage (at its peak).

Questions and Problems

Electricity at Rest

1. An electric cord for an appliance consists of two pieces of copper wire surrounded by a plastic covering. Explain the functions of the copper and the plastic.
2. Mounted on the undercarriages of some trucks that transport gasoline and other highly combustible substances is a chain that comes in contact with the ground. What primary purpose does such a chain serve?
3. If you have acquired a positive charge from sliding across a car seat on a cold winter day, what can you say about the charge you left on the seat? What can you say about the charge induced on the metal door handle of the car just before you touch it?

Coulomb's Law

4. In a classroom demonstration, two small spheres are given equal charges of $+0.150\ \mu C$ ($+0.150 \times 10^{-6}$ C). If they are separated by a distance of 0.115 m, what is the force between them? Is the force attractive or repulsive?
5. If you charge a rubber balloon by rubbing it with wool, the balloon will stick to a wall or a wooden door. Why? Would you expect it to stick to a bare metal surface? Why or why not?

The Charge on the Electron

6. Ordinary salt is made up of equal numbers of sodium and chlorine atoms. Each chlorine atom attracts and holds one of the electrons originally attached to a sodium atom. Each chlorine atom thus has an excess of one electron, while each sodium atom has a deficiency of one electron. (a) What is the charge on each sodium atom? What is the charge on each chlorine atom? Is the electrical force between sodium atoms and chlorine atoms one of attraction or repulsion? Explain. (b) Is a crystal of ordinary salt, which contains millions of sodium and chlorine atoms in equal numbers, neutral or charged? Explain.

Electric Fields

7. (a) A charge of $+5.00\ \mu C$ is placed at a location and is found to have an upward force of 0.250 N on it. What is the direction of the electric field in which this charge is located? What is the magnitude (size) of the electric field?

(b) The same charge is moved to another location and it is found to now have an upward force of 0.500 N on it. How does the electric field at the second location compare with that at the first location?

Electric Potential and Electric Charge in Motion

8. It is found that it takes 600 J of work to move a charge of +5.00 C from point A to point B. (a) What is the electric potential (in volts) of point B relative to that of point A? (b) If the charge is allowed to return to point A, how much work will be available as the charge returns?

Electric Current

9. What is the electric current (in amperes) if 4.50 C of charge flow from one point in a circuit to another in a time of 2.00 s?

10. A starter motor in a car draws a current of 58.2 A for 4.25 s. How much charge passed through the starter motor?

Electromotive Force

11. A battery in a flashlight supplies 5.00 J of energy to the bulb when 0.600 C of charge pass through the bulb. What is the emf $\mathcal{E}$ of the battery in volts?

12. How much electric charge moves in a circuit that carries a current of 20.0 A for a period of 10.0 s? If the potential furnished to this circuit by a source is 12.0 V, what is the power supplied to the circuit?

Power Supplied by an Electromotive Force

13. A storage battery with a potential of 12.0 V is used on a boat to furnish the voltage for a spotlight that draws a current of 6.50 A. How much power is supplied by the battery to the spotlight?

Resistance and Ohm's Law

14. Why are large-diameter rather than relatively small-diameter copper wires used to conduct electric current throughout our homes?

15. How can birds perch on high-voltage wires without danger of being electrocuted? If the birds were so close that a bird's wings touched two wires simultaneously, would the bird be in danger?

16. How much current is used to operate a hair dryer, the resistance of which is 16 Ω, and which is plugged into a 120-V outlet?

17. A student using a small heater to warm her dormitory room is distressed when the heating element "burns out." She disconnects the heater, finds the place that has burned through, and shortens the heating element (a coiled wire) by twisting the burned ends together with a pair of pliers (definitely not recommended). Is the resistance of the heater after the "repair" more or less than it was before, assuming there is negligible resistance between the twisted ends? Will the heater now draw more or less current than before if it is "plugged into" the same socket?

18. If the electrical resistance of a mechanic's body is 1000 Ω, how much current would flow through it if he touched both terminals of a 12.0-V battery?

19. The electrical resistance of a 100-W light bulb designed to be operated from a 120-V source is 144 Ω. How much current passes through the bulb when it is operating? (Note: You do not need all the numbers given in this problem.)

20. A wire conductor 100 m long has a resistance of 90.0 Ω. (a) What will be the resistance of 300 m of the wire? (b) How much current would flow through the longer wire if it were directly attached to a 12.0-V battery?

Theories of Magnetism

21. Each piece of iron has domains which may be thought of as tiny magnets. Why, then, is not every piece of iron a magnet?

22. Two rods identical in appearance are handed to you. One rod is a magnet; the other is not. How can you determine which rod is the magnet, using only observations of one rod's effect on the other?

23. What factors might determine the strength of a permanent magnet?

Electromagnetism

24. Describe how you would make an electromagnet if you were given a battery, a piece of iron, and a long piece of wire.

Alternating Current

25. What is the effective value of alternating current produced by an electrical generator if the maximum value is noted to be 190 V?

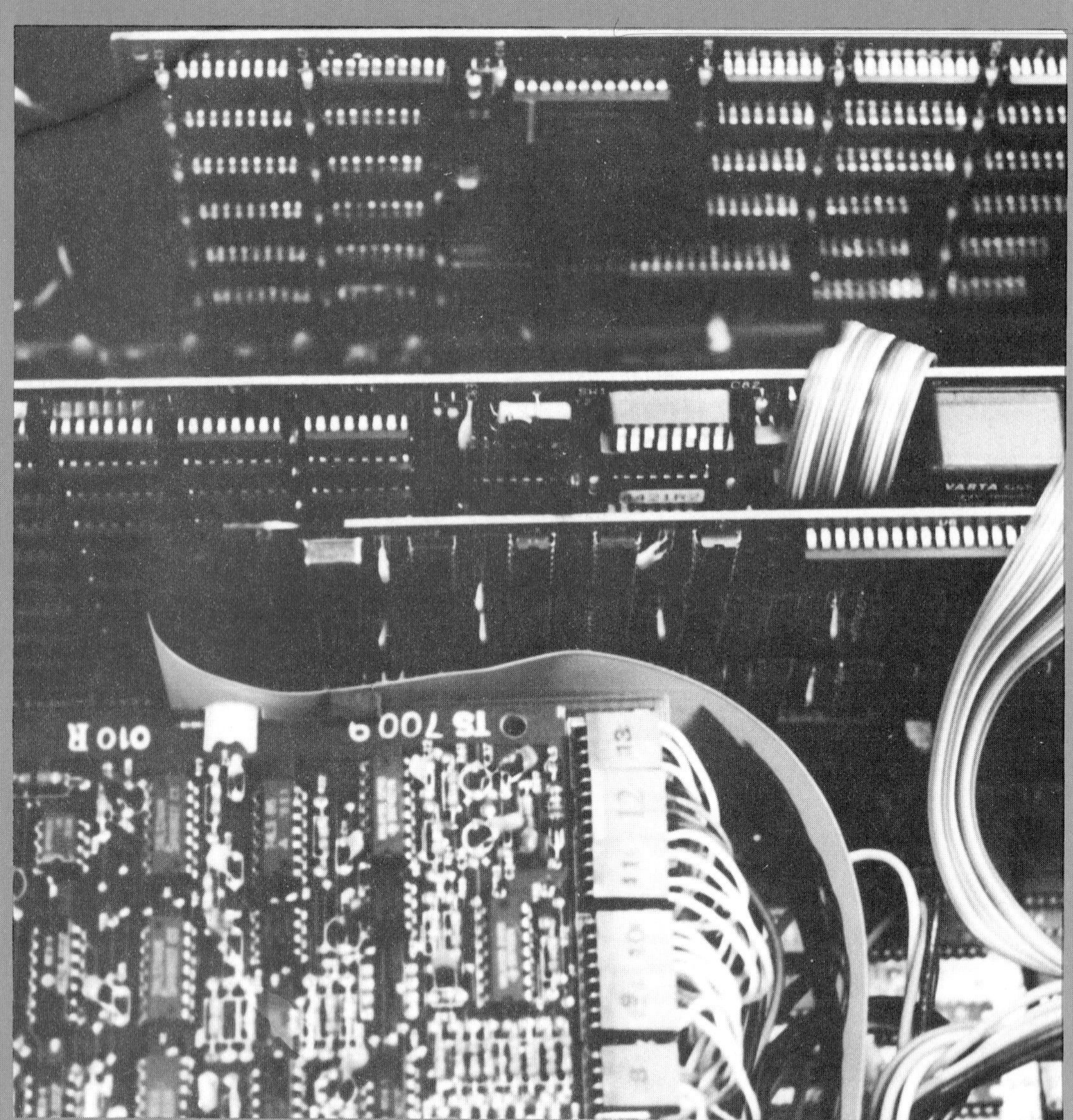

9 Applied Electricity and Electronics

Chapter outline

Key terms/concepts

The microcomputer exemplifies the applications of electricity. It not only utilizes the principles of electricity, but at the same time may be used as a tool to further improve applications.

The flow of electric charge in a circuit is an electric current. The amount of charge that moves from the negative pole to the positive pole of the source per second (coulombs/second) gives the magnitude of the current in amperes (as discussed in Chapter 8). If *the charge flows in one direction only,* it is called **direct current.** If the charge *continuously changes direction in the conductor,* the flow is called **alternating current.**

The flow of electric charge is caused by some electric generating device that provides the work required to push the charge through the circuit. This charge travels from the negative pole of the generator through a path in the circuit to the positive pole of the generator. The charge changes direction only as the polarity of the generator reverses.

There is a similarity between what occurs when electric charge flows in a circuit and what occurs when water flows in a pipe. The water pressure in the pipe results from the greater potential energy of the water at the reservoir than at the other end of the pipe. Charge flows in a conductor because there is a difference in electric potential across the ends of the conductor. The flow of water or of charge persists until both ends of the pipe or of the conductor have a common potential. If the pipe is small, less water flows, even though the difference in potential energy remains constant. In electricity, the amount of charge that flows is less in a small wire than in one of larger diameter, even though the electric potential remains unchanged. If water in the pipe is forced to turn a paddle wheel placed in its path, less water also flows, since there is resistance in addition to that related to the size of the pipe. The amount of electric current that flows in a wire or the amount of water that flows in a pipe is directly related to the potential energy provided and is inversely related to the total resistance that the flow encounters.

Energy and Power in Electric Circuits

The three basic quantities in electricity—current (I), electric potential (V), and resistance (R)—are related according to **Ohm's law,** $V = IR$. V is measured in volts, I is measured in amperes, and R is measured in ohms (see Chapter 8). An electric potential of 1 volt causes a current of 1 ampere if the electrical circuit has a resistance of 1 ohm.

These three quantities also govern the amount of heat produced in a circuit. A voltage difference between two points joined by a conductor forces electric charge to flow through the conductor. This charge meets resistance to its motion, requiring work be done on the charge by an electromotive force to maintain its motion. The work done on these charges increases their potential energy, which is then converted into heat by collisions within the conductor as the charges pass through the circuit. For a fixed amount of current, the greater the resistance, and the greater the amount of heat produced. The heat developed by a current generally causes the temperature of the conductor to rise. This increase in temperature depends not only on the amount of heat, but also on how rapidly this heat is dissipated. In an electric iron, for instance, the heat emitted produces no visible light. However, in an electric toaster the heating element reaches such a high temperature that visible light is produced. The wire glows red-hot. The filament in a light bulb becomes so hot that it emits "white" light. In the iron and the toaster, the amount of heat generated is much more than in a typical light bulb, but the heat is dissipated more rapidly, resulting in a lower temperature.

The energy dissipated by the charges moving through a conductor is usually supplied by an electromotive force, typically within a battery or generator. Recall that the voltage is a measure of the energy (E, in joules) of the charge per unit of charge (q, in coulombs). Thus, when a positive charge moves to a lower voltage (or a negative charge to a higher voltage), it loses an amount of potential energy equal to the product of the voltage and the charge, Vq. In an electric circuit, it is more convenient to measure current than charge. Current has been defined in Chapter 8 to be the rate of flow of charge, or $I = q/t$, so that 1 ampere = 1 coulomb/second. Then $q = It$, giving

$$E = Vq = VIt. \qquad \text{ELECTRIC ENERGY}$$

When electric lights and appliances are used, each of us pays the power company for the energy used by these devices in doing work. For example, if a light bulb operating at 120 volts has a current of 1 ampere passing through it for 60 seconds, the energy used is 120·1·60 J, or 7200 J (joules of work are done). The joule is an inconveniently small unit of energy for the power company to use. The larger unit that is used by power companies, the kilowatt-hour, will soon be discussed.

The *rate of energy usage* is the **power** (P) in **watts;** that is, $P = E/t$, so 1 watt (W) = 1 J/s. Taking the previous equation for electric energy, power is computed from

$$P = \frac{E}{t} = \frac{VI\cancel{t}}{\cancel{t}};$$

thus,

$$P = VI. \qquad \text{ELECTRIC POWER}$$

(Note that this is in accordance with the law of conservation of energy when compared with the power supplied by an emf, as discussed in Chapter 8.) Remember that if this is the voltage drop across a conductor and the current through the conductor, the work done on the charges is dissipated in the form of heat (called Joule heating). In a conductor, the voltage, current, and resistance are related by Ohm's law, $V = IR$ (p. 157). Thus, two other equations

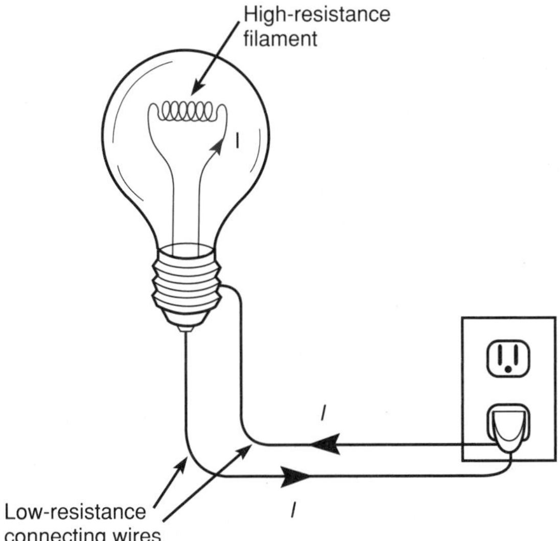

Figure 9.1 A simple lamp circuit illustrates the effect of resistance on parts of a circuit through which the current is the same. The resistance of the connecting wires is made small so that little heat will be produced in them, while the filament of the lamp has a high resistance so that it will be heated to incandescence.

for electric power in a conductor can be found, either by substituting for V in the power equation from Ohm's law, or by solving Ohm's law for I ($I = V/R$), and substituting for I.

Substituting IR for V gives

$$P = (IR)\,(I) = I^2R, \qquad \text{ELECTRIC POWER}$$

or substituting V/R for I provides yet another expression,

$$P = (V)\left(\frac{V}{R}\right) = \frac{V^2}{R}. \qquad \text{ELECTRIC POWER}$$

These last two equations show how power depends on the resistance of the conductor. If the current through a conductor is increased, the power increases as the current squared, so that doubling the current produces four times as much heat. If the current is kept the same, the heat generated is in direct proportion to the resistance; that is, twice the resistance gives twice the heat. Actually, most emf's give approximately a constant voltage, so that the current changes when the resistance changes. In this case, the last power equation shows that if the voltage is constant, decreasing the resistance increases the amount of power dissipated in the wire in the form of heat. For constant voltage, doubling the resistance will result in one-half as much heat being produced.

An important use of the equation $P = I^2R$ is in the design of appliances used in our everyday lives. Consider the simple lamp circuit shown in Figure 9.1. The same current, I, must pass through the connecting wires and the filament of the lamp. The resistance of the connecting wires is made as low as is practical by using large wires of highly conductive metal so that they will not become hot, but the resistance of the filament of the bulb is made high so that it will become hot. Similarly, house wiring is made with very low resistance to prevent overheating and possible fire. Any ordinary wire will become hot if enough current passes through it, so circuit breakers and fuses are designed to limit the current to safe levels, should a short circuit occur or should the circuit be overloaded by an excessive number of appliances.

Example 9.1

An electric heater with a resistance of 8.00 ohms is connected to a 120-volt wall outlet in a home. A circuit breaker or fuse is designed so that if the current exceeds a safe value it will interrupt the current. Find the smallest value for a circuit breaker or fuse that will protect the circuit but allow operation of the heater, assuming there are no other appliances connected to the same circuit. What is the power rating of the heater?

Solution

The first part of the problem requires computation of the current in amperes that will pass through the heater. Ohm's law, $V = IR$, can be used in the form $I = V/R$, where $V = 120$ V and $R = 8.00\ \Omega$. Then

$$\begin{aligned} I &= \frac{V}{R} \\ &= \frac{120\text{ V}}{8.00\ \Omega} \\ &= 15.0\text{ A}. \end{aligned}$$

So the circuit requires at least a 15-A circuit breaker or fuse.

The second part of the problem can be solved by using any of the three equations given for computing electric power, since $V = 120$ V, $R = 8\ \Omega$, and $I = 15.0$ A are all known. Thus,

$$\begin{aligned} P &= VI \\ &= (120\text{ V})\,(15.0\text{ A}) \\ &= 1800\text{ W}, \end{aligned}$$

or

$$\begin{aligned} P &= I^2R \\ &= (15.0\text{ A})^2(8.00\ \Omega) \\ &= 1800\text{ W}, \end{aligned}$$

or

$$\begin{aligned} P &= \frac{V^2}{R} \\ &= \frac{(120\text{ V})^2}{8.00\ \Omega} \\ &= 1800\text{ W}. \end{aligned}$$

So the heater has a power consumption of 1800 W.

The energy expended in a wire may also be expressed in units of thermal energy, Q (calories), as well as joules. It is found experimentally that 4.184 joules = 1 calorie, so that the equation for electric energy can be put in the form

$$Q = (VIt)\left(\frac{1 \text{ calorie}}{4.184 \text{ J}}\right)$$

$$Q = (0.239 \text{ cal/J})\,(VIt),$$

or

$$Q = (0.239 \text{ cal/J})(I^2Rt), \qquad \text{HEAT ENERGY}$$

where Q is now in calories. This last expression is known as **Joule's law.** (If kilocalories are desired, the multiplying factor is 2.39×10^{-4}.) As an example, the heater described in Example 9.1 gives off an amount of heat energy in 1.00 second equal to (0.239 cal/J)(120 V)(15.0 A) (1.00 s), or 430 calories.

The typical household depends on electricity for the operation of most home appliances. The power requirement for each home amounts to thousands of watts, since many of these devices are used simultaneously. Since the amount is so great, a multiple unit of power is ordinarily used along with a time factor to determine electric energy consumption instead of using the small unit of energy, the joule. This combination is called the **kilowatt-hour** (kWh). The measure, as the name implies, is equivalent to 1000 watts being used for a time of one hour. It is a simple matter to show that 1 kWh = 3.6×10^6 J.

Example 9.2

A television set that requires 600 watts of electrical power is continuously operated for 10.0 hours. How many kilowatt-hours of energy are used? If the electrical energy costs 11.0¢ per kilowatt-hour, what is the cost of operating the television set for this period of time?

Solution

The energy usage is found by multiplying the power by the time. Since kilowatt-hours are desired and the power is given in watts, a conversion is necessary. It can be made before or after the multiplication. Carrying out the conversion before the multiplication gives

$$P = (600 \text{ W})\left(\frac{1 \text{ kW}}{1000 \text{ W}}\right)$$
$$= 0.600 \text{ kW}.$$

Then the energy in kilowatt-hours is found to be

$$E = (0.600 \text{ kW})\,(10.0 \text{ h})$$
$$= 6.00 \text{ kWh}.$$

So 6.00 kWh of electrical energy were used to operate the television set.

The cost is found by multiplying the energy by the cost per unit of energy, 11.0¢/kWh, giving

$$\text{cost} = (6.00 \text{ kWh})\left(\frac{11.0¢}{\text{kWh}}\right)$$
$$= 66¢.$$

It is found that the cost is 66¢ to operate the television set for 10 h.

Figure 9.2 The amount of energy consumed by each household is measured by the kilowatt-hour meter.

The company that supplies electricity to the consumer uses a meter similar to that illustrated in Figure 9.2. The meter is connected so that all electrical energy consumed must pass through it before the energy enters the home. The device measures the product of three variables—voltage, amperage, and time—and constantly adds the result to the dial readings. The difference in monthly dial readings determines the monthly consumption, and the consumer is billed accordingly.

Series and Parallel Circuits

Many useful electrical circuits contain devices connected in combinations called series and parallel. If devices are connected between two points in a circuit such that to get from one point to the other requires *passage through all of the devices,* they are said to be connected in **series.** On the other hand, **parallel** combinations

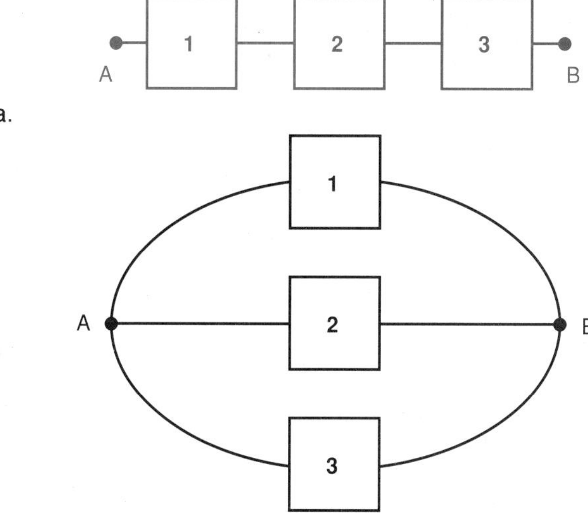

Figure 9.3 In (*a*), the elements are in series between points A and B, since to get from A to B requires passage through all elements. If there is a choice of paths between two points, as in (*b*), the elements are said to be in parallel between A and B. A and B represent two points within a circuit, the remainder of the circuit not being shown in this diagram, while the elements 1, 2, and 3 represent various circuit elements, such as batteries or resistors.

permit *passage along alternate paths.* Figure 9.3 illustrates series and parallel combinations.

A single dry cell, such as that used in a standard flashlight, has an emf of 1.5 volts, whereas one cell of an automobile storage battery has an emf of 2 volts. The voltage between the terminals of a cell is slightly less than the emf when a current is passing through the cell because the cell itself has a small amount of resistance, called the internal resistance. Batteries made up of several single-cell units can deliver more voltage than an individual cell. An increase in voltage above 1.5 or 2 volts, respectively, indicates that the cells are arranged and connected so that the voltage of the cells is additive and that they are, therefore, part of a series connection. This arrangement calls for the negative terminal of one cell to be connected to the positive terminal of another. The total voltage produced in this way is equal to the sum of the single-cell voltages, a concept illustrated in Figure 9.4. Mathematically, the result of connecting cells in series is given by the expression $V_t = V_1 + V_2 + V_3 + \cdots + V_n$, where V_t is the total voltage and V_1, V_2, and so on, are the voltages of the individual cells.

The electrical system of the automobile is generally a 6-volt or a 12-volt system. The 6-volt system uses three cells of 2 volts each, connected in series. The 12-volt system uses a series of six cells with an electric potential of 2 volts each. Both these arrangements are illustrated in Figure 9.5.

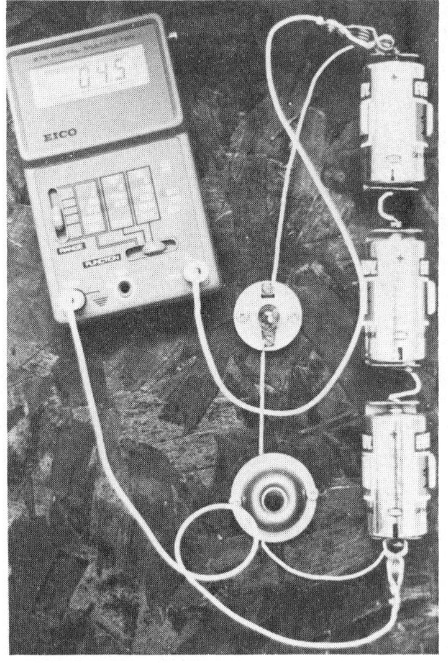

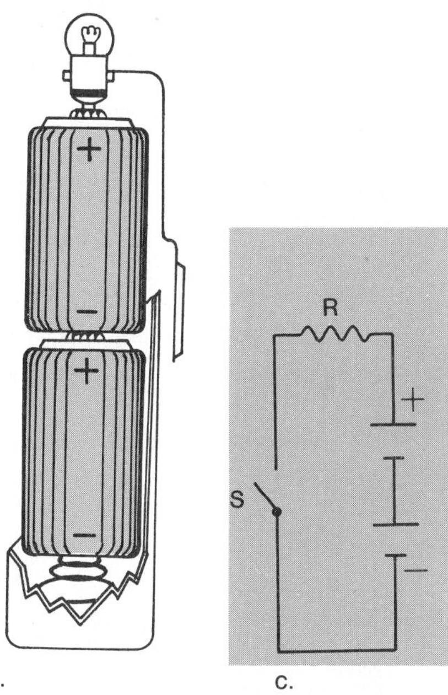

Figure 9.4 (*a*) The method of connecting three 1.5–V dry cells in series to provide 4.5 V. (*b*) A pair of 1.5–V dry cells arranged to provide 3 V, a series arrangement typical in the flashlight. (*c*) A schematic of the arrangement in (*b*).

Figure 9.5 Typical arrangements of cells to produce voltages required for automobile electrical circuits. The caps on top of each battery provide a means to add water to each 2-V cell.

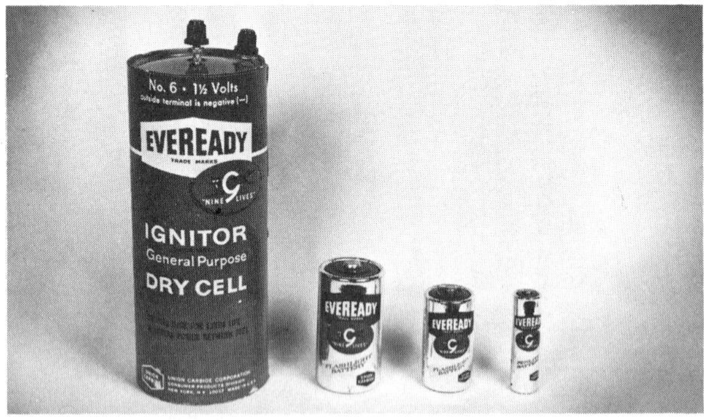

Figure 9.6 The variation in size of dry cells of equal voltage.

a.

b.

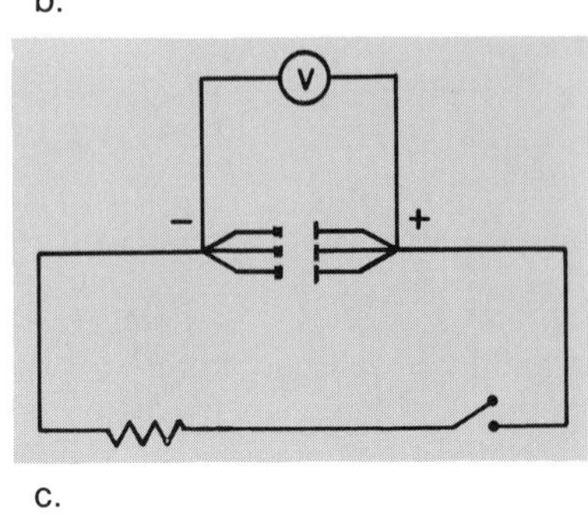

c.

Figure 9.7 (*a*) The result of connecting individual dry cells in parallel. All positive poles are connected jointly as are all negative poles, a method that produces a potential difference of 1.5 V (note the accompanying schematic). (*b*) This special battery is composed of four 1.5-V dry cells. Through parallel connections, the total voltage of the battery is 1.5 V. (*c*) The schematic of the circuit in (*a*).

Generally speaking, the dry cell has a relatively short useful life because the amount of chemical energy that can be stored in the battery and converted to electrical energy is limited. In addition, the internal resistance increases with age and use. The typical dry cell cannot have this energy replenished as the storage cell (Fig. 9.5) can. However, rechargeable (in the sense that the energy is replenished) dry cells are gaining in popularity and undoubtedly will replace, to a great extent, the standard nonrechargeable dry cell. Although all the standard cells deliver 1.5 volts, dry cells are made in many shapes and sizes. Each cell shown in Figure 9.6 delivers 1.5 volts. Naturally, the larger cells will last longer, since they can supply more total electrical energy than the smaller cells can.

Sometimes the cells on hand are not large enough to offer the amount of current required by a given circuit. The solution to the situation is obtained by connecting a number of 1.5-volt cells in parallel; that is, by connecting all positive terminals to each other and all negative terminals to each other. The method of connecting several dry cells in parallel is illustrated in Figure 9.7a. A large dry cell composed of smaller cells connected in parallel is shown in Figure 9.7b. Parallel circuits composed of many dry cells have a resultant voltage equal to the highest voltage of any of the cells connected in parallel. Normally, all the cells connected in parallel have the same voltage, so that $V_t = V_1 = V_2 = \cdots = V_n$, where V_t is the total voltage and V_1, V_2, and so on, are the voltages of the individual cells. If the cells are not of equal voltage, the cell of highest voltage will tend to "charge" the cells of lower voltage, resulting in a drain of energy from the high-voltage cell. (Sometimes this is desirable, as will be discussed shortly for automobile batteries.) The purpose of connecting dry cells of equal voltage in parallel is to make a greater amount of electrical energy available. The electrical energy that can be furnished by dry cells wired in parallel is the sum of the electrical energy that the individual dry cells can provide. If four cells, each with an electric potential of 1.5 volts, are connected in parallel, the electrical energy made available to the circuit is four times the amount available from one 1.5-volt cell. In addition, the four cells will last significantly longer when they are connected in parallel than each cell would last if used individually, due to several factors beyond the scope of the current discussion.

Automobile owners occasionally forget to turn off their headlights or parking lights. To their dismay, they return to discover the battery no longer has sufficient electrical energy to start the automobile. With a set of "jumper" cables, the weak battery may be attached in parallel to a strong battery; that is, the positive terminal of the weak battery is connected by means of the cable clamps to the positive terminal of the strong battery. Then the negative terminal of the strong battery is connected to a negatively "grounded" component on the engine as generally described in the owner's manual. The circuit should be completed in this manner to guard against the possibility of a spark igniting hydrogen gas released by the weak battery as it converts electrical energy to chemical energy.

Cells are not the only electrical components that are ordinarily wired into a circuit in either parallel or series. Various types of resistances are also connected in both ways for specific effects. When several electrical devices are connected in series, the resistance (R) of the combination is equal to the sum of the resistances of each component. In a symbolic fashion, the expression is written as $R_t = R_1 + R_2 + R_3 + \cdots + R_n$. Components called resistors are included in electronic circuits to govern the current that flows through a circuit.

Example 9.3

Three resistors with resistances of 5.00 ohms, 3.00 ohms, and 1.00 ohm are connected in series in a circuit that contains a battery, the electric potential of which is 18.0 volts. What is the total resistance that the resistors supply the circuit, and what is the current through the circuit?

Solution

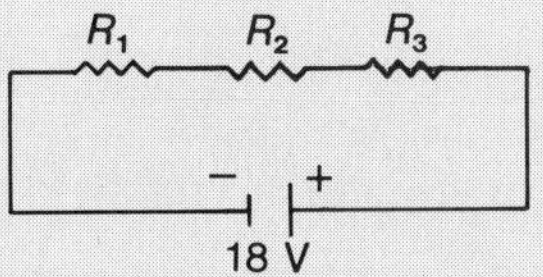

The resistance is found from

$$\begin{aligned} R_t &= R_1 + R_2 + R_3 \\ &= 5.00\ \Omega + 3.00\ \Omega + 1.00\ \Omega \\ &= 9.00\ \Omega. \end{aligned}$$

SERIES RESISTORS

Then, using Ohm's law,

$$\begin{aligned} I &= \frac{V}{R} \\ &= \frac{18.0\ \text{V}}{9.00\ \Omega} \\ &= 2.00\ \text{A}. \end{aligned}$$

In reference to the previous analogy between the flow of water and electricity, consider the effect on the flow of water if three identical pipes were placed parallel to each other and all connected to the water source. How much of the total volume of water that flowed would be carried by each pipe? How would pipes of various diameters placed parallel affect the amount of water that flowed through each of them? Likewise, would wiring resistors in parallel affect the current through a circuit? The total effect of wiring resistors in parallel is obtained from the expression $1/R_t = 1/R_1 + 1/R_2 + 1/R_3 + \cdots + 1/R_n$. If a resistor wired in parallel with other resistors fails to perform, voltage is still present in the balance of the circuit. Also, total resistance can be varied with parallel resistors to provide a control of current.

Example 9.4

Three resistors are wired in parallel and included in an electrical circuit. If $R_1 = 12.0$ ohms, $R_2 = 8.00$ ohms, and $R_3 = 24.0$ ohms, what is their total resistance? If the circuit includes a 120-volt source of AC, how much electrical power is used?

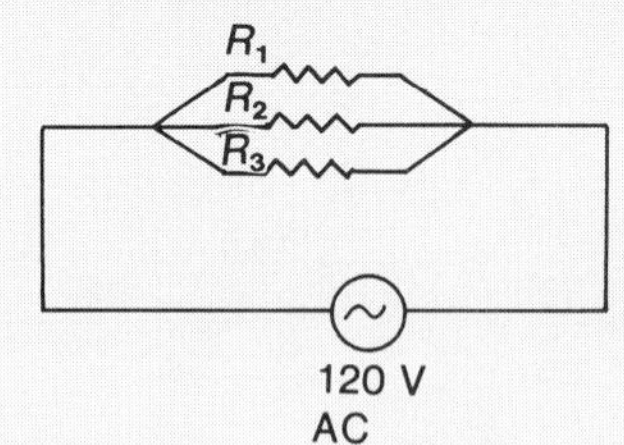

$$\frac{1}{R_t} = \frac{1}{R_1} + \frac{1}{R_2} + \frac{1}{R_3} \qquad \text{PARALLEL RESISTORS}$$

$$= \frac{1}{12.0\ \Omega} + \frac{1}{8.00\ \Omega} + \frac{1}{24.0\ \Omega}$$

$$= \frac{2.00}{24.0\ \Omega} + \frac{3.00}{24.0\ \Omega} + \frac{1.00}{24.0\ \Omega}$$

$$= \frac{6.00}{24.0\ \Omega}.$$

Cross-multiplying,

$$6.00\ R_t = 24.0\ \Omega$$

$$R_t = 4.00\ \Omega.$$

Then, using Ohm's law,

$$I = \frac{V}{R}$$

$$= \frac{120\ \text{V}}{4.00\ \Omega}$$

$$= 30.0\ \text{A}.$$

Finally, from the expression for power,

$$P = VI = (120\ \text{V})(30.0\ \text{A}) = 3600\ \text{W},$$

or

$$P = \frac{V^2}{R} = \frac{(120\ \text{V})(120\ \text{V})}{4.00\ \Omega} = 3600\ \text{W}.$$

Lights used for holiday decorations are ideal examples of parallel and series circuits. If the bulbs in a set of lights are wired in parallel and one or more of the bulbs burn out, the remaining bulbs will continue to glow. The incandescent lamps in a light fixture or in a room with multiple fixtures all controlled by a single switch are wired in parallel. The switch, however, is wired in series with the light fixtures, as is the fuse or circuit breaker in the circuit. The method of wiring multiple light fixtures on the same circuit is illustrated in Figure 9.8a. In series circuits, if a single lamp is burned out, the entire set of lamps fails to glow. If the filaments in all lamps are intact, the brightness of each lamp wired in the series is considerably less than its rated intensity. A series circuit is displayed in Figure 9.8b. Lamps of the same-rated intensity as those used in Figure 9.8a are used in this series circuit.

Alternating Current and the Transformer

Electricity now furnishes the energy for almost every task undertaken in the home and at work. It has furnished us energy with which to illuminate our homes and streets, to heat our homes, to communicate with one another, and to operate our household appliances.

Batteries remain our most common source of direct current; however, direct current can be produced more economically with a mechanically driven generator. In this device, a mechanical force is used to rotate a series of wire loops in a magnetic field, generating alternating current in the wire loops. The generator brushes—electrical conductors of copper or carbon that make sliding contact between a stationary and a moving part of the generator—receive only part of each AC cycle produced by the generator and carry the pulsating type of direct current that results to the consuming device or storage battery.

Prior to the 1960s, automobiles used DC generators to provide the electrical power needed and to charge the cars' storage batteries. The AC generator, commonly called the alternator, was found to be more advantageous and has replaced the DC generator in modern automobiles. The alternator has the capability of delivering more current at low speeds than a DC generator can usually provide; this is important in meeting large electrical demands in congested city traffic. Today's automobiles contain motors, conventional storage batteries, radios, tape players, CD players, ignition-system components, and electrical gauges that still must be furnished with DC voltages. To provide the proper type of voltage, a *rectifier,* a device that allows current to flow in only one direction, converts the alternating current into direct current.

Electricity supplied to homes and commercial installations comes from large generating stations that generate alternating current. The voltage of alternating current can be easily controlled with a **transformer** (Fig. 9.9), a device that can raise ("step up") or lower ("step down") the voltage using coils of wire wound on an iron core. Transformers require the use of alternating current. Recall from Chapter 8 that when a magnet is pushed into or withdrawn from a coil of wire, a current is produced in the coil. If the

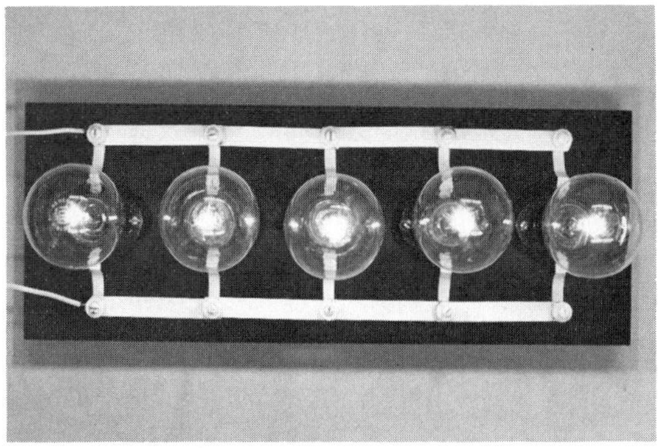

a.

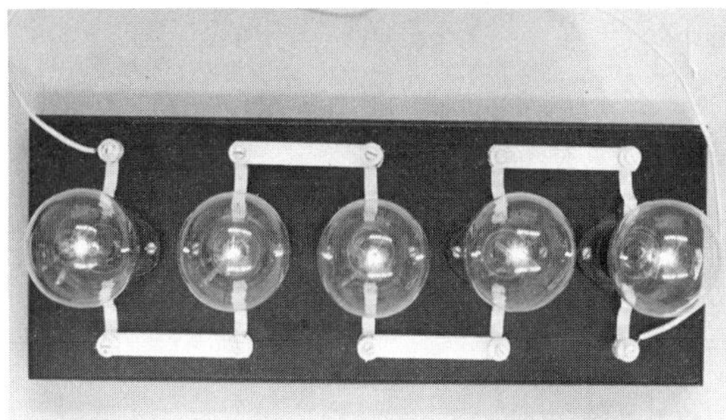

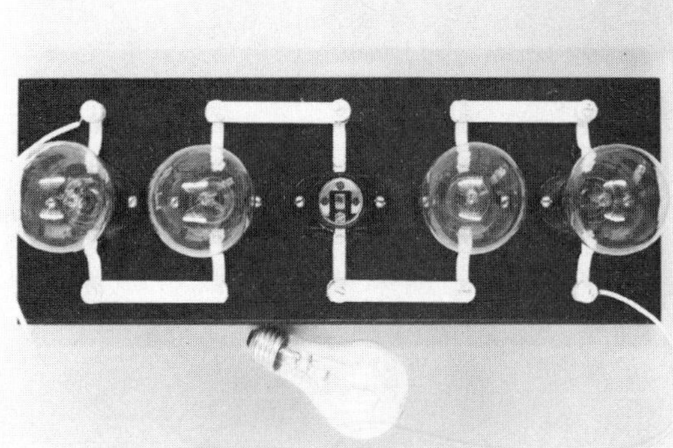

b.

Figure 9.8 (*a*) In parallel circuits, if one or more lamps burn out or are removed, the others continue to glow with no change in intensity. (*b*) In a series circuit, all lamps would fail to glow if one or more were burned out or removed.

magnet is held stationary, whether it is in the coil or not, no current is produced. The moving magnet changes the magnetic field in the coil; it is this *change* that produces the current. When a transformer is connected to an alternating current source, alternating current in one coil of wire, the "primary," produces a continuously changing magnetic field that passes through a second coil, the "secondary," inducing a voltage in the secondary. The iron core is made from "soft" iron, which can be easily magnetized and demagnetized. The core provides a path so that the magnetic field from the primary coil can be passed through the secondary coil. The voltage of the secondary depends on three factors: the voltage across the primary, the number of turns of wire on the primary coil, and the number of turns of wire on the secondary coil. The relationship between primary and secondary voltage is

$$\frac{\text{secondary voltage}}{\text{primary voltage}} = \frac{\text{\# of turns on secondary coil}}{\text{\# of turns on primary coil}}.$$

Example 9.5

A step-up transformer has its primary coil attached to a 120-volt AC source. The primary coil has 200 turns of wire and the secondary coil has 1400 turns. Calculate the secondary voltage.

Solution

Using the basic relationship

$$\frac{\text{secondary voltage}}{\text{primary voltage}} = \frac{\text{\# of turns on secondary coil}}{\text{\# of turns on primary coil}},$$

$$\frac{\text{secondary voltage}}{120\text{ V}} = \frac{1400\text{ turns}}{200\text{ turns}}.$$

Multiplying both sides by 120 V,

$$\text{secondary voltage} = \frac{(120\text{ V})(1400\text{ turns})}{200\text{ turns}}$$
$$= 840\text{ V}.$$

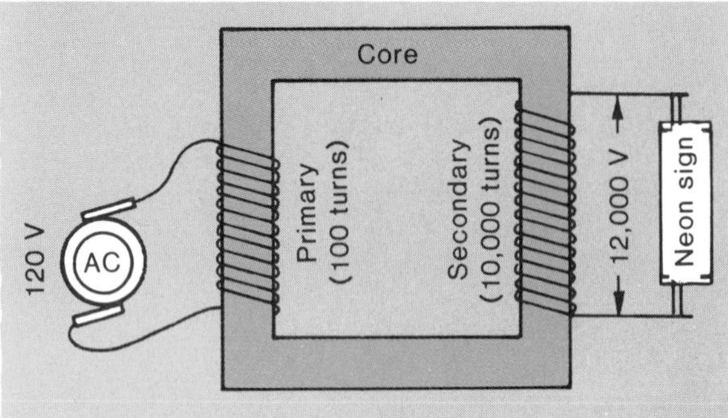

a.

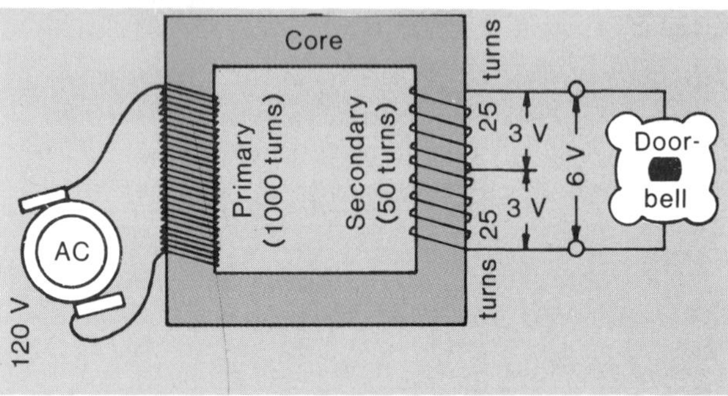

b.

(*a*) The step-up transformer has more turns of wire in the secondary coil than in the primary coil. The photograph shows an electrical discharge between the secondary terminals of a step-up transformer. (*b*) The step-down transformer has fewer turns of wire in the secondary coil than in the primary coil. The photograph shows a disassembled step-down transformer.

It would seem, at first glance, that a transformer offered something for nothing; the voltage can be increased by a transformer without any additional power source. However, the power input to a device or its power output is equal to the product of the voltage and the current. If a transformer were 100 percent efficient, the product of the voltage input and the current input would equal the product of the voltage output and the current output. If the voltage is stepped up by a transformer, the current must be stepped down proportionately. For example, if the output voltage is 10 times the input voltage, the output current will be 1/10 the input current, etc. The equation relating the secondary current to the primary current is

$$\frac{\text{secondary current}}{\text{primary current}} = \frac{\text{\# of turns on primary coil}}{\text{\# of turns on secondary coil}}.$$

(Note that the ratio on the right side of the equation is inverted when compared to the voltage equation for transformers given earlier.)

Example 9.6

A transformer has 100 turns on its primary coil and 500 turns on its secondary coil. If the current into the primary coil is 12.0 amperes, determine the value of the secondary current produced.

Solution

Using the relationship

$$\frac{\text{secondary current}}{\text{primary current}} = \frac{\text{\# of turns on primary coil}}{\text{\# of turns on secondary coil}},$$

$$\frac{\text{secondary current}}{12.0\ \text{A}} = \frac{100\ \text{turns}}{500\ \text{turns}}.$$

Multiplying both sides by 12.0 A gives

$$\text{secondary current} = \frac{(12.0\ \text{A})(100\ \text{turns})}{500\ \text{turns}}$$
$$= 2.40\ \text{A}.$$

One of the many uses of transformers is stepping up the voltage output of generators so that power may be distributed at high voltages using low currents. The advantage of using low currents is that less power is wasted in the transmission lines. Power converted into heat in a power line (Joule heating) is proportional to the square of the current, so that lowering the current reduces the amount of power wasted. As illustrated in Figure 9.10, voltage from a generator is stepped up for distribution on large power lines, then stepped down for distribution inside cities, and stepped down even further before it enters homes, industrial plants, office buildings, and so forth. The lower voltages are, of course, safer. The transformers used in power-distribution systems are pictured in Figure 9.11.

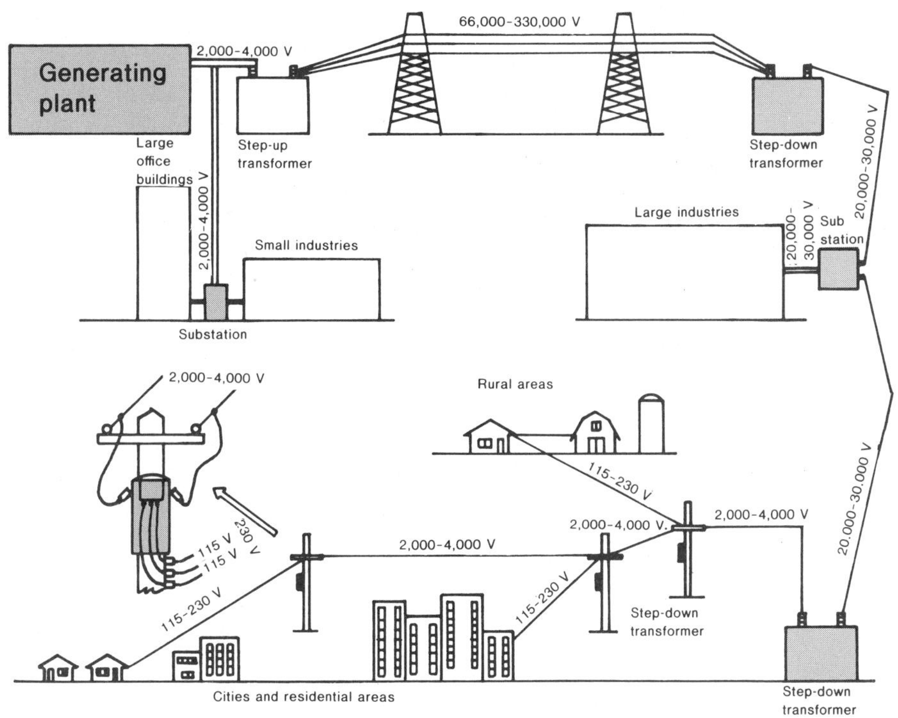

Figure 9.10 Electric energy travels many paths as it leaves the generating plant.

a.

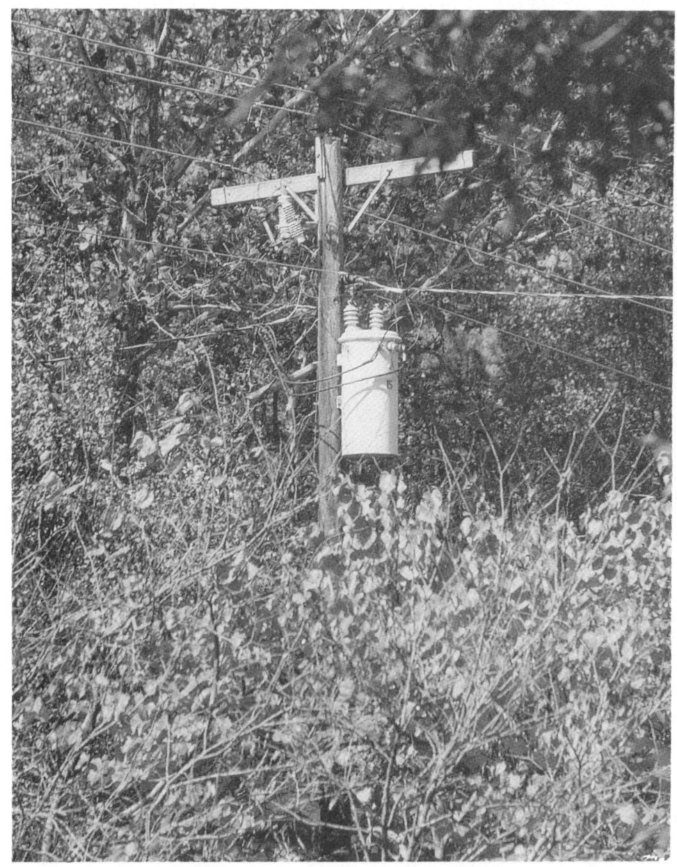

b.

Figure 9.11 (*a*) An electrical "substation" is simply a collection of transformers to step the voltage down to a safe level for local distribution. The large dark gray transformers reduce the voltage supplied by the lines coming in from the upper right in the picture from about 25,000 V to about 2500 V, which is distributed by the lines going out at the upper left. (*b*) The lines along a street have a voltage of about 2500 V, which is reduced to voltages of 110 V and 220 V by the familiar transformer seen on power line poles.

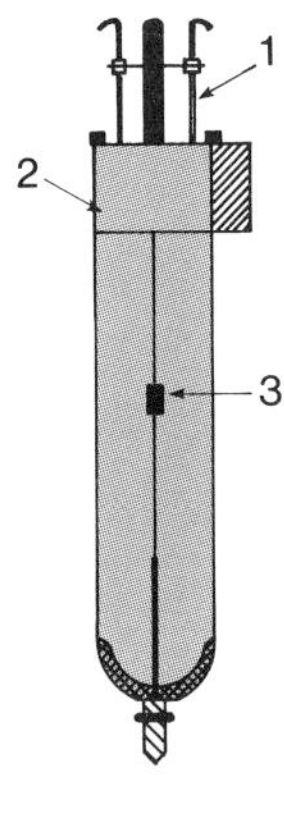

a.

b.

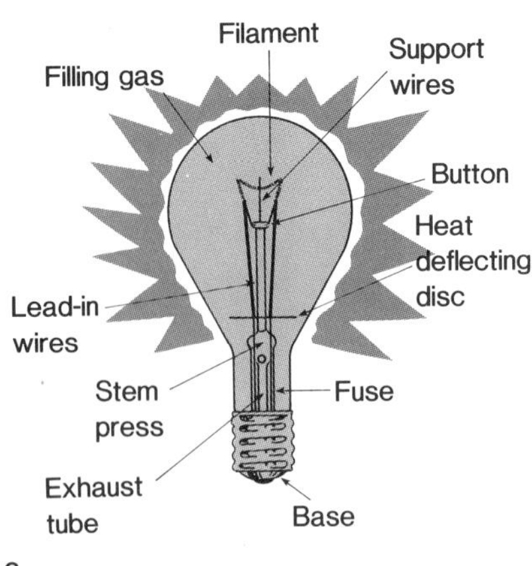

c.

Figure 9.12 (*a*) An illustration of the open arc lamp of about 1878. The lifetime of this model was quite short, and the lamp had little practical use. (*b*) A version of Edison's first lamp. The evacuated lamp became one of the first practical applications of electricity. (*c*) The modern incandescent lamp has many components.

Individual devices often have transformers to step voltage up or down. For example, a television set may have a transformer to step the voltage up to 25,000 volts to operate the picture tube, whereas a doorbell or an electric train may have a transformer to step the voltage down to 6 volts.

Household Appliances

Electricity serves us in a multitude of ways at home and at work. One of its earliest applications was as a source of artificial illumination. The invention of the carbon filament lamp by Thomas Edison in 1879 introduced electricity as a practical source of energy. The Chicago World's Fair of 1893 was illuminated by some 20,000 electric lamps and served as the first commercial display of the illuminating potential of electricity. The vacuum-type enclosure that contains a tungsten filament, our most commonly used source of artificial illumination, was first put to practical use at the New York World's Fair in 1939.

If the current through a conductor encounters a high enough resistance, **incandescence** will occur; that is, *a body at high temperature will emit visible light.* The high temperature of the filament in a lamp is brought about by its resistance to current, and it becomes white-hot. For this reason the ordinary household lamp is called an *incandescent lamp.* The filament is made from the element tungsten, a metal with both a high electrical resistance and a higher melting point than any other metal. Illustrations of the development of filament bulbs appear in Figure 9.12. The proper mixture of various inert gases causes the filament in a modern lamp to last for thousands of hours without burning into fragments, whereas Edison's first lamp, made public in 1879, lasted only some forty hours. The life of the filament is inversely related to the rate at which the filament combines with oxygen; thus, the more expensive bulbs are more carefully manufactured to prevent the leakage of air into the evacuated bulb. The incandescent bulb, by its nature, has a low efficiency and produces great amounts of heat. Some filaments reach temperatures as high as 2650°C, even though they may be no thicker than a human hair.

The *fluorescent lamp* differs from the incandescent lamp in that light is emitted from it through the discharge of electrical energy from two points of significantly different electric potential. The electrodes in fluorescent lamps, such as the types illustrated in Figure 9.13, are slightly heated by the friction produced from the electric current that passes through them. The hot electrodes emit electrons that bombard the atoms of argon and mercury vapor that are used to fill the tube. The collisions produce an electric discharge that creates a faint bluish light as well as invisible ultraviolet radiation. These emissions, in turn, strike a powdered coating of a *phosphor,* generally a metallic sulfide, on the inside of the tube and cause the coating to undergo **fluorescence,** or *to glow with light that is in the visible range.* Various combinations of phosphors used as a coating can create such visible colors as yellow, pink, and daylight blue, as well as others. Once the process is underway, the hot electrodes are no longer operationally necessary, and the starter automatically disconnects them, thus extending their lifetime.

The fluorescent lamp has proved superior to incandescent lighting in many different ways. Fluorescent lamps offer greater lighting efficiency per watt than do incandescent lamps. For instance, a 40-watt fluorescent lamp furnishes as much light as a 150-watt incandescent lamp. Also, the life of the fluorescent lamp is about 8000 hours; it is limited by the length of time that the coating on the filament can last and thus furnish electrons to start

Thomas Alva Edison: The Inventor's Inventor

Thomas Alva Edison (1847–1931) was born in Milan, Ohio. Since his progress in school was slow, he received most of his education at home with the aid of his mother. Edison was an ambitious youth, and he developed a flourishing business selling newspapers, candy, and tobacco aboard trains. He traveled frequently and had many jobs, including the position of telegraph operator. In his spare time he studied chemistry, and he patented his first invention—an electrographic vote recorder—in 1868.

Edison had considerable influence on the formation of the American scientific tradition. Henry Ford proclaimed that Edison definitely ended the distinction between the theoretical scientist and the practical scientist. As a result of Edison's work, scientific discoveries are now usually evaluated as to their potential application in society.

Edison's "invention factory" was built at Menlo Park, New Jersey, in 1876 and became the prototype for the large research laboratories that were soon to follow. At this new facility Edison invented the first phonograph and developed an improved model of the telephone, which had been invented earlier by Alexander Graham Bell. He also is credited with improving Thomas Armat's cinematic projector. Edison moved his research laboratories and his many salaried aides to West Orange, New Jersey, in 1887. Thereafter, Edison spent little time as a practical inventor but became involved in public relations and management.

Although Edison was credited with over a thousand patents during his lifetime, many were for things that he did not actually invent. His real genius was in taking an impractical device and making it functional. As an example, experiments had been made on the incandescent lamp in 1840, and a prototype that worked briefly had been constructed in 1860. However, it remained for Edison in 1879 to make a practical lamp by using a carbon filament. His major contribution to theoretical science came from the announcement of his discovery in 1883 that the lamp could be used as a valve, receiving negative but not positive electricity. This observation, known as the *Edison effect,* is the result of electrons being ejected from the filament by thermal vibrations. The phenomenon involved serves as the basis for the electron tube, an invention credited to Ambrose Fleming, and patented in 1904.

Edison worked with remarkable concentration. He required very little sleep and often lived in his laboratory. By the end of his life, "The Wizard of Menlo Park" had been honored by numerous countries around the world ■

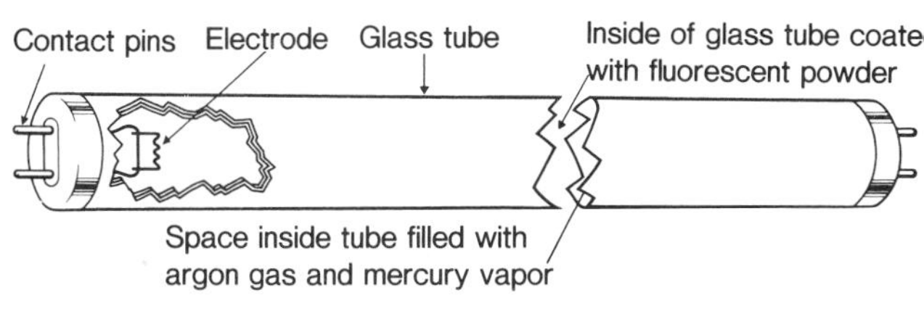

a.

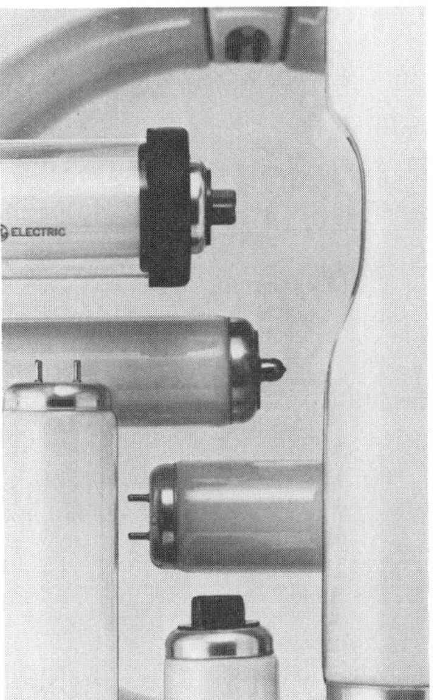

b.

Figure 9.13 (*a*) The fluorescent lamp emits wavelengths that closely resemble daylight; hence, such lamps are used in kitchens and in dressing rooms where light more closely resembling light from the sun is preferred. (*b*) The shape of the modern fluorescent lamp has changed considerably from earlier models. Most design changes are to increase efficiency and convenience.

Figure 9.14 The tunnel on the right is equipped with a fluorescent lighting system. Notice the difference in brightness between the fluorescent lighting and the incandescent lighting used in the tunnel on the left.

the fluorescent process. Fluorescent lamps, though, are not suitable for use as spotlights and do not typically function well in temperatures that vary outside the range from about 15 to 32°C; therefore, in many climates they are not very useful outdoors. Research by various lighting divisions has overcome this deficiency to some degree, and some additional applications have been accomplished. For example, Figure 9.14 shows a successful application of fluorescent lighting in tunnels. This figure illustrates the contrast between incandescent lighting, installed in the westbound Tuscarora Mountain tunnel on the Pennsylvania Turnpike in 1940, and the fluorescent lighting that replaced incandescent lighting in the eastbound tunnel.

The development of efficient, energy-saving light sources is an ongoing concern for major lighting industries. The 90-watt Sylvania Capyslite flood lamp displayed in Figure 9.15a serves as a prime example. This unit produces the same illumination as the standard 150-watt flood or spot lamp. Figure 9.15b shows a new energy-saving fluorescent lamp that can substitute for the standard household incandescent lamp. The fluorescent conversion unit fits the standard incandescent socket and produces the same illumination as a 75-watt bulb, but lasts many times longer than an incandescent lamp at considerably less operating expense. The miniature high-pressure lamps shown in Figure 9.15c increase light output and consume less power than comparable incandescent and mercury vapor light sources. The compact 70-watt Lumalux lamp delivers the same amount of light as a 300-watt incandescent lamp, yet has a 24,000-hour life.

a.

b.

c.

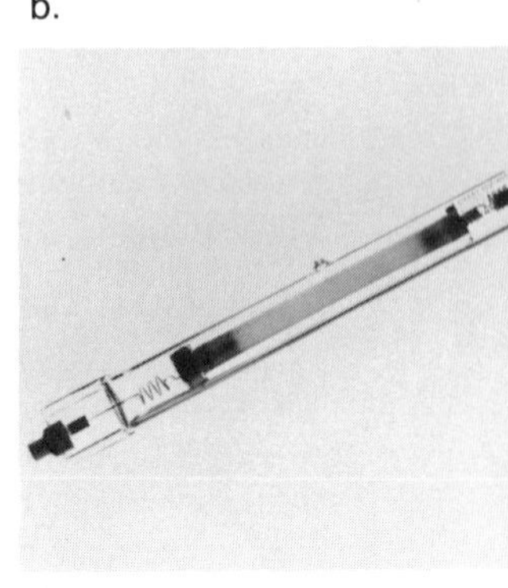

d.

Figure 9.15 (*a*) The Sylvania Capyslite flood lamp cuts consumer cost about 40 percent. (*b*) The energy saving Sylvania fluorescent lamp can be used in any incandescent lamp socket. (*c*) A compact 70–W Sylvania sodium Lumalux lamp can deliver the same illumination as a 300–W incandescent lamp. (*d*) The high-pressure sodium lamp requires about one-fourth the cost to operate, compared to the tungsten halogen lamp it replaces.

The light source in Figure 9.15d was developed for use in sodium fixtures as an energy-saving replacement for tungsten halogen lamps of higher wattage. The rated life of the lamp pictured is 24,000 hours, twelve times longer than the unit it can replace. Such lamps can be used in factories, warehouses, garages, and at exit areas along interstate highways.

A second application of electricity in the home involves its heating ability. Various appliances, such as toasters, irons, dryers, electric blankets, coffee makers, and stoves, use a resistance wire to produce heat. The resistance wire is commonly composed of an alloy of nickel and chromium, called *nichrome*. The wire is wound into a coil or a screenlike grid pattern so that a long length of it

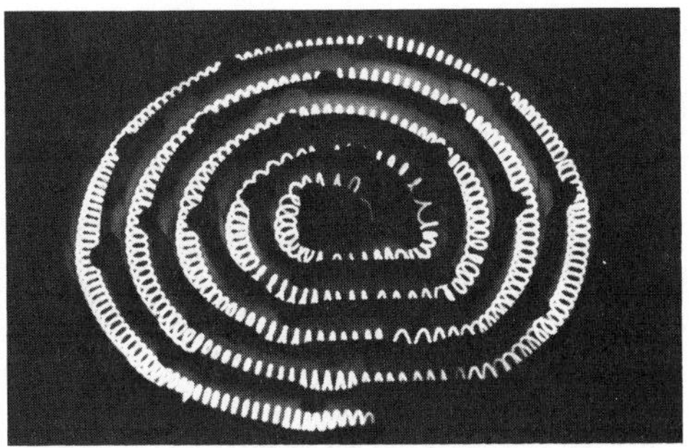

a.

b.

c.

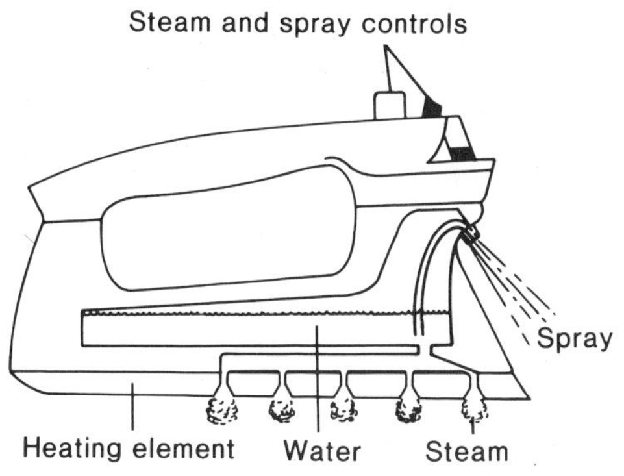

d.

Figure 9.16 (*a*) The coiled-wire element offers greater resistance in a confined space to current flow than would a straight-wire element because of its greater length; thus, greater heat is produced. The dark areas on the spring are clips that hold the spring in place. (*b*) The enclosed electrical element of a hot plate. (*c*) The element is often molded into the base of an electric iron. (*d*) Water is converted into steam by the heat of the element in the steam iron.

can be installed in a small space (see Fig. 9.16a). The *element,* as the wire is often called, is electrically insulated from the rest of the appliance by ceramic materials or by sheets of mica. Other elements, such as those illustrated in Figure 19.16b and 19.16c, are enclosed in a metal tube to prevent damage to them from air and moisture.

The typical electric iron is capable of converting water into steam (see Fig. 9.16d). Temperature control of each of the appliances is accomplished by varying the percentage of time that power is applied to the heating element. The heat regulator, called a *thermostat,* opens and closes the electrical circuit according to various settings of the calibrated temperature control.

Wall-mounted thermostats, such as the model shown in Figure 9.17, provide the means by which to control the heating and cooling units in a home. These devices contain a *bimetallic strip,* a component constructed of two metals fused together to function as a single strip. The metals used have significantly different coefficients of linear expansion (see Chapter 6) and are good electrical conductors. Thus, as the temperature of a room increases and reaches the predetermined setting, the bimetallic strip bends away from the point of contact with the balance of the circuit; hence, the electrical circuit is broken and the furnace is shut off. If cooler air is desired, the thermostat is so set as to close or complete the circuit that activates the air conditioning unit. As one might surmise, the circuit that turns on the cooling system is on the opposite side of the bimetallic strip from the circuit that controls the heating of the home.

A third major application of electricity is in converting electrical energy to mechanical energy using the electric motor. The motor is opposite in principle to the generator in that the generator converts mechanical energy to electrical energy. The generator produces electricity by rotating a coil of wire in a permanent magnetic field. If an electric current is connected to the coil of wire, known as the *armature,* the coil becomes an electromagnet, and the generator is converted into a motor. The north pole of the armature is repelled by the north pole of the permanent magnet, just as the south pole of the armature is repelled by the south pole of the permanent magnet. The armature is caused to rotate by the

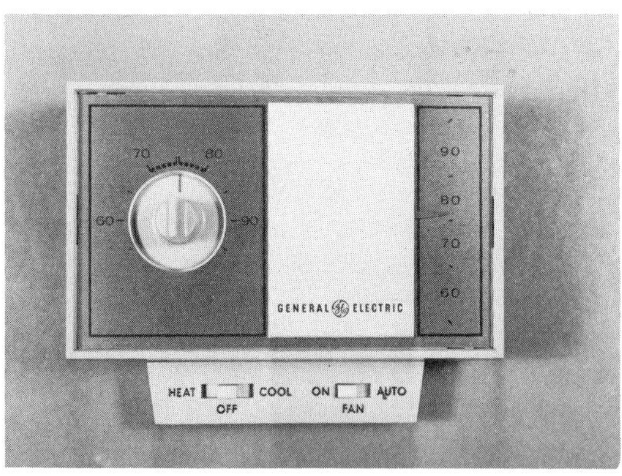

a.

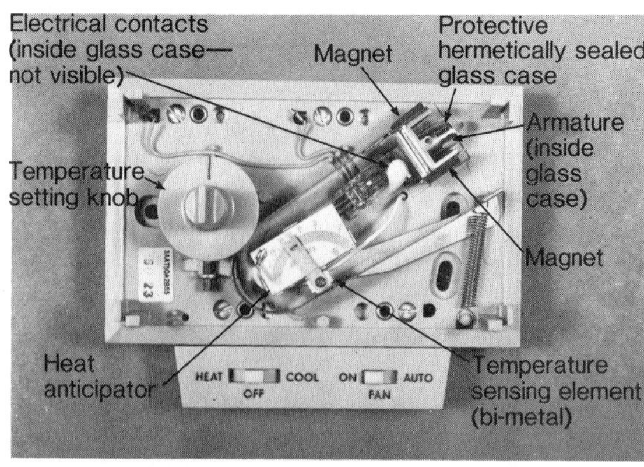

b.

Figure 9.17 (*a*) A deluxe heating-cooling thermostat designed to provide close control of temperature. (*b*) The internal components in a typical heating-cooling thermostat.

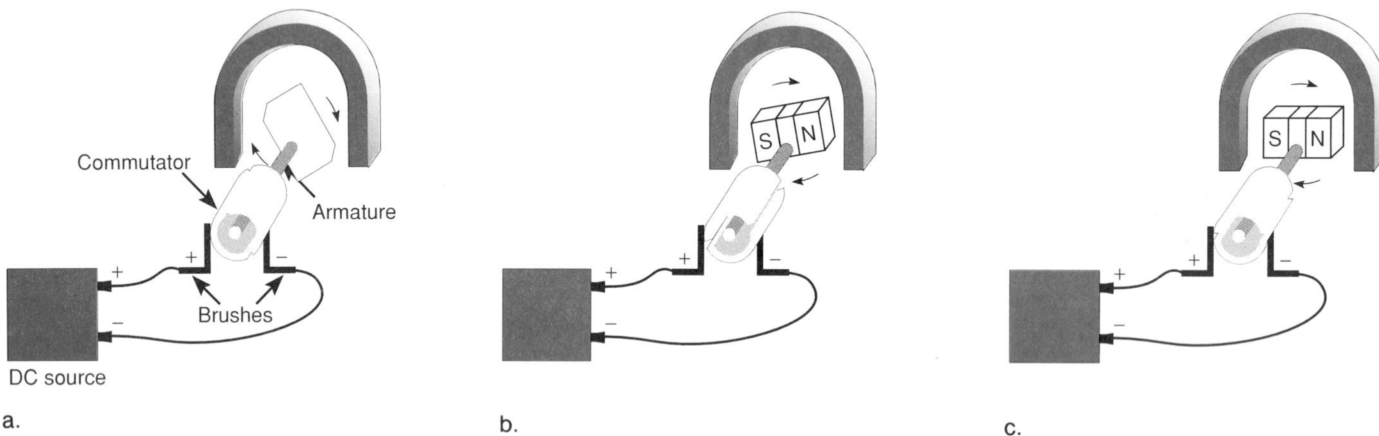

Figure 9.18 (*a*) The current through the armature gives the ends of the coil N and S poles as described in Chapter 8. These poles are repelled by the like poles of the permanent magnet and (*b*) attracted to its opposite poles making the coil and commutator rotate. (*c*) When the coil rotates so that the brushes touch the opposite contacts on the commutator, the current in the coil is reversed, making what was the S pole a N pole and vice versa, thus continuing the rotation.

repulsion of the magnetic fields, but the fields of the armature are reversed by design before the armature completes one full rotation. The reversal of fields of the armature forces the armature to continue to rotate, as is shown in the sequence in Figure 9.18. Then, by simple conversion, the rotational energy is eventually transformed into more useful forms of mechanical energy or is used in its original form to do work.

Electronic Components

The separation of electronics from the concepts of electricity is extremely difficult. The field of electricity is generally concerned with magnetism, lighting, heating, and the production of electricity by generators and chemical action. Electronics usually involves the application of electricity in communication, as in radios, televisions, and other devices where transistors and vacuum tubes are employed. The vacuum tube (or radio tube) is considered one of the most important inventions of recent times. It was discovered in principle by Edison shortly after he invented the incandescent bulb. He learned, by accident, that if an additional wire were attached to the outer surface of the bulb and then connected to the circuit, an electrical current would be produced in the wire, even though it was not connected to the filament. The vacuum tube, as the device came to be known, has all but been replaced by the transistor in electronic circuits that do not require large amounts of electrical power.

Electronics has undergone a major transformation from the days of the development of the vacuum tube. Much of this change was brought about by scientists who sought to learn more about the properties of matter—particularly, crystalline solids.

a.

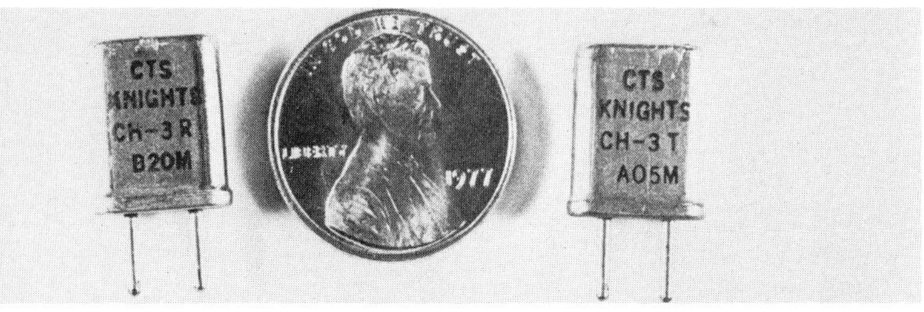

b.

Figure 9.19 (*a*) The atoms in a crystalline material have a definite geometrical arrangement. The three models show variations of cubic crystalline structures, with the model on the far right having the structure of common table salt. (*b*) Crystals enclosed within the protective containers have natural vibrational frequencies. These natural frequencies permit the transmission and reception of signals on the various channels of a CB set.

Figure 9.20 Liquid crystal display devices have become common in calculators and portable computers.

The crystalline nature of most solids is evident whether one studies snowflakes or examines the minerals that form Earth's crust. Atoms in most solids are arranged in orderly three-dimensional patterns called *lattices.* These specific inner arrangements are often reflected by the outer shapes of such substances as table salt (see Figs. 9.19a and 10.15). Most crystalline solids, however, are conglomerations of small crystals and do not reveal their inner orders by their external appearance.

Some crystals, because of their specific atomic structures, have unique effects on electricity and light. Other crystals can withstand large amounts of distortion because of the arrangements of their atoms. Quartz crystals are caused to vibrate (or oscillate) by surges of electric current. Thin wafers of this natural component of Earth's crust are used to permit radio transmitters to broadcast on assigned frequencies (see Fig. 9.19b). Other quartz crystals are used in accurate timepieces. The quartz crystal that furnishes the time base for one design of modern watch vibrates at a rate of 32,768 hertz. The manufacturers of today's timepieces are constantly searching for ways to provide more accurate time, and they realize that the accuracy of a watch or other timing device depends on the frequency of vibration of certain components of the timepiece. For example, the tuning fork used in some watches oscillates at 360 hertz, and the typical balance wheel used in other types of watches has a frequency of 5 hertz.

Conversely, if quartz crystals are twisted or compressed, electric current is produced by the crystals. The elements selenium and silicon have proven valuable in our space age. These crystalline substances, when exposed to sunlight and sources of artificial light, emit electric current. The electric current is produced as the electrons in the atoms that form the crystal lattice are forced into excited states by the light to which they are exposed, resulting in the creation of an electric potential within the crystal. Both the solar cells that power orbiting satellite components and the light meters in most cameras contain one of the two essential elements. Synthetic ruby crystals in the shape of rods are placed inside a coil that contains the rare gas xenon. When light emitted from an electrical discharge in the xenon gas is absorbed by the ruby crystal, the electrons in it become excited and generate an intense beam of light. The laser, as this device is called, is discussed in detail in Chapter 11.

Researchers have discovered that certain properties of solid crystals are retained by the substances in their liquid state. A **liquid crystal** (LC) is *a liquid that has some degree of molecular orderliness.* (Some scientists consider these ordered liquids to be a fifth state of matter.) These liquids, which include some cholesterol derivatives and other carbon compounds, assume the shape of their containers as other liquids do, but have an inherently weak crystalline structure. A small force can cause realignment of this crystalline structure and change its optical properties. In a *liquid crystal display* (LCD) a liquid crystal is placed between two electrodes, at least one of which is partially transparent. A small voltage difference between the electrodes has the capability to reorder the crystalline structure, thus altering its optical properties. LCDs have become a standard display device in calculators, computers, and small flat-screen television receivers (Fig. 9.20). The optical properties of liquid crystals are discussed further in Chapter 11.

Semiconductors are *materials that lie between insulators and conductors in their resistance to charge flow.* A good insulator may have a resistance more than 10^{22} times the resistance of a good conductor of the same dimensions. That is, if a conductor

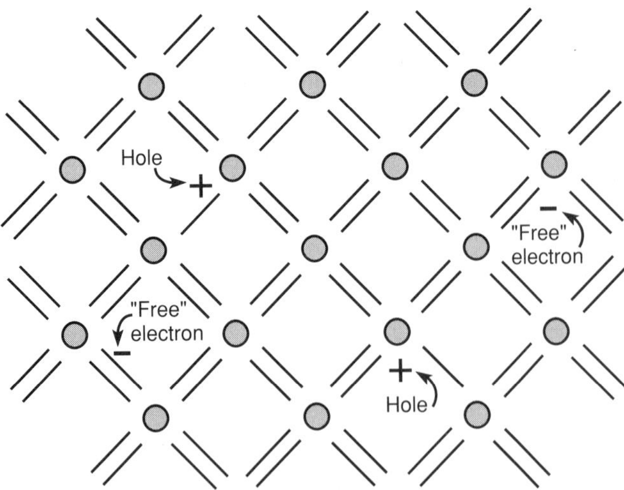

Figure 9.21 A symbolic representation of electron bonding in silicon and germanium. Circles represent the sites of the atoms and lines represent electrons bonding the atoms into the crystal. Each atom has four neighboring atoms. Normally, two electrons are bonded between each atom, but thermal vibrations are sufficient to break some bonds, freeing the electrons. The vacant sites, called "holes," behave in the manner of positive charge carriers.

had a resistance of 1 ohm, the resistance of an insulator of the same dimensions could be greater than 10,000 billion billion ohms! This is why insulators effectively prevent the flow of charge. On the other hand, a semiconductor of the same dimensions as the conductor (with a resistance of 1 ohm) might have a resistance of only (!) 30 billion ohms. The most common semiconductors are the crystalline materials silicon and germanium. Both materials have relatively weak interatomic bonds that are continuously being broken by thermal vibrations at room temperature. Such broken bonds result in free electrons, and they move from the site where the break occurred (see Fig. 9.21). At the site of the break, a net positive charge results, since the atom loses its electrically neutral balance. The positive region attracts other electrons from surrounding atoms with sufficient force to break the bonds that hold them in place. These electrons, then, are attracted to the positive site, neutralizing it, but creating a new positively charged site in the region they vacated. The process repeats and the free electrons may be accelerated toward an externally applied positive electric potential, thus contributing to electron flow. The vacated site of the broken bond, called a *hole* or an *electron hole,* behaves as if it were a positively charged particle and moves toward the externally applied negative electric potential, adding to the total current (Fig. 9.22). The positive charges attracted to a negative electrode appear to have the same effect in an external circuit as do electrons that are moving toward the positive electrode.

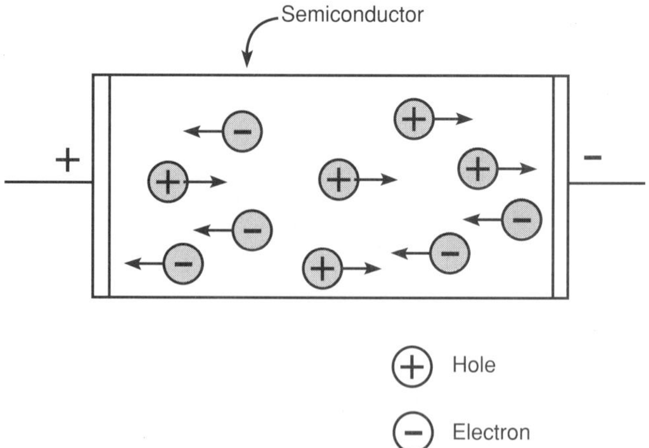

Figure 9.22 The electrons and holes move in opposite directions in a semiconductor. The total current is found by summing the contributions from the positive carriers (holes) and the negative carriers (electrons).

A pure crystal of a semiconductor will have an equal number of free electrons and "holes," the number of which depends on the temperature of the material as well as its composition and size. If a very small amount of a suitable impurity is added to the crystal, a process known as doping, the "doped" crystal may have either an excess of electrons or an excess of "holes"; hence, more *electron current* or *hole current,* respectively. A semiconductor with an excess of free electrons is called an n-type (negative) semiconductor. At the positions of the atoms added to the crystal that serve as the impurity, such as those of the element arsenic, there is a net positive charge; therefore, the crystal remains electrically neutral overall. If an element such as boron replaces some of the atoms in the silicon or germanium crystal lattice, there are not sufficient electrons to satisfy all bonds. Although the crystal conducts primarily by means of "hole current," overall it remains electrically neutral. This type of crystal is known as a p-type (positive) semiconductor.

When p- and n-type semiconductor materials are joined together, the electronic component is called a semiconductor **diode.** If an external potential, a voltage, is applied to the diode in one direction, the diode conducts an appreciable current, but when the electric potential is reversed, the diode conducts very little current. Thus, the diode acts as a one-way valve allowing current to flow in one direction only. The diode may be used for **rectification,** which is the *conversion of alternating current into direct current.* (The direct current produced is not constant, but may be made so by additional circuit elements.) Diodes are also used for other purposes in radio receivers, voltage regulators, and other electronic circuits. The diode and its schematic symbol are shown in Figure 9.23.

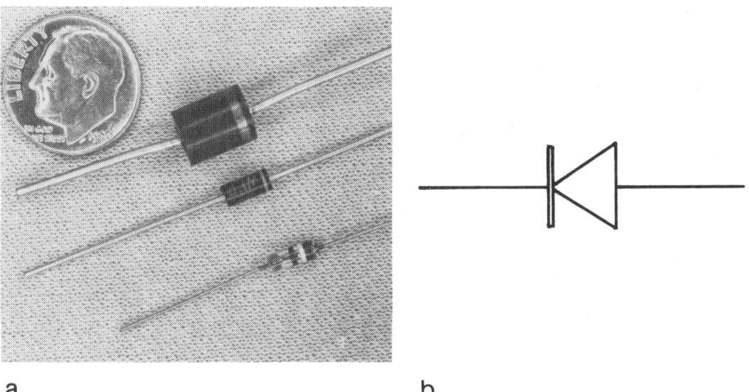

Figure 9.23 (*a*) Three semiconductor diodes are shown. Electric current will flow in only one direction through these devices. (*b*) The symbol for a diode used in circuit diagrams. The arrowhead points in the direction in which conventional current would flow. Electrons flow in the opposite direction.

Figure 9.24 Light-emitting diodes (LEDs) are used in timing devices and radios, such as those found in automobiles, office buildings, and homes.

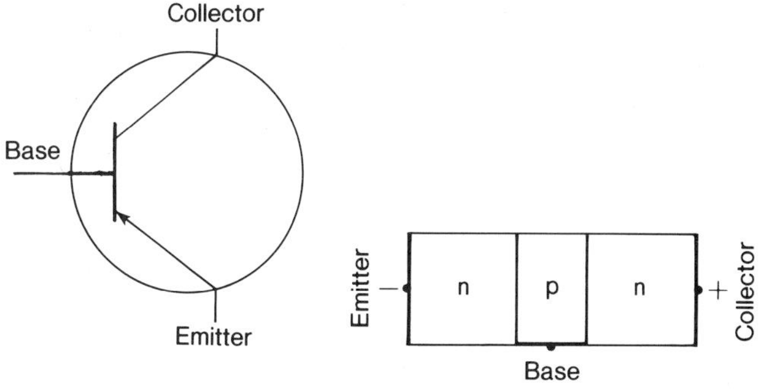

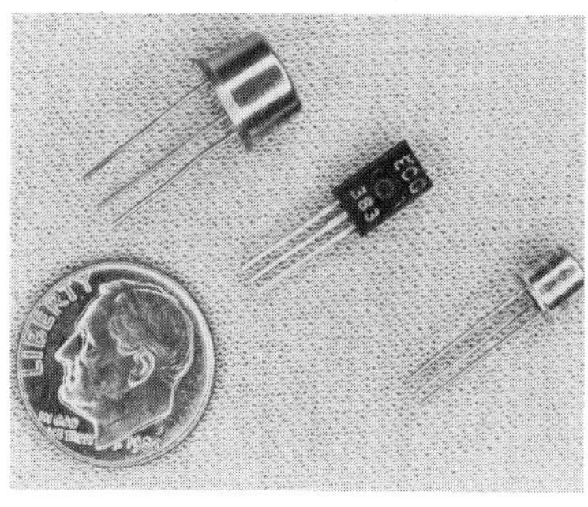

Figure 9.25 (*a*) Symbols used for transistors in circuit diagrams. Transistors are semiconductor devices that may be used to control or amplify signals. (*b*) A photograph of some ordinary transistors. Note that each has three leads, allowing connection to the emitter, base, and collector. When a fixed voltage is established between the emitter and the collector, small changes between the emitter and the base can control the current through the transistor.

The *tunnel diode* has proven to be a valuable contribution to electronic circuitry. This device operates on a negligible amount of electric current and at an electric potential of less than 0.5 volt. The tunnel diode plays an essential role in the development of wrist telephones and CB sets, pocket-size television sets, ring radios, minicomputers, and other such miniature electronic devices.

Another type of diode that has proved very useful is the *light emitting diode* (LED). This electronic component may be used as an indicator lamp, a signaling device, an alphanumeric display on calculators, or a display for digital timing devices. The LED is made from crystals of the element gallium or silicon that contain minute amounts of phosphorus, carbon, or arsenic. The LED can be a continuous light emitter, as in electric digital clocks or, as used in a battery-powered wristwatch, the LED can be activated and made to emit light by a button or by the touch-sensitive case in which it is housed. The LED is used in various timing devices (Fig. 9.24) and in other instruments where digital displays are advantageous.

A **transistor** is in reality *a diode with an additional "doped" region;* thus, the diode with a third component becomes a **triode.** For example, a p-type region is sandwiched between two n-type regions. One of the n-type regions is called the *emitter,* and the second n-type region is called the *collector.* The p-type area is known as the *base.* This type of transistor is called an n-p-n transistor. A simple diagram of a transistor appears in Figure 9.25a, and pictures of several are shown in Figure 9.25b. The effect a transistor has on a circuit depends on the relative voltages applied to its three regions. Generally, a voltage applied between the emitter and collector results in a weak current through the transistor, but a small additional voltage applied between the emitter and the base can result in a large current from the emitter to the collector. Thus, the transistor acts as a sort of electrical valve, where small changes in one voltage (and its corresponding current) can be made to produce large changes in another current (and its corresponding voltage)—a process called *amplification.* The well-designed transistors of today permit practically all the free electrons to move through the base and allow for a large current from the emitter to the collector. The n-p-n transistor and its counterpart, the p-n-p transistor, represent only two of the types of transistors in use today. The transistor, according to the method of circuit wiring, can also be used as a diode or as an *oscillator,* an electronic component, the circuit of which converts direct current to a periodically varying electric output and serves as an amplifier stage.

After the development of transistors, scientific and technological progress in this field of physics continued at an astonishing pace, and soon an electronic marvel, the **integrated circuit** (IC), was developed. This microminiature device *contains an entire electronic circuit of transistors and related components on a single silicon chip.* As advancement continued, an integrated circuit chip less than 1 square centimeter in area outmoded the most sophisticated electronic devices of the 1950s. Further development of the integrated circuit continues, and circuits with over hundreds of thousands of elements have been produced. The great storage capacity of these tiny integrated circuit chips has made possible the rapid development of microcomputer systems. Presently, IC chips with 50 or more components are so small that they can easily pass through the eye of a sewing needle. An integrated circuit, as well as its two accepted symbols, is shown in Figure 9.26.

In order to smooth the pulsating current in circuits with diodes and transistors, a *capacitor* (occasionally called a condenser) is added to the circuit. This device consists of two conductors that are insulated from one another by an insulating sheet of material called a *dielectric.* The source of electric potential is connected to the capacitor in such a way that a positive charge is put on one of the conducting plates and a negative charge on the other. These charges attract each other, and so they remain on the inner surface of the plates and represent a stored (static) charge that may be used to compensate for a sudden decrease in electric potential, an effect that causes the pulsating DC to resemble more closely the DC supplied by a battery. A capacitor, along with its schematic symbol, is shown in Figure 9.27.

The *resistor* is an electronic component used to adjust the relationship between voltage and current in a circuit. The value of a resistor is known as its resistance and is measured in ohms, as presented in Examples 8.7 through 8.9.

When current passes through a resistance, power, measured in watts, is dissipated as heat. In order to avoid damage to a resistor, its size and construction must be capable of withstanding a known amount of power. A variety of resistors is required to meet demands that range from those suitable for dissipating a fraction of a watt as used in transistor radios to those that must withstand several kilowatts of power as used in industrial plants.

Many resistors are made of inexpensive materials and can be mass-produced. The resistive component consists primarily of finely powdered carbon spread evenly throughout an inert substance such as fireclay. The two materials are mixed in various proportions with a liquid adhesive, then pressed into cylindrical shape and molded into insulated sleeves that have treated copper leads attached to them. After further processing, bands of colors that signify the rated resistance of the component are added. Several

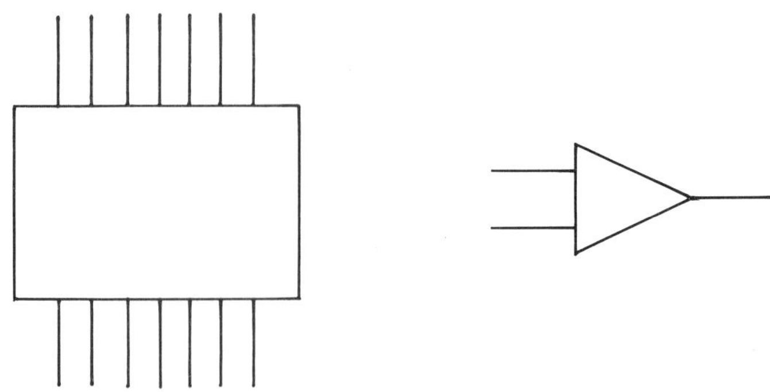

a.

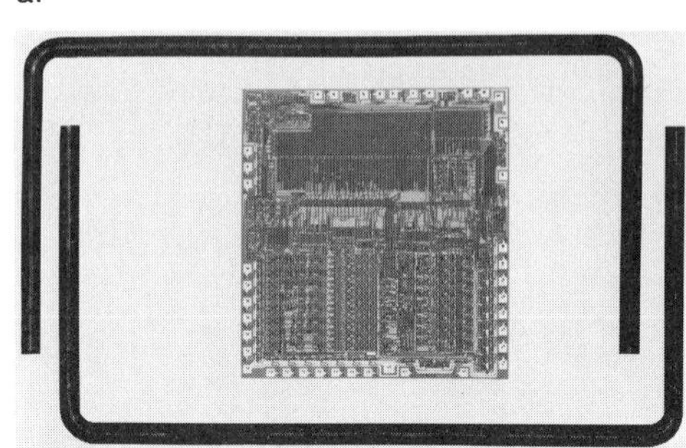

b.

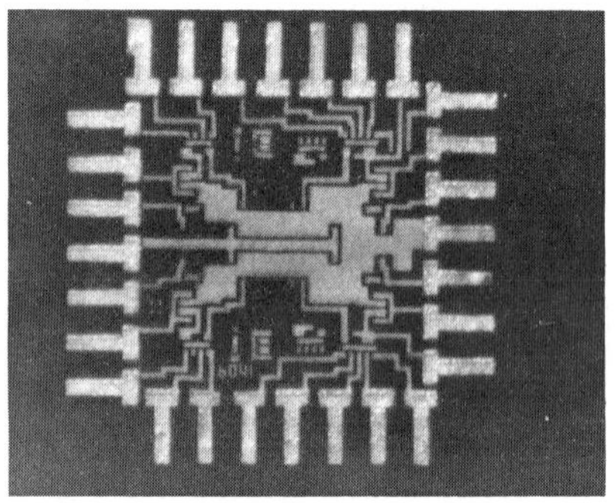

c.

d.

Figure 9.26 (*a*) Symbols used in circuit diagrams for integrated circuits. (*b*) The IC chip shown is smaller than the standard office staples that surround it. (*c*) The integrated circuit chip in this microscopic photo contains many transistors and resistors. The tiny chip of silicon has an area of less than one square millimeter. (*d*) This microcircuit component, an IC chip called a static RAM, operates over a temperature range from −55°C to 125°C.

Transistors

Advancements in electronics during World War II resulted in vast improvements in the crystal detectors used in communications. Various elements used in relatively pure crystalline form were found to limit the flow of electric charge in much the same manner a turnstile at the entrance to a sports event controls the flow and enforces one-way traffic of fans.

In 1948, scientists at Bell Telephone Laboratories made a discovery that revealed the true worth of the metal germanium. They attached two fine wires to a crystal of pure germanium with the ends of the wire about 0.01 centimeters apart and pressed the opposite side of the crystal to a metal plate. They found that current flowed between one of the wires and the plate and that the amount of current between these two electrodes could be adjusted by varying the voltage of the second wire. The latter wire, then, acted like a valve and the germanium crystal not only could serve as a rectifier but as an amplifier, too. The three components, along with the crystal, form the device now known as a transistor ■

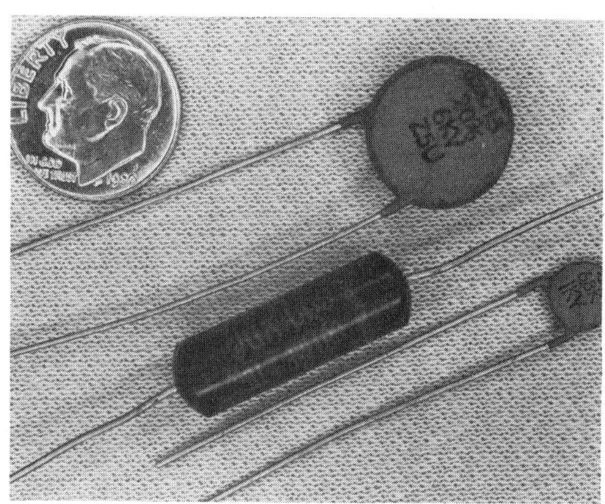

a.

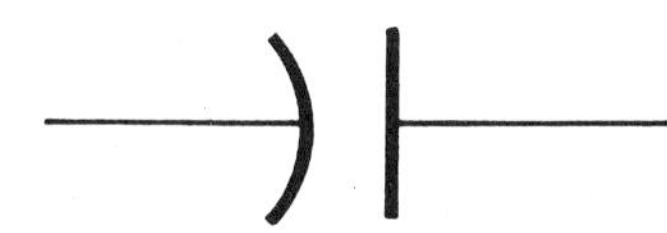

b.

Figure 9.27 (*a*) Three capacitors are shown. Capacitors store electric charge by using two conductors near each other separated by a dielectric (insulator). The flat capacitors have two disk-shaped conductors separated by a dielectric, while the cylindrical capacitor has two pieces of foil rolled in a spiral with a thin piece of insulating material between them. (*b*) The capacitor symbol used in circuit diagrams.

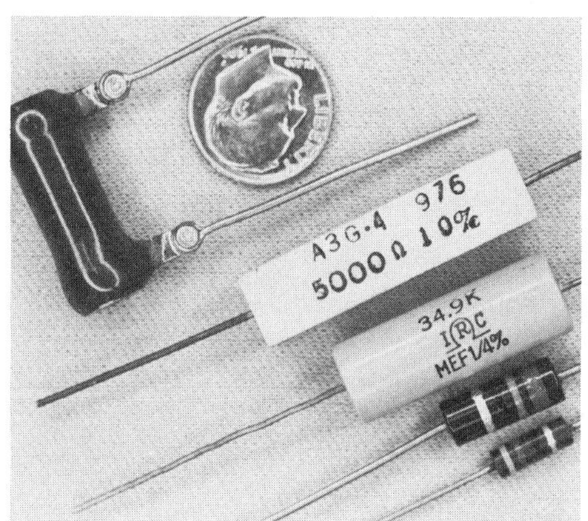

a.

b.

Figure 9.28 (*a*) Resistors used in electric circuits are elements with specific resistances placed in circuits to limit currents or to produce voltages when currents pass through them. The resistance in ohms and the amount of power must be taken into account when designing circuits. (*b*) The resistor symbol used in circuit diagrams.

resistors, along with the symbol for a resistor, are shown in Figure 9.28.

Electronic Systems

Electronics serves as a means of transmission and reception of signals via radiated electromagnetic waves—*radio waves.* The radiations are produced by an alternating current of very high frequency and are transmitted in all directions. (Generation and transmission of radio waves are discussed in Chapter 10.) Sounds are converted into low-frequency electrical impulses by a *microphone* and are then strengthened by the action of an *amplifier.* An *oscillator,* the third component of the circuit, generates high-frequency alternating current that is blended with the electrical signals from the microphone in a *mixer.* The mixer uses the microphone output to control the high-frequency alternating current by producing signals that either *vary in amplitude,* called **amplitude modulation** or signals that *vary in frequency,* called **frequency modulation.** The resultant high-frequency signal is then radiated from an antenna. The receiving set, or radio, is tuned to a transmitting station by selecting the high-frequency signal of the desired station. Each station within range of the radio uses a different high frequency so that many stations may be received by the same radio. Once the signal has been received by the radio, it passes

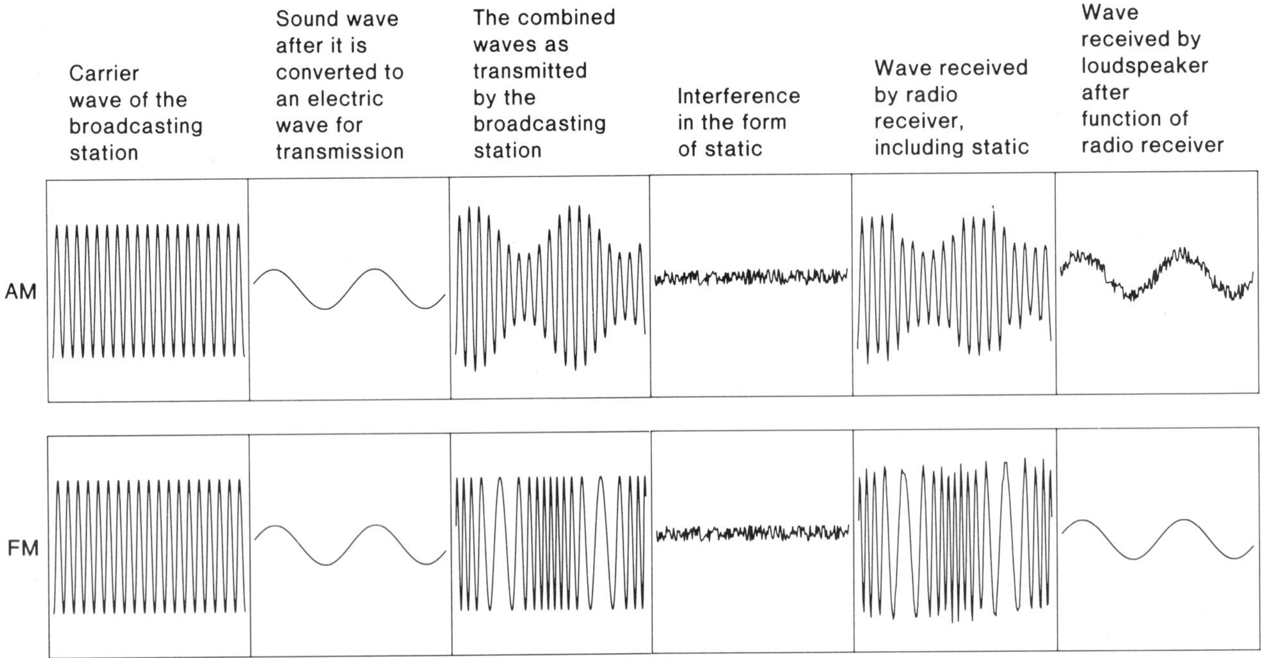

Figure 9.29 A comparison of amplitude modulation (AM) with frequency modulation (FM). The signal to be transmitted is impressed upon a carrier wave, either by changing its amplitude (AM) or by changing its frequency (FM). Since the amplitude of the wave is strongly affected by static, AM radio receivers cannot discriminate between the desired signal and the static. However, because static has almost no effect on the frequency of the waves, FM radio receivers are mostly unaffected by it.

through a *detector* which "discards" the high frequency and recovers the signal impressed upon it, using either amplitude of frequency modulation. An *amplifier* in the radio amplifies the recovered signal and sends it to a *loudspeaker* which converts the electrical signal back into sound waves.

Amplitude modulation is susceptible to interference from electrical storms, electrical appliances, and power lines because any disturbance that changes the size of the signal is amplified by the radio receiver. This interference is heard by the listener as static. Frequency modulation, on the other hand, is relatively static-free since most sources of interference do not affect the frequency of the wave received by the radio. Amplitude- and frequency-modulated waves are illustrated in Figure 9.29, and a schematic diagram of a simple radio receiver appears in Figure 9.30.

Either method of modulation can be used to produce stereo radio broadcasts, although FM is more common. A program source with two separate channels is required. The source may be a stereophonic recording or, for live broadcasts, two microphones. The two signals are combined in the transmitter and broadcast at slightly different frequencies. The receiver separates the two signals from the carrier wave, and each signal is sent through a separate amplifier to different speakers. The resulting *stereophonic sound* allows the listener to detect the direction from which the sounds are coming, giving a "space effect."

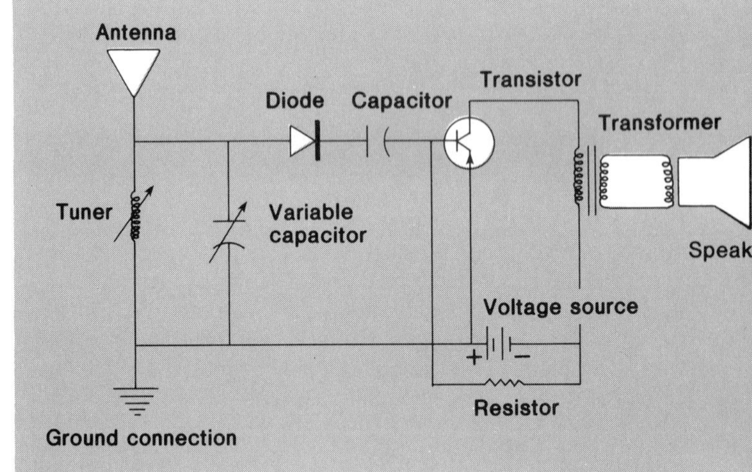

Figure 9.30 A schematic of a simple radio receiver. Such a diagram is necessary for a technician to trace any difficulties in the circuit so that repairs can be made.

Not long after the transmission of sound over long distances was successfully accomplished using radio waves, a concentrated search was begun for a practical way to transmit pictures. The resulting technique is, of course, *television.* Light is converted into electrical impulses, transmitted in the form of electromagnetic waves, and the signal is then converted back into light at the receiver. Light may be converted into electricity by a phenomenon called the **photoelectric effect,** the property present in some materials that causes them to *emit electrons when their surfaces are struck with light,* or the conversion may take place by allowing the light to strike a small semiconductor with thousands of "cells" arranged in a gridlike pattern, which is called a *charge-coupled device,* or *CCD.* In a CCD, light striking the semiconductor causes the cells to acquire a charge proportional to the intensity of the light. The charge on each cell can be "read" electronically and converted into the desired signal. CCDs have the advantages of being smaller, requiring less power, being more rugged, and having greater sensitivity to light than devices using the photoelectric effect.

The television camera initially creates an optical image of the scene to be televised with the aid of a zoom lens. The image so formed is focused on the "target" component of the camera, either a tube coated with photoemissive material or a CCD. If a photoemissive material is used, a beam of electrons is swept across the tube from an "electron gun" located inside the tube, generating a current proportional to the intensity of the light. This current is converted by resistors and amplifiers to voltages suitable for transmission. Each time the electron beam sweeps across the tube, a narrow part of the scene to be televised, called a scan line, is converted. As the electron beam begins its next sweep, it is moved slightly downward, scanning another part of the scene. In the United States, the entire picture is built up of 525 of these scan lines.

When the camera uses a CCD, the entire image is focused on the cells of the device, but the cells are "read out" a row at a time, giving the same effect as scanning one line with the electron beam and making the two methods compatible with each other.

The picture signal is transmitted as radiated waves from the antenna of the television station either directly to the receiving antenna (which may be a community antenna), or to a satellite from which it is retransmitted to a satellite dish. The televised scene is then reproduced in the receiver by the picture tube (Fig. 9.31), a special form of *cathode ray tube* (CRT). An electron beam from the electron gun located in the CRT scans the CRT screen in precisely the same pattern as the camera-tube beam, with the intensity of the electron current controlled by voltages produced in the camera. The picture-tube screen is coated with a sulfide phosphor that emits light when struck by an electron beam, the intensity of the light being proportional to intensity of the current in the electron beam. Thus (in the United States) the picture is made up of 525 lines of varying brightness as the electron beam sweeps across the face of the tube. High-definition television (HDT), now being developed, will use many more lines and will also allow the intensity along each line to vary more rapidly.

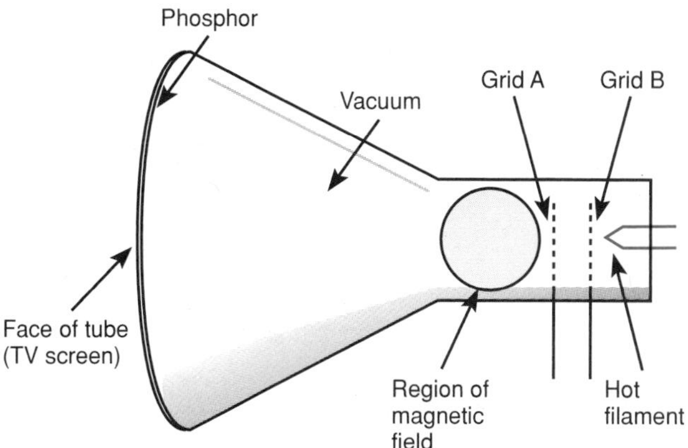

Figure 9.31 A diagram of the principal parts of a cathode ray tube (CRT). Cathode rays (electrons) are emitted by a hot filament and accelerated toward the screen by a high positive voltage on grid A. The number of electrons, and thus, the intensity of the spot on the screen, is controlled by a voltage applied to grid B. The beam is steered to different points on the screen by magnetic fields in the shaded region from coils outside the tube.

A scene that is to be transmitted in color must be created with a special color television camera. The image is split into its red, green, and blue components. In a camera utilizing a tube, special mirrors that reflect only certain colors are employed, each mirror reflecting light to a separate tube. In a camera having a CCD as its active element, small color filters can be incorporated directly on the semiconductor chip, eliminating the need for the mirrors. Each of the red, green, and blue images is transmitted. The receiver has three electron guns and three colors of phosphor on the inside of the tube. Each electron gun in the CRT is aimed so as to strike only one color of phosphor, allowing each color to be separately displayed on the screen, but in an area so small that from a normal viewing distance they blend together to produce the beautiful hues we see. Close examination of the face of a CRT, especially with a small magnifying glass, clearly reveals the individual lines and phosphor colors making up the picture.

The television pictures we view are transmitted at a rate of 30 per second, so rapidly that the human eye cannot distinguish the individual pictures. Instead, the eye interprets the images as continuous motion. This is, of course, the same effect as that of a motion picture, where the images are flashed on the screen at a rate of 24 per second.

The high reliability of microminiature electronic systems has led to their application in a wide variety of fields. For instance, the list of relatively new devices available for the diagnosis, treatment, and monitoring of medical patients that incorporate these tiny components continues to grow (see Fig. 9.32). Electronic pacers and hearing aids have been greatly improved with recent electronic discoveries. Cordless telephones and wristwatch-size television sets are also prime examples of technological advances. The utilization of charge-coupled devices in place of film has led to

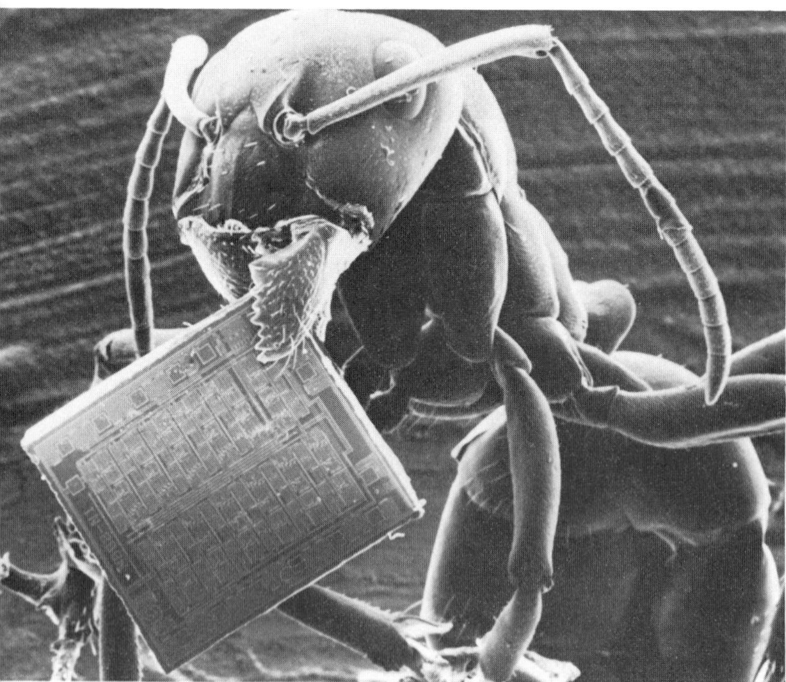

Figure 9.32 The world of miniaturization in electronics is upon us. Even an ant has little difficulty holding the integrated chip used in numerous miniature electronic circuits.

greatly improved detection of light at low levels, allowing astronomers to study very faint and distant objects.

The progress being made in the field of electronics and related areas may be further enhanced by recent research involving the development of new plastics. Since about 1977, researchers have known that the process of impregnating various plastics with certain doping materials causes the treated plastics to behave in the manner of electrical conductors. In metals, the outermost orbital electrons move about somewhat freely, whereas in plastics and other insulating materials these electrons are more firmly held in place. Some researchers feel that the impurities either capture some of these electrons, as in p-type doping, or furnish additional electrons, as in n-type doping. In either case, electrons are permitted to move about freely; the greater the amount of dopants added, the greater the electron movement, and hence the greater the conductivity. The development of the treated plastics, known as plastic metals, may ultimately result in the availability of conducting wires that need no insulation, as well as of plastic computer chips and solar cells, both cut to some convenient size from a roll. Plastic batteries are also under development and may be available this decade. The lifetime of conventional rechargeable wet-cell batteries is limited, since the electrodes are subject to pitting by the action of the acid solution, eventually rendering them useless. Plastic metal electrodes perform much like a sponge, furnishing electrons when power is required and taking them back as the battery is recharged. Since no chemical reaction occurs, the lifetime for such a battery could be a long one. Plastic batteries could be used in numerous ways, such as to furnish heat in the lining of clothing and sleeping bags or to provide electric power for electric cars. They could be most useful in providing electrical energy for computers, so that the memory would be retained in case of conventional power failure. Then the batteries could be recharged when power was restored.

Problems must be solved before the plastic metal, with its seemingly unlimited applications, becomes available to the consumer, but research will continue.

Summary

Electric current is the flow of electric charges, usually electrons. Even though electrons usually flow from the negative pole of a source to its positive pole, most scientists consider the direction in which positive charges would flow and call this the direction of conventional current. Ohm's law describes the relationship of current, electric potential, and resistance in a circuit. The current (measured in amperes) is equal to the electric potential (in volts) divided by the resistance (in ohms). The joule is used as a unit of work and energy in electricity. The power consumed or provided by a circuit is measured in watts (joules per second). Joule's law, $Q = 0.239\ I^2Rt$, gives the heat produced in calories per unit of electrical energy consumed. The electrical energy consumed by homes and industry is measured in kilowatt-hours.

Electrical circuits may be constructed with elements either in parallel, in series, or in combinations of parallel and series wiring. Batteries are composed of a series of cells so connected that the voltages of the cells are additive. House circuits are primarily wired in parallel; however, fuses are placed in series with lights and receptacles within a circuit. Sources of electric potential and resistances may be wired in parallel or in series.

Batteries produce direct current. Alternating current is produced by AC generators. Devices that increase or decrease the voltage of the alternating current produced by a generator are called transformers. Transformers are composed of at least two coils, a primary coil and a secondary coil. Step-down transformers decrease voltages; step-up transformers increase voltages. When the voltage is stepped up the current in the secondary is stepped down proportionately, and vice versa, so that the product of voltage and current in the primary and secondary is the same. Many household appliances contain transformers that provide proper voltages for the necessary electrical circuitry.

The electrical energy used in our homes is often converted into heat and light. Incandescent and fluorescent lamps provide artificial lighting. Each type of lamp has advantages over the other type; therefore, most homes contain both types of lamps. Thermal energy produced from electrical energy often heats houses and water, controls room temperature, and cooks food. Some household appliances convert electrical energy into mechanical energy. The conversion is accomplished by a motor, a device that contains a coil of wire that is forced to rotate in a magnetic field by utilizing the laws of repulsion and attraction of magnetic poles.

Electronics is the branch of science that deals with the emission, properties, and effects of electrons as well as with the devices that employ these characteristics. Major components of various electronic circuits are the capacitor, diode, resistor, transformer, transistor, and voltage source. Because of their small size, accuracy, low power consumption, and reliability, semiconductors—including diodes and transistors—have replaced the vacuum tube in most fields of electronics. Integrated circuits have miniaturized the electronic world and have enabled many new devices for the consumer.

Equation Summary

Electric energy:

$$E = VIt,$$

where the energy E (joules) is equal to the applied potential V (volts) times the current I (amperes) times the time t (seconds).

Electric power:

$$P = VI,$$

where the power P (watts) is equal to the applied potential V (volts) times the current I (amperes). For devices that obey Ohm's law, it also holds that

$$P = I^2 R$$

and that

$$P = \frac{V^2}{R},$$

where R (ohms) is the resistance of the device.

Heat energy:

$$Q = (0.239 \text{ cal/J}) I^2 Rt,$$

where the heat energy Q (calories) is equal to the constant times the square of the current I (amperes) times the resistance R (ohms) times the time t (seconds).

Cells in series and parallel:

$$V_t = V_1 + V_2 + V_3 + \cdots + V_n,$$

where the total voltage V_t (volts) is the sum of the voltages of each cell in series. In parallel, the cells usually have the same voltage so that

$$V_t = V_1 = V_2 = V_3 = \cdots = V_n$$

and the total voltage is equal to the voltages of the individual cells.

Resistors in series and parallel:

For resistors in series, the total resistance R_t (ohms) is equal to the sum of the resistances, so

$$R_t = R_1 + R_2 + R_3 + \cdots + R_n.$$

For resistors in parallel, the total resistance (ohms) is found by adding the reciprocals of the individual resistances and then taking the reciprocal. Thus,

$$\frac{1}{R_t} = \frac{1}{R_1} + \frac{1}{R_2} + \frac{1}{R_3} + \cdots + \frac{1}{R_n}.$$

Transformers:

$$\frac{\text{secondary voltage}}{\text{primary voltage}} = \frac{\text{\# of turns on secondary coil}}{\text{\# of turns on primary coil}},$$

where the secondary (output) voltage (volts) may be made higher or lower than the primary (input) voltage (volts) by using different ratios of turns on the secondary and primary coils. Also,

$$\frac{\text{secondary current}}{\text{primary current}} = \frac{\text{\# of turns on primary coil}}{\text{\# of turns on secondary coil}}.$$

The ratios of the currents (amperes) is the reciprocal of the ratio of the voltages; that is, the right side of the second transformer equation is inverted relative to the first transformer equation.

Questions and Problems

Energy and Power in Electric Circuits

1. Why does the filament of an electric lamp glow "white-hot" when it is connected to a source of the proper voltage? Why don't the wires connecting it to the source glow "white-hot"?
2. (a) If a certain piece of wire is carrying a current, how does the power dissipated as heat change if the current through the piece of wire is doubled? (b) How does it change if the current is reduced to half of its original value?
3. A light bulb designed to operate from a 120-V wall outlet has a resistance of 360 Ω. What is the power rating of the bulb?
4. A clothes dryer and a television set operate at 4000 W and 300 W, respectively. Find the cost to operate each for 1.00 h if electrical energy costs 11.0¢ per kWh.
5. (a) Determine the electrical power consumed by a hair dryer that draws a current of 10.0 A when plugged into a 120-V outlet. (b) If all the energy of the hair dryer is converted into heat, determine the amount of heat energy, in calories, provided by the hair dryer in 1.00 min.

Series and Parallel Circuits

6. What is the purpose of connecting batteries of equal voltage (a) in parallel and (b) in series?
7. Cite several examples of series and parallel circuits commonly used in our homes.
8. Determine the current through a circuit that contains two resistors, one having a resistance of 10.0 Ω and the other having a resistance of 30.0 Ω (a) when they are wired in series and (b) when they are wired in parallel to a voltage source of 12.0 V.
9. Explain why lights wired in series all "go out" if one becomes defective and why if lights are wired in parallel only the defective one "goes out."

Alternating Current and the Transformer

10. What is a rectifier and what is its purpose?
11. State the purpose of a transformer and list its essential parts.
12. Why is the voltage usually "stepped up" to a high value as it leaves a generating station? Why is it stepped down before it enters our homes?
13. The transformer for a doorbell has 500 turns on its primary and 50 turns on its secondary. What is its secondary voltage if its primary is connected to a 120-V source?
14. A transformer is used at a generating station to step up the voltage from the generator for the transmission line. The transformer has 150 turns on its primary and 30,000 turns on its secondary. If the current in the primary is 1500 A, what is the current the secondary supplies to the transmission line?

Household Appliances

15. List the advantages of fluorescent lighting over incandescent lighting as used in the home. What are the disadvantages of fluorescent lighting in the home?
16. Why is the armature of an electric motor usually composed of a soft-iron core rather than one of hardened steel?
17. How does a thermostat control the operating temperature of an appliance?
18. What are the three major household uses of electricity?

Electronic Components

19. What is a crystal lattice?
20. Describe the characteristic that makes semiconductors different from conductors and insulators. What are the two most common semiconducting materials? How do semiconductors differ in structure from insulating materials?
21. How does a solar cell produce electricity?
22. What are liquid crystals and how may they be used in electronic devices?

Electronic Systems

23. State the main components used in converting sound waves into radio waves and explain the use of each component. How do radio waves differ from sound waves?
24. What is the difference between amplitude modulation and frequency modulation of radio waves?
25. Describe how a cathode ray tube (CRT) produces a television picture.

10 Electromagnetic Radiation

Chapter outline

Key terms/concepts

The sunburst is created when the light from the sun passes through small openings between the leaves deep within the forest. Electromagnetic radiation in the visible region of the spectrum is only a small portion of the total radiation emitted by our nearest star.

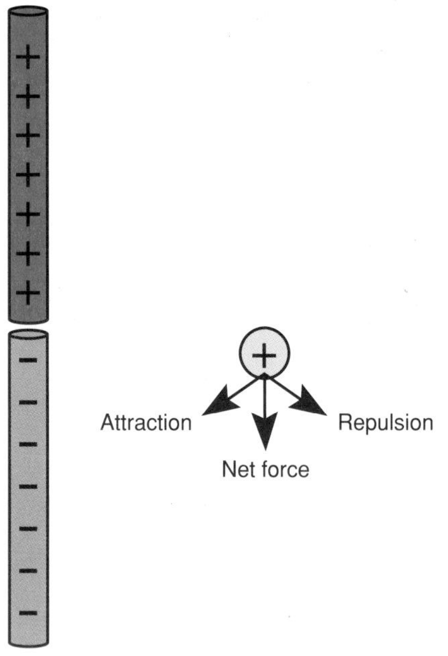

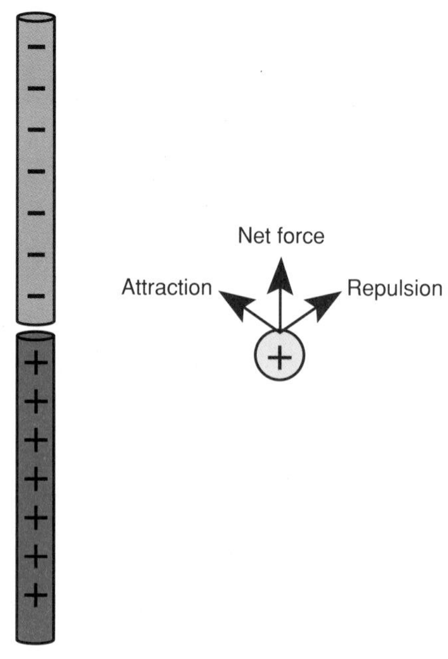

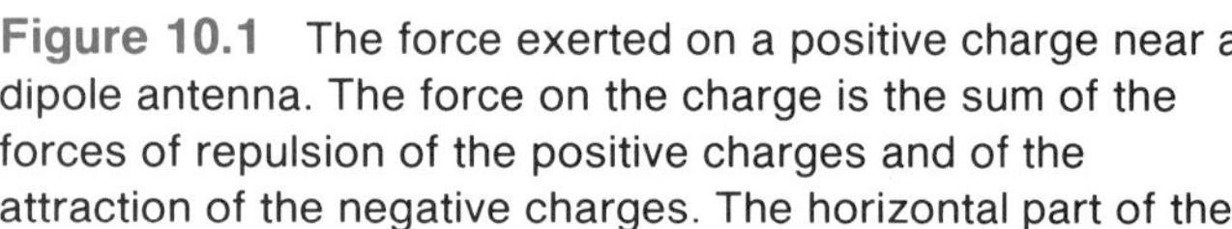

a. b.

Figure 10.1 The force exerted on a positive charge near a dipole antenna. The force on the charge is the sum of the forces of repulsion of the positive charges and of the attraction of the negative charges. The horizontal part of the forces cancel out, leaving only a vertical force. Compare this with the electric field pattern of the two charges in Figure 8.13b.

Electromagnetic waves were postulated by James Clerk Maxwell (1831–1879), a brilliant theoretical physicist from Scotland. He found the equations of electricity and magnetism to be incomplete, and by including a term to account for the effects of a changing electric field, he predicted completely new phenomena. It had long been known that changing a magnetic field produced forces on charged particles, as we discussed in Chapter 9 in conjunction with transformers and generators. In other words, the changing magnetic field "induces" an electric field. Maxwell's revised equations showed that a changing electric field conversely induces a magnetic field. This interaction between the changing fields allows for the production of electromagnetic waves, and the equations permit calculation of the speed with which these waves will propagate through space or any material medium. Maxwell, in 1864, calculated the speed of these electromagnetic waves and found it to be the same as the experimentally determined speed of light. Further work with the equations showed that almost all of the experimentally determined properties of light could be explained by assuming that light is an electromagnetic wave.

At the time of Maxwell's death, there was no known method for generating and detecting electromagnetic waves using electrical and magnetic apparatus, so Maxwell's hypothesis could not be verified experimentally. In 1888, Heinrich Hertz, a German physicist, performed experiments that demonstrated that electrical disturbances could be transmitted through space in the form of waves. Maxwell's equations were thus experimentally validated. At one time they were considered fundamental laws of the universe in the manner of Newton's laws of motion and gravitation, but they are now known to be inadequate for phenomena that require explanation in terms of quantum theory (discussed later in this chapter, in the section entitled "Light—Wave or Particle?"). Thus, they are found to have a limited range of validity, just as Newton's laws. Nonetheless, they are still extremely useful for most ordinary electromagnetic phenomena and, indeed, they form the basis for most of electrical engineering.

Electromagnetic Waves

To better comprehend the nature of electromagnetic waves, consider two pieces of wire charged as shown in Figure 10.1a. Such an arrangement is called a *dipole antenna.* The force exerted on a positive charge located near the antenna is shown. Now, suppose that the positions of the positive and negative charges on the antenna could be rapidly switched. The force on the positive charge would now be in the opposite direction, as shown in Figure 10.1b. It would, however, take a short time for the positive charge to "sense" the change in the charges on the antenna. In other words, the electric field of the antenna would be disturbed, but the disturbance would take a finite amount of time to move the distance from the antenna to the positive charge.

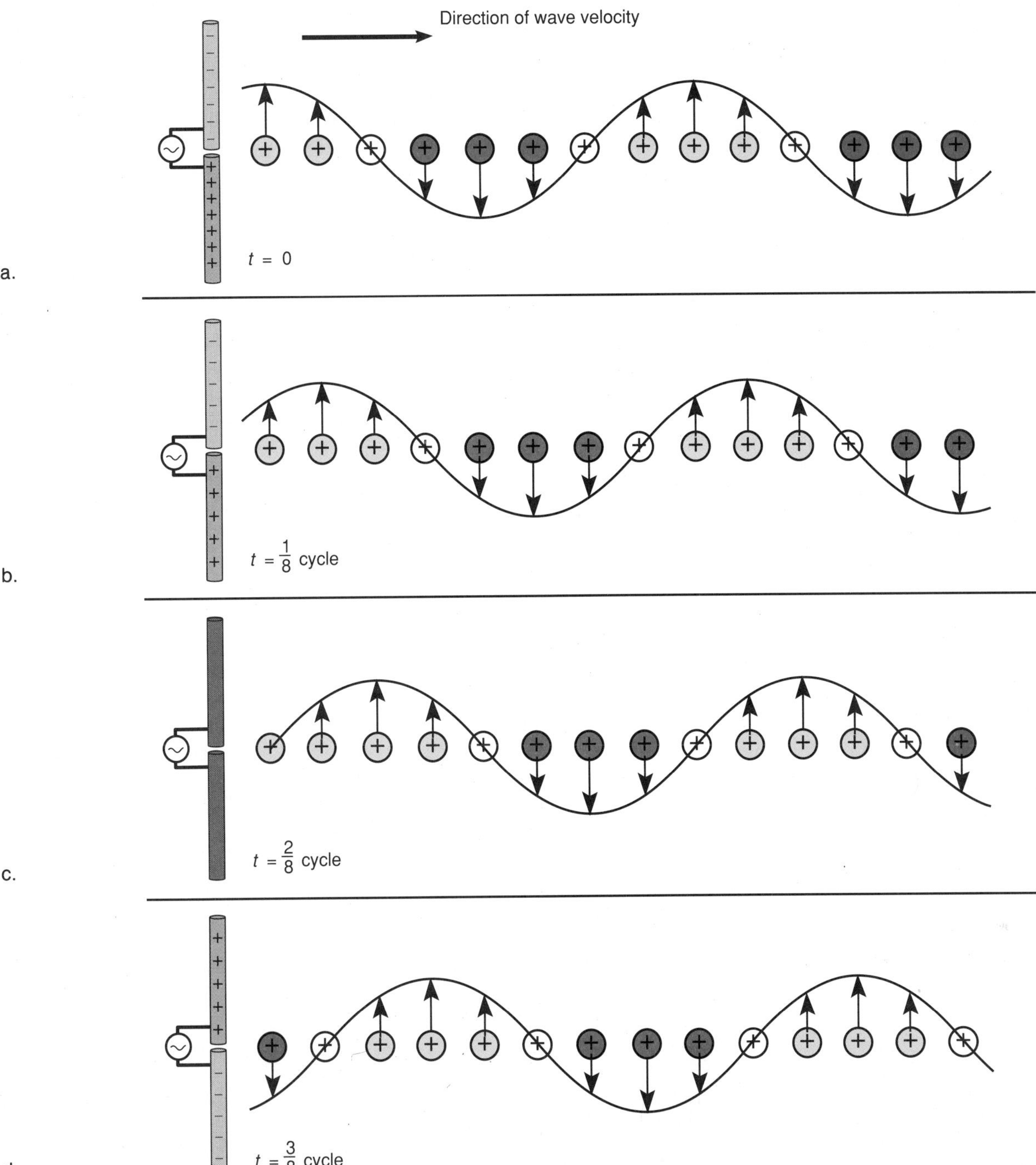

Figure 10.2 The wave generated by a dipole antenna attached to an AC source. The charge on the two sections of the antenna continually changes, causing the forces on the charges to continually change. Note that when the source has completed 1/8 cycle, the wave has moved 1/8 wavelength.

If the charges on the antenna were continually changed by an alternating voltage source, instead of just once, the disturbance would move outward from the antenna, requiring more time for the signal to reach more-distant points. A charge at any given location would experience a continually changing force, tending to make the charge oscillate upward and downward. Charges farther from the antenna would also be experiencing this oscillatory force, but since the signal would take longer to reach them, they would lag behind in their response. The situation for a line of charges is shown in Figure 10.2a. Note that the force on some of the charges is of the same magnitude and in the same direction. These charges are far enough apart that the antenna has gone through one or more complete oscillations while the signal has moved a distance equal to that between the charges.

The line connecting the tips of the force vectors on all the charges shows that the electric field at each point in space has a magnitude that varies in the shape of a wave. Further, this wave is moving outward away from the antenna at the speed of light, so that at later times the disturbance appears as shown in Figures 10.2b–d. Not shown in any of the diagrams is the magnetic field

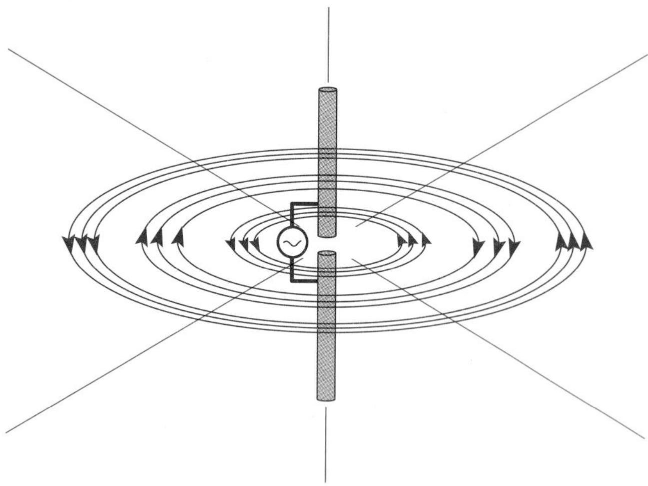

Figure 10.3 The magnetic field around a dipole antenna is induced by the current flowing in the antenna. As the current reverses direction, the field reverses direction, with the signal moving outward at the speed of light.

surrounding the antenna. Recall that a magnetic field circles any current, and that while the charges are flowing in the antenna, as they must in order to continually change the charge on it, there will be a magnetic field pattern circling the antenna, as shown in Figure 10.3. Note that this magnetic field is perpendicular in direction to the electric field.

Near the antenna, the electric and magnetic fields are "out of phase"; that is, the electric field reaches a maximum and the magnetic field a minimum at the same time, and vice versa (to be more exact, the fields are 90 degrees out of phase). It turns out that near the antenna, all of the energy contained in the fields is returned to the antenna when the fields themselves return to zero every one-half cycle, unless some of this energy is absorbed by charges near the antenna. The energy contained in these fields can be detected with an ordinary AM radio if the radio is placed near a strong oscillating source. The buzzing sound of a car radio when it passes under a high-voltage transmission line is evidence of this oscillating field.

Far from the antenna, the electric and magnetic fields are "in phase," and this is the electromagnetic field *radiated* from the antenna, which can be received at long distances. This radiated wave is the one predicted by Maxwell's equations and is shown in Figure 10.4. The diagram actually shows the strengths of the electric and magnetic fields along one line and the direction in which the wave is moving. Note that the electric and magnetic fields are oscillating in a direction perpendicular to the direction in which the wave is traveling; hence, electromagnetic waves are *transverse waves.* (The waves generated on a horizontal string that is shaken up and down

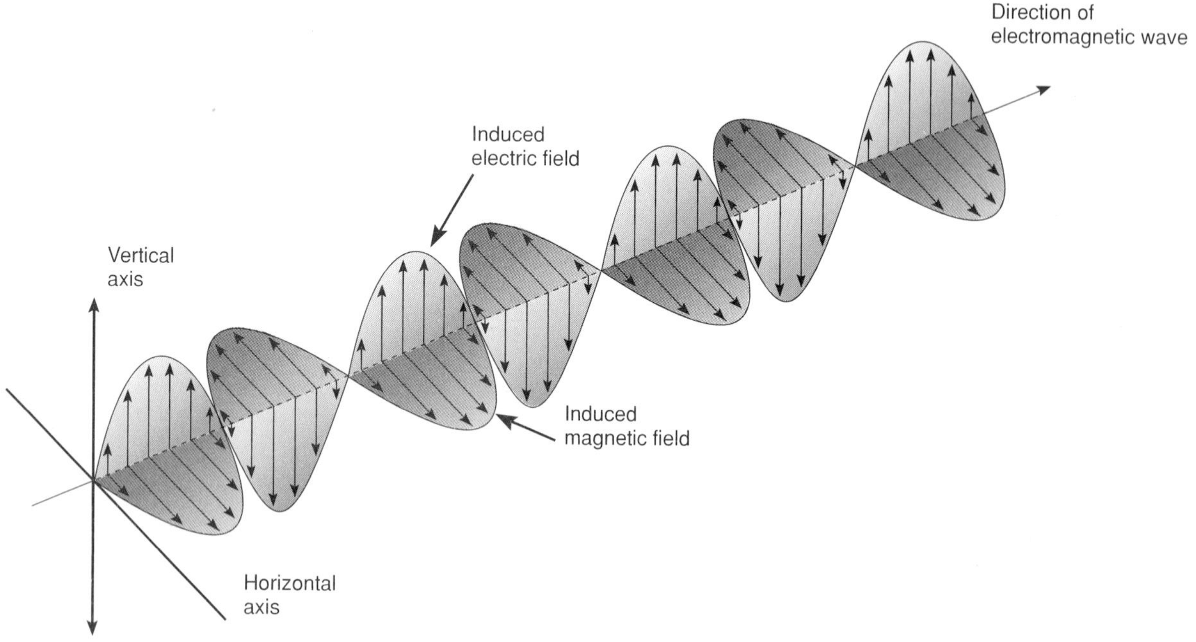

Figure 10.4 Far from an antenna, the electric and magnetic fields are in phase, so they reach their maximum values at the same point. The electric and magnetic field are transverse; that is, perpendicular to the direction in which the wave is traveling.

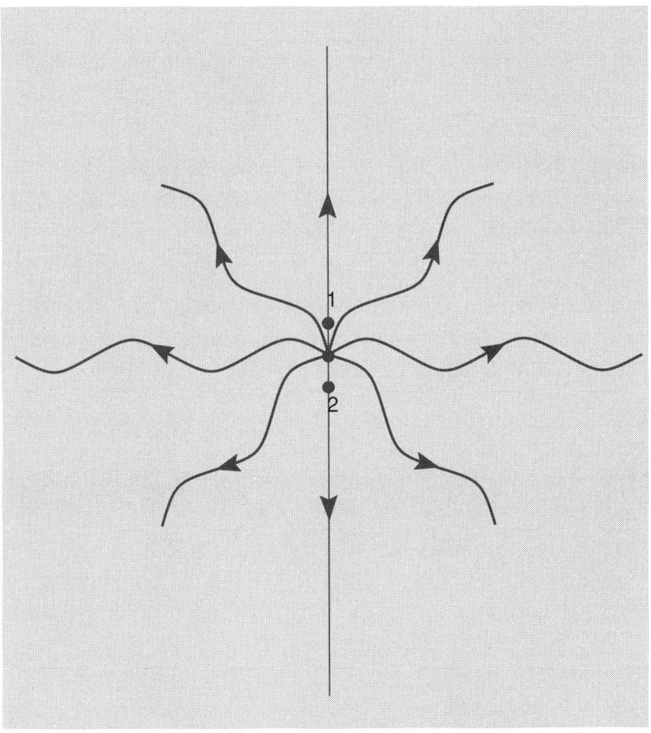

Figure 10.5 As a positive charge oscillates between points 1 and 2, its electric field is distorted. In this diagram the charge is midway between the points. Far from the charge, only the oscillating part of the field has appreciable strength; that is, one could subtract the field shown in Figure 8.13a and find the "wave part" of the field. Note that no wave is emitted in the direction along which the charge is oscillating, only perpendicular to it.

at one end are a familiar example of transverse waves. See the section entitled "Kinds of Waves" in Chapter 7.) A wave is said to be **polarized** if *the oscillations are in a particular direction.* This wave is said to have vertical polarization, since the vibrations of the electric field are all in the vertical direction. (The direction of vibration of the electric field is usually taken to be the direction of polarization, but the magnetic field could be said to be horizontally polarized.)

Before completing the discussion of the nature of electromagnetic waves, note that any vibrating charge can send out electromagnetic waves. In Figure 10.5 the oscillating charge causes the electric and magnetic fields to change, thus sending out an electromagnetic wave. If the oscillating charge is an electron, proton, or the charge on an atom, the oscillations may be extremely rapid, allowing the generation of high-frequency waves.

The Speed of Light

As has been previously mentioned, the speed of propagation of electromagnetic waves can be computed from theoretical constants in Maxwell's equations and is equal to the measured speed of light. The first successful measurements of the speed of light were based on the data of Danish astronomer Olaus Roemer (1644–1710). While undertaking a long-term study to determine the time it takes Jupiter's moons to complete their orbits, he found his measurements indicated shorter times when Earth was moving toward Jupiter and longer times when Earth was moving away from Jupiter. He concluded that the difference was caused by the finite speed of light, with the signal reaching Earth sooner as it moved closer to Jupiter and later as it moved farther away (Fig. 10.6).

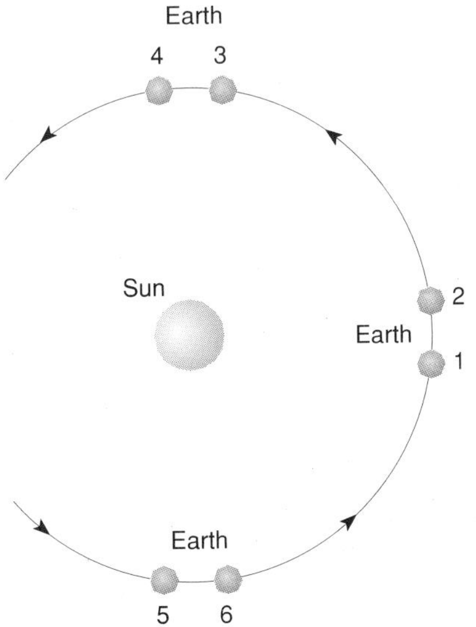

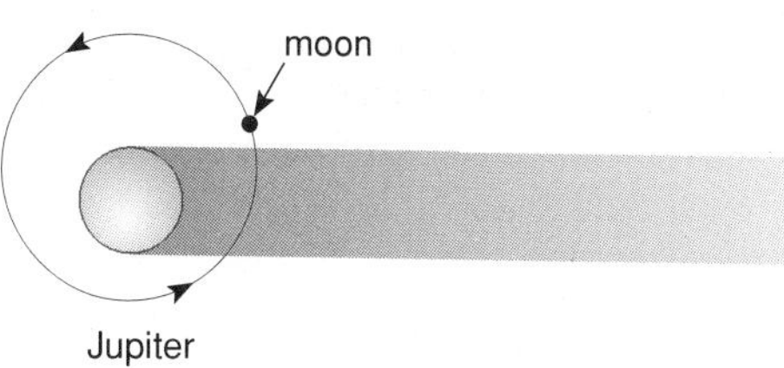

Figure 10.6 Roemer measured the time for Jupiter's moons to complete one orbit by observing when they reappeared from Jupiter's shadow. He found that they seemed to take longer when Earth moved from position 3 to 4 than from 1 to 2. This is because position 4 is farther from Jupiter than position 3, and it takes longer for the light to reach Earth at position 4 than at 3. On the other hand, positions 1 and 2 are the same distance from Jupiter. Conversely, the observed time is less when moving from 5 to 6 than when moving from 1 to 2, since the second signal reaches Earth sooner at 6 than at 5.

The speed of light through space, as based on Roemer's measurements, is about 230,000,000 meters per second (140,000 mi/s). As measurements have become more and more accurate, the speed of light in a vacuum has been determined to be 299,792,458 meters per second. (Actually, this is now the *defined* value of the speed of light in a vacuum, since in 1983 the following definition was adopted for the meter: The meter is the length of the path traveled by light in a vacuum during a time interval of 1/299,792,458 of a second.)

The speed of light in a vacuum has been found to be a fundamental constant of nature, relating mass and energy and providing a sort of universal speed limit for everything in the universe as detailed in the special theory of relativity developed by Albert Einstein in 1905. (The special theory of relativity shows that the speed of ordinary particles is always less than the speed of light, but it does not preclude the existence of particles created at speeds exceeding the speed of light. An experimental search for such particles, called tachyons, has yielded only negative results.) Since it is used so widely, a lowercase c is the universally accepted symbol for the speed of light. The speed of light is lower in materials than in a vacuum, the consequences of which will be discussed in the section entitled "Refraction" later in this chapter.

The oscillations causing electromagnetic waves vary over a wide range in frequency, but in a vacuum (or, approximately, in air) the speed with which the waves move away from the source is the constant value 3.00×10^8 meters per second (to three significant figures), regardless of the frequency of the oscillations. As discussed in Chapter 7, the **wavelength,** λ, and the **frequency,** f, of oscillation are related to the speed of the waves. The distance between successive crests or troughs (places where the field reaches a maximum value in one direction or the other) is the wavelength of an electromagnetic wave. Consider the diagram shown in Figure 10.7. During the time required for the source to complete exactly one cycle of its oscillation, the wave must have traveled exactly one wavelength, since now the source will have returned to the same condition as at the start of the previous cycle. (For example, if it had been generating a crest it will now be generating the next crest.) The *time for one cycle* is the **period,** denoted by T. Since the speed is the distance divided by the time, it follows that $c = \lambda/T$. A more useful form of this equation is found by using the frequency, f, instead of the period T. The frequency in hertz (1 Hz/s) is equal to the reciprocal of the period T. For example, if a source were oscillating with a period of 0.1 second it would complete 10 oscillations per second, and its frequency would be 10 hertz. Therefore, $f = 1/T$, and the equation relating wavelength, frequency, and speed in a vacuum becomes

$$c = f\lambda. \qquad \text{SPEED, FREQUENCY, WAVELENGTH (in vacuum)}$$

If the electromagnetic wave is traveling in some medium, the speed is less than c, so the speed in that medium must be used. The frequency of the wave is not changed when passing from one

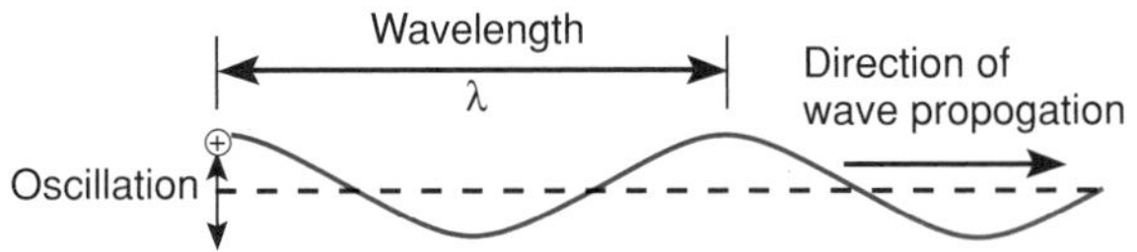

Figure 10.7 As the source vibrates, waves propagate outward. During the time for one complete vibration (or cycle), the wave must move exactly one wavelength.

medium to another (otherwise the waves would "pile up" at the boundary or new waves would be created there). Thus, if the speed in the medium is v, the equation becomes

$$v = f\lambda. \qquad \text{SPEED, FREQUENCY, WAVELENGTH (in medium)}$$

Example 10.1

Determine the frequency of an electromagnetic wave, the wavelength of which is 2.50×10^{-7} meter as it travels through a vacuum.

Solution

Using the relationship

$$\begin{aligned} c &= f\lambda, \\ f &= \frac{c}{\lambda} \qquad \text{FREQUENCY} \\ &= \frac{3.00 \times 10^8 \text{ m/s}}{2.50 \times 10^{-7} \text{ m}} \\ &= 1.20 \times 10^{15} \text{ Hz.} \end{aligned}$$

Example 10.2

The broadcasting frequency of a local radio station is 90.3 megahertz. What is the wavelength of the radio wave it transmits? Assume the velocity of the wave to be 3.00×10^8 meters per second.

Solution

The solution is found using the formula

$$\begin{aligned} \lambda &= \frac{c}{f} \qquad \text{WAVELENGTH} \\ &= \frac{3.00 \times 10^8 \text{ m/s}}{90.3 \times 10^6 \text{ Hz [1/s]}} \\ &= 0.0332 \times 10^2 \text{ m} \\ &= 3.32\text{m.} \end{aligned}$$

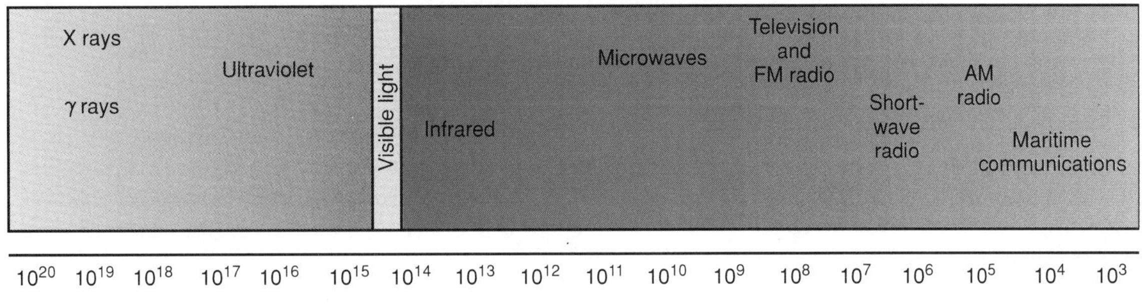

Figure 10.8 The electromagnetic spectrum. Note that the product of the exponents of each frequency and the wavelength of that specific frequency (which appears immediately above it) are consistently in the order of 10^8, the proper exponent for the velocity of electromagnetic waves in air and in most other mediums when measured in meters per second.

The Spectrum

The previous examples show that wavelengths and frequencies of electromagnetic waves have a tremendous range. If an electric charge oscillated only once per second (1.00 Hz), the wavelength of the electromagnetic wave would be 300,000,000 (3.00×10^8) meters, whereas if one oscillated with a frequency of 5.00×10^{14} Hz, its wavelength would be 6.00×10^{-7} meter. Wavelengths generally range between about 100,000 meters and 10^{-12} meter, this range encompassing the electromagnetic **spectrum.** Figure 10.8 shows the general range of frequencies and wavelengths of the spectrum, as well as the descriptive names used for each region. The term *spectrum* is also used to describe the range of wavelengths (or frequencies) emitted by a particular source. The sun, for example, emits a broad range of wavelengths, and one speaks of this range as the solar spectrum.

The spectrum is limited on both ends by natural constraints. At the low-frequency, or long-wavelength, end of the spectrum, the size of antennas becomes a limiting factor. An antenna radiates energy efficiently only if it has a length approximately equal to one-half of a wavelength. So, at low frequencies, such as power-distribution frequencies (60 Hz), the waves are so long (5 million meters) that not even long transmission lines radiate appreciable amounts of power. Wavelengths ranging from 187 to 544 meters are used for AM radio transmissions, necessitating tall transmitting antennas, whereas FM radio broadcasts use frequencies with wavelengths between 2.7 and 3.4 meters, with antennas that are much smaller. (FM and television stations put their antennas on higher towers because the signals do not follow the curvature of Earth, as to be discussed.) The shortest wavelengths come from cosmic rays, and are limited by the energy source available for generation.

The frequencies used for AM radio broadcasts, then, are between 535 and 1605 kilohertz (kHz), whereas FM radio uses frequencies from 88 to 108 megahertz (MHz). Television uses frequencies of from 52 to 216 megahertz (called *very high frequencies,* or VHF); channels 14 to 83 transmit at frequencies of from 470 to 890 megahertz (called *ultrahigh frequencies,* or UHF). Each TV station is allotted a bandwidth of 6 megahertz, whereas the various AM broadcasting stations operate at a bandwidth of 10 kilohertz. The greater bandwidth for TV broadcasting is necessary because a television transmitter must send out the information required to produce a complete picture and its accompanying sound 30 times per second, whereas a radio station, sending out much less information per unit of time, requires a correspondingly smaller bandwidth.

In order to obtain the frequencies used for radio waves (550 to 1600 kHz), a new method of producing electromagnetic waves had to be developed, since the induction coil could not reverse the electric field produced rapidly enough to provide a major portion of the total energy spectrum. A small crystal was found to reverse the direction of the electric field more rapidly. Various crystals vibrate naturally at a constant high frequency directly related to their respective sizes and composition, in much the same way that a bell or chime has its own natural frequency. Each radio station has its own wafer of crystalline quartz, about the size of a nickel, that constantly vibrates at the assigned frequency of the station. Smaller crystals are designed and employed to produce the transmitting signal of FM radio and television stations. Other crystals, microscopic in size, vibrate at higher frequencies known as infrared, or heat waves. The observation can be made, therefore, that infrared waves are produced by vibrating atoms and molecules that carry an electric charge.

Frequencies emitted in the range of visible light, 4.3×10^{14} to 7×10^{14} hertz, are obtained through the vibration of (negatively charged) electrons. If the electron is caused to vibrate above 7×10^{14} hertz, the wave produced may be an ultraviolet ray or an X ray, depending on the specific frequency that results. Beyond the X-ray range lies the gamma-ray region. The gamma ray is produced in the nucleus of the atom through a disturbance of the

Early Theories of the Nature of Light

One of the earliest attempts to explain the nature of light is attributed to the Greek philosopher, Plato, ca. 428–348 (or 347) B.C. Plato maintained that light consists of threadlike streamers emitted by the eye. According to his theory, seeing occurs when these emissions make contact with an object. Euclid, another Greek scholar who lived shortly after Plato, pointed out that we seldom notice a small object, such as a needle on the floor, unless our eyes are caused to seek it, an observation that appeared to support Plato's contention. Even as late as the seventeenth century A.D., Descartes, a French mathematician-philosopher, published the streamer theory in his writings.

However, not all of the ancient philosophers and other scholars expressed a belief in the streamer theory of light. Scholars affiliated with the School of Pythagoras, another Greek philosopher who lived before Plato, insisted that light travels from a luminous object in the form of tiny particles to the eye. About the same time during that century, still another Greek philosopher, Empedocles, argued that light travels from objects to the eye in some form of wave. The wave-particle controversy about light, then, was conceived long before the rise of modern science and continues to offer apparently conflicting views as to light's true identity even today. We do, however, recognize the dual nature of electromagnetic waves and have established the existence of both inherent characters through experimentation ■

electric field of the nucleus. As was previously discussed, all electromagnetic waves travel at the same velocity, 299,792,458 meters per second in a vacuum, but they differ in frequency and, hence, wavelength. As the frequency of the wave increases, its wavelength decreases, since the velocity of the wave in a vacuum is constant (recall $c = f\lambda$).

The velocity of electromagnetic waves varies with the medium through which they travel. Visible light travels through water at a velocity of 2.25×10^8 meters per second and through glass at about 2×10^8 meters per second. In no medium does light travel faster than it does in a vacuum. Several properties of electromagnetic waves are unique. First, as visible light passes through glass, it slows down instantly from 3×10^8 meters per second, its velocity in air, to 2×10^8 meters per second. But the light regains its original velocity as it leaves the glass; that is, the light immediately reassumes a velocity of 3×10^8 meters per second, which causes the emitted ray (the one that leaves the glass) to be parallel to the incident ray (the one that entered).

Electromagnetic waves also differ in their ability to penetrate matter. Such high-frequency waves as radio waves can penetrate the air but are absorbed by thin layers of metals. Low-frequency radio waves are reflected from the ionic sublayer of the thermosphere, which is composed of charged particles, whereas high-frequency radio waves pass through the ionosphere into outer space. Radio reception is better at low frequencies at night than during the day because the ionosphere is higher above Earth's surface and is thus able to reflect more-distant sources. Television frequencies, which are higher than those of radio waves, penetrate the ionosphere; hence, they are not reflected. The TV frequencies (about 10^8 Hz) can be received only in straight-line distances from the transmitting station. Because of the curvature of Earth, the taller a transmitting antenna is (Fig. 10.9), the farther away from the TV station a suitable signal can be received. This relationship may be expressed mathematically as $s = 3.59\sqrt{h}$, where s is the distance of transmission in kilometers and h is the height of the antenna in meters. The distance in the English system, when h is expressed in feet and s in miles, is determined from $s = 1.23\sqrt{h}$.

Example 10.3

What is the effective range of a television station, the transmitting antenna of which is located on the peak of a mountain 3600 meters (11,800 ft) above sea level?

Solution

The range is found using the relationship

$$\begin{aligned} s &= 3.59\sqrt{h} \qquad \text{TRANSMITTING RANGE} \\ &= (3.59)(\sqrt{3600}) \\ &= (3.59)(60) \\ &= 215 \text{ km, or } 134 \text{ mi.} \end{aligned}$$

Infrared (or heat) waves can penetrate dry air but are stopped by humid air. The human body also absorbs waves in the infrared region, which accounts for the sun's ability to warm our bodies. Infrared rays, whose wavelengths are slightly longer than those at the red end of the visible band, also affect special photographic film, as illustrated in Figure 10.10.

Electromagnetic waves shorter in wavelength than those of the violet end of the visible band are unusual in many ways. Ultraviolet radiation (UV), like visible light, can cause photochemical reactions in which radiant energy is converted into chemical energy. Ultraviolet radiation produces ozone in the upper atmosphere and produces melanin in the human skin. The production of melanin results in what we call a "suntan." X rays penetrate denser materials than do electromagnetic waves of lesser frequencies; however, some of the X rays are absorbed by matter. Any substance, depending on its structure and its density, can absorb, reflect, or transmit the various electromagnetic waves selectively.

Figure 10.9 Television transmitting antennas are placed on tall towers and at elevated locations, if possible, because the high-frequency waves essentially follow the line of sight. Due to this lack of bending, the curvature of Earth limits their range.

Figure 10.10 Film sensitive in the infrared region reveals the emission and reflection of otherwise invisible light from the iron bottoms and its reflection from the statue.

The Behavior of Electromagnetic Waves

Many properties of electromagnetic waves make them useful in various applications. Even though these properties are unified by Maxwell's theory, some of them depend on the wavelength (or frequency) of the radiation. For example, radio waves are reflected just as visible light is reflected, but the strength of the reflected wave relative to the incident wave may be very different for radio waves and light. Additionally, the speed of the waves in various materials depends somewhat upon the wavelength. Variations in speed generally cause the waves to bend, thus the wavelength has an effect on the magnitude of the bending. Another important property that is affected by the frequency of the radiation is absorption. Electromagnetic waves of certain frequencies are strongly absorbed by some materials, while those of other frequencies are hardly absorbed at all. Thus, it is natural when studying certain characteristics of electromagnetic waves to separate the study into small regions of the electromagnetic spectrum. Since the visible region of the spectrum can be detected by the human eye, it is the most widely studied portion of the spectrum.

Atoms may absorb energy in various ways, such as from an electrical discharge, from heating, or by the absorption of electromagnetic radiation. For example, as discussed in Chapter 9 in connection with fluorescent lamps, fluorescence is the absorption of ultraviolet light and the subsequent release of visible light by an atom. (Absorption and emission of light by atoms is discussed more fully in the last section, "Light—Wave or Particle?" of this chapter.) Fluorescent substances are widely used; for example, in paints and in the inside of fluorescent lamps. A "black light" is simply a source of ultraviolet light that, while not visible to our eyes, causes fluorescent paint to "glow." (Actually, most "black lights" emit some violet light, which is visible.) Detergents that supposedly clean clothes whiter than other detergents often contain fluorescent substances that cause the materials to absorb ultraviolet light from the sun. These substances then weakly fluoresce, emitting blue light, causing the clothes to appear whiter than ordinary white clothes.

All substances, when heated sufficiently, emit electromagnetic waves, with some of the frequencies falling in the visible region of the spectrum. Individual atoms in vapors and gases, which may consist of molecules, emit frequencies that are unique to the particular substance being heated. These frequencies can be used to identify the substance. For example, the sodium vapor lamp emits yellow light, and the mercury vapor lamp emits various frequencies that produce blue-white light. Neon lights are red, and argon lights are purple. (The specific colors of light emitted by various lamps are shown in Figure 10.11.) Other substances do not emit light, yet they have color because they reflect light rather than emit it. But what is color? The scientist considers color to be a physiological phenomenon that depends on the frequency of light emitted or reflected by a substance. We perceive color by virtue of variances in frequency, the lowest visible frequency being red, the highest, violet; the other colors and hues come from frequencies in between. Some substances can reflect only the specific wavelength that gives them the color that an observer sees as discussed later in this section.

The visible part of the electromagnetic spectrum is that portion to which the human eye is sensitive, as illustrated in Figure 10.11. The wavelengths of visible light are usually expressed in *namometers* (nm), with 1 nanometer equal to one-billionth of a meter. As can be seen in Figure 10.11, the shortest visible wavelength is about 400 nanometers and corresponds to violet light while the longest visible wavelength is about 700 nanometers and corresponds to red light.

Ultraviolet radiation (200–400 nm) and infrared radiation (700–1200 nm) are usually included as portions of the spectrum that are normally associated with our light sources, but they are not visible to the human eye.

Sunlight and some other sources of white light include practically all wavelengths between 400 and 700 nanometers. But white light may also be produced by mixing or superimposing red, blue-violet, and green light, as shown in Figure 10.12. The three colors that produce white light are known as the *primary colors*—red, blue-violet, and green.

A concentrated effort to obtain quantitative measurements of various colors was made by the International Commission on Illumination (ICI) in 1931. The ICI at that time adopted the three primary colors of light as we ordinarily know them today. However, close examination points out that specific identification of the three primaries that will provide all the spectral colors may be an impossible task. In order to duplicate a specific shade, the intensity of the primaries also must be taken into consideration.

In order to match the color of a given sample of light, one useful technique involves projecting the light source onto a screen. The three primary colors of light, as perhaps obtained by a given set of filters, are then superimposed near the sample, and the intensity of each primary is adjusted in an attempt to duplicate the sample. If a match is obtained, the color of the sample may be uniquely specified by the measured intensity of the three primaries used. These intensities are dependent on the actual wavelengths that represent the primaries. For instance, suppose the primaries were noted to be the spectral colors red (700 nm), green (530 nm), and blue (450 nm). The intensities of each color to produce the color of the specific sample may vary if other wavelengths, such as 680, 540, and 430 nanometers, respectively, are used. One should note, then, that identifying any of the three primary colors of light by a narrow band of wavelength, much less by specific wavelength, is an insurmountable task. The primary colors as specified are only arbitrary representations, sanctioned by the ICI in much the same way that standardization of many units of measurement is accomplished.

As illustrated in Figure 10.12, two specific colors of light can combine to produce white light and are known as *complimentary colors,* such as blue-violet and yellow light or specific shades of red and blue. Note also that the various shades of purple are missing. Purple is referred to as a nonspectral color and is obtained by mixing red and violet.

Elementary school students learn that a mixture of blue pigment and yellow pigment produces the color green and that a mixture of red and yellow pigment yields the color orange. But although a mixture of the three primary colors of light produces white light, a mixture of the pigments of these colors produces the color brown. The process of mixing pigments creates a color opposite that obtained from adding colored lights to produce white light. This difference is because the mixing of pigments is actually a *subtractive* process, and an observer sees only the color that is reflected after all other colors are absorbed (see Fig. 10.13). If every color in the white light that illuminates a surface painted with a mixture of the three primary colors were absorbed, black, rather than brown, would result.

The manner in which light interacts with matter accounts for the colors of the objects we see. When an electromagnetic wave is incident on an object the negatively charged electrons and the positively charged nuclei in the atoms and molecules vibrate at the same frequency as the electromagnetic wave. Vibrating particles have certain natural frequencies of vibration (just as a mass hanging on a spring), and when the electromagnetic wave corresponds to one of these natural frequencies it is said to be in resonance. (This is exactly the type of resonance phenomenon discussed in Chapter 7 in connection with sound waves.) Resonance causes the oscillations of the electrons or nuclei to absorb energy from the electromagnetic wave, with the energy then usually being converted into heat. The importance of the process as far as color is concerned is that the frequency (hence, the wavelength) at which the resonance occurs is absorbed and thus removed from the light.

Pigments in paints and dyes have resonances at certain wavelengths that lead to *preferential absorption.* For example, a blue sweater has been prepared with a dye that has resonances in the red and green regions of the spectrum which are then absorbed leaving the blue region to be reflected. If the blue sweater is illuminated with only red light, it will appear black because the red light will be absorbed and no light will be reflected (Fig. 10.14). This phenomenon is the major reason that objects appear to be one color in the sunlight but other colors under incandescent lamps, mercury vapor lamps, or fluorescent lamps. Each source differs in the range and intensity of visible wavelengths that it provides. For example, the incandescent lamp provides greatest intensity in the red region of the visible spectrum; thus, red objects appear brighter under incandescent lighting than under fluorescent lighting. Fluorescent lighting, on the other hand, enhances blue objects, and the appearance of brown, blue, and green objects in fluorescent light closely resembles their appearance in sunlight. A black object has resonances at essentially all wavelengths in the visible region of the spectrum; hence, all the light is absorbed.

The conversion of light into heat upon absorption can be demonstrated with a simple experiment. If two glasses of water are

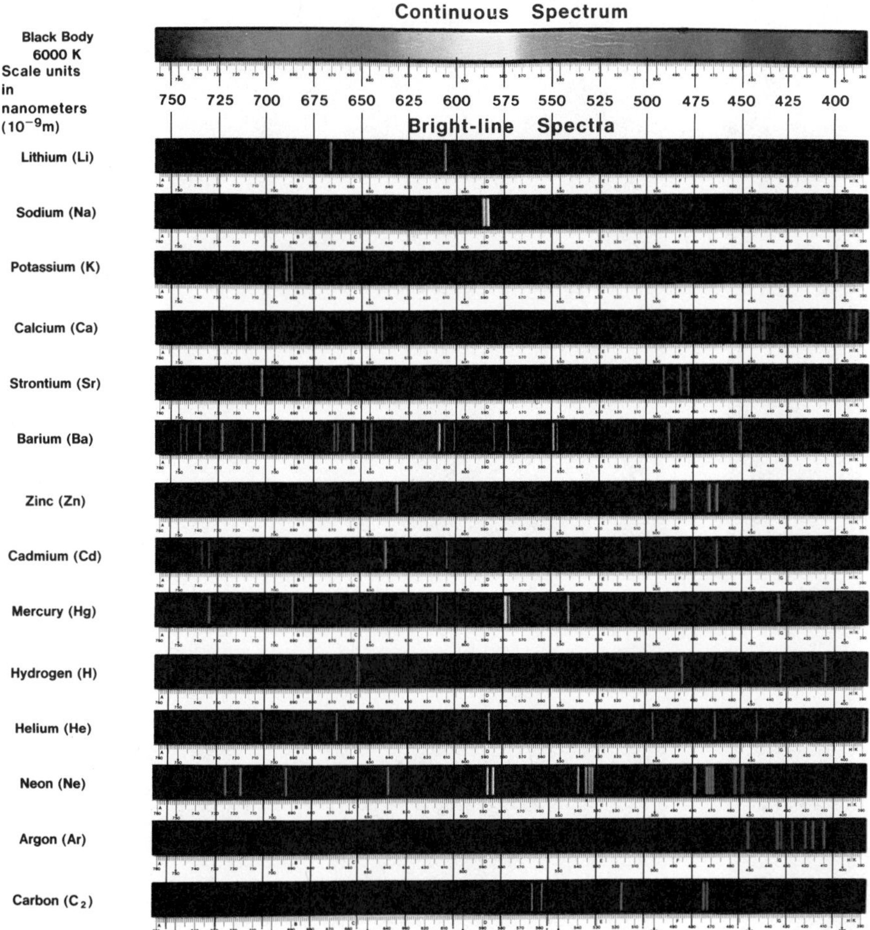

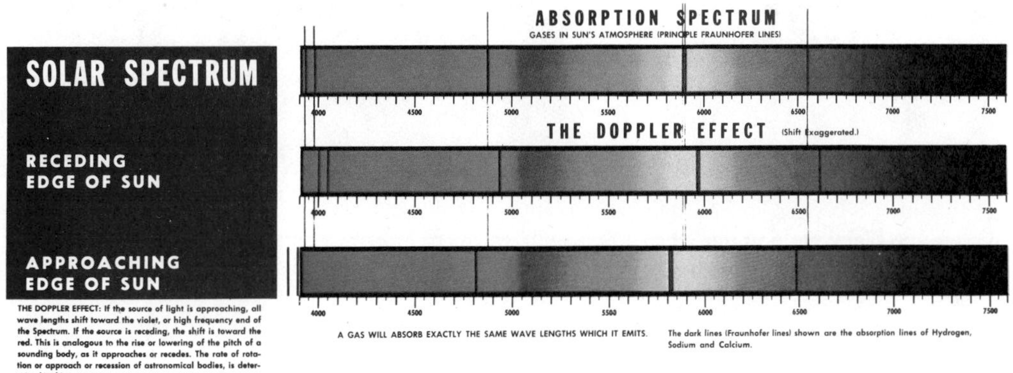

SOLAR SPECTRUM
ABSORPTION SPECTRUM
GASES IN SUN'S ATMOSPHERE (PRINCIPLE FRAUNHOFER LINES)
4000 4500 5000 5500 6000 6500 7000 7500
THE DOPPLER EFFECT (Shift Exaggerated.)
RECEDING EDGE OF SUN
APPROACHING EDGE OF SUN
THE DOPPLER EFFECT: If the source of light is approaching, all wave lengths shift toward the violet, or high frequency end of the Spectrum. If the source is receding, the shift is toward the red. This is analogous to the rise or lowering of the pitch of a sounding body, as it approaches or recedes. The rate of rotation or approach or recession of astronomical bodies, is determined in this manner.
A GAS WILL ABSORB EXACTLY THE SAME WAVE LENGTHS WHICH IT EMITS.
The dark lines (Fraunhofer lines) shown are the absorption lines of Hydrogen, Sodium and Calcium.

Figure 10.11 The visible region of the spectrum extends from about 400 nm (400×10^{-9} m) on the violet end of the spectrum to a little over 700 nm (700×10^{-9} m) on the red end of the spectrum. The continuous spectrum (the top bar) is the type of spectrum emitted by a hot solid object or a hot gaseous object at high pressure at a temperature of 6000 K (the surface temperature of the sun). The bright-line spectra are emitted by various gases at low pressure when excited by an electrical discharge (for example, as in a neon sign). Absorption spectra occur when an absorbing medium is between the observer and a source emitting a continuous spectrum. (The spectrum of the sun is an absorption spectrum, with the body of the sun emitting a continuous spectrum and the cooler gases in the sun's atmosphere absorbing certain wavelengths.)

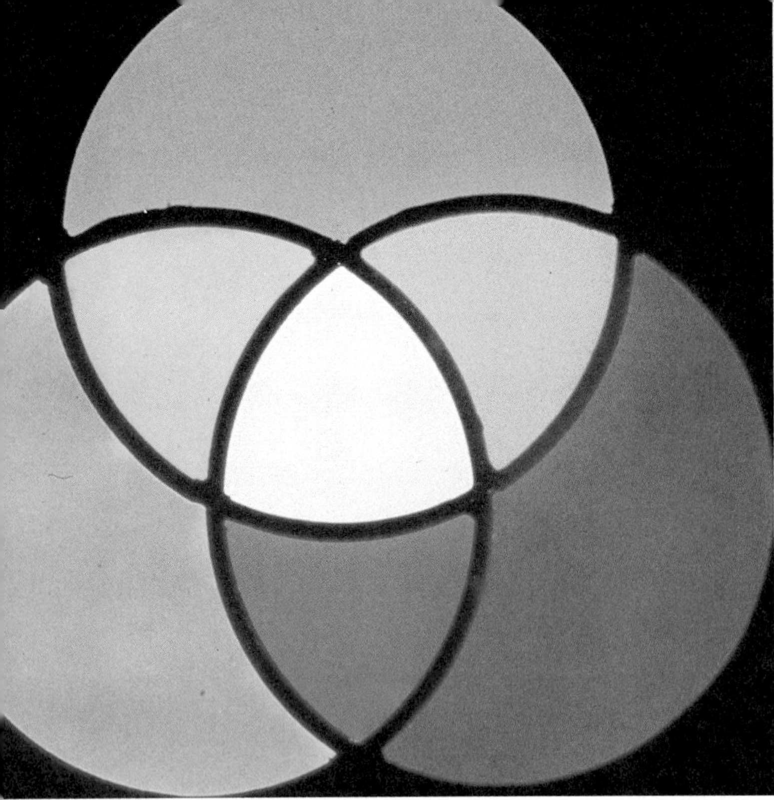

Figure 10.12 White light is produced where red, green, and blue-violet light overlap on a common surface. Other colors are produced where two of the three primary colors of light overlap.

Figure 10.13 The combination of red, yellow, and blue pigments (such as dyes or paints) creates brownish-black. Other colors and hues result if various shades of only two of the three primary colors are mixed.

a.

b.

Figure 10.14 (*a*) Pieces of blue and black cloth under white light appear their "true" colors. (*b*) When illuminated with red light, the blue cloth appears black.

placed in sunlight, one wrapped in a black cloth and the other in a white cloth, the temperature of the water in the glass wrapped in the black cloth will rise to a higher value than that wrapped in the white cloth, showing that more energy is being absorbed and converted to heat. This is why light-colored clothing is preferred in summer and dark-colored clothing preferred in winter.

Transparent materials, such as glass and water, have resonances outside of the visible spectrum. Most objects that appear white are the result of light scattering from very small particles of material that are actually transparent. Consider, as examples, water and salt. When water droplets are finely dispersed in air, as in a cloud or fog, the multiple scattering from the droplets makes the droplets appear white. The same is true for salt. Individual crystals of salt are transparent, but when many small salt crystals scatter light, all wavelengths are scattered and the salt appears white (Fig. 10.15). It is easy to demonstrate that all wavelengths are reflected from a white object by illuminating it with light of a single color at a time. Salt will appear red if illuminated with red light, green if illuminated with green light, and so forth.

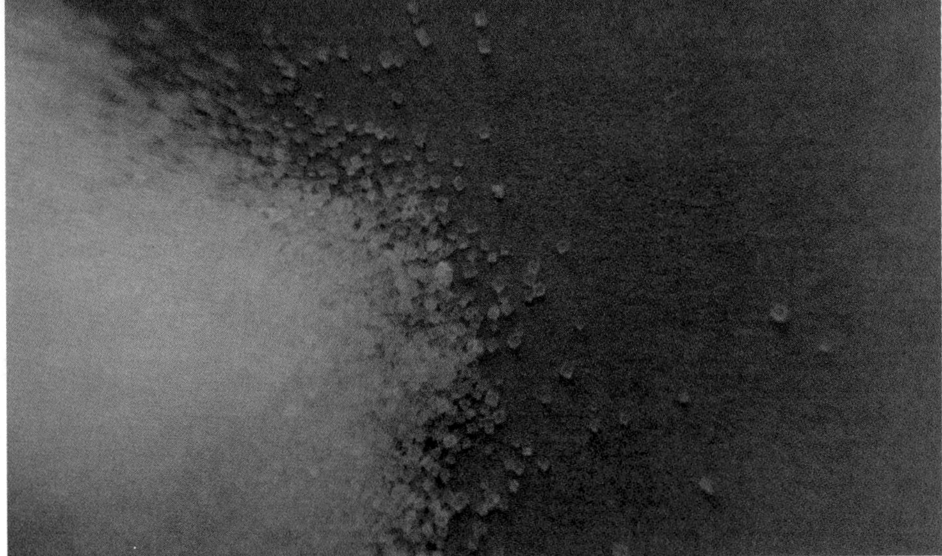

Figure 10.15 Individual crystals of salt are transparent, but the multiple scattering of all wavelengths from many small crystals makes the pile of salt appear white.

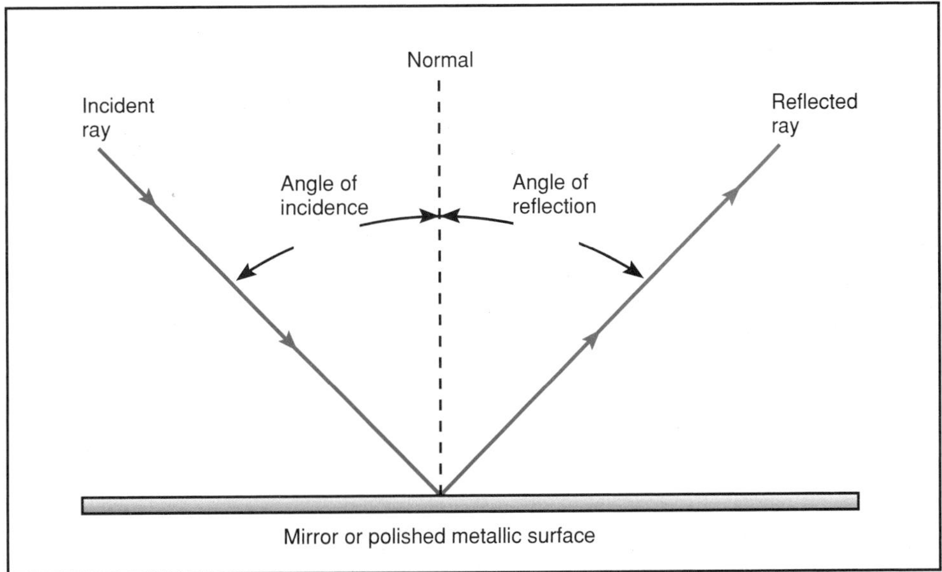

Figure 10.16 The angle of incidence equals the angle of reflection for each ray striking a reflecting surface. If the surface is rough or curved, it is necessary to consider the changing direction of the normal for every individual ray.

Reflection, Diffusion, and Scattering

We can see an object only when a source of light is present. Some objects, such as the sun, a burning match, or an incandescent lamp, are *luminous;* that is, they emit visible light. Most objects, however, are visible because they reemit, and thus reflect, light that reaches them from a luminous body.

Window glass and clear plastic permit light to pass through them in relatively straight lines and are said to be *transparent. Translucent* objects, such as frosted glass and waxed paper, allow the passage of light, but scatter it in many directions so that one cannot see clearly through them. Most objects, however, are *opaque;* that is, they do not permit light to pass through them except when the objects are extremely thin.

Light travels in straight lines, but its path may be affected by various substances. For instance, when light strikes a surface, part or all of it may be reflected, depending on the various characteristics of the surface and the angle with which the light strikes the surface. Almost all *incident light,* the light that strikes a surface, is reflected from a clean and polished metallic surface. The path along which light travels is called a *ray.* The law of **reflection,** illustrated in Figure 10.16 states that the *angle of incidence,* measured from the **normal** (a line perpendicular to the plane of the reflecting surface at the point of incidence), equals the *angle of reflection,* also measured from the normal. This law applies to each individual ray reflected from a surface.

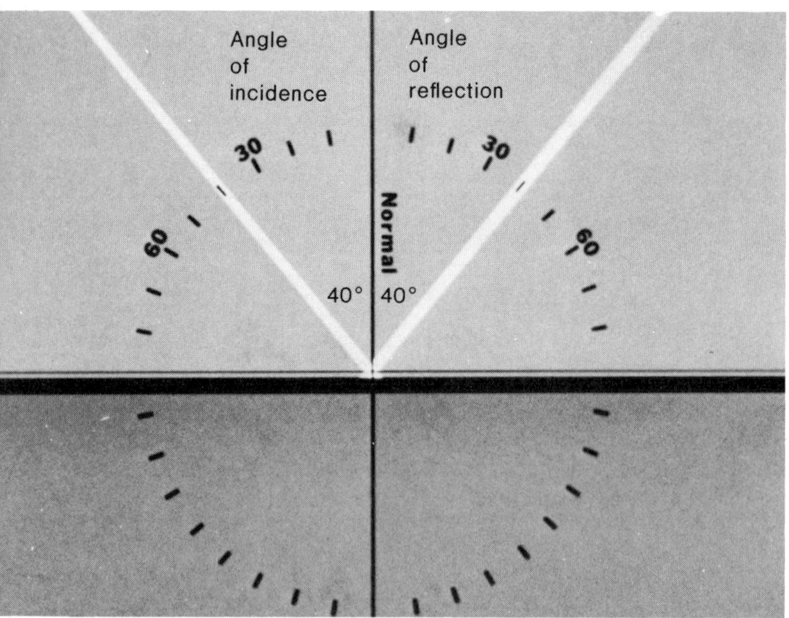

Figure 10.17 A light-ray incident on a plane glass mirror and the resulting reflected ray. Note the angle of the incident ray on the left with that of the reflected ray.

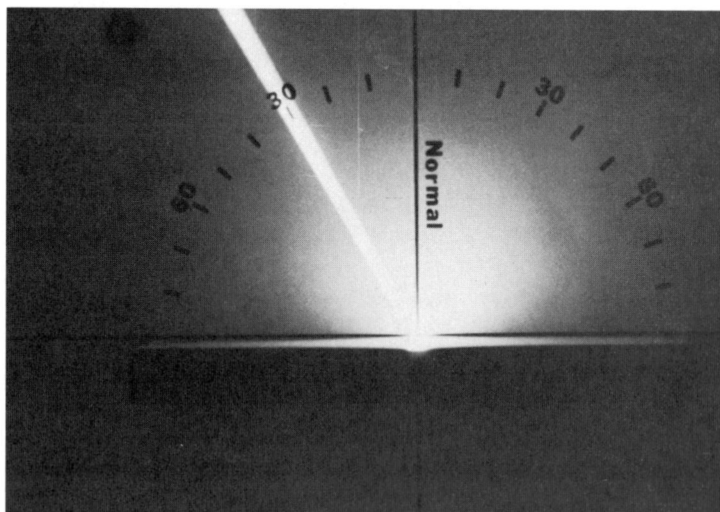

Figure 10.18 An incident light ray that strikes an irregular surface is reflected at many angles.

A mirrored surface forms an image of things that reflect light onto it. The image occurs because the pattern of light rays remains the same, except that it is reversed from right to left. Mirrors are generally composed of smooth glass with a thin coating of a shiny metal such as silver or aluminum bonded to the reverse side, although front-surfaced mirrors are used in the study of light in the laboratory. The reflection of an incident ray of light from a plane glass mirror is shown in Figure 10.17.

To be perceived by an observer as shiny, a surface must be quite smooth. For example, a wall with a coat of high-gloss enamel is extremely smooth, and therefore, shiny. A shoe with a well-buffed surface is also very smooth and, like the enameled wall, reflects light exceptionally well. A rough surface, such as that of a suede shoe, reflects little incident light. The light that is reflected leaves the surface at many different angles; therefore, the object does not have a high gloss (see Fig. 10.18). A flat enamel, then, leaves the surface of a wall to which it is applied rough and uneven and causes incident rays to be either absorbed or reflected at many different angles. The reflection of incident light rays at various angles results from their respective normals being nonparallel. This reflected light is said to be diffused. The *diffusion* of light makes a newspaper more comfortable to read than a magazine that has a glossy surface, since less light from the source of illumination of the newspaper is reflected into our eyes. The glass envelope of "soft white" light bulbs scatters light from the filament more uniformly than ordinary frosted bulbs, resulting in less glare.

A similar phenomenon, *scattering,* occurs when atoms absorb electromagnetic waves and release them in many directions. Higher frequencies are more readily scattered by small particles than are lower frequencies. The sky appears blue because the atoms and molecules in our atmosphere scatter blue light slightly more

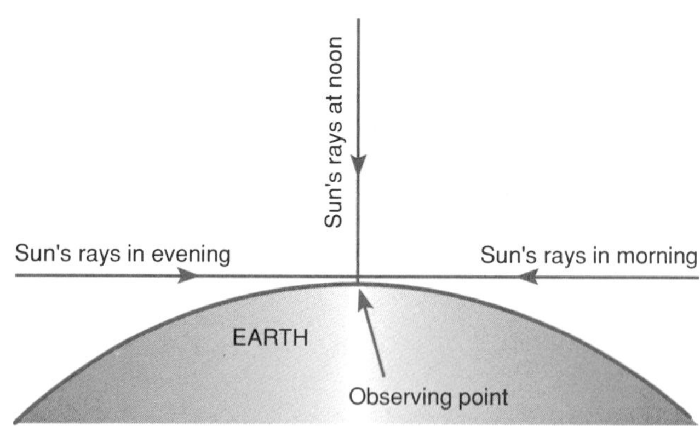

Figure 10.19 During sunrise and at sunset, the rays of light from the sun must travel farther through our atmosphere than at noon. The additional distance allows more scattering of the shorter blue and intermediate green wavelengths, leaving a much more reddened beam and producing the beautiful red and orange skies of dawn and evening.

efficiently than red light. Without this scattering effect, the sky would appear black. Clouds appear almost white because the larger water droplets do not exhibit the selective scattering as strongly, although it may be noticed near the bottoms of thick clouds.

Atmospheric scattering of light becomes most effective when sunlight must travel through the thickest portion of our atmosphere, such as at sunrise and sunset. When the sun is directly overhead, the scattered blue light is subtracted from the white light emitted by the sun. As a result, we see the sun as a white ball with a very slight yellowish tinge. In the late afternoon or early morning, the sun's rays skim across the surface of Earth and thus must travel through a much thicker layer of air than at noon (see Fig. 10.19). Because of this increased path length, not only the blue light, but even the more weakly scattered, longer wavelength green light, suffers an appreciable scattering from the beam. The beam

a.

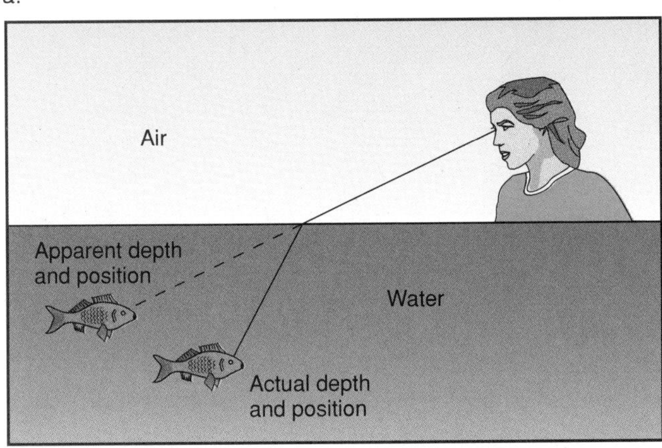

b.

Figure 10.20 (*a*) Refraction creates many phenomena when objects are viewed in two mediums in which the velocity of light differs. Note how the spoon appears broken below the surface of the water and also bent when viewed through the top surface of the water. (*b*) Because of refraction, the fish appears to be closer to the surface than it actually is.

becomes progressively redder as it traverses a longer and longer path through the atmosphere, permitting us to see the sun in a spectacular array of red and orange colors. The effect is enhanced in the evening as the turbulence and dust in the atmosphere create additional scattering.

Refraction

Undoubtedly all of us have noticed at an early age that objects immersed in water or some other liquid seem to be bent or broken at the point where they enter the liquid (Fig. 10.20). For example, a spoon used to stir a glass of iced tea may appear to be bent, but upon removal from the tea is found to be straight. These effects are a result of **refraction,** *the bending of a ray of light when it passes from one material, or medium, into another in which the speed is different.* This bending occurs if the light strikes the boundary between the two materials at an angle other than 90 degrees. The bending increases as the difference in the speed of light between the two materials increases.

Since the bending depends on the relative speed of light in the two materials, it is useful to define a quantity that is related to the speed of light in a material relative to the speed in a vacuum. The **index of refraction,** *n,* is equal to *the speed of light in a vacuum divided by the speed of light in the material;* that is,

$$n = \frac{\text{speed of light in vacuum}}{\text{speed of light in material}},$$

$$n = \frac{c}{v}, \qquad \text{INDEX OF REFRACTION}$$

where $c = 3.00 \times 10^8$ meters per second and v is the speed of light in the material. Since the speed in any material is always less than c, the index of refraction is always greater than 1, the larger values signifying the slower speeds.

For example, the speed of light in a vacuum is 2.42 times the speed of light in a diamond; therefore, the index of refraction of a diamond is 2.42. The indices of refraction of various mediums, along with the speed of light in each substance, are listed in Table 10.1.

Table 10.1

The index of refraction and the speed of light in various mediums. Note that all values except for air and vacuum are given to three significant figures.

Medium	Speed (m/s)	Index of refraction (n)
Vacuum	299,792,458 (c)	1.0000
Air	299,700,000	1.0003
Diamond	124,000,000	2.42
Glass (light crown)	199,000,000	1.51
Glass (flint)	180,000,000	1.67
Ice	229,000,000	1.31
Water	225,000,000	1.33

Example 10.4

Determine the index of refraction of a transparent substance in which light travels 175,000,000 meters per second.

Solution

$$n = \frac{c}{v}$$
$$= \frac{3.00 \times 10^8 \text{ m/s}}{1.75 \times 10^8 \text{ m/s}}$$
$$= 1.71.$$

According to Table 10.1, the specimen may be a form of dense flint glass.

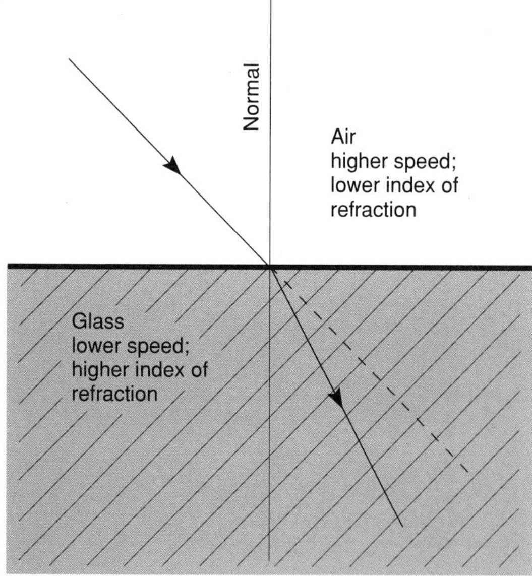

a.

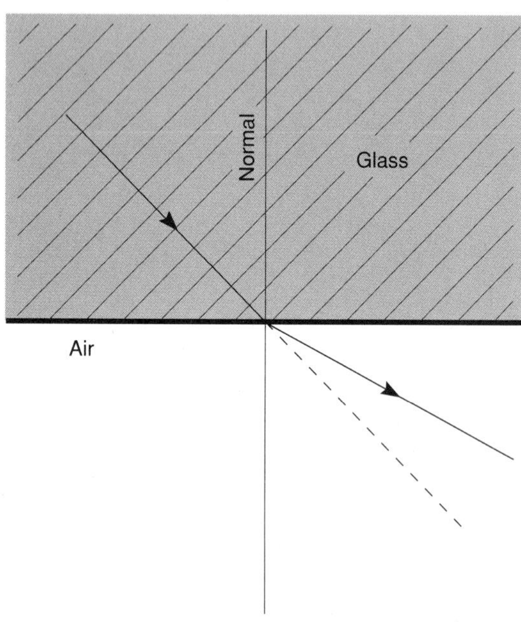

b.

Figure 10.21 When light passes from one material to another material in which the speed is different, the ray bends at the interface. When the light slows down, it bends toward the normal, as in (*a*), whereas when it speeds up, it bends away from the normal, as in (*b*). Actually, (*b*) is exactly like (*a*) except turned upside-down and with reversed arrows.

When a ray of light passes from one medium into another in which the speed is slower, the ray is bent toward the normal; but if the light has a greater speed in the second material, the ray is bent away from the normal (Fig. 10.21). Figure 10.22 shows a ray of light passing from air into a clear block of plastic (Lucite). Note that the ray is bent toward the normal, showing that the speed of light in the plastic is lower than the speed in air. When the ray passes back into the air upon exiting from the block, it bends away from the normal. If the two sides of the block are parallel, the ray leaving the block is parallel to the ray entering

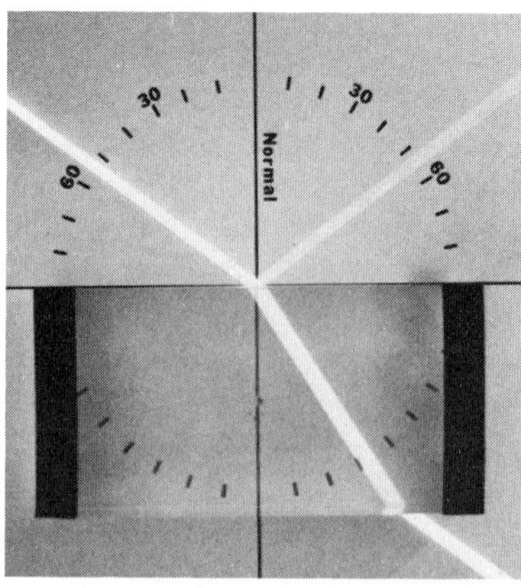

Figure 10.22 A beam of light incident on the surface of a Lucite block, other than perpendicular to it, is refracted, with the ray bending toward the normal. When the ray exits the other side of the block, it is refracted in the opposite direction. Since the sides of the block are parallel, the ray emerging from the block is parallel to the ray that entered it. The weak reflected ray obeys the law of reflection.

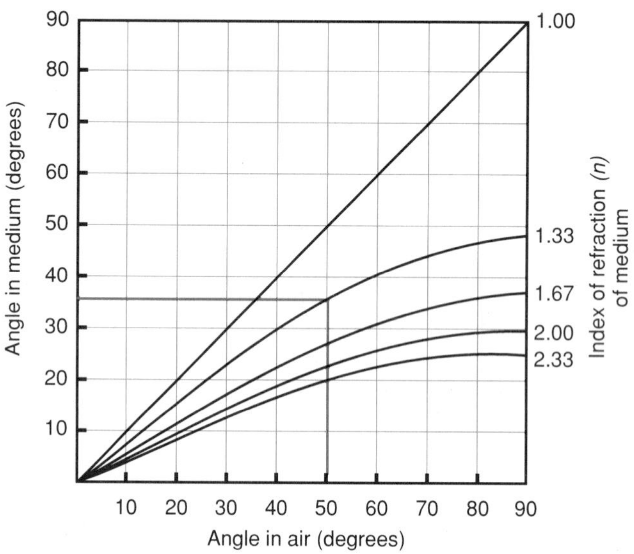

Figure 10.23 The graph shows the angles, relative to the normal, of rays in air and different mediums with given indices of refraction. For example, water has an index of refraction of 1.33, so a ray of light with an angle of 50° in air would have an angle of about 35° in water (colored lines). The ray may be passing either from the air into the water or from the water into the air. Note, again for water, that the greatest angle in the water is about 49°, making this the critical angle. For any angle in water greater than this, all light is reflected back into the water by total internal reflection.

the block. A reflected ray is also seen (obeying the law of reflection) as the ray enters the block.

Determination of the exact amount of the bending requires the use of trigonometry and is beyond the scope of this text; however, the graph in Figure 10.23 allows an approximate determination of the amount of bending. Each line on the graph represents a different index of refraction, and the angles are computed assuming the object is immersed in air. The graph works equally well whether the light ray is passing from air into the medium or from the medium into air.

Example 10.5

A certain type of glass, flint glass, has an index of refraction of 1.67. If a ray of light passing from air into the glass strikes the glass at an angle of 65 degrees to the normal, what is the angle of the ray in the glass (relative to the normal)?

Solution

Since the graph has a curve representing an index of refraction of 1.67, the necessary data are on the graph. The 65-degree angle is located on the "angle in air" axis and a line followed upward to the point of intersection with the curve marked 1.67 (for index of refraction). Then a line is followed horizontally from this point over to the axis titled "angle in medium." The angle in the glass can then be estimated to be about 33 degrees.

One of the most interesting things about the graph of Figure 10.23 is that all of the lines end where the angle in air is 90 degrees. Since light bends away from the normal when passing from a material into air, and the greatest angle the ray can have relative to the normal (and still pass into the air) is something less than 90 degrees, the question arises as to what happens if the angle in the material exceeds what is shown on the graph. In this case, light is no longer refracted into the air, but instead *all of the light is reflected back into the material,* a phenomenon known as **total internal reflection.** Note that for this to occur, the light must be traveling in the medium with the higher index of refraction. This ray, reflected without loss, obeys the law of reflection and is utilized in optical devices such as fiber optics light guides discussed in Chapter 11. *The smallest angle at which total internal reflection occurs* is called the **critical angle,** and is smaller for larger indices of refraction as can be determined from the graph. Total internal reflection is illustrated in Figure 10.24 for a glass-air interface.

Total internal reflection has been observed by most of us when looking at an aquarium. If the eyes are just below the level of the water, the underside of the water seems to be silvered, reflecting perfectly the fish and other contents of the aquarium.

The critical angle for diamond, which has one of the highest indices of refraction of any substance, is about 24 degrees. This condition causes the diamond to appear to sparkle as light entering the top of the diamond is nearly all reflected, a feature enhanced by the jeweler when he cuts the diamond at the proper angles to take full advantage of the small critical angle.

a.

b.

Figure 10.24 (*a*) An incident ray of light is shown as it strikes a transparent surface at slightly less than the critical angle. (*b*) When the angle of the incident ray is increased slightly beyond the critical angle there is no refracted ray, and all the light is reflected back into the material.

Example 10.6

Using the graph of Figure 10.23, find the critical angle for the flint glass of Example 10.5, which has an index of refraction of 1.67.

Solution

The critical angle is the angle in the medium that would give a refracted ray in air an angle of 90 degrees. Moving upward along the right edge of the graph (which represents 90 degrees in air) until the line representing an index of refraction of 1.67 is reached shows that the angle in the material (in other words, the critical angle) is about 37 degrees.

When light is compared to other types of energy, one of its properties seems particularly unusual. This characteristic has to do with the transmission of light through various mediums. A bullet, for example, slows down as it penetrates a block of wood, because it loses energy as it collides with the particles of wood in its path. As it emerges from the wood, it does so at a lesser speed than that at which it entered the block. Light enters a second medium from air at speed *c* (for practical purposes), slows down as it travels through the second medium but, unlike the bullet, resumes its original velocity as it emerges. The light ray that exits from a piece of transparent substance is parallel to the incident ray if the opposite sides of the medium are parallel. This behavior is explained most readily if light is viewed as individual *photons,* or bundles of energy, rather than as a solid beam—as light is generally represented.

As a photon collides (thus, interacts) with the particles that constitute the second medium (the atom and its orbital electrons), the similarity of light to other electromagnetic waves is more readily recognized. The photon that enters the surface of clear glass is absorbed by an atom and causes an electron of that atom to oscillate at a frequency equal to that of the photon. This vibration in turn causes the emission of a second photon of the same frequency as that of the incident photon. The second photon also travels at *c* until it is absorbed by another atom. The absorption causes the electrons to oscillate and a third photon to be released at velocity *c.* The process continues until a photon reaches the surface, where it continues through the original medium, air, at velocity *c.* All the absorptions and re-emissions of the photons require time; this observation explains why light travels at a lesser velocity than *c* in any medium other than a vacuum. Therefore, the photon of light energy that emerges from the second medium is not the same photon that entered it initially, but one that originated inside the medium. (It should also be noted that some of the photons absorbed will be converted to heat energy, thus reducing the intensity of the light.) The origin of photons is discussed more fully in the last section of this chapter, "Light—Wave or Particle?".

Dispersion

Sir Isaac Newton was extremely interested not only in the science of mechanics, for which he is well known, but also in the study of the properties of light. He once caused a narrow beam of white light to be transmitted through glass of various shapes with nonparallel sides, including the sphere and the triangle. The results of his efforts have been applied in many useful ways, but the spectrum of color that he produced from "white" light has proven the most important result of his investigations. The most common display of such a spectrum found in nature is that produced by raindrops as sunlight passes through them (Fig. 10.25a).

When a spectrum is produced by a prism such as that illustrated in Figure 10.25b, all the colors of the "rainbow" are formed, from red through orange, yellow, green, blue, and violet. Newton recognized that white light is a mixture of all colors and that the prism simply serves as a mechanism to separate the colors from each other. *The separation of light into colors arranged according to their frequencies* is known as **dispersion.**

Dispersion comes about because the index of refraction is slightly different for different wavelengths of light in the same material; that is, the speed of light in a material depends slightly on the color of the light. This causes each color to be bent a slightly different amount when passing from air into the material. (In a vacuum, the speed is the same for all wavelengths, *c.*) For example, the speed of violet light in crown glass is 195,000,000 meters per second while that of red light is 197,000,000 meters per second.

a.

b.

Figure 10.25 (*a*) A beam of light from the lower left strikes a glass disk to simulate rays of the sun, striking a raindrop and creating a rainbow. Notice that the two weak beams of light emerging from both sides of the disk in the direction from which the original beam came are faintly colored by dispersion in the glass. (*b*) A beam of white light passing through a prism is separated into the spectrum by dispersion.

Dispersion by prisms is one of the principal methods of separating the wavelengths of light, allowing the study of light emitted by different sources and also permitting a determination of absorption of various colors by different materials (see the section entitled "Diffraction and Interference" in Chapter 11). Figure 10.11 shows how different spectra appear when analyzed with a prism.

Diffraction

Electromagnetic waves—indeed, any waves—have a tendency to "spread out" as they travel. Sound waves, for example, can easily bend around corners, so that people talking in an adjoining room with the door open can be heard even when they are out of sight of the listener. The ease with which waves bend around corners is strongly dependent upon their wavelengths. Long wavelengths (low frequencies) bend much more readily than short wavelengths. Again, consider sound waves. The low-frequency (long-wavelength) waves emanating from a stereo can be heard clearly if a chair is placed in front of the speakers, but the high-frequency (short-wavelength) waves are blocked by the chair.

The deviation of electromagnetic waves from their original direction due to an obstruction in their path is called **diffraction.** Waves from AM radio stations are very long, and in addition to being reflected by the ionosphere, can bend enough to follow the curvature of Earth. On the other hand, waves from FM and television stations are much shorter and do not bend appreciably. For this reason, hills, buildings, and the curvature of Earth block these

Figure 10.26 Narrow fringe lines caused by diffraction are visible around the shadow of the paper clip.

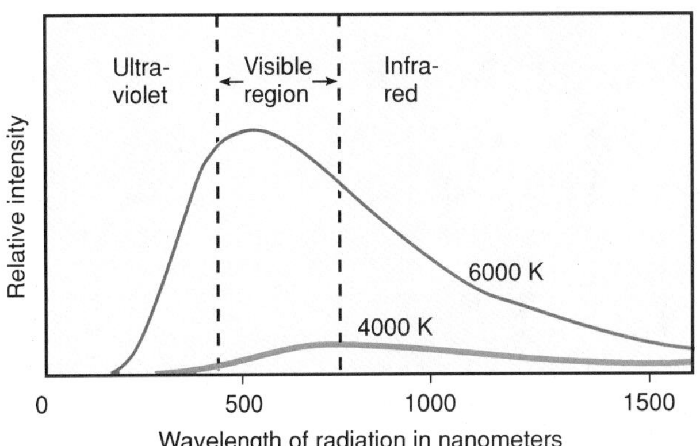

Figure 10.27 The intensity of radiation from a blackbody at a certain wavelength depends on its temperature. Cooler bodies have lower overall intensities, the maximum of intensity peaking at longer wavelengths than for hotter bodies. The 4000-K object would appear reddish because the intensity is greater on the red end of the visible portion of the spectrum. The 6000-K object (about the temperature of the sun) would appear close to white because there is not much variation in the relative intensities of all visible colors. The shapes of these curves led Planck to adopt the idea of quanta.

shorter wavelengths, requiring taller towers for transmission, as previously discussed in the section "Electromagnetic Waves."

Light waves have a very short wavelength, and casual observation seems to show no bending at all. However, the diffraction of light can be observed if one looks very carefully at shadows cast by objects in a beam of light. This slight bending of light causes the edges of the shadows produced to be blurred (diffuse) rather than sharp, as are the sides of the object producing the shadow. (The edge of a shadow produced by a source of light that is not a point is also diffuse. For example, frosted light bulbs have more diffuse shadows than clear bulbs because the filament of a clear bulb more nearly approximates a point. This is different from diffraction.) Diffraction of light waves is most easily observed using a source that produces only one wavelength (one color) of light and that has all its rays emanating from a single point or all parallel to each other. The shadow of a paper clip made using a laser, which makes an excellent source, is shown in Figure 10.26. (The laser is discussed in Chapter 11.) The shadow is found to be bounded at the edges by narrow bands or fringes of light. Diffraction was initially explained by Sir Isaac Newton, who assumed that light is composed of small particles that obey the laws of mechanical motion and are radiated from a source of light at high speeds in all directions. The "stream" of particles bends around an object because, if both the particles and the object have mass, there is a gravitational attraction between them, and the particles thus respond accordingly. Newton's particle, or *corpuscular,* theory of light required that the light travel faster in a medium than in air. This was later found by experiment not to be true. The theory was replaced by the hypothesis that light is composed of many waves of extremely short wavelengths, which explained not only diffraction, but also such characteristics as reflection, refraction, and polarization.

Diffraction, like many other effects, has both useful and troublesome aspects. For example, diffraction allows direct measurement of the wavelength of light waves, but it limits the sharpness of images in optical instruments. These effects are discussed in Chapter 11.

Light—Wave or Particle?

At the beginning of the twentieth century, Max Planck (1858–1947), a German physicist, advanced a theory to account for electromagnetic radiation from blackbodies. A *blackbody* is an ideal substance with a totally black surface that absorbs all electromagnetic radiation that strikes it and emits radiation in specific ways depending only on the temperature of the body. Although the search for such a substance has not been entirely successful, some materials approximate this condition closely enough to support the blackbody theory. For instance, observations made of available substances point out that blackbodies do not emit all wavelengths in equal amounts. Instead, specific wavelengths are emitted more readily than others. As the temperature of the blackbody increases, the wavelengths that it emits become shorter (see Fig. 10.27). Specifically, the wavelength of maximum emission varies inversely with the absolute temperature of the blackbody. Planck developed his theory by suggesting that matter can only emit *energy in discrete amounts* called **quanta.**

Although Planck's studies concentrated on the energies of vibrating atoms, the quantum specifies the discreteness of any system. If some representative quantity of a system occurs only in discontinuous and discrete amounts, that quantity is said to be *quantized.*

Albert Einstein

Albert Einstein, who helped revolutionize a major portion of scientific thinking, was born in Ulm, Germany, March 14, 1879, of Jewish parents. As a young lad, Einstein appeared to be anything but a genius. He was very slow to learn to speak, failed to grasp the French language, and was such a poor student that his teachers considered him incapable of advancing very far in his studies. He did, however, show rapid progress in the development of his manipulatory skills. Young Einstein was given a magnetic compass at the age of five with which he is reputed to have visualized the mysterious invisible lines that existed in space and attracted the compass needle. A Euclidian geometry book he read as a youth helped him express his beliefs about the consistency of nature.

Einstein, having a total distaste for the rote learning he was expected to accomplish, dropped out of formal school at the age of sixteen. After several unsuccessful attempts to enroll in other schools of instruction, he was admitted to a progressive school in Aarau and permitted to investigate concepts and devices in which he showed interest and to use his somewhat unusual tendency to think in concrete images. Einstein's unique talents were recognized by his instructors, and he was urged to pursue his endeavors.

The great scientist was left to his own thoughts, his own observations of nature, and his simple but meaningful experiments. In 1902, he became an examiner in the Swiss patent office at Bern. This position gave him time for further study and he published four important physics papers in 1905. One of his contributing papers explained the Brownian motion theory, the movement of particles in suspension. A second paper laid the foundation for the photon, or quantum, theory of light through his explanation of the photoelectric effect. Another of his publications contained his special theory of relativity, a contribution for which he was awarded the Ph.D. degree from the University of Zurich and a professorship at the institution. His fourth paper included his revolutionary expression of the relationship between mass and energy.

In 1912, Einstein undertook the task of solving various problems in astronomy and applying his thoughts about relativity. His general theory of relativity was published in 1916, and in 1922 he was awarded the Nobel Prize in physics for his explanation of the photoelectric effect. During the next few years, Einstein traveled abroad, lecturing on physics and on Zionism, the international movement for reestablishing the Jewish national state in Palestine. With the rise of Nazism, his doctrines were denounced, his property confiscated, and his books publicly burned. He moved to the United States in 1933 and was appointed a life member of the Institute for Advanced Study at Princeton, New Jersey. Einstein became a United States citizen in 1940 and resided in his adopted country until his death in 1955■

In 1905 Albert Einstein expanded the quantum concept in his explanation of the photoelectric effect (see the section entitled "Electronic Systems" in Chapter 9). As light strikes certain metals, electrons in these elements escape and form an electric current. The energy of the escaping electrons is independent of the amount of radiation that strikes the metal. The maximum energy was noted to be related to the frequency of the incident light rays. Einstein theorized that matter absorbs or emits energy in precise amounts (quanta), and that the photoelectric effect can be explained by assuming that all electromagnetic waves occur in bundles. This explanation gave new life to the particle theory of the electromagnetic wave, even though it was quite different from the original particle theory. Later experiments supported the concept that light energy travels in quanta. *A single "particle" of light possesses one quantum of energy* and is called a **photon.**

The manner in which matter can serve as a light source is explained by the **quantum theory.** When specific elements are heated, they emit light of certain wavelengths. When the light from a given element is viewed with a spectroscope, discrete bright lines are noted. Each element provides its own unique spectrum, and the wavelengths of these bright lines can be accurately measured. This set of wavelengths becomes a positive identifying characteristic of that substance.

Scientists have attempted to discover why atoms of a given element, although provided a wide range of energies by the intense heat of a hot gas flame, give off only the specific energies inherent in their spectra. The modern theory of atomic structure provides the most acceptable explanation. The atom is composed of massive positive particles (protons), equally heavy neutral particles (neutrons), and very light negatively charged particles (electrons) that orbit the heavy particles that constitute the nucleus. The electrons can assume only certain fixed energy values, called *energy levels,* with one another and with the nucleus of the atom. The electrons in like (similar) positions in atoms of the same element have the same energy levels. The electrons must occupy one of the discrete energy levels; that is, they cannot possess energies between the specific energy levels.

When an electron is stimulated by energy, and thus caused to move to a higher orbit (energy level), its respective atom is said to be *excited.* The farther from its original orbit the electron moves, the greater the degree of excitation the atom has experienced. The potential energy an electron gains is restricted to distinct values; that is, an electron can move outward in only discrete energy jumps. The excited states of an atom are unstable. It regains stability as the electron returns to its original level, or *ground state,* directly or in intermediate steps through the release of energy in discrete

The Discrete Nature of Energy, Matter, and Charge—Quantization

The idea that energy in light waves comes in discrete "bundles" was a novel idea that came into use starting in 1900 with the work of the German physicist Max Planck. Planck did not originally understand the significance of the work he had done to explain radiation from hot bodies until others had extended the concept of quantized radiation; that is, photons. Albert Einstein's paper on the photoelectric effect in 1905 gave validity to the idea and was an important contribution to the foundations of quantum theory.

Radical is the proper word to use when describing this departure from the ideas of Newton and Maxwell, on which were based the sciences of mechanics and electromagnetic theory. The kinetic and potential energies of objects in mechanics can have any value, as can the energy carried in a "classical" electromagnetic wave.

The concept that matter itself is quantized is an ancient one, however. The Greek philosopher Democritus postulated around 400 B.C. that a given kind of matter could be subdivided only so far and still retain the properties of the material. His idea was resurrected and expanded in 1808 by the English chemist John Dalton who used the atomic concept to explain chemical properties of materials (see Chapter 12).

The model of the atom as composed primarily of electrons, protons, and neutrons has evolved during this century. The smallest electric charge that has ever been found is that carried by the electron and the proton, so that as far as is now known electric charge comes in discrete units also, and since atoms are made up of definite numbers of electrons, protons, and neutrons, the mass is also "quantized." (The nature of these particles is an area of current research.)

In everyday life, the discreteness, or quantized nature, of energy, matter, and charge is not noticed because the units are so small. When we purchase a certain number of gallons of gasoline for our automobiles the measurement is coarse enough that a few molecules more or less will not be noticed. Likewise, when we pay the utility company for electrical energy, the energy is so "fine grained" that its quantized nature is of no practical significance to us. If, however, we wish to understand the nature of matter and energy at a fundamental level, its discreteness is an essential consideration ■

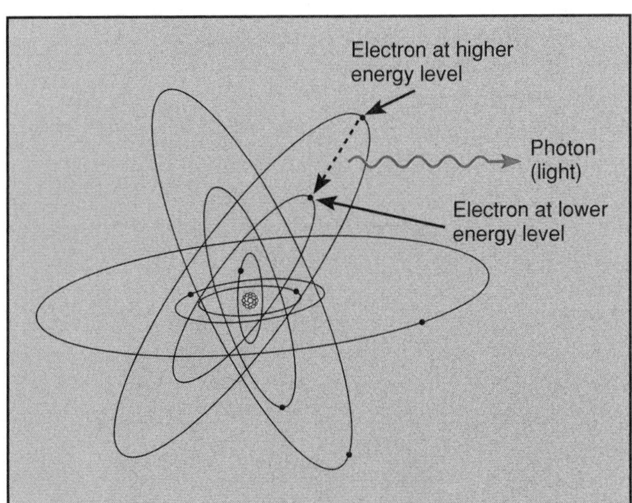

Figure 10.28 An electron may jump to a lower energy level by releasing the energy difference between the levels in the form of electromagnetic radiation. The "packet" of energy released is called a photon.

bunches called *quanta.* If the quanta are within a certain energy range, visible light of a specific color results (see Fig. 10.28).

When an atom is given energy, perhaps by heating or by an electrical discharge, one or more of its electrons may be raised to a higher energy level. However, the electron usually returns to a lower level and gives off an electromagnetic wave (a photon) equal in energy to the difference between the two energy levels (see Fig. 10.28). When this energy is in the visible band of the electromagnetic energy spectrum, it shows up as a bright line in the spectrum of the element. The spectrum of each element is different because each has a different number of electrons and different energy levels in which the electrons can potentially relocate (see Fig. 10.11).

Spectral analysis of the hydrogen atom reveals a single bright line in the red region, one in the green, one in the blue, and one in the violet part of the visible spectrum. Many lines are also known to appear in the ultraviolet and infrared regions. The distance between bright lines becomes progressively less toward the violet region, for reasons beyond the scope of this discussion.

The atomic theory did not explain the dual identity of light. Physicists considered light as a particle *or* as a wave, depending on which theory was more useful to explain the particular phenomenon under study. The paradox was further contemplated in 1924 by Louis de Broglie (1892–1987), a French physicist. He postulated that matter, generally considered as a collection of particles, has a wave aspect as well. Many experiments, including those that involve the electron, reveal the wave nature of all substances. The circumference of the innermost orbit of an electron is equal to one wavelength of the electron wave. The second orbit has a circumference of two electron wavelengths, the third has three, and so on. Since the circumference of electron orbits is discrete, the radii of these orbits—and hence the energy levels—are also discrete.

As electrons jump from the higher energy levels of the outermost orbits to lower energy levels near the nucleus, they emit energy in the form of photons. The energies of the photons are equal to the differences between the two energy levels and obey the quantum relationship, $E = hf$. Planck's constant, h, is a fundamental constant of nature that represents the ratio of the energy

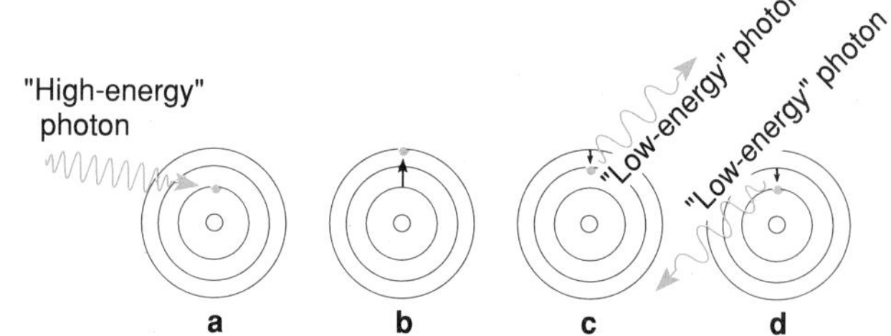

Figure 10.29 (*a*) A photon with an energy equal to the energy difference an electron has in two "orbits" approaches an atom. (*b*) The photon is absorbed, raising the energy of the electron. (*c*) and (*d*) The electron returns to the lowest energy in two steps, emitting photons of lower energy, and hence longer wavelengths. In an actual atom, the electrons are not confined to "neat" orbits as shown in this diagram, but the energies of each configuration are nearly exact. Also, quantum theory allows only the computation of the probability that certain photons will be absorbed or emitted.

of a photon to its frequency and has an experimentally determined value of 6.63×10^{-34} joule-second. As the units indicate, the energy of photons is determined in joules. Through the application of this quantum numbering system, light from a source is noted to be emitted in units of hf, rather than in a continuous fashion as sound is produced by a vibrating string. One might note that the photon of energy that excited the atom need not be the photon of energy released as the atom leaves the excited state. The degree of excitation may be lost by an atom through the release of energy photons in step-by-step increments (quanta) or as a single event (see Fig. 10.29).

Example 10.7

Calculate the energy of a photon, the frequency of which is 2450 megahertz.

Solution

The frequency of 2450 MHz can be written in scientific notation as 2.45×10^9 Hz (1/s). The energy can be calculated from

$$\begin{aligned} E &= hf \qquad \text{ENERGY} \\ &= (6.63 \times 10^{-34}\ \text{J}\cdot\text{s})(2.45 \times 10^9\ \text{vib/s}) \\ &= 1.62 \times 10^{-24}\ \text{J}. \end{aligned}$$

Summary

Electromagnetic waves are a changing of electric and magnetic fields produced by oscillating charges. The fields vibrate in a direction perpendicular to the direction in which the wave is traveling; hence, they are transverse waves. The wave is polarized if the fields are vibrating in only certain directions. All electromagnetic waves travel at approximately the same speed through a given medium such as air, although they differ in frequency and in wavelength. Electromagnetic waves are usually identified according to the way in which they are generated or used.

Radio waves, visible light, and all other electromagnetic waves have common properties. All such waves may be reflected, absorbed, or transmitted by matter. Electromagnetic radiation is visible if its wavelength is between 400 and 700 nanometers. White light is produced by the presence of all wavelengths in the visible band. Other combinations, such as the primary colors—red, blue-violet, and green—produce white light. The mixing of light is an additive process, whereas the mixing of pigments is a subtractive process.

A substance that transmits light readily without significant scattering is considered transparent. However, when light strikes the boundary between two different materials, some of the light is usually reflected. The angle of reflection equals the angle of incidence. Metals reflect most waves. If the surface is not smooth, light is diffused. Some substances scatter incident light; that is, they absorb photons and release them in all directions. Some of the photons released in scattering may be at lower frequencies than those absorbed. The amount of light reflected from an object, then, depends on several features of the object: its composition, its surface characteristics, and its position relative to the incident beam.

Refraction is the bending of electromagnetic waves as they pass obliquely from one medium into another having a different index of refraction. The speed with which a given electromagnetic wave travels through a medium other than a vacuum depends slightly on its wavelength, with violet light traveling more slowly than red light, giving rise to dispersion. As light passes into a material in which its speed is slower, it bends toward the normal, whereas if its speed in the second medium is greater it bends away from the normal. The angle in a slower medium that gives an angle of refraction of 90 degrees in a faster medium is called the critical angle, and any ray in the slower medium that strikes the surface at an angle greater than this will be totally internally reflected.

Light is bent from its straight-line path when it passes through openings or encounters obstacles. This bending of light rays, with the resultant spreading of the transmitted waves, is known as diffraction.

Light displays properties of both particles and waves, with longer wavelengths tending to display more wavelike characteristics and shorter wavelengths being predisposed toward characteristics of particles. The difficulties with the dual identity of light were generalized by Planck, Einstein, and others. Energy is radiated in definite units called quanta, or photons. All material particles have wave properties.

Equation Summary

Speed, frequency, and wavelength:

In a vacuum (and approximately in air)

$$c = f\lambda,$$

where the speed of light c (3.00×10^8 meters per second) equals the frequency of the radiation f (hertz) times the wavelength λ (meters). This equation can be rearranged to give

$$f = \frac{c}{\lambda}$$

and

$$\lambda = \frac{c}{f}.$$

In a medium, the speed is symbolized by v (meters per second) (v is always less than c) and the preceding equations become

$$v = f\lambda,$$
$$f = \frac{v}{\lambda},$$

and

$$\lambda = \frac{v}{f}.$$

Transmitting range:

The transmitting range of an antenna emitting short-wavelength waves (limited by the curvature of Earth) is

$$s = 3.59\ \sqrt{h},$$

where the distance s (kilometers) the station can be received equals 3.59 times the square root of the height h (meters) of the antenna.

Index of refraction:

The index of refraction of a medium n (no units) equals the speed of light in a vacuum c (meters per second) divided by the speed of light in the medium v (meters per second); that is,

$$n = \frac{c}{v}.$$

The index of refraction is always greater than 1.

Energy of a photon:

The energy of a photon E (joules) equals Planck's constant h (6.63×10^{-34} joule-second) times the frequency f (hertz):

$$E = hf.$$

Questions and Problems

Electromagnetic Waves

1. What is "vibrating" in an electromagnetic wave?
2. Why are electromagnetic waves transverse waves?
3. What is meant by "polarization" of an electromagnetic wave?

The Speed of Light

4. The longest wavelengths of visible light are those we call red, intermediate wavelengths of visible light are green, and the shortest wavelengths of visible light are violet. How do the frequencies of these three wavelengths compare; that is, which has the highest frequency, which the lowest, and which has a frequency in between?
5. A certain blue light has a wavelength of 475 nm (475×10^{-9} m). What is the frequency of this light?
6. What is the wavelength of waves transmitted by a radio station operating at a frequency of 840 kHz?

The Spectrum

7. State the approximate ranges of infrared radiation, visible light, and ultraviolet radiation in terms of wavelengths and frequencies.
8. Calculate the frequency of green light as it travels through a vacuum.
9. Determine the effective range of a television station that transmits from an antenna 400 m high in a region where there are no hills. (The actual range is somewhat greater than this because the receiving antennas are also elevated.)

The Behavior of Electromagnetic Waves

10. Before purchasing an article of clothing that you consider to be your favorite color, why should you view it in direct sunlight rather than in articifical light?
11. What is selective absorption? What color would a red dress appear in blue light?
12. Why does mixing all primary colors of light produce white light, whereas when all colors of paint are mixed, a very dark brown (theoretically black) color results?
13. Many streetlights and security lights make use of an electrical discharge through mercury vapor as their source of light. Using Figure 10.11, list the colors of light emitted by a mercury lamp. Knowing that the color sensed by our eyes is bluish, which of these colors do you think is most intense? Why do lips and blemishes on the skin that appear reddish in sunlight appear purplish under mercury lights?

Reflection, Diffusion, and Scattering

14. What causes the difference in the reflection from a smooth surface, such as a piece of glass, and that from a piece of notebook paper?
15. Explain why the sun appears so red at sunset on a hot summer day.

Refraction

16. The speed of light is said to be constant. Does this statement indicate that the speed of light is the same in all materials? Defend your response.
17. What does the index of refraction have to do with the amount of bending that occurs when light passes from air into a material?

18. When a ray of light passes from air into water, in what direction does it bend? When a ray of light passes from water into air, in what direction does it bend?
19. The index of refraction of a substance is found to be 2.00. Calculate the speed of light in this substance in meters per second.
20. Using Table 10.1 and the graph of Figure 10.23 estimate the critical angle for light crown glass. What happens to a ray of light in the glass if it strikes the surface of the glass from within at an angle greater than the angle you estimated? What happens if a ray in air encounters the surface of the glass at an angle greater than this angle?

Dispersion

21. Why does a glass prism separate white light into various colors? Suppose a beam of yellow light passes through a prism and the beam is found to be made of two colors. According to the additive primaries illustrated in Figure 10.12 what would be the two colors?

Diffraction

22. What is diffraction? How can diffraction be observed?

Light—Wave or Particle?

23. Keeping in mind conservation of energy and the idea that only certain "orbits" are allowed for electrons in an atom, explain why light emitted by a particular atom can have only certain frequencies; hence, certain wavelengths.
24. Calculate the energy of a photon, the frequency of which is 4.75×10^{14} Hz, which corresponds to red light from a laser.
25. X rays are produced in X-ray machines when electrons that have been accelerated to high speeds strike a metal target and are brought to rest. The energy of the electrons is converted to electromagnetic radiation in the collision. If the energy lost by an electron is about 1000 times the energy released when visible light is emitted by an atom, how would the wavelength of the X rays emitted compare to that of visible light?

11 Applications of Electromagnetic Radiation

Chapter outline

Key terms/concepts

The efficient reception of electromagnetic radiation depends on its wavelength, which is inversely related to its frequency. A satellite dish and its associated antenna are efficient for waves having wavelengths on the order of centimeters, while the eye is efficient for wavelengths 100,000 times smaller. The satellite dish above is located on the north shore of Oahu, Hawaii.

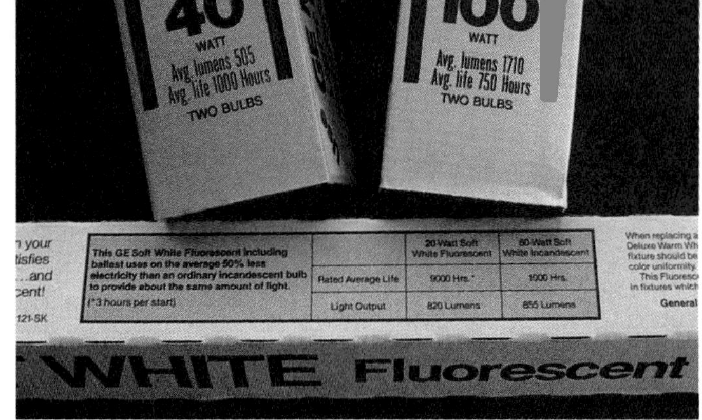

Figure 11.1 Light bulb jackets indicate not only power consumption in watts, but also light output in lumens.

Electromagnetic waves transmit energy from one place to another. Generally, energy is transmitted for two purposes: to transfer energy from one place to another or, by controlling the energy in different ways, to convey information from one place to another.

Energy transmitted by the sun across the distance of 93,000,000 miles to Earth is indispensable to maintaining life on Earth. Electromagnetic waves are also radiated from heating and lighting devices. A light bulb illuminates a scene by the electromagnetic radiation it emits and a laser drills a hole in a piece of steel because of the intensity of its radiation.

Most of the information we receive comes either directly or indirectly in the form of electromagnetic waves. Our senses can detect electromagnetic waves directly in the form of light or heat, and much of what we hear is transmitted to us indirectly through television and radio which, of course, utilize electromagnetic waves.

Such information may be transmitted, broadly speaking, in three ways. First, the intensity of the energy in a transmitted beam may be varied as the beam is transmitted, with examples being AM radio waves and, on a simpler level, a blinking traffic warning light located at a dangerous intersection. Second, variations in the frequency of the wave may be used to carry information, as in FM radio or a traffic light using different colors to convey messages to a driver. Third, instead of variations in intensity or frequency with time, there may be variations of intensity and frequency in space. When this book is read, the intensity and color of the light varies spatially (that is, from place to place), and this variation carries information.

Since our senses can directly detect visible light, this small region of the electromagnetic spectrum assumes an importance far out of proportion to its small breadth in the complete spectrum. A considerable portion of this chapter is devoted to *the study of visible light,* called **optics.** Numerous optical instruments have been designed to control light; hence, to extend the capabilities of the human eye. These instruments mainly depend on principles discussed in Chapter 10: reflection, refraction, diffraction, and polarization.

Light Intensity

As a person looks at a scene, the amount of light available for viewing depends on several physical factors, including the energy radiated by the source of light illuminating the scene, the distance from the source, and the angles between the source and the objects being viewed. Another important factor for humans is the wavelength of the light emitted by the source. Our eyes, of course, respond only to those wavelengths in the visible region of the spectrum, so that other wavelengths are not useful to us for vision. Even in the visible spectrum the response is not uniform, being greatest in the yellow region and dropping off toward both the violet and red ends of the spectrum. For this reason, a unit is needed to take into account the response of the eye when referring to the intensity of light.

The total amount of energy emitted in the visible region of the spectrum per unit of time by a source, the **luminous flux,** is given in a unit called the **lumen** (lm), and is denoted by Φ (uppercase Greek phi). (The SI base unit is actually the candela, which gives the visible radiation in a certain direction based on a wavelength of 555 nanometers in air. For comparison, 1 lumen = 12.57 candela. See Appendix A and the glossary for more detailed information.) A 100-watt incandescent light bulb typically emits about 1700 or 1800 lumens, while a fluorescent tube of the same wattage emits about three or four times as much visible light (Fig. 11.1).

An ordinary light bulb's radiated light is among the simplest to study because, if the bulb has no reflector, the light is radiated approximately uniformly in all directions. Figure 11.2 shows the way in which light spreads out around a source. Even though the source of light in the diagram has been idealized to a "point source," actual practice demonstrates that if the source is small compared to the distance from the source where the intensity is measured, the assumption of idealization works well. The figure illustrates the well-known fact that the farther one is from a source of light, the dimmer the light appears.

Over ordinary distances, air does not absorb an appreciable amount of light, so if a series of concentric spheres is drawn around a light source, the same total amount of light must pass through all the spheres. Since the spheres farther from the source are larger, the same amount of light passes through larger and larger areas, "diluting" the light or, in other words, making it appear dimmer. The actual quantity of interest when using a source to illuminate a scene is the **illumination,** E, which is *the number of lumens falling on each area* within the scene, measured in lumens per square meter (lm/m²). The area of a sphere is given by $A = 4\pi r^2$, so that the illumination, E, of a surface from a point source emitting Φ lumens is given by

$$E = \frac{\Phi}{4\pi r^2}. \qquad \text{ILLUMINATION}$$

Since the illumination is proportional to the reciprocal of the distance squared, this is an inverse square law. (Some other inverse square laws are universal gravitation, Coulomb's law, and the intensity of sound radiated by a small source.)

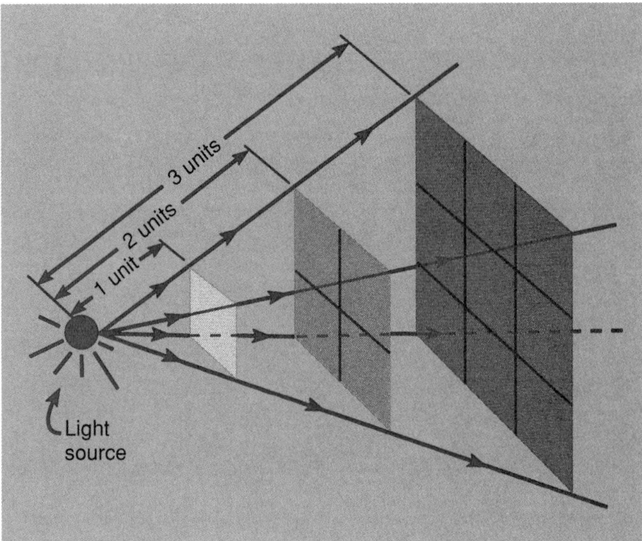

Figure 11.2 Illumination from a small light source is inversely proportional to the distance from the source squared. The same total amount of light is spread over larger areas as the distance from the light source is increased.

The minimum illumination level recommended for reading printed material is about 320 lumens per square meter, while for general study it is approximately 750 lumens per square meter. For comparison, the illumination provided by the sun on a bright day may reach 100,000 lumens per square meter.

The unit of illumination called the *lux* is equal to 1 lumen per square meter. The lux is often used to specify the lowest level of illumination that will provide satisfactory operation of devices such as video cameras.

Example 11.1

A student studying in her room uses a 60-watt light bulb that emits 855 lumens and is 60.0 centimeters (0.600 m) from her physical science textbook. What is the illumination incident on the book?

Solution

If the light bulb is assumed to be a point source, this is a straightforward case of using the previously stated equation for illumination. Using $\Phi = 855$ lm and $r = 0.600$ m gives

$$E = \frac{\Phi}{4\pi r^2} = \frac{855 \text{ lm}}{4\pi(0.600 \text{ m})^2} = 189 \text{ lm/m}^2.$$

The illumination on the book is thus 189 lm/m^2, level somewhat below the recommended level for reading printed material and far below the level recommended for general study areas.

Actually, this example illustrates the fallacy of using simple laws to solve complex problems. The student probably has her light bulb in some type of fixture that directs more of the light in a certain direction, so the assumption of light being uniformly emitted in all directions is not applicable. In addition, light is scattered from the objects in the room, increasing the actual amount of light striking the page. Illumination engineers consider all these factors when designing lighting systems. In addition, the adequacy of existing lighting may be determined with light meters. Unless the student is an illumination engineer, the best way for her to determine the proper bulb size is to rely on past experience and to have several bulb sizes available!

Mirrors

Diffuse reflection of light from somewhat rough surfaces is the most common application of light reflection. Most surfaces around us diffusely reflect light. (See Fig. 10.18 in Chapter 10.) The next most common use of reflection to control light is the **plane mirror,** which is simply *a smooth, flat reflecting surface coated with a thin metallic layer.*

Consider a small source of light placed near a plane mirror, as shown in Figure 11.3a. Light rays spread out in all directions, with only a few being shown in the diagram. Using the law of reflection and a little geometry, it is easy to see that all of the rays that strike the mirror are reflected as if they were coming from a point behind the mirror. A **virtual image** is *formed at the point from which the rays appear to be coming.* Since there are no actual rays of light at the location of the virtual image, a screen placed there will have no image on it. Any actual object can be thought of as a collection of small sources and, as shown in Figure 11.3b, the entire virtual image of an extended object can be found by looking at the images of each point.

Simple symmetry shows that the image appears the same distance behind the mirror as the object is in front of the mirror, or in the notation of the figure, $d_o = d_i$. **Magnification** is *the ratio of the image size to the object size,* and for the plane mirror the ratio equals 1, so that the image and object are the same size. The image is also an **upright image.** (Thank goodness! It would be hard to shave or put on makeup if the image were upside-down.)

Curved mirrors are also used extensively. Makeup mirrors that magnify or wide-angle mirrors used as rearview mirrors on cars or to see large areas in department stores are common examples. Most of these mirrors are *small portions of a spherical surface,* and are called **spherical mirrors** (Fig. 11.4). The center of the sphere (C) is called the *center of curvature.* Mirrors using the *inside surface of the sphere* are **concave mirrors,** while those using the *outside surface* are **convex mirrors.**

Concave mirrors are also called **converging mirrors** because parallel rays of light reflected from concave mirrors converge to *a point approximately halfway between the center of curvature and the center of the mirror (the vertex)* called the **focal point,** (F). (See Fig. 11.5.) The *distance from the center of the mirror to the focal point* is the **focal length,** f, where $f = R/2$ with R being the radius of the spherical surface of which the mirror is a part.

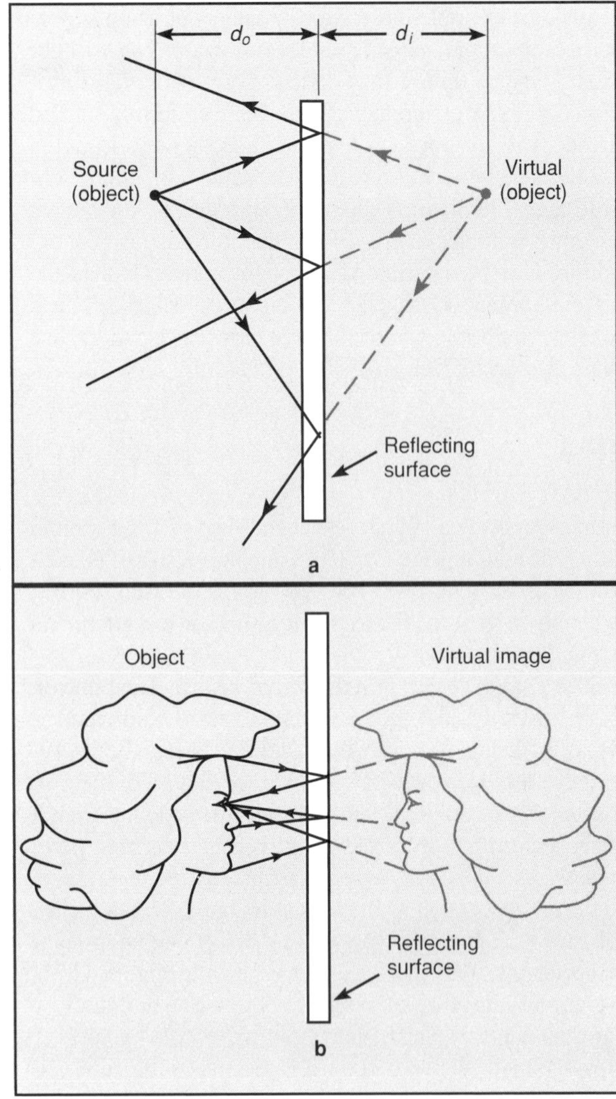

Figure 11.3 Light reflected from a plane mirror gives a virtual image the same distance behind the mirror as the object is in front of the mirror. The image is upright and the same size as the object.

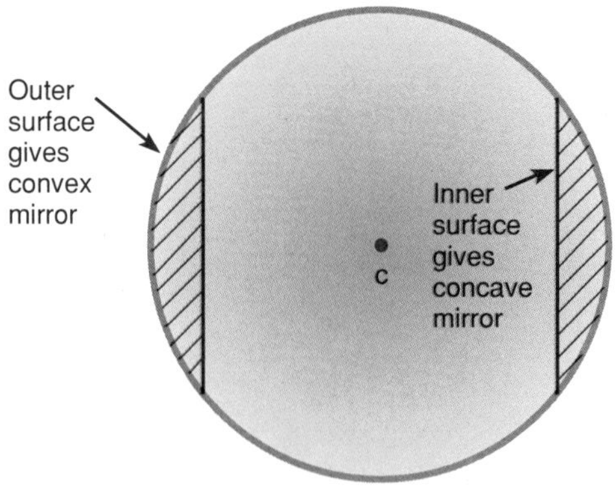

Figure 11.4 Spherical mirrors are small portions of a spherical surface. Using the outer surface gives a convex (diverging) mirror, while the inner surface gives a concave (converging) mirror.

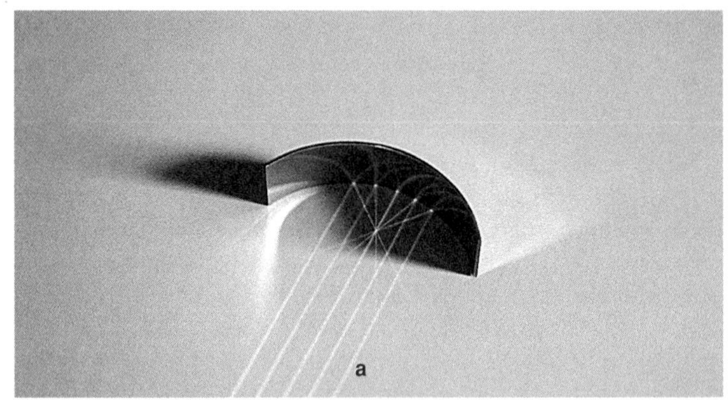

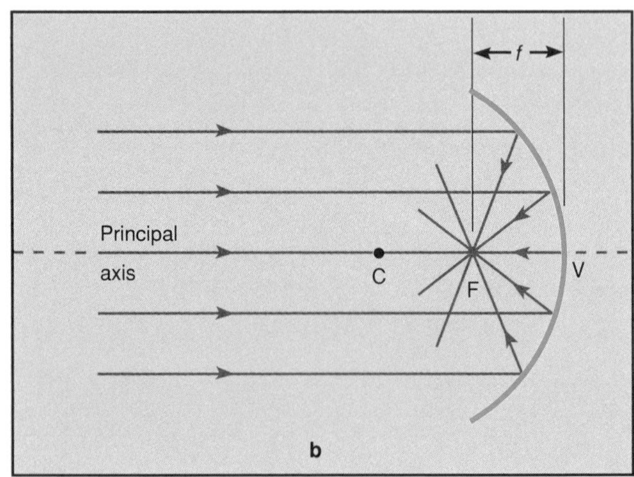

Figure 11.5 (*a*) Parallel light rays are brought to a focus at the focal point of a concave mirror. This mirror has curvature in only one direction, but most mirrors used in image-forming systems are "bowl"-shaped. (*b*) If the mirror is spherical, the focal point (F) is halfway between the center of curvature (C) and the vertex (V) of the mirror.

The line through the center of curvature, the focal point, and the vertex is the principal axis. In most applications, the rays of light do not come in parallel to the principal axis but instead are diverging (spreading out) from some source. The type of image formed by a concave mirror depends on whether the source of light is nearer to or farther from the mirror than the focal length.

The location and type of image formed can be determined graphically by a *principal-ray diagram* (Fig. 11.6). The paths of three of the rays leaving any point on the object and striking the mirror can be easily determined by knowing the locations of the focal point and the center of curvature. The mirror is represented by a vertical line. Ray 1 is parallel to the axis and must be reflected toward the focal point. A second ray (ray 2) passes through the focal point before striking the mirror and is reflected parallel to

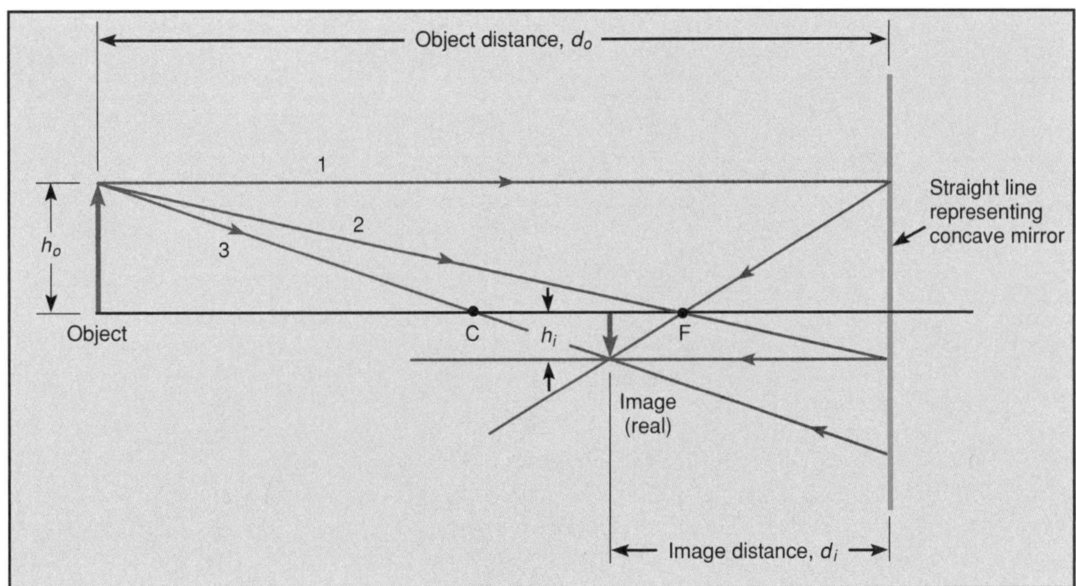

Figure 11.6 The location of the image of an object, its type (real or virtual), its orientation, and its size can be graphically determined by following the three principal rays in a principal-ray diagram. Even though the mirror is curved, it is represented by a straight line in a principal-ray diagram.

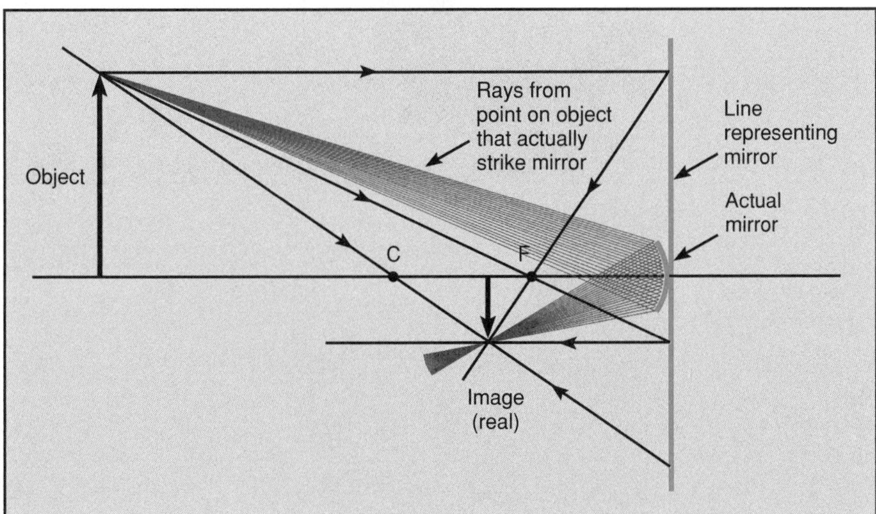

Figure 11.7 The principal-ray diagram is simply a geometrical construction for determining image properties. The actual rays may not coincide with any of the principal rays.

the axis (recall that the direction of the arrows on rays may be reversed). Finally, a ray that strikes the mirror coming from the direction of the center of curvature will be reflected back on itself since it lies along the normal (ray 3). A **real image** is formed *when the rays converge,* as in this diagram. This real image would show up on a screen placed at the position to which the rays are converging. Actual objects, of course, have many points on them, each scattering or emitting light. The three principal rays could be drawn for each point, showing how the image is "built up" of these points, but usually (for simple objects) only one point is needed to locate the image. If there is any doubt about the location of other points in the image, the principal rays are followed for these points.

Before proceeding further, the nature of the principal-ray diagram needs to be discussed. The principal-ray diagram is merely a geometrical tool used to locate images formed by mirrors (and lenses). Consider a mirror with a diameter of 10 centimeters used to form an image of an object of height 50 centimeters, as shown in Figure 11.7. The principal rays from the top of the object are shown, but it can be seen that none of these rays actually strike the mirror. Instead, the rays within a narrow cone, represented by the red shaded region in the figure, strike the mirror, coming to a

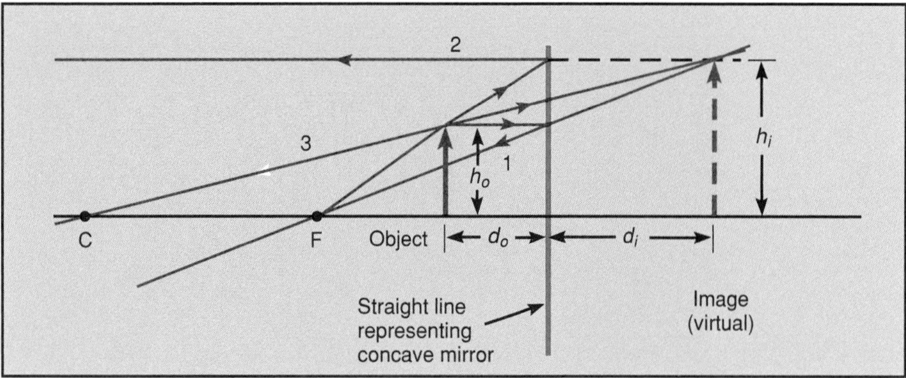

Figure 11.8 The principal-ray diagram for a converging mirror when the object is between the focal point and the mirror shows that the image is virtual, upright, and larger than the object.

focus at the point indicated by the principal-ray diagram, and represented by the lighter region that converges to the image. Thus, the actual rays forming the image are not necessarily any of the principal rays.

Returning to Figure 11.6, it is seen that the image is an **inverted image;** that is, *upside-down relative to the object.* It is also smaller than the object, so that the magnification is less than 1. Since all the rays are reversible, this diagram would also work if the object and image were interchanged, with the direction of travel of all rays reversed. The magnification would then be greater than 1, since the image would be larger than the object. If the object is farther from the mirror than the center of curvature, the image lies between the center of curvature and the focal point and is smaller than the object. If the object lies between the center of curvature and the focal point, the image is beyond the center of curvature and larger than the object. If the object is at a distance equal to that of the center of curvature, so is the image, with both the object and the image being the same size.

Consider now the case where the object is closer to the mirror than the focal length, as shown in Figure 11.8. Ray 1 is the same as before, but rays 2 and 3 are drawn in the directions they would have if they came from the focal point and center of curvature, respectively (only the direction is important when they strike the mirror). It is seen that these rays are still diverging after striking the mirror, seeming to come from some point behind the mirror, and thus indicating that a virtual image is formed behind the mirror. Note that this image is upright and larger than the object. This is the principle used in shaving and makeup mirrors, which magnify images.

Convex mirrors cause parallel rays striking them to diverge (Fig. 11.9) and so are called **diverging mirrors.** The diverging rays seem to emanate from a point behind the mirror, halfway between the mirror and its center of curvature, called the *virtual focus.* The sign convention used for focal length requires that for diverging mirrors f must be negative, so that $f = -R/2$.

The principal-ray diagram for a convex mirror (Fig. 11.10) shows how to locate an image formed by such a mirror. Ray 1 leaves the object and travels parallel to the axis, being reflected

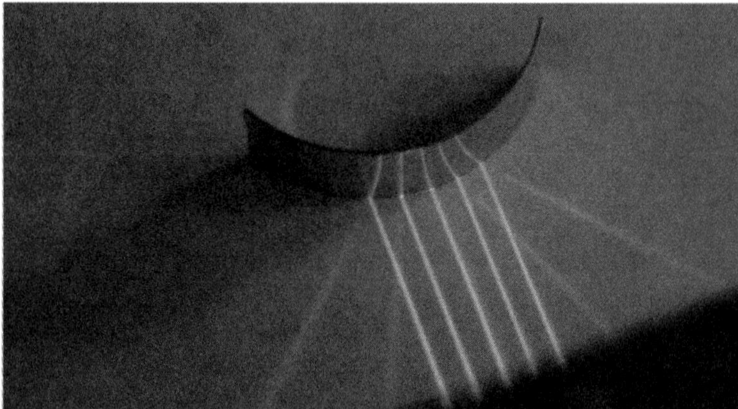

a

V
F
C

b

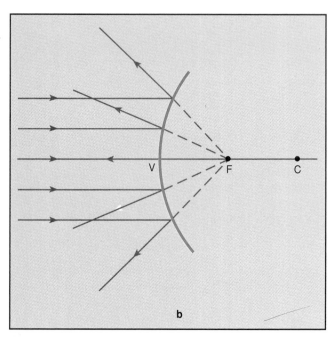

Figure 11.9 (*a*) Parallel light rays striking a convex mirror seem to diverge from a point behind the mirror called the virtual focus. (*b*) For a spherical mirror, the virtual focus (F) is halfway between the vertex (V) of the mirror and the center of curvature (C).

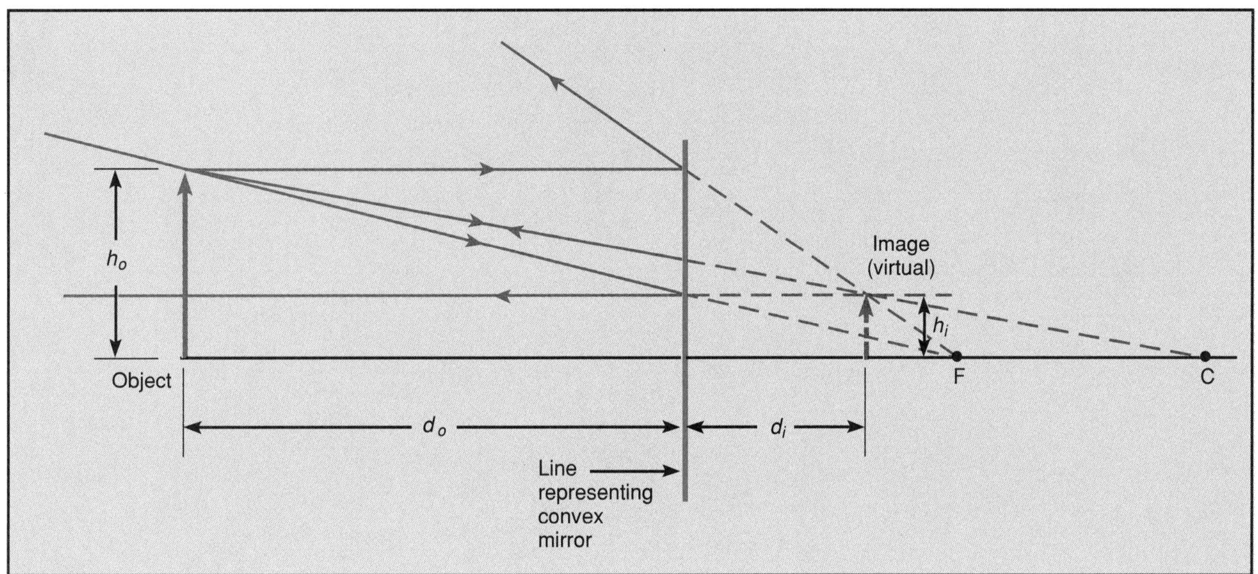

Figure 11.10 The principal-ray diagram shows that for a convex mirror the image is always virtual, upright, and smaller than the object, if there are no other optical elements in the system.

away from the virtual focus. Ray 2 travels toward the virtual focus and is reflected parallel to the axis, with ray 3 traveling toward the center of curvature and being reflected back on itself. The image formed by a diverging mirror is seen to be virtual, upright, and smaller than the object. These conditions hold, no matter how near or far the object may be, but the magnification depends on the distances. (In systems with more than one optical element; i.e., several mirrors or lenses, it is possible to have a real image formed by a diverging mirror.)

Since objects appear larger when they are closer, objects viewed in a diverging mirror often "fool" the viewer into thinking they are farther away than they actually are, hence the warning found on automobile rearview mirrors designed to give a wide field of view: "Objects in mirror are closer than they appear."

The distances of object and image from the mirror, as well as the magnification, can also be computed mathematically. The object distance, d_o, the image distance, d_i, and the focal length, f, are related by the mirror equation.

$$\frac{1}{d_o} + \frac{1}{d_i} = \frac{1}{f} \qquad \text{MIRROR EQUATION}$$

In this equation, f must be positive if the mirror is converging and negative if the mirror is diverging. The image is real if d_i is determined to be positive, and it is virtual if it is computed to be negative.

The magnification, m, is the ratio of the image size to the object size, h_i/h_o. The magnification, the image and object sizes, and the image and object distances are related to each other by the equation

$$m = \frac{h_i}{h_o} = -\frac{d_i}{d_o} \qquad \text{MAGNIFICATION}$$

The minus sign before the image and object distances is inserted so that when the magnification is negative, the image is inverted and when positive, the image is upright.

Example 11.2

A light bulb with a height of 10 centimeters is placed 0.300 meter in front of a converging mirror with a focal length of 0.250 meter. Determine the location of the image and fully describe it (with respect to the image being real or virtual, upright or inverted, and size of the image).

Solution

The location of the image can easily be found by using the mirror equation. The distance from the mirror to the object is the object distance, so $d_o = 0.300$ m. Then with $f = 0.250$ m, the mirror equation becomes

$$\frac{1}{0.300\ \text{m}} + \frac{1}{d_i} = \frac{1}{0.250\ \text{m}}.$$

Solving for d_i gives

$$d_i = 1.50\ \text{m},$$

showing that the image is formed 1.50 m from the mirror. Since the image distance is positive, the image is real. The magnification is found to be

$$\begin{aligned} m &= -\frac{d_i}{d_o} \\ &= -\frac{1.50\ \text{m}}{0.300\ \text{m}} \\ &= -5.00. \end{aligned}$$

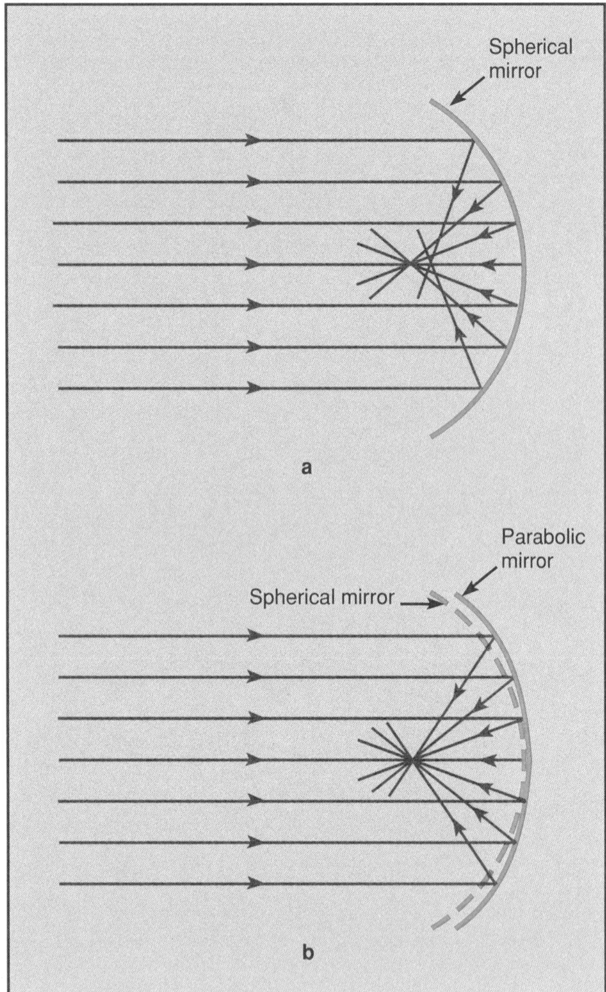

Figure 11.11 (*a*) Spherical mirrors suffer from spherical aberration, which causes poor image quality. (*b*) Parabolic mirrors allow sharper images if the object lies far from the mirror and near the principal axis.

The magnification of −5.00 means that the image is five times as large as the object, so that it must be 5.00 × 10.0 cm or 50.0 cm tall, and that it is inverted (the negative sign). From this calculation, it is concluded that a screen placed 1.50 m from the mirror would have a 50.0-cm-high inverted image of the light bulb on it.

Spherical mirrors are relatively easy to construct, but they do suffer one serious problem when used in image-forming systems. Rays of light coming in parallel to the axis of the mirror tend to come to a focus nearer the mirror the farther they are from the axis (Fig. 11.11a). This *spherical aberration* is not appreciable for relatively "shallow" mirrors, but it results in poor image quality for more deeply curved mirrors. Spherical aberration can be corrected by slightly flattening the edges of the mirror, giving the mirror a *parabolic* cross section instead of a spherical one

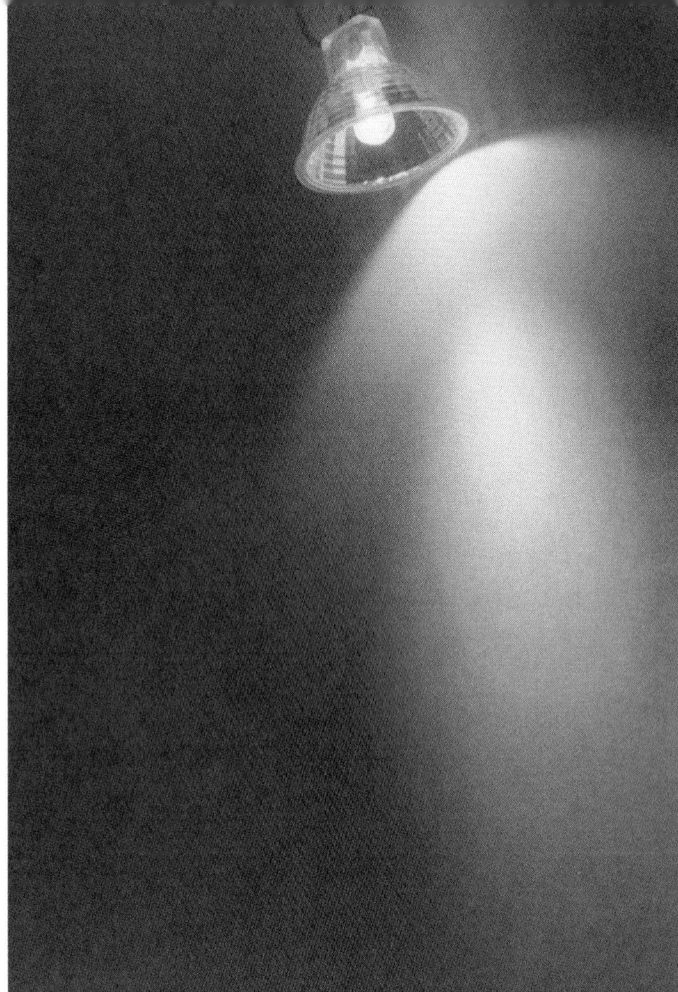

Figure 11.12 Parabolic mirrors are used to direct beams of light. Note that the "cone" of light farther from the lamp is more intense than the cone near the lamp because of the focusing effect of the reflector.

(Fig. 11.11b). Most of the reflecting telescopes used in astronomical work have parabolic mirrors rather than spherical mirrors because of the superior image quality.

Many instruments, other than image-forming devices, use curved mirrors to control light. Parabolic mirrors are used to produce high-intensity beams of light, with the rays parallel to one another, by placing a source of light at the focal point (Fig. 11.12). Lamps used in slide and movie projectors, automobile headlamps, and the dentist's lamp used to illuminate the inside of a patient's mouth are a few common examples.

Lenses

The two main types of lenses, converging and diverging, have the same characteristics as converging and diverging mirrors.

Converging lenses are characterized by being *thicker at the center than at the edges,* causing parallel rays of light to be brought to a focus at the focal point because of refraction of light (see Chapter 10) at the surfaces (Fig. 11.13). There is a focal point on each side of the lens and, if the lens is reasonably thin, the distances from the lens to these focal points is the same whether or not the lens is symmetrical (Fig. 11.14).

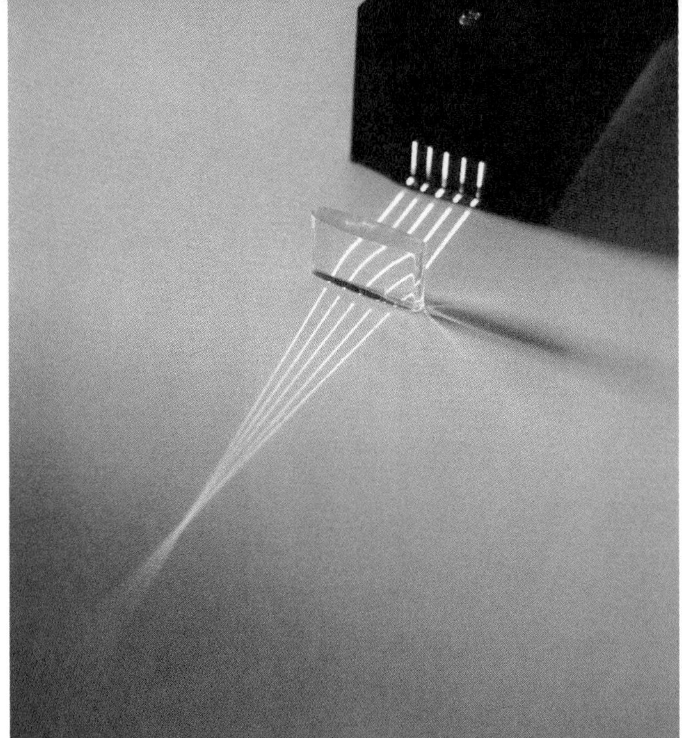

a.

b.

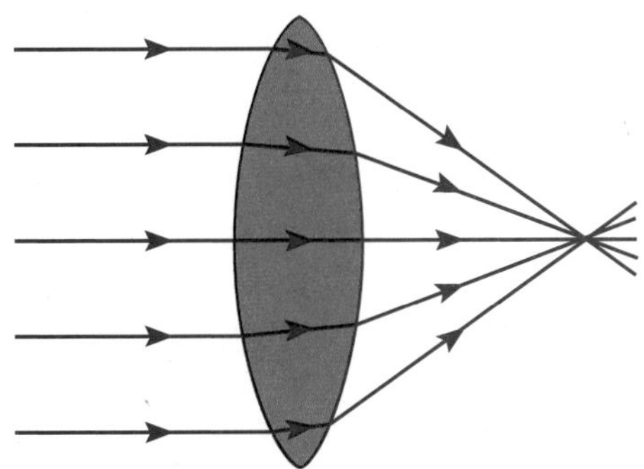

c.

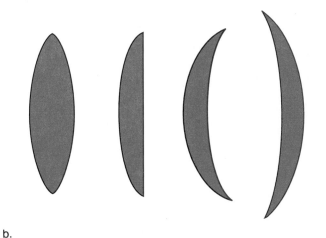

Figure 11.13 (*a*) Parallel rays are brought to a focus by a converging lens. (*b*) Converging lenses are characterized by being thicker at the center than at the edges, with inexpensive lenses having spherical surfaces. (*c*) Rays parallel to the axis of a convex lens are bent towards the axis by refraction. The amount of bending increases as the distance the ray strikes the lens from its center increases.

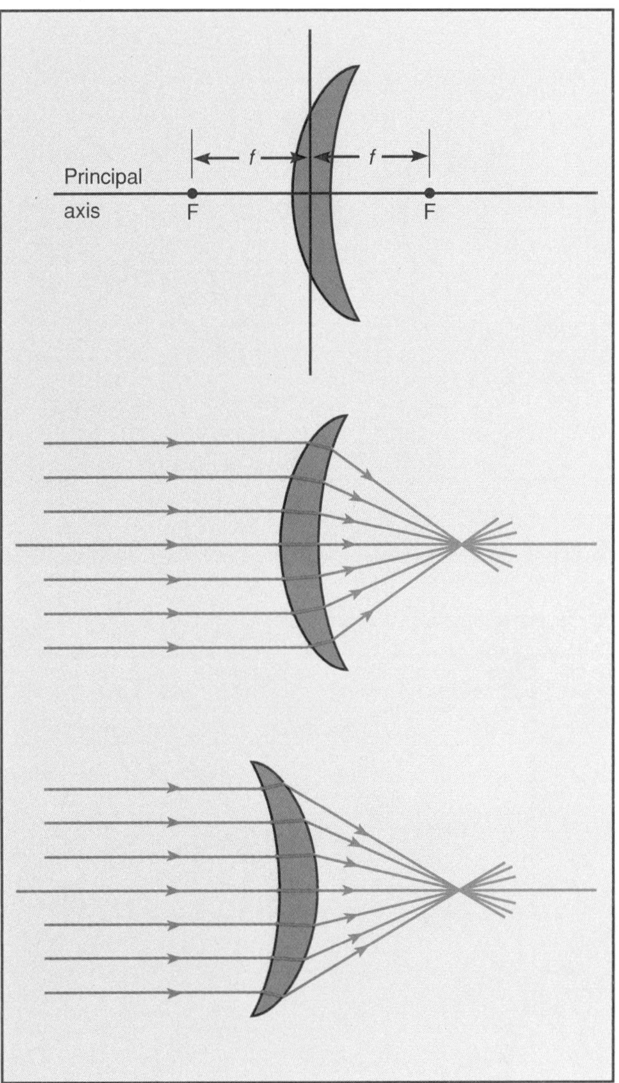

Figure 11.14 Parallel rays will be brought to a focus at the same distance on either side of a converging lens if the lens is thin, even if the lens is not symmetrical. The focal length, *f*, depends on the index of refraction of the glass as well as on the curvature of the surfaces.

A principal-ray diagram may be used to graphically determine the location and type of image formed by a lens just as for a mirror, but with a few modifications. Figure 11.15 shows the three "principal" rays for a lens. Ray 1 strikes the lens parallel to the axis and is refracted through the far focal point. Ray 2 passes straight through the center of the lens and is unbent. (At the middle of the lens the opposite faces are parallel, the situation shown in Fig. 10.22 in Chapter 10, except the lens is much thinner.) Ray 3 passes through the near focal point and emerges parallel to the axis. The image is seen to be real and inverted, which is the case so long as the object is farther from the lens than the focal length. As the object approaches the focal point, the image moves farther from the lens, and the magnification is greater. If the object is moved farther from the lens, the image moves nearer the lens with

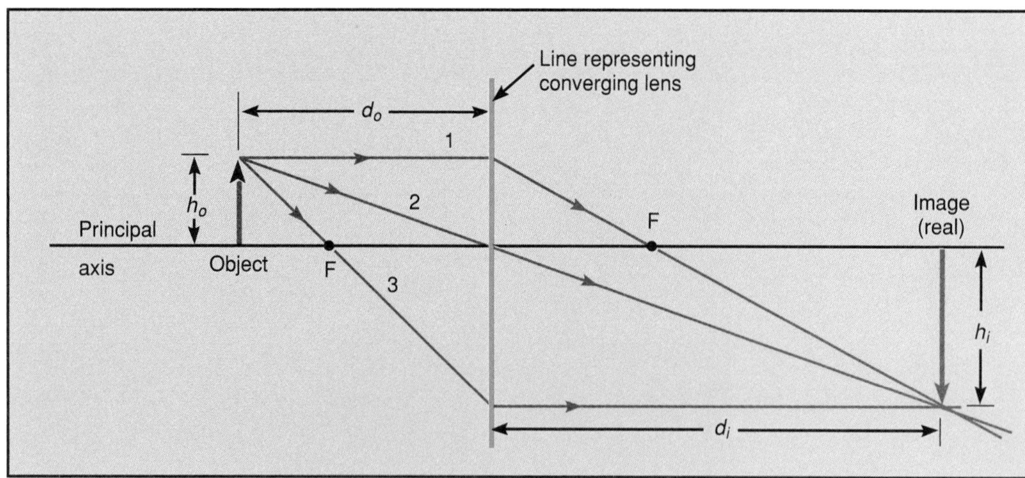

Figure 11.15 The principal-ray diagram for a converging lens utilizes rays through both focal points, as well as an unbent ray through the center of the lens. Image characteristics can be determined from the diagram.

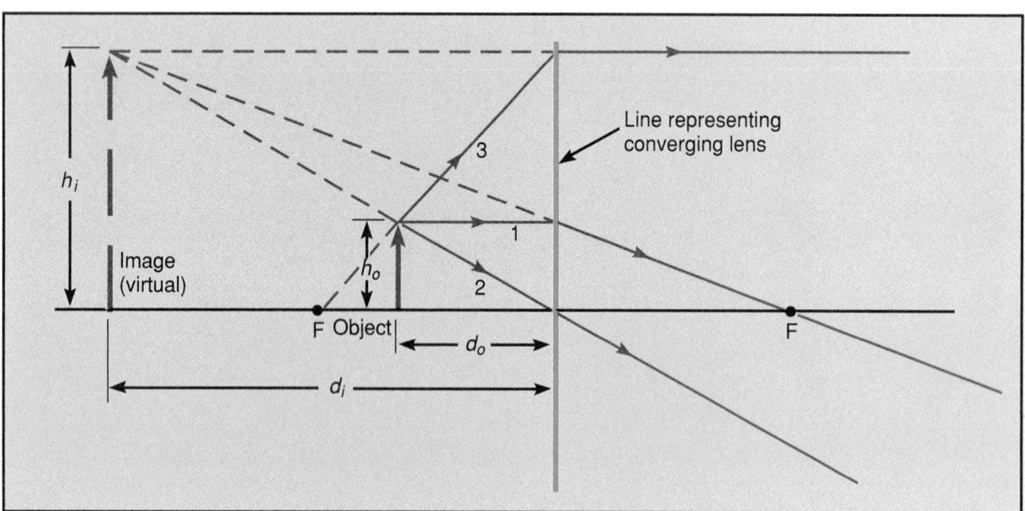

Figure 11.16 When the object is between the focal length and the lens, a converging lens becomes a "magnifying glass." The image is virtual, upright, and larger than the object.

a smaller magnification. Large magnifications would be used in devices such as projectors, whereas cameras generally employ small magnifications.

If the object is placed nearer to the lens than the focal length, the principal-ray diagram (Fig. 11.16) shows that the rays diverge from a point on the same side of the lens as the object. This virtual image is upright and larger than the object, in other words, the lens is being used as a "magnifying glass."

If the lens is *thinner at the center than the edges,* a **diverging lens** results (Figs. 11.17a and 11.17b). There is a virtual focus equidistant on each side of the lens. The three principal rays are (1) parallel to the axis, diverging from the near focal point after passing through the lens; (2) toward the far focal point, then parallel to the axis after passing through the lens; and (3) straight through the center of the lens. The image is virtual, upright, and smaller than the object, no matter how far from the lens the object is positioned (Fig. 11.17c).

The same equations used to describe the distances and magnification for mirrors are also applicable to lenses. Thus,

$$\frac{1}{d_o} + \frac{1}{d_i} = \frac{1}{f}. \qquad \text{THIN-LENS EQUATION}$$

works for lenses as well as mirrors. The focal length is positive for a converging lens and negative for a diverging lens, just as for converging and diverging mirrors. The magnification equation is also the same; that is,

$$m = \frac{h_o}{h_i} = -\frac{d_o}{d_i}. \qquad \text{MAGNIFICATION}$$

a.

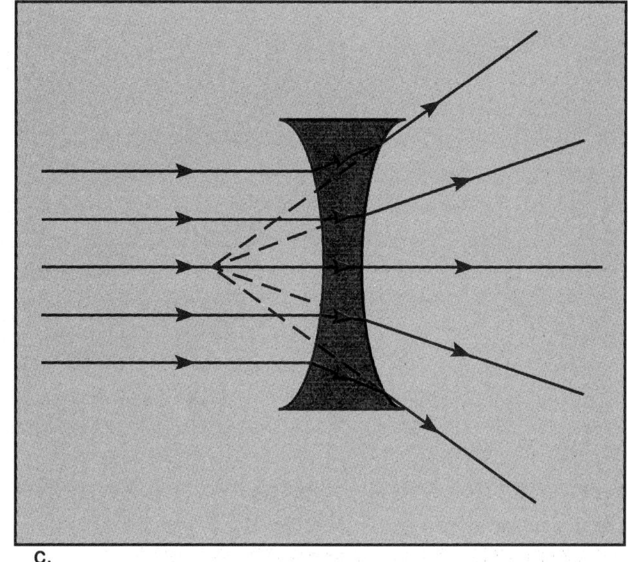
c.

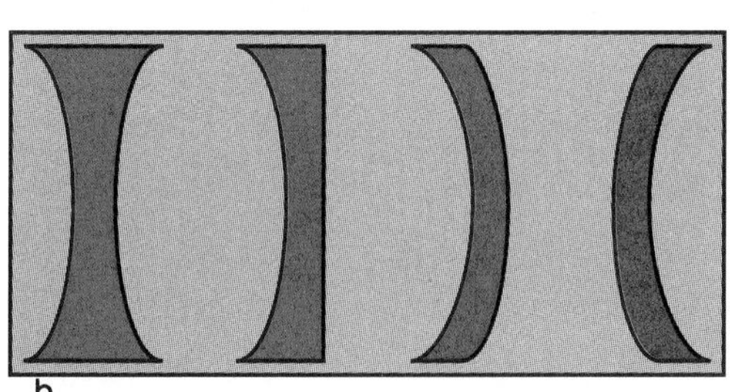
b.

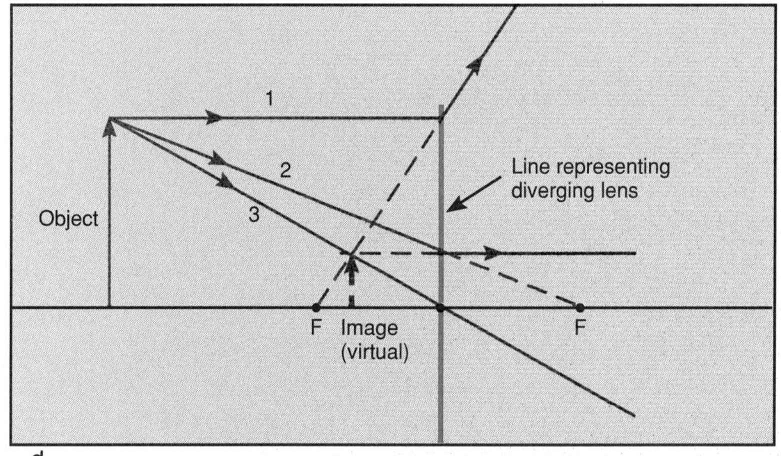

d.

Figure 11.17 (*a*) Parallel rays passing through a diverging lens seem to diverge from the virtual focus. (*b*) Diverging lenses are thinner at the center than at the edges. (*c*) Rays parallel to the axis of a concave lens are bent away from the axis by refraction. The amount of bending increases as the distance the ray strikes the lens from its center increases. (*d*) The principal-ray diagram for diverging lenses shows that the image is virtual, upright, and smaller than the object.

The sign of the image distance, d_i, is again positive for real images and negative for virtual images. Positive magnification means an upright image and negative magnification an inverted one.

Lenses made with spherical surfaces suffer from spherical aberration just as mirrors do, with rays near the edge focusing closer to the lens. This problem can be corrected by making the surfaces nonspherical (and more expensive). Lenses also exhibit *chromatic (color) aberration,* caused by the edges of the lens acting as prisms. Dispersion bends the violet light more than the red, causing images to display a colored "halo." Good-quality lenses must be made of several pieces of glass (hence, raising their cost) to limit chromatic aberration to an acceptable level.

An important use of converging lenses is in cameras. The essential elements of a camera (Fig. 11.18) are a lens, a lighttight box, and some type of light-sensitive material (film) or device such as a CCD (discussed in the "Electronic Systems" section in Chapter 9). If the camera is for pictures made on film, some type of shutter is also needed to control the flow of light. The distance from the lens to the detector must equal the image distance for the image to be in focus. In lower-cost cameras used for "snapshots," the lens is usually fixed at a distance approximately equal to its focal length, but better cameras either allow the operator to adjust this distance or automatically adjust the distance for the operator.

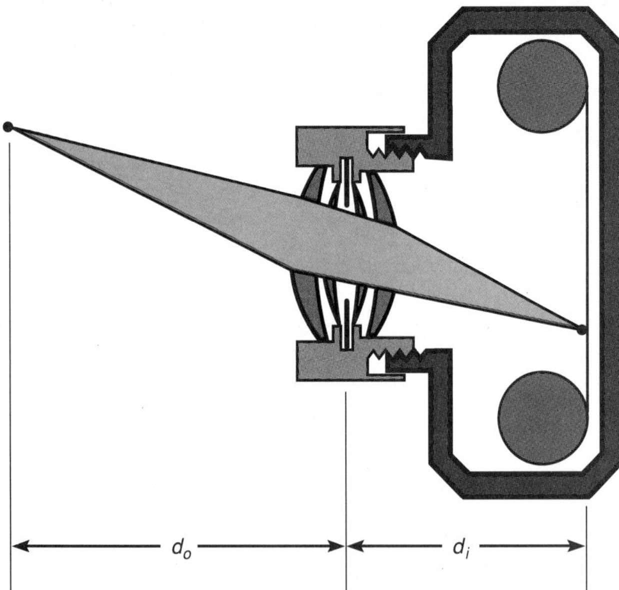

Figure 11.18 The principal elements of a photographic camera are a converging lens to focus the image on the film plane, a shutter (not shown) to control the length of time light is allowed to strike the film, and a light-tight box to enclose the film. Usually the object distance, d_o, is much greater than the image distance, d_i, making the image distance approximately equal to the focal length, *f*, of the lens.

Example 11.3

A camera that uses 35-mm film has a lens with a focal length of 50.0 millimeters. If a photographer takes a picture of his child from a distance of 2.00 meters, how far from the film should the lens be positioned?

Solution

The distance from the lens to the film must equal the image distance for the picture to be in focus. The lens equation with $d_o = 2.00$ m and $f = 50.0$ mm $= 0.0500$ m becomes

$$\frac{1}{2.00\text{ m}} + \frac{1}{d_i} = \frac{1}{0.0500\text{ m}}.$$

Solving for d_i gives

$$\begin{aligned} d_i &= 0.0513\text{ m} \\ &= 51.3\text{ mm}, \end{aligned}$$

so that the lens should be 51.3 mm from the film.

Example 11.4

A geology student is using a converging lens as a magnifying glass to study a small crystalline mineral sample. The lens has a focal length of 5.00 centimeters, and the sample is placed 4.00 centimeters from the lens. Find the location of the image, its type, and its magnification. What does the sign of the magnification mean?

Solution

The location of the image is found from the thin-lens equation, with $d_o = 4.00$ cm and $f = 5.00$ cm. Then

$$\frac{1}{4.00\text{ cm}} + \frac{1}{d_o} = \frac{1}{5.00\text{ cm}}.$$

Solving this for d_o gives

$$d_o = -20.0\text{ cm}.$$

The minus sign means that the image is virtual, so that the student must look through the lens to see the image. The magnification is found from

$$\begin{aligned} m &= -\frac{d_i}{d_o} \\ &= -\frac{(-20.0\text{ cm})}{4.00\text{ cm}} \\ &= 5.00 \end{aligned}$$

Thus the image is five times as large as the object. Since the magnification is positive, the image is upright. Summarizing, the student looks into the lens and sees a virtual upright image five times as large as the object 20.0 cm "behind" the lens.

The most common optical instrument in the world is the eye (over 10 billion are currently in use by humans alone!). The human eye operates much as a camera when focusing images, except that the focal length of the lens itself is changed for focusing on objects at different distances. This accommodation, as it is called, is brought about by a muscle squeezing the small lens near the front of the eye, adjusting the curvature of its surfaces. Even though most of the focusing is accomplished by the cornea, the fixed bulge of transparent material on the front of the eye, accommodation allows the normal eye to clearly focus on objects that are at any distance greater than about 25 centimeters from the eye. Age generally causes the small lens to stiffen and the muscle to weaken, making it difficult to focus on nearer objects (farsightedness, or presbyopia). When this occurs, extra converging power is needed, and eyeglasses with converging lenses are used (Fig. 11.19a). If, on the other hand, the converging power of the eye lens is too great when the muscle is completely relaxed, distant objects cannot be focused clearly (nearsightedness, or myopia). In this case, a diverging lens must be used to allow the image to be focused on the retina (Fig. 11.19b). If the eye is incapable of focusing on either near or far objects, bifocal lenses are used, being constructed with one focal length over a portion of the area and another focal length over the rest of the area. By looking through one portion of the lens, the wearer can see distant objects clearly, while nearby objects are in clear focus through the other portion of the lens.

Another eye problem that can be corrected with lenses is astigmatism. This occurs when the eye lens lacks symmetry about its axis. Correction requires another lens lacking symmetry about

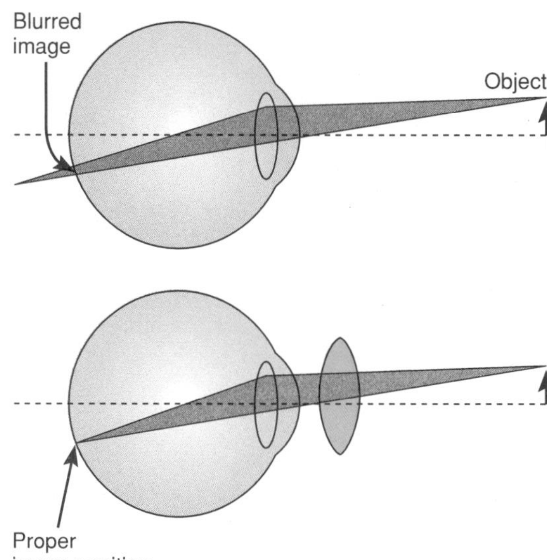

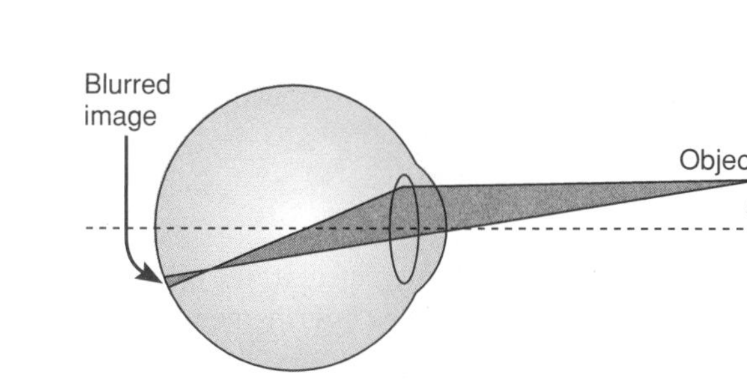

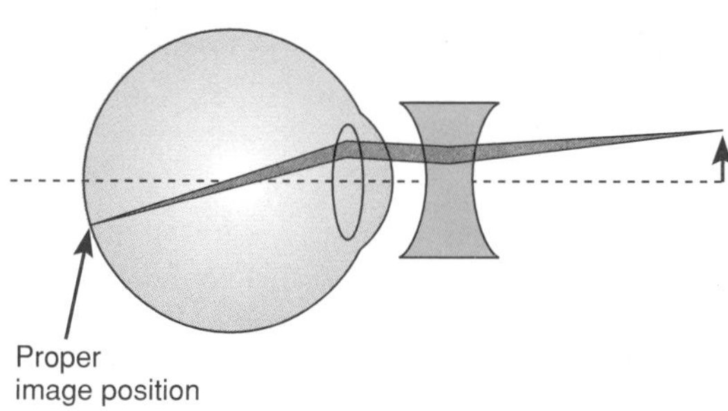

Figure 11.19 (*a*) The way in which a convex (converging) lens corrects farsightedness. (*b*) Nearsightedness is corrected by means of a concave (diverging) lens.

its axis, but, of course, with opposing features. This type of lens must be worn with the correct orientation relative to the eye to work properly.

Complicated optical instruments, such as telescopes and microscopes, use combinations of lenses to produce images. In such devices, the image produced by the first lens or mirror becomes the object for the second lens or mirror, and so forth, until the final image is formed by the last optical element. For example, in an ordinary compound microscope a lens of short focal length, the objective, placed near the specimen forms a real image of the specimen that is then viewed with an eyepiece acting as a "magnifying glass." The total magnification is the product of the two separate magnifications. The maximum useful magnification is limited by the wave nature of light itself, as discussed later in the section on diffraction and interference.

Figure 11.20 Two lenses of the same diameter but of differing focal lengths produce images that are of different sizes and different brightnesses. The longer focal-length lens has an *f*-number twice that of the shorter focal-length lens and its larger image is one-fourth as bright.

The ability of lenses and mirrors to form sharp images is critical in most applications, but another vital, and sometimes equally important, function is that of "gathering" light. If a camera can focus an image perfectly, but the image is not bright enough to expose the film, the camera is not useful. If the intensity of light in the image reaching the retinas of our eyes is insufficient, we cannot see.

The **intensity of the image** produced by a lens is dependent upon its **focal ratio,** also called the *f*-stop or *f*-number. This value is determined by the ratio of the focal length of the lens to its aperture; that is,

$$\text{focal ratio} = \frac{\text{focal length}}{\text{aperture}}. \qquad \text{FOCAL RATIO}$$

The *aperture* is the effective diameter of the mirror or lens, which is often varied by an adjustable diaphragm. The pupil of the eye is the aperture of the eye lens, with the diameter being automatically regulated by the iris in response to the light falling on the retina. The focal ratio of the eye ranges from about 2.3 in dim light to about 8 in bright light (written $f/2.3$ or $f/8$).

The intensity of the image, assuming the source has a fixed intensity, is inversely proportional to the square of the *f*-number. The amount of light passing through a lens (or mirror) depends on the area of the lens, with the area being $\pi(\text{diameter})^2/4$. Thus, the amount of light passing through the lens is proportional to the diameter squared. Lenses with longer focal lengths have larger images, spreading the light out over a larger area. Consider two lenses with the same diameter, but with one having a focal length twice that of the other (Fig. 11.20). The image from the lens with the longer focal length will be formed (in normal usage) approximately twice as far from the lens. The magnification will be twice as much (in each direction), giving the image four times the area, and hence making the image one-fourth as bright. Combining these two ideas shows that the image brightness is proportional to $1/(f\text{-number})^2$.

Example 11.5

If the human eye could adjust to changes in light intensity only by changing the diameter of the pupil, what range of brightnesses could be observed; that is, how many times brighter would the brightest observable scene be relative to the dimmest observable scene?

Solution

Since the focal ratio of the eye ranges from $f/2.3$ to $f/8$, and since the intensity of an image is inversely proportional to the focal ratio squared, under equal lighting conditions the images formed would have brightnesses proportional to $1/(2.3)^2$ and $1/(8)^2$. Squaring the numbers gives $1/5.29$ and $1/64$. The ratio of the brightnesses of the two images is thus $64/5.29$, or 12.1. Conversely, if one scene is 12.1 times as bright as another, adjustment of the diameter of the pupil is capable of producing images of equal intensity on the retina. This range of brightnesses is not nearly great enough to cope with ordinary changes in lighting, so most of the adaptation of the eye to differing brightness levels takes place in the retina.

The "standard" f-stops on a camera are chosen so that a change of one stop either doubles or halves the intensity of the image. Suppose that a photographer is "shooting" a picture with the lens set at $f/8$ but finds that the shutter speed is too slow to "stop" the action in the picture. A change to a speed of $f/5.6$ will allow a shutter speed twice as fast, since 8^2 is approximately twice as much as 5.6^2. Cameras with extremely low f-numbers permit picture taking in low light levels or with short shutter times, but aberrations become more important for these lenses, making them more difficult to construct.

Covering part of a lens affects the amount of light that passes through the lens and, consequently, affects the intensity of the image. Consider the opaque piece of material placed over part of the lens shown in Figure 11.21. This material blocks light from the lower half of the lens, but it still allows all the rays that pass through the upper half of the lens to form the image. The result is an image that is half as bright as the unobstructed lens, but one that still shows the entire scene.

Figure 11.21 When half of a lens is covered, the image is half as bright, but the image still shows the entire object.

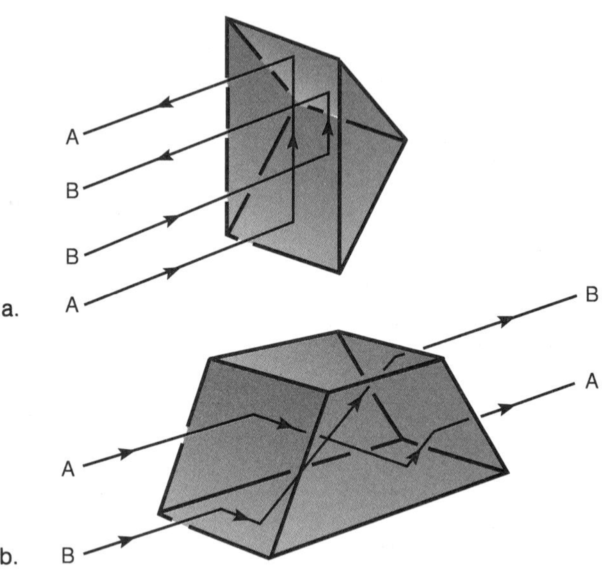

Figure 11.22 Total internal reflection is often utilized in prisms. These prisms are used to invert images in optical instruments such as binoculars.

Total Internal Reflection

Total internal reflection has many practical uses. Recall from Chapter 10 that when a ray of light in a material with a high index of refraction strikes a boundary with a lower index material at an angle greater than the critical angle, all of the light is reflected back into the material with the higher index. This property allows for the construction of mirrors that reflect light with no loss in light intensity.

Consider the prisms shown in Figure 11.22. These types of prisms are used in binoculars to invert images. The glass from which these inverting prisms are made usually has a critical angle of about 42 degrees, so that the reflecting faces need not be silvered. Both types of prism reverse the upper and lower rays (or left and right if they are rotated 90 degrees). The familiar offset in some binoculars occurs because two prisms of the type shown in Figure 11.22a at right angles to each other are needed in both barrels of the binoculars. More compact binoculars can be constructed by using the prism shown in Figure 11.22b.

Another area in which extensive use of total internal reflection has recently been made is in the rapidly growing area of **fiber optics.** Consider the two diagrams of glass fibers in Figure 11.23. The diagrams show a light ray entering the ends of the fibers. If the ray strikes the side of a fiber at an angle greater than the critical angle, the ray will reflect back into the fiber, bouncing back and forth as it traverses the length of the fiber. If the fiber is bent, the ray will strike at a slightly different angle, but if the fiber is thin enough it will still be greater than the critical angle. By using extremely thin fibers, light can be made to flow along a path just as electricity follows a conducting wire (Fig. 11.24). Even though the light ray is reflected from the sides of the fiber millions of times,

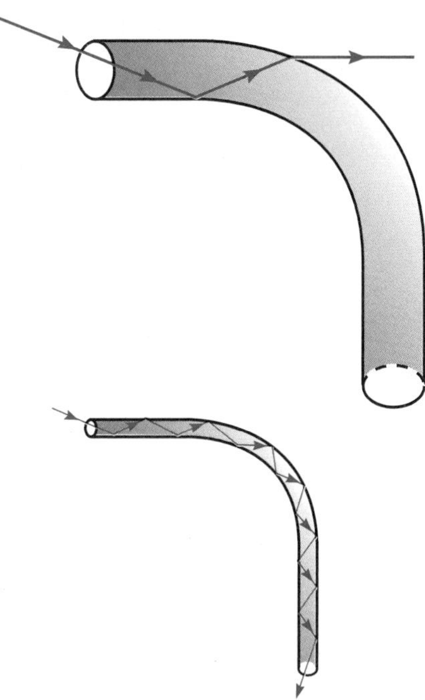

Figure 11.23 Fiber-optic light guides are made extremely thin so that when they are bent, the light will still strike the interface at an angle greater than the critical angle, thus allowing no light to escape.

Figure 11.24 Multistranded optical fibers transmit light and other electromagnetic waves with little loss of intensity.

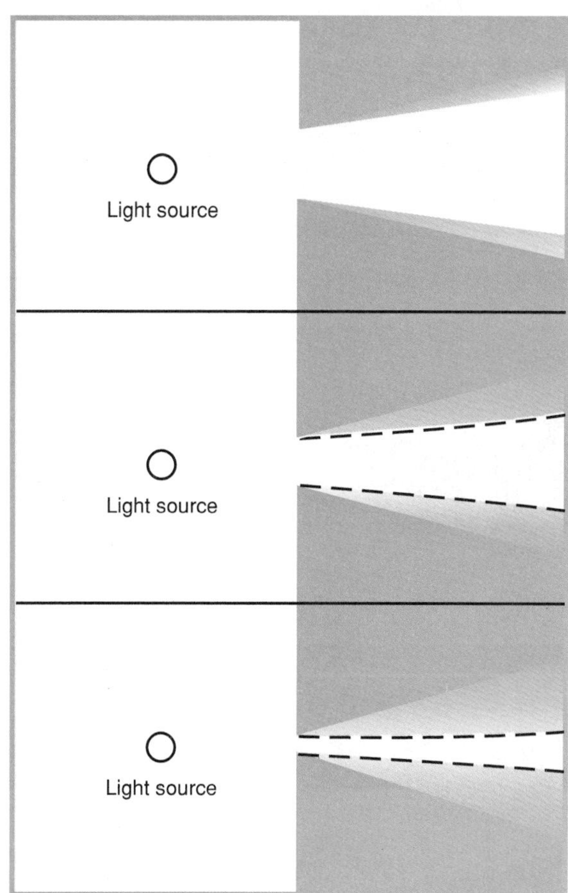

Figure 11.25 Shadows from a light source are always slightly "fuzzy." If the opening through which the light is passing is made very small, the light spreads appreciably because of diffraction.

its intensity remains high because of total internal reflection. However, the glass used in optical fibers must be of exceptional clarity so that absorption losses in the glass do not diminish the intensity of the transmitted light below an acceptable level over long distances.

Optical fiber networks have been constructed to transmit information. When information is transmitted using electromagnetic waves, customarily a "carrier wave," with a frequency much higher than that of the information being transmitted, is modulated (see the discussion of amplitude and frequency modulation in Chapter 9). Each channel of information requires a range of frequencies, called its "bandwidth." For example, ordinary telephone signals have a bandwidth of 5000 hertz. The higher the frequency of the carrier wave, the more channels that can be transmitted, since each channel must have a range of available frequencies equal to its bandwidth. The extremely high frequency of light permits simultaneous transmission of many channels of information, allowing for a reduction in the number of lines needed to transmit, and a consequent reduction in installation and maintenance costs. Bundles of optical fibers can also transmit images from previously inaccessible locations, such as inside the human body, as well as provide a path for intense light that can be used, for example, for surgical procedures.

Diffraction and Interference

The spreading of light waves when they encounter obstacles, called diffraction, produces effects encountered in almost every area of optics. One of the negative effects of this spreading is a limitation on the sharpness of images in optical systems, and one of the positive effects is that the spreading allows different parts of the same beam to interact.

Consider a beam of light passing through an opening, as in Figure 11.25. When the opening is large compared to the wavelength of the light, the edges of the shadow appear relatively sharp,

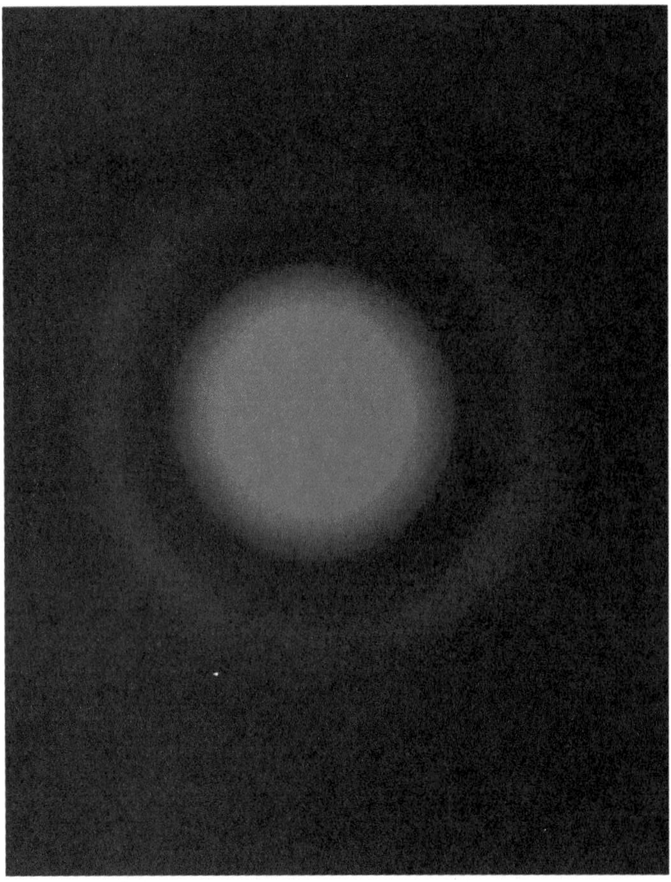

a.

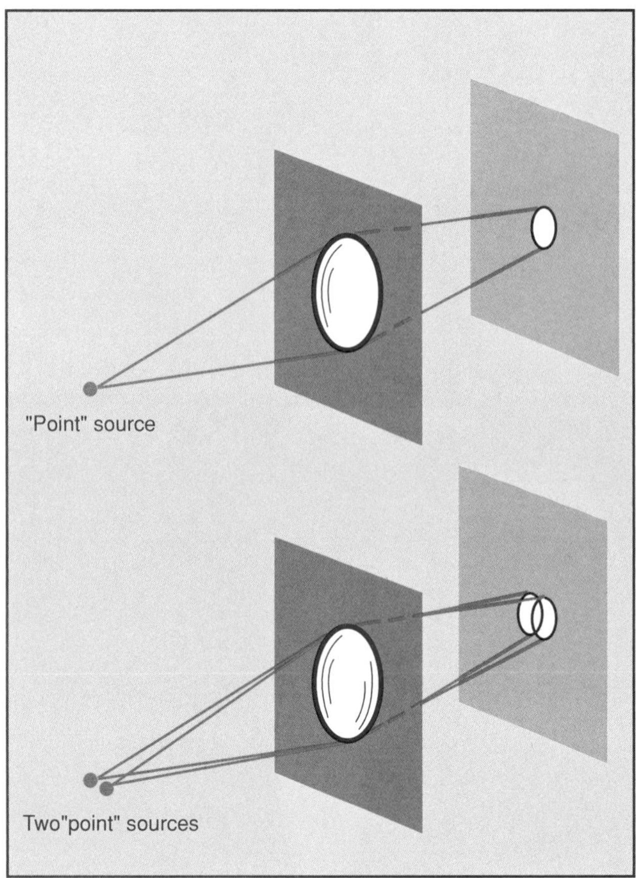

b.

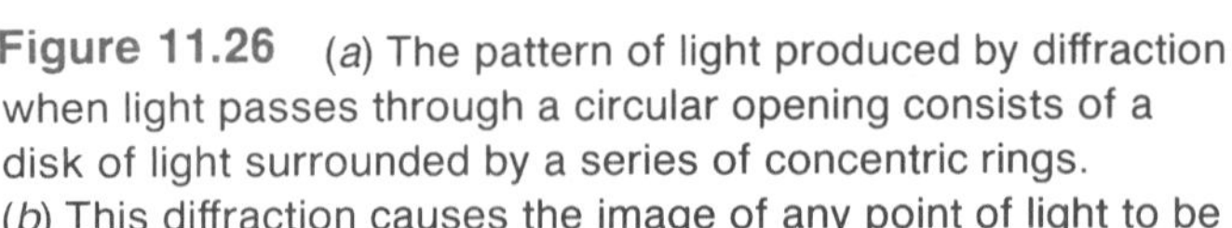

Figure 11.26 (*a*) The pattern of light produced by diffraction when light passes through a circular opening consists of a disk of light surrounded by a series of concentric rings. (*b*) This diffraction causes the image of any point of light to be spread into a small disk when imaged by a lens. Nearby points on objects will have disks of light that overlap, blurring the image.

but close inspection will reveal a slight "fuzziness." If the opening is made very small, a point will be reached where the size of the spot of light on the screen will begin to increase because of diffraction. Of course, the intensity will be very low since the opening is so small, and the light will be difficult to observe. Consider the case in which the opening contains a perfect lens (Fig. 11.26a), and the light source is a perfect point of light. The spreading caused by the opening will still be present, and even with the lens in place the beam will spread slightly. This will cause the image to appear as a small disk of light, and not as a point. The larger the lens, the smaller the disk will be. Every point on an object that is reflecting or emitting light will have an image that is a small disk of light, with neighboring disks overlapping and blurring the image (Fig. 11.26b), even if the lens is perfect. This blurring, while not serious in ordinary imaging, becomes the limiting factor when the images are enlarged in microscopes. The same problem is present in telescope lenses and mirrors, with larger lenses and mirrors giving sharper images until the distorting effects of the atmosphere become dominant. If the Hubble space telescope were limited only by diffraction, its primary mirror, with a diameter of 96 inches, would be able to discern detail about five times finer than can any Earth-based telescope. Unfortunately, because of an error in the grinding of the mirror, its capability to resolve detail is only about twice that of Earth-based telescopes.

The spectrum of a light source can be analyzed by utilizing not only the spreading effect just discussed, but also the fact that when more than one wave passes through a region of space, the total wave is the sum of the individual waves (Fig. 11.27). When the crests and the troughs of two waves of equal amplitude (size) coincide, the resultant is a wave with twice the amplitude of the component waves. If the crest of one wave coincides with the trough of the other wave, the waves "cancel" each other out. These effects are collectively called interference, the former being **constructive interference** and the latter **destructive interference.** For light waves, when destructive interference occurs, the result is darkness; when constructive interference occurs, the result is a brighter light than either of the waves produces individually.

When light of a single wavelength (monochromatic light) passes through a narrow slit, the crests and troughs will spread outward beyond the slit in semicircular arcs (Fig. 11.28a). If

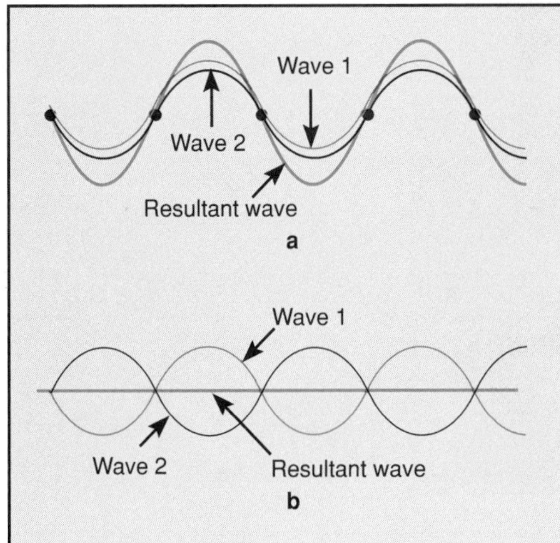

Figure 11.27 Interference occurs when two or more waves pass through the same region at the same time. If two waves of equal amplitude (size) are in phase, as in (*a*), the resultant wave has twice the amplitude of the two component waves. If the waves are out of phase (180°), as in (*b*), the waves "cancel" each other out, leaving no resultant wave.

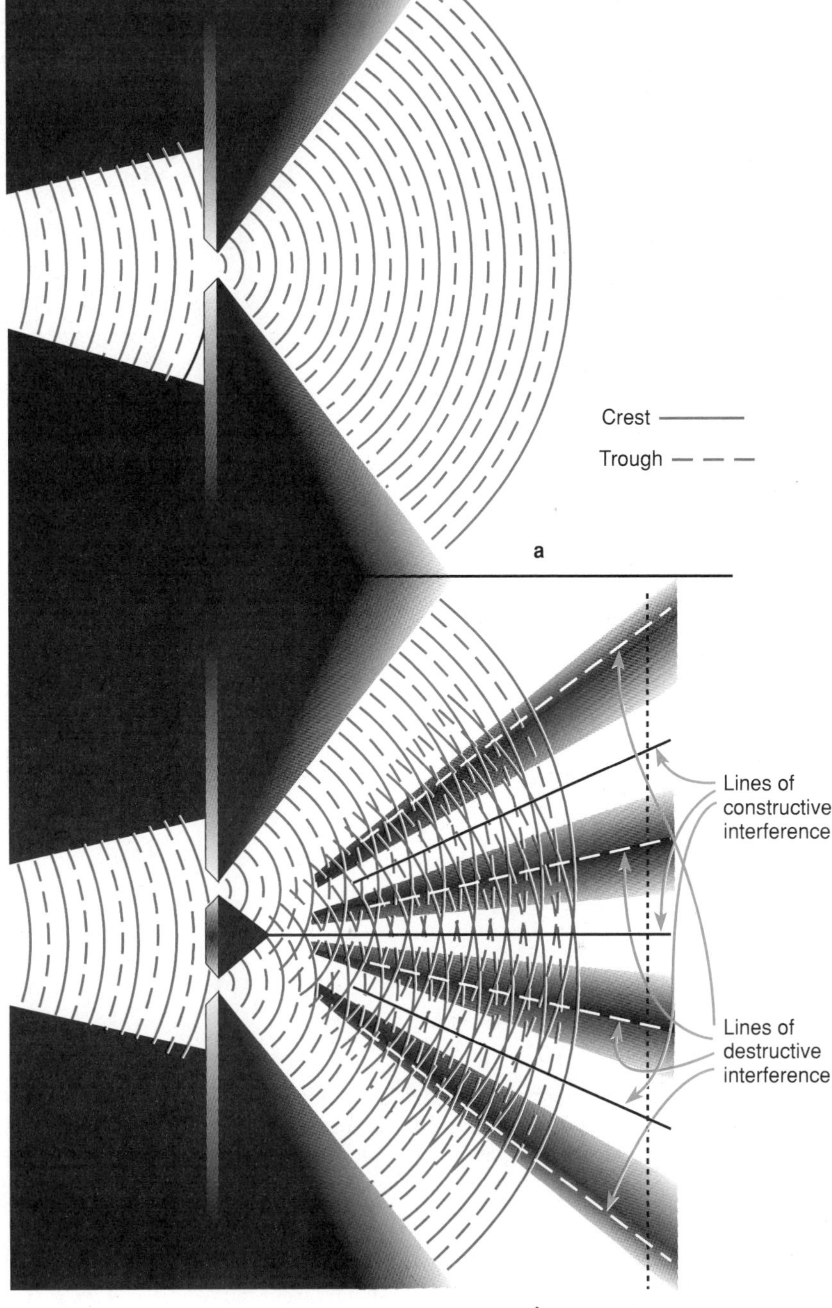

Figure 11.28 (*a*) Light of a single wavelength passes through a narrow slit, spreading because of diffraction. (*b*) If the light passes through two slits, the waves from the two slits will overlap, causing interference. Along lines of constructive interference, crests and troughs are in phase, resulting in a large amplitude wave. Along lines of destructive interference, the waves are out of phase, with the result that the waves cancel out.

another narrow slit exists near the first, there will be a region of space through which two waves are passing, creating interference between the waves (Fig. 11.28b). The points where the crests and troughs are together (they are said to be "in phase") are seen to fall along certain lines, as do the points where the crests and troughs are opposite ("out of phase"). It should be remembered that the waves are moving away from the openings, generally toward the right in the diagram. An observer stationed along one of the lines where the waves are in phase would see the electromagnetic field oscillating vigorously; in other words, the light would be bright along these lines. Conversely, darkness would prevail along the lines where the waves are out of phase. As indicated, a screen placed near the right edge of the diagram would exhibit a series of fringes, alternating from bright to dark. A photograph showing actual fringes from a double slit is shown in Figure 11.29. This experiment is called Young's double-slit experiment because it was demonstrated and explained in the early nineteenth century by Thomas Young, an English physician. The experiment showed that light has properties of waves.

Figure 11.29 A photograph showing fringes produced at the location of the dotted line in Figure 11.28. The intensity varies gradually from a maximum to a minimum. The photograph was made using a laser as a source of light.

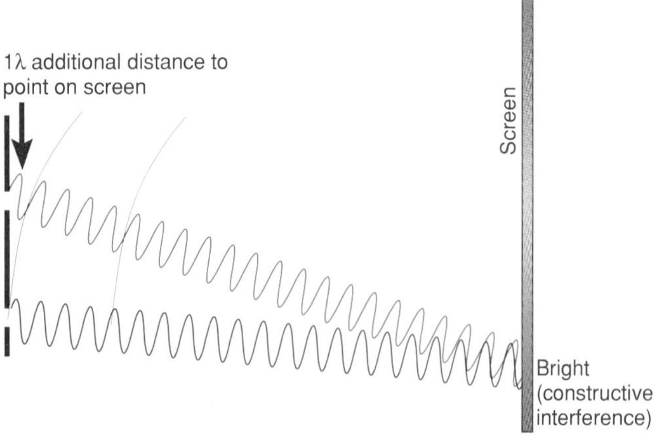

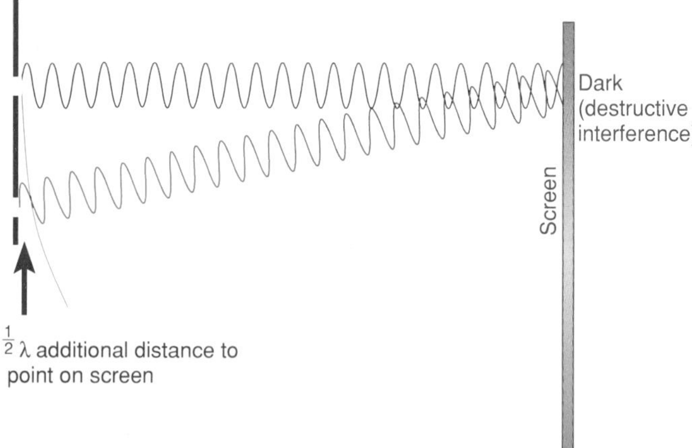

Figure 11.30 A construction to show how to locate points on a screen where waves are in phase (bright) and out of phase (dark). (*a*) If the difference in path lengths is a whole number of wavelengths, the waves arrive in phase and the point is bright. (*b*) If, however, the difference is an odd number of half wavelengths, the waves arrive out of phase and the point is dark.

An easy way to locate the bright and dark fringes in Young's experiment is shown in Figure 11.30. If the distances from the two slits to a point on the screen differ by a whole number of wavelengths; for example, 0λ, 1λ, 2λ, 3λ, . . . , $n\lambda$, the waves will always reach that point in phase, causing a bright place on the screen. If the distances differ by an odd number of half wavelengths; for example, $(1/2)\lambda$, $(3/2)\lambda$, $(5/2)\lambda$, . . . , $(n/2)\lambda$, the waves will always arrive out of phase and the point will be dark. The bright fringes are numbered outward from the central fringe according to how many wavelengths difference there is between the two distances. So the first bright fringe on either side of the central one is called the first-order fringe, the second is called the second-order fringe, and so on.

Consider now a *very large number of uniformly spaced parallel slits,* a **diffraction grating** (Fig. 11.31). On a screen located far from the diffraction grating, light from a monochromatic (single-color, single-wavelength) source that passed through the grating would cause a series of sharp bright fringes with large dark areas between the fringes. The fringes are created by constructive interference, as in the double slit. When the difference in path length between two adjacent slits to a point on the screen is a whole number of wavelengths, the light from all the slits arrives in phase (Fig. 11.32a). If, however, it deviates only slightly from this, it will be found that the intensity of the light drops rapidly to zero. To understand this rapid drop in intensity, consider two adjacent slits where the difference in distances is 1.01λ (Fig. 11.32b). The light

a.

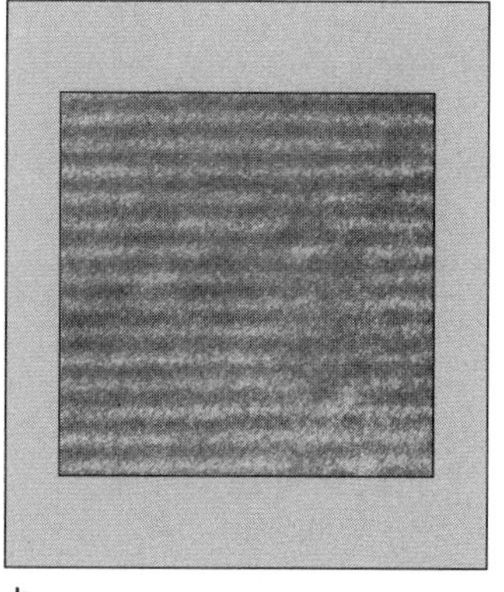

b.

Figure 11.31 (*a*) The transmission diffraction grating consists of parallel lines that produce fringe areas. (*b*) Microscopic views of diffraction gratings reveal that they are produced by scratching parallel lines on a glass surface or occur through the natural characteristics of various crystalline substances. The diffraction grating shown is of a crystalline nature.

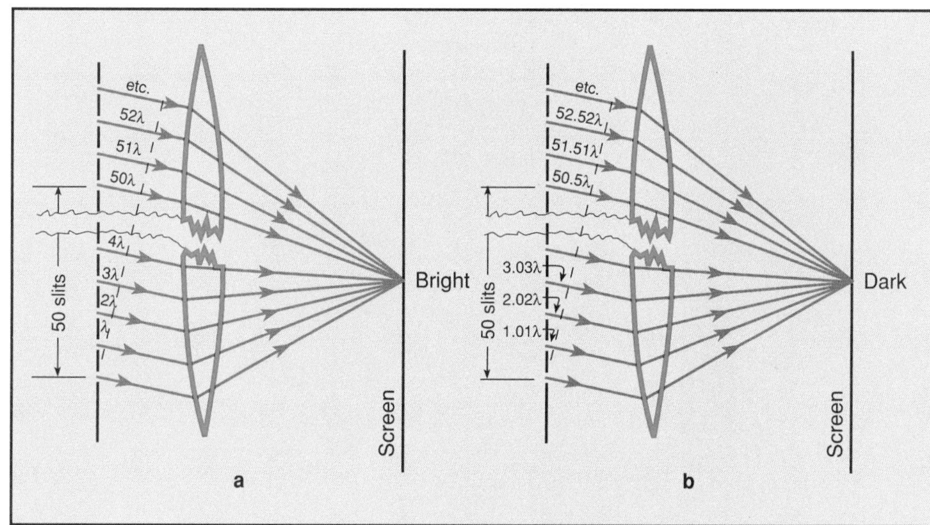

c

Figure 11.32 (*a*) Light passing through all the slits of a diffraction grating will be in phase if the difference in paths for light from adjacent slits is a whole number of wavelengths (one wavelength difference is illustrated). (*b*) If the angle is changed only slightly, say 0.01λ, then a slit can be found that will cancel out the light going in that direction, making the fringes very sharp (*c*). Compare these "fringes" to those in Figure 11.29.

from these two slits is "almost" in phase, but remember that there are hundreds of slits. The light from the fiftieth slit upward from the starting slit will travel a path that differs from the original slit by 50.50λ and hence will "cancel it out." Similarly, the light from slits 1 and 51 will "cancel out," and so forth, meaning that no light will strike the screen. The difference in direction to the screen caused by a change of only 0.01λ is very small, so that the fringes are very sharp (Fig. 11.32c).

If light of more than one wavelength is passed through a diffraction grating, the light will be spread out into a spectrum just as if it had passed through a prism, since each color will have a particular direction for which constructive interference occurs. The diffraction grating has several advantages over a prism. First, since the spacing between the adjacent slits can be precisely controlled and determined, the construction of gratings with different angles of spread between different colors is possible, also allowing for a

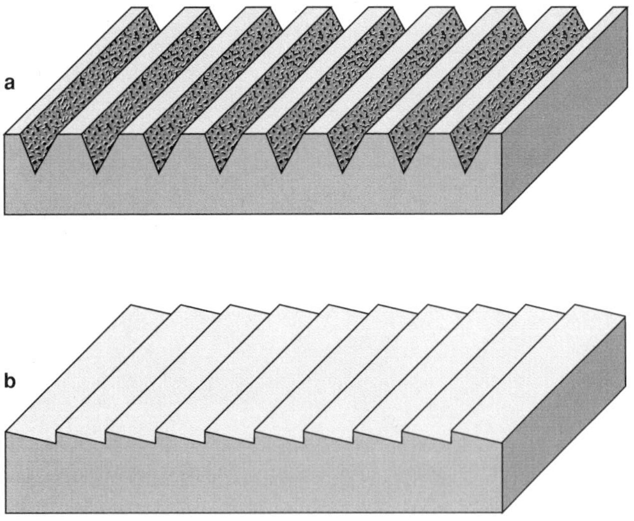

Figure 11.33 (*a*) Diffraction gratings may be mass-produced by scratching fine lines on a piece of soft metal that is then used as a mold for clear plastic. The smooth areas act as the slits. (*b*) Mirror surfaces can also be made to act as diffraction gratings. The angles of the facets can intensify the constructive interference at a certain angle, making the light much brighter.

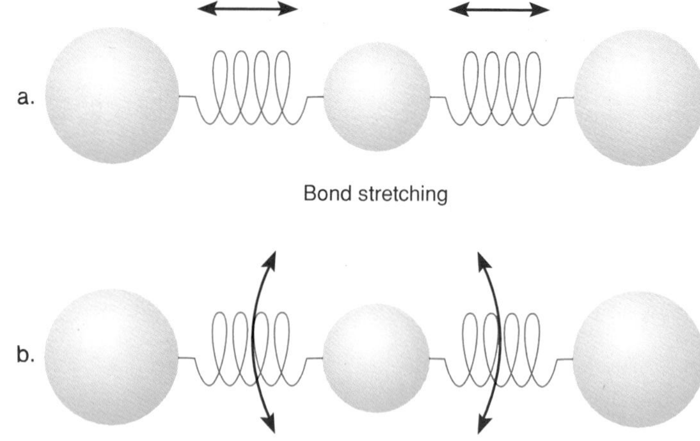

Figure 11.34 A simple model of a triatomic molecule. The frequencies of vibration depend on the masses of the atoms (spheres), the strengths of the bonds (springs), and the manner of vibration (stretching or bending). Resonance occurs when the frequency of the radiation impinging upon the molecule matches one of these frequencies. At resonance, the amplitude of the vibration increases, increasing the energy of the molecule and removing energy from the radiating beam.

direct computation of the wavelength of the light. Second, the gratings can be mass produced, with the same characteristics for all gratings, so that instruments made with the gratings will all have the same characteristics. Third, the gratings are flat and lightweight. There are also some disadvantages associated with diffraction gratings, the main one being that most of the light passes straight through the grating, making them less useful when analyzing dim sources of light (though this can be overcome to some extent by using special gratings made on mirrors).

Diffraction gratings may be produced by scratching fine lines close together (about five hundred lines per millimeter) on a soft piece of metal, coating the metal with a thin layer of clear plastic, and peeling the plastic off the metal (Fig. 11.33a). The thin layer of plastic will then have smooth "windows" that act as the slits. They can easily be mass-produced and have even become popular in bumper stickers and costume jewelery. The colors seen when light is reflected from a compact disc are the result of the very closely spaced tracks acting as a diffraction grating.

An alternative method of making gratings is to make a "stepped" mirror, as shown in cross section in Figure 11.33b. The light waves striking different levels of the mirror travel different distances and arrive at the destination either in or out of phase. Gratings made from mirrors are especially useful when working in the ultraviolet and infrared portions of the spectrum, since absorption of waves is not a problem.

The ability to spread light into its component colors has led to important uses in the analysis of compounds. Consider the model of a simple triatomic molecule shown in Figure 11.34. The springs are one way of modeling the bonds holding the molecule together. Each spring will have a certain stiffness, and that, along with the masses of the atoms, will determine the frequency at which the atoms will vibrate. If light is passed by these molecules, it will be absorbed if its frequency corresponds to one of the vibrational frequencies (it will be in resonance). Since different molecules will have different sets of vibrational frequencies, the absorption spectrum will depend on the compound. Instruments that analyze blood or urine for different compounds, or those that are used to determine pollutants, often rely on this technique (Fig. 11.35). The absorption spectrum of dyes used in dyeing cloth also can be analyzed in this manner.

Polarization

It will be recalled from Chapter 10 that when radio waves are generated by charges moving in antennas, the waves are polarized; that is, the vibrations are in a certain direction. Ordinary sources of light are not polarized, because the atoms that emit the individual photons of light do so independently of each other. Thus, in ordinary light the direction of the electric field randomly and rapidly changes direction. Several methods are available for producing polarized light, including crystalline separation, absorption, reflection, and scattering.

If a beam of light is passed through certain materials that are not uniform in all directions, the light vibrating in one direction will behave differently from the light vibrating in a perpendicular direction. Certain plastics display this effect if they are

On Polarizing Light

Polaroid, the innovative material that prevents most glare and has other optical effects on light, was developed by Edwin H. Land in 1938. The Polaroid process stretches large plastic sheets to several times their original length. The plastic sheets contain long chains of various molecules, primarily carbon and hydrogen atoms with special bonds. The stretching is continued until the molecules line up with each other. Then the sheet is attached to a rigid transparent backing and submerged in an iodine solution. The molecules react with the iodine, forming long chains of carbon atoms with iodine atoms attached. This fine light-conducting network limits the directions in which light can pass through it, and thus polarizes light.

Land served as head of a research group concerned with the improvement of military optics and the further development of Polaroid materials. During a holiday season, his daughter photographed numerous family activities and expressed a concern over the seemingly excessive length of time before she could see the developed pictures. Land decided to investigate various ways to produce photographs other than through the normal development procedure. The research he conducted culminated in 1947 in the 60-second production of a single-color print that had greatly improved sharpness and quality over other films of similar sensitivity. Land's association with the Polaroid Corporation led to the film and the Land camera he invented being known by the name "Polaroid," although there is no association between the camera and the polarizing materials he produced ■

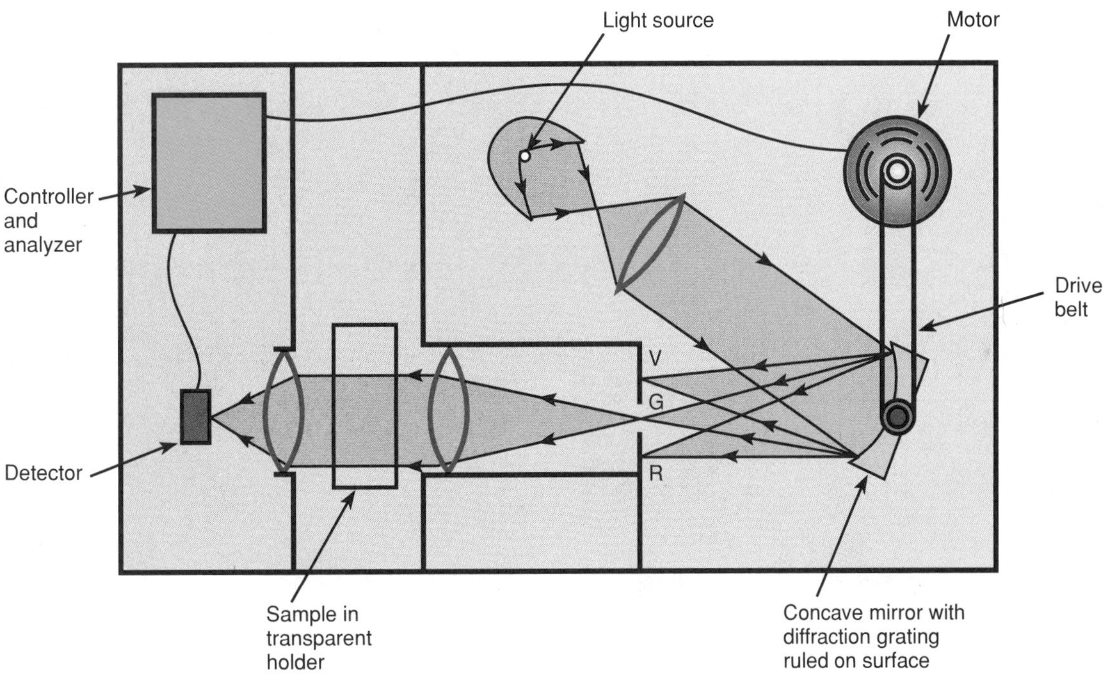

Figure 11.35 A diagram of a spectrograph. Light from the source strikes a diffraction grating ruled on the surface of a mirror. By rotating the grating, different colors of light are made to pass through the sample. The controller runs the motor to rotate the grating, and the detector senses how much light is absorbed for each wavelength, allowing an identification of compounds in the sample.

manufactured so that their long molecules are aligned. The effect is also evidenced in some crystals. When light is passed through these materials, all light vibrating in one direction is absorbed, leaving the transmitted beam with vibrations only in the perpendicular direction (Fig. 11.36a). Two disks of polarizing material are shown in Figures 11.36b–d. As shown, if the directions of polarization are parallel, light will be transmitted, but if the directions of polarization are perpendicular, the second polarizer will absorb the light that has passed through the first polarizer.

Some materials have the capability of rotating the plane of polarization. When placed between two polarizers, the amount of rotation can be determined by turning the second polarizer (called the analyzer) to allow passage of the maximum amount of light. The strength of sugar solutions from grapes, for example, can be analyzed in this way. Stress in certain transparent materials also affects polarization, allowing models of structures to indicate points of great stress (Fig. 11.37).

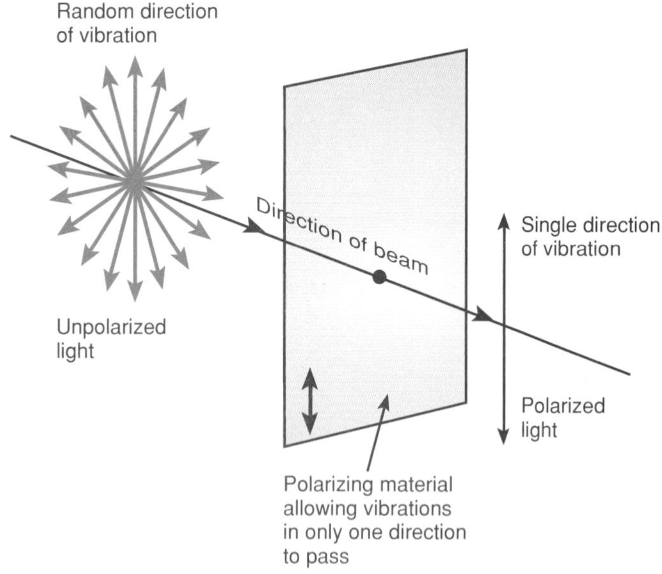

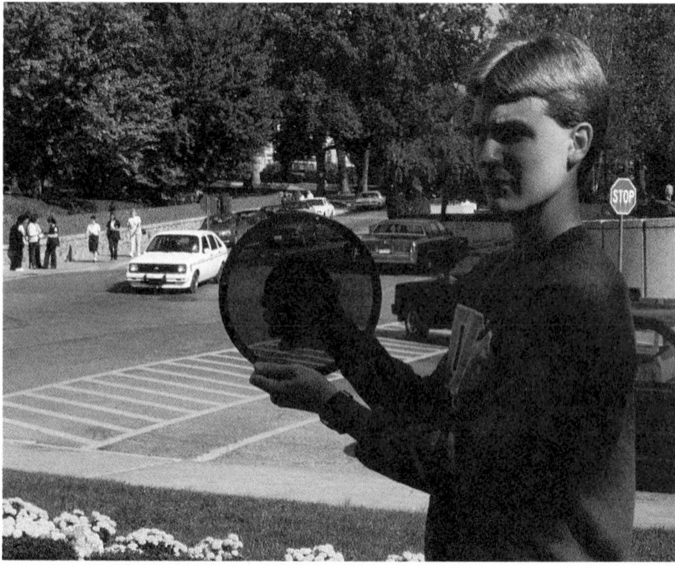

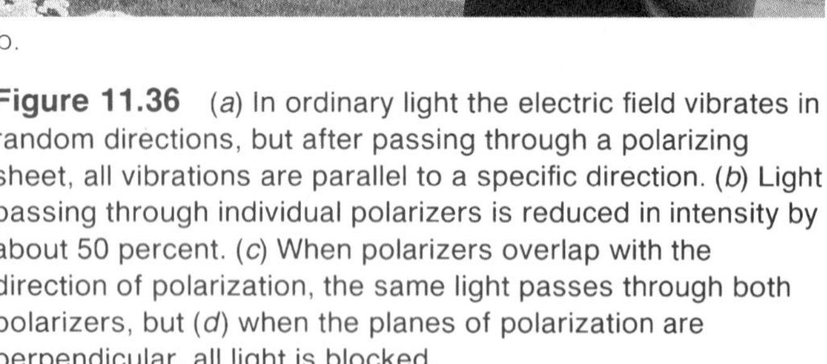
Figure 11.36 (*a*) In ordinary light the electric field vibrates in random directions, but after passing through a polarizing sheet, all vibrations are parallel to a specific direction. (*b*) Light passing through individual polarizers is reduced in intensity by about 50 percent. (*c*) When polarizers overlap with the direction of polarization, the same light passes through both polarizers, but (*d*) when the planes of polarization are perpendicular, all light is blocked.

a.

b.

c.

d.

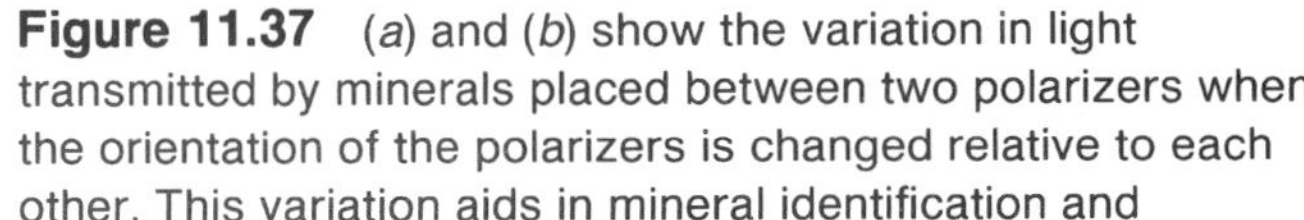

Figure 11.37 (*a*) and (*b*) show the variation in light transmitted by minerals placed between two polarizers when the orientation of the polarizers is changed relative to each other. This variation aids in mineral identification and classification. The small plastic model in (*c*), when placed between cross polarizers (*d*), shows the material's stress pattern.

Liquid crystal displays used in calculators, computers, and flat-screen television sets utilize the principle of crossed polaroids with an intervening material that affects the polarization. Certain organic molecules have the capability to rotate the plane of polarization if they are aligned properly. The alignment of these long molecules can be accomplished by applying a weak electric field in the desired direction of alignment. Figure 11.38 shows a typical arrangement, with a "sandwich" made of crossed polaroids, a solution of the organic molecules, and two partially silvered conductors used to apply the electric field. When a voltage is applied between two parts of the conductors, the electric field aligns the molecules, rotating the direction of polarization of the light, and allowing it to pass through the second polarizing sheet. When the voltage is removed, thermal vibrations cause the molecules to assume random orientations, destroying the rotation and preventing the passage of the light. A mirror on the back face would allow viewing in reflected light.

A second common method of polarization is by reflection. When light is reflected from a nonmetallic surface, it is at least partially polarized (if the angle between the reflected ray and the refracted ray in the material from which the light is reflected is 90 degrees, the polarization is complete), with the vibrations of the electric field being parallel to the surface. Light reflected from horizontal surfaces will be horizontally polarized, so that if a sheet of polarizing material is oriented to allow only vertical vibrations, the reflected glare will be greatly reduced (Fig. 11.39). Polarizing sunglasses take advantage of this effect. If the head is tilted away from the vertical when wearing polarizing sunglasses, the increase in reflected light is usually very pronounced.

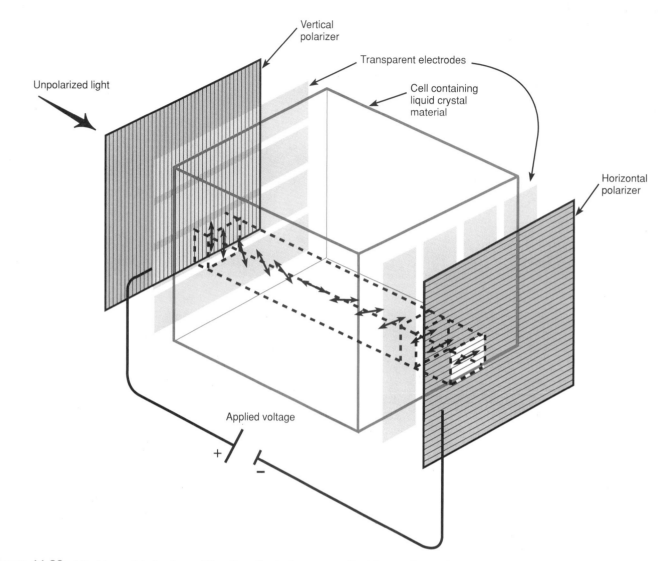

Vertical polarizer

Transparent electrodes

Unpolarized light

Cell containing liquid crystal material

Horizontal polarizer

Applied voltage

+

–

Figure 11.38 Liquid crystal displays utilize the effect of rotation of the plane of polarization of light by certain organic compounds when their molecules are aligned with an electric field. Transparent electrodes in the form of perpendicular strips allow an electric field to be applied to a portion of a liquid crystal, rotating the plane of polarization and allowing the light in this area to pass through the crossed polaroids. Using a mirror or silvering one set of the electrodes allows light to be reflected back through the display.

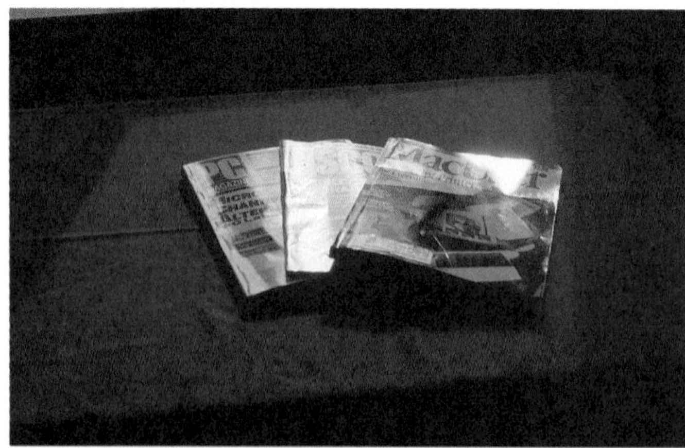

a.

b.

Figure 11.39 The scenes in (*a*) and (*b*) are made under identical lighting conditions, the difference being that the polarizing sheet in (*b*) is rotated to block horizontally polarized light while in (*a*) the horizontally polarized light is allowed to pass.

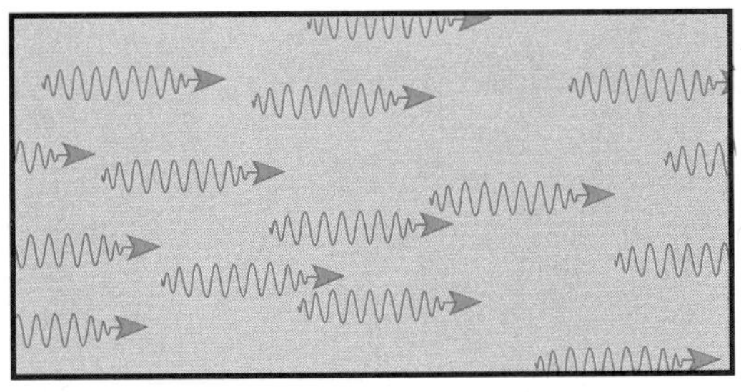

a.

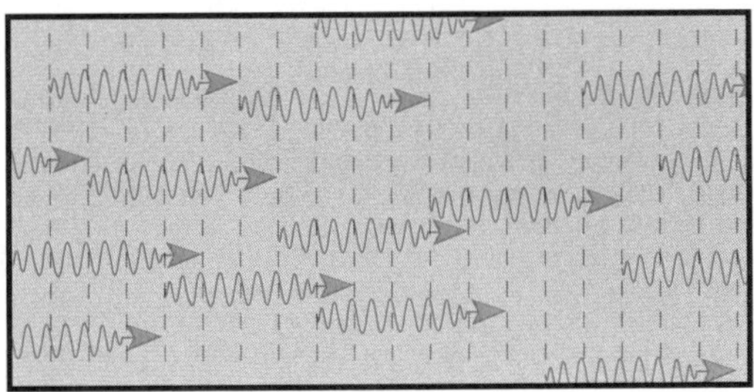

b.

Figure 11.40 (*a*) An ordinary light beam, even if it is monochromatic, consists of many individual photons. The photons are emitted randomly by the atoms of the source; hence, they have no definite phase relationship to one other. (*b*) Coherent light consists of photons that are all in phase.

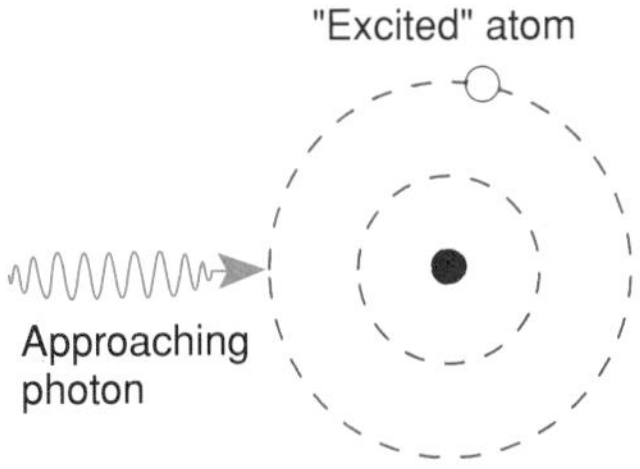

a.

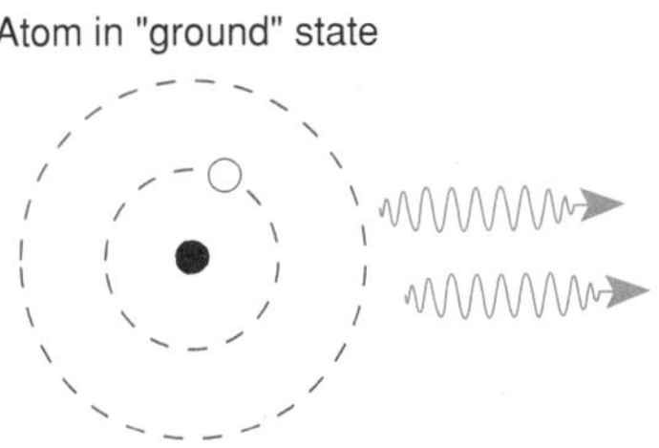

b.

Figure 11.41 When a photon passes near an atom in an excited state (*a*), it may stimulate the atom into jumping to a lower state (*b*). The original photon must have an energy exactly equal to the energy difference in the two levels of the atom. The two photons will be in phase and traveling in the same direction.

The Laser and the Maser: Stimulated Emission

Ordinary light sources emit a mixture of colors. Since each color has a different wavelength, there is no way to keep the waves in phase with each other. Even in sources that emit only a single wavelength, the waves usually do not maintain a definite phase relationship to each other, because different parts of the source act independently of each other. Photons are usually emitted from atoms by **spontaneous emission,** whereby *an atom with an electron in an excited state spontaneously "jumps" to a lower state, emitting a photon in the process* (see the section "Light—Wave or Particle" in Chapter 10). Since the jumps are spontaneous, each photon is emitted independent of the other photons, and even though the wavelength of the light of each photon may be the same, *the locations of the crests and troughs of one photon bear no relationship to those of another photon* (Fig. 11.40a). Such light is referred to as **incoherent light.** If *photons of the same wavelength are emitted so that all the crests and troughs are in phase* (Fig. 11.40b), the light is referred to as **coherent light.** Coherent light has properties similar to the radio waves discussed in Chapter 10, making it much easier than ordinary light to modulate for purposes of transmitting information. In addition, the light is much more intense and, because of the method of generation, all photons in the beam follow essentially parallel paths with little spreading of the beam over long distances.

Coherent light is generated by the process called **stimulated emission.** Consider an atom in which an electron is not in the innermost orbit (Fig. 11.41a). The electron tends to make a transition to the innermost orbit by emitting a photon, the photon having an energy equal to the energy difference between the orbits and a wavelength inversely proportional to the energy. If the electron jumps to the inner orbit on its own, spontaneous emission occurs. If, however, a photon having an energy equal to the energy difference between the two orbits passes near the atom before the electron can jump into the inner orbit, the electron may be stimulated to jump sooner than it would on its own (Fig. 11.41b). This stimulation causes the emission of a photon, just as if the electron had jumped on its own, except *the photon emitted will be moving in exactly the same direction as the "stimulating" photon and will be exactly in phase with it.* Thus, stimulation is a method of amplifying the light, because the two photons now moving together can stimulate the emission of even more photons, and so on (Fig. 11.42).

Einstein speculated about the possibility of stimulated emission in 1917. In the 1950s American and Soviet scientists independently worked out theories as to how this process actually might be used to amplify light (microwave amplification had been achieved in 1953 and will be discussed later in this section). The first working device was developed by Theodore H. Maiman in 1960 at Hughes Research Laboratories in Miami, Florida, even though a graduate student at Columbia University, Gordon Gould, conceived a working device in 1957.

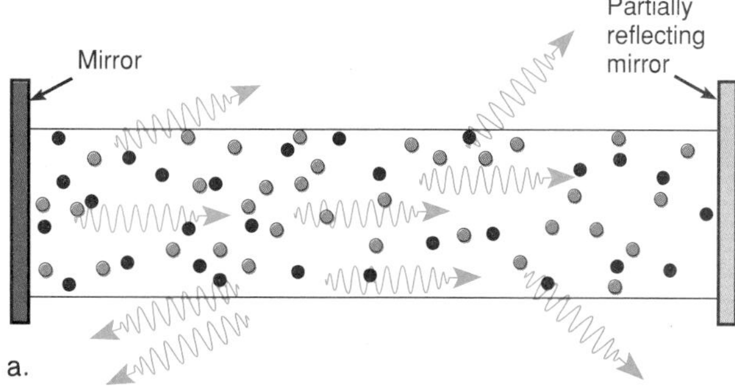

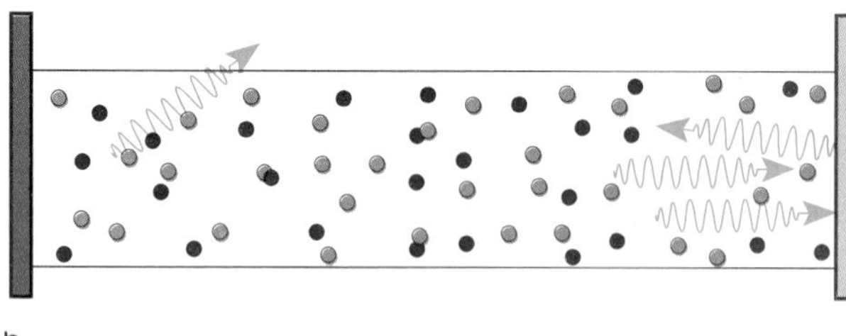

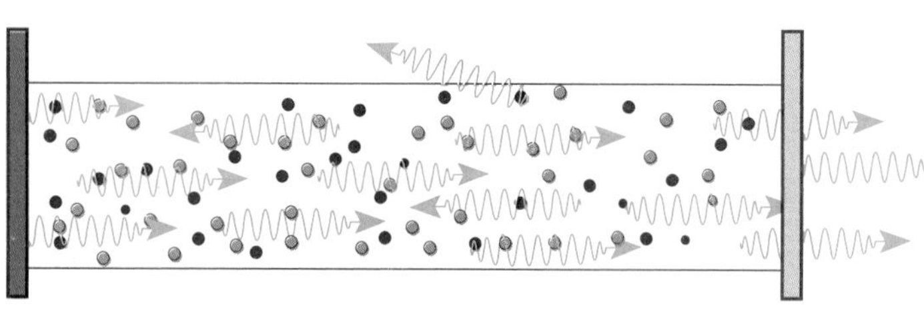

Figure 11.42 A schematic of the operation of a gas laser. Open circles represent atoms in an excited state (usually from an electrical discharge or collisions with other atoms). Spontaneous emission causes photons to be emitted randomly from atoms (*a*), but when a photon is emitted parallel to the axis of the tube, the mirrors at each end reflect it (*b*), causing it to pass near many atoms. Soon (*c*), light bouncing back and forth between the mirrors consists of many photons, each with the capability of stimulating emission from other excited atoms. Atoms are continually excited into higher energy levels so that the process is continuous, with some light leaving the tube through one of the mirrors, which is only partially reflecting.

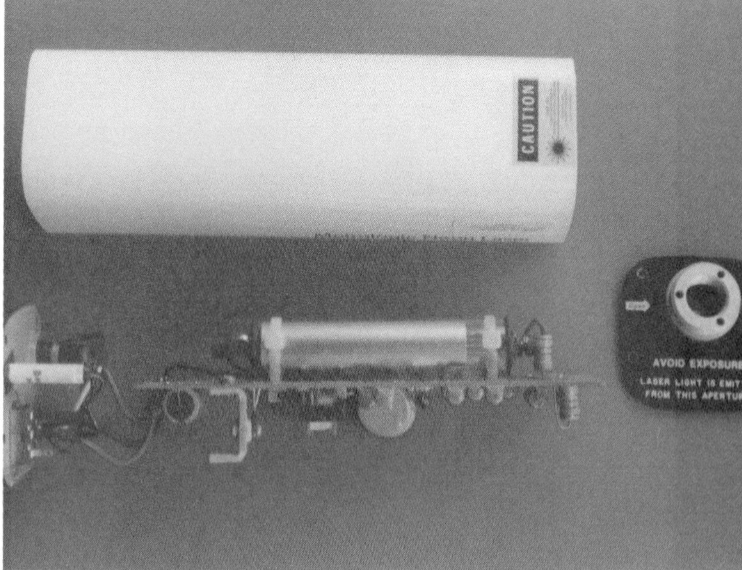

Figure 11.43 The basic components of a laser. Federal regulations require that lasers carry warning labels as seen on the end cap and the laser housing.

A **laser** is *a device used for amplifying light by stimulated emission.* The word laser is an acronym for ***l**ight **a**mplification by **s**timulated **e**mission of **r**adiation.* In a typical gas laser (Fig. 11.43), light is reflected back and forth between two parallel mirrors through a gas in which the atoms have been "excited" by an electrical discharge and subsequent collisions. As the light bounces back and forth, it is amplified by stimulated emission. One of the mirrors is designed to allow some of the light to penetrate the mirror, with typically about 1 percent of the light continuously escaping, meaning that for steady-state operation the amplification during each round-trip must be about 1 percent. Other types of lasers operate in a pulsed mode, usually with the pulses being emitted so rapidly that they appear to have a continuous output.

The outstanding properties of laser light, namely the high intensity, coherence, single wavelength, and single direction, have led to an increasing number of applications in many fields of research and industry, and other uses are certain to be discovered as research continues.

For instance, optical lenses can be used to bring a laser beam into very sharp focus. The intensity of the tiny beam produced is sufficient to rapidly vaporize almost any substance. Since there is no physical contact with the material, very small holes may be produced without affecting the surrounding material, a decided advantage over other techniques. The same feature permits the trained surgeon to "weld" together sensitive tissue, such as a detached retina in the eye, with no need for actual surgery. This procedure is accomplished by aiming the tiny beam from the surgical laser onto the transparent pupil of the eye, permitting the laser light to interact with the more opaque tissue located at the back of the eye, where localized heating fuses the tissues together. Tumors that were formerly considered inoperable, as well as some types of cataracts in the eye, are routinely destroyed with laser light. The laser is also of great value in gynecological treatments and other surgery, partly due to its ability to vaporize tissue and to cauterize blood vessels. Many specialists who perform surgery with the laser contend that laser-produced incisions heal faster and with less scar tissue than those made by conventional methods. Dentists also find the laser beneficial in removing tooth decay. The intensity of the tiny laser beam is also employed to cut cloth with great precision for commercially made garments.

The straight path of the laser is used to align long sections of pipe as well as to ensure vertical alignment during the construction

Figure 11.44 A researcher operates a gas-flow laser that uses a mixture of carbon dioxide and nitrogen.

of skyscrapers. The laser has also been used to study the surface of the moon; with it the depths of lunar craters and the heights of mountain ranges have been determined with great accuracy. The lunar explorations of the 1970s included experiments with lasers; sensitive laser reflectors were set up on the surface of the moon to permit studies of Earth's atmospheric conditions, of the relative motion of the moon and Earth, of rotational characteristics of Earth, and even to measure the rate at which the continents are moving relative to each other.

The capability to focus laser beams to extremely small points of light has led to their use in information storage and retrieval systems. The familiar compact audiodisc and the laser videodisc are examples of this relatively new technology. The discs consist of tracks with millions of microscopic pits, each pit representing a single bit of information. The pits can be packed so densely that it is estimated that all the books ever written could be stored on an area of only 6 square feet. The discs are now making their way into the personal computer market as an alternative to "floppy" disk storage.

Bell Telephone Laboratories has patented a carbon dioxide laser that can be used to burn holes through steel plates and to weld ceramic materials together. This laser operates at a wavelength of 10,000 nanometers and thus is in the infrared region of the spectrum (Fig. 11.44). Research into other high-power lasers leads some researchers to envision them as tremendous defensive weapons, though the technical problems seem almost insurmountable to other researchers.

One of the most spectacular uses of laser light, and one that has captured the imagination of the public, is the production of three-dimensional pictures. A **hologram** is *a three-dimensional image captured on a photographic plate* by utilizing the interference of waves. Consider the diagram shown in Figure 11.45. Laser light is passed through a beam-splitter (a piece of glass), with part of the light being reflected directly to a photographic plate (the reference beam) and the other part striking the object (the object beam). The lenses are used to expand the narrow beam of the laser, but they play no part in focusing an image as in an ordinary photograph. A portion of the light that strikes the object will be scattered to the photographic plate. If the light scattered from some point on the object reaches the photographic plate "in phase" with the light in the reference beam, the light will be "bright" at that point, exposing the photographic plate. If the light from some point arrives "out of phase," there will be a dark place, and the plate will not be exposed. The photographic plate will thus be a record of the wave pattern of the object, not an actual picture of the object. Note also that the light used to make the picture must be coherent, so that the interference pattern of the waves will not change during the time the plate is being exposed.

If, after the photographic plate is exposed, a beam of laser light (or well-collimated monochromatic light) is passed through the plate in the direction of the original reference beam, an interference pattern will now be created that will recreate the original scene. This recreated scene will actually possess the wave pattern of the original scene, and thus will appear three-dimensional to the observer, just as if the observer were looking through a window at the object. If the observer moves his or her head, the object will be seen from different directions as long as it remains in view "through the window."

Holograms that may be viewed in ordinary light (so called white-light holograms) have recently become very common. They appear on credit cards, magazine covers, and in novelty shops. They are made by taking an ordinary hologram and allowing laser light to pass through a slit placed over the hologram, making a second hologram from the first. This second hologram will have three-dimensional information in only one direction (parallel to the slit), but now may be viewed in ordinary light without blurring. Researchers are actively working on new techniques for producing and viewing holograms. True three-dimensional television and movies may soon be a reality.

The research into properties of laser light, methods of generation, and uses in other areas of research continues at an ever-accelerating pace. Tremendous advancement is likely to continue, leading to better understanding of materials and to even more applications (Fig. 11.46). Considering that the first laser was invented in 1960, it is likely that we have just begun to exploit its possibilities.

As mentioned previously, ***microwave amplification by stimulated emission of radiation,*** the **maser,** was invented in 1953. The principle is the same as for the laser; that is, microwaves pass near excited molecules which then give up their energy in the form of

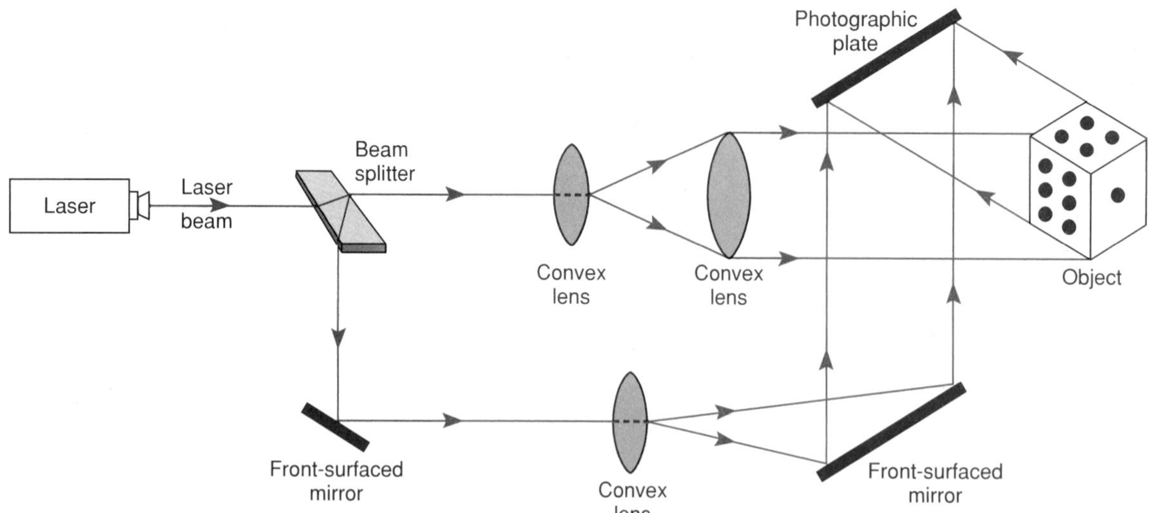

Figure 11.45 The means by which three-dimensional holograms are produced (as viewed from above). A beam-splitter and a series of lenses cause the laser light to strike two sides of the object to be photographed. The light that is reflected to the sensitive photographic plate then produces an image that is viewed with laser light or filtered light as from a slide projector.

photons, amplifying the original beam. Microwaves have wavelengths on the order of centimeters, with correspondingly lower frequencies than visible light. Molecular vibrations are used for microwave generation, instead of electrons making transitions from one level to another. With the aid of the maser, the science of communications has progressed to such a degree that it is now possible to simultaneously send many telephone conversations, TV programs, and other forms of communication over the same radio carrier, thus minimizing costs.

The maser is also very useful in the field of radio astronomy. With this device, astronomers have been able to determine the rotational properties of planets having surfaces that are not optically visible. They have been able to apply their observations to the Doppler shift as it relates to astronomy, an application that has led to a greater comprehension of the nature of the universe. (The Doppler shift and the Doppler effect are discussed in Chapters 16 through 19 about astronomy and in Chapter 7 on sound.)

An apparatus that has partially replaced the maser in its role as a communications device is the cooled parametric amplifier, abbreviated *paramp.* General Telephone and Electronics has successfully maintained ground stations for satellite communications throughout the world. The installation and implementation of these paramp units at strategic locations has established a satellite communication system that provides worldwide telephone service as well as live television broadcasting.

Another useful application of electromagnetic waves in the microwave region (that is, waves having wavelengths in the order of centimeters) is **radar, ra***dio detection and ranging.* The microwaves travel in more nearly straight lines than do the longer radio waves and are readily reflected by objects such as airplanes, mountains, and areas of rain. The location of objects relative to the antenna is determined by the direction and distance from the antenna. The direction of the beam is determined by the direction in which the antenna is pointed. The distance to the object reflecting the waves is determined by sending out a short pulse of waves and timing how long it takes the reflected wave to return to the antenna. Since microwaves travel at the speed of light, the distance to the object, s, is equal to the speed of light, c, times the time, t, divided by 2 (since the waves must make a round-trip). Thus the distance can be computed from the equation $s = ct/2$. The timing is done electronically, with the result usually being displayed on a cathode ray tube (Fig. 11.47).

Example 11.6

A signal sent out from a radar antenna is reflected by a rainstorm, returning to the antenna 35.0 microseconds (35.0×10^{-6} s) after it was sent. How far is the rainstorm from the antenna?

Solution

Since the time is known, substituting the numbers into the equation

$$s = \frac{ct}{2}$$

gives

$$s = \frac{(3.00 \times 10^{8}\ \text{m/s})\ (35.0 \times 10^{-6}\ \text{s})}{2}$$
$$= 10500\ \text{m} = 10.5\ \text{km}.$$

Thus the rainstorm is 10.5 km from the radar site.

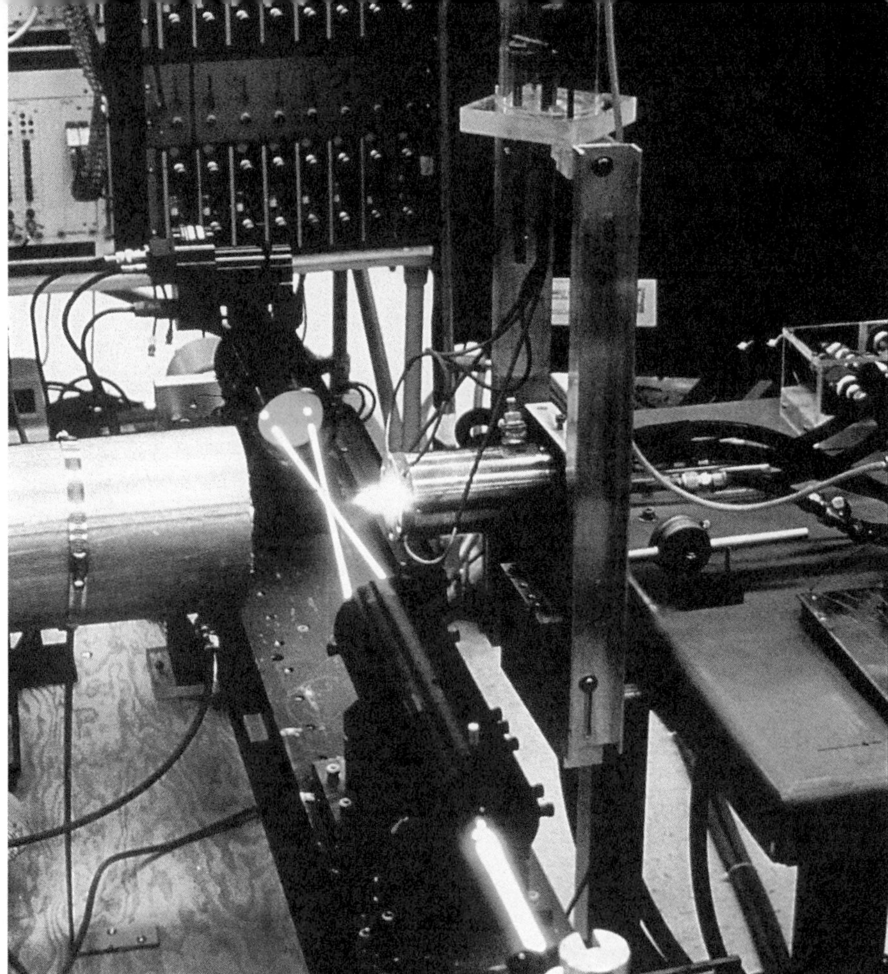

a.

b.

Figure 11.46 (*a*) Laser Raman spectroscopy is used by the National Bureau of Mines to study the chemical reactions occurring inside a plasma arc. In Raman spectroscopy, the light scattered from a monochromatic (single-wavelength) source has characteristics that enable identification of the scattering atoms and molecules. Plasmas may prove useful in extracting metals from raw ore, but the chemical and physical processes occurring in the plasma need to be better understood. (*b*) Laser on a penny. As sufficient electrical power is applied to it, the semiconductor laser is tuned electronically to transmit information through pulses of electromagnetic radiation over glass fibers. Such is the manner in which many phone calls are completed between Boston and Washington.

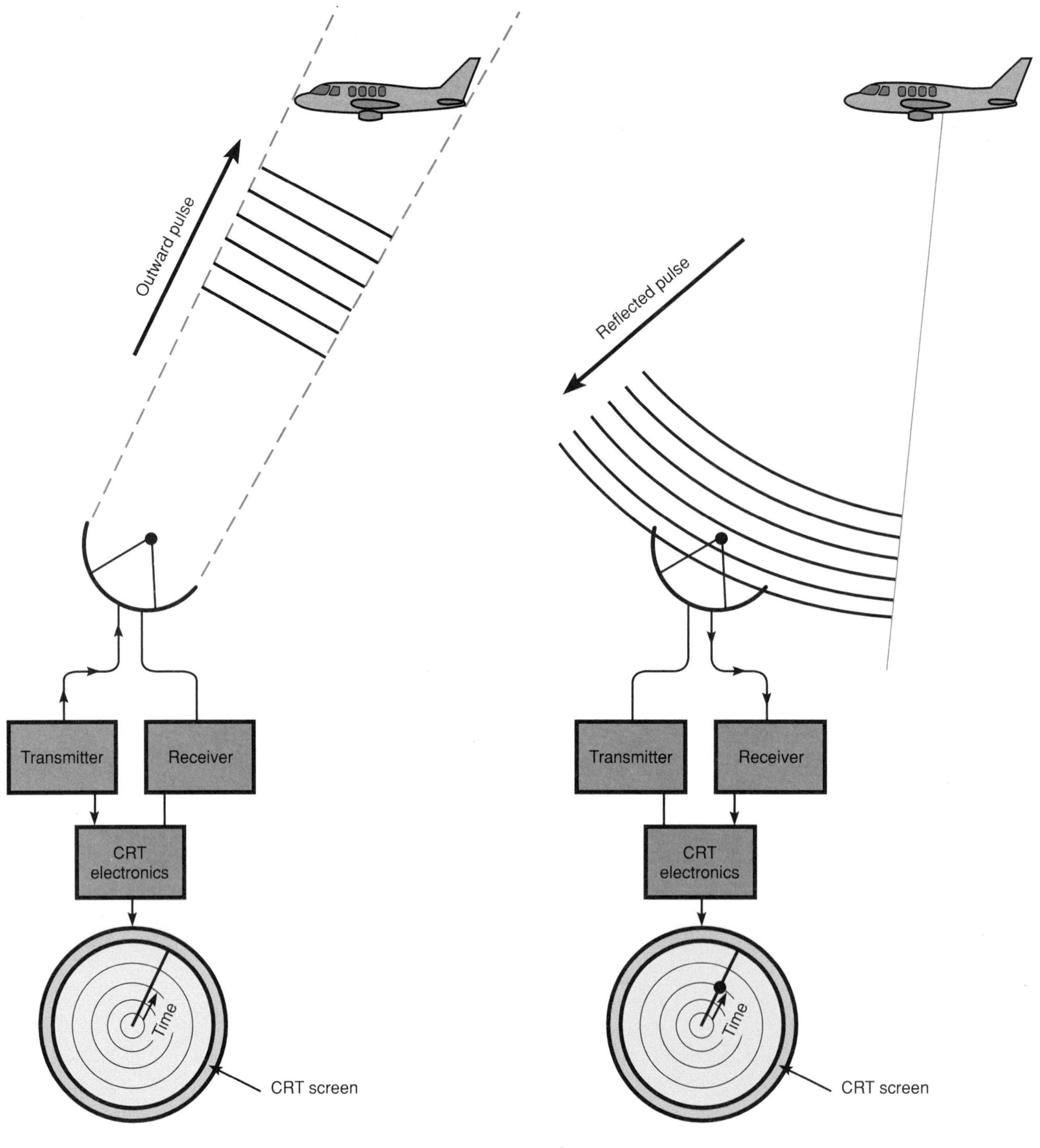

Figure 11.47 (*a*) A short burst of microwaves is emitted from the antenna, and at the same time a beam of electrons is started outward from the center of a cathode ray tube (CRT). (*b*) When the reflected wave is received from an object, the intensity of the electron beam is increased, resulting in a bright spot on the CRT. The distance to the object is determined by how far the spot is from the center of the CRT, and the direction of the object is determined by moving the electron beam outward in a direction corresponding to the direction of the antenna.

The familiar weather radar seen on television weathercasts utilizes a microwave frequency that reflects particularly well from water. When the microwaves sent out by the antenna encounter raindrops, a small portion of the wave is reflected, allowing most of the wave to continue onward. Thus, the radar beam can penetrate the rainstorm, but now the reflected signal will persist as long as the wave remains within the rainstorm. Again, by timing when the echoes start and stop, the extent of the area covered by the rain can be determined. Snow is a much poorer reflector of microwaves than rain, making it harder to detect, especially if the snowfall is light.

Standard radar gives the position of an object, but to use this radar to determine the object's speed, it would be necessary to determine its position at least twice so that the change in position

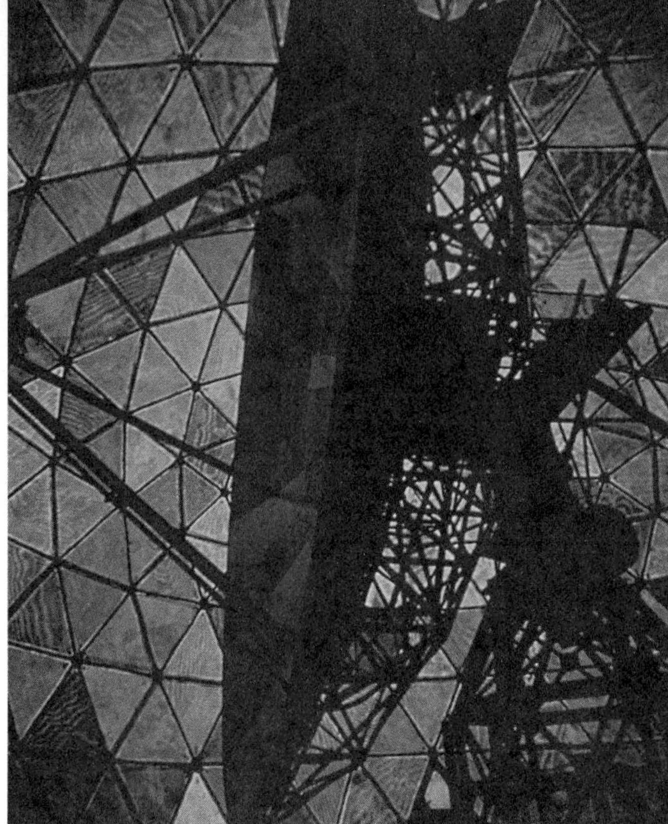

Figure 11.48 A Doppler radar unit at the National Severe Storms Laboratory at Norman, Oklahoma, has greatly improved the means by which meteorologists can monitor dangerous weather patterns.

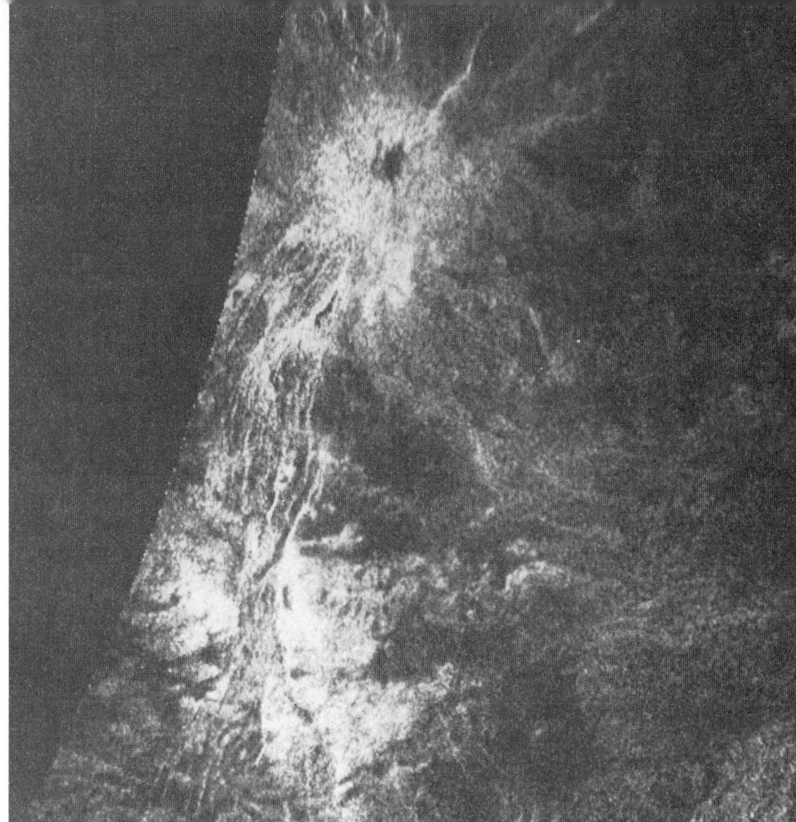

Figure 11.49 A high-resolution radar image of central Beta Region on the planet Venus, obtained with the 12.6–cm wavelength radar system at Arecibo Observatory in Puerto Rico. The oval-shaped highland region shown is presumed to be volcanic; tectonic features extend about 2500 km across the planet's surface.

could be divided by the time. The speed of a moving object can be determined much more quickly with **Doppler radar.** Recall that the Doppler effect is a shift in frequency observed if the source of a wave and the observer are moving relative to each other. A radar wave sent out at a particular frequency will be reflected at a different frequency if the reflecting object is moving. If the object is moving toward the transmitting and receiving antenna, the reflected wave will have a higher frequency, and if the object is moving away from the source, the reflected wave will have a lower frequency. Devices known as frequency counters can precisely determine the frequency shift, which is almost instantaneously converted to speed by a microcomputer built into the radar unit. The speed may then be displayed in different ways, depending on the purpose of the particular radar unit. For example, if the radar unit is being used by the police to determine the speed of vehicles on the highway, the display is usually in digital form, allowing the officer to quickly read the speed of an individual vehicle. If the radar is being used to track a storm front or measure wind speeds in a tornado, the frequency shifts may be translated into colors displayed on a video screen, so that the locations and speeds may be simultaneously displayed. Doppler radar makes the issuance of earlier and more accurate storm warnings than possible with other types of radar facilities. The intensity of an observed storm is constantly monitored by sweeping radar dishes, such as the unit located at the National Severe Storms Laboratory in Norman, Oklahoma (Fig. 11.48). This unit, 9 meters in diameter, sweeps the Oklahoma prairies with narrow pencil-like beams of microwave pulses that have a wavelength of 10 centimeters.

Radar technology continues to advance, and additional applications are being investigated in commercial, military, and scientific fields. For instance, NASA's Pioneer Venus spacecraft and Earth-based radar have recently revealed the surface topography of the planet Venus, the dense atmosphere of which has prevented even the world's most powerful telescopes from disclosing the rugged surface features of this closest planet to Earth (Fig. 11.49).

Microwaves are electromagnetic waves, just as are AM and FM radio, VHF and UHF television, infrared and visible light, ultraviolet rays, X rays, and gamma rays. We hear of radiation and identify it with X rays and gamma rays, both potentially hazardous to living organisms. Color TV sets that lack proper shielding emit low-level X-radiation and could potentially damage biological tissue just as other sources of X rays can. Therefore, government regulations set limits on permissible radiation levels, with which TV manufacturers must comply. A significant portion of the frequency range of electromagnetic waves is believed capable of damaging tissue, and the effect is considered both cumulative and irreversible. However, microwave radiation is not capable of ionizing the molecules of biological tissue because this frequency, like those of infrared and radio waves, lacks sufficient energy and resonating frequencies. An analysis of related research reveals that damage done to biological tissue by microwaves is in the form of thermal effects; however, governmental and independent researchers are continuing to study potential hazards from devices that emit microwaves.

Some time ago a scientist for Raytheon Corporation discovered that his hand became uncomfortably warm almost immediately when exposed to a beam of microwaves emitted from a radar device he was testing. Pursuing this discovery, he is reported to have placed a bag of popping corn into the path of the beam. He discovered that the corn popped in a matter of seconds and that the paper bag did not burn. This discovery led to the development of the microwave oven, a popular commercial and household appliance that receives its energy from the household AC that is converted into high-frequency waves by electronic circuitry. The energy travels back and forth in the food and causes a frenzy of molecular motion that exemplifies intense heat. The heat is, therefore, spread throughout the food, a feature that results in shorter cooking time. For example, hamburgers are cooked in about two minutes, bacon and raw potatoes in four minutes, and various frozen foods in about five minutes. The cooking may be done in glass baking dishes, plastics, or even on paper plates. (Recall that metals reflect microwaves, so conventional pots, dishes with metallic trim, and metal foils cannot be used safely.)

Specifically, the microwave oven makes use of the principle of preferential absorption discussed in the section "Behavior of Electromagnetic Waves" in Chapter 10. The central element of the appliance transmits electromagnetic radiation, the frequency of which corresponds to the frequency that water molecules can readily absorb. The water molecules rotate in a tumbling fashion in resonance with the frequency of the microwaves and, as they tumble, they reemit microwaves that are scattered to other parts of the food, affecting yet other water molecules. The excited molecules of water interact with surrounding molecules, and the resulting friction between molecules generates the heat that cooks the food. The air within the microwave oven does not get hot as does the air inside a conventional oven, since the air is relatively transparent to the microwave region. The container is warmed slightly, mostly by the transfer of heat by conduction from the cooked food itself.

Research has revealed that microwaves at a frequency of 2450 megahertz are in resonance with water and other molecules that have structural similarities with water; therefore, this frequency has been assigned by the Federal Communications Commission for this purpose alone. Since it is contained inside the oven, this frequency does not ordinarily cause interference with television transmission or radio broadcasting. A popular microwave oven is pictured in Figure 11.50 along with a simplified drawing of its components.

Microwave applications are still an important subject of contemporary research and are new enough for us to predict with confidence that other uses will be developed for this range of electromagnetic radiation.

a.

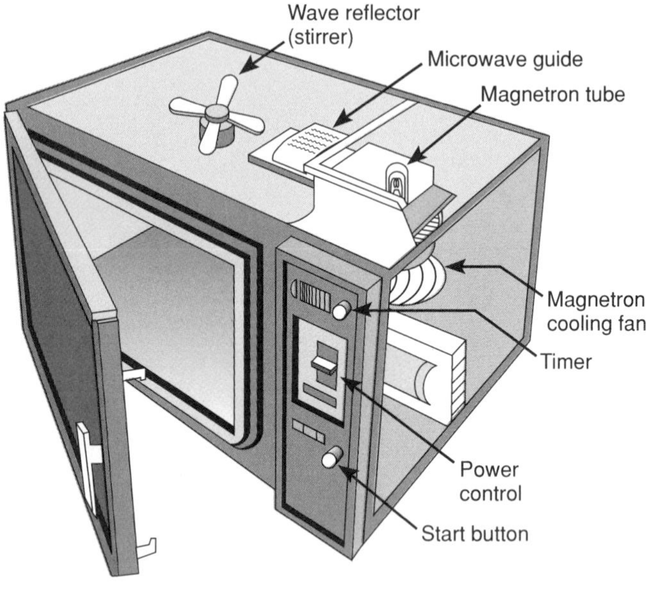

b.

Figure 11.50 (*a*) Microwave ovens are used in many households to cook complete meals or to warm leftovers. The appliance cooks foods in much less time than does the conventional oven. (*b*) Electrons generated from the heated cathode in the magnetron tube move under the combined force of an electric field and a magnetic field in such a way as to provide a continuous source of microwaves. The wave guide directs the microwave radiation toward a slowly rotating stirrer, a component that reflects the microwaves evenly throughout the cooking area.

Summary

Electromagnetic radiation transfers energy from one place to another. The process may involve simply a transfer of energy, or it may be controlled in some manner to transmit information.

The amount of electromagnetic energy radiated from a source depends on many factors, with usually only a small portion being radiated in the visible region of the spectrum. The rate at which light is radiated in the visible region of the spectrum is measured in lumens, and the illumination of a scene is measured in lumens per square meter. If the source of light radiates uniformly in all directions and is small compared to the distance from the source to an object being illuminated, the illumination is inversely proportional to the distance squared.

Mirrors and lenses are used to control light by reflection and refraction, with one of the primary uses being the formation of images. Real images, formed when rays of light converge to some location, may be focused on screens. Virtual images cannot be focused on screens because the rays of light only appear to diverge from the location at which the virtual image is formed. Mirrors and lenses that cause parallel rays to converge are called converging lenses, the point of convergence being the focal point. Diverging lenses and mirrors cause parallel rays to appear to diverge from the virtual focus. The location, size, and orientation of images may be determined graphically or algebraically. Spherical surfaces used in the manufacture of lenses and mirrors are relatively inexpensive, but these surfaces lead to distortions of the images called aberrations. Parabolic mirrors and nonspherical lenses are used in more exacting applications. Converging mirrors and lenses are also used to intensify beams of light.

Total internal reflection is used in fiber-optic light guides to transmit light along a certain path with virtually no loss. Vast fiber-optic networks have replaced traditional wiring in many information systems because a single fiber can transmit much more information than a wire, making them less expensive. Total internal reflection is also used in more traditional optical devices to provide mirrors that reflect without loss of light.

Diffraction limits the theoretical performance of optical instruments, causing the light waves to spread out when they pass through an opening, thus smearing images. This same spreading, combined with interference of waves, is used to advantage in the diffraction grating, a device giving scientists the capability of accurately measuring the wavelength of light and allowing for the study of spectra.

Ordinary light is unpolarized; that is, its vibrations occur in random directions. Light can be polarized by reflection or by transmission through suitable materials. Once light is polarized, a second polarizer may be used to control the beam by virtue of its orientation relative to the direction of polarization. The beam may thus be blocked, or the effects of objects placed in the beam on its polarization may be studied.

Equation Summary

Illumination:

The illumination from a point source E (lumens per square meter) is equal to the ouput Φ (lumens) divided by 4π times the square of the radius r (meters):

$$E = \frac{\Phi}{4\pi r^2}.$$

Mirrors and lenses:

The focal length f (meters) of a spherical mirror is equal to half the radius R (meters) of the sphere:

$$f = \frac{R}{2}.$$

The focal length f is positive for converging (concave) mirrors and negative for diverging (convex) mirrors.

The lens and mirror equation is

$$\frac{1}{d_o} + \frac{1}{d_i} = \frac{1}{f},$$

where d_o is the object distance, d_i is the image distance, and f is the focal length. The distances may all be in centimeters or all be in meters. The value of f is positive for converging lenses or mirrors and negative for diverging lenses or mirrors. A positive image distance indicates a real image and a negative image distance indicates a virtual image.

The magnification m (dimensionless) is given by

$$m = \frac{h_i}{h_o} = -\frac{d_i}{d_o},$$

where h_i is the image height, h_o is the object height, d_i is the image distance, and d_o is the object distance. The distances in each of the fractions must be in consistent units. A negative magnification indicates an inverted image.

Focal ratio:

The focal ratio $f\#$ of a lens (dimensionless) is given by

$$\text{focal ratio} = f\# = \frac{\text{focal length}}{\text{aperture}},$$

where the focal length and aperture (diameter of lens) are in the same units. The intensity of an image is proportional to the reciprocal of the focal ratio $f\#$ squared; that is,

$$\text{intensity of image} = \text{constant}\frac{1}{f\#^2},$$

where the constant depends on the amount of light striking the lens.

Radar distance determination:

The distance s (meters) of an object from a radar site equals the speed of light ($c = 3.00 \times 10^8$ meters per second) times the time t (seconds) divided by 2; that is,

$$s = \frac{ct}{2}.$$

Questions and Problems

Light Intensity

1. (a) Which is more important for visual purposes when purchasing a light bulb, its wattage or its rating in lumens? Why? (b) Do incandescent bulbs (ordinary light bulbs) usually give more or fewer lumens for the same wattage rating than fluorescent bulbs?

2. Suppose that a 100-W light bulb radiates 1750 lm and that a 50-W bulb radiates 700 lm. Would it be better to illuminate a scene with one 100-W bulb or two 50-W bulbs? In your answer discuss total illumination, shadows, and other such considerations.
3. A streetlight consists of a bare 300-W light bulb radiating 6000 lm. If the bulb is suspended from a pole 6.00 m high, what is the level of illumination on the ground directly beneath the bulb? Neglect any scattered light.

Mirrors

4. (a) When you look into a plane (flat) mirror and see an image of your face, what kind of image is it? (b) If your face is 50.0 cm in front of the mirror, where is the image located? How large is the image compared to your face?
5. What is a spherical mirror? Does a concave spherical mirror cause parallel rays of light to converge or diverge? If the radius of curvature of this mirror is 40.0 cm, what is the focal length of the mirror?
6. A student with a makeup mirror in her dormitory room notices that if she holds the mirror 1.20 m from a light in the ceiling a real image of the light is formed on the ceiling right next to the light; that is, the image and object distances for this arrangement are both 1.20 m. Is this mirror convex or concave? What is the focal length of this mirror?
7. Where have you seen convex mirrors used? State two places and explain what the mirrors were used for in each case.

Lenses

8. What type of lens is used in a camera, converging or diverging? What type of image is formed by the camera lens, real or virtual? Would the image formed by a camera lens be upright or inverted?
9. (a) Using a principal-ray diagram, locate the image of a 3-cm-high object placed 15.0 cm in front of a converging lens of focal length 6.00 cm. Describe the image, giving its size, type, and orientation. (b) Repeat for the same lens and object, but with the object placed 4.00 cm from the lens.
10. Using the lens equation, calculate the image distances for the two cases given in Problem 9. If you have completed Problem 9, compare your numerical values with those measured off your principal-ray diagram.
11. (a) What is nearsightedness (myopia) and how is it corrected? (b) What is farsightedness (presbyopia) and how is it corrected?
12. A photographer using a camera to take pictures of a basketball game with a lens of focal ratio $f/2$ finds that for proper exposure of the film the fastest shutter speed he can use is 1/100 s. When he develops the pictures, he notices that the players are moving too fast for the shutter speed he has used, causing the images to be smeared on the film. He decides to purchase a "faster" lens. Should he buy the $f/1.4$ lens or the $f/4$ lens he finds at the camera store?

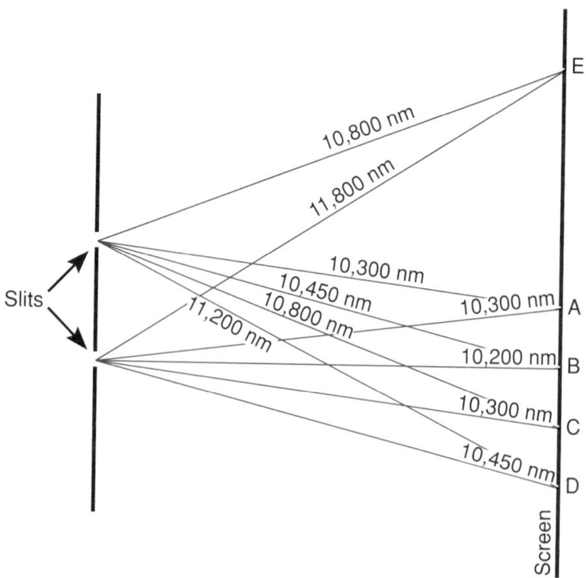

Figure 11.51 Light is incident on the two slits, traveling from left to right. A fringe pattern is formed on the screen.

When he returns to the same gymnasium for the next basketball game, what is the shortest time he can use for the shutter speed with his new lens?

Total Internal Reflection

13. What is the advantage of total internal reflection over reflection from a mirror? Using a diagram, show how a prism may be used to change the direction of light 90 degrees from its original direction by means of total internal reflection.
14. What is "fiber optics"? What is an advantage of using light to transmit information using fiber optics?

Diffraction and Interference

15. The maximum "useful" magnification for a telescope is determined by the diameter of the objective; that is, the main lens or mirror. Why? A general guide to find the maximum "useful" magnification is to multiply the diameter of the objective lens or mirror (given in centimeters) by 24. If a catalog advertising telescopes with lenses 6.00 cm in diameter claims that they have a 500-power capability, would this be a good buy? What would be the maximum "useful" power for this telescope? (Note: The light-gathering ability of telescopes is generally more important than maximum magnification. Extremely large magnifications on small telescopes yield fuzzy images.)
16. Light of 500-nm wavelength is passing through two narrow slits, as shown in Figure 11.51. For each of the points labeled A through E, state whether the light waves will arrive "in phase" or "out of phase." Also state which points on the screen would be illuminated and which would be dark.

17. State several advantages of a diffraction grating over a prism when they are used to analyze a spectrum.

Polarization

18. What is the difference between polarized and unpolarized light?
19. Explain why polaroid sunglasses and polarizing filters used on cameras are effective in cutting reflected glare.

The Laser and the Maser: Stimulated Emission

20. What is coherent light and how does it differ from incoherent light?
21. What is the difference between spontaneous emission and stimulated emission of light from an atom?
22. What is a hologram? Why is it necessary to have a "reference beam" when making a hologram? Why is coherent light used to make holograms?
23. What is Doppler radar? How does it differ from radar used to locate objects?
24. A radar antenna used to track airplanes approaching an airport sends out a pulse of microwaves. An airplane reflects the pulse, and the reflected signal returns to the antenna 133 μs (133×10^{-6}s) later. How far from the antenna (in kilometers) is the plane?
25. (a) Microwave ovens operate at a frequency of 2450 MHz. Why? What is the wavelength of the microwaves used in a microwave oven? Give your answer in millimeters (mm). (b) Electromagnetic waves will not pass through a sheet of conducting material that has openings much smaller than the wavelength of the waves. What importance does this have in the design of microwave oven doors, which have small holes so that one can see inside the oven while it is operating?

PART THREE

The physical attributes of nature have been categorized into both practical and theoretical domains, among which is the general area of chemistry. This branch of the physical sciences deals with the general study of matter, the substances of which the material universe is composed. The science of chemistry separates from this heterogeneous assemblage of matter the pertinent information that identifies individual composition and properties of various substances.

As one attempts to trace the origin and development of chemical knowledge, it must be realized that this science was originally based on very obscure principles. The history of chemistry dates back through the numerous centuries of alchemy (the study of chemistry, astrology, philosophy, mysticism, magic, and religion in the Middle Ages) into the ancient times of primitive superstitions and religions.

Ancient civilizations, particularly of the Middle East, gained useful knowledge of metals and their alloys, developed methods to produce fermented liquors, and discovered ways to make glass, leather, soap, stoneware, and many other products. There is no evidence, however, that they possessed any appreciable knowledge of what is presently considered to be chemistry. Their discoveries undoubtedly were accomplished by trial and error or by accident.

Perhaps the earliest imaginative concept of the physical sciences worthy of being identified as a theory was that of the "four elements" and their qualities, ascribed to Aristotle but also reportedly proposed by Indian and Egyptian scholars about 1500 B.C. The transmutation of one of these elements into another perhaps initiated an idea that carried over into later times when alchemists attempted to change common metals into gold.

With the exception of chance discoveries, alchemy contributed little to scientific advancement until the close of the fifteenth century. Early in the next century, alchemy greatly increased in scope to ally itself with medicine. The earlier goals of the alchemists were slowly replaced with this new field of interest to all.

Chemistry: Order Among the Atoms

Since the sixteenth century, chemistry has made steady and often spectacular advances, both experimentally and theoretically. True research chemists, whose discoveries and writings stimulated this field of the physical sciences, came into being. The scientific procedure of quantitative analysis was introduced and the atomic concept was expanded. The rise of chemistry as a modern science was brought about by the development of a sensitive balance and other precise instruments that permitted a study of chemical change on a quantitative as well as on a qualitative basis.

Scientists into the seventeenth century lacked essential knowledge of gases and a suitable experimental technique of collecting them; thus, misinterpretations about chemical theory were perpetuated. Combustion was considered to be brought about by the presence of a constituent in some substances called phlogiston. Burning was thus equivalent to a loss of phlogiston, and a metal was regarded as a compound of its calyx (oxide) with phlogiston. It was believed that metals lost weight when sufficiently heated, as a piece of wood or a candle did upon burning, since the gaseous products of combustion could not be readily collected or accurately weighed. However, there was existing experimental evidence that tin and several other metals increased in weight when heated in the presence of air and that confined air contracted when a substance was burned in it. The phlogiston theory, then, was not without its flaws and, therefore, its critics. Various chemists set about to challenge its accuracy, and eventually, near the end of the eighteenth century, chemical theory was renovated with phlogistic concepts totally abandoned. Combustion was correctly identified as the chemical combination of the substance with oxygen.

The abolition of old theories and the rapid accumulation of accurate data from experimentation led to the formulation of the atomic theory around 1800. The law of definite proportions markedly influenced the growth of chemistry into the exact science as we perceive it today. Chemistry, like all other natural sciences, is now established on the firm basis of logic and experimentation.

12 Atoms, Molecules, and Chemical Changes

Chapter outline

Key terms/concepts

The explosion of ether and alcohol is an example of a chemical reaction in which chemical energy is transformed into heat and light energy.

Chemistry was first practiced when human beings learned how to perform the relatively high temperature oxidation of cellulose products—that is, when they learned to build a fire. Although they were not aware, of course, of the chemical reactions occurring, they nonetheless created the necessary conditions for the reaction of oxygen with molecules of cellulose and other molecules present in wood. Later in time, people accomplished a similar reaction using black "rocks" instead of just wood. Because these people had seen fires started by lightning, they surely set out determined to learn how to start a fire from wood. Then they probably learned by sheer accident about the black "rocks" that burn. Imagine their astonishment when the black "rocks" on which the fire had been built ignited and burned with a heat more intense than that of the burning wood. Thus, coal was discovered as a source of heat and light. Quite often in science, one discovery leads to another. Such was the case with the discovery of how to make fire, which inevitably led to a number of other discoveries, such as how to use chemical reactions. These reactions were the partial thermal decomposition of various proteins, carbohydrates, fats, and so forth, combined in living or freshly killed organisms—or the cooking of meat.

There is no way to know when heat was first used to dehydrate certain clays to form pottery and bricks, but burned clay artifacts dating from 30,000 to 20,000 B.C. have been found. Later, about 5000 B.C., humans learned that when certain green stones (the mineral malachite—a copper carbonate hydroxide) were heated in a hot fire, the metal copper was produced. By 2500 B.C. bronze had been produced from copper and tin. This discovery ushered in the Bronze Age, with bronze fighting instruments proving their superiority to any stone weapons. By 1600 B.C. humans had learned to smelt iron ore, to burn the excess carbon out of the resulting soft iron to produce a fairly hard grade of iron, and to considerably increase the hardness of iron by plunging it into water after heating it to a dull red heat.

For thousands of years after the beginning of the Iron Age, humans continued to improve on the discoveries of the ancients and expanded the existing chemical technology. From about A.D. 500 to 1700, the development of chemistry was in the hands of the alchemists, whose chief aims were to transmute such base metals as iron, copper, lead, tin, and zinc into gold by a magical instrument called the *philosopher's stone.* Since a green rock could be converted into copper and a reddish earth material changed into iron, why could not lead be transmuted into gold? During the Middle Ages many monarchs appointed court alchemists whose chief job was to find the philosopher's stone. Most of the work of the alchemists involved a trail-and-error technique that involved considerable mysticism and secrecy. Consequently, very few significant advances in chemistry were made during this period.

A great number of charlatans flourished during the alchemical stage of the development of chemistry. One scheme in particular, which occurred in England, gives insight into the ways in which an enterprising alchemist would use his secret knowledge of chemistry to swindle whole villages out of considerable sums of money. The alchemist was aware that when iron was placed into a solution of copper sulfate, the iron would slowly go into solution and the copper ions in solution plate onto the iron to give a copper-plated piece of iron. The alchemist would go into a village and promise the villagers that for a certain sum of money he would show them how to transmute their iron utensils into gold. He would then have them scrupulously clean their iron utensils and place them in a nearby stream that he knew contained copper sulfate. During the next day he would chant various incantations over the submerged utensils. By nightfall it was obvious to the villagers that their utensils were becoming gold-colored, and they assumed that the iron was being changed to gold. About this time, the deceiving alchemist would collect the villagers' money and ride off into the night.

Since early in the seventeenth century, chemisty has made steady and even spectacular advances, both experimentally and theoretically. For instance, near the turn of the century, Jan Baptista van Helmont (1577–1644) introduced the scientific concept of quantitative analysis. Then, Joseph Proust (1754–1826) proposed the law of definite proportions. Next, John Dalton (1766–1844) resurrected and refined the atomic concept attributed to Democritus, the Greek philosopher of the fourth century B.C. According to various authorities, however, modern chemistry began in 1774, when Lavoisier proposed the law of conservation of mass and correctly explained the process of combustion.

The Atomic Concept

Our universe consists of an immense void in which there is scattered a comparatively minute amount of matter. It is, however, the matter in our universe with which we are concerned. It is matter that makes up our galaxies, our planets, and our human bodies. Understandably then, we need to comprehend the secrets of matter, for unless we have an understanding of what matter is and how it responds to various conditions, there is no hope of our ever understanding the planet on which we live, much less our universe.

The inhabitants of our planet have always been confronted with matter in a bewildering array of sizes, shapes, and colors. The ancients reasoned that there must be some fundamental principle holding diverse forms of matter together. If one were to split a rock in half and split each half in half again and so on, would there eventually be an infinitesimal portion of the rock that could not be split and still retain the properties of what is called rock? Are there pieces of matter too small for the eye to see that combine with other pieces of similar matter to form a rock or a drop of water? Careful thinking about just such questions caused the Greek

Table 12.1
The various properties of the typical elementary particles found in the atom.

Particle name	Symbol	Charge	Relative mass	Mass (u)	Actual mass
electron	e^- ($_{-1}^{0}e$)	−1	1	0.000,548	0.000,91 × 10^{-24} g
proton	p^+ ($_{+1}^{1}p$)	+1	1838	1.007,277	1.62 × 10^{-24} g
neutron	n^o ($_{0}^{1}n$)	0	1840	1.008,665	1.675 × 10^{-24} g

philosopher Democritus (460–362 B.C.) to declare that all matter was composed of minute particles, too small to be seen with the naked eye, that he called atoms. Democritus maintained, then, that matter is essentially discontinuous, even though it appears to the eye to be continuous and unbroken. However, neither Plato nor Aristotle accepted the atomic theory of matter that Democritus taught, and it was not until about 2000 years later that the atomic concept of matter was reintroduced into scientific thought. In 1808 the English chemist John Dalton expanded the atomic concept to explain the action of gases as well as chemical reactions. Dalton's atomic theory can essentially be summed up in five statements:

1. Matter consists of indivisible minute particles called atoms.
2. All the atoms of any particular element are exactly alike in shape and mass.
3. The atoms of different elements differ from one another in their masses.
4. Atoms chemically combine in definite whole-number ratios to form chemical compounds.
5. Atoms can neither be created nor destroyed in chemical reactions.

Chemists found Dalton's atomic concept extremely useful in correlating and explaining known facts and in correctly predicting the results of new experiments. Today he is called the father of the modern atomic theory of matter. It should be noted, however, that Dalton did not know about isotopes. In view of the present knowledge of isotopes, statements 2 and 3 above are not strictly correct.

The Anatomy of Atoms

Atoms are unbelievably small in size. For instance, one gram of hydrogen contains approximately 600,000,000,000,000,000,000,000 (six-hundred-thousand-billion-billion or 6 × 10^{23}) atoms. Since Dalton's time, it has been learned that atoms are not indivisible units, but consist of even smaller (elementary) particles. With the exception of the most common atom of the element hydrogen, all atoms contain three subatomic particles. The names, symbols, and certain physical properties of these particles are listed in Table 12.1.

The masses of these subatomic particles are usually stated in terms of atomic mass units (u). One **atomic mass unit** equals 1.66 × 10^{-24} gram. All atoms, except those of the element hydrogen, consist of a very small and very dense nucleus of protons and neutrons surrounded by one or more electrons grouped into shells, or energy levels. The atom with the simplest possible structure is an atom of the element hydrogen, designated by the symbol H. In the preceding diagrams, the proton has been represented by the letter *p,* and the electron and its orbit have been represented by a large black dot on a circle. The positively charged proton is electrically balanced by the negatively charged electron; that is, the hydrogen atom is electrically neutral. The next simplest atom is an atom of the element helium (He), which contains two protons, two neutrons (represented by the letter *n*), and two electrons. The element lithium (Li) contains three protons, three electrons, and four neutrons. Notice that the three electrons of lithium are not placed together, but that two electrons are shown in the first electron shell surrounding the nucleus, and the third electron is shown

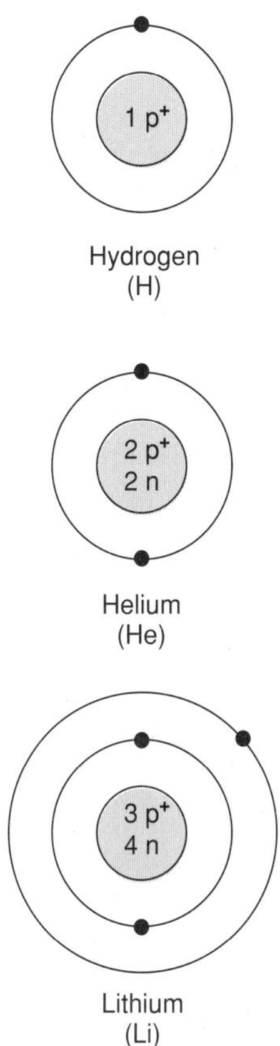

in a second electron shell surrounding the first. No atom contains more than two electrons in the first electron shell—this limit is a fundamental law of nature. As we shall see, the second electron shell can contain no more than eight electrons. In fact, there is a maximum number of electrons that an atom can have in any of its shells. The third electron shell can hold up to eighteen electrons and the fourth shell can contain as many as thirty-two electrons. No element has yet been prepared with a filled fifth electron shell.

Shell	*Maximum number of electrons*
first	2
second	8
third	18
fourth	32

The expression $2n^2$ (where "n" denotes the energy level) gives the maximum number of electrons that can occupy any given energy level.

An element is a substance, the atoms of which all have the same number of protons. An element can neither be decomposed nor transformed by ordinary chemical or physical reactions (see the section "Chemical and Physical Changes" in Chapter 13). The simplified structures of three common elements are shown in the following illustrations.

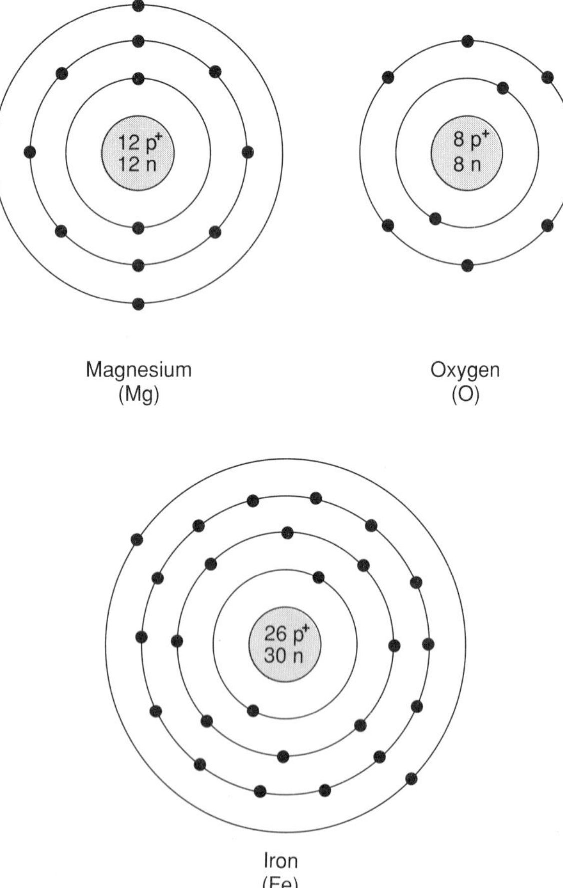

Magnesium (Mg)

Oxygen (O)

Iron (Fe)

Table 12.2

Some common elements and their uses.

Element	Symbol	Uses
Aluminum	Al	Cans, pans, boats, foil, flashbulbs
Chlorine	Cl	Bleach, aerosols, water purification
Copper	Cu	Electric wire, brass, plumbing
Fluorine	F	Anti-tooth-decay medicine, insecticides, special rocket fuel
Gold	Au	Jewelry, coins, anti-infrared coatings
Iodine	I	Iodized salt, antiseptic, iodine lamps
Iron	Fe	Vehicles, steel structures, machines, metal equipment
Magnesium	Mg	Airplanes, flashbulbs, flares
Mercury	Hg	Thermometers, fluorescent lights
Nitrogen	N	Fertilizers, explosives
Oxygen	O	Breathing apparatus, welding, rocket fuel
Phosphorus	P	Matches, fertilizers, soft-drink additives, incendiary devices
Silver	Ag	Silverware, coins, jewelry, photographic films
Silicon	Si	Semiconductors, transistors, integrated circuit chips
Sodium	Na	Salt, baking powders
Sulfur	S	Matches, rubber vulcanization
Tungsten	W	Incandescent light filaments, X-ray tubes
Zinc	Zn	Dry cells (batteries), protective coating for iron, steel, and brass

To indicate the elements, their names are abbreviated in a "shorthand" method. Such abbreviations are called *symbols*. Each symbol consists of one or two letters. The first letter of the symbol for an element is always capitalized and the second letter, if there is one, is lowercased.

Some of the most commonly used elements are listed in Table 12.2. Table 12.3 lists 11 of the various elements, combined with other elements to form compounds, that are most prevalent in the human body. Only 15 of the many elements found in trace quantities are listed.

Many of the names of elements are derived from Latin, but some names have their roots in German, French, Arabic, and English (see Table 12.4). Some elements are named after countries or large geographic areas such as the Americas, France, Germany, and Scandinavia. Other elements are named after states and cities (California and Berkeley, for example). A few elements, particularly the ones more recently found or synthesized, are named after great scientists—Einstein, Fermi, Mendeleev, and Lawrence.

In 1913, H. G. Moseley studied the X rays that are emitted when elements are bombarded with high-energy cathode rays. He found that the pattern of X rays provided a measure of the positive electric charge of an atom's nucleus. Since the positive charge within a nucleus is attributed to the nucleus's protons, Moseley was able to determine the number of protons present in the nuclei of atoms.

Symbols for the Elements

Most symbols for chemical elements are derived from the first letter or the first two letters of the element's name. Thus, H is the symbol for hydrogen, O is the symbol for oxygen, U is the symbol for uranium, and I is the symbol for iodine. If the names of different elements have the same first letter, then two letters are used for the symbols of the elements to prevent confusion. For example, the names of the elements carbon, calcium, and chromium all begin with the letter "C"; therefore, the symbol for carbon is "C," whereas calcium is represented by "Ca" and chromium by "Cr."

There are some symbols that do not appear to relate to the names of the elements they represent. The reason for the apparent inconsistency is that scientists in this country accepted the symbol for an element that scientists in another country proposed, but did not accept the proposed name for the element.

Thus, scientists in our country did not accept the name natrium that German scientists proposed (we call the element "sodium"), but we did accept the symbol "Na" that represents the first two letters of the name natrium. Likewise, we did not accept the name aurum (we call the element "gold"), but we did accept the symbol "Au" ■

Table 12.3
Elements present in the human body.

Element	Symbol	Percentage of mass (approximate)
Oxygen	O	65
Carbon	C	18
Hydrogen	H	10
Nitrogen	N	3
Calcium	Ca	1.5
Phosphorus	P	1
Potassium	K	0.4
Sulfur	S	0.3
Sodium	Na	0.2
Chlorine	Cl	0.1
Magnesium	Mg	0.1
Arsenic	As	Trace
Cobalt	Co	
Copper	Cu	
Chromium	Cr	
Fluorine	F	
Iodine	I	
Iron	Fe	
Manganese	Mn	
Molybdenum	Mo	
Nickel	Ni	
Selenium	Se	
Silicon	Si	
Tin	Sn	
Vanadium	V	
Zinc	Zn	

Note that only 8 elements account for more than 99 percent of the mass of the human body. The same is true for the matter of all living organisms. Although only 8 elements account for over 99 percent of the mass of living organisms, at least 29 other elements, 21 of which are metals, are essential to life. These elements are needed in only trace amounts, but they are very necessary for normal life functions. For example, vanadium must be present in a person's diet only to the extent of 0.1 part per million for normal growth; without vanadium in the diet, a 30 percent growth retardation of a child will result.

Table 12.4
Origins of the names of some elements.

Symbol	Original name	Name derivation	Present name
Ag	Argentum	Latin	Silver
Au	Aurum	Latin	Gold
Cu	Cuprum	Latin	Copper
Fe	Ferrum	Latin	Iron
K	Kalium	Arabic	Potassium
Na	Natrium	Latin	Sodium
Pb	Plumbum	Latin	Lead
Sn	Stannum	Latin	Tin
W	Wolfram	German	Tungsten

The *number of protons an atom has* is called the **atomic number,** Z, of the atom. Thus, the atomic number of the element oxygen (O) is 8, and the atomic number of the element iron (Fe) is 26. Since all atoms are electrically neutral, the atomic number also designates the number of electrons surrounding the nucleus. The atomic numbers of elements are shown as a left-hand subscript of the element's symbol; for example, ${}_{8}O$ and ${}_{26}Fe$.

The *number of neutrons* (electrically neutral particles) *in the nucleus* is referred to as the **neutron number,** N. The *atomic number,* Z, when *added to the neutron number,* N, gives the **mass number,** A, of an atom:

$$Z + N = A.$$

The mass number is shown as a left-hand superscript of the symbol of an element as shown:

$${}^{16}_{8}O \qquad {}^{56}_{26}Fe \qquad {}^{108}_{47}Ag \qquad {}^{235}_{92}U.$$

Isotopes and Atomic Mass

It is the number of protons in the nucleus of an atom that determines its identity. If an atom has only one proton in its nucleus, the atom must be an atom of hydrogen. If there are eight protons present in the nucleus, the atom is an atom of the element oxygen. Even though all the atoms of a certain element must have the same number of protons present in their nuclei, the number of neutrons may vary. *Atoms of a single element, the nuclei of which contain different numbers of neutrons* are called **isotopes.** Even though isotopes of an element contain different numbers of neutrons, they contain the same number of protons and have the same *electronic configuration* (the same number and arrangement of electrons in their electron shells).

Since the mass of an atom is basically due to the mass of the protons and neutrons present in its nucleus (the mass of the electrons is comparatively neglible), one would expect the mass (or weight) of an atom to be a whole number multiple of the masses of its protons and neutrons. Under such conditions the atomic mass of elements, in atomic mass units, u, would be close to a whole number. In fact, there are but few exceptions to this observation. One of the most common examples of those elements that do not follow the general rule is the element chlorine, the atomic mass of which is cited as 35.453, the average of the atomic masses of all chlorine isotopes.

The explanation for the atomic mass of chlorine is that atoms of the element chlorine exist as isotopes of varying abundance. Many elements exist as a collection of numerous isotopes, but chlorine has two isotopes, chlorine-35 and chlorine-37, as shown in the diagram below. The "35" and "37" are referred to as the mass numbers of the isotopes, and is shown as a left-hand superscript of the symbol for the element (^{35}Cl and ^{37}Cl), or simply as Cl-35 and Cl-37.

The isotope of chlorine-35 has an atomic mass of 34.97 u and has 17 protons and 18 neutrons in its nucleus. The chlorine-37 isotope has an atomic mass of 36.97 u, and has 17 protons and 20 neutrons in its nucleus. Today, the relative weight of atoms is based on assigning the carbon-12 isotope a mass of exactly 12 u. Therefore, an atom with a mass twice that of a carbon-12 atom would be assigned a mass of 24 u, and an atom with one-third the mass of carbon-12 would have a mass of 4 u.

Table 12.5

Data on the isotopes of sulfur. Sulfur has an average atomic mass of 32.06.

Isotope	Mass number	Isotopic mass	Percentage of natural abundance
$^{32}_{16}S$	32	31.97	95.00
$^{33}_{16}S$	33	32.97	0.76
$^{34}_{16}S$	34	33.97	4.23
$^{36}_{16}S$	36	35.97	0.01
			100.00

In nature, 75.77 percent of the chlorine atoms have been experimentally found to be chlorine-35 and 24.23 percent are chlorine-37. If the isotopic masses and the percentages of the two masses of these isotopes are "rounded off," then 76 percent will have a mass of 35 and 24 percent will have a mass of 37. The average atomic mass of chlorine can be calculated as follows:

$$
\begin{aligned}
76 \text{ atoms of Cl} \times 35 \text{ u} &= 2660 \text{ u} \\
24 \text{ atoms of Cl} \times 37 \text{ u} &= \underline{888 \text{ u}} \\
100 \text{ atoms of Cl} &= 3548 \text{ u.}
\end{aligned}
$$

It follows, then, that the average atomic mass of naturally occurring chlorine atoms is

$$\frac{3548 \text{ u per 100 chlorine atoms}}{100 \text{ chlorine atoms}} = 35.48 \text{ u per chlorine atom.}$$

This value is quite close to the 35.45 listed for one chlorine atom in the periodic chart. The value of 35.48 u corresponds to a weighted average that reflects the natural abundances and atomic masses of the Cl-35 and Cl-37 isotopes. Chemists normally refer to this mass as the *atomic weight* of chlorine. The atomic weight of any element is the average of the atomic masses of its isotopes as they are found in nature. Chemists realize that the terms "mass" and "weight" have different scientific meanings, but long-standing tradition has established the common usage of the term "atomic weight." Table 12.5 lists data from which the value for the atomic weight of the element sulfur can be determined.

17 p+
18 n

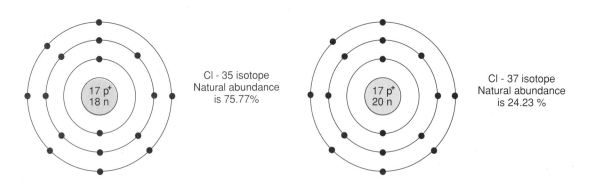

Cl - 35 isotope
Natural abundance
is 75.77%

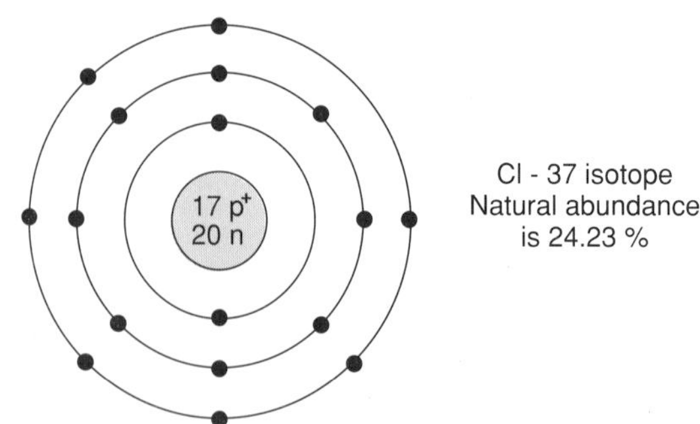

The Reality of Atoms and Molecules

Until 1970, all the evidence for the existence of atoms and molecules (two or more atoms bound together) was strictly indirect. Do not get the impression, however, that scientists doubted the existence of atoms. Many observed phenomena could be explained only if scientists assumed the validity of the atomic theory of matter formulated around 1800. Also, guided by the theory, they were able to correctly predict the physical properties of some elements that had not yet been discovered, as discussed later in the next section. Still, no one had ever seen an atom. Why, then, for more than 150 years were scientists convinced that matter consisted of atoms?

Consider what happens when one places a very small amount of a dye in a glass of water. The color of the dye slowly spreads, and eventually the color permeates the whole glass of water. How can one explain the diffusion of the dye unless one assumes that the water and the dye are both made up of very minute particles that intermingle to form a homogeneous mixture of water particles and dye particles? You may also have observed, in shafts of light, the erratic and zigzag motion of smoke particles or dust particles in a room in which the air is still. This random and irregular motion of smoke and dust particles, called **Brownian motion,** can be explained only if one assumes that the dust or smoke particles are being bombarded by minute particles of air (see Fig. 6.7).

In 1968 two intertwined deoxyribonucleic acid (DNA) molecules were photographed with a special electron microscope that magnified the two molecules more than 7 million times. In 1970 single atoms of thorium and uranium were "seen," as shown in Figure 12.1. Since this time, other types of atoms have been imaged by this and other techniques. We have finally observed what we logically "knew" must exist.

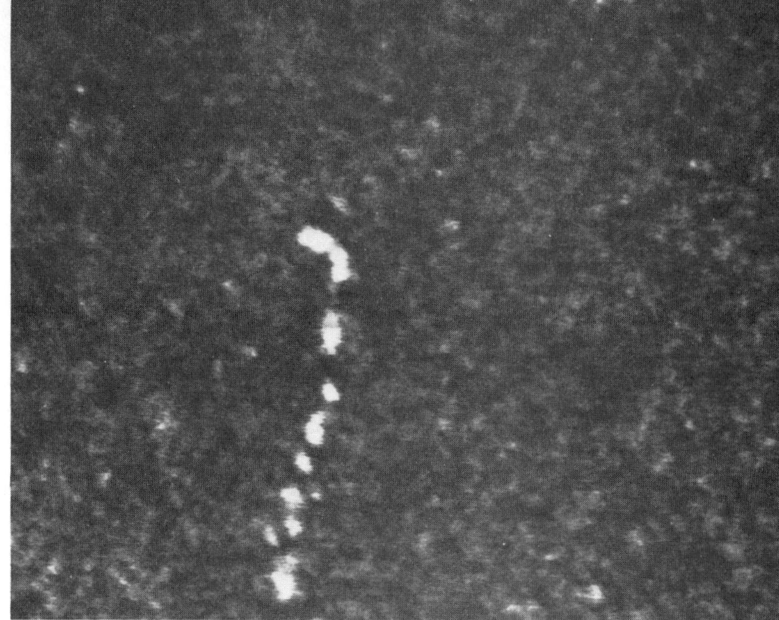

Figure 12.1 The first view of atoms (thorium) ever obtained. This photograph of a chain of thorium atoms (the white dots), magnified approximately 5 million times, was made with a scanning electron microscope by Professor Albert Crewe of the University of Chicago.

A Chart of Periodic Properties

If one lists the elements in order of increasing atomic number, such as

H	He	Li	Be	B	C	N	O	F	Ne	Na	Mg	Al	Si
1	2	3	4	5	6	7	8	9	10	11	12	13	14
P	S	Cl	Ar	K	Ca								
15	16	17	18	19	20	etc.,							

a predictable periodic repetition of physical and chemical properties can be noted. For example, helium ($_{2}He$), neon ($_{10}Ne$), and argon ($_{18}Ar$) are all gases and are all chemically very unreactive. Note that $_{10}Ne$ is the eighth element after $_{2}He$ and that $_{18}Ar$ is the eighth element after $_{10}Ne$. Lithium ($_{3}Li$), sodium ($_{11}Na$), and potassium ($_{19}K$) are all silver-colored solids and extremely reactive elements. Potassium is the eighth element after sodium, which is the eighth element after lithium. From the preceding information, one would expect, then, that if from the list of elements above, any two elements were picked that were separated by seven other elements, the two elements would have similar chemical and physical properties. This is indeed the case, with the exception of hydrogen.

It is convenient to list the elements with similar properties and still maintain the order of increasing atomic number. This can be done in the following manner:

(a)	H							He
(b)	Li	Be	B	C	N	O	F	Ne
(c)	Na	Mg	Al	Si	P	S	Cl	Ar
(d)	K	Ca						

Each element in row (c) has properties similar to the element directly above it in row (b), and to the element directly below it in row (d). The same relationship is valid for elements in other rows.

About 1869, Dmitry Mendeleev (1834–1907), a Russian chemistry professor at the University of St. Petersburg (formerly called Leningrad) noted the periodic repetition of the properties of each seventh element (the noble gases were not yet discovered) and arranged those elements known in his day in the manner indicated above. In 1872 Mendeleev published his Periodic Classification of the Elements, part of which is as follows:

Li	Be	B	C	N	O	F
Na	Mg	Al	Si	P	S	Cl
K	Ca	□	Ti	V	Cr	Mn
Cu	Zn	□	□	As	Se	Br

The three blank spaces were left empty because at that time no elements were known that had properties similar to the properties that Mendeleev felt should be exhibited by elements in the blank positions. It is not sufficient for a scientific theory or concept to merely correlate known data; the real test of the validity of a new concept is to correctly predict the outcome of future tests or discoveries. Convinced that his classification of elements was a valid one, Mendeleev predicted that the three elements missing in his chart would be discovered and that they would fit into his chart in the blank spaces. He even forecast the physical and chemical properties that the unknown elements would have. Within twelve years all three elements had been found, and their properties turned out to be extremely close to those predicted.

The three elements that Mendeleev predicted would be found were scandium (first found in and named for Scandinavia), gallium (discovered in France and named after Gaul—an ancient region including France), and germanium (first discovered in Germany). Some predicted and eventually observed properties of gallium and germanium are shown in Table 12.6.

Properties and Electron Arrangements

Why should every seventh element have similar chemical properties? The explanation lies in the number of electrons in the outermost shell of the atoms of each of the elements. To illustrate, consider the electron arrangement in the atoms of the Li-Na-K column, referred to as group I and IA on the accompanying periodic chart (see Fig. 12.3).

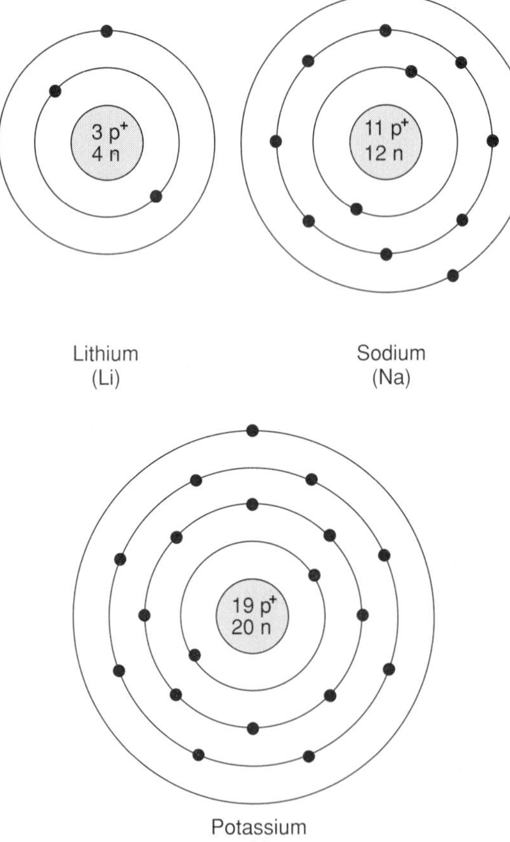

Table 12.6
Some properties of gallium and germanium.

Properties	**Gallium** *Predicted values*	*Observed values*
Atomic weight	68	69.7
Density	4.9	5.9
Melting point	low	29.8°C
Boiling point	high	1983°C
Formula of oxide	Ga_2O_3	Ga_2O_3
Properties	**Germanium** *Predicted values*	*Observed values*
Atomic weight	72	72.6
Density	5.5	5.4
Formula of oxide	Ge_2O_2 (GeO)	GeO and GeO_2

Note that each of these elements has only one electron in its outer electron shell.

On the other hand, fluorine (F) and chlorine (Cl) both have seven electrons in the outer electron shell. Similarly, beryllium (Be), magnesium (Mg), and calcium (Ca) all have two electrons in the outer shell; boron (B) and aluminum (Al) both have three electrons in the outer shell, and so on.

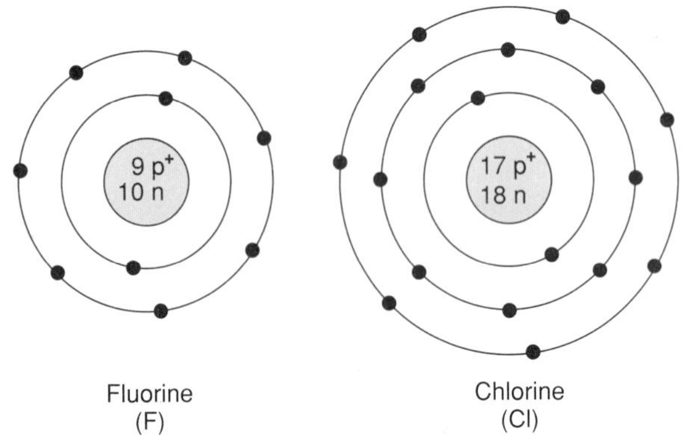

It would be helpful, then, to indicate the number of electrons in the outer shell of all the atoms in a particular column. To do this, Roman numerals are used:

I	II	III	IV	V	VI	VII	VIII
H							He
Li	Be	B	C	N	O	F	Ne
Na	Mg	Al	Si	P	S	Cl	Ar
K	Ca						

It is obvious, then, that atoms of those elements that have the same number of electrons in the outer shell have similar properties. Since sodium and potassium, for instance, each have one electron

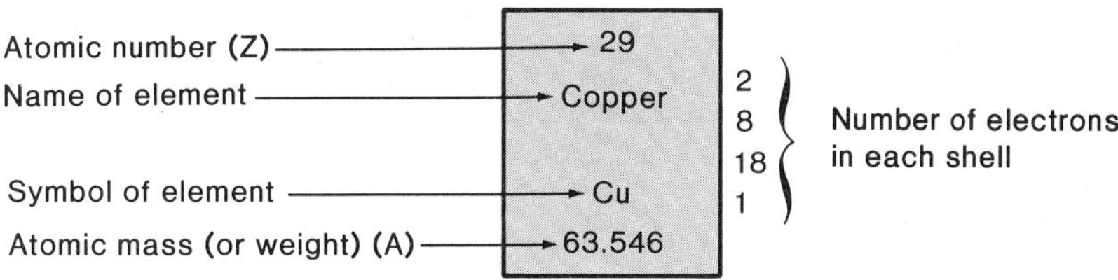

Figure 12.2 The data available in each box of the periodic chart indicate numerous characteristics of the specific element.

in their outer shell, they both have only one electron that can be involved in electron interaction with other atoms. Consequently, they both react in a similar manner.

As pointed out earlier, Ne and Ar are chemically rather unreactive; they have eight electrons in the outer shell. Indeed, all atoms with eight electrons in their outer shell are chemically very stable and unreactive.

The Modern Periodic Chart

Scientists now believe that all the elements that exist naturally on Earth have been discovered. Over 90 elements have been found on Earth, and numerous others have been synthesized in the laboratory.

Although Mendeleev arranged the elements in his chart according to increasing atomic weight, modern charts have the elements arranged according to atomic number, as stated in the **periodic law:** *When the elements are arranged in order of increasing atomic number, there occurs a periodic repetition of physical and chemical properties.* The seven horizontal rows of the periodic chart, arranged in numerical sequence according to increasing atomic number, are called *periods.* Period numbers (one through seven) indicate the highest occupied electron shell (energy level) of the atoms that make up a particular period. Or, said differently, the period number indicates the total number of partially or fully occupied electron shells of atoms of elements in any given period.

Vertical columns, called *groups* or families, contain elements that have the same outer electron configuration. For instance, atoms of the elements in group IIA (Be, Mg, Ca, Sr, Ba, and Rn) each have two electrons in their outermost shell.

Elements in the periodic table can be classified as belonging to basically three subdivisions: *representative elements, transition elements,* and *noble gases.* The representative elements are labeled according to the number of electrons in their outermost electron shells, and are all contained in the groups IA, IIA, IIIB, IVB, VB, VIB, and VIIB. Thus, the representative elements of period four are K, Ca, Ga, Ge, As, Se, and Br, and they have, respectively, one, two, three, four, five, six, and seven electrons in their outermost electron shells.

Transition elements include those elements in which inner electron shells are not completely filled. The inner transition elements, which include the lanthanide and actinide series placed at the bottom of the periodic chart, also have unfilled inner electron shells. There are a few exceptions to the preceding two statements, but they are usually not discussed at this level.

Each period ends with a noble gas. Atoms of all noble gases have eight electrons in their outermost orbit.

If the lanthanide and actinide series were placed in the periodic chart according to the atomic numbers of their elements, the width of the chart would be significantly greater than the length, and the chart could not easily be contained to a single page. A possible solution would be to reduce the size of the chart, but then the information would be difficult to read.

The periodic chart (or table) provides a considerable amount of information (see Fig. 12.2), and the amount of knowledge that can be obtained from it continues to grow as one gains a greater comprehension of its instrinsic design. The chart is presented in Figure 12.3; further information concerning the elements is provided in the following list:

1. All elements to the left of the dark staircase line are metals (including the lanthanide and actinide series), and all elements to the right of the staircase line are nonmetals. Note that a majority of the elements that make up our planet are metals. Some physical properties of metals and nonmetals are given in Table 12.7.
2. The elements boron (B), silicon (Si), germanium (Ge), arsenic (As), selenium (Se), antimony (Sb), and tellurium (Te), which are on the border between metals and nonmetals, are the *semiconducting elements.* These elements physically resemble metals, but behave chemically in the manner of nonmetals.
3. The number above the name of each element represents the number of protons in the nucleus and is called the atomic number, *Z,* of that element. This number also indicates the number of electrons surrounding the nucleus.
4. The number below the symbol for each element represents the average atomic mass for that element and also denotes its atomic weight. Values in parentheses indicate the mass of the isotope of that element having the longest known half-life.
5. The vertical columns are called groups, or chemical families. Each column is designated by a Roman numeral with a letter attached. Some groups have special family names such as *alkali metals* (IA), *alkaline earths* (IIA), *halogens* (VIIB), and *noble gases* (farthest to the right).

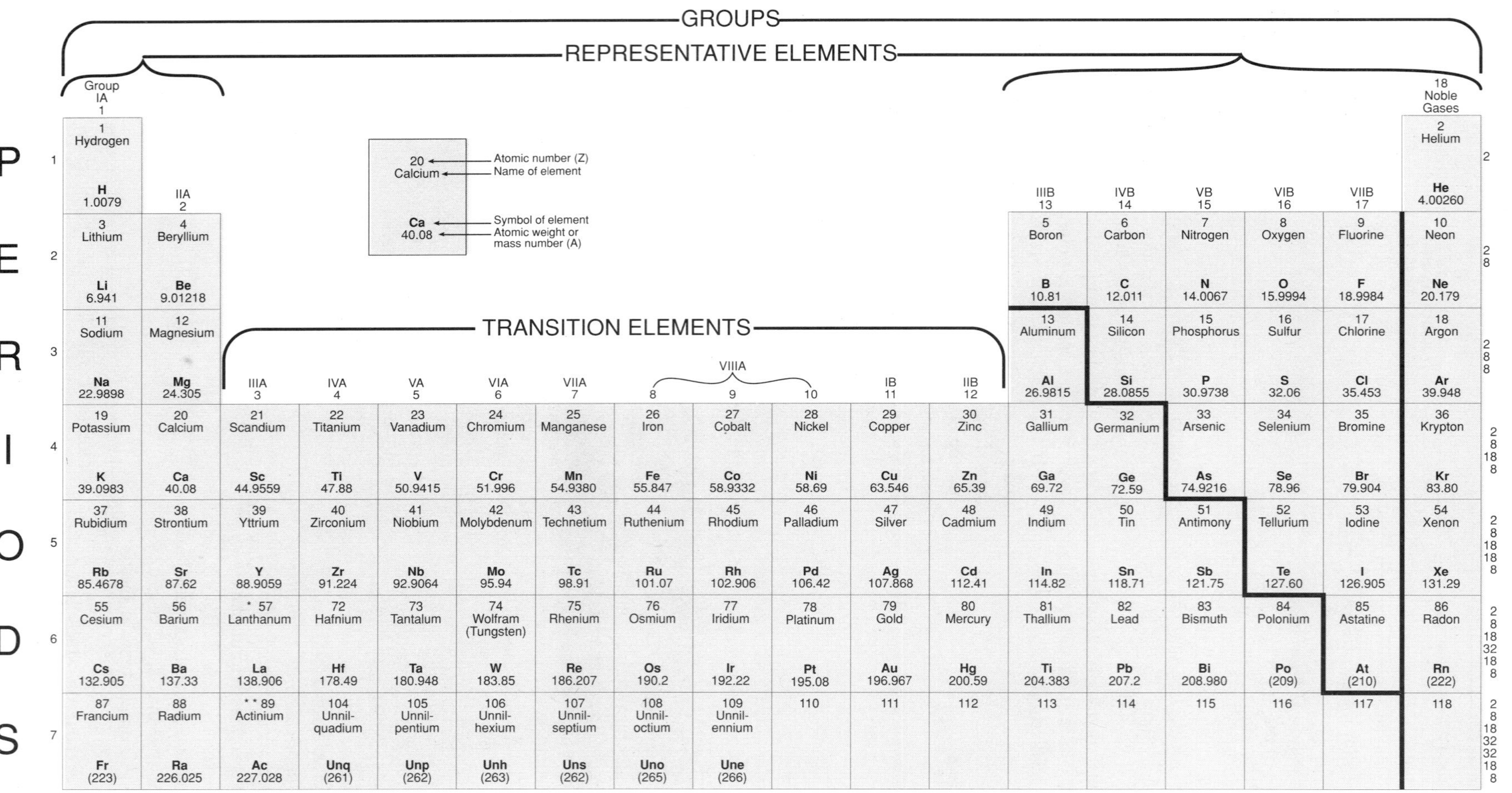

Period	Group IA 1	IIA 2	IIIA 3	IVA 4	VA 5	VIA 6	VIIA 7	VIIIA 8	9	10	IB 11	IIB 12	IIIB 13	IVB 14	VB 15	VIB 16	VIIB 17	18 Noble Gases	
1	1 Hydrogen H 1.0079																	2 Helium He 4.00260	2
2	3 Lithium Li 6.941	4 Beryllium Be 9.01218											5 Boron B 10.81	6 Carbon C 12.011	7 Nitrogen N 14.0067	8 Oxygen O 15.9994	9 Fluorine F 18.9984	10 Neon Ne 20.179	2 8
3	11 Sodium Na 22.9898	12 Magnesium Mg 24.305											13 Aluminum Al 26.9815	14 Silicon Si 28.0855	15 Phosphorus P 30.9738	16 Sulfur S 32.06	17 Chlorine Cl 35.453	18 Argon Ar 39.948	2 8 8
4	19 Potassium K 39.0983	20 Calcium Ca 40.08	21 Scandium Sc 44.9559	22 Titanium Ti 47.88	23 Vanadium V 50.9415	24 Chromium Cr 51.996	25 Manganese Mn 54.9380	26 Iron Fe 55.847	27 Cobalt Co 58.9332	28 Nickel Ni 58.69	29 Copper Cu 63.546	30 Zinc Zn 65.39	31 Gallium Ga 69.72	32 Germanium Ge 72.59	33 Arsenic As 74.9216	34 Selenium Se 78.96	35 Bromine Br 79.904	36 Krypton Kr 83.80	2 8 18 8
5	37 Rubidium Rb 85.4678	38 Strontium Sr 87.62	39 Yttrium Y 88.9059	40 Zirconium Zr 91.224	41 Niobium Nb 92.9064	42 Molybdenum Mo 95.94	43 Technetium Tc 98.91	44 Ruthenium Ru 101.07	45 Rhodium Rh 102.906	46 Palladium Pd 106.42	47 Silver Ag 107.868	48 Cadmium Cd 112.41	49 Indium In 114.82	50 Tin Sn 118.71	51 Antimony Sb 121.75	52 Tellurium Te 127.60	53 Iodine I 126.905	54 Xenon Xe 131.29	2 8 18 18 8
6	55 Cesium Cs 132.905	56 Barium Ba 137.33	* 57 Lanthanum La 138.906	72 Hafnium Hf 178.49	73 Tantalum Ta 180.948	74 Wolfram (Tungsten) W 183.85	75 Rhenium Re 186.207	76 Osmium Os 190.2	77 Iridium Ir 192.22	78 Platinum Pt 195.08	79 Gold Au 196.967	80 Mercury Hg 200.59	81 Thallium Tl 204.383	82 Lead Pb 207.2	83 Bismuth Bi 208.980	84 Polonium Po (209)	85 Astatine At (210)	86 Radon Rn (222)	2 8 18 32 18 8
7	87 Francium Fr (223)	88 Radium Ra 226.025	** 89 Actinium Ac 227.028	104 Unnil-quadium Unq (261)	105 Unnil-pentium Unp (262)	106 Unnil-hexium Unh (263)	107 Unnil-septium Uns (262)	108 Unnil-octium Uno (265)	109 Unnil-ennium Une (266)	110	111	112	113	114	115	116	117	118	2 8 18 32 32 18 8

Lanthanide Series 6 *	58 Cerium Ce 140.12	59 Praseo-dymium Pr 140.908	60 Neodymium Nd 144.24	61 Promethium Pm (145)	62 Samarium Sm 150.36	63 Europium Eu 151.96	64 Gadolinium Gd 157.25	65 Terbium Tb 158.925	66 Dysprosium Dy 162.50	67 Holmium Ho 164.930	68 Erbium Er 167.26	69 Thulium Tm 168.934	70 Ytterbium Yb 173.04	71 Lutetium Lu 174.967	2 8 18 32 9 2
Actinide Series 7 * *	90 Thorium Th 232.038	91 Protactinium Pa 231.03	92 Uranium U 238.029	93 Neptunium Np 237.048	94 Plutonium Pu 239	95 Americium Am (243)	96 Curium Cm (247)	97 Berkelium Bk (247)	98 Californium Cf (252)	99 Einsteinium Es (252)	100 Fermium Fm (257)	101 Mendelevium Md (256)	102 Nobelium No (259)	103 Lawrencium Lr (260)	2 8 18 32 32 9 2

Figure 12.3 The periodic chart or periodic table represents a convenient means by which to determine the relative and exact properties of the elements. Scientists predict that elements 110 to 118 would extend period 7, as illustrated. The *groups* are identified by Roman numerals established by the International Union of Pure and Applied Chemists (IUPAC) and by the new 18–column format developed by the same authority.

The Periodic Chart of the Future

With the aid of the linear accelerator, an apparatus that accelerates charged particles such as electrons, protons, or heavy ions, about a dozen transuranium elements have been produced. Although elements 106 and 109 have been created, supposedly by both Russian and American scientists, universal acceptance of their discovery has been delayed until other laboratories are successful in producing similar results.

The search for heavier elements continues. A periodic chart of the future has been designed by scientists that predicts the positions of the elements as high as $Z = 168$. Starting about $Z = 110$, the elements to be discovered should have isotopes that are less unstable and special properties unlike those of any elements known today ■

Table 12.7
Properties of metals and nonmetals.

Metals	Nonmetals
Most elements	Only 21 elements
All solids at room temperature except for Hg	At room temperature, 10 are gases, 10 are solids, and 1 is a liquid
Good conductors of heat	Poor conductors of heat
Good conductors of electricity	Poor conductors of electricity, except for carbon in the form of graphite
Hard and strong but malleable	Brittle (for solids)
Ductile (can be drawn into wire)	Nonductile
Shiny appearance (reflect light at all wavelengths) except for gold and copper	Not shiny

The Roman numeral above a column indicates the number of outermost electrons in the outer shell of each atom of a particular representative element. Thus, the alkali metals (IA) all have one electron in their outermost shell, the alkaline earths (IIA) all have two electrons, and the halogens (VIIB) all have seven electrons.

6. The last period, period seven, is incomplete. Elements 94 through 109 do not occur naturally but have been synthesized in various scientific laboratories throughout the world.

Other information of value, but which does not appear in the periodic chart, includes the following:

1. The size of the atoms in the chart increases from the right side to the left side and from the top to the bottom.
2. Metallic properties of the elements decrease from the left side to the right side and from the bottom to the top of the chart.
3. The **electronegativity** (discussed in the next section) of the elements increases from the left side to the right side and from the bottom to the top of the chart.

Why Some Atoms React—The Formation of Ions

Some atoms have an affinity for one or more electrons in spite of the fact that the atoms of an element are all electrically neutral; that is, they have the same number of positive protons as negative electrons. This attraction for extra electrons varies with different elements and is highest for elements in the upper right-hand corner of the periodic chart (excluding the noble gases) and lowest for elements in the lower left-hand corner of the chart. The relative ability of a bonded atom or a group of atoms to attract one or more extra electrons to itself is called electronegativity. As a consequence of the electronegativity of fluorine, for example, the following reaction takes place. The fluorine atom, which has the highest electronegativity of all elements, attacts one electron and, as a result, then has ten electrons in its electron shells but still has only the original nine protons in its nucleus. Consequently, the structure then has a charge of -1 associated with it. *An atom with a charge, either positive or negative,* is called an **ion,** and the charge present on the ion is designated as a right-hand superscript (F^{-1}). Positive ions have the same name as the atoms the ions were made from, but the suffix *-ide* is used with the name of the element to indicate a negative ion. Thus, the ion of the element sodium, which has a $+1$ charge (Na^{+1}) is called the sodium ion, but the ion of the element fluorine (F^{-1}) is called the fluoride ion, the ion of the element oxygen (O^{-2}) is called the oxide ion, and the ion of the element sulfur (S^{-2}) is called the sulfide ion.

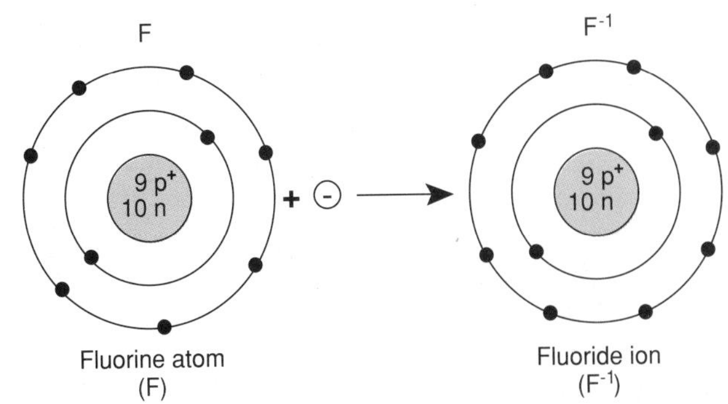

Table 12.8

Group number and ion formation relationship of representative elements.

Group Number	IA	IIA	IIIB	IVB	VB	VIB	VIIB	Noble gases
Number of outermost electrons	1	2	3	4	5	6	7	8
Electrons lost or gained	1 lost	2 lost	3 lost	0 to 4 *	3 gained	2 gained	1 gained	0
Charge on the resulting ion	+1	+2	+3	+2 and +4 *	−3	−2	−1	0

*Carbon forms a few −4 ions but no others. Silicon does not form ions.

In the case of oxygen, two electrons are captured. Note that the fluorine atom and the oxygen atom both form ions with eight electrons in the outermost electron shell. We will find that atoms that capture electrons will capture just the right number to result in an ion with eight electrons in its outer shell.

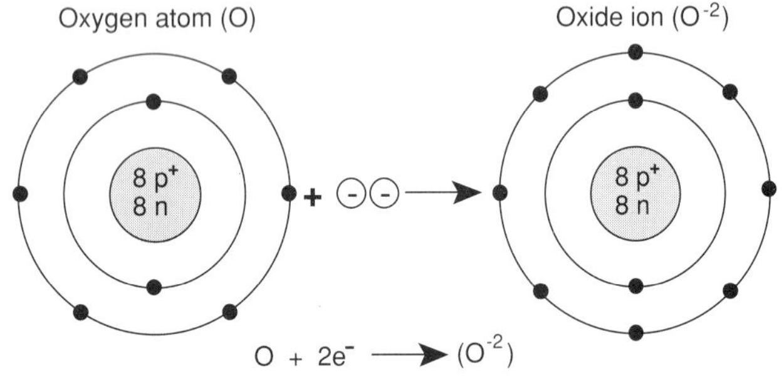

In general, atoms on the right side of the periodic chart (groups VB, VIB, and VIIB) are sufficiently electronegative to attract electrons, whereas atoms on the left side of the chart (groups IA, IIA, and IIIA) have such low electronegativity that they actually lose electrons to the more electronegative atoms on the right side of the chart.

Those elements that readily accept electrons are the ones that need only a few electrons to attain the noble gas electronic configuration of a completed octet. Those elements that readily lose electrons are the ones that by loss of a few electrons expose their completed octet below their outermost electrons. Atoms of group IVB have four electrons in their outermost shell, and they have little tendency to lose or gain electrons. Table 12.8 indicates the type of ions that can be formed from some groups in the periodic table.

The system used to name chemical substances is called *chemical nomenclature.* The name of a compound should indicate the composition of the substance and the number and type of atoms that compose the substance. To determine the correct name for a compound, you must know its formula. For ionic structures, the formula must be written so that *polyatomic ions* (ions containing more than one atom) are identified as such.

Quite often, metals in groups IIIA to IIB form more than one positive ion. Thus, iron, (column VIIA) forms Fe^{+2} and Fe^{+3} and mercury (column IIB) forms Hg^{+1} and Hg^{+2}. These ions are named by using a Roman numeral to indicate the charge on the ion. For example, Fe^{+2} is iron II, and Fe^{+3} is iron III, Hg^{+1} is mercury I, and Hg^{+2} is mercury II.

In naming compounds containing only nonmetals, we will consider only compounds with two nonmetals. Such binary compounds have the first element in the formula named as usual, but only the stem of the second element is used. To the stem is attached the suffix *-ide*; for example, hydrogen chloride for HCl and iodine chloride for ICl. The number of atoms of each element in the compound is indicated by prefixes, such as

mono- (one atom)
di- (two atoms)
tri- (three atoms)
tetra- (four atoms)
penta- (five atoms).

Examples are phosphorus tribromide (PBr_3), sulfur dichloride (SCl_2), nitrogen dioxide (NO_2), dinitrogen pentoxide (N_2O_5), and carbon monoxide (CO).

The Octet Rule

It is significant that the ions produced when sodium combines with chlorine, magnesium with oxygen, and aluminum with fluorine all have eight electrons in their outer shells, once again showing the unusual stability of an outer shell of eight electrons. Sodium, in group IA, loses one electron; magnesium, in group IIA, loses two electrons; and aluminum, in group IIIA, loses three electrons. Note that Na^{+1}, Mg^{+2}, and Al^{+3} all have eight electrons in their outer shell. On the other hand, chlorine and fluorine, in group VIIB, gain one electron each to give them the needed total of eight electrons. Oxygen, in group VIB, gains two electrons to complete its outer shell of eight electrons. The *octet rule* states that an atom that attains eight electrons (an octet) in its outermost shell by transfer or sharing of electrons will acquire unusual stability, such as that of a noble gas atom. Once again, hydrogen is an exception since it attains a complete outermost shell of two when it gains one electron. Two electrons in the outer hydrogen shell give it the configuration of the noble gas helium (He).

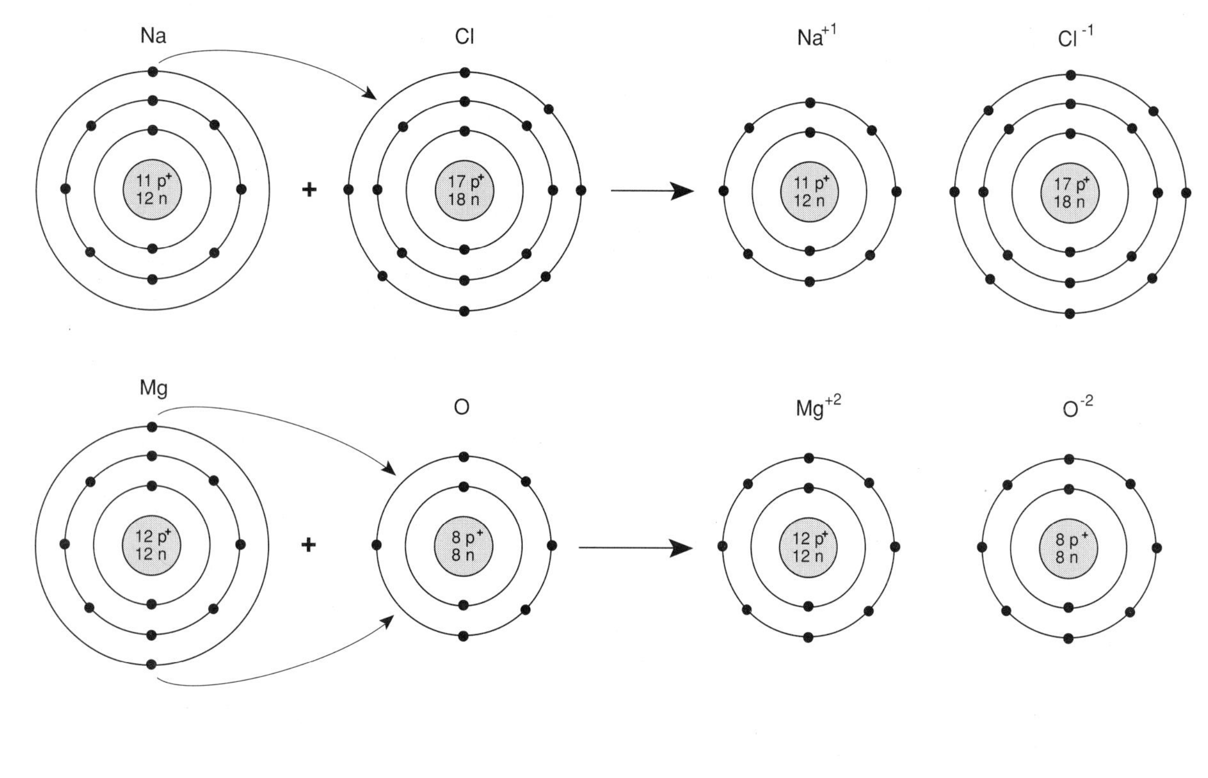

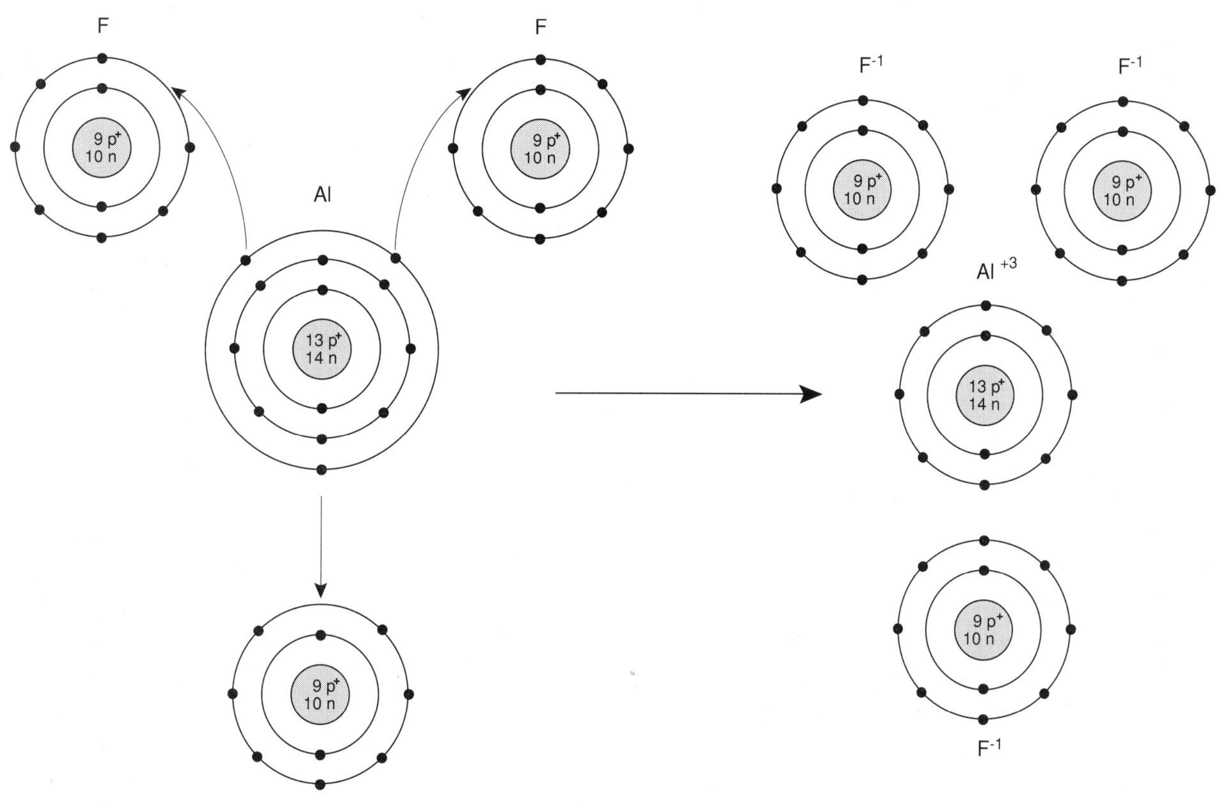

The three reactions may be written as follows:

$Na + Cl \rightarrow NaCl$ (One Na atom reacts with one Cl atom) sodium + chlorine → sodium chloride

$Mg + O \rightarrow MgO$ (One Mg atom reacts with one O atom) magnesium + oxygen → magnesium oxide

$Al + 3F \rightarrow AlF_3$ (One Al atom reacts with three F atoms) aluminum + fluorine → aluminum fluoride

NaCl is the formula for the compound sodium chloride (common table salt) and shows that the compound is made up of

Table 12.9
Some simple ions.

Symbol	Name
Cl^{-1}	chloride ion
Cu^{+1}	copper I ion
H^{+1}	hydrogen ion
I^{-1}	iodide ion
K^{+1}	potassium ion
Na^{+1}	sodium ion
Ca^{+2}	calcium ion
Mg^{+2}	magnesium ion
O^{-2}	oxide ion
S^{-2}	sulfide ion
Sn^{+2}	tin II ion
Al^{+3}	aluminum ion

equal numbers of sodium and chloride ions. The formula for aluminum fluoride is AlF_3 and shows that the compound is composed of three fluoride ions bound to one aluminum ion. Although the three compounds formed consist of ions, each unit of the compounds is electrically neutral:

$Na^{+1}Cl^{-1}$ (one + 1 and one − 1)
$Mg^{+2}O^{-2}$ (one + 2 and one − 2)
$Al^{+3}F_3^{-1}$ (one + 3 and three − 1).

As you will note, a compound consists of two or more different atoms or ions that are chemically combined.

The simplest unit indicated by the formula of an ionic compound is called a **formula unit,** which indicates the smallest sample of an ionic compound that has the characteristics of the compound. A solid ionic compound consists of positive and negative ions arranged in a three-dimensional structure in which no single positive ion can be said to be attached to any one negative ion (see Fig. 13.2). The same lack of identification of one specific ion to another specific ion is also true for an ionic compound in the liquid state.

A formula unit of an ionic compound represents, then, the types and relative number of positive ions and negative ions necessary for an algebraic balance of ionic charges present in the compound. Thus, the formula unit for sodium chloride is NaCl; that for aluminum fluoride is AlF_3.

The sum of all the ions present in a formula unit of an ionic compound is called the unit's formula weight, which is obtained by adding the atomic weights of all the positive and negative ions present in the formula unit.

In each formed compound (NaCl, MgO, and AlF_3), the positive and negative ions are held together by mutual attraction since unlike charges attract each other. Thus, the three − 1 fluoride ions are attracted to the single +3 aluminum ion and vice versa. This *bonding between oppositely charged ions* is known as **ionic bonding.**

Some simple ions are listed in Table 12.9. Some that are very important in physiological functions are listed in Table 12.10. Many ions have more than one atom and are called *polyatomic ions* (see Table 12.11).

Table 12.10
Physiological functions of important ions.

Ion	Function
HCO_3^{-1}	Maintains balance of certain substances in blood.
Cl^{-1}	Needed for gastric HCl and is involved in blood transport of O_2 and CO_2.
HPO_4^{-2}	Needed for construction of bones.
K^{+1}	Maintains pressure in cells; needed for nerve and muscle activity.
Na^{+1}	Maintains water balance and osmotic pressure of blood; needed for nerve and muscle activity.
Ca^{+2}	Needed for construction of bones and teeth and for muscle activity.

Table 12.11
Some common polyatomic ions.

Name	Formula	Name	Formula
Ammonium	NH_4^{+1}	Nitrate	NO_3^{-1}
Bicarbonate	HCO_3^{-1}	Nitrite	NO_2^{-1}
Cyanide	CN^{-1}	Permanganate	MnO_4^{-1}
Hydroxide	OH^{-1}	Phosphate	PO_4^{-3}
		Sulfate	SO_4^{-2}
		Sulfite	SO_3^{-2}

The formulas of compounds containing polyatomic ions are written in a manner to stress the fact that the ions are independent units. Thus, the formula for sodium nitrate, which contains the nitrate ion (NO_3^{-1}), is written as $NaNO_3$ and not as, say, $NNaO_3$. If a compound contains two or more of the same polyatomic ion units, the ion is enclosed in parentheses with a subscript outside the parentheses to indicate the number of ion units in the compound. For instance, calcium bicarbonate has the formula $Ca(HCO_3)_2$ and ammonium sulfide is $(NH_4)_2S$.

The properties of ions differ greatly from the properties of the participating atoms. Sodium is a silver-colored metal that reacts very vigorously with water and chlorine is a greenish-yellow gas that also reacts with water. When sodium and chlorine react with each other, sodium chloride is produced. Sodium chloride is a white solid that does not react with water or hardly anything else. Some of the most common ionic compounds are listed in Table 12.12.

Molecules and Equations

The equations presented at the start of the preceding section, "The Octet Rule," are not written as a chemist would write them, since the gases chlorine, oxygen, and fluorine should be shown as Cl_2, O_2, and F_2, respectively. The three gases used in the equations exist as diatomic molecules, that is, with two atoms to each molecule.

Table 12.12
Examples of ionic compounds.

Ammonium carbonate	$(NH_4)_2CO_3$	smelling salts
Barium sulfate	$BaSO_4$	X-ray examination of internal organs
Calcium carbonate	$CaCO_3$	antacids
Magnesium sulfate	$MgSO_4$	laxatives
Silver chloride	AgCl	photographic film
Sodium bicarbonate	$NaHCO_3$	baking soda, antacids
Sodium chloride	NaCl	table salt
Sodium fluoride	NaF	dental decay preventive

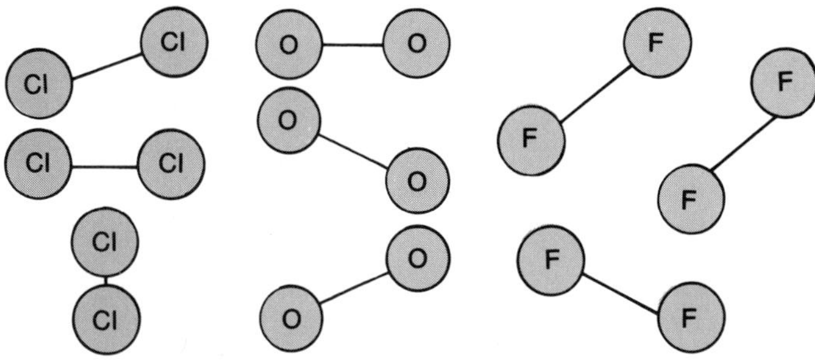

On the other hand, atoms of sodium, magnesium, and aluminum exist as individual atoms.

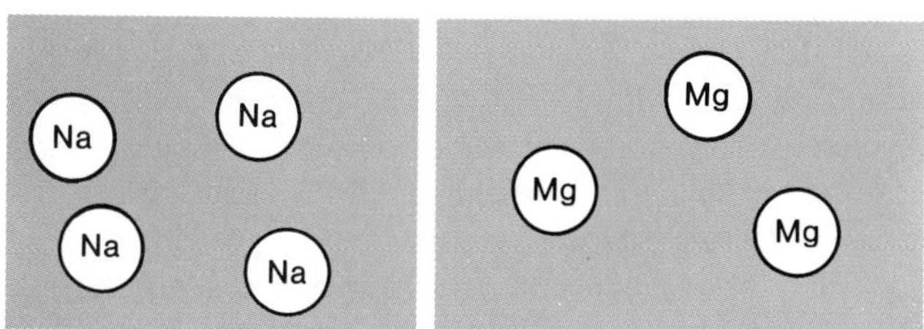

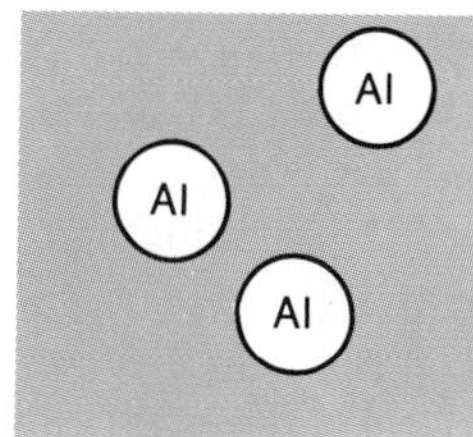

One should be very careful in distinguishing between atoms and molecules. A **molecule** *is composed of two or more atoms.*

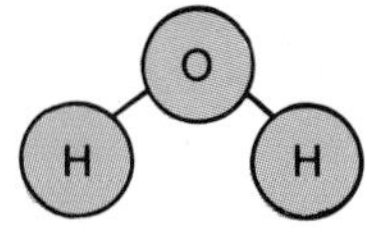

Thus, water consists of molecules containing one oxygen atom and two hydrogen atoms. Atoms join together to form molecules. Molecules are the smallest particles of a compound that can exist, and all the molecules of a compound have the same kind and number of atoms.

In general, atoms are the basic units of elements, and molecules are the basic units of covalent compounds. The most obvious exceptions to this statement are those elements that react with themselves to form diatomic molecules containing atoms of only one element. Examples of these elements are as follows:

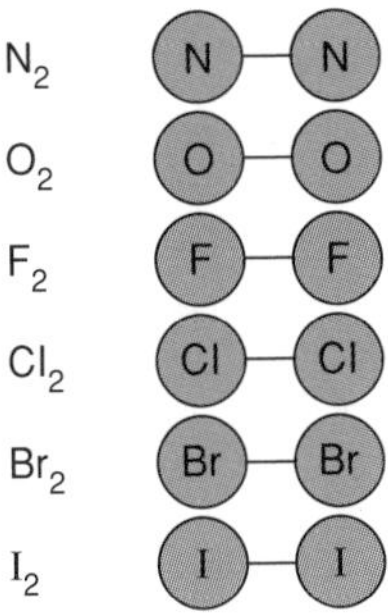

The kind and number of atoms present in a molecule or compound are given by its *formula.* The right-hand subscript that follows the symbol for an atom indicates the number of atoms of that specific element present in one molecule. Thus, O_2 shows that there are two atoms of oxygen in one molecule of oxygen, and H_2O shows that there are two hydrogen atoms and one oxygen atom in one molecule of water. If an atom in a molecule does not have a right-hand subscript, it is assumed to be a single atom and is not written as H_2O_1 but simply as H_2O. If one wishes to indicate more than one molecule of a substance, a left-hand coefficient is used. For example, $3H_2O$ shows that there are three molecules of water and that each molecule is made up of two hydrogen atoms and one oxygen atom.

Now, at last, we are able to write the three equations that we first started several pages ago as a chemist would record them:

$$2Na + Cl_2 \longrightarrow 2\ NaCl$$
$$2Mg + O_2 \longrightarrow 2\ MgO$$
$$2Al + 3F_2 \longrightarrow 2\ AlF_3$$

In these balanced equations, the ratio of Na atoms reacting with Cl atoms is 1:1, the ratio of Mg atoms reacting with O atoms is 1:1, and the ratio of Al atoms reacting with F atoms is 1:3. These are the same ratios employed in the original equations. The equations are called *balanced equations* because the number of atoms of each element is the same on both sides of the equation arrow.

The Language of Chemical Equations

Chemical equations are a type of shorthand notation used to show a chemical reaction. An equation indicates what substances are reacting (the reactants) and what substances are produced (the products). The reactants are on the left side of the equation and the products are on the right side. The reactants and products are separated by an arrow that points from the reactants to the products. The equation,

$$Mg + S \longrightarrow MgS$$

should be "read" as "magnesium plus sulfur yields the compound magnesium sulfide."

As with the periodic chart, a large amount of information can be obtained from chemical equations by someone skilled in reading them. Consider the reaction of carbon with oxygen in the presence of heat; that is, the combustion of carbon:

$$C + O_2 \xrightarrow{\text{heat}} CO_2$$

Note that the number of carbon atoms (one) on the left is the same as the number of carbon atoms (one) on the right and the number of oxygen atoms (two) on the left is the same as the number of oxygen atoms (two) on the right. Since the number of atoms of each element is the same before and after the reaction, the reaction is said to be *balanced.*

For the equation

$$Fe + O_2 \longrightarrow Fe_2O_3$$

the equation is clearly not balanced since there is one atom of Fe on the left but two atoms on the right. Also, there are two atoms of oxygen on the left but three atoms on the right. The equation indicates this relationship between its reactants and its products:

Reactants	*Products*
One atom Fe	Two atoms Fe
Two atoms O	Three atoms O

The equation can be balanced by the proper use of coefficients:

$$4Fe + 3O_2 \longrightarrow 2Fe_2O_3$$

Now the following relationship exists:

Reactants	*Products*
Four atoms Fe	Four atoms Fe
Six atoms O	Six atoms O

First, in a qualitative sense, the equation that shows the reaction of carbon with oxygen tells us that carbon burns in the presence of oxygen. The carbon may be in the form of coal, charcoal, graphite, lampblack or, if one is rich and foolish, diamond. We can also consider the equation from a quantitative viewpoint. Although there are no coefficients shown in the equation, it is understood that the equation should be read as:

$$1C + 1O_2 \xrightarrow{\text{heat}} 1CO_2$$

Now consider the different ways in which this equation can be read:

	$1C$	$+$	$1O_2$	$\longrightarrow$	$1CO_2$
(1)	1 atom of carbon	plus	2 atoms of oxygen	yields	1 atom of carbon and 2 atoms of oxygen combined as carbon dioxide
(2)	1 atom of carbon	plus	1 molecule of oxygen	yields	1 molecule of carbon dioxide
(3)	1 atomic mass (or weight) of carbon	plus	2 atomic masses (or weights) of oxygen	yields	1 atomic mass (or weight) of carbon + 2 atomic masses (or weights) of oxygen
(4)	1 atomic weight of carbon	plus	1 molecular weight of oxygen	yields	1 molecular weight of carbon dioxide
(5)	12.011 grams of carbon	plus	31.998 grams of oxygen	yields	44.009 grams of carbon dioxide

Once the equation is written out and balanced, the ratio of the weight of carbon to the weight of oxygen that will react to form carbon dioxide is determined by using the atomic weights of the atoms present in the carbon dioxide molecule. Thus, one atomic weight of carbon is 12.011u, and two atomic weights of oxygen are 2×15.999u, or 31.998u. The **molecular weight** of carbon dioxide is one atomic weight of carbon plus two atomic weights of oxygen: 12.011u + 31.998u = 44.009u. Thus, 12.011 weights of carbon will react with 31.998 weights of oxygen to form 44.009 weights of carbon dioxide. The molecular weight of any substance is the sum of the atomic weights of all the atoms present in one molecule of the substance.

The atomic weight of an element expressed in grams (instead of atomic mass units) is called the **gram atomic weight** of the element, and *the molecular weight of a compound expressed in grams* is called the **gram molecular weight** of the compound. Thus, the gram atomic weight of hydrogen (H) is 1.008 grams, the gram atomic weight of carbon (C) is 12.001 grams, the gram molecular weight of oxygen (O_2) is 31.998 grams, and the gram molecular weight of carbon dioxide (CO_2) is 44.009 grams. The number of gram atomic weights or gram molecular weights of a substance can be found by dividing the number of grams of the substance by its gram atomic or gram molecular weight. For example, 30 grams of magnesium contains $30/24.31 = 1.23$ gram atomic weights of magnesium and 45 grams of water (H_2O) contains $45/18.01 = 2.50$ gram molecular weights of water. The *formula* weight for an ionic compound expressed in grams is called the **gram formula weight** of the compound.

The concept of gram atomic weight and gram molecular weight is particularly valuable in chemistry since a gram atomic weight of all elements contains the same number of atoms—6.02×10^{23} atoms. Also, a gram molecular weight of all molecules

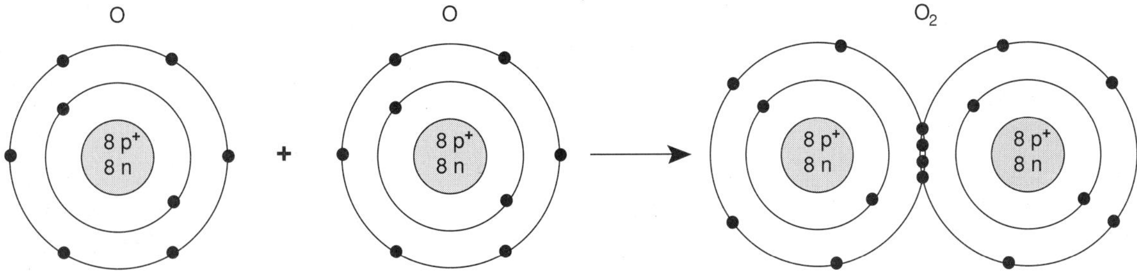

Figure 12.4

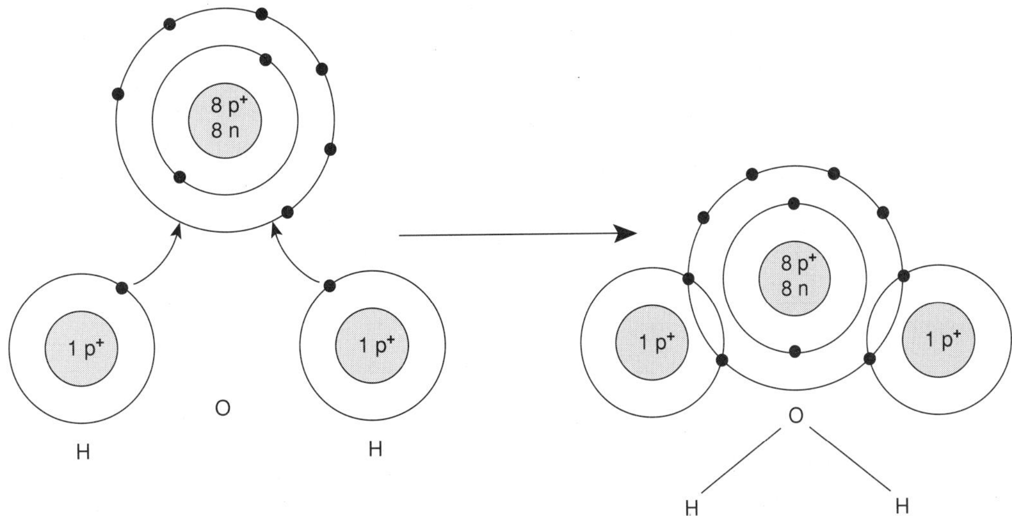

Figure 12.5

contains the same number of molecules—6.02×10^{23} molecules. The number 6.02×10^{23} is called **Avogadro's number,** after the Italian chemist who discovered it.

A special unit, the mole, is defined as Avogadro's number of particles, whether the particles are atoms, ions, or molecules. A mole of carbon (C) has an atomic mass of 12.01u, a gram atomic weight of 12.01 grams, and contains 6×10^{23} carbon atoms. A mole of chlorine (Cl_2) has a molecular mass of 70.91u, a gram molecular weight of 70.91 grams, and contains 6×10^{23} chlorine molecules.

Another Way Atoms React—Covalent Bonds

Not all atoms react by losing or gaining electrons due to differences in electronegativity. Atoms of one element may have almost the same attraction for electrons as atoms of a different element; that is, they have similar electronegativities. In such a case neither atom can capture the other's electrons. They can, however, share electrons mutually.

In the case of the O_2 molecule (Fig. 12.4), the two oxygen atoms are sharing the four electrons shown between them. The four electrons surround the nuclei of both oxygen atoms. This sharing of electrons produces two bonds (a double bond) that hold the two oxygen atoms together. *Bonds formed by sharing pairs of electrons* are known as **covalent bonds.** After the reaction occurs, both oxygen atoms have eight electrons in their outer shells because of the electron sharing. (This observation represents another example of the octet rule.) Covalent bonds exist in the vast majority of chemical compounds. Note, for example, the electron structure of water.

The sharing of a pair of electrons between two atoms; that is, a covalent bond, is represented by a dash. Thus, the molecules in Figure 12.5 can be written as

O=O (a double bond)	and	H—O—H (two single covalent bonds).

By using a shorthand method of presenting the outermost electrons of atoms without having to show all the electrons of an atom, less space is used to show only atoms and bonds:

$$\mathrm{H}\cdot + \cdot\ddot{\underset{..}{\mathrm{Cl}}}: \longrightarrow \mathrm{H}:\ddot{\underset{..}{\mathrm{Cl}}}: \text{ or } \mathrm{H}{-}\mathrm{Cl}$$

In the compound HCl, hydrogen has two electrons in its outermost shell; that is, it now has the noble gas configuration of He. The chlorine atom in the molecule now has eight electrons in its outermost shell and the configuration of the noble gas Ar:

$$:\ddot{\underset{..}{\mathrm{Cl}}}\cdot + \cdot\ddot{\underset{..}{\mathrm{Cl}}}: \longrightarrow :\ddot{\underset{..}{\mathrm{Cl}}}:\ddot{\underset{..}{\mathrm{Cl}}}: \text{ or } \mathrm{Cl}{-}\mathrm{Cl}$$

Linus Pauling

Linus Pauling published a book in 1939 titled *The Nature of the Chemical Bond* in which he described the essential character of ionic and covalent bonding. In 1954, Pauling was awarded the Nobel Prize in chemistry for his research into the nature of the forces that bind atoms together.

Pauling's concern about the harmful effects of radiation from atmospheric testing of nuclear weapons on heredity led him to present a petition to the United Nations requesting a ban on exploding nuclear devices in the atmosphere. The petition bore the signatures of over 900 scientists. As a result of his efforts, a ban on atmospheric testing was signed by more than 100 countries. For his accomplishments, Pauling was awarded the Nobel Peace Prize in 1962. Only Linus Pauling, Madame Marie Curie, and John Bardeen have ever been awarded two Nobel prizes ■

Each chlorine atom in the Cl_2 molecule has the Ar electron configuration.

$$:\dot{\underset{\cdot}{N}}\cdot \; + \; \cdot\dot{\underset{\cdot}{N}}: \longrightarrow \; :N \vdots N: \; \text{or } N \equiv N$$

(a triple bond). Each nitrogen atom in the N_2 molecule has the Ne electron configuration.

The electronegativity of various atoms determines whether atoms react with one another to form ionic bonds or covalent bonds. The greater the difference in electronegativity between atoms, the greater the probability that the atoms will react with one another to form ionic bonds. The smaller the difference in electronegativity, the greater the probability that the atoms will react to form covalent bonds. Compounds containing only nonmetallic elements (including hydrogen) are always covalent structures. The relationship between the group number of some representative elements and the number of covalent bonds they generally form is indicated in Table 12.13.

Molecules that contain covalent bonds can be represented in a shortened form as O_2, H_2O, Cl_2, H_2, N_2, and so on. The same type of notation can be used for ionic compounds, as in NaCl, MgO, and AlF_3. A comparison of some properties of covalent and ionic compounds is presented in Table 12.14.

In a way, the most unusual type of bonding occurs in metals. Since atoms of metals generally have rather low electronegativity, the outer electrons of metal atoms are bound to the nucleus by rather weak forces. Indeed, the outer electrons of the atoms of a metal can move within the metal rather freely, so that it would be incorrect to say that a particular electron belongs to a specific nucleus; the electrons may be considered "delocalized." One way to look at metals is to consider the positive metal ions in a geometric pattern. These ions are completely surrounded by a "cloud" of freely moving electrons. It is this freedom of movement of electrons in a metal that makes metals such excellent conductors of electricity and heat.

Covalent bonding is not very important between metal atoms, but the force that results from the nuclei of metal atoms being submerged in a cloud of electrons is very strong. *This force that binds the metal atoms together* is known as the **metallic bond.**

Table 12.13

Relationship of group number and number of covalent bonds formed.

Group	Number of covalent bonds
IIIB	3
IVB	4
VB	3
VIB	2
VIIB	1
Noble gases	0 (varies)

Table 12.14

Properties of compounds with covalent and ionic bonds.

Property	Covalent compounds	Ionic compounds
Melting point	rather low (often liquids or gases)	high (always solids at room temp.)
Boiling point	low	very high
Electrical conduction	very poor (even when melted)	poor (good when melted)
Water solubility	most are not soluble	many are soluble

The Noble Gases

The vertical column at the right side of the periodic chart contains the elements known as the noble gases: helium (He), neon (Ne), argon (Ar), krypton (Kr), xenon (Xe), and radon (Rn). The noble gases used to be called *inert gases* because they were considered completely unreactive. The name *argon,* from the Greek *argos* meaning "inert" or "idle," clearly indicates its lack of chemical reactivity. Since 1962, however, all the elements except helium have been reacted with certain powerful reagents to form reasonably stable compounds.

The long time that it took—almost 150 years after the beginnings of modern chemistry—for all the noble gases to be discovered is suggested by the Greek names of three of the elements. *Neon* means "new," *krypton* means "hidden," and *xenon* means "stranger." Helium, named after Helios, the Greek sun god, was first discovered in a spectrum of sunlight, and only later found to be present on Earth.

Radon, one of the products of the decay of radium and uranium nuclei, is itself radioactive. Rock strata containing uranium slowly give off radon gas, which eventually works its way to the surface and can collect under houses and in basements. The danger for people living in such houses is of much concern today.

The chemical inertness of the noble gases is used to advantage in certain industrial applications. Argon is used as a shielding gas in welding. Regular incandescent light bulbs and photoflash bulbs are filled with a mixture of approximately 90 percent argon and 10 percent nitrogen. Noble gases are also used to provide the inert atmospheres needed in the preparation of crystals of semiconductors. Xenon is an excellent general anesthetic.

Some noble gases are used in making "neon" signs. When an electric current is passed through a tube of a noble gas under low pressure, a bright glow results. Different gases produce different colors. Neon produces a red glow; helium, an off-white color; and argon, a blue color. A mixture of helium and argon produces an orange light, whereas a neon and argon mixture yields a lavender color. A combination of various noble gases, sometimes with a small amount of mercury added, can produce almost any desired color.

Summary

When the elements are arranged in order of their atomic numbers, they exhibit similar chemical and physical properties at regular intervals. This similarity of chemical and physical properties is the result of the similarity of the arrangement of the electrons of the atoms of these elements. Thus, the atoms of elements with only two electrons in the outermost shell all have similar properties.

The symbols of most elements are derived from the first letter or the first two letters of the element's name.

The atomic number, Z, of an atom plus its neutron number, N, gives the mass number, A, of the atom. The atomic weight of an element is the average of the atomic masses of its isotopes as they are found in nature.

In the periodic table the vertical columns are called groups or chemical families and the horizontal rows are called periods.

Atoms with fewer than four electrons in their outermost shell tend to lose these electrons to become positively charged ions. Atoms with more than four electrons in the outermost shell tend to gain enough electrons to give a total of eight electrons in their outermost shell and become negatively charged ions. Atoms with eight electrons in their outermost shell, such as those of the noble gases, are relatively unreactive chemically because of the stability of this electron arrangement. The atoms of some elements neither lose nor gain electrons but share their electrons with other atoms to form covalent chemical bonds.

An ionic bond results from the electrostatic attraction between positively and negatively charged ions. The ions are the result of the transfer of one or more electrons from one atom to another. A covalent bond is produced when a pair of electrons is shared by two atoms.

An equation is "balanced" when the number and kind of atoms on the right side of the equation are the same as on the left side of the equation.

The gram atomic weight of an element is the atomic weight of the element expressed in grams and the gram molecular weight of a compound is the molecular weight of the compound expressed in grams. A gram atomic weight of an element and a gram molecular weight of a compound both contain 6×10^{23} atoms (or molecules).

Questions and Problems

The Atomic Concept

1. What line of reasoning probably caused the Greek philosopher Democritus to assume that matter was discontinuous, or atomic, in nature?
2. List at least four concepts of matter that John Dalton included in his atomic theory.

The Anatomy of Atoms

3. How does the mass of an electron compare with the mass of a proton?
4. What is the maximum number of electrons that can occupy the second and fourth shells (energy levels) of any atom?
5. Define the atomic number and the neutron number of a nucleus.
6. What are the symbols for the elements oxygen, aluminum, copper, and calcium?

Isotopes and Atomic Mass

7. What particle determines the chemical identity of an atom?
8. Define an isotope in your own words. What major characteristics do all isotopes of a given element have in common?
9. From the data in Table 12.5, calculate the value of the atomic weight of naturally occurring sulfur. How does your value correspond to the value listed on the periodic chart included in the chapter?

The Reality of Atoms and Molecules

10. How does Brownian motion support the atomic concept?

A Chart of Periodic Properties

11. What determines the length of the rows in the periodic chart?

12. Why was Dmitry Mendeleev able to predict some of the physical and chemical properties of some elements that were yet to be discovered?

Properties and Electron Arrangements

13. Why do all the elements of group IA in the periodic chart have similar chemical properties?
14. How many electrons are in the outer shell of atoms of oxygen and sulfur?

The Modern Periodic Chart

15. Which two elements should most resemble silver (element 47) in chemical and physical properties?
16. Which representative elements have the same number of electrons in their outer shells as atoms of calcium have?
17. What is a periodic chart "group"? How does one group differ from another?
    How many electrons surround the nucleus of the following?
    (a) Sr
    (b) Fe^{+2}
    (c) Au
    (d) S^{-2}
18. How do the following properties of the elements vary in the periodic chart?
    (a) size of atoms
    (b) metallic properties
    (c) electronegativity

Why Some Atoms React—The Formation of Ions

19. Show the reaction between potassium (element 19) and bromine (element 35). Indicate any loss and gain of electrons by the atoms.
20. How do ions differ from atoms?
21. Write the equations for the reaction of the following elements:
    (a) potassium and sulfur
    (b) aluminum and iodine
    (c) magnesium and fluorine
22. Name the following ions:
    (a) O^{-2}
    (b) Mg^{+2}
    (c) HCO_3^{-1}
    (d) SO_4^{-2}
    (e) OH^{-1}
23. What is the formula of the compound produced when boron (B) reacts with fluorine (F)?
24. What charge would you expect to be associated with the ions of the elements Ba, S, Ga, and Cs?

The Octet Rule

25. Except for the noble gas elements, how does an atom obtain eight electrons in its outer shell?

Molecules and Equations

26. What is a "balanced" chemical equation?
27. Balance the following reactions:
    (a) $Li + S \longrightarrow Li_2S$
    (b) $H_2 + Cl_2 \longrightarrow HCl$
    (c) $H_2SO_4 + NaOH \longrightarrow Na_2SO_4 + H_2O$
    (d) $Na + H_2O \longrightarrow NaOH + H_2$
    (e) $KClO_3 \xrightarrow{\text{heat}} KCl + O_2$
    (f) $CH_4 + O_2 \longrightarrow CO_2 + H_2O$
28. Complete and balance the following reactions:
    (a) $Al + S \longrightarrow$
    (b) $Na + O_2 \longrightarrow$
    (c) $Ca + Cl_2 \longrightarrow$
    (d) $Li + Br_2 \longrightarrow$
    (e) $Mg + Br_2 \longrightarrow$
29. Name the following compounds:
    (a) P_2O_5
    (b) CO
    (c) Na_2SO_4
    (d) $FeCl_3$
    (e) $SnCl_2$
    (f) NH_4NO_3

The Language of Chemical Equations

30. A molecule contains two atoms of chlorine (Cl) and one atom of sulfur (S). What is the gram molecular weight of the compound?
31. What is the molecular weight of $C_2H_4O_2$?
32. Define Avogadro's number. What role does it play in chemical reactions?

Another Way Atoms React—Covalent Bonds

33. How does a covalent bond differ from an ionic bond?
34. Explain the type of bonding present in a molecule such as Cl_2 or O_2.

The Noble Gases

35. Why were the noble gases at one time called the inert gases?
36. What elements other than neon are used to make "neon" signs?

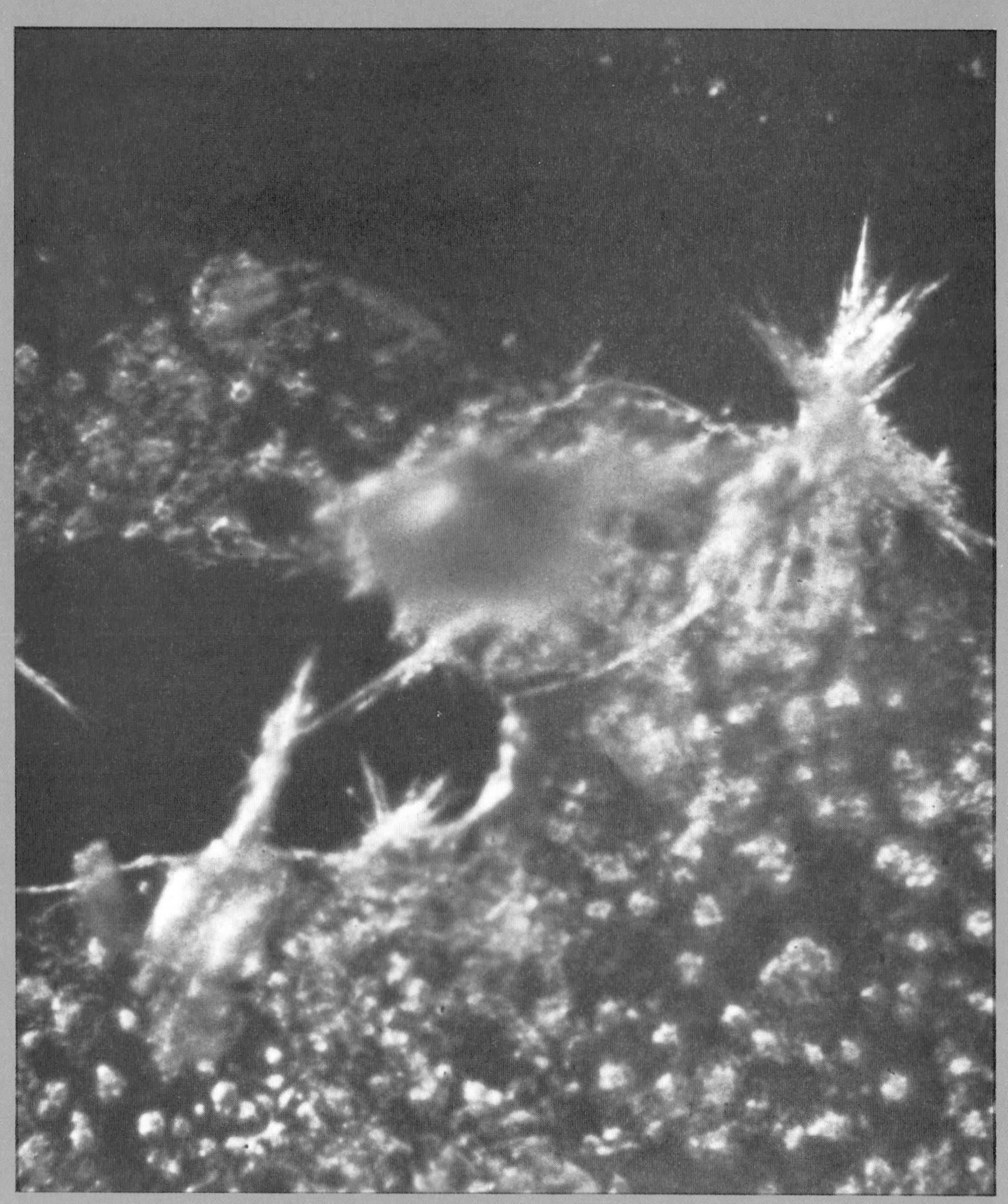

13 Chemical Reactions and Equilibrium

Chapter outline

Key terms/concepts

The precipitation of a solid from a water solution.

The science of chemistry has steadily developed since prehistoric peoples first burned wood, cooked meat, and made pottery. Although chemical technology grew relatively slowly for many centuries, the development of bronze, iron, glass, and other basic substances, along with increasingly sophisticated techniques for experimentation, has considerably expanded our understanding of chemistry. This progress was impeded somewhat during the Middle Ages, when alchemy flourished, but chemistry has made continual advances since the seventeenth century. Today chemistry, in the manner of all sciences, builds on a firm foundation of tested rational principles.

Chemical and Physical Changes

What is the difference between the melting of wax and the burning of wax and between the dissolving of sugar in water and the heating of sugar until it chars? The answers lie in the difference between physical changes and chemical changes. *Physical changes* do not alter the chemical composition of a substance, but they do alter one or more of its physical properties, such as color, structure, density, transparency, and hardness. Physical changes would not, however, alter the temperature at which a substance melts or boils. Examples of physical changes are the melting of wax, the dissolving of sugar, and the freezing of water.

All samples of a given substance will show identical physical properties under similar conditions, and as a result, physical properties can be used to identify unknown substances. If one had samples of the silver-colored metals cadmium, silver, and platinum, and wished to determine the identity of each, the physical property of density would be sufficient to distinguish one from another.

Substance	*Density (g/cm³)*
cadmium	8.6
silver	10.5
platinum	21.5

Other examples of the use of physical properties to distinguish one substance from another are as listed.

Substances	*Physical property used to distinguish*
sugar and salt	taste
copper and cadmium	color
oxygen and hydrogen sulfide	odor

Chemical changes result in the formation of one or more new substances with physical properties that differ from those of the original substances. Thus, when iron reacts with oxygen to produce rust (iron oxide), the rust formed is red-brown in color, has no structural strength, and has a density of 5.2 grams per cubic centimeter, whereas the original iron is gray in color, has considerable structural strength, and has a density of 7.9 grams per cubic centimeter. Also, iron is attracted to a magnet whereas rust is not.

In order to better understand chemical changes, consider the reaction illustrated below. The new substance, iron II sulfide (FeS), formed from the chemical reaction of iron with sulfur, has physical properties that are different from the physical properties of the reactants.

A comparison of the chemical and physical properties of three substances—sulfur, sugar, and iron—is shown in Table 13.1.

Table 13.1

Different substances have different properties.

Substance	Chemical properties	Physical properties
Sulfur	Burns in air. Reacts with metals when heated. No reaction with acid.	Yellow in color. Soluble in CS_2. Insoluble in water.
Sugar	Chars on heating. Reacts with acid.	White solid. Soluble in water. Tastes sweet.
Iron	Reacts with acid. Reacts with moist oxygen to form iron oxide.	Gray-colored solid. High strength. High melting point.

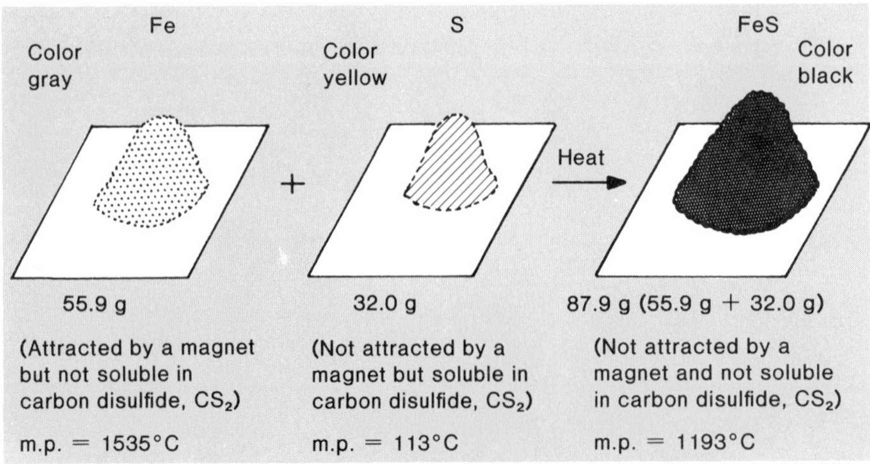

Mixtures and Solutions

There are very few naturally occurring substances that are pure elements or compounds. Nature has produced small quantities of pure copper, silver, gold, platinum, and carbon (in the form of diamonds), but most substances found in nature are mixtures. A **mixture,** which may be homogeneous or heterogeneous, *contains two or more compounds or elements.* A heterogeneous mixture, such as salt-pepper, is a mixture of individual salt grains and individual pepper grains. A homogeneous mixture, such as salt water, is a mixture at the *molecular* level—salt molecules intimately mixed with water molecules. Some examples of heterogeneous mixtures are salt and pepper, sugar and flour, concrete and dirt. The components of a heterogeneous mixture can exist in any proportions; for example, salt and sugar can be mixed in any ratio. However, in many cases the components of a homogeneous mixture can vary only within certain limits; thus, only so much salt can be dissolved in water. *Homogeneous mixtures* are given a special name: **solutions.** Some examples of solutions, or homogeneous mixtures, are as follows:

Example	*Type of solution*	*Components of solution*
sugar-water	solid in liquid	sugar, water
air	gas in gas	nitrogen, oxygen
carbonated drink	gas in liquid	carbon dioxide, water
alcohol-water	liquid in liquid	alcohol, water
brass	solid in solid	copper, zinc

The various types of matter may be indicated diagrammatically:

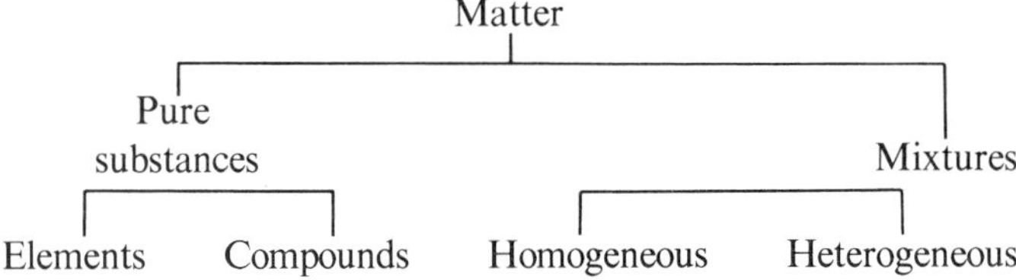

We are very familiar with solid-liquid solutions: salt in water, sugar in water, iodine in alcohol (a "tincture"), and citric acid in water (present in fruit juices). However, we are much less familiar with solid-solid solutions, some of which are listed in Table 13.2.

How does a mixture differ from a compound? The various elements in a compound are chemically bonded to each other in definite proportions by weight. This is the *law of definite proportions.* Thus, in the compound H_2O all molecules of water contain one oxygen and two hydrogen atoms; stated another way, 2.016 (or 2×1.008) grams of hydrogen always combine with 15.999 grams of oxygen. On the other hand, a mixture can consist of elements or compounds in variable weight ratios. For instance, one can mix salt and sugar in any proportions.

Solutions are actually mixtures at the molecular level; that is, the components of the mixture are molecules rather than large or small particles. When salt is dissolved in water, the salt is called the *solute* and the water is called the *solvent.* In general, the solvent is understood to be the medium in which the solute is dissolved, and it usually is not too difficult to tell the role that each substance plays. Thus, when a small amount of alcohol is dissolved in a lot of water, the alcohol is the solute and the water is the solvent. Similarly, if a small amount of water is dissolved in a lot of alcohol, the water is the solute and the alcohol is the solvent. But if one has a 50-50 mixture of alcohol and water, which is the solute and which is the solvent? In this case, either choice is correct.

Table 13.2

Solution (alloy)	Components
Amalgam	Mercury, silver (plus 4% Cu and 1% Zn)
Brass	Copper, zinc
Bronze	Copper, tin
Pewter	Tin, copper, bismuth
Solder	Lead, tin
Steel	Iron, carbon
Silver, sterling	Silver, copper

The amount of solute that can be dissolved in a certain amount of solvent varies with the identities of the solute and solvent and with the temperature of the solution. For example, at 20°C approximately 200 grams of sugar will dissolve in 100 grams of water (whereas sugar is insoluble in gasoline), but only 36 grams of salt (sodium chloride) and almost no barium sulfate will dissolve in 100 grams of water. Also, whereas 200 grams of sugar will dissolve in 100 grams of water at 20°C, about 360 grams of sugar will dissolve in 100 grams of water at 80°C. In general, the reverse is true for gases. An increase in temperature of a solution results in a decrease in the solubility of a gas in a solvent. Almost everyone who has opened a warm can of a carbonated beverage can attest to the accuracy of this observation. The spewing of the soft drink is a result of the sudden release of the carbon dioxide used in the manufacturing of the beverage.

In general terms, if a solution contains only a small quantity of solute, the solution is said to be a *dilute solution,* and if the solute is present in large quantities, the solution is a *concentrated solution.* More specifically, if there is less solute dissolved in the solvent than can be dissolved at the existing temperature of the solution, the solution is an *unsaturated solution.* If the solution contains the maximum amount of solute that can be dissolved in the solvent at a given temperature, the solution is a *saturated solution* (see Fig. 13.1).

Under controlled conditions it is possible to dissolve more solute in a certain quantity of solution than is normally possible at a given temperature. Such a solution is called a *supersaturated solution.* A supersaturated solution is unstable—the solution will eventually force the excess solute out of the solvent to become a saturated solution. It is relatively easy to form supersaturated solutions of sugar and water.

Dry Ice

In 1835, a French chemist, C. S. A. Thilorier, discovered that if the pressure were sufficiently increased on a confined volume of carbon dioxide, the gas would liquefy. He also noted that the liquid, when released, quickly evaporated and that its temperature dropped rapidly. Upon further investigation he found that the liquid, in fact, froze upon contact with some materials, such as wool, and formed a white solid. A most unusual property of the compound was revealed as the compound in the solid state seemed to disappear before his very eyes. (Some substances sublime; that is, they change from a solid state directly into a gaseous state.) This unusual characteristic applies to carbon dioxide under ordinary atmospheric pressure.

Solid carbon dioxide, which is white, does not ordinarily appear in the liquid state and is known as dry ice. Carbon dioxide in the solid state is very useful as a refrigerant to preserve delicate foods during shipment and delivery. Dry ice sublimes at −78.7°C ■

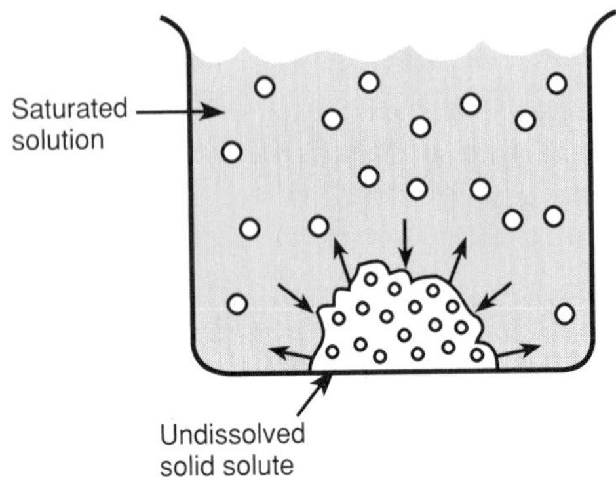

Figure 13.1 A saturated solution. The rate at which the solid solute is dissolving equals the rate at which solute molecules are precipitating back out of solution.

The Structure of Solids

Just as most of the elements that exist on Earth are solids, so are most of the compounds found in nature or synthesized in the laboratory. With the exception of water, we see relatively few liquids, and the ones that we do see are mostly solutions or suspensions of water or oil. We seldom see gases, for most gases are colorless and therefore cannot be seen. Even though we are completely immersed in the gaseous solution called air, we are not usually aware of the air's presence.

Many solids such as wood, rocks, and iron are not particularly attractive in appearance. Other solids such as a snowflake or ice crystal, gold, a diamond, or colored glass are extremely attractive. Everyone is familiar with the general properties of solids—solids resist a change in shape and are relatively firm and compact. This resistance varies considerably with different solids. Some solids, such as glass, are extremely brittle because they are amorphous and have no definite structural order. Therefore, they break under the slightest deformation. Other solids, such as gold, can be beaten into foil that is less than a micrometer thick or drawn into wire so thin that 1.6 kilometers of the gold wire will have a mass of only about half a gram.

Why are solids firm, compact, and resistant to a change in shape? These characteristics are the direct result of the attractive forces between the ions or molecules in a solid and their three-dimensional *ionic crystal* arrangement. Figure 13.2 shows the regular geometric arrangements of Na^{+1} and Cl^{-1} in solid sodium chloride that constitutes the ionic crystal lattice of the solid. Since the illustration shows only a portion of the sodium chloride lattice, we should be aware that the Na^{+1} and Cl^{-1} would be continued in all three directions. With this understanding, we notice that each sodium ion has six neighboring chloride ions and each chloride ion has six neighboring sodium ions. In short, the sodium chloride crystal consists of Na^{+1} and Cl^{-1} arranged in such a manner as to produce maximum mutual attraction between the positive and negative ions. This maximum mutual attraction results in a compact, relatively rigid solid that resists deformation.

A diamond is an example of a solid consisting of atoms rather than ions. Since atoms are electrically neutral, what then holds the atoms in place? Each atom of carbon in the diamond crystal shares a pair of electrons (covalent bond) with four other carbon atoms. This covalent bonding of every carbon atom in the diamond crystal with four other carbon atoms results in a very hard solid.

In a way, the most unusual arrangement of *atomic crystal lattices* occurs in metallic crystals. Since atoms of metals generally have rather low electronegativity, the outer electrons of metal atoms are bound to the nucleus by rather weak forces. In fact, the outer electrons of the atoms of a metal can move within the metal rather freely, so that it would be incorrect to say that a particular electron belongs to a specific nucleus. One way to look at the metal crystal is to consider the positive metal ions in a geometric pattern. These ions are completely surrounded by a cloud of freely moving electrons. It is this freedom of movement of electrons in a metal that makes metals such excellent conductors of electricity.

Many common substances consist of *molecular crystal lattices,* in which only molecules are present. Some examples are ice, sugars, dry ice (solid carbon dioxide), and certain plastics.

Combustion

Fire afforded early humans a means by which to protect themselves against wild animals, to warm themselves during periods of cold weather, to cook food, and, eventually, to smelt metal ores to produce metals. Combustion is no less important today than it was

Third Finger, Left Hand

One of nature's most glamorous accomplishments is the formation of diamonds. These precious stones are specimens of crystallized carbon and are very similar to graphite, but over one and a half times as dense. The chemical nature of diamonds was discovered in 1772 by the famous French scientist, Lavoisier. He convinced several of his close friends to assist him in the purchase of a diamond so that they could learn how to duplicate it. After several days of unfruitful testing, the scientists proceeded to heat the diamond to such a high temperature that it vaporized before their eyes. The diamond, they discovered, had been converted into carbon dioxide gas and they abruptly discontinued their investigation. Additional attempts to discover nature's secrets about diamonds were conducted by other scientists nearer the turn of the century. Their results proved that diamonds were pure carbon and that a diamond could be converted into a practically worthless lump of graphite. But could carbon dioxide or graphite be used to synthesize diamonds? Many efforts were made, and there were numerous reports of success. Among the most noteworthy was the claim by the French chemist, Moissan, in 1893. He dissolved graphite in molten cast iron and reportedly found small diamonds present as the mass cooled. His widely publicized feat was never duplicated, however. Although various grades of synthetic diamonds have been produced, none closely compares to those of nature. However, the attempts of one team of investigators led to the development of silicon carbide, now known as carborundum, the effective abrasive used for grinding and polishing.

Industrial diamonds, made from graphite under a pressure of 200,000 times atmospheric pressure and at temperatures of over 5000°C, are produced today. They are presently too small and too impure to be of significant monetary value as gems; however, synthetic diamonds are most effective as special abrasives and as edges of cutting tools ■

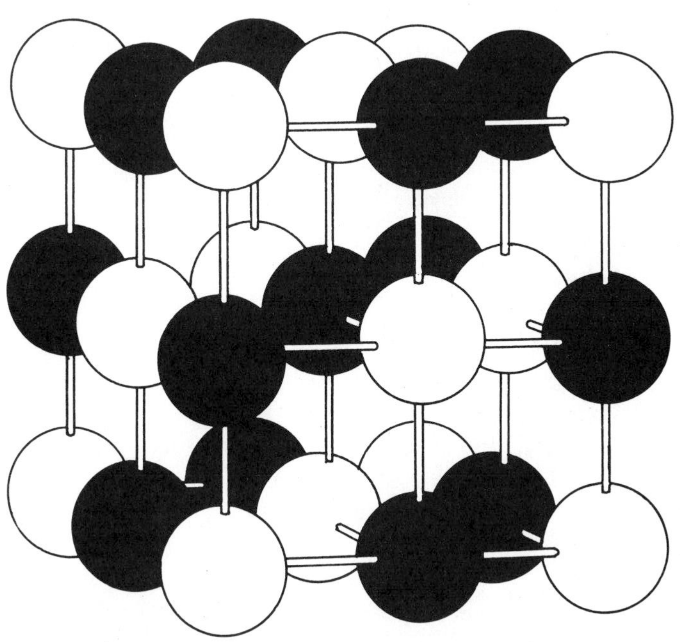

a. Sodium chloride

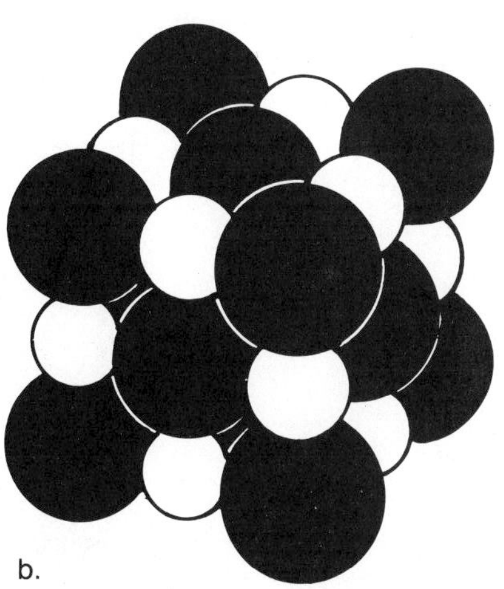

b.

Figure 13.2 The ionic crystal lattice of sodium chloride. (*a*) The black balls represent Cl^- and the white balls represent Na^+. (*b*) A more realistic representation of solid NaCl as a close-packed structure.

The Chemistry of Rocket Fuel

Many of the fuels used to propel rockets into orbit or toward rendezvous in outer space are liquefied gases. The space shuttle uses two solid rocket boosters in addition to liquid propellant. The various fuel mixtures must very rapidly react chemically to provide the necessary thrust for "lift-off" and for the following period of acceleration. A liquid combustible, such as alcohol or kerosene, reacts with an oxidizing agent, such as oxygen, to provide the energy for some rockets. The oxygen or other oxidizer must be carried aboard the rocket because of the absence of air beyond Earth's outer atmosphere. The element is liquefied before being pumped into the fuel cylinders because liquids are more dense than gases; therefore, a greater quantity of the oxidant can be stored in a given space than if the oxygen were gaseous.

One of the most popular fuel combinations is liquid oxygen and liquid hydrogen. The product of the very efficient reaction is, of course, water. The advantages of this reaction compared to other possible combinations are numerous.

A fuel of the future may be liquid fluorine as an oxidizing agent and diborane (a boron-hydrogen compound) as a reducing agent. Before these two substances can be used, however, materials must be developed that will not become highly eroded when fluorine is pumped from holding tanks. It will also be necessary to find a means of lowering the cost of production of diborane. Fluorine is the most powerful oxidizing agent known, and diborane releases about twice as much heat when oxidized than does kerosene ■

in the distant past, and for many of the same reasons. Heat from the combustion process is still used to heat most of our homes, to run our cars, to smelt metal ores, and to generate electricity.

Early peoples did not understand the chemical processes going on when substances burned. The Greeks were the first to attempt an explanation of combustion that did not assume the mystical workings of a god. The Greeks explained that all objects had both earth and fire in them and that during the combustion process the fire was released and the Earth, as ashes, remained. It was not until about 2000 years later, near the end of the eighteenth century, that the chemical explanation of combustion was discovered.

Essentially, *combustion* is the rather rapid combination of oxygen from the air with various materials such as oil, gas, wood, and coal. For example, the natural gas that is piped into many homes is primarily methane (CH_4), and it burns according to the following equation:

$$\underset{\text{methane}}{CH_4} + \underset{\text{oxygen}}{2O_2} \xrightarrow{\text{flame}} \underset{\text{carbon dioxide}}{CO_2} + \underset{\text{water}}{2H_2O} + \text{heat}$$

Wood and coal burn in much the same way, but considering the complexity of the compounds in wood, more than a single equation is necessary to show the combustion of wood.

Oxidation and Reduction

Actually, combustion is a specific example of the chemical process of oxidation. One example of combustion with which we are all familiar is the burning of magnesium or aluminum ribbon in a photographic flashbulb. The flashbulb contains a mass of fine magnesium or aluminum ribbon in an oxygen atmosphere. When the camera is operated, an electric current is sent through the ribbon. The passage of current through the ribbon heats up the ribbon until it begins to burn in the oxygen. The magnesium or aluminum burns with such an intense flame that for a short period of time a brilliant light is produced. This *reaction of oxygen with an element or compound* is known as **oxidation.** In the case of the combustion of magnesium in oxygen, the magnesium is said to be oxidized to magnesium oxide.

We find, however, that combustion may occur in the absence of oxygen if another reactive gas is present. Magnesium ribbon also burns quite easily in the presence of chlorine gas, and in doing so produces a very intense flame:

$$\underset{\text{magnesium}}{Mg} + \underset{\text{chlorine}}{Cl_2} \longrightarrow \underset{\text{magnesium chloride}}{MgCl_2} + \text{heat}$$

Since some material will burn in gases other than oxygen, the term oxidation is best defined to include more reactions than combustion in oxygen. To see whether there is an underlying principle that can be used to more generally define the term "oxidation," we can take a more careful look at the equations for the burning of magnesium in oxygen and the burning of magnesium in chlorine.

In this reaction, magnesium loses two electrons, and oxygen gains two electrons. When magnesium reacts with chlorine, magnesium loses two electrons, and each chlorine atom gains one electron. In both reactions magnesium is oxidized and loses electrons. The important change that magnesium undergoes in both oxidations is the loss of electrons. Consequently, oxidation is broadly defined as the loss of electrons. *The gain of electrons* is called **reduction.** In a more restricted sense, the loss of oxygen by a compound is also called reduction. If carbon is heated to a very high temperature with magnesium oxide, the magnesium oxide is reduced to metallic magnesium:

$$\underset{\text{magnesium oxide}}{MgO} + \underset{\text{carbon}}{C} \xrightarrow{2000^\circ C} \underset{\text{magnesium}}{Mg} + \underset{\text{carbon monoxide}}{CO}$$

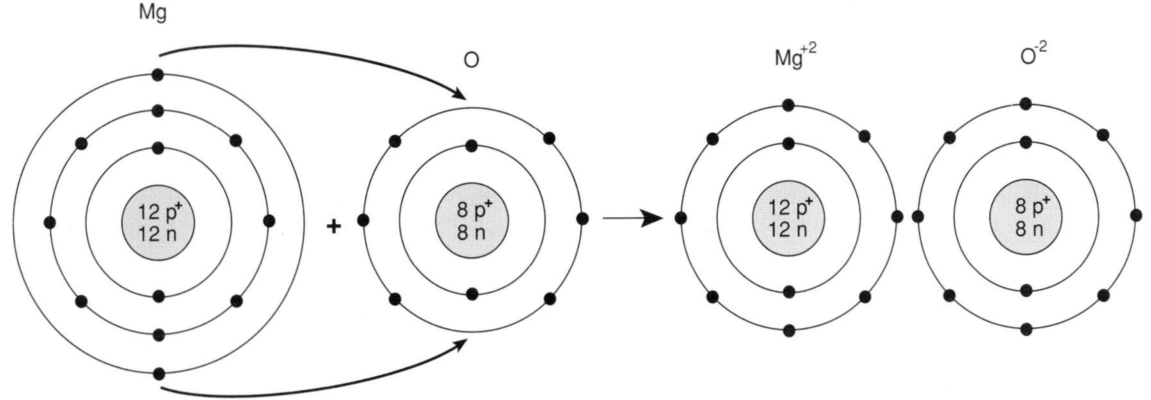
Mg
O
Mg+2
O-2
12 p+
12 n
8 p+
8 n
+
12 p+
12 n
8 p+
8 n

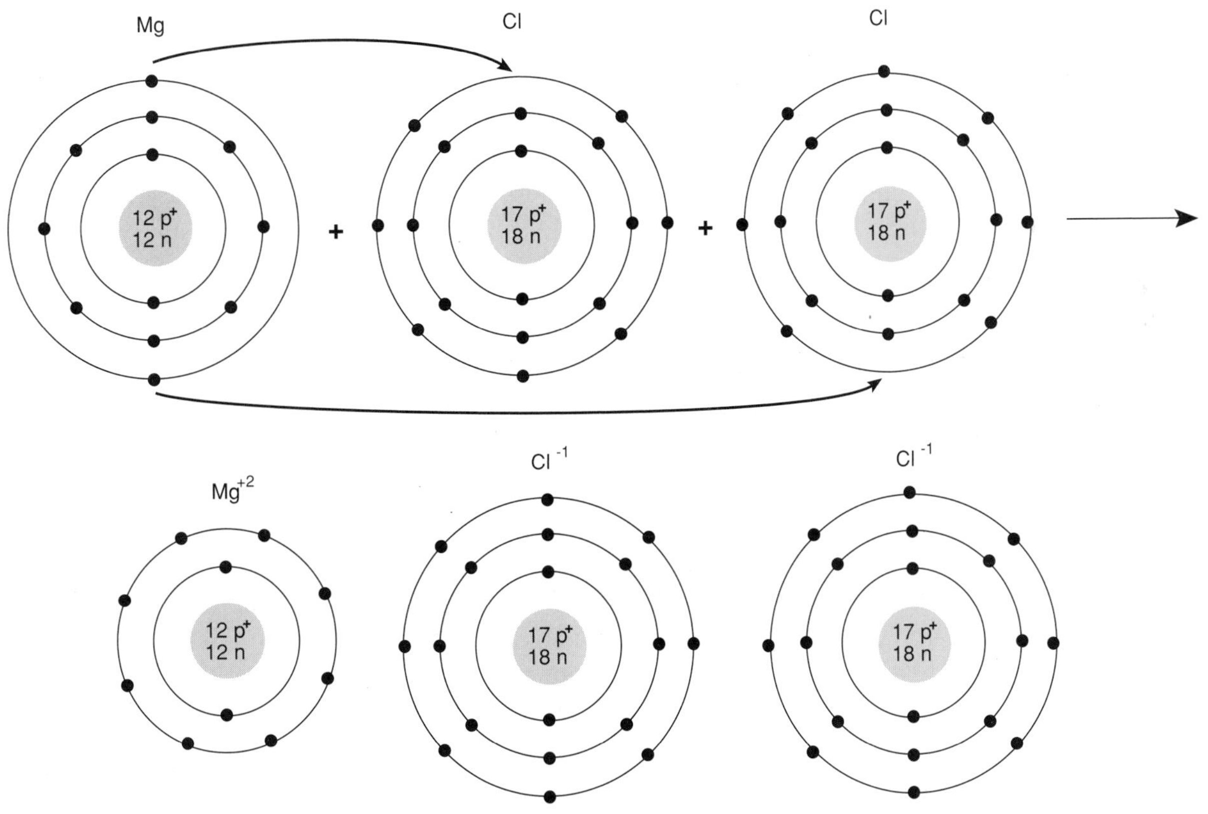
Mg
Cl
Cl
12 p+
12 n
+
17 p+
18 n
+
17 p+
18 n
Mg+2
Cl -1
Cl -1
12 p+
12 n
17 p+
18 n
17 p+
18 n

The magnesium oxide loses oxygen and, therefore, is reduced. The magnesium ion of magnesium oxide also gains electrons (reduction), as follows:

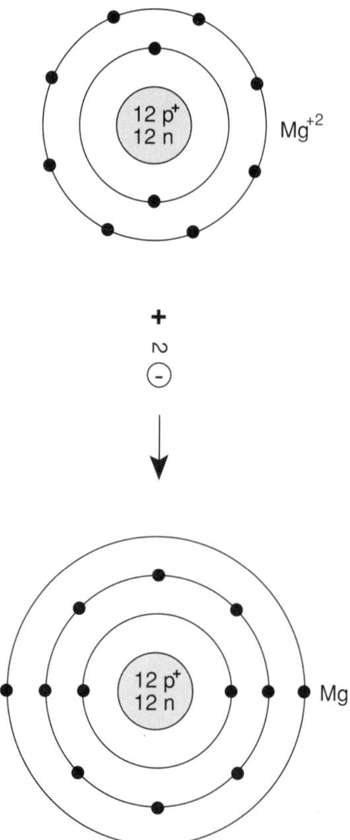

An oxidation-reduction reaction very important to our civilization is the reaction of iron ore (Fe_2O_3) with coke (C) in a blast furnace to produce iron:

$$2Fe_2O_3 + 3C \xrightarrow{\text{heat}} 4Fe + 3CO_2$$

In this reaction the Fe^{+3} ion of the Fe_2O_3 has been reduced because it has lost oxygen and also because it has gained electrons. The carbon, having lost electrons, has been oxidized.

When reduction occurs, electrons are gained by atoms or ions. In order for electrons to be gained, they must first be lost by the process of oxidation. Likewise, when electrons are lost in an oxidation process, the lost electrons are always taken up (gained) by other atoms or ions present in the reaction. It can be said, then, that oxidation is always accompanied by reduction and that reduction is always accompanied by oxidation. If in a chemical reaction one substance is oxidized, another substance must be reduced and vice versa.

Equilibrium

A chemical reaction usually is the result of the collision of atoms or molecules. When the particles collide with sufficient energy, existing bonds are broken and new bonds are formed. In the reaction of hydrogen molecules with iodine molecules, if the collisions are energetic enough, the chemical bonds between the atoms of the hydrogen molecules and the atoms of the iodine molecules are broken. New chemical bonds form between hydrogen atoms and iodine atoms to yield hydrogen iodide:

$$H_2 + I_2 \longrightarrow 2\ H\text{–}I$$

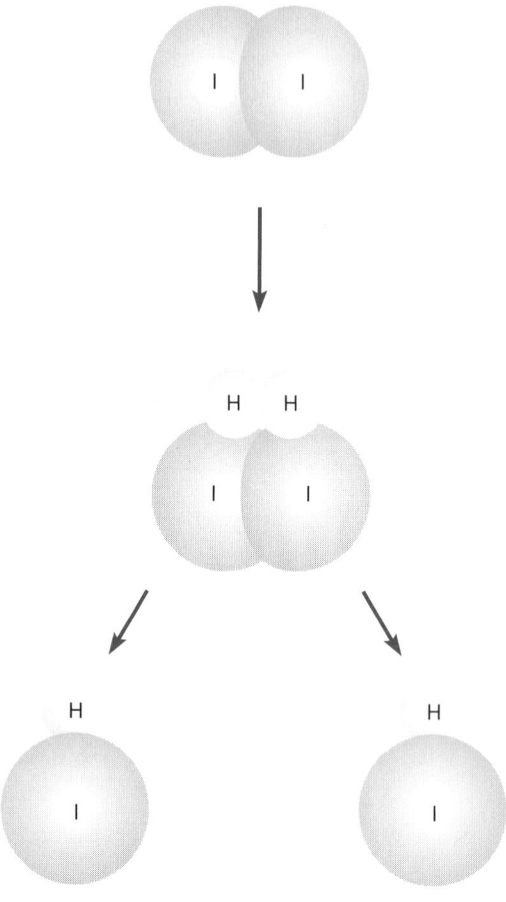

Once the hydrogen iodide molecules are formed, they in turn can collide with each other to break the H–I bonds and reform the H–H and I–I bonds. As a result, the reaction of hydrogen with iodine does not go to completion, where all the reactants are converted to products. Instead, a **reversible reaction** results between hydrogen molecules and iodine molecules on the one hand, and between two hydrogen iodide molecules on the other. Such a reversible reaction is indicated by means of a double arrow:

$$H_2 + I_2 \rightleftharpoons HI$$

At first, the forward reaction predominates, but as time passes and the concentration of the product increases, the rate of decomposition of the HI molecules increases until hydrogen iodide is decomposing into hydrogen and iodine as fast as hydrogen iodide is being formed. That is, the rate (speed) of formation of hydrogen iodide is the same as the rate of its reconversion to hydrogen and iodine. When *reversible reactions occur at the same rate,* the two reactions are said to be in **equilibrium.** Once the reaction reaches equilibrium, the concentrations of H_2, I_2, and HI no longer change, but it should not be assumed that the reactions are no longer occurring. The equilibrium appears to be static because by simple observation no changes are detected to be taking place, but the reactions are continuing to occur. On the molecular level, H_2 and I_2 continue to react together and HI molecules continue to decompose. Such a system is a dynamic system, and the system is in a state of *dynamic equilibrium;* that is, a state in which the forward and reverse reactions have the same rate. If you started with 1.0 mole of H_2 and 1.0 mole of I_2 at a reaction temperature of 450°C, at equilibrium there would be 0.2 mole of H_2, 0.2 mole of I_2, and 1.6 moles of HI, all still continually undergoing reactions.

Reaction Rates

The rate of a chemical reaction is expressed in terms of the change in the concentration of one component per unit time. The rate can be expressed as a decrease in the concentration of a reactant (or an increase in the concentration of a product) in a certain period of time. The **reaction rate** is usually described as the change in the number of moles of a reactant or product per second. Rates of reactions vary from very large to very small. When dynamite molecules decompose in a fraction of a second, the rate of reaction is very large and an explosion results, but when iron rusts because of the presence of oxygen in the air, the number of iron atoms undergoing a reaction with oxygen per second of time is very small.

Reaction rates are determined by a number of factors, such as the nature of reactants, concentration of reactants, particle size of reactants, temperature of the reaction, and presence of a catalyst. The most important factor affecting reaction rates is the nature of the reactants. Some substances react very rapidly whereas other substances react very slowly, if at all. Sodium metal reacts very vigorously with water, but zinc reacts with water quite slowly.

The effect of a change in the concentration of reactants can be quite pronounced. Iron wool (fine wire) will burn slowly when placed in a flame (due to the oxygen in the air), but the same iron wool burns with an almost blinding white flame when placed in 100 percent oxygen. The increase in the concentration of one of the reactants, oxygen, from 20 percent in air to 100 percent in pure oxygen, accounts for the difference in the rate of the reaction.

The particle size (state of subdivision) of the reactants also has considerable influence on reaction rates. For liquids reacting with liquids or gases reacting with gases, the state of subdivision is as small as it can possibly be, since atoms or molecules are reacting with atoms or molecules. For solids, however, only the outer surface of the solid particles can react. If the total surface area of the reactants can be increased, the rate at which the reaction will occur should increase. Consider a solid cube 1 centimeter on a side. The surface area of the cube is 1 cm × 1 cm × 6, or 6 cm^2. If each side of the cube is divided by 10 to form 1000 smaller cubes, each cube will have a surface area of 0.1 cm × 0.1 cm × 6, or 0.06 cm^2. Since there are 1000 small cubes, the total surface area of the cubes is 60 cm^2, or 10 times the surface area of the original cube. If the original cube were divided into 1 million small cubes, the resultant total surface area would be 100 times that of the original 1-cm^3 cube. The increase in the rate of reaction that results with an increase in state of subdivision is the reason coal is powdered before being used in industrial furnaces. In a plant that produces flour by grinding grain, the flour that is suspended in the air must be constantly kept to a minimum, or else the finely divided flour that mixes with the oxygen of the air could react with the oxygen of the air in the presence of a spark to produce a devastating explosion. In general, doubling the surface area of reactants doubles the rate of reaction.

Another very important factor affecting the rate of a reaction is temperature. Reaction rates that are only modest at room temperature may be extremely large at much higher temperatures. As a rule of thumb, a 10°C increase in the temperature of a reaction will result in a doubling of the reaction rate. There are many exceptions to this rule, since the rate of some reactions increases as much as 350-fold with a 10°C increase. As the temperature of a reaction is increased, the average kinetic energy of the colliding atoms or molecules is increased proportionally.

If the only factor involved in bringing about a reaction were the collisions of the reactants, the reaction would all be over in an incredibly short time, since in some reactions 10^{30} collisions occur every second! However, when molecules approach very closely, their outer electrons mutually repulse the molecules. This repulsion must be overcome and the molecules activated before they can react upon collision. The amount of available energy from the kinetic energy of the molecules must be greater than the minimum amount of energy necessary to overcome the repulsion forces. This *quantity of energy necessary to initiate a chemical reaction* is known as the **activation energy** of the reactants. For most reactions, the activation energy may be considered a barrier to the reaction. Every chemical reaction has a particular activation energy. The greater the activation energy, the slower the rate of the reaction.

As previously mentioned, a chemical reaction causes bonds between the atoms of molecules of reactants to be broken, whereupon new bonds are formed to produce molecules of products. Energy is absorbed to break bonds, and energy is released when bonds are formed. If more energy is produced by the formation of bonds in product molecules than is used to break bonds in reactant molecules, the extra energy is given off in the form of heat (or sometimes as light, sound, or electricity). *A chemical reaction in which energy is released* is called an **exothermic reaction.** An example of an exothermic reaction is the oxidation of aluminum:

$$4Al + 3O_2 \longrightarrow 2Al_2O_3 + \text{energy}$$

If more energy is used to break bonds in reactant molecules than is given off when bonds are formed in product molecules, energy is absorbed during the reaction. *A chemical reaction in which energy is absorbed* is called an **endothermic reaction.** An example of an endothermic reaction is the decomposition of calcium carbonate (limestone):

$$CaCO_3 + \text{energy} \longrightarrow CaO + CO_2$$

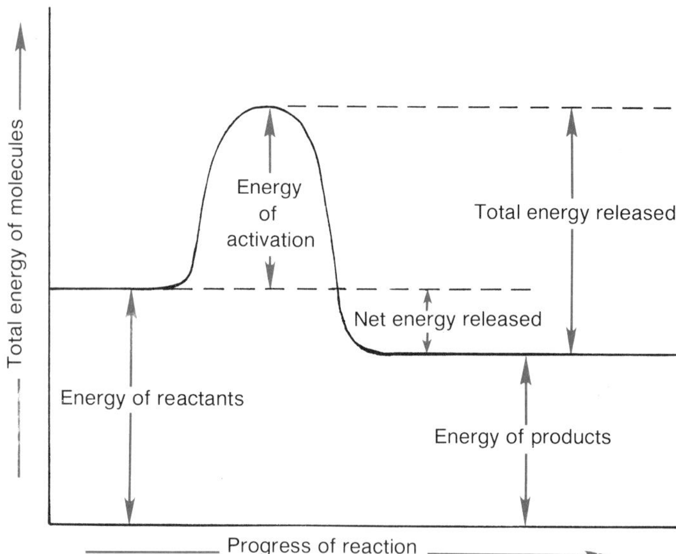

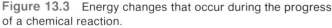

Figure 13.3 Energy changes that occur during the progress of a chemical reaction.

Consider the exothermic reaction of the burning of methane (natural gas):

$$CH_4 + 2O_2 \longrightarrow CO_2 + 2H_2O + \text{energy}$$

In this reaction, more energy is released in the formation of the carbon-oxygen and hydrogen-oxygen bonds of the products than is absorbed in the breaking of the carbon-hydrogen and oxygen-oxygen bonds of the reactants. The products, carbon dioxide and water, possess less energy than the reactants, methane and oxygen. Although extra energy was needed to activate the reaction, that activation energy was returned during product formation. The relationship between the energy of reactants, the energy of products, and the energy of activation for this reaction is shown in Figure 13.3.

In many reactions, it is possible to alter the activation energy of the atoms or molecules by adding a **catalyst.** A catalyst is *a substance that speeds up or slows down the rate of a reaction without being consumed in the reaction.* The word catalyst comes from the Greek words *kata* and *lyein,* meaning "to loosen or release." A catalyst changes the rate of a reaction by entering into the reaction and altering the activation energy of the reaction, but the catalyst emerges from the reaction unchanged. A *positive catalyst* increases the rate of a reaction; a *negative catalyst* decreases the rate of a reaction. A catalyst has no effect on the position of equilibrium but simply changes the rate of the reaction in both the forward and reverse directions. A positive catalyst, then, results in a system reaching equilibrium in a shorter period of time than would be the case without the catalyst.

One well-known use of positive catalysts is in the catalytic converters on cars. The catalysts are platinum and palladium deposited as a thin layer on ceramic beads that have a very large surface area. The catalysts convert carbon monoxide and unburned hydrocarbons (see Chapter 14 on organic chemistry) to carbon dioxide and water vapor by the action of oxygen from the air.

The hydrogen peroxide (H_2O_2) present in some antiseptics and bleaching agents contains a small amount of a special compound that acts as a negative catalyst to slow the natural decomposition of H_2O_2.

Water—Our Most Important Chemical

Water is the chemical compound that exists in overwhelmingly greater volume than any other compound on our planet. There are sixteen hundred thousand million billion (16×10^{20}) liters of water in the ocean basins of Earth and fifty thousand million billion (50×10^{18}) liters of frozen water in the polar caps and ice sheets of our globe. This ubiquitous substance covers about 71 percent of Earth's surface in the form of marine water and approximately 10 percent in the form of huge ice sheets covering vast stretches of land and water. About 97 percent of all water on Earth is marine water. Water is present in gaseous solution in our atmosphere and is physically trapped and chemically combined within the rocks of Earth's crust and mantle. Water falls from the skies as soft delicate snowflakes, as hail, or as rain. Of course, most of the water that falls upon the land finds its way back to the great ocean basins of our planet.

Most naturally occurring substances exist in varying concentrations in solution in our oceans, which are approximately a 3.5 percent solution of dissolved materials. These dissolved substances are mostly present in the form of ions. The ions that constitute a major portion of the dissolved matter are Na^{+1}, Mg^{+2}, Ca^{+2}, K^{+1}, Cl^{-1}, SO_4^{-2}, and HCO_3^{-1}. Surely water is the compound closest to being a universal solvent.

We are quite dependent on water for our very existence. We must drink water daily in order to live. We must use huge quantities to maintain the vast industrial structure of society, and we must have water available for sanitation purposes. It is this very dependence on water that makes the pollution of rivers, lakes, and oceans so dangerous in terms of our future on this planet. Our Earth is the only "water" planet in the solar system, and we must protect it if we are to survive.

The Action of Water on Ions and Molecules

The electron-pair bonds between the hydrogen atoms and the oxygen atom of the water molecule are not evenly distributed between the hydrogen and the oxygen atoms. Because the electronegativity of oxygen is much greater than that of hydrogen, the electron pairs are pulled closer to the oxygen atom:

```
        O
     ·.   .·
   H        H
```

This unequal charge distribution results in a high electron density around the oxygen atom and a low electron density around the hydrogen atoms. This unequal charge distribution produces what is called a *polar molecule* and can be represented as follows:

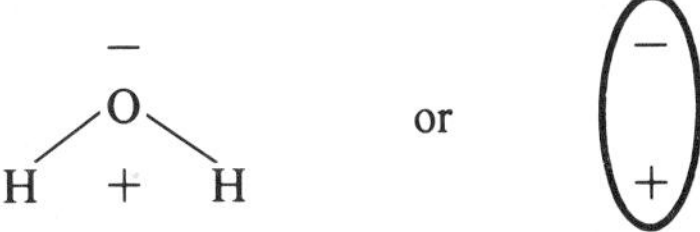

Water molecules align themselves so that the high electron density of the oxygen end of one water molecule is next to the low electron density of the hydrogen atoms of an adjacent water molecule as illustrated:

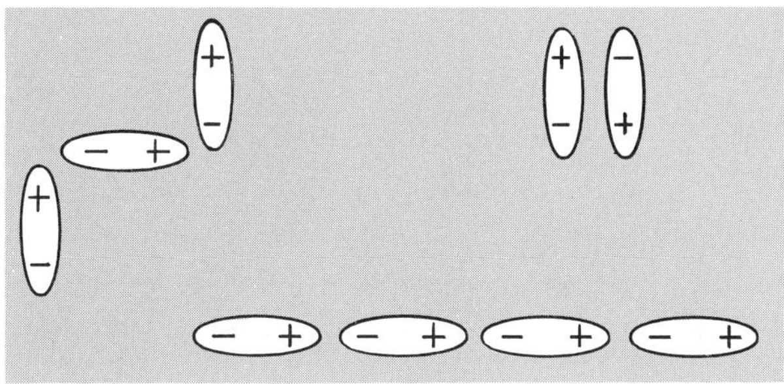

If water molecules are grouped together because of this polarity, certainly they should group around an ion that is in water solution. Consider what happens when a grain of salt is dissolved in water. In the solid crystalline state, Na^{+1} and Cl^{-1} are held together by mutual attraction. When the solid NaCl is placed in water, the positive sodium ions on the outside of the solid attract the high-electron-density end of the water molecules, and the negative chloride ions attract the low-electron-density end of the water molecules. The result is that the Na^{+1} and Cl^{-1} are surrounded by the appropriate ends of water molecules. This arrangement of water molecules weakens the attraction of Na^{+1} for Cl^{-1}, with the result that eventually the Na^{+1} and Cl^{-1} bonds break. The water molecules continue to align themselves around the ions until an envelope of water molecules completely encases the separate ions as illustrated:

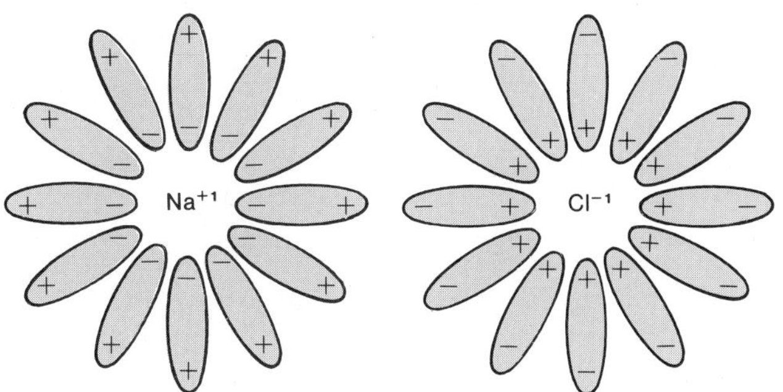

It turns out, then, that negative and positive ions in solution are not held rigidly together by ionic bonds as they are in the solid state, but they are completely free to move about at random among the water molecules. The ions do not exist as individual ions but carry with them a complete envelope of water molecules.

The behavior of molecules that are not ionic in nature when dissolved in water is very similar to the behavior of ions in water solution. Thus, although sugar molecules are not composed of ions, the sugar molecules, like water molecules, have an unequal distribution of electrons and are, therefore, polar in nature. Consequently, when sugar molecules are dissolved in water, water molecules align themselves in appropriate fashion to encase each sugar molecule in a water envelope (a process called *solvation*). Indeed, it is this ability of water to align itself around ions or polar molecules that accounts for the solubility of so many substances in water.

Nonionic Molecules That Produce Ions

The vinegar on the kitchen shelf is approximately a 5 percent solution of acetic acid in water. This vinegar will conduct an electric current. If, however, one tests pure acetic acid with no water present, the pure acetic acid will not conduct an electric current. We also know that pure water does not normally conduct an electric current, whereas salt water does. We might assume, then, that since salt water contains sodium and chloride ions, it is the presence of ions in the salt water that is responsible for the conduction of an electric current. Since acetic acid molecules themselves do not consist of ions, how do water molecules produce ions from the nonionic acetic acid molecules?

The formula for acetic acid is

```
    H   O
    |   ||
H — C — C — O — H
    |
    H
```

The hydrogen-oxygen bond of acetic acid, like the hydrogen-oxygen bond of water, is polar and, therefore, water molecules align themselves around the acetic acid hydrogen-oxygen bond as illustrated:

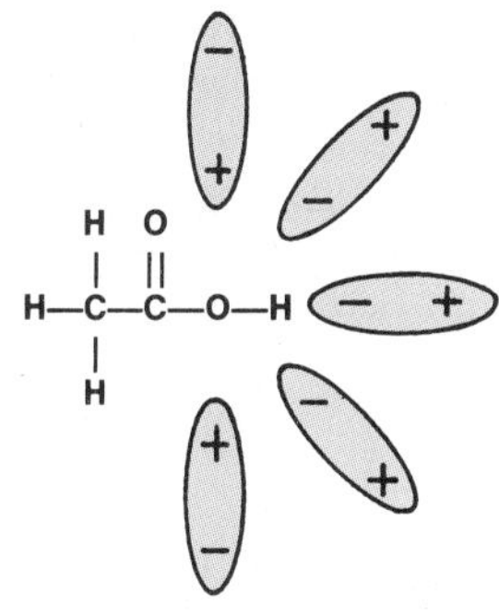

This arrangement of water molecules around the covalent hydrogen-oxygen bond weakens the bond to such an extent that the bond ruptures and produces a positively charged hydrogen ion and a negatively charged acetate ion:

$$HC_2H_3O_2 \xrightarrow{\text{(in water)}} C_2H_3O_2^{-1} + H^{+1}$$

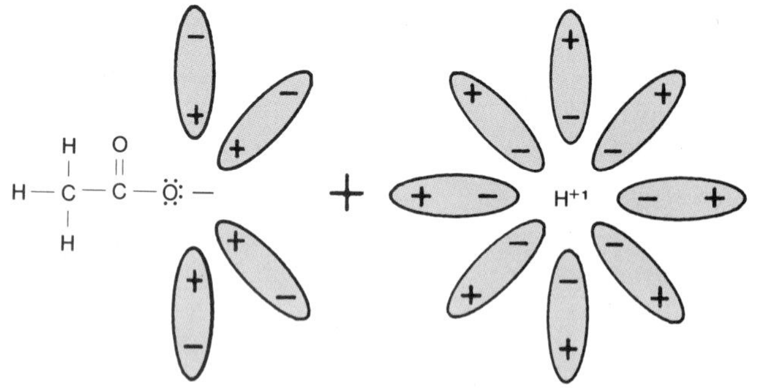

Acetic acid undergoes solvation by water molecules, causing the covalent acetic acid structure to form ions. This process of ion formation from molecules is called *ionization.* It is now easy to see why pure acetic acid does not conduct an electric current, whereas acetic acid dissolved in water does.

Acids

An **acid** *has a sour taste and changes the color of the natural dye litmus from blue to pink.* Other properties of acids include their reaction with certain metals, their ability to neutralize bases, and their conduction of an electric current when dissolved in water. Some acids are strong and some are weak. Sulfuric acid (H_2SO_4) is a strong acid and attacks skin, clothing, metal, and most other substances. On the other hand, carbonic acid (H_2CO_3), made by dissolving carbon dioxide in water, is a very weak acid and is so harmless that it is used in many beverages. In fact, the term *carbonated beverage* stems from the fact that the beverage contains carbon dioxide and, therefore, carbonic acid. Hydrochloric acid (HCl), present in our stomachs, is a strong acid, but our stomachs contain only a dilute solution of HCl. We are conscious of the acid content of certain citrus fruits, such as the grapefruit and the lemon, because of their sour taste. Although we are not as aware of the acid content of the orange or tangerine, all four of these citrus fruits contain about the same concentration of acid (citric acid). The orange and the tangerine have enough natural sugar to mask the sour taste of the citric acid so that it is not as noticeable. For other examples of acids, see Table 13.3 and Figure 13.4.

Why are acids sour-tasting, and why do they turn blue litmus to pink? The reason is that in water solution acids ionize to produce hydrogen ions (H^{+1}). It is the hydrogen ions that are responsible for the acidic properties of acids.

Actually, the hydrogen ions produced by acids do not remain by themselves in water solution. The hydrogen ions react with a neutral water molecule to form what is called the *hydronium ion:*

$$H^{+1} + H_2O \longrightarrow H_3O^{+1}$$

Table 13.3
Some common acids.

Formula	Name		Example or application
HCl	Hydrochloric acid	Strong acid	Present in the stomachs of mammals.
H_2SO_4	Sulfuric acid	Strong acid	"Battery acid."
HNO_3	Nitric acid	Strong acid	Used to make nitroglycerine.
$HC_2H_3O_2$	Acetic acid	Weak acid	Main ingredient in vinegar.
H_2CO_3	Carbonic acid	Weak acid	Present in carbonated beverages.
$HC_6H_7O_7$	Citric acid	Weak acid	Present in citrus fruits.

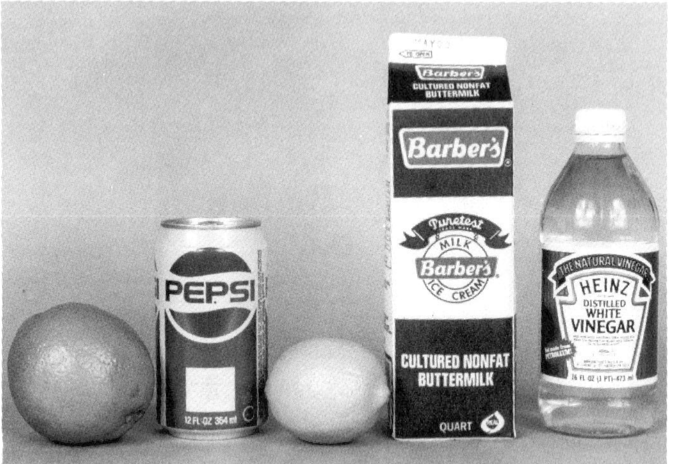

Figure 13.4 Examples of common household acids.

The reaction of a hydrogen ion with water is not very unusual, considering that the hydrogen ion is actually a lone proton. The +1 charge of the proton is concentrated in an incredibly small volume and, therefore, has a very high charge concentration in its small volume. It is natural that this bare proton, with its very high charge density, will be attracted to the high electron density of the outer electrons of the oxygen atom of water. To be more precise in our equations showing the reaction of nonionic acids with water, we should show the formation of hydronium ions rather than hydrogen ions:

$$HCl + H_2O \longrightarrow H_3O^{+1} + Cl^{-1}$$
$$H_2SO_4 + 2H_2O \longrightarrow 2H_3O^{+1} + SO_4^{-2}$$

Bases

Bases are compounds that *have a bitter taste, a soapy feel, and change the color of litmus from pink to blue.* Bases also neutralize acids and conduct an electric current when dissolved in water. Whereas most common acids are nonionic liquid compounds, most bases are ionic solid compounds. As evidence of the ionic nature of bases, a pure base that has been melted will conduct an electric current. Some bases, such as sodium hydroxide (NaOH), are strong bases; others, such as magnesium hydroxide ($Mg(OH)_2$, milk of

Table 13.4
Some common bases.

Formula	Name	Example or application
NaOH	Sodium hydroxide	Used in preparations to unstop clogged drains.
NH_4OH	Ammonium hydroxide	Household ammonia. Used for removing grease.
$Ca(OH)_2$	Calcium hydroxide	"Slaked lime" used in plaster.
$Mg(OH)_2$	Magnesium hydroxide	"Milk of magnesia" used as a laxative.

magnesia), are relatively weak. Strong bases used in various commercial products to unstop the drains of sinks consist mostly of the base sodium hydroxide. Many liquid preparations used to remove wax from floors contain the weak base ammonium hydroxide.

When bases are dissolved in water, they all yield one or more hydroxide ions as shown. For other examples of bases, see Table 13.4 and Figure 13.5.

$$NaOH \longrightarrow Na^{+1} + OH^{-1}$$
$$Ca(OH)_2 \longrightarrow Ca^{+2} + 2OH^{-1}$$

Some bases, such as ammonia gas (NH_3), are not ionic and do not contain hydroxide ions in the pure state; however, when ammonia is dissolved in water, it chemically reacts with the water molecules to produce hydroxide ions:

$$NH_3 + H_2O \longrightarrow NH_4^{+1} + OH^{-1}$$

Neutralization and Salts

One characteristic of acids and bases is that they react with each other; and if equal amounts of acid and base are combined, the resulting solution is neither acidic nor basic, but neutral. *The reaction of an acid with a base* is known as **neutralization.** Thus, sodium hydroxide is neutralized by hydrochloric acid to produce sodium chloride and water:

$$NaOH + HCl \longrightarrow NaCl + H_2O$$

As indicated, this type of reaction is quite general:

$Ca(OH)_2$	+	H_2SO_4	$\longrightarrow$	$CaSO_4$	+	$2H_2O$
calcium hydroxide		sulfuric acid		calcium sulfate		water
$Mg(OH)_2$	+	2HCl	$\longrightarrow$	$MgCl_2$	+	$2H_2O$
magnesium hydroxide		hydrochloric acid		magnesium chloride		water

Although we use the term *salt* as the common name for NaCl, **salt** in the wider sense refers to a whole family of compounds. For our purposes here, we can consider a salt as *the ionic substance produced when an acid and a base react.* Thus, calcium sulfate ($CaSO_4$), formed from the reaction of sulfuric acid (H_2SO_4) with calcium hydroxide ($Ca(OH)_2$), and magnesium chloride ($MgCl_2$), formed from the reaction of magnesium hydroxide ($Mg(OH)_2$) with hydrochloric acid (HCl), are both salts. Salts are characterized as ionic solids with extremely high melting points. For other examples of salts, see Table 13.5.

Figure 13.5 Examples of common household bases.

The Activity Series

One type of oxidation-reduction reaction is the displacement reaction. In a **displacement reaction,** *atoms or ions of one substance take the place of other atoms or ions in a compound.* The element hydrogen was first prepared in 1671 by the displacement of hydrogen from sulfuric acid by iron:

$$Fe + H_2SO_4 \longrightarrow FeSO_4 + H_2$$

A number of metals can displace hydrogen from acids, and some metals can displace hydrogen from water:

$$Ni + 2HCl \longrightarrow NiCl_2 + H_2$$
$$2K + 2H_2O \longrightarrow 2KOH + H_2$$

The metals in the preceding reaction have lost electrons (oxidation), and hydrogen ions have gained electrons (reduction).

A metal more reactive than another metal will displace the less-reactive element from its salt:

$$Fe + CuSO_4 \longrightarrow FeSO_4 + Cu$$
$$Zn + CuCl_2 \longrightarrow ZnCl_2 + Cu$$

Zinc and iron atoms lose their electrons more easily than do copper atoms; hence, they lose electrons to copper ions to produce copper atoms.

The relative reactivity of metals with water or acids is shown in the *activity series* presented in Table 13.6. Metals located above hydrogen in the activity series will produce hydrogen when reacted with an acid or, in some cases, with water. Metals located below hydrogen will not. Any metal in the series will replace any metal below it from a solution of the salt of the less-reactive metal.

Table 13.5
Some common salts.

Formula	Name	Example or application
$NaCl$	Sodium chloride	A flavoring agent and a necessary constituent of body fluids.
$NaHCO_3$	Sodium hydrogen carbonate (sodium bicarbonate)	"Baking soda." An antacid and an ingredient in baking powders.
$NaNO_2$	Sodium nitrite	A food preservative. Used to retard spoilage and discoloration of some meats.
NH_4NO_3	Ammonium nitrate	A fertilizer and an explosive.
SnF_2	Stannous fluoride	A toothpaste additive used to prevent dental caries (tooth cavities).
$AgCl$	Silver chloride	A coating for photographic film.
$MgSO_4$	Magnesium sulfate	"Epsom salts." Used as a laxative and to relax tired feet.
$BaSO_4$	Barium sulfate	Used to obtain X rays of the gastrointestinal tract.
$CaCO_3$	Calcium carbonate	"Limestone."

Table 13.6
Activity series of metals.

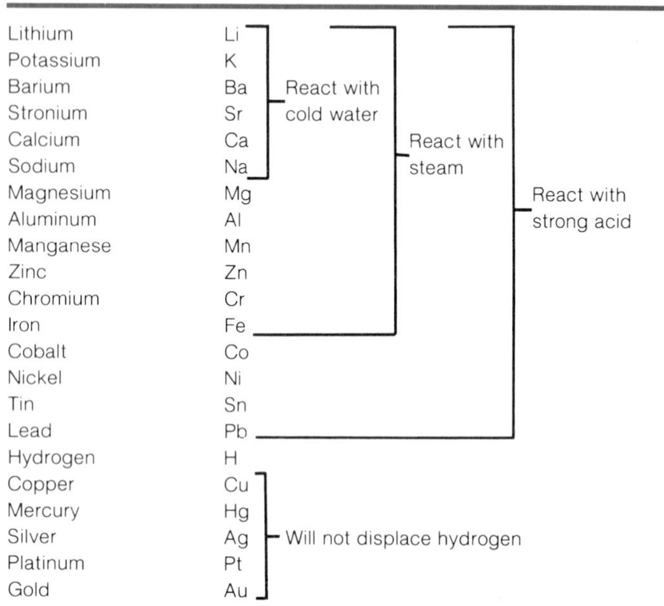

Lithium	Li	React with cold water (Li–Na); react with steam (Li–Fe); react with strong acid (Li–Pb)
Potassium	K	
Barium	Ba	
Stronium	Sr	
Calcium	Ca	
Sodium	Na	
Magnesium	Mg	React with steam (Li–Fe); react with strong acid (Li–Pb)
Aluminum	Al	
Manganese	Mn	
Zinc	Zn	
Chromium	Cr	
Iron	Fe	
Cobalt	Co	React with strong acid (Li–Pb)
Nickel	Ni	
Tin	Sn	
Lead	Pb	
Hydrogen	H	
Copper	Cu	Will not displace hydrogen (Cu–Au)
Mercury	Hg	
Silver	Ag	
Platinum	Pt	
Gold	Au	

The most reactive metals are at the top of the activity series. The activity of the metals decreases down the series. For example, iron (Fe) is more reactive than copper (Cu), and magnesium (Mg) is more reactive than both copper and iron. Silver (Ag), platinum (Pt), and gold (Au) are at the bottom of the series, indicating that they are quite unreactive. For this reason, as well as for their beautiful luster and color, they are used in jewelry, ornaments, and tableware.

The pH Scale

Water molecules, whether in a solution or in pure water, ionize to a limited extent:

$$H_2O \longrightarrow H^{+1} + OH^{-1}$$

The concentration of hydrogen ions (H^{+1}) and hydroxide ions (OH^{-1}) is measured in *moles per liter.* In pure water the concentration of H^{+1} equals the concentration of OH^{-1} ions, and both are present in a concentration of 10^{-7} mole per liter.

In an acidic solution the H^{+1} concentration is higher than 10^{-7} mole per liter, and the OH^{-1} concentration is lower than 10^{-7} mole per liter. In a basic solution the OH^{-1} concentration is higher than 10^{-7} mole per liter, and the H^{+1} concentration is lower than 10^{-7} mole per liter.

Since in pure water the concentration of H^{+1} and OH^{-1} is the same, pure water is considered neutral, as is any solution in which the H^{+1} and OH^{-1} concentrations are equal. If a solution contains a higher concentration of H^{+1} than OH^{-1}, the solution is acidic, and if a solution contains a higher concentration of OH^{-1} ions than H^{+1}, the solution is basic.

A standard method of indicating the H^{+1} concentration of solutions has been developed and is universally used in dealing with acidic and basic solutions. That method expresses the *hydrogen ion concentration* as a quantity known as **pH.** The pH of a solution is the negative power to which 10 must be raised to equal the H^{+1} concentration. Thus,

$$H^{+1} = 1 \times 10^{-pH}$$

Since the H^{+1} in pure water (at 25°C) is 1×10^{-7} mole per liter, the pH of pure water is 7. If the concentration of H^{+1} in a solution is 1×10^{-8}, then the pH is 8. For a concentration of H^{+1} of 1×10^{-3}, the pH is 3; for 1×10^{-10}, the pH is 10.

The H^{+1} concentration of a solution times the OH^{-1} concentration of the solution always equals 1×10^{-14}. Therefore, if the H^{+1} concentration is 1×10^{-3}, then the OH^{-1} concentration must be $10^{-14}/10^{-3} = 1 \times 10^{-11}$. Similarly, if the OH^{-1} ion concentration of a solution is 1×10^{-5}, the H^{+1} concentration must be $10^{-14}/10^{-5} = 1 \times 10^{-9}$.

In terms of pH, then, an acidic solution has a pH of less than 7, a neutral solution has a pH of 7, and a basic solution has a pH greater than 7. Figure 13.6 shows the pH scale.

Although it might seem that a difference in pH of 1, say from 6 to 7 or 7 to 8, should represent a very small difference, a difference of 1 in a pH value represents a tenfold difference in

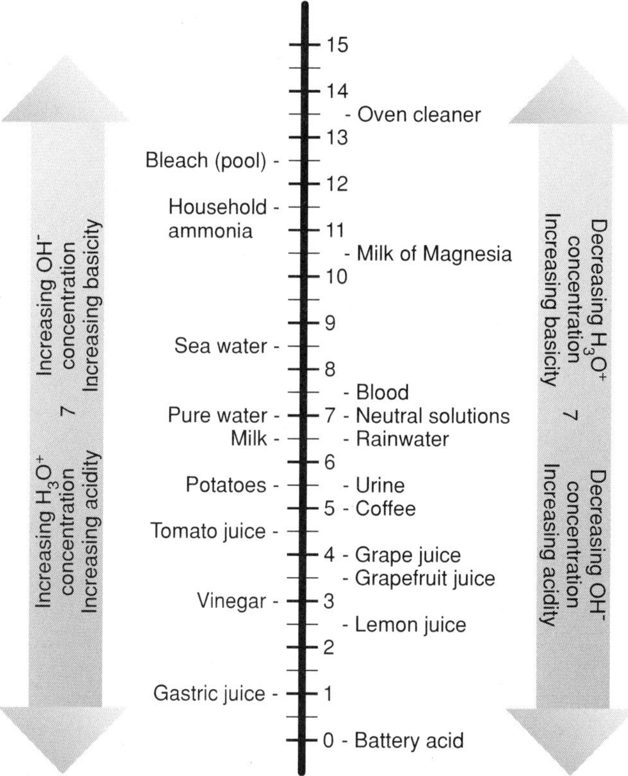

Figure 13.6 The pH scale from 0 to 15.

strength. Thus, an acid with a pH of 2.0 is 10 times as "strong" (has 10 times the H^{+1} concentration) as an acid with a pH of 3.0. An acid with a pH of 1.0 is 10 times as strong as an acid with a pH of 2.0 and 100 times (10 × 10) as strong as an acid with a pH of 3.0. A base of pH 12.4 is 10 times as strong as a base with a pH of 11.4, 100 times as strong as a base with a pH of 10.4, and 1000 times as strong as a base with a pH of 9.4. Consequently, a relatively small change in the pH of a solution will represent a rather large change in the acidity or basicity of a solution. The average pH of some common foods is given in Table 13.7.

The pOH of a solution is the negative power to which 10 must be raised to equal the OH^{-1} concentration. That is,

$$OH^{-1} = 1 \times 10^{-pOH}$$

and

$$pH + pOH = 14$$

Summary

The different elements present in a compound are chemically bound to one another in definite proportions by weight. This is known as the law of definite proportions. A mixture is a combination of two or more substances that are not chemically united and that do not exist in a fixed proportion to each other. A solution is a mixture on the molecular level.

The atoms or molecules of a solid are held together by various forces so as to give the solid a definite shape and to resist deformation.

Table 13.7
The average pH of some common foods.

Food	Average pH
Soft drinks (pop)	2.0–4.0
Lemon	2.5
Wine	3.0
Vinegar	3.0
Grapefruit or grapefruit juice	3.2
Oranges	3.5
Peaches	3.5
Grape juice	4.0
Tomato juice	4.5
Bananas	4.5
Coffee, black	5.0
Milk (cow or human)	6.5

The chemical combination of oxygen with another element is known as oxidation. Oxidation is defined in more general terms as the loss of electrons by an element or compound. The rapid oxidation of a substance is known as combustion. When an oxygen-containing compound chemically loses its oxygen, the process is called reduction. A more general definition of reduction is the gain of electrons.

Equilibrium is a dynamic state in which the rate at which reactants combine or decompose equals the rate at which products combine or decompose to reform the reactants, so that no net change occurs in the system with time.

The rate at which substances are used up or are formed during a chemical reaction is known as the rate of the reaction. The rate of the reaction is determined by such factors as the nature of the reactants, concentration of the reactants, particle size of the reactants, temperature of the reaction, and presence of a catalyst. The amount of energy necessary to start a chemical reaction is known as the reaction's activation energy.

Water is a polar substance and therefore dissolves other molecules that have unequal charge distributions, such as ionic and polar covalent compounds. In the dissolution of a compound in water, water molecules align themselves so that the low-electron-density end of the water molecule is near the high-electron-density end of the dissolving polar molecule or ion and vice versa. Consequently, the dissolving of a substance in water results in the molecules of the substance being completely encased in water molecules.

When a compound has a bond that is sufficiently polar, water molecules align themselves around the polar bond, and the alignment of the water molecules causes the bond to rupture and to produce positive and negative ions in solution; that is, ionization occurs.

Exothermic reactions release energy and endothermic reactions absorb energy.

In a displacement reaction, atoms or ions of one substance take the place of other atoms or ions in a compound.

An acid produces hydrogen ions (H^{+1}) in water solution. Once the hydrogen ion is present, it reacts with a water molecule to form a hydronium ion (H_3O^{+1}). Bases are compounds that produce a hydroxide ion (OH^{-1}) in water solution. When an acid reacts with a base to produce a salt and water, the process is termed neutralization. Salts are ionic crystalline solids produced by the reaction of an acid with a base.

The standard method of indicating the hydrogen ion concentration of a solution is to determine the negative power to which 10 must be raised to equal the H^{+1} concentration. The resulting number is known as the pH of the solution. Thus, a solution that has a molar concentration of hydrogen ions of 1×10^{-5} has a pH of 5. Given two solutions, a difference of 1 in the numerical value of the pH represents a difference of a factor of 10 in the hydrogen ion concentrations of the solutions. A solution that has a pH of less than 7 is acidic, and a solution that has a pH of greater than 7 is basic. A neutral solution, such as pure water, has a pH of 7.

Questions and Problems

Chemical and Physical Changes

1. What is the difference between a physical change and a chemical change?

2. How could one separate a mixture of powdered iron and powdered sulfur?

3. What is the difference between a mixture and a solution?

4. Explain the difference between an element, a compound, and a mixture.

Mixtures and Solutions

5. How does a mixture differ from a solution?

6. Which of the following are solutions?

(a) bronze
(b) iron
(c) air
(d) steel
(e) gold

7. What is a supersaturated solution?

The Structure of Solids

8. Why do solids resist deformation?

9. A chemist has two solid compounds. One is powdered sodium carbonate and one is powdered sugar. Both solids are white and they look alike. How can the chemist determine which solid is which compound?

10. What is a molecular crystal?

Combustion and Oxidation and Reduction

11. What is the principle behind quenching flames with a foam-type fire extinguisher?

12. Explain the oxidation that occurs when methane burns in air.

13. Explain the oxidation that occurs when iron reacts with sulfur. Write equations showing the process of oxidation and reduction in this reaction.

Equilibrium

14. How does a catalyst affect the equilibrium of a chemical reaction?

15. What is meant by "a reversible reaction"?

16. Define "chemical equilibrium."

Reaction Rates

17. List the factors that influence the rate of a reaction.

18. What is meant by "activation energy" of a reaction?

19. Define "rate of a reaction."

20. How does the state of subdivision of reactants influence the rate of a reaction?

Water—Our Most Important Chemical

21. Why is water so important to us as individuals?

The Action of Water on Ions and Molecules

22. Why are polar molecules surrounded by water molecules in an aqueous solution?

23. How does unequal electron charge distribution aid in the water solubility of compounds?

Nonionic Molecules That Produce Ions

24. Explain what happens when HCl dissolves in water.

25. If pure liquid HCl (at −85°C) does not conduct an electric current, why does HCl dissolved in water conduct an electric current?

Acids

26. How could one prepare a solution of solvated protons?

27. What is the pH of a solution that contains 1×10^{-8} moles per liter of OH^{-1}?

28. Why are curdled milk, grapefruit juice, and vinegar all sour tasting?

29. How can an orange have a pH of 3.5 but not have a sour taste?

30. What is the difference in a hydrogen ion and a hydronium ion?

Bases

31. Explain the difference between an acid and a base.

Neutralization and Salts

32. Write an equation for the neutralization of magnesium hydroxide with hydrochloric acid.

33. How would you define a salt?

The Activity Series

34. Which of the following elements will liberate hydrogen from cold water?
(a) Cu
(b) Mg
(c) Ni
(d) Na
(e) Pb

35. Which of the following elements will produce Cu from a solution of $CuSO_4$?
(a) Zn
(b) Ni
(c) Hg
(d) Pt
(e) Al

The pH Scale

36. How much difference is there in the strength of solutions with pH levels of 6 and 4?

37. What is the pOH of a solution that has a hydrogen ion concentration of 10^{-4}?

14 Organic Chemistry

Chapter outline

Key terms/concepts

A model of the DNA structure—the fundamental molecule of life.

f the approximately 7 to 8 million chemical compounds known to exist, about 2½ million contain the element carbon. The only other element that is so ubiquitous in chemical compounds is hydrogen. Chemical compounds are broadly divided into two classes: *organic* and *inorganic* substances. Organic compounds contain one or more carbon atoms, whereas inorganic compounds, with a few exceptions, do not contain carbon atoms. Most of the chemicals constituting the cells, and therefore the bodies, of all living things are organic.

Early scientists used the term *organic compounds* to describe compounds closely related to living processes. Since, for instance, urea was found only in the urine of animals, and sugar was found only in living plants, and vinegar and ethyl alcohol were produced only by the fermentation of various parts of living plants, it was natural to assume that these organic compounds were different in some mysterious way from inorganic compounds from which nonliving matter was constructed. However, in 1828 the German chemist Friedrich Wöhler (1800–1882) found that the supposedly inorganic compound ammonium cyanate could be converted by heat into the organic compound urea. Since that time a vast number of compounds present in living organisms have been synthesized in the laboratory. Today the term **organic compound** is used to indicate *any compound containing carbon* (other than a few simple compounds such as carbon dioxide), rather than just compounds found in living cells.

The variety of organic compounds synthesized in laboratories and produced by industry is incredible. Some manufactured organic compounds include medicines, bleaches, detergents, fertilizers, synthetic fibers, food additives, fuels, pesticides, plastics, and many others.

The major natural sources of organic compounds are petroleum, coal, and natural gas. Minor sources of organic compounds are plants and animals.

The Central Role of Carbon

Why should the element carbon have the unique ability to form such a vast array of different compounds, whereas other elements form comparatively few compounds? It is because carbon, unlike most other elements, can combine with itself to form chains consisting of carbon atoms. In fact, diamond consists solely of carbon atoms bonded to one another. Silicon can also combine with itself to form chains, but not with the versatility exhibited by carbon.

Carbon has four electrons in its outer shell. Each of these electrons combines with an electron from another atom to form a total of four electron-pair covalent bonds. There are many elements with which carbon combines, such as hydrogen, oxygen, sulfur, chlorine, nitrogen, and iodine, but carbon also very readily combines with itself. The following example is of a simple organic compound. (Only the electrons actually involved in covalent bonding are shown for each atom.)

$$\begin{array}{ccccccc} & & \mathrm{H} & & \mathrm{H} & & \\ & & \cdot\cdot & & \cdot\cdot & & \\ \mathrm{H} & : & \mathrm{C} & : & \mathrm{C} & : & \mathrm{H} \\ & & \cdot\cdot & & \cdot\cdot & & \\ & & \mathrm{H} & & \mathrm{H} & & \end{array}$$

ethane

Generally, the covalent bonds are presented by dashes as illustrated:

$$\begin{array}{ccccccc} & & \mathrm{H} & & \mathrm{H} & & \\ & & | & & | & & \\ \mathrm{H} & - & \mathrm{C} & - & \mathrm{C} & - & \mathrm{H} \\ & & | & & | & & \\ & & \mathrm{H} & & \mathrm{H} & & \end{array}$$

ethane

This formula is sometimes written as CH_3—CH_3 and occasionally condensed into the molecular formula C_2H_6. A structural formula shows which atoms are bonded to which, whereas a molecular formula shows only the number of atoms of each element in the molecule.

Hydrocarbons

Organic compounds that *contain only hydrogen and carbon* are consequently called **hydrocarbons.** Without this class of organic compounds, our mechanized society could not exist as we know it, since the hydrocarbons include various gasolines and oils essential as fuels and lubricants for our internal-combustion engines. From an economic point of view, finding petroleum—a complex mixture of hydrocarbons—has captured the popular imagination second only to finding gold as a quick way to accumulate wealth.

On the basis of the bonding between carbon atoms, hydrocarbons may be divided into four main classes: *alkanes, alkenes, alkynes,* and *aromatic hydrocarbons.*

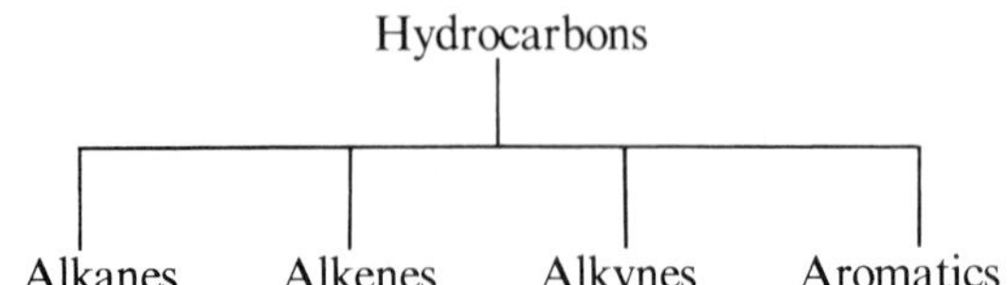

Alkanes

The alkanes are hydrocarbons that contain only single bonds. The suffix of alkanes is *-ane* and, except for the first four members, the first part of the name is derived from the Greek prefix that indicates the number of carbon atoms present in the molecule. Thus, *pent-* indicates five carbon atoms, *hex-* indicates six, and so forth. (See Table 14.1 for the relationship of Greek prefixes to the number of carbon atoms present in alkanes.)

If you examine the formulas of the alkanes in Table 14.1, you will see that each member differs from the member before and after it by a —CH_2— unit (a *methylene* group); thus,

$$\begin{array}{lr} \text{Methane} \;—\!\!— & CH_4 \\ + & \underline{CH_2} \end{array}$$

The First Organic Synthesis

During the nineteenth century, Friedrich Wöhler was one of many leaders in the field of chemistry. His prime interest was inorganic chemistry, and he made several discoveries in that area that were quite significant. His greatest contribution, however, was in the chemistry of organic compounds. Wöhler, reportedly by accident, performed what was presumed to be an impossible task at the time—he synthesized an organic substance out of an inorganic one without the involvement of living tissue. To his excitement, he discovered that, upon heating, ammonium cyanate formed crystals of urea. This substance, the major nitrogenous waste product of all mammals, is present in urine and is organic.

The importance of Wöhler's 1828 discovery was overrated, according to the opinions of scientists who succeeded him. (Even today, some authoritative sources consider ammonium cyanate an organic compound.) In any event, Wöhler's efforts did inspire others to seek means of synthesizing organic compounds from inorganic ones.

The level of success of chemists who followed Wöhler is evident when one considers the hundreds of thousands of organic compounds that have been synthesized, and imagines the countless others yet to be produced in the chemists' laboratories ■

Table 14.1

Relationship between prefixes and number of carbon atoms.

Name	Formula	Greek prefix	Number of carbons
Methane	CH_4	—	1
Ethane	C_2H_6	—	2
Propane	C_3H_8	—	3
Butane	C_4H_{10}	—	4
Pentane	C_5H_{12}	Pent-	5
Hexane	C_6H_{14}	Hex-	6
Heptane	C_7H_{16}	Hept-	7
Octane	C_8H_{18}	Oct-	8
Nonane	C_9H_{20}	Non-	9
Decane	$C_{10}H_{22}$	Dec-	10

Ethane —— C_2H_6
+ CH_2

Propane —— C_3H_8
+ CH_2

Butane —— C_4H_{10}

A series of compounds in which each member differs from the next member by a constant amount is called a **homologous series.** Members of the series are called homologs. Also, you might notice that in the formula of each member of the series the number of hydrogen atoms equals two more than twice the number of carbon atoms, which leads to the general formula for alkanes: C_nH_{2n+2}. The use of this formula is shown in Table 14.2.

The simplest alkane, the hydrocarbon methane, contains one carbon atom and four hydrogen atoms. The manner in which the hydrogen atoms attach to the carbon atom is illustrated in Figure 14.1. The four hydrogen atoms are bonded to the carbon atom of methane by covalent bonds; that is, by the sharing of electron-pair bonds. As seen in Figure 14.1, the methane molecule has a tetrahedral arrangement. The four bonds of carbon are directed toward the vertices of the tetrahedron. The angles between the four carbon-hydrogen bonds are all 109.5 degrees.

Table 14.2

Use of general formula of alkanes to determine specific formulas.

Alkane	Value of n	General formula	Specific formula
Methane	1	$C_1H_{2(1)+2}$	CH_4
Propane	3	$C_3H_{2(3)+2}$	C_3H_8
Pentane	5	$C_5H_{2(5)+2}$	C_5H_{12}
Decane	10	$C_{10}H_{2(10)+2}$	$C_{10}H_{22}$

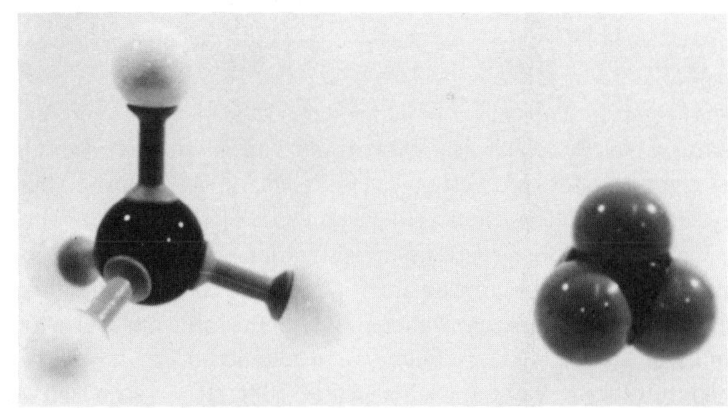

Figure 14.1 The three-dimensional structure of the methane molecule. The "ball-and-stick" representation on the left correctly shows the carbon-hydrogen bond angles. The "space-filling" illustration on the right more closely represents the compactness of the molecule.

Methane is normally written as a "flat" structure since it is easier to draw than a tetrahedron.

```
    H
    |
H — C — H
    |
    H
```

methane

Methane is the main constituent (up to 97 percent) of natural gas, which is used to heat homes and supply energy for industries. Methane is constantly being released into the atmosphere in small quantities by the anaerobic (without oxygen) decay of organic matter from the bottoms of swamps. Bubbles of methane gas, often called *marsh gas,* can be seen rising to the surface of swamps and marshes. Also, the deadly gas called *fire damp* that causes explosions in coal mines is methane. Although there are only trace amounts of methane in our atmosphere today, methane was one of the major constituents of Earth's primitive atmosphere, along with ammonia, water, and hydrogen. Stanley Miller, a Ph.D. student of Professor Harold C. Urey, found that an electric discharge will convert a mixture of methane, ammonia, water, and hydrogen into complex organic compounds of biochemical importance to life forms (see "The Chemical Origin of Life" section in Chapter 23).

Propane and butane are normally written as follows:

```
    H   H   H           H   H   H   H
    |   |   |           |   |   |   |
H — C — C — C — H   H — C — C — C — C — H
    |   |   |           |   |   |   |
    H   H   H           H   H   H   H
     propane                butane
```

Since all the bond angles of alkanes are 109.5 degrees, it would be better to represent these molecules as illustrated, with the lines indicating bonds 109.5 degrees apart. The hydrogen atoms of the molecules are not shown.

```
                              —109.5°
   C                    C   C
C    C              C     C
      109.5°               109.5°
```

The principal sources of alkanes are crude oil (petroleum) and natural gas. Petroleum is separated into various fractions (groups of compounds with a certain boiling-point range) by distillation. A list of petroleum fractions is given in Table 14.3.

When scientists were first investigating the chemistry of hydrocarbons, they became aware of an initially confusing situation. They found two different compounds with the same molecular formula of butane (C_4H_{10}). One type of butane boiled at $-0.6°C$, and the other type of butane boiled at $-10°C$. It was soon learned that the two butanes differed in the way in which their carbon chains were constructed, as indicated:

```
    H   H   H   H               H   H   H
    |   |   |   |               |   |   |
H — C — C — C — C — H       H — C — C — C — H
    |   |   |   |               |   |   |
    H   H   H   H               H   |   H
                                H — C — H
       C4H10                        |
     b.p. −12°C                     H
      n-butane
                                  C4H10
                                b.p. −0.5°C
                                 isobutane
```

Table 14.3

Petroleum fractions.

Number of carbon atoms	Approximate boiling-point range in degrees Celsius	Name	Uses
1	−162°C	Natural gas	Heating
2–5	Under 40°C	LP gas	Heating, bottled gas, cigarette lighters
5–10	40–200°C	Gasoline	Motor fuel
10–16	200–300°C	Kerosene	Jet and motor fuel
15–18	250–375°C	Fuel oil	Diesel fuel
16–30	Above 310°C	Heavy gas oils	Heating oil
28–38	Nonvolatile	Heavy oils	Lubricating oil
Over 40	Nonvolatile	Asphalt	Road and parking area construction

Different compounds that have identical molecular formulas are called **isomers.** The word isomer comes from two Greek words, *isos* (equal) and *meros* (parts).

The simplest isomer of an alkane is called the *normal* isomer, abbreviated by n-. Two normal alkanes, shown without their hydrogen atoms, are

```
C — C — C — C          C — C — C — C — C
   n-butane                n-pentane
```

and their simplest branched isomers, indicated by the prefix iso-, are as shown:

```
C — C — C        C — C — C — C
    |                |
    C                C
 isobutane       isopentane
```

Thus, n-butane and isobutane are isomers.

An n-isomer has all of its carbon atoms joined in a line, whereas an iso- structure has a single one-carbon branch (a *methyl* group, CH_3-) on the next to the first (or last) carbon atom of a chain. The existence of different organic compounds that have identical molecular formulas is common and is called *isomerism.* Different isomers have different physical and often chemical properties. Whereas butane has two isomers, pentane (C_5H_{12}) has three:

```
CH3 — CH2 — CH2 — CH2 — CH3        CH3 — CH — CH2 — CH3
          n-pentane                       |
                                         CH3
            CH3                       isopentane
             |
      CH3 — C — CH3
             |
            CH3

         neopentane
```

The hydrocarbon hexane (C_6H_{14}) has five isomers, and a hydrocarbon with 20 carbon atoms has approximately a third of a million isomers! As the number of carbon atoms in an alkane increases, the possible number of isomers increases at even a faster rate (see Table 14.4). The phenomenon of isomerism leads to the possibility of an infinite number of organic compounds. The three-dimensional molecular structure of the alkanes n-propane, n-butane, and isobutane is shown in the illustrations that follow.

The hydrocarbon formulas given so far in this chapter show only *a single covalent bond between any two carbon atoms.* Hydrocarbons having this particular feature are referred to as **saturated hydrocarbons,** or alkanes.

Alkanes are relatively inert. Under normal conditions they do not react with acids, bases, oxidizing agents, or reducing agents. At a high temperature, such as in a flame, alkanes will undergo oxidation with oxygen:

$$CH_4 + 2O_2 \xrightarrow{\text{flame}} CO_2 + 2H_2O + \text{heat}$$

Although the burning of methane, or natural gas, produces carbon dioxide and water, the most important product produced is heat. If a space heater in a house is not properly adjusted so that sufficient oxygen from the air can mix with the methane, incomplete combustion results:

$$2CH_4 + 3O_2 \xrightarrow{\text{flame}} \underset{\text{carbon monoxide}}{2CO} + 4H_2O.$$

Since the inhalation of carbon monoxide can be fatal, the proper adjustment of space heaters is very important.

With certain gasolines the smooth burning of the fuel-air mixture in a car's cylinders is replaced by a series of explosions that occur because ignition in the system is premature. The explosions produce a "knocking" in the engine that greatly reduces its efficiency. When pure n-heptane is burned, knocking in the engine is pronounced; but when isooctane is burned, very little knocking occurs. An arbitrary scale has been established with isooctane given an *octane number* of 100 and n-heptane an octane number of 0. A mixture of 90 percent isooctane and 10 percent n-heptane has an octane rating of 90. Isooctane reacts with oxygen in the air in the same way as methane:

$$2C_8H_{18} + 25O_2 \longrightarrow 16CO_2 + 18H_2O + \text{heat}$$

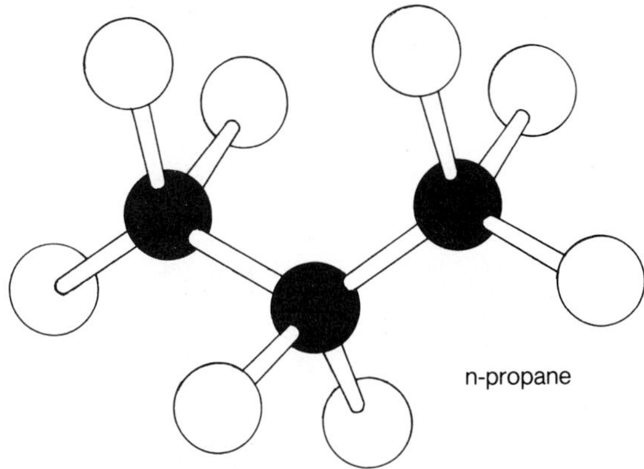

Table 14.4

The possible number of isomers increases rapidly for alkanes that contain over 15 carbon atoms.

Number of carbon atoms in an alkane	Number of possible isomers of the alkane
1	1
2	1
3	1
4	2
5	3
6	5
7	9
8	18
9	35
10	75
15	over 4000
20	over 300,000
30	over 4 billion
40	over 62 trillion

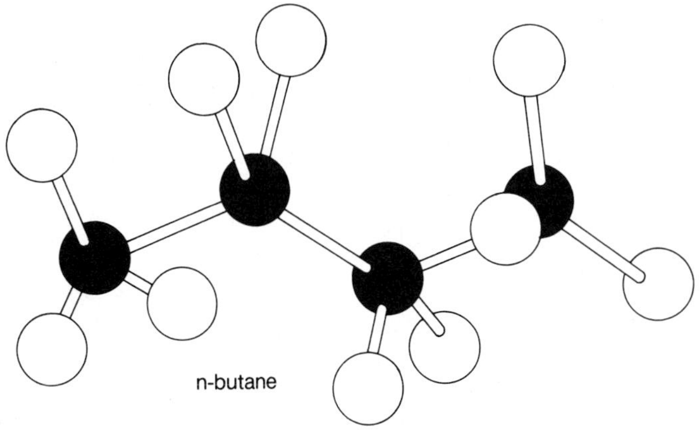

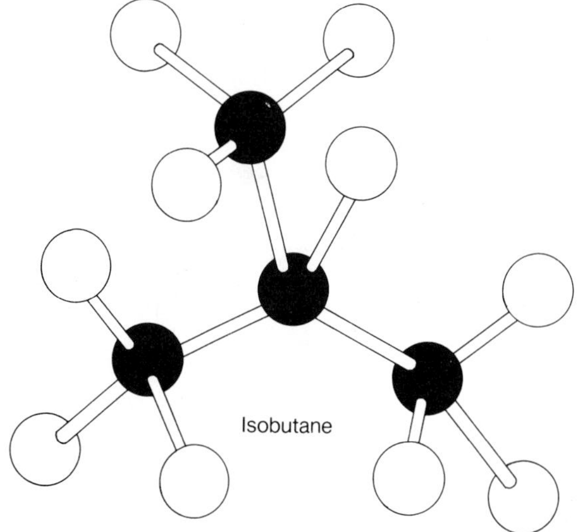

Under the influence of ultraviolet light or elevated temperatures, alkanes react with chlorine or other halogens to form haloalkanes, or *alkyl halides*. One of the products of the reaction of methane with chlorine is methyl chloride:

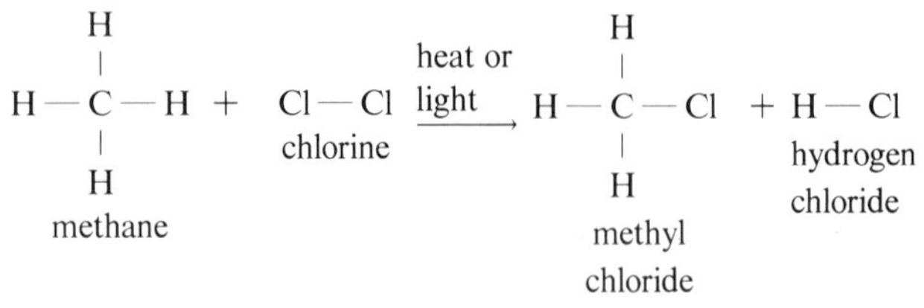

$$\underset{\text{methane}}{H-\overset{\displaystyle H}{\underset{\displaystyle H}{C}}-H} + \underset{\text{chlorine}}{Cl-Cl} \xrightarrow[\text{light}]{\text{heat or}} \underset{\text{methyl chloride}}{H-\overset{\displaystyle H}{\underset{\displaystyle H}{C}}-Cl} + \underset{\text{hydrogen chloride}}{H-Cl}$$

One hydrogen atom has been substituted by a chlorine atom. Such a reaction is a **substitution reaction.** Actually, when methane reacts with chlorine, more than one substitution occurs for some of the molecules, and a mixture of haloalkanes is obtained:

$$CH_4 + Cl_2 \xrightarrow[\text{light}]{\text{heat or}} \underset{\text{methyl chloride}}{CH_3Cl} \text{ and } \underset{\text{dichloromethane}}{CH_2Cl_2} \text{ and } \underset{\text{trichloromethane, or chloroform}}{CHCl_3} \text{ and } \underset{\text{carbon tetrachloride}}{CCl_4}.$$

Methyl chloride and dichloromethane are used as solvents, chloroform is an anesthetic (but it is no longer used on humans), and carbon tetrachloride is a nonflammable cleaning agent and is also used in some types of fire extinguishers.

Certain fluoroalkanes are of considerable industrial importance. One such substance, Teflon, will be taken up in the discussion on polymers in the next section. Another widely used fluorocompound is dichlorodifluoromethane, CCl_2F_2, known as Freon 12. Freon 12 is used as a refrigerant in refrigerators and air conditioners, as a plastic foam, and as an aerosol propellant in some spray cans.

Cycloalkanes are a special type of alkane and exist as cyclic structures:

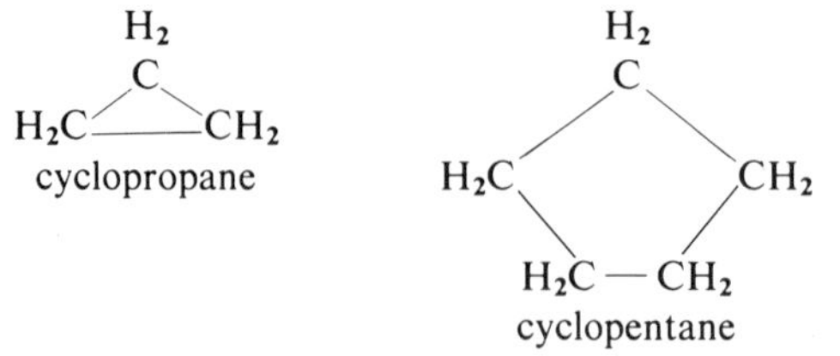

cyclopropane: H_2C (top) bonded to H_2C and CH_2, which are bonded to each other.

cyclopentane: H_2C (top) bonded to H_2C and CH_2, which are bonded to H_2C-CH_2.

The general formula for cycloalkanes is C_nH_{2n}. Cycloalkanes are named by adding the prefix *cyclo-* to the name of the parent alkane.

Alkenes and Alkynes

Hydrocarbons that have more than one bond between two carbon atoms are called **unsaturated hydrocarbons.** Unsaturated hydrocarbons with one or more double bonds between carbon atoms are called *alkenes;* those containing one or more triple bonds are called *alkynes.*

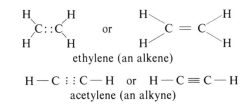

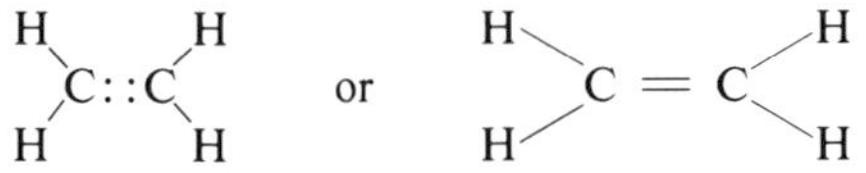

$$\text{H}_2\text{C::CH}_2 \quad \text{or} \quad H_2C=CH_2$$

ethylene (an alkene)

$$H-C\,\vdots\vdots\,C-H \quad \text{or} \quad H-C\equiv C-H$$

acetylene (an alkyne)

Alkenes with one double bond have the general formula C_nH_{2n}. Ethylene's formula is $C_2H_{2(2)}$, or C_2H_4. Alkynes with one triple bond have the general formula C_nH_{2n-2}. Acetylene's formula is $C_2H_{2(2)-2}$, or C_2H_2. Ethylene is a flat molecule and acetylene is linear.

Unsaturated compounds can react with themselves to produce long chains made up of many units of the starting material. The starting compound is called a *monomer* and the long chain product is called a **polymer** (from the Greek *poly,* meaning "many," and *meres,* meaning "parts").

The process by which a monomer is converted into a polymer is known as *polymerization.* Thus, ethylene polymerizes to produce the well-known polymer polyethylene, which is widely used as a transparent packaging material. Two molecules of ethylene at first react as follows:

$$H_2C=CH_2 + H_2C=CH_2 \longrightarrow -CH_2-CH_2-CH_2-CH_2-$$

Other ethylene molecules react in turn to build up a saturated hydrocarbon consisting of a long chain:

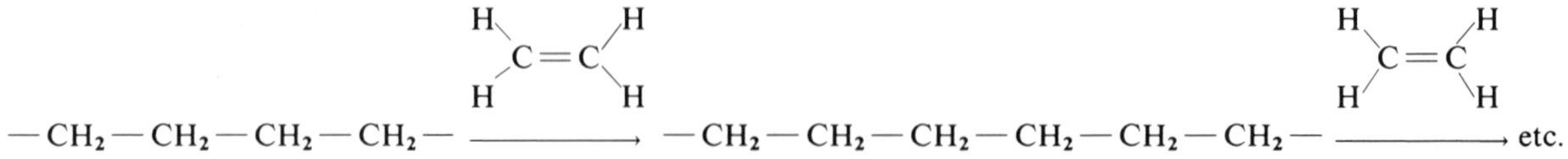

$$-CH_2-CH_2-CH_2-CH_2- \xrightarrow{H_2C=CH_2} -CH_2-CH_2-CH_2-CH_2-CH_2-CH_2- \xrightarrow{H_2C=CH_2} \text{etc.}$$

Table 14.5
Some important addition polymers.

Monomer name	Polymer name	Monomer formula	Polymer repeating unit	Uses
Ethylene	Polyethylene	$CH_2{=}CH_2$	$(CH_2-CH_2)_n$	films, bags, plastic bottles
Isobutylene	Polyisobutylene	$CH_2{=}C(CH_3)-CH_3$	$(CH_2-C(CH_3)_2-)_n$	special rubbers
Tetrafluoroethylene	Polytetrafluoroethylene ("Telfon")	$CF_2{=}CF_2$	$(CF_2-CF_2)_n$	Teflon coatings
Vinyl chloride	Polyvinyl chloride	$CH_2{=}CH-Cl$	$(CH_2-CH(Cl))_n$	films, water pipes, raincoats

$$\underset{\text{vinyl chloride}}{H_2C{=}CHCl} \longrightarrow \underset{\text{polyvinyl chloride}}{-CH_2-\underset{Cl}{CH}-CH_2-\underset{Cl}{CH}-CH_2-\underset{Cl}{CH}-}$$

In a way similar to the formation of polyethylene, vinyl chloride polymerizes to form polyvinyl chloride, which has been used to make phonograph records and waterproof wearing apparel.

The reactions presented in this section are known as *addition polymerization* reactions, since the polymers formed by the reactions are the result of the addition of one molecule of monomer at a time to the ends of a growing polymer chain. Some addition polymers are given in Table 14.5. Approximately 25 million tons of synthetic organic polymers are produced each year in the United States.

An example of an **addition reaction** of alkenes other than polymerization is the addition of a halogen to the carbon-carbon double bond:

$$\underset{\text{propylene}}{CH_2{=}CH-CH_3} + Cl_2 \longrightarrow \underset{\text{1,2—dichloropropane}}{\underset{Cl}{CH_2}-\underset{Cl}{CH}-CH_3}$$

It is necessary to number the positions of the chlorine atoms in the name of the resultant alkyl halide so that it is understood not to be another possible dichloropropane isomer such as one of the following:

$$\underset{\text{1,3—dichloropropane}}{\overset{1}{\underset{Cl}{CH_2}}-\overset{2}{CH_2}-\overset{3}{\underset{Cl}{CH_2}}} \qquad \underset{\text{1,1—dichloropropane}}{\overset{1}{Cl_2CH}-\overset{2}{CH_2}-\overset{3}{CH_3}}$$

$$\underset{\text{2,2—dichloropropane}}{\overset{1}{CH_3}-\overset{2}{C}Cl_2-\overset{3}{CH_3}}$$

Aromatics

A special type of unsaturated hydrocarbon that has unique chemical properties is benzene:

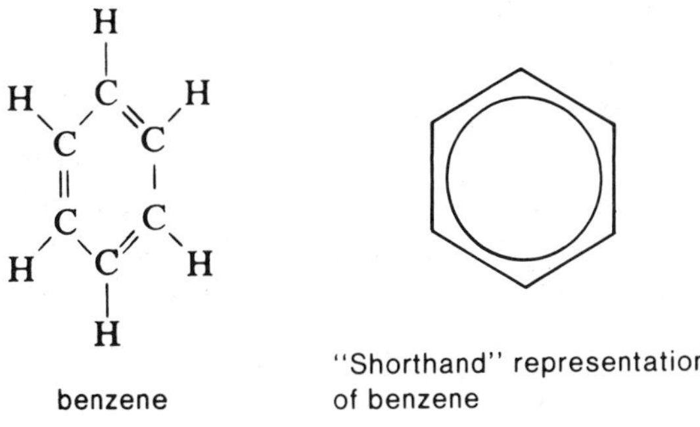
benzene

"Shorthand" representation of benzene

In view of its unusual chemistry, a more meaningful representation of the benzene molecule is portrayed in the "shorthand" structure on the right.

Benzene compounds are known as **aromatic hydrocarbons** and do not add other molecules to their double bonds, nor do they polymerize with themselves. Benzene's unusual properties are the result of the alternation of its double bonds with single bonds in a cyclic structure. This particular arrangement allows some of the electrons in the double bonds of benzene to circulate around the benzene molecule, with the result that the electrons are not available to react as do those of normal unsaturated hydrocarbons. Instead, benzene undergoes substitution reactions similar to the reactions of saturated hydrocarbons in which other atoms or groups of atoms are substituted for the hydrogens of benzene. For instance, toluene reacts with nitric acid to form trinitrotoluene (TNT):

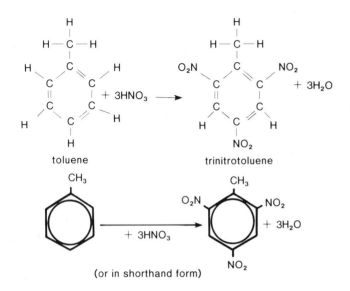

```
          H                               H
          |                               |
       H—C—H                           H—C—H
          |                               |
  H       C       H          O2N          C          NO2
    \   /   \\   /               \     /   \\    /
     C         C                   C         C        + 3H2O
     ||        |  + 3HNO3  ——→     ||        |
     C         C                   C         C
    /   \   //  \                 /   \   //   \
  H       C       H             H       C        H
          |                             |
          H                            NO2
       toluene                    trinitrotoluene
```

```
     CH3                          CH3
      |                     O2N    |    NO2
  (benzene ring)  ————————→   (benzene ring)
                  + 3HNO3                     + 3H2O
                                   |
                                  NO2
              (or in shorthand form)
```

All organic compounds, with the exception of hydrocarbons, have **functional groups** attached to a carbon chain or ring. The reactive nonalkane portion of the molecule is the functional group. This group determines the chemical properties of the molecule. Selected functional groups are listed in Table 14.6.

Alcohols and Ethers

Organic compounds that have a *hydroxyl* group (—O—H) in place of one of the hydrogen atoms of a hydrocarbon are called *alcohols*. Methyl alcohol (wood alcohol) is produced by heating wood in the absence of air to yield charcoal, methyl alcohol, and other chemicals. (A *methyl* group is CH_3—.) Large quantities of methyl alcohol are also made by reacting carbon monoxide with hydrogen. Methyl alcohol is used mainly as an antifreeze and as a solvent. Like ethyl alcohol, methyl alcohol is an intoxicant, but it is a very poisonous chemical to ingest. Methyl alcohol attacks the optic nerve of the eyes; unwise consumption could result in blindness, or possibly even death.

```
     H
     |
  H—C—O—H
     |
     H
methyl alcohol
```

Perhaps the most widely known alcohol is ethyl alcohol, the alcohol present in all alcoholic beverages and also widely used for industrial purposes.

```
     H  H
     |  |
  H—C—C—O—H
     |  |
     H  H
 ethyl alcohol
```

Both methyl and ethyl alcohols are mixed with gasoline in some regions of the United States. A few countries, such as Brazil, use ethyl alcohol undiluted with gasoline as their main source of fuel.

Ethyl alcohol is often called grain alcohol since it is produced in large quantities by the fermentation of the starch present in various grains or from the sugars present in sugar cane or sugar beets.

$$\underset{\text{sugar}}{C_{12}H_{22}O_{11}} + H_2O \xrightarrow{\text{enzyme}} \underset{\text{ethyl alcohol}}{4CH_3CH_2OH} + 4CO_2$$

Enzymes are *special organic catalysts that are produced by living organisms.* Enzymes can also be synthesized by chemists. The enzyme used in the previous reaction is present in yeast.

The concentration of ethyl alcohol in such beverages as beer, wine, and whiskey is given by the "proof" of the beverage. Proof is twice the alcohol content on a percentage basis. Pure ethyl alcohol (100 percent) is defined as 200 proof, and a bottle marked 90 proof is 45 percent ethyl alcohol by volume.

Isopropyl alcohol, also known as rubbing alcohol, is used extensively as a solvent. Isopropyl alcohol, as well as higher-molecular-weight alcohols, is not intoxicating.

```
     H     H
     |     |
  H—C———C—O—H
     |     |
     H  H—C—H
           |
           H
  isopropyl alcohol
```

Ethylene glycol, a molecule containing two hydroxyl groups, is the main ingredient in so-called permanent antifreezes.

```
        H  H
        |  |
  H—O—C—C—O—H
        |  |
        H  H
   ethylene glycol
```

Glycerol, which contains three hydroxyl groups, finds widespread use in tobacco products and cosmetics because of its moisture-retaining property.

```
        H  H  H
        |  |  |
  H—O—C—C—C—O—H
        |  |  |
        H  O  H
           |
           H
       glycerol
```

Table 14.6

Some of the functional groups that attach to a carbon ring or chain.

General structure	General name	Example	Specific name
$\begin{array}{cccc} & \mid & & \\ - & \mathrm{C} & -\mathrm{O} & -\mathrm{H} \\ & \mid & & \end{array}$	an alcohol	$\begin{array}{ccccc} & \mathrm{H} & & \mathrm{H} & \\ & \mid & & \mid & \\ \mathrm{H}- & \mathrm{C} & - & \mathrm{C} & -\mathrm{O}-\mathrm{H} \\ & \mid & & \mid & \\ & \mathrm{H} & & \mathrm{H} & \end{array}$	ethyl alcohol
$\begin{array}{ccccc} & \mid & & \mid & \\ - & \mathrm{C} & -\mathrm{O}- & \mathrm{C} & - \\ & \mid & & \mid & \end{array}$	an ether	$\begin{array}{ccccccc} & \mathrm{H} & & \mathrm{H} & & \mathrm{H} & \\ & \mid & & \mid & & \mid & \\ \mathrm{H}- & \mathrm{C} & -\mathrm{O}- & \mathrm{C} & - & \mathrm{C} & -\mathrm{H} \\ & \mid & & \mid & & \mid & \\ & \mathrm{H} & & \mathrm{H} & & \mathrm{H} & \end{array}$	methyl ethyl ether
$\begin{array}{cccc} & \mathrm{O} & & \\ & \Vert & & \\ - & \mathrm{C} & -\mathrm{O} & -\mathrm{H} \end{array}$	a carboxylic acid (an organic acid)	$\begin{array}{ccccc} & \mathrm{H} & & \mathrm{O} & \\ & \mid & & \Vert & \\ \mathrm{H}- & \mathrm{C} & - & \mathrm{C} & -\mathrm{O}-\mathrm{H} \\ & \mid & & & \\ & \mathrm{H} & & & \end{array}$	acetic acid
$\begin{array}{ccccc} & \mathrm{O} & & & \\ & \Vert & & \mid & \\ - & \mathrm{C} & -\mathrm{O}- & \mathrm{C} & - \\ & & & \mid & \end{array}$	an ester	$\begin{array}{ccccccc} & \mathrm{H} & & \mathrm{O} & & \mathrm{H} & \\ & \mid & & \Vert & & \mid & \\ \mathrm{H}- & \mathrm{C} & - & \mathrm{C} & -\mathrm{O}- & \mathrm{C} & -\mathrm{H} \\ & \mid & & & & \mid & \\ & \mathrm{H} & & & & \mathrm{H} & \end{array}$	methyl acetate
$\begin{array}{ccc} & \mathrm{O} & \\ & \Vert & \\ - & \mathrm{C} & -\mathrm{H} \end{array}$	an aldehyde	$\begin{array}{ccccc} & \mathrm{H} & & \mathrm{O} & \\ & \mid & & \Vert & \\ \mathrm{H}- & \mathrm{C} & - & \mathrm{C} & -\mathrm{H} \\ & \mid & & & \\ & \mathrm{H} & & & \end{array}$	acetaldehyde
$\begin{array}{ccccccc} & & & \mathrm{O} & & & \\ & \mid & & \Vert & & \mid & \\ - & \mathrm{C} & - & \mathrm{C} & - & \mathrm{C} & - \\ & \mid & & & & \mid & \end{array}$	a ketone	$\begin{array}{ccccccc} & \mathrm{H} & & \mathrm{O} & & \mathrm{H} & \\ & \mid & & \Vert & & \mid & \\ \mathrm{H}- & \mathrm{C} & - & \mathrm{C} & - & \mathrm{C} & -\mathrm{H} \\ & \mid & & & & \mid & \\ & \mathrm{H} & & & & \mathrm{H} & \end{array}$	acetone
$\begin{array}{ccc} & \mid & \\ - & \mathrm{C} & -\mathrm{NH_2} \\ & \mid & \end{array}$	an amine	$\begin{array}{ccc} & \mathrm{H} & \\ & \mid & \\ \mathrm{H}- & \mathrm{C} & -\mathrm{NH_2} \\ & \mid & \\ & \mathrm{H} & \end{array}$	methylamine
$\begin{array}{ccccc} & & & \mathrm{O} & \\ & \mid & & \Vert & \\ - & \mathrm{C} & - & \mathrm{C} & -\mathrm{O}-\mathrm{H} \\ & \mid & & & \\ & \mathrm{NH_2} & & & \end{array}$	an amino acid	$\begin{array}{ccccc} & \mathrm{H} & & \mathrm{O} & \\ & \mid & & \Vert & \\ \mathrm{H}- & \mathrm{C} & - & \mathrm{C} & -\mathrm{O}-\mathrm{H} \\ & \mid & & & \\ & \mathrm{NH_2} & & & \end{array}$	glycine

Glycerol reacts with nitric acid to produce nitroglycerine, from which dynamite is made:

$$\begin{array}{ccc} & \mathrm{H} & \\ & \mid & \\ \mathrm{H}- & \mathrm{C} & -\mathrm{O}-\mathrm{H} \\ & \mid & \\ \mathrm{H}- & \mathrm{C} & -\mathrm{O}-\mathrm{H} \\ & \mid & \\ \mathrm{H}- & \mathrm{C} & -\mathrm{O}-\mathrm{H} \\ & \mid & \\ & \mathrm{H} & \\ & \text{glycerol} & \end{array} + 3\mathrm{HNO_3} \longrightarrow \begin{array}{ccc} & \mathrm{H} & \\ & \mid & \\ \mathrm{H}- & \mathrm{C} & -\mathrm{O}-\mathrm{NO_2} \\ & \mid & \\ \mathrm{H}- & \mathrm{C} & -\mathrm{O}-\mathrm{NO_2} \\ & \mid & \\ \mathrm{H}- & \mathrm{C} & -\mathrm{O}-\mathrm{NO_2} \\ & \mid & \\ & \mathrm{H} & \\ & \text{nitroglycerine} & \end{array} + 3\mathrm{H_2O}$$

Nitroglycerine alone is too easily detonated by shock to be used safely as an explosive, but Alfred Nobel, the Swedish industrialist who established the Nobel prizes, found that mixing nitroglycerine with inert substances such as sawdust produces a product (dynamite) that can be transported and handled safely.

Ants and Nature's Acid Rain

The rain that falls in cities is not like the rain that falls in remote areas devoid of manufacturing plants and heavy vehicle traffic. City rains have harmful amounts of nitric acid and sulfuric acid and may have a pH as low as 3.1. Both city rain and "normal" (country) rain have small quantities of carbonic acid produced from the reaction of the carbon dioxide and water present in all air and some nitric acid produced by the reaction of atmospheric nitrogen and oxygen caused by lightning discharges. But rain, particularly country rain with a pH of around 5.3, also has some formic acid present. If no acids were present in the air, rain would have a pH of 7.0.

Where does the formic acid in rain come from? About 50 percent of it comes from ants! Formic acid is the noxious chemical that ants, especially red ants, inject into a person's skin when they sting. Formic acid gets its name from the Latin *formica,* meaning "ant." The other 50 percent of formic acid comes from the chemical reactions in the air that convert formaldehyde and methane into formic acid.

One may wonder how very small ants can possibly produce enough formic acid to appreciably affect rain. Even though each individual ant will produce only a small quantity of formic acid, there are estimated to be 10 million billion ants in the world. Since approximately 15 percent of formicine ants' body weight is formic acid, ants release about 150 million kilograms of formic acid into the atmosphere each year.

Instead of sound, ants use formic acid, other organic acids of high molecular weight, esters, and proteins to communicate. Such chemical messengers are called pheromones. Ants also use formic acid for defense.

Most fire ants, when they bite, also inject cyclic organic compounds containing nitrogen, as well as certain proteins ■

In ether compounds, two carbon atoms are joined to a single oxygen atom. Diethyl ether, usually just called *ether,* was used for over a century as a general anesthetic in hospitals.

```
    H   H       H   H
    |   |       |   |
H — C — C — O — C — C — H
    |   |       |   |
    H   H       H   H
```

diethyl ether ("ether")

Organic Acids

Molecules of organic acids all contain the *carboxyl* group:

```
  O
  ||
— C — O — H
```

Like an inorganic acid, an *organic acid* (a carboxylic acid) is a covalent compound that ionizes in water to produce a hydronium ion. A carboxylic acid does not give up its proton as readily as an inorganic acid and is, therefore, a weaker acid than hydrochloric or sulfuric acid.

The simplest carboxylic acid is formic acid, which was first isolated from the distillation of red ants. In fact, the pain produced by the sting of a red ant is the result of the ant's giving a hypodermic injection of formic acid to its victim.

```
    O
    ||
H — C — O — H
```

formic acid

Acetic acid derived its name from the Latin *acetum,* meaning "vinegar," since vinegar is a 5 percent solution of acetic acid in water.

```
    H   O
    |   ||
H — C — C — O — H
    |
    H
```

acetic acid

Butyric acid has a very unpleasant odor. The objectionable odors of perspiration, rancid butter, and certain strong cheeses are due to the presence of butyric acid.

```
    H   H   H   O
    |   |   |   ||
H — C — C — C — C — O — H
    |   |   |
    H   H   H
```

butyric acid

Capric acid is named from the Latin *capricornus,* meaning "goat," since it is this acid in the skin secretions of goats that gives them an unpleasant smell.

```
    H   H   H   H   H   H   H   H   H   O
    |   |   |   |   |   |   |   |   |   ||
H — C — C — C — C — C — C — C — C — C — C — O — H
    |   |   |   |   |   |   |   |   |
    H   H   H   H   H   H   H   H   H
```

capric acid

It might be interesting to blue-cheese and roquefort-cheese lovers that the rank smell of these cheeses is due mainly to the presence of the "goat acid."

Citric acid occurs in many plants and animals but is most prominent in certain citrus fruits (from which citric acid gets its name) in which the acid tartness is easily detected. Artificial lemon juice is a 7 percent solution of citric acid. Because of citric acid's tartness and pleasant flavor, it is widely used in soft drinks and foods.

Table 14.7
Some organic acids.

Name	Formula	Present in
carbonic (carbonated water)	$[H_2CO_3]$	soft drinks
acetic	$HC_2H_3O_2$	vinegar
oxalic	$H_2C_2O_4$	rhubarb
lactic	$HC_3H_5O_3$	buttermilk (sour milk)
tartaric	$H_2C_4H_4O_6$	grapes
citric	$H_3C_6H_5O_7$	citrus fruits
acetylsalicylic	$HC_9H_7O_4$	aspirin

Probably the most commonly used drug is the carboxylic acid (and ester) acetylsalicylic acid, known as aspirin. Aspirin is used for reducing pain (its analgesic effect) and for reducing fever (its antipyretic effect).

```
                  O
                  ‖
                  C–O–H
          O       |
          ‖      / \
   CH3–C–O–|  ○  |
                 \ /
```

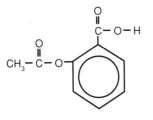

Table 14.7 lists some organic acids; some commonly used products containing organic acids are pictured in Figure 14.2.

Neutralization of long-chain organic acids with bases forms salts that find various uses in our everyday lives. For example, sodium and potassium salts of long-chain carboxylic acids are the main ingredients of bath soaps, and the calcium salt of propionic acid, called calcium propionate, is added to bakery products and processed cheeses as a fungicide to retard spoilage.

```
    H  H  O                  O  H  H
    |  |  ‖                  ‖  |  |
 H—C—C—C—O⁻¹   Ca⁺²   O⁻¹—C—C—C—H
    |  |                        |  |
    H  H                        H  H
```
calcium propionate

Esters, Fats, and Oils

A carboxylic acid reacts with an alcohol to produce a compound called an *ester:*

```
    H  O                        H  H
    |  ‖    ┌─────────────┐     |  |
 H—C—C─┤O—H + H├─O—C—C—H ———→
    |       └─────────────┘     |  |
    H                           H  H
 acetic acid               ethyl alcohol
```

Figure 14.2 A few commercially available organic compounds. All products shown, except the alcohol and aspirin, are alkanes.

```
    H  O       H  H
    |  ‖       |  |
 H—C—C—O—C—C—H + H2O
    |          |  |
    H          H  H
```
ethyl acetate (an ester)

An ester, then, is a compound in which the proton of a carboxylic acid has been replaced with a group containing one or more carbon atoms. Esters are widespread in nature and are mainly responsible for the pleasant fragrances of flowers and the delightful aromas and savory flavors of fruits. The aromatic bouquet of brandies and wines is also produced by esters. Although flowers and fruits generally contain a complex mixture of esters, quite often a single predominant ester is mainly responsible for their fragrances. For instance, there are four esters that contribute to the aroma and taste of pineapple; however, ethyl butyrate has been detected in the fruit in overwhelming abundance. Other esters are manufactured in the laboratory to be used as artificial fruit essences and in perfumes. Such is the case for ethyl acetate, prepared according to the preceding reaction. The characteristic smell and flavor of bananas are due to the presence of isoamyl acetate and butyl acetate. The structures of these two esters are as follows:

```
                          H
                          |
    H  O       H  H  H—C—H   H
    |  ‖       |  |       |      |
 H—C—C—O—C—C————C————C—H
    |          |  |       |      |
    H          H  H       H      H
```
isoamyl acetate

```
    H  O       H  H  H  H
    |  ‖       |  |  |  |
 H—C—C—O—C—C—C—C—H
    |          |  |  |  |
    H          H  H  H  H
```
butyl acetate

Pears contain the ester amyl acetate:

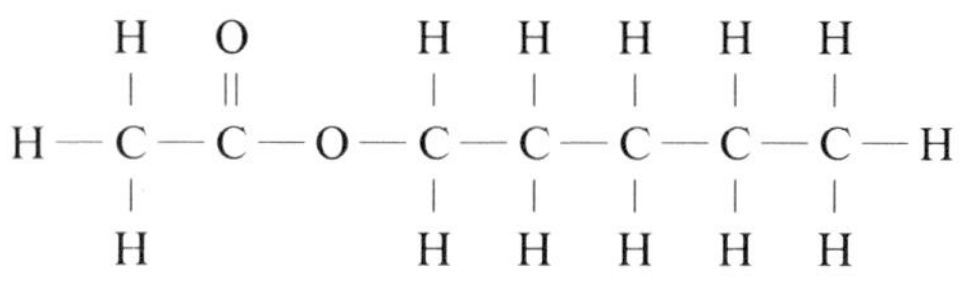

$$\begin{array}{ccccccccccccccccccc} & & H & & O & & & & H & & H & & H & & H & & H & & \\ & & \mid & & \| & & & & \mid & & \mid & & \mid & & \mid & & \mid & & \\ H & - & C & - & C & - & O & - & C & - & C & - & C & - & C & - & C & - & H \\ & & \mid & & & & & & \mid & & \mid & & \mid & & \mid & & \mid & & \\ & & H & & & & & & H & & H & & H & & H & & H & & \end{array}$$

amyl acetate

In view of how different pears and bananas smell and taste, it is amazing how similar in structure are the esters responsible for the odor and flavor of the two fruits. Even the delicate fragrance of roses is due to a bewildering mixture of many esters, although the main component is amyl undecanoate.

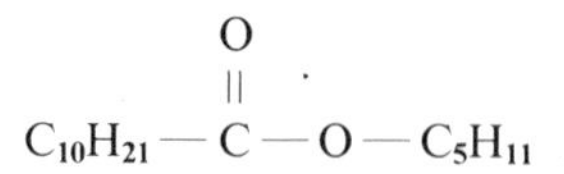

$$\begin{array}{ccccccc} & & O & & & & \\ & & \| & & & & \\ C_{10}H_{21} & - & C & - & O & - & C_5H_{11} \end{array}$$

amyl undecanoate

Since esters may be synthesized very readily in the laboratory, they are used in various mixtures in the preparation of perfumes and artificial food flavors. Such perfumes and flavors are often much less expensive than the natural products.

Fats and *oils* are special types of esters formed from the reaction of carboxylic acids with an alcohol (glycerol) containing three hydroxyl groups. Shown below is the reaction between three molecules of a carboxylic acid and one molecule of glycerol. The R in the carboxylic acid molecules represents a long hydrocarbon chain.

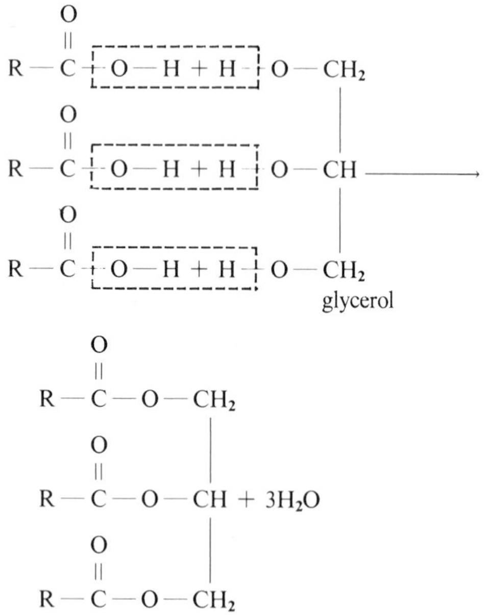

$$\begin{array}{ccccccccccc} & & O & & & & & & & & \\ & & \| & & & & & & & & \\ R & - & C & - & O & - & H + H & - & O & - & CH_2 \\ & & O & & & & & & & & \mid \\ & & \| & & & & & & & & \mid \\ R & - & C & - & O & - & H + H & - & O & - & CH \\ & & O & & & & & & & & \mid \\ & & \| & & & & & & & & \mid \\ R & - & C & - & O & - & H + H & - & O & - & CH_2 \end{array} \longrightarrow$$

glycerol

$$\begin{array}{ccccccc} & & O & & & & \\ & & \| & & & & \\ R & - & C & - & O & - & CH_2 \\ & & O & & & & \mid \\ & & \| & & & & \mid \\ R & - & C & - & O & - & CH + 3H_2O \\ & & O & & & & \mid \\ & & \| & & & & \mid \\ R & - & C & - & O & - & CH_2 \end{array}$$

a fat or oil

The difference between fats and oils is simply that at room temperature fats are solids and oils are liquids. Fats and oils may be of animal origin (such as beef tallow or lard) or of vegetable origin (such as peanut oil or soybean oil). Vegetable oils contain a greater number of unsaturated hydrocarbon chains than animal fats or oils do.

Many of the waxes found in nature are high-molecular-weight esters. For example, beeswax is mostly myricyl palmitate, one molecule of which contains a total of 47 carbon atoms.

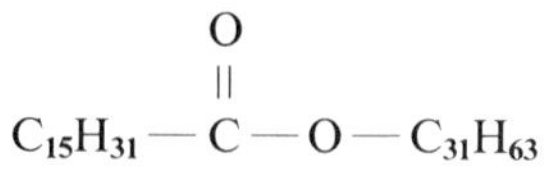

$$\begin{array}{ccccccc} & & O & & & & \\ & & \| & & & & \\ C_{15}H_{31} & - & C & - & O & - & C_{31}H_{63} \end{array}$$

myricyl palmitate

Aldehydes and Ketones

If the hydroxyl group of a carboxylic acid is replaced by a hydrogen atom, the resulting compound is called an *aldehyde*. When the hydroxyl group is replaced by a hydrocarbon group, the resulting compound is called a *ketone*.

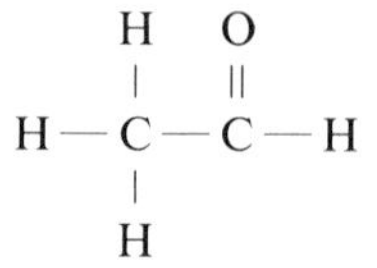

$$\begin{array}{ccccccc} & & H & & O & & \\ & & \mid & & \| & & \\ H & - & C & - & C & - & H \\ & & \mid & & & & \\ & & H & & & & \end{array}$$

acetaldehyde
(an aldehyde)

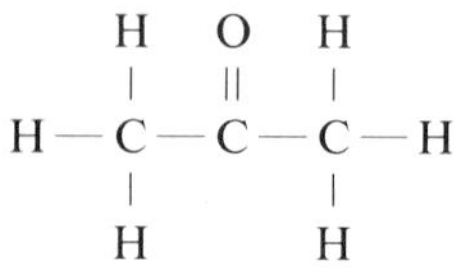

$$\begin{array}{ccccccccc} & & H & & O & & H & & \\ & & \mid & & \| & & \mid & & \\ H & - & C & - & C & - & C & - & H \\ & & \mid & & & & \mid & & \\ & & H & & & & H & & \end{array}$$

acetone
(a ketone)

As can be seen from their formulas, aldehydes and ketones are characterized by the presence of

$$\begin{array}{ccc} & O & \\ & \| & \\ - & C & - \end{array}$$

which is known as the *carbonyl* group. Aldehydes have at least one hydrogen atom attached to the carbonyl group, whereas ketones have only carbon atoms attached to the carbonyl group. Aldehydes and ketones generally have rather pleasant odors, and those aldehydes having approximately 10 carbon atoms in their molecular chains are used in perfume formulations. The simplest aldehyde is formaldehyde.

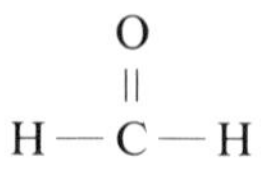

$$\begin{array}{ccccc} & & O & & \\ & & \| & & \\ H & - & C & - & H \end{array}$$

formaldehyde

Formaldehyde has a very irritating odor and is used as a disinfectant. A water solution of formaldehyde, called *formalin,* is used as a preservative for animal specimens. Ketones are widely used as solvents. One of the ketones, methyl amyl ketone, is partially responsible, along with carboxylic acids, for the distinctive odor of blue cheese. The characteristic flavor of butter is due to the presence of the following two ketones:

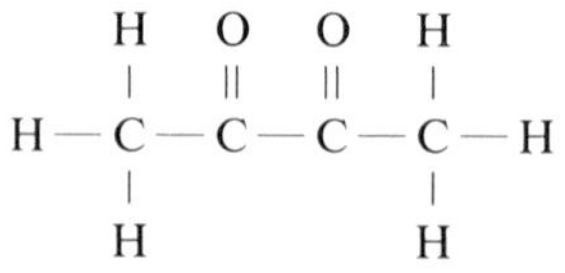

$$\begin{array}{ccccccccccc} & & H & & O & & O & & H & & \\ & & \mid & & \| & & \| & & \mid & & \\ H & - & C & - & C & - & C & - & C & - & H \\ & & \mid & & & & & & \mid & & \\ & & H & & & & & & H & & \end{array}$$

diacetyl

$$\begin{array}{ccccccccccc} & & & & H & & & & & & \\ & & & & / & & & & & & \\ & & H & & O & & O & & H & & \\ & & \mid & & \mid & & \| & & \mid & & \\ H & - & C & - & C & - & C & - & C & - & H \\ & & \mid & & \mid & & & & \mid & & \\ & & H & & H & & & & H & & \end{array}$$

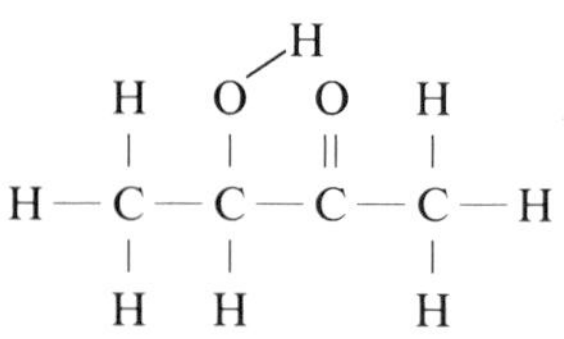

3-hydroxybutanone

Carbohydrates

What do wood, cotton, starch, sugar, and the exoskeletons of insects and crabs all have in common? They all are made up to a large extent of a class of organic compounds called carbohydrates. *Carbohydrates* are molecules containing many hydroxyl groups (alcohol groups) plus usually an aldehyde or a ketone group. The simplest carbohydrates all exhibit varying degrees of sweetness, and for this reason carbohydrates in general are referred to as saccharides, from the Latin *saccharum,* meaning "sugar." We will deal briefly with monosaccharides, disaccharides (carbohydrate molecules containing two monosaccharide units), and polysaccharides (carbohydrate molecules containing many monosaccharide units).

The most important monosaccharide is glucose, which, either as free glucose or combined glucose in disaccharides and polysaccharides, is the most abundant organic compound on Earth. Glucose is the monomer in the natural polymers starch and cellulose and is also present in sugar, milk, honey, syrups, flower nectars, and many fruits. The structure of glucose is typical of many of the monosaccharide structures. Note that glucose is a polyalcohol with an aldehyde group, whereas fructose is a polyalcohol with a ketone group. Fructose is found along with glucose in honey and fruits. Fructose is the sweetest of all carbohydrates, being sweeter even than cane sugar (sucrose).

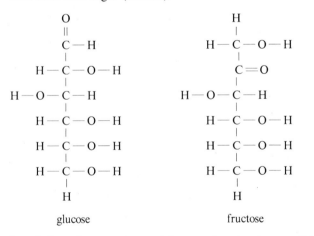

```
        O                       H
        ||                      |
        C—H                 H—C—O—H
        |                       |
    H—C—O—H                     C=O
        |                       |
H—O—C—H                 H—O—C—H
        |                       |
    H—C—O—H                 H—C—O—H
        |                       |
    H—C—O—H                 H—C—O—H
        |                       |
    H—C—O—H                 H—C—O—H
        |                       |
        H                       H
     glucose                fructose
```

In solution, glucose exists mainly as a ring structure, as follows:

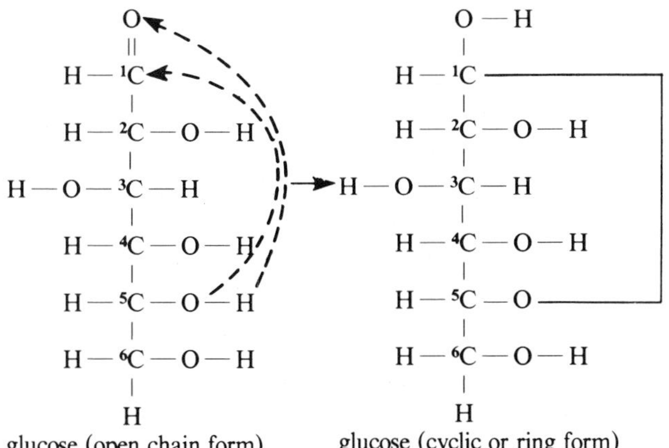

glucose (open chain form) glucose (cyclic or ring form)

The cyclic form is usually represented as a Haworth formula:

```
              ⁶CH₂OH
               |
              ⁵C———————O
             / |        \
        H   /  H         \   H
         \ /              \ /
         ⁴C                ¹C
         /  \             /  \
     H—O     \   O—H   H /    O—H
              \  |     |/
              ³C———————²C
               |        |
               H        O—H
```

glucose
(Haworth projection formula)

```
        O                       O
        ||                      ||
        C—H                     C—H
        |                       |
    H—C—O—H                 H—C—H
        |                       |
    H—C—O—H                 H—C—O—H
        |                       |
    H—C—O—H                 H—C—O—H
        |                       |
    H—C—O—H                 H—C—O—H
        |                       |
        H                       H
      ribose               deoxyribose
```

The organic compounds ribose and deoxyribose are present in various organic catalysts called enzymes, and deoxyribose is a vital constituent of *deoxyribonucleic acid* (DNA), the double helix structure that carries the genetic code for the chromosomes of cells.

The sugar sucrose was probably the first organic compound to be isolated in a pure state. The first recorded purification of sucrose from sugar-producing plants occurred in India about A.D. 300.

Sucrose is formed by chemically combining the monosaccharides glucose and fructose. Sucrose is, therefore, a disaccharide. Sucrose is more widely known as *table sugar* and is extracted from the juices of sugar cane and sugar beets. Another common disaccharide, lactose, contains the monosaccharides glucose and galactose. Lactose is known as *milk sugar,* since it is present in the milk of all mammals.

The polysaccharides of greatest significance and distribution in nature are starch, glycogen, and cellulose. All three of these polysaccharides are natural polymers composed of the monomer glucose. The difference between starch and cellulose is essentially the way in which the glucose units are chemically bound. Starch and glycogen have the same chemical bonds but differ in size and shape of their molecules.

Starch is a reserve food supply that plants store in various parts of their structures, such as their roots, seeds, and fruit. The number of glucose units present in a starch molecule varies from 200 to a million.

Glycogen, like starch, is made up of glucose units, but in general the number of units per glycogen molecule is greater by a factor of 10 than for starch. Glycogen, sometimes called *animal starch,* serves a function for animals similar to the function that starch serves for plants. Glycogen acts as a reserve supply of glucose units in various animal tissues. When the level of glucose in the blood drops below the critical value, glycogen from the liver is broken down into glucose. The glucose then enters the bloodstream to maintain the necessary blood glucose level.

The fibrous parts of plants are made up of cellulose; dry wood is approximately 50 percent cellulose and cotton is 98 percent cellulose. It is the cellulose part of wood that is utilized in the production of paper. Whereas starch is quite easily digested by the enzyme systems of animals, cellulose cannot be directly decomposed by most animal enzymes into glucose in the way that starch is. This inability of the enzyme systems is a direct result of the difference in the way in which the glucose units in cellulose and starch are held together. Herbivores do eat vegetation composed mainly of cellulose, but these animals' stomachs contain microorganisms that convert the cellulose into glucose, which the herbivores can digest.

Amines, Amino Acids, and Proteins

Amines are compounds that have anywhere from one to three carbon atoms attached to a nitrogen atom. Most amines have objectionable odors. The decay of flesh, for example, is accompanied by the liberation of various amines. For example, part of the unpleasant odor of decaying fish is due to trimethylamine,

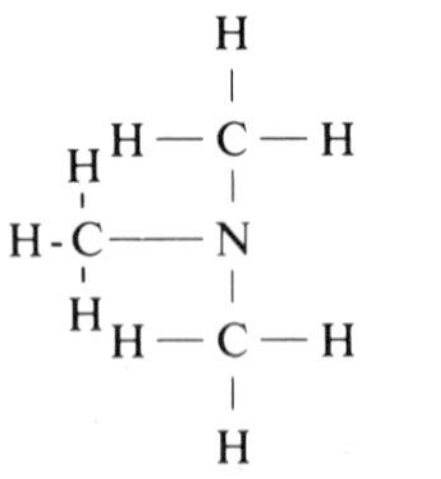

$$N(CH_3)_3$$

trimethylamine

and much of the disagreeable smell of other types of putrefying flesh is due to the presence of two amines (*putrescine* and *cadaverine*), each of which contains two *amino* groups ($-NH_2$). The names of these two amines obviously are intended to reflect their offensive odors.

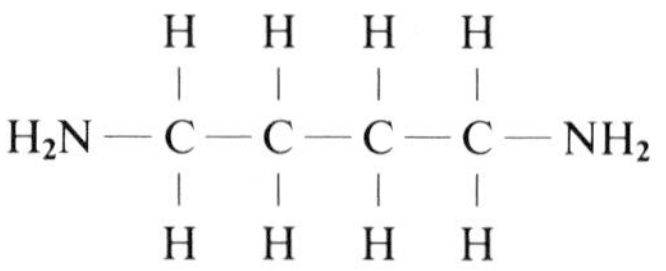

$$H_2N-CH_2-CH_2-CH_2-CH_2-NH_2$$

putrescine

$$H_2N-CH_2-CH_2-CH_2-CH_2-CH_2-NH_2$$

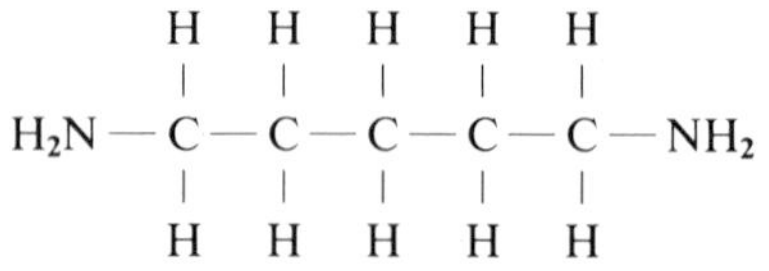

cadaverine

Amines are organic bases and react with carboxylic acids to form compounds called *amides.*

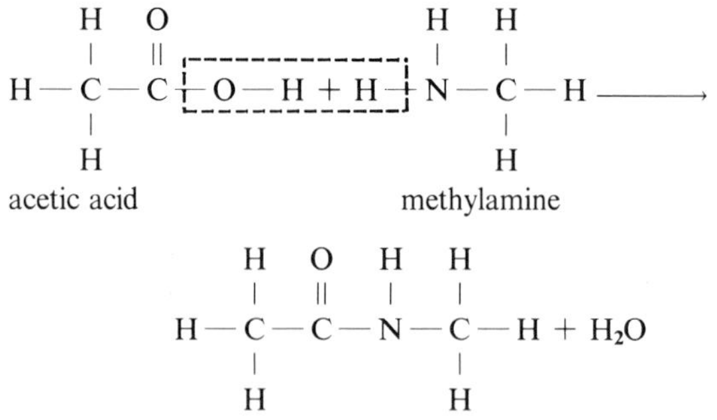

$$CH_3-C(=O)-O-H + H-N(H)-CH_3 \longrightarrow$$

acetic acid; methylamine

$$CH_3-C(=O)-N(H)-CH_3 + H_2O$$

an amide
(N-methylacetamide)

Nylon is a polymer containing many amide units and has the general formula

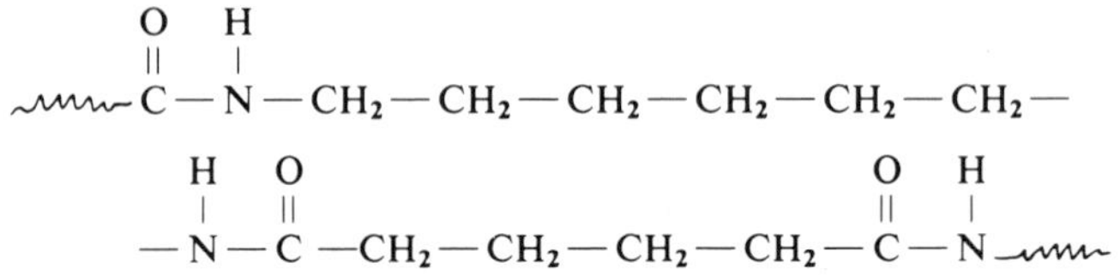

$$\sim C(=O)-N(H)-CH_2-CH_2-CH_2-CH_2-CH_2-CH_2-$$
$$-N(H)-C(=O)-CH_2-CH_2-CH_2-CH_2-C(=O)-N(H)\sim$$

Amino acids are compounds that contain both an amino group and a carboxylic acid group. Although the total number of amino acids that can be synthesized in the laboratory is virtually unlimited, there are only about 100 that occur in nature. Of these, about 25 occur in overwhelming abundance. The formula of amino acids can, in general, be represented as

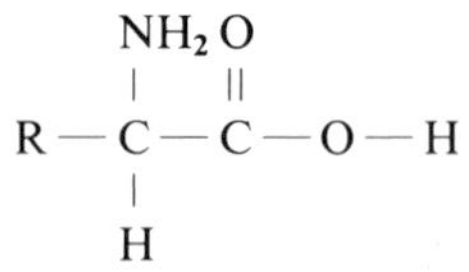

$$R-CH(NH_2)-C(=O)-O-H$$

where R represents a large variety of structures. If the R-group of an amino acid is a hydrogen atom, the resulting molecule is the simplest possible amino acid, glycine:

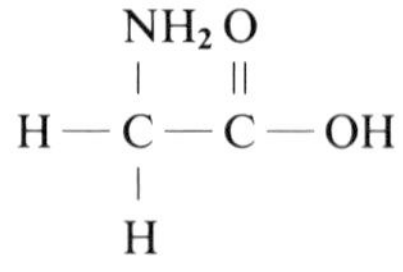

$$H-CH(NH_2)-C(=O)-OH$$

glycine

If the R-Group is

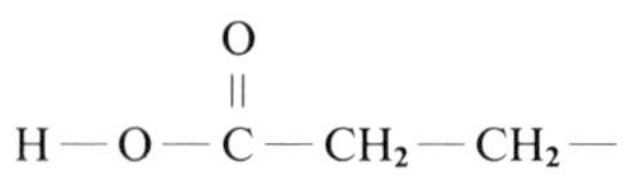

$$H-O-C(=O)-CH_2-CH_2-$$

the resulting molecule is glutamic acid:

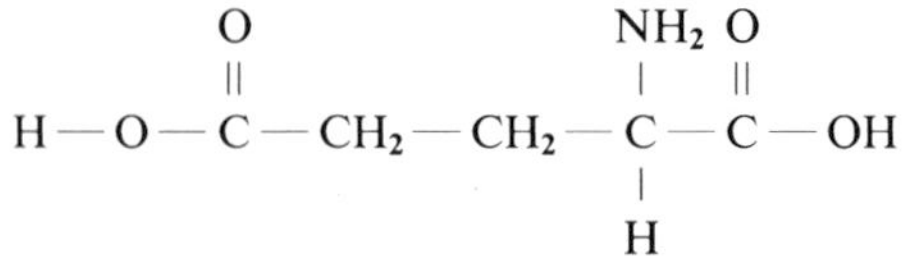

$$H-O-C(=O)-CH_2-CH_2-CH(NH_2)-C(=O)-OH$$

glutamic acid

The single-sodium salt of glutamic acid is called monosodium glutamate (MSG), which is used extensively in foods to increase the sensitivity of the taste buds, with the result that flavors naturally present in foods are considerably enhanced. Aspartame (trade names: Equal and NutraSweet) is an amino acid derivative that is 160 times sweeter than table sugar (sucrose), but it contains very few calories.

All living things require the presence of amino acids to carry out various biochemical functions necessary to life. Plants can synthesize all the amino acids they require from inorganic substances present in the air and soil. Animals, however, can only synthesize about half of the amino acids that they require and must obtain the remaining amino acids that they need by eating plants or by eating animals that have consumed plants.

In essentially the same way in which molecules of ethylene react to form polyethylene and molecules of amines and carboxylic acids react to form the polymer nylon, amino acids can react to form large polymeric molecules known as *proteins.* Proteins have the following general formula:

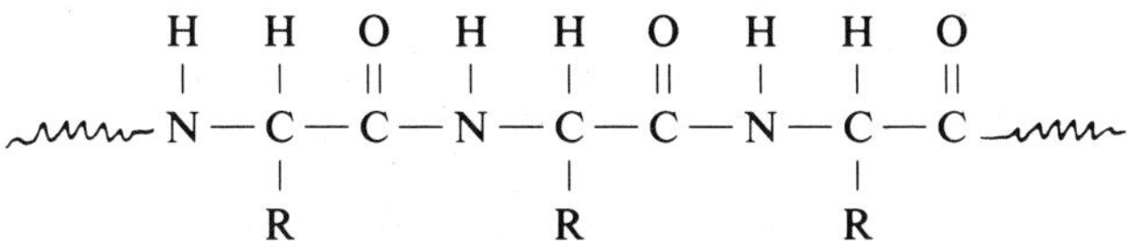

```
        H   H   O   H   H   O   H   H   O
        |   |   ||  |   |   ||  |   |   ||
 ~~~~ N — C — C — N — C — C — N — C — C ~~~~
            |           |           |
            R           R           R
```

Since proteins are of such fundamental importance to life forms (the word "protein" is from the Greek, *proteios,* "first in importance"), and since proteins are made up of amino acids, amino acids are frequently called the building blocks of life. Proteins may contain as few as 51 amino acid molecules, as in the hormone insulin (a deficiency of which causes diabetes), or as many as millions of amino acid molecules, strung like beads on a string, as in virus proteins.

Although all parts of living organisms contain varying amounts of proteins, some parts contain much higher concentrations than others. The flesh, skin, hair, fingernails, toenails, horns, and hoofs of animals and the seeds of plants have high protein concentrations. Proteins make up about 75 percent of the dry weight of animals and are the structural components of animal bodies. All biochemical reactions that occur in living cells are controlled by enzymes.

Literally thousands upon thousands of different protein molecules exist in nature, but only about 25 different amino acids are prevalent. How is it possible for such a great variety of separate protein molecules to be produced from only 25 different amino acids? In order to answer this question, we must first answer the question, How do the proteins differ from one another? The difference in various proteins is a result of the difference in the sequence in which the various amino acids are arranged. Thus, if we let letters of the alphabet represent different amino acids in a protein molecule, one protein may have an amino acid sequence of ACAGGHACY . . . , whereas another protein may have a sequence of ACCHAAYGG. . . . It is because of this difference in possible sequences of the amino acids in protein molecules that there can be such a great variety of protein molecules built up from the same few amino acids. If only five different amino acids existed from which to make protein molecules, they could be arranged in various sequences to produce 120 different molecules. Double the number of different amino acids available and 3,628,800 different molecules could be constructed. Triple the number (15) of starting amino acids and over 100 billion different molecules could be prepared! With 25 amino acids available in nature, we should not be surprised at the diversity of protein molecules found in different life forms.

DNA—The Fundamental Molecule of Life

Inheritance is controlled by genes that reside in the chromosomes of the nucleus of all cells. Each gene carries a single characteristic, such as hair color or color blindness. It is now known that genes are simple DNA (deoxyribonucleic acid) molecules or segments of DNA molecules. Each organic chemical reaction occurring in cells is controlled by an enzyme (organic catalyst). Each particular hereditary trait in a cell's chromosomes is the result of the action of a specific enzyme.

The DNA molecule is a very large polymer made up of three different essential units: (1) a carbohydrate sugar, deoxyribose; (2) an inorganic acid, phosphoric acid (H_3PO_4); and (3) an organic nitrogen base. The nitrogen-containing bases that are present in the DNA molecule are adenine, cytosine, guanine, and thymine. The structure of one of the nitrogen bases in DNA is as follows:

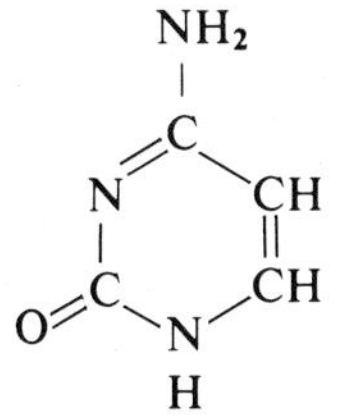

```
          NH2
           |
           C
         //  \
        N     CH
        |     ||
        C     CH
      // \   /
     O    N
          |
          H
```

cytosine

The DNA molecule consists of chemicals arranged, for example, in the following manner.

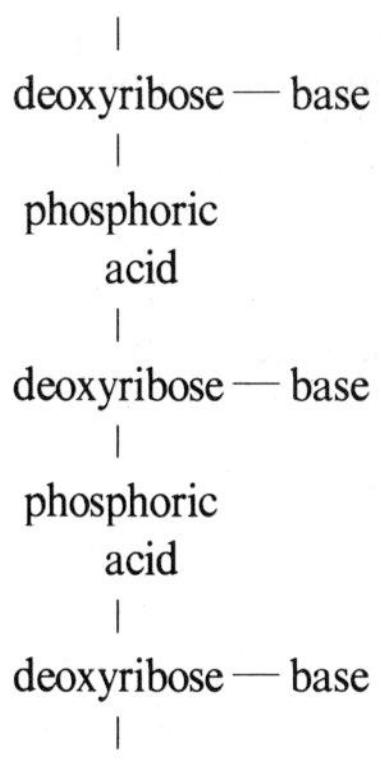

```
       |
  deoxyribose — base
       |
   phosphoric
      acid
       |
  deoxyribose — base
       |
   phosphoric
      acid
       |
  deoxyribose — base
       |
```

Actually, however, DNA exists as two separate molecules held together by weak bonds (shown as small dashes in Fig. 14.3). The two molecules twist together to form a double helix. The *genetic*

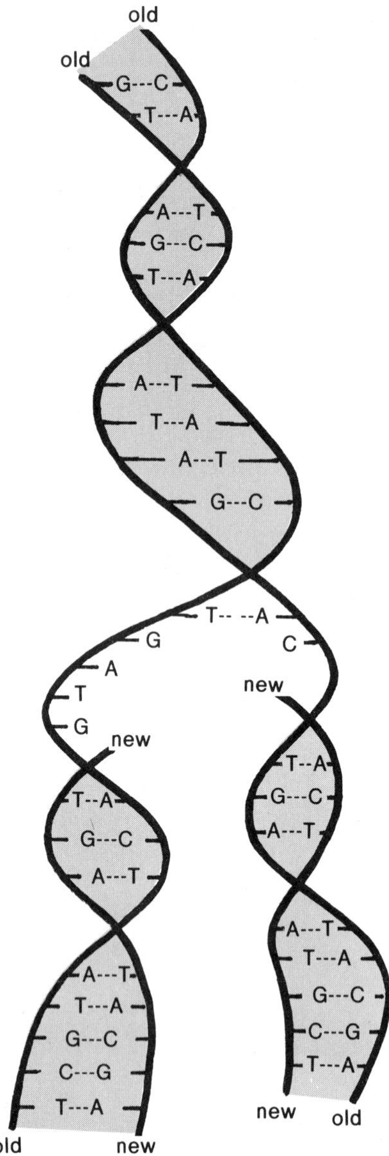

Figure 14.3 A parent DNA molecule duplicates itself to produce two identical daughter DNA molecules. The letters A, C, G, and T represent adenine, cytosine, guanine, and thymine, respectively. The heavy lines represent the deoxyribose-phosphoric acid "backbone."

code is the result of the sequence in which the organic nitrogen bases are arranged in a DNA molecule. That is, DNA molecules contain a "blueprint" by which new proteins are synthesized. Thus, in the synthesis of a protein molecule, each amino acid present in the molecule is predetermined by a single code unit of three organic nitrogen bases.

By varying the sequence of the single code units and by varying the organic nitrogen bases within each code unit, the specific amino acids present in a protein molecule, as well as the particular position of each amino acid in the molecule, are precisely controlled. Bear in mind that there are only 4 different organic nitrogen bases in DNA and that a set of only 3 of the 4 in a definite sequence determines the code for a single amino acid. There are 64 ways in which the 4 bases can be grouped into units of 3 bases. Since only 20 to 25 amino acids are used in synthesizing proteins, this "triplet" code system is more than adequate for the synthesis of all known proteins. Consequently, if a protein molecule containing 100 amino acids arranged in a definite sequence is synthesized, the synthesis is controlled by a portion of a DNA molecule consisting of 100 single code units or 300 individual organic nitrogen bases.

The arrangements of the triplet units of bases in each single code unit constitute a coded message that is "read" by the cell. The cell in turn translates the code into various proteins. Some proteins produced are incorporated into the structural portions of the cell. Other proteins that are produced are enzymes that control the many processes of cell operation. These operations include the repair of damaged proteins and the production of energy by the breakdown of carbohydrate sugars.

DNA molecules are able to duplicate themselves exactly, so that when a cell divides into two daughter cells, each daughter cell is genetically identical to the parent cell. The parent double-helix DNA molecule accomplishes this replication by separating its constituent DNA strands. Each strand, as it unwinds, synthesizes a replica of the other strand, so that by the time the parent double helix is completely unwound, two new and exact copies of the parent DNA molecule are produced. Then, when the cell divides, each daughter cell has an exact copy of the DNA molecule of the parent cell. In this way the hereditary genetic code is passed on to all new cells (see Fig. 14.3).

When a cell divides into two daughter cells, changes (mistakes) can occur in the duplication of the cell's DNA molecule. Such changes produce mutations (variant forms). Most mutations are disadvantageous but some may be harmless or even beneficial. Some mutant forms are better equipped for obtaining needed resources or responding to predators. If such is the case, the new forms will more likely survive and reproduce, and so will their offspring. Such a scenario is the essence of the concepts of *natural selection* and *evolution.*

Summary

Organic chemistry is the chemisty of carbon compounds. The fact that carbon can form millions of different compounds is the result of the ability of carbon atoms to react with other carbon atoms to form carbon chains and rings of all shapes and sizes. Organic compounds consist mainly of carbon atoms covalently bound to other carbon atoms and to various other elements. Carbon atoms can form single, double, and triple covalent bonds.

Isomers are compounds that have the same molecular formulas but different structural formulas; consequently, they have different chemical and physical properties.

A series of compounds in which each member differs from the next member by a constant amount is called a homologous series. The formula of each member of the series of alkanes differs from the member before and after it by a —CH_2— unit.

A carbon atom in a saturated organic compound has its four covalent bonds directed to the vertices of a tetrahedron. The carbon's bond angles are 109.5 degrees.

Saturated alkanes react with chlorine to produce alkyl halides that are very useful as solvents, cleaning agents, and refrigerants.

The reactive nonalkane portion of an organic molecule is called a function group. Some typical functional groups are

$$-\mathrm{O}-\mathrm{H},\quad -\overset{\mathrm{O}}{\overset{\|}{\mathrm{C}}}-\mathrm{H},\quad -\overset{\mathrm{O}}{\overset{\|}{\mathrm{C}}}-\mathrm{O}-\mathrm{H},\ \text{and}\ -\mathrm{NH_2}.$$

The process whereby many small molecules are reacted together to produce a very long chain (a polymer) is called polymerization. Chemists produce many different polymers, such as polyethylene, rayon, and Dacron, in the laboratory. Nature also produces many polymers such as proteins, starch, and cellulose.

Deoxyribonucleic acid (DNA) is a large natural polymer made up of a carbohydrate sugar (deoxyribose), an acid (phosphoric acid), and four different nitrogen bases. DNA molecules carry the genetic code of life that is passed on from one generation to the next.

Questions and Problems

The Central Role of Carbon

1. Why are there so many more compounds containing carbon than there are compounds containing other elements?

2. What is the difference between organic compounds and inorganic compounds?

3. What types of bonds are present in organic compounds?

Hydrocarbons

4. Why are hydrocarbons so important to our country's economy?

5. How does a hydrocarbon differ from other organic compounds?

Alkanes

6. What is the main source of methane, as found in coal mines and garbage dumps?

7. What is an isomer?

8. What is the molecular formula for a 16-carbon alkane?

9. Draw the structure for isononane.

10. Draw the structure of cyclohexane.

11. What is meant by octane number?

12. Name three alkyl halides and describe their uses.

13. Why is it important that a practical substitute for Freon 12 be found?

Alkenes and Alkynes

14. How does a monomer differ from a polymer?

15. What is an addition polymer?

16. Show the formation of Teflon from its monomer.

Aromatics

17. How do aromatic hydrocarbons differ from unsaturated hydrocarbons?

Alcohols and Ethers

18. In which class of organic compounds would you expect to find chemical properties similar to those of water?

19. What is the distinguishing structural feature of an alcohol?

20. What is special about the "functional group" of a molecule?

Organic Acids

21. What is the distinguishing structural feature of organic acids?

22. Name four foods or medicines in your home that contain organic acids. What acids are present in the substances?

Esters, Fats, and Oils

23. What do esters, fats, oils, and some waxes all have in common?

24. Write an example of a structural formula for each of the following compounds:

(a) isohexane
(b) polyvinyl bromide
(c) neopentyl alcohol
(d) methyl isopropyl ether
(e) ethyl butyrate
(f) 1,2-dichloropentane

Aldehydes and Ketones

25. What is the distinguishing structural feature of an aldehyde?

Carbohydrates

26. From what two simple sugars is sucrose (table sugar) formed?

27. What carbohydrate is an important constituent of DNA? Draw the structure of the carbohydrate.

28. What compounds other than sucrose are sweet-tasting?

29. What do alcohols, sugars, starch, and cellulose all have in common?

Amines, Amino Acids, and Proteins

30. In what way does one type of protein differ from another type of protein?

31. According to Table 14.6, how does an ester differ from an ether? An amine from an amino acid?

32. What are the identifying characteristics of a protein in terms of chemical composition and basic chemical unit?

DNA—The Fundamental Molecule of Life

33. Define in your own words "genetic code."

34. What different types of organic compounds exist in a molecule of DNA?

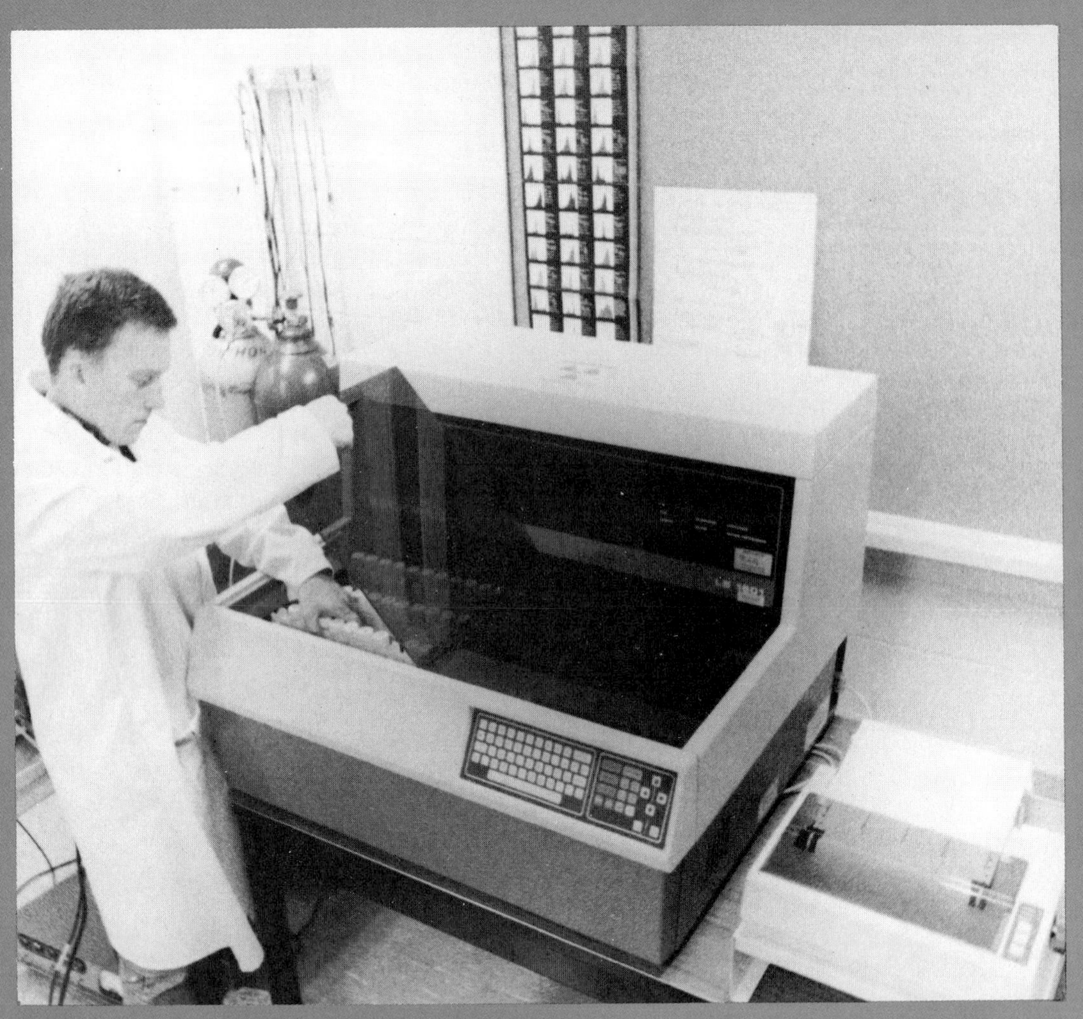

15 Radioactivity and the Nucleus

Chapter outline

Key terms/concepts

The modern nuclear science laboratory includes, as an integral part, an efficient beta spectrometer for analysis of radionuclides that are beta particle emitters.

We are at a time in history when the energy of the atom has been harnessed and serves as a means by which work is done for us. Our formal education includes studying the nature of the atom, its components, and some of its characteristics. But for most students, very little time (if any) is spent in developing an understanding of why atoms spontaneously release mass and/or energy. Often our views of the effects of unstable atoms are misconceptions that arise more from mass media, such as TV programs and movies, than from scientific sources. This chapter is designed to accurately explain radioactivity and describe its many applications.

The atom is the basic building block of all matter. Its diameter is on the order of 10^{-10} meter, and the diameter of the nucleus is on the order of 10^{-14} meter, a comparison that indicates that the nucleus occupies about one-trillionth the volume of the atom. The atom, like our solar system and the universe, then, is primarily composed of empty space. Also, as in the case of the sun and the relation of its mass to that of the rest of the solar system, most of the mass of the atom is found in its nucleus. If an atom of gold, for example, were to be enlarged to about 40 meters in diameter, the nucleus of this atom would be the size of a BB. Yet, if this atom were to have a mass of 1000 kilograms, the nucleus would account for 999.5 kilograms of its mass.

The electron that orbits the nucleus of the atom is considered indivisible; hence, it is referred to as an *elementary particle,* or one that is fundamental in nature. The nucleus of an atom, with a single exception (hydrogen−1), is composed of one or more neutrons and protons, which are not classified as elementary by most scientists. Other particles that are considered elementary may be present in the nucleus of various atoms, and these will be discussed later in this chapter.

The discovery in 1896 of instability in some atoms caused scientists to realize that the atom had a nucleus. They were also able to almost immediately determine some of the properties of this central part of all atoms. The original study and the interpretation of radioactivity in terms of the structure of the atom are credited to the British physicist Ernest Rutherford (1871–1937). Other great discoveries concerning the atom occurred almost simultaneously, including the discovery of X rays by Germany's Wilhelm Roentgen (1845–1923), the discovery of radioactivity by France's Henri Becquerel (1852–1908), and the proof of the existence of the electron by England's J. J. Thomson (1856–1940).

During the following decade, curiosity about the fundamental nature of radioactivity also contributed to the new science. The element thorium was found to be radioactive. Then, Marie and Pierre Curie were successful in isolating two elements, polonium and radium, both of which were found to be radioactive. Eventually, three families of elements—those of uranium, thorium, and actinium—were noted to contain most of the 70 radioactive nuclides from the list of 354 nuclides that have been found in nature. (A nuclide is a species of atom characterized by its number of protons, its number of neutrons, and energy content in its nucleus.) A partial listing of the nuclides for various elements known to exist or to have been artifically produced appears in Table 15.1. For the elements above atomic number 83, no stable nuclides, either manufactured or naturally occurring, appear to exist.

Table 15.1

Nuclides presently known to exist for some selected elements.

Element	Atomic number	Stable isotopes	Radioactive isotopes
Hydrogen	1	^{1}H,^{2}H	^{3}H
Carbon	6	^{12}C,^{13}C	^{9}C,^{10}C,^{11}C,^{14}C,^{15}C,^{16}C
Nitrogen	7	^{14}N,^{15}N	^{12}N,^{13}N,^{16}N,^{17}N,^{18}N
Oxygen	8	^{16}O,^{17}O,^{18}O	^{13}O,^{14}O,^{15}O,^{19}O,^{20}O
Radium	88	None	^{213}Ra,^{219}Ra through ^{230}Ra
Uranium	92	None	^{227}U through ^{240}U
Nobelium	102	None	^{254}No,^{255}No,^{256}No
Lawrencium	103	None	^{256}Lw,^{257}Lw

The Particles and the Rays

At this point, some of the concepts concerning the atom and its characteristics presented in earlier chapters will be reviewed in a manner that should enhance your understanding.

The term *atom* (Greek, *atomos*) literally means "cannot be broken into pieces" or "indivisible." The atom can, however, be separated into pieces by chemical processes in which the atom is *ionized;* that is, the number of electrons that orbit the nucleus can be increased or decreased during chemical reactions. The nuclei of certain atoms are found to be unstable, so these atoms are said to be radioactive. The term, **radioactivity,** is now used to describe *a particular type of radiation,* such as alpha, beta, or gamma emission by a radioactive substance. The expression was apparently coined by one of Becquerel's graduate students who observed a sample of radium as it glowed in the dark. It is assumed she derived the term by combining the words "radiance" and "activity." *The spontaneous emission of particles and/or energy from the nucleus of an atom* is known as **radioactive decay.** The atom is transformed from one element into another, if the particle emitted is electrically charged.

Most *radioactive isotopes,* called radioactive nuclides or **radionuclides,** emit one or more distinct types of particles or rays, called alpha (α), beta (β), and gamma (γ) radiation (after the first three letters of the Greek alphabet). **Alpha particles** are *positively charged,* **beta particles** are *negatively charged,* and **gamma rays** have *no net charge* at all. Each particle responds to a magnetic or electric field in such a manner that either type of field can be used

The Elusive Quark

Scientists consider the universe to be an orderly place. They believe that the best explanation for a natural phenomenon is ordinarily the simplest. Simplicity as well as symmetry are to the scientists what beauty is to artists. As to the *number* of fundamental particles, the fewer the better. Nuclear physicists assumed this attitude about the middle of the twentieth century as they sought the few discrete entities of which they thought matter was composed.

During the 1960s scientists performed a multitude of sophisticated investigations in search of the identity of strongly interactive particles that had markedly different properties. As their studies continued, they found that a degree of order existed between particles and that the particles could be grouped into families according to certain characteristics. The mystery centered around why certain similarities existed and other suspected similarities did not. A hypothesis that explains the family structure indicates that certain "elementary particles," including neutrons and protons, are composed of more fundamental particles, called *quarks.* Those particles not composed of quarks are called *leptons,* which include the electron, the positron, and the neutrino. The quark was named by physicist Murray Gell-Mann, who adopted the term from James Joyce's novel *Finnegan's Wake.* The five known varieties of quarks are called *up, down, strange, charm,* and *bottom.* A proposed sixth type, the *top* quark, continues to evade researchers. Each supposedly has an antiparticle; hence, if a proton is made up of two up quarks and a down quark, an antiproton is composed of two antiup quarks and an antidown quark. Photons and leptons are considered completely elementary; therefore, they are not further divisible. In theory, however, particles such as neutrons, protons, and mesons are composed of more elementary particles. The neutron, for instance, has no electric charge when viewed as a whole, but its interior consists of net positive and negative charges.

An international team of physicists in late 1979 uncovered strong evidence for the existence of the *gluon,* the particle that presumably holds the nucleus of an atom together. This discovery lends additional support to the theory that neutrons, protons, and other so-called elementary particles (except leptons) are indeed composed of quarks, and it won a Nobel prize for the team's leader, Carlo Rubbia ■

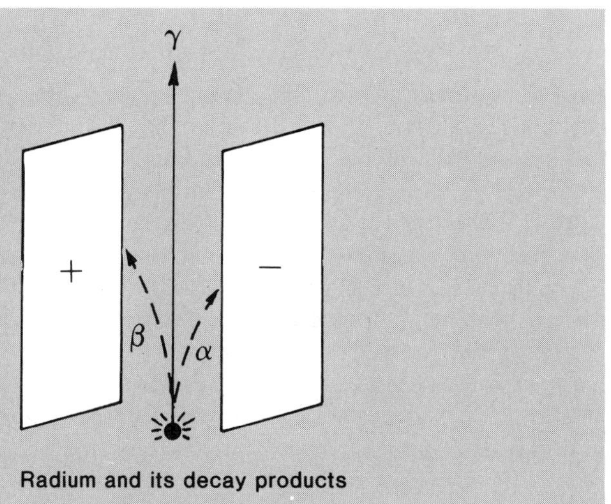

Figure 15.1 An electric field permits separation of the three primary types of radiation: alpha, beta, gamma.

to separate one of the types of radiation from the others. Note in Figure 15.1 the use of an electric field to separate the three types of radiation.

The total electrical charge of the nucleus, as explained in Chapter 8, is equal to the number of positively charged particles located in it, the sum of which is known as the *atomic number, Z.* The positively charged particle was discovered by Rutherford in 1920 during experiments with positive "rays" and was named the proton, from the Greek *protos,* "first." The proton has been recognized for some time as identical to the nucleus of the common hydrogen atom. A second particle, the neutral neutron, is also part of the nucleus of all atoms, except atoms of common hydrogen. It was suggested by Rutherford in 1920 and ultimately identified by English physicist Sir James Chadwick (1891–1974) in 1932. The *neutron number, N,* when added to the atomic number yields the *mass number, A,* of the atom; that is, $A = Z + N$.

The electron was adopted as a basic unit of electricity near the end of the nineteenth century. The charge of the electron is equal in magnitude but opposite in sign to the positively charged proton. The mass of the electron is often used as a unit of mass in measures that concern atomic particles and has a value of 9.106×10^{-28} gram. Scientists have established that the mass of the proton is 1838 times the mass of the electron, and the mass of the neutron is equivalent to 1840 electron masses. Similarly, the charge of the electron is accepted as the basic unit of charge and is said to be -1. The proton's charge, therefore; is $+1$. The neutron is neutral and has a net charge of 0. These three particles—the electron, the neutron, and the proton—constitute the so-called elementary particles of each atom. They, along with other particles, are divided into families according to similarities in mass. Even though over 200 particles are currently classified as elementary, scientists continue to search for internal structure among them. Excluding protons, current studies strongly suggest that mesons and baryons decay into more fundamental particles yet to be identified. (The predictions of certain theories have led some scientists to believe that protons, too, are unstable, and they have been searching for proof since 1983. These investigators propose that half of a given number of protons would undergo decay in 10^{31}

Table 15.2

Families of various elementary particles, as categorized according to similarities in mass. The table well illustrates the true complexity of the atom.

Family	Particle name	Symbol	Charge, in *e* units	Rest energy, in MeV	Antiparticle	Half-life, in seconds
Leptons	Electron	e^-	−1	0.511	e^+	Stable
	Muon	μ^-	−1	105.7	μ^+	1.5×10^{-6}
	Electron neutrino	ν_e	0	0	$\bar{\nu}_e$	Stable
	Muon neutrino	ν_μ	0	0	$\bar{\nu}_\mu$	Stable
Mesons	Pion	$\pi^\pm$	±1	139.6	$\pi^\mp$	1.8×10^{-8}
		π^0	0	135.0	Self	5.8×10^{-17}
	Kaon	K^+	+1	493.7	K^-	Stable
		K^0	0	497.7	$\bar{K}_0$	Stable
	Eta	η	0	548.8	Self	5.3×10^{-19}
	Eta′	η'	0	957.6	Self	1.7×10^{-21}
Baryons	Proton	p	+1	938.3	$\bar{p}$	Stable
	Neutron	n	0	939.6	$\bar{n}$	720
	Lambda	Λ^0	0	1116	$\bar{\Lambda}^0$	1.8×10^{-10}
	Sigma plus	Σ^+	+1	1189	$\bar{\Sigma}^+$	5.5×10^{-11}
	Sigma zero	Σ^0	0	1193	$\bar{\Sigma}^0$	4.0×10^{-20}
	Sigma minus	Σ^-	−1	1197	$\bar{\Sigma}^-$	1.0×10^{-10}
	Delta star	Δ^*	+2, +1, 0, −1	1232	$\bar{\Delta}^*$	4×10^{-24}
	Xi zero	Ξ^0	0	1315	$\bar{\Xi}^0$	2.0×10^{-10}
	Xi minus	Ξ^-	−1	1321	$\bar{\Xi}^-$	1.1×10^{-10}
	Sigma star	Σ^*	+1, 0, −1	1385	$\bar{\Sigma}^*$	1.4×10^{-23}
	Xi star	Ξ^*	−1, 0	1530	$\bar{\Xi}^*$	4×10^{-23}
	Omega minus	Ω^-	−1	1672	$\bar{\Omega}^-$	5.7×10^{-11}

years.) A brief discussion of three families follows, and particle members of these families, along with selected properties, are presented in Table 15.2.

1. *Leptons* are particles, the masses of which are less than that of the proton. These particles experience no strong interactions.
2. *Mesons* are particles, the masses of which are greater than that of the lepton but less than that of the proton. These lightweight particles radioactively decay into electrons. They are unstable, strongly interacting particles and occur in many varieties.
3. *Baryons* are particles, the masses of which are equal to or greater than that of the proton. Each member has the same angular momentum and radioactively decays into protons.

The closely compacted nucleus is primarily composed of positively charged protons and uncharged neutrons. The repulsive force between protons is a relatively strong one and should cause the nucleus to tear apart, but the nucleus of an atom is so tightly bound and compacted that there is no other mass on Earth as dense. Estimates indicate that one cubic centimeter of mass composed of nothing but nuclei would approach 10^{14} grams (100 trillion grams), whereas like volumes of water or gold would have masses of 1 gram and 19 grams, respectively.

Particles in the nucleus are bound together by several forces. The *nucleons,* neutrons or protons found in the nucleus of an atom, attract each other because of the gravitational force between their respective masses. The gravitational force is relatively inconsequential, however, when compared to the *strong nuclear force.* This attractive force, like the gravitational force, acts on nucleons regardless of electrical charge. It is the effective force that binds the nucleus together. The gravitational force follows the inverse square law regarding force and distance. The strong force, however, is effective only at distances of less than 3×10^{-15} meter. This maximum range of effectiveness indicates the nuclear force is of great consequence only in the nucleus itself and only between nucleons that are immediately adjacent to one another, since the diameter of neutrons and protons is about 2.4×10^{-15} meter. (The role of other elementary particles, such as mesons and neutrinos, in affecting stability in a nucleus is beyond the scope of this text.) In comparison, within its range of effectiveness, the strong force is over 100 times as strong as the force of repulsion between two adjacent protons. In many nuclei, the sum of the attractive forces appears to be more than sufficient to overcome the repulsive force between the positively charged and closely compacted protons; hence, these nuclei are considered to be stable.

There appears to be a limit to the maximum number of protons that a stable nucleus can contain regardless of the number of neutrons present. All nuclei that contain 83 or more protons are unstable. That is, from $Z = 83$ to $Z = 109$, no known stable nuclei have been found to exist nor have any been artifically produced by scientists. Indirect evidence of the previous existence of superheavy elements 113 to 115 in the solar system has also been revealed by several scientists through studies of meteorites. Proof of existence of such massive atoms would have decided effects on modern atomic theory.

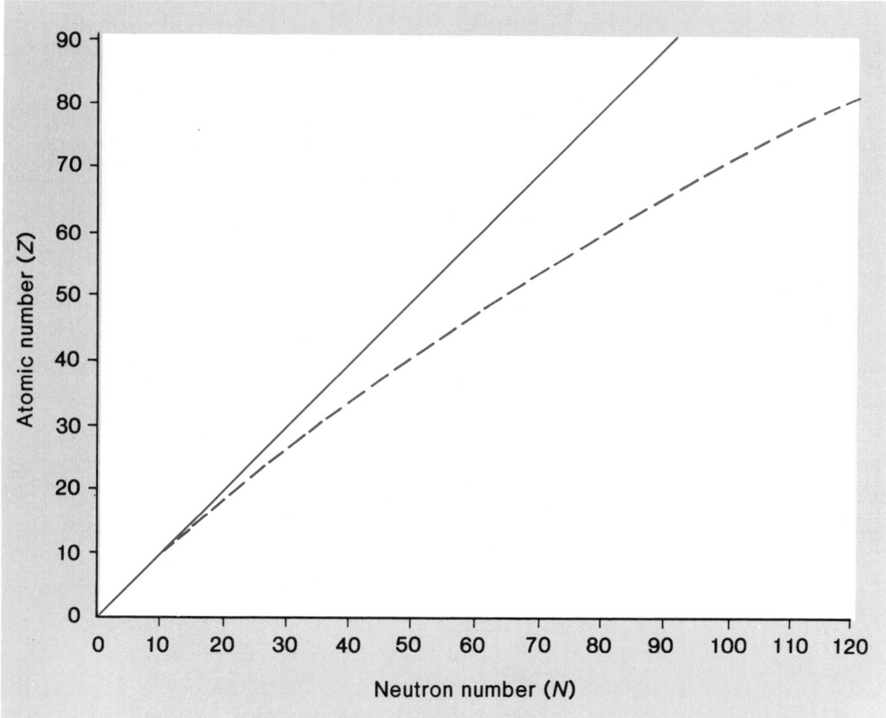

Figure 15.2 The stability ratio Z/N. The solid line identifies a ratio of $Z/N = 1$, and the dashed line indicates the Z/N ratio of the stable isotopes.

In summary, some combinations of Z number and N number produce stable nuclei, whereas other combinations do not. The Z/N ratio is equal to, or approximately equal to, unity for the light elements. In the case of helium-4, lithium-6, boron-10, carbon-12, nitrogen-14, and oxygen-16, $Z = N$, and the ratio Z/N is 1. Above calcium-40, where $Z = 20$ and $N = 20$, no stable nuclei exist with equal numbers of protons and neutrons. As Z increases, the repulsive electrical forces of the positive nuclei become much greater than the attractive forces in the nucleus. The loss of stability of forces creates the potential for particle and/or energy emission, and a radioactive nucleus is created. The increase of additional neutrons in the nucleus serves to increase the distance between protons and thus to decrease the amount of repulsive force that tends to rip the nucleus apart. The stable nuclei of the heavier elements reveal an N number equal to about one and a half times the value of the Z number. The ratio of stability, Z/N, is portrayed graphically in Figure 15.2.

The 279 stable nuclides known to exist can be categorized in four groups, according to whether their A, Z, and N numbers are odd or even. Table 15.3 lists the number of stable nuclides present in each category. As the table indicates, A, Z, and N are even numbers in the majority of stable nuclides. The lowest frequency occurs when both Z and N are odd numbers; thus, A must be even, since it is the sum of Z and N. The distribution also points out that the strong nuclear force between two nucleons is not related to their identities; that is, whether the nucleons are neutrons or protons.

Table 15.3

The distribution of stable nuclides included in this table shows that the greatest number of stable ones exists when Z and N are even numbers. If Z and N are even, A must be, also. A is small for the four odd-odd stable nuclides noted in the table.

A	*Z*	*N*	Number of stable nuclides
even	even	even	168
odd	even	odd	57
odd	odd	even	50
even	odd	odd	4
			279

Nuclides, then, are said to be "radioactive" due to their emission of mass and/or energy in an attempt to reach stability. The most significant types of radiation are alpha, beta, and gamma radiation.

Alpha (α) *radiation* is actually a stream of alpha particles. The alpha particle is identical to the nucleus of a helium atom. It lacks only the two orbiting electrons to make it an electrically neutral helium atom. The common helium atom has an atomic number (Z) of 2 and a mass number (A) of 4. Therefore, the helium nucleus is composed of two protons and $A - Z$, or two, neutrons. As the alpha particle (α) loses speed, and thus its energy, from collisions with atoms that it encounters in the atmosphere, it attracts two free electrons that ultimately are drawn toward the particle and become orbital electrons. At this instant it assumes the identity of a helium atom in all aspects. Uranium-227 decays by alpha emission according to the following reaction:

$$^{227}_{92}U \longrightarrow ^{223}_{90}Th + ^{4}_{2}\alpha.$$

Beta (β) *radiation* is actually a stream of high-speed electrons that are emitted from the nuclei of atoms. They are of two types: the negative electron, or negatron (e^-), and the positive electron, or positron (e^+). The masses of the negatron and positron are equal, but the particles are opposite in charge. The positron is, in effect, the antiparticle of the negatron. As the negatron loses energy in its trip through space, it has the capability of becoming the orbital electron of an atom that has previously been positively ionized. The positron generally collides with matter and often comes in contact with a negatron, with the result that both are totally annihilated; that is, converted into energy. When the ratio of neutrons to protons is too great, the nucleus is unstable. Stability is reached through the decay of a neutron into a proton, an electron, and an antineutrino within the nucleus:

$$^{1}_{0}n \longrightarrow ^{1}_{+1}p + ^{0}_{-1}e + ^{0}_{0}\bar{\nu}. \qquad \text{NEUTRON DECAY}$$

As an example:

$$^{35}_{16}S \longrightarrow ^{35}_{17}Cl + ^{0}_{-1}\beta + ^{0}_{0}\bar{\nu}.$$

On the other hand, the proton (p^+), which remains in the nucleus, adds to the Z number one unit of positive charge, while the neutron number, N, is reduced one unit. The A number remains constant, but the element changes to the next higher element on the periodic chart in all physical and chemical properties. The electron (e^-) is emitted from the nucleus and is commonly referred to as beta (β^-) radiation. The antineutrino ($\bar{\nu}$) is a neutral particle with little or no mass, but, like the photon, it possesses energy. Note the conservation of both charge and mass in the reaction.

If the Z number is excessively greater than the N number, stability is reached through the reaction

$$^{1}_{+1}p \longrightarrow ^{1}_{0}n + ^{0}_{+1}e + ^{0}_{0}\nu. \qquad \text{PROTON DECAY}$$

Carbon-11 undergoes proton decay according to the following reaction:

$$^{11}_{6}C \longrightarrow ^{11}_{5}B + ^{0}_{+1}\beta + ^{0}_{0}\nu.$$

The proton decays into a neutron and a positron (β^+), the latter of which is ejected. Additional energy loss that is noted is attributed to the neutrino (ν). As the name and symbol imply, the antineutrino is the antiparticle of the neutrino. Most authorities believe that the spontaneous decay of free protons may not occur, since such a reaction appears to be energetically impossible. The mass of the neutron itself is greater than the mass of the proton. But inside a nucleus, energy from surrounding nucleons can be converted to mass, a topic discussed in the section on the transmutation of elements, so a proton within the nucleus can decay under certain conditions.

One should realize that this discussion should not be taken to mean that a neutron is composed of a proton and an electron or that a proton is composed of a neutron and a positive electron. The neutron disintegrates into a proton and an electron just as a tree can be reduced to toothpicks. The tree, however, is not composed of toothpicks.

In addition to β^- and β^+ emission, there is a third related process. **Electron capture** occurs when *a nucleus absorbs one of its orbiting electrons.* Beryllium-7 decays by this mechanism and becomes lithium-7:

$$^{7}_{4}Be + ^{0}_{-1}e \longrightarrow ^{7}_{3}Li + ^{0}_{0}\nu. \qquad \text{ELECTRON CAPTURE}$$

Generally, it is an electron in the innermost (K) shell of an atom that is captured. In this case, the process is often known as K-capture. The electron is captured by a proton in the nucleus of the atom and the pair assumes the identity of a neutron. A neutrino accompanies the process of electron capture.

Gamma (γ) *radiation* is a stream of photons of electromagnetic radiation, generally higher in frequency than X rays. The photon is particulate (appears to be a particle) in the manner in which it behaves as a result of collisions with matter, but it has no rest mass (see also Chapter 10). Therefore, this bundle of energy is basically different from alpha and beta radiation. Gamma radiation often accompanies the emission of alpha particles and beta particles from the various nuclei. Its emission does not alter the nucleus except in the degree of excitation present. There exists among the unstable nuclides pure gamma-emitters, such as manganese-54 and cadmium-109. Note the following reactions that involve gamma emission:

$$^{60}_{27}Co \longrightarrow ^{60}_{28}Ni + ^{0}_{-1}\beta + ^{0}_{0}\gamma + ^{0}_{0}\bar{\nu}, \qquad \text{GAMMA EMISSION}$$

$$^{54}_{25}Mn \longrightarrow ^{54}_{25}Mn + ^{0}_{0}\gamma.$$

The speed of gamma radiation, like that of the rest of the electromagnetic spectrum, is a constant (c), 3.00×10^8 meters per second. X rays differ from gamma rays in that X-radiation is electromagnetic radiation released by energy disturbances of orbital electrons, whereas gamma radiation is energy that results from disturbances in the nucleus. The frequencies may overlap, however, and these two types of radiation are indistinguishable except for their respective origins. There are many examples of this slight difference between waves of various energies that compose the electromagnetic spectrum. The amount of energy released as gamma, X ray, radio, television, ultraviolet, or some other type of electromagnetic radiation is directly related to the frequency the oscillating object produces. This phenomenon, discovered by Max Planck in 1900 and discussed in detail in Chapter 10, can be expressed as

$$E = nhf, \qquad \text{ELECTROMAGNETIC ENERGY}$$

where E represents energy, n signifies the number of photons released, and f is the frequency of the radiation. The factor h is Planck's constant, the accepted value of which is 6.63×10^{-34} joule-second. The only way that energy from an electromagnetic wave can be increased, then, since n and h do not vary, is by increasing the frequency of the wave (an increase in frequency results in a decrease in wavelength, from $v = f\lambda$). The energy unit of *all* types of radiation—including alpha, beta, and gamma—is the *electron volt* (eV), the energy that an electron would acquire in accelerating through an electric potential of 1 volt.

Some years ago, Louis de Broglie suggested that since light and other electromagnetic radiation seemed to have properties causing it to act in the manner of waves as well as of particles (photons, or quanta), there was logic in the conclusion that each elementary particle had an associated wavelength. The hypothesis was strengthened by increased knowledge of atomic structure, and consequent experimental measurements led to the collection of ample data to afford proof of the theory. For instance, an electron accelerated through a potential of 100 volts (thus with an energy gain of 100 eV) reaches a speed of 5.90×10^6 meters per second with a wavelength of 1.22×10^{-10} meter. An alpha particle with energy of 100 eV, however, reaches a speed of only 6.90×10^4 meters per second and a wavelength of 1.44×10^{-12} meter. An alpha particle of energy equal to 1,000,000 eV (1 MeV) has a speed of 6.90×10^6 meters per second and a wavelength of 1.44×10^{-14} meter, a figure representative of wavelengths of electromagnetic waves classed as gamma radiation.

In the 1920s scientists became aware that atomic nuclei contained vast amounts of energy. The ability to detect the number of protons and neutrons that an atom has in its nucleus, in conjunction with the development of the mass spectrograph, led investigators to a startling conclusion. The total mass of the individual nucleons, the protons and neutrons in the nucleus, amounted to more than the actual mass of the nucleus. *With the exception of the common hydrogen nucleus,* where $A = 1$, *all nuclei have less mass than the sum of their particle masses,* a condition referred to as the **mass defect** of nuclei. On the modern atomic scale, recall, the mass of the particles is expressed in terms of atomic mass units (u). The masses of various particles are listed in Table 15.4, and the atomic masses of various nuclides are shown in Table 15.5.

Consider the mass defect in the oxygen atom, the A number of which is 16. The chart of the nuclides, published by Knolls Atomic Power Laboratory, lists the atomic mass of $^{16}_{8}O$ as 15.994,915. The Z number and N number are both equal to 8. According to the masses of the three elementary particles present in the oxygen-16 atom, the exact mass of the atom can be obtained by the following calculation:

Particle	*Number present*	*Mass per particle*	*Total*
Proton	8	1.007,277 u	8.058,216 u
Neutron	8	1.008,665 u	8.069,320 u
Electron	8	0.000,548 u	0.004,384 u
	Total mass of elementary particles		16.131,920 u

The difference between the total mass of the elementary particles and the value obtained from a chart of the nuclides is the mass assumed to be converted to energy for the oxygen-16 atom:

$$16.131{,}920 \text{ u} - 15.994{,}915 \text{ u} = 0.137{,}005 \text{ u} \qquad \text{MASS DEFECT}$$

The mass of one atomic mass unit, recall, is 1.66×10^{-27} kilogram. The mass defect for oxygen-16 can be expressed as $(1.66 \times 10^{-27}\text{ kg/u})(1.37 \times 10^{-5}\text{ u}) = 2.27 \times 10^{-32}$ kg. Then, from the equation to relate mass to energy, a relationship discussed in considerable detail in the "Mass-Energy Equivalence" section later in this chapter, the energy derived from the mass defect can be determined. Energy, measured in joules, equals the product of mass and the square of a constant equivalent to the speed of light. That is,

$$\begin{aligned} E &= mc^2 \text{ (recall, } c = 3.00 \times 10^8 \text{ m/s)} \\ &= (2.27 \times 10^{-32}\text{ kg})(9.00 \times 10^{16}\text{ m}^2/\text{s}^2) \\ &= 2.04 \times 10^{-15}\text{ joule.} \end{aligned}$$

In comparison, one atomic mass unit, expressed in energy equivalence, would equal 1.49×10^{-10} joule. As pointed out earlier in this section, the scientist finds it convenient to express the energy of particle emissions in electron volts, and even more useful in millions of electron volts (MeV). In order to relate the MeV to joules, the following conversion may be used:

$$1.00 \text{ MeV} = 1.60 \times 10^{-13}\text{ joule.}$$

Thus,

$$(1.49 \times 10^{-10}\text{ joule}) \times \frac{1.00 \text{ MeV}}{1.60 \times 10^{-13}\text{ joule}} = 931.48 \text{ MeV.}$$

In other words, one atomic mass unit = 931.48 MeV. The mass defect of the oxygen-16 nuclide, then, can be expressed as 127.62 MeV. This amount of energy is known as the binding energy for

Table 15.4

Comparative masses of various particles in atomic mass units. The mass of each particle has been carefully determined.

Particle	Atomic mass units (u)
Alpha	4.001,506
Electron	0.000,548
Neutron	1.008,665
Proton	1.007,277

Table 15.5

The atomic mass of selected nuclides, indicated in atomic mass units.

Element	*Z* number	*A* number	Atomic mass (u)
Hydrogen	1	1	1.007,825
Hydrogen	1	2	2.014,102
Helium	2	4	4.002,603
Carbon	6	12	12.000,000
Oxygen	8	16	15.994,915
Chlorine	17	35	34.968,853
Chlorine	17	37	36.965,903
Iron	26	56	55.934,939
Gold	79	197	196.966,560
Uranium	92	235	235.043,925
Uranium	92	238	238.050,786

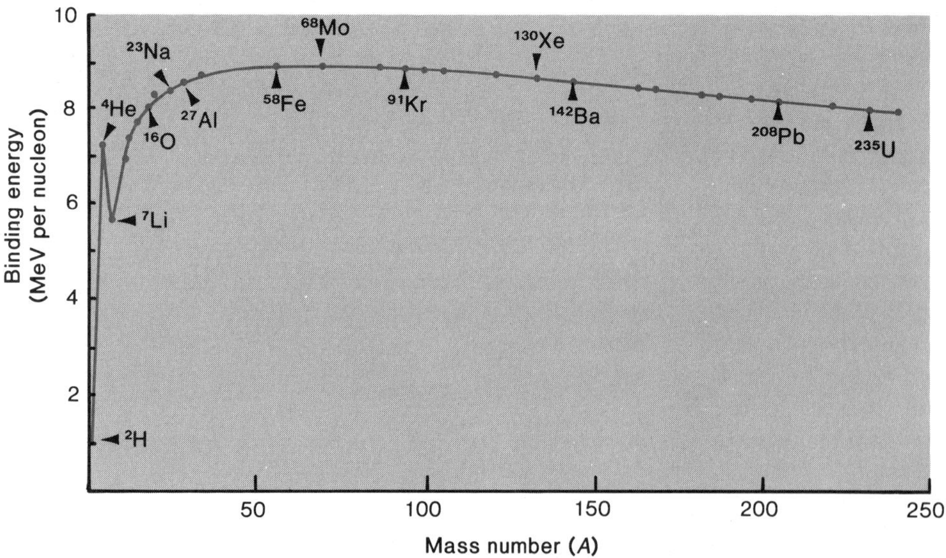

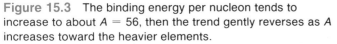

Figure 15.3 The binding energy per nucleon tends to increase to about $A = 56$, then the trend gently reverses as A increases toward the heavier elements.

that nuclide and is, in effect, the energy with which the particles are held together in the oxygen-16 nucleus. The binding energy per nucleon for oxygen-16 would be 127.62 MeV/16 = 7.98 MeV.

Example 15.1

The atomic mass of an isotope of iron, $^{58}_{26}Fe$, is 57.933,28. Determine the mass defect of the isotope and the binding energy per nucleon.

Solution

$$\text{mass defect} = 26(1.007{,}277\ \text{u}) + 32(1.008{,}665\ \text{u}) + 26(0.000{,}548\ \text{u}) - 57.933{,}28\ \text{u} = 0.547{,}45\ \text{u}$$

BINDING ENERGY

$$\frac{\text{binding energy}}{\text{per nucleon}} = \frac{(931.48\ \text{MeV/u})\ (0.547{,}45\ \text{u})}{58} = 8.792\ \text{MeV}$$

Binding energy may also be defined as the difference between the sum of the rest energies of a specific atom's components and the rest energy of the specific atom. A generalized graph of binding energy per nucleon with respect to atomic mass appears in Figure 15.3. As the graph indicates, nuclei that fall along the plateau, about mass number 60, have the highest binding energy per nucleon; therefore, these nuclei are the most stable ones among the elements. Nuclei with mass numbers outside of this specific group have nucleons not as strongly bound. For this reason, nuclei of mass numbers significantly greater or less than 60 will more readily release energy during various nuclear reactions.

Natural Radioactivity

The growth of knowledge in new frontiers of science is usually a chain reaction, since one great discovery generally leads to others. For example, Roentgen discovered X rays in 1895, and Becquerel discovered radioactivity the following year. Becquerel revealed that radiations from uranium salts created silhouettes of individual crystals on photographic plates. He also found that uranium compounds can cause electrically charged objects to discharge, a discovery that led to a quantitative method of measuring radiation and to the development of the goldleaf electroscope and the ionization chamber. Later, Rutherford used the discoveries of Becquerel to study the penetrating power of radiation from uranium salts and found that the radiation consisted of two types: alpha radiation, which is readily absorbed by matter, and a more penetrating type, beta radiation.

Not long after Rutherford's discovery, the properties of alpha and beta radiation were under concentrated study. A radioactive source was placed in a lead container that had a small hole through which the particles emitted could escape. The beam of particles was directed to travel between the charged plates of a capacitor onto a fluorescent screen, creating three distinct bright spots on the screen. The beta particles were bent toward the positive plate of the capacitor, the alpha particles were bent toward the negative plate, and another beam was unaffected, indicating the particle had no net charge. Hence, the third particle, a high-energy photon, was discovered and named the gamma ray, after the third letter in the Greek alphabet (recall Fig. 15.1).

Marie and Pierre Curie discovered that the activity of uranium salts was directly related to the mass of the uranium in the salts. Through their concentrated efforts they demonstrated the atomic nature of radioactivity. They also discovered two other naturally occurring radioactive elements, radium and polonium, for which Marie Curie was awarded her second Nobel prize. (In 1903, the Curies shared the Nobel prize in physics with Becquerel for their work in radioactivity, but Pierre was killed by a horse-drawn carriage in 1906, and Marie alone accepted the Nobel prize in chemistry in 1911.) Further research by Marie Curie proved that radium has over a million times the activity as the same mass of uranium.

Many more radioactive nuclides that occur in nature have since been discovered. Plutonium, $Z = 94$, represents the element with the highest atomic number found in nature. Traces of this element have been found in ancient African uranium deposits. From the list of over 2000 nuclides that have been identified, approximately 1700 have been found to be radioactive; about 50 of these radionuclides occur naturally. The remainder of the known radionuclides have been produced in laboratories. As noted in Table 15.3, there are at least 279 stable nuclides in the interior and atmosphere of Earth, about 250 of which are present in relatively abundant amounts.

Transmutation of the Elements

Transmutation, also known as nuclear transformation, is *a process in which the atomic nucleus of one element is converted into an atomic nucleus of another element.* This process involves a gain or a loss in the number of protons by the original nucleus; therefore, it may be brought about by the ejection of an alpha, a beta, or other charged particle, including the recently discovered proton reaction in which these charged particles are ejected in pairs. (Note that X-ray and gamma emission do not cause transmutation to occur, since their charge is zero.) A carbon nucleus, for instance, transforms into a nitrogen nucleus upon emission of a beta particle and the increase in the number of protons that results from beta emission. Carbon-14, a beta emitter, has 6 protons, since $Z = 6$, and a total of 14 nucleons, since $A = 14$; thus, the number of neutrons in its nucleus is $A - Z$, or 8. Since the emission of a beta particle indicates that a neutron has disintegrated into a proton, a beta particle, and an antineutrino, consider the following reaction:

$$^{14}_{6}\text{C} \longrightarrow {}^{14}_{7}\text{N} + {}^{0}_{-1}\beta + {}^{0}_{0}\bar{\nu}. \qquad \text{BETA EMISSION}$$

In a similar manner, the instant an alpha particle leaves the nucleus of a uranium atom, the atom is no longer uranium; its physical and chemical properties become those of the element thorium. Note the following reaction in which A decreases 4 and Z decreases 2 due to alpha emission:

$$^{238}_{92}\text{U} \longrightarrow {}^{234}_{90}\text{Th} + {}^{4}_{2}\alpha. \qquad \text{ALPHA EMISSION}$$

In both reactions, the mass numbers (A) of both sides balance, as do the atomic numbers (Z). Since the reactions are nuclear in character, the number of orbiting electrons may be ignored. The thorium atom produced is also radioactive and undergoes transmutation according to the reaction that follows. Typically, when one writes such reactions, uncharged particles, such as antineutrinos, are disregarded.

$$^{234}_{90}\text{Th} \longrightarrow {}^{234}_{91}\text{Pa} + {}^{0}_{-1}\beta.$$

Note again that both the A and Z numbers are balanced. The transmutation of the uranium atom continues through the emission of charged particles until the resulting element is lead ($Z = 82$, $A = 206$). Other elements produced through the natural transmutation of uranium include bismuth, polonium, protactinium, radium, radon, and thallium. The radioactive decay scheme of uranium is only one of several transmutation series that occur in nature. Figure 15.4 is a graphical representation of the total decay of uranium-238. Note the variations in time between the formation of one radionuclide and the formation of others in the decay scheme. Note also that radium-226 is a "daughter" (decay product) of uranium-238. The half-lives of the various members of the uranium-238 decay scheme are listed, and the concept of "half-life" is discussed in detail in the next section. Such a decay series suggests why so many nuclides of various elements are present in nature.

The ancient alchemists and their predecessors spent 2000 years attempting to change common elements into gold, silver, and other valuable elements. Of course, none succeeded, for their most determined efforts were all centered around chemical reactions. The transmutation of one element into another requires a nuclear reaction. Regardless of the violence of the chemical reaction, the nucleus of an atom is protected from its immediate environment by the protective shield of the atom's orbiting electrons. To change an element into another, the number of charged particles in the nucleus must be altered. For example, to change lead into gold, three positive charges must be removed from the lead nucleus since the Z number of lead is 82 and that of gold is 79.

The answer to the alchemists' dilemma was at hand, but their scientific knowledge was insufficient to realize the solution. The constant decay of radium, uranium, and other available elements would have given them a mechanism, for transmutation is possible by the addition of charged particles to a nucleus as well as by the emission of charged particles from a nucleus. Transmutation can also be effected by the interactions of some nuclides with some uncharged particles, such as neutrons—a technique discussed in the section on radionuclides. The alpha particle ejected from uranium is an excellent particle with which to create new elements. This particle enables transmutation to occur, since its energy is sufficient to pass through the protective shield of the orbiting electrons with a degree of difficulty comparable to that of a bullet passing between the rotating blades of an electric fan.

In 1919 Rutherford bombarded nitrogen nuclei with alpha particles from a naturally occurring radioactive source and successfully transmuted nitrogen into oxygen. The nuclear reaction was as follows:

$$^{14}_{7}\text{N} + {}^{4}_{2}\alpha \longrightarrow {}^{17}_{8}\text{O} + {}^{1}_{1}\text{p}. \qquad \text{ARTIFICIAL TRANSMUTATION}$$

It is somewhat discomforting to note that there appears to be no worthwhile means to obtain stable gold from other elements. However, artificial transmutation was accomplished in the laboratory before some of the naturally occurring elements we use today were discovered.

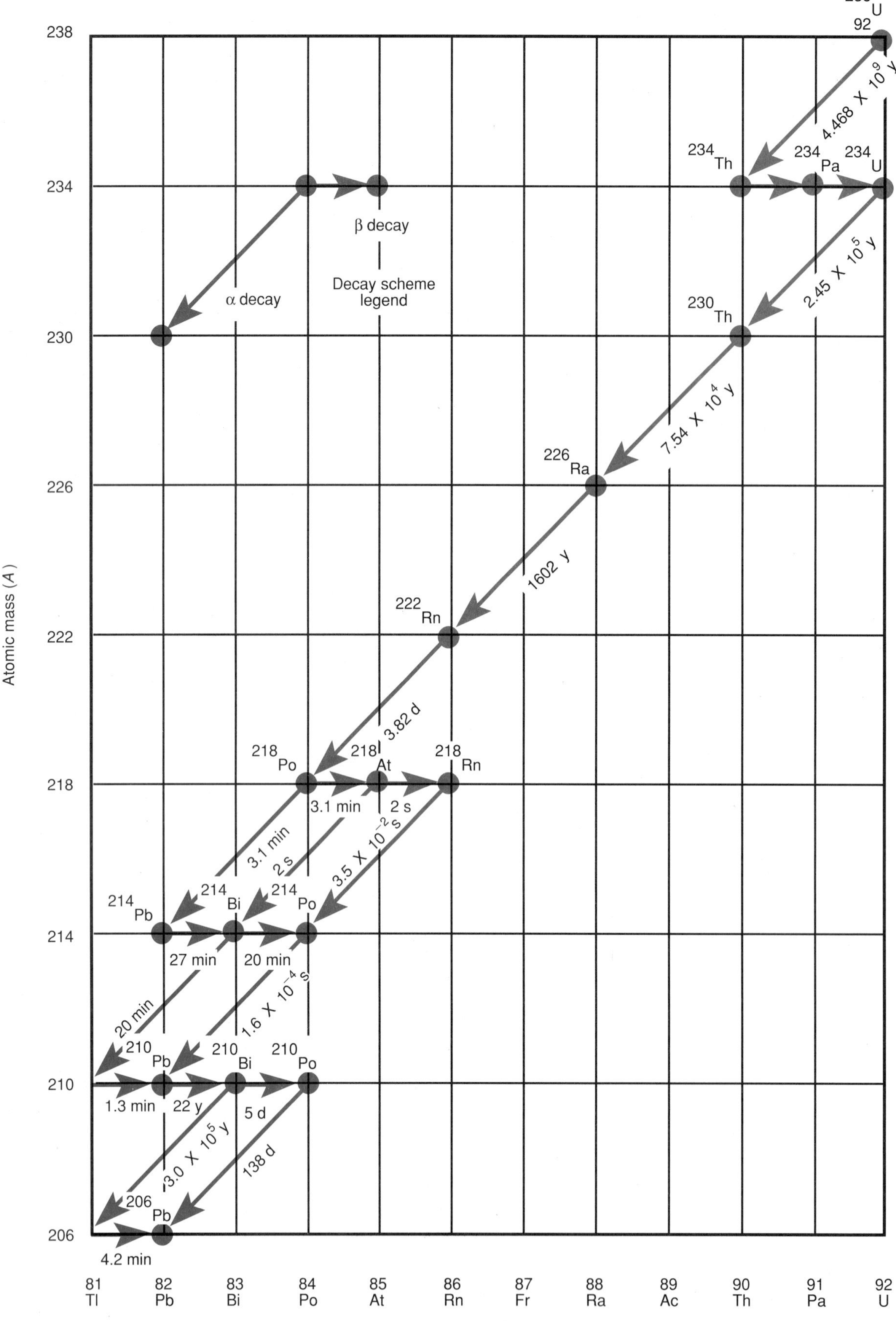

Figure 15.4 A graphical representation of the decay scheme of uranium-238. Note the alternate modes of decay from polonium-218 until the stable nuclide, lead-206, is formed.

The Discovery of Radioactivity

Scientific discoveries in a particular field follow one another very rapidly, as stated earlier in the chapter. Scientists learn to rely on the work of others and to relate reported research to their own endeavors. For instance, Antoine Henri Becquerel (1852–1908), a French physicist, discovered radioactivity less than a year after Roentgen announced the discovery of X rays in 1895. The discovery of radioactivity seemed to confirm what chemists had suspected for many years; namely, that the elements were inherently related. The French scientists Pierre (1859–1906) and Marie (1867–1934) Curie succeeded in isolating polonium and radium in 1898. (Polonium was named by Marie Curie in honor of her native country, Poland.) They discovered that these elements would decay spontaneously into other elements, which in turn would decay into still lighter elements.

Lord Ernest Rutherford (1871–1937), a British physicist, applied his experimental genius to the problem of unraveling the mysteries of the Becquerel rays, the various types of radiation emitted from the element uranium. In 1911, he announced that an atom consisted of a small positively charged nucleus containing most of the mass of the atom and tiny electrons that orbited the nucleus of the atom in the same way that planets orbit the sun. He also noted the similarity of X rays and gamma rays. Rutherford is given credit for offering conclusive evidence that the alpha particle is identical to the nucleus of the helium atom. He is also credited with the discovery of the proton ■

Half-Life and Activity

The nucleons that compose the nucleus of an atom contain large amounts of energy as a result of electrical and nuclear forces. Some of the nuclei have excess energy, which causes them to exist in an excited state. If the amount of excess energy is high, alpha particles are ejected. Excited nuclei with insufficient energy to expel an alpha particle may reach stability through beta or gamma emission.

For any given nucleus there is no known way to predict when the particle or energy will be emitted. The nucleus may disintegrate in the next second or at some future time. Yet in a large mass of the isotope, a constant rate of disintegration can be determined. This rate is commonly expressed in terms of *activity,* the number of disintegrations per second. So that activity could be expressed in a more meaningful way, radium-226 was chosen as the standard. Close examination revealed that one gram of radium-226 emits 3.7×10^{10} alpha particles per second. The standard was called the *curie* (Ci) in honor of the discoverers of the element radium. (As more accurate techniques of detection and measurement were developed, however, 3.7×10^{10} disintegrations per second (dis/s) were noted to emanate from 0.97 gram, rather than from 1.0 gram of radium, but the unit has remained as initially defined.) A curie of iodine-131 has a mass of about 0.000,000,1 gram, and a curie of uranium-238 has a mass of 1000 kilograms. So that the unit can be adapted to any radionuclide, the unit now represents 3.7×10^{10} disintegrations per second of any given nuclei and thereby encompasses all types of radiation. (In various scientific circles the curie has been replaced by the *becquerel* [Bq], a derived SI unit representing events per second, which can be combined with the metric base units. One curie is equal to 3.7×10^{10} becquerels.)

The rate of disintegration of a given mass of a radionuclide is expressed in yet another term—**half-life.** Half-life ($t_{½}$) is *the time required for a given mass of a radionuclide to disintegrate to half its original activity.* For instance, half of a given amount of cobalt-60 will decay into nickel-60 in 5.26 years. At the end of 10.52 years (two half-lives), only one-fourth of the original amount of cobalt-60 will remain. The ratio of cobalt-60 to nickel-60 through three half-lives is shown in the graphs presented as Figure 15.5. Half-lives of the naturally occurring radionuclides are in thousands or millions of years, for shorter-lived radionuclides would have decayed to activities well below detection long ago. Those radionuclides produced in the laboratory have much shorter half-lives that vary from a few thousand years to millionths of seconds. Some radionuclides and their respective half-lives are listed in Table 15.6.

Example 15.2

A sample of cobalt-60 was found to have an activity of 2.00×10^5 becquerels. What would be its activity in 10.5 years?

HALF-LIFE

Solution

According to Table 15.6, the half-life of cobalt-60 is 5.26 y. Therefore, 10.5 y represents two half-lives for the radionuclide. Hence, its activity would be

$$\frac{2.00 \times 10^5}{2 \times 2} \text{ Bq or } 5.00 \times 10^4 \text{ Bq}$$

at the end of the time specified.

The concept of half-life can be further illustrated by means of bar graphs. Note that Figure 15.5b points out the ratio of cobalt-60 to nickel-60 (cobalt-60's daughter) at the end of one, two, and three half-lives.

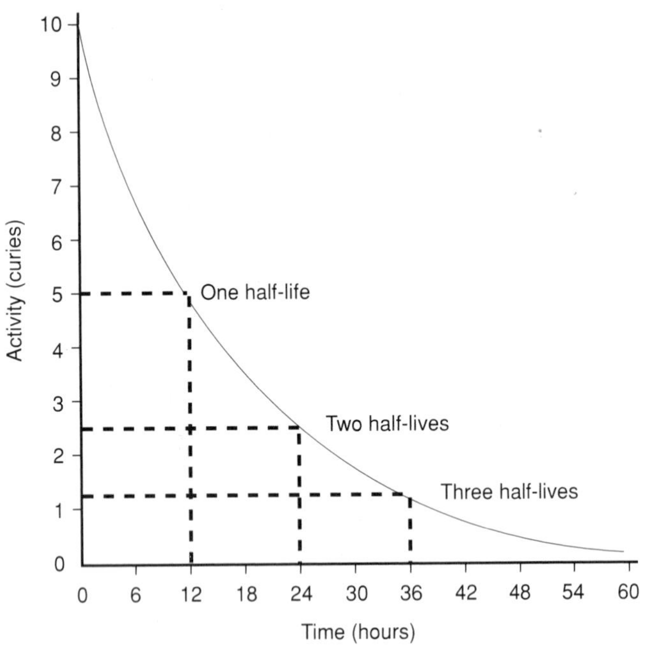

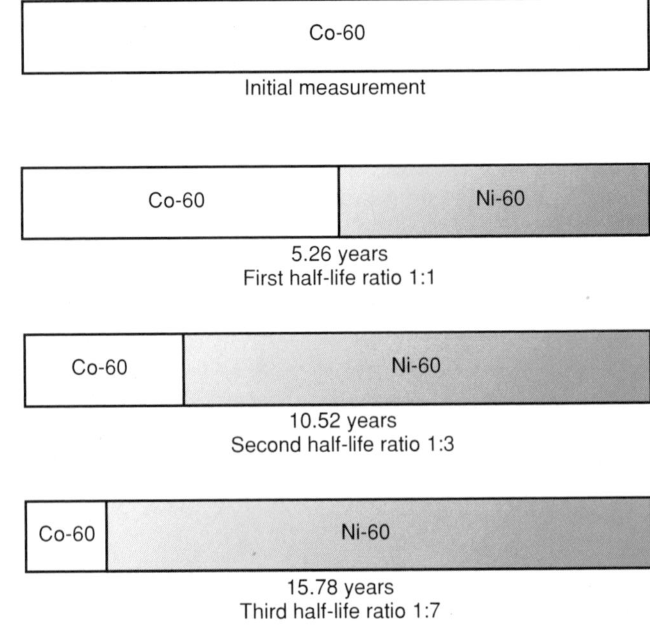

Figure 15.5 (*a*) The activity curve of a radionuclide, the half-life of which is 12 h. The original activity of the sample was 10 curies. (*b*) Half of a given amount of cobalt–60 will decay into nickel–60 in 5.26 y. At the end of 10.52 y, only one-fourth of the original amount of cobalt–60 will remain. The ratio of cobalt–60 to nickel–60 is shown in the bar graph through three half-lives of this radioactive nuclide.

Table 15.6
Half-lives of some commonly used radionuclides.

Elements	Atomic number (*Z*)	Mass number (*A*)	Particle(s) emitted	Half-life ($t_{1/2}$)
Calcium	20	45	Beta	162.7 d
Californium	98	252	Alpha, gamma	2.55 y
Carbon	6	14	Beta	5730 y
Chlorine	17	36	Beta	3.07×10^5 y
Cobalt	27	60	Beta, gamma	5.26 y
Copper	29	64	Beta, gamma	12.7 h
Copper	29	66	Beta	5.1 min
Hydrogen	1	3	Beta	12.33 y
Indium	49	114	Beta	71.9 s
Iodine	53	131	Beta, gamma	8.065 d
Neptunium	93	237	Alpha, gamma	2.14×10^6 y
Nitrogen	7	16	Beta, gamma	7.1 s
Phosphorus	15	32	Beta	14.3 d
Plutonium	94	239	Alpha, gamma	24,360 y
Potassium	19	40	Beta, gamma	1.25×10^9 y
Radium	88	226	Alpha	1602 y
Silver	47	108	Beta, gamma	2.4 min
Silver	47	110	Beta, gamma	24.6 s
Uranium	92	238	Alpha, gamma	4.51×10^9 y

Dating with Radioactivity

Radioactivity has many useful applications, including age determination of the rock layers that form Earth's crust. When molten material solidifies, the elements in the deposit become fixed in position. The uranium in the crust, along with the daughter nuclei, provide the scientist with an accurate indication of the age of the various rock layers that contain them. As was noted in Figure 15.4, the decay scheme of uranium-238 produces many radionuclides before stable lead-206 is eventually reached. By measuring the amount of uranium-238 remaining in the rock layer relative to the amount of various daughter nuclei, the geologic time at which that layer solidified can be calculated. Radioactive dating with uranium-238 indicates that the age of the oldest rocks found on Earth is about 4 billion years old. The age of rocks in which the oldest fossilized organisms have been detected is approximately 3 billion years.

Another radionuclide that has proved valuable to geologists is potassium-40. This isotope of potassium, the half-life of which is 1.25 billion years, spontaneously converts into argon-40 by electron capture. In this decay mode, recall that a proton captures an orbital electron and converts into a neutron; thus, the Z number is decreased by one while the A number remains constant. Although argon is a gas, the element is trapped in the crystalline structure of the rock and is rather easy to detect and measure. If a rock layer contained both uranium and potassium when it formed, a comparison of the percentage of decay products present during the analysis affords the scientist with a very reliable means to date the rock layer, since two different "clocks" are at work. Other radionuclides and their decay products (daughters) have been studied to substantiate the age of Earth's crust. This method of analysis has been exceedingly valuable as geologists attempt to reconstruct Earth's past history and how biological organisms with various characteristics developed.

One further application of radioactive dating involves the method used to determine the age of plants and animals that existed up to about 40,000 years ago. The radionuclide, carbon-14, decays by beta emission and has a half-life of 5730 years. All living plants obtain carbon dioxide from the air, utilizing the carbon and expelling the oxygen. The vast majority of carbon atoms present in the atmosphere are the stable isotope carbon-12. However, a very small fraction, about one part per trillion (1×10^{12}) carbon atoms, are carbon-14. This same ratio is assumed to have existed for many thousands of years, even though the half-life of carbon-14 is relatively short when compared with those of uranium-238 and potassium-40. This assumption about the consistent ratio of carbon-14 to carbon-12 seems valid because neutrons in Earth's cosmic radiation bombard atoms in the planet's atmosphere. Specifically, neutron's collisions with nitrogen atoms produce the following nuclear reaction:

$$^{1}_{0}n + ^{14}_{7}N \longrightarrow ^{14}_{6}C + ^{1}_{1}p.$$

That is, as a neutron collides and is absorbed by the nucleus of a nitrogen atom, a proton is expelled and the remaining nucleus is that of a carbon-14 atom. The constant rate of production practically balances the rate of decay, so as long as a plant, such as a tree, survives, it continually absorbs carbon dioxide to produce new cells and replace the old cells. As animals eat the plants they, too, absorb a constant supply of carbon to build new tissue. Since the ratio of carbon-14 to carbon-12 stays practically constant in the atmosphere, the ratio in either form of life also remains constant. But when either a plant or an animal dies, the intake or release of carbon dioxide is halted, and because the carbon-14 decays into nitrogen, the ratio of carbon-14 to carbon-12 decreases with time. In 5730 years after the death of the plant or animal, the ratio has dropped to half, or approximately one carbon-14 atom per 2×10^{12} carbon-12 atoms. After another 5730 years, the ratio decreases by half again, to one carbon-14 atom per 4×10^{12} carbon-12 atoms, and so forth. The actual ratio of carbon-14 to carbon-12 in a specific sample from an ancient life-form compared to the ratio in living tissue provides the scientist with the length of lapsed time since the organism was alive. The carbon-14 dating procedure has been exceptionally useful in determining the precise age of such organic remains as those found at archeological "digs," including human and animal bones, seeds, plants, insects, and microscopic organisms. The process has also been helpful in correlating events that took place in the past. Radioactive carbon dating has been used to establish when various civilizations existed, what foods they ate, and how they migrated from one area to another. The study of the cultures discussed in Chapter 2 was enhanced with this method of dating. The painstaking manner in which the specimens must be handled and analyzed requires everyone involved to have a high level of expertise, but the results have helped unravel many mysteries from the past.

Actually, as the available technology has become more sophisticated, scientists have had to make minor corrections in radioactive carbon dating because alternative procedures, such as counting the annual rings in very old trees, have indicated that a slight fluctuation has occured over the centuries in the carbon-14-to-carbon-12 ratio in the atmosphere. Even greater proof of need to adjust for fluctuations in the carbon-14 environment over the past 30,000 years has become evident as researchers probe various coral reefs. Their studies of drilling samples taken from such deposits as those off the coast of Barbados indicate that relative amounts of uranium and its decay product thorium detected do not agree with the accepted carbon-14 dating scale. The thorium concentration, if correct, will force scientists to revise many milestones of prehistory by increasing their estimated ages by perhaps 3500 years. For instance, the last ice age may have reached its peak about 21,000 years ago, rather than 18,000 years ago, as previously thought. Such adjustments, however, will not affect the order of occurrence of geologic events.

Radionuclides—Their Other Uses and Their Detection

A very valuable application of controlled nuclear reactions is producing radionuclides of various elements, also called "radioisotopes." The production takes place in several ways. Many radionuclides are produced as by-products of the fission process, since the nuclei of the fissionable material are split into fragments that are actually lighter nuclei, some of which are radioactive. (Fission is discussed in detail in its own section later in the chapter.) Other radionuclides are created when the neutrons produced by the fission process collide with and are absorbed by some of the nuclei of stable atoms introduced into their paths as "targets." Some of these new nuclides are unstable, and thus radioactive. Many commonly used radionuclides are prepared by this technique in a nuclear reactor, in a process known as *neutron bombardment* (see Figs. 15.6 and 15.7). (Also note the nuclear reactors depicted in Chapter 26.)

Figure 15.6 The experimental floor of Brookhaven National Laboratory's High Flux Beam Reactor (HFBR). The 60–MW HFBR is one of the most advanced research reactors in the world. It produces intense beams of neutrons that are used to study the structure of atoms and molecules as well as to produce various radionuclides.

Figure 15.7 Beamline at Brookhaven National Laboratory's High Flux Beam Reactor through which neutrons bombard materials placed in their path.

Every element known has at least one isotope that is radioactive. The radioisotope does not differ chemically from stable isotopes of the same element. All nuclides of an element, stable or radioactive, have the same Z number (the same number of protons) but a different A number. The element tin ($Z = 50$), for example, has 25 isotopes with A numbers from 108 to 132; all are identical in chemical properties, but each isotope has a different number of neutrons. Some tin isotopes are stable, whereas others are radioactive. In fact, natural tin, $Z = 50$, is composed of 10 stable isotopes, ranging from $A = 112$ to $A = 124$. But how can we decide which atoms are stable? We have learned that radionuclides can be detected in numerous ways with sophisticated instrumentation, as well by indirect methods. Atoms are constantly colliding with others in gases or liquids, but not enough energy is involved in the typical collision to disturb the atoms to the point where they will lose some orbiting electrons as a result of the collision. Alpha and beta particles, however, are energetic enough to cause one or more atoms with which they collide to lose their orbiting electrons and thus their electrical balance. An avalanche of electrons often results from collisions of the energetic alpha particles with the atoms in gases. The atoms that lose the electrons are said to be *ionized;* in effect, they become positively charged particles rather than electrically neutral atoms. The ionization that results becomes a mechanism by which radioactivity can be detected.

The *Geiger-Müller counter* is commonly used to detect both beta and gamma radiation. This device, pictured in Figure 15.8, uses a Geiger-Müller (GM) tube, which detects the presence of radiation. (Alpha particles are not detected by most GM systems, since these cumbersome particles cannot penetrate the GM tube and, hence, be counted.) The tube (see Fig. 15.9) consists of a central wire mounted in a hollow cylinder made of (or coated with) metal. The tube is filled with one of several available gases, and the radioactive substance is placed beneath the vertically mounted tube. The various particles enter the "window" of the GM tube and ionize the gas. The ionized gas in turn is attracted to the charged central wire or the metal in the tube, according to the charge on the components involved. A pulse of electrical current results, which is interpreted as an "event" by the electronic counter. The GM tube varies in sensitivity to alpha, beta, and gamma radiation; it is most sensitive to beta particles. No measure of particle energy is directly available from the Geiger-Müller system; however, a trained investigator can establish the particle's energy by determining the amount of radiation absorbed by aluminum, lead, or paper sheets placed between the radioactive source and the GM tube.

Unlike the range of energies of beta particles emitted from atoms of a given radionuclide, energies of gamma photons from any gamma-emitting radionuclide are consistently of some specific value; thus, a measurement of their energies becomes an accurate means of identification of the radionuclide. For instance, cobalt-60 decays by beta emission to become nickel-60, a stable nuclide. The energy of the beta particles ejected from the radioactive cobalt varies from slightly above zero to a peak of about 0.31 million electron volt. The beta radiation is accompanied by two gamma photons, the energies of which are 1.17 and 1.33 million electron volts, respectively. An instrument called a multichannel analyzer or a *gamma ray spectrometer* utilizes various crystals in the

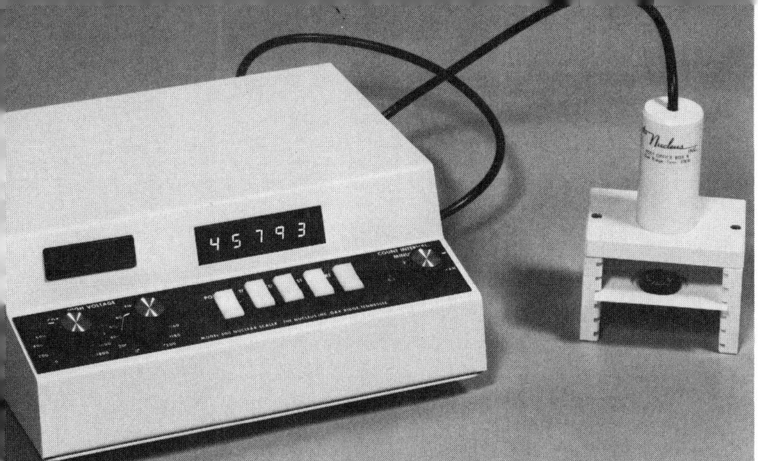

a.

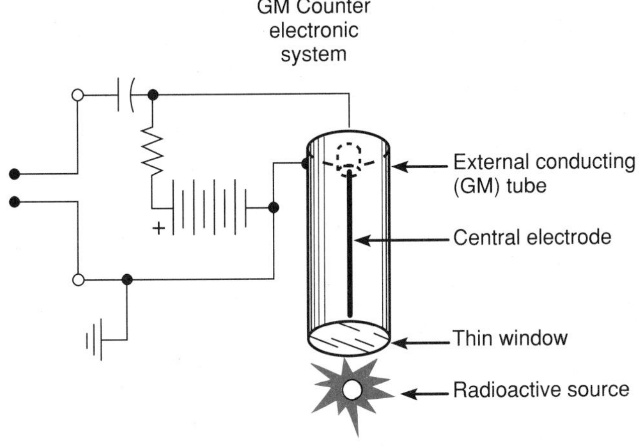

b.

Figure 15.8 (*a*) Beta or gamma radiation is detected by the Geiger-Müller counter. Samples of radioactive substances are placed on the tray in the tube mount under the GM tube to be measured for activity. (*b*) A simple schematic diagram of a Geiger-Müller system as shown in (*a*).

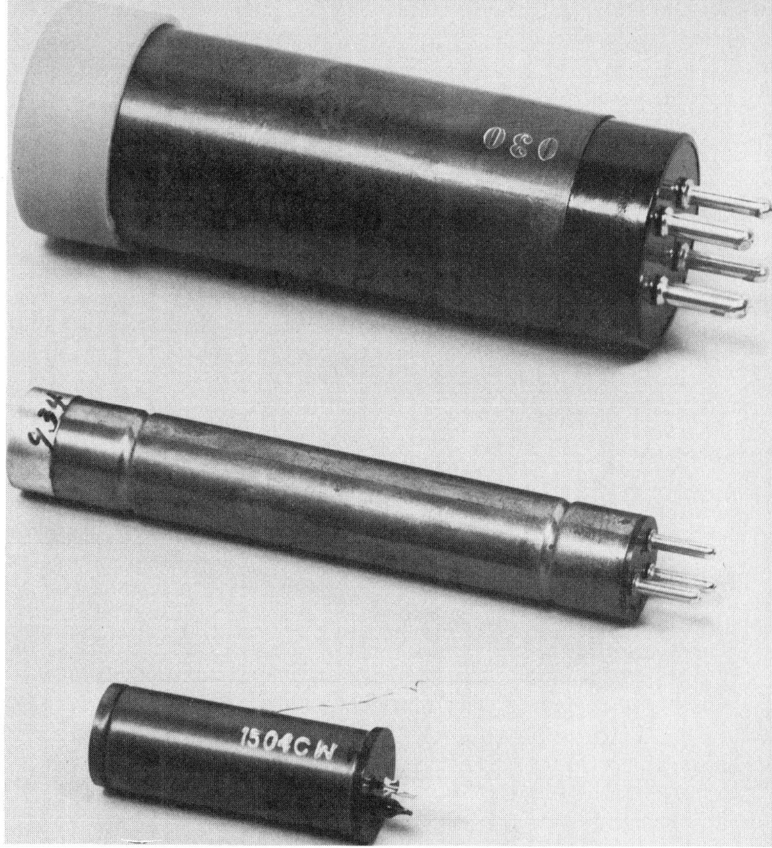

Figure 15.9 The Geiger-Müller tube is usually made of metal by design and is filled with a mixture of gases. The radiation enters the tube through a thin "window" located at the side or, more typically, the end of the tube. The particle or ray is detected as it interacts with the components of the tube.

detector and is designed to analyze gamma energy (see Fig. 15.10). As gamma radiation enters a crystal, the alignment of its atoms is disrupted as the atoms absorb energy; the excited atoms in turn emit a flash of light as they move back to their original energy state. The flash of light that results is called a *scintillation.* The brilliance of the scintillation is directly related to the energy of the incident gamma ray. Other parts of the instrument measure the intensity of the flash and interpret the energy of the gamma ray accordingly. The interpretation of data from a given nuclide permits the identification of the nuclide with much greater accuracy than is possible with most methods of chemical analysis.

A third instrument used to detect the presence of radioactivity is the *cloud chamber.* As a charged particle is introduced into a closed chamber where a saturated condition has been created by water vapor or alcohol vapor and a "cloud," or "fog," has formed, the particle creates a visible path similar to that produced by a high-flying jet plane. The length of the path that the charged particles produce, as well as the degree of deflection caused by an electric or magnetic field brought near the chamber, permits a scientist to determine the particle's mass, charge, and energy. Many of the previously discussed particles were first discovered during the operation of a cloud chamber. Electromagnetic radiation, such as gamma and X rays, is also detected in the cloud chamber, but the mechanisms involved are beyond the scope of this text.

The *bubble chamber,* a fourth type of detecting and measuring device, permits particle trails in its liquid to be observed. The trails are actually gas bubbles created by particles in the liquid. Often the liquid used is hydrogen that is maintained in a pressurized chamber at a temperature near boiling. The pressure is suddenly released in the chamber, and an incoming particle causes the hydrogen to boil along the particle's path. The path dissipates almost instantly; therefore, time photography is the only practical means of observing properties of the particle. The mass, charge, and energy of the incident particle can be measured with the bubble chamber. A photograph of particle interaction in a bubble chamber is shown in Figure 15.11. In the encircled area, the spiral patterns provide evidence of particle interactions in which various particles are destroyed and others are created. The less significant tracks outside the encircled area indicate that other particles have lost their kinetic energies to the liquid.

These detecting and measuring devices are similar in principle to other devices currently in use. As investigations continue, more properties of the elementary particles will be revealed. Instrumentation will undoubtedly become more sophisticated and efficient as each investigator makes contributions to the storehouse of knowledge concerning elementary particles.

The Role of the Health Physicist

When Roentgen discovered X rays in 1895, he paved the way for one of the most useful diagnostic and therapeutic techniques known to scientists. However, problems were soon to arise as a result of the application of his discovery. Within four months, three investigators who worked with X rays suffered radiation burns to their skin. As soon as the use of X rays became more widely accepted, numerous accidental exposures followed. The scientific community did not become particularly concerned with the hazards of radiation until the Manhattan Project and its development of the atomic bomb in 1942. The protection of the Project workers and the public was assigned to a group of eight scientists, who were to be known as *health physicists.*

As a research scientist, the health physicist (HP) must constantly study how radiation interacts with matter and how it affects us, as well as our environment. As a consultant, the HP advises those who use radioactive materials or ionizing radiation, including dentists, physicians, researchers, and industrial users. The HP is responsible for enforcing the safety regulations that deal with radiation. These scientists monitor hospital equipment, nuclear medicine applications, microwave ovens, and all other possible devices that may emit harmful ionizing radiation due to a malfunction of some safety device.

The HP must develop a variety of academic skills, including the acquisition of considerable knowledge in biology, chemistry, nuclear engineering, and physics, as well as in other disciplines. Programs offered by various colleges and universities lead to an associate degree or even to a doctorate. A certificate awarded by the American Board of Health Physics is not mandated by law, but it serves as recognition by colleagues of a high achievement level. The Health Physics Society contends that a career in health physics holds a bright future in the employment picture for years to come ■

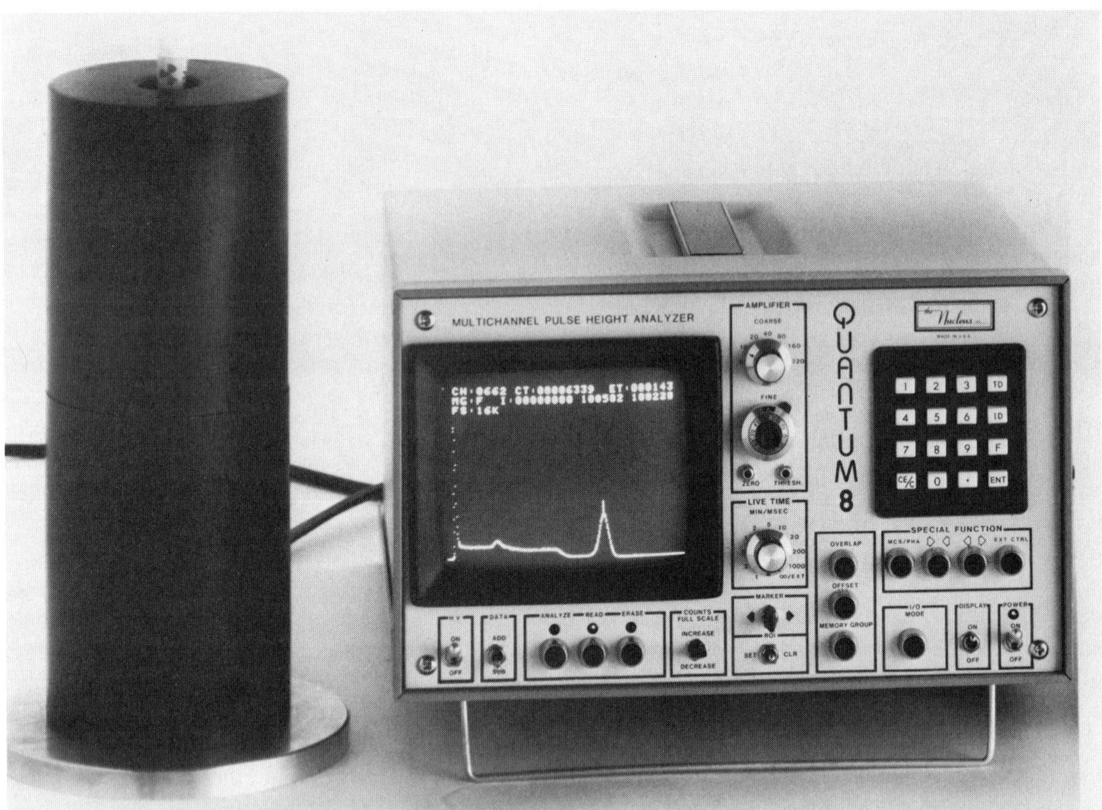

Figure 15.10 An analysis of the data obtained from a gamma-emitting radionuclide reveals that the sample placed under the detector contains cesium–137. The major peak in the spectrum shown was caused by gamma photons, the energy of which is 0.662 MeV.

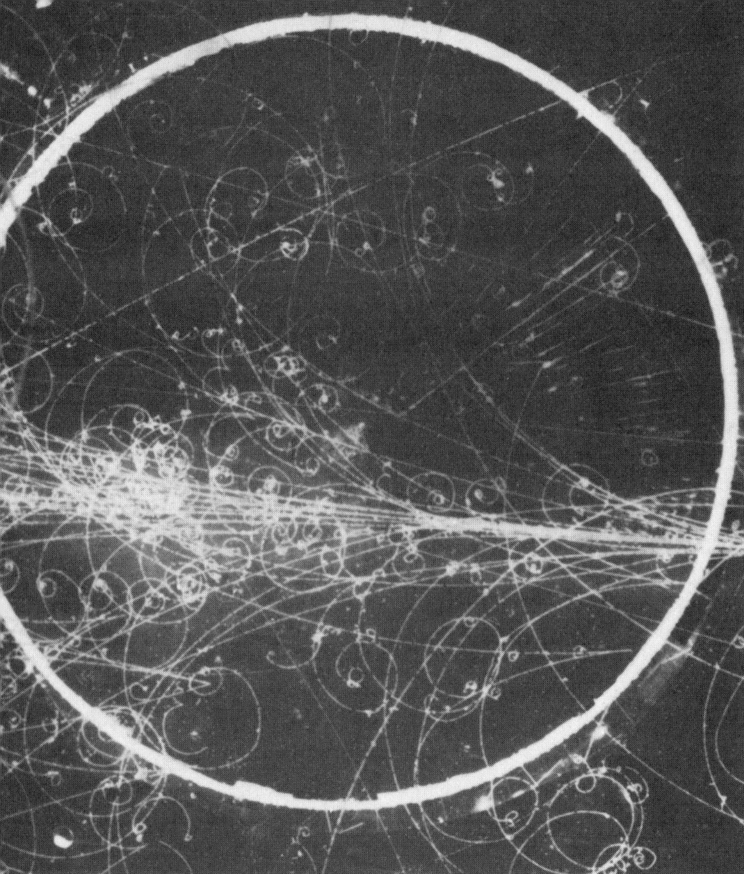

Figure 15.11 The interaction encircled in the photograph was detected with the 15-ft bubble chamber at the Fermilab in Batavia, Illinois. The analysis of such particle interactions has led to the discovery of numerous elementary particles.

The continued refinement of the process of inducing radioactivity by neutron bombardment has led to hundreds of applications that involve minute amounts of the radionuclides produced. For example, essentially all the hairs on an individual's head contain the same amounts of metallic elements. This constant quantity can be determined with such accuracy that a single hair can be traced to a given individual. In the 1960s, the analysis of a single hair from Napoleon's head reportedly indicated that the statesman died of arsenic poisoning, and there was speculation at the time that the doses were self-administered for the stimulating effect arsenic produces. Later, more sophisticated studies found that the original results overestimated the amount of arsenic present. According to the researchers involved, perhaps the arsenic could be traced to the wallpaper found in Napoleon's place of exile on the island of St. Helens.

Specks of dirt and grease too small to be seen by the naked eye can be identified and related to similar specks. This type of study, along with a determination of the amount of gunpowder on a suspect's hands, has led to the solution of many crimes.

Similar applications of this technique, known in scientific fields as *activation analysis,* have been of great value in lunar studies, animal and human physiology, geology, and biology. Industry has applied the principle in many different ways, such as in determining the amount of automobile piston wear under various operating conditions, the optimum mixing time of paints, and in countless studies of fluids in motion.

The following reactions reveal how, through neutron bombardment, very small quantities of some elements can be identified:

NEUTRON ACTIVATION

$$\underset{\text{stable}}{{}^{55}_{25}\text{Mn}} + {}^{1}_{0}\text{n} \longrightarrow \underset{\text{unstable}}{{}^{56}_{25}\text{Mn}} \longrightarrow \underset{\text{stable}}{{}^{56}_{26}\text{Fe}} + {}^{0}_{-1}\beta, \quad t_{1/2} = 2.58 \text{ h}$$

$$\underset{\text{stable}}{{}^{35}_{17}\text{Cl}} + {}^{1}_{0}\text{n} \longrightarrow \underset{\text{unstable}}{{}^{36}_{17}\text{Cl}} \longrightarrow \underset{\text{stable}}{{}^{36}_{18}\text{Ar}} + {}^{0}_{-1}\beta, \quad t_{1/2} = 3.00 \times 10^{5} \text{ y}$$

$$\underset{\text{stable}}{{}^{108}_{48}\text{Cd}} + {}^{1}_{0}\text{n} \longrightarrow \underset{\text{unstable*}}{{}^{109}_{48}\text{Cd*}} \longrightarrow \underset{\text{stable}}{{}^{109}_{48}\text{Cd}} + {}^{0}_{0}\gamma, \quad t_{1/2} = 453 \text{ d.}$$

Radionuclides, particularly those that emit gamma rays, such as cobalt-60, are sometimes used instead of the costly X-ray devices and the rare and naturally occurring element radium. In the medical field, the energy from radionuclides has been used to retard the growth of unhealthy tissue and to sterilize surgical instruments. Other products being sterilized by radiation include baby bottle nipples, disposable diapers, milk cartons, feminine hygiene products, cosmetics, and glass products. The energy of the particle ejected from the radionuclide has been used to preserve agricultural products and to sterilize seeds to prevent their germination. Radionuclides that emit gamma radiation are potential sources of penetrating energy with which radiographs, the successors to X-ray photographs, are made. Industry has also made use of radionuclides in thickness gauging and in producing structural changes in wood, plastics, and other materials, thereby making them more useful.

Radionuclides can also be used as invisible tracers. Minute amounts can be mixed unnoticeably with varying proportions of a stable isotope of the same element and traced throughout a closed system. For instance, the length of time required for a chemical reaction to occur can be measured with great precision. Also, a small quantity of a radionuclide can be placed inside a capsule that is to be inserted into a pipe, then followed by a detecting device as the liquid carries the capsule along. This technique may be used to locate a blockage inside the pipe or to determine where the pipe goes. If poured directly into the liquid carried in the pipe, a concentration of activity from the radionuclide readily marks the area of small leaks that might otherwise be undetected.

The Biological Effects of Radiation

Many of us have cringed at the "clicking" of a Geiger counter as used in movies and TV productions. We have learned to associate this sound with the presence of space aliens, "monsters" with unusual physical abilities resulting from exposure to radioactive materials, "mad scientists," or impending danger for the ingenuous characters who dared venture into a tomb of the ancients. It is little wonder that many people have developed an inherent fear of the often misunderstood natural phenomenon of radiation.

As has been previously discussed, a significant amount of radioactivity in our immediate environment occurs naturally. There are radionuclides that emit alpha particles, some that emit beta particles, and others that emit gamma rays. Radionuclides that simultaneously emit more than one of the three types are also found in nature. Another source of radiation that adds to our degree of exposure is that radiation emitted by heavenly objects, referred to as cosmic radiation. The combination of the various sources provides our natural background radiation.

As noted, radioactivity is commonly expressed in curies or in becquerels. With respect to safety, other types of measure must be considered, particularly in the case of ionizing radiation. Exposure to X-radiation or gamma radiation is ordinarily expressed in *roentgens* (R). This unit is defined as the amount of X- or gamma radiation that will produce 2.08×10^9 ion pairs in 1 cubic centimeter of dry air. In other terms, 1 roentgen is capable of the production of 1.61×10^{12} ion pairs per gram of air or of an ion charge per unit mass of 2.58×10^{-4} coulomb per kilogram. The expression in ion charge per unit mass is the SI standard of measure. Many laboratory instruments or those used to monitor radioactive sources are calibrated in submultiples of the roentgen per unit time, such as milliroentgens per hour (mR/h). In addition to the GM survey instrument discussed earlier in the chapter, radiation exposure is monitored by pocket-sized ion chambers called *dosimeters* and by radiation-sensitive film sealed in lightproof paper and worn as *film badges* by laboratory investigators or technicians.

Also important among the measures of exposure levels is the unit called the *rad,* derived from an indication of *r*adiation *a*bsorbed *d*ose. It is defined as the amount of radiation necessary to deposit energy equal to 1.00×10^{-5} joule in 0.00,100 kilogram of material. An exposure of 1 R absorbed by various tissue deposits doses ranging from about 0.83 to 0.93 rad, depending on the energy of the X- or gamma radiation involved. The SI unit of absorbed radiation dose is the *gray* (Gy), the amount of radiation that will deposit 1 joule of energy in 1 kilogram of material. In terms of equivalence:

$$1 \text{ rad} = 1.00 \times 10^{-2} \text{ J/kg} = 1.00 \times 10^{-2} \text{ Gy}.$$

The biological effects of radiation are directly related to the rate at which energy is deposited in living tissue. Different types of radiation lose energy through the ionization they create along their paths. Faster moving particles and higher energy X- or gamma radiation ionize less effectively than slower moving particles or radiation of lower energy. The unit of effective dose is called the *rem,* or *r*oentgen *e*quivalent *m*an. The rem and millirem, although popular among radiation technicians, essentially have been replaced in scientific literature by the SI unit known as the *sievert* (Sv). One rem is equal to 0.001 Sv. The average person annually receives an effective dose of about 210 millirem from such sources as cosmic rays, building materials, medical and dental diagnoses, color televisions, food, and the air we breathe. Also included in this amount is the exposure from the radionuclides present in our bodies, about half of which results from decay of the present potassium-40 and the balance from uranium and its decay daughters. The exposure from global fallout created by nuclear-weapons testing is not as severe as in past decades because of the international ban on this activity; it has decreased from a peak of about 4 mrem to a present value of less than 0.5 mrem annually.

According to the United States Nuclear Regulatory Commission, there was an annual average of 10,000 cases of measurable radiation exposure during the 1980s. Primarily, these individuals were employees at commercial nuclear power plants, and their level of exposure, for the most part, was considered to be minimal. Scientists are continuing their efforts to determine the health risks associated with even low levels of exposure. Their data also come from the ongoing study of those who survived the detonation of atomic bombs during World War II and those who were exposed to low-level radiation as a result of the Chernobyl nuclear reactor incident in Russia (see Chapter 27). Human beings who are exposed to radiation doses over their entire bodies in excess of 250 mrem in a very brief period of time, such as in a matter of seconds, have been found to suffer from varying degrees of radiation sickness. Without proper treatment, half the people who absorb a single whole-body dose of 350 rem (350,000 mrem) in a brief period of time are expected to die within a few days to several weeks; hence, 350 rem is called the LD50 (lethal dose for 50 percent of the recipients). However, there are many factors that influence the response of an individual to various amounts of whole-body irradiation. For instance, the very young and the very old apparently are more radiosensitive than the middle-aged or young adults. Also, the female, in general, appears to have a greater tolerance to radiation exposure than does the male. In addition, the biological effect on the human body has been found to depend on the type of radiation involved. For instance, exposure to high-energy neutrons or alpha particles, as measured in rads, is twenty times as damaging as exposure to the same number of rads of beta particles, X rays, or gamma rays. The survivors who receive no prescribed medical treatment, such as bone marrow transplants or transfusions of blood platelets and red blood cells, may recover, but apparently have a shortened life span. A brief exposure of 900 rem or more is essentially fatal, so 900 rem is presently considered as LD100. Radiation exposures to various parts of the body, such as to the hands or the feet, may exceed what would be considered a lethal whole-body dose without apparent threat to life.

When ionizing radiation passes through living tissue, the hazardous effects result from the ionization of molecules within the various cells. The ionized molecules may interfere with the cell's ability to function. While many cells are able to repair themselves, others cannot; hence, they die without reproducing new, healthy cells. Still other damaged cells live long enough to reproduce new cells that simulate the altered parent cell. A defective cell, then, may trigger a genetic change, and may bring about a cancerous growth or an unwanted mutation in the organ that contains it.

Genetic defects have been produced and studied in detail in many species of plants, in insects such as fruit flies, as well as in mice and other animals. The dosages for such studies number into the hundreds of rem for animals and into the thousands of rem for insects and plants. However, no inherited genetic effects have been noted among humans, even among those who survived the atomic explosions at Hiroshima and Nagasaki, Japan, near the close of World War II. The radiation exposure among the survivors is estimated at over 200 rem. (There is, however, significant statistical evidence that links high radiation doses to cancer

The Atomic Pile

Numerous contributions to our knowledge about the atom are attributed to scientists who fled Europe to escape persecution associated with Fascism. Among those great researchers was Enrico Fermi (1901–1954), an Italian physicist. His scientific investigations were to have a great impact toward bringing World War II to a close.

Fermi had closely followed the research conducted by Sir James Chadwick, an English scientist who, along with other physicists, had discovered in 1932 that alpha particles could knock neutral particles (neutrons) from beryllium nuclei. Fermi found that these free neutrons could initiate nuclear reactions, particularly after they were passed through water or paraffin, a procedure that slowed them down so they could be more readily absorbed by nuclei. He received the Nobel prize in 1938 for his investigations with slow (thermal) neutrons in neutron bombardment. Immediately after the conclusion of the awards ceremony held in Stockholm, Sweden, Fermi and his family fled to the United States where he became a professor of physics at Columbia University.

Fermi was appointed a senior scientist in the Manhattan Project, the effort that resulted in the means to produce the chain reaction that led to the development of the atomic bomb, although he maintained his Italian citizenship until 1944. He designed a structure composed of graphite blocks that slowed the neutrons, along with blocks of uranium and uranium oxide that served as a source of neutrons, alpha particles, and fissioning material. Rods made of the element cadmium were raised or lowered among the blocks to control the rate of nuclear reaction by absorbing the thermal neutrons. This pile of blocks, known as an atomic pile, was the world's first nuclear reactor. Located at the University of Chicago, it became operational in 1942. Two and a half years later, the first bombs that employed an uncontrolled, thus explosive, fission reaction leveled two Japanese cities and World War II abruptly ended. Fermi died from cancer before he could witness the many benefits obtained from his accomplishments. In the year following his death, a new element ($Z = 100$) was discovered; it was named fermium (Fm) in his honor ■

frequency.) Despite the relatively few people who have been exposed to obviously harmful radiation levels, the biological effects of radiation and the potential hazards that exposure may bring about have been studied in great detail. In fact, each case has been so thoroughly monitored that, with the exception of exposure to certain chemicals, scientists have more information about radiation dangers than about any other biological hazard, including pesticides and pollutants introduced into our environment.

The radiologist who administers radiation treatments to destroy abnormal tissue growth (cancer) must carefully consider the danger to the patient and compare it to the benefit expected from the exposure to various radiation levels. In diagnostic medicine, various radionuclides are used. For example, fluorine-18 is used in bone imaging, iodine-131 has proved invaluable in thyroid imaging and treatment, and gold-198, along with indium-113, is used in diagnosing suspected liver problems. In order to destroy cancerous cells, the physician may implant an encapsulated needle that contains cobalt-60, radium-226, iridium-192, or another beta-gamma emitter directly into the tumor. The ionizing radiation kills the cancerous cells (and, unfortunately, also the healthy cells) that surround the radioactive source. Externally, cobalt-60 and cesium-137 are used in therapy radiation to treat cancerous growths in deep-seated organs. The gamma rays emitted from either source can be essentially focused, as visible light is done with convex lenses.

New high-energy devices that generate beams of X rays more penetrating than the gamma rays emitted from the cobalt and cesium radionuclides are now available. Medical researchers continue to study the effects of high-energy beams of alpha particles and other heavy ions in hopes that greater efficiency can be obtained in treating malignancies without increasing the damage to surrounding healthy tissue. It appears that the number of people per year who have had their life span significantly extended in their bouts with cancer as a result of some nuclear medicine treatment will continue to increase as new technological advances are made.

Fission among the Atoms

The work of early twentieth-century scientists revealed that if two light nuclei combine to form a heavier one, the new nucleus has *less* mass than the sum of the masses of the original ones, an observation that holds true for elements lighter than iron. If an iron-56 nucleus collides with another nucleus and a nuclear reaction results, the iron nucleus splits apart rather than forms a heavier nucleus. (Recall the great binding energy per nucleon of iron-56. The nuclides beyond iron-56 apparently are formed through successive captures of neutrons and the resultant beta decays.) Accordingly, if heavy nuclei are divided into parts, energy is released and the sum of the masses of the fragments is less than that of the original nucleus. Thus, a given amount of matter is lost—converted to energy. The application of this source of energy is the focus of the field of *atomic energy.* Splitting nuclei and combining nuclei are more commonly known as *fission* and *fusion,* respectively.

Fission is *the process by which the nuclei of various large atoms split into two approximately equal nuclei.* The basic material of the fission process is generally an isotope of the element uranium. The process was discovered in 1939 by two German scientists, Otto Hahn and Fritz Strassman, as they were bombarding a uranium target with neutrons in an attempt to produce heavier elements. They found that barium, an element with a mass about half that of uranium, was produced by the neutron bombardment. Later, the process was called fission after its similarity to biological cell division (see Fig. 15.12).

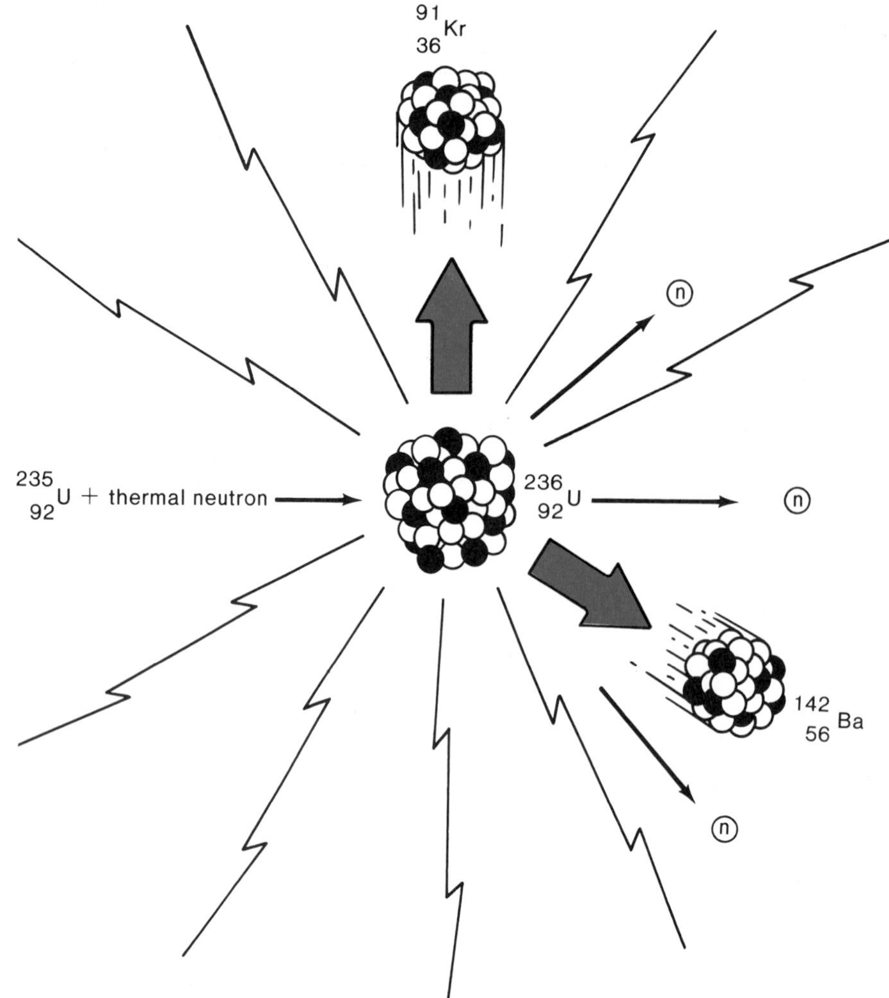

Figure 15.12 The fission reaction depicted in this illustration is only one of many possible combinations of new nuclei that can result. The released neutrons can bring about other fission reactions as they interact with other fissionable nuclei.

The potential for heavy nuclei to split into approximately equal pieces depends on the delicate balance of various forces within the nucleus. The repulsive electrical forces in the nucleus cause it to expand from its typical spherical shape into an elongated shape, similar to that of a peanut shell. The forces of attraction are overcome by the repulsive forces, and the nucleus splits, accompanied by the release of a tremendous amount of energy—over 6 million times the energy released by a molecule of TNT.

A neutron that undergoes an inelastic collision with a uranium-235 atom apparently adds sufficient energy to cause elongation of the uranium nucleus. The resultant fission of the nucleus may produce many combinations of lighter nuclei. A typical reaction that results is

FISSION REACTION

$$^{1}_{0}n + ^{235}_{92}U \longrightarrow ^{236}_{92}U \longrightarrow ^{91}_{36}Kr + ^{142}_{56}Ba + 3(^{1}_{0}n).$$

As in the case of this reaction, the products of fission ordinarily have an excessively low Z/N ratio; consequently, they have very short half-lives. The new atoms reach stability only through a chain of spontaneous emissions of alpha and beta particles.

The energy released from the fissioning of uranium atoms is immense. For example, the energy released from the controlled fissioning of 1 kilogram of uranium-235 would be equivalent to that obtained from the burning of over 150 massive truckloads of coal! An appreciable amount of energy is also released from the fissioning of certain isotopes of plutonium, thorium, and californium.

Unlike the other elements mentioned, thorium is often overlooked as a nuclear fuel, but thorium-232 undergoes the process of fissioning after absorbing a neutron, undergoing two consecutive beta emissions, and then absorbing a second nuetron. The first reaction after a neutron is absorbed by the thorium nucleus is

$$^{232}_{90}Th + ^{1}_{0}n \longrightarrow ^{233}_{90}Th.$$

This isotope of thorium, Th-233, is very unstable, with a half-life of about 22 minutes. The newly formed radionuclide decays by beta emission; that is,

$$^{233}_{90}Th \longrightarrow ^{233}_{91}Pa + ^{0}_{-1}\beta.$$

Protactinium-233 (Pa) is also radioactive, with a half-life of 27 days. The element undergoes another beta decay and becomes a radionuclide of uranium with a half-life of 159,000 years:

$$^{233}_{91}Pa \longrightarrow ^{233}_{92}U + ^{0}_{-1}\beta.$$

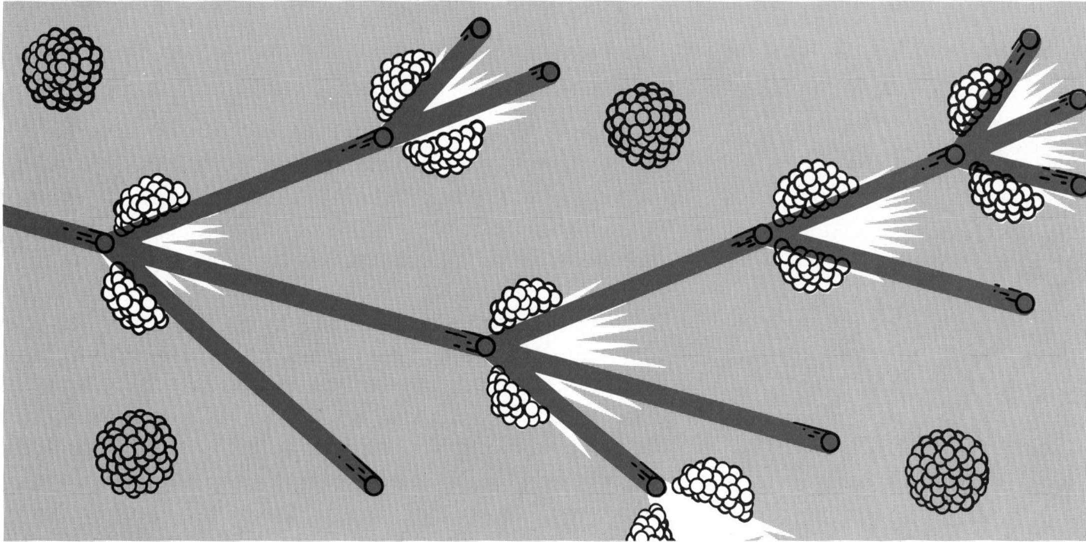

Figure 15.13 A thermal or slow neutron can create a chain reaction among certain nuclides of uranium or plutonium. The typical interaction creates an average of 2.5 neutrons per fission event to carry on the reaction. Many new radionuclides are produced, depending on the manner in which the nuclei split.

The uranium-233 nucleus so formed undergoes fissioning upon absorption of another neutron, a process shown by this equation:

$$^{233}_{92}U + ^{1}_{0}n \longrightarrow ^{234}_{92}U \longrightarrow \text{fission produces} + \text{energy.}$$

There is only an established probability that a U-233 or a U-235 atom will split and release nuclear energy when struck with a neutron. Just as some baseball players prefer to hit fast-thrown balls and others prefer to be pitched slower-moving balls, neutrons of various kinetic energies are capable of initiating the process of fissioning through inelastic collisions with specific nuclei. For instance, some nuclei interact effectively with and absorb only the slow-moving or *thermal neutrons.* These neutrons have a velocity of about that of a bullet from a target rifle. Fast neutrons have velocities thousands of times greater than thermal neutrons and, as one might expect, there are categories between these extremes. Thermal neutrons are more easily absorbed by U-235 nuclei than by U-238 nuclei. Nuclear power reactors (see Chapter 26) use a fuel mixture of many times more U-238 atoms than U-235 atoms, and the preferential absorption of the relatively slow neutrons by the nuclei of U-235 atoms, rather than by those of U-238, offers a mechanism by which the rate of fissioning can be delicately controlled. Figure 15.12 graphically portrays the fissioning of U-236.

As the equation for the fission of a uranium-236 atom previously presented in this section indicates, two or three neutrons are usually ejected as individual particles. These specific neutrons have the potential of causing other atoms of U-235 to split, with the consequence that more neutrons are released to cause additional atoms to fission. The total process is, then, a *chain reaction* that can continue to release neutrons to split other atoms at a constant rate if the *mass* is of proper, or *critical,* size. If subcritical segregated masses of the fissionable material are suddenly caused to collide, such as with an explosive device, tremendous amounts of energy are released as billions upon billions of atoms fission in a few millionths of a second. This technique is used as the primary mechanism of an atomic (nuclear fission) bomb. The critical mass of the atomic bomb dropped over Hiroshima on August 6, 1945, was less than 2 kilograms and produced a blast equal to about 20,000 tons of the explosive TNT. Bombs available to the various world powers today are rated in the equivalents of megatons of TNT. The fissioning of 1 kilogram of U-236 in this manner yields about 24 million kilowatt-hours, or as much energy as provided by about 3300 tons of coal or 440,000 gallons of gasoline. A chain reaction is depicted in Figure 15.13.

A chain reaction does not take place with naturally occurring uranium, since the particular nucleus (^{235}U or U-235) that fissions upon absorbing a neutron represents only 0.7 percent of all uranium. The common nucleus (^{238}U or U-238) absorbs neutrons without splitting; by its presence it actually prevents the fission process from occurring. The separation of the two radionuclides is accomplished at various strategic plants within the United States as well as in several foreign countries.

The Fusion of Atoms

Fusion generally involves *the inelastic collision of light nuclei to form a heavier nucleus.* For fusion to occur, the various nuclei must collide with one another at sufficiently high velocities for the mutual electrical repulsion of the nuclei to be overcome in order for the nuclei to form a single more massive nucleus. The velocities involved correspond to the velocities of nuclei caused by the exceedingly high temperatures present in stars. In a star similar to our sun, hydrogen atoms fuse to form helium atoms, and a tremendous amount of mass is converted into energy. In the sun an estimated 5.97×10^{11} kilograms of hydrogen are converted to 5.94×10^{11} kilograms of helium each second. The difference in mass (3.00×10^{9} kg) is converted to energy and assumes the form of light, heat, and other types of electromagnetic radiation. Astronomers refer to this fusion as "hydrogen burning." The nuclear reaction of fusion produces new stable atoms or excited atoms that, through a series of particle and energy emissions, eventually form very compact and stable nuclei.

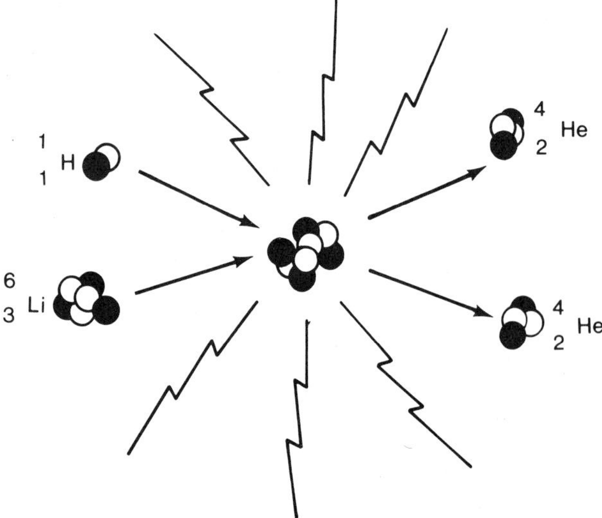

Figure 15.14 The fusion reaction between light nuclei results in a significant amount of mass that is converted to energy. The by-product of this reaction is simply helium.

Nuclei significantly heavier than the nuclei of helium can be caused to fuse at thermonuclear temperatures, but considerably less energy is produced per reaction. Older stars that have "burned" most of their hydrogen must rely on the fusion of heavier elements to maintain their energy needs; otherwise, they will undergo gravitational collapse and ultimate destruction.

The energy released by the fusion of hydrogen atoms to form helium atoms is tremendous, but it varies with the isotopes of hydrogen that are involved. A single reaction could release 20 million electron volts of energy. On the average, in a gallon of ordinary water there is the equivalent of about 20 drops of the isotope of hydrogen called *deuterium* ($^{2}_{1}H$). This rare hydrogen atom is composed of an electron, a proton, and a neutron. The amount of energy that would result from fusion of these 20 drops of water could yield as much energy as that produced by the combustion of 300 gallons of gasoline or 10 tons of TNT.

Two possible fusion reactions that involve deuterium, (hydrogen-2), along with the energy released as a result of each reaction, are:

$$^{2}_{1}H + ^{2}_{1}H \longrightarrow ^{3}_{2}He + ^{1}_{0}n + 3.27\ MeV$$

and

$$^{2}_{1}H + ^{2}_{1}H \longrightarrow ^{3}_{1}H + ^{1}_{1}H + 4.03\ MeV.$$

Figure 15.14 illustrates a possible fusion reaction in which the nucleus of a lithium-6 atom interacts with a hydrogen-2 nucleus. The unstable beryllium-8 nucleus that results from the fusion reaction disintegrates almost instantly into two helium-4 nuclei (alpha particles) and a vast amount of energy is released. The alpha particles shortly capture two electrons each and become harmless helium-4 atoms.

As has been previously discussed, fusion reactions under controlled conditions require thermonuclear temperatures that reach into the millions of degrees. To provide and maintain such high temperatures with any measure of success has demanded great expenditures of funds and time on the part of researchers. In electric arcs temperatures have been reached that are sufficiently high to convert gases into **plasma,** *matter that has become disrupted so that it consists merely of electrically charged particles.* (Matter in the form of plasma is considered so abundant in stars that 99 percent of all substances in the universe is presumed to exist in this fourth state of matter.) The temperature of a plasma is increased further by beams of high-energy electrons or with laser beams. But what matter can be used to contain a plasma? At about 4000°C all known substances melt and vaporize, so no conventional receptacle can be used. A solution to the problem came from the continued research on magnetic fields. A "magnetic bottle," a magnetic field that is not affected by high temperatures, contains particles set in motion by the great forces it can provide. As the magnetic field is increased, it contains and compresses the plasma as well as produces the thermonuclear temperatures necessary for fusion to occur. The total thermal energy present in the magnetic field is very low, even at thermonuclear temperatures, since the amount of matter remaining in this region approximates that in a vacuum.

In order for fusion to be self-sustaining, stellarlike temperatures apparently are necessary. In addition, new hydrogen atoms must be introduced into the target area to replace those that have undergone fusion. Although fusion has been achieved with various techniques, the length of time over which it can presently be maintained is not sufficient to make it a practical major source of energy.

When the various technical problems have been surmounted and sufficient production facilities constructed, a search will have ended for the ultimate answer to our energy problems. Typically there are no radioactive by-products associated with the fusion of hydrogen atoms, only the release of helium. This light gas is useful to industry, and its release in fusion reactions will be welcomed as a cheaper, more accessible supply of this valuable element. The fusion process cannot get out of control, and thus does not present a potential hazard. Hydrogen, the fuel for fusion, is the most abundant element in the universe; thus, the amount of energy that can be produced through controlled fusion is far beyond any of our future needs. When fusion-produced electric power becomes a reality and is cost-competitive, we will be able to use our remaining fossil fuel supply to extract from it the valuable and necessary raw materials it contains.

The harnessing of the energy provided by the processes of nuclear fission and fusion in a fashion most beneficial to us is accomplished by the facilities called *nuclear reactors.* Their characteristics and uses are discussed in detail in Chapter 26. The structure of a fusion reactor is depicted in Figure 15.15.

Mass-Energy Equivalence

If a force sufficient to overcome the state of rest of an object is constantly applied, the object will accelerate to an increasingly higher speed. If the force remains constant, so will the acceleration; thus, the object will steadily increase in speed. There is, however, a maximum speed an object can attain, and that value is the speed that light in a vacuum travels, *c.* In fact, an object could not

Figure 15.15 This research fusion reactor exemplifies the large investment made toward obtaining energy from nuclear fusion.

be accelerated to reach this speed. Recall from Newton's second law of motion that the acceleration of an object is dependent on the mass of an object, as well as on the acting force. According to a theory proposed by Einstein, as work is done to increase the speed of an object, the mass of the object also increases. Therefore, the force impressed on the object produces consistently less acceleration as the speed of the object increases. The relationship between mass and speed is expressed as

$$m = \frac{m_0}{\sqrt{1 - \frac{v^2}{c^2}}}.$$

In this equation, m_0 represents the mass of an object at rest, and m is the relativistic mass of the object as it moves at speed v. According to the equation, as v approaches c (the speed of light in a vacuum—3.00×10^8 meters per second), m approaches infinity. If the mass became infinite, then the acceleration of the object would be zero unless the force were also infinite. The effect on mass as speed approaches c is presented in Table 15.7. A particle accelerated until it reached the speed of light would have an infinite mass and could only gain the speed c if an infinite force acted on it. Such a force would be impossible to attain; thus no particle could be accelerated to the speed of light, although charged particles of various identities have been accelerated by electric or magnetic fields in particle accelerators to speeds of over $0.99c$. (The next section of this chapter will take a closer look at particle acceleration.) Electrons reportedly have been accelerated to such a high speed that the relativistic mass of these negatively charged particles has been measured at 40,000 times the rest mass of the electron. This relativistic mass corresponds to a speed that approaches within a fraction of the speed of light. The accuracy of the relativistic mass of the electron determined in the experiment is under close scrutiny as the precise value of c is further investigated (as was discussed in Chapter 10).

Table 15.7

Comparison of relativistic mass at various speeds.

Speed ratio v/c (percent)	Speed (m/s)	Relativistic mass (m/m_o)
1	2,997,925	1.000
10	29,979,250	1.005
50	149,896,230	1.15
86	257,821,520	2.00
90	269,813,210	2.30
94	281,804,910	3.00
96	287,800,760	4.00
99.5	298,293,350	10.0
99.9	299,492,670	22.3
100	299,792,458	Infinity

Example 15.3

Determine the relativistic mass of a 50.0-kilogram object, the speed of which is 1.73×10^8 meters per second.

Solution

RELATIVISTIC MASS

$$\text{relativistic mass} = \frac{50.0 \text{ kg}}{\sqrt{1 - \frac{(1.73 \times 10^8 \text{ m/s})^2}{(3.00 \times 10^8 \text{ m/s})^2}}}$$

$$m = \frac{50.0 \text{ kg}}{\sqrt{1 - \frac{(3.00 \times 10^{16})}{(9.00 \times 10^{16})}}}$$

$$= \frac{50.0 \text{ kg}}{\sqrt{1 - \frac{1}{3}}}$$

$$= \frac{50.0 \text{ kg}}{\sqrt{0.667}}$$

$$= \frac{50.0 \text{ kg}}{0.817}$$

$$= 61.2 \text{ kg}.$$

For centuries scientists considered matter and energy to be two completely separate entities. However, early in the twentieth century Albert Einstein concluded that mass and energy were really different aspects of the same thing. Energy can assume the form of mass, and mass can be converted into energy. Einstein described the relationship between energy and mass quantitatively in the well-known equation

$$E = mc^2.$$

In this equation, E is a measure of energy, m stands for mass, and c^2 represents the square of the speed of light. The equivalence of mass and energy is a major part of the **special theory of relativity.** The conversion of mass to energy is quite common in our

The World's Premier Atom Smasher

Congress has recently approved funds to initiate the construction of a *Superconducting Super Collider* (SSC). This high-energy facility is to be located on about 17,000 acres in Waxahachie, a farming community located in northeast Texas.

In tests completed at Fermilab in Batavia, Illinois, the prototype magnet developed for the SSC successfully carried 7500 amperes of electrical current without any appreciable energy loss. The next step is the production of about 8000 such giant superconducting magnets, each almost 17 meters long, that will be needed for the main collider ring. The magnets will be used to direct proton beams so that they will follow the collider's circular tunnel to the target area. The tunnel will have a circumference of almost 90 kilometers.

The project represents an investment that may surpass $8 billion dollars and is billed as "the most expensive physics experiment ever conducted." However, it will ensure that the United States remains in the forefront of innovation and technological advances in this field for decades to come ■

immediate surroundings. As a match is struck and a chemical change occurs, a slight portion of the mass of the atoms involved in the reaction is converted into energy. This point could be made for any chemical equation, regardless of the length of time involved. Also, the decrease in the mass of the sun as nuclear reactions produce electromagnetic radiation has been estimated to be in excess of 4 million tons per second. The mass-energy equivalence is not restricted to chemical and nuclear reactions; in fact, any change in energy corresponds to a change in mass.

The expression $E = mc^2$ does not indicate that matter is converted into energy when it is traveling at the speed of light. It points out that as matter increases in speed, its mass increases accordingly. As matter approaches the speed of light, its mass and hence its energy approach infinity. In fact, c^2 is a *proportionality constant* that relates energy to mass in a correct fashion. This constant comes from the relativistic expression for mass.

The accuracy of Einstein's conceptual view of mass-energy equivalence was significantly corroborated in 1932. C. D. Anderson, an American physicist who was studying cosmic radiation, discovered that a gamma ray had interacted with a photographic emulsion in such a manner that it was converted into two particles. The identity of the pair of newly formed particles was found to be that of a negative electron, and its antiparticle, the **positron.** This discovery also confirmed the existence of antiparticles, which reportedly were predicted by British physicist P. A. M. Dirac several years before. Further investigations proved that an electromagnetic wave of sufficient energy may, on interacting with a charged particle such as a nucleus, bring about the conversion of energy into mass by a process called *pair production.* Other particles and their antiparticles have been produced by the interactions of higher energy photons with nuclei.

The inverse process, the annihilation of a particle and its antiparticle, has also been observed. The interaction between an electron and a positron results in total annihilation of the two particles and the creation of two photons, each with an energy of at least 0.511 million electron volts. (The rest mass of an electron or positron is equivalent to an energy of 0.511 MeV.) The photons propagate in opposite directions so that both energy and momentum are conserved. Pair production and annihilation represent dramatic evidence of the mass-energy equivalence predicted by Einstein's relativistic theory.

The Subatomic World of Nature

We have wondered about the submicroscopic nature of things since the days of Aristotle and his contemporaries, and our inquisitiveness continues to grow. As discussed in the section on radionuclides, our method of studying the world too small to see is by means of particle interactions, including the annihilation of a particle as it collides with its antiparticle. Various devices, other than those discussed in this chapter, have been constructed to study particle properties. They have such descriptive names as cyclotrons, synchotrons, beam reactors, and, simply, particle accelerators. The first such machines were constructed in the 1930s to accelerate particles to the energies considered necessary to penetrate to the nucleus of the atom. Intense electric fields, along with their accompanying magnetic fields, became the mechanisms by which charged particles were accelerated in a manner basically the same as that employed in a TV picture tube.

The accelerator types are classified into two general categories: circular and linear. The circular accelerators, as the name implies, repeatedly accelerate a beam of charged particles, such as electrons or protons, in a circular path until the desired particle energy is reached. Circular accelerators can maintain the particle beam at a specified energy, in a sense "storing" it for further study. Linear accelerators hurl charged particles, such as electrons, protons, and their antiparticles, at a target area chiefly by means of an electric field. The field, however, is not produced by a voltage potential difference; rather, it is provided by electromagnetic radiation in the microwave region (recall Fig. 10.8). The Two Mile Linear Accelerator, shown in Figures 15.16 and 15.17, is located at the Stanford Linear Accelerator Center (SLAC) in California. The beam of charged particles produced by this facility is diverted into numerous paths so that several experiments can be carried on simultaneously. The beam of microwaves transports the injected electrons and constantly accelerates them along the length of the facility.

Figure 15.16 Aerial photograph of the Stanford Linear Accelerator Center. Electrons and positrons are accelerated to high energy under the 2-mi structure at center and brought into collision inside a huge particle detector housed in the large building at lower right.

An obvious means to make available an intense concentration of energy is through particle-antiparticle interactions. The rate of annihilation of electrons and positrons is directly related to the energy of the two particles. Again, both circular and linear colliding-beam machines were developed to accelerate particles and their antiparticles in opposite directions so that the head-on collisons would provide much more energy than would simple linear or circular acceleration with a relatively stationary target. As a result of the high-energy collisions, new particles are created and are available for increasing our knowledge of the submicroscopic world. A linear collider facility, such as the Stanford Linear Collider (SLC), uses electron and positron beams obtained from the SLAC linear accelerator.

The SLAC has produced its first Z particle, an important contribution to ongoing attempts toward understanding the structure of matter. The feat has also been accomplished at the Large Electron-Positron (LEP) collider in Geneva, Switzerland. At both facilities, an electron is accelerated almost to the speed of light and made to collide with an equally fast positron to produce this elusive particle. It decays almost instantly into leptons or into a

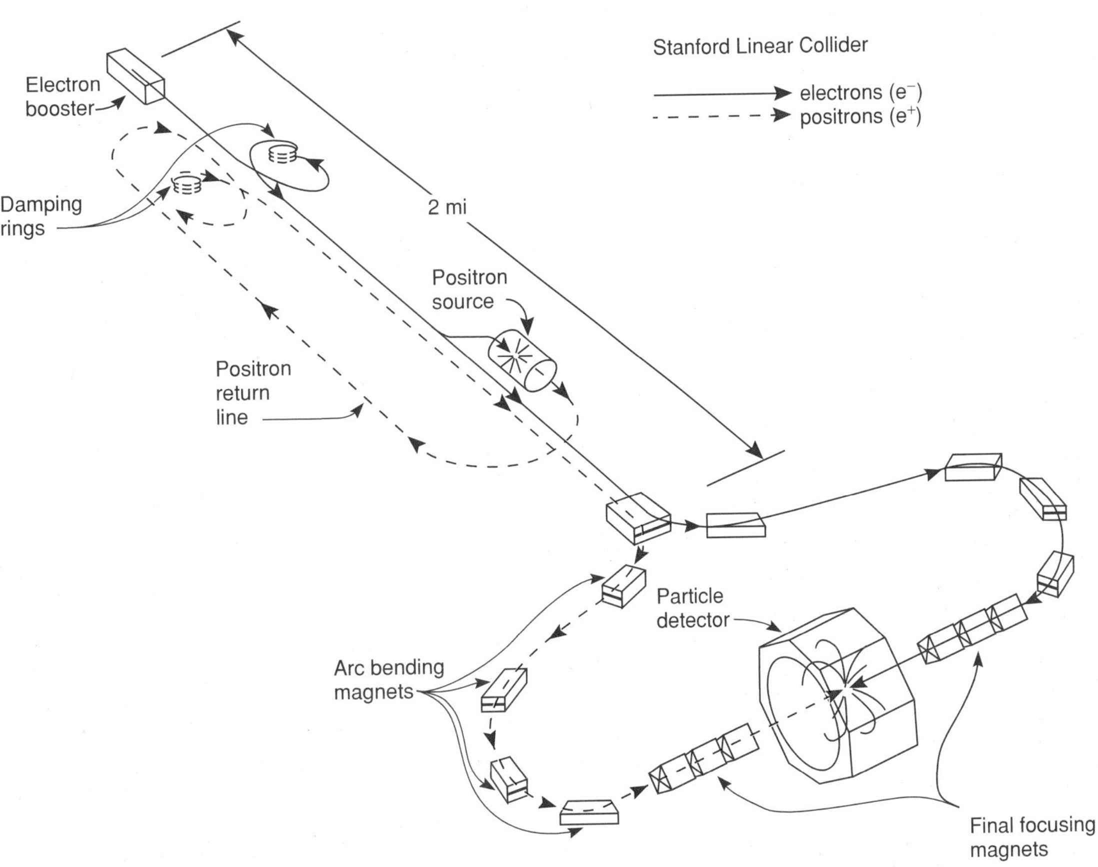

Figure 15.17 An artist's conception of the Stanford Linear Collider and its basic operations. High-energy electrons and positrons annihilate each other inside a large particle detector, producing massive *Z*-particles that subsequently decay into showers of other particles. Courtesy of SLAC.

quark-antiquark pair that disintegrates into a spray of debris. The *Z particle* is one of three carriers of the weak nuclear force that explains certain kinds of radioactive decay.

Researchers first had to solve the problem of eliminating muons, particles similar to electrons but with a mass 200 times as large (recall Table 15.2), created when stray electrons and positrons interacted with the collider's walls. Special magnets were installed to disperse any muons present, keeping them out of the target area. The modifications were successful in improving the collider's efficiency, thus permitting the scientists to detect more Z particles.

The combined efforts of scientists at the collider facilities currently lead them to surmise that there are but three families of fundamental particles known as leptons (charged and neutral) and quarks that are the building blocks of matter. The search for other members of the quark and lepton families is currently under way. In addition, the basic subatomic forces that are involved in the atom are being studied. Many investigators consider the determination of the characteristics of these forces to be as important to the understanding of the world within the atom as is the search for other subatomic particles and the determination of their properties.

Summary

The atom is primarily composed of three elementary particles: electrons, neutrons, and protons. Various nuclear configurations of each element are unstable, and stability is reached by spontaneous ejection of elementary particles and/or energy, a phenomenon known as radioactivity. Primary radiation released by unstable atoms is known as alpha, beta, and gamma radiation. Alpha and beta radiation consists, in reality, of streams of certain charged particles. Other particles released are classified into various families, among which are leptons, mesons, and baryons. The ratio of protons to neutrons in an atom's nucleus seems to govern the stability of an atom. Upon the emission of a charged particle, the nucleus of an unstable atom assumes a new identity instantly. Alpha particles are composed of two protons and two neutrons; therefore, upon alpha emission, the Z number of an atom decreases two units, the A number decreases four units, and the identity of the atom is immediately changed accordingly. A beta particle is negatively charged and has negligible mass; therefore, upon beta emission, the Z number of the radioactive atom increases one unit and the A number remains the same. Gamma-ray emission does not change the identity of the unstable atom, since it has no electrical charge or detectable mass. The total energy of all radiation is expressed in terms of the electron volt. The nucleus of an atom is held together by mass that has been converted to energy, a measure of which is called binding energy. In order for a stable nucleus to release its particles, energy must be added to it to overcome the strong nuclear force.

Becquerel and others discovered that about 70 nuclides of the elements found in nature are radioactive. Most naturally occurring radionuclides are among the heavier elements, such as radium and uranium.

The rate at which an unstable nuclide (radionuclide) disintegrates is called activity, a measure generally calculated in disintegrations per second. The curie is the activity of essentially 1 gram of radium, 3.7×10^{10} becquerels (disintegrations) per second, and is an accepted unit of measure of activity for all types of radiation.

Half-life is the time required for the activity of a given mass of a radionuclide to decrease to one-half its original value. The loss of mass from radioactive decay among the heavier elements is negligible. The statistical reliability of the time required for a given radionuclide to undergo radioactive decay is such that scientists can accurately predict the age of certain rock layers or of the ancient life-forms found in them. Uranium-238, potassium-40, and carbon-14 are among the relatively few radionuclides that have proven to be of greatest use in dating procedures.

The biological effects from exposure to radiation from various sources are directly related to the rate at which energy is deposited in living tissue. Plants and animals are affected by radiation exposure in relationship to their degrees of biological complexity; hence, cockroaches and flies can withstand higher dosages without suffering harmful effects than can human beings. The major concern regarding low levels of exposure experienced by human beings is the damaging effects that may occur in future generations.

Fusion is the combining of relatively light nuclei to form heavier nuclei. The resultant atom assumes an identity according to its new Z and A numbers. Fission is the splitting of a nucleus into two nuclei of approximately equal Z and A numbers. The mechanism by which stars "burn" is fusion, primarily that of hydrogen atoms to form helium atoms. The process of fission is responsible for the tremendous energy supplied by nuclear reactors.

Elements are transformed into other elements both in the laboratory and in nature. Uranium, a naturally occurring radioactive element, is eventually transmuted to lead through a series of alpha- and beta-particle emissions. Artificial transmutation of one element into another has been possible for several decades, and new techniques are being developed. All elements with Z numbers from 93 to 109 have been produced by the transmutation of lighter elements. The stability of a target atom is upset by the bombardment of its nucleus by particles or by electromagnetic energy.

Scientists have accelerated various atomic particles, such as electrons and protons, to speeds more than 99 percent the speed of light ($0.99c$). Their masses have been found to increase as predicted by the formula to determine relativistic mass. The charged particles are accelerated by the attractive and repulsive forces produced by various combinations of alternating and constant magnetic and electric fields in linear accelerators, cyclotrons, and bevatrons. These devices have permitted scientists to identify many of the subatomic particles and to determine many characteristics of each.

The initial proof that energy could be converted to mass came in 1932 as an American physicist, C. D. Anderson (b. 1905), found that a photon of gamma radiation entered the emulsion of a photographic plate and changed into two separate particles. The pair of particles turned out to be two electrons equal in mass but opposite in electric charge. The positive electron, or positron, is the antiparticle of the more common negatively charged electron. When a photon is converted from energy to mass, a pair of particles results. The antiparticle faces a brief existence, for at the instant that it encounters a particle identical to it except for charge, both particles are annihilated and two photons are released.

Radionuclides are unstable isotopes of various elements. At least 70 radionuclides occur in nature. Many others are artificially produced during fission or by bombardment of a stable element with particles or energy. Every element has at least one radionuclide. The radionuclide does not chemically or physically differ (except in mass) from a stable isotope of the same element. The charged particles and energy emitted during radioactive decay are detected by Geiger-Müller counters, scintillation detectors, cloud chambers, and bubble chambers. In addition, other instruments similar in principle have been developed. Radionuclides are used as tracers in many scientific fields. Gamma radiation from some radionuclides has partially replaced X-ray devices. The particles and energy from radionuclides have been found to affect various substances in some ways beneficial to society.

Scientists are continuing to probe the submicroscopic world in search of more knowledge about nature. The list of particles that belong to the quark and lepton families is constantly growing. Most researchers think that all ordinary matter is composed of various combinations of charged leptons, neutral leptons, and quarks. However, the search is on for a fourth particle, a massive neutrino. Circular and linear accelerators are used to give beams of charged particles high kinetic energies. Various colliders are used to produce head-on collisions of the energized charged particles, in hopes of learning more about the building blocks of nature and the forces that interact with them.

Equation Summary

Neutron decay:

$$^{1}_{0}\text{n} \rightarrow {}^{1}_{+1}\text{p} + {}^{0}_{-1}\text{e} + {}^{0}_{0}\bar{\nu}.$$

In neutron decay, a neutron n decays into a proton p that remains in the nucleus, and an electron $^{0}_{-1}\text{e}$ (or beta $^{0}_{-1}\beta$) is ejected. Accompanying the reaction is an antineutrino $^{0}_{0}\bar{\nu}$.

Proton decay:

$$^{1}_{+1}\text{p} \rightarrow {}^{1}_{0}\text{n} + {}^{0}_{+1}\text{e} + {}^{0}_{0}\nu.$$

In proton decay, a proton p disintegrates into a neutron n that remains in the nucleus, and a positron $^{0}_{+1}\text{e}$ (or beta $^{0}_{+1}\beta$) is ejected. A neutrino $^{0}_{0}\nu$ is also produced.

Electromagnetic energy:

$$E = nhf,$$

where energy E (electron volts) is released and is directly related to the number n of photons released and the frequency f of the photons. The constant h is known as Planck's constant and has a value of 6.63×10^{-34} joule-second.

Mass defect:

mass defect = total particle mass − mass of nuclide.

The difference between the total mass of the elementary particles of a nuclide and the actual mass of the nuclide is known as the mass defect.

Binding energy:

binding energy per nucleon = (931.48 MeV/u)(mass defect)/mass number

The binding energy of a given nuclide is determined by multiplying the energy released through the conversion of one mass unit of energy, 931.48 MeV, by the mass defect in atomic mass units (u). By dividing this product by the mass number of the nuclide, the binding energy per nucleon is obtained.

Relativistic mass:

$$m = \frac{m_0}{\sqrt{1 - \frac{v^2}{c^2}}}.$$

The relativistic mass m of a moving object is determined by the comparison of its mass at rest m_0 and the speed v at a given instant.

Mass-energy equivalence:

$$E = mc^2.$$

The equation is used to calculate the energy E obtained when a given mass m is converted into it. The constant c, 3.00×10^8 meters per second, is derived from the relativistic expression for mass.

Questions and Problems

The Particles and the Rays

1. How are X rays and gamma rays similar? How do they differ?
2. What forces tend to cause a nucleus to maintain its stability? What forces apparently cause a nucleus to be unstable, thus radioactive?
3. According to Figure 15.2, what is the approximate ratio for stability of Z to N if (a) $Z = 40$ and (b) $Z = 70$?
4. Determine the mass defect and binding energy of $^{238}_{92}\text{U}$. The mass number of uranium-238 is 238.0508. Maintain significant digits, as illustrated in Example 15.1.

5. Is the neutron composed of a proton and an electron, and the proton composed of a neutron and a positive electron? Support your answer.
6. Complete the following reactions:
   (a) ${}^{45}_{20}Ca + {}^{1}_{0}n \longrightarrow$
   (b) ${}^{225}_{88}Ra + {}^{1}_{0}n \longrightarrow$
   (c) ${}^{252}_{98}Cf + {}^{10}_{5}B \longrightarrow$
   (See the periodic chart on the inside front cover.)

Natural Radioactivity

7. According to the graph in Figure 15.3, what two values of mass number have a binding energy/nucleon of 8 MeV? With respect to the same graph, what is the *total* approximate binding energy of ${}^{4}_{2}He$? Of ${}^{56}_{26}Fe$?

Transmutation of the Elements

8. List and briefly discuss the quantities that must be conserved when a radionuclide decays. What quantities, if any, are not conserved?
9. How many atomic mass units do atoms lose when the following types of radiation are released? How is the *Z* number of the nuclide affected in each case?
   (a) alpha emission
   (b) beta emission
   (c) gamma emission
   (d) X-ray emission
   (e) neutron emission
   (f) positron emission

Half-Life and Activity

10. Complete the following nuclear reactions with the aid of Table 15.6 and the periodic chart on the inside front cover.
    (a) ${}^{32}_{15}P \longrightarrow$
    (b) ${}^{238}_{92}U \longrightarrow$
    (c) ${}^{60}_{27}Co \longrightarrow$
11. A radioactive sample of iodine-131 was determined to have an activity of 6.40×10^{10} Bq. What would be the activity from the remaining radioisotope after two half-lives had lapsed? (See Table 15.6.)
12. Note the half-life of carbon-14 as listed in Table 15.6. As a general rule, after a period of time passes equal to five half-lives, a small sample of a radionuclide is rendered harmless, unless the daughter is also radioactive. (a) How much time is required before the given sample of carbon-14 is considered safe? (b) How long must calcium-45 remain isolated under the same conditions?
13. A sample of radium-226 has a mass of 32.0 g. How many years must lapse until only 1.00 g of radium remains? (See Table 15.6.)
14. Determine the activity in becquerels emitted from 26.5 g of radium, according to the original definition of the curie.

Dating with Radioactivity

15. An animal bone was discovered that contained one-eighth the level of carbon-14 of a similar living specimen. Approximately how old is the bone?
16. Why would a researcher attempt to date a rock layer assumed to be over a billion years old by means of its uranium-238 or potassium-40 content rather than with the amount of carbon-14 present?

Radionuclides—Their Other Uses and Their Detection

17. Why can't the standard GM system distinguish between beta- or gamma-emitting radionuclides that have the same activity?

The Biological Effects of Radiation

18. Does external exposure to alpha, beta, or gamma radiation cause a person to become radioactive? Defend your answer.
19. How does ionizing radiation affect living tissue? Why is it used to treat cancer patients?

Fission among the Atoms

20. Is there a serious concern that a stray neutron could cause the uranium in one of Earth's uranium deposits to fission, bringing about the total destruction of our planet? Defend your answer.
21. Neutrons are more commonly used to bombard targets to produce radioactivity than are alpha or beta particles, even though both particles are readily available. Why?
22. Why are thermal neutrons more likely to bring about fissioning of heavy nuclei than fast neutrons?

The Fusion of Atoms

23. Where does the energy come from when light nuclei undergo fusion?

Mass-Energy Equivalence

24. During certain nuclear reactions, mass is lost. Where does this mass go? In other nuclear reactions, mass is gained. What is the source of this additional mass?
25. Calculate the relativistic mass of a 100-kg object that is moving at 2.45×10^{8} m/s.

The Subatomic World of Nature

26. How do various accelerators cause charged particles such as protons to reach high speeds?
27. How do scientists determine the properties of subatomic particles if the particles are too small to observe directly?

PART FOUR

Humans have been fascinated with the night skies since prehistoric times. All early civilizations studied the stars and the motions of the planets, and their application of these observations was among the earliest sciences practiced. The ancients made use of their observations to tell time, to find their way about, to predict the coming seasons, to explain eclipses, and to reinforce their religious beliefs.

Undoubtedly, one of the most intriguing features of the celestial bodies is their inaccessibility, disregarding our recent space ventures. A seemingly endless amount of scientific effort has been involved in the attempt to unveil the secrets of the heavens.

In order to determine the dimensions of the sun and the moon, as well as the distances to them, some gifted individual had first to establish the size of Earth with a high degree of accuracy. This measurement is credited to the Greek scholar Eratosthenes (276–194 B.C.). Eventually, the practice of studying and measuring the heavens with the unaided eye for the most part gave way to a more detailed view available with a telescope. Galileo, contrary to popular belief, did not invent the first telescope. He did, however, construct a three-power telescope in 1604 (the first of many he devised), and according to the evidence available, he was the first person to use the instrument to study the planets and the stars. Galileo found, among other things, that the Milky Way was composed of countless stars and that the orbit of the planet Venus lay between Earth and the sun. He also was the first to observe some of the moons of Jupiter.

Today, artificial satellites bearing telescopes orbit Earth and improve our view of our neighboring planets and their moons, along with various nebular and stellar objects. In some instances, telescopes aboard space vehicles have been sent

The Universe beyond Earth

forth to orbit various planets and their moons. (Numerous photographs in the following chapters are evidence of these outstanding accomplishments.) Other telescopes aboard satellites have studied comets and their tails of cosmic fluff that sometimes lead and sometimes follow these visitors, perhaps from the mysterious outer reaches of our solar system and beyond.

About 1640 it was discovered that our sun is just another star, and eventually scientists understood the mechanism by which our sun and other such stellar objects produce their brilliant energy. Scientists currently believe that after a period of billions of years, an average star experiences a sharp decline in energy production followed by one of several routes that lead to its demise.

Nature would seem to prefer collections of objects rather than individual ones. Many visible stars are members of a binary, triple, or even more complex system. In general, clusters of stars range from systems that contain only a few members to arrangements of hundreds of thousands of stars. Some galaxies contain trillions of stars, and most galaxies are, themselves, members of clusters.

Various theories attempt to explain how the universe was formed, each school of thought having its dedicated researchers who constantly seek decisive proof to support their ideas. The most widely held cosmological evolutionary theory postulates that approximately 18 billion years ago the universe was born from a titanic explosion, and the energy and matter from that event have been expanding ever since. The analysis of light spectra received from various objects in space offers strong evidence for the validity of this theory, although scientists are currently testing alternative concepts in an attempt to reconcile new evidence uncovered in recent galactic surveys.

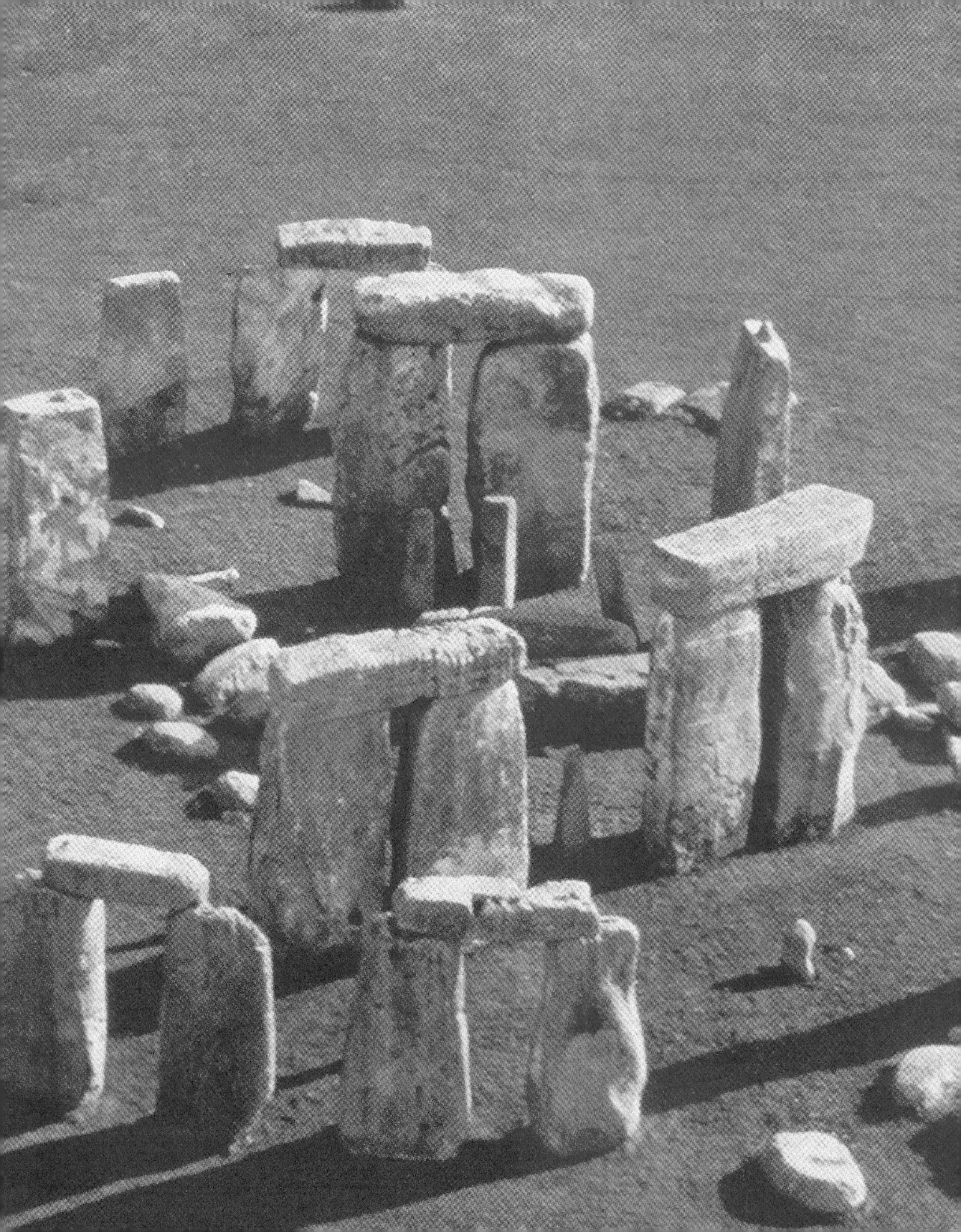

16 Ancient Astronomy

Chapter outline

Key terms/concepts

Stonehenge is thought to have been an astronomical observatory and shrine of the Druids of western England.

Members of the earliest civilizations, particularly those who were assigned the tasks of keeping constant vigil over their villages and livestock, could not have helped but wonder about the beauty of the night skies. Even now, on a dark evening relatively free of atmospheric contamination and light pollution, we inevitably look upward toward the splendor created by nature, and our thoughts must parallel those of the ancient observers. We can readily understand why the early scientists and philosophers sought answers to questions about the observable universe—either questions of their own, or those posed by individuals who preceded them.

The science of astronomy points out very vividly that so-called scientific truths are always open to improvement and revision. Consider the early confrontation between the various authorities and scientists with regard to the rejection of the concept of a stationary Earth. A similar conflict continues, as geologists and other scientists maintain that Earth is about 4.5 billion years old, based on mounting geological evidence, in contrast with literal interpretations of ancient religious writings.

This introductory chapter on astronomy points out how the ancients determined, with amazing accuracy, some of the measurements we can scientifically verify today, but only through the application of much more sophisticated instrumentation than was available to them.

The Essence of Astronomy

Astronomy is *the science concerned with celestial objects along with the observation and the interpretation of the radiation from the component parts of the universe detected on Earth.* It is the scheme of nature that derives the properties of celestial objects, and from these properties we deduce the laws that describe the way the universe responds.

Astronomy is presumably the oldest science, since observations of celestial bodies were made and noted as early as any other phase of recorded history. It is common knowledge that the skies were observed in careful detail by the ancients. It was they who first noted the specific motions of various starlike objects. In earliest recorded time, the practical motivation for studying astronomy was the desire to predict certain significant events, such as the changing of the seasons, based on the knowledge of various cyclic phenomena. It was important to the ancient cultures to know when to make the necessary provisions for the approaching winter or other season. They also had to know when to prepare for floods or droughts. Some cultures had to know when the time had arrived for them to move to more suitable areas, where shelter, hunting, and the gathering of other food and water would be more to their liking. Their knowledge of the positions of celestial objects let them venture long and far across deserts, seas, and unfamiliar lands to their chosen places of refuge for that time of year, then to return to their original locations as the seasons changed once again. They also traveled to other villages to trade wares or to certain regions to gather the various raw materials for their pottery, weapons, and the like.

Practical knowledge about the observable universe was also put to use in both spiritual and whimsical applications. The ancients were convinced that events in the heavens exerted considerable influence over their lives. Hence, many of the early astronomers practiced what is now known as **astrology,** *the ancient belief that the positions of the sun, moon, and planets* with respect to the zodiac *influenced earthly affairs and human lives.* The *zodiac* is an imaginary band across the sky that is wide enough to encompass the paths of all planets visible to the naked eye. Each monarch arranged for the services of an astronomer who could forecast the future success of strategies that involved war, love, investments, and politics. This person came to be known as an astrologer. During this period, and amid various cultures, astronomy was closely tied to religion, and astronomers were often held in as high esteem as the religious leaders of the time.

Modern astronomy, as a result of the contributions of the ancients and of such individuals as Copernicus, Brahe, Kepler, Galileo, and Newton, is founded on a firm physical basis. The practical aspects abound as we seek to expand the available knowledge of the universe, to strengthen or refute existing theories, and to develop new and better theories. Simply consider the technological advances brought about by the space programs since the first artificial satellite was launched into orbit about Earth in 1957. Further applications of related achievements have had a major impact on our everyday lives.

The Priest-Astronomers

In ancient Egypt the astronomers were temple priests. These priest-astronomers found it advantageous to be able to predict when eclipses of the sun and moon would occur, since this ability was interpreted by their rulers as evidence of their understanding of and communication with various gods, thus enhancing their own prestige and power. (An **eclipse** is *the total or partial obscuring of one celestial object by another.*)

Once each year the Nile River overflowed its banks and renewed the fertility of Egypt's farmlands with its silt-laden waters. Since much preparation was needed in order for the people to reap maximum advantages from this event, it was of great economic importance to accurately predict this flooding.

Astrology vs. Astronomy

Astrology began in ancient Sumer, Babylon, and Egypt and was used to predict or explain events surrounding kings, pharaohs, and empires. The topmost terraces of the ziggurats of Babylon and other Mesopotamian cities were used by priests for observations of the sun, the moon, and the five planets known to them. Astrology was and is based on the belief that human events are influenced by conjunctions of the seven heavenly bodies in relation to the background of immutable stars. Once it was accepted that various celestial bodies influenced happenings on Earth, then if a disastrous flood occurred while the planets Mars and Venus were in conjunction, people assumed such a conjunction was the cause of the great flood.

It was believed that Earth was ruled by the gods of the sky (the sun, moon, and known planets), and that the gods provided omens of good and of evil for those who knew how to read the omens. It was also assumed that the evil intentions of the gods might be mitigated or nullified by performing certain religious rites. Thus, the astrologer-priests were consulted to predict the destiny of planned military adventures, affairs of state, and the future greatness or demise of kings. Since such predictions were of inestimable importance, it was necessary for the astrologers to know precisely the movements of the seven celestial gods. Their studies resulted in the accumulation of considerable knowledge of the movements of all celestial objects. From the systematic study of the sky by ancient civilizations, the science of astronomy was born.

A scientific study should enable tests of its theories by repeatable observations and experiments. Astrology does not meet this criterion, but astronomy does.

Today's astrology, like palmistry and crystal-ball gazing, is concerned mainly with predicting what events and people will bring good fortune or adversity to a person's future. Most astrological predictions concerning an individual's future are based on the positions of the planets at the time of the person's birth. Such a concept assumes predestination and eliminates the possibility of free will. Many people associate astronomy with astrology, even though astrology is not a science and has no scientific validity. The development of astronomy, however, owes a great debt to ancient astrology ■

By 2700 B.C., predating the building of the great pyramid of Cheops at Giza, the Egyptian priest-astronomers had observed that the Nile overflowed its banks within a few days after the bright star Sothis (called Sirius today) peeped above the horizon just a little south of and simultaneously with the rising of the sun. They had also noted that after this simultaneous rising of Sothis and the sun, Sothis rose slightly earlier than the sun each day. After approximately 91 days Sothis was high in the night sky, halfway between rising and setting when the sun was just beginning to rise. After 182 days Sothis would be setting as the sun was rising. Then after 365 days Sothis and the sun would once again rise simultaneously, followed in a few days by the overflowing of the Nile River. The whole scenario would be repeated in another 365 days. Thus, the Egyptians established a fairly accurate yearly calendar. They found that it actually took a period of four years plus one day, or 1461 days for the sun and Sothis to rise together four times. Therefore, the length of the year was in reality 1461 ÷ 4, or 365.25 days. We know the extra day the Egyptians added every four years as February 29, which occurs every leap year.

Vegetation growth cycles are obviously keyed to yearly seasonal cycles, and agricultural planning and practices are, in turn, keyed to the vegetation growth cycles. Since agriculture provides the necessary food for the existence of any civilization, it was important for ancient civilizations to know with some precision when the various seasons were to begin. In order to obtain the needed information, the priests of early civilizations developed various calendars, based on astronomical events, that allowed them to anticipate impending seasonal changes.

From a cultural point of view, a calendar also enabled priests to accurately determine when certain feasts, sacrifices, and holidays should be celebrated.

The Egyptian priest-astronomers also pursued the study of the stars in order to add to the information that they thought necessary to predict human events through astrology. This belief in astrology, a sort of astronomical magic, was generally held by the ancients to be based on truths as revealed by various sky gods.

Over 5000 years ago, the Babylonians were using a calendar divided into 12 months of 30 days each. The number of days in a month was based on the time necessary for the moon to go through all its phases—about 29.5 days. The Babylonians also devised the week as a unit of time and named the seven days of the week after the sun, the moon, and the five planets known to them—Mars, Mercury, Jupiter, Venus, and Saturn.

The ancients noticed that different groups of stars were visible at different seasons of the year. Certain *groups or clusters of stars,* called **constellations,** assumed fanciful shapes in the minds of some observers. These constellations became associated with certain mythological gods and goddesses and were eventually named in

Early Observatories

We all feel proud (as we should) of our modern technological world. We have landed men on the moon, landed automated space vehicles on Venus and Mars, and sent space probes to encounter Jupiter, Saturn, Uranus, and Neptune. One of our space vehicles, *Pioneer 10,* has left the farthest reaches of our solar system and is heading for the distant stars. The spacecraft is well over 4 billion miles from Earth, yet a long journey lies ahead to another heavenly body.

At the same time we should acknowledge the marvelous intellectual achievements in observational astronomy that the ancients accomplished thousands of years ago.

There are the remains of ancient astronomical observatories at Stonehenge, England, the Big Horn Medicine Wheel in Wyoming, and the once great cities of the Aztecs of Mexico and the Mayans of Central America. Even more well known are the astronomical observatories at the great cities of ancient Babylon and Alexandria, Egypt.

These ancient observatories recorded the passing of the equinoxes and the solstices (see Chapter 21), the changing positions of the planets, and in the case of Egyptians, the simultaneous rising of the sun and the star Sothis (Sirius), which resulted in the establishment of the length of time we call a year ■

their honor. The Babylonian priest-astronomers compiled lists indicating the positions of their constellations. By 250 B.C., they had produced a compilation of lunar eclipses, both past and future, and described the calculations necessary to predict other eclipses.

Many ancient societies, as well as more modern but equally nontechnical ones, have relied upon the stars and the sun for long-distance navigation. In Homer's epic poem the *Odyssey* (eighth century B.C.), the goddess Calypso gave Odysseus directions on how to sail from what must have been the Madeira Islands to the southwest coast of Spain. These directions included using three constellations—the Pleiades, Boötees, and the Bear (the Big Dipper)—as navigational steering aids. The Polynesians voyaged many hundreds of miles across portions of the uncharted Pacific Ocean by noting which stars and constellations were rising or setting on the horizon at a point near the island to which they were headed. During the day, they closely observed the position of the sun in order to estimate their approximate latitude.

The ancient civilizations of the Orient, although cut off from the influences of other cultures, made numerous contributions worthy of mention. For instance, the Chinese developed a practical calendar perhaps as early as the fourteenth century B.C. Some scholars maintain that the Chinese had determined the length of the year to be 365.25 days as early as during the twelfth century B.C. About 350 B.C., the astronomer Shih Shen developed what appears to be the earliest star catalogue. His impressive works contained about 800 entries. As early as 700 B.C., the Chinese also maintained quite accurate records of comets and meteors they observed. Recordings were also made of sunspots visible to the naked eye and of "guest" stars that normally were too dim to be seen with the naked eye but increased in brightness for periods of weeks or even months. The most significant of these unusual observations undoubtedly was the supernova of A.D. 1054, a brilliant display caused by the star that exploded and created the heavenly object we know as the Crab nebula.

The Greek Philosopher-Astronomers

In general, physical science began to flourish only after Greek scientists began to use experimental techniques that involved specially designed equipment, yet some earlier Greeks made significant contributions to the physical sciences. For instance, numerous achievements are attributed to one of the earliest natural philosophers, Thales of Miletus (ca. 625–545 B.C.).

It is thought that Thales traveled to Egypt and Mesopotamia, where he studied astronomy and mathematics. This Ionian Greek is credited with accurately predicting a solar eclipse that occurred two decades after his death, although various scholars of ancient literature discount the authenticity of this claim, based upon data that would have been available to him at this time in history. His practical applications of geometry were numerous, including some ideas that were undoubtedly original and others that he presumably heard about during his travels. Most authorities feel his works eventually led to the development of the theoretical structure of this field of mathematics. Regardless of the accuracy of any of the claims regarding Thales, his logical explanations, based on observation, helped set him aside from others among the ancients. For example, he considered water to be the primary constituent for all that existed. According to his belief, Earth was a large flat disc that floated on water. Though simple and naive, his view would have appeared logical to anyone of his time who had journeyed to Egypt and witnessed life abound from the barren regions surrounding the Nile after the annual floods. Thales, as one would note, sought natural physical explanations (a scientific view), rather than relying on gods or goddesses to explain the occurrence of events. With this approach, he explained that earthquakes were caused by hot water from somewhere on Earth that gushed into the oceans with such fury that the event caused Earth to tremble. Thales, then, is among the first of the ancient philosophers who

The Seven Gods of the Week

Before civilization had even gotten a foothold in most of the world, the Babylonian priest-astronomers had divided the year into 12 months and the week into seven days. They were aware that there were seven celestial objects that had the freedom to move among the immutable background stars. The stars held their positions relative to each other, but the seven brilliant lights that stirred the imagination of the Babylonians were obviously not bound by the laws that governed the stars. Because of their seemingly special powers, the seven celestial lights were thought to be gods and goddesses.

The largest (therefore, the most important) was the sun. It not only marked the time period "day" but also provided daytime light. Its warmth and light caused plants and trees to grow. When the sun's path was lowest in the heavens, the days were cold; when its path was highest, the days were hot. The people of Babylon realized that the sun quite literally gave them the greatest gift a god could give humans—life. The moon, although it did not have the same life-giving power, lighted one's way at night and was a celestial adornment of unquestionable beauty. The changes the moon went through, from full to new and back to full, marked the time of one month. The remaining five objects in the magnificent scheme of seven gods and goddesses were the five planets (Greek, *planetes,* wanderer) known to the priest-astronomers.

The Babylonians throughout their history were constantly besieged by plagues, invading armies, and floods. As with all early civilizations, they thought that such catastrophic events were caused by evil spirits or by displeased gods; hence, they sought the protection and forgiveness of their gods. Since it was assumed that the seven celestial gods were invincible and controlled the lives and fortunes of people and nations, wisdom dictated that the gods should be worshiped in a manner befitting their omnipotence.

To curry the favor of the gods, the priest-astronomers set aside seven days (one week) to honor them. The sun was honored on the first day of the week, Sunday, the day to praise the Sun. The moon was given homage on the second day of the week, "Moonday" (our Monday). Tuesday (Old English, *Tiwesdaeg,* day of Tiu—a Germanic god) was named after the red planet. The ancients assumed a connection between the red color of the planet and the color of blood. The red planet was the Babylonian god of war. Today we call the planet Mars (Roman god of war). Wednesday (Old English, *Woodnesdaeg,* day of Woden—the Teutonic god) was named to honor the planet we call Mercury (Roman messenger of the gods). Thursday (Old Norse, *Thorsdagr,* god of thunder) was the day to show reverence for the planet we know as Jupiter (the supreme Roman god). Friday (Old Norse, *Frigg,* goddess of love) was the day to worship the planet we call Venus (Roman goddess of love and beauty). Saturday (Old English, *Saeterdaeg*), or "Saturnday," was the day to venerate the planet we know as Saturn (Roman God of agriculture).

The gods worshipped 5000 to 6000 years ago in old Babylon give us today the seven days of our week. Had there been 10 moving celestial objects witnessed by the priest-astronomers, the "week" as we know it would have been of different duration ■

attempted to provide natural, rather than supernatural, explanations for the world as he witnessed it.

In contrast, Pythagoras (ca. 560–500 B.C.), a very influential Ionian Greek religious leader and legendary scientist-mathematician, established a brotherhood that advocated a simple, religious, and ethical life of poverty, content with aristocratic rule. He and his followers were persecuted to the point that they left their homes on the island of Samos, close to Miletus, and settled in Crotona, located in southern Italy, where they founded the Pythagorean academy. Eventually, their doctrines were deemed unacceptable, and again Pythagoras and many of his followers were forced to flee—this time to a region on the Gulf of Taranto, which forms the heel of Italy's "boot." Like the Babylonians, the Pythagoreans considered the celestial objects to be divine and the planets (the celestial objects that moved) to be at different distances from Earth than the stars. Their interest in numbers brought about the establishment of an order of distance at which various celestial objects seemed to orbit about Earth. After some refinement, the ascending order they adopted was the moon, the sun, Mercury, Venus, Mars, Jupiter, and Saturn. The Pythagoreans' appreciation of symmetry led them to reach conclusions of some significance. The orbiting objects moved regularly about Earth in circular patterns, an observation that was to have a decided effect on Greek astronomy and that of other civilizations. Also, they maintained that Earth was spherical, as was the balance of the heavens. Although their rationale for these beliefs is not really known, most scholars of ancient literature feel that they considered a spherical heaven to have greater elegance than any other shape, especially the hemisphere described in the writings of Homer. There is some speculation that Pythagoras witnessed a ship sail from sight over the horizon and, hence, concluded Earth was spherical. However, his conception of a spherical Earth rather than an irregularly shaped one is generally attributed to his faith in beauty, rather than to any rational explanation he might have conceived.

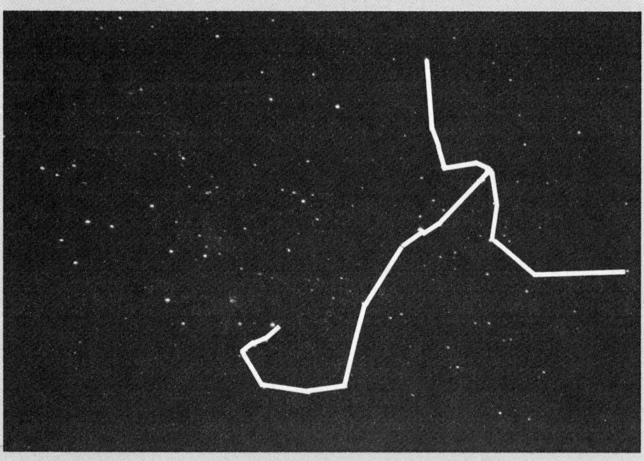

a.

b.

Figure 16.1 Two constellations closely resembling that for which they are named: (*a*) Scorpius, the scorpion and (*b*) The Big Dipper (part of the constellation Ursa Major, the Greater Bear).

Myths and Constellations

From legends dating back 6000 and possibly 10,000 years, we know that observers of the heavens filled the night sky with familiar deities, ordinary mortals, animals, and mere objects they fancied resembled certain groups of stars. The star groups are called constellations (Latin *constelatio,* group of stars). The most ancient constellations occupied only the northern sky from about 25 degrees to 45 degrees north latitude, whereas a large portion of the Southern Hemisphere was without constellations. Such an orientation suggests that the constellation-makers could not see the Southern Hemisphere from their location.

Considerations from archaeology, astronomy, and ancient history all point to the Minoan and Babylonian (or Akkadian) civilizations (or precivilizations) as the birthplace of the constellations. Of course, the exact origin of many of the constellations is shrouded in the dimness of antiquity. Undoubtedly, many of the objects represented by the constellations have been renamed or changed over the millennia. Many of the constellation names come to us from Greek and Latin rather than from the language of their origins. On the other hand, the names we use for most stars have their source in Arabic.

Although some constellations bear a striking resemblance to what they are said to represent (see Figs. 16.1, 16.2, and 16.3), some do not. The constellations Corona Borealis, Cygnus, Draco, and Leo do approximate a crown, swan, dragon, and lion, respectively. It is more difficult, if not impossible, to relate the geometrical patterns of the star groups Cassiopeia, Pegasus, and Canis Major to a chained lady, a flying horse, and

Around 450 B.C., several outstanding concepts were offered for all to consider. Anaxagoras proposed the actual cause of solar and lunar eclipses, and he also maintained that the moon shone by reflected sunlight. Philolaus, a dedicated disciple of the late Pythagoras, introduced the very controversial view that Earth, indeed, did move like other planets among the stars and was not, as all before him had thought, a stationary body. And the philosopher Democritus advanced the concept that all matter is made up of indivisible particles too small to be seen that are held together by some unknown force.

The Greek philosopher Aristotle (384–322 B.C.) reasoned that Earth must be spherical, offering as proof his observations of ships and other objects on the horizon, along with his comprehension of lunar eclipses. From the latter he emphasized that the shape of Earth's shadow on the moon represented part of a circle. He had made relatively short journeys in many directions from his home, yet of sufficient distance so that different stars appeared above the horizon at some given time; thus, Earth was really not very large, according to Aristotle's reasoning.

Aristarchus (310–230 B.C.), the noted Greek astronomer at Alexandria, Egypt, was the first person to teach the **heliocentric concept,** which assumed that *Earth and all the other planets revolve around the sun* (see Fig. 16.4). He also believed that Earth rotates on its axis daily. The heliocentric concept ("heliocentric" comes from two Greek words—*helios,* meaning "sun," and *kentron,* meaning "center"), although ably defended by Aristarchus, was not generally accepted. Aristotle had considered the heliocentric concept as a possibility but had rejected it in favor of the **geocentric concept** (from the Greek *ge,* meaning "earth," and *kentron,* meaning "center"), which assumed that *the sun, the moon, and all other planets revolve around Earth* (see Fig. 16.5). Aristotle obviously committed a grave error in judgement for supporting the

Figure 16.2 The stars that make up the constellation of Orion. Note the middle three stars of Orion's belt and the stars directly under them that represent his sword. The giant reddish Betelgeuse is the star of the upper left, while brilliant white Rigel is at the lower right.

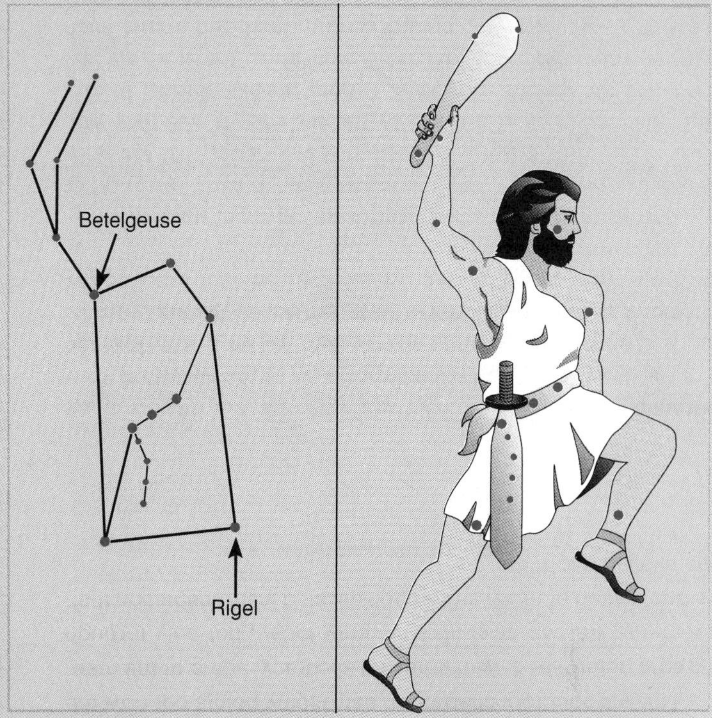

Figure 16.3 Compare this mythological figure with the photograph of the stars in the constellation of Orion in Figure 16.2.

a great dog. The most well-known constellation is generally called the Big Dipper (Fig. 16.1b) and does, indeed, look like a dipper. The constellation is actually named Ursa Major (Latin, *ursa,* bear) or the Big Bear. Two of the stars in the Big Dipper point to Polaris, which is our north pole star. Since Polaris' position in the sky is approximately in line with Earth's spin axis, it is sometimes called "the star that never moves."

The splendid constellations shine in the heavens for all to see and enjoy. It has been said that they make up "the oldest picture book of all" ■

geocentric model without careful consideration. Being primarily a philosopher, and although he had access to the numerous observations recorded by the Babylonian and Egyptian astronomers, he chose to focus his attention on general questions of such magnitude that mere observations were of little value. He did agree that the stars and other celestial objects moved in a circular path, but the theory of Earth as the center of the universe apparently satisfied his basic beliefs. Then, too, if Earth really moved after all, would there not be many futile and irrelevant questions that would have to be considered before the pertinent ones could be culled from among them? Also, why, if Earth moved, did he not detect any shifting of its motion among the stars?

Because of Aristotle's great reputation, the heliocentric concept was accepted by only a small number of philosophers and astronomers during Aristarchus's life. Furthermore, another great Greek-Egyptian astronomer who studied and taught at Alexandria, Claudius Ptolemaeus (A.D. 85–165), usually called Ptolemy, sided with Aristotle in the heliocentric versus geocentric argument and managed to deal a death-blow to the heliocentric concept. It should be pointed out that Ptolemy's geocentric representation of the motion of the sun, the moon, and the planets did rather accurately account for the observations made during Ptolemy's life. One of the most baffling of these observations was the apparent change in direction of the outer planets, Mars, Jupiter, and Saturn, as they were observed to move through the nighttime sky over a period of several months. This **retrograde motion,** as it is called, can be viewed as *an illusion caused by Earth's faster motion around the sun* (Fig. 16.6). However, Ptolemy's model of the solar system also accounted for the retrograde motion, and the accuracy of predictions according to his ingenious geocentric concept were the best available at the time.

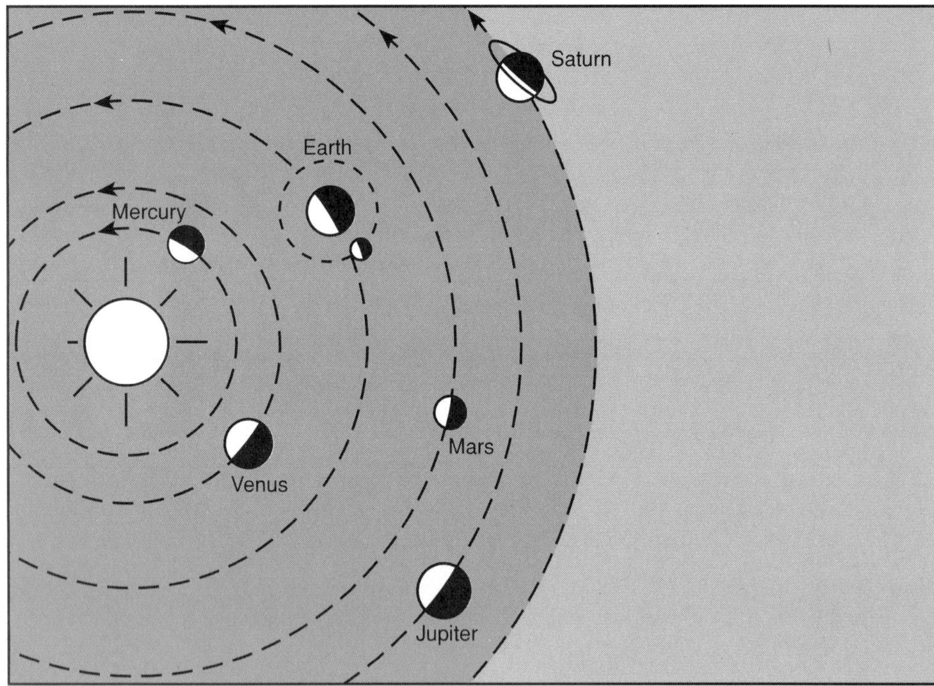

Figure 16.4 Copernicus's heliocentric view of the solar system.

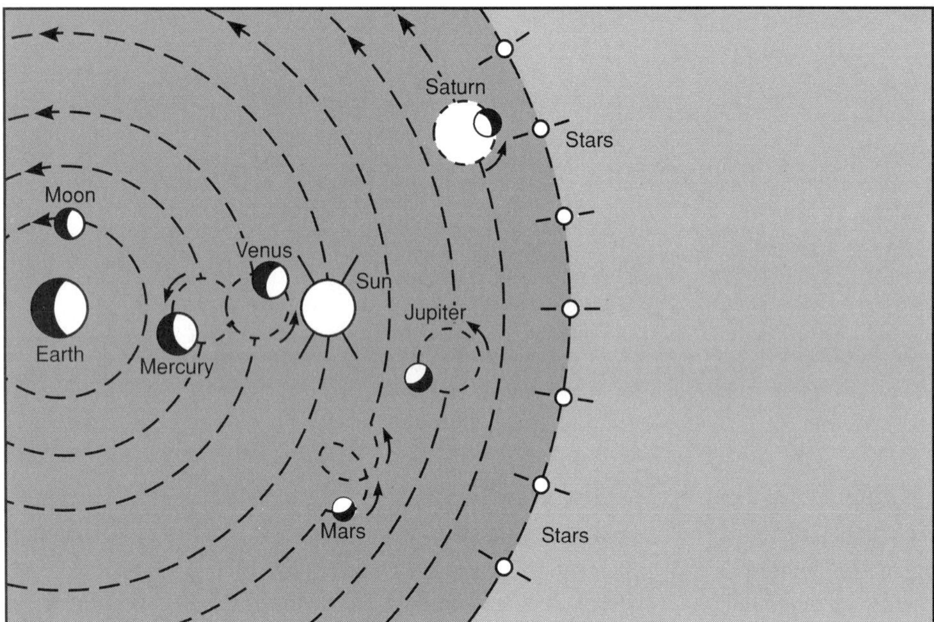

Figure 16.5 Ptolemy's geocentric view of the solar system, in which each planet moves in a circular epicycle. The center of each epicycle moves in a circular deferent, which also characterizes the motion of the sun and moon.

The heliocentric concept was not seriously reintroduced into science until A.D. 1530 (about 1800 years after the concept was first introduced by Aristarchus) when Nicolaus Copernicus (1473–1543), a Polish astronomer, published a book in which he pointed out the logical reasons why the heliocentric concept must be the correct interpretation of planetary motion. (Aristarchus is often called the Copernicus of antiquity.) Galileo (1564–1642) vigorously defended the heliocentric concept as outlined by Copernicus.

Measuring the Size of Earth

The first reasonable measurement of the size of our planet was accomplished by Eratosthenes (276–194 B.C.), who was the Greek librarian for the museum at Alexandria, Egypt. In order to understand the principles that Eratosthenes used in his measurements, we must realize that rays of light reaching us from a source very, very far away travel approximately in parallel lines.

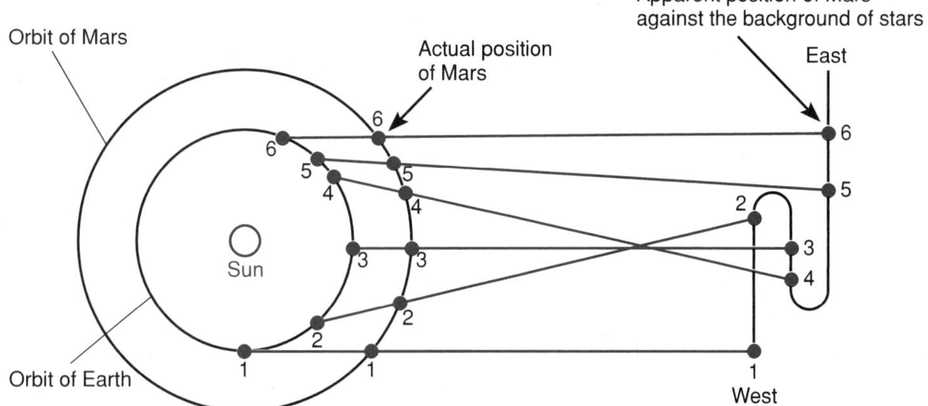

Figure 16.6 The retrograde motion of Mars. In the Ptolemaic system, epicycles were used to explain the retrograde motion of the planets. In general, the planets move from west to east in relation to the background stars. Earth, however, travels faster in its orbit than does Mars. As Earth overtakes Mars, the line of sight from Earth to Mars moves from east to west relative to the stars. As Earth passes Mars the line of sight begins once again to move from west to east.

Eratosthenes made the essentially correct assumption that the sun is sufficiently far from Earth for light from the sun to reach Earth in virtually parallel lines.

Eratosthenes noticed one day at Syene (a town on the Nile River approximately where the Aswan Dam stands today) that the sun at noon shone directly down a well, so that the bottom of the well was completely illuminated. This observation could only mean that the sun was directly overhead at the time. On the same day in the following year, Eratosthenes was at Alexandria, which is approximately due north of Syene. At noon he measured the angle between his **zenith,** *a point directly overhead,* and the center of the sun, as shown in Figure 16.7. The angle measured slightly more than 7 degrees (more precisely, 7.2 degrees). He reasoned that the light ray that illuminated the bottom of the well at Syene and the light ray that formed the same angle with the vertical at Alexandria were parallel. Since these two parallel lines would be intersected by the straight line *CZ,* angle *B* had to equal angle *A.* The accepted standard for determining length in Greece at that time was the longest dimension of the famous Greek sports stadium, and Eratosthenes had a pacer, a person trained to determine the distance between different locations by walking with a definite stride, to determine the number of stadium units, or *stadia,* between the cities of Syene and Alexandria. The pacer reported a distance of 5000 stadia. Since 7.2 degrees is 1/50 of a circle (7.2 degrees/360 degrees), the distance from Syene to Alexandria had to be 1/50 of a great circle around Earth; that is, Earth's circumference. Through the application of an accepted geometric relationship (theorem) that applies to circles:

$$\frac{\text{length of arc}}{\text{central angle}} = \frac{\text{circumference}}{360°}.$$

Therefore,

$$\frac{5000 \text{ stadia}}{7.2°} = \frac{\text{distance around Earth}}{360°}$$

and

$$\text{distance around Earth} = \frac{(5000 \text{ stadia})\,(360°)}{7.2°} = 250{,}000 \text{ stadia}.$$

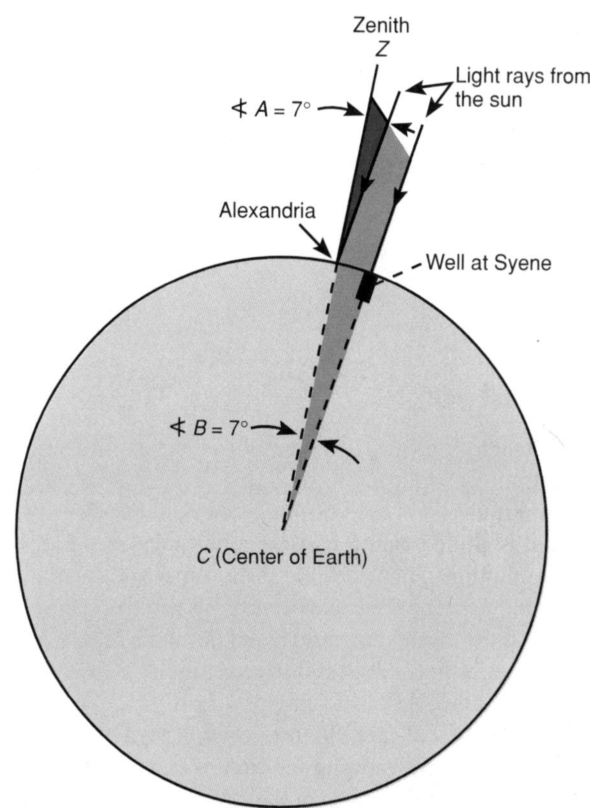

Figure 16.7 Eratosthenes's method of determining the circumference of Earth.

If the generally accepted conversion between the unit stadium and the *mile* is applied; that is, one stadium = 0.1 mile, the accuracy of Eratosthenes's method of determining Earth's polar circumference was spectacular. The measurement he obtained was within 0.5 percent of the value determined by modern satellite technology (25,000 miles compared to the more precise 24,901 miles, as considered accurate today)! To determine the polar diameter of Earth,

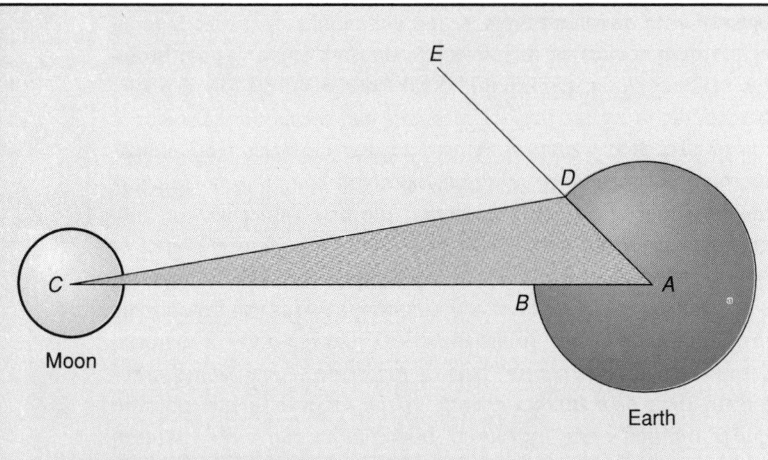

Figure 16.8 Ptolemy's geometrical method of determining the distance between Earth and the moon.

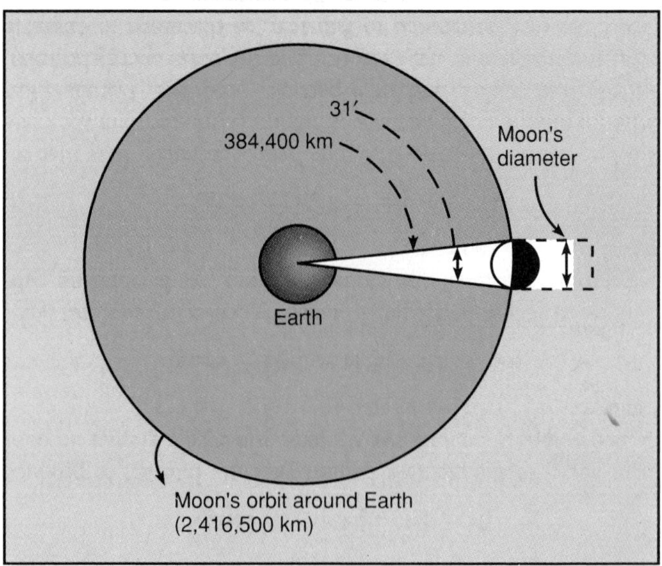

Figure 16.9 A method of determining the diameter of the moon.

the ancients who accepted this result simply had to divide the circumference by π, or 3.14:

$$\frac{250{,}000 \text{ stadia}}{3.14} = 79{,}618 \text{ stadia (7962 mi)}$$

(The actual value has been determined to average 7930 miles [12,770 km].)

Measuring the Distance to the Moon

One of Ptolemy's greatest accomplishments was measuring the distance from Earth to the moon (see Fig. 16.8). Ptolemy knew that on a certain day of the year the moon would be directly overhead (that is, at his zenith) at Alexandria (point *B* in Fig. 16.8). At that same time he had a fellow astronomer stationed at point *D* on Earth determine the angle between his zenith at point *D* and the center of the moon; that is, he found the angle *EDC*. Since the angle of straight line *EA* is 180 degrees, the angle *CDA* equals 180 degrees—angle *EDC*. Ptolemy had the distance on Earth between points *B* and *D* carefully measured. Knowing the distance between *B* and *D* and knowing the circumference of Earth, Ptolemy determined what fraction of the circumference of Earth was represented by the distance *BD*. It follows that the angle *BAD* must be the same fraction of the total degrees of a circle (360 degrees). For instance, if the distance *BD* were 25,000 stadia, *BD* would be one-tenth of the 250,000-stadia circumference of Earth, and the angle *BAD* would be one-tenth of 360 degrees, or 36 degrees.

From Eratosthenes's measurement of the circumference of Earth, Ptolemy simply calculated the radius of Earth, line *AD*. Now Ptolemy knew the angle *CDA*, the angle *BAD*, and the distance *AD* (two angles and the included side of the triangle *CAD*). With this information he solved the triangle for the distance *CA*, the distance from the center of Earth to the center of the moon. Ptolemy obtained a value for *CA* of 233,050 miles, which is only 2.4 percent less than the mean distance of 238,713 miles (384,400 km)! Ptolemy's result was an outstanding accomplishment, considering the instruments available at the time. The determination of the distance to the moon also allowed Ptolemy to calculate the distance that the moon travels to complete one orbit around Earth. The circumference of the moon's orbit would be equal to the distance of the moon from Earth times 2π, since the circumference of any circle equals its radius times 2π.

Measuring the Diameter of the Moon

From Eratosthenes's measurement of the circumference of Earth, Earth's diameter and radius could be calculated. Knowing the radius of Earth, Ptolemy was able to determine the distance to the moon. From this information the circumference of the moon's orbit could be calculated. With knowledge of the circumference of the moon's orbit, the diameter of the moon can be determined, and so it is with science. Each new discovery stands upon the shoulders of previous ones.

The angle made by drawing a line from one edge of the moon to an observer on Earth and back to the other edge of the moon is called the *angular diameter* of the moon. This angle is slightly more than 0.5 degrees (31 min and 5 s of arc to be more precise, usually written 31′5″). Because the moon is relatively small compared to the circle made by the moon's orbit around Earth, the small arc of the moon's orbit passing through the moon (see Fig. 16.9) is essentially the moon's diameter. Even though this arc is curved, a small portion of a large circle is essentially a straight line. The solution to the problem of the diameter of the moon is basically the reverse of Eratosthenes's method of determining the circumference of Earth. If we determine the fraction of the total degrees in a circle represented by about 31 minutes, then the diameter of the moon should be the same fraction of the total circumference of the moon's orbit. Since there are 360 degrees in a circle and 60 minutes in a degree, there are 360 × 60, or 21,600 minutes in a circle. The angular diameter of the moon is then 31 minutes/21,600 minutes, or 0.001,435 of the total degrees in a circle. Therefore, the moon's diameter should be 0.001,435 of the total circumference of the moon's orbit around Earth. The circumference of the moon's orbit is equal to the radius of its orbit (about 384,400 km) times 2π, which gives a value of 2,415,000 kilometers. The diameter of the moon, then, is 0.001,435 × 2,415,000, or 3466 kilometers. Very accurate measurements of the moon's diameter give a value of 3476 kilometers.

Summary

Astronomy is the science that is concerned with celestial objects as well as with the observation and the interpretation of the radiation detected by astronomers from the component parts of the universe.

The ancients made many observations and recordings of the night skies. Studies of their literature provide us with insight as to their lives and beliefs. They considered some of the celestial objects to be deities, a belief that led to the practice of astrology.

Astronomy is considered the oldest of the physical sciences. The various ancient cultures used the information they gathered to predict conditions that told them of the approaching seasonal changes. They depended on their views of the stars for the time to plant their crops, when to harvest them, and when to prepare for annual flooding or extreme cold weather. Their knowledge of the stars permitted them to travel over land or sea to distant places, and then to safely return home.

Many individuals devoted most of their lives to the study of the visible universe. The locations of the stars and planets permitted these astrologers to make predictions about events that were to occur. These individuals are referred to as priest-astronomers by those who study the ancient writings. Each community had at least one person who assumed the role of predicting the future, and this person was held, spiritually and otherwise, in high esteem.

Groups or clusters of stars became known as the constellations, and many were named after the gods or goddesses of the given culture. The ancients also used the constellations to locate particular stars or other celestial objects, much as astronomers do today.

Questions and Problems

The Essence of Astronomy

1. What were the original basic assumptions of astrology?

2. What is the historical connection between astrology and astronomy?

The Priest-Astronomers

3. Why do some astronomers still learn the characteristics and locations of the constellations conceived by the ancients?

4. How did the economy of Egypt influence the growth of astronomy?

5. Why are there 12 months in a year, seven days in a week, and 365.25 days in a year?

6. According to various authorities, how were the names of each day of the week derived?

7. Why were vegetation and growth cycles so important to ancient civilization?

8. How was the number of days in a month determined?

The Greek Philosopher-Astronomers

9. How did the practical application of the concept of the circle affect ancient civilizations?

10. Why was there so much opposition to the heliocentric concept?

11. List the major contributions made to astronomy by Thales of Miletus.

12. List and briefly discuss the important concept attributed to Philolaus.

13. Explain the geocentric concept in your own words. For what major reasons did the theory become so widely accepted in ancient times?

14. What was Anaxogoras's major contribution to science?

15. Who was Pythagoras? List three major contributions he or his followers made to ancient astronomy.

Measuring the Size of Earth

16. Suggest a method for determining the distance from Earth to the sun.

17. Could Ptolemy have made his determination of the distance to the moon without knowledge of Eratosthenes's determination of the circumference of Earth?

18. What is the major significance of a certain water well in the town of Syene (near what is known today as Aswan, Egypt)?

Measuring the Distance to the Moon

19. For what major scientific contribution is the ancient Greek Eratosthenes remembered?

Measuring the Diameter of the Moon

20. In what way was Ptolemy's determination of the diameter of the moon related to Eratosthenes's measurement of the circumference of Earth?

17 The Solar System

Chapter outline

Key terms/concepts

A montage of images of the Saturnian system prepared from Voyager 1 photos taken in November, 1980. Clockwise, the moons are as follows: Dione (in front of Saturn), Enceladus, Rhea, Titan, Mimas, and Tethys. A star background has been added by an artist.

Our solar system consists of a central star, the sun; its nine planets (the search is on for a tenth planet); no less than 62 *satellites*, or moons, that orbit them; tens of thousands of asteroids; innumerable comets; and myriads of meteors. Consider, however, what intragalactic visitors would observe as they approached our solar system. Until they were quite close to our system, they would notice only our sun. Even if they came closer, our sun would still reign supreme against the black void of space. By comparison with the sun, the sun's planets, comets, and asteroids would appear as mere specks, since the sun has an estimated 99.86 percent of all the mass in our solar system. On closer inspection, Jupiter, Saturn, and perhaps Uranus would appear reasonably impressive, but the rest of the planets, including our Earth, would seem rather inconsequential and be difficult to locate. Unless some of the comets happened to be close to the sun, they would be almost impossible to detect, since at great distances from the sun they have no tails and reflect barely any of the sun's light. The great variation in the sizes of the sun's planets might be startling to our space visitors. They would observe that the largest planet, Jupiter, is so massive (larger than the mass of all the other planets combined) that one of its moons is larger than either Mercury or Pluto. (Saturn also has a satellite larger than either of these small planets.)

After the visitors had had time to study in detail and to catalog the various components of our solar system, they would notice that as they looked down on our solar system from the vicinity of the North Star, all the planets revolve around the sun in a counterclockwise direction and, but for Venus and Uranus, spin on their axes in a counterclockwise direction. With a few exceptions, the numerous satellites, or moons, of the planets also revolve around their mother planets in a counterclockwise manner. The visitors would further notice that the planets circle the sun in approximately the same plane, whereas the comets orbit the sun at all possible angles to the plane of the orbits of the planets.

In the Beginning

Many scientists theorize that our solar system originated from a nebulosity of gas and dust. About 5 billion years ago a tremendous formless cloud, possibly encompassing about 2×10^{41} cubic kilometers of space, started to contract due to the mutual gravitational attraction of the atoms, molecules, and particles present. Each particle exerted a small gravitational attraction on each of its neighbors, causing all the particles to be pulled closer together. As this condensation occurred throughout the vast cloud, a gradual contraction toward the center resulted. The contraction was a relatively slow process at first. It probably took millions of years for the cloud to contract from a diameter of perhaps 5 trillion kilometers to a diameter of 16 billion kilometers. During this initial contraction, the particles were so far apart that mutual gravitational attraction was extremely weak. As a result, the particles would have "fallen" very slowly toward the center of the cloud.

After the diameter of the cloud reached approximately 16 billion kilometers, the speed of the particles as they moved toward the center of the cloud was greatly accelerated, until the particles eventually attained speeds estimated at thousands of kilometers per hour. As the primeval nebula continued to contract and the particles moved even faster toward the center, the deep interior of the cloud began to heat up as the gravitational potential energy was slowly changed into heat energy. During the process of contraction, the whole cloud rotated slowly in a direction that would appear counterclockwise to a stellar observer in the vicinity of the North Star. As the mass of material increased its movement toward the center, various eddy currents, or whirlpools, of matter in the swirling cloud began to build up much smaller concentrations at varying distances from the center of the contracting cloud. It was these smaller contracting masses produced by eddy currents in the rotating cloud that eventually became the planets.

As the heat liberated in the center of the mass continued to increase, the temperature rose until the dense center became incandescent. This brilliant birthshine of the new star heralded the beginning of life for our sun. However, although our sun had begun its lifework of pouring out untold quantities of warmth, light, and other forms of energy, it was still not ensured a long life. This long life, probably about 10 billion years, could only be ensured if the internal temperature of the new star continued to increase until it reached 10 million degrees Celsius.

Approximately 10 million years after the star began to shine, the temperature of the sun's interior rose to the critical value of 10 million degrees Celsius by virtue of its continued contraction and conversion of gravitational potential energy into heat energy. When this temperature was reached, various nuclei of the light elements in the sun underwent nuclear fusion. This process, known as a *thermonuclear reaction,* is the mechanism that furnishes the long-term energy for stars. It was described in the studies of Hans Bethe in 1938. The great abundance of hydrogen in stars such as our sun makes it the major constituent in the various stellar reactions that occur. The second most abundant substance is helium-4, ${}^{4}_{2}\text{He}$. When the nuclei of such light elements as helium undergo fusion, or "burn," to form heavier nuclei, the reaction is often referred to as *nucleosynthesis.*

As the critical temperature of 10 million degrees Celsius was reached, hydrogen nuclei (which are actually protons) collided with such violence that they came close enough to cause the strong nuclear force to "overcome" the repulsive force between like electrical charges, and the charged particles fused to form heavier nuclei accompanied by the release of a certain amount of energy. The actual nuclear reaction between two nuclei of hydrogen-1 produces hydrogen-2 (deuterium), a positron (${}^{0}_{+1}\text{e}$), a neutrino, and energy. That is,

$${}^{1}_{1}\text{H} + {}^{1}_{1}\text{H} \rightarrow {}^{2}_{1}\text{H} + {}^{0}_{+1}\text{e} + \nu + \text{energy.}$$

Solar System Formation—The Case for a Violent Beginning

Hydrogen and helium, the most abundant elements in the universe, are believed to have been produced when the universe began. It is theorized that all the other elements were formed in stellar interiors, with the formation of the most massive elements requiring the extreme conditions present in exploding stars (see the discussion of supernovae in Chapter 18). Thus, the sun and planets must have formed from materials that were once present in the interiors of stars that have long since ceased to exist.

Recent evidence suggests that the formation of the solar system may have been even more directly influenced by nearby dying stars. The abundance of certain radioactive isotopes found in meteorites suggests that they formed between 1 million and 20 million years after a nearby exploding star showered the cloud from which our solar system was formed with debris. This observation also implies that these rocky fragments formed relatively quickly on an astronomical time scale.

There is also evidence that the shock wave from a second stellar explosion may have triggered the collapse of the pre-solar-system cloud. Shock waves are produced when an object exceeds the speed of sound in a material. The explosion of a star hurls material outward at tremendous speeds, generating shock waves in the thin clouds of gas in interstellar space. These large gas clouds are relatively stable, but the passage of a shock wave through the cloud causes localized heating which can create instability in the cloud. This instability may have started the collapse of the cloud from which the solar system formed.

Even though most astronomers agree that the solar system formed from the collapse of a large interstellar cloud, it can be seen from this brief discussion that the details of the collapse are still unresolved. Research is continuing in an attempt to unravel the mysteries of the formation of the solar system. New telescopes in space and larger Earth-based telescopes under construction will permit astronomers to make better observations of nearby young stars. Supercomputers permit astronomers to test mathematical models of complexity only dreamed of a few years ago. Perhaps this combination of enhanced observations and improved models will yield the answers ■

In turn, the new deuterium nuclei, themselves traveling at awesome speeds, collide and fuse with other hydrogen nuclei to form nuclei of the element helium (He) and release gamma radiation, along with considerably more energy:

$$^{2}_{1}\text{H} + {}^{1}_{1}\text{H} \rightarrow {}^{3}_{2}\text{H} + \gamma + \text{energy.}$$

Then, nuclei of the element helium violently interact to produce another isotope of helium, two protons, and even a greater amount of energy:

$$^{3}_{2}\text{He} + {}^{3}_{2}\text{He} \rightarrow {}^{4}_{2}\text{He} + 2\,{}^{1}_{1}\text{H} + \text{energy.}$$

So, in essence, four hydrogen-1 nuclei, by consecutive interactions, fuse into a nucleus of the element helium, while emitting subatomic particles and energy:

$$4\,{}^{1}_{1}\text{H} \rightarrow {}^{4}_{2}\text{He} + 2\,{}^{\;0}_{+1}\text{e} + \text{energy.}$$

The sum of the masses of the four hydrogen nuclei that fuse together to form a single nucleus of helium is greater than the mass of the helium nucleus and other particles produced. The mass lost is transformed into energy according to the mass-energy relationship, $E = mc^2$, attributed to Einstein, as discussed in detail in Chapter 15. The mass lost from the conversion amounts to about 0.0286u. From $E = mc^2$, the energy produced by the nuclear processes is determined as follows:

$$\begin{aligned} E &= mc^2 \\ &= (0.0286\text{u})\,(1.66 \times 10^{-27}\,\text{kg/u})\,(9.00 \times 10^{16}\text{m}^2/\text{s}^2) \\ &= 4.27 \times 10^{-12}\,\text{joule(J)}. \end{aligned}$$

This nuclear reaction, whereby hydrogen nuclei are fused into helium, is essentially identical to the process that is used on a much smaller scale in the nuclear weapon known as the hydrogen bomb. Although the amount of mass that is converted into energy when each helium nucleus is created appears extremely small, the total number of helium nuclei being produced in the sun in a short period of time, such as a second, generates a phenomenal amount of energy.

Meanwhile, the eddy currents present in the contracting, slowly rotating cloud inexorably pulled more and more material toward their own smaller centers at the expense of the material rushing toward the central part of the cloud which would eventually become our sun. In the same way that the mass of material was collected to produce our sun, material on a far less grand scale slowly collected in these whirlpools of dust and gas at different distances from the main cloud and eventually formed bodies of material that slowly circled around the new sun. Like the sun, material condensing to eventually become planets was probably mostly hydrogen, a small amount of helium, and a relatively smaller percentage of all the heavier elements and compounds present in the original cloud. Eventually, many comparatively small bodies of material congealed in the slowly rotating cloud. These **protoplanets** were distinguishable as the beginnings of future planets. Their volumes were very large in terms of present-day planetary volumes, and probably they encompassed a volume of space thousands of times larger than their present volumes. At the same time and in the same way that the planets were formed, the moons of the planets were forming as still smaller condensations circling their parent planets (see Fig. 17.1).

By the time that our new sun became sufficiently hot to reach incandescence, the material forming the planets was fairly well consolidated. Even so, the denser material was concentrated in the

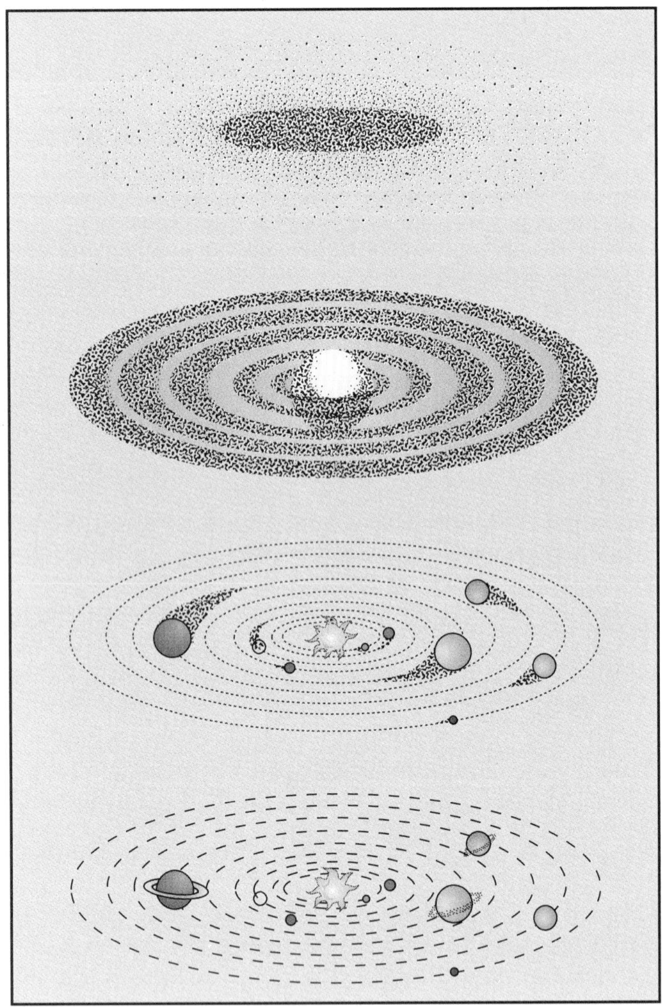

Figure 17.1 An artist's conception of the solar system during various stages of its formation.

center of the mass forming the protoplanets, whereas the extremely light hydrogen and helium, forming the greater bulk (90 percent) of the total protoplanet masses, enveloped the denser centers in a cocoon extending outward to great distances. The mass of proto-Earth must have been 1000 times greater than that of Earth today, and its diameter may have approached 16 million kilometers.

As the new star began to shine more and more brilliantly, releasing an increasingly greater amount of heat, the hydrogen and helium molecules surrounding those protoplanets closest to the new star reached the speeds that permitted them to escape the protoplanets' gravitational force. Also, the stream of ionized particles, mainly electrons and protons, which are continually ejected by the sun to form the so-called solar winds, physically bombarded and pushed away the light hydrogen and helium molecules. The protoplanets were thus denuded of their hydrogen and helium atmospheres.

The closest protoplanet, Mercury, being so near the sun, received the most heat and solar particles; consequently, it was the first protoplanet to lose its envelope of hydrogen and helium. What remained was only a small fraction of the total mass of material of which it had once been composed. On the other hand, the elements and molecules left behind were essentially the dense material that we are familiar with as constituting the mass of the present-day inner planets—Mercury, Venus, Earth, and Mars. These four planets all have similar densities (about five times as heavy as an equal volume of water), as one would expect if they all had formed under the same conditions. Also, the inner planets, because of their relatively small sizes, exerted correspondingly modest gravitational attractions on the hydrogen and helium gases surrounding them. (The higher kinetic energy of the gaseous particles that orbited the inner planets made escape even more feasible than among the outer ones.)

However, those planets farther removed from the sun (the outer planets)—Jupiter, Saturn, Uranus, and Neptune—were so far away and, because of their huge masses, exerted such large gravitational attractions for the gases that surrounded them that even the prodigious outpouring of heat and charged particles from the sun was not sufficient to heat up and brush away more than a fraction of the vast quantities of hydrogen and helium that initially condensed and encircled them. Consequently, these outer planets today have comparatively very low densities (from 0.7 to 2.3 times as dense as water), and a major portion of their masses is due to the relatively lightweight hydrogen and helium remaining from the initial formation. Saturn, for example, is only 100 times the mass of Earth, but is almost 800 times the volume of our planet. Thus, Saturn is less dense than water, and if there were an ocean large enough to hold the planet, a major portion of this celestial body would float.

As the slowly rotating original cloud condensed and its radius shrank, the cloud began to spin faster and faster. Why the speed of rotation of the cloud must have increased as the cloud shrank is easily illustrated by the example of an ice skater, such as in Figure 4.14. When an ice skater starts spinning, the skater's arms are outstretched, but as he or she pulls the arms in closer to the body, the skater spins more rapidly. This is an example of the conservation of angular momentum. In order for the angular momentum of a contracting nebula to remain constant, as it must, the speed of rotation of the nebula increases.

As a consequence of the increased speed of rotation of the contracting cloud, it is estimated that the newborn sun rotated on its axis approximately 50 times faster than it does today. How then can we explain the present speed of rotation of our sun (one rotation in 26 Earth days) in view of its original rotation once in 12 Earth hours? As the rapidly rotating new star shone upon the infant protoplanets, the intense bombardment of charged particles and light from the star ionized the envelope of gases relatively near the sun. Interaction took place between the sun's magnetic field and the now-ionized rotating envelope of gaseous material. This interaction evidently resulted in a decrease of the rotation of our sun in the course of time and a corresponding increase in the speed of rotation of the rest of the solar system material.

As the powerful radiation from the sun pushed away the miscellaneous gases and materials left from the formation of the sun and the planets of our solar system, not all the material was

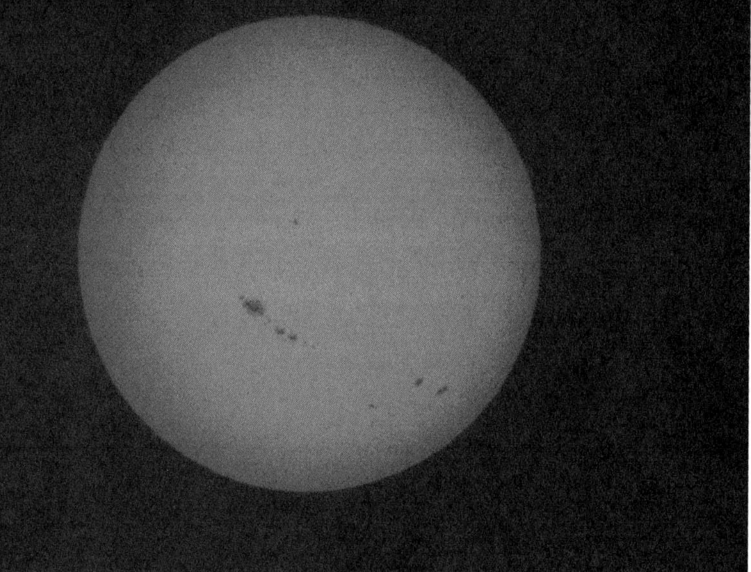

Figure 17.2 The photosphere of the sun. Several sunspots are also visible.

Figure 17.3 The corona of the sun as seen during a total solar eclipse.

completely ejected from the solar system into interstellar space. At distances beyond the orbit of Pluto, some of the debris condensed into relatively small bodies that circle the sun at the very outposts of our solar system—probably some extend a trillion kilometers from the sun. Occasionally these cold dark objects come close enough to the sun to shine by reflection and fluorescence and to become the spectacularly brilliant objects we know today as comets.

The Sun

Our sun is the nearest star. Like other such celestial objects, it is a gigantic incandescent globe of highly ionized gases known as *plasma,* a substance composed of an approximately equal number of positive ions and electrons. The interior temperature of the sun is thought to be between 15 million and 20 million degrees Celsius. Its diameter is calculated to be 1,390,000 kilometers, a value equal to 3.6 times the distance from Earth to the moon. (Still, our sun is rated as a dwarf among other stars.) The mass of the sun is estimated to be about 2×10^{30} kilograms, or 330,000 times more massive than Earth. Spectral analysis of the electromagnetic radiation received from the sun concludes that this brilliant celestial object is composed of about 75 percent hydrogen and about 23 percent helium. (The element helium was actually first detected by spectral analysis of the sun's light before it was found to exist on Earth, and its name is derived from the Greek word *helios,* meaning "sun.") All the remaining 65 elements identified and those yet to find, then, constitute about 2 percent of the sun's mass.

The *surface of the sun that is visible to us* is called the **photosphere** (Greek for "light sphere"), shown in Figure 17.2. The temperature of this region reportedly averages about 6000°C (approximately 5700 K), ranging from 4300°C to 6800°C. Above the photosphere the gases are practically transparent, whereas below it they are so dense as to be opaque. *Surrounding the photosphere and extending outward for several thousand kilometers* is a region called the **chromosphere** (Greek for "color sphere"). This layer can be seen as a red glow for a very brief period before and after a total solar eclipse, when the photosphere is blocked from the observer's view by the moon. (The term "layer" should be interpreted very generally, in view of the fact that none of the layers of the sun are well defined; rather, they merge into one another.) The lower portion of the chromosphere is often called the *reversing layer,* since it is the major source of the absorption lines in the solar spectrum (refer back to Fig. 10.11). The luminosity of the sun is primarily due to the conversion of hydrogen to helium, providing energy at the rate of 4×10^{26} joules per second. The shining ivory halo that is visible when viewing a totally eclipsed sun is called the *corona* (see Fig. 17.3). This region surrounds the chromosphere and extends outward for millions of kilometers. Among the most spectacular of the sun's phenomena are the **prominences,** *gigantic, arching columns of red-hot gases that are ejected upward perhaps at 1300 kilometers per second to heights of over 150,000 kilometers above the chromosphere,* thus, well into the corona. Many of these disturbances form a fountain of extremely hot gases that arch toward the sun's surface. Solar prominences, such as those depicted in Figures 17.4a and 17.4b, are visible along the sun's edge (a region referred to as the limb) during a total solar eclipse or through special photographic techniques.

Perhaps the most outstanding, though short-lived, phenomena on the sun are the *solar flares.* Such displays represent very bright spots in the photosphere that reach a maximum intensity in just a matter of several minutes, then fade beyond detection within the hour. Flare temperatures are much higher than the balance of the photosphere, and they emit large amounts of energy of various sorts, including ultraviolet radiation. The calm, smooth appearance of the sun's surface we can view at sunrise or during sunset belies the violent turbulence that actually exists, as shown in the series of photographs of Figure 17.4.

Apparent dark spots have been observed on the sun's disk for over three centuries. The occurrence of spots yielded evidence to Galileo that the sun was not the perfect, flawless object that most scholars of his time believed it to be. **Sunspots,** as the dark spots are called, appear as *great dark blotches on the sun's surface;* at any given time only a few (or none) may be visible with a telescope, or they may appear in large numbers, singly or in sizeable

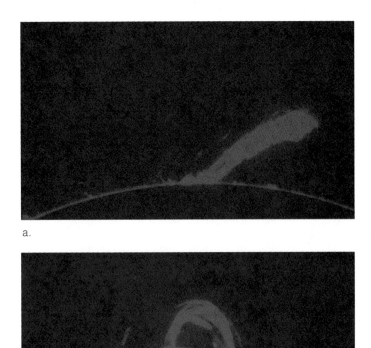

a.

b.

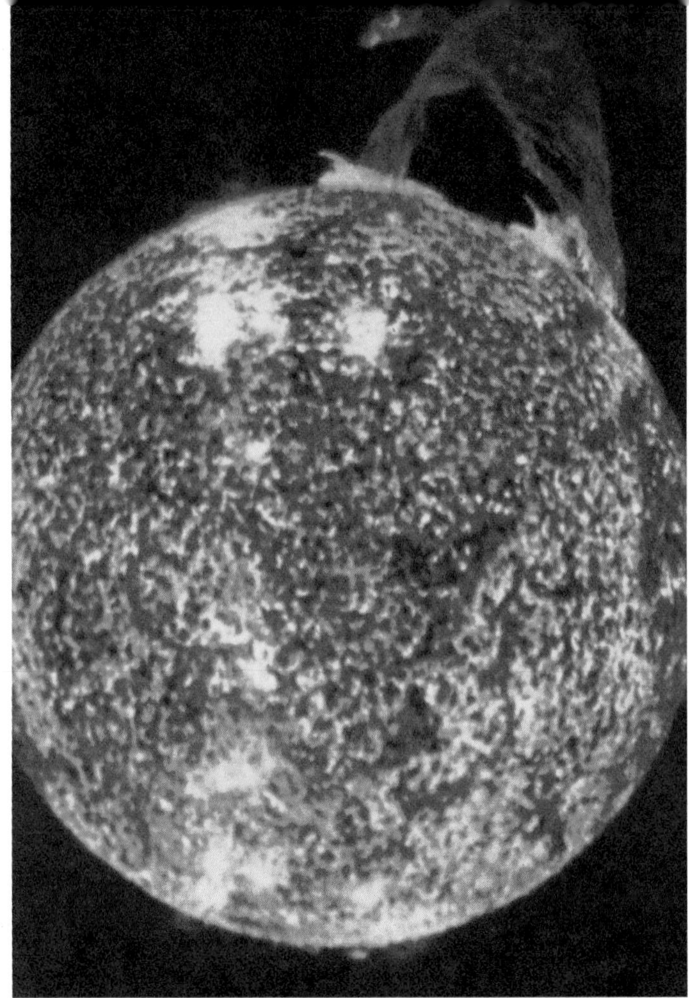

c.

Figure 17.4 The tumultuous surface of the sun. Photos (*a*) and (*b*) show the various stages in solar prominences. Photo (*c*) reveals the tremendous turmoil present on the surface. The lighter regions are presumed to be significantly hotter than the surrounding portions.

groups. (Special precautions to prevent serious eye damage must be taken in attempting to observe sunspots, particularly with binoculars or telescopes. Consult your instructor before undertaking such a venture.) At periods of maximum activity, it may be possible to view several dozen spots at a time. They are almost always located between 5 degrees and 35 degrees of latitude in each hemisphere; thus, they seem to be absent from a narrow zone near the equator and from the sun's polar regions.

Sunspot activity, as determined through observations and magnetic field measurements, infers a *solar cycle.* Since 1851, observers of the sun have noted a rough periodicity, with an average peak of approximately 11 years. A graph produced from the comparison of sunspot numbers to time shows a relatively rapid rise from minimum to maximum number of sunspots in a period of 4 to 5 years, then a slow decline to the next minimum. However, recent data analysis has caused some astronomers to doubt that the sun as a whole is in a specific phase of a cycle. The data reveal that solar activity of a given cycle starts near one of the poles and migrates toward the equator. New zones may form each 11 years, but the migration from the vicinity of one of the poles to the region near the equator may take about 18 years; consequently, successive cycles may overlap. This observation points out that one portion of the sun may be in a different phase of activity than another. The pattern of sunspot activity appears to be a complex one, indeed.

Continued observations of the sunspots have revealed numerous characteristics of the sun. This brilliant object in our solar system, like all others observed in our night skies, rotates on its axis. Unlike Earth, the moon, and other such objects, all parts of the sun do not rotate in the same period of time. For instance, regions near the equator rotate once in about 25 days, as determined by the appearance and disappearance of sunspots around the edge of the sun near this imaginary line. Nearer the poles, the period of rotation seems to be about 31 days. Obviously, we are simply observing the manner in which the outer layer of the sun responds to the way that gases, including the clouds in our atmosphere, are affected by a rotating body they surround. Sunspots are shown in Figure 17.2, and a close-up view is presented in Figure 17.5. Sunspots appear as dark blotches on the sun's surface only by comparison with the brighter photosphere. They are actually about 800 Celsius degrees lower in temperature than the layer in which they occur; hence, they are still considerably brighter than the flame of a welding torch. These regions emit electromagnetic radiation in the radio frequency region and can create considerable static interference with radio transmissions here on Earth, particularly at or near maximum levels of activity.

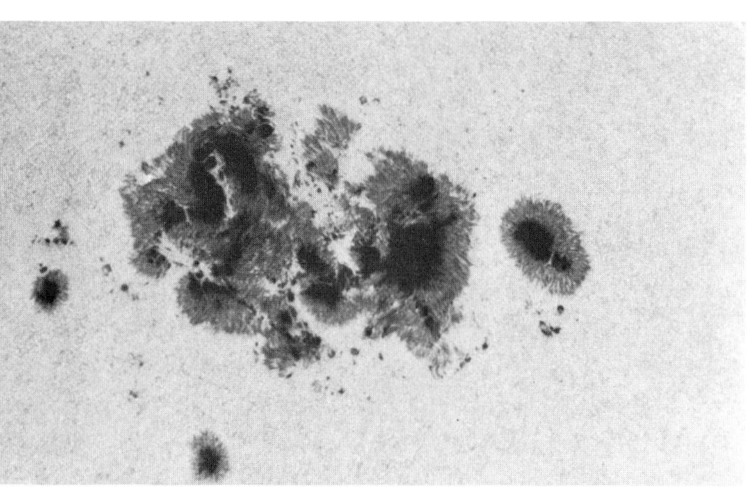

Figure 17.5 A close-up view shows the characteristics of sunspots. The darker regions reveal portions of the sun's surface that are several hundred degrees below the temperatures of the other areas.

In addition to the sunspot cycle, there are other complicated time scales that have been studied. For instance, scientists have noted an apparent periodicity associated with solar flare activity. Some studies have indicated that the number of major flares reaches a maximum every 155 days, whereas other data show a recurrence rate of activity each 51 days. Further data analysis will indicate the actual frequency of maximum flare occurrence, if indeed one exists. Scientists have noted that most solar flares occur in the vicinity of sunspots and that their rate of occurrence is related to the degree of sunspot activity. In any event, the number of charged particles emitted by the sun appears to increase with the level of sunspot and solar flare activity. These charged particles, mostly *electrons and protons that reach speeds up to about 2200 kilometers per second, along with the moving electric current associated with them,* basically form what is called the **solar wind.** Since the speed that this stream of particles reaches is considerably less than the speed of light (300,000 km/s), over 4 days pass after the outburst of a solar flare before its effects are noticed on Earth. It is this solar wind, along with the interactions of photons (bundles of electromagnetic energy), that forces the tails of comets to assume a constant direction away from the sun as they journey around it. A portion of the deluge of charged particles is trapped by Earth's magnetic field to produce the *Van Allen radiation belts* that surround Earth. The remaining charged particles bombard the atoms and molecules in the upper reaches of our atmosphere to produce the brilliant displays of light sometimes present about Earth's polar regions. The **auroras** (the northern and southern lights), often known as *aurora borealis* and *aurora australis,* respectively, are produced when the solar wind increases in intensity perhaps 1000 times beyond its normal value when there is increased activity on the surface of the sun. The colorful auroras often appear as flickering streaks or curtains of light; at other times they may consist of steady colored lights, resembling draperies that change brightness and color (such as from pale pink to blue to green), often in a matter of hours. Most displays are located between 80 and 160 kilometers above Earth's surface. An example of a beautiful aurora is presented in Figure 17.6.

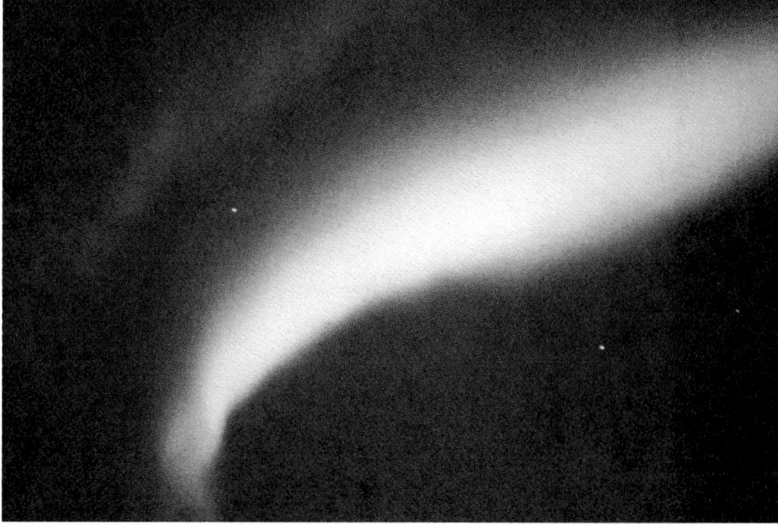

Figure 17.6 The aurora borealis (northern lights). In the Southern Hemisphere, this phenomenon is known as aurora australis, or southern lights.

Besides visible light and charged particles, the sun essentially emits frequencies that sweep along the entire electromagnetic spectrum, including radio waves, infrared rays, ultraviolet rays, X rays, and gamma rays. A large percentage of the radio waves headed toward Earth is reflected into space by the ionized gases in our upper atmosphere.

The damaging high-frequency ultraviolet radiation emitted by our sun and other stars is mostly absorbed by Earth's ozone layer, a region that is composed of triatomic molecules of oxygen (O_3), located high in our atmosphere. There is a serious concern among scientists that without this protective layer, various forms of life on our planet would perish. The thickness of the layer is diminishing, particularly over the polar regions. Many scientists think that current industrial practices are destroying the ozone layer, a contention that is being carefully examined.

The Planets

The planets and their motions have been studied in a dedicated manner at least since the days of Claudius Ptolemy (A.D. 85–165), the last notable astronomer of antiquity. His accomplishments were discussed in detail in Chapter 16. Nicholas Copernicus (1473–1543), the Polish astronomer-mathematician, is credited with determining that the orbits of the planets are essentially perfect circles. His conceptual view of planetary motion was strongly supported by the Italian scientist Galileo Galilei (1564–1642). However, after 20 years of study and the reliance on the observations made by the Danish investigator, Tycho Brahe (1546–1601), who recorded the positions of the planets known to him for over two decades, the German-born Johannes Kepler (1571–1630), a contemporary of Galileo, published in 1609 (and again in 1619) what are known as **Kepler's laws of planetary motion.** Kepler's first law of planetary motion—*the law of ellipses,* maintains that *the orbit of each planet is an ellipse with the sun at one of the foci.* Such a figure can be drawn by inserting two pins or thumbtacks, representing fixed points or foci, into a piece of cardboard. A piece

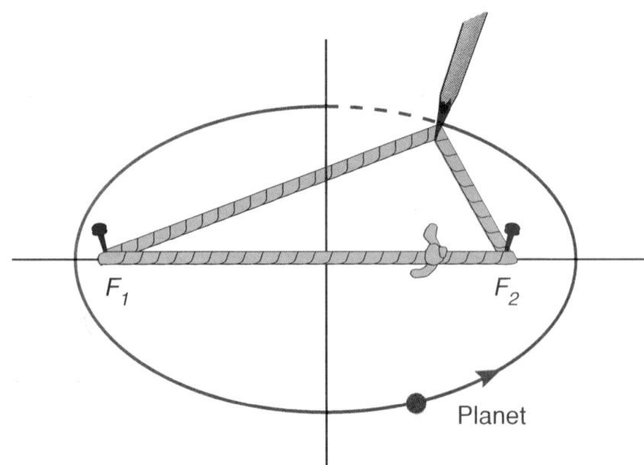

Figure 17.7 A string tied into a loop and placed about the two pins, as shown, will yield a figure called an ellipse. Planetary orbits are ellipses, with the sun at one focus (for instance, F_1), but the orbits actually are more nearly circular than the ellipse shown in this figure.

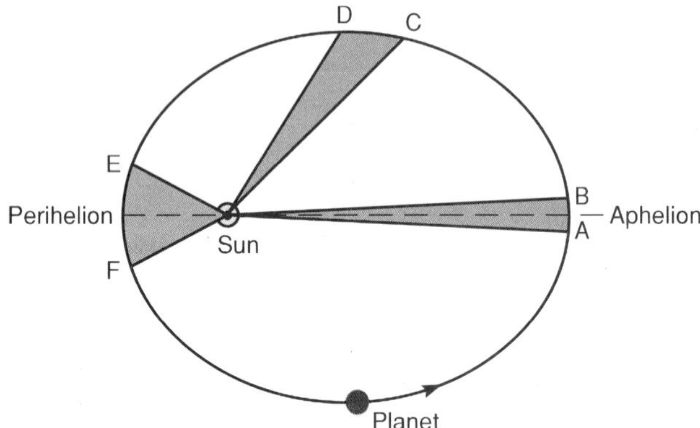

Figure 17.8 According to Kepler's second law, the shaded areas encompassed by points A and B, C and D, and E and F are equal, being swept out in the same length of time. One should note that this valid observation indicates that a planet would travel considerably farther in its orbit between E and F than at any other points in its orbit in the same time interval.

of string is looped over the two pins or tacks, and then a pencil or pen is used to keep the string taut as it is traced around the pins. The model assumes the sun is at the position of one of the foci (see Fig. 17.7). This law indicates that the distance between a planet and the sun will vary depending on the position of the planet in its orbit. When the planet is *nearest the sun,* it is said to be at **perihelion,** and when it is *farthest from the sun,* it is at **aphelion.** (Earth is at perihelion on or about January 4 and at aphelion on or about July 5 each year.)

The second law of planetary motion attributed to Kepler, known as *the law of areas,* notes that *the radius vector to a planet sweeps out equal areas in equal intervals of time.* A straight line joining the sun and the planet would sweep out equal areas in equal intervals of time, as noted in Figure 17.8. A literal interpretation of this law indicates that the closer a planet is to the sun, the faster it moves in its orbit, and, the farther it is from the sun, the slower it moves. Thus, a planet increases in speed as it approaches perihelion and slows down as it approaches aphelion.

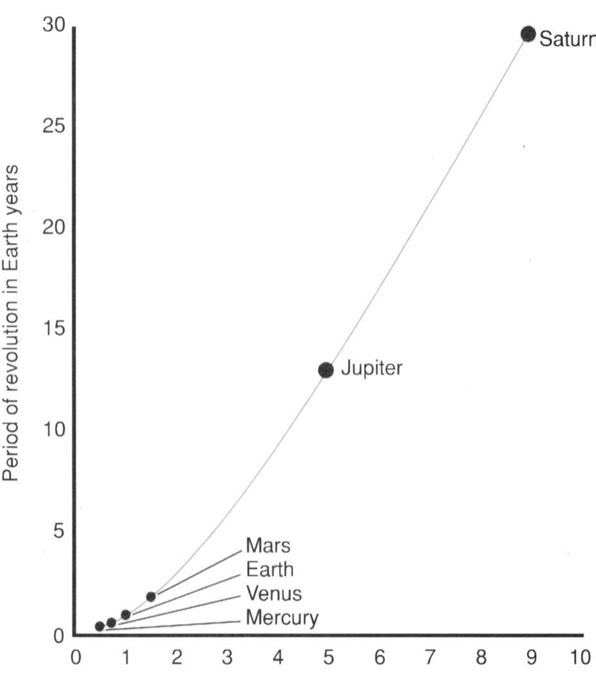

Figure 17.9 Kepler's third law, when expressed in a graphical manner as shown, would permit an observer to determine the period of revolution of a given object if its distance from the sun were known. Also, if the orbital period of the object had been determined, then the distance could be obtained from the same graph.

Kepler's third law of planetary motion is often referred to as *the harmonic law.* The law states that *the square of the period of revolution of a planet is directly proportional to the cube of the planet's average distance from the sun.* That is,

$$\frac{\text{period}^2}{\text{distance}^3} = \text{a constant that applies to all planets, or } \frac{T^2}{s^3} = k.$$

If T is expressed in terms of Earth's period, that is, one year, and s is represented in terms of Earth's orbiting distance, designated as one *astronomical unit,* 1 AU, then $k = 1$, and thus $T^2 = s^3$.

Jupiter averages a distance of 5.20 AU (5.20 times Earth's orbiting distance, or approximately 778,000,000 km) from the sun. Its period of revolution can be calculated as

$$\begin{aligned} T^2 &= s^3 \\ &= (5.20)^3 \\ &= 141 \\ T &= \sqrt{141} = 11.9 \text{ y.} \end{aligned}$$

PERIOD OF A PLANET

A graph that illustrates the relationship of the distance of a planet from the sun and its period of revolution is shown in Figure 17.9.

The four planets closest to the sun, called the *inner planets,* or *earth-type planets,* are all relatively small in size and have relatively high densities, that is, from 4 to 5.5 grams per cubic centimeter (compared to the density of water, which is 1 g/cm^3). These inner planets (Mercury, Venus, Earth, and Mars) have either little

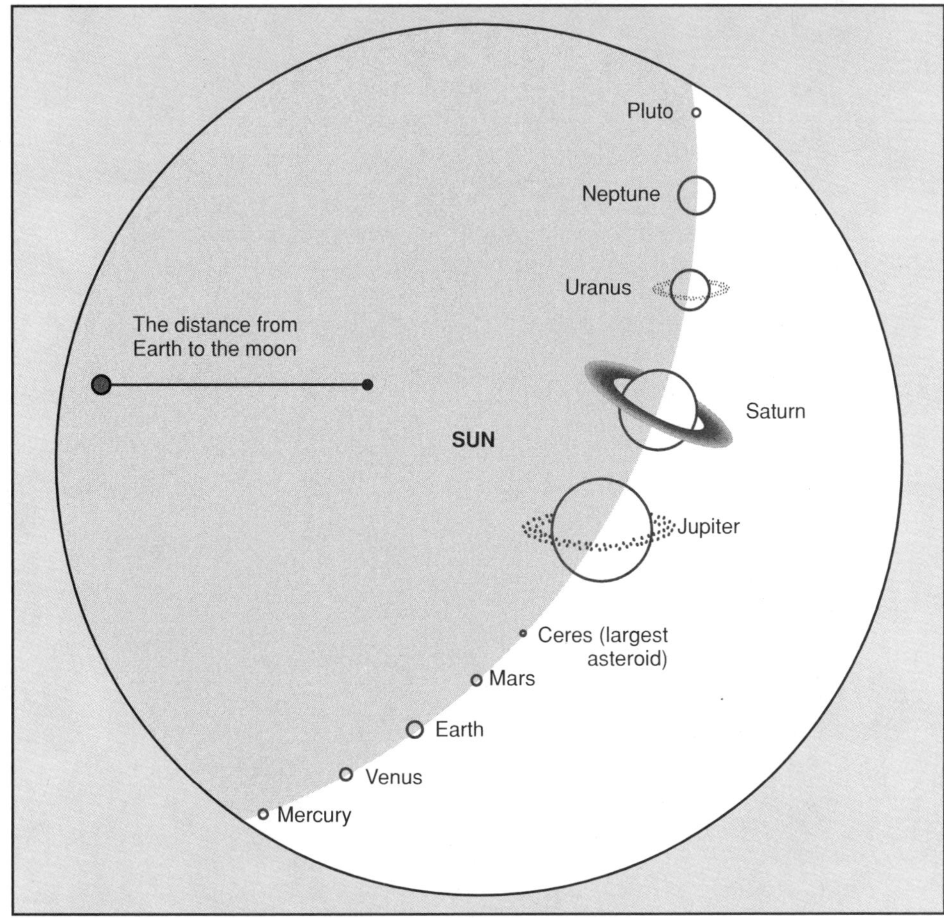

Figure 17.10 The relative sizes of the planets as compared to the size of the sun.

or no atmosphere, such as Mercury, or oxidizing atmospheres somewhat similar to Earth's. An oxidizing atmosphere is one that in general contains varying amounts of oxygen, carbon dioxide, and nitrogen, whereas reducing atmospheres, such as those of the outer planets, contain varying amounts of hydrogen, helium, methane, and ammonia. Because of the closeness of the inner planets to the sun, they are all either fairly warm or hot bodies. Mercury and Venus have daytime temperatures of about 340°C and 480°C, respectively, and Earth and Mars have daytime temperatures of about 21°C and 4°C, respectively. Of course, these temperatures are average temperatures of the planets' sunlit sides. The nightside temperatures of Mercury and Mars become quite low, because neither their atmospheres nor their surfaces store as much heat as Earth. Only Earth and Mars have temperatures that allow water to exist as a liquid, and in the case of Mars this temperature exists only during the daytime. The two planets Mercury and Venus, the orbits of which lie between those of Earth and the sun, are called *inferior planets,* and the remaining planets—Mars, Jupiter, Saturn, Uranus, Neptune, and Pluto—the orbits of which are farther from the sun than Earth's orbit, are called *superior planets.*

Jupiter, Saturn, Uranus, Neptune, and Pluto are called the *outer planets,* although the planet Pluto much more closely resembles the smaller inner planets. The other outer planets are quite alike in chemical composition, size (in that they are quite large compared to the inner planets), temperature, and density (from 0.7 to 2 grams per cubic centimeter). The planets Jupiter, Saturn, Uranus, and Neptune all have chemical compositions similar to those of the sun, other stars, and nebulas. Because of their high escape velocities (a result of their large masses) and low temperatures (because of their great distances from the sun), they have not lost much of their atmospheres and, consequently, still have approximately the same chemical makeup that they had at their birth when the solar system condensed out of a dark nebula 4.5 billion years ago. On the other hand, the inner planets have lost their atmospheres of hydrogen, helium, methane, and ammonia. If the sun, planets, satellites, asteroids, meteors, and comets did condense simultaneously from the same primordial dust cloud, then they should all be of the same approximate age. Scientists have determined from the analysis of rock samples that Earth and its moon, Mars, and the meteorites are all between 4.5 and 4.6 billion years old.

Of the nine planets in our solar system, only Mercury and Venus have no known moons. The relative sizes of the planets is illustrated in Figure 17.10, and a comparison of some of the planets to various moons in our solar system is shown in Figure 17.11. Figure 17.12 illustrates the order in which the planets orbit the sun. Further information about the planets and Earth's moon is provided in Table 17.1 (p. 377).

Figure 17.11 This montage of photographs displays the smaller planets and larger moons of the solar system at the same scale.

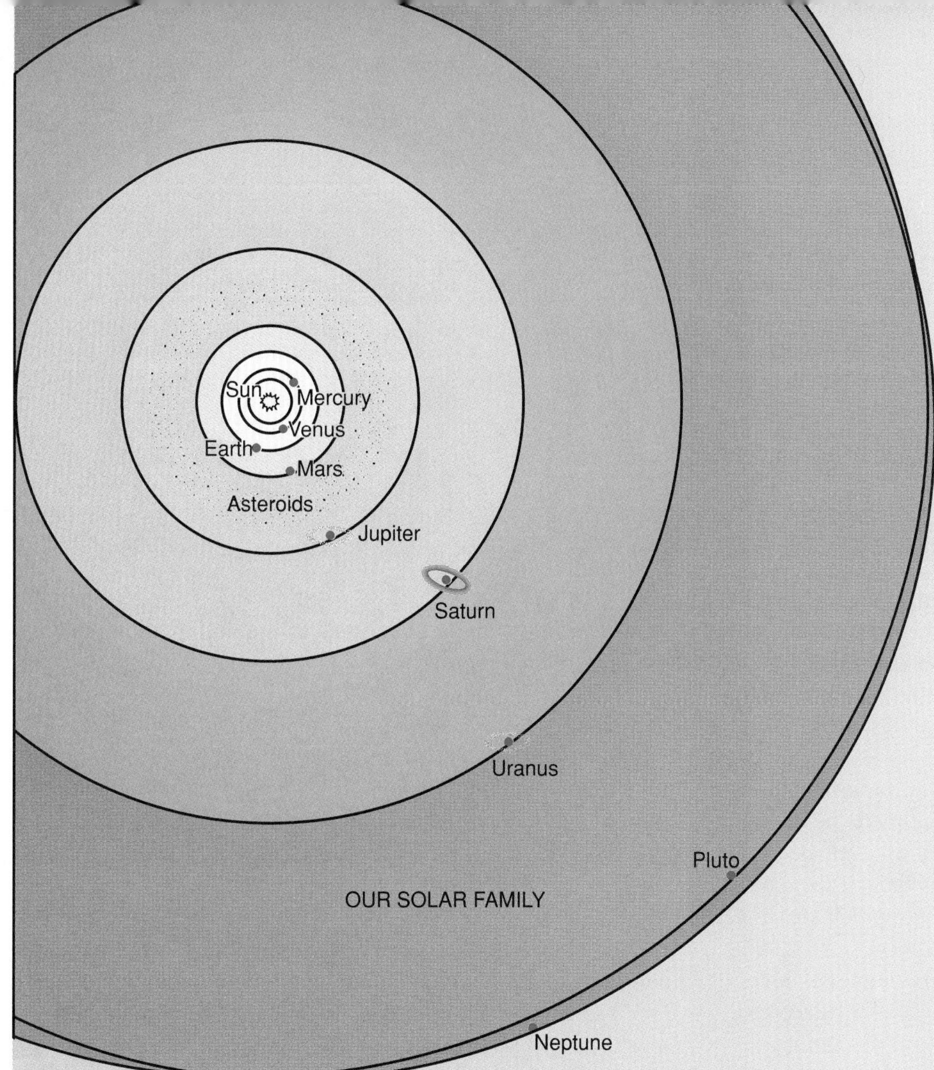

Figure 17.12 Mercury orbits closest to the sun. Ordinarily, Pluto's orbit carries this planet farthest from the sun, but for the next decade or so, Pluto's eccentric path brings it inside the projected orbit of Neptune.

Table 17.1

Selected Data on our solar system.

Object	Diameter*	Mass*	Mean distance from sun (AU)	Known number of satellites	Revolution period (years)	Rotation period (days)	Escape velocity (km/s)	Surface gravity*
Sun	109	333,000		9		~25	624	28
Moon	0.27	0.1	~1	0	0.075	27	2.4	0.16
Mercury	0.38	0.06	0.4	0	0.24	59	4.2	0.5
Venus	0.95	0.8	0.7	0	0.62	243†	10.3	~1
Earth	1	1	1	1	1	1	11.3	1
Mars	0.53	0.1	1.5	2	1.88	~1	5	0.5
Jupiter	11	318	5.2	17	12	0.41	61	2.6
Saturn	9.4	95	9.5	18–19	29.7	0.44	36	1.2
Uranus	4.1	15	19.2	15	83.7	0.72†	22.1	~1
Neptune	3.8	17	30.3	8	165	0.67	24.7	1.5
Pluto**	0.18	0.002	39.5	1	248.4	6.4	1.2	006

*Earth's = 1
†Exhibits retrograde (reverse) rotation
**Most data about Pluto are somewhat uncertain at this time.

Mercury

The planet Mercury is the closest planet to the sun. Like Pluto, Mercury has a highly eccentric (elliptical) orbit. Its distance from the sun varies from about 45 million kilometers at perihelion to approximately 69 million kilometers at aphelion. Mercury was named after the swift, wing-footed Roman god who served as the messenger for all the gods. This name is appropriate since Mercury, being the closest planet to the sun, moves at a higher speed than any of the other planets. Since Mercury's orbit is so close to the sun, we can see the planet for only a short period of time after sunset and before sunrise. In fact, there are only about two weeks of each year during which Mercury is easily seen, since most of the time it is lost in the solar glare. As a result, only those observers who have kept a constant vigil have ever seen Mercury. The planet is the second smallest in the solar system, with a diameter of only 4878 kilometers. The Earth's moon has a diameter two-thirds that of Mercury, and Jupiter's largest moon and Saturn's largest moon are both larger than the planet Mercury.

For a long time it was thought that Mercury always kept the same face to the sun in the same way that our moon keeps the same face toward Earth. In order for Mercury to keep one side perpetually facing the sun, its period of rotation on its axis would have to equal its period of revolution around the sun. Mercury's period of revolution has been known for centuries to be about 88 days, but in 1965 it was shown by radar observations that Mercury rotates on its axis in 59 days. It does not, therefore, keep the same portion of its surface pointed toward the sun. The sunlit side of Mercury may be as hot as 425°C, depending on the planet's orbital distance from the sun. This temperature is high enough to melt various metals, including lead, tin, and zinc. In contrast, its nightside, the portion turned away from the sun, has a temperature of about −120°C. This comparison represents a temperature difference of about 545 °C between the two sides. The planet has no satellites and, according to the *Mariner 10* space probe and recent ground-based observations, it has a tenuous atmosphere composed primarily of sodium, with lesser amounts of helium and hydrogen (see Fig. 17.13).

Mariner 10 also discovered that Mercury has a cratered surface similar to Earth's moon with no indication of recent geologic activity. In addition, instruments aboard the spacecraft determined that the planet has a substantial magnetic field. Planetary scientists, however, have not been able to detect a molten core considered necessary to produce a magnetic field. Radio maps made of the planet indicate that all of the heat it releases is from reradiated sunlight.

Venus

With the exception of our moon, Venus is the most brilliant and beautiful object in the night sky. At its brightest it can even be seen during the day if one knows where to look for it, and at night its light can cast shadows. The shining splendor of Venus

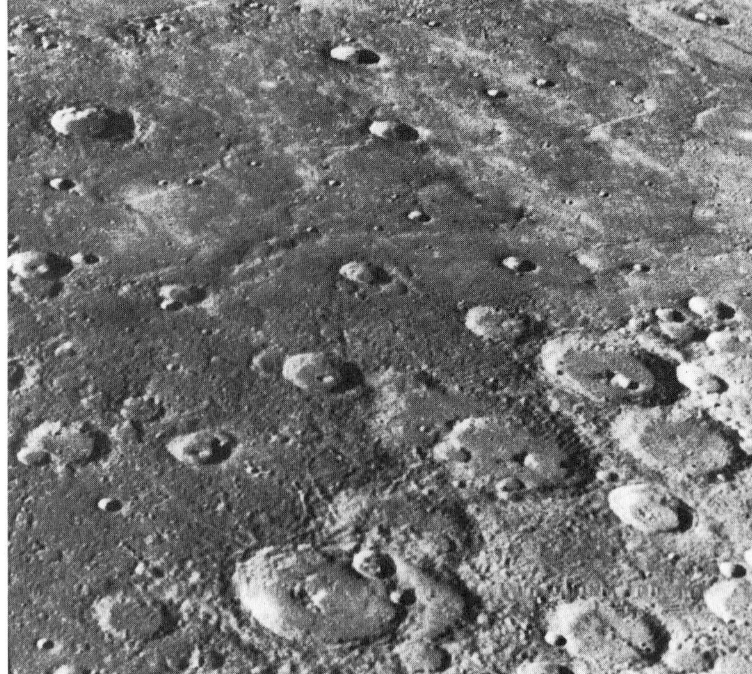

Figure 17.13 A photograph of a portion of the surface of Mercury. Numerous craters are clearly visible.

occasionally results in reports that the star of Bethlehem has reappeared. Venus has been fittingly named after the Roman goddess of beauty.

Venus is often called Earth's twin, since its diameter is 12,100 kilometers, or about 650 kilometers less than that of Earth. In addition to size, Earth and Venus are very similar in mass and average density. However, numerous U.S. and Soviet space probes that have encountered Venus, particularly the *Mariner 10* mission, have detected a magnetic field considerably less than the strength of Earth's. This discovery confirms the belief that the planet has an interior structure, including a molten iron core, much like the center of Earth.

The orbit of Venus about the sun is the most circular of all planets, and it completes one revolution in 225 Earth days. Various studies have revealed that Venus, like Uranus, rotates on its axis in a clockwise fashion; hence, the two planets rotate in a direction opposite that of their counterparts. A planet that rotates backward is said to undergo *retrograde rotation.* The sun, then, rises in the west and sets in the east on Venus. One "day" on Venus, the length of time from one sunrise to the next, is almost 117 Earth days. However, as a result of its retrograde motion, the rotation period of Venus, as viewed from Earth, has been measured at 243 days. The rate of rotation of the planet is apparently too slow to produce any appreciable currents among the charged particles in its molten core which would induce a detectable magnetic field.

At closest approach, 40 million kilometers, Venus is closer to Earth than any other planet. Since Venus is an inferior planet, it, like Mercury, goes through phases similar to the phases shown by our moon. Venus is brighter when it is only a crescent than when it is more fully lit because it is closer to the Earth at crescent phase.

The veil of beautiful white clouds that encircles and conceals the surface of Venus is mainly responsible for the extraordinarily

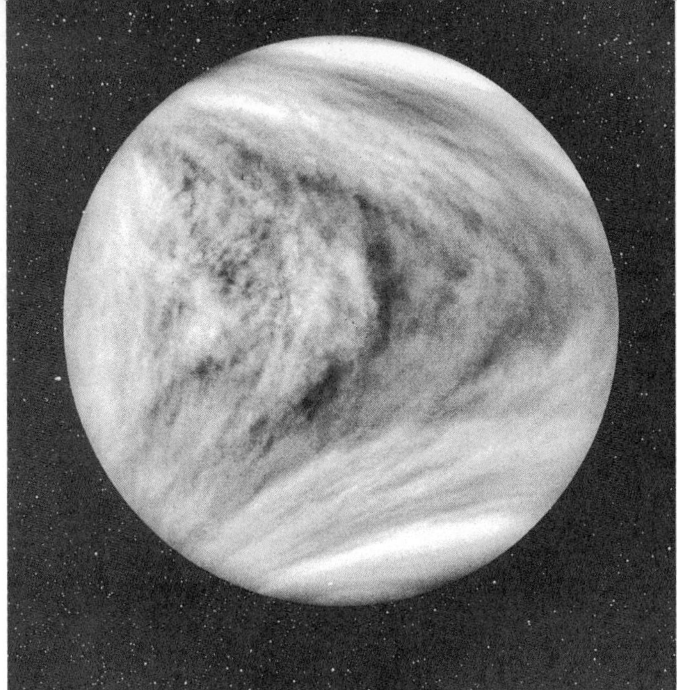

Figure 17.14 This photograph of Venus taken from a *Pioneer* Venus probe reveals the structure and motion of the planet's atmosphere. The clouds form two broad layers and revolve about Venus in about four days. (The star background has been added by an artist.)

high surface temperature of this planet (see Fig. 17.14). Venus's surface temperature of 480°C is the direct result of the high (96 percent) carbon dioxide content of its atmosphere. The high concentration of carbon dioxide produces a **greenhouse effect** *by trapping the infrared radiation of the sun* in the same way that greenhouses on Earth trap the sun's warmth. The atmosphere of Venus also contains water vapor and nitrogen (3 percent), plus various amounts of an assortment of acids. A compound present in considerable abundance in the Venusian clouds appears to be sulfuric acid, formed by the reaction of sulfur dioxide and water. (On Earth, sulfur dioxide released into our environment reacts with the water in our atmosphere to produce a portion of our acid rain.) Because of the high reflectivity of the Venusian clouds, light from the sunlit side of Venus is probably reflected into all portions of the nightside so that instead of having dark moonless nights, Venus enjoys a perpetual twilight.

The air pressure on the surface of Venus is over 90 times that of Earth's; on Earth one would have to descend to an ocean depth of over 900 meters in order to experience a pressure equal to Venus's surface air pressure. The dense atmosphere of Venus shrouded the surface of the planet from our view until radar mapping and other scientific strategies were employed. The radar image shown in Figure 17.15a reveals the topography of the planet, consisting of mountainous regions and relatively level plains. The highly corrosive atmosphere apparently causes an intense amount of weathering of Venus's surface. A close-up taken from the surface of the rugged planet by the *Venera 13* spacecraft appears in Figure 17.15b.

The *Magellan* spacecraft, launched from the space shuttle *Atlantis* in May 1989, entered an orbit of 3.26 hours about Venus in August 1990 after completing a 948-million-mile journey that included one and one-half trips about the sun. The radar imaging

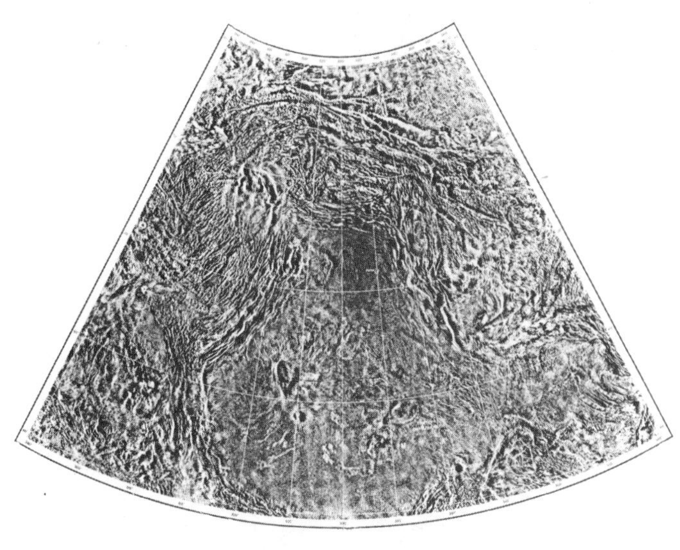

a.

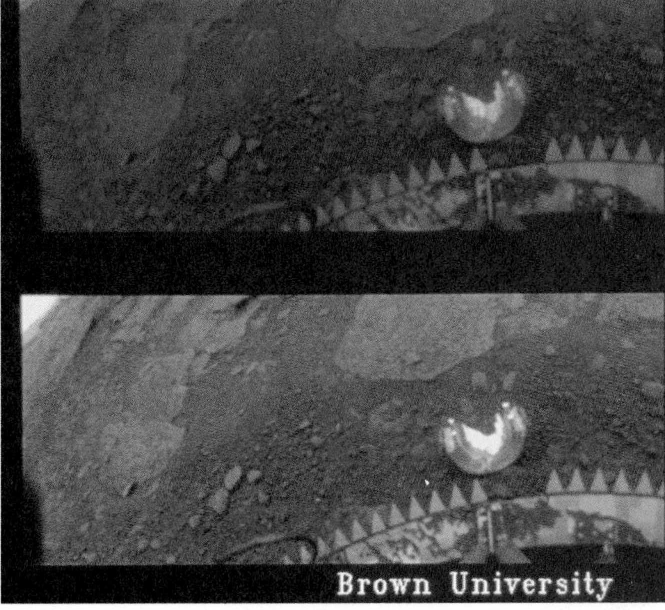

b.

Figure 17.15 (*a*) A radar image of a portion of the planet Venus. Note the rugged appearance of the planet's surface. Soviet Academy of Sciences/Brown University. (*b*) *Venera 13* color panorama of the surface of Venus. The upper panel is a reproduction similar to the actual appearance at visible wavelengths. The orange hue is caused by diffuse incident radiation when blue radiation has been removed by the planet's surface. The lower panel is the same region reprocessed to remove the effects of the strongly colored radiation; hence, as Venus would appear in white light without atmospheric interference. The "teeth" and other protrusions in both panels are parts of the spacecraft.

system aboard the spacecraft yielded pictures and maps of most of the surface of the planet with a level of detail significantly better than the images obtained from the Soviet *Venera* spacecraft in the 1980s. Pictures received from *Magellan* were taken within 250 kilometers of the planet. They included a belt of mountains and valleys that appear to have been stretched apart by a tremendous

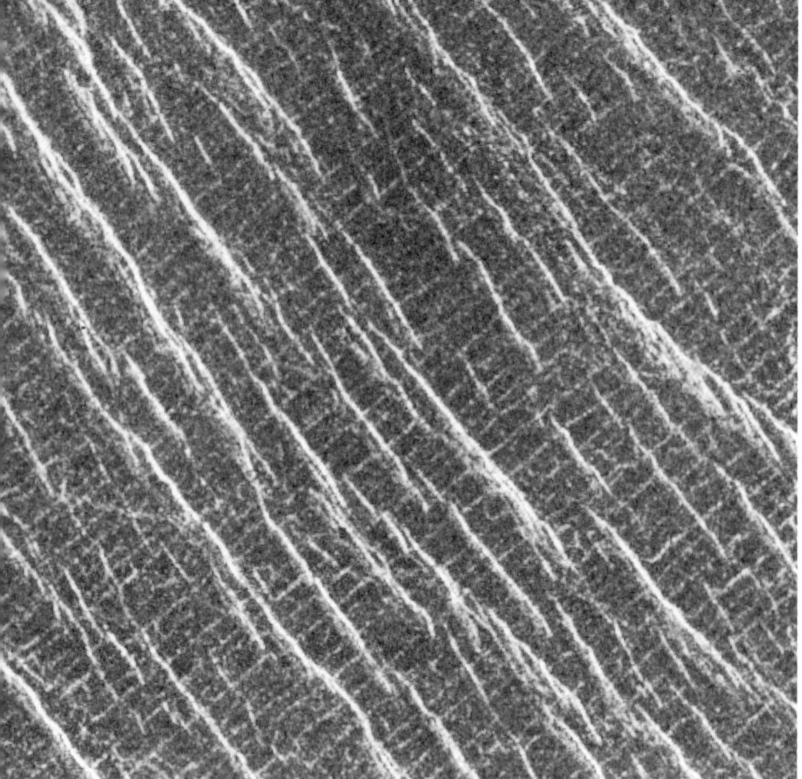

a.

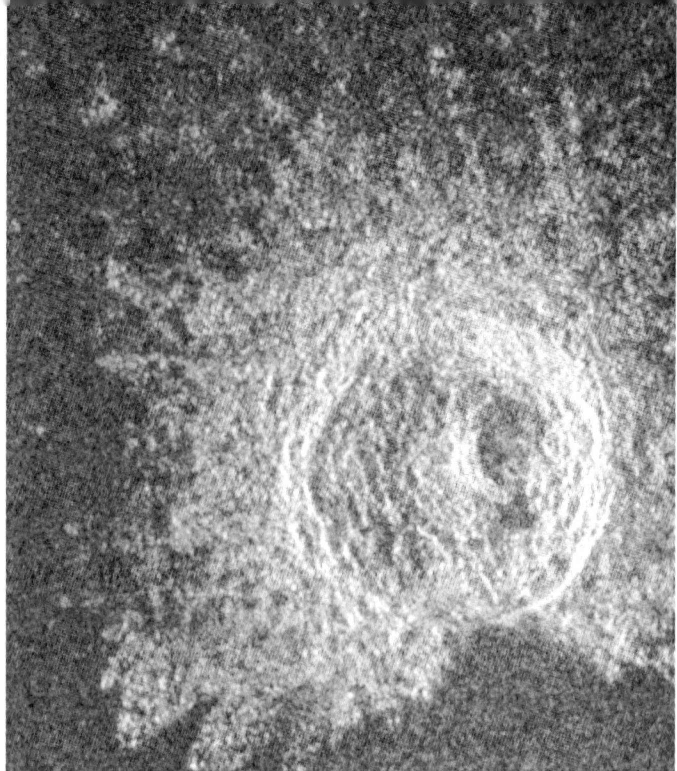

b.

Figure 17.16 (*a*) A *Magellan* radar image mosaic of a portion of Venus, acquired in September 1990. The fainter parallel lineations are about 1-km apart and intersect with the brighter ones almost at right angles. This type of terrain, perhaps representing faults or fractures, had not been detected before on any object in the solar system, including Venus, prior to the *Magellan* mission. (*b*) A radar image mosaic of an impact crater taken during the *Magellan* mission. The crater has a rim diameter of 12.5 km. According to the surrounding ejecta, the crater was probably caused by an object moving toward the north (top of photo) at a shallow angle to the surface before collision.

landslide as highlands gradually collapsed onto surrounding lowlands. Also included was a picture of a unique meteorite impact crater that indicated a meteorite struck Venus at a low angle, rather than from directly above (the oblong crater is 8 kilometers wide and 11 kilometers long). *Magellan* also photographed thousands of volcanoes on Venus from which lava had gently flowed. Petal-shaped lava flows were located about circular pits and linear fissures. In some cases, various narrow, river-like channels were formed where extremely hot lava flowed down a steep grade. However, the landscape near other volcanoes was found to be covered with volcanic ash, indicating that some volcanoes had undergone explosive eruptions similar to many of those on Earth. Some of the surface features photographed during the *Magellan* mission may be seen in Figure 17.16. According to NASA officials, the mission to Venus represents the beginning of a new era of planetary exploration.

Earth

Our planet is unique in its beauty. It is the only known planet that presently has large quantities of water on its surface and is the only one with lush vegetation. These two features give us the blue of our oceans and the green of our hills. The other solid planets in our solar system, in contrast, appear stark and bleak.

Earth and its lone satellite orbit the sun in a slightly eccentric orbit between the orbits of Venus and Mars. A spectacular view of Earth, as seen from the remoteness of outer space by the astronauts, appears in Figure 17.17. Our planet, as photographed by the astronauts in one of their landings on the surface of the moon, is shown in Figure 17.18a. A view of the moon's surface obtained during the *Apollo 17* mission (Fig. 17.18b) shows the actual ruggedness of its surface. Pertinent data about Earth are provided in Table 17.1. In order to more readily interpret the table's information, one should note that Earth's equatorial diameter is considered to be 12,760 kilometers, its mass has been calculated at 6.0×10^{24} kilograms, and its average orbital distance from the sun is estimated at 1.5×10^{8} kilometers. Chapters 20 through 27 discuss our planet in considerable detail.

Mars

Mars is one of the planets readily visible with the naked eye. Its actual ruddy red-orange color appeared bloodlike to the ancient Babylonians, who called it the Star of Death. Its present name is no less a suggestion of bloodshed and death. Mars is named after the Roman god of war, and its moons, Phobos and Deimos, are named after the two sons of Ares, the god of war in Greek mythology who personified terror and fear. The shape of Phobos, shown in Figure 17.19 is reminiscent of a giant potato. It races along in its orbit about Mars so rapidly that it completes one revolution in a mere 7.5 hours. As a result of its speed, it can possibly rise in the west and set in the east twice in one night. On the other hand, Deimos takes considerably longer than a Martian day to make one complete revolution and, consequently, rises in the east and remains in the sky two nights before setting in the west. The

Figure 17.17 As *Apollo 11* closes in on planet Earth, the cloud-whorled blue planet reveals its predominance of water, as seen in this view of the Pacific Ocean. It is late afternoon in western North America (upper right), and near midday the following day in Australia (lower left). At the top, the north polar cap gleams white.

a.

b.

Figure 17.18 (*a*) Astronauts during one of their visits to the surface of the moon. (*b*) Scientist-astronaut Harrison H. Schmitt is standing next to a huge split lunar boulder at the time of an *Apollo 17* landing on the moon. The region is known as the Taurus-Littrow landing site.

moons of Mars are among the smallest of the satellites detected in the solar system. Phobos, with its oblong shape, is about 26 kilometers long and 21 kilometers wide, whereas Deimos resembles a sphere about 13 kilometers in diameter.

The first telescopic observation of Mars was accomplished by Galileo early in the seventeenth century. About 50 years later, Huygens sketched the surface features of the red planet he observed through his telescope. Additional surface characteristics of Mars were mapped by other observers, including Father Secchi, a Roman Catholic priest who, in the early 1870s, believed he detected the presence of long straight lines on the planet. He referred to these linear features as *canali,* an Italian word referring to natural channels capable of transporting water. Then, in 1877, Giovanni Schiaparelli, director of the Milan observatory, reported that he possibly observed a network of the features described by Father Secchi. Other observers confirmed his detection of the canali, and soon the name began to be interpreted as canals—wide ditches constructed to carry water from the polar regions. Percival Lowell, a Boston aristocrat, further peaked public curiosity with his 1895 book suggesting the possible existence of a highly intelligent Martian civilization that had constructed the canals to irrigate the regions of Mars more conducive to life near the planet's equator. The belief in the existence of a Martian civilization became fixed in the minds of the public, and even today various science fiction books and movies, including H. G. Wells's leading production of *The War of the Worlds,* relate to this belief. However, later photographs taken with Earth-based telescopes, such as the photograph shown in Figure 17.20, and close-up views furnished by the *Mariner* series revealed no evidence of a network of canals. Interestingly enough, other photographs taken from satellites while orbiting the planet, such as shown in Figure 17.21a point out sinuous channels that probably represent old riverbeds carved out in times past by rushing water. On the other hand, the desolate regions in other portions of the same figure indicate that barren wastelands abound similar to some desert regions on planet Earth. Also noted is evidence of crater impact and past volcanic activity. However, most researchers contend that the outer portion of Mars is considerably different from Earth in that the Martian crust is composed of broken blocks that do not move relative to each other.

Although the possibility of some forms of life on Mars has not been totally ruled out, instruments aboard the *Viking 1* and

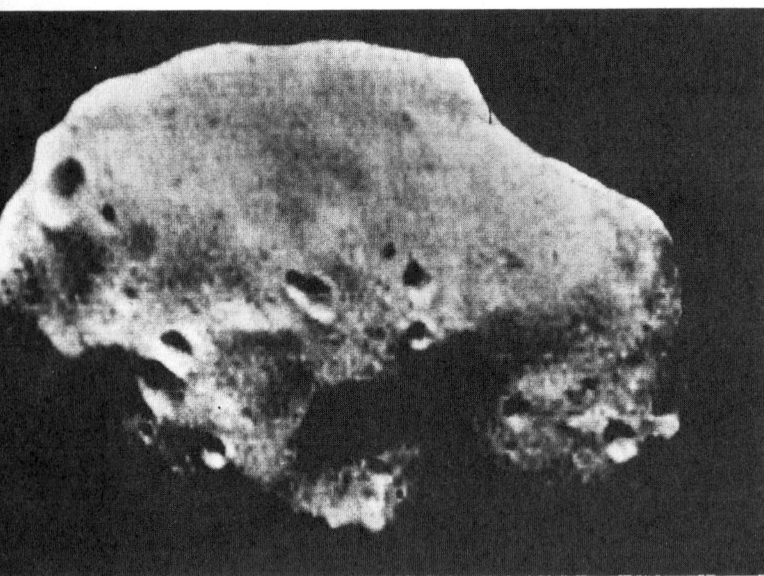

Figure 17.19 Phobos, the innermost satellite of Mars, is about 26 km long and 21 km wide.

Figure 17.20 A spectacular photograph of the Martian surface as photographed from Earth. The vast rift canyon, called *Valles Marineris,* stretches nearly 4000 km along the planet's equator. A sinuous valley, apparently created by the erosive action of moving water, appears in the Northern Hemisphere and several large craters, perhaps of volcanic origin, are visible near the center of the planet. During various times of its year, Mars has polar icecaps composed of atmospheric carbon dioxide (dry ice).

Viking 2 landers have been unable to detect any supporting evidence for this supposition on the Martian surface. The two Viking landers discovered an even lower water content on the planet than had been previously believed. Various photographs these spacecraft transmitted to Earth reveal why Mars is correctly identified as the "Red Planet" (see Fig. 17.22).

Mars rotates on its axis in 24.6 hours; hence, the Martian day and night are about the same length as those of Earth. The length of a Martian year is 687 days, or almost twice as long as an Earth year. The four seasons on Mars are thus almost six months long. Mars has a diameter of about 6788 kilometers, or slightly more than half the diameter of Earth.

The most prominent feature of Mars is its polar caps. During the coldest of Martian seasons, the winter polar cap may reach halfway to the Martian equator, but during the summer the same polar cap may completely disappear. Since these caps shrink rapidly, they must be very thin, perhaps no more than a meter thick in most places. The winter Martian polar cap is probably similar to the winter snows on Earth which may cover an area that extends almost halfway to our equator but which quickly melt off during our warm spring.

Large dust storms, as they blow across the Martian surface, have been observed for many years. Their presence offers definite proof that Mars has an atmosphere. However, it was not until the *Mariner IV* space probe flew past Mars in 1965 that it was realized how tenuous and thin its atmosphere really is. The spacecraft reported that the atmosphere of Mars is only about 2 percent as dense as the atmosphere of Earth and consists almost entirely of carbon dioxide, with but traces of argon, oxygen, nitrogen, and water vapor. Because of the thinness of the planet's atmosphere, the temperature changes the Martian surface experiences are rather extreme. For instance, at the 1976 *Viking 1* landing site, about 23°N latitude, the temperature was found to range from a daytime high of −40°C to a nighttime low of −80°C. Other regions of the planet were estimated to reach −17°C, whereas still others experienced temperatures as low as −120°C. The layer of watery ice that coated the rocks and soil near the landing sites was found to be about 1 millimeter thick. Obviously, the conditions that exist on Mars today are not at all supportive of life-forms by Earth standards.

The Asteroids

Before the discoveries of Uranus, Neptune, and Pluto, astronomers wondered about the regularity of the spacing of planets about the sun. In 1772, Johann Bode, the director of the Berlin Observatory, published a paper in which he popularized the earlier work of Johann Titius. The German astronomer Titius had noted an emperical relationship between the various distances from the sun and the known planets in the solar system. Since Bode reemphasized the little realized connection Titius had noted, it came to be known as *Bode's law.* However, as the outer planets were later discovered, the relationship failed to predict their accurate distances; hence, the accepted law proved not to be totally correct in

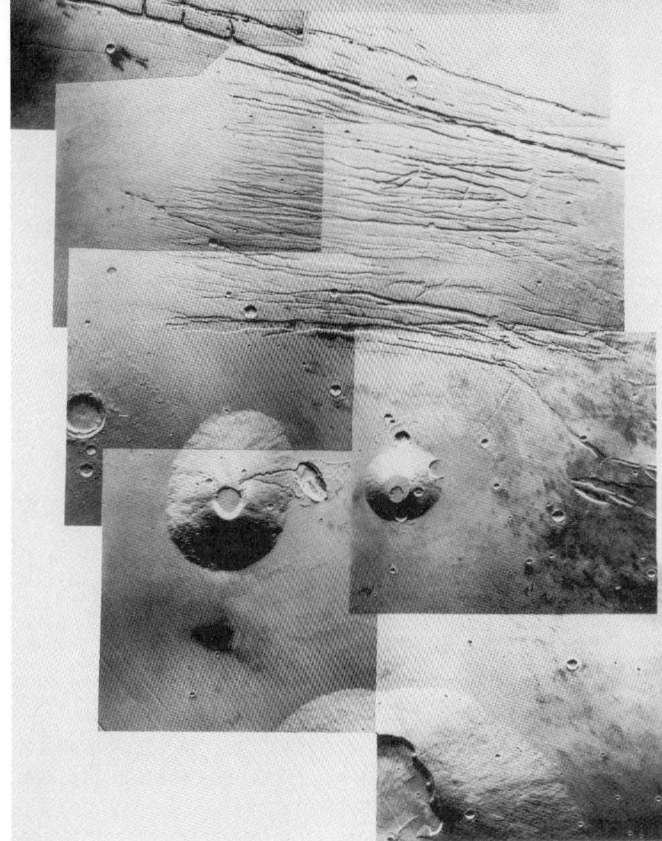

a.

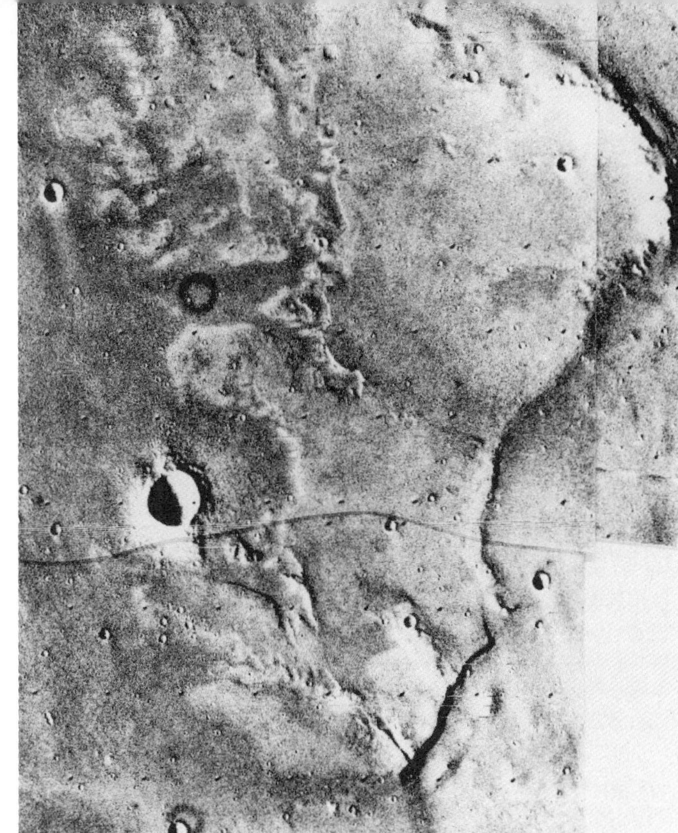

b.

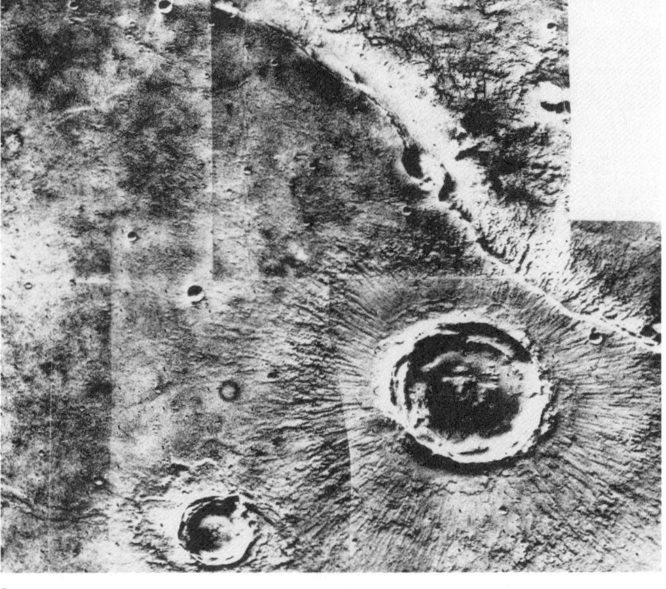

c.

d.

Figure 17.21 (*a*) *Viking Orbiter 1* photographed this mosaic of the northeast region of Mars in 1977. The area is the youngest volcanic region on the red planet. Note the areas pockmarked with impact craters. (*b*) *Viking 1* photographed this plateau near the original primary landing site. The lower area was shaped by the erosion of running water. (*c*) A fresh young crater located near the "Western Front" landing site photographed by *Viking 1*. The crater is about 30 km in diameter. (*d*) A spectacular photograph of the Martian landscape by the *Viking 1* lander. Note the dune field with features similar to those of desert scenes on Earth. The structure in the lower right portion of the photograph is part of the *Viking 1* landing craft.

Speculation about the Asteroids

There is some speculation among scientists that billions of years ago a planet did exist between the orbits of Mars and Jupiter, but that it was destroyed by impact with some celestial object or torn apart by the gravitational pull of the massive planet Jupiter to form the asteroids. The theory is reinforced by the presence of irregularly shaped asteroids that orbit counterclockwise about the sun. Some of these asteroids are earthy or stony; others are very dark, like many of Earth's igneous rocks; and still others have a high iron and nickel content. Assuming only a relatively small amount of fragments escaped, the unfortunate planet would have been of extremely small size, since it is estimated that all the asteroids combined account for no more than 10 percent of the mass of the smallest planet, Pluto. Indeed, the assumed planet would not have been large enough to be considered a respectable satellite, since it would have been only 5 percent the mass of our moon ■

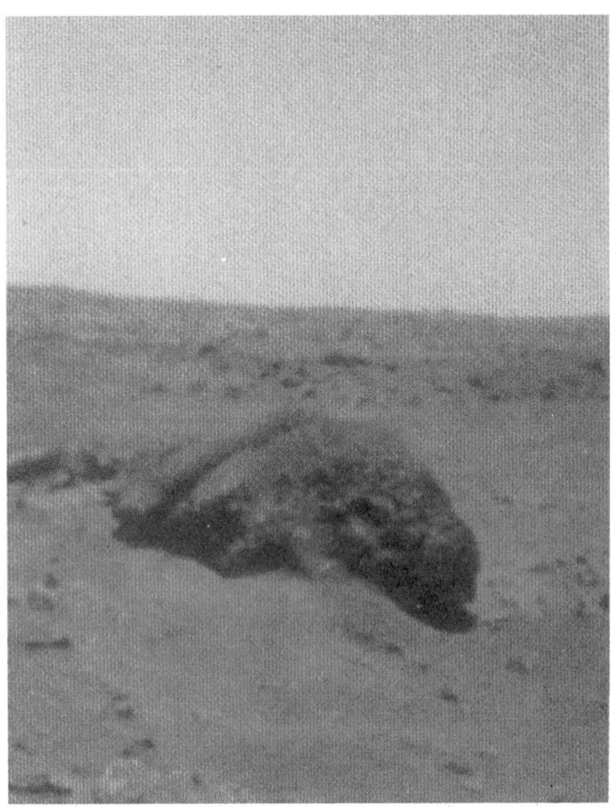

Figure 17.22 "Big Joe" (alias "Big Bertha" and "Big John") represents the largest rock located near the Viking landing sites. This specimen, probably composed of basaltic lava, is over 2 m long and about 1 m high. The giant boulder may have been hurled to its present position by the impact of a meteorite several kilometers away. The specimen is dark in color, but it is covered by reddish Martian soil, offering evidence of the red dust carried by large Martian dust storms.

all cases and was eventually discounted. Since the correlation appeared to be valid at the time, it predicted that a planet should occupy an orbit between the orbits of Mars and Jupiter, 2.8 astronomical units (AU) from the sun. (Recall that an astronomical unit is the distance from Earth to the sun.) However, no planet had been observed at this distance. In 1801, the Sicilian astronomer Piazzi found a planetlike object that was very close (2.9 AU) to the predicted orbit. Piazzi named the new object *Ceres,* after a goddess of his native Sicily. A more careful investigation of the newly discovered object showed that it was only about 800 kilometers in diameter. Could an object that small be considered a planet when the smallest planet previously known in the solar system was about 3200 kilometers in diameter? Surely not. The question became academic when a year later a second diminutive planet was discovered with the same general orbit as Ceres; it was named *Pallas,* after Athena, the Greek goddess of wisdom. The diameter of Pallas has been determined to be 450 kilometers. In 1804, a third object was found and named *Juno,* after the wife of the Roman god Jupiter. Today, thousands of these "minor planets," known as **asteroids** or *planetoids,* have been found occupying orbits around the sun that correspond to the predicted orbit of the "missing planet." There are about 10 known asteroids that have a diameter from 150 to 300 kilometers, over 100 that average from 80 to 150 kilometers in diameter, and approximately 500 that average from 40 to 80 kilometers across. There are probably untold thousands of asteroids, too small to be detected at the present time. Most astronomers and other scientists think that the asteroids represent debris that remained after the other celestial objects in the solar system were formed. The effect of the enormous mass of Jupiter prevented small masses from accreting into a larger object.

Even though the asteroids basically orbit between Mars and Jupiter, some stray into the path of Earth. Our planet is assumed by scientists to have been constantly bombarded by numerous members of this ring of celestial objects. According to various teams of researchers, exciting evidence is mounting that Earth collided millions of years ago with large asteroids (and perhaps with comets). The areas of study offering promising results include the coastal regions of Cuba and Haiti, as well as the oceanic crust between North and South America. These collisions are thought to have brought about a series of significant environmental changes. The scientists' conclusions are based upon the presence of "shocked" mineral grains discovered in unusually thick layers of clay and deep-sea sediments collected since the 1970s. Other evidence of impact includes the discovery of a circular depression 300 kilometers in diameter, located in the Colombian basin near thick clay deposits under study.

A significant number of celestial objects from perhaps the outer portion of the solar system have been observed as they passed through the plane of Earth's orbit in recent times. For instance, in 1989, an asteroid estimated to be the size of an aircraft carrier, and traveling at a speed of 74,000 kilometers per hour, passed within 640,000 kilometers of Earth, thus avoiding a collision by only six hours. According to the American Institute of Aeronautics and Astronautics, significantly large foreign objects pass through Earth's orbit every two to three years. Technology is reported to be available to detect, track, and possibly divert or even destroy such objects if an impact with Earth appears imminent.

Jupiter

As the solar system beyond the asteroid belt is examined, one finds that the physical characteristics of the outer planets are distinctly different from those of the inner planets. The four inner planets are relatively small and rocky, with solid surfaces and average densities about five times the density of water. These inner planets are called **terrestrial planets,** meaning earthlike. The four giant planets, Jupiter, Saturn, Uranus, and Neptune, in contrast, are large, gaseous bodies without solid surfaces and average densities about equal to that of water. They are the **Jovian planets,** meaning jupiterlike. (The present available knowledge about Pluto does not permit astronomers to neatly fit it into either category.) Even though their densities are low, the Jovian planets, because they are so large, contain about 200 times the mass of all the other bodies in the solar system, with the exception of the sun. The Jovian planets are composed mostly of hydrogen, but they have thick atmospheres with colorful cloud patterns created by compounds with low boiling points.

Jupiter is the giant among all planets in our solar system; thus, it is aptly named after the King of the Roman gods. Jupiter has a diameter of over 140,000 kilometers, about 11 times that of Earth, giving Jupiter a volume over 1300 times as great as Earth's. The mass, about 318 times the mass of Earth, is about 2.5 times as great as all the other planets combined. Jupiter, along with several of its moons, is shown in Figure 17.23.

Figure 17.23 The swirling atmosphere of Jupiter becomes the background for various moons of the giant planet. The innermost large satellite, Io, can be seen against Jupiter's disk. The darkest satellite, Callisto, although twice as bright as Earth's moon, is barely visible at the bottom left of the photograph. Jupiter's colorfully banded atmosphere displays various patterns highlighted by the Great Red Spot, a large circulating atmospheric disturbance.

As viewed from Earth, this giant planet is generally the second brightest in appearance, being outshone only by Venus. The brilliance of a planet depends on its distance from the sun, its distance from Earth, its surface area, and the fraction of light reflected from its surface. *The fraction of light reflected from the surface of a planet* is called its **albedo.** Venus has the highest albedo of any planet, 0.76, meaning that 76 percent of the light striking the surface is reflected. This, combined with its size and proximity to both the sun and Earth (see the discussion of the inverse square law of intensity in Chapter 11) makes it the most brilliant planet. Jupiter, with an albedo of 0.51 and a tremendous surface area, is brightest when it is at **opposition;** that is, *in a direction exactly opposite to that of the sun as viewed from Earth.* In opposition, Jupiter not only appears in the full phase, but it is the closest to Earth that the two planets' orbits permit. (The albedo of Mercury is only 0.056 and that of Mars, only 0.15. The moon, with a very low albedo of 0.073, appears bright because of its relative closeness to Earth.)

One might intuitively conclude that a magnificent planet such as Jupiter would be rather sluggish and would spin slowly on its axis. However, Jupiter is the fastest rotating planet in the solar system. The features close to the planet's equator complete one rotation in 9 hours, 50 minutes, while those near the poles require about 9 hours, 56 minutes. Thus, the great planet exhibits **differential rotation.** (Recall that the sun also rotates differentially because of its fluid nature.) The exceedingly high rate of rotation causes Jupiter to be quite oblate; that is, the diameter through its poles is considerably smaller than the diameter through its equator (a partially deflated basketball pressed between two boards would be oblate). The diameter through the poles is 1/15 smaller than the diameter through the equator, whereas for Earth the comparable figure is 1/298. The high rate of rotation indicates that objects at Jupiter's equator travel about the planet's center at approximately 45,000 kilometers per hour. Compare this with the speed of objects at Earth's equator that travel at about 1600 kilometers per hour.

This high rate of rotation is also believed to be responsible in large part for Jupiter's turbulent atmosphere. Jupiter is about 75 percent hydrogen, but the atmosphere contains significant quantities of helium (He), methane (CH_4), and ammonia (NH_3). Although the planet's makeup certainly must include water, the low temperature in the upper atmosphere (about $-150°C$) would result in the water existing only in the form of ice. (Deeper in the atmosphere it may be warm enough for liquid water to exist.) The

Figure 17.24 This *Voyager 1* photograph of the Great Red Spot shows a white oval with its "wake" of counterrotating vortices. The distance from top to bottom of the area is about 24,000 km.

Figure 17.25 Saturn and its satellites Tethys (outer left), Enceladus (inner left), and Mimas (right of rings) are seen in this mosaic of images taken by NASA's *Voyager 1* on October 30, 1980, from a distance of 18 million km. The projected width of the rings at the center of the disk is 10,000 km.

overall chemical composition of Jupiter, as well as that of Saturn, Uranus, and Neptune, is very similar to the chemical composition of the sun, other stars, and gaseous nebulas from which our solar system was formed. This similarity of composition is the result of the great masses of these planets (with the consequence of large gravities) and their extreme distances from the sun (which results in very low temperatures). Because of its high gravitational field and very low temperature, Jupiter has not lost very much of its original composition.

The most spectacular feature on Jupiter is the Great Red Spot, shown in Figure 17.24 (as photographed during the *Voyager* missions) and also in the lower left portion of Figure 17.23. This gigantic mass of circulating fluid, located in the planet's thick atmosphere, is about 50,000 kilometers long and an estimated 15,000 kilometers across. Such an area would be capable of holding four planets the size of Earth within its perimeter and have ample space remaining to place several objects the size of our moon. The Great Red Spot, despite the chaotic fluid flow that surrounds it, seems to be situated in a shear zone, rolling much like a giant ball between a westward current above it (to the north) and an eastward current below it. The rest of the planet is encased by white, red, yellow, and blue-green bands parallel to the equator. These bands, like the Great Red Spot, appear to be permanent features of Jupiter's atmosphere; however, their colors and bandwidths slowly change with time.

Jupiter's internal source of heat is apparently such that the planet radiates about twice as much total energy into space as that it receives from the sun. Perhaps this observation is a pale reminder that the planetary temperature of this planet was much higher billions of years ago than the currently estimated core temperature of 30,000°C.

Jupiter is surrounded by huge radiation belts similar to Earth's Van Allen radiation belts. Its radiation belts mean, of course, that the planet must have a very strong magnetic field that has trapped large quantities of charged particles. Radio telescopes detect strong radio emissions from the region of Jupiter's radiation belts. These radio emissions are undoubtedly the result of the charged particles trapped in the radiation belts being accelerated as they spiral through the magnetic lines of force of Jupiter's magnetic field. Radio emission caused by the acceleration of charged particles in a magnetic field is known as *synchrotron radiation.*

Jupiter has a large metallic (that is, solid) hydrogen core as a result of the extremely high pressures existing near the center of the planet. There is no solid surface anywhere near the outer layers of Jupiter. Its density simply increases as the distance from the "surface" toward the center increases. Some of Jupiter's satellites have atmospheres. For instance, Ganymede, Jupiter's largest moon, has an atmosphere of methane, and Io has a slight atmosphere that contains sodium and sulfur. Loki, the huge volcano on Io, Jupiter's most famous and active moon, may be ejecting sulfur dioxide, a gas that accounts for the snowlike appearance of the satellite. Like Saturn and Uranus, a ring of debris surrounds Jupiter, although its ring pattern is quite small compared to those of the other two planets.

Saturn's Rings and the Roche Limit

Saturn is the most beautiful and spectacular telescopic object in the solar system, mainly because of the splendid set of radiant rings that encircle it. Saturn was named after the ancient Roman Titan god of agriculture who was the father of Jupiter. This ringed planet, as seen in Figure 17.25, is the second largest planet in the solar system, with a diameter of over 120,000 kilometers. Like Jupiter, the planet presents a banded view, but these atmospheric features are less distinct and not as colorful as those of Jupiter. Saturn, too, has colored "cloud spots" in its atmosphere similar to the Great Red Spot on Jupiter, although they are small and are not as well-defined. However, in October 1990, astronomers discovered a giant white spot that suddenly developed in the clouds above Saturn. The oval spot was detected first by Earth-bound telescopes, then closely observed by the orbiting Hubble space telescope. The oval-shaped white spot eventually evolved into a

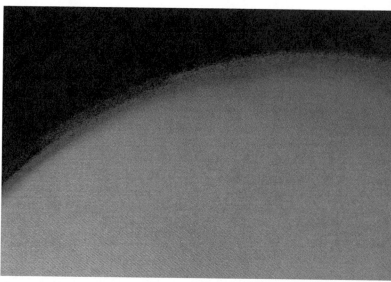

a.

b.

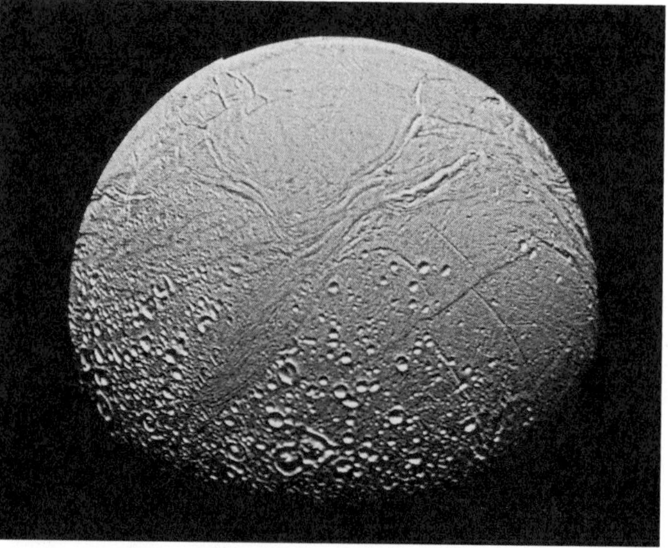
c.

d.

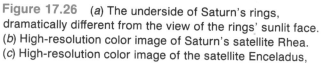

Figure 17.26 (*a*) The underside of Saturn's rings, dramatically different from the view of the rings' sunlit face. (*b*) High-resolution color image of Saturn's satellite Rhea. (*c*) High-resolution color image of the satellite Enceladus, showing surface detail similar to Jupiter's moon Ganymede. (*d*) Saturn's largest moon, Titan, enveloped in a thick haze layer, as shown in this enhanced color image. All photos were taken during the two *Voyager* missions.

gargantuan storm over 20,000 kilometers long and about 5000 kilometers wide, remaining visible for several months. A comparable event on Saturn had not been observed by scientists since 1933, when two white spots that lasted for several weeks were observed. Although Saturn's period of rotation is slightly less than that of Jupiter (see Table 17.1), the oblateness of the gaseous giant is the largest among all planets in our solar system. (The polar diameter of Saturn is 1/11 or 9 percent smaller than its equatorial diameter.) The analysis of recent data received from the *Voyager* spacecraft reveals that Saturn, like Jupiter, radiates about twice as much energy as it receives from the sun.

Although Uranus and Jupiter are known to have rings of debris in orbit about them, Saturn unquestionably has the largest, most magnificent, and most intricate ring system in the solar system. Saturn's rings are very thin (perhaps no more than 2 kilometers thick), and even disappear from view when both the sun and Earth are in the ring plane, much like the effect of the same view one might obtain of a person's hat. The planet's rings seem to consist of particles that range in size from those of dust particles to the dimensions of large houses. Close-up views of the planet, its rings, and some of its satellites are shown in Figure 17.26, a beautiful montage obtained from the *Voyager* missions.

Figure 17.27 *Voyager 1* looked back at Saturn on November 16, 1980, four days after the spacecraft flew past the planet, to observe its appearance from this unique perspective.

The outermost ring of Saturn extends to perhaps 200,000 kilometers above the planet's surface. There are apparently seven major ring systems in all, and the rings are designated (starting with Saturn's innermost ring) by the letters D, C, B, A, E, F, and G. There is a "gap" between the B ring and the A ring known as the *Cassini division,* in which the concentration of particles is much less than in the regular rings. Each major system is composed of rings within rings within rings. The outer portion of the A ring appears rippled, suggesting the presence of more small satellites yet to be detected. (A newly detected ring ripple suggests the discovery of the nineteenth moon.)

The particles of rings A, E, F, and G are apparently kept within their ring boundaries by some of Saturn's satellites. These satellites act as gravitational "sheepdogs" that "shepherd" the ring particles and keep them from straying outside their ring boundaries. The rings are subdivided into a multitude of narrow ringlets. Superimposed on the rings is a pattern of radial spokes. These spokes move around the rings in the direction of the rings' rotation, and are probably composed of electrically charged fine particles. There are actually thousands of "ringlets" within the major ring systems of Saturn. A farewell view of the planet is provided by *Voyager I* in Figure 17.27 as the spacecraft continues on its journey to the outer reaches of our solar system and beyond.

Saturn has at least 18 satellites that vary in size from the huge Titan, which is slightly larger than the planet Mercury, to S-10, which is only about 2 percent the size of Titan. Many of Saturn's satellites' presumably rocky cores are surrounded by ice. The mix of these two components ranges from about 50 percent to 85 percent ice.

Titan, with a diameter of 5118 kilometers, has an atmosphere that is 1.5 times more dense than Earth's atmosphere. Titan's atmosphere is composed primarily of nitrogen, small quantities of methane and carbon monoxide, and traces of hydrocarbons with high molecular weights. This dense atmosphere does not allow even modern cameras or telescopes to examine Titan's surface. The orange appearance of Titan's atmosphere may be due to bands of methane, along with the carbon monoxide gas recently detected.

The existence of Saturn's rings had puzzled astronomers for more than 200 years, but in the middle of the nineteenth century calculations made by French astronomer Edouard A. Roche (1820–1883) indicated that if a satellite came closer to its mother planet than 2.4 times the planet's radius, the orbiting object would be totally broken apart. This critical distance came to be known as the **Roche limit.** Scientists since have concluded that there is not a single Roche limit; rather, the limit for a small moon is closer to the planet than for a larger moon. Also, the greater the density of the satellite, the closer it can orbit without being destroyed. Most important, the limit applies only to objects of significant size held together mainly by gravitational forces. If such an object enters this limit, it would be broken apart into particles that would form rings about the planet. (Satellites launched from Earth are held together by the cohesive strength of their materials as well as by gravitational forces; thus, the Roche limit does not apply to them.) Likewise, a ring of solid particles within the Roche limit would indefinitely be prevented from merging to form a single satellite.

Although a portion of Saturn's ring system is beyond the Roche limit for the planet (an estimated 145,000 kilometers), gravitational disturbances caused by some of Saturn's existing satellites prevent the various particles from coalescing. Also, certain satellites of the giant planet orbit within the Roche limit but are so small that they can apparently withstand the gravitational tidal forces that are present.

Saturn has the lowest density of all the planets, 0.7 times the density of water. With this low density Saturn would quite easily float. The planet's atmosphere has a temperature of $-176°C$. Saturn rotates on its axis in 10 hours and 40 minutes.

An orbiting body has been detected between the paths of Saturn and Uranus, far beyond the asteroid belt. This object, formerly called Kowal, was discovered in 1977 and has recently been named Chiron. The diameter of this object has been estimated to be about 400 kilometers (250 mi) and it has an orbital period calculated to be about 50.7 years. The similarity of Chiron to a large asteroid may point out that not all asteroids orbit the sun within the asteroid belt located between Mars and Jupiter.

Uranus

The discovery in 1781 of Uranus, the seventh planet from the sun, is attributed to William Herschel, an English amateur astronomer who was destined to become one of the leading astronomers of the eighteenth century. At first sighting with his homemade telescope, Herschel believed the planet to be a comet. Other astronomers and various mathematicians, however, concluded from its motion that it was not a comet. Two months after its detection, Uranus was decreed to be a planet, the first discovered with the aid of a telescope. Uranus was named after the legendary Greek god of heaven who was the father of Saturn and the Titans, as well as grandfather of Jupiter. (Reportedly, Herschel wanted the new planet to be named after King George III of England, but his suggestion was never widely accepted among scientific circles.)

Uranus is one of the four gaseous giants, differing from Jupiter and Saturn in relative size and atmospheric composition, but quite similar to Neptune, its closest planetary neighbor. It presents a hazy, pale blue-green appearance due to the methane in its upper

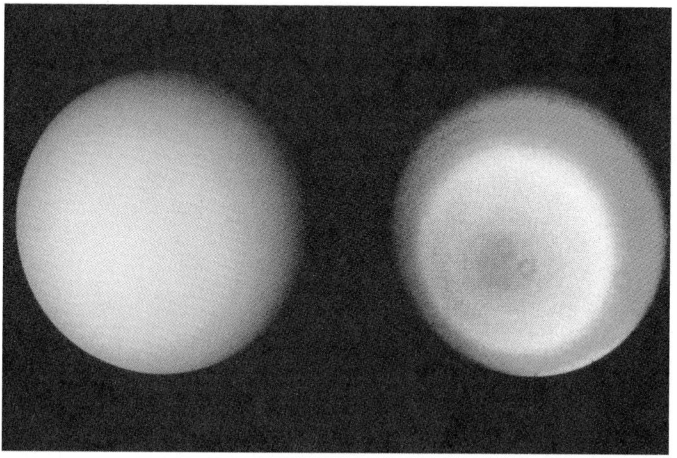

a.

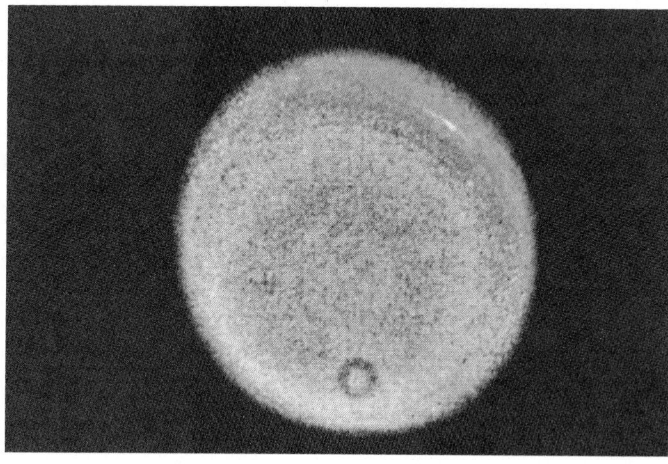

b.

Figure 17.28 (*a*) *Voyager 2*'s views of the planet Uranus. The photograph on the left represents the actual color of the planet, whereas the view on the right is a false-color image used to study the composition of the planet's atmosphere. (*b*) A view of the clouds that surround the planet. They are composed primarily of hydrogen and helium. Methane present in the upper atmosphere absorbs red light, giving Uranus its blue-green color.

atmosphere that absorbs sunlight in the red portion of the spectrum. The planet has a substantial magnetic field, roughly comparable in strength to the field that surrounds Earth, although the field about Uranus varies much more from point to point because of its large offset from the center of the planet. Magnetic fields are usually thought to be generated by electrical currents produced in a planet's molten core. Uranus is assumed to be partially molten deep within its interior, since its temperature increases steadily with depth to thousands of degrees. However, the region is not large enough to generate, by itself, the magnetic field strength detected by *Voyager's* instruments. The source of the magnetic field about Uranus is presently unknown; the electrically conductive, superpressurized ocean of water and ammonia once thought to be present between the core and the atmosphere now appears to be nonexistent.

The diameter of Uranus has been difficult to establish because of the haziness near its surface, but is presently estimated at about 52,000 kilometers. Photographs of the planet and its hazy atmosphere are presented in Figure 17.28. Uranus undergoes retrograde rotation, somewhat less pronounced than that of Venus. Basically, the spin axis of most planets is perpendicular to the plane of the solar system. However, the axis of rotation of Uranus is tilted to within 8 degrees of pointing directly at the sun. Consequently, one Uranian pole spends one-half of the period of revolution, or 42 years, in total darkness while the other polar region is in continuous daylight. The period we know as dusk on Earth would be brighter than noon on Uranus. In fact, Uranus receives about 1/400 of the sunlight that illuminates Earth. The period of rotation of Uranus, as viewed from Earth, is about 17 hours. Instruments aboard the 1986 *Voyager 2* space probe revealed an average temperature for the planet of $-212°C$ and very thin cloud formations that moved in the direction of the planet's rotation, Uranian west, at 160 kilometers per hour. (On Earth, the jet streams that so influence our weather patterns average 180 kilometers per hour.) A unique blanket of hydrogen and helium was found in the outer atmosphere. Immediately beneath this layer were detected clouds of methane ice, and closer to the core, scientists suspect dense clouds of ammonia, as well as the water previously discussed.

The *Voyager 2* mission also verified that Uranus has at least 11 dark, narrow rings of sparse, boulder-sized debris, perhaps the remnants of small moons that penetrated the planet's Roche limit. The spacecraft also located 10 previously undetected satellites, some shepherding the particles that form the rings in the manner various small satellites tend the rings of Saturn. This detection of other moons adds to the total of 5 previously discovered Uranian satellites that includes the exotic moon, Miranda, with its canyons, grooves, and innumerable impact craters similar to those observed on Mars and Mercury as well as on Jupiter's largest moon, Ganymede. Various views of Uranus and some of its moons appear in Figure 17.29.

Neptune

In the years that followed the discovery of Uranus, astronomers periodically noted its positions in the night skies and from this information were able to calculate the exact orbit that Uranus should take in its path around the sun. Before very many years had passed, it became obvious that Uranus was deviating from its calculated orbit. (The phenomenon is known as *perturbation* in the language of astronomers.) Many of the astronomers of that day began to question whether perhaps Newton's universal law of gravitation was invalid at such great distances or perhaps was correct only theoretically. Other astronomers wondered whether Uranus was being pulled from its predicted path by the gravitational influence of an unknown planet that was circling the sun beyond the orbit of Uranus.

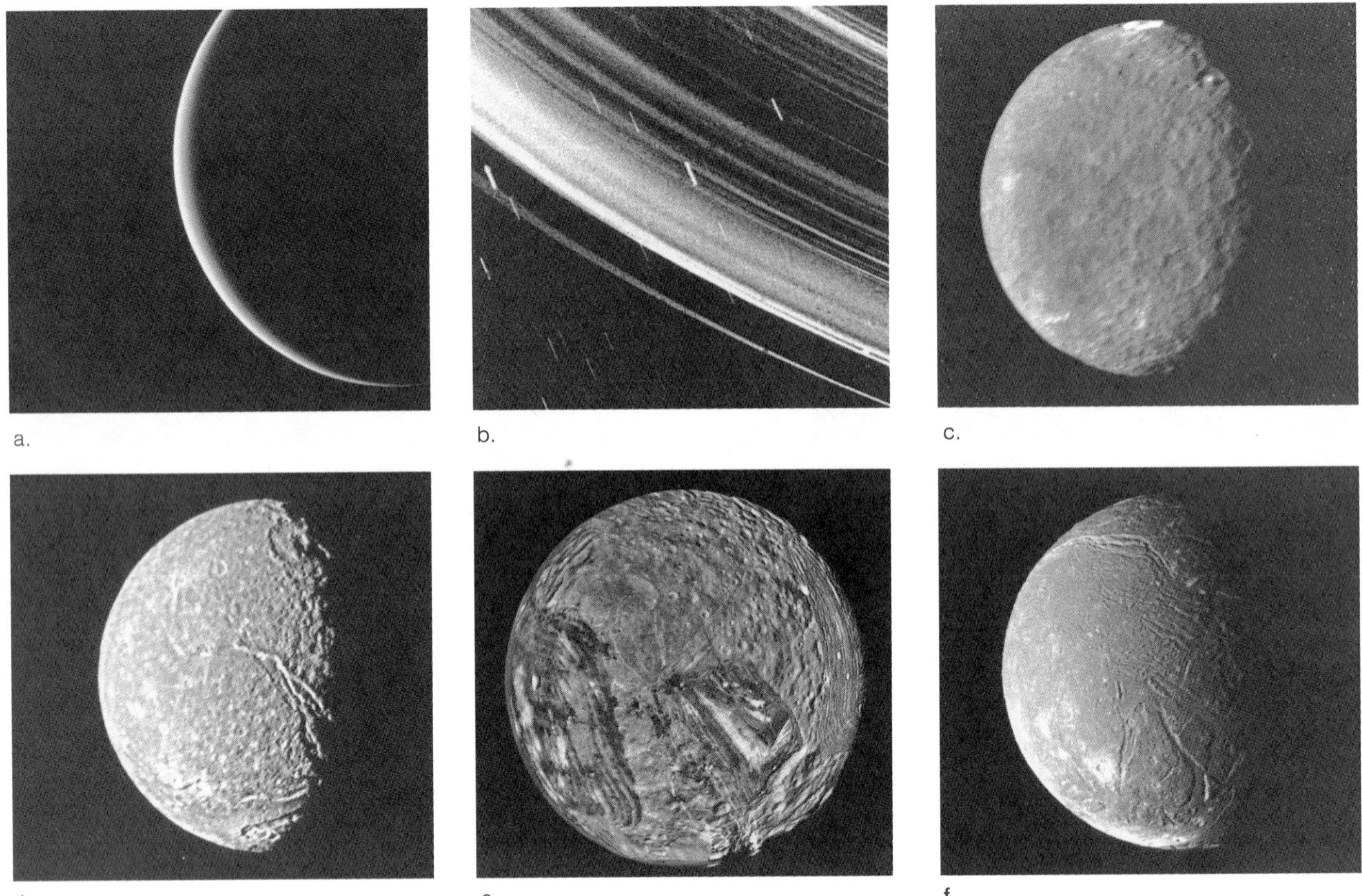

Figure 17.29 A montage from *Voyager 2*'s Uranus encounter. (*a*) A farewell view of Uranus as the spacecraft speeds toward Neptune. (*b*) The rings of Uranus with backlit dust particles. (*c*) A high-resolution photograph of Uranus's moon Umbriel. (*d*) A high-resolution shot of the moon Titania. (*e*) A mosaic of the moon Miranda. (*f*) A high-resolution photograph of Uranus's Ariel.

In 1845, the 26-year-old English astronomer John Adams calculated exactly where an unknown planet should be in order to disturb the orbit of Uranus by the observed amount. Adams sent his calculations to Sir George Airy, the Astronomer Royal of England. Sir George apparently did not have much faith in so young an astronomer and did not bother to have the telescopes at his command assigned the task of looking for the new planet. The following year the French astronomer Leverrier, who did not know of Adams's work, also calculated the position of the unknown planet and sent his result to the astronomer Galle at the Berlin Observatory. Galle looked for and found the new planet on the night of the same day that he received Leverrier's letter. Today, both Adams and Leverrier equally share the honor of having predicted the position of the new planet. This newly discovered planet was named Neptune, after the Roman god of the sea.

Neptune has an estimated diameter of 48,000 kilometers. Like other planets, including Earth, Jupiter, and Uranus, Neptune generates a magnetic field, detected only recently by astronomers at the University of California at Berkeley. Two satellites of this remote planet were initially discovered with Earth-based telescopes. The August 1989 *Voyager 2* encounter detected six others as it approached within 4950 kilometers of the planet's north pole, including those that aid the planet in retaining a ring of debris like that of Jupiter and Saturn. The ring was reportedly discovered by a team of astronomers who announced its existence in the mid-1980s, based on light fluctuations they detected in a background star. (However, during a more recent **occultation,** *the event in which one celestial body blocks out the view of another,* astronomers failed to locate a ring pattern about the planet.) Still, radio emissions detected from Neptune indicated that a radiation belt surrounds the giant planet.

The chemical composition and structure of Neptune appears to be very similar to that of Uranus. Since Neptune is so distant from the sun, about 4.5 billion kilometers, the planet requires about 165 years to complete a single revolution about the sun.

The most obvious feature of Neptune, according to *Voyager* photos, is its blue color, as shown in Figure 17.30a. The color is the result of the relatively high methane content in its atmosphere.

a.

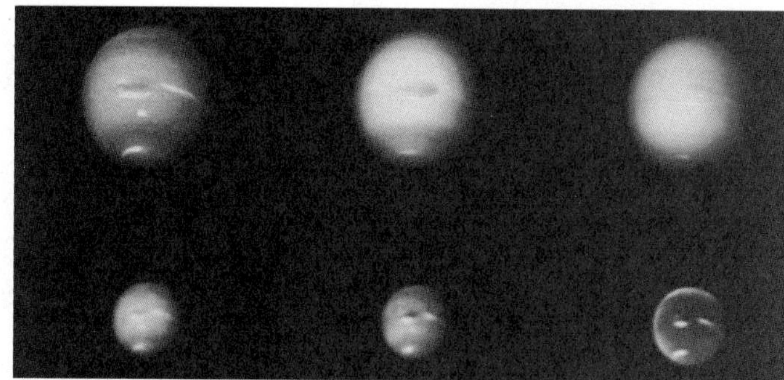

b.

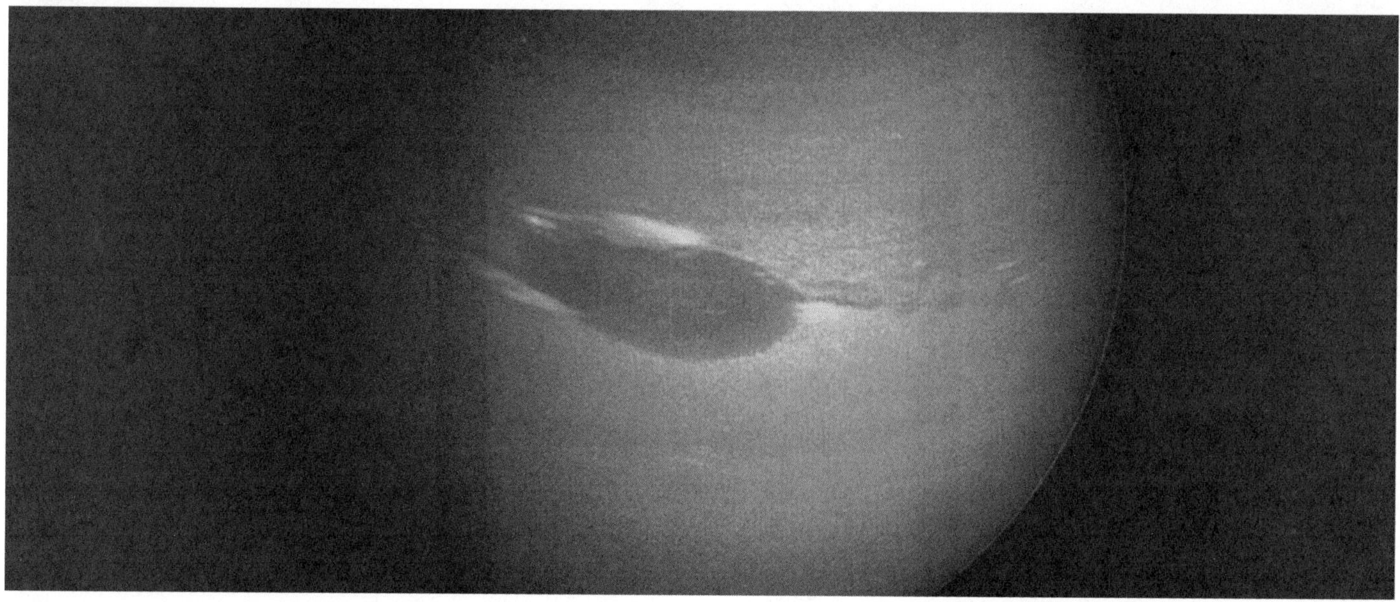

c.

Figure 17.30 (*a*) A false-color image of Neptune. The red areas represent a semitransparent haze that covers the planet. (*b*) Neptune, as it appears through various camera filters. These views also reveal altitude data on cloud features. (*c*) This false-color image shows detail of the Great Dark Spot and variations in cloud composition and structure at different altitudes.

This light gas preferentially absorbs the longer wavelengths of sunlight (those near the red end of the spectrum). The appearance of Neptune changes considerably when the planet is photographed through various camera filters, as seen in Figure 17.30b.

Neptune is a dynamic planet, although it receives only 3 percent as much sunlight as the planet Jupiter. Several large dark spots somewhat resemble Jupiter's hurricanelike storms. In fact, one of the regions is large enough to encompass Earth. This feature, which appears to be an anticyclone like Jupiter's Great Red Spot, has been designated the "Great Dark Spot" by its discoverers. Most of the winds on Neptune blow in a westward direction, thus retrograde, or opposite, to the rotation of the planet. Winds near the Great Dark Spot reach speeds of over 2400 kilometers per hour as they rotate about a center of high pressure; they represent the strongest winds measured on any planet. The Great Dark Spot is visible in Figure 17.30c. A more detailed view of this region of tremendously high disturbance is shown in Figure 17.31a. The *Voyager 2* mission also confirmed the presence of Neptune's ring system, as seen in Figure 17.31b.

Of the eight known moons that orbit Neptune, Triton is the largest. This satellite has a diameter of about 2700 kilometers and orbits Neptune in about six days at an average distance of 330,000 kilometers. Its relatively high density (twice as dense as water) indicates that Triton contains a greater percentage of rock in its interior than the icy satellites of Saturn and Uranus. The unusually high density and the retrograde orbit offer strong evidence that this satellite did not originate near Neptune, but is a captured object. Triton appears to have the same general size, density, temperature, and chemical composition as Pluto, the only planet not closely encountered by any spacecraft; thus, this satellite of Neptune will serve as our best model of Pluto for many years to come. Several views of Triton from the *Voyager* mission appear in portions of Figure 17.31.

a. b. c. d. e. f.

Figure 17.31 A montage of *Voyager 2*'s encounter with Neptune. (*a*) The planet's Great Dark Spot, as photographed by the narrow-angle camera of the spacecraft. The image shows feathery white clouds that overlie the boundary of the dark and light blue regions. (*b*) Neptune's ring system, shown in two exposures that lasted nearly 10 min each. (*c*) The bright southern hemisphere of Triton. (*d*) A high-resolution color mosaic of Triton. (*e*) Neptune and Triton as they appeared three days after flyby. Triton is the smaller crescent and is closer to the viewer. (*f*) Satellite 1989N1, discovered first by *Voyager 2*.

Little knowledge has been gained about Nereid, the second known moon of Neptune, since its discovery with Earth-based telescopes in 1948. This tiny satellite has the most eccentric orbit of any object in the solar system, ranging from about 1.35 million kilometers to almost 10 million kilometers as it revolves about Neptune. *Voyager's* best photos of Nereid were taken at a distance of about 4.7 million kilometers and show that its surface reflects about 14 percent of the sunlight that strikes the satellite, making it somewhat more reflective than Earth's moon. Nereid is slightly smaller than the moon of Neptune known as 1989N1. This third moon, first discovered by *Voyager 2,* is shown in Figure 17.31f. The last of the series of moons of this remote planet first viewed by *Voyager* has been designated 1989N6. These newly discovered satellites will likely be named at a later date by the International Astronomical Union, a task specifically delegated to this organization.

Pluto

The method that led to the discovery of the planet Neptune was applied to locate the planet Pluto. Shortly after Neptune's discovery, the planet was noted to deviate from its prescribed orbital path about the sun. Scientists immediately initiated a search for a planet farther from the sun than Neptune. In 1930 this outermost planet was discovered by Clyde W. Tombaugh, an assistant astronomer at the Lowell Observatory. He compared photographs taken over a period of years and identified a pinpoint of light in the constellation Gemini that had shifted among the thousands of stars that formed the background on separate photographic plates. This celestial object was named Pluto, for the ancient Greek god of the dark and lonely underworld (and possibly for the initials of Percival Lowell whose Lowell Observatory made a major contribution to the discovery of the planet).

Exploring the Solar System

Mariner II rocketed by the planet Venus in December 1962, the first in a long series of United States' robot space probes destined to unlock many mysteries of our solar system.

From 1964 to 1968 came the flights of *Rangers* and *Surveyors,* all robot space probes that photographed and/or landed on the moon's surface. Then followed the *Apollo* manned landings on the moon that occurred between 1969 and 1972.

Pioneer and *Explorer* probes analyzed the solar winds, while *Viking* probes photographed the entire surface of Mars. Robot space vehicles landed on Mars in 1976 and conducted an unsuccessful search for extraterrestrial life.

Voyager 1 and *2* spacecraft have taken "close-up" photographs of the turbulent surface of Jupiter, detecting active volcanoes on at least one of its satellites. They sent back beautiful photographs of the magnificent rings within rings of Saturn. After almost 8.5 years in flight, *Voyager 2* flew within 81,500 kilometers of Uranus's cloud tops in 1986. In 1989 the spacecraft reached Neptune, its final planetary target, passing within 4950 kilometers of the planet's north pole. Both spacecraft, as they leave our solar system, will continue to study ultraviolet sources among the stars, and will send back additional information about interstellar space for decades to come.

Astronauts have walked on the surface of our moon, and our unoccupied spaceships have sampled the solar winds, have flown through the hot sulfuric acid clouds of Venus, and have performed exobiological experiments on the cold, red, sandswept surface of Mars. The planetary mission of the space probe *Magellan* to Venus in 1990 provided fine resolution of the planet's mountains and plains. After perhaps years of study of the information transmitted to Earth by the space vehicle, scientists will determine whether Venus, like Earth, has been sculptured by drifting continental plates or by internal forces compressing the surface horizontally, buckling the crust.

The scheduled revisits to Mars and Jupiter will be accomplished by the space probes *Mars Observer* and *Galileo* in 1993 and 1995, respectively. *Mars Observer* will examine the surface geology and atmosphere from a polar orbit of Mars throughout the Martian year (687 days). *Galileo* will provide scientists with their first detailed look at Jupiter and some of its moons as it orbits the giant gaseous planet and sends a spinning probe toward the planet's surface. The probe, although it will be crushed as it penetrates Jupiter's intense atmosphere, will enable scientists to learn more about the planet's intense atmospheric composition, pressure, and structure. The prevailing belief is that Jupiter is composed of the same gases and dust as was the nebula from which the planets formed 4.6 billion years ago.

The fifth spacecraft sent toward Jupiter was launched by a NASA space shuttle in October 1990. This probe, called *Ulysses,* was developed by the European Space Agency (ESA). Its planned path takes it close to Jupiter in 1992 so that it can use the planet's strong gravitational field in an attempt to attain a path that would send it into a polar orbit about its primary objective, the sun, in 1994. This procedure is necessary, since no available rocket is powerful enough to accelerate a spacecraft out of a plane near the sun's equator. Though there is no camera aboard, the craft will attempt to determine how the sun's poles differ from the lower latitudes investigated earlier by German, Russian, and American probes. Instruments aboard the space vehicle also will study such characteristics as solar flares, the sun's magnetic field, and the speed and composition of solar wind, as technical difficulties permit.

By the end of the century, the United States intends to complete a manned space station and to establish a permanent lunar station by the year 2010. There are definite plans to land astronauts on the planet Mars by 2019. Such projects hold out the promise that the United States will maintain leadership in space exploration ■

Although various procedures have been used to determine the diameter of Pluto, the results are still not definitive. For instance, data collected during a near occultation (when Pluto passed almost in front of a star) indicated that the planet is about 6800 kilometers in diameter. On the other hand, the analysis of data collected and corrected for turbulence in Earth's atmosphere by scientists at the Hale Observatory yielded a diameter that ranged from 3000 to 3600 kilometers. More recent indications are that the diameter of Pluto is about 2300 kilometers, making the planet considerably smaller than Earth's moon. These last data were obtained as scientists observed Pluto's lone satellite, Charon, as it eclipsed the planet that is only about twice its diameter. (The alignment of Earth, Pluto, and Charon is such that the eclipses were studied for a five-year period that ended in 1990. Charon orbits the planet Pluto once each 6.4 days, the approximate length of time as Pluto's period of rotation.)

The companion of Pluto was discovered in 1978 and named after the ferryman who carried the souls of the deceased across the River Styx to Hades, the kingdom of Pluto. Both celestial objects are shown in Figure 17.32. Together they have assumed a highly eccentric orbit about the sun, varying from 29.7 AU to 49.3 AU. The orbit of Pluto is also unusual in that it is out of alignment with the orbits of the other planets ("inclined to the ecliptic") by about 17 degrees. The eccentricity of the planet's orbit is such that the planet comes closer to the sun at times during each revolution than does the planet Neptune. Pluto actually sweeps inside the

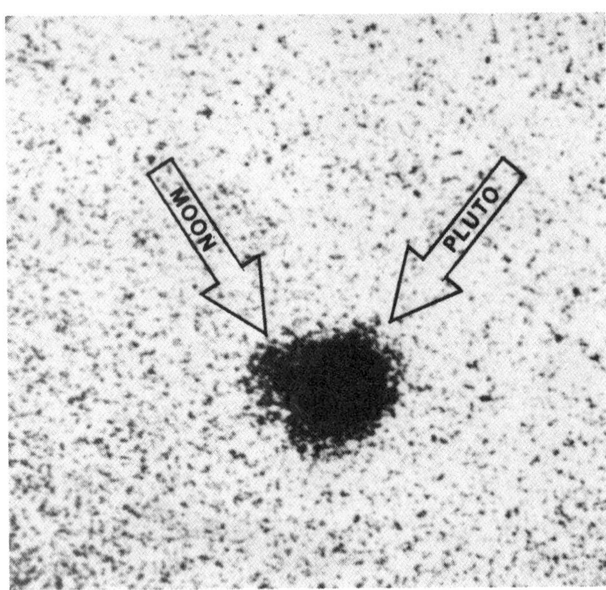

Figure 17.32 Pluto's lone satellite, Charon, was discovered in 1978, less than 50 years after the detection of the planet itself. As it orbits within 20,000 km of the tiny planet, Charon appears as a bump in the image of the two objects.

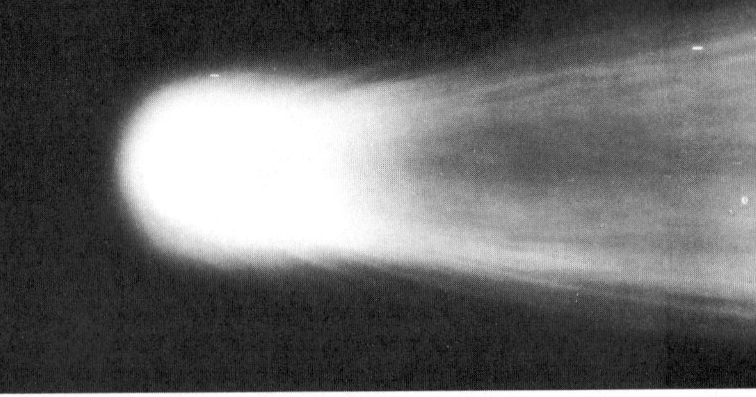

Figure 17.33 The head (or nucleus), along with a portion of the tail of Halley's comet, is shown as it appeared on May 8, 1910. Unfortunately, the 1986 encounter with the famous comet was such that a comparable view was only possible from the Earth's Southern Hemisphere.

orbit of Neptune and remains so positioned for 20 years. In fact, Pluto is presently closer to the sun than Neptune, and it reached perihelion in 1989. The planet's orbital speed has been determined to average about 4.7 kilometers per second as it completes one revolution about the sun each 248 years.

Infrared spectral analysis shows that frozen methane coats a portion of Pluto's surface, indicating that the surface temperature is no higher than about −233°C. At this cold temperature, only such elements as helium, hydrogen, or neon would exist as gases and none have been detected; thus, it is generally accepted that Pluto has no appreciable atmosphere. Further spectral analyses, from data obtained when both objects are visible, as well as when Pluto blocks the scientists' view of Charon, have failed to indicate that methane is present on Charon. It also has been noted that the albedo of Pluto is significantly higher than that of its satellite. (However, various astronomers still maintain that Pluto and Charon share a common thin atmosphere of methane; thus, they doubt the accuracy of the estimated temperature for the two celestial bodies.) Charon also may have water ice on its surface, according to the preliminary results of several ongoing studies that were initiated in 1986.

Scientists now contend that Pluto is not large enough to have any appreciable effect on the orbit of Neptune; thus, perturbation should not have played a role in Pluto's discovery. Clyde Tombaugh's discovery of Pluto about 60 years ago was but a mere coincidence, acccording to the opinions of numerous experts. The search for a tenth planet, closer to the sun than Pluto, that would explain the small perturbations in Neptune's orbit is still under way.

Comets

From ancient times to the Middle Ages, the appearance of a **comet** was considered an apparition that portended grave disasters for individuals and for nations. Comets were thought to predict the defeat of armies and the coming of pestilence and plague. For example, a comet appeared in 1453 when Constantinople fell to the Turks. People of that time were convinced that the comet had been sent to mark the end of the 1000-year-old Byzantine empire. Today we realize that comets are natural members of our solar system. As recently as 1910, however, considerable fear was generated when Earth passed through the tail of Halley's comet. Scientists had pointed out that, among other things, comets contain methane, ammonia, and hydrogen cyanide. Knowing that the gases present in comets are deadly in high enough concentrations, many people prophesied that life on our world would end when we passed through the lethal gases present in the tail of Halley's comet. (Actually, no significant effects have been detected as a result of Earth's last two encounters with a comet's tail.) The head, or nucleus, along with a portion of the tail of Halley's comet appears in Figure 17.33. The latest view of this comet, not as detailed as the view obtained in 1910, is presented in Figure 17.34.

A large comet is a beautiful sight to behold. The comet appears as an elongated filmy patch of soft light that may cover a larger portion of the night sky than any other astronomical object. The tenuous nature of the comet is obvious from the fact that background stars shine through the tail with very little diminution of their brightness. The name *comet* is very well chosen from the Greek term meaning "long hair." Thus, "comet" implies an apparition of soft and glowing heavenly hair. (One of the less famous comets is shown in Fig. 17.35.)

Comets are most spectacular when they are at or near perihelion and consist of a head and a tail. The head, which can be from 15,000 kilometers to over 2.5 million kilometers in diameter, consists of a cloud of gas surrounding a collection of loose particles. After closer examination made possible by various space probes, the heads of comets were found to contain highly reflective specks of matter that form a nucleus. This central region is surrounded

Figure 17.34 Halley's comet on the morning of March 8, 1986. The comet had reached perihelion 27 days earlier and reappeared in February, displaying both a blue tail (glowing carbon monoxide gas) and a pale yellow tail (sunlight reflected from the dust particles trailing behind the comet).

Figure 17.35 Comet Arend-Roland, April 30, 1957.

by a globular, cloudlike mass known as a *coma.* The tails of comets have been estimated to reach over 150 million kilometers in length.

When a comet is far from the sun, it is mainly a collection of frozen water, ammonia, and methane in which rocklike particles are embedded. The diameters of comets under these conditions probably range from a few kilometers to a few thousand kilometers. As a comet approaches the sun, it becomes warmer, and a portion of the frozen pristine material melts, vaporizes, and escapes through vents in the crust of the nucleus to produce the coma and tail of the comet. The tail grows in length, getting longer the closer the comet is to perihelion, and always pointing away from the sun, as illustrated in Figure 17.36. The growth of the tail, as well as its direction, is the direct result of the solar wind and radiation pressure of the sun. As the comet recedes from the sun, the tail decreases in length until it finally disappears, and the head gradually shrinks. The gaseous material of the head slowly condenses and solidifies until once again the comet is a frozen mass of material that is no longer visible even through telescopes. Comets are presently considered to be debris remaining from the formation of the solar system.

Whereas the planets all revolve around the sun in approximately the same plane, comets revolve around the sun at all possible angles. Thus, the orbits of some comets are in the same plane as the orbits of the planets, others are perpendicular to the orbits of the planets, and still others are between these extremes. The orbits of comets are highly elliptical (also shown in Fig. 17.36), and there is a great variety in the aphelion distances of various comets. Some comets have an aphelion distance from the sun fairly near the aphelion distance of the planet Jupiter, and are actually gravitationally influenced by Jupiter. These comets are called Jupiter's family of comets. The most famous comet, and a member of Jupiter's family of comets, is Halley's comet, whose every passage around the sun since 240 B.C. has been faithfully recorded. Halley's comet returns every 76 years and was last observed in 1986, as it passed within 63,000,000 kilometers of our planet. In March 1986 two Soviet spacecraft, known as *Vega 1* and *Vega 2,* were launched toward the comet Halley. They both passed within less than 9000 kilometers of the head of the comet. Also, the European-sponsored vehicle *Giotto* was programmed to approach within 600 kilometers of the comet's head. All three vehicles yielded valuable information about the regular visitor from the outer

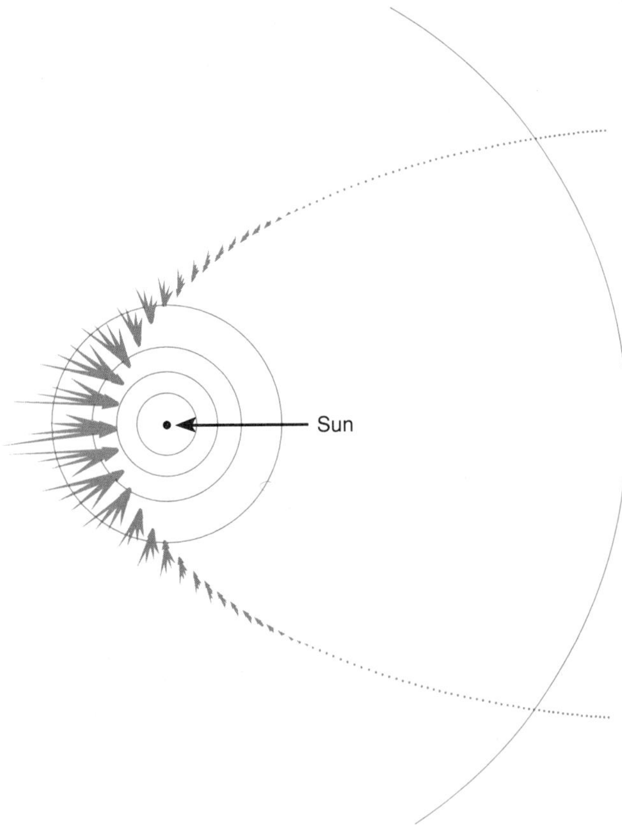

Figure 17.36 An illustration of the position of Halley's comet at 10-day intervals. The dot represents the sun and the four concentric circles denote the orbits of the inner planets. A portion of Jupiter's orbit appears to scale. Note the distance between positions and the length of the tail as the comet approaches the sun, then continues its journey toward the outer part of the solar system.

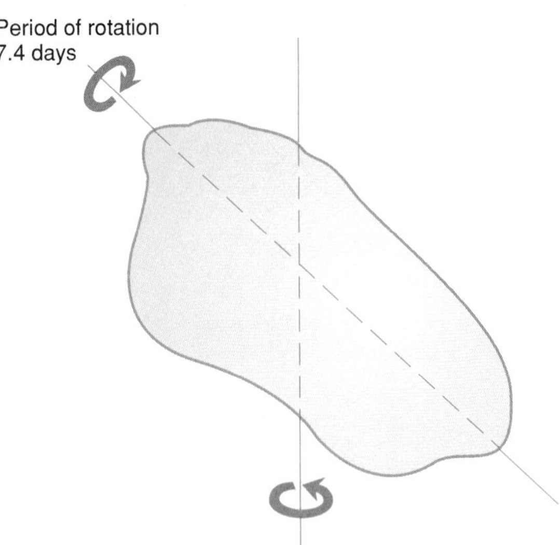

Figure 17.37 An artist's rendition of the probable shape of Halley's nucleus. The irregular shape somewhat resembles Phobos, a satellite of Mars. Its complex system of rotation is not totally resolved, but its period along the major axis seems to be 7.4 days.

reaches of our solar system. The images received from Soviet space probes indicated that the nucleus resembles a giant avocado, about 16 kilometers long and 8 kilometers wide. An interpretation of the shape and the complicated pattern of rotation of the nucleus of Halley's comet is illustrated in Figure 17.37. The instruments aboard the Soviet spacecraft found that the surface of the nucleus is very dark (it reflects only about 4 percent of the light that strikes it). The collective data analysis from the three vehicles indicated that the surface of the nucleus contains particles similar to finely ground charcoal. *Giotto's* images indicated that the nucleus might be even more irregular in shape. All three missions yielded information that discounted the once popular "dirty snowball" theory and verified the idea that the nucleus emits jets of gas and dust. In effect, they determined that the nucleus and coma of such a comet are more likely to be composed of various kinds of residue that do not readily evaporate. The data received from the *Giotto* flight also indicated that Halley's nucleus is covered with a nonvolatile insulated crust composed of some unusually dark material.

Some astronomers who conduct detailed studies of our solar system feel that a great number of comets reach 100,000 AU at aphelion as they orbit the sun. According to these astronomers, there may exist a vast spherical mass of cometary nuclei composed of icy debris. These cometary nuclei, numbering perhaps 7 trillion, are known as the *Oort cloud,* named after the Dutch astronomer Jan Oort, who initially proposed its existence in the late 1950s. The total mass of the cloud is currently estimated to be about fifty times the mass of Earth—not nearly as large as previously imagined. Comet Wilson, discovered in August 1986, may represent the first known celestial object detected from the cloud of comets. Such comets may vary in aphelion from about 6 billion kilometers to several trillion kilometers. Although the Oort cloud of comets has never been observed, various astronomers suspect that it is occasionally disrupted by passing stars, an effect that sends comets hurtling into the inner portions of the solar system. Recent theories suggest the presence of an inner Oort cloud within the larger sphere of comets, 10,000 to 20,000 AU from the sun, that slowly feeds other members into the more distant and major Oort cloud. Still others propose the existence of a concentration of comets that orbit the sun at a distance of about 40 AU, known as the *Kuiper belt.* Two of NASA's spacecraft, *Pioneer 10* and *11,* launched in the early 1970s, are scheduled to examine this region in 1992 to search for the proposed band of comets. The interest in comets continues, since many authorities think that these icy objects primarily represent the other solar system members as they originally formed.

Meteors

Countless numbers of particles called *meteoroids* plummet through our surrounding space. As these particles enter Earth's atmosphere, they are heated by friction to a point of incandescence. The "shooting stars" that result are known as **meteors,** and they produce bright streaks of light that ordinarily are visible for but a brief instant at the rate of about 10 per hour. Many of the meteors that we see are no larger than the head of a pin, yet most such objects that invade our atmosphere are too small to give any sign of their presence. It has been estimated that Earth gains about 3.6×10^8 kilograms annually from meteoroids that strike it. Assuming this value has remained constant since Earth's formation

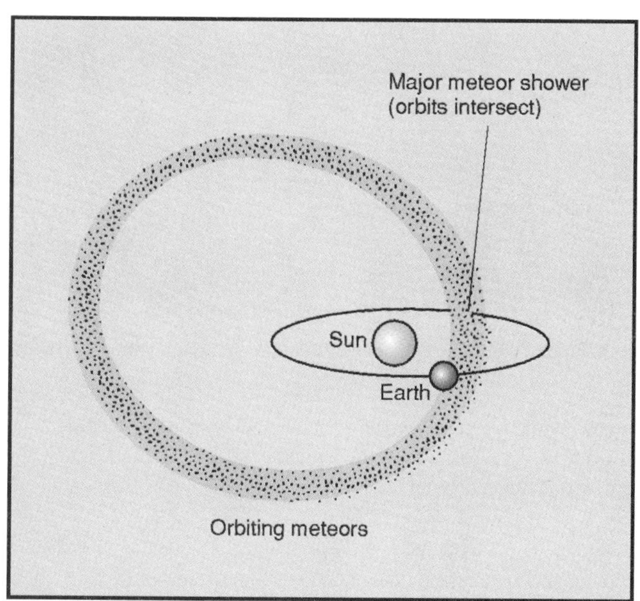

Figure 17.38 Annual meteor showers occur when debris from dead comets is dispersed throughout a given orbit. Generally, the orbit of the remaining particles is tilted so that Earth's orbit intersects at only one point. The orbits of other streams of particles, such as those that create the Perseid shower each August, are closer to alignment to Earth's orbit; thus, there may be a minimal meteor shower at another time of year.

Figure 17.39 A meteorite found in Antarctica by a research team composed of scientists from the National Science Foundation, the Smithsonian Institute, and NASA. The specimen had a mass of 32 grams and is basically black and gray with white and gray fragments imbedded throughout. Its characteristics are very similar to the rocks collected from the lunar highlands by *Apollo* astronauts.

4.5 billion years ago, the mass of Earth has increased some 1.6×10^{18} kilograms as a result of meteoric bombardment. The total added mass, however, is negligible compared to that of Earth.

A few times each year an unusually large number of meteors are seen. These exceptional meteoric displays, called *meteor showers,* are sometimes the result of Earth's orbit intersecting the orbit of a "dead" comet. Each time a comet makes a close approach to the sun, some of its gaseous material is swept away because of the action of radiation pressure and solar wind. Perhaps after hundreds of revolutions about the sun, comets lose their compactness and separate into large fragments or possibly disintegrate, leaving behind only small particles of rocks and metals. (The nucleus of Comet West was photographed in 1976 as it broke apart into four pieces after its orbit took it near the sun.) When Earth intercepts these small particles, a stream of beautiful and spectacular streaks of visible light results (note Fig. 17.38). Such an event occurs annually on or about August 11, and is known as the *Perseid shower,* so named for the constellation Perseus, its apparent source.

If a meteoroid strikes Earth before totally vaporizing, it is called a **meteorite.** Many hundreds of such meteorites have been found, but most meteoroids are totally destroyed by friction during their passage through the air and never reach Earth's surface. Of those that do survive, more than two-thirds plummet into the great oceans of our planet and are never found. Most of the meteorites that are discovered are relatively small; however, a 45,000-kilogram meteorite was found in southwest Africa, and a 32,000-kilogram meteorite found in Greenland is on display at the American Museum of Natural History. There is only one reliable report of someone having been struck by a meteorite. A meteorite crashed through the roof and into the living room of an Alabama home and injured a woman. Another meteorite recently destroyed a mailbox in Georgia, as witnesses can attest.

Continuing studies of meteorites indicate that a few have come from the moon, ejected into space from a lunar crater that was perhaps formed by the collision with other celestial objects. The Lunar and Planetary Science Conference of 1983 studied a plum-sized meteorite discovered in Antarctica, and practically every scientist attending agreed that it had a lunar origin. The search that was initiated in 1970 by teams of scientists in Antarctica has yielded thousands of meteorites, including five other smaller ones that strongly resemble the moon's composition (see Fig. 17.39). Analysis of the largest of the suspected lunar fragments indicated that the meteorite was blasted away from Earth's satellite about 100,000 years ago as the result of an asteroid impact. Another meteorite, according to the analyses of the nitrogen isotopes and the relatively rare gases it contained, could well be a fragment of the planet Mars. The specimen has been studied by numerous scientists since its discovery in 1979.

Meteorites are normally too small to have any visible effect on impact with the ground, but there have been some spectacular exceptions. In 1908 a meteorite (perhaps a large fragment of a comet's nucleus) estimated to have had a mass of 90,000,000 kilograms collided with Earth, causing a spectacular explosion above the Tunguska region of Siberia. The explosion, essentially equivalent to that resulting from the detonation of a powerful nuclear missile, destroyed hundreds of square kilometers of forestland. The

a.

b.

Figure 17.40 (*a*) An aerial view of the Barringer Meteorite Crater near Winslow, Arizona. The meteorite struck Earth about 50,000 years ago and created a crater that now averages 1186 m across and 167 m deep with a rim crest about 50 m above the surrounding plain. (*b*) A view of the same crater from within.

explosion was heard over 1000 kilometers away, and the shock wave was recorded on seismographs at distant stations throughout Europe.

One of Earth's youngest and the largest confirmed meteoric impact crater discovered in the United States is the Barringer Meteorite Crater, located near Winslow, Arizona (see Fig. 17.40). The meteorite apparently exploded on impact, and small pieces primarily composed of the element iron have been found over a radius of 7 kilometers about the crater. Over 23,000 kilograms (about 25 tons) of iron meteorite fragments have been found. Drillings beneath the crater floor have not located the main portion of the meteorite, if one exists. The crater is estimated to have been formed about 22,000 years ago.

An even larger circular depression was discovered in 1950 in Quebec, Canada. This suspected impact crater is over 3 kilometers in diameter; thus, it is twice the size of the Barringer Crater. No traces of meteorite fragments have been found in the area, however. There are over a dozen other craters in Earth's crust that are considered of meteoric origin. Undoubtedly, many craters formed thousands of years ago have been leveled by the agents of erosion and have escaped detection.

For instance, analysis of samples from deep oil wells drilled in 1988 near Sioux City, Iowa, indicate a crater with a diameter of over 30 kilometers was created by an object over 3 kilometers in diameter that collided with Earth about 66 million years ago. After further investigations are completed, this impact crater may prove to be the largest detected in the United States.

The first crater in the ocean floor that is attributed to a meteorite was discovered in 1987 off the coast of Nova Scotia. The huge depression lies at an ocean depth of over 110 meters, but it undoubtedly will be thoroughly investigated when scientists decide on the best method to undertake the study.

Summary

Our solar system consists of a central star—the sun—with its attendant nine known planets and their approximately 67 satellites (or moons), thousands upon thousands of asteroids (planetoids), innumerable comets, and myriads of meteors of various origins.

Our sun is an average star (actually, it is considered a dwarf when compared to certain other stars) with a diameter of 1,391,000 kilometers, a "surface" temperature of 6000°C, and a core temperature estimated to be about 15,000,000°C. Sunspots, frequent disturbances on the outer portion of the sun, are regions that become temporarily cooler in temperature than the balance of the sun's photosphere. A steady stream of charged subatomic particles is emitted in all directions by the sun and produces the solar wind that permeates much of our solar system. The rapidly moving charged particles are trapped by the magnetic fields of Earth to form the doughnut-shaped Van Allen radiation belts.

The planets, in order of their distances from the sun, are Mercury, Venus, Earth, Mars, the asteroids, Jupiter, Saturn, Uranus, Neptune, and Pluto. (At the present time, Pluto's highly eccentric orbit takes it inside the orbit of Neptune.) Pluto and Mercury are the smallest planets and Jupiter is the largest. Only Mercury and Venus have no satellites—at least none have been detected at the present time. Jupiter has 17 satellites that have been located, and Saturn has at least 18. Venus and Uranus rotate in the opposite direction of the other planets; hence, they are said to undergo retrograde rotation. Saturn has a very elaborate and readily observable ring system. Jupiter and Uranus also have rings that have been detected by the NASA space probes. The existence of a ring of debris about Neptune has also been recently confirmed.

Between the orbits of Mars and Jupiter are a staggering number of orbiting rocky bodies known as asteroids or planetoids (minor planets). These objects, because of the tremendous gravitational influence of Jupiter, supposedly were never able to accrete into larger objects, such as a planet or even moons. They represent some of the debris left over from the formation of the balance of the solar system. The asteroids vary in diameter from 800 kilometers to that of small pebbles. The combined mass of the asteroids, however, could only form an object considerably smaller than Earth's moon.

Comets are relatively small masses that revolve about the sun in highly elliptical orbits oriented in all possible directions to the plane of the orbits of the planets. When a comet approaches perihelion, it forms a tail that always points away from the sun. The tail may attain a length of millions of kilometers.

Meteors, commonly called "shooting stars," are meteoroids that continually enter Earth's atmosphere. If they survive the frictional effects of the air that surrounds Earth and strike our planet, they are known as meteorites.

Equation Summary

Period of a planet:

$$T^2 = s^3.$$

The square of a planet's period of revolution T expressed in Earth years is directly related to the cube of its distance s from the sun, measured in astronomical units (AU).

Questions and Problems

In the Beginning

1. According to the most commonly accepted scientific theory, how did the solar system originate?
2. Why do the outer planets, with the possible exception of Pluto, have densities considerably lower than the inner ones?

The Sun

3. Why do sunspots, although they are considered to be hotter than the flame of a welding torch, appear as dark regions on the sun?
4. What phenomenon apparently creates the auroras? During what two seasons are the northern displays most noticeable?
5. What is solar wind?

The Planets

6. Cite three major differences between the terrestrial and the Jovian planets.
7. How does the speed of a planet vary as it orbits the sun?
8. An object, previously undetected, is noted to orbit the sun at a distance of 4 AU. According to Kepler's third law of planetary motion, what would be its period of revolution?
9. What is meant by a planet's aphelion? What is meant by its perihelion?
10. According to Kepler's first law, the eccentric orbit of a planet about the sun is essentially a natural phenomenon. Therefore, what shape would a moon's orbit about a planet be expected to resemble?

Mercury

11. How does Mercury's period of revolution compare to its period of rotation?
12. What is the composition of Mercury's thin atmosphere?

Venus

13. Although Mercury is the planet closest to the sun, why does the planet Venus, rather than Mercury, have the highest surface temperature?
14. Why is Venus often considered Earth's twin?
15. Why is the atmospheric pressure on Venus so high?

Earth

16. How does the diameter of Earth compare to the largest planet? To the smallest planet? (See Table 17.1.)

Mars

17. Why is Mars known as the "Red Planet"?
18. Why did observers in the late 1800s think that there was intelligent life on Mars?

The Asteroids

19. What are the asteroids? Where are they primarily located?
20. How do astronomers think the asteroids originated?

Jupiter

21. What conditions apparently created Jupiter's Great Red Spot? What characteristics help to maintain it?
22. How does the planet Jupiter compare to Earth in terms of diameter, mass, period of rotation, and atmosphere?

Saturn's Rings and the Roche Limit

23. How might the Roche limit be involved in explaining the beautiful ring structure about the planet Saturn?
24. According to astronomers, why are some of the moons in our solar system spherical whereas others are jagged or otherwise irregularly shaped?

Uranus

25. How does the axis of rotation of Uranus compare to that of the other planets?
26. Why does Uranus have a pale blue-green appearance?

Neptune

27. What is a perturbation, as it applies to a planet's predicted path?
28. Briefly describe the Great Dark Spot discovered by the *Voyager* spacecraft on the planet Neptune.

Pluto

29. How was the planet Pluto discovered?
30. How does the diameter of Pluto compare to Earth's moon?

Comets

31. Why do the tails of comets invariably point away from the sun?
32. According to past observations, during what year will Halley's comet be expected to return to our part of the solar system?

Meteors

33. How are various meteors related to comets? What proof of relationship is accepted by most scientists?
34. Why do major meteor showers occur only at certain times of the year?

18 The Stars

Chapter outline

Key terms/concepts

The magnificent constellation Orion can be seen from almost anywhere in the world on a dark clear evening between November and April. Two bright stars near the top mark the hunter's shoulders, the left one being the red giant Betelgeuse. Rigel, a hot blue star on the lower right, is one of two marking his legs. Below the three stars, in a row near the center marking the belt of Orion, hangs the sword. The center object in the sword is a nebula containing many young hot stars.

The stars have fascinated people through the ages, but early observers must have felt that they would never be able to discover the secrets of these mysterious points of light in the night sky. However, once it was realized that our own sun was, in fact, a star, and that the billions of other stars in the universe must be much like our sun, new areas for scientific investigation were opened.

Our Closest Star: The Sun

The sun is the only star close enough to us for its surface features to be directly observed (Fig. 18.1). Telescopic photographs of the sun reveal the turmoil that exists in the outer layers. Gigantic explosions and upheavals are constantly occurring on the "surface" of the sun to produce the magnificent flaming loops and arches of star material that sometimes are hurled a half million kilometers into space. These violent events, leading to *sunspots* and *solar flares,* have been discussed in Chapter 17. Yet in spite of its enormity and importance to Earth, the sun is a fairly ordinary star.

With a diameter of 1.39 million kilometers (864,000 mi), this sphere, consisting mostly of hydrogen, has a volume greater than a million Earths, and a mass over 300,000 times greater than Earth's. A study of its electromagnetic spectrum shows that the sun radiates at all wavelengths—from the radio region of the spectrum through infrared, visible, ultraviolet, and into the X-ray region. The intensity of the spectrum at each wavelength permits astronomers to determine the composition of the sun's surface, its surface temperature, and even the "states" of the atoms on its surface and in its atmosphere. The sun also emits a steady stream of particles, mostly protons and electrons.

Study of the sun, then, is the first step toward understanding the stars. Studying the radiation emitted by these points of light, along with an understanding of physical laws, allows astronomers to deduce not only the mechanisms by which stars shine, but also how they are formed, how they will spend their lives, and how they will die. It should be noted that, since it is impossible to bring a star into the laboratory for study, there is debate concerning the validity of some of the theories about stars; that is, there are alternate theories that explain the same phenomena, and there are puzzling discrepancies yet to be resolved. As new and better observations are made and computing methods are advanced, theories will be refined or discarded, and scientists will need to make objective judgments, as free of personal bias as possible.

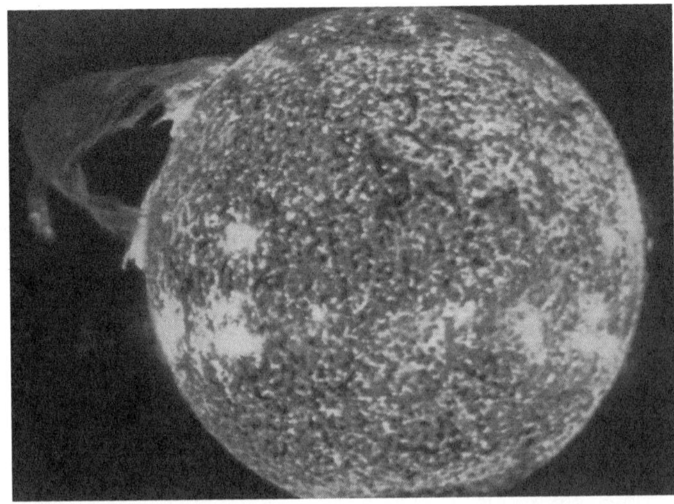

a.

b.

Figure 18.1 (*a*) The sun is the only star close enough to allow details on its surface and in its atmosphere to be seen with telescopes. (*b*) All stars other than the sun are so distant that even in the largest telescopes they appear only as points of light. In this telescopic view, the different apparent sizes of stars are caused by the photographic process, not by actual differences in the sizes of the stars.

What Makes the Stars Shine?

Observations of the sun's outer layers show that hydrogen is the dominant element, making up almost 79 percent of the mass in these layers, with helium making up a little less than 20 percent. Traces of the other elements complete the outer layers. The composition of the interior of the sun, as well as its interior temperature, must be inferred from models, primarily using observations of the outer layers and a knowledge of nuclear physics and hydrodynamics (the study of fluid motion). Current models indicate that the sun's mass is about 73 percent hydrogen and 24.5 percent helium with a temperature at the center of approximately 15 million kelvins.

At the temperature of about 15 million kelvins, the electrons are stripped from the hydrogen atoms, leaving bare their nuclei, which normally consist of a single proton. The protons collide with such high speeds that they can occasionally overcome the force of repulsion caused by their positive charge, approaching closely enough to fuse together. Thus, *the primary source of energy within a star similar to the sun* is **nuclear fusion.** The main fusion reaction at this temperature is the **proton-proton chain,** sometimes

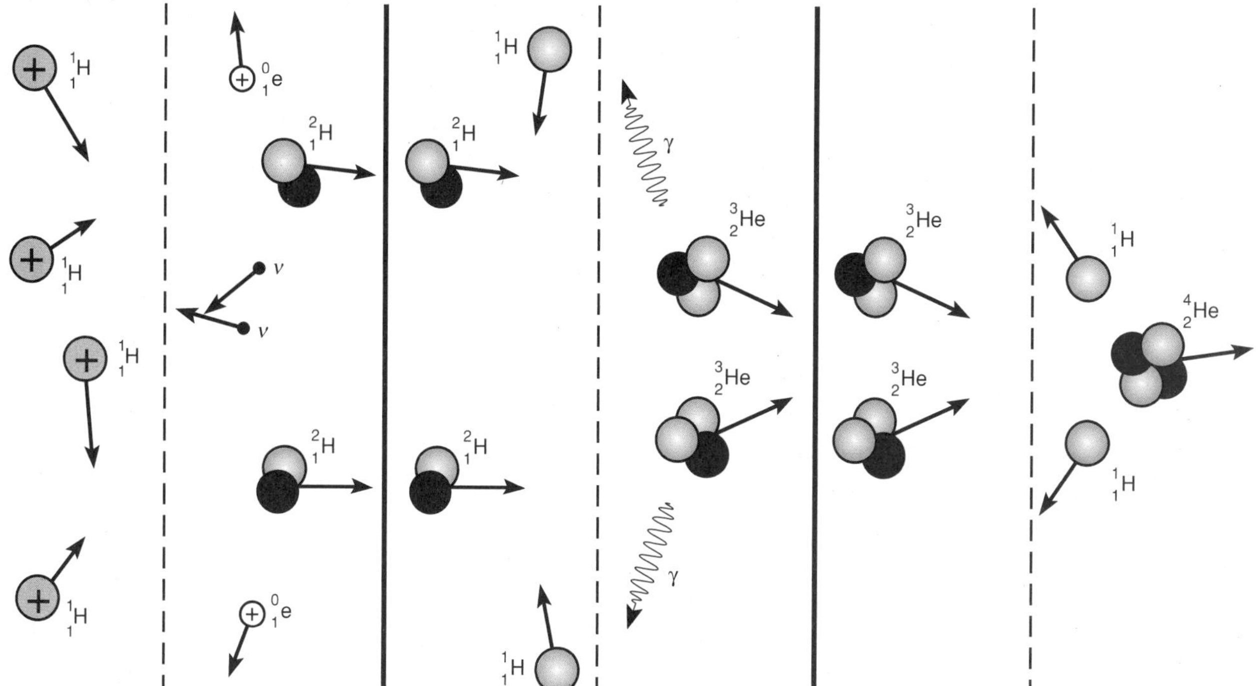

Figure 18.2 The proton-proton chain is the primary energy source for stars when their interior temperature is between about 10 and 15 million kelvins. In (*a*), four protons fuse to produce "heavy" hydrogen nuclei (deuterium); neutrinos (ν), which escape from the star; and positrons (^0_1e), which interact with electrons to form gamma rays. The reaction proceeds with two more hydrogen atoms combining with the "heavy" hydrogen to produce a light form of helium and releasing gamma rays (*b*). The reaction is completed (*c*) when the two light helium nuclei fuse to form one ordinary helium nucleus and two ordinary hydrogen nuclei. Energy is released in every step of the reaction, which has the net result of combining four protons (hydrogen nuclei) to form one helium nucleus.

referred to as "hydrogen burning," although it is certainly not burning in the traditional sense. The net result (leaving out two intermediate reactions) of the proton-proton chain is:

$$4\,^1_1H \longrightarrow \,^4_2He + 2\,^{\;0}_{+1}e + 2\nu + 2\gamma + \text{energy.}$$

In other words, *four hydrogen atoms fuse to form one helium atom, two positrons, two neutrinos, two gamma rays, and energy is liberated.* The positrons almost immediately combine with electrons (annihilating each other), releasing two more gamma rays, which are just high-energy photons. The mass of the four hydrogen atoms is greater than the mass of the helium atom, thus mass is converted into energy. The amount of energy liberated from the conversion of mass is equal to mc^2, according to Einstein's theory of relativity (see Chapter 15). The gamma rays also contribute to the energy, but the neutrinos escape from the star. The energy is mainly in the form of kinetic energy of the particles, and their random motions maintain the star's interior temperature. The complete set of reactions involved in the proton-proton chain is illustrated in Figure 18.2.

Stars more massive than the sun must maintain a higher core temperature to prevent collapse. The increased weight of the overlying layers of the star must be balanced by an increased pressure, which is generated by a greater density of particles and a higher temperature. If the temperature of the core rises above 15 million kelvins, the **carbon cycle** becomes a major source of energy for the star. The net result of the carbon cycle is the same as that of the proton-proton chain; that is, *four hydrogen nuclei are converted to one helium nucleus, some other particles, and energy is released.* The carbon cycle has a different set of intermediate steps, with the carbon nucleus acting as a *catalyst,* a substance that is involved in the reaction, but in the end is unchanged.

During the "old age" of a star, a considerable percentage of the star's hydrogen will have been converted into helium, and the core temperature of the star may reach a temperature of 100 million kelvins. At this temperature, a nuclear fusion reaction known as the **triple-alpha process** becomes an important source of energy for the star. This process amounts to *the conversion of three helium nuclei into one carbon nucleus:*

$$3\,^4_2He \longrightarrow \,^{12}_{\;6}C + \text{energy.}$$

The triple-alpha process is so named because a helium nucleus is also known as an alpha particle and three of them are involved in the reaction.

While the triple-alpha process is occurring, the carbon cycle and the proton-proton chain are also still producing energy for the star. Other nuclear reactions are also carried on at very high temperatures. Theories of the beginning of the universe postulate that hydrogen was the only element present in the beginning (it is still by far the dominant element). It is believed that all the heavier elements were "created" by nuclear reactions that took place in

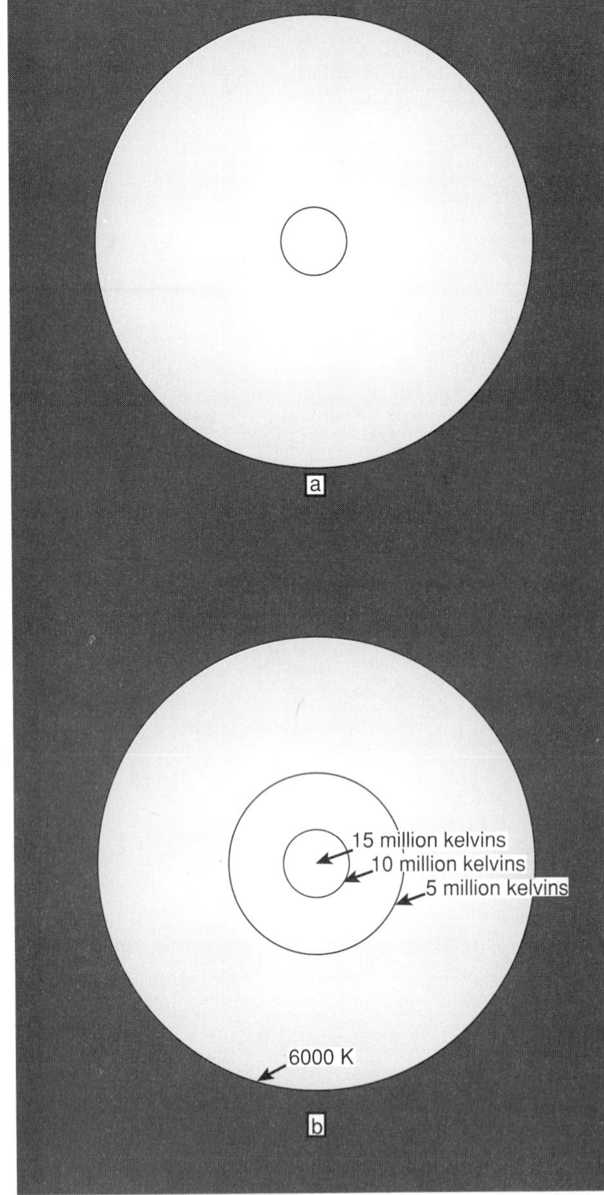

Figure 18.3 (*a*) About 80 percent of the energy generated in a star is produced within the inner 15 percent of its radius, a small volume equal to about 1/300 of the total volume of the star. (*b*) The energy flows outward from the hotter core to the relatively cool surface, with the temperature dropping rapidly as the distance from the center increases.

stellar interiors. As discussed in the "Life Cycles of the Stars" section, this material is ejected into space by cataclysmic events that take place in the life cycle of stars.

Most of the tremendous energy generation by fusion occurs near the core of the star (Fig. 18.3a). As this energy is transported from the interior of the star to its surface, the temperature drops from the 15-million-kelvin core temperature to the surface temperature of the star, which can range from about 3000 kelvins to over 100,000 kelvins (Fig. 18.3b). Since the star is "isolated" in the vacuum of space, most of the energy is radiated from the star in the form of electromagnetic radiation, with wavelengths covering nearly the entire spectrum.

The **luminosity** of a star is *a measure of the star's total electromagnetic energy output*. The amount of radiated energy mainly

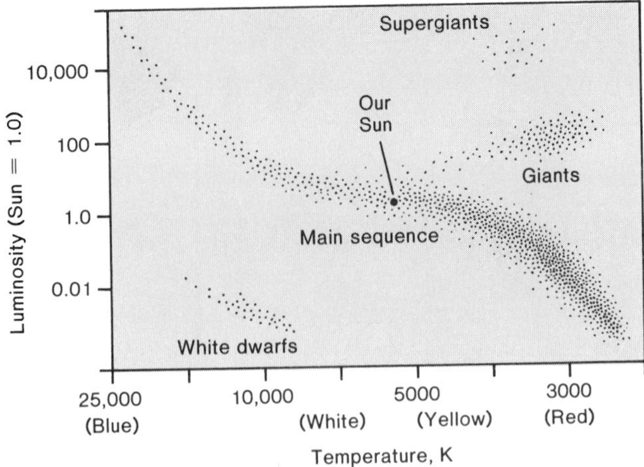

Figure 18.4 On a Hertzsprung-Russell diagram, stars progress from bright at the top to dim at the bottom and from hot at the left to cool at the right. The color of a star depends on its temperature (blue stars being the hottest and red stars the coolest), whereas the luminosity (brightness) depends on both the temperature and the size of the star. Most stars fall near the band labeled "main sequence," but large relatively cool stars (the red giants and supergiants) appear in the upper right region and small hot stars (the white dwarfs) appear in the lower left region.

depends on two factors: the star's surface temperature and surface area. For two stars with the same surface temperature, the larger star will, of course, be more luminous. Similarly, for two stars of the same size, the "hotter" star will be more luminous. Hotter stars also radiate more of their energy at short wavelengths, making them appear bluer, whereas cooler stars radiate more of their energy at long wavelengths, giving them a redder appearance (see Fig. 10.27).

The surface temperature and surface area of normal stars are related to each other. Larger stars must have higher core temperatures to prevent collapse; therefore, they must generate more energy. However, they also have larger surface areas from which to radiate the energy. Establishment of equilibrium requires that the energy generated and the energy radiated by a star be the same. A star with a mass 10 times that of the sun generates about 1000 times as much energy, but the more massive star's surface area is not 1000 times as great; therefore, it must have a higher surface temperature to rid itself of the greater energy.

Types of Stars

The concepts of the preceding two paragraphs can be graphically depicted by showing *the luminosity of stars versus their temperatures* (Fig. 18.4). Two astronomers, Ejnar Hertzsprung of Denmark and Henry Norris Russell of the United States, independently plotted such diagrams about 1911 and 1913, respectively, and these diagrams are now called **Hertzsprung-Russell** (or just H-R) **diagrams.** Each point on the diagram represents a star. The luminosities, plotted in the vertical direction, are relative to the sun; that is, the luminosity of the sun is taken to be 1.0. A point

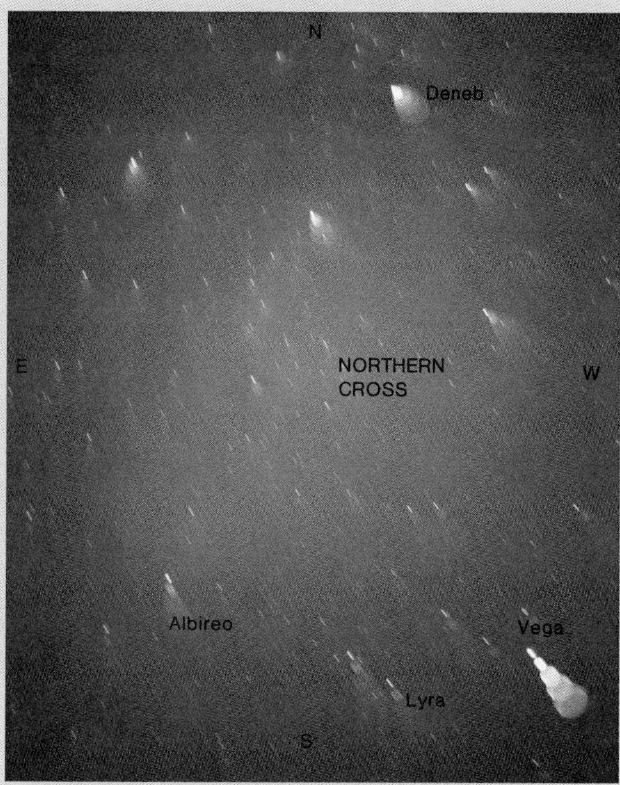

a.

N
Deneb
NORTHERN
CROSS
E
W
Vega
Albireo
LYRA
S

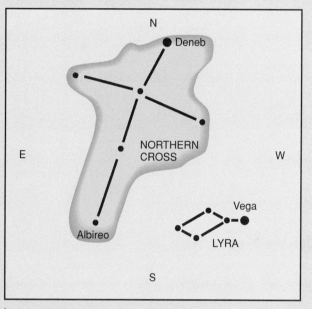

b.

Figure 18.5 (*a*) The colors of the stars are apparent in this exposure in which the stars are allowed to trail and the camera is deliberately defocused. The colors of the brighter stars appear when the image is largest and those of the fainter stars when the image is smallest. The region shown includes the Northern Cross, part of the constellation Cygnus (the Swan), and Lyra (the Harp), as well as the bright stars Vega and Deneb and the dimmer star Albireo (*b*). The upright of the Northern Cross lies almost exactly along the plane of the Milky Way, which causes the haze across the center of the picture.

Kiss a Star

Astronomers have developed a system by which to arrange the various stellar spectra into an orderly lettered sequence for quick reference. After numerous revisions, the spectral classification scheme presently employed places the stars into categories of O, B, A, F, G, K, and M, along with a relatively few stars classified as R, N, and S. The correct order of letters may be recalled by the use of the following mnemonic exclamation; *O*h, *B*e *A* *F*ine *G*irl, *K*iss *M*e *R*ight *N*ow (*S*mack)! O-type stars are hot, blue stellar objects (perhaps 30,000 K or more) with dominant hydrogen and helium spectral lines; G-type stars, like our sun, are yellow (5000 to 6000 K), and have strong metallic-related spectral lines, including the element calcium; and M-type stars are red (2000 to 3400 K), and cool enough to have spectral features attributed to compounds as well as to various elements (see Fig. 18.5)

representing a bright star would be near the top, a dim one near the bottom. Note that the scale is nonlinear, encompassing a tremendous range of luminosities. The temperature is plotted in the horizontal direction, seemingly backward (the original diagrams were plotted using a measure called the color index). Hot stars are to the left and cool stars to the right. The color is dependent on the temperature, and a few colors are listed in parentheses under the temperatures. The "normal" stars lie in a narrow band, called the **main sequence,** running more or less diagonally from the upper left to the lower right. The sun, explicitly pointed out, is seen to be a member of the main sequence. Stars near the upper left corner of the diagram are called **blue giants.** They are *large, hot stars,* pouring out copious amounts of radiation. At the other end of the main sequence are the *small, cool stars,* the **red dwarfs.** (The sun is a white star of average size. Its spectral output peaks in the green region of the spectrum, but the particular mixture of light makes it appear white to our eyes.)

It is apparent from the diagram that not all stars fall within the main sequence. Stars near the upper right corner of the diagram are *extremely bright,* even though they *have cool surfaces.* These stars, which must have extremely large surface areas to emit so much energy, are classed as **red giants** or **red supergiants.** At the opposite extreme, the stars in the lower left corner, despite their *high surface temperatures,* are *very dim.* They must be small stars and are called **white dwarfs.** These unusual star types represent different stages in the development of stars, as discussed in the section on stars' life cycles later in the chapter.

The ranges in the sizes and densities of the stars is striking. The red supergiants may have diameters up to 400 times the diameter of the sun, so that if one were placed in our solar system with its center at the sun, all the planets out to and including Mars would be within the star. These supergiants, however, may have a mass only about 20 times the mass of the sun, making their outer layers the equivalent of a respectable vacuum on Earth. On the other hand, white dwarfs have diameters about equal to that of Earth, but masses about the same as the sun. This material is so dense that a piece the size of a baseball would have a mass of over 2 million kilograms!

Multiple Stars and Star Clusters

In the year 1650 the Italian astronomer Giovanni Baptista Riccoli discovered that the star Mizar in the handle of the Big Dipper, which appears to the naked eye as a single star (Fig. 18.6), was actually two stars. Present information indicates that 55 to 60 percent of all stars are not single stars, but actually multiple-star systems.

Two stars gravitationally bound together are called a **binary system.** In a binary system, the stars revolve about their common center of mass (called the barycenter) with speeds that depend on the masses of the stars and the distance between them. If the distances to the stars are known, it is often possible to compute the individual masses of the stars, allowing for a check on stellar theory.

Binary systems, in which one of the components is too dark to be seen, can often be detected by studying the "wobble" of a star as it is orbited by its unseen companion, or by detecting a periodic shift in the frequency of the star's light (see the "Doppler Effect" section in Chapter 7). In principle, these same techniques allow astronomers to detect planet-sized objects orbiting stars. However, the small masses of planets relative to the masses of stars makes interpretation of the data extremely difficult, and reports of the detection of planet-sized objects orbiting stars other than the sun is still a subject of controversy among astronomers. As of this writing, the observed motion of several nearby stars suggests the presence of planet-sized companions, but more refined data is needed for confirmation.

Occasionally, Earth lies in the plane formed by the two revolving members of a binary system. During each revolution, *one star passes in front of the other star with the consequence that each star at least partially eclipses the other.* Such a system is called an **eclipsing binary system.** This rhythmic eclipsing results in a periodic fluctuation in the light intensity coming from the binary system. When one star eclipses the other, the light is reduced, since the nearer star blocks some or all of the light from the eclipsed star. Figure 18.7 shows how an eclipsing binary appears at maximum and minimum brightness. The most famous eclipsing binary is Algol, in the constellation Perseus. Figure 18.8 shows the orbital motions involved in the Algol system and how the light intensity varies. The variation in light intensity with time allows astronomers to calculate the sizes of the stars relative to their orbits, the brightness of each star, and their relative masses. If the distance to the eclipsing binary is known, the actual mass and luminosity of each star can be computed.

a.

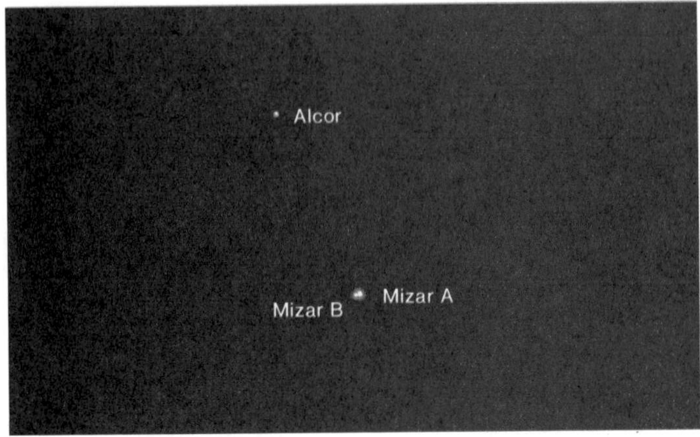

b.

Figure 18.6 (*a*) The brighter of the stars at the bend of the handle of the Big Dipper is Mizar, the dimmer star being Alcor. (*b*) Upon examination with a telescope, Mizar is seen to be a double star. Further observation shows that each component of Mizar is a binary star, with the two components orbiting their common center of mass.

A small percentage of the stars in multiple systems contain three or more stars. Polaris, the North Star, is actually a triple-star system. Two major types of star clusters, gravitationally bound together, also can be located. The **open clusters** usually contain from a few dozen to a few hundred stars (Fig. 18.9a). All the stars in an open cluster lie about the same distance from Earth and are about the same age, giving astronomers the opportunity to analyze the effect of mass on the evolution of stars. Because they are all about the same distance from Earth, comparisons in luminosity within the group may be made without such complicating effects as the inverse square law or absorption of light by interstellar dust

a.

b.

Figure 18.7 The eclipsing binary star U Cephei in the constellation Cepheus. In (*a*), U Cephei is in its normal state; in (*b*), it is much dimmer. The dimming is caused by a relatively dark star eclipsing its brighter companion.

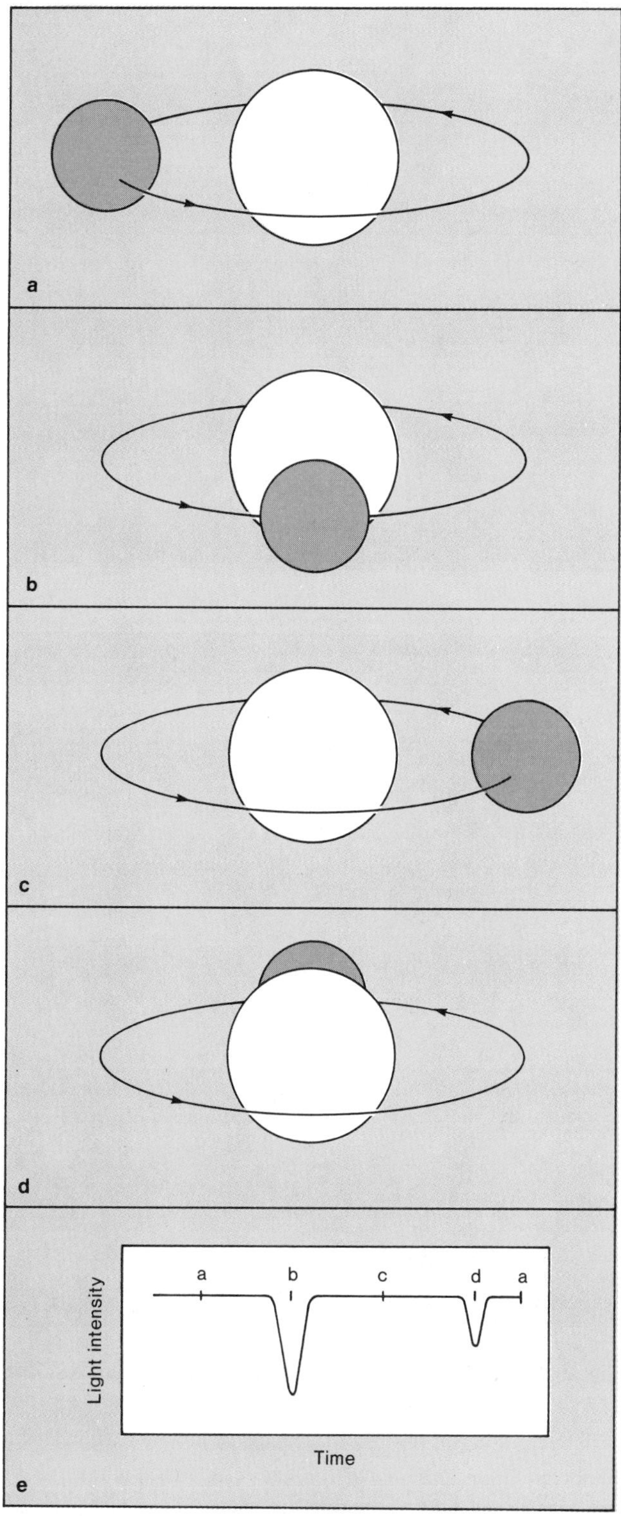

Figure 18.8 The orbital motions and light intensity of Algol, best known eclipsing binary. In (*a*), light from both stars reaches an observer on Earth, but as the darker star moves across the face of the brighter star (*b*), the observed light intensity is reduced. When the darker star moves past the brighter (*c*), the observed intensity returns to the level in (*a*). When the darker star passes behind the brighter (*d*), the intensity is reduced slightly. The cycle then repeats. The observed light intensity is shown in (*e*).

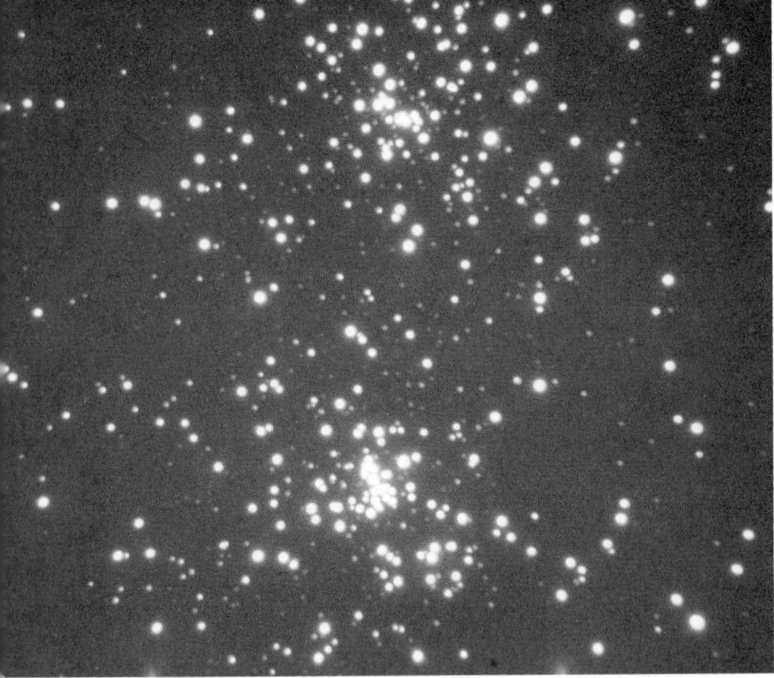

a.

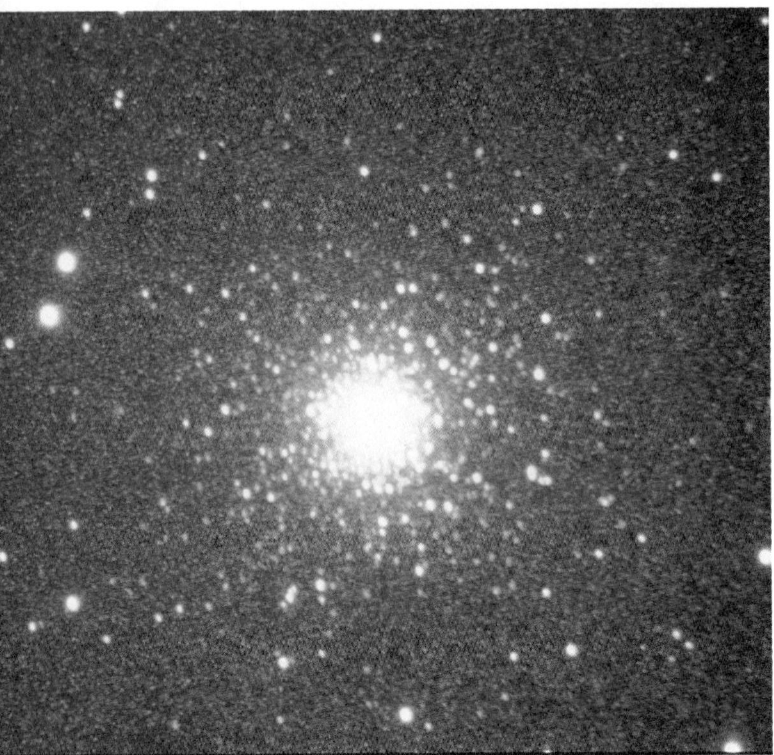

b.

Figure 18.9 Gravity binds stars together in two types of star clusters. Open clusters (*a*) usually contain a few hundred stars or less and tend to lie within the plane of the Milky Way, whereas globular clusters (*b*) may contain hundreds of thousands or millions of stars and are distributed fairly uniformly around the center of the galaxy. The open cluster is the "double cluster" in the constellation Perseus; the globular cluster is M53 in the constellation Coma Berenices.

and gas. Another type of cluster, the **globular cluster** (Fig. 18.9b), may consist of a few thousand to about a million stars. They, too, allow for a study of the evolution of stars and, as indicated in Chapter 19, are useful in galactic study. By way of contrast, the open clusters are usually composed of young, hot stars and lie in the plane of our galaxy, while the globular clusters contain older stars and form a "halo" around the center of the galaxy.

Many open clusters, for example, the Big Dipper and the Pleiades, may be easily seen with the naked eye. With a small telescope, beautiful views of many globular clusters can be seen.

Variable Stars

The discussion of individual stars, up to this point, has implied that they all exist in a state of equilibrium. Various degrees of disequilibrium exist, especially when the star is forming or when it begins to exhaust its nuclear fuel.

A star that exhibits a periodic variation in the intensity of its light (not caused by eclipses in a binary system) is called a **pulsating variable** (Fig. 18.10). Pulsating variables are stars that periodically expand and contract. Just as a mass hanging on a spring oscillates because its inertia carries it past the equilibrium point, the star oscillates in size. When the star expands, it cools, lowering its internal pressure and allowing the outer layers to fall inward. The inward fall heats the star, increasing the internal pressure, eventually stopping and reversing the inward fall, but not until inertia has carried the inwardly falling layers past the equilibrium point. The star expands, again past the equilibrium point, and again cools, repeating the cycle. The time for one cycle ranges from less than a day to over a year. As might be expected, larger stars generally take longer to complete their cycles. Since the size of a star is related to its luminosity, the relationship between the star's brightness as it appears from Earth and the time of one of its one expansion-and-contraction cycles is an important tool in determining the distances to these stars.

The most spectacular stars that inhabit our universe are the **nova** and **supernova** stars. *Nova* comes from Latin, meaning "new," but a nova is actually *an existing star, the brightness of which sharply increases by a factor of thousands.* Current theory suggests that the outburst of energy from a nova comes from a binary star system, one of the members being a white dwarf and the other being a larger star that is losing mass from its outer layers (Fig. 18.11). The mass lost by the larger star is attracted to the small, dense white dwarf where it crashes down with such violence that tremendous energy is released. The release of energy is accompanied by an ejection of material (Fig. 18.12). After the outburst of energy, the nova generally returns to its original state, seemingly unchanged. The brightest naked-eye nova observed recently was in the constellation Cygnus (the Swan) in 1975 (Fig. 18.13).

A supernova can be millions of times brighter than its precursor star. A supernova represents the explosive destruction of a star as it consumes most of its nuclear fuel. The star literally blows itself apart, showering the surrounding region of space with debris. The remnants of such stars form many of the various *clouds of*

a.

b.

Figure 18.10 The long-period variable Mira (often called "The Wonderful") in the constellation Cetus. Mira brightens from its minimum, as seen in (*a*), to its maximum, as seen in (*b*), about every 11 months, making it a long-period variable. At minimum intensity, Mira is too dim to be seen by the naked eye, but at maximum it is a conspicuous star. The greater than 100 fold increase in brightness from minimum to maximum of this giant star is a much greater increase than for most variable stars.

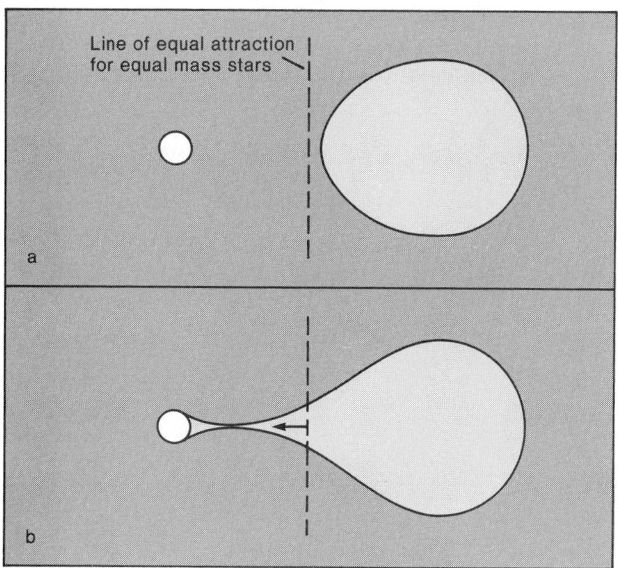

Figure 18.11 The most probable mechanism for a nova is believed to be the transfer of material from a giant star to a white dwarf in a binary system. (*a*) The outer layers of the giant star are distorted by the gravitational pull of the white dwarf. As the giant star expands (*b*) some material reaches the point at which the gravitational pull of the white dwarf exceeds that of the giant. Tremendous energy is released when this material crashes down on the surface of the white dwarf, causing a tremendous increase in its brightness.

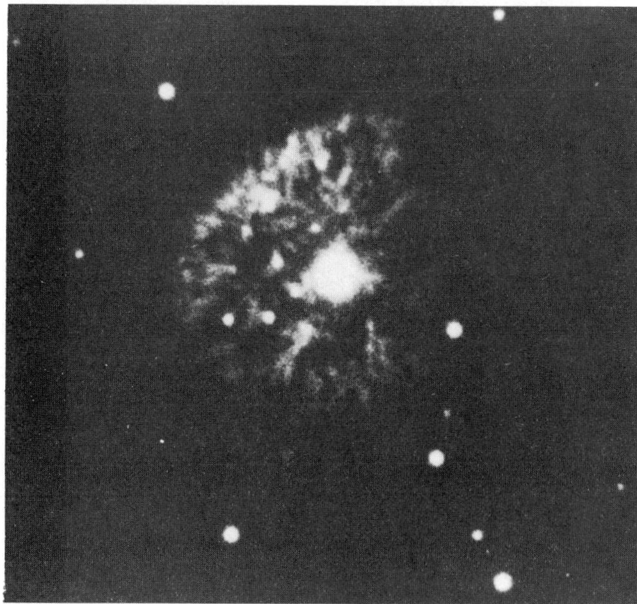

Figure 18.12 Material ejected from Nova Persei, which occurred in 1901, can be seen in this recent photograph.

interstellar gas or dust known as **nebulas.** The Veil nebula, shown in Figure 18.14, presumably represents the remnants of an ancient supernova explosion. The Crab nebula, depicted in Figure 18.15, is the remnant of the supernova explosion of 1054.

There is much more to the story of the Crab nebula than has been addressed so far. Careful observation shows that in the center of the nebula there is a star that acts like a giant celestial beacon, its light flashing on and off 30 times a second. More careful scrutiny reveals that this star is sending out pulses from the X-ray,

Figure 18.13 The rich star field in the constellation Cygnus (the Swan) was punctuated in the autumn of 1975 with a brilliant nova. The brightest star in the photograph was far too faint to see with the naked eye on the evening of August 28, 1975, but by the morning of August 29 it had become a prominent object. The star's greatest brilliancy, reached on August 30, was followed by rapid fading, with the star falling below the limits of visibility on September 8.

Figure 18.14 Veil nebula in Cygnus. The nebula is the remnant of a supernova outburst that occurred about 50,000 years ago.

Figure 18.15 The Crab nebula in the constellation Taurus is the remnant of a supernova explosion recorded in 1054.

Supernova

The brightest supernova in recorded history was the "guest" star that appeared in the constellation Taurus, recorded by Chinese and Japanese astronomers in 1054. At the height of the "guest" star's brilliance, it outshone the planet Venus and was visible during the daytime for three weeks. Some supernovas have been known to increase in brightness over 100 million times. A supernova explosion throws out so much stellar material that an expanding shell of material surrounds the star. The expanding material may travel away from the star at speeds of up to 1000 miles per second. Since the Taurus supernova burst forth more than 900 years ago, the expanding material has covered a diameter of 20 trillion miles and is today called the Crab nebula (Fig. 18.15). Since the Crab nebula is about 6500 light years from Earth, light from the explosion was first seen here in A.D. 1054, although the explosion occurred about 5500 B.C. A light year is the distance light travels in one year, almost 10 trillion kilometers

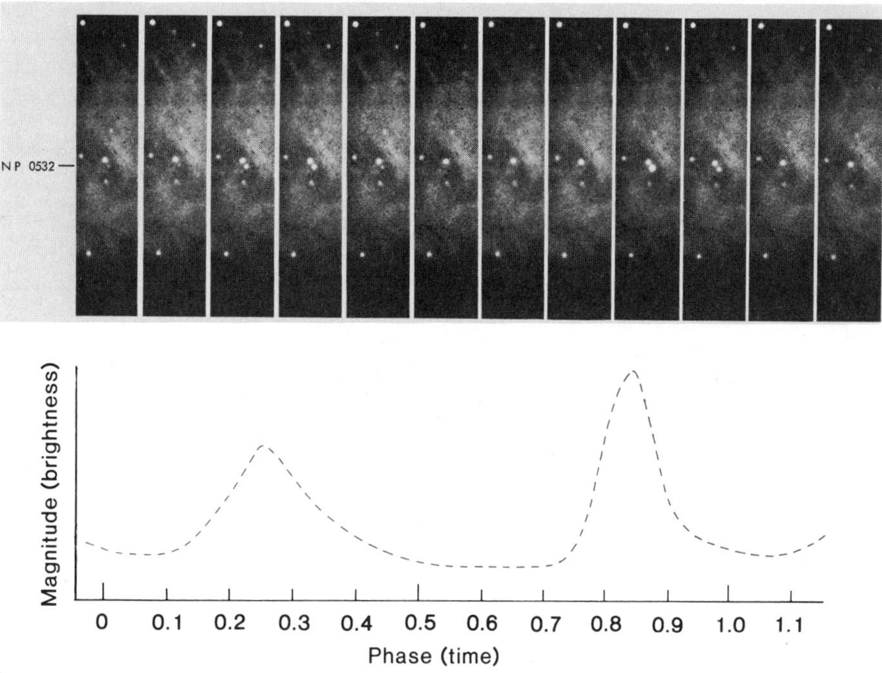

Figure 18.16 The pulsar NP 0532, identified by the line drawn on the series of photographs, is located at the center of the Crab nebula. Note that in some of the scenes the object is not visible, as indicated on the corresponding graph by a decrease in brightness.

infrared, and radio-wave regions of the spectrum that are synchronized with those in the visible region. There are many other such stars in our galaxy, in addition to the central star of the Crab nebula. *Stars that emit electromagnetic radiation in definite pulses* are called **pulsars.** Pulsars are believed to be rapidly rotating and highly magnetic neutron stars—the crushed core of a supernova (see Fig. 18.16).

Life Cycles of Stars

Stars are formed by the gravitational condensation of dust and gas in interstellar space (see Fig. 18.17). The life cycle of most stars is characterized by a relatively short period during which the star is formed; a long period spent as a "normal" star (i.e., a main-sequence star); another relatively short period when the star undergoes tremendous changes; and then by various fates depending on the mass of the star. That most of the stars observed are main-sequence stars is a consequence of the relative lengths of time spent in each phase and how long since the formation of the universe stars have had to evolve.

During the birth of a star, gravitational potential energy is converted into kinetic energy (heat), and part of the kinetic energy is converted into light. This increase in heat results in an increase in pressure within the star. Eventually, the internal pressure of the star is high enough to stop the continued gravitational collapse of the star's material. An equilibrium is then established between the internal pressure of the star, which tends to expand the star, and the inward gravitational pull of the star's matter, which tends to contract the star.

a.

b.

Figure 18.17 (*a*) Small dark knots of material in the "Rosette nebula" are believed to be concentrations of material from which new stars are formed. (*b*) The Great Nebula in Orion, which is clearly visible in the late autumn and early winter sky, contains many young hot stars.

If the initial mass of the material that gravitationally collapses to form a star is less than about one-twelfth the mass of our sun, then the internal temperature of the star when it reaches equilibrium is not high enough to start nuclear fusion reactions, and the star shines for only a few hundred million years by the conversion of gravitational energy into heat and light. The star eventually cools and becomes a "brown dwarf."

The initial rate of gravitational collapse strongly depends on the amount of mass in the collapsing nebula. Large masses, even though they occupy larger volumes of space, have a stronger force pulling the particles inward and collapse faster. As the material heats during the phase of initial collapse, the embryonic star begins to glow. The low temperature and large size of the cloud would place it to the right on an H-R diagram (Fig. 18.18), although it may not be visible because of surrounding clouds of gas and dust. The time for a star to contract and reach the main-sequence phase ranges from approximately 100 million years for stars with one-tenth the sun's mass to about 10,000 years for stars 100 times as massive as the sun.

Once a star has reached equilibrium and becomes a main-sequence star, the star remains in a stable state for a length of time, again dependent upon its mass. With nuclear fusion supplying their energy, stars the size of the sun have enough fuel to remain stable for approximately 10 billion years. Larger stars, in spite of their greater fuel supply, exhaust their stores in as little as 1 million years, whereas smaller stars may linger on the main sequence for over 100 billion years.

Eventually, the depletion of the hydrogen fuel supply in the core slows energy production, and gravity causes the central core of the star to begin collapse. Gravitational energy, rapidly released during the collapse, heats the core region of the star to a higher temperature, with the additional energy being absorbed by the outer layers of the star, causing it to swell. The expansion precipitates cooling of these layers, thus the star evolves into a red giant. A star the size of the sun will require about 1 billion years for this evolution, with larger stars evolving faster and smaller ones evolving more slowly. During the expansion, the core continues to heat, accelerating the fusion of hydrogen in the layers surrounding the core, increasing the star's total luminosity.

When the temperature of the core reaches the fantastic temperature of 100 million kelvins, the triple-alpha process begins to convert the helium in the core into carbon. This reaction proceeds very rapidly, and soon the star has a carbon core. By continuation of nuclear fusion processes similar to those responsible for converting a hydrogen core into a helium core and a helium core into a carbon core, all the elements known in the universe have probably been forged in these stellar furnaces. (Fusing nuclei releases energy until the iron nucleus is reached, so elements more massive than iron probably are generated in supernova explosions. The solar system may owe its existence to a nearby supernova. See Chapter 17.)

The star is now ready to relinquish its role as a red-giant star, and it begins a process of ridding itself of a portion of its mass. For stars of the sun's mass and of less, the following scenario is considered the most likely. The central core of the star continues to pour out energy that is absorbed by the outer layers, forcing them farther and farther from the center. The weakening pull of gravity eventually allows the outer layers to separate from the core. Thus, there is a shell of gas with perhaps 0.1 or 0.2 of the total mass of the star surrounding the bare core, a white-hot sphere about the diameter of Earth with its mass still nearly equal to the original mass of the star. Observations show many of these *gaseous shells,* called **planetary nebulas,** *surrounding dim, hot stars* (see Fig. 18.19). (Planetary nebulas have nothing to do with the planets, other than the fact that some of them have the pale greenish, disklike look of the outer planets as viewed with small

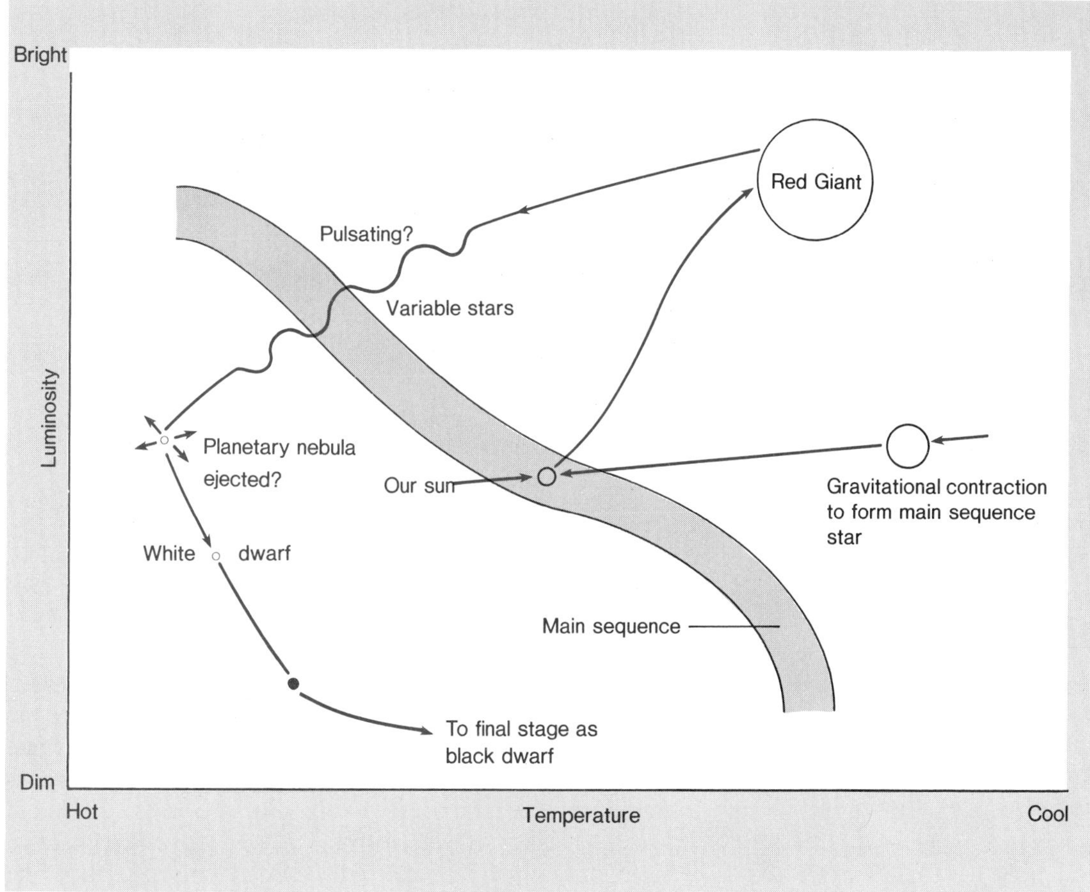

Figure 18.18 The evolution of our sun shown on the Hertzsprung-Russell diagram.

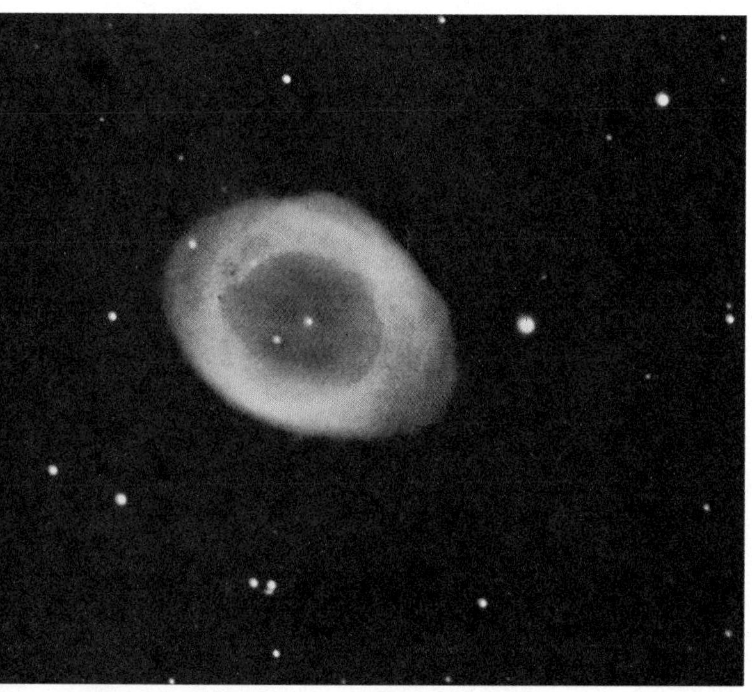

Figure 18.19 The Ring nebula in Lyra. The star in the center of the "halo" ejected the expanding gas.

telescopes. They are far beyond the boundaries of the solar system.) The dim, hot stars are white dwarfs, discussed in the "Types of Stars" section. The white-dwarf stage represents "old age" for most stars. This stage of life turns out to be the longest of any of the stages. The white dwarf, even though it has burned its nuclear fuel, has such a small surface area that it takes many billions of years for it to lose its residual heat. It will eventually cool, however, becoming a **black dwarf,** a *burned-out cinder of a star.* Black dwarfs have not been observed, perhaps because they are too faint to see or, more probably, because there has not been enough time since the creation of the universe, estimated to have been about 18 billion years ago, for any white dwarfs to have sufficiently cooled.

Stars considerably more massive than the sun are believed to end their lives in a much more spectacular way—by undergoing the cataclysmic explosion that produces a supernova. When the carbon is exhausted in the core of an extremely massive star, the situation is similar to the exhaustion of hydrogen, except it is much more intense. Lacking a source of energy, the core begins to collapse from gravitational forces, but now the extra mass is so great that the collapse continues even when more nuclear fusion processes begin (perhaps even aided by some fusion processes that require energy). The collapse runs out of control, releasing gravitational energy so rapidly that new heavier elements are formed, and the star literally blows itself apart, showering the surrounding region of space with debris (see Fig. 18.14) and emitting copious quantities of X rays and neutrinos in addition to the visible light.

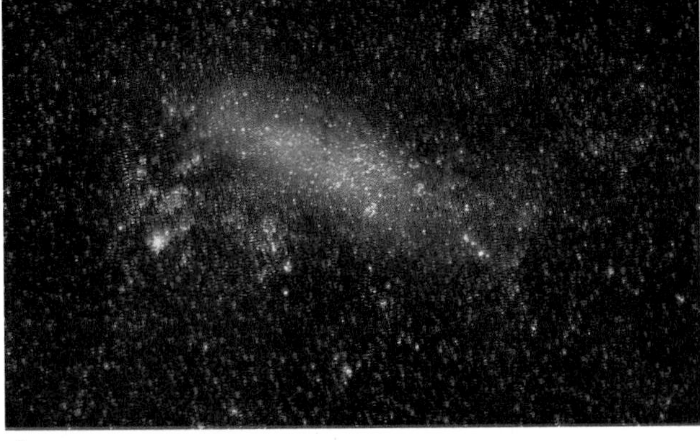
a.

b.

Figure 18.20 (*a*) The Large Magellanic Cloud, an irregular galaxy visible to the naked eye in the Southern Hemisphere, is one of the nearest neighbors of the Milky Way at a distance of about 160,000 LY. (*b*) On the night of February 24, 1987, the first naked-eye supernova in 383 years was observed. Supernova 1987A is the bright star marked with the arrow. Compare this photograph with (*a*).

Remaining is *a small core in which the electrons have been forced to combine with the protons, leaving only neutrons.* The slow rotation of the larger parent star would speed up, just as an ice skater pulling in his or her arms, and this **neutron star,** with a diameter of perhaps 10 miles, would emit rapid pulses of electromagnetic energy, becoming a pulsar.

Supernovas rarely occur in our own galaxy—once about every 500 years. They are often observed in nearby galaxies, but their great distances make them difficult to study. The report that a "nearby" supernova had been observed on the night of February 24, 1987 by Ian Shelton, working at the Las Campanas Observatory in Chile, was received with great delight in the astronomical community. Supernova 1987A, as it was called, occurred in the Large Magellanic Cloud, a very close neighbor of the Milky Way galaxy (Fig. 18.20). Although the increase in brightness was only (!) about 4000 times, the fact that the progenitor star could be identified and the entire range of the radiation it emitted could be studied with readily available equipment made this astronomical event one of the most valuable of recent years. It provided the data scientists needed to substantiate or disconfirm certain theories about supernova explosions. One of the most surprising findings was that the star that exploded was a blue giant instead of the expected red giant. As of this writing, some astronomers are speculating that the chemical composition of the star may have accounted for this unexpected finding. Nevertheless, X-ray and neutrino observations support the general theory of core collapse, indicating that the core of the star probably collapsed in a time period on the order of one second!

Black Holes

The contraction of the core of a star, as has been seen, usually leads to formation of a white dwarf or a neutron star. In both these types of stars, the tremendous inward pull of gravity is balanced by outward forces, but it is possible to envision a body so dense that no known force could prevent its total collapse. Theory speculates that no known force can prevent the total collapse of a body remaining after a supernova explosion if the mass is about 20 or more times the mass of the sun. When this ultimate stellar event occurs, a **black hole** is produced. Since the diameter of the resulting mass of the star is, for all practical purposes, negligible, the density of the matter at the center of a black hole is unimaginably large.

The black hole derives its name from the fact that nothing, including light, emitted from any point within a certain distance of the center can escape. Recall that for an object to escape the gravitational pull of a body, it must have an upward velocity equal to the escape velocity for that body. According to Einstein's general theory of relativity, if the escape velocity exceeds the speed of light, even light itself will be unable to escape. Of course, since no object can exceed the speed of light, no material objects will be able to escape. Thus, any matter that is drawn into the black hole will remain trapped forever, causing some people to compare black holes to giant cosmic vacuum sweepers, devouring whatever comes near them. There has been no confirmed observational evidence of a black hole, although circumstantial evidence is mounting in several cases. Material drawn into a black hole would undergo tremendous acceleration, emitting X rays. Several suspicious X-ray sources are still under investigation. Also, there seems to be a tremendous unobservable mass concentrated near the center of our galaxy.

Summary

Our sun and other stars produce energy by the very high temperature fusion of four hydrogen nuclei into one helium nucleus, two positrons, and energy. The difference in the mass of four hydrogen nuclei and one helium nucleus, along with two positrons, shows up as energy. At very high temperatures hydrogen is converted into helium by the carbon cycle, and at even higher temperatures these helium nuclei are fused into a carbon nucleus by the triple-alpha process.

Some yellow and red supergiant stars have diameters that exceed 500 million kilometers, whereas a star such as our sun has a diameter of only about a million kilometers. White dwarfs, no larger than about one-half the size of Earth, are still huge in comparison to neutron stars, which have a diameter of only a few kilometers. The ultimate minimum-size star is the so-called black hole, which has, for all practical purposes, a negligible diameter and a density high enough to produce an escape velocity exceeding the speed of light.

Approximately one-third of all the stars we can see from Earth are binary systems, and a very small percentage are systems of three or more stars.

Some stars fluctuate in brightness. There are three main reasons for light-intensity fluctuations of stars. A star's photosphere may expand and contract, with a resulting pulsation of its light output. Other pulsating stars are the result of eclipsing binary systems, in which the pulsations result from the eclipsing of one star by another as they revolve around a common gravity center. Some stars, such as the star in the center of the Crab nebula, are highly magnetic neutron stars that rotate very rapidly, with the result that very short bursts of light and radio energy are emitted. Such stars are called pulsars.

A star is formed out of the stardust of space when a large cloud of dust and gas slowly coalesces due to the mutual gravitational attraction of the dust and gas particles. During the contraction, gravitational potential energy is converted into sufficient quantities of heat energy to cause the star to shine. When the internal temperature of the new star reaches many millions of degrees, the fusion of hydrogen atoms into helium atoms begins with subsequent conversion of mass into energy. The average star then has a relatively long life of about 10 billion years, after which it expands to become a red giant. Larger stars during this period of expansion may expand with explosive violence to form a supernova. After the explosion of a supernova, contraction of the supernova remnant produces a neutron star or a black hole. The average star, however, after it expands to the red-giant stage, slowly contracts to become a white-dwarf star, possibly losing its outer layers in the formation of a planetary nebula. The white dwarf slowly cools over many billions of years to produce a burned-out cinder of a star called a black dwarf.

Questions and Problems

Our Closest Star: The Sun

1. Why is the sun unique as a star to astronomers on Earth?
2. Since the surface of the sun is very hot, the gaseous materials, such as hydrogen, must be moving at very high average speeds. Why, then, have not hydrogen and the other elements escaped into outer space?

What Makes the Stars Shine?

3. Why do old stars have a greater percentage of helium than young stars?
4. What are the two most important energy processes in typical stars?
5. What are the two major factors that govern the luminosity of stars?

Types of Stars

6. What is the difference between a red-giant star and a blue-giant star?
7. What type of stars are the most luminous? The least luminous?
8. With reference to Figure 18.14, respond to the following:
   (a) Describe the luminosity and temperature of the average red-giant star.
   (b) Describe the luminosity and temperature of the average white-dwarf star.
   (c) What appears to be the average luminosity of stars on the main sequence that have a surface temperature of about 10,000 K? Of about 20,000 K?

Multiple Stars and Star Clusters

9. What is an eclipsing binary? What can astronomers learn about stars by studying eclipsing binary systems?
10. What is an open cluster? How does it differ from a globular cluster?

Variable Stars

11. Describe what causes the variation in light received from a pulsating variable star as opposed to that received from an eclipsing binary system.
12. The term *nova* means "new star." What, in reality, is a nova?
13. What is a supernova?
14. With respect to most other nebulas, why is the Crab nebula unique?
15. How do pulsars differ from other classifications of stars?

Life Cycles of Stars

16. How are stars "born"?
17. Why does the fate of the solar system rest with the aging process of the sun?
18. What processes in action identify a star as being an old star? A relatively young star?
19. Do planetary nebulas represent the birthplace of planets? Explain your response.
20. How do black-dwarf stars differ from white-dwarf stars?
21. What is the major determining factor as to whether a star becomes a white dwarf or a neutron star?

Black Holes

22. How is a black hole believed to form?
23. What is the origin of the term "black hole"?

19 Galaxies and the Universe

Chapter outline

Key terms/concepts

A beautiful spiral galaxy in the constellation Pegasus. Galaxies typically contain hundreds of billions of stars and there are billions of galaxies within the observable universe. The individual stars in the picture are part of our own galaxy, the Milky Way, and are very much closer than the galaxy shown.

To the eye, the universe appears static, as if all the objects were frozen, much as objects in a photograph. Nothing could be further from the truth. The universe is a dynamic collection of objects in constant motion. It is only because of the great distances that separate us from all the objects in the universe (with the exception of our solar system) that we experience the illusion of motionlessness.

The detail by which we understand the universe has been greatly enhanced by the telescopes that now orbit our planet, well beyond the frustrating artificial light in even the most remote areas and the atmospheric conditions that have bothered astronomers since the days of Galileo. The images compiled by the IRAS (*I*nfra*R*ed *A*stronomical *S*atellite) are revealing a universe heretofore beyond the reach of even the most powerful optical devices on Earth. The faintest emanations from outer space in the infrared portion of the electromagnetic spectrum are transmitted to Earth by the satellite, analyzed by computers, and studied by astronomers from many nations. The IRAS has detected an infrared universe in which new stars constantly form among the dust-filled regions of space. It has also discovered cool dust encircling several stars, including the star Vega. The telescope, however, lacks sufficient resolution to distinguish between spherical shells that accompany the nova stage of stars and flat disks that might suggest planets in the process of formation. Data analysis has promoted further study of these stars, chiefly performed with the 2.5-meter reflecting telescope at Las Campanas Observatory in Chile. This investigation has revealed that the star Beta Pictoris, 50 light years distant, is surrounded by a massive disk of dust and rocklike objects that orbits so as to provide us an edge-on view. The image of the star and perhaps an incipient planetary system was obtained through the application of various imaging and processing techniques. Undoubtedly, as data are obtained and analyzed, our present theories about space within and beyond our solar system will need to be modified to provide a more accurate model of our entire universe. The revamping of our view of the universe is definitely an ongoing adventure that will continue as long as scientific inquiry is pursued.

Nebulas

All of us have savored the beauty of the stars on a clear, dark night. About 6000 stars are visible with the naked eye under ideal conditions, making the half that can be seen at one time number about 3000. All of these stars are but a small fraction of those which, along with the sun, are bound together gravitationally to form our **galaxy.**

Scattered throughout the vast reaches of our galaxy, in addition to the stars, are immense regions containing gas and dust particles. Some sections of the spiral arms of our galaxy, the Milky Way, contain relatively high concentrations of these gases and solid particles. The distribution of the dust particles in our galaxy is not uniform, however, with the result that some regions are more dense than others and have a cloudlike appearance. These shadowy patches, extending over enormous regions, are called **nebulas,** from the Latin meaning "cloud."

Even though the concentration of material in the nebulas is thousands of times higher than in interstellar space in general, the nebulas, consisting of gas and dust particles, constitute a better vacuum than we can create in our laboratories.

There are three types of nebulas that we can see in the vast stretches of the universe: dark nebulas and two types of bright nebulas. Although both the dark and the bright nebulas are composed of essentially the same material, the bright are close enough to large, hot stars to be illuminated either by reflection or by fluorescence. On the other hand, the dark nebulas lie so far from bright stars that no appreciable amount of starlight is reflected by them. The light that shines from stars and bright nebulas is sometimes obscured by the material of the dark nebulas. This absorption of light by dark nebulas is what makes them appear as dark patches blotting out background stars. Two of the most famous dark nebulas are the so-called Great Dark Rift, which appears to split the Milky Way, and the Horsehead, which conceals part of the bright nebula in the constellation Orion (see Figs. 19.1 and 19.2). Although the density of nebulas is very low, some extend over such vast distances that the total mass of material contained in them is unimaginably large.

How were these nebulas formed? Their formation is part of the endless cycle of cosmic events that includes the birth, life, and death of stars in our galaxy and throughout the universe. Nebulas are formed as the result of the violent explosions of unstable stars that suddenly and explosively eject part of their mass as a rapidly expanding shell of superhot gaseous material. These unstable stars hurl part of themselves into the far reaches of the space between them. This ejected "stardust" accumulates over aeons to form the nebulas of our universe. The Dumbbell nebula, a cloud of ejected stellar material that assumed an unusual shape, is shown in Figure 19.3.

What is the composition of this matter that stars "suicidally" eject into space? It depends a great deal on the size and age of the star. If the star is of average size and age, it will have slowly converted parts of its hydrogen into helium and heavier elements by nuclear fusion reactions. The material ejected by a star contains hydrogen, helium, and varying amounts of other elements. Of course, the different elements can chemically combine once they escape from the fantastically high temperatures of a star. One would expect at least all the nonradioactive elements possible in nature to be present in the debris that exists in the space between stars, but most of the elements have not yet been detected. Some of the various elements and compounds found in various nebulas are hydrogen, helium, carbon, oxygen, nitrogen, sodium, calcium, sulfur, argon, titanium, iron, methane, water, ammonia, hydrogen cyanide, formaldehyde, methyl alcohol, and chlorophyll-type compounds. These are the types of elements and compounds that constitute the gases and dust particles of nebulas.

Figure 19.1 This "fish-eye" photograph captures the beauty of the entire sky, with the Milky Way stretching from the upper left (northward) to the lower right. Made in Australia on the morning of April 12, 1986, many star clouds and regions of nebulosity are visible. Halley's comet is just above the Milky Way at right, while the streak is that of a meteor. Jupiter is the bright image within the wedge of light that rises out of the dawn's glow, while Mercury is closer to the horizon. Mars is the bright image slightly left of the picture's center, while Saturn is above and to the right of center, slightly brighter than the star Antares, which is to its lower right.

The condensation of the material of these nebulas gives birth to second- and later-generation stars. Then, these second- and later-generation stars eventually become unstable and in turn eject part of their matter into the space from which it came, and in the process give birth to new nebulas that potentially can become stars of the next generation.

The Milky Way

One of the great beauties of the universe that can be seen without the aid of telescopes is the archway of soft gentle light that meanders across the vault of our night sky from horizon to horizon—the **Milky Way.** This "lighted passage to heaven" dominates the moonless night sky. When one looks at the Milky Way through a telescope, what appears to the naked eye as a soft diffused glow is revealed to be an incredible multitude of individual stars and patches of nebulas. Obviously, the Milky Way is actually an aggregation of billions of stars.

Our sun is one of perhaps 250 billion stars that make up the Milky Way, our home galaxy. If viewed from the outside, the more luminous stars of the Milky Way would appear as a lens-shaped grouping with a bulge at the center. A convenient unit of distance

The Skies of Messier

There have been numerous projects established to catalog nebulous objects. The most popular system, particularly among many amateur astronomers, was completed in 1781 by French astronomer Charles Messier (1730–1817). Messier reportedly was obsessed with discovering a comet to which his name could be attached, as was the case with several other astronomers of his time. To impose orderliness on his search, he recorded 103 objects that were frequently mistaken for distant comets. Messier's long list of objects contains a portion of the most outstanding objects within our view, including globular and open clusters, nebulas, and galaxies of various shapes. Messier assigned a number to each object; in fact many are known today by the M-number, rather than by any specific name. The most famous objects he classified include the Crab nebula, M 1; the Lagoon nebula, M 8; the Dumbbell planetary nebula, M 27; the Orion nebula, M 42; the Andromeda galaxy, M 31; and the "Whirlpool" spiral galaxy in Canes Venatici, M 51. A considerable number of Messier's objects are pictured in this text, several of them in their beautiful colors, a feature not available in his time. Even though Messier did discover many comets, none bear his name. However, his popular system of registration has won for him an important position among astronomers of all time. Other systems of cataloging were developed in the nineteenth and twentieth centuries; these have been amended to include considerably more stellar objects than the 103 recorded by Messier ■

Figure 19.2 The Horsehead nebula is a combination of bright and dark nebulas. The dark nebula lies between us and the bright nebula. Note the relative scarcity of stars in the dark region, indicating that the dark cloud is obscuring the light from all luminous objects in that direction.

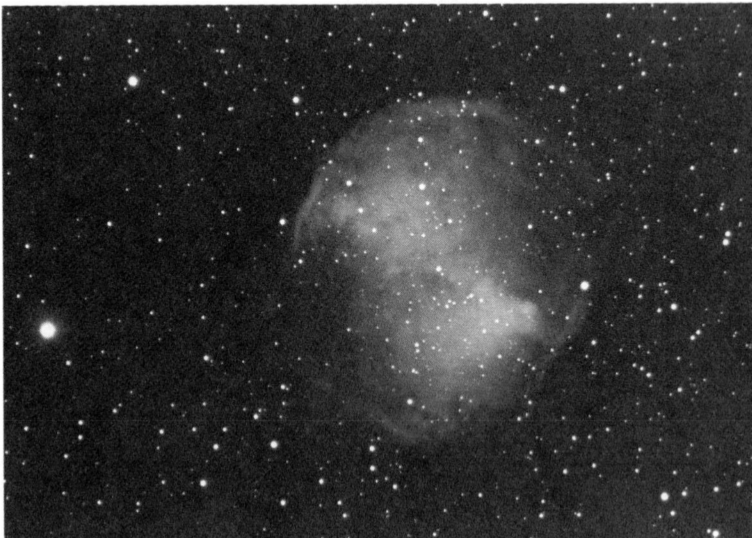

Figure 19.3 The Dumbbell nebula is a planetary nebula. This is an example of a dying star returning gas and dust to the interstellar medium

to state the size of the galaxy is the light year (LY), the distance light travels in one year. This distance is over 9 trillion kilometers (about 6 trillion miles). The unit is discussed in more detail later, in the section "Astronomic Distances." The distance across the brighter part of the galaxy is about 100,000 light years, with the thickness of the central bulge, the nucleus, being about 10,000 or 15,000 light years. Figure 19.4 suggests what our galaxy probably would look like if viewed from the outside. When we look up at night from our vantage point within the galaxy and see the luminous path of light called the Milky Way, we are looking edge-on through our galaxy (see Fig. 19.1). If we look 90 degrees away from the plane of the galaxy, there are relatively few stars, for in this view we are looking in the "thin" direction of the galaxy.

The stars making up our galaxy rotate around its nucleus, and the flattened shape of our galaxy is the result of this rotation. Our sun revolves around the galactic center at a speed of 800,000 kilometers per hour, and even at this fantastic speed it takes our sun 200 million years to complete one revolution. That the galactic "year" is 200 million Earth years is an indication of how immense our galaxy is. In fact, our galaxy has made only 90 complete revolutions since its birth about 18 billion years ago. Our sun is about two-thirds of the way out from the galactic center (about 30,000 LY) in one of the galaxy's spiral arms (see Fig. 19.5).

Stars at different distances from the galactic center revolve around it at different speeds, similar to the way the planets revolve around the sun at different speeds. The speeds of the individual

a.

b.

Figure 19.4 (*a*) The spiral galaxy NGC 253, which is similar in shape to our Milky Way galaxy. (*b*) An edge-on view of the spiral galaxy in Coma Berenices. Our Milky Way galaxy would have a similar appearance if viewed from the side.

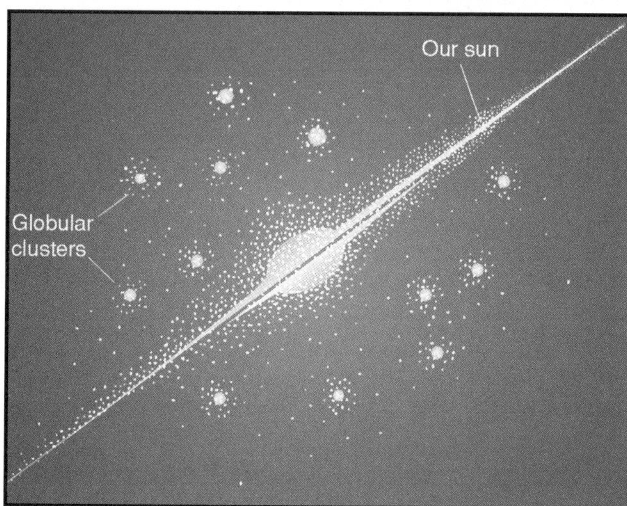

Figure 19.5 An artist's conception of an edge-on view of our galaxy, showing the disk, nucleus, location of the sun, and the corona containing a thin "veil" of stars and the globular clusters. The stars in the corona would be too faint to see with the naked eye from the vantage point indicated.

Figure 19.6 The globular cluster M 80. Astronomers use the distribution of globular clusters to help locate the center of the galaxy.

stars are more complicated than the motions of the planets, however. This is because the galaxy is not concentrated at a single place, and the gravitational effects of nearby stars and gas clouds affect the motions of the individual stars. The motion of a star toward or away from the sun can be determined from the *Doppler effect;* that is, by studying the shift in the color of light emitted by elements in the star's atmosphere (see Chapter 7 and Fig 10.11). Some nearby stars also can be observed to move slowly across the sky relative to the more distant stars. This motion may be measured by comparing two photographs of the same area of the sky taken at different times (generally years or decades apart). The **space velocity,** *the velocity of a star relative to the sun,* can be computed by combining these velocities using vector addition as described in Chapter 3. Values of the space velocity range from a few kilometers per second up to hundreds of kilometers per second, with typical values of 20 to 25 kilometers per second. If we could travel this fast on Earth, it would take us only about three minutes to cross the United States. In spite of these high speeds, the stars are so far apart that little danger of collision exists.

The spiral arms of the galaxy contain much cosmic dust and gas and relatively young and bright stars, called *population I stars.* These relatively young stars were formed from the gas and dust lying near the central plane of the galaxy. The nucleus of the galaxy contains older, redder stars called *population II stars.* The Milky Way's nucleus is not visible to us, but it can be studied with radio and infrared telescopes. The younger population I stars, including our sun, contain a greater abundance of heavier elements than do population II stars, indicating that at least some of the material from which they formed was "cooked up" in stars that have long since ceased to exist. (See the section on variable stars in Chapter 18.)

Our flattened galactic disk is embedded in a spherical veil made up of beautiful symmetrical clusters of stars, the *globular clusters* (Fig. 19.6), discussed in Chapter 18. The globular clusters surrounding the spiral-armed flattened disk of our galaxy, as well as a "mist" of individual stars surrounding the nucleus of the disk, make up the *corona* of our galaxy. The principal part of the corona extends about 50,000 light years above and below the disk of the galaxy. The stars in the corona, including the globular clusters, are population II stars.

The difficulty of observing faint stars has complicated the task of defining the rather inexact "boundaries" of our galaxy. Recent observations with radio telescopes, infrared telescopes (both earth-bound and aboard spacecraft), orbiting X-ray telescopes, and more sensitive detectors on optical telescopes (see CCDs, Chapter 9) have led astronomers to conclude that the galaxy extends far beyond the bright disk and nucleus. Its extremes may reach a distance of over 250,000 light years from the center, the visible disk occupying a small portion of the total volume. The nature of the material in the far reaches of the galaxy is still uncertain and may include planet- or asteroid-sized objects in addition to gas. The total mass of our galaxy may be more than 500 billion times the mass of the sun.

Figure 19.7 A barred spiral galaxy in the constellation Eridanus, photographed with a 5-m telescope.

Other "Island Universes"

When we consider the immensity of our Milky Way galaxy (light traveling at 300,000 kilometers per second takes at least 100,000 years to go from one edge of our galaxy to the other), it is difficult to imagine how any structure could be larger and more impressive. Comparatively speaking, however, our galaxy is but a speck of dust on a voyage in a sea of empty space. There are tens of billions of other galaxies much like ours scattered throughout the limits of our cosmos. The reference to galaxies as "island universes" aptly describes their lonely isolation in the void of space.

What do the other galaxies in our universe look like? Would they resemble the Milky Way galaxy to an intergalactic traveler? Is our galaxy of average size? There are a number of different types of galaxies based on differences in galactic shapes. Perhaps we should first consider the type of galaxy that characterizes our Milky Way—that is, the **spiral galaxy.** The considerable rotation of a spiral galaxy causes flattening of a major portion of the galaxy into a disk or lens shape. In the nucleus of the galaxy, stars were formed from the cosmic dust of the original nebula at an early stage; therefore, the nucleus today contains old stars in general. Because of the rotation of the galaxy, the cosmic dust and gas outside the nucleus have sufficient motion to slow the formation of stars by gravitational accretion. Consequently, the spiral arms of such a galaxy contain younger stars and even stars presently in the process of formation. The stars in the globular clusters surrounding a spiral galaxy consist of old stars that probably formed about the same time the stars of the galactic nucleus were formed. The nuclei of some spiral galaxies are elongated into a bar shape, and these spiral galaxies are consequently referred to as *barred spiral galaxies.* An example of a barred spiral galaxy is shown in Figure 19.7. Perhaps the most famous and most spectacular spiral galaxy other than our own Milky Way is the great *Andromeda galaxy* (see Fig. 19.8). The number of stars in spiral galaxies ranges from 1 billion to over 1 trillion.

Figure 19.8 The Andromeda galaxy, M 31. This beautiful spiral galaxy is a neighbor to our own Milky Way galaxy, lying "only" 2 million LY away. It is similar in size and shape to our galaxy. Notice the two small elliptical companion galaxies.

The most common type of galaxy in the universe is the **elliptical galaxy.** The shape of elliptical galaxies varies from spherical to near lens-shaped. The difference in the shapes of elliptical galaxies is the direct result of the difference in the speed of rotation of these galaxies. If a galaxy does not rotate to any appreciable extent, its shape is essentially spherical, but if the galaxy has a reasonable rotation, then it is flattened out and approaches the shape of the faster-rotating spiral galaxies. Elliptical galaxies consist almost entirely of old stars, the ages of which approximate the ages of stars in the nuclei of spiral galaxies and in globular clusters. The largest galaxies in the universe are elliptical galaxies (see Fig. 19.9).

Not all galaxies have symmetrical shapes. Members of a third class of galaxies, called **irregular galaxies,** lack any definite form, even though the billions of stars in them are bound together gravitationally. An example of an irregular galaxy located in the constellation Ursa Major near the Big Dipper (although, of course, much farther away) is shown in Figure 19.10.

Just as the stars cluster in galaxies, on a much larger scale the galaxies themselves tend to form **clusters** (Fig. 19.11). The Milky Way is one galaxy of about two dozen galaxies in a cluster

a.

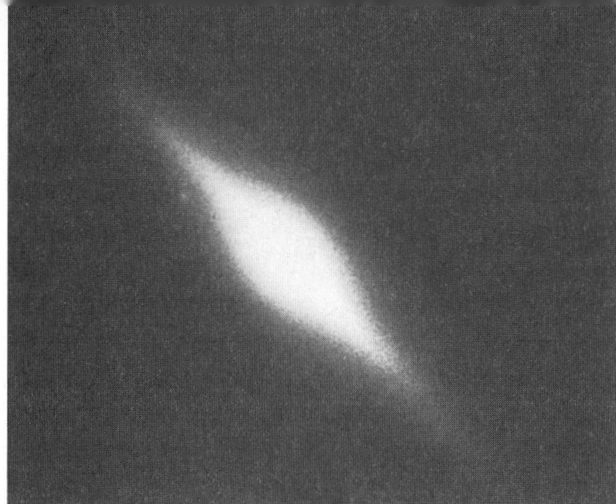

b.

Figure 19.9 (*a*) An elliptical galaxy with a relatively slow rate of rotation. (*b*) This elliptical galaxy has a considerably higher rate of rotation than does (*a*).

Figure 19.10 The irregular galaxy M 82 located in the constellation Ursa Major lies at a distance of about 7 million LY.

Figure 19.11 A cluster of galaxies in the constellation Hercules. Each of the "fuzzy" patches of light is an entire galaxy consisting of billions of stars.

called the *local group.* The galaxies in the local group lie in a roughly disklike region about 2.5 million light years in diameter, with no other galaxies within about 3.5 million light years. Clusters may have only a few members or they may have thousands. Clusters themselves tend to form even larger groups called **superclusters.** Recent evidence from a small section of the sky seems to suggest that the clusters and superclusters are not distributed randomly through space. Rather, the evidence points to great voids in the universe, with the clusters being concentrated on the "surfaces" of these voids. This scheme is termed the "bubble" structure of the universe, since, if the evidence is confirmed by future studies, the universe could be likened to a large bowl of many soap bubbles, with galaxies lying on the surfaces of the bubbles.

There is increasing evidence that astronomers may be currently witnessing the formation of a new galaxy in a cluster of galaxies as it occurred 240 million years ago. A jet of gaseous material emanating from a galaxy has been discovered as it interacts with a massive cloud of stellar gas and dust, generating detectable radio waves as a result of the impingement. The intense turbulence the jet of gases creates as it penetrates deeper into the dense cloud is suspected to be a mechanism that initiates gravitational contraction of steller material into stars.

Astronomic Distances

There is no way for the unaided eye to detect the difference in distances from Earth of various celestial objects. The sun and the moon appear to be about the same size to an observer on Earth, yet the sun is about 390 times farther from Earth than the moon. Some stars appear brighter than other stars, which may lead us to believe that the brighter stars are closer, but this is not necessarily true. Indeed, a very bright star may be much farther away than a faint star.

Since the distance to astronomical objects is so very large, the distance unit of a mile or a kilometer is simply too small to use. Consequently, astronomical distance units have been developed. *The mean distance between Earth and the sun,* 149,500,000 kilometers (92,900,000 mi), is called an **astronomical unit** (AU). The astronomical unit is used in reference to the distances of planets from the sun, distances between planets, and even distances to some of the nearest stars. The **light year** (LY) is another, much larger, distance unit used in astronomy. A light year is *the distance that light travels in one year.* Since light travels at a velocity of about 300,000 kilometers (186,000 mi) per second, and since there are about 31.6 million seconds in a year, then there are approximately 9.46×10^{12} kilometers (5.88×10^{12} mi) in 1 light year. There are also 63,100 AU in 1 LY.

A unit often preferred by astronomers when referring to vast distances in the universe is known as the **parsec** (pc). The parsec is *equal to approximately 30.9 trillion kilometers or about 3.27 light years.* The unit is discussed in more detail as to its derivation in Chapter 21. One megaparsec equals 3,270,000 light years.

For the remainder of this discussion on astronomic distances, the length of time it takes light to travel to various celestial objects is used for the sake of consistent comparisons.

In order to get some feeling for how far away we are from various objects in our universe, consider the length of time that it takes light to reach us from different bodies within the universe. At a velocity of 300,000 kilometers per second, light would take only a little more than 0.1 second to travel around Earth. Light from the moon, about 385,000 kilometers (239,000 mi) away, requires a mere 1.3 seconds to reach us. The sun sends its light from an average distance of about 150,000,000 kilometers (93,000,000 mi) in 500 seconds (8.3 min). Pluto, the lonely sentinel at the outpost of our solar system, sends us the sun's reflected light in 5.3 hours, over an average distance of 5,746,000,000 kilometers (3,570,000,000 mi).

As an example of the solitary isolation of stars even within our galaxy, the star nearest Earth, Proxima Centauri of the Alpha Centauri group, is about 40.7 trillion kilometers away. Light from that lonely beacon in space must travel 4.3 years before it reaches us. If Proxima Centauri were annihilated in a great cataclysmic explosion, more than four years would have to pass before we could know that the star no longer existed. Light from the closest galaxy to our own Milky Way, the Large Magellanic Cloud, must travel for 163,000 years before reaching our planet. Thus, the light that we see today from the Large Magellanic Cloud must have left that galaxy about the time the early Neanderthal lived.

The spectacular Andromeda galaxy, shining like a spiral of pearls set in the darkness of space, is so distant from us that light from Andromeda must travel for more than 2 million years before it reaches Earth. When we look at the Andromeda galaxy today, we are seeing light that left Andromeda tens of thousands of years before the beginning of the great Ice Age, in which immense sheets of ice covered most of the northern portion of the United States.

How far into the depths of space can we look and still find galaxies? This question has no viable answer. As larger and larger telescopes have been built, we have been able to see farther and farther toward the edges of the universe, and each new telescope reveals even more galaxies. A "summary" of these distances is presented in Figure 19.12.

A telescope designed to study celestial objects is rated according to the size or diameter of its *aperture,* the lens or mirror used to gather light from the distant object. The larger aperture intercepts more light from a star, galaxy, or nebula than does a smaller one; hence, the larger telescope enables the astronomer to resolve finer details of the image and to photograph the object in shorter exposure time. The largest telescope in use today has an aperture of 6 meters (236 in) and is located on Mount Pastuklov in southeastern Russia. The 5-meter (200-in) Hale telescope at Mount Palomar in southern California is second largest. A 10-meter (approximately 400-in) telescope under construction at a Hawaiian Island site by the University of California might enable the deepest penetration into the unknown possible with modern technology.

The farther into space our telescopes allow us to probe, the farther back into the dim past we can see. It has been said that astronomy is the scientific study of the past history of our universe.

What is the universe like today? We don't know! Since the light from our nearest neighboring galaxy left that galaxy about 160,000 years ago and the light from one of the farthest known galaxies left about 12 billion years ago, the universe that we are seeing "today" is actually the universe as it existed hundreds of thousands or billions of years ago. We can only assume that the current universe is doing the same sorts of things as the incredibly ancient universe that is finally visible to us.

Cosmology and the Redshifts

As has been discussed in this and the previous chapter, the illusion of an unchanging universe exists only because of the great distances to celestial objects outside the solar system and the immense times required for most changes. With the lifetimes of stars measured in hundreds of millions or billions of years and the distances to nearby stars being tens of trillions of miles, a few thousand years of human observation, for the most part with the naked eye, are not sufficient to notice many changes. However, the universe is a dynamic place, with stars and presumably planets forming, moving, and dying. Galaxies, too, have their motions and life cycles. **Cosmology** is *the study of the origin and structure of the universe.* (Formally, the term *cosmogony* applies to theories of the origin of the universe, but in general usage, *cosmology* has been extended to include the origin in relation to current structure.)

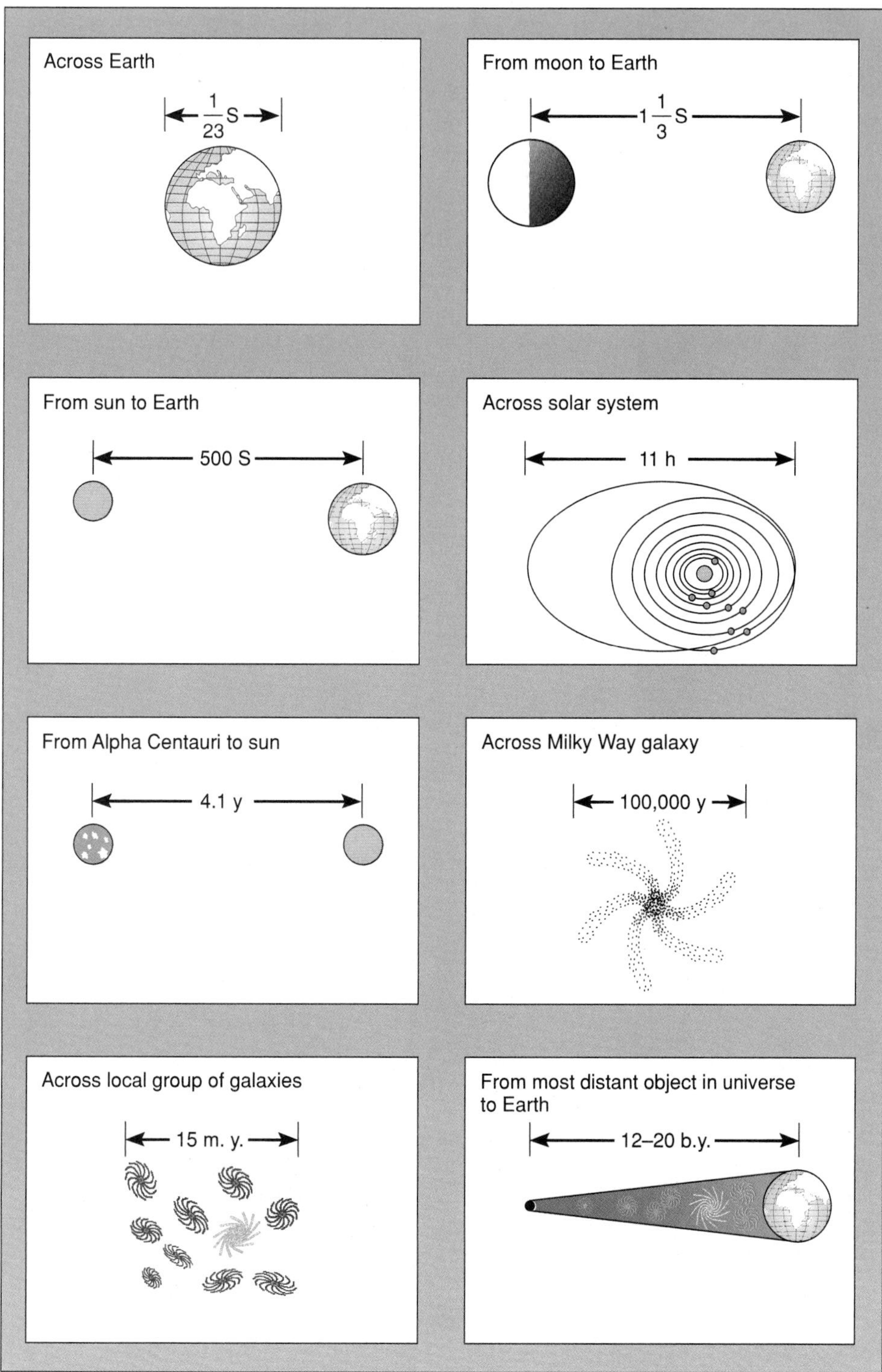

Figure 19.12 The times required for light to travel different distances is shown. Even using this convenient "measuring stick," it is difficult to comprehend the vastness of space.

Galaxies—the Farther, the Faster

For over three centuries after Galileo first used the telescope to study the heavens, astronomers were convinced that the numerous small, hazy, stellarlike objects they observed were simply nebulous stars or dense clouds of gaseous materials. However, in 1845, Ireland's Lord Rosse announced that he had observed a nebula that had a spiral structure. In 1918, the 2.5-meter (100 in.) reflecting-type telescope was installed at Mount Wilson, California, initially to study the structure of the Milky Way galaxy. In 1924, Edwin P. Hubble used the instrument to study distant objects and succeeded in resolving various nebulas into individual stars, including some supergiants. He was also able to show that numerous stellar bodies were not objects in our galaxy, but were complete galaxies in themselves, similar to the Milky Way. Hubble also noted that the brightness of nebulas close to our galaxy fell within a specific range and that the distances of fainter nebulas could be calculated according to their apparent brightness.

Hubble, in 1929, discovered that clusters of galaxies all exhibited systematic redshifts, the amount of which is a function of distance. The farther the galaxy is from our Milky Way galaxy, the greater its velocity of recession. The constant of proportionality, known as the *Hubble constant,* has undergone various revisions as technological advances have been made. The value generally accepted today is about 100 kilometers (62 mi) per second per megaparsec, although recent studies indicate the constant may be only one-half this value. An object one megaparsec distant (3.27×10^6 LY) has a recessional velocity of 100 kilometers per second. The Virgo-cluster galaxies, 11 megaparsecs from our galaxy, are receding at 1100 kilometers per second.

About two dozen galaxies near the Milky Way galaxy do not show a general recessional velocity as do the more distant galaxies. These galaxies are called the "local group." The local group is bound together gravitationally, with some of the galaxies approaching the Milky Way galaxy and some receding from it ■

When the light from certain stars in various galaxies is analyzed by a spectrograph, it is found that the spectral lines representing certain gaseous elements present in the stars are in a different position than that observed when light from these same elements is passed through a spectrograph in the laboratory. Instead, the spectral lines present in the light of galaxies are shifted toward longer wavelengths (and lower frequencies); that is, shifted toward the red end of the spectrum (see Chapter 10). This **redshift** is due to the Doppler effect (note Fig. 10.11) on the light sent to us from these receding galaxies.

If a source of light is moving toward an observer, or if the observer is moving toward a source of light, the observer sees the wave front of the light "bunched" closer together than is normal, and spectral lines are shifted toward shorter wavelengths (higher frequencies) and, therefore, toward the blue end of the spectrum. If a source of light is traveling away from an observer, or if an observer is traveling away from a source of light, the observer sees wave fronts of the light "spread" apart more than is normal, and spectral lines are shifted toward longer wavelengths (lower frequencies); that is, shifted toward the red end of the spectrum.

With the exception of only a few galaxies close to the Milky Way that are members of what is called the local group, all galaxies show a shift of spectral lines toward the red end of the spectrum. This redshift can only mean that all the galaxies that we can see (except a few members of our local group) are receding from our Milky Way galaxy. Interestingly, there is a direct relationship between the recessional speed of a galaxy and its distance from us. Thus, the more distant a galaxy is, the faster it is receding from us. This distance-velocity relationship is known as the *law of redshifts*. For example, if a distant galaxy that has been checked for a redshift shows a recessional speed of 270 million miles per hour (41 percent of the speed of light), the law of redshifts indicates that the galaxy must be more than 5 billion light years away. Some galaxies have been photographed that are 12 billion light years from Earth.

It is rather disconcerting to realize that all the galaxies in the universe are rushing away from us. Does this mean that we are at the center of the universe and that all other galaxies are expanding away from the center? Actually, observers on any other galaxy in the universe would have the illusion that all the other galaxies in the universe were rushing away from their galaxy. How can this illusion be? A very simple analogy can explain this apparent anomaly. Consider a balloon that has dots painted on it. As the balloon is blown up, the distance between the dots increases. An ant on any one of the dots would observe that all the other dots were receding from it as the balloon continued to increase in size (see Fig. 19.13).

We are forced, then, to conclude that all the galaxies in the universe are involved in a general expansion from some central site in our universe. Calculations indicate that the expansion started about 18 billion years ago. We can assume then that some 18 billion years ago (time zero for our universe) all the matter in the universe was concentrated into a colossal primeval "egg." This original egg must have undergone an explosion, the likes of which we cannot even conceive, with the result that matter was sent hurtling outwards at unimaginable speeds. During the continued expansion of this matter, galaxies slowly formed and the distances between these galaxies continued to increase. This theory of the

Figure 19.13 If the balloon on the left were inflated to the size of the balloon on the right, all the dots would move away from each other. Imagine that each dot represents a cluster of galaxies, and note that the general expansion looks the same from any dot.

birth of the universe is commonly known as the **big-bang theory.** At present there is no way to know if the universal expansion will continue or, as some scientists speculate, the rate of expansion will decrease. According to their theory, the expansion may cease, and the universe will begin to contract due to mutual gravitational attraction between the galaxies. Then all the galaxies of the universe will move back toward the place of their "birth" at ever increasing velocities until they once again fuse together to form a new "egg." We have, therefore, a picture of a limited universe with a definite maximum diameter. If, after the formation of the new egg, the egg again explodes to begin a new expansion-and-contraction cycle, the universe can be considered cyclic and never-ending, a scheme generally referred to as the **oscillating-universe theory.** Is the cycle we are presently in the first one, or have there been untold numbers of cycles before this one? Perhaps even from a philosophical point of view it makes no difference whether this is the first cycle or not, because the start of the present cycle was the "beginning" of the universe as far as we will ever know.

Another theory proposed in 1948 by scientists at Cambridge University is known as the **steady-state theory.** Although most research astronomers discount the theory in its entirety, it has gained some support. According to this proposal, there was no beginning to the universe, nor will there be an end. It has appeared always the same, with but minor variations, and it always will. Matter is constantly being created in the form of hydrogen atoms that eventually will become part of the new galaxies. New galaxies replace those that move away; thus, the density of galaxies in any given region remains a constant. According to this theory, there was no cataclysmic explosion, and hence no cosmic fireball that would have created a detectable background radiation level, as predicted by other theories. On the basis of this observation, the theory has been mostly discounted, since measurements from all directions have revealed spectra radiated from objects that are consistently in the thermodynamic range of 3 K. The existence of this level of radiation is prime evidence in support of a cosmic explosion.

How large is our present universe? Since the distance and speed of various galaxies are known, one can calculate how long these galaxies, at their present speeds, must have taken to travel from the initial egg to the positions they presently occupy. Such a calculation gives a value of about 18 billion years. With this age of the universe, we can calculate the maximum distance that light could have traveled in a straight line from the center of the explosion 18 billion years ago to the present time. This distance, then, would be the maximum theoretical radius of the universe. Since light travels 9.46×10^{12} kilometers per year, the maximum value of the radius can be calculated from

$$\begin{aligned} \text{radius} &= (9.46 \times 10^{12}\ \text{km/y})(18 \times 10^{9}\ \text{y}) \\ &= 1.7 \times 10^{23}\ \text{km}\ [1.1 \times 10^{23}\ \text{mi}]. \end{aligned}$$

Since matter cannot travel at the speed of light, the actual radius of the matter portion of the present universe must be something less than this value.

Quasars

In 1960, two faint starlike objects were found to be emitting enormous quantities of radio waves. An analysis of these two objects revealed that they had characteristics quite different from those of regular stars. Consequently, they were named quasi-stellar radio sources, which was soon shortened to **quasars.** More recent studies reportedly indicate that there are millions of quasars in the universe. Quasars have been found to exhibit extraordinarily large redshifts. The redshifts of quasars generally greatly exceed the redshifts of galaxies so far investigated. One quasar appears to be receding from us at the fantastic rate of 282,000 kilometers per second, or about 94 percent the velocity of light. Even though the exact scale of the universe is unknown, this quasar is estimated to be at least 14 billion light years away. As astonishing as the recessional velocities of quasars are, these celestial puzzles have another property that has stimulated major interest in them among scientists. It has been determined that quasars eject incredible quantities of electromagnetic energy into space. Even though they are the most distant objects, they are among the brightest. The most distant object ever observed, presently considered to be a quasar, is also the most luminous object observed. Although quasars in general are only about as large as our solar system, they pour out energy at the rate of 25 trillion suns, or more energy than about 100 Milky Way galaxies. So far astronomers have been unable to unravel the riddle of the paradoxical quasars. A theory under consideration perceives quasars as galaxies, the cores of which have become overly active compared to those of other galaxies. The source of the vast amount of energy emitted from a quasar appears to be the core itself.

Not all quasars emit large quantities of radio energy. In fact, only 1 quasar in 300 is an extraordinary radio emitter. In spite of the rarity of quasars that are very active in the emission of radio waves, the name quasars, implying a radio source, has been retained.

Summary

Our galaxy, the Milky Way, is a lens-shaped collection of over 250 billion stars isolated in the almost endless space of our universe by vast distances from other galaxies. Our sun is approximately two-thirds away from the center of our galaxy. At night we can see a portion of our galaxy by looking at the Milky Way, which meanders from horizon to horizon across the sky.

The flattened lens-shaped galaxies, such as the Milky Way, are called spiral galaxies. Elliptical galaxies vary in shape from spherical to nearly lens-shaped and are the most common type of galaxy in the universe. Other galaxies have no particular symmetrical shape and are referred to as irregular galaxies.

Currently, the most widely accepted theory of the history of the universe is the big bang theory. About 18 billion years ago all the matter in the universe was collected into a gigantic single mass that exploded with titanic violence. All the matter in the universe is still expanding outward from that initial explosion. Light from other galaxies shows a Doppler redshift that indicates that the universe is still expanding today. Perhaps the expansion will eventually slow to a stop and the universe will begin to contract. The contraction will continue until once again all the matter will be compressed into a new single mass. From this mass a new cycle of explosion, expansion, and contraction will occur.

Questions and Problems

Nebulas

1. Why are some nebulas dark and others light in appearance?
2. How are nebulas formed? What is their major chemical composition?
3. What is meant by a second-generation star? How is the composition of these stars different from first-generation stars?

The Milky Way

4. Describe the dimensions of the Milky Way galaxy in light years, the general shape of the galaxy, and the sun's approximate location in the galaxy.
5. What are the major components of a galaxy?
6. Where are globular clusters primarily found?
7. This chapter refers to the sun as one of *perhaps* 250 billion stars in our galaxy. Other estimates put the number of stars at 400 billion, and the current estimate of the mass of our galaxy is 500 billion times the mass of the sun. Why do you think the number of stars in our galaxy is so uncertain? And why is the mass of the galaxy considerably greater than the total mass of the stars in the galaxy?
8. What is the major difference between the stars in the nucleus of the galaxy and those in the spiral arms?

Other "Island Universes"

9. What are the three major classifications of galaxies? Which type is most common?
10. Why are some elliptical galaxies lens-shaped while others are quite spherical?
11. What is the range in the number of stars in typical spiral galaxies?
12. What is the "local group" of galaxies?

Astronomic Distances

13. Define a light year. How is its value determined?
14. Why has it been said that astronomy is the scientific study of the past history of the universe.
15. Why do astronomers constantly seek to build telescopes with larger apertures?

Cosmology and the Redshifts

16. What is cosmology?
17. What is the "redshift" as it is applied in astronomy? How is the redshift interpreted?
18. What is the "big-bang" theory of the universe? How does this theory fit in with the "oscillating-universe" theory? Cite some evidence that supports the big-bang theory.
19. Cite some evidence that seems incompatible with the "steady-state" theory of the universe.

Quasars

20. What are quasars? What are two unusual characteristics they exhibit?

PART FIVE

The origin of Earth, its age, its variety of features, and its diverse forms of life have been under study ever since intelligent beings appeared on the planet. Knowledge about our planet has been made available through the efforts of inquisitive people who approached problem solving through the collection and interpretation of pertinent data.

Our knowledge of Earth may be divided into several closely related sciences, including geology, oceanography, and meteorology. And since Earth is a planet, a portion of the accumulated knowledge belongs in the field of astronomy. The study of Earth may also include the various resources it can provide, how we obtain them, and how we affect our environment when we use them.

We commonly think of geology as the study of terrifying volcanic eruptions, devastating earthquakes, or destructive tidal waves, but this area of the physical sciences actually involves the study of Earth and *all* its changes, regardless of whether the changes occur quickly or slowly. The study of geology is traditionally separated into Earth's physical and historical aspects. Physical geology focuses mainly on the materials that make up our planet and the changes they undergo. Historical geology is more concerned with Earth's origin and development and their impact on the evolution of life.

For centuries, the science we know today as geology simply existed as scattered facts and observations with no central theme. Miners and others concerned with the extraction of valuable ores from Earth had limited knowledge about the solid portion of the planet, while natural philosophers formulated various theories, usually undocumented by systematic investigation.

Various natural resources, such as metals, coal, clay, and salts, were considered living matter that grew and replenished themselves. It was a somewhat common practice to close mines periodically to permit the ores to regenerate. Fossils were generally considered to be failed attempts by nature to produce viable plants and animals. Although there were some individuals in the seventeenth century who did believe that fossils were the remains of ancient life, even they generally thought that the life-forms had been destroyed by great floods and washed from the mountains to become fossilized.

The action of such natural catastrophes aided in the formation of rock layers and their fossil content, according to some eighteenth-century investigators, but others emphasized the role that heat and volcanic activity played in raising the mountains and dry land above the level of the seas. These

Earth Sciences: Our Blue Planet

theories, known as Neptunism and Vulcanism, respectively, created their shares of controversy. Theorists argued the two concepts well into the nineteenth century, when the origin and meaning of fossils, along with the roles of water, heat, and volcanic activity in the development of Earth, finally became clarified.

Of all the sciences, geology may be the one most tied to the dimension of time, since geologists must learn not only what geologic processes are occurring today, but whether those same processes have been occurring throughout all geologic history. The present and the past are inescapably entwined, and the geologist strongly relies on the premise that the various geologic processes that operate in and on Earth today are the same ones that acted in and on Earth in the past. Only present geologic forces should be used to explain past formation of the rocks, according to these investigators. Today, geologists primarily use sophisticated instruments to study the degradation of mountains, the drift of continents, and the upheaval of ocean floors.

Meteorologists study Earth's atmosphere. They are interested not only in the physics, chemistry, and dynamics of the air that surrounds our planet, but also in the atmosphere's direct effects on Earth's surface, the oceans, and on life in general. Among their goals is to develop a deeper comprehension, a more accurate means of prediction, and a better means of artificially controlling atmospheric conditions. Their investigations, then, involve more than simply discovering and describing the physical conditions that create our weather.

The task of conserving our dwindling natural resources, such as coal and oil, is shared by scientists from all fields. The search for alternative means to produce our needed energy may well represent the most outstanding opportunity in history for scientists to work toward a common goal that is crucial to the well-being of society in the immediate future. Scientists from many disciplines and practically all nations have become actively involved. Some scientists have expressed concern about how Earth's overall environment has been detrimentally affected by the human race and are making concerted attempts to ensure that further, possibly irreparable, damage does not occur. It is the desire of concerned citizens the world over to see that the balance of nature on our fragile Earth remains intact.

20 Planet Earth

Chapter outline

Key terms/concepts

The Earth, observed by an ocean-observing satellite, showing our planet's curvature against the sky.

After most of the hydrogen and helium that surrounded proto-Earth had disappeared, there remained but a bleak shrunken core of material held together by mutual gravitational attraction. These gases had escaped into the reaches of outer space from the new planet because of the relatively low speed required to do so and because of the high kinetic energy, thus speed, the gases had attained during Earth's early stages of formation. The aggregate of atoms, molecules, and unsorted chunks of material that remained was destined to become that which composes our Earth today.

As Earth was forming, so was Earth's lone satellite, the moon. The two bodies were probably formed together as gravitationally bound twin condensations during the creation of most members of the solar system.

The Infant Earth

There still remained in the general vicinity of Earth, as well as in the vicinity of the other planets, debris from the creation. This debris ranged in size from dust particles to huge mountains of rocklike material circling the sun in the general orbit of Earth. With the passage of perhaps a billion years, these myriads of debris were slowly captured by the more massive Earth until eventually Earth's path around the sun was almost swept clean.

The overall process of gravitational contraction had caused warming of the infant planet. Also, as Earth attracted the debris in its vicinity, the fragments of material would crash into Earth at very high speeds and the kinetic energies of the fragments would be converted into heat. This continual release of heat energy, produced by constant bombardment of Earth, resulted in further warming of the planet. In addition, another heat-releasing process was taking place. Uranium, thorium, radium, and other unstable elements were slowly liberating heat from radioactive processes of decay that are an intrinsic part of the life processes of radioactive elements. Over millions and millions of years, these radioactive elements liberated vast quantities of heat. Between the heat liberated by radioactive decay and that generated by debris crashing into Earth, Earth became very hot. It probably never became hot enough to completely melt but it did reach a sufficiently high temperature to allow various elements and compounds to segregate and migrate according to their chemical and physical similarities and affinities.

It was during this period, and perhaps extending up to the present day, that the very dense iron from various parts of Earth slowly separated and migrated under the influence of gravity toward the center of the planet. Since certain types of meteorites, which can be considered relics of the solar system's creation, contain iron mixed with 5 to 15 percent nickel, it is reasonable to assume that nickel also migrated with the iron.

Eventually, at least part of the iron-nickel core of Earth melted. Although some scientists do not agree, it is possible that the outer portion of Earth surrounding the core melted also. One reason for this assumption is that although Earth is at least 4.6 billion years old, no one has so far found any rocks definitively shown to be more than 4 billion years old, nor any minerals older than 4.3 billion years old. However, a team of geologists has reported the discovery of rocks believed to be in excess of 4 billion years old in western Australia, increasing the age of the oldest known rocks by 300 million years. If their system of age determination proves valid, there may still exist some of Earth's original crust. It is of interest to note that samples of soils and certain rocks retrieved by astronauts from the moon have an estimated age of 4.6 billion years.

The iron-nickel alloy that makes up Earth's core started melting when the temperature of the core reached about 1500°C (iron melts at 1535°C), but it is estimated that today the core has a temperature of approximately 3500°C.

It is because of the molten portion of the iron-nickel core that Earth has a strong external magnetic field, resulting in north and south magnetic poles as well as the Van Allen radiation belts that surround our planet from 3230 kilometers to 8060 kilometers from the planet's surface. Mercury, Venus, and Mars have no appreciable magnetic fields and, hence, probably no molten iron-nickel core.

Earth's magnetic field is produced by Earth's rotation, which sets up eddies that circle from west to east within the liquid outer core. The result is an electric current moving in the same direction. Just as a magnetic field surrounds any conductor carrying an electric current, the electric current in the core acts as a huge electromagnet that produces a magnetic field external to our planet with lines of force extending north and south. (Note that Earth's magnetic field is oriented at right angles to the core's electric field and is roughly parallel to the axis of Earth's rotation.)

Besides the elements and compounds initially present in the primeval cloud, many other substances were formed during the formation of the solar system. On proto-Earth the oxygen present reacted with hydrogen, silicon, magnesium, iron, aluminum, calcium, carbon, sodium, potassium, and so on, to produce their corresponding oxides: H_2O, SiO_2, MgO, FeO and Fe_2O_3, Al_2O_3, CaO, CO and CO_2, Na_2O, K_2O, and so forth. Some of these oxides in turn reacted with each other in the presence of differing amounts of heat to produce minerals of other classifications (a mineral is a naturally occurring inorganic substance having distinctive chemical and physical properties). Minerals are often divided into *sial* and *sima* materials. The sial minerals are rich in silicon and aluminum, whereas the sima materials are rich in silicon, magnesium, and iron.

The Shape of Earth

As early as the fourth century B.C., some people realized that Earth must be round. Aristotle reasoned that the world must be round because as one travels north (or south), new stars and constellations become visible that were not visible before, while known stars and constellations slowly disappear below the horizon. Much later in history, when ships with very tall masts sailed the seas, people noticed that as a ship sailed into the distance, it did not simply

Old Ideas Die Hard

Earth is flat! So contend members of the Flat Earth Research Society of Lancaster, California. They assume that Earth is flat and circular like a phonograph record. Members of the society also believe that the sun and the moon are both only 52 kilometers in diameter and that the stars are no more than 4030 kilometers from Earth (note that the moon and sun, according to scientists, are approximately 384,400 kilometers and 15 million kilometers from Earth, respectively).

Members of the society also claim that the moon landings were staged in Hollywood studios. They consider all the publicity concerning the space shuttle flights to be ludicrous since one cannot "orbit" a flat Earth.

Only a handful of people in the United States believe that Earth is flat, but even so it is startling that there exists a Flat Earth Research Society in this age of space flights to the moon and other planets.

When Christopher Columbus set sail, he knew that Earth was spherical. He expected to sail from Portugal to India around the curvature of Earth. He did not know, of course, that the continent of North America lay between Europe and India. Since at first Columbus did not realize that he had not reached India, he assumed the people he initially encountered were inhabitants of India; consequently, the natives of North America are called Indians.

Indeed, as early as 340 B.C., Aristotle reasoned that Earth must be spherical because Earth's shadow on the moon during a lunar eclipse was part of a circle (see chapter 21). Aristotle also thought, as did most ancients, that the sphere was the perfect geometrical shape, and that God would not have made an imperfect Earth ■

appear to grow smaller and smaller until it disappeared. Instead, the hull of the ship soon disappeared over the curve of Earth while the tops of the masts remained visible—another indication of Earth's spherical shape.

Earth, however, is not a perfect sphere but an oblate spheroid—which means that the diameter of Earth at the equator is greater than it is at the poles. (If Earth were longer through the poles than through the equator, it would be a prolate spheroid.) This flattening at the poles, with the subsequent bulging at the equator, is the result of Earth's rotation on its axis. As a result of Earth's rotation, an object at the equator is traveling approximately 1700 kilometers per hour. The difference in Earth's equatorial diameter, 12,785 kilometers, and its polar diameter, 12,742 kilometers, is quite small, only 43 kilometers; consequently, as astronauts approach Earth from outer space, they describe Earth's appearance as a perfect sphere.

Since the outward force exerted on a body at the equator, due to Earth's rotation, opposes the force of gravitational attraction for the body, there is a slight decrease in the weight of the body. At the poles, where the force due to rotation is zero, the weight of a body is exactly what would be expected by the attractive force of gravity. Thus, a person who weighed 136.4 kilograms (1336 N) at one of Earth's poles would weigh only 135.9 kilograms (1332 N) at the equator. Actually, there is a second factor that affects the weight of objects at the equator and at the poles. Since Earth's rotation flattens Earth in the polar regions, objects at the poles are 21.8 kilometers closer to the center of Earth than are objects at the equator. Since an object weighs less the farther away it is from Earth's center, objects at the equator weigh less than objects at the poles. It is both factors—Earth's rotation and Earth's oblate shape—that are responsible for the decrease in the weight of objects at the equator compared with the weight of objects at the poles.

The Structure and Composition of Earth

After the new sun swept away the mist surrounding proto-Earth, there remained the rock mass that was infant Earth. The infant Earth now had a diameter of about 12,765 kilometers, a volume of 1.1×10^{21} cubic meters, a mass of 6.0×10^{24} kilograms, and a density of 5.5 grams per cubic centimeter (5.5 times the density of water). It is difficult to comment on the overall inner structure of the infant Earth, but it is useful to assume that after a period of a few hundred million years the structure of infant Earth was essentially similar to its structure today. The interior of present Earth is structured in zones, as shown in Figure 20.1.

Knowledge of the different zones within Earth is deduced largely from a study of the speed and nature of earthquake waves as they pass through different portions of our planet. In a manner of speaking, scientists use earthquakes to "X-ray" Earth.

The boundary between the crust and the mantle was discovered by recording the speed with which earthquake waves travel through Earth. Earthquake waves travel at a speed of 6.5 kilometers per second in the crust, but abruptly increase to 8 kilometers per second on entering the upper part of the mantle. The boundary between the two zones is called the *Mohorovičić discontinuity,* after its discoverer, Andrija Mohorovičić, a Yugoslavian geophysicist (see Fig. 20.4). This variation in velocity is accounted for by assuming that the rocks of the mantle and the crust have different compositions.

The geoscience of earthquakes and the properties of Earth deduced from earthquake waves, both natural and man-made, are known as *seismology,* from the Greek work *seismos,* "shock" or "quake."

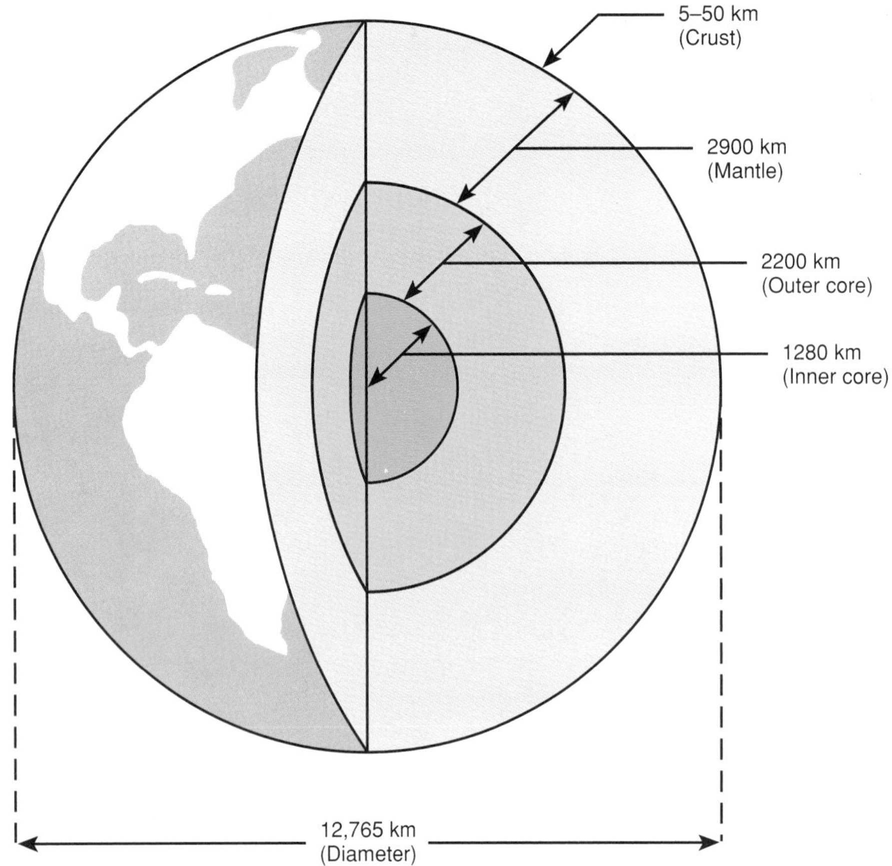

Figure 20.1 The structure of Earth and the approximate thickness of its respective layers.

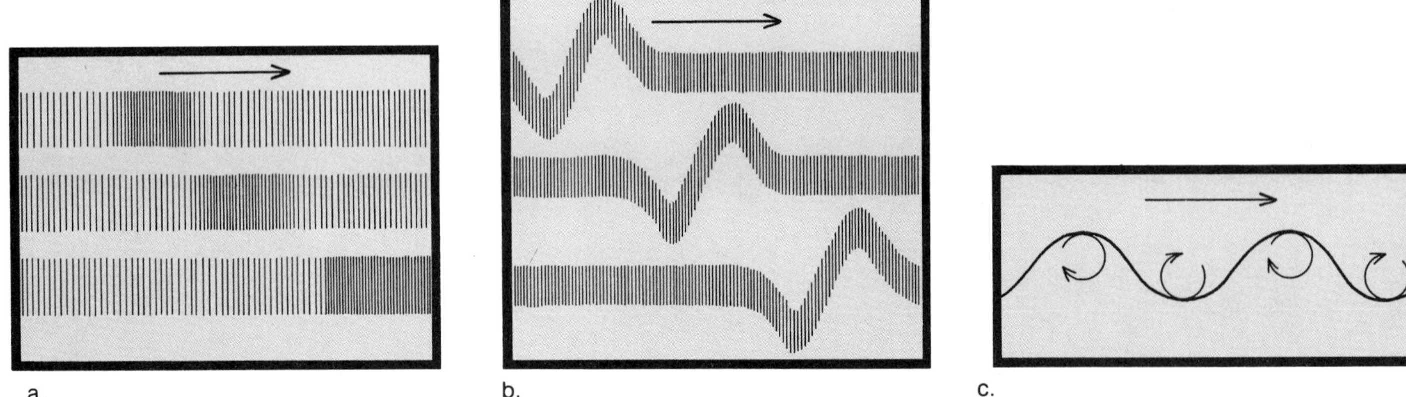

Figure 20.2 Earthquake waves. (*a*) P (primary) waves are compressional waves. (*b*) S (secondary) waves are shear waves. (*c*) L (long) waves, like surface-water waves, move in a sine wave pattern that is superimposed on the small circular motion of solid particles or water molecules making the L wave.

An earthquake produces three kinds of seismic waves:

1. P (primary) **waves**—*longitudinal* and *compression waves.* P waves, like sound waves, consist of particles that vibrate in the direction in which the waves move. The way in which P waves move through Earth is similar to a domino effect. As the first standing domino topples, it disturbs the second domino, and so forth (see Fig. 20.2a). P waves can travel through both solids and liquids.
2. S (secondary) **waves**—*transverse* or *shear-type waves.* S waves travel at right angles to the direction in which the waves are moving, just as the waves produced in a rope that is jerked sideways at one end (see Fig. 20.2b). S waves cannot travel through a liquid.
3. L (long) **waves**—*surface waves.* L waves do not travel through Earth, but are confined to Earth's crust. These waves are similar in pattern to ocean waves (see Fig. 20.2c).

A startling thing happens to S waves from an earthquake when they reach a depth of 2900 kilometers below the surface, which is about halfway to the center of Earth. They are stopped completely.

The "Year" of the Scientist

What has been called the greatest coordinated scientific endeavor of all time took place from July 1, 1957 to December 31, 1958. This period is officially known as the *International Geophysical Year (I.G.Y.)*. Approximately 30,000 scientists from 66 countries were involved. Many of the investigators conducted their studies at or near more than 2000 established stations. An almost overwhelming amount of data about our planet and outer space was collected, evaluated, and classified.

In October 1957, the USSR launched into orbit Earth's first artificial salellite. With the aid of balloons, rockets, and the earliest of orbiting satellites, an amazing number of secrets of Earth's atmosphere were unveiled, including the presence of two intense bands of radiation, now known as the Van Allen radiation belts. The slightly pear-shaped appearance of Earth also was detected. Cosmic and X-ray bombardment of Earth's upper atmosphere was measured with such sensitivity that a change in latitude of less than 10 kilometers was detectable.

The I.G.Y. was initiated in conjunction with the projected peak of activity on the sun. Several violent upheavals on its surface were observed and studied during this time. Solar disturbances were noted to relate to radio communication interference, such as static, as well as to the auroras seen from Earth.

Earth's surface and interior were investigated in detail during the I.G.Y. Teams of scientists concentrated their studies on the polar ice caps and the oceans, along with the life-forms each supports. Much was learned about the characteristics of earthquakes and the variations in Earth's gravitational force. The first recorded surface crossing of Antarctica occurred during the I.G.Y. to obtain a profile of ice thickness and to uncover clues to Earth's past hidden in the depths of the ice deposits. Ships, equipped with complete laboratories, measured water temperatures, tidal actions, and ocean-floor conditions. The rich nutrients found in the depths of the oceans revealed the importance of the cyclic ocean currents to the survival of all sea life.

The wealth of information collected during the I.G.Y. was sorted, indexed, and supplied to each of three major centers located in the United States, Russia, and western Europe. Scientists are still reaping major benefits from this priceless information ■

This can mean only that the S waves have reached a liquid zone, which must be molten core material.

Note from Figure 20.1 that the center portion of Earth consists of an inner core and an outer core. Although the outer core is liquid, the inner core is solid. The solidity of the inner core is implied because P waves travel through the outer liquid core at a set speed (slower than their speed through a solid), but at a distance of 1290 kilometers from Earth's center the P waves suddenly increase in speed, indicating they are now traveling through solid material (Fig. 20.3).

It is estimated that the chemical composition of our planet is as follows:

Section of Earth	*Substance*	*Percentage (by weight)*
Crust	SiO_2 miscellaneous	1–2
Mantle	SiO_2	32
	MgO	23
	FeO	6–7
	Fe_2O_3	6–7
	Al_2O_3	2
	CaO	2
	Na_2O	1
Core	Fe	24
	Ni	4
	Si	4

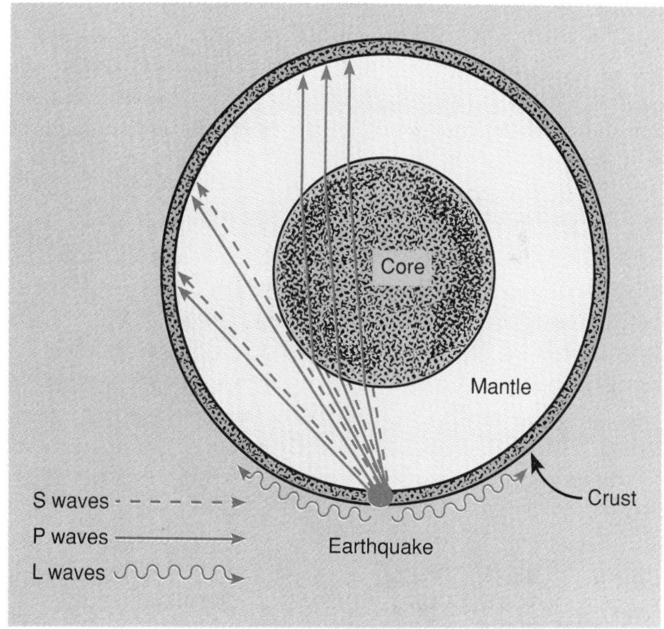

Figure 20.3 The bending of certain earthquake waves gives valuable information about the various zones of Earth's interior. The L waves also shown cause widespread devastation.

These percentages are very close to the percentages obtained for the chemical composition of certain meteorites. This similarity is not unexpected, since we assume a common origin for planets and meteorites.

The core accounts for 32 percent of the mass of our planet but only 22 percent of its volume. Obviously, the core must be made of very heavy material. The core's density is calculated to be 15 grams per cubic centimeter in order for Earth as a whole to have an overall density of 5.5 grams per cubic centimeter. The assumption that the core is an iron-nickel alloy is consistent with the calculated density.

It has already been pointed out that the motion of the molten iron-nickel outer core is responsible for Earth's magnetic fields. It is interesting to note that, for some as yet unexplained reason, Earth reverses the polarity of its magnetic poles with fair regularity. The north and south magnetic poles are reversed approximately every million years. The present magnetic field polarity has lasted for the past 700,000 years.

Between the core and the crust is a zone approximately 2900 kilometers in thickness called the **mantle.** The mantle, which contains slightly more than two-thirds of Earth's mass, must be somewhat solid since it transmits S waves during an earthquake. It appears that the mantle probably is composed of **peridotite,** a rock made up mainly of the iron-and-magnesium-rich mineral olivine (olivine = 2 parts magnesium + 2 parts iron + 1 part silicon + 4 parts oxygen).

The high temperatures and high pressures that exist in the mantle make it slightly plastic in nature. Consequently, there are huge convection currents operating below the crust, the slow but constant movements of which greatly affect the surface features of our planet.

The thin outermost shell of Earth, which accounts for only about 1 percent of Earth's mass, is called the crust. This is the portion of Earth that we live on and are most familiar with; it is, therefore, that portion of our planet about which we have the most information. The crust varies in thickness from about 8 kilometers to 48 kilometers, being thickest under the continents and thinnest under the oceans. In most places on our planet, the crust can be divided into an upper and lower part. The lower part has properties similar to the rock layer underlying the oceans and consists mostly of **basalt.** Chemically, the crust is mostly oxygen and silicon, as noted earlier in this section. Interestingly, only eight elements make up approximately 98.6 percent of the crust, with oxygen and silicon constituting almost three-fourths of the total. The crust is illustrated in Figure 20.4.

The infant Earth must have had only a modest amount of water on its surface. A considerable amount of the water that exists today in our ocean basins was produced by "outgassing" (the squeezing out of gases and liquids by high pressures, high temperatures, and crystallization of minerals) from the mantle and crust of our planet. This outgassing took place through volcanoes, hot springs, and fumaroles over aeons and, along with the water initially present, produced the only "water" planet in our solar system. Outgassing yielded not only water but carbon dioxide, which resulted in a fairly constant deposition rate of limestone (calcium carbonate) in the seas.

The Plastic Mantle and Isostasy

We have all been awed by the lofty mountains that exist on Earth, but we would be even more impressed if we realized that nine-tenths of the mass of a mountain exists beneath the surface crust that we stand upon. Indeed, an analogy may be drawn between a mountain and an iceberg. Approximately nine-tenths of the mass of an iceberg is below the surface of the water. This means, of course, that nine-tenths of the iceberg must displace its weight of water to give the necessary buoyancy to float the whole mass. The same is required for our immense mountains. Some mountains have "roots" that extend down into the mantle for 65 kilometers and there float in the denser mantle. Mountains and the roots are, however, part of the crust of Earth. As the massive mountains are eroded by wind, water, ice, and earthquake action, the buoyancy of the monolithic mass slowly pushes the mountains up, so that their towering heights are maintained much longer than would otherwise be possible.

This upward movement, in response to the decreased weight of a mountain (caused by erosional agents), is similar to the response of a heavily ladened ship that sits very low in the water but, while being unloaded, slowly rises in the water in response to its decreased weight.

How can the mantle of our Earth be sufficiently fluid to "float" an immensely heavy mountain range? If the mantle is rigid (as it must be in order to transmit S waves from earthquakes), it is difficult to imagine how it can be fluid enough to act as a liquid and buoy massive structures such as mountains in the same fashion as the ocean floats icebergs. In order to understand this apparent anomaly of the mantle's physical properties, consider the common candle. A candle will most certainly break if struck a sharp blow, but a long candle supported horizontally by one end for a long period of time will slowly but inevitably bend under the force of its own weight. The same types of physical characteristics are exhibited by the mantle of Earth. The mantle is sufficiently rigid to break under stresses and strains to cause earthquakes and transmit S waves, but under tremendous pressures and temperatures over long periods of time the mantle acts as a plastic "fluid" material that will float huge masses existing on the crust of our planet. *This condition of equilibrium, whereby tall heavy landmasses are supported by deep roots in the denser plastic mantle,* is known as **isostasy,** from two Greek words: *isos,* "equal," and *stasis,* "standing still."

Composition of Earth's Crust	
Element	*Percentage (by weight)*
Oxygen	46.6 } 74.3%
Silicon	27.7
Aluminum	8.1
Iron	5.0
Calcium	3.6
Sodium	2.8
Potassium	2.6
Magnesium	2.1
All other elements	1.5

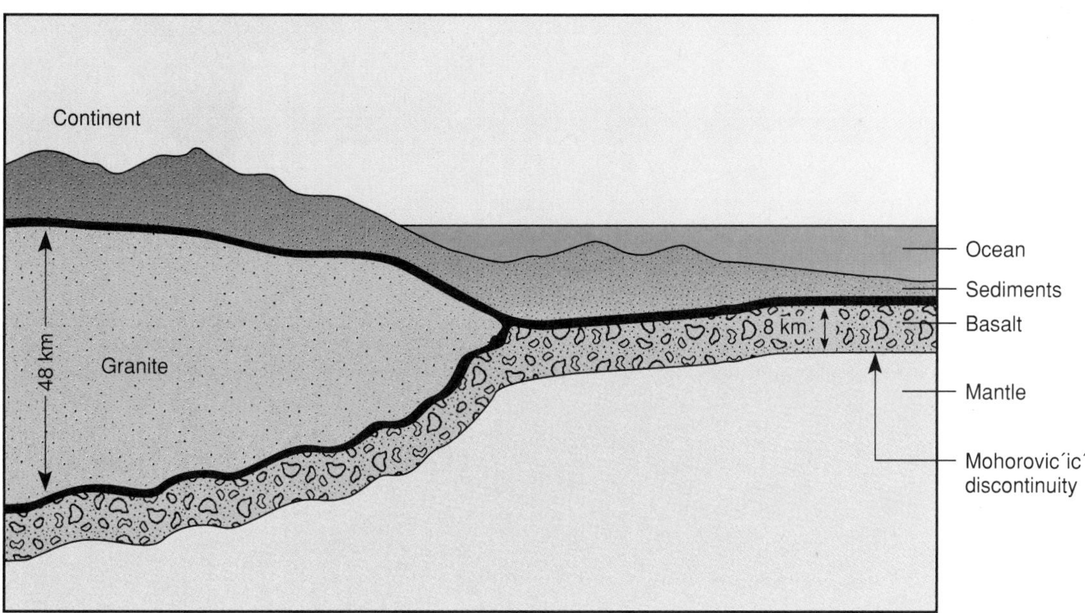

Figure 20.4 The crust of Earth, showing the basalt layer under the oceans and under the granite of the continents.

Continents That Drift—Plate Tectonics

While looking at maps of the world, you may have noticed that the opposing shorelines of the continents of South America and Africa appear as though they would fit as two pieces of a jigsaw puzzle (see Fig. 20.5). The surprising fact is that at one time in the distant past, these two continents, as well as the others, were a single landmass. Slowly, over hundreds of millions of years, they drifted apart to reach their present positions.

For at least half a billion years collections and fragments of present continents slowly wandered at random on Earth's surface. The consolidation of most of these landmasses into two great masses called *Laurasia* and *Gondwanaland* was complete about 350 million years ago. The present areas of North America, Europe, Asia, and Greenland were welded together to form Laurasia, which was positioned in the Northern Hemisphere. Gondwanaland, located in the Southern Hemisphere, consisted of South America, Africa, India, Australia, and Antarctica. Approximately 250 million years ago, Laurasia and Gondwanaland collided. The impact from the collision welded the two huge land masses into one supercontinent called **Pangaea** (from Green: *pan,* "all"; *gaea,* "land" or "earth"). This solitary, behemoth landmass, created by the colossal forces of nature, sat alone in a single world-encircling sea.

About 200 million years ago, when dinosaurs reigned over the lands of Earth, ruptures appeared in the Pangaea supercontinent, and the fragments that formed drifted slowly away from each other at an average speed of approximately 2 centimeters per year. The North and South American continents drifted westward, and India broke away from Antarctica and drifted a considerable distance northward to eventually crash into the Asian continent. This collision resulted in the formation of the Himalaya Mountains

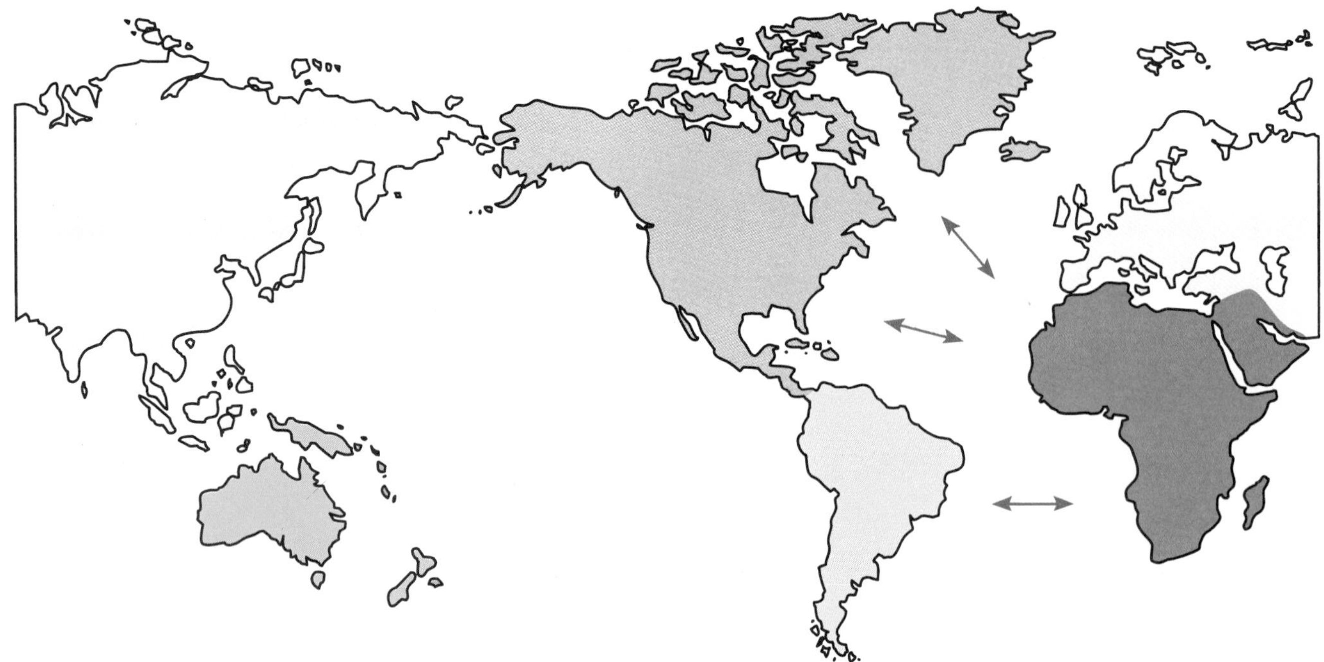

Figure 20.5 The shorelines of the American continents would fit the shorelines of Europe and Africa like pieces of a jigsaw puzzle.

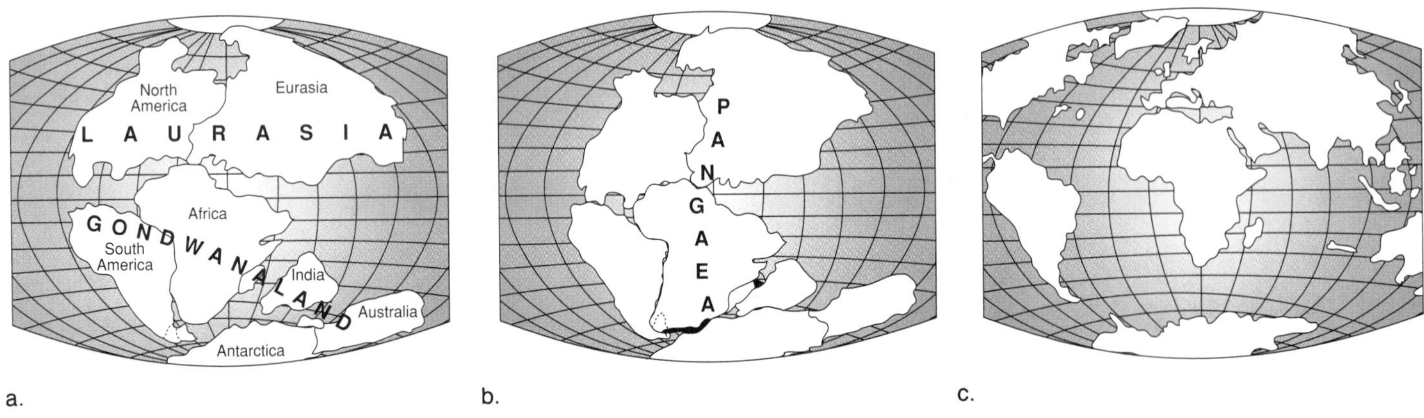

Figure 20.6 The landmasses of Earth: (*a*) about 350 million years ago—Laurasia and Gondwanaland; (*b*) about 250 million years ago—Pangaea; and (*c*) today.

between Tibet and India. Australia drifted eastward, and Antarctica drifted southward to its present position at the South Pole. Figure 20.6 shows the changes in the positions of Earth's landmasses over the past 350 million years.

What evidence is there to suggest that Antarctica at one time was part of Gondwanaland and was, therefore, attached to the present African continent? Geologic studies of the Antarctic continent reveal that during the Permian period (about 250 million years ago) large portions of the Antarctic continent were covered by lush tropical forests and swamps, and *Lystrosaurus,* an early dinosaur type, roamed Antarctica. How could this situation have existed if Antarctica had always been in its present position at the frigid South Pole? The answer, of course, is that it has *not* always been in its present position. Rather, 200 million years ago Antarctica was attached to Africa only 15 to 30 degrees south of the equator. It was during this period that *Lystrosaurus* and the hot,

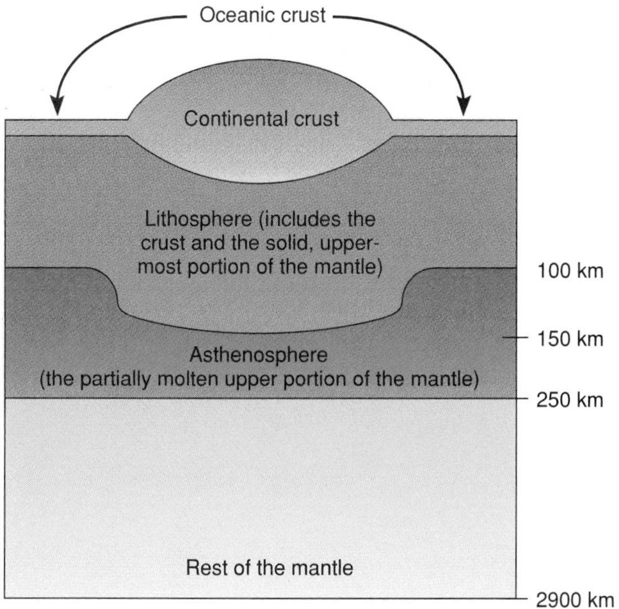

Figure 20.7 The outermost layer of Earth is called the lithosphere. The lithosphere is a rigid solid layer that floats on the partially molten asthenosphere. The top portion of the lithosphere is a thin crust of basalt under the oceans and a much thicker continental crust of granite. The astenosphere is the somewhat molten portion of the mantle.

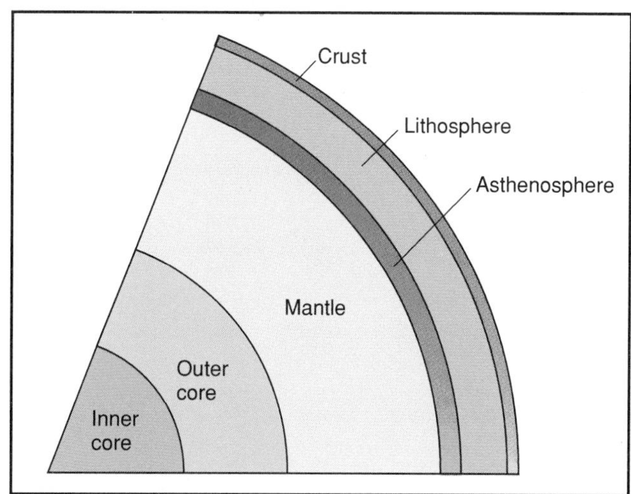

Figure 20.8 The lithosphere and asthenosphere portions of the mantle.

moist climate conducive to flourishing vegetation existed. Moreover, fossils of plants and animals that lived about 225 million years ago, including *Lystrosaurus,* found in Antarctica, have also been found in the southern parts of South America, Africa, and India. In addition, some mineral deposits found along the west coast of Africa continue on the east coast of South America. Since *Lystrosaurus,* as well as the flora present during the same period of time, could not have survived an ocean trip between these continents, one must assume that Antarctica, South America, Africa, and India were joined together during the Permian period.

These "continental islands," floating on a sea of heavy mantle materials, drifted to the various corners of the globe. The North and South Atlantic oceans were slowly formed by the westward drift of the North and South American continents, until today a distance of approximately 2400 kilometers separates these two continents from their original positions against Europe and Africa.

The obvious question arises: Are the continents still drifting today? The answer is yes. The continued movement of continents is being measured and studied by scientists from many nations at the present time. The concept of continental drift, first proposed by Alfred Wegener in the early nineteen hundreds, was initially greeted with general skepticism, but is now accepted as a scientific fact.

In the 1970s, the unifying theory of **plate tectonics** emerged to explain the idea of drifting continents and seafloor spreading and to provide a possible mechanism for the underlying cause of the continents' motion. The central idea of plate tectonics depicts Earth as being divided as shown in Figures 20.7 and 20.16.

The **lithosphere,** from the Greek *litho,* meaning "rock," is *a strong, solid outermost shell* about 7 kilometers thick under the oceans and 100 to 150 kilometers thick under the continents. The lithosphere includes the crust and the upper part of the mantle. The lithospheric shell rides on the weak, partly molten **asthenosphere,** from the Greek *astheno,* meaning "weak" (also a portion of the mantle), which ends at a depth of about 250 kilometers.

The crust of Earth is divided into eight large lithospheric plates and many smaller plates which move independently of each other (see Figs. 20.9 and 20.16). The lithospheric (or crustal) plates that include the crust move relatively to each other. The plates apparently move in various directions depending on the direction of the movements of the convection currents within the mantle of Earth. Some plates are sliding past each other, others are moving away from each other, and still others are colliding with each other (see Fig. 20.16).

The movements of the crustal plates cause great geological stresses between the moving rock masses, resulting in many earthquakes and volcanoes. The close relationship between geologic processes occurring between plate boundaries and the general distribution of earthquakes is shown in Figure 20.10.

Seafloor Spreading

Separating, divergent plate boundaries typically produce very large structures in the oceanic crust called mid-ocean ridges. The ridges encircle Earth to form the largest single geologic structure on our planet. This behemoth undersea mountain range encircles the continents of Earth and extends for more than 75,000 kilometers (see

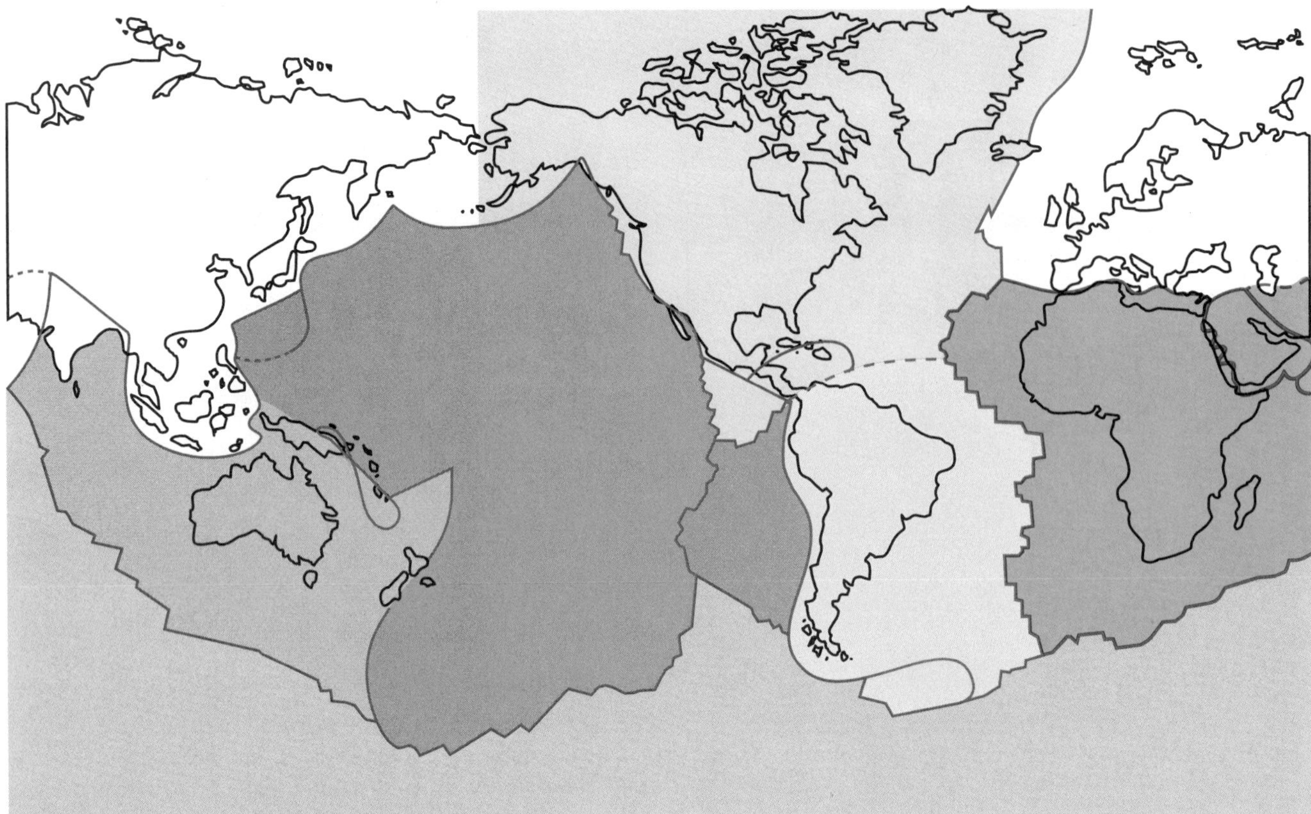

Figure 20.9 The crustal plates of Earth.

Fig. 20.11). It is at these ridges that material from the hot mantle issues forth and spreads out along both sides of the ridges. This outpouring produces spreading of the seafloors. Figure 20.12 shows the convection currents in the mantle and Figure 20.13 illustrates seafloor spreading at a mid-ocean ridge.

The plastic upper mantle, the asthenosphere, being lighter than the cooler lithosphere, wells up and creates new oceanic crust. This spreading causes the ocean floor to move away from the ridges and toward the continents. As new seafloor is formed at a ridge, the seafloor splits lengthwise, and each newly formed strip moves in opposite directions away from the ridge. The youngest seafloor is at the mid-ocean ridges, becoming increasingly older with increasing distance from the ridges. The oldest sections of the ocean floor in the North Atlantic are regions near the North American continent and Europe. These sections are 200 million years old and were formed when the North American continent started separating from Pangaea and commenced its long journey westward.

The mid-ocean ridges are the scars remaining from deep planetary wounds from which molten mantle material constantly issues forth. These are the same wounds that opened up under the ancient continent, Pangaea, to cause the monolithic structure to fragment into smaller continental masses.

The concepts of seafloor spreading and plate tectonics were significantly reinforced during the 1960s when magnetic fields along the ocean floors were studied in detail. Evidence initially collected in the 1930s was added to newer data and scientists concluded that Earth's magnetic field has reversed its polarity in a rather periodic fashion. During *magnetic reversal,* Earth's north magnetic pole and south magnetic pole exchange positions. The preponderance of evidence for magnetic reversals originated from the study of lava flows that occurred on the continents. At various sites on Earth, magnetic particles in solid lava were found to point in a direction that differs from the present-day north and south directions, indicating a change of magnetic orientation. The magnetic particles in the lava are very small crystals of the mineral magnetite (Fe_3O_4), called lodestone in the form of larger crystals or rocks.

Numerous periods of normal and reverse magnetization are noted in continental lava flows. Magnetic reversals in the oceans near the mid-ocean ridges offer additional evidence to the supposition that Earth's seafloor is indeed spreading. As spreading lava from mid-ocean ridges cools, magnetic particles within the lava line up with the magnetic poles of Earth. Once the lava hardens, the direction of magnetism in the lava rock is frozen in

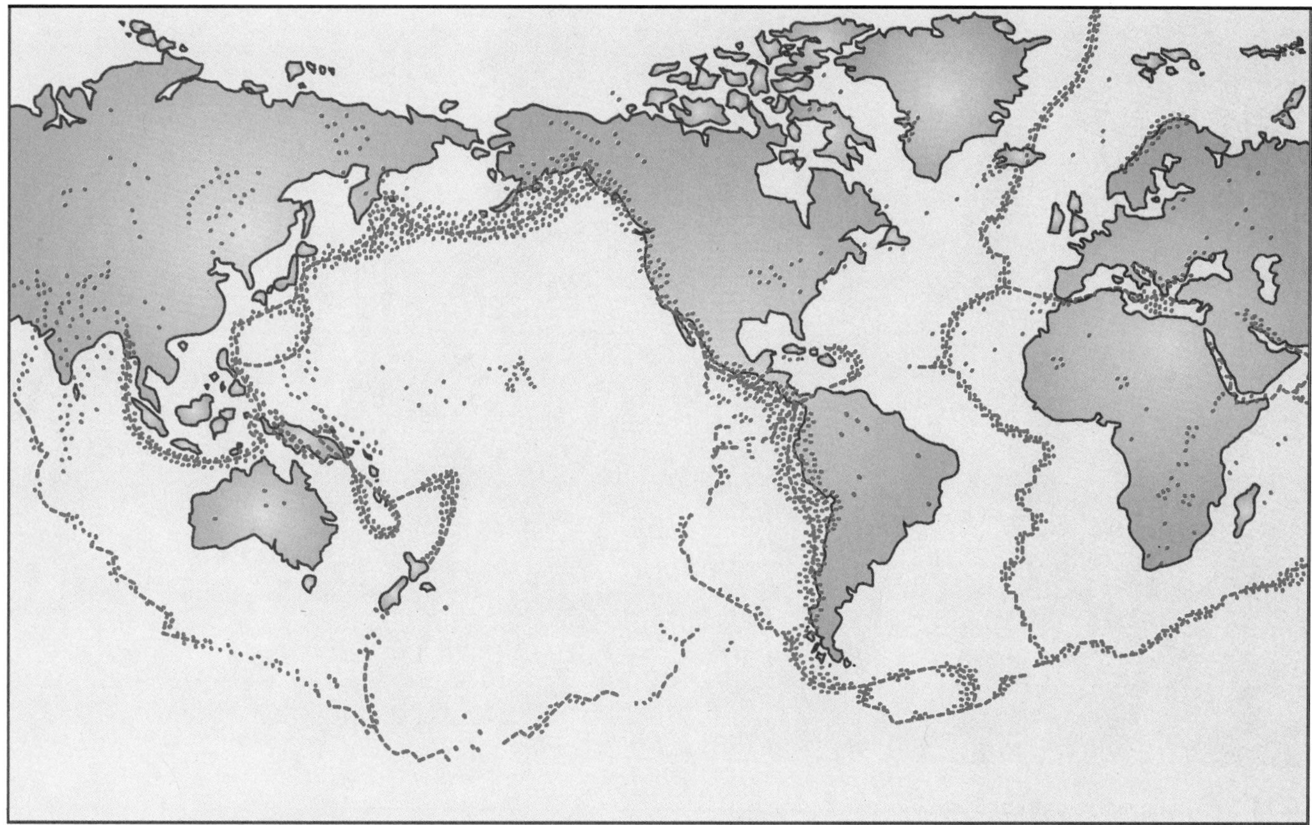

Figure 20.10 The relationship between plate boundaries and earthquakes.

place and reflects the polarity of Earth's magnetic field at the time of solidification (see Fig. 20.14). The direction of Earth's magnetic field at the time of solidification is shown by the direction the arrows point in each strip of new seafloor. Strip 1 is 0 to 5 million years old, strips 2 may be 5 to 20 million years old, strips 3 may be 20 to 35 million years old, and strips 5 may be 60 to 120 million years old.

The continuous formation of new seafloor along the ocean ridges has been confirmed by the *Glomar Challenger,* a research vessel that has been drilling into the ocean floor for over two decades. As the drilling continues, more information about Earth's structure and processes is revealed.

If a continent is attached to a seafloor that is spreading, the continent will move with the seafloor. Thus, seafloor spreading is the agent responsible for continental drift. Near the continental slopes, the ocean floor dips down under the continents and is mixed with and melted by the hot plastic mantle.

The creation of new crust material must be balanced by the destruction of old crust, or Earth would be continually expanding. However, there is no evidence that Earth is expanding. Instead, what happens is that at some plate boundaries a rigid lithospheric plate; for example, the Nazca plate, converges with another plate, such as the western edge of the South American plate, and descends deep into the asthenosphere below the other plate. The subducted lithospheric plate is ultimately destroyed by melting. Regions where two plates converge and one is subducted beneath the other are called *subduction zones* (see Figs. 20.15 and 20.16). The most powerful earthquakes occur at subduction zones.

Strong earthquakes also occur in regions along faults where crustal plates are grinding past one another. In the United States the most well-known such contact between plates occurs along the San Andreas fault.

An excellent example of a sea presently being formed by the spreading of two landmasses is the Red Sea. The width of the Red Sea is slowly increasing as the Arabian Peninsula gradually separates from Ethiopia, Sudan, and Egypt. There is also evidence that the Mediterranean Sea is decreasing in width as a result of the motion of the landmasses that enclose it.

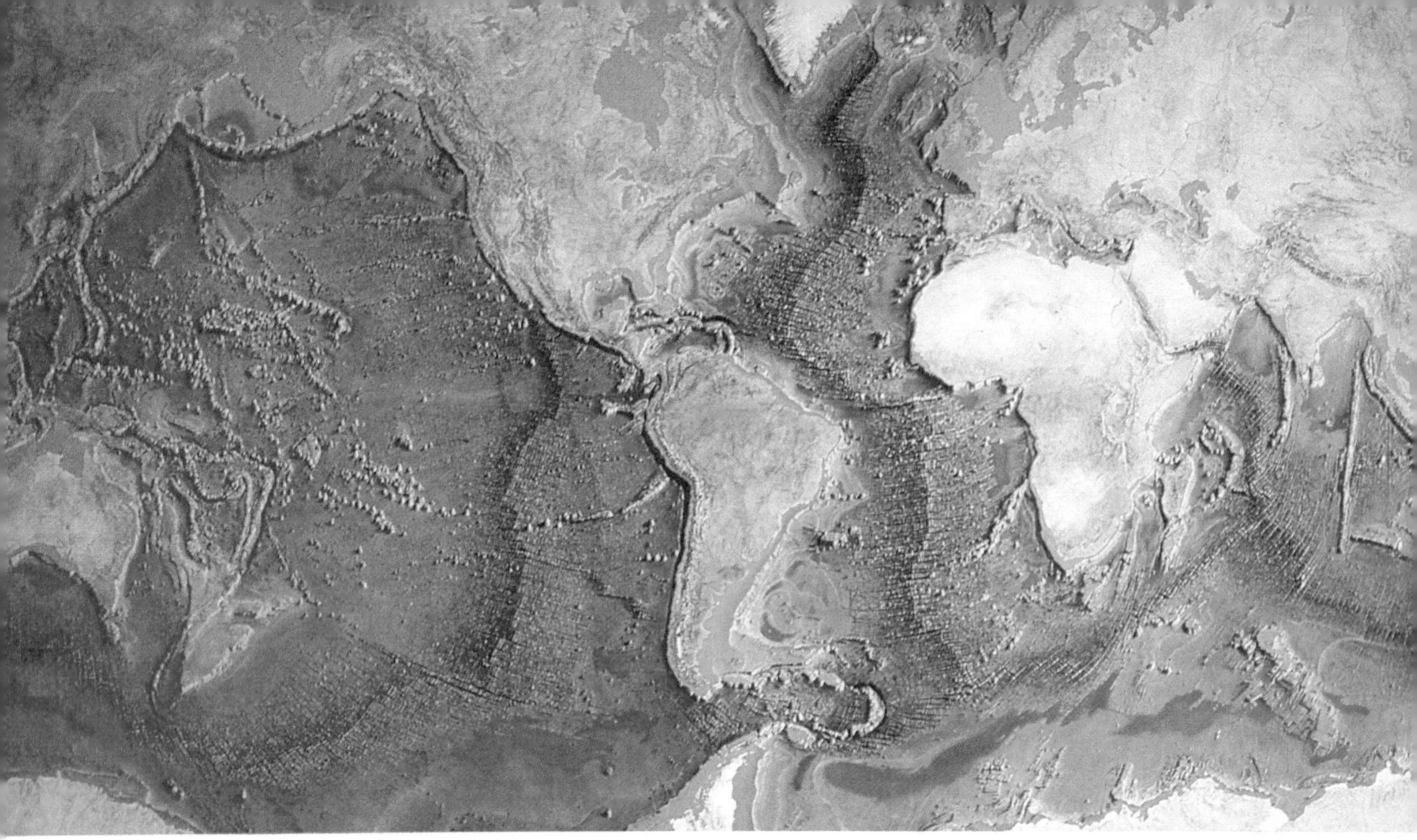

Figure 20.11 Physiographic map of Earth's ocean floors. It is at the mid-ocean ridges that seafloor spreading occurs.

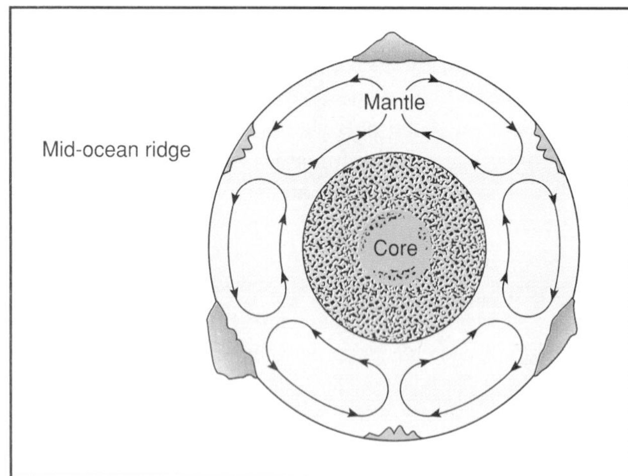

Figure 20.12 Convection currents within the mantle.

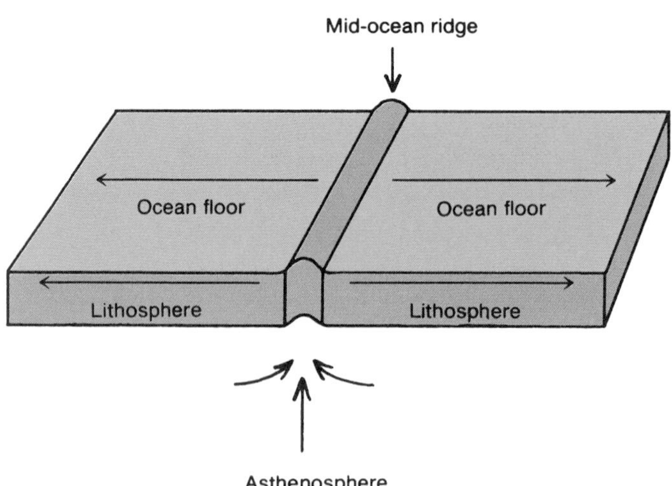

Figure 20.13 Mid-ocean ridge seafloor spreading.

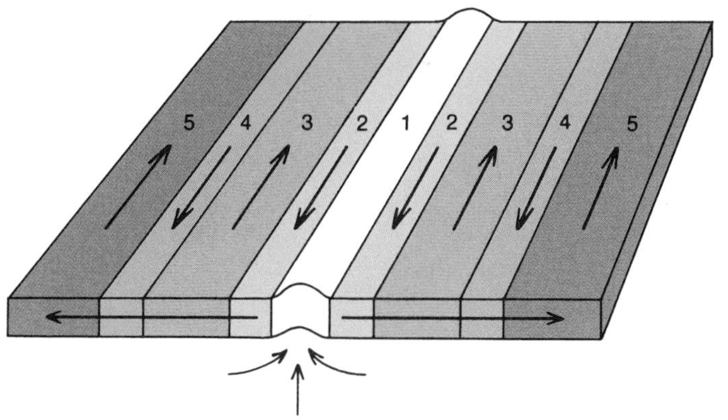

Figure 20.14 Magnetic reversals occurring in the seafloor along mid-ocean ridges.

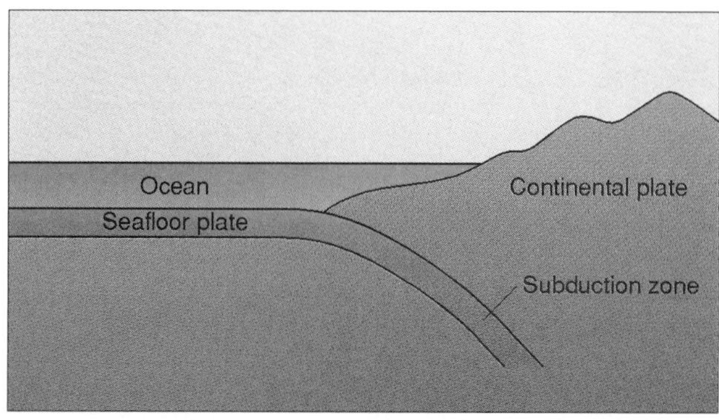

Figure 20.15 A subduction zone, where the rigid lithosphere descends deep into the asthenosphere at the boundary between the seafloor and a continental plate.

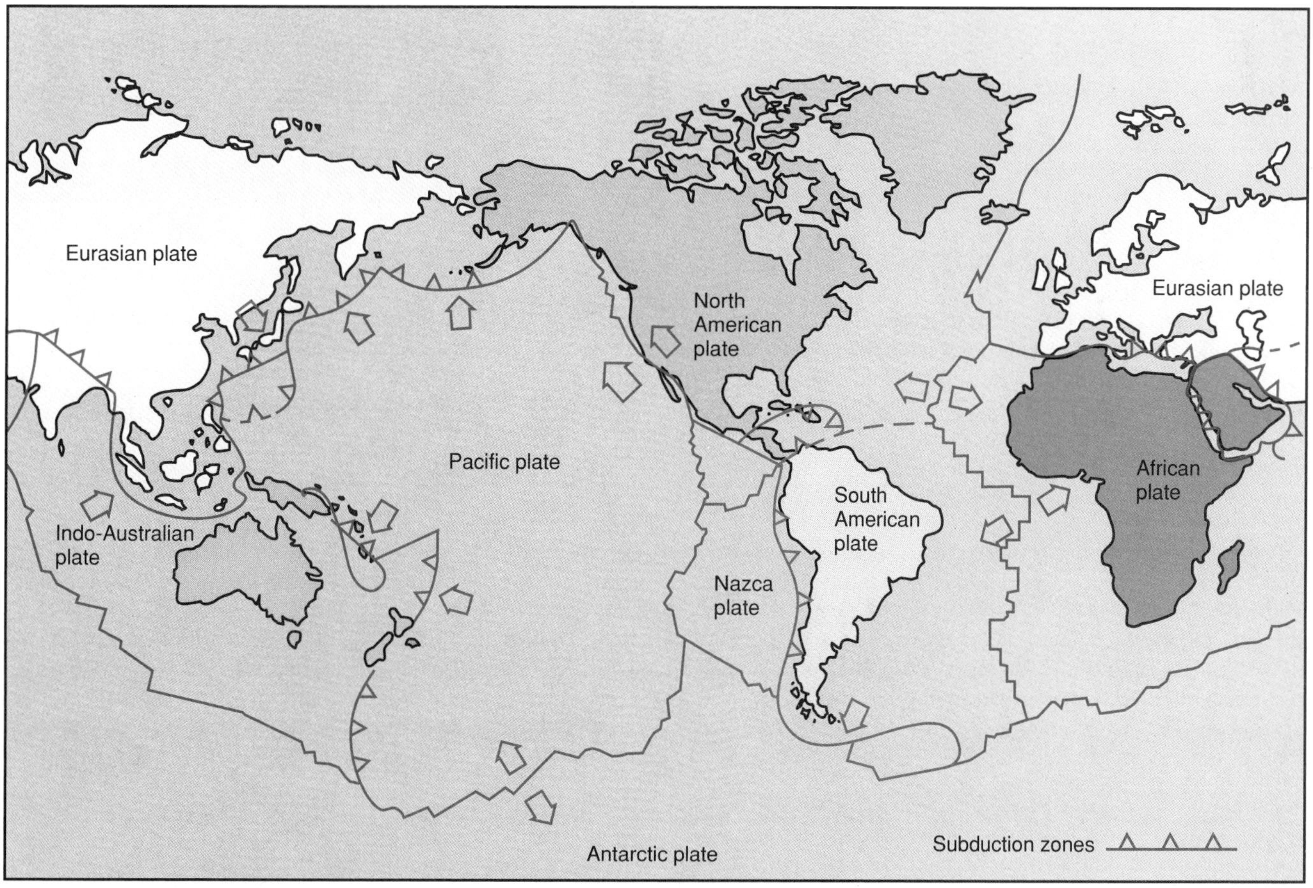

Figure 20.16 The three main types of plate boundaries. Subduction zones are also shown.

Oceans of Minerals

Not only are the mid-ocean ridges important in terms of seafloor spreading, they are also important as potential sources of mineral wealth. At the mid-ocean ridges there is much hydrothermal (hot-water) activity. Water, at temperatures of 300°C to 350°C (hot enough to melt lead), issues from chimneylike vents. The superheated water does not boil because of the tremendous pressure at the depths of the ridges. Some of the hot water comes from the lava oozing forth at the ridges, and some is produced by the cold water of the ocean floor flowing over the hot lava. The water of the ocean's abyssal plain surrounding the ridges has a temperature of only 2°C. The water from the lava contains numerous minerals dissolved from the asthenosphere. As the superhot water mixes with the cold water of the ocean floor, the dissolved minerals precipitate and form the many vents present at the ridges. Most of the minerals are metal sulfides formed from the fairly high concentration of hydrogen sulfide present in the hot water. Nearly all of the vents are of the black-smoker type, since the effluent waters that jet out of the chimneys are clouded with black particles of metal sulfides. The minerals being deposited at the ridges are mainly iron, copper, and zinc sulfides with smaller amounts of silver and gold sulfides. Mining companies are presently interested in the possibilities of mining the accumulated minerals of the rifts, but in many areas the minerals are at too great a depth and in too small a quantity for commercial purposes. There are some areas, however, that do hold considerable promise. For example, along a section of the East Pacific Rise off the coast of Oregon and Washington there are extremely rich deposits of zinc, iron, and silver sulfides (see Fig. 20.17).

Figure 20.17 Sulfide minerals pour from a black-smoker chimney on the East Pacific Rise (a section of the Pacific mid-ocean ridge).

A considerable portion of the ocean floor is dotted with black nodules of manganese oxides and hydroxides. The nodules vary from pea-size up to 25 centimeters in diameter. Besides manganese, the nodules also contain lesser quantities of iron, nickel, copper, cobalt, and zinc. In some areas (for instance, in the North Pacific near Hawaii), the manganese nodules are plentiful enough for possible commercial utilization. Exploratory dredging has been carried out to determine cost of mining operations. As land deposits of the metals present in the sea nodules become more costly to mine and the technology of deep-sea mining advances, the mining of the seas will unquestionably increase.

The most promising location for a successful mining operation is in the Red Sea. Initial mining has revealed very rich deposits of various metals in the muds around the spreading rift zone at the bottom of the Red Sea. The muds contain a mineral bonanza of silver, gold, lead, copper, and zinc ores.

Besides the more esoteric and expensive metals present in our seas and oceans, there is an unbelievable quantity of more prosaic substances dissolved in these waters. In fact, the seas and oceans of our planet contain enough salts to cover all the land masses of Earth to a depth of 150 meters. Half of the world's supply of sodium chloride (table salt) is harvested each year by the evaporation of seawater. We should not, however, assume that because there are such huge quantities of a certain salt or mineral present in seawater that it would be cost-effective to extract the material. Although there are approximately 500 billion kilograms of silver and 10 billion kilograms of gold salts in the ocean, it would cost many times the value of the silver and gold to extract them. There are only two elements of the nearly 55 that occur in easily measurable quantities that are presently being commercially extracted from seawater—bromine and magnesium.

Where did all the salts of the oceans come from? They came from two sources. One source involves the hydrologic cycle (see p. 472) in which rain or snow falls on the land and during its journey to the sea dissolves various minerals and salts from the land. The waters of the seas and oceans are continually undergoing evaporation, leaving behind the dissolved solids. Another source is the dissolved minerals and salts present in the superheated water issuing from the mid-ocean ridges and from the dissolution of minerals and salts from the lava by circulating ocean waters. It is assumed by most scientists that both sources contribute about equally to the mineral and salt content of our seas and oceans ■

Summary

The outermost layer of rock on Earth, the crust, is composed mainly of oxygen and silicon. The crust varies in thickness from about 8 kilometers under the oceans to about 48 kilometers on the continents.

The core of Earth has a diameter of about 7100 kilometers and, except for the inner portion of the core, consists of molten iron and nickel. The small inner core is solid iron and nickel. The section of Earth between the core and the crust is the mantle. The mantle is solid but has plastic characteristics in the part known as the asthenosphere. The outermost layer of Earth is called the lithosphere. The lithosphere, which floats on the partially molten asthenosphere, includes the oceanic and continental crust and the upper part of the mantle.

Because of Earth's rotation, the molten iron-nickel outer core generates the magnetic field that surrounds Earth. Earth's magnetic field has reversed its polarity many times during geologic history. The last known time of a magnetic reversal was about 700,000 years ago.

Earth is not spherical but is an oblate spheroid. Because of Earth's rotation its equatorial diameter is 43 kilometers greater than its polar diameter. Earthquake waves are used by scientists to study the interior of Earth.

About 200 million years ago the supercontinent, Pangaea, ruptured into many fragments or smaller continental plates that have slowly drifted apart to their present positions. The drifting of the continents is the result of seafloor spreading that occurs at the mid-ocean ridges. At the mid-ocean ridges, hot mantle material issues forth to cause spreading of the seafloors.

The crust of Earth is divided into eight large lithospheric plates and many smaller plates that move independently of each other. Regions where two plates converge and one is submerged beneath the other are called subduction zones. Most earthquakes are the result of the movements of lithospheric plates.

Questions and Problems

The Infant Earth

1. What would be the effect on Earth's magnetic field if the planet were to cease its rotation?

2. What processes caused Earth to be much hotter at one time in the planet's history than at present?

3. How closely related do scientists consider Earth and the moon? Earth and the other planets?

The Shape of Earth

4. Why is our planet Earth not a perfect sphere?

5. Why does an object weigh slightly less at the equator than at the North or South Pole?

The Structure and Composition of Earth

6. From the mass and volume of Earth provided in this chapter, verify the value given for Earth's average density. Recall $d = m/v$.

7. What evidence is offered to support the belief that a portion of Earth's core is liquid?

8. How were the oceans of Earth formed? What appears to be the major source of the limestone they contain?

9. What type of earthquake waves can travel only through solids? Why?

The Plastic Mantle and Isostasy

10. Earthquake waves indicate that Earth's mantle is solid, but it is believed that the continents "float" on part of the mantle. Explain.

11. What two elements constitute over 70 percent of the chemical composition of Earth's crust?

Continents That Drift—Plate Tectonics

12. What evidence suggests that Antarctica was at one time connected to the African continent?

13. What theory explains the concepts of drifting continents and seafloor spreading?

14. How were the Himalaya Mountains formed? What evidence is provided to support this accepted theory?

15. What supportive evidence, other than shape, indicates that Africa and South America were at one time attached?

Seafloor Spreading

16. What is magnetic reversal? How is it determined?

17. How does the phenomenon of seafloor spreading support continental drift?

18. What are crustal plates? Where are they located?

19. How do volcanoes and earthquakes relate to crustal plates?

20. Describe the appearance of the ocean floors, as indicated by the illustrations and accompanying discussion in this chapter.

21. How are the continents of Africa and Europe presently moving in relation to each other? What proof of this relative motion does the scientist have to offer?

22. Where did all the salts of the oceans come from?

21 The Motions of Earth

Chapter outline

Key terms/concepts

Earth is shown rising over the lunar surface of its moon with the sun flare on the edge of Earth's limb.

Although Jupiter's moon Ganymede is the largest satellite (approximately 5300 kilometers in diameter) in our solar system, Earth's moon is among the largest in comparison with its parent planet. In fact, the Earth-moon system is frequently referred to as the "double planet." Looking "down" on Earth from the surface of the moon, we see the most beautiful and colorful planet in the entire solar system. Brilliant, multihued Earth hangs in the black, airless sky in striking contrast to the moon's desolate landscape of dark shadows and shining white light. This chapter's introductory photograph is one of the text's most impressive illustrations. Look at it carefully to experience its full impact.

If we leave the moon and travel toward Earth, a journey of about 382,000 kilometers, we find at a distance of slightly over 24,000 kilometers from Earth the first of the Van Allen radiation belts that encircle our planet (see Fig. 21.1). The belts contain charged particles, mainly electrons and protons, captured from the solar wind and held fast by Earth's magnetic field. At approximately 8000 kilometers we encounter the second belt and, as we did with the first belt, hurry through so as to expose ourselves to the least possible number of X rays generated by the high-speed electrons and protons striking the walls of our space vehicle. As we approach Earth, we see soft white clouds gently floating in an azure sky above the surface of a planet-encircling blue-green sea. At each pole are large, cold white caps of ice and snow. The continents and islands scattered around the planet contain lush dark-green forests, eye-searing white-sand deserts, and lofty mountain ranges that span whole continents.

As far out as thousands of kilometers from Earth, we would notice faint and tenuous vestiges of Earth's atmosphere, and from about 10,000 kilometers down to roughly 160 kilometers, there may be visible a brilliant display of glowing curtains of light that silently move in a contortive dance of color and beauty. These features are the *auroras,* the northern and southern lights. As noted in Chapter 17, the auroras are caused by charged particles from the sun bombarding the atoms in the polar regions of Earth's upper atmosphere. Spaceborne visual sensors aboard a weather satellite detected a spectacular auroral display over Canada in 1974, shown in Figure 21.2. Contrast it with the photograph of Figure 17.6, which was taken from Earth.

During our trip from the outer reaches of the atmosphere down to the surface, we notice a continuous increase in the concentration of air molecules with an accompanying increase in pressure. The gravitational attraction of Earth for the air molecules pulls them toward the surface. If this gravitational attraction were the only force operating on the air molecules, they would eventually settle into a thin, hardpacked layer encasing Earth. However, since the solar-heated molecules have a certain kinetic energy that results in a constant random motion, the air molecules are closer together nearer the surface and increasingly farther apart at greater distances from Earth. As a

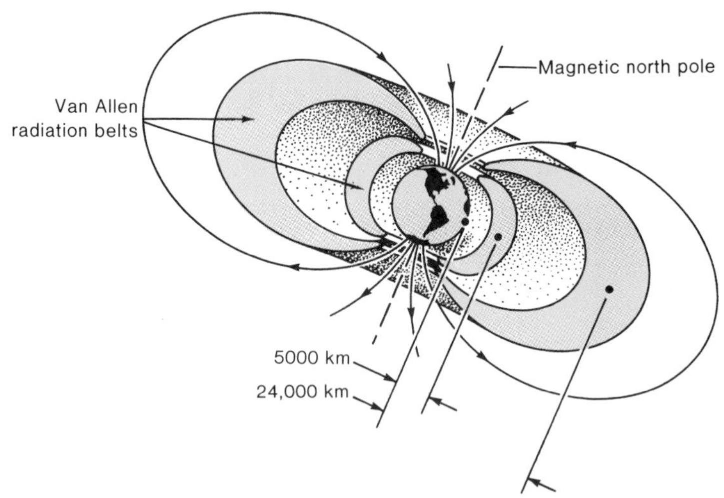

Figure 21.1 Earth's magnetic field and Van Allen radiation belts.

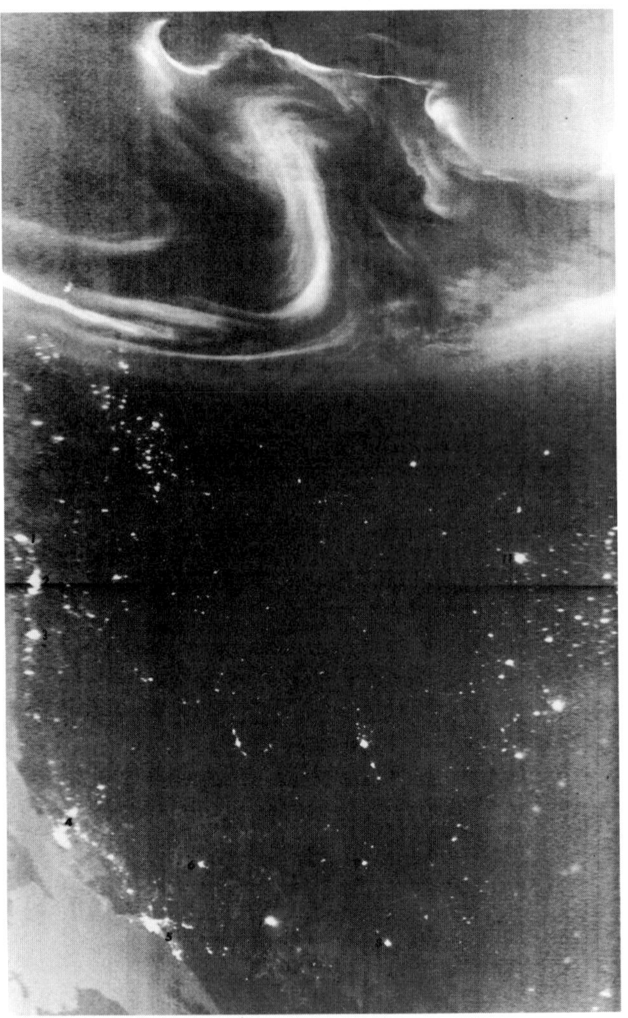

Figure 21.2 A spectacular display of the northern lights detected over Canada in 1974. Lights from various cities, ranging from Seattle to El Paso to Minneapolis-St. Paul, are also displayed in the photograph.

The Proof Is in the Swing

Jean Foucault (1819–1868) was destined to become one of the most famous experimental scientists of his generation. As a child he was bored with the routine of school, particularly with the rigorous study of the classics that were emphasized at that time. He preferred to construct his own toys, including a model steam engine, and he displayed a high degree of dexterity in addition to an outstanding scientific instinct. His intelligence and craftsmanship were put to use in the design of a pendulum named after him. Results of his best-known experiment were published in 1850. The pendulum he constructed and set into motion, reportedly from the tall ceiling of the Pantheon in Paris, consisted of an approximately 30-kilogram mass suspended by a wire some 60 meters long. It was capable of sustaining the vibrational mode for hours at a time. Upon release, the pendulum would traverse a line marked on the floor beneath it. The device maintained its vibrations in the same plane due to its inertia, while Earth's rotation slowly turned the reference line in a somewhat distorted elliptical figure. The fact that the pendulum changed its direction of swing with respect to Earth is proof that Earth does indeed rotate. In the Northern Hemisphere the apparent motion of the pendulum with respect to Earth would be clockwise, and in the Southern Hemisphere the motion would be exactly opposite. At a point along the equator, the plane of rotation of Earth is such that the swing of the pendulum would not appear to be altered. At either pole, the time for the pendulum to sweep out a complete path would be equal to Earth's period of rotation, so the path the pendulum would make in approximately 24 hours would be a complete circle. This shape of figure would indicate that the directional period of such a pendulum would be equal to Earth's period of rotation, whereas at various distances from the poles, the directional period would be longer. For instance, in New York City, the directional period of the pendulum would be about 40 hours. Various renditions of the Foucault pendulum are on display at many museums throughout the world.

Foucault's diverse scientific interest led to other noteworthy achievements. For instance, in 1852, he published plans for a gyroscope, a rotating wheel that strongly resists any attempt to change its direction. The gyroscope has many uses because of its rotational inertia. His other great accomplishment was to develop a rotating mirror with which he measured the velocity of light to within 1 percent of today's accepted value. The Foucault test he developed for detecting microscopic-sized flaws in the concave mirrors used in telescopes is still considered a practical test by various manufacturers■

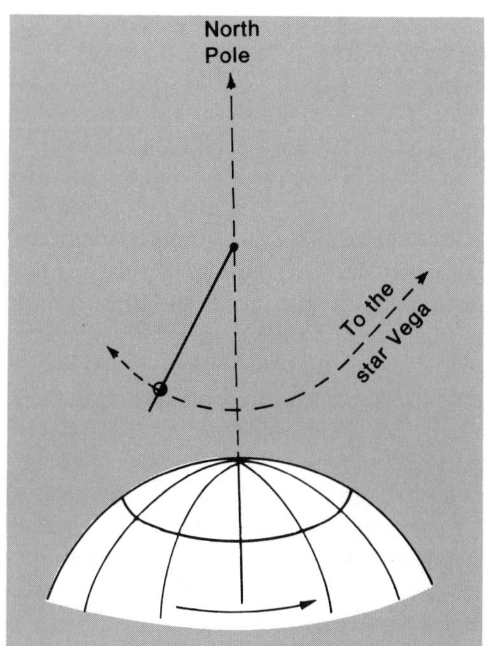

Figure 21.3 A Foucault pendulum experiment at the North Pole.

result of the opposing forces of gravity and molecular motion, 90 percent of all of Earth's atmosphere is contained within 16 kilometers of the surface, and 50 percent is contained within 5.6 kilometers of the surface.

Analysis of the Earth's atmosphere reveals that it consists of 78 percent nitrogen, 21 percent oxygen, 0.9 percent argon, 0.03 percent carbon dioxide, 0.01 percent water vapor, and traces of helium, neon, hydrogen, methane, and other gases. (This analysis does not include the smog and noxious gases that prevail around large cities.) The atmosphere is discussed in considerable detail in Chapter 22.

The Foucault Pendulum Experiment

How do we know that Earth rotates on its axis? For centuries the ancient Egyptians, Greeks, Babylonians, and Romans wondered whether Earth rotated and the stars, sun, and moon were stationary or whether Earth was stationary and the heavens moved. Although there were advocates for both views, the ancients generally believed that Earth stood still and the heavens moved. It was not until 1851 that this question was finally settled beyond all doubt by an experiment devised by the French physicist Jean Foucault. The famous Foucault pendulum experiment was performed in Paris, but it will be easier to understand if we imagine it being performed at one of Earth's poles.

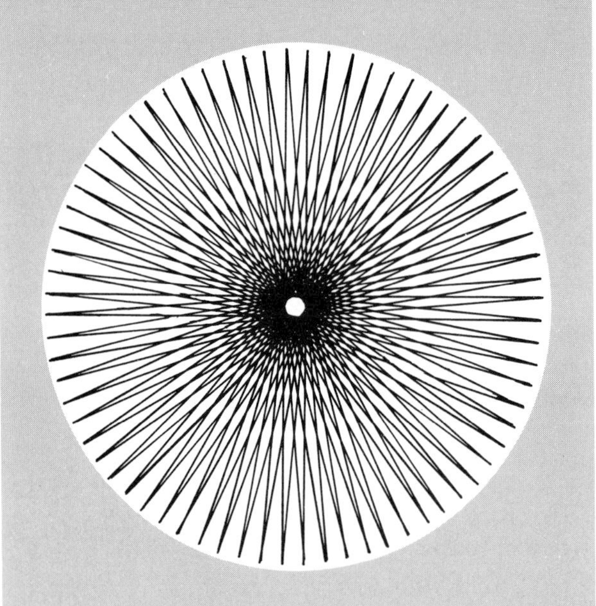

Figure 21.4 The type of pattern that would be obtained if the Foucault pendulum experiment were carried out at the North Pole.

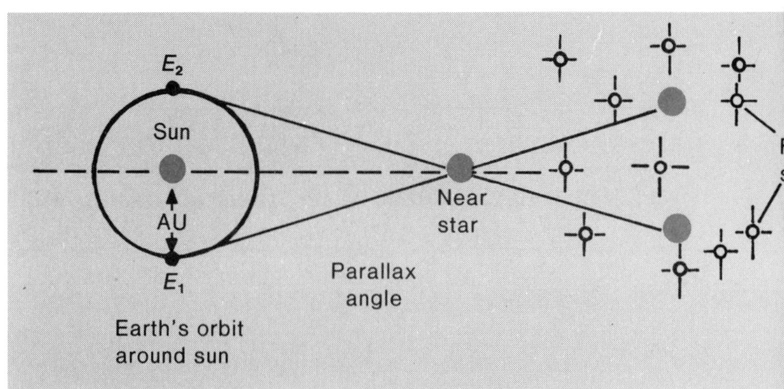

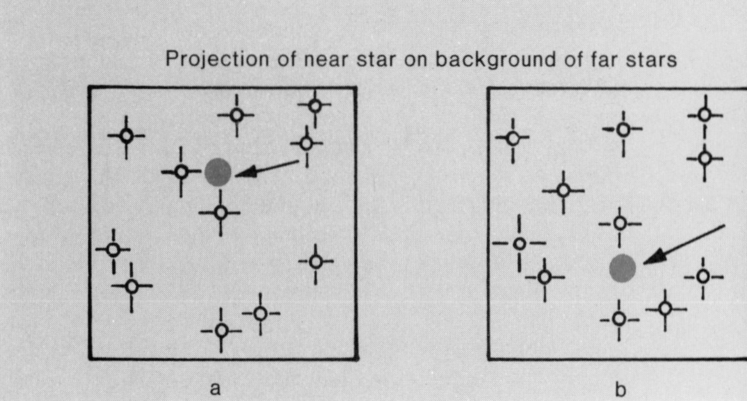

Figure 21.5 The observation of stellar parallax. When Earth is at E_1, the near star would appear to be positioned as shown in (*a*) in relation to background stars. Six months later, when Earth is at E_2, the near star would appear to be positioned as shown in (*b*) in relation to background stars.

Consider setting up a tall tripod, the center of which is directly over the North Pole. From the top of the tripod a steel ball is suspended by a fine steel wire (see Fig. 21.3). At the bottom of the steel ball is a steel point of such length that as the ball swings like a pendulum, the point will trace the ball's path in a layer of sand. If the pendulum is set into motion, we find that the point does not make the same line in the sand over and over. Instead, after a few hours many marks are made by the pointer, all passing through a common point. The reason for this occurrence is that the pendulum, once set in motion, continues to swing in the same direction while Earth rotates underneath it. Indeed, if the experiment had been carried out at night and if the pendulum had been set in motion swinging toward and away from the star Vega in the constellation Lyra, it would have continued swinging toward and away from Vega throughout the whole experiment. Obviously, Earth must be turning beneath the swinging pendulum; hence, a line made by the pointer is displaced a small amount by the rotation of Earth before the pendulum can retrace its previous mark. If we continued the experiment for 23 hours and 56 minutes (the length of Earth's day with respect to the stars), we would see the pattern shown in Figure 21.4, in which, eventually, the first trace made by the pendulum's pointer would be retraced, since Earth would have completed one rotation on its axis. Foucault did not obtain a perfect pattern in his experiment because he was at Paris, France, instead of at one of the poles. As this experiment is performed at different places on a line from one of Earth's poles to the equator, the circular pattern becomes more and more distorted until at the equator only a single straight line is obtained since the part of the Earth directly underneath the pendulum is moving in a single direction. At one of the poles, however, Earth is rotating 360 degrees beneath the pendulum.

You may wonder why, since the tripod of this experiment is attached to Earth and the steel ball is attached to the tripod by a wire, the direction of the swing of the pendulum does not change as Earth rotates. Rest assured that it does not. You can verify this fact by swinging an object on the end of a string. If you twist the string between your fingers while it is swinging, the direction of the swing will not change.

Evidence of Annual Motion

As early as about 315 B.C. and as late as A.D. 1600, the philosopher Aristotle and the Danish astronomer Tycho Brahe, respectively, dismissed the possibility that Earth revolves around the sun because they could observe no shifting back and forth of nearby stars with respect to very distant stars during a year. They correctly assumed that if we did revolve around the sun, our motion would cause the position of some stars to shift in relation to other stars as Earth sped along in its orbit (see Fig. 21.5) in the same way that a nearby telephone pole is seen to block out different distant objects as one moves past the pole. Since they were not able to observe any *apparent shifting of position of stars,* called **stellar parallax,** they concluded that either all stars are unimaginably distant from Earth (and, therefore, the parallactic shift is too small to see) or that Earth is stationary.

The largest parallax of any star is only 0.75 of a second of arc (there are 60 seconds of arc in 1 minute, and 60 minutes in 1 degree of arc), and since the eye is not capable of resolving anything less than 1 minute of arc, the parallactic shift of the nearest stars (in the Alpha Centauri system) is too small to be observed with the naked eye. Small wonder, then, that Aristotle and Brahe rejected the heliocentric concept. They were wrong—but for a good reason!

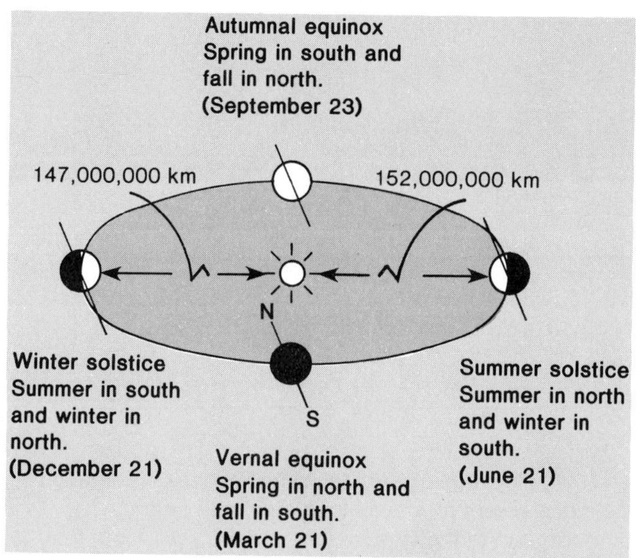

Figure 21.6 The seasons are caused by the inclination (66.5°) of Earth's axis of rotation to the plane of Earth's orbit, not by its changing distance from the sun.

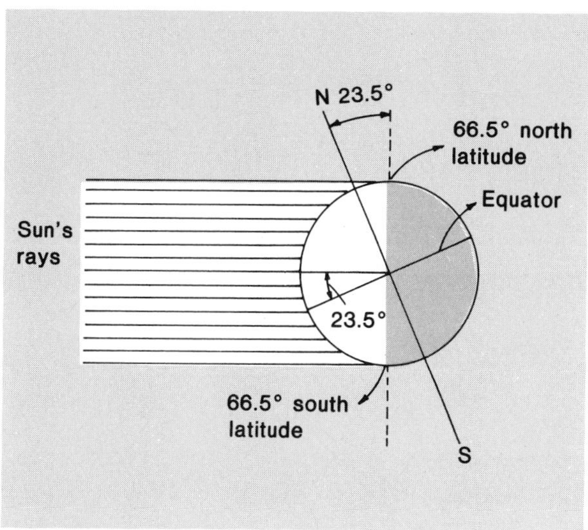

Figure 21.7 Earth during the summer solstice (June 21). The sun is at zenith (directly overhead) 23.5° north latitude. The sun's rays reach 23.5° beyond the North Pole, or to 66.5° north latitude. Note that the sun's rays are more slanted, or oblique, in the Southern Hemisphere than in the Northern Hemisphere.

A star with a parallax of 1 second of arc (1 arc s) is located at a distance of 2×10^5 astronomical units from the sun, or 3×10^{13} kilometers. As was noted in Chapter 19, it is customary among astronomers to refer to this distance as the *parsec* (pc) and to apply this unit for major distances in the universe. The parsec is equal to 3.27 light years.

In one year the sun appears to make a complete journey about the **celestial sphere,** *the imaginary sphere of the heavens that is half-visible at any time when viewed from Earth,* supposedly its central figure. The concept of a celestial sphere found early use by the ancient Babylonians, Chinese, Egyptians, and Greeks; it was a view that strongly supported the geocentric concept of Aristotle and others. The apparent route the sun takes against the background of stars within the celestial sphere is known as the *ecliptic,* since eclipses can occur only when the moon's apparent position is located on or near the sun's path. In reality, the motion of the sun along the ecliptic is a false perception, created by the revolution of Earth about the sun. If we could look past the sun on each day of the year and see the shining objects behind it, we would be aware of the constant changes in the sun's star-studded background due to Earth's motion.

Further evidence of Earth's revolution is offered by various measurements of light from the stars, initially conducted by the English astronomer James Bradley in 1729. He mounted a telescope in a vertical position, reportedly by attaching it to the chimney of his home. He attempted to measure the parallax of the stars in his field of view as Earth's rotation caused them to move across a scaled measuring device he had installed. The results of his experiment contradicted his expectations. He could detect no apparent shift of position due to parallax with his telescope. (The sensitivity of his equipment was undoubtedly inadequate for the purpose.) Instead, he discovered that every star he examined shifted its direction during a year by 20 seconds of arc on either side of its average position, a phenomenon now called *aberration.* Bradley is credited with offering an explanation to the perplexing results of his study. He cited a similar situation when one observes the direction the flag assumes aboard ship when the vessel is considered to be moving in the same direction as the wind. He concluded that aberration of light from the nearer stars was caused by the motion of Earth. The effect is maximized when Earth moves at right angles to the stars under study. No effect is noted when Earth moves directly away from or toward the specific stars. In the latter case, a star that is located on the ecliptic appears to shift back and forth on a straight line since, through a portion of the year, Earth moves in one direction relative to the star, and the balance of the year, Earth moves in the opposite direction. Stars between the ecliptic and located at right angles to Earth's orbit appear to shift directions by the 20 seconds of arc, as noted by the English astronomer. Since the sun is constantly within our view year after year, it is evident that the motion of Earth must be one of revolution about the sun.

The Seasons

Earth follows an elliptical path in its annual journey around the sun, so that at times it is closer to the sun and at other times it is farther from the sun. Actually, Earth is closest to the sun (*perihelion*) during the Northern Hemisphere's winter and farthest from the sun (*aphelion*) during summer. Earth is at maximum perihelion on about January 4 and at maximum aphelion on about July 5. Obviously, then, we must seek an explanation for the seasons elsewhere. Earth's axis, determined by Earth's rotation, makes an angle with the plane of Earth's revolution around the sun of 66.5 degrees—more commonly expressed as 23.5 degrees from a line perpendicular to the plane of Earth's orbit. Because of this tilt of the axis, the Northern Hemisphere receives more sunshine than the rest of the planet during part of the year, thus producing the Northern Hemisphere's summer.

As shown in Figures 21.6 and 21.7, the sun's rays strike the Northern Hemisphere almost perpendicularly during summer and the Southern Hemisphere at an extreme slant. This difference means that more sunlight falls on each square meter of Earth in

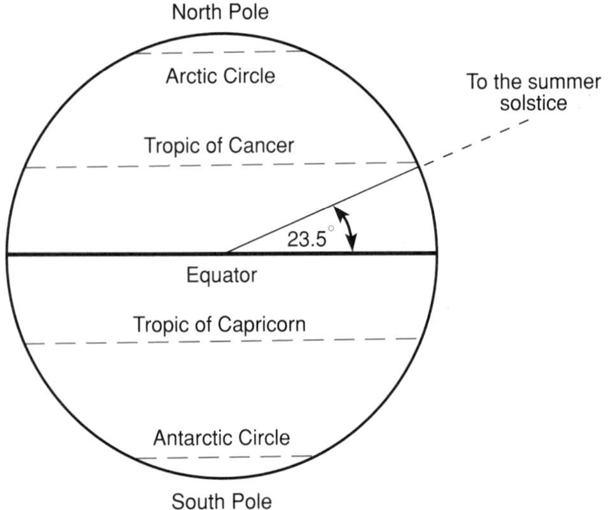

Figure 21.8 The latitude zones of Earth. At the summer solstice, as shown, the sun is directly overhead at a latitude of 23.5° N. This point marks the most northerly position the sun's path makes, and defines the Tropic of Cancer. Approximately six months later, at the winter solstice, the sun is overhead at the Tropic of Capricorn at a latitude of 23.5° S.

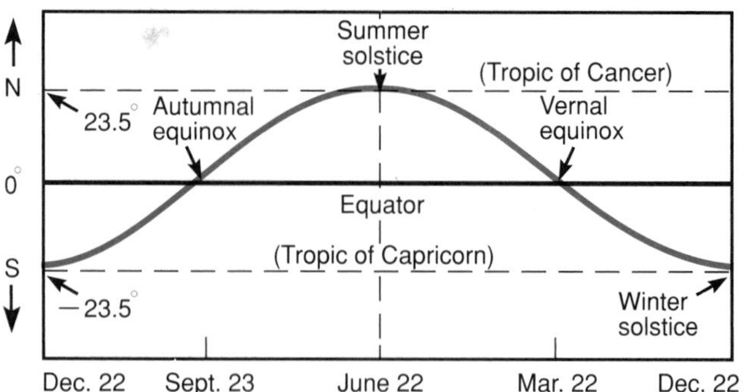

Figure 21.9 Because of Earth's orbital motion and its tilted axis, the sun appears to move along a smooth path from south to north during the various times of the year. Note that the path always stays between 23.5° N and 23.5° S.

the Northern Hemisphere than in the Southern Hemisphere. Thus, while the Northern Hemisphere is having summer, the Southern Hemisphere is experiencing winter. Six months later the situation is reversed, and the Southern Hemisphere receives the greater amount of sunlight per square meter. On or about June 21, the **summer solstice,** the sun is at zenith (directly overhead) on 23.5 degrees north latitude, called the **Tropic of Cancer** (see Fig. 21.8). On the same day, the sun shines 23.5 degrees past the North Pole toward the equator, or to 66.5 degrees north latitude, called the *Arctic Circle;* thus, on or about June 21, all points within the Arctic Circle bask in 24 hours of daylight. At the same time, the sun reaches only to 66.5 degrees south latitude (23.5 degrees from the South Pole), called the *Antarctic Circle.* On June 21, people at latitudes above the Antarctic Circle face 24 hours of total darkness. Six months later, on or about December 21, the **winter solstice,** the sun is at zenith on 23.5 degrees south latitude, called the **Tropic of Capricorn.** On or about March 21 and September 23, Earth is halfway between the winter solstice and the summer solstice and is said to be at the **equinoxes** (from the Latin meaning "equal night"). During these two days, Earth's axis neither slants toward nor away from the sun but instead is "parallel" to the sun, as shown in Figure 21.6. Consequently, Earth receives 12 hours of sunshine and would experience 12 hours of night were it not for the fact that diffraction and scattering of the sun's rays considerably extend the period of daylight. On about March 21 is the *vernal* (spring) *equinox,* and approximately September 23 is the *autumnal equinox.* During the equinoxes, the sun is at zenith on the equator. The apparent smooth path the sun makes from south to north during Earth's year is depicted in Figure 21.9.

Solar and Sidereal Time

Over a year, the average time from noon (the precise time when the sun crosses the observer's meridian) to noon of the next day is 24 hours, the length of the **solar day.** However, an observer in

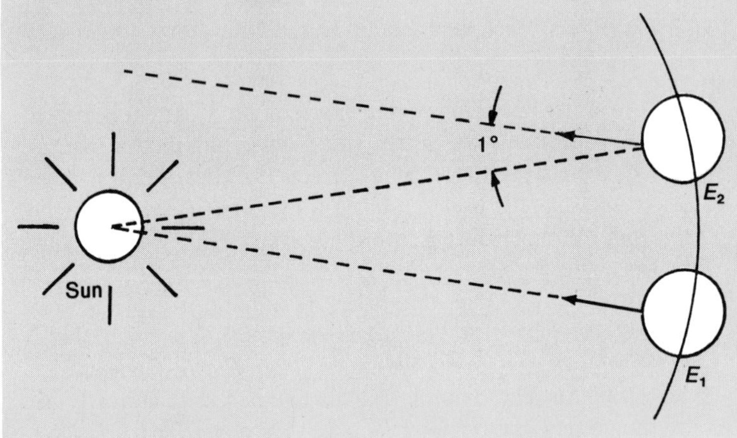

Figure 21.10 In one sidereal day Earth makes one complete rotation (360°) on its axis. Earth must rotate 361°, however, in order to complete one solar day.

outer space would find that it takes only 23 hours and 56 minutes for Earth to make one rotation on its axis, and this amount of time is the length of the **sidereal day.** Obviously, the solar day is 4 minutes longer than the sidereal day. How can we account for this difference?

As shown in Figure 21.10, while Earth is making one complete rotation on its axis, it is also traveling in its orbit around the sun (about 2,580,000 kilometers per day). Consider a large arrow so constructed that it points directly at the center of the sun. During the time that Earth makes one complete rotation on its axis; that is, one sidereal day, Earth will have moved from position E_1 to position E_2. At position E_2, after Earth has made one rotation on its axis, the arrow will be pointing not at the center of the sun but in the same direction that it had pointed while at position E_1. In order for the arrow to once again be pointing directly at the center of the sun, Earth must turn an average of 1 degree more on its axis. It takes Earth about 4 minutes to rotate 1 degree. Since Earth must rotate 361 degrees for a solar day but only 360 degrees for a sidereal day, the solar day is 4 minutes longer than a sidereal day.

Figure 21.11 Precession of the axis of a top.

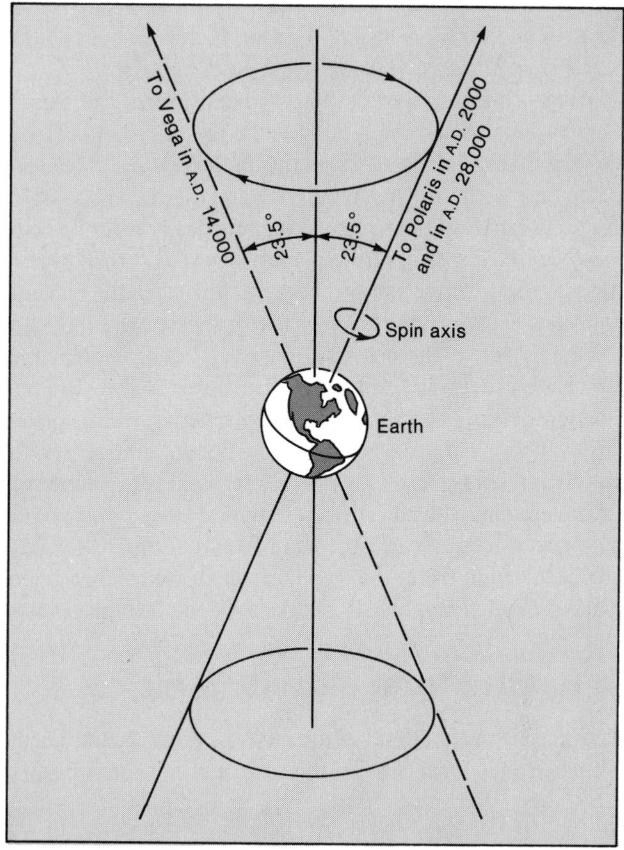

Figure 21.12 Precession of Earth's rotational axis. The planet completes one full precession every 26,000 years. About A.D. 28,000, Earth's axis will once again point to the star Polaris.

Precession of Earth's Axis

If you ever watched a spinning top carefully, you would have noticed that as the top spun on its axis, it did not stand upright or perpendicular to the floor but, instead, had a tilted motion that caused its axis to sweep out a cone in space, as shown in Figure 21.11.

This *conical motion of the axis of a spinning object* is called **precession.** The precession of a spinning top is caused by the gravitational attraction of Earth for the top. If a top is held rigidly perpendicular to the surface of Earth, the center of gravitational attraction on the top is along the top's axis and no precession occurs. If the top wobbles, however, the direction of the force of Earth's attraction for the top is not in line with the axis of the spinning top. This nonalignment of the direction of gravitational force with the axis of the spinning top produces the observed precession of the axis of the spinning top.

Consider Earth as a huge spinning top. The sun exerts on Earth a gravitational attraction that is not in line with Earth's spin axis. Consequently, Earth's axis precesses like the axis of a toy top. Earth's axis maintains its 23.5-degree tilt from a line perpendicular to the plane of Earth's orbit around the sun, but the axis precesses and cuts out a cone in space. It takes 26,000 years for Earth's axis to make one complete precession (see Fig. 21.12).

Earth's axis today is pointing approximately to the star Polaris, the North Star. Because of precession, however, Earth's axis is slowly pointing farther and farther away from Polaris, and in the future Polaris will not be our pole, or north, star. Indeed, for a thousand years before and well over a thousand years after the birth of Jesus, Earth had no north star. As shown in Figure 21.13, by the year A.D. 3500 there will be no star close enough to the direction in which Earth's axis points to be called a pole star. By the year A.D. 7000, a star in the constellation Cepheus will assume the position as pole star; and, by A.D. 11,500, a star in the constellation Cygnus will be fairly close to the spin axis of Earth. By about the year A.D. 14,000, the bright star Vega, in the constellation Lyra, will be our pole star. Then we must wait about 9000 years before there will be another pole star, for it will not be until A.D. 23,500 that a star in the constellation Draco will be in the

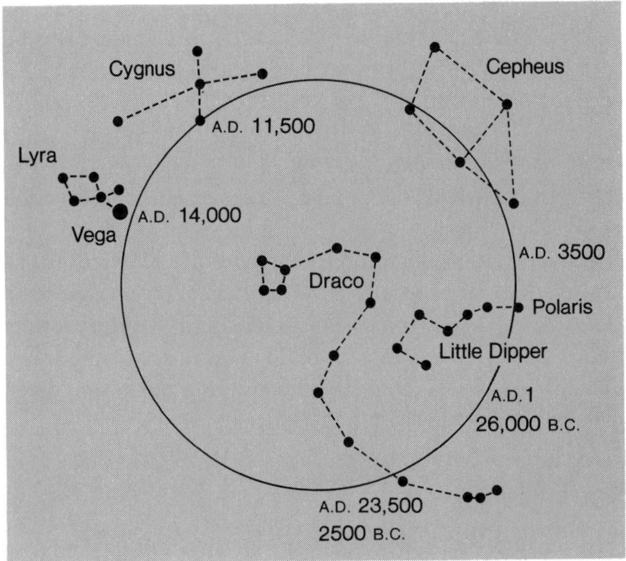

Figure 21.13 The path among the stars followed by Earth's north axis over a 26,000-y period. This path is the result of the precession of Earth's axis.

approximate position. This star will be the same pole star seen in 2500 B.C. by the Babylonians and Egyptians. Then, about 26,000 years from now, Polaris will once again be our pole star.

As Earth undergoes precession, its hemispheres will experience significant seasonal changes, since the planet's rotational axis will sweep out an angle close to the maximum of 47 degrees with respect to the sun's location between now and the year A.D. 14,000. By this time, Earth will have completed practically half of a complete precession. Earth's Northern Hemisphere will then receive the sun's direct rays (the sun will be at zenith) when the planet is at perihelion. Therefore, the Northern Hemisphere will experience a considerably hotter summer than at present, but during the time of year when it currently undergoes the winter season. Winters will be more severe in the Northern Hemisphere since the planet will be at aphelion and will occur during the present summer season. The Southern Hemisphere will assume a seasonal pattern opposite that of the present, with cooler midyear summers and milder winters that occur about six months later. Even the time when past ice ages occurred in the Northern Hemisphere seems to correlate well with the inclination of Earth's axis as a result of precession.

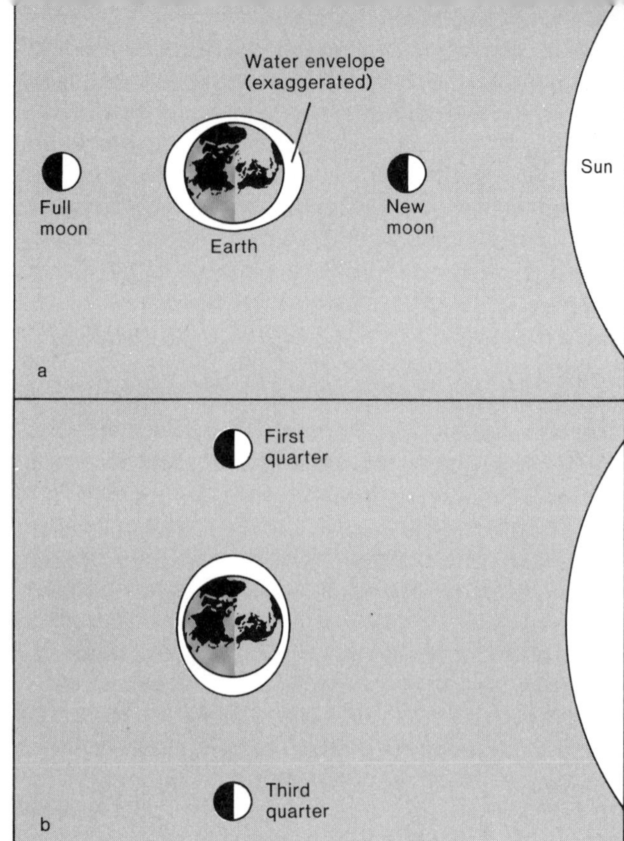

Figure 21.14 Different types of tides produced by different positions of the moon; (*a*) spring tides, where the moon is either full or new, and (*b*) neap tides, where the moon is either in its first or third quarter.

The Earth-Moon Gravity Couple

We have all said or thought that the moon revolves around Earth, and although this appears to be the case, it is not the full story. Actually, any two orbiting masses, such as Earth and the moon, revolve around a common point on a line connecting their centers. *The point around which the system revolves* is called the center of gravity, or **barycenter,** of the system.

If the bodies are of equal mass, such as is the case in some binary systems, the two bodies revolve around a barycenter that is halfway between their centers, on a line connecting them. What about a system such as the Earth-moon system in which Earth is considerably more massive than the moon? If a very large child and a very small child want to seesaw, the larger child must sit in the center of half of the seesaw, whereas the smaller must sit as close to the outer edge of the other half of the seesaw as possible in order for the seesaw to be balanced. The Earth-moon system must balance itself similarly. Because of the greater mass of Earth, the barycenter is very close to Earth's center; indeed, it lies about 1600 kilometers below the surface of Earth. The moon and Earth both revolve around this center of gravity. If one were looking "down" on the solar system from the direction of Polaris, one would see Earth wobble as it speeds along on its yearly journey around the sun. This wobble would be the result of Earth's revolving around the Earth-moon barycenter in addition to the Earth-sun system revolving about a point within the sun.

The Tides

Many ancient societies worshiped the globe of soft light that we know as the moon. They thought that the moon directly affected their lives as well as other animate and inanimate things on Earth. Today, however, while there is no evidence that the moon directly affects our personal lives, we do know that it does cause physical changes in the atmosphere, oceans, and landmasses of our planet.

Anyone who has spent at least five or six consecutive hours at the seashore has noticed that the sea level rises and falls in a rhythmic pattern. This coming in and going out of the tides can cause typical variations from mean sea level ranging from 0.1 meter to 1 meter. In many geographic areas, the rise in sea level occurs about every 12 hours and 25 minutes. The level of the sea reaches a minimum midway between consecutive rises. The average difference between the highest level and the lowest level is about 0.58 meter. The gravitational pull of the moon causes the water of the oceans to be heaped up at certain places and to be lowered at other places. Even the huge, seemingly rigid, continental masses cannot resist this unrelenting attractive force of the moon, and they develop tidal bulges up to 0.23 meter. The moon also causes tides in our atmosphere.

Even though the sun is much more massive than the moon, the sun is so much farther away that its tide-raising force is only about one-third that of the moon's. During the full moon and the new moon (Fig. 21.14), the moon and sun are lined up so that their tide-raising forces are additive and they produce tides that are higher than normal. These *highest tides* are called **spring tides,** a term that has absolutely nothing to do with the spring season. The concerted effort of the moon and sun, particularly during certain spring tides, could be a potential cause of major earthquakes along fault zones, such as those located near continental coasts.

Figure 21.15 Boat owners and their crews must be prepared to move their vessels to deeper water during low tides such as experienced in the Bay of Fundy and shown in these photographs. Veteran seamen may prefer to secure their boats in upright positions to await the return of high tide.

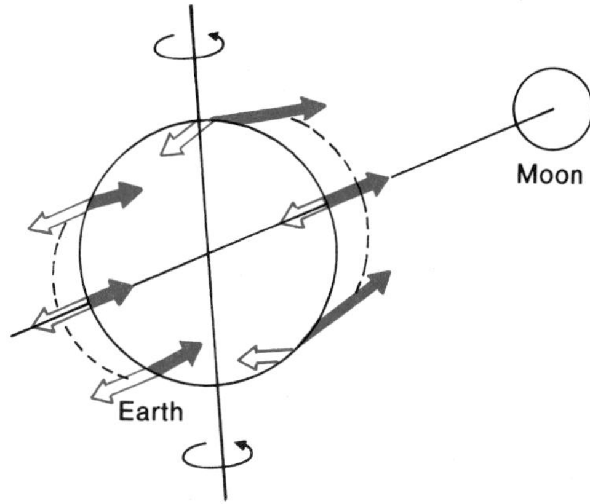

Figure 21.16 The forces that cause the formation of lunar tides on Earth. The gravitational pull of the moon on Earth helps create high tides on the side of Earth opposite the moon as the planet is pulled away from its waters.

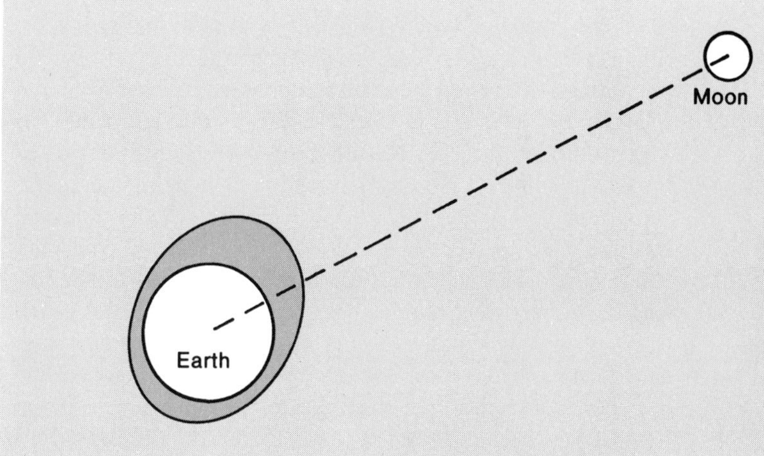

Figure 21.17 The tidal bulge caused by the moon actually points ahead, or east of the moon, because Earth's rotation eastward pulls the bulge ahead of the moon.

When the moon is in its first and third quarters, the pull of the moon and the pull of the sun are at right angles to each other, and the tides produced are correspondingly lower. These *lower tides* are called **neap tides.** The variation from mean sea level between high and low tides may be no more than about 0.1 meter along some coasts.

Tidal differences in the oceans offshore rarely exceed 1 meter; however, the tidal current of the Amazon River may vary from low tide to high tide by as much as 8 meters. Various parts of the Bay of Fundy, the body of water that essentially separates the Canadian provinces of Nova Scotia and New Brunswick, experience spring tides that may cause the water level to rise over 15 meters. The record vertical tidal range for the Bay of Fundy is said to have been as great as 20 meters. Crews of fishing vessels and others must be prepared to adjust for tidal effects on water depths, as illustrated in Figure 21.15.

A complete accounting of all the forces involved in the causation of Earth's tides would be quite complex; however, Figure 21.16 provides an explanation adequate for our purposes. The dark arrows represent the forces due to the moon's gravitational attraction. The light arrows represent the centrifugal forces due to Earth's movement around the Earth-moon center of gravity (barycenter) as well as the centrifugal forces caused by Earth's axial rotation. The resultants of the gravitational forces and the centrifugal forces are causes of the tides. These resultant forces cause the waters of the oceans to flow toward a point on Earth's surface that is directly between Earth and the moon on the side toward the moon and to a point on the opposite side of Earth that is in line with the moon. In addition, the centrifugal force caused by the planet's rotation, combined with the gravitational attraction of the moon as it pulls the solid portion of Earth away from its waters, creates a lesser high tide on the side of Earth opposite the moon.

The tidal bulge does not point toward the moon as one might expect, but instead points ahead of the moon in an easterly direction. The tidal bulge is pulled ahead of the moon by the eastward rotation of Earth on its axis, since Earth rotates in a much shorter period of time than it takes the moon to revolve around Earth (see Fig. 21.17).

There is considerable friction produced as the tides roll over the seafloors, particularly where huge quantities of water are funneled through very shallow and narrow passages. Energy must be used to overcome these frictional forces, and energy is made available at the expense of Earth's rotation on its axis. The consequence

Coincidence or Consequence?

Did you ever wonder why the sun and the moon appear to be about the same size, or how our small moon can block out our view of the sun during a total solar eclipse? These appearances are simply happenstance; the sun's greater diameter is almost exactly counterbalanced by its average distance from Earth. An object, when moved twice as far from an observer, appears half as large. The sun's and moon's diameters and present distances from Earth make these two celestial bodies appear to be practically the same size from our vantage point.

Several thousand years from now, however, Earth's inhabitants will not experience the same view. The moon is constantly receding from Earth because of the effect of Earth's ocean tides on it. Sometime in the future, Earth's lone natural satellite will reach such a distance from our planet that it can no longer mask the view of the sun. Then there will never again be a total eclipse of the sun visible from Earth ■

of this loss of rotational energy is that Earth is slowly decreasing its speed of rotation. This decrease means that the length of our day is correspondingly increasing. Even though the increase in the length of the day is very small, approximately 1/1000 of a second per century, this increase can be an appreciable length of time if one considers the many millions of centuries that our planet has existed. In fact, the U.S. Naval Observatory officially increased the day of June 30, 1983 by 1 second in order to maintain synchronization of our atomic clocks with Earth's rotation. A second was also added to December 31, 1990 to correct the synchronization even further. It has not been established beyond dispute how long Earth's period of rotation has been on the increase, although one would assume that the slowing effect has been occurring since our planet has had oceans. Frictional forces in the atmosphere and in Earth's core add to the effect created by our major bodies of water. It has been determined, however, that the decrease in the rate of Earth's rotation has been going on at least since the Devonian period, about 400 million years ago. Geological evidence shows that during the Devonian period Earth's year consisted of 400 days, which means that each Devonian day was only about 22 hours long.

If the forces previously discussed cause the crust of our planet to bulge out as much as 0.23 meter, what must be the tidal effect of the much more massive Earth on the moon's crust? It is probable that the surface of the moon bulges out at least 1 meter due to the tidal action of Earth.

Solar and Lunar Eclipses

As the moon slowly comes between Earth and the sun and eventually blocks our view of the sun, the sky slowly darkens and a chill pervades the air. When the sun's golden disk is completely blotted out by the moon, birds fly to their nests to roost and stars can be seen shining brightly at midday. It is small wonder that primitive societies were stricken with fear by the awesome spectacle of a **solar eclipse.** Their life-giving sun was apparently being swallowed up by some god of darkness.

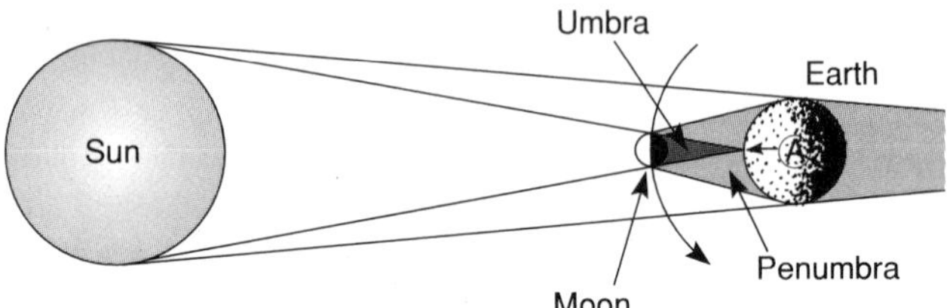

Figure 21.18 Earth's umbra and penumbra during a solar eclipse. The total eclipse would be viewed only at point A at that particular time of the occurrence.

As shown in Figure 21.18, Earth always casts a long conical shadow, called the **umbra,** within which there is no light from the sun. Surrounding the umbra is a less dark region, called the **penumbra,** in which only part of the sun's light is visible.

Earth's umbra is about 1,371,000 kilometers in length. Since the moon is only 383,000 kilometers from Earth, we would expect the moon to pass within Earth's shadow and be eclipsed. From this discussion one might quite reasonably expect that the moon should be eclipsed once each 29.5 days (from new moon to the next new-moon phase); this would indeed be the case if the plane of the moon's orbit around Earth coincided with the plane of Earth's orbit around the sun. The plane of the moon's orbit is inclined 5 degrees away from the plane of Earth's orbit; consequently, a **lunar eclipse** can occur only where the two planes intersect and only when the line of intersection includes the moon, Earth, and the sun, in that order (see Fig. 21.19). This special arrangement of the moon, Earth, and the sun occurs but twice each year; therefore, we can expect to experience two lunar eclipses annually.

Like lunar eclipses, solar eclipses can occur only when Earth, the moon, and the sun are aligned, in that order, along the line of intersection of the plane of the moon's orbit and the plane of Earth's orbit. Consequently, a solar eclipse can occur only about every six months. Interestingly enough, the next total solar eclipse visible in the United States will not occur until the year 2024.

It is pure happenstance that we on Earth can observe a total solar eclipse. In the first place, if the moon were slightly farther

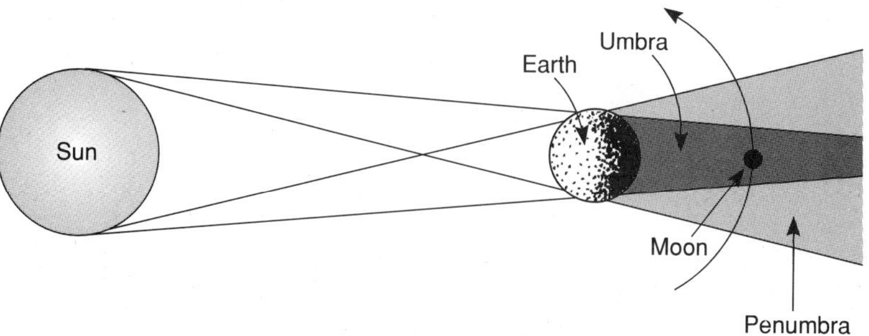

Figure 21.19 Earth's umbra and penumbra during a lunar eclipse.

from Earth, the moon's shadow, or umbra, would not be long enough to fall upon Earth's surface. In fact, there are numerous times each decade when the alignment of Earth, the moon, and the sun allow for a total eclipse, but the moon's umbra is not long enough to reach Earth. The average length of the moon's umbra is only 375,000 kilometers, whereas the mean distance of the moon from Earth is approximately 384,000 kilometers. Therefore, it is only when the moon is near perigee (closest approach to Earth) that a total solar eclipse can occur, since at this distance the moon is only about 354,000 kilometers from Earth. Secondly, it appears to be pure coincidence that the apparent size of the moon's disk is approximately the same as the apparent size of the sun's disk, so that the sun can be completely covered by the moon. Although the sun has a diameter 400 times the diameter of the moon, the sun is approximately 400 times farther from Earth than is the moon. (Recall, the diameter of the moon is 3476 km.) As a result, the apparent size of the moon's disk is large enough to cover the sun. Within our solar system there are many moons that are close enough to their parent planets to produce eclipses. Figure 21.20 shows a solar eclipse occurring on a small portion of the planet Jupiter.

Figure 21.20 A solar eclipse on the planet Jupiter. Jupiter's satellite Ganymede casts its shadow above Jupiter's Great Red Spot.

The Face of the Moon

From Earth the moon appears as a magnificent celestial object that bathes the night Earth in a soft, pale light. This, however, belies the severe conditions that actually exist on the surface of the moon, where a harsh scorching sun blazes down on a very inhospitable landscape (note Fig. 21.21). During the day the surface temperature soars to above the boiling point of water and during the night plunges to $-175°$ C.

Only recently have photographs and observations been made of the far side of the moon. Until the advent of space travel, we had to be content with seeing only one side of the moon. The reason that the moon keeps one side perpetually facing Earth is that the rotation of the moon on its axis is synchronized with the revolution of the moon about Earth. The moon has not always had its rotation and revolution synchronized, however. The tidal bulge on the moon, created by the Earth's gravitational pull, has slowly reduced the speed of the moon's axial rotation to its present rate. The moon's present rate of rotation is now locked by gravitational forces to its revolution around Earth.

Figure 21.21 *Apollo 17* astronaut-geologist Harrison Schmidt examines a large boulder on the surface of the moon.

Figure 21.22 This gibbous (more than half full) Earth lights up the nightside of the moon with "earthlight" that is about five times brighter than the nightside of Earth bathed in moonlight from a gibbous moon.

Earthlight

When the moon is at crescent phase, we see the sun shining on only a thin, sickle-shaped slice of the moon's surface, whereas the rest of the moon that we faintly see is the nightside. The light of the sun does not touch this nightside, yet we see this region faintly illuminated. What is the source of this illumination? It is "earthlight," Earth's equivalent of moonlight. The nightside of the crescent moon is facing the daylight side of Earth with the result that earthlight shines upon the nightside of the moon in the same way that moonlight shines upon the nightside of Earth. An astronaut on the nightside of the moon would be bathed in earthlight from a "full Earth" that would be five times brighter than the moonlight we see from a full moon. From Figure 21.22 we can imagine the brilliance of the sunlit portion of Earth. The reflection of light from its surface can be seen in Figure 21.23 as a dark segment of the moon is softly illuminated by earthlight.

The Future of Earth

Earth has existed for 4.6 billion years and will probably exist in close to its present state for another 4.6 to 5 billion years, assuming of course that the planet is not significantly altered by human intervention. After another 5 billion years, however, our sun will enter the red-giant stage and expand until its outer edges engulf the planet Mercury, Venus, and perhaps beyond. Then the surface of our planet will be so hot that the water of the oceans and ice of the polar caps will be turned into superheated steam. All life forms will have perished, leaving a barren, inhospitable planet. Eventually, the sun will begin to shrink and will continue to do so until it reaches the stage of a white dwarf. As the sun reaches the white-dwarf stage, Earth will become a bare frozen rock encased in continental ice sheets and frozen oceans. Ultimately, after many hundreds of billions of years, the nuclear fires of our sun will be completely exhausted, and the sun will become a cold, burned-out cinder (a black dwarf), which, with its retinue of dead planets, will continue its lonely journey through space.

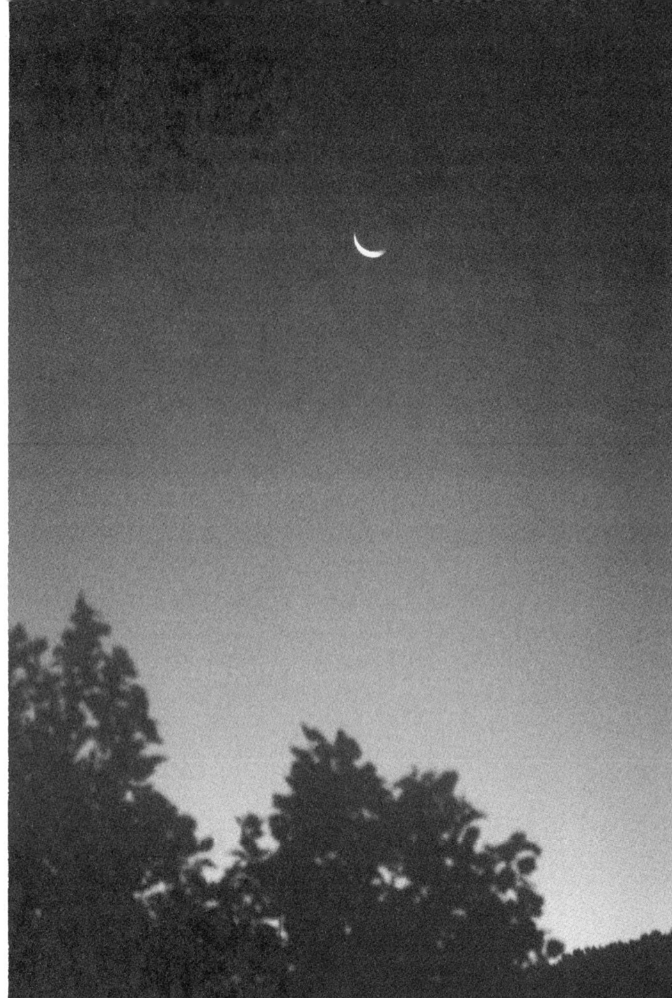

Figure 21.23 Even though only a portion of the moon is directly illuminated by sunlight, the balance of the curvature of the moon is faintly visible as light reflected from Earth's surface strikes it.

Summary

The Foucault pendulum experiment proves that Earth rotates on its axis.

On or about June 21, the summer solstice, the sun is at zenith on the Tropic of Cancer at 23.5 degrees north latitude. Then land above the Arctic Circle at 66.5 degrees north latitude has 24 hours of sunshine and land above the Antarctic Circle at 66.5 degrees south latitude has 24 hours of darkness. On or about December 21, the winter solstice, the sun is at zenith on the Tropic of Capricorn at 23.5 degrees south latitude. On that day, land above the Arctic Circle has 24 hours of darkness and land above the Antarctic Circle has 24 hours of sunshine.

On or about March 21, the vernal equinox, and on or about September 23, the autumnal equinox, the sun is overhead at the equator.

Because Earth is a rotating object influenced by the gravitational fields of the sun and the moon, Earth's spin axis precesses once every 26,000 years.

An eclipse of the sun occurs whenever part of Earth passes through the umbra, or shadow, of the moon. An eclipse of the moon occurs whenever part of the moon passes into the umbra of Earth.

The gravitational attraction of the sun and the moon for Earth produces tides in Earth's atmosphere, oceans, and solid crust. Only the tides produced in the oceans and seas of Earth are easily discernible. The tides in the crust of Earth may slightly exceed 0.2 meter.

Questions and Problems

The Foucault Pendulum Experiment

1. How does a Foucault pendulum prove that Earth rotates on its axis?
2. How does the pattern made by the Foucault pendulum change as the pendulum is moved from one of Earth's poles toward the equator?

Evidence of Annual Motion

3. How does stellar parallax prove that Earth revolves about the sun?
4. Why do some stars appear to move against a background of other stars during Earth's journey around the sun?
5. What is the ecliptic and where did the name originate?
6. What is the celestial sphere? How is our view of it affected during the course of a year?

The Seasons

7. Where on Earth will the sun be directly overhead (at zenith) on or about December 21? Where will the sun be at zenith on or about June 21?
8. Where is the Tropic of Cancer located? Where is the Arctic Circle?
9. According to Figure 21.6, during what season does the North Pole of Earth receive the most sunlight? The least sunlight?
10. During what season in the Northern Hemisphere is Earth closest to the sun (at perihelion)? When is the Earth farthest from the sun (at aphelion)?
11. If not for Earth's tilted axis, what season of the year would be the hottest in the Northern Hemisphere?

Solar and Sidereal Time

12. Why is a solar day four minutes longer than a sidereal day?
13. Why doesn't sunrise occur at the same time each day?

Precession of Earth's Axis

14. What is the cause of Earth's precession? How long is its period of precession?
15. What effect does precession have on Earth's seasonal pattern?

The Earth-Moon Gravity Couple

16. What is the barycenter and where is it located?
17. Why is Earth's orbital path about the sun a wobbly one?

The Tides

18. What causes tides? How are Earth's atmosphere, oceans, and continental masses affected by these forces?
19. How do spring tides differ from neap tides? On the average, how much do high tides and low tides differ from mean sea level?
20. Why is the rate of rotation of Earth slowly decreasing? How does the decrease in rotational speed of Earth affect the moon?

Solar and Lunar Eclipses

21. Why don't we experience a total eclipse of the sun each month?
22. Describe the positions of the sun and moon during (a) a lunar eclipse and (b) a solar eclipse. (c) During what phase of the moon could a solar eclipse occur? (d) During what phase could a lunar eclipse occur?

The Face of the Moon

23. Why do we always see the same side of the moon?

Earthlight

24. What is earthlight and during what stage of the moon is it most obvious?
25. How does the brightness of a fully illuminated Earth compare to the brightness of a full moon?

The Future of Earth

26. How will the projected evolutionary stages of the sun affect Earth and the other planets?

22 The Atmosphere of Earth

Chapter outline

Key terms/concepts

A mixture of blue skies and dark clouds warn small craft of a pending storm over coastal waters.

Although we obtain our food and water from the surface of our planet, our survival is directly dependent on the envelope of air we call the atmosphere. This blanket of air gives us oxygen to breathe, fresh water to drink, and enables the plants and animals we eat to live. Because of the atmosphere, the temperature of Earth is moderated so that we do not freeze or burn to death. The atmosphere blocks out enough of the ultraviolet and X-radiation from the sun to prevent sterilization of our landmasses. The atmosphere allows certain types of radio waves to be received worldwide. The atmosphere, thousands of miles deep, gives us our colorful auroras and the beautiful clouds that provide a perfect contrast to the planet's blue skies and magnificent red sunsets.

Composition

The atmosphere is a mixture (more precisely, a solution) of many different gases bound to our planet by gravity (see Table 22.1). A volume of dry air at sea level consists of 78 percent nitrogen and 21 percent oxygen; the remaining 1 percent is mainly argon. Carbon dioxide, so important to the life of plants, constitutes only 0.033 percent of the atmosphere but, due to the widespread combustion of fossil fuels, is apparently increasing. All the remaining ingredients of the atmosphere account for only 0.01 percent of the total.

The balance between the oxygen and the carbon dioxide content of the air is extremely important for our survival. For instance, if the oxygen content should increase substantially, a fire, once started, would be extremely difficult to extinguish. If the carbon dioxide content should increase substantially, the resulting greenhouse effect would raise Earth's surface temperature to the point that extensive melting of arctic and antarctic ice would result in the submergence of many of the world's coastlines and coastal cities.

Table 22.1 does not list such atmospheric constituents as water vapor or dust particles, both of which are very important to our weather and climate. The atmospheric water content (humidity) and dust-particle content vary widely from place to place and from time to time and, therefore, cannot be quantitatively listed in any meaningful way in a table of the composition of the atmosphere. Even though the total atmosphere of our planet contains approximately 12,500 cubic kilometers of water as water vapor, the daily water content of the atmosphere at various geographical locations varies from a low of about 0.02 percent in desert regions to about 4 percent in humid tropical regions.

The continual exchange of water between the atmosphere and our planet's surface is paramount in determining weather conditions, since during the evaporation process large quantities of heat are absorbed and during the condensation process equally large quantities are released.

The nitrogen, oxygen, water, and carbon dioxide in the atmosphere are constantly being used, but also constantly being replenished (see Table 22.2).

Table 22.1
The composition of dry air.

Substance	Percentage of volume
Nitrogen (N_2)	78.08
Oxygen (O_2)	20.95
Argon (Ar)	0.93
Carbon dioxide (CO_2)	0.03
Neon (Ne)	trace amounts
Helium (He)	trace amounts
Methane (CH_4)	trace amounts
Krypton (Kr)	trace amounts
Xenon (Xe)	trace amounts
Hydrogen (H_2)	trace amounts
Nitrous oxide (N_2O)	trace amounts

Table 22.2
Utilization and replenishment of major air components.

Substance	Used for	Replenished by
Nitrogen	Plant nutrients formed by nitrogen-fixing microorganisms	Denitrifying bacteria using fixed nitrogen compounds
Oxygen	Respiration by animals	Photosynthesis by plants
Water	Rain and snow	Evaporation from bodies of water
Carbon dioxide	Photosynthesis in plants to produce carbohydrates	Respiration of animals and the decomposition of dead organisms

The Layered Atmosphere

The atmosphere can be considered to consist of five main layers: (1) the *troposphere,* (2) the *stratosphere,* (3) the *mesosphere,* (4) the *thermosphere,* and (5) the *exosphere* (see Fig. 22.1). These subdivisions are based on differences in the thermal characteristics of each layer.

The **troposphere** is *the layer of air closest to Earth's surface.* It extends from sea level to a height of 16 kilometers over the equatorial region and to 8 kilometers over the polar regions. By contrast, the top of Mount Everest is 8.58 kilometers above sea level. The troposphere contains almost all the atmosphere's water vapor and dust, and it is within the troposphere that all of our weather (air movements, clouds, rain, and storms) occurs. Approximately 90 percent of the total mass of the atmosphere is found in the troposphere. The temperature of the troposphere gradually decreases with increasing altitude. This *temperature decrease* with altitude is known as the **lapse rate** and for the troposphere averages about 6 C° per kilometer. Air is most dense near the bottom of the troposphere because the weight of the air molecules above presses the lower-lying air molecules together.

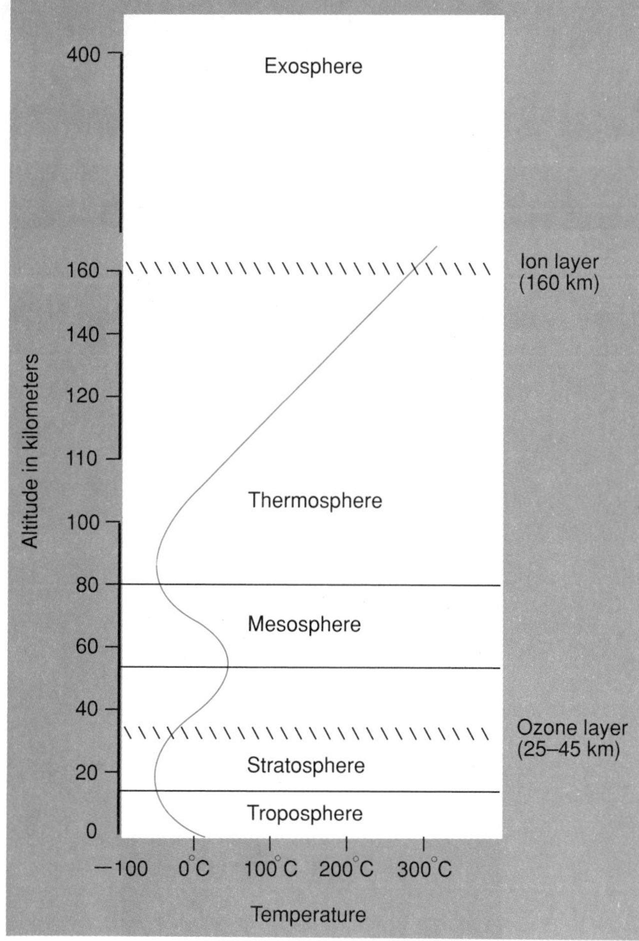

Figure 22.1 Change in atmospheric temperature with altitude.

At the upper limit of the troposphere, the lapse rate abruptly changes. Above this altitude the temperature increases with increasing altitude. The layer above the troposphere is the **stratosphere.** Where the troposphere ends and the stratosphere begins, the temperature is approximately −60°C, and this temperature increases until at the top of the stratosphere, about 55 kilometers up, the temperature reaches about 0°C. The stratosphere is characterized by the lack of clouds or weather.

Within the stratosphere is a sublayer (extending upward from about 25 to 45 kilometers) where oxygen molecules (O_2) are broken up into oxygen atoms (O) by ultraviolet radiation from the sun. Many of these oxygen atoms then combine with unaffected oxygen molecules to produce ozone (O_3):

$$O_2 + \text{UV raditation} \longrightarrow O + O$$
$$O_2 + O \longrightarrow O_3$$

The ozone produced exists as a trace element of only about ten parts per million at its highest concentration. Even at this low concentration the ozone very efficiently absorbs ultraviolet radiation and as a result, dissociates into oxygen atoms and molecules. Fortunately, this whole process results in a large amount of ultraviolet radiation absorption. If the thin *ozone layer* surrounding Earth did not exist, the surface of Earth would be bombarded with the full blast of the sun's ultraviolet radiation. Exposed bacteria would die, and the outer tissues of unprotected animals would be severely damaged with the resulting destruction of many life-forms. Exposure of fair-skinned humans who were not completely covered would be disastrous. Various means of protecting this atmospheric layer have been undertaken by most major countries.

Above the stratosphere lies the **mesosphere,** which extends upward from 55 kilometers to about 80 kilometers. In the mesosphere the temperature again decreases with increasing altitude, falling from about 0°C at an altitude of 55 kilometers to about −90°C at 80 kilometers.

Above the mesosphere is located the **thermosphere,** which extends from about 80 kilometers to 400 kilometers above Earth. The oxygen atoms and nitrogen atoms in the thermosphere are subjected to rather intense solar X rays and ultraviolet radiation, which results in ionization reactions:

$$O + \text{radiation} \longrightarrow O^+ + e^-$$
$$N + \text{radiation} \longrightarrow N^+ + e^-$$

Even though the temperature in the thermosphere does become very high (about 1000°C in the upper thermosphere) in the sense that the particles present are traveling at high speeds (high kinetic energies), the density is so low that there is very little heat content or total heat energy present.

The free electrons and positive ions produced in the thermosphere reflect radio waves in the AM band back to Earth. The reflection allows us to be able to hear radio broadcasts from around the world. If the ion layer, sometimes called the *ionosphere,* of the thermosphere did not exist, we would have only line-of-sight radio reception.

Above the thermosphere, at about 400 kilometers above Earth's surface, is the **exosphere,** which extends outward from Earth for thousands of kilometers. The exosphere is the transition layer between Earth's atmosphere and outer space.

Although the temperature changes dramatically at times from zone to zone in the atmosphere, there is no radical change in the pressure with increasing altitude. Our atmosphere is held to Earth by gravity. But if gravity were the only force acting on the air, all the air would be pulled down to form an extremely thin layer on the surface of our planet. However, all the atoms and molecules in our atmosphere, by virtue of their kinetic energy, are in constant motion. Their kinetic energy content results in atoms and molecules moving in all possible directions and at a multitude of different speeds. The movement of the atoms and molecules opposes the pull of gravity to give us the atmosphere as we know it. The atoms and molecules of the air that have relatively low kinetic energies cannot overcome the pull of gravity as well as those with high kinetic energies. Consequently, we find that air pressure in the atmosphere decreases with increasing altitude. For example, at a height of 5.5 kilometers air pressure is about 50 percent that at sea level, and at 18 kilometers air pressure is about 10 percent that at sea level.

Our Fragile Ozone Layer

About 25 kilometers up in our atmosphere, a layer of ozone surrounds Earth. Ozone, a molecule composed of three oxygen atoms, O_3, absorbs ultraviolet light. Our ozone layer blocks much of the ultraviolet radiation that Earth receives from the sun. If the amount of ozone in the layer were diminished to any appreciable extent, the increase in ultraviolet radiation reaching the surface would greatly increase the incidence of human skin cancers and probably the number of eye cataracts. It is estimated that for every 1 percent decline in concentration of the ozone layer, an eventual 5 percent increase in skin cancer will occur. If the ozone layer ceased to exist, many life-forms would be exterminated and others would be considerably diminished.

In the past few decades the amount of ozone in this very important protective layer has decreased about 5 percent worldwide. In 1985, it was found that a "hole" in the layer exists over the south polar region. During September and October, the ozone is depleted over Antarctica to about 50 percent of its normal concentration. The concentration increases during the rest of the year, but this drastic annual formation of the Antarctic ozone hole poses very serious consequences for life on Earth. In years of highest depletion, such as 1987 and 1989, the region of ozone depletion grew to twice the size of the Antarctic continent. Fortunately, the hole shows no sign of dramatically increasing in size. A swirling vortex of winds, called the polar vortex, circles the Antarctic region at about 66 degrees south latitude; consequently, the hole maintains a relatively constant size. A smaller but equally disturbing hole in the ozone layer exists in the north polar region. Besides the adverse effect of an increase in ultraviolet radiation on humans, the production of phytoplankton (photosynthetic microorganisms) in the antarctic ocean region is reduced. These phytoplankton constitute the bottom of the food chain that supports all polar animal life.

Most scientists believe that the ozone in the ozone layer is being depleted mainly by chlorofluorocarbons (CFCs), of which Freon 12 is the main culprit. The CFCs decompose in the upper atmosphere to release chlorine that destroys the ozone molecules. Through a chain reaction, a single chlorine atom can cause the destruction of tens of thousands of ozone molecules. In 1987, 40 nations met under the sponsorship of the United Nations to find how to best limit the production of chlorofluorocarbons in the world■

The Earth-Sun Energy Balance

Earth is constantly being bathed in the radiant energy given off by the sun. Without this absorbed radiant energy, life on our planet could not exist. Fortunately, though, not all the energy that falls on Earth is absorbed, since total absorption and retention of the sun's radiant energy would result in the steady increase of Earth's temperature until life could no longer exist.

Actually only about two-thirds of the sun's energy that is emitted toward Earth enters our atmosphere; the rest is reflected into outer space. About one-third of the energy that penetrates the atmosphere is absorbed by water vapor and carbon dioxide; the remaining energy is absorbed by Earth's surface. A sizable portion of the absorbed energy is reradiated at longer wavelengths, to be absorbed by our atmosphere or reflected again toward Earth.

Dust particles in the air play a very important role in influencing weather conditions. Airborne particles from dust, smoke from forest fires or industrial plants, pollen grains, volcanic ash, and salt from sea spray act as condensation nuclei for the formation of water droplets and ice crystals to produce clouds, rain, snow, or hail. Whereas most hailstones are pea-sized or smaller, some are large enough to cause very serious damage to property and humans (see Fig. 22.2).

Figure 22.2 Hailstones larger than baseballs may form, with the potential to cause considerable physical damage. Their crystalline structures can appear blue when they are photographed in sunlight.

Dust particles also act as multifaceted surfaces that reflect sunlight out into space and, therefore, reduce the amount of heat that Earth receives from the sun. As an example of how much influence suspended dust particles have on the amount of heat absorbed by Earth, the explosion in 1816 of the volcano Tambora in Indonesia blew so much particulate material into the air that the average temperature of the entire planet was reduced by 6°C. In

Halos and Coronas

Varying conditions that periodically exist in Earth's atmosphere create unusual optical phenomena. A halo is one of the large class of atmospheric occurrences that appear as colored or white rings and arcs about the sun or moon. Halos are caused by the presence of ice-crystal clouds or by falling ice crystals. The halos with prismatic coloration are produced by refraction of light by the crystals, whereas those of whitish luminosity are caused by the reflection of light from crystal faces.

Coronas differ from halos in that coronas are produced by the diffraction and reflection of light from water droplets. A colored halo may be distinguished from a corona in that the halo has the red nearest the sun or moon, whereas the corona has red in its exterior rings. On rare occasions, the sky will be filled with a spectacular display of four or five simultaneous occurrences. The ancients fashioned much supernatural lore around these multiple displays ■

fact, the year following the eruption of Tambora is known as "the year without a summer" because of the unusually low summer temperatures that resulted from the concentration of dust particles in the air.

Clouds are especially important in absorbing energy from the sun, as well as the energy radiated from Earth. Much of the energy absorbed by Earth would otherwise be radiated directly into space. On a cloudless night, the temperature of Earth drops much more than it would if the night were cloudy.

A considerable amount of the solar energy that reaches Earth's surface is absorbed by water. The solar energy causes water to be converted from the liquid state to the gaseous state. This conversion consumes at least 540 calories for each gram of water converted (heat of vaporization). The water vapor so produced rises in the air, is transported by winds to other locations, and eventually condenses back into water to form clouds or to produce rain. As each gram of water vapor is converted to liquid water, 540 calories are released (heat of condensation). The result of these processes is that large quantities of heat at one location are transferred to other regions on Earth's surface.

If Earth received more energy from the sun than it reradiated into space, the temperature of Earth would steadily increase; conversely, if Earth received less energy from the sun than it reradiated into space, the temperature of Earth would steadily decrease. Since Earth, over the years, has neither become exceedingly hot nor exceedingly cold (actually, the average temperature of Earth is remarkably constant), Earth must be reradiating into space the same amount of energy that it absorbs from the sun; that is, the heat input and heat output of Earth are balanced.

The Greenhouse Effect

As long as there have been an atmosphere and oceans on our planet, there have been changes in atmospheric and oceanic temperatures. A graph of global temperature versus time shows a continuous series of peaks and troughs. Global temperature changes occur over billions of years and also over only a few years. About 10,000 years separate us from the end of the most recent Ice Age, which began about 2 million years ago and during which almost one-third of Earth's landmass was covered with ice. Obviously, the Ice Age was not caused by human activity; however, there is concern about the effect of recent human activity (over the last 150 years) on the global temperatures of the future.

The industrial revolution brought with it not only an abundance of all kinds of consumer goods but also extensive air pollution. Most air pollution is the result of the burning of fossil fuels (oil, coal, and natural gas). Besides particulate matter, industrial plants and automobiles release carbon dioxide, carbon monoxide, oxides of nitrogen, sulfur dioxide, and hydrocarbons. Although some of these gases are harmful in terms of acid rain and depletion of the ozone layer (principally by freon), they all contribute to the heat-retaining property of the atmosphere. The retention of atmospheric heat by various gases is referred to as the *greenhouse effect* (see Fig. 22.3). The name derives from the use of structures (or houses) covered with glass or plastic to maintain a higher temperature for growing plants inside them than the temperature of the surrounding outside air. Sunlight enters a greenhouse through the glass or plastic and is absorbed by the soil, plants, and objects inside. They then radiate some of the energy they have gained as infrared radition. The infrared radiation cannot pass through the glass or plastic as easily as did the sunlight; hence, it is trapped inside the greenhouse. The same effect is responsible for part of the sunlight's warming of the inside of a closed car on a cold day.

In the atmosphere, water vapor and carbon dioxide are mainly responsible for the absorption of infrared rays radiated from Earth. Although water vapor is the most important absorber, absorption by carbon dioxide is very important since human activities have a considerable effect on the concentration of atmospheric carbon dioxide. There has been an increase in the concentration of atmospheric carbon dioxide of about 15 percent since the beginning of the Industrial Age. Whereas the concentration of water in the atmosphere is usually controlled by the balance between the evaporation of ocean water and the reduction of atmospheric water by rain and snow, excess carbon dioxide is not so easily regulated. Some carbon dioxide is removed by the oceans through conversion to calcium carbonate sediments, but this process is not sufficient to counter present-day carbon dioxide buildup.

Because of the considerable increase in the CO_2 content of the atmosphere predicted for the future, Earth is expected to experience a corresponding reduction in heat loss by radiation into

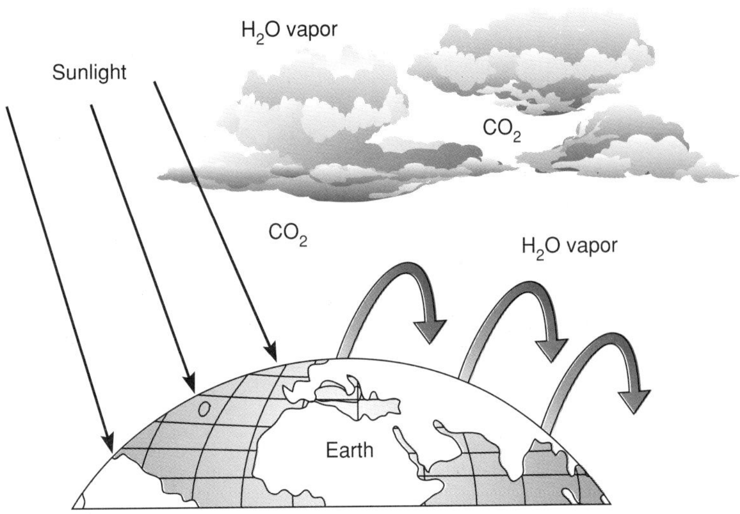

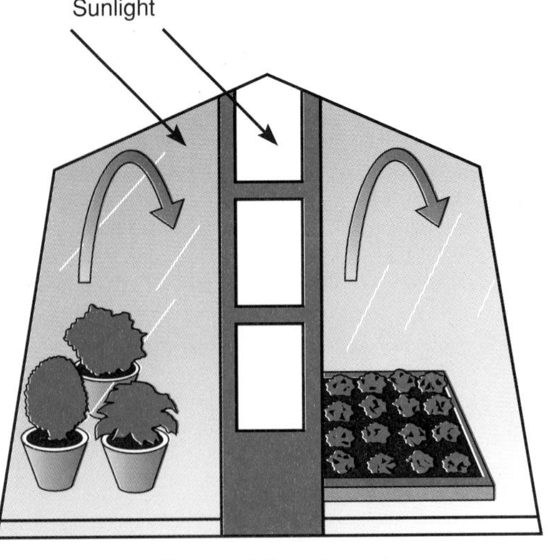

Figure 22.3 The greenhouse effect. Sunlight easily passes through the air, as well as through glass or plastic sheets, and warms Earth or the contents of a greenhouse. The energy absorbed from sunlight is then radiated as infrared energy. The water vapor and carbon dioxide of the atmosphere, like glass or plastic, prevent the infrared radiation from escaping into space, thus effecting an increase in the temperature of the atmosphere or the greenhouse.

space. The resulting increase in atmospheric temperature could have enormous climatic consequences. Any appreciable warming would produce a partial melting of present ice sheets that would cause serious flooding of many ports and cities along our coasts.

The prediction of Earth's future atmospheric temperature is extremely complex, however, because temperature variations are not determined only by humans. From 1920 to 1940 Earth's average atmospheric temperature increased by about 0.35 C°, but from 1940 to 1960 the temperature decreased by about 0.22 C° and from 1960 to 1990 again increased by about 0.25 C°. Ironically, as stricter antipollution laws result in decreased atmospheric pollution by particulate matter, the atmosphere will reflect less sunlight, a condition which will produce an increase in the temperature of Earth's atmosphere. On the other hand, some scientists speculate that an increase in Earth's temperature would result in an increase in the rate of evaporation of water from the oceans. Such an increase could produce more clouds that would reflect more of our sun's light into space, resulting in a cooling of Earth's atmosphere. More study over a longer period of time will be necessary before the future of our atmosphere can be determined with any degree of certainty.

Air Circulation and the Coriolis Effect

The surface of Earth is not heated evenly by the sun. Whereas grass- and tree-covered areas absorb about 85 percent of the radiation that falls upon them, snow-covered regions absorb only about 15 percent of the radiation and reflect about 85 percent. Also, since sunlight strikes the equatorial regions almost perpendicularly, the amount of solar energy received per square meter of surface area at the equator is much greater than that received in the polar regions, where the sun's rays strike Earth at more acute angles. For instance, at 60 degrees north or south latitude only about 50 percent as much solar radiation is received as at the equator. In the equatorial regions of our planet (between 10 degrees north and south latitude), the relatively large amount of heat absorbed by the land and the oceans heats the air near Earth's surface. The heated air has a lower density than it had before it was heated and, consequently, expands and rises to produce a low-pressure area. As the air rises, it displaces air above it and pushes the displaced air upward and outward toward the north and south. Cooler and denser air moves along Earth's surface toward the low-pressure area created by the rising warm air. Thus, the unequal heating of different parts of Earth's surface produces low- and high-pressure areas. As air from a high-pressure area flows toward a low-pressure area, the horizontally moving cooler air from the high-pressure area is called **wind.**

As the warm air moving north and south continues to rise, the colder temperature at the higher altitudes eventually cools the air to the point that the cool, now denser, air begins to sink back to Earth's surface. The air movements produced in this manner are called *convection currents* (see Fig. 22.4). If there were no other effects to be considered, convection currents would cause warm air from the equatorial regions to rise, flow north and south at high altitudes, become cool, and sink near the poles. Cool air from the polar regions would flow along the surface toward the equator, where it would be heated and then rise and flow north and south, thus starting another cycle. Actually, the pattern of planetwide air circulation is much more complex.

Assume that you are standing on the rim of a rotating disk. The distance from you to the center of the disk is 20 meters. You throw a ball at an object 10 meters from you toward the center of the disk. The object is traveling at a much slower speed than

Figure 22.4 The atmospheric circulation that would occur around Earth if the planet did not rotate and if its axis were not tilted.

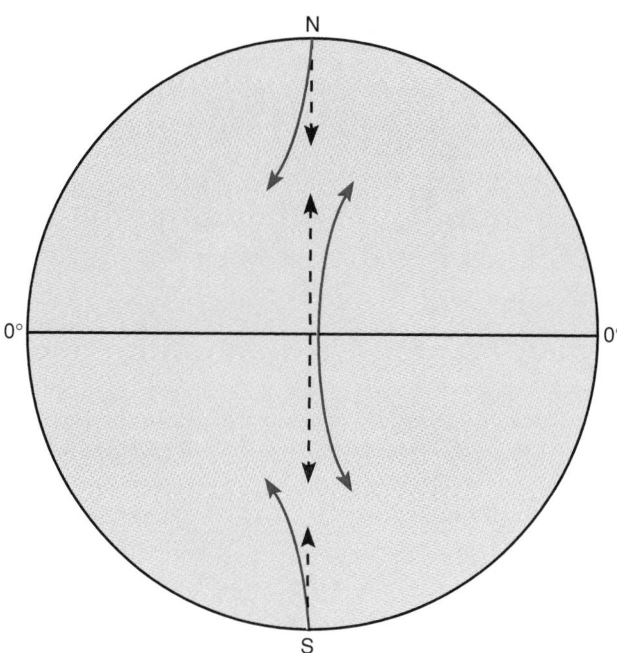

Figure 22.5 The Coriolis effect created by Earth's rotation causes winds to be deflected to the right in the Northern Hemisphere and to the left in the Southern Hemisphere.

you are, since the circumference of the disk at the position of the object is considerably smaller than the circumference at your position. In a given length of time, you will be traveling a greater distance than the object. The instant the ball leaves your hand, the ball is traveling at your speed. As the ball approaches the object, the ball passes over points that are traveling at increasingly slower speeds. You will see the ball pass to the right of the object in the direction of rotation of the disk. To an observer above the disk the ball will appear to curve in the direction of rotation as it leaves your hand and to pass in front of the object even though the ball is actually traveling in a straight line. For similar reasons, if a person near the object throws a ball toward you, the ball will also appear to curve to the right and pass behind you. Likewise, because of Earth's rotation, air that is traveling north or south in the Northern Hemisphere is deflected to the right, and air that is traveling north or south in the Southern Hemisphere is deflected to the left, as shown in Figure 22.5. Another way of stating this concept is that air that is moving toward the equator veers to the west and air that is moving away from the equator veers to the east. *The deflection that results because of Earth's rotation* is known as the **Coriolis effect.**

Air starting its journey north from the equator also is traveling eastward in the same direction as Earth's rotation since it shares in Earth's rotational motion. However, at 30 degrees north or south latitude the circumference of Earth is considerably smaller than its circumference at the equator. Any place on the equator will make one complete 360-degree rotation in 24 hours (actually 23 hours and 56 minutes, as discussed in Chapter 21). Since the circumference of Earth at the equator is 40,250 kilometers, and since a spot on the equator makes one entire revolution in 24 hours, the spot is traveling eastward at a speed of 1677 kilometers per hour. However, at 30 degrees north or south latitude the circumference of Earth is only about 34,560 kilometers, and therefore, a place on Earth at that latitude is traveling eastward at a speed of only 1440 kilometers per hour. Directly over Earth's poles the circumference is zero, so that a person at one of the poles would appear stationary, simply completing one rotation during a 24-hour period.

Other conditions also affect the pattern of air movements. Where warm air is rising, a low-pressure area exists, and where cooler air is descending, a high-pressure area exists. Naturally, air will flow from an area of high pressure to an area of low pressure. In the Northern Hemisphere, air flowing in a northerly direction from a high-pressure area will be deflected eastward, and air flowing in a southerly direction will be deflected westward. The result of these deflections is that around a high (a high-pressure area) air currents (wind) circulate in a clockwise direction. By the same reasoning, around a low (a low-pressure area) winds circulate in a counterclockwise direction (see Fig. 22.6).

Air motion at the equator is mainly upward, with very little horizontal movement. Therefore, only light winds are present at the equator most of the time, or there are no winds at all. These equatorial areas are called **doldrums.** As the air moving upward at the equator flows north and south, it cools and descends at about 30 degrees north or south latitude. Again, since the air movements here are vertical, only occasional light winds are found. The areas around 30 degrees north or south latitude are called the **horse latitudes.** Air that flows northward from the horse latitude in the Northern Hemisphere is deflected eastward, and these winds are

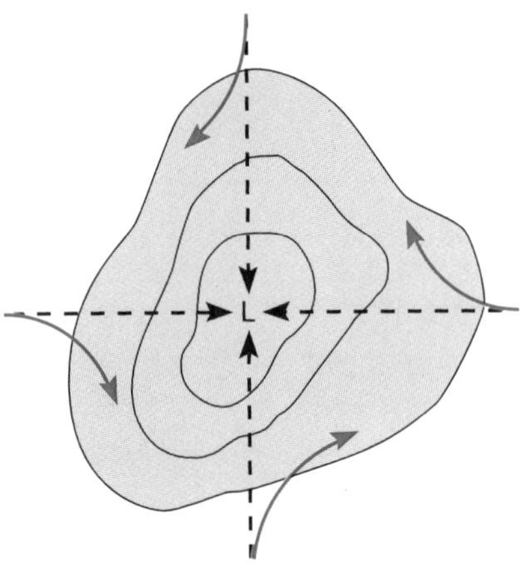

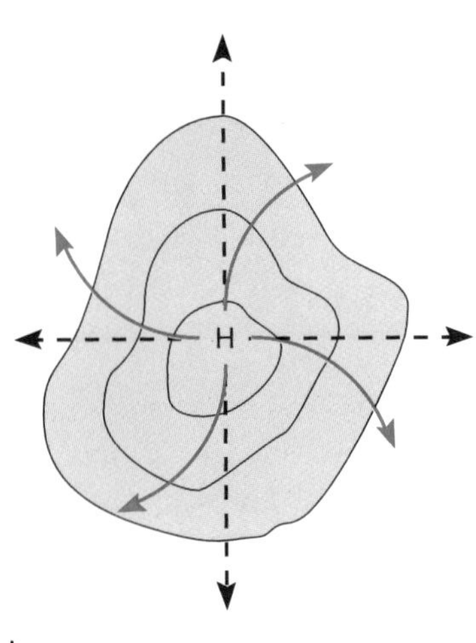

Figure 22.6 Because of the Coriolis effect, (*a*) air flow around a low (L) is counterclockwise and (*b*) air flow around a high (H) is clockwise. The reverse is true in the Southern Hemisphere.

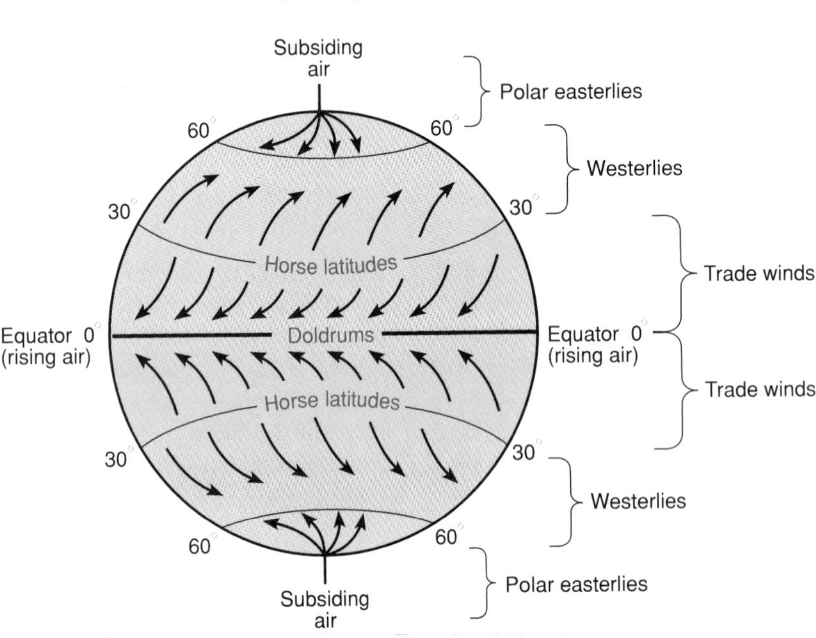

Figure 22.7 The general pattern of prevailing surface winds caused by Earth's rotation.

known as the *prevailing westerlies* (by convention, wind direction is considered the direction from which the wind is blowing). Air that flows from the horse latitude southward toward the equator is deflected westward, and these winds are called the *trade winds* (see Fig. 22.7).

Above 60 degrees north or south latitude are the *polar easterlies.* These winds are the result of cold polar air masses moving toward the equator. Due to the Coriolis effect, the polar easterlies are deflected toward the west.

In general, global wind patterns are the result of a combination of distance from the equator (and, therefore, differences in the heating of Earth's surface), and the Coriolis effect. In practice, though, global wind patterns are modified by local conditions. For instance, a regular pattern of wind direction often is found along seashores. During the day a constant onshore breeze occurs, and at night a constant offshore breeze. (Note Figs. 22.8a and 22.8b.)

The specific heat of water is approximately five times greater than that for land. Consequently, when water and land receive the same amount of heat from the sun, the land becomes hotter during the day than does the water. The air above the land heats up, expands, and rises. In comparison, the air over the water is relatively cool and dense. The density differential results in a convection cell that forces the air above the water to flow inland, producing an onshore breeze. Since water cools more slowly than land, the

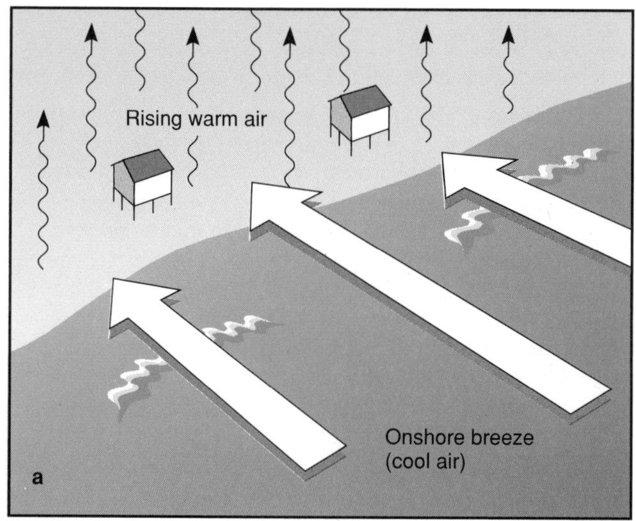

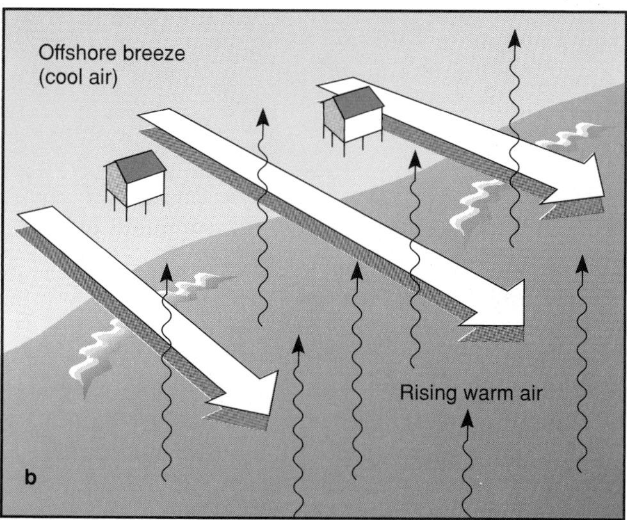

Figure 22.8 Onshore and offshore breezes. (*a*) During the day the sun heats the land more quickly than the water, causing an onshore breeze. (*b*) During the night the land cools faster than the water, causing an offshore breeze.

situation is reversed at night. At night the air over the water rises and the air over the land flows toward the water to produce an offshore breeze. In certain locations there are exceptions to this generalization.

Jet Streams

In the upper troposphere there are some *long, narrow, meandering air currents* called **jet streams.** Jet streams, which occur at heights of 10 to 50 kilometers, result from pressure gradients caused by temperature differences in the troposphere. They can be thousands of kilometers long, hundreds of kilometers wide, and several kilometers thick. Wind speeds near the centerline of a jet stream average about 100 kilometers per hour, but sometimes attain speeds of 500 kilometers per hour. Since jet streams occur at such high altitudes, the direction of their winds is little influenced by the terrain below them. The winds generally blow directly from west to east, with variations in the jet stream's direction of movement resulting mainly from deflections due to the Coriolis force and to temperature changes between the equator and the poles.

The jet stream that most affects the weather of the United States is the polar jet stream that occurs over our middle latitudes. The position of the polar jet stream varies from about 30 degrees latitude in the winter to about 50 degrees latitude in the summer. In the winter a southward bulge of the polar jet stream brings a cold air mass to the southern part of the country, and in the summer a northward bulge brings a warm air mass to the northern part of the country. Thus, the jet stream brings to the northern part of the United States occasional heat waves as well as winter blizzards (see Fig. 22.9).

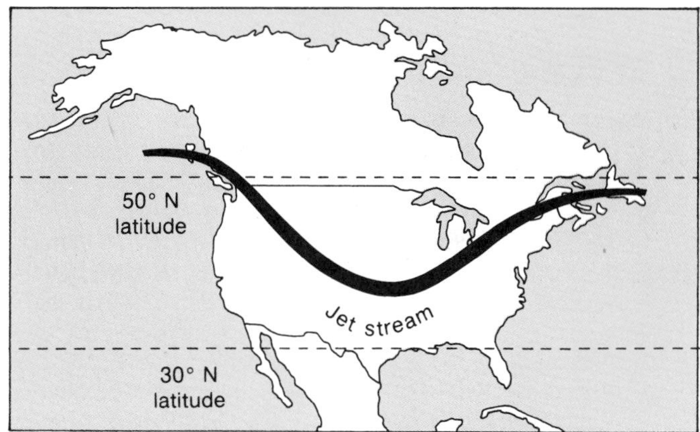

Figure 22.9 Average position of the jet stream across the United States. During the year the path of the jet stream moves between 30° and 50° N latitude.

Particles lifted by winds from the ground and oceans, volcanic debris, and radioactive matter from nuclear tests and accidents are carried by jet streams and eventually deposited at places all over the globe. For that reason, a shovelfull of dirt from any spot on Earth probably would contain at least some particles and molecules from almost every place on our planet.

Planes flying from west to east take advantage of jet streams when possible, and planes flying from east to west are careful to avoid them. A plane can add from 100 to 500 kilometers per hour to its ground speed while flying with the winds of a jet stream.

Humidity and the Hydrologic Cycle

Humidity is the water vapor content of air. Most of the water vapor in the air comes from the evaporation of water from oceans, lakes, and moist land, but a fair amount is from the *transpiration* of plants. (Plants absorb water through their root systems and transport the water to their leaves, where much of it escapes into the atmosphere by evaporation.) An average forest in the southern United States releases to the atmosphere, by transpiration, approximately 30,000 liters of water per acre per day.

Air is said to be *unsaturated* when the amount of water vapor in the air is less than the maximum amount the air can hold at a

El Niño Grande

El Niño, Spanish for "the child," is a short-term global climatic change that usually starts about Christmas time; hence, the reference to "Christ child." El Niños vary in their severity and usually occur several times per decade. They normally begin with changing wind conditions and a warming of the Pacific waters off the coast of Peru. Massive climatic change was initiated by the El Niño of 1982–1983, the greatest oceanic-atmospheric disturbance recorded in over a century. Record torrential downpours and massive avalanches and mudslides occurred in Ecuador and Peru during 1983, whereas devastating droughts simultaneously struck portions of Indonesia, Australia, South America, and Asia. The western portion of the United States experienced a high rate of precipitation in the form of rain and snow during the same period. Coastal regions of California concurrently suffered damage from numerous severe storms, and the number of hurricanes in the Caribbean Sea and the Gulf of Mexico was considerably reduced.

The temperature of the area of the Pacific Ocean involved rose about 6°C above normal, retarding the usual upwelling of cold, nutrient water from the ocean depths. As a result, the plankton upon which small forms of marine life depend perished, and the food chain for all ocean life in the region was seriously interrupted. The results were devastating. Great quantities of anchovies and other types of fish died of starvation as did coastal sea birds that depended on them as food.

El Niño of 1983 not only created chaotic weather patterns early in the year, but also caused a retarding effect on Earth's rate of rotation, increasing the length of a day. A million years from now this rate of retardation will have a decided effect on Earth's day and the climate of our planet will change accordingly. Although there have been other El Niños in the past ten years, the 1982–1983 El Niño led to the death of 1000 people and caused an estimated $8 billion in worldwide damages ■

given temperature and *saturated* when the air has all the water vapor it can hold at a given temperature. **Relative humidity** is *the amount of water vapor present in a sample of air compared with the amount of water vapor the air would hold if it were saturated at a given temperature.* Relative humidity is usually expressed as a percentage. For instance, if a certain volume of air contains 25.5 grams of water vapor but at saturation could hold 51.0 grams of water vapor, the relative humidity of the air sample is determined as follows:

$$\text{relative humidity} = \frac{25.5 \text{ g water vapor}}{51.0 \text{ g water vapor}} \times 100\% = 50\%.$$

Humidity greatly influences our weather. Depending on the humidity and the temperature of air, some areas on Earth experience little or no rain, snow, or sleet. Furthermore, during the day the humidity that air receives by evaporation (a process that absorbs heat) keeps daytime temperatures from getting too high, and at night when cooling causes water vapor to condense out (a process that releases heat) as clouds, dew, rain, or snow, the nighttime temperature is higher than it would be without condensation.

The amount of water vapor that air can hold increases with temperature. Thus, if air is saturated at one temperature and the air is heated to a higher temperature, the air is no longer saturated but can now hold more water vapor. On the other hand, if the air is cooled to a low enough temperature, it will become saturated. Continued cooling results in liquid water condensing out of the air. The temperature at which air becomes saturated is the *dew point.* Below the dew point, water vapor in the air is changed to liquid water that adheres to condensation nuclei to produce *dew* or *fog.* Fog occurs when a layer of low-lying air, cooled below the dew point, produces small droplets of water. Fog occurs mainly on calm, clear nights. Dew results when small droplets of water are deposited on grass, shrubs, or land and usually occurs on clear nights. When the dew point is below freezing, frost instead of dew is deposited.

When large masses of moist air are pushed upward, they cool below the dew point and some of the water vapor present is converted to droplets of water. When the small water droplets bump into each other, larger drops of water are formed. When the drops of water become too large to remain suspended in the air, they fall to Earth in the form of precipitation (rain or snow). The rainwater journeys to the oceans or lakes by runoff, river flow, or underground water flow. Water from the oceans and lakes evaporates to replace the moisture lost by rain and snow. The natural system of circulating water from the oceans to the air, to the ground, to the oceans, and then back into the atmosphere is called the *water cycle,* or the **hydrologic cycle.**

The source of energy for the water cycle is the sun. Because of the sun's nearly inexhaustible supply of energy, the water cycle has continued to operate since our oceans were first formed and will remain operative until our sun begins its decline. The water cycle starts when energy from the sun causes evaporation of the ocean and other waters. The ocean water is literally distilled by the sun's heat and converted from a salt solution into gaseous water in the atmosphere. The atmospheric gaseous water collects into small droplets that form clouds. These clouds then release their hoard of water on the sea and the land in the form of rain or snow. The water that falls upon the land slowly finds its way back into the sea, and the cycle starts over again (see Fig. 22.10). The water cycle clearly illustrates how important the oceans of Earth are, not only to the life that abounds in them but also to the plants and animals that live on the land. Without oceans and the water cycle, the land areas of Earth might be devoid of life.

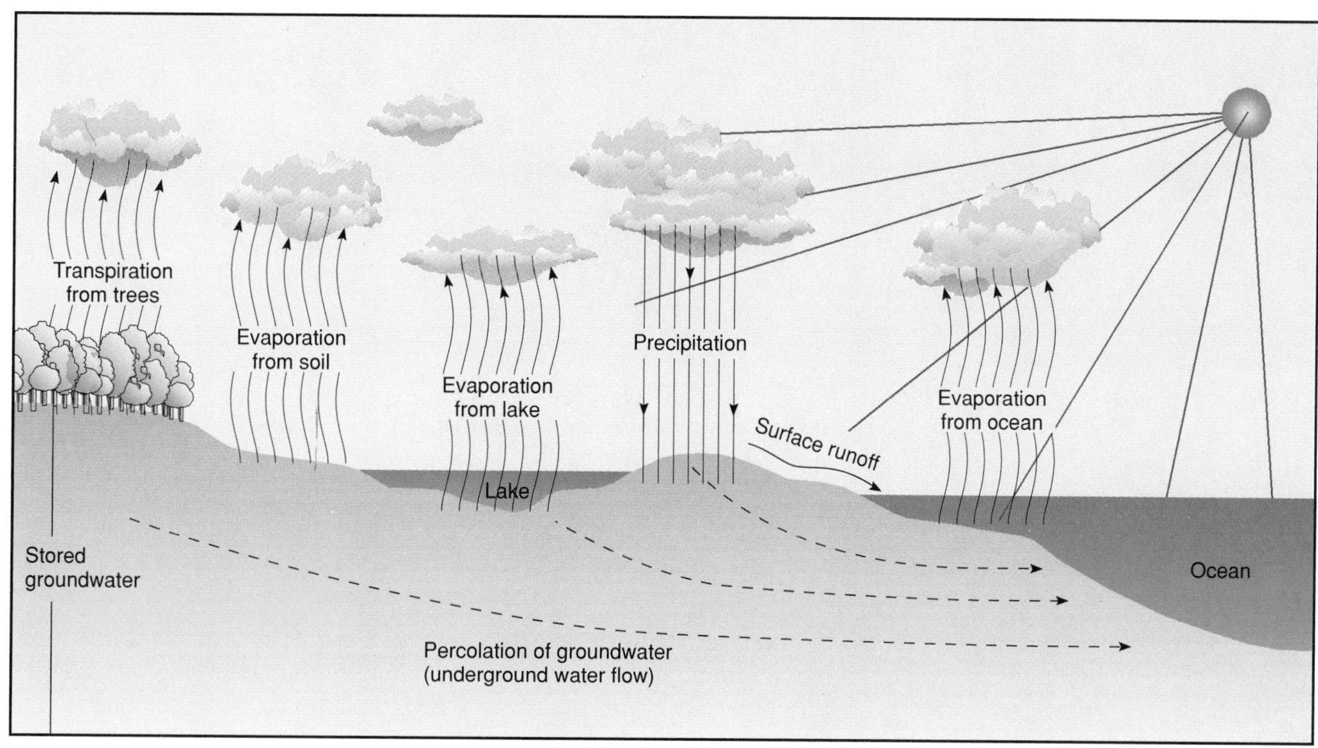

Figure 22.10 The hydrologic or water cycle. Water moves from the sea into the air, to the land, and back to the sea.

Clouds

Clouds are masses of tiny water droplets, ice crystals, or a mixture of both. In general, clouds higher than 5 kilometers are composed of ice crystals since the temperature of the troposphere above 5 kilometers is usually below freezing. Clouds are formed when a rising warm air mass containing water vapor expands into the cooler surrounding air. The ascending air mass cools since the heat energy necessary to cause expansion comes from the expanding air mass's internal energy.

Clouds are classified according to their shape or appearance and altitude. There are four root names used to describe clouds. *Cirrus* means "curl" in Latin, and cirrus clouds are wispy and featherlike. Because of their high altitudes, they are usually composed of minute ice crystals (see Figs. 22.11 and 22.21). Cirrus clouds are often likened to a mare's tail or an artist's brush strokes. *Cumulus* is Latin for "heap," and these clouds resemble billowy, cottonlike puffs (Fig. 22.19). *Stratus* means stratified or layered. Stratus clouds are usually seen in horizontal layers or as fog, as shown in Figure 22.16. *Nimbus* is from the Latin meaning "rain" or "rain cloud." A nimbus cloud is a dark cloud, threatening or producing rain, snow, or sleet.

Clouds are also named according to a combination of features they may possess. Thus, nimbostratus clouds combine features of stratus and nimbus clouds, and cirrocumulus clouds have the appearance of both cirrus and cumulus clouds.

Clouds are also classified according to the altitude at which they normally appear as summarized in Table 22.3. Examples of cloud types are shown in Figures 22.11 through 22.21.

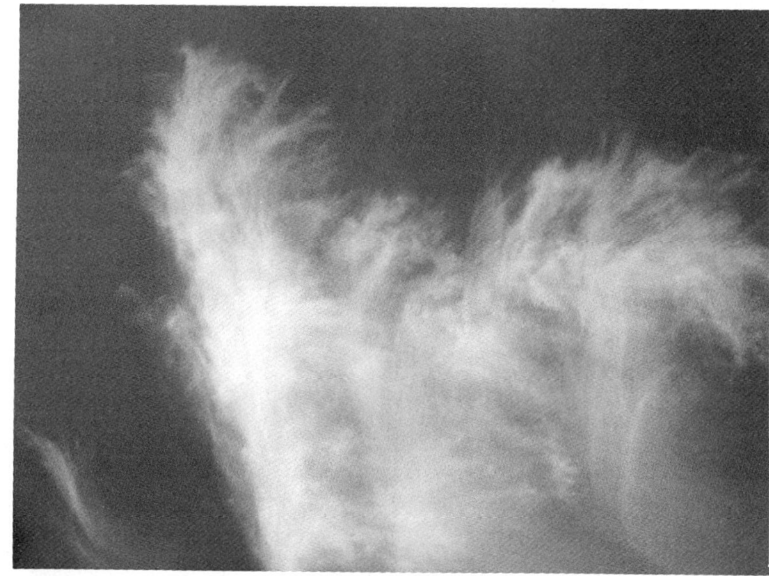

Figure 22.11 Cirrus cloud. Note the wispy structure of the cloud.

Figure 22.12 Cirrocumulus clouds. These high-altitude patches of clouds are the result of considerable vertical movement.

Figure 22.13 Cirrostratus clouds. The cirrostratus clouds shown here have a whitish foglike appearance and consist mainly of ice crystals.

Figure 22.14 Altocumulus clouds. Middle-height clouds of cotton puffs and rolls make up this formation.

Figure 22.15 Altostratus clouds. This type of cloud is semitransparent; the sun is reminiscent of a Seurat or other postimpressionist painting.

Figure 22.16 Stratus clouds. This particular stratus cloud is rather formless, extending almost to the ground.

Figure 22.17 Stratocumulus clouds. This cloud formation consists of masses of cottonlike billows.

Figure 22.18 Nimbostratus clouds. The low dark clouds shown here are producing rain.

Figure 22.19 Cumulus clouds. These billowy white clouds are usually seen on a day of fair weather.

Table 22.3

Cloud Classification.

Altitude of cloud (km)	Type cloud
High clouds (occur at altitudes between 5 and 13 km)	Cirrus Cirrocumulus Cirrostratus
Middle clouds (occur at altitudes between 2 and 7 km)	Altocumulus Altostratus
Low clouds (occur at altitudes between 0 and 2 km)	Stratus Stratocumulus Nimbostratus
Vertical clouds (~0.5–20 km in height)	Cumulus Cumulonimbus

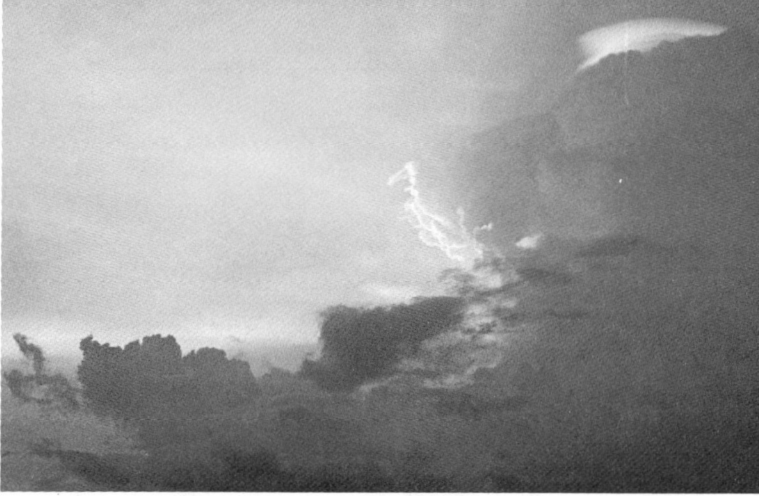

Figure 22.20 Cumulonimbus clouds. When cumulus clouds become dark and threaten to produce a storm, they are called cumulonimbus clouds or thunderheads. Here we see an excellent example of vertical development.

Figure 22.21 Cirrus clouds. In Latin, *cirrus* means "curl."

Weather

Weather refers to *the condition of the atmosphere for a designated geographic area over a short period of time.* It is characterized in terms of cloudiness, humidity, precipitation, pressure, temperature, and wind. Whereas weather refers to specific short-term conditions of the atmosphere of a region, *climate* refers to average long-term atmospheric conditions of a region. The weather that you are experiencing now may have had its beginning many hundreds of kilometers away. Much of the weather we experience is the result of the meeting and interaction of different *air masses.*

An air mass is a large body of air that stays over a certain area of Earth's surface long enough to acquire the humidity and temperature of that region. It has relatively distinct boundaries and properties. The area from which an air mass acquires its humidity and temperature is known as its *source region.* The properties of density, humidity, and temperature vary for different air masses depending on their origin.

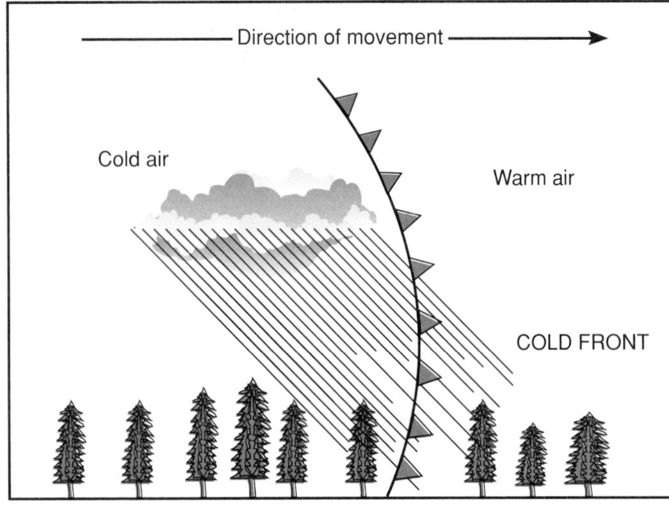

a

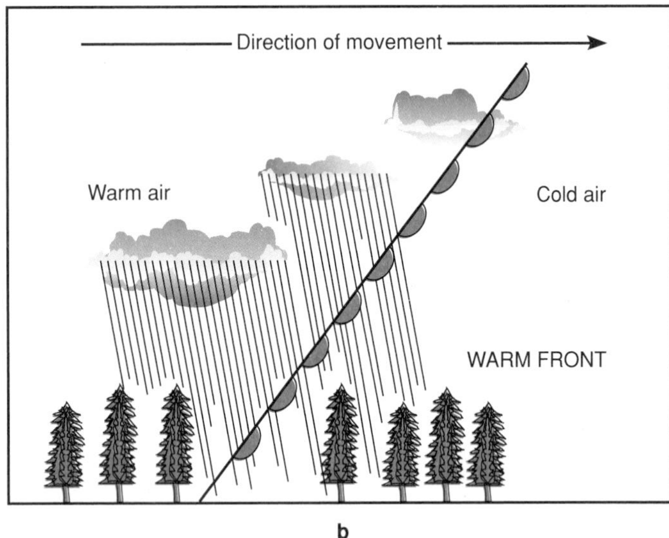

b

Figure 22.22 Rain and cloud formation along (*a*) a cold front and (*b*) a warm front.

Table 22.4

Some symbols used on weather maps.

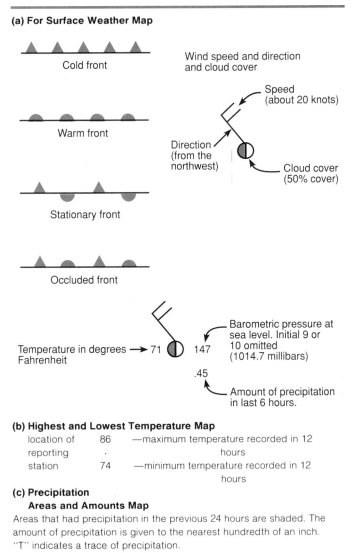

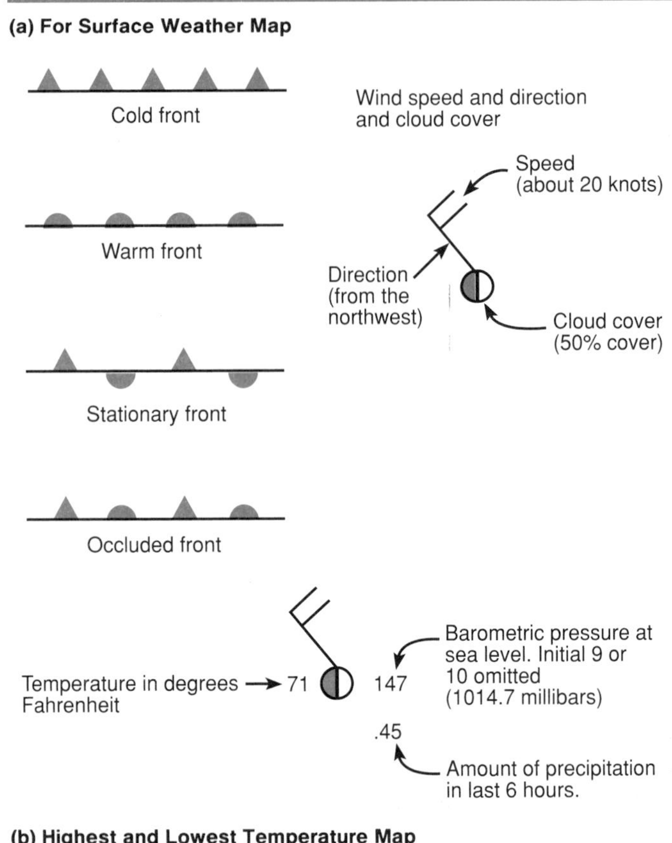

(b) Highest and Lowest Temperature Map

location of	86	—maximum temperature recorded in 12 hours
reporting	·	
station	74	—minimum temperature recorded in 12 hours

(c) Precipitation Areas and Amounts Map

Areas that had precipitation in the previous 24 hours are shaded. The amount of precipitation is given to the nearest hundredth of an inch. "T" indicates a trace of precipitation.

Warm and moist air masses, called *maritime tropical air masses,* form over the warm waters of the Atlantic Ocean, the Caribbean Sea, the Gulf of Mexico, and the north Pacific Ocean between 10° and 40° north latitude. These air masses are responsible for much of the precipitation that occurs over large areas of the eastern and southwestern coastal portions of the United States. A warm and dry air mass, called a *continental tropical air mass,* forms in the summer over northern Mexico and the southwestern United States.

Cool and moist air masses, called *maritime polar* (or *arctic*) *air masses,* form over the north Pacific Ocean and influence the weather along our western coast. Cold and dry air masses, called continental *polar* (or *arctic*) *air masses,* form over the northern portions of Canada and the north polar region. These air masses greatly influence the weather of the north central and northeastern areas of the United States.

In the tropic and polar regions, weather conditions and wind directions are reasonably stable, but in the middle latitudes (the regions of the westerlies), the weather is subject to considerable day-to-day variation. The reason for the variability of weather in the regions of the westerlies lies in the flow of cold air masses from the polar regions to the equatorial regions and the flow of warm air masses from the tropical regions to the polar regions.

The contact zone between air masses is called a **front,** which is usually accompanied by clouds, rain, and storms (see Fig. 22.22). If a cold air mass is moving into territory occupied by a warm air mass and horizontally replacing the warm air (since warm air is less dense than cold air, the warm air is forced upward over the colder air mass), the contact zone between the two air masses is called a *cold front.* If a warm air mass is overriding a cold air mass, the front between the two air masses is called a *warm front.* If neither of the two air masses is moving, the contact zone is called a *stationary front.* An *occluded front* occurs when an advancing cold front overtakes a retreating cold air mass. The warm air mass that did separate the two cold air masses is pushed upward, and the warm air is occluded from Earth's surface. (Symbols for the various types of fronts are included in Table 22.4). A cold-front occlusion occurs when the advancing cold air is colder than the

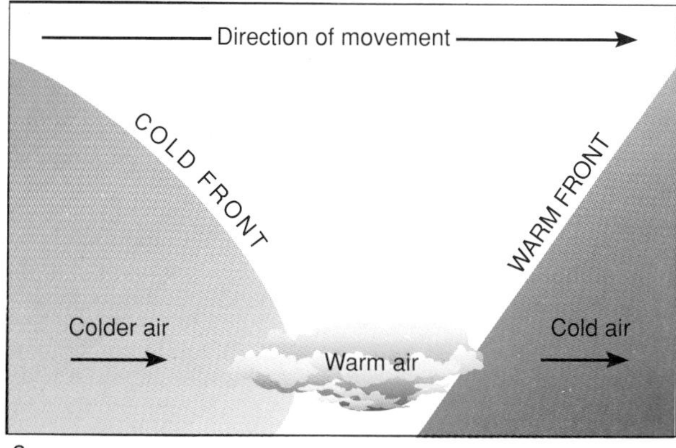

a.

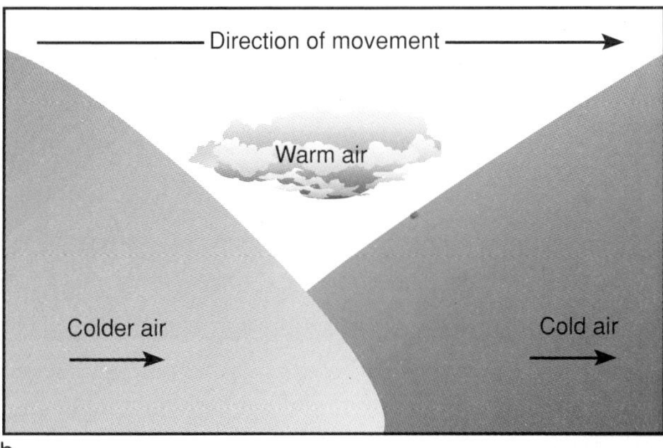

b.

Figure 22.23 An occluded front. (*a*) Advancing cold front overtaking a retreating cold-air mass. (*b*) Resulting warm-air occlusion.

cold air it is overtaking (see Fig. 22.23). If the cold air ahead is colder than the cold air advancing on it, the front is called a warm-front occlusion.

Data taken at various weather stations in the United States are collated and analyzed to prepare **weather maps.** Weather maps, prepared by the Climate Analysis Center, are issued daily by the United States National Weather Service. The Surface Weather Map shows major frontal systems, high- and low-pressure areas, and numerous data such as barometric pressure, temperature, precipitation, wind speed and direction, cloud cover, and so forth (see Table 22.4). Other, more specific maps are prepared that give information on high and low temperatures for the past 12 hours and amounts of precipitation for the past 24 hours. For pictures of some of the weather maps available each day, see Figure 22.24.

Most local TV news broadcasts present the mold spore count and the pollen count for the day and predict the same information for the following day at certain times of the year. Although such data is not a part of weather forecasting, it is of considerable interest to many people who watch the weather report. A fair percentage of the population suffers from allergic reactions to some extent.

To predict tomorrow's or next week's weather, weather forecasters must consider the probable movements of various fronts

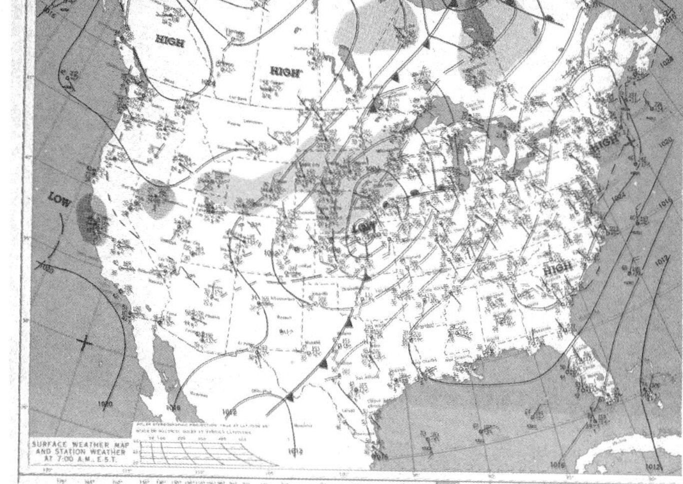

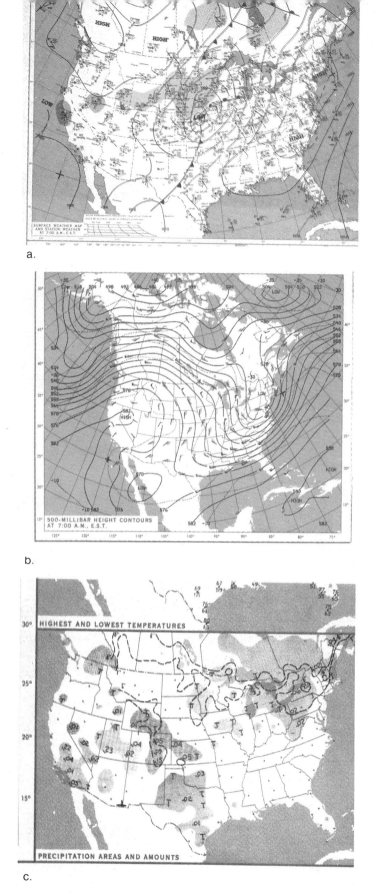

a.

b.

c.

Figure 22.24 (*a*) A typical surface weather map, as provided by the National Weather Service. (*b*) A millibar contour map. The lines indicate connecting geographic positions of equal pressure. (*c*) Precipitation areas and the amounts received.

and pressure areas. Forecasters are aided in their analyses by the knowledge that, in general, weather moves across the United States in a west to east direction. Thus, for a specific location, as a low-pressure area approaches from the west there is a decrease in barometric pressure (the barometer falls) and the winds blow from the east toward the low. As the low area passes, the barometric pressure rises and the winds begin coming from the west. In actual practice, however, the forecasting of the weather is much more complex. If one considers all the data that must be studied before a meaningful prediction can be made, the difficulty of accurate forecasting becomes obvious. Certainly, much more is involved than relying on such old adages as

Red sky in the morning, sailors take warning;
Red sky at night, sailors' delight.

Chill Factor and Heat Index

Quite often, people who are suddenly exposed to increased wind velocity suffer serious frostbite. Hunters and skiers are among those particularly vulnerable. The faster air moves over exposed skin, the faster evaporation of skin moisture occurs. The evaporation removes about 540 calories of body heat for each gram of water evaporated. This condition brings about extreme discomfort, even though the individual may be dressed properly for exposure to the temperature indicated on the thermometer. A study of this phenomenon has led to the introduction of a new concept in the recording of weather observations. The **chill factor** (or *windchill index*) is an attempt to describe an atmospheric condition by combining observations of temperature and wind speed. Since winter outdoor recreation is rapidly growing in popularity, an awareness of the chill factor is becoming increasingly important.

In Figure 22.25a the effect of wind speed on our comfort is displayed in terms of equivalent temperature. From this, one can ascertain the degree of danger of frostbite to the face, hands, and other exposed areas that results from any combination of wind speed and temperature. As an example, at a football game when the temperature is 15°F with gusts of wind that reach 25 miles per hour, the chart shows that the spectators and players are exposed to a temperature equivalent of 22 degrees below zero! There is considerable danger from frostbite to all exposed areas of the body as well as to those areas that are insufficiently protected. If an individual were exposed to these conditions while riding on a snowmobile or while skiing, the equivalent temperature would have to be adjusted to compensate for the additional wind speed to which he or she were exposed. If the snowmobile were moving into the wind at a velocity of 10 miles per hour, the equivalent temperature would drop from −22°F to −28°F, and the probability of being frostbitten would increase accordingly. Conversely, if the snowmobile were moving in the same direction as the wind, the chill factor would be less because of decreased resultant speed. A knowledge of the dangers associated with the chill factor can prevent a pleasurable outing from turning into a disaster.

The summer equivalent of the winter chill factor is called the **heat index.** The heat index *takes into account how we react to a given temperature at different relative humidities.* As air humidity increases, the rate of evaporation of skin moisture decreases. In Figure 22.25b the effect of relative humidity on our comfort is indicated by how our body perceives the actual temperature.

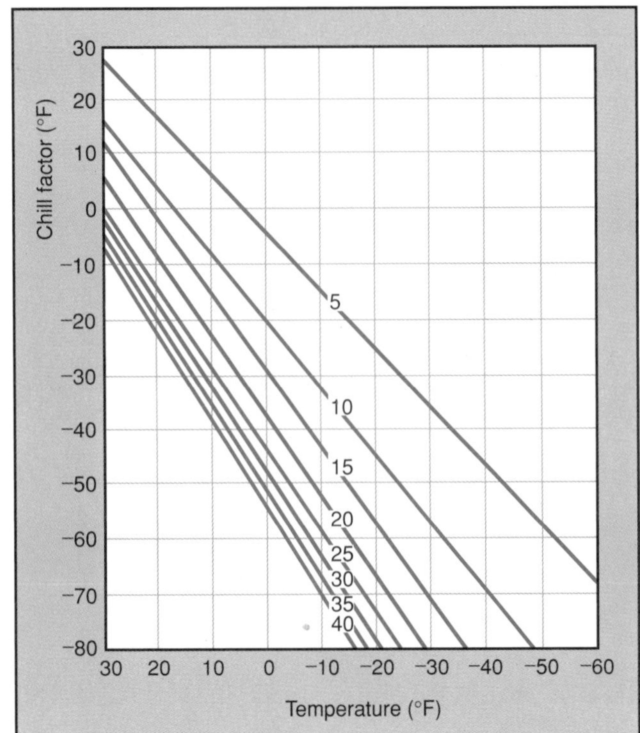

a.

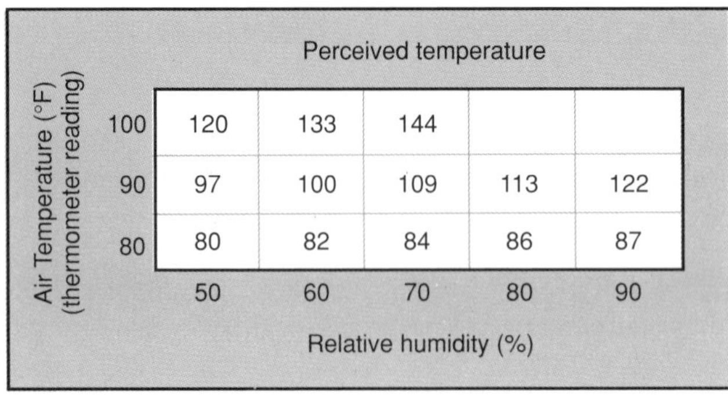

Perceived temperature

Air Temperature (°F) (thermometer reading)					
100	120	133	144		
90	97	100	109	113	122
80	80	82	84	86	87
Relative humidity (%)	50	60	70	80	90

b.

Figure 22.25 (*a*) Chill factor: The effect of wind speed on exposed body tissues. The diagonal lines represent wind speeds in miles per hour. (*b*) Heat index: A chart indicating how our body perceives air temperatures at different relative humidities.

Hurricanes

The most violent and destructive of all storms are *hurricanes* and *tornadoes*. A **hurricane** is a tropical cyclone, which is called a *hurricane* in the Atlantic Ocean and the Gulf of Mexico, a *typhoon* in Asian Pacific waters, and a *willy-willy* off the Australian coast.

Figure 22.26 A hurricane as seen by satellites. The violent storm is located along our eastern coast.

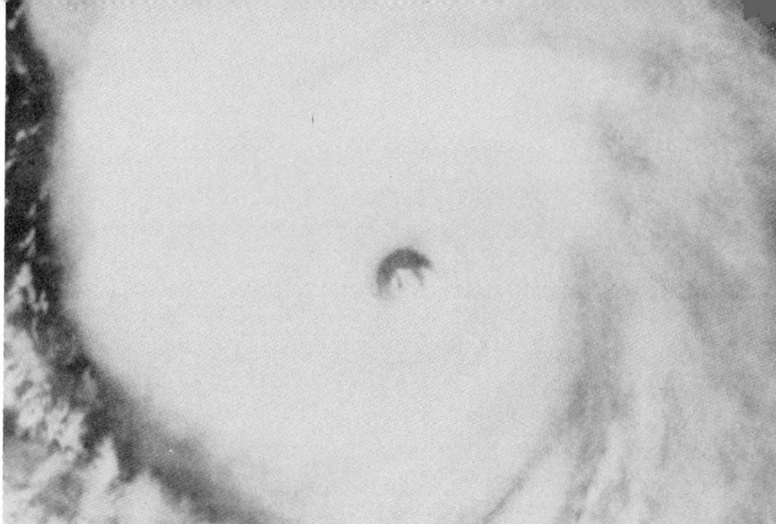

Figure 22.27 The "eye" of a hurricane.

A tropical storm becomes a hurricane when its wind speed exceeds 120 kilometers per hour (75 mi/h).

Hurricanes are huge whorls of fast-moving air that form over tropical oceans where, in the late summer, the sun heats large masses of moist air. The moist air rises and is cooled to the point that condensation occurs. For every gram of condensed water formed, 540 calories of heat are released and the air temperature increases accordingly. The Coriolis force is responsible for beginning the cyclonic motion (see Fig. 22.26).

Within the doughnut-shaped storm is a calm central core, or eye, where almost no wind blows and where no rain falls, as shown in Figure 22.27. The eye of a hurricane is usually about 8 to 40 kilometers in diameter, whereas the overall diameter of the hurricane itself varies from about 150 to 500 kilometers. Within the storm, wind speeds reach from 120 to 320 kilometers per hour. Although the winds inside a hurricane move at high speeds, the hurricane itself travels at a very modest 8 to 40 kilometers per hour.

Once formed, hurricanes are carried westward by the trade winds and then deflected northward (in the Northern Hemisphere) by the Coriolis effect and eventually eastward by both the Coriolis effect and the eastward-blowing westerlies. Consequently, coastal areas in both the south and the northeast of the United States can be ravaged by the same hurricane. As warm humid air continues to flow into the storm region and rises, clouds and torrential rains are produced. The energy for the hurricane is primarily provided by the heat liberated when water vapor in the air is converted to liquid water.

Wind damage from a hurricane can be extremely great when the winds reach velocities of 200 to 300 kilometers per hour. The pressure on buildings (due to the wind's kinetic energy) varies directly with the square of the wind's speed. For example, a doubling of the wind's speed results in a fourfold increase in pressure. Extensive damage also results from flooding. A hurricane can dump as much rain on an area in a matter of hours as the area normally would receive in a full year (see Fig. 22.28).

The sea wave, or storm surge, is the most destructive part of a cyclone or other tropical storm in a low coastal area. Among the most devastating tropical storms in the past 300 years were those that hit Calcutta, India, in 1737 and Haiphong, Vietnam, in 1881.

Figure 22.28 The destructive force of a hurricane in action.

The storm surges that resulted produced 12-meter walls of water that inundated these two cities and drowned about 300,000 people in each case. Perhaps the greatest death toll, however, resulted from the rotating storm that struck Bangladesh in 1970, killing 500,000 people. This cyclone was considerably weaker than the one that hit the low-lying islands and southern coast of Bangladesh in April 1991. As a result of warnings issued prior to the arrival of the latter storm, three million people were evacuated, yet an estimated 139,000 people perished and two million were left homeless. Many casualties resulted from lack of communication with some residents of remote islands and from those who did not heed the early warnings.

Many hurricanes have struck the United States in the past century. Such destructive storms have caused numerous deaths, as noted in Table 22.5. Casualties as a result of hurricanes have been almost nonexistent in the United States since 1972 due to better early warning systems developed by such agencies as the United States Weather Service.

Table 22.5

Some killer-hurricanes in the United States (50 or more deaths).

Location	Year	Deaths
Texas	1900	6000
Florida	1906	164
Alabama, Florida, Mississippi	1906	134
Louisiana	1909	350
Louisiana	1915	275
Texas	1915	275
Florida	1919	~750*
Florida	1926	243
Florida	1928	1836
Florida	1935	408
New England	1938	600
Northeast United States	1944	390*
Florida, Louisiana, Mississippi	1947	51
South Carolina, North Carolina	1954	95
Northeast United States	1954	60
Northeast United States	1955	184
Louisiana, Texas	1957	390
Florida, Louisiana	1965	75
Louisiana, Mississippi	1969	256
Northeast United States	1972	122

*Most of these deaths occurred from boat sinkings off the coast.

Tornadoes

Tornadoes, although very small in size compared to other storms, are extremely violent and destructive. They have much less total energy than a hurricane, but the energy of a tornado is concentrated in an extremely small region. Tornadoes, sometimes called *twisters,* have the appearance of a long funnel extending downward from a very dark cloud to the ground. The funnel of a tornado is usually about 150 to 300 meters wide, but the size can vary. A tornado is shown in Figure 22.29.

The great destructive power of a tornado is the result of two factors. One is the high speeds of the rapidly revolving winds—up to 640 kilometers per hour. The second factor is the very low pressure area produced within the funnel, caused by the air inside the funnel moving upward at such high speeds—about 160 kilometers per hour. As a tornado passes over a building, the sudden drop in pressure outside a building causes the air inside to push violently outward. The rapid reduction in normal air pressure, due to the upward movement of air inside the funnel, can be as much as 10 percent. This much reduction in total air pressure will exert an outward force of 1 newton per square meter (1.5 lb/in^2). Consider the outward force on all the walls of a house as well as its roof, and it is not surprising that a house with tightly closed windows and doors is frequently destroyed as if by a charge of explosives.

When tornadoes occur over water, they are called *waterspouts* since the reduced air pressure within the funnel draws large quantities of water into the air.

Figure 22.29 The dark funnel of a tornado. The funnel's destructive power is difficult to comprehend.

Although tornadoes occur occasionally in many nations of the world, they are much more frequent in the United States and Australia. An average of 650 tornadoes are reported in the United States each year.

Summary

Our atmosphere consists of 78 percent nitrogen, 21 percent oxygen, almost 1 percent argon, 0.03 percent carbon dioxide, and traces of other elements and compounds.

The continual exchange of water between the planet's surface and the atmosphere is a major factor in determining weather conditions. Solid particles suspended in the atmosphere also play an important role in determining weather.

Our atmosphere is divided into layers. The main difference between the various layers is the way in which the temperature changes from layer to layer.

The troposphere is the layer closest to Earth's surface; it extends upward to a height of 16 kilometers over the equatorial region and to 8 kilometers over the polar regions. It is in the troposphere that all our weather occurs. The temperature of the troposphere decreases an average of 6°C per kilometer above Earth's surface. This decrease in temperature with altitude is known as the lapse rate.

The layer above the troposphere is the stratosphere, which reaches to a height of 55 kilometers. The temperature in the stratosphere increases from about −60°C at the bottom to about 0°C at the top. Within the stratosphere is the ozone layer that absorbs much of the ultraviolet light from the sun.

Above the stratosphere is the mesosphere, which extends upward to about 80 kilometers. In the mesosphere the temperature decreases from about 0°C at the bottom to about −90°C at the top.

The layer above the mesosphere is the thermosphere, which extends upward to 500 kilometers. Within the thermosphere is a layer of ions that reflects most of the radio waves transmitted from Earth.

Above the thermosphere is the exosphere, which extends outward for thousands of kilometers. The exosphere is the transition between Earth's atmosphere and outer space.

Approximately two-thirds of the sun's radiant energy that reaches Earth enters our atmosphere, and the remainder is reflected into space. Of the energy that enters our atmosphere, about one-third is absorbed by the atmosphere, and the balance is absorbed by Earth's surface. The burning of great quantities of fossil fuels has added such compounds as carbon dioxide, carbon monoxide, sulfur dioxide, and oxides of nitrogen to the air. These compounds selectively absorb long-wavelength radiation from the sun. This extra absorption of sunlight could result in a disastrous increase in the temperature of our atmosphere. The retention of atmospheric heat by various gases is known as the greenhouse effect.

Because of Earth's rotation, air in the Northern Hemisphere is deflected to the right, and air in the Southern Hemisphere is deflected to the left. These deflections are the result of the Coriolis effect.

Air motion near the equator is mostly upward with very little horizontal movement, and the equatorial areas are referred to as the doldrums. Air movement at about 30 degrees north or south latitude is also mostly vertical (a downward movement). These latitudes are known as the horse latitudes. Air that flows northward from the horse latitude in the Northern Hemisphere is deflected eastward to produce winds known as the prevailing westerlies. Air that flows southward from the horse latitude in the Northern Hemisphere is deflected westward to produce winds known as the trade winds. Long, narrow air currents, called jet streams, exist in the upper troposphere. The jet streams travel at an average speed of about 100 kilometers per hour.

Air is called unsaturated when the amount of water vapor in the air is less than the air can hold at a given temperature and saturated when the air has all the water vapor it can hold at a given temperature. The temperature at which air becomes saturated is its dew point. Low-lying air, cooled below its dew point, produces fog or frost, depending on the temperature.

The natural system of circulating water from the oceans to the air, to the ground, then to the oceans, and then back into the atmosphere is called the water or hydrologic cycle.

Weather is the condition of the atmosphere over a short period of time and for a designated geographic area. A large body of air that stays over a certain area of Earth's surface long enough to acquire the humidity and temperature of that region is known as an air mass.

The contact zone between air masses is called a front. If a cold air mass is moving into territory occupied by a warm air mass, the contact zone between them is called a cold front. If warm air moves into an area occupied by cold air, a warm front is the result. If neither of the air masses is moving, the contact zone is called a stationary front. The fronts between air masses are usually accompanied by clouds, rain, and storms.

An atmospheric condition that results from a combination of air temperature and wind speed is called the chill factor. A temperature equivalent for a given air temperature and wind speed indicates possible dangers to exposed areas of the body. The physiological effect of a combination of air temperature and relative humidity is called the heat index.

A surface weather map shows major frontal systems, high- and low-pressure areas, and numerous data such as barometric pressure, temperature, precipitation, wind speed and direction, and cloud cover.

A hurricane is a tropical cyclone and acquires its circular motion because of the Coriolis effect. A tornado is similar to a hurricane, but a tornado's funnel may be less than 300 meters in diameter, whereas the huge whorls of air of a hurricane extend for hundreds of kilometers.

Questions and Problems

Composition

1. What might be the effect of an appreciable increase in the carbon dioxide content of our atmosphere?
2. On what basis do scientists divide Earth's atmosphere into layers?

The Layered Atmosphere

3. How does the thin ozone layer protect Earth?
4. List three important characteristics of the troposphere.
5. What is our major use of the ion layer located in the thermosphere?
6. How does the "hole" in the ozone layer over Earth's south polar region pose a threat to our food chain?

The Earth-Sun Energy Balance

7. How do scientists determine that Earth radiates as much energy as it receives from the sun?
8. How does the concentration of dust particles in our atmosphere affect the climate of Earth?

The Greenhouse Effect

9. Why is there a great concern about the amount of carbon dioxide that burning fuel adds to the atmosphere?
10. What different processes remove carbon dioxide from the atmosphere?

Air Circulation and the Coriolis Effect

11. In what way are the doldrums and the horse latitudes similar?
12. How does the Coriolis effect determine the general direction of the winds?

13. With regard to Earth's atmosphere, what is meant by the term *convection currents?*
14. Why are weather conditions in the tropics and near the poles quite stable, whereas such conditions vary at other locations on Earth?
15. How does a coastal city's location on the Gulf of Mexico help moderate its summer temperatures?

Jet Streams

16. The jet streams occur at about what altitude?
17. What is the general direction of movement of the jet streams?
18. How do pilots use jet streams to their advantage?

Humidity and the Hydrologic Cycle

19. Why does the continual exchange of water between the atmosphere and the surface of our planet have such a great effect on weather conditions?
20. What conditions must occur for a portion of the water in our oceans to be converted into water vapor?
21. What is relative humidity? What event would bring about a decrease in its value? An increase?
22. What is meant by the term *water* (hydrologic) *cycle?*
23. When air reaches its dew point, what is its relative humidity?

Clouds

24. How do cirrus clouds differ in appearance from cumulus clouds?
25. How do clouds form?
26. What type of cloud exists at altitudes from 2 to 7 km?

Weather

27. What is a front? How does the type of front that is invading a region influence the immediate weather?
28. How does a cold front that moves into a region increase the probability for precipitation in the immediate area?
29. What is a stationary front? What is an occluded front?

Chill Factor and Heat Index

30. According to Figure 22.25, what would be the value of the chill factor (a) if the temperature were 0°F and the wind speed were 20 mi/h, (b) if the temperature were −10°F and the wind speed were 25 mi/h, and (c) if you were to ride a motorbike at 30 mi/h into a gentle wind of 10 mi/h when the air temperature was 10°F?
31. How does relative humidity affect our perception of temperature?

Hurricanes

32. What is the wind speed in the eye of a hurricane?
33. Why do hurricanes in the Northern Hemisphere eventually move eastward?

Tornadoes

34. What characteristics make a tornado so destructive?
35. What is the term for tornadoes that occur over water?

23 The Record of Life in the Rocks

Chapter outline

Key terms/concepts

Fossils represent various forms of life from prehistoric times. Clockwise from upper left: trilobite, crincid, fern leaf, and skull of hominid child.

The composition of the atmosphere, the contents of the oceans, and the variety of surface features that characterized primitive Earth certainly were different from what we find today. Our planet has undergone, and is still undergoing, geological as well as biological and chemical evolution.

The atmosphere of our Earth has changed from one containing mostly hydrogen and helium with lesser amounts of methane, water, ammonia, and oxides of carbon to an atmosphere of 78 percent nitrogen, 21 percent oxygen, and small amounts of argon, carbon dioxide, and water.

Various reactions converted the chemical constituents of Earth's original primeval atmosphere into complex organic compounds that were the precursors of primitive life on our planet. These organic compounds, over untold millennia, slowly increased in complexity until, by chemical evolution, the first self-duplicating forms of life were evolved. Our seas and oceans have changed from almost freshwater bodies containing a thin organic "soup" to a 3.5 percent solution of mainly inorganic salts that support a staggering tonnage of life in an unbelievable diversity of forms.

What the distant future holds for humans in terms of evolution we cannot say with certainty; we may be fortunate that this knowledge is not available. According to Greek mythology, Prometheus, the Titan who made and loved humankind, took away our knowledge of the future and in its place established hope. It was this hope, the ancient Greeks believed, that had always sustained the individual in the face of disasters and ill fortunes. Humans may stand now, not at the end of nature's evolutionary path, but at the beginning of the evolution of intelligent, thinking people. It is quite possible that humans in the distant future may evolve in ways we cannot now conceive, even in our wildest imaginings. Although we have come a long way from the mere sentient beings considered our ancestors, the path that humankind is to follow may appear to be endless.

Life on Earth has evolved for approximately 4 billion years. Although life supposedly originated in the seas, only on land have highly intelligent life-forms (humans) evolved. Life began with single-cell organisms and progressed through simple multi-celled systems to the very complex mammals that culminated as humans. As of this writing, Earth is the only place where life of any sort is known to exist, but who can say how many worlds harbor life in various forms? History has taught us not to underestimate the discoveries the future holds.

The Chemical Origin of Life

At the end of the formation of our planet, Earth circled its star with its eight companion planets. Its once overwhelming mass of atmosphere had been bombarded by the sun's outpouring of radiant energy and charged particles until only a remnant of its initial gaseous envelope remained. The atmosphere that remained consisted mainly of hydrogen (H), ammonia (NH_3), and methane (CH_4), with smaller amounts of carbon monoxide and dioxide (CO and CO_2), water (H_2O), helium (He), hydrogen cyanide (HCN), and free nitrogen. Notice the conspicuous absence of free oxygen (O_2 or O_3) from the constituents of the primeval atmosphere. This absence is certainly not due to any lack of oxygen in Earth itself, since the crust alone is made up of minerals, the weight of which is approximately 45 percent combined oxygen. This high percentage indicates what happened to the original free oxygen; it reacted chemically during the early stages of Earth formation to form the compounds and minerals that make up the crust and mantle of our planet. There simply was no appreciable amount of free oxygen left.

In support of this assumed chemical composition of Earth's primitive atmosphere are the facts that nebulas, such as the one that formed our solar system, have been shown to contain these compounds and that comets, the leftover debris from our solar system's birth, are made of methane, ammonia, and water, among other compounds.

What happened to change the atmosphere of Earth from a primitive mixture of hydrogen, helium, ammonia, and methane to the present-day nitrogen and oxygen mixture? With the passage of aeons, the hydrogen and helium escaped into interplanetary space because Earth's gravitational field was not strong enough to retain them, while the ammonia was oxidized to nitrogen and water and the methane was oxidized to carbon dioxide and water. The oxygen for these conversions was available only after plants began to produce large quantities of oxygen by photosynthesis. Much of the carbon dioxide produced was slowly incorporated into deposits of limestone.

The argument for the loss of most of the original hydrogen and helium in Earth's atmosphere because of Earth's relatively small gravitational field is borne out when one considers the present atmospheres of the other planets of our solar system. The massive planets Jupiter, Saturn, Uranus, and Neptune have large gravitational fields; consequently, they have retained their original concentrations of hydrogen and helium.

Although a vast array of organic compounds needed for the emergence of life on our planet were synthesized in the atmosphere of our world during the first 500 million years after the birth of our sun, many of the compounds were initially present in the primodial mixture of gas molecules and dust particles from which our solar system was formed. Dozens of organic compounds, some extremely complex, have been detected in nebulas throughout the universe. It is from just such nebulas that solar systems are formed. Contrary to what people used to believe, the formation of complex organic compounds from atoms is not a matter of sheer chance but, instead, the result of compound formation by natural chemical laws.

It was probably during the earlier part of the 500-million-year period when Earth still had a large amount of hydrogen in its atmosphere that the temperature of Earth reached its highest point. Toward the end of this period, the temperature of Earth decreased significantly. Once the temperature of the land fell below the boiling point of water, the torrential rains that were falling upon Earth collected in depressions and crevices and, over a long period of time, slowly filled the depressions to produce oceans, seas, lakes, ponds, and rivers. The rains falling upon the hot land were in turn warmed, and the warm waters flowed by the force of gravity to the now existing bodies of water. During the time that the water was upon the land, it dissolved small quantities of various chemicals, mostly inorganic substances. The unending rains continued to leach out these salt minerals from the land and carry them in solution to the seas. Thus, the buildup of the salt in the seas and oceans began.

When the dissolved chemicals and suspended particles carried by swift streams were carried into the still waters of ponds, lakes, and oceans, the solids began to settle out of suspension and some of those in solution precipitated. Thus, these compounds came to rest on the sides and the bottoms of lakes and seas. Some of the compounds were natural catalysts of various types.

The desolate and barren Earth continued to receive endless rains. The blazing sun, occasionally obscured by dark clouds of methane and ammonia, bathed Earth in light, in the warmth of infrared rays, and in the searing energy of ultraviolet rays. Lightning flashed across the energy-filled skies in great electrical discharges that ruptured chemical bonds of molecules in the atmosphere and caused new bonds to be formed. Water, passing over subterranean molten rock, rose to the surface to produce hot springs and pools containing water at various temperatures.

In most places on the planet, the seas of the land were warm, but at higher altitudes land and water were much cooler. From the subterranean fires there were occasional erupting volcanoes that spewed forth great lava flows. At this stage of development of Earth, temperatures ranged from below the freezing point of water to the temperature of molten lava. Radioactive elements, existing as compounds dissolved in the waters and deposited on the land, silently and incessantly emitted high-energy radiation. Debris from the solar system formation crashed into Earth at high speeds to produce shock waves.

We have, then the following picture of the primitive Earth:

1. Various compounds such as H_2, NH_3, CH_4, H_2O, N_2, and CO_2 are present in the atmosphere.
2. Many energy sources are available for the breaking of existing chemical bonds. These energy sources are electric discharge (lightning), ultraviolet radiation from the sun, high-energy particles from radioactive substances, heat, and shock waves.
3. Many different catalysts are present on the land and in the waters.
4. Temperatures vary from below the freezing point of water to over a 1000° C.

What may have been the consequence of all the materials, catalysts, and energy forms coexisting on our infant planet? Laboratory experiments and theoretical discussions by many scientists have provided some answers to this question. It is assumed that the beginnings of primitive life-forms were produced, which in turn evolved into self-duplicating, life-sustaining cells.

When a mixture of methane, ammonia, and water is subjected to electric discharges simulating lightning in the primitive atmosphere, subjected to high-energy electrons acting as the natural radioactivity present on the infant Earth, or simply heated together at various temperatures, a generous mixture of amino acids is created:

$$CH_4 + NH_3 + H_2O \xrightarrow{\text{various energy forms}} \text{amino acids.}$$

Simple heating of the resulting amino acids produces some protein materials similar to natural proteins that form cell membranes. When hydrogen is added to the initial mixture (or one in which methane and ammonia have been replaced by carbon dioxide and nitrogen) and the same conditions exist, other compounds as well as amino acids are formed:

$$H_2 + CH_4 + NH_3 + H_2O \xrightarrow{\text{various energy forms}} \text{aldehydes, amino acids, cyanides, hydrocarbons, and so on.}$$

From these relatively simple organic structures, many substances of biochemical significance to life-forms can be produced.

The same energy forces that created the amino acids and other substances can produce all the various chemical units present in deoxyribonucleic acid (DNA), the genetic material that carries the genetic code. Different sites on the DNA control the production of different enzymes (chemical catalysts in cells). The enzymes produced by the DNA in turn direct the construction of carbohydrates, proteins, and fats in the cells, tissues, and organs that make up living organisms. Most of the chemical units that are present in DNA have been identified in meteorites. The amino acids, proteins, sugars, and DNA precursors rained upon and were washed into the ponds, lakes, and oceans of our new world to produce a warm dilute organic "soup" that contained the chemical building blocks of life itself.

Chemical and Biological Evolution

Many scientists theorize that during the passage of untold aeons, large protein molecules were formed. Because of the large number of sites of unequal charge distribution in protein molecules, they aligned themselves together in water to form geometrical patterns similar to the structures in present-day living cells. Sidney Fox, a biochemist, and other scientists have produced microspheres of synthetic protein material that are membrane-type spherules reminiscent of living cells. Photomicrographs of the proteinoid microspheres show that the spheres increase in number by "budding" and increase in size in concentrated protein solutions (see Fig. 23.1). Furthermore, these protocells exhibit enzyme activity similar to that of enzymes from living systems. It is not too difficult to visualize a natural chemical evolution of the microspheres to cells that can undergo life processes. It seems that the formation of protein substances and DNA must have occurred rather early on primitive Earth.

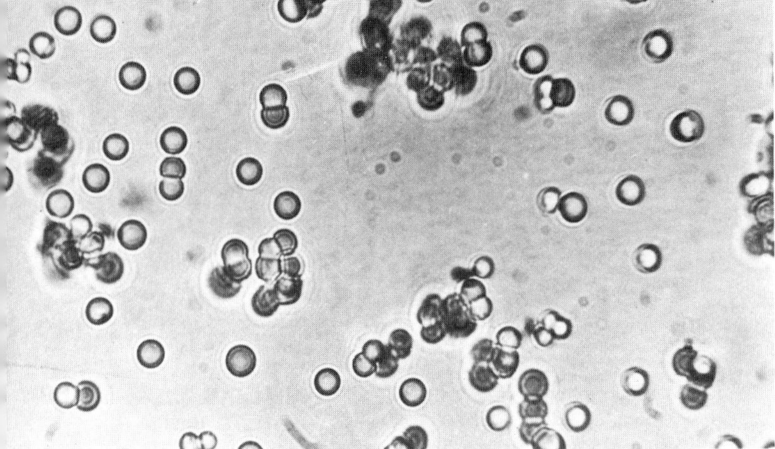

Figure 23.1 Microspheres of synthetic protein material.

After the transition from the membrane-type spherules or protocells to regular-reproducing primordial single-cell forms, what did the new life eat? Fortunately, there was available in the ocean a whole menu of nutritious organic compounds of the exact type needed by the new organisms. If the primordial cells had evolved no further, however, the newly emerged life-forms would have simply consumed the thin organic soup that surrounded them and multiplied, until eventually all the organic material produced through hundreds of millions of years would have been consumed. When all the organic "food" had disappeared, life on Earth would have perished from starvation.

An organism's acquisition of needed energy by the simple breakdown of organic compounds present in its environment is a very inefficient way to obtain energy. However, there are many different primitive organisms that sustain themselves in this manner even today (essentially, the process is that of *fermentation*). If, however, life was to continue to evolve, a more efficient way of obtaining energy was imperative. The manner in which the original life-forms managed this transition was by evolving into organisms that could acquire their energy from the sun in a direct manner; that is, through the process of photosynthesis.

Photosynthesis depends on the three-dimensional molecular structure of chlorophyll, which holds an ion of magnesium in a central position in the structure, as illustrated in Figure 23.2. This particular arrangement of organic molecule and inorganic metal ion *allows an organism to trap the energy of sunlight directly and use that energy to synthesize from carbon dioxide, water, and trace amounts of other materials the food that it requires for its growth and reproduction.* The process of photosynthesis, while consuming carbon dioxide and water, produces glucose (a sugar) and releases free oxygen. Therefore, with the advent of photosynthesis by living organisms, oxygen was slowly released into the atmosphere of Earth:

$$6CO_2 + 6H_2O + \text{energy} \longrightarrow C_6H_{12}O_6 + 6O_2.$$

There already were very small quantities of oxygen in the atmosphere, produced by the photodissociation of water vapor into hydrogen and oxygen (the dissociation of water molecules into hydrogen and oxygen atoms by ultraviolet light). As life-utilizing photosynthesis flourished in the waters of our planet, the concentration of oxygen released into the atmosphere continuously increased. As the amount of oxygen slowly built up, the sun's radiation changed some of the oxygen into molecules of ozone:

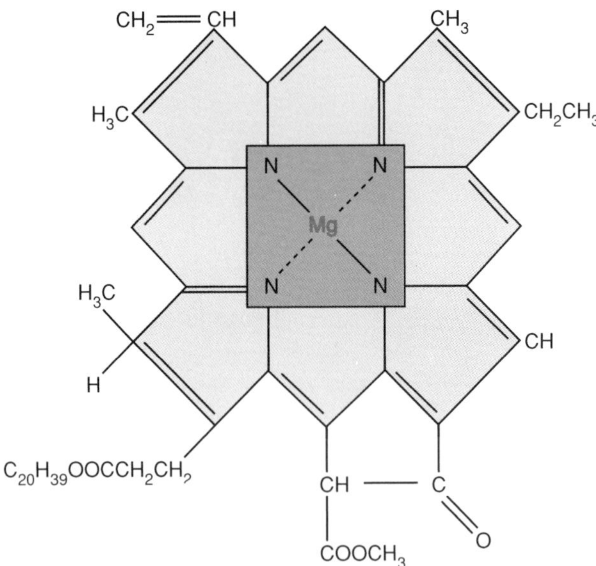

Figure 23.2 The structure of chlorophyll a.

$$3O_2 \xrightarrow{\text{energy}} 2O_3 \text{ (ozone)}.$$

The ozone was formed about 10 miles up in the atmosphere, so that after many centuries had passed, a layer of ozone completely surrounded our planet (as it still does today). After the ozone layer was formed, the amount of ultraviolet radiation reaching the surface of the planet was greatly reduced.

As it turned out, this development was absolutely necessary for life to emerge from the waters and invade and multiply upon the solid Earth. Life-forms as we know them cannot live in the presence of high concentrations of ultraviolet radiation. Consequently, until the ozone layer was formed, the land remained sterile and devoid of life. Life was able to continue to thrive in the seas, however, since the upper layers of water absorbed much of the ultraviolet radiation.

Experiments verify that in an atmosphere consisting of hydrogen, methane, ammonia, and water, various energy forms will convert these small molecules into the building blocks of life, including amino acids, peptides, proteins, carbohydrates, and DNA. One should realize, however, that living things do not use proteins, carbohydrates, and fats only because they are necessary for various life-form functions, but because these are the compounds that were available and plentiful on primitive Earth. Therefore, it is natural that life systems are composed of and utilize these particular building blocks as food-energy sources.

A summary of the concepts included in the most widely accepted view of the origin of life is as follows:

1. Organic compounds formed from inorganic compounds.
2. Interaction of organic compounds occurred to form more-complex structures and, eventually, enzyme systems.
3. The complex structures with enzymes became primitive, self-reproducing heterotrophs, (which exist by using preformed organic compounds but which cannot synthesize needed organic molecules. Heterotrophic organisms feed on each other, any available organic compounds, or an autotrophs).

4. The heterotrophs evolved lipid-protein membranes that shielded them from the surrounding environment.
5. Heterotrophs evolved into autotrophs (able to make their own organic molecules). Most autotrophs are photosynthetic organisms that harness energy from sunlight (photosynthesis) to build their needed organic compounds from carbon dioxide. A few autotrophs are chemosynthetic in nature, using various inorganic substances such as hydrogen sulfide, ammonia, or iron compounds to obtain energy to build their needed organic compounds from carbon dioxide.

Actually, then, given a water-type planet with an atmosphere of compounds similar to those that existed on primitive Earth, and given normal energy forms such as heat, ultraviolet radiation, and electric discharge by lightning, many scientists theorize that it is inevitable that life systems will be produced if sufficient time lapses.

Radioactive "Clocks"

Scientific evidence indicates that Earth is at least 4.6 billion years old, that primitive animals and plants left their homes in the seas and invaded the lands of Earth 400 million years ago, and that dinosaurs dominated our planet 200 million years ago. But how can scientists possibly know when various events occurred or when certain animals lived in the dim, distant past? In order to answer this question, we must first understand the principles that underlie the radioactive decay of certain elements. Consider the radioactive decay of the element ^{238}U. Uranium spontaneously transmutes into other elements until eventually ^{238}U becomes the element lead (^{206}Pb). There is no way to predict exactly when a specific atom of uranium will decay. An atom may decay instantaneously, or it may wait several million years before decaying. Although we do not know when a specific atom of uranium or any other radioactive element will decay, the length of time it takes for a certain fraction of a large group of atoms to release their particles and/or energy can be calculated with great precision. Analogously, it is impossible for an insurance company to say when a certain person will die, but the company can very accurately calculate exactly how many deaths will occur each year in a very large group of people.

As shown in Figure 23.3, if one kilogram of ^{238}U had been isolated at the formation of Earth, today (4.6 billion years later) there would be 0.5 kilogram of ^{238}U left. Thus, one-half of all the uranium atoms initially present would have commenced the series of transformations that end when they become stable lead atoms. Likewise, 4.5 billion years from now (9 billion years from when the kilogram of uranium initially existed) one-half of the 0.5 kilogram (or 0.25 kg) of uranium would remain, and so on.

In other words, with the passage of each 4.5 billion years, the amount of uranium present will decrease by one-half. The length of time it takes for one-half of a radioactive substance to decay, as discussed in detail in Chapter 15, is referred to as the half-life of the substance. If a rock containing uranium is analyzed and the ratio of uranium to lead is determined, then one can calculate how long it must have taken for that particular ratio of uranium to lead to have resulted. Thus, the approximate age of rocks that contain uranium can be determined with reasonable accuracy.

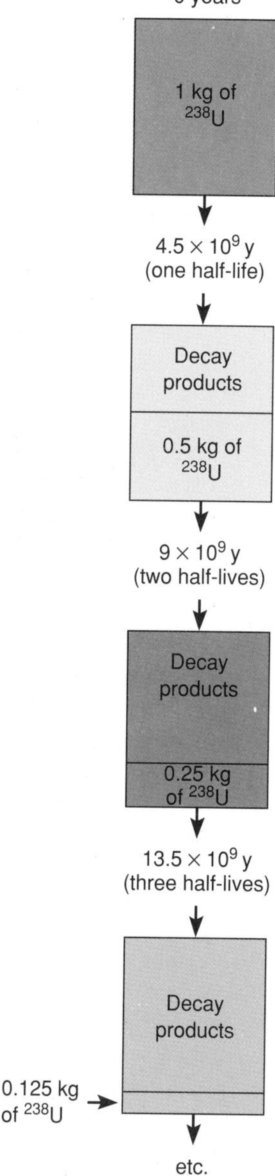

Figure 23.3 Radioactive decay of uranium-238 into its decay products, including stable lead-206. The mass of the decay products is almost equal to the mass of uranium-238 that decays; only a small portion of the mass is converted to energy.

Several other radioactive decay series besides uranium to lead are used to date rocks and fossils; for example the decay of K-40 to Ar-40 (potassium-argon dating) and that of Rb-87 to Sr-87 (rubidium-strontium dating). It is by the use of these various radioactive "clocks" that scientists can tell when in the past certain animals and plants lived. If the remains of a saber-toothed tiger are found buried in layers of rocks that, from various dating methods, are shown to be 25 million years old, then the saber-toothed tiger must have lived and died 25 million years ago.

Telling Time by Rock Layers

Unfortunately, not all the layers of rock that exist in the crust of our planet contain the various radioactive materials necessary to determine the rocks' ages. However, other methods can be used to obtain, if not absolute ages of formation, at least the relative ages of rock formation. For instance, as shown in Figure 23.4, if layer A has been dated by radioactive clock techniques as having been formed 100 million years ago and if layer E has been dated as having been formed 80 million years ago, then layers B through D must have been formed between 80 million and 100 million years ago. One can further state that since layer B is below layer C, layer B must be older than layer C.

This conclusion can be generalized by pointing out that each layer or bed of rock is younger than the beds below it and older than the beds above it, if the bed sequence has not been overturned by geologic processes—a statement of the **law of superposition.** Thus, in reference to Figure 23.4, although we cannot tell the absolute age of each of the beds B through D, we do know that beds B through D are between 80 million and 100 million years old and that bed D is closer than the other beds to being 80 million years old and that bed B is closer than the other beds to being 100 million years old. Bed C would be intermediate in age between beds D and B.

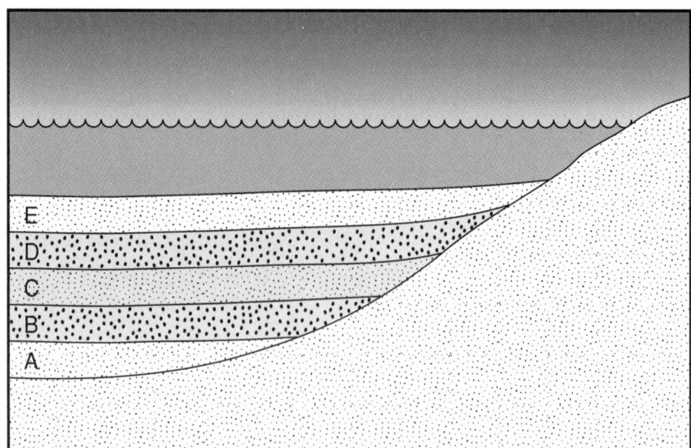

Figure 23.4 An illustration of the law of superposition. Layer A is older than layer B, which is older than layer C, and so on.

Ancient Life—Layer by Layer

As early as about 500 B.C., the Greek philosopher Xenophanes postulated that the rocks that were formed in the bottom of seas and lakes were later made part of the dry land. He reasoned that the transition from sea bottom to dry land was the result of the sea's receding or of the sea bottom's being pushed up above the surrounding water. His reasoning was based on the fact that shells of certain mollusks were found great distances inland and even in rocks high up in mountains. Further, he had observed in broken rocks impressions of many different kinds of fishes and other marine life. The indications of past life that Xenophanes observed are today called **fossils.** A fossil may be defined as *any indication of past life.* (By convention, indications of life that existed after the beginning of recorded history are not included.) These indications of past life may be the bones or shells of once-living creatures or replicas of bones or shells in which the original bones or shells have been replaced by various minerals in the same way in which petrified forests are produced. Fossils also may be only impressions of past life, such as the imprint of a leaf or the footprint of a dinosaur.

Rarely are complete flesh-and-blood carcasses of ancient animals preserved. Occasionally insects are found embedded in amber, and frozen carcasses of huge woolly mammoths also have been found on occasion. The mammoths apparently had fallen into glacial crevasses and were hurriedly deep-frozen. After more than 20,000 years, those portions of glaciers melted to reveal the undecayed flesh of these huge ancient animals.

When fossils are found embedded in layers of rock, it is possible to use the law of superposition to determine which layer of rock contains the oldest fossils and which layer of rock contains the youngest fossils. Thus, fossils present in different layers of rock present a record of the succession in time of different life-forms. Logically, the fossils in the uppermost layer are a record of the most recent life. The fossils in the succession of rock layers from the bottom to the top show that the organisms that formed the fossils in the bottom layer were the oldest; that the organisms that formed the fossils in the next higher layer were more recent than those in the bottom layer, but older than the remains of life in the layer immediately above them; and so on. This orderly succession of ancient through recent life forms recorded in the rocks is referred to as the **law of faunal succession.** The word *fauna* is actually a term referring to animals, but it is understood that the law of faunal succession applies to plant fossils as well as to animal fossils.

The Succession of Life on Earth

As one studies the fossils that exist in rock layers, the gradual increase in the complexity of life from the lower, more ancient beds to the higher, more recent beds becomes apparent. In fact, the rock layers offer a picture book of the succession of life from lower to higher forms.

This succession is worldwide—changes that were taking place in one part of the world were, in general, also taking place in other parts of the world. Consequently, if certain types of fossils are found in a particular bed in Europe, then a bed in the United States that contains the same fossils should be of the same general age as the European bed. In fact, the ages of different beds are actually dated in this manner by paleontologists. Fossils, then, are the Rosetta stone of the development of life on Earth.

Even the environment in which animals and plants lived can be deduced from the fossils and the mineral composition of the rocks in which the fossils are embedded. If a geologist finds oyster shells, coral, or fossilized fish in a bed, it is safe to assume that the bed was formed at the bottom of an ancient sea. On the other hand, if the fossils are of four-legged animals, one would postulate

Table 23.1

The geologic time scale and important biological events.

Age in millions of years	Era	Period	Epoch	Est. age in millions of years	Major evolutionary changes	Age of
	Cenozoic	Quaternary	Holocene (recent)	0.01	Man dominates Earth	Man
			Pleistocene	1.8	Ice ages	
		Tertiary	Pliocene	5	Flowering plants widespread	
			Miocene	22.5	Great diversity of mammals	
			Oligocene	37.5	Grasses abundant; first saber-toothed tigers	Mammals
			Eocene	53.5	Horses, rhinoceroses, whales appear	
65			Paleocene	65	First primates	
	Mesozoic	Cretaceous		136	Dinosaurs become extinct; flowering plants appear	
		Jurassic		195	Dinosaurs rule Earth; first mammals, first toothed birds	Reptiles
225		Triassic		225	First dinosaurs; many mammal-like reptiles	
	Paleozoic	Permian		280	Rise of reptiles	
		Pennsylvanian		320	Extensive coal-forming swamps; reptiles appear; giant insects	
		Mississippian		345	Great diversity of fish; insects evolve wings	
		Devonian		395	Amphibians appear on land; first insects	Amphibians
		Silurian		435	First air-breathing animals; first land plants	Fishes
		Ordovician		500	First vertebrates appear—primitive fish	
570		Cambrian		570	Marine invertebrate animals and algae abundant	Invertebrates
800	Precambrian	Precambrian			Variety of simple marine plants and soft-bodied animals	
1600					Diverse cellular plant life-forms	
2500					Single-cell photosynthetic organisms similar to present-day blue-green algae	Photosynthesis
3600					First primordial life-form	Life
4000(?)					Complex organic compounds present in Earth's seas	Chemical evolution
4500					Formation of Earth	

The time scale is divided into four main time segments called **eras.** These eras are

1. *Cenozoic*—from 65 million years ago to the present. *Cenozoic* means "recent life."
2. *Mesozoic*—from 225 million years ago to 65 million years ago. *Mesozoic* means "middle life."
3. *Paleozoic*—from 570 million years ago to 225 million years ago. *Paleozoic* means "ancient life."
4. *Precambrian*—from the time of the formation of Earth to 570 million years ago. The term *Precambrian life* is used to indicate any life-forms that existed before the beginning of the Paleozoic era.

In turn, each era, with the exception of the Precambrian, is subdivided into smaller segments of time called **periods.** The periods of the Cenozoic are in turn subdivided into even smaller time segments called **epochs.**

Note: Since geology is ultimately based on the study of rocks, then geologic time should start with the oldest known rocks, presently dated at 4 billion years. The span of time from 4 billion to 4.6 billion years would more correctly be called astronomic time. Source: Some data from the U.S. Department of the Interior.

that the rock was formed on land or in a shallow stream or sea. If a bed containing fossils also contains salt, then the fossils probably lived in a desert climate in which a shallow sea evaporated to produce the salt deposit. A bed of coal suggests a very warm climate with a generous rainfall that would produce the lush vegetation necessary for the eventual formation of coal.

The Geologic Time Scale

The geologic time scale, shown in Table 23.1, indicates some of the main events that heralded the changes in life-forms on Earth with the passage of time.

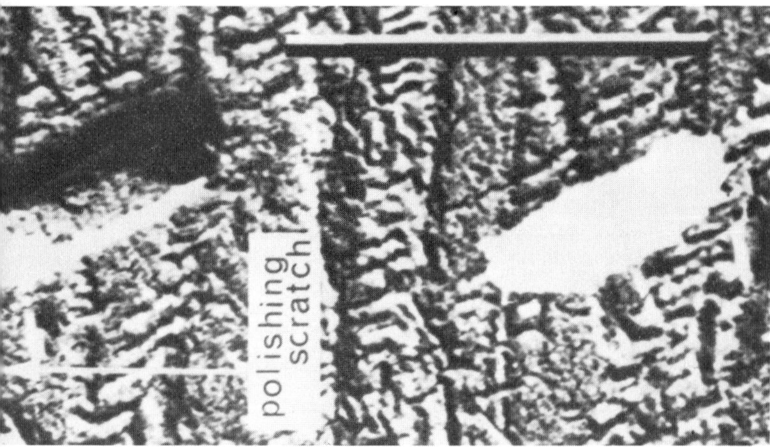

Figure 23.5 Fossil remains of microorganism (left portion of picture) 3.4 billion years old, from the Precambrian era of South Africa.

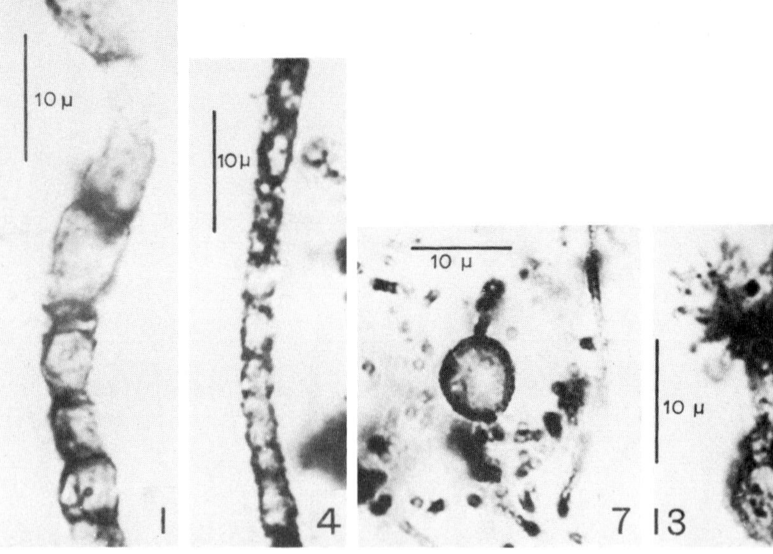

Figure 23.6 Fossil remains of photosynthetic cellular organisms that are 2 billion years old. They were found in Ontario, Canada.

Precambrian Life

The very ancient Precambrian rocks contain few fossils in comparison to the great number and variety of fossils found in rocks of the Paleozoic, Mesozoic, and Cenozoic eras. The scarcity of Precambrian fossils is mainly due to the following factors:

1. Most of the Precambrian animal life consisted of soft-bodied organisms without hard parts easily preserved as fossils.
2. Life was not as abundant and widespread as in later geologic eras.
3. Most Precambrian rocks have been subjected to intense heat and pressure; consequently, fossil remains that may have been present at one time have been destroyed.
4. Precambrian rocks have not been studied extensively by paleontologists due to the preceding factors.

In any case, Precambrian fossils are relatively scarce. There have been, however, some very notable fossil finds in Precambrian rocks in the last 30 years. Enough information has now been collected to allow scientists to sketchily reconstruct the succession of life from earliest Precambrian times.

Surely the most momentous event in life's evolutionary climb during the **Precambrian era** was the first step—the appearance of the first primordial self-replicating life system. This primordial form slowly evolved over thousands of centuries into single-cell organisms similar to present-day blue-green algae. Fossil remains of these single-cell organisms have been found in South Africa by Elso Barghoorn and William Schopf in rock that is 3.4 billion years old (Fig. 23.5). (Other researchers reportedly have found evidence of previous life, but confirmation is not irrefutably complete at this time.)

Eventually life developed photosynthesis. The ability of life to obtain much of its energy by the photosynthetic process was a great leap forward. Prior to the development of photosynthetic ability, cells had to acquire nourishment from the organic nutrients of the shallow seas in which they lived—a method much less efficient than photosynthesis.

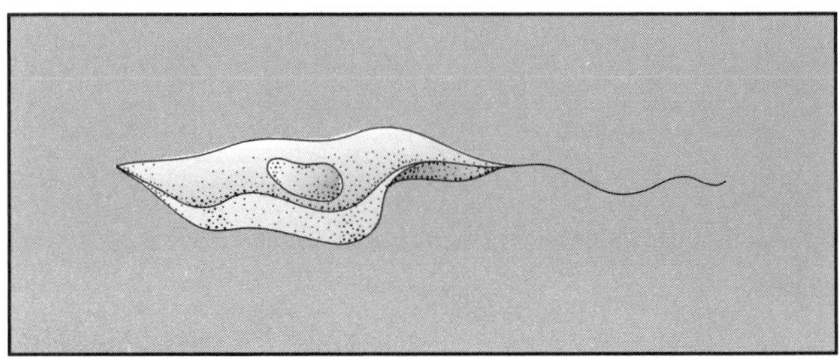

Figure 23.7 A drawing of a present-day mastigophoran.

The next major step in the evolving of life was the development of a cell containing a nucleus. With the appearance of nucleated cells, sexual reproduction became possible. Sexual reproduction opened the door to genetic variability that resulted, by natural selection techniques, in increased complexity of multicellular organisms. Eventually, a single-cell plant-animal form, similar to present-day Mastigophora, developed. The mastigophoran-type organism could live strictly as a plant by obtaining its energy by photosynthesis or could live strictly as an animal and in the absence of sunlight obtain its energy by ingestion of organic nutrients from plants (see Figs. 23.6 and 23.7).

By at least 1 billion years ago multicellular forms had developed in which different cells of the organism had different specific functions necessary for the organism's normal life operations. The relatively simple multicellular organisms evolved into the more complex plants and soft-bodied animals represented by the fossils found in the **Ediacara Formation** in Australia (see Fig. 23.8).

The two major milestones in the evolution of life on Earth were

1. the origin of cells, which occurred about 3.4 billion years ago; and
2. the development of multicellular plants and animals, which occurred about one billion years ago.

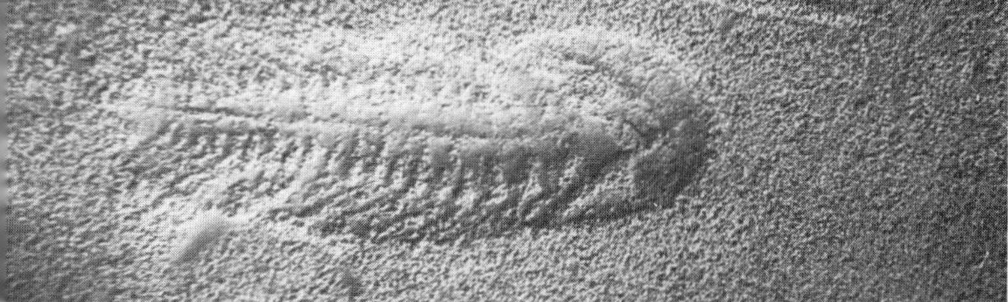

Figure 23.8 Fossil remains of organisms that lived about 700 million years ago. These fossils were found in the Ediacara Formation in Australia. Top to bottom: *Spriggina floundersi, Tribrachidium meraldicum, Dickinsonia costata.*

It took about 22 percent of Earth's span of existence as a planet to produce single-cell organisms and about 76 percent of its span of existence to produce multicellular forms. It required 71 percent of the total time frame of history of life on our planet to evolve from single cells to multicellular forms, and 52 percent of the entire span of existence of our planet for the transition!

The Paleozoic Era

At the close of the Precambrian era and during the whole of the Paleozoic era, great inland seas inundated the continents of Earth.

Figure 23.9 Life in the Cambrian seas. Jellyfish, trilobites, trilobitomorphs, shrimplike arthropods, and organ-pipe and pin-cushion sponges are mixed with filamentous algae.

These ancient intracontinental marine seas teemed with life. The ancient life left its fossil remains deep within the interior of continents for us to see and ponder many hundreds of millions of years later.

By the Cambrian period of the Paleozoic era, 570 million years ago, the shallow seas of Earth swarmed with invertebrate animal life; that is, life without internal skeletons but with hard external shells. Some of the soft-bodied animals in the latter part of the Precambrian era had acquired body armor that protected them from predators and harmful doses of ultraviolet radiation. The dominant animal in the Cambrian seas was the trilobite, which existed in a great variety of shapes. Other important animals present in the Cambrian seas were brachiopods, sponges, snails, worms, and jellyfish (see Fig. 23.9). Plant life consisted mostly of blue-green algae. Although life flourished in the Cambrian seas, the stark desolation of the Cambrian landmasses was mute testimony to the fact that evolution had not yet succeeded in producing life that was viable enough to live out of the life-protective seas.

The Ordovician period is remarkable for the great diversity and abundance of marine animal life that existed. As an indication of the profuse animal sea life of the Ordovician, the oldest and largest deposits of oil and gas come from chemically changed animal bodies from this geologic period. The most important development of the Ordovician period was the appearance of the first vertebrates, which were primitive fish. The largest animals that had yet existed on Earth were the cephalopods of the Ordovician. Some of the cephalopods attained a length of over 4.5 meters and were about 25 centimeters in diameter at their widest part (see Fig. 23.10).

The Silurian period marks a double milestone in the history of life on our planet. The first appearance of land plants and of air-breathing animals occurred in the Silurian. This invasion surely was a long and difficult struggle. Before plants could successfully survive on land, they had to develop a protective outer layer to prevent loss of water. Furthermore, the plants had to evolve a vascular system for transporting water from the soil to the rest of the plant. The first air-breathing animals were apparently sea scorpions, some of which attained a length of over 2.7 meters. Silurian life is shown in Figures 23.11 and 23.12.

Figure 23.10 Life in the Ordovician seas. Shown are straight-shelled nautiloid cephalopods (striped), colonial coral, crinoid, bryozoans, brachiopods, and trilobites.

Figure 23.11 Life in the Silurian seas. Sea scorpions Pterygotus (center, left foreground, and right background) live with worms and "shrimp" (translucent).

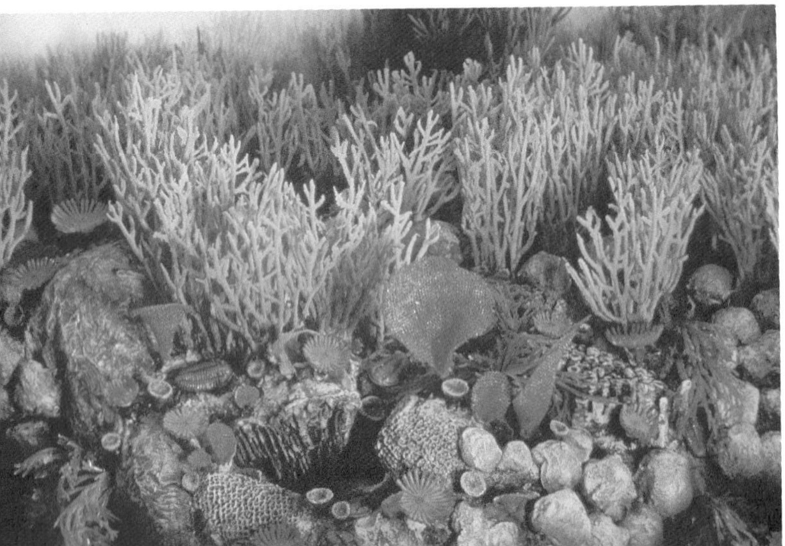

Figure 23.12 Other Silurian life included chain coral (foreground), organ-pipe coral (right side), bryozoans, flowerlike cystids, straight-shelled nautiloid cephalopods, and trilobites.

Figure 23.13 Life in the Devonian seas. Lacy bryozoan (funnels) and budlike blastoids mix with colonial and organ-pipe coral, crinoids (sea lilies), snails, trilobites, and straight nautiloid cephalopods.

Figure 23.14 The *Diplovertebron,* an early amphibian. The first land animals had distinct walking legs as opposed to the stubby fleshlike fins of the first fish that laboriously crawled upon land. Courtesy of Dr. Charles A. Payne; Susan Heulsing, artist.

The outstanding events of the Devonian period were the appearance of insects and the transition from fish that used gills to fish that used lungs. The lungfishes, descendants of which have survived up to the present time, evolved into amphibians that slowly developed legs and are represented today by salamanders. Thus did the first land vertebrates appear (see Figs. 23.13 and 23.14).

During the Mississippian, Pennsylvanian, and Permian periods of the Paleozoic era, the climate of Earth was mildly tropical and, therefore, very favorable to plant growth (note Fig. 23.15). Consequently, the latter part of the Paleozoic era saw the development throughout the world of great forests and lush vegetation that today yield most of the world's coal. Because coal formed during the Mississippian and Pennsylvanian periods is so plentiful

Figure 23.15 Life in the Mississippian seas. Crinoids share the seas with blastoids and starfish (on bottom).

Figure 23.16 A coal forest during the Carboniferous period.

Figure 23.17 The *Brontosaurus*, a Mesozoic sauropod (herbivorous) dinosaur. The giants reached a length of 24 m and had a mass of about 45,000 kg. They represented the largest land animals that ever existed.

and widespread, the Mississippian and Pennsylvanian periods are often referred to collectively as the *Carboniferous period* (see Fig. 23.16).

During the Mississippian period, insects developed wings, and there was a great diversity of fishes. By the Pennsylvanian period, some giant insects had evolved a wingspread of up to 75 centimeters, and the swish and rustle of their wings filled the air above the luxuriant tropical vegetation with their great vibrations. Of all animals that have ever lived on Earth, the insects and the fishes have been more carefully tailored by nature to survive capricious environmental changes. Insects and fishes have flourished with only modest changes for about 400 million years.

The **amphibians** of the Pennsylvanian period slowly evolved legs and began laying eggs with a tough outer shell that prevented dehydration. This adaptation allowed the animals to lay their eggs on the land rather than in the water. This transition from marine animals to land animals resulted in the first reptiles. The reptiles, continuing to evolve during the Permian period, must have been well suited to their environment because they established a dynasty that ruled the Earth, uncontested, for 200 million years.

The Mesozoic Era

During the early part of the Mesozoic era, the South American continent was still attached to Africa as part of Gondwanaland. Since the oldest rocks in the floor of the Indian Ocean were formed no later than the Cretaceous period, the landmass of India must also have been attached to Africa in early Mesozoic time. About the middle of the Mesozoic era, South America and India broke away from the African continent and began their slow journeys to their present positions.

The North American continent during the early Mesozoic era was mostly free of the great inland seas that had been present during the latter part of the Paleozoic era. Toward the middle of the Mesozoic era, however, the seas began once again to encroach upon the land, and by the Cretaceous period an extensive marine sea extended from the Arctic Ocean to the Gulf of Mexico. The middle of the North American continent was thus completely inundated, and North America appeared as two large elongated islands.

The Mesozoic era is called the **Age of Reptiles,** for during this era the reptiles were by far the dominant animals on this planet. During the Triassic period of the Mesozoic era, the most famous of all ancient life forms appeared—the dinosaur. The word *dinosaur* comes from the Greek meaning "terrible lizard." The first dinosaurs were modest-sized animals, but by the Jurassic and Cretaceous periods evolution had produced great reptilian monarchs such as *Brontosaurus, Triceratops,* and *Tyrannosaurus rex* (see Figs. 23.17 and 23.18). *Brontosaurus,* one of the largest animals ever to walk on Earth, attained a length of over 24 meters and a mass of 45,000 kilograms (a weight of 50 T).

The Jurassic period saw winged reptiles invade the skies and eventually become larger than any animal that ever learned the secrets of flight. Some pterosaurs developed a wingspread of over 8 meters (see Figs. 23.19 through 23.21).

The fishlike or dolphinlike ichthyosaurs were common in the Triassic period and reached a Jurassic peak. From the Triassic nothosaurs evolved the chiefly Jurassic-Cretaceous sea-serpent-like plesiosaurs (see Fig. 23.22).

Even while the great reptilian lords were thundering over the Earth, however, the animals that would one day usurp their sovereign rule were scurrying about almost unnoticed. During the

Figure 23.18 *Tyrannosaurus rex,* the "king tyrant lizard," a Mesozoic theropod (carnivorous) dinosaur. This fearsome animal was 6 m high, 15 m long, and had a mass of over 8000 kg. The moderate-sized *Triceratops,* shown at the left of the drawing, probably used the wide, flat bony frill that extended over its neck region to protect vital and vulnerable areas from attack.

Figure 23.19 The Age of Reptiles.

Figure 23.20 During the Jurassic period some reptiles invaded the skies. The *Archaeopteryx* represents the oldest known fossil bird. It had several reptilian features, including jaws with teeth and a long, jointed reptilian tail with feeble but well-feathered wings and was about the size of the modern crow. An artist's representation of the creature is illustrated in e figure.

Figure 23.21 Giant sea lizards, massive turtles, and flying reptiles *(pteranodons)* were in considerable abundance near the close of the Mesozoic era, particularly during the Cretaceous period.

Figure 23.22 Three plesiosaurs and two ichthyosaurs in a Cretaceous sea. Some plesiosaurs reached lengths of over 16 m.

Triassic period mammals had slowly evolved from small reptiles. The unpretentious appearance of these mouse-sized mammals belied their future greatness.

During the Cretaceous period flowering plants appeared. At the close of the Cretaceous period (and, therefore, the close of the Mesozoic era), dinosaurs became extinct throughout the world. The end of the reptilian reign appropriately marks the close of an era.

The Cenozoic Era and the Rise of Human Beings

The Cenozoic era opens with the appearance of the first primates (any mammal of the order Primates, including humans, apes, and monkeys). Possibly it was these initial primates that were destined to give birth, approximately 40 million years later, to *Homo sapiens*. Mammals had been transformed from rather inconspicuous animals thinly scattered over Earth to the dominant animal life-forms.

During the Tertiary period there was a great diversity of mammals. Some mammals—bats—evolved wings and sought their food in the air. Other mammals—whales and porpoises—reentered the seas and oceans that had been the home of their ancestors.

During the middle of the Tertiary period, grasslands became widespread, with the result that grazing animals evolved. Herds of horses, camels, and rhinoceroses were constantly preyed upon by fierce dire wolves and saber-toothed tigers (Fig. 23.23). Flightless birds of prey about 2.5 meters tall, belonging to the order of Diatrymiformes, roamed the continents in search of food. The largest land mammal that ever existed flourished during the middle of the Tertiary period. This mammal, *Baluchiterium,* an ancestor of the modern rhinoceros, stood 5.5 meters high at the shoulders and was over 9 meters long. Its skull alone, according to fossilized remains, measured well over a meter in length.

The Pleistocene epoch of the Quaternary period of the Cenozoic era is usually called the *Ice Age.* During the Pleistocene Ice Age, great continental glaciers covered as much as one-fourth of the land. Since the water used to produce these massive ice sheets came from evaporation of ocean water, the level of the world's oceans was more than 100 meters lower than it is today.

Figure 23.23 A saber-toothed tiger *(Smilodon)* that roamed North America during the Tertiary period.

Four times did the vast continental glaciers spread over much of North America, Europe, and Asia, and four times did the climate warm up and the glaciers retreat. The glaciers did not completely disappear, however, since remnants of the Pleistocene glaciers still exist as the Greenland and Antarctic ice sheets.

During the early Pleistocene era, about 1.8 million years ago, creatures that had a mass of no more than 45 kilograms and that were no taller than 1.5 meters roamed southern Africa. These creatures were only slightly more humanlike than they were apelike. These human-apes, called ***Australopithecus africanus*** ("southern apes of Africa"), did not make tools or weapons. They did, however, use bone and wooden clubs that by happenstance had the desired shape for weapons. They did not modify these weapons but used them as they found them. At the same time that the southern apes of Africa were roaming their domain, there appeared in southeast Africa a slightly more humanlike ape-creature called ***Homo habilis*** ("man with ability"). *Homo habilis,* who lived at least 2 million years ago, was a pygmy by our standards and was very apelike in appearance. These creatures, *Australopithecus africanus* and *Homo habilis,* had a common ancestor,

Mass Extinctions on Earth

Major and relatively abrupt disappearances of extremely large numbers of species of animals and plants have occurred at certain times in the history of life on Earth. About 435 million years ago, during the Ordovician period, nearly 30 percent of Earth's species became extinct, and in the course of the Devonian period (370 million years ago) slightly over 20 percent of nature's species disappeared. The most extraordinary example of widespread extinctions occurred during the Permian period at the Paleozoic-Mesozoic boundary when approximately 50 percent of all life-forms (95 percent of all marine species) vanished! This mass extinction was the most critical turning point in the history of life on Earth.

The most recent widespread extinction took place around 65 million years ago at the border between the Cretaceous and Tertiary periods of the Mesozoic and Cenozoic eras. The Cretaceous-Tertiary (CT) extinctions resulted in the loss of about 17 percent of land animals. Loss of marine animals was not nearly so extensive. Besides a general reduction in various life-forms, the CT boundary witnessed the beginning of the end for Earth's most renowned species (with the exception of *Homo sapiens*) when the death of the dinosaurs was set in motion. Most scientists believe that the devastation of the dinosaurs began when an asteroid almost 10 kilometers in diameter slammed into Earth. If one assumes the asteroid was traveling at 25 kilometers per second, about 74 times the speed of sound, the impact with Earth would have released energy equivalent to 250,000 times the energy produced during the 1980 Mount St. Helens' eruption. Mighty shock waves and earthquakes radiated out from the impact site, and winds stronger than any hurricanes or tornadoes that Earth had ever experienced ravaged the land, destroying all living things within hundreds of kilometers. Massive amounts of dust and debris were ejected high into the air, and the ejected material was joined by great quantities of soot produced by raging forest and plant fires ignited by the intense heat of the impact. Tens of thousands of square kilometers of land were incinerated. The dust, debris, and soot, blown by Earth's winds, produced dark globe-encircling clouds that scattered, absorbed, and obscured the life-giving rays of the sun for approximately three months.

Plant life was decimated without its normal ability to carry on photosynthesis for the production of needed biochemical compounds, and much of our tree and other plant life had to wait for the following year's spores and seeds to germinate. With the great reduction in plant life, most plant-eating dinosaurs and dinosaurs that ate plant-eating dinosaurs perished. A few species of saurischian and ornithischian dinosaurs did survive the CT catastrophe, but they soon died out. The energy released by the impact caused a general increase in temperature of 10 to 15°C throughout Earth's atmosphere. At the site of impact, however, the 1000 cubic kilometers of material thrown upward and outward was superheated to 2000°C. This extremely high temperature resulted in the reaction of nitrogen and oxygen molecules to produce large quantities of nitrogen oxides. The nitrogen oxides in turn caused considerable destruction of the ozone layer and produced intense acid rains.

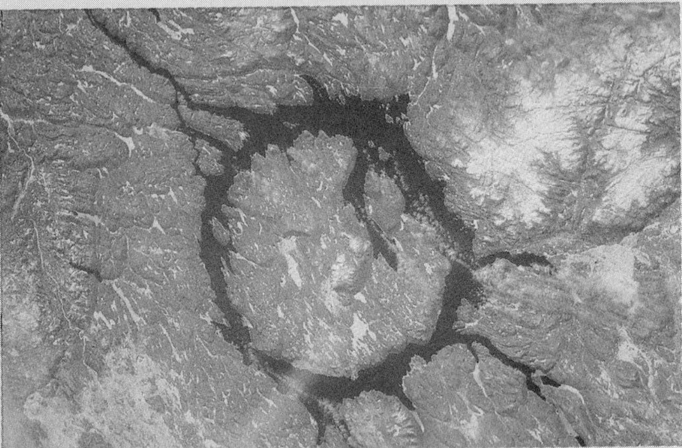

Figure 23.24 The Manicouagan crater in Quebec. A 100-km-diameter ring lake presently marks the spot where an asteroid (meteor) hit Earth about 200 million years ago.

Many times in Earth's history, meteor impacts have caused extinctions on a smaller scale. Although some impact craters can be seen today (see Fig. 23.24), most have been erased by erosion processes.

What is the sceintific evidence to support the asteroid-impact theory of mass extinction at the CT boundary? In many places on Earth there is a very thin seam of clay, laid down 65 million years ago, that contains iridium in concentrations 200 to 10,000 times higher than normally found in the crust. Iridium is about 1000 times more abundant in meteorites than in Earth's crust. This same layer also contains shock-metamorphosed quartz of the kind that would be produced by an asteroid impact and a carbon-rich sediment (thought to be from soot) of the same age.

For almost 150 million years dinosaurs ruled over the lands of Earth only to be struck down in a geologic blink of time by an astronomic accident ■

Figure 23.25 Neanderthals, presumably as they lived in France about 130,000 years ago. Courtesy of Dr. Charles A. Payne; Susan Heulsing, artist.

Figure 23.26 Cro-Magnon artists painted hunting scenes similar to this one.

Australopithecus afarensis (the best specimen of which is widely known as "Lucy"), that existed about 3.5 million years ago during the Pliocene epoch. By 1 million years ago, *Homo habilis* had evolved into ***Homo erectus,*** who made tools consisting of large, shaped hand axes of flint. Remains of *Homo erectus* are now known in eastern and northern Africa, Asia, and Europe. *Homo erectus* was more culturally advanced than *Homo habilis* but, more importantly, *Homo erectus* learned how to use and control fire.

About 130,000 years ago the most famous of all extinct human species was spread throughout Europe and the Near East. The Neanderthals (***Homo neanderthalensis***), although they resembled present humans more than had their ancestors, did have prominent brow ridges, slightly bowed legs, and great barrel chests. They did not have double curves to their spines as we have and, therefore, could not hold their heads up as high as we can. Neanderthals averaged only 1.6 meters in height and covered themselves from the cold of the glacial period with animal fur. We should be careful not to assume a patronizing attitude toward the Neanderthals, however, because their brain cases were as large as the average brain case of modern humans, and their cunning and strength allowed them to kill the great mammoths. A great advantage that Neanderthals had over their predecessors was better made weapons. The use of fire, along with better tools, gave them a decided advantage in survival, allowing Neanderthals to dominate Europe and the Near East for 100,000 years (see Fig. 23.25). Neanderthals, as far as is now known, were the first animals to bury some of their dead. They also occasionally marked a grave with a tombstone. These attitudes about death suggest the Neanderthals may have developed a belief in a life after death.

It is not known what caused the decline of the Neanderthals. Perhaps it was the scarcity of food brought about by the advance of glaciers. In any case, they became extinct or were assimilated by the Cro-Magnons, who appeared during the last ice age about 35,000 years ago.

Modern human beings, after approximately 3.5 billion years of evolution of life-forms, had arrived. (Cro-Magnons were the first members of the species known as ***Homo sapiens.***) Cro-Magnons were magnificent human specimens. They stood slightly over 1.8 meters tall and, because they had a double curvature of the spine, could hold their heads high. They had a high forehead, a highly developed chin, and an absence of heavy brow ridges.

The tools that the Cro-Magnons made were considerably more sophisticated than those of the Neanderthals. They also fashioned a greater variety of tools and weapons. Besides the more common stone axes, knives, clubs, and spears, Cro-Magnons produced bone needles, fishhooks, hunting darts and, eventually, bows and arrows. They cooked their food and wore fur clothes sewn together with bone needles.

Cro-Magnons, from whom we descended, were considerably more advanced than Neanderthals intellectually and culturally. In fact, Cro-Magnon culture was advanced to the point that within a particular community there were skilled craftsmen, cooks, painters, and religious leaders who were at least partially supported by a highly organized work force. The people of a Cro-Magnon community decorated themselves with beads and bracelets made from ivory and seashells much as people do today.

Perhaps the greatest accomplishment of the Cro-Magnons was their art. The skilled Cro-Magnon painters produced superb paintings, as well as reasonably delicate carvings and sculpture. Members of this civilization not only recognized the beauty and wonder of nature but could create their own beauty, giving expression to their feelings. An ancient artist at work is depicted in Figure 23.26.

If the first primordial life forms did appear on Earth about 3.6 billion years ago, it required about 3.598 billion years for the evolutionary process to produce humans. We, then, have been on this planet for only a very, very brief part of the time that life ha existed on Earth (see Fig. 23.27).

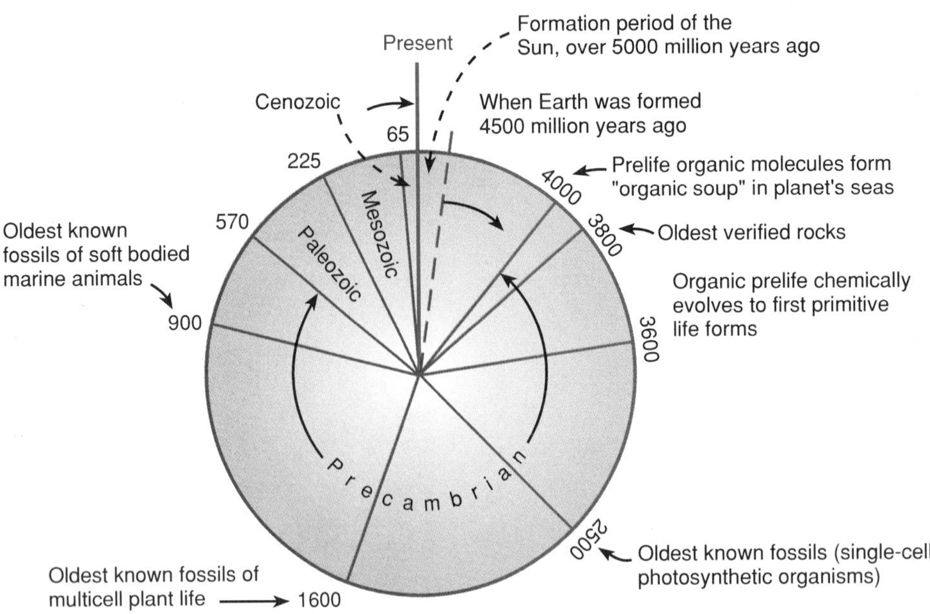

Figure 23.27 A geologic clock. Humans have been on Earth for such a relatively short period that it is impossible to show their existence on this clock. Numbers around the clock indicate millions of years.

Life without Light

At 90 degrees west longitude on the equator, near the Galápagos Islands (one thousand kilometers west of Ecuador), the deep submersible, the *Alvin,* dived to a depth of 2.5 kilometers. The scientists aboard the *Alvin* knew that there would be scalding hot, magma-heated seawater issuing from vents at this site of the Galápagos Rift. What they didn't expect was the incredible quantity and diversity of sea life that they found on the ocean floor where sunlight never penetrates. There were large foot-long clams, giant red-plumed tube worms (some reaching a length of over 3 meters), as well as white crabs, eyeless shrimp, mussels, and eel-like fish (see Fig. 23.28). Since the 1977 discovery at the Galápagos Rift, numerous other hydrothermal vent ecosystems have been found. Such ocean-floor colonies exist on the East Pacific Rise near Easter Island (approximately halfway between Tahiti and Chile); about 240 kilometers south of the tip of Baja California Sur, Mexico; off the northwestern coast of the United States; about 3000 kilometers east of Miami on the Mid-Altantic Ridge; and possibly at many other sites along the mid-ocean ridges.

Almost all of the animals and plants in our oceans live within about 250 meters of the surface. This is because plants need light to produce their organic compounds of photosynthesis and animals must feed on plants or on animals that feed on plants. But the great depths of ocean seafloors (2.5 to 3.5 kilometers deep) are dark and barren abyssal plains that no sunlight for photosynthesis can reach. The oases of life that do exist on the otherwise desolate oors manage to survive because the bacteria present in the nineral-rich waters surrounding the hot vents are auto- teria and are the basis of the food chain for these life- teria are chemosynthetic in nature, utilizing the d carbon dioxide in the vent waters to obtain sary for them to manufacture their needed

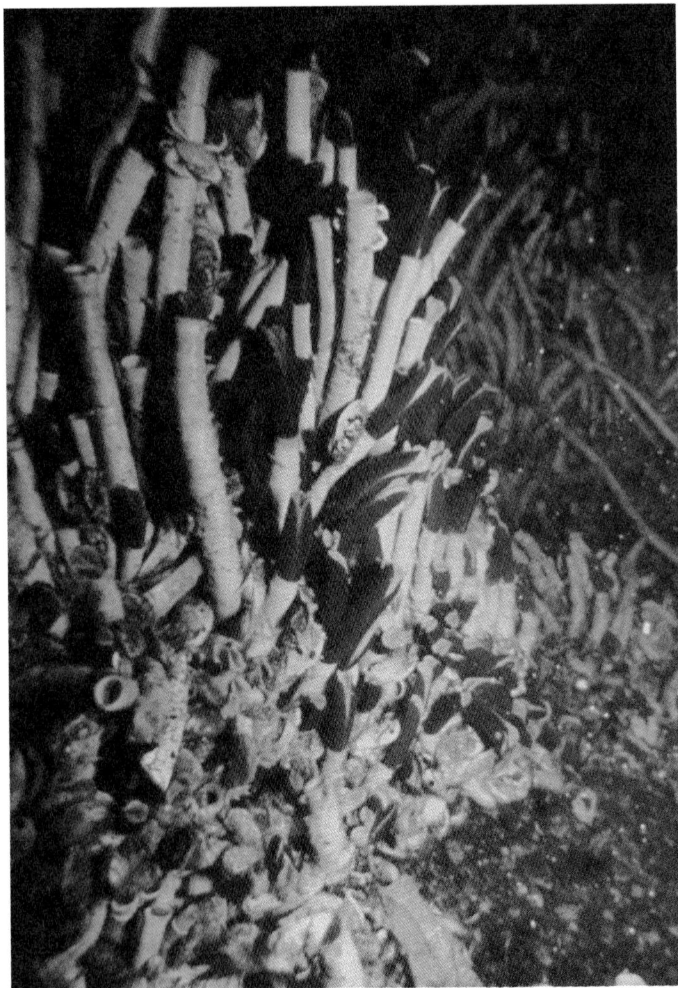

Figure 23.28 Hydrothermal vent life at the Galápagos Rift.

A "Tick" of the Clock

Suppose we were to plot the entire existence of the planet Earth on a scale represented by one calendar year. That is, we let one Earth year equal the time from the formation of Earth until the present. Using this scale, we can date selected major events. For example, on January 2, the crust of Earth formed from a melt. On March 6, the first living things appeared. During October, plants and animals of various sorts developed. In November, our coal supply started to form. On December 1, dinosaurs were abundant, but disappeared on December 20. This day also marked the arrival of various mammals. And, on December 31, human species appeared. To put other significant events into the same perspective, at 11:59:30 P.M. on the last day, the Roman Empire was at its peak. The New World was discovered with eight seconds left on the clock. On the same scale, our nation assumed a major role in the scientific endeavor less than a "tick" of the clock ago ■

organic compounds. The colonies of bizarre life-forms on our ocean seafloors are an excellent example of nature's way of providing for new ecological niches.

The sea life flourishing in the special enclaves depend on the superheated water circulating around the mid-ocean vents. The temperature of the water at the ridges quickly drops to the 2°C temperature of the water of the surrounding ocean floor. Thus, the waters that support life at these ridges varies all the way from 350°C to 2°C. The energy to produce the molten lava at the vents comes from the nuclear decay occurring in the interior of Earth; hence, energy formed by nuclear reactions in the sun is not needed! One cannot help but wonder whether on some planets and satellites, too far from the sun for normal photosynthetic reactions, there might be life-forms that obtain their energy and food by similar nonphotosynthetic means.

Life on Other Worlds?

If one accepts the hypothesis that life systems will "spontaneously" be created if the necessary compounds and conditions exist, then the question arises, Is there life on planets in other solar systems in our galaxy and in other galaxies of the universe? In order to answer this question, we must have knowledge of the existence of other solar systems. In 1968 Peter van de Kamp analyzed data collected over a period of 51 years on the motion of the star called Barnard's star. The results of his study caused him to conclude that this star has at least two companion planets.

Barnard's star may have many planets revolving around it, as our sun does. At present, however, our instruments are not sufficiently sensitive to detect the presence of small planets with the mass of, say, our Earth. Also, since planets do not emit their own light, and any light received from them is reflected from their "sun," it is not possible to actually "see" planets outside our solar system. In 1983, scientists at the U.S. Naval Observatory discovered a "wobble" in the motion of a star known as Van Biesbroeck-8 (VB 8). The deviation in the predicted path of VB8 indicated that an object several times the mass of Jupiter was orbiting the star. The star and its orbiting mass are located some 21 light years from our solar system. The object, called VB 8B, is considered to be a substellar companion of VB 8, since it supposedly lacks the mass considered necessary to initiate the "hydrogen-burning" process in its interior. (Some scientists theorize that a mass eight times the mass of Jupiter is necessary to support stellar fusion.) Although the discovery has not been confirmed, the report created new interest in the process called *speckle interferometry,* a technique that adjusts for the turbulence in the Earth's atmosphere.

Similar applications of the infrared speckle technique were completed on various other stars suspected of supporting orbiting companions, including VB 10, Stein 2051, and CC 1228. No bright objects similar to VB 8B were detected in orbit about the stars that were investigated. Thus, it was inferred that any companions orbiting these stars must be very small, perhaps the size of the planets in our solar system, or are exceptionally dim.

Still, Barnard's star, as well as others, may have numerous planets revolving about them. In fact, careful observations made by astronomers at the Las Campanas Observatory in Chile and reported in 1985 revealed a swarm of particles present in a disk-shaped nebula that surrounds Beta Pictoris, a star located some 50 light years from Earth. The chemical composition of the particles appears to be identical to that from which Earth and other members of our solar system formed. It was assumed in the past that stars with planets were a very rare occurrence in nature, but now scientists believe that it may be the norm rather than the exception for stars to have planets.

Certain conditions other than the mere existence of planets must be met, however, in order for life to exist on them. A few such conditions are as follows:

1. The star (sun) must not fluctuate to any appreciable extent in energy output.
2. Only planets at certain distances from their sun (called the ecosphere or habitable zone) can sustain life. If they are too far from their sun, the planets are so cold that water stays frozen, and life-forms as we know them must use liquid water in their cells. If the planets are too close to their sun, the resulting high temperatures destroy life. In our solar system only Earth and Mars in a 24-hour period have surface temperatures between the freezing and boiling points of water. Pulsating stars would not provide the narrow temperature range necessary for life on their planets. Double- or triple-star systems would also probably lack planets that could sustain life since the planets' orbits would be very complex, resulting in wide temperature variations.

3. The planets' orbits must not be too eccentric or they will receive too much heat when at perihelion (closest approach to their sun) and too little heat at aphelion (farthest distance of planetary orbit from their sun). Obviously, a nearly circular orbit is the safest for life-forms.
4. The planets must be "water" planets so that the protoplasm of cells can function.
5. The primitive atmosphere of planets must be the type of atmosphere that Earth once enjoyed. This is the easiest requirement to fulfill since it is probably a universally existing condition.
6. The rotation of the planets on their axes must not be synchronized with their revolution around their sun. If this synchronization occurred, then the planet would keep the same face to the sun and the same hemisphere away from the sun, with the result that all water on the planet would eventually be collected and frozen on the dark side, and the lighted side would become a searing, inhospitable desert.
7. The masses of the planets must be sufficiently great. If the mass were too small, all the planetary atmosphere would escape.
8. The speed of the planets' rotation should not be too fast in relation to their masses. For instance, if our planet rotated on its axis about 18 times faster than it does, the centrifugal force at our equator would slightly exceed the gravitational force and our atmosphere would all escape in a short time.

If our solar system is a reliable gauge, then one in nine planets can readily support life.

In view of these restrictions, assume that only one star in 10 million (10^7) has a planet with prerequisites to produce and sustain life. Since there are about 300 billion stars in our galaxy, there may be $(3 \times 10^{11})/10^7$, or 3×10^4, stars with planets that probably have life-forms present in our galaxy alone. Even more stunning in its implication is the fact that since there are 10^{21} observable stars in the universe, there may approximately be $10^{21}/10^7$, or 10^{14}, stars in the universe that we can see that have planets on which life exists.

Summary

The original atmosphere of Earth contained mainly hydrogen, helium, methane, ammonia, and water. Various forms of energy available on primitive Earth caused the elements and compounds present in Earth's atmosphere to react and produce amino acids, sugars, organic nitrogen bases, and many other organic com-
unds. The amino acids in turn reacted to form proteins. Further
ns of proteins, DNA, and other compounds may have pro-
first primordial life cells.
reproduced for untold generations and eventually
lity to utilize chlorophyll to obtain their energy
to synthesize their needed food.

Evidently, as a result of the photosynthetic process, life slowly changed the primitive atmosphere of Earth by releasing oxygen from photosynthesis into the air. Over untold aeons, the oxygen reacted with the compounds in the air of Earth until the present atmosphere of nitrogen and oxygen prevailed.

Comparing the ratio of a radioactive substance to its decay products present in a rock allows one to determine the rock's age and the age of fossils present in the rock.

The order in which fossils occur in rock layers allows one to deduce the sequence in which various ancient life-forms occurred on Earth. It is possible, therefore, to estimate the chronology of life-forms from the simplest single-cell organisms of billions of years ago through many intermediate life forms to the eventual evolution of humans.

Geologic time is divided into the following time segments called eras: Cenozoic (from 65 million years ago to the present), Mesozoic (from 225 million years ago to 65 million years ago), Paleozoic (from 570 million years ago to 225 million years ago), and Precambrian (from the time of Earth's formation to 570 million years ago). Eras are subdivided into time spans called periods, and periods are further subdivided into epochs.

Life existed on Earth at least 3.6 billion years ago, when single-cell organisms lived in the shallow seas of Earth. At least a billion years ago multicelled organisms developed, and approximately 800 million years ago complex soft-bodied animals were gathering food in the seas of our planet. About 570 million years ago, the seas of Earth swarmed with thriving invertebrate animal life, many with hard external shells. Life-forms continued to change until the evolutionary process culminated in *Homo sapiens*.

Since there are other stars in the universe with accompanying planets, scientists estimate that there may be trillions of planets in the universe on which life exists.

Questions and Problems

The Chemical Origin of Life

1. How was our present atmosphere apparently formed?
2. Considering Earth's present atmosphere, if all forms of life on Earth today were to perish, could life as we know it originate again? Explain your response.
3. List five compounds that were apparently abundant in Earth's primitive atmosphere.
4. What elements and compounds are assumed to have made up the original atmosphere of Earth?

Chemical and Biological Evolution

5. Does it seem likely that the chemical evolution that occurred on other planetary bodies gave rise to life-forms?
6. How does fermentation differ from photosynthesis?
7. How does a heterotroph differ from an autotroph?

Radioactive "Clocks"

8. Briefly discuss three possible explanations as to why dinosaurs suddenly disappeared from Earth's rock record.

9. What radioactive elements other than uranium are used to date rocks and fossils?

Telling Time by Rock Layers

10. Explain the logical assumptions that support the law of superposition.

11. How are fossils used to correlate rock layers separated by great distances, such as along the coasts of Africa and South America?

Ancient Life—Layer by Layer

12. What is a fossil?

13. How is the law of faunal succession applied to the determination of the age of rock layers?

The Succession of Life on Earth

14. How can the environment that existed when an animal or plant died be deduced by the type of rocks surrounding the animal or plant fossil?

15. How can the succession of ancient life-forms be determined by the relative positions of fossils in rock layers?

The Geologic Time Scale

16. What is the age of the most recent geologic era in millions of years?

17. In which geologic period did fish first appear?

18. Which era is considered the age of mammals?

19. For what percentage of the time that life has been present on Earth has humankind existed?

20. How long was life present in the seas of Earth before evidence of life on land appeared?

Precambrian Life

21. Why are soft-bodied animals such as jellyfish usually not preserved as fossils among the rock layers?

22. How can the climatic conditions present many millions of years ago be deduced by the types of fossils found in a given area?

23. During which era did photosynthesis develop?

24. Cite two reasons why Precambrian fossils are scarce.

25. When did life originate on Earth, according to geologists? What evidence do they have to offer for this conclusion?

The Paleozoic Era

26. During which period did animals first walk upon the land?

27. When did the first vertebrates appear?

28. During which periods did most of the world's coal form?

The Mesozoic Era

29. Why is the Mesozoic era referred to as the "Age of Reptiles"?

30. During which period did birds first appear? Approximately how many years ago?

The Cenozoic Era and the Rise of Human Beings

31. How did the abilities of *Homo habilis* differ from those of *Homo erectus* and *Homo neanderthalensis?*

32. What are some of the main differences between *Homo neanderthalensis* and *Homo sapiens?*

Life without Light

33. Why are the life-forms found at various hot-water vents on the ocean floor different from the rest of life-forms on Earth?

34. How can plant life exist in the absence of photosynthesis or fermentation?

Life on Other Worlds

35. What are some of the reasons that most scientists believe that some form of life exists on other worlds?

36. What planetary conditions would probably preclude the development of life-forms?

24 The Physical Aspects of Geology

Chapter outline

Key terms/concepts

Sedimentary rock layers in the Grand Canyon, as exposed by the continued processes of degradation.

The origin of Earth is of considerable interest to geologists; however, geology usually concentrates on the study of Earth after solidification of the surface occurred. Various details about Earth's composition, shape, and size have been previously discussed, but a brief reconsideration of these items helps set the stage for the topic at hand.

Earth is considered an oblate spheroid—almost spherical but somewhat flattened at the poles and slightly bulged about the equator, both characteristics caused by the planet's rate of rotation. The diameter of this planet through the equator is about 43 kilometers greater than the same measure through the poles. The difference between the two measures amounts to about 0.34 percent, an indication of the flattening of Earth.

The two major features of our planet are the continents and the ocean basins. The total area of Earth is about 510 million square kilometers, 29 percent of which is land; the remaining area is covered by water. Most of Earth's exposed surface appears as part of the continental masses that are unevenly distributed over the planet, with two-thirds located north of the equator. There is also a significant difference in elevation of the landmasses. The highest point on Earth's surface is Mount Everest, 8850 meters above sea level. The two lowest points are located on the ocean floor, both of which are in excess of 11,000 meters below sea level. One region is located near the Philippine Island of Mindanao; the other point occurs in the Mariana Trench south of the island of Guam. The total relief of 20 kilometers amounts to a variation of about 0.3 percent when compared to Earth's radius. The continents average about 800 meters above sea level. Asia, some 975 meters above sea level, is the highest, and Europe, elevated slightly less than 300 meters, is the lowest. The North American continent averages about 730 meters above sea level.

Uniformitarianism

The outer layer of Earth is called the crust. This layer is the only part of Earth accessible to our direct observation, although only about 30 percent of it is exposed to our view because of the amount of the surface covered by lakes, rivers, and oceans. Classical geology developed from the study of the continents. One of the most fundamental observations that concerns the outer layer of Earth is that of **uniformitarianism,** a modern view of Earth credited to James Hutton (1726–1797), a Scottish geologist. This concept suggests that the present is the key to the past; any structure in old rocks must have been formed by constant or catastrophic processes identical or similar to those that presently occur on Earth. We note the processes of erosion, soil formation, and the effects of earthquakes and realize that geologic development, while uniformitarian, encompasses differences in the speed and intensity by which geologic processes occur. After seeing a deep valley such as the Grand Canyon, we can understand why early observers believed that it was created by great earthquakes. The fact that water flows in the valley was readily explained by earlier scholars of geology through comparison of the low area with a higher surrounding land. We now take a closer look at nature and note the material carried by muddy rivers and streams and conclude from the amount of accumulated sediment that rivers eroded the deep valleys and formed the present beds. The main process occurring today on Earth's surface is clearly erosion, as evidenced by the sediment deposits located at the mouths of rivers and streams. Earth is constantly being worn down and, theoretically, all topography could be removed in 44 million years. However, erosion has been going on for billions of years. The work of the rivers, easily the most important erosion agent, is countered by the uplift of the continents. Thus, much of the science of geology is centered on the struggle between the forces of erosion and the forces of uplift.

The Geologic Processes

There are numerous processes that may operate in and upon Earth's crust as well as deeper in the lithosphere. **Gradation** includes those processes responsible for the relatively smooth surface of our planet. Among the group is *degradation,* the wearing down of rocks by water, wind, and ice. The opposing process is called *aggradation,* the building up of low spots on Earth's surface by the accumulation of sediment deposited by the action of water, wind, and ice.

Deposits of rock composed of the sediment produced primarily by gradation are more than 10 kilometers thick in the mountain belts of the United States, whereas such layers extend only to a depth of from 1 to 2 kilometers in the central plains. In the eastern portion of Canada known as the Precambrian Shield, no sedimentary layers exist, whereas some sedimentary layers elsewhere reach to over 15 kilometers into Earth's lithosphere.

Many of Earth's mountain ranges are composed primarily of sedimentary deposits. Prior to the acceptance of the *plate-tectonic theory,* described in detail in Chapter 20, these mountain ranges were perceived to have resulted from deformed *geosynclines*—elongated regions of shallow ocean basins that appeared to be constantly sinking, then were continually replenished with sediment. The geosynclinal theory has been essentially discounted in favor of the plate-tectonic theory. Geologists now realize that such mountain systems, some thousands of kilometers long, are closely associated with crustal plate boundaries, as illustrated in Figure 24.1. (Compare to Fig. 20.9.) Most of these systems we observe today contain at least second-generation mountains. Evidence strongly points out that these mountains were leveled to no more than hills by the process of degradation, then they rose at least once again in later periods of uplift.

In the plate-tectonic model, a process responsible for most of Earth's major mountain systems is known as **orogenesis.** It involves *the mountain-building processes that result in displacement (faulting) or deformation (folding) of Earth's crust.* Other concepts involved in the theory include *subduction zones* and *seafloor spreading,* both discussed in Chapter 20. Also associated with orogenesis are the concepts of *igneous activity* and *metamorphic*

Panorama of the Ocean Floor

Innovative techniques, including those applied during various space shuttle missions, provide a detailed view of the ocean basins. Seafloor mapping is not a relatively recent development; however, the accuracy and detail now available point out underwater surface features as diverse as any observed on the seven continents. Radar images that penetrate the ocean floor reveal many hidden features well beneath the bottom sediment.

Computerized applications, such as color graphics, have added valuable details to our comprehension of the forces in action along the boundaries of Earth's plates. Color and shading have added a whole new dimension, bringing out features that were formerly obscure. Computer graphics has proven valuable in determining the age and the rate of seafloor motion and in comparing regions where the magnetic fields have reversed ■

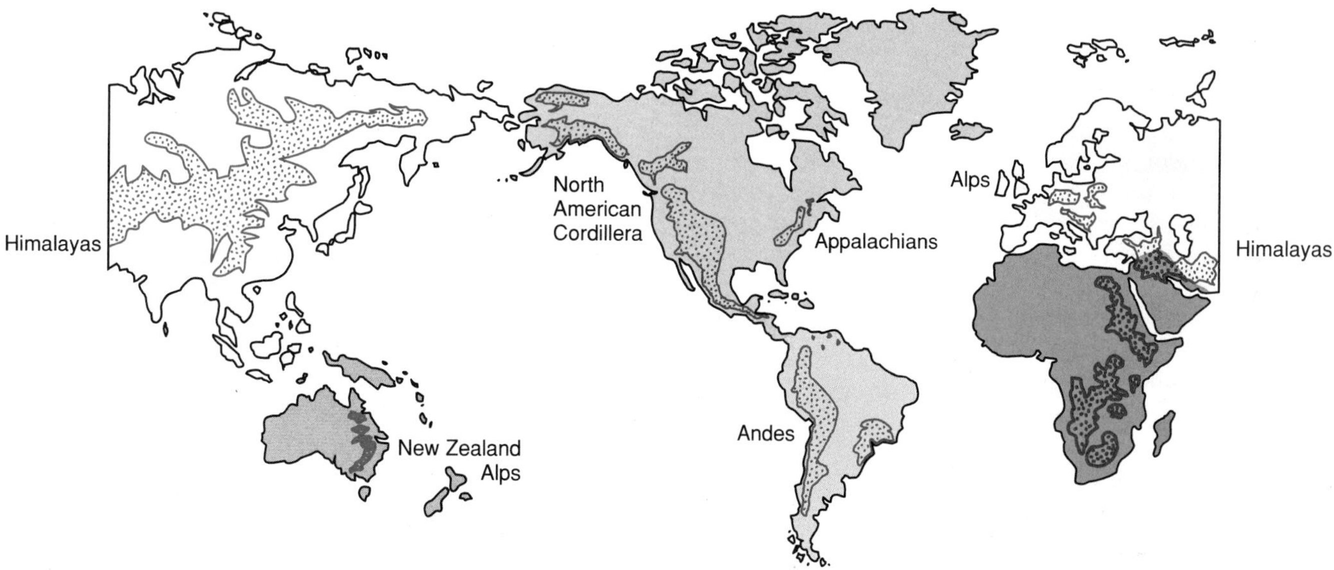

Figure 24.1 Earth's major mountain ranges are closely associated with crustal plate boundaries (see Fig. 20.9).

processes, both to be discussed later in this chapter. *A sequence of events occurring within a specific time frame and generally leading to intense deformation of rock layers* is known as an **orogeny.**

Faulting is the process in which the fracturing of Earth's crust is accompanied by a shifting of rock masses in a direction parallel to the fracture. The crust of Earth is considered solid, but rocks that compose the crust may also be plastic under great pressure. When the pressure and the accompanying high temperature exceed the limit that the rock layers can withstand, the rocks are prone to bend. Slight bending over a long area is called *warping,* a process directly involved in early stages of mountain building. *Folding* is much like warping but is more intensive. It is generally caused by forces that mostly act horizontally in Earth's crust. Such forces are in action along the zones where the seafloor is sliding beneath a continent or island arc. Much of the knowledge about Earth is derived from the mountains that rise above Earth's surface, since some of Earth's lower crust is exposed through uplift and subsequent erosion. Folded mountains occur in systems of ranges that appear as a series of ridges that alternate with valleys. This type of mountain range extends far across the continents. The Appalachian Mountains, which extend from northeast to southwest across the eastern United States, are an example of folded mountains.

Igneous activity refers to *all movements of molten rock and the formation of solid rock layers from the molten state.* This process is divided into two categories: *volcanism,* surface deposition or lava flow, and *plutonism,* deep-seated activity that takes place about 1000 meters or deeper inside Earth. The *molten rock that is present under the surface of Earth* is called **magma.** It is mainly composed of silicates. About 11 percent of its composition is steam and other gases dissolved under pressure in addition to crystals that were previously formed in the inner regions. The temperature of magma ranges from 500°C to 1400°C. The magma, according to its chemical composition, may form such rocks as basalt and granite or separate from a homogeneous melt by various processes. One such method is *magmatic differentiation,* a layering effect in which the heaviest of the first crystallized constituents of the melt concentrate in the lower zone of the rock mass.

Magma is often lighter and naturally more fluid than the solid rock that surrounds it. As a result of lesser density, it tends to rise in the lithosphere of Earth, additionally supported by the pressures acting upon it. During the formation of mountains, magma is squeezed from its deep reservoirs upward into areas of lower pressure. It expands and releases some of its corrosive gases, which flow upward into surrounding rock layers. As the hot fluid reaches shallow regions where cracks may exist, it moves more rapidly upward and emerges as *lava.* The molten rock cools rapidly on contact with the air and water and retains the small vacant spaces from which the trapped gases are released.

The molten rock that cools deep within Earth's lithosphere takes on an entirely different appearance. The structure of the cooled magma is that of tiny discrete particles, mostly crystalline in nature. The size of the crystals varies, but the crystals generally grow larger if the rate of cooling is slow.

Igneous rocks, those formed from magma, are of two kinds: extrusive and intrusive. *Extrusive rocks* are those formed from cooling lava or from other volcanic materials spewed out onto Earth's surface. *Intrusive rocks* are those formed from magma that was forced between existing rocks or rock layers within Earth's lithosphere. Intrusive rock formations are classified according to shape, size, position, and location.

Several of the intrusive rock bodies are common enough to warrant a brief discussion. If an igneous intrusive body of rock is so large that more than 100 square kilometers of its surface are exposed, geologists refer to it as a *batholith.* Batholiths originated during an orogeny that also created many mountains on Earth's continents and on the ocean floors. Because of their very large size, these batholiths cooled very slowly and produced coarse-textured rock.

Those intrusive bodies with a surface exposure of less than 100 square kilometers are known as *stocks.* Other flows of magma that are partially exposed by erosion pass downward continuously and form pipelike bodies called *volcanic necks.* Still other flows fill cracks or fissures in the outer crust and form *dikes. Sills* resemble dikes in that they are flat and thin, but they intrude upon other parallel rock layers. If the masses of igneous rock that intrude between the sedimentary layers cause a bulging of the overlying layers, the deposits are called *laccoliths.* The various types of igneous intrusions are illustrated in Figure 24.2.

A vast amount of knowledge can be obtained about igneous processes from volcanoes. Volcanoes offer the only direct evidence about the existence of magma within the crust, for they are central vents through which heated rock and magma are carried to the surface. Most of the main volatiles in magma are nothing other than water that escapes in the form of steam, along with some carbon dioxide and sulfur gases. If the magma erupts very suddenly, it may form a frothy substance that solidifies into a light rock that will float in water. The formation of *pumice,* as the light rock is known, is much the same as the formation of froth produced by opening a warm bottle of soft drink. As the lava and solid volcanic material are deposited on Earth's surface, they form a conical hill or mountain with a funnel-shaped center called a *crater.* On the other hand, the fluid release of magma, along with

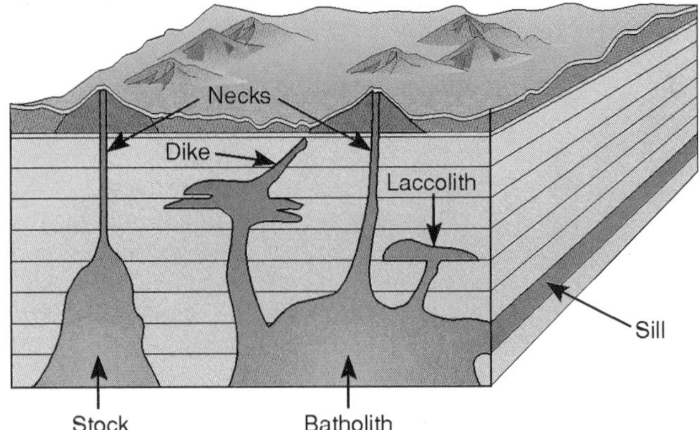

Figure 24.2 Common intrusions that often make their way through Earth's crust.

exceedingly hot gases, ashes, and other materials, may be quite violent and cause total destruction over large areas. For example, in 1902 an eruption of Mount Pelée on Martinique destroyed the city of Saint Pierre and reportedly killed all but one of the city's 30,000 inhabitants. The lone survivor was a prisoner locked behind the strong, thick walls of the prison.

Many mountains formed by volcanoes are sites of beauty because the volcano is *extinct* (that is, inactive); its former vents are closed. A *dormant volcano* is one in which activity is only temporarily suspended. The threat of renewed eruption is always present and warnings are often brief. For instance, in 1991, Mount Pinatubo in the Philippines, after a series of tremors, erupted after 600 years of dormancy. The week-long eruptions left 200,000 people homeless, including U.S. military personnel and their dependents stationed at Clark Air Base and Subic Bay Naval Station. The residents are always alert for the earthquakes and tremors that usually accompany the awakening movement of magma under Earth's crust.

One of the most spectacular eruptions in modern history occurred the morning of May 18, 1980, with a release of *ejecta* (ash and rock) that rivaled the most famous eruption of all times—Mount Vesuvius, a holocaust that buried Pompeii and Herculaneum, two thriving Italian cities, in A.D. 70. In 1980, an area of 400 square kilometers (150 mi^2) was literally devastated as Mount St. Helens, a volcano that had been dormant since 1831, violently erupted. The famous peak, located in the beautiful Cascades of southwestern Washington state, blew itself apart in one of the most cataclysmic explosions ever recorded. The eruption hurled a major portion of the mountain into the air with a surge of energy surpassing that of an atomic bomb. After the ejection of *pyroclastics* or igneous rock fragments of various sizes, the release of accompanying gases, and the resulting shock waves, the eruption triggered the largest landslide and release of scalding mud ever witnessed. Spirit Lake, located at the base of Mount St. Helens and shown in Figure 24.3a, became a boiling pool of mud and was filled with thousands of shattered trees. The hot mud eventually flowed into the mighty Columbia River, raising its temperature and causing millions of fish, primarily salmon, to perish. The searing

a.

b.

c.

d.

Figure 24.3 (*a*) Spirit Lake reflected the beauty of Mt. St. Helens. The outstanding body of water owed its name to Indian lore about canoeists who mysteriously disappeared on it. (*b*) The eruption's heat melted vast amounts of ice and snow, and mud, ash, and debris of all sorts were carried into the valley below. (*c*) Searing gases, ash, and rock from the eruption spread to form an ominous cloud of death and destruction. (*d*) Mature trees over a vast area were uprooted by the forces released by the eruption.

cloud of ash and gases that accompanied the eruption suffocated about 55 people who survived the initial explosion as it shot outward at over 90 meters per second. The vast clouds of ejecta fell on major cities as far as 1200 kilometers from the eruption.

The eruption was not without warning, however. Following a series of tremors on March 27, 1980, modest bursts of ash and steam were ejected at irregular intervals. On April 23, an ominous bulge began to appear near the summit of the mountain. For the next three weeks the bulge increased in size to an area of over 1.3 square kilometers and rose constantly until it extended outward over 120 meters. On Sunday morning, May 18, 1980, Mount St. Helens was shaken by two strong earthquakes. The bulge quivered violently, creating a great avalanche of dirt, mud, and rocks that roared down the peak, destroying everything in its path (see Fig. 24.3b). A large mushroom cloud appeared that rose 20 kilometers into the atmosphere, and the fury continued until it left a gigantic crater about 3 kilometers long, 2.4 kilometers wide, and 1.6 kilometers deep in the mountain's north side (see Fig. 24.3c). The surrounding forests were knocked down and destroyed by the accompanying intense heat, as shown in Figure 24.3d. A magnificent display of lightning ignited forest fires well beyond the area of devastation. The air became so electrically charged that electric sparks reportedly encircled the ice axes of a party of mountain climbers on Mount Adams, almost 50 kilometers away. Even at this distance, the survivors who witnessed the event felt the intense heat of the eruption and were bombarded by ashes, rocks, and singed pinecones.

Mount St. Helens marks but one of over a dozen volcanoes of varying ages and degrees of activity that are located in the Pacific Northwest. Mount Hood, another towering Cascade peak near

Mount St. Helens and about 60 kilometers from Portland, Oregon, was the site of over 50 small tremors in July, 1980. It, along with several dormant volcanoes in Mexico and Chile, are under constant surveillance by geologists and other scientists.

El Chichón, a volcano that remained dormant for over a century, came to life in 1982. The eruption devastated a wide area in southeastern Mexico and killed 22 people. At least 10 times as much material was ejected from El Chichón as was ejected from the Mount St. Helens eruption of 1980. In fact, a cloud of material that entered the atmosphere from the Mexican volcano within a 3-month period created a cloud that extended over one-fourth of Earth's surface. The overall effect of the eruption on the planet's weather patterns is still under study. In the case of El Chichón, the material was ejected vertically, whereas the eruption of Mount St. Helens was from a lateral vent.

The eruption of Mount St. Helens was but a minor event compared to eruptions of the past, as measured by the volume of airborne debris. In fact, the eruptions of Mount St. Helens in 1900 B.C. and in A.D. 1500 were much larger than the 1980 explosion, which released about a cubic kilometer of ejecta. Indonesia's Tambora erupted in 1815, killing 12,000 people and releasing over 80 cubic kilometers of molten material, solid fragments, and gases into the atmosphere. The airborne ash and its effect on the atmosphere created "the year without a summer" in 1816. In 4600 B.C. the eruption of Mount Mazama in Oregon released about 40 cubic kilometers of ejecta. After the major eruptive activity subsided, Mount Mazama collapsed into the underlying chamber that the ejecta vacated, forming a *caldera,* a greatly enlarged volcanic crater. The 8-kilometer-wide crater eventually filled with water to form the beautiful Crater Lake.

The number of active volcanoes on Earth is estimated to be over 400. They are believed to occur in regions where Earth's crust is either sinking or rising along the plate boundaries, as discussed in Chapter 20. Most active or intermittently active volcanoes are on islands in the Pacific Ocean or along its coasts. In early 1990, Kilauea volcano, in the Hawaiian Islands, destroyed numerous homes as the intense heat from the flowing lava ignited them. A night scene during an eruption of this famous volcano is shown in Figure 24.4. Another volcano, Redoubt in Alaska, also underwent a series of eruptions about the same time after 25 years of dormancy. The eruptions from this volcano formed a plume of steam and gritty volcanic ash that reached an altitude of over 12,000 meters, disrupting air traffic in the Anchorage area at various intervals.

One of the most devastating volcanic eruptions in the past century occurred in November 1985. Colombia's Nevado del Ruiz volcano, a 5500-meter peak near Armero, woke from a dormancy of 390 years and destroyed several western towns located in the surrounding Andes mountain valley. The torrent of melted snow and mud that resulted buried practically everything in its path, leaving behind a death toll of over 20,000 people.

A major portion of the volcanic activity in Earth's lithosphere is directly associated with movement in the crustal plates. The violence of eruptions apparently depends upon the rate of crustal-plate movement and the chemical composition of the magma (or lava) at that location. Of equal importance, the violence is directly related to the amount of gases present in the magma, and how readily these gases can escape through the vents that form.

Figure 24.4 The Kilauea volcano spews forth steam and molten lava from deep within its interior.

Great tremors in Earth's inner region and across its surface have also provided many clues about the composition and structure of Earth's interior. An **earthquake,** as the disturbance is called, is the result of fundamental geological processes, never the actual cause. Locally a quake may set off landslides, avalanches, explosions, and the seemingly inextinguishable fires that usually accompany such major calamities in populated areas. The actual cause, as in the case of Mount St. Helens, may be the eruption of a volcano and the resulting great tremors that the vast release of energy produces, or, more typically, a sudden movement along a **fault**—*a fracture in the rock layers along which lateral (horizontal) or vertical displacement has taken place.* Many major faults have a constant rate of lateral or vertical displacement that measures over 2 meters per century. One of the most publicized faults in North America is the *San Andreas fault* that stretches along about 1000 kilometers of the western portion of California. The fault continues through the San Francisco area and then northwest under the Pacific Ocean. During the earthquake of 1906 that struck San Francisco, rock layers along the San Andreas fault underwent a sudden lateral displacement of over 3 meters. Also, the outer crust in other areas along the fault was ripped open by the major tremor (see Fig. 24.5). Movement along this fault in October 1989 triggered the earthquake that claimed scores of lives and damaged major highways, including a double-decked freeway that runs near Oakland, shown in Figure 24.6a. The destructive quake rocked the crowded ballpark nearby shortly before a World Series game was to be played. It was considerably stronger than several other major tremors that occurred in the same region within a two-year period.

The destruction caused by an earthquake depends primarily on such factors as origin, suddenness, frequency, duration, intensity, and population density of the immediate area. The major

a.

b.

Figure 24.5 (*a*) In the California earthquake of 1906, fences were offset about 3 m by the lateral movement along a fault. (*b*) The ground near Olema, California, was ripped open by the 1906 earthquake.

earthquakes occur along the plate boundaries. The more violent ones cause significant changes in Earth's topography. For instance, large fissures or cracks open Earth's surface, and massive downhill movements of mud, rocks, and soil engulf widespread areas. The great compressions caused by the movement of crustal plates toward or away from each other cause ridges or depressions to appear in Earth's surface. At other times, vertical faults several stories high are formed, created by uplifted fractures. Gigantic icebergs are set free as the tremors separate them from the glaciers to which they are attached. Among the most vivid results of an earthquake are large bodies of water, such as Reelfoot Lake along the Mississippi River in Tennessee, that are formed when the floodplain of a major river sinks.

Most of the widespread destruction from earthquakes results from the seismic sea waves, called *tsunami,* that are caused by violent disturbances under the ocean floor. The long, low waves travel at high speeds and suddenly break on the shores of islands and mainlands. Vast destruction accompanies the single surge of the waves, being most severe at the water's edge. The waves may also be disastrous to aquatic life in the nearshore environment. Extremely destructive tsunami struck Japan in 1703 and 1896 and the Aleutians at the close of World War II.

As many as 3000 earthquakes per day occur, about 450 of which may be felt by people near the *epicenter,* a point directly above the underground source, or *focus,* of disturbance. An estimated 100 earthquakes per year are severe enough to cause significant damage, again depending on the population density near the epicenter as well as on other factors.

One of the major destructive earthquakes of modern times, in terms of lives taken and property destroyed, occurred in late November 1980. This devastating tremor struck the densely populated area west of Naples, Italy. Over 3000 people were killed, and hundreds of thousands were left homeless. The ruins of the ancient city of Pompeii, buried by the eruption of Mount Vesuvius in A.D. 79 and excavated in the eighteenth century, suffered extensive damage from this modern quake. Numerous aftershocks further damaged the ancient ruins, as well as other structures that survived the initial devastating earthquake.

In late October 1983, a devastating earthquake struck a mountainous province of Turkey, destroying 50 villages and killing over 1200 people. At about the same time, an earthquake occurred in Idaho, almost equal in magnitude to the quake in Turkey. Tremors were felt over eight states and into Canada from the Idaho earthquake, yet damage was moderate and only two deaths resulted, since the area near the epicenter was sparsely populated. The quake did, however, create a scarp (cliff) 5 meters high, 40 kilometers long, and 5 meters wide. It was the strongest quake the United States had experienced in the three decades prior to the San Francisco-area earthquake of 1989. Several major quakes leveled portions of Mexico City in the fall of 1985, although the epicenters were about 400 kilometers west of the world's largest metropolis. The death toll reached 9500 in the dust-shrouded stricken area as over 400 buildings collapsed upon the residents.

The magnitude of earthquakes is measured with the *Richter scale.* The rating is derived from the amplitude of an earthquake wave and is a measure of the amount of energy released at the source of the earthquake. The scale ordinarily describes magnitude on a logarithmic base from 1 to 10. A disturbance of 7, for example, is ten times as great as one of 6 and one hundred times as severe as one of 5. However, the scale is open-ended; thus, it can measure the magnitude of earthquakes less than 0 or greater than 10. An earthquake with a magnitude as low as 2.5 on the Richter scale may be detected by persons nearby. A magnitude of 7 or greater indicates a major earthquake capable of causing extensive damage. The 1983 earthquakes measured approximately 7 on the Richter scale. The earthquake in southern Italy in 1980 was also rated at 7. An earthquake of this intensity, although deadly, is far from being as intense as others that have occurred in recent times. Unfortunately, the point of disturbance of the 1980 quake was near a populated area, and the type of construction typical of the region could not withstand the distortions caused by a major earthquake. In comparison, the 1985 earthquakes in Mexico had a maximum intensity of about 8.1. The 1964 earthquake that struck Anchorage, Alaska, had a reading of 8.6 on the Richter scale and was estimated to be twice as severe as the destructive earthquake at San Francisco in 1906. The resulting tsunami caused damage to areas as high as 30 meters above sea level, as seen in Figure 24.6b. Many structures in this earthquake-prone region of Alaska were built on the bedrock, and they did not succumb to the great tremors to the degree of those built on less secure foundations. In fact, no loss of life was reported as a direct consequence of the Alaskan earthquake, yet the uplift and subsidence of Earth's surface exceeded 10 meters and affected about 88,000 square kilometers of Alaska's south-central region. The proper selection of building sites, along with selective construction materials and design, appears to minimize the damage

a.

b.

Figure 24.6 (*a*) A tragic scene caused by the earthquake that struck the San Francisco Bay area in October 1989. The freeway collapse occurred in the Oakland area shortly before the scheduled start of a World Series baseball game. (*b*) The Alaska earthquake of March 27, 1964. Tsunamis washed many vessels and parts of harbor structures into the heart of Kodiak, Alaska.

brought about by the Earth's sudden tremors. Also, the terrain's characteristics and conditions are significant factors in determining how much damage will be done. On October 10, 1986, San Salvador, a city in El Salvador, was struck by an earthquake that measured but 5.4 on the Richter scale, yet mudslides from the hills saturated by violent rainstorms and toppled buildings killed approximately 1500 people and left perhaps 250,000 homeless. An earthquake of 6.8 under similar conditions caused mudslides that buried 1000 people in Equador on March 7, 1987. Still another such disaster occurred in August 1988 during the monsoon season along the India-Nepal border region. An earthquake, rated at 6.5 on the Richter scale, triggered a series of landslides and floods that claimed about 1000 lives and destroyed over 3000 homes.

A devastating earthquake in Soviet Armenia in December 1988 leveled numerous cities and claimed the lives of about 25,000 people, leaving hundreds of thousands homeless. The quake, rated at 6.9 on the Richter scale, had an epicenter near Spitak, a city of about 16,000, which was totally devastated.

An even more destructive earthquake, estimated at 7.7, struck northern Iran in June 1990, demolishing scores of villages. The landslides set off by the great tremor, along with the aftershocks that followed in conjunction with torrential rains, hindered rescue attempts for days. The death toll reached over 50,000 with an estimated 500,000 left homeless. This tragedy marks the largest loss of life from earthquakes during the twentieth century.

An earthquake of equal magnitude struck a somewhat remote mountain resort in the Philippines in July 1990. About 1000 people lost their lives even though miners brought to the scene worked frantically to locate survivors in the remains of homes and buildings. Again, aftershocks slowed rescue operations, but one person survived for two weeks in the rubble of eight hotels that collapsed. A comparison of the mortality rates of this earthquake and the one in Iran again illustrates how population density and other factors affect the seriousness of such an uncontrollable natural event.

Sunspots, tides, planetary positions, and weather conditions are under observation as possible causes of Earth's internal disturbances and the tremors that result. Also, according to the theory of plate tectonics, some plates slip and slide laterally against one another rather than vertically, a phenomenon that may cause major tremors. Such lateral movement is reportedly occurring today at the mouths of the Red Sea and the Persian Gulf and is suspected of being the major cause of earthquakes and volcanic eruptions along the plate boundaries in this part of the world.

Although seismographs can record the shocks and slight ground movements that usually precede a major earthquake, scientists have not yet been able to predict with great accuracy the actual time when a tremor of significance will occur. However, the recent NASA satellite project LAGEOS is designed to establish an early warning system and may provide the information needed to make accurate predictions of forthcoming major earthquakes. Also under investigation is the reported increase in the electrical resistance of rocks under great stress and the decrease in seismic-wave velocity as the shock waves move through them. Both effects seem to be related to *dilatancy,* the opening of microscopic cracks in rocks that are exposed to tremendous pressure. Other efforts at earthquake prediction include a comparative study of the concentration of radioactive radon gas in Earth's interior, as well as that of anomalous soil-gas levels, including the helium formed from alpha particles released by the decay of naturally occurring uranium in rock. Scientists think that if crustal stress drops before a shock, the pores in rocks could dilate, allowing more gases to escape. Some researchers have also reported that they have developed a monitoring technique that has detected sudden changes in a vertical electric field prior to several earthquakes. These pulses of electromagnetic radiation are assumed to be associated with microfracturing of rock layers. Still other scientists are using lasers to complete a precise land survey that can be compared with similar surveys in the future. A significant change in the relative positions of landmarks may signify an impending tremor. The efforts of the scientific community to develop reliable earthquake prediction techniques continue.

Some people question the benefits of earthquake prediction and fear that if the forecasts are made public, the panic, looting, and traffic jams that would result could prove more costly than the earthquake itself. Others strongly oppose this point of view. Most casualties, they note, are caused by collapsing buildings. There would also be great loss of life in low areas if large dams were caused to rupture. These people generally propose stronger building codes and recommend that funds be provided to reinforce existing structures in areas of high earthquake activity.

Some scientific study is being directed toward ways to control earthquakes. A possible means was suggested in the late 1960s, when it was observed that small tremors accompanied the injection of water under pressure into oil wells to increase their productivity. The small quakes occurred for the four months during which the project was underway and ceased almost immediately as the water and oil were pumped out. This technique applied along converging crustal plates may prevent major earthquakes through the release of stresses in the form of small tremors. Also, the surrounding rock would effectively be strengthened and the fault along the plates presumably would be locked in place.

Although earthquakes have had a decided effect on the topography of Earth's surface, they are of minor significance compared to the slow, gentle processes of erosion and folding that have played the major roles in sculpturing our lands as we know them.

Minerals

Atoms, elements, molecules, and compounds were discussed in detail in Chapters 12 through 15. In nature, atoms of the various elements are combined to form most of the *minerals.* A **mineral** is *a naturally occurring solid element or inorganic compound that has an ordered internal structure and a chemical composition that can vary only within narrowly defined limits.* Each mineral has a set of physical properties that are also fixed within limits. A substance must meet all criteria to be properly classed as a mineral. Substances such as coal, gas, and oil are classed as *mineral fuels* rather than as minerals since these substances contain predominantly organic materials.

Table 24.1

Various common minerals can be classified into chemical groups.

Chemical group	Mineral name	Chemical composition
Oxides	Hematite	Fe_2O_3
	Magnetite	$FeFe_2O_4$
Sulfides	Pyrite	FeS_2
	Galena	PbS
Sulfates	Gypsum	$Ca(SO_4) \cdot 2H_2O$
	Anhydrite	$Ca(SO_4)$
Carbonates	Calcite	$Ca(CO_3)$
	Dolomite	$CaMg(CO_3)_2$
Halides	Halite	$NaCl$
	Fluorite	CaF_2
Silicates	Quartz	SiO_2
	Olivine	$(Mg,Fe)_2SiO_4$
(Micas)	Muscovite	$KAl_2(AlSi_3O_{10})(OH)_2$
	Biotite	$K(Mg,Fe)_3(AlSi_3O_{10})(OH)_2$
	Chlorite	$(MgFeAl)_6(AlSi_4O_{10})(OH)_8$
	Talc	$Mg_3(Si_4O_{10})(OH)_2$
	Kaolinite	$Al_2(Si_2O_5)(OH)_4$
(Feldspars)	Orthoclase	$K(AlSi_3O_8)$
	Plagioclase	Ca and Na Al silicate
	Albite	$Na(AlSi_3O_8)$
	Anorthite	$Ca(Al_2Si_2O_8)$

Geologists have collected and identified over 2600 minerals, and they discover about 10 new ones each year. About 50 of the known minerals are elements; the rest are compounds that contain two or more chemically combined elements. Most of the minerals can be classified into major chemical groups, such as those shown in Table 24.1.

Most minerals are *crystalline solids,* a feature that indicates that these specimens have an ordered internal structure. Such familiar minerals as diamond, mica, and quartz have a well-defined internal structure, and the angles between corresponding faces are constant regardless of specimen size. Other minerals, such as limonite and opal, are classified as noncrystalline or *amorphous solids.* Their atoms are randomly arranged; hence, they do not have crystal faces or the other directional properties of crystalline solids. In addition, some minerals form very fine-grained crystals, and others have such poorly crystallized forms that crystal faces are not evident. The precise internal structure of a mineral can be determined by the use of **X-ray diffraction,** *a technique that determines the manner in which X rays are spread or bent by a crystal lattice and indicates the orderliness of the spacing between atoms or ions in the crystal lattice of the specimen.*

The most abundant elements in Earth's crust are listed in Table 24.2. By weight, more than 98 percent of the rocks and minerals of the crust are formed from the eight elements listed, and the same elements account for all but a very small amount of the crust in terms of volume. Obviously, most of the minerals that form rocks are composed predominantly of oxygen and silicon. In addition, aluminum and one of the other common elements, such as sodium or potassium, are generally present. Minerals that form rocks contain about 90 percent oxygen by volume. This condition exists not only because of the relative abundance of oxygen, but also because of the size of the oxygen atom (ion) compared to that of silicon, iron, and the other major elements. The possible combinations in which oxygen can arrange itself along with other atoms depend on the respective radii of the atoms that attach to the oxygen atom. For instance, the radius ratio of silicon to oxygen is such that four oxygen atoms can attach to it (see Fig. 24.7). This unit, SiO_4, is the building block of the vastly abundant family of silicate minerals. In all the silicate minerals, silicon and oxygen are joined in a tetrahedral (four-sided) arrangement. Oxygen frequently reacts with atoms much smaller than the silicon atom.

Table 24.2

The major elements present in Earth's crust.

Element	Abundance by weight (percent)	Abundance by volume (percent)
Oxygen	46.60	93.77
Silicon	27.72	0.86
Aluminum	8.13	0.47
Iron	5.00	0.43
Calcium	3.63	1.03
Sodium	2.83	1.32
Potassium	2.59	1.83
Magnesium	2.09	0.29
All others	1.51	Negligible

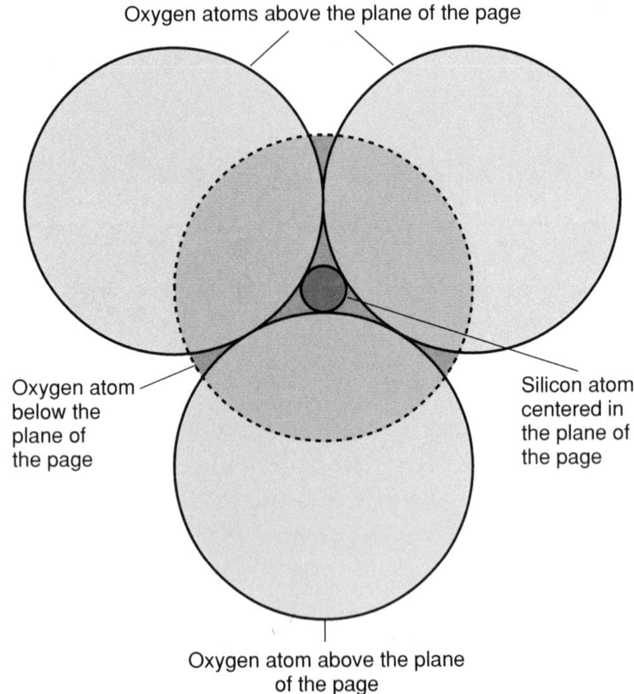

Figure 24.7 The tetrahedral arrangement of a silicon atom and four oxygen atoms (SiO_4) is a common occurrence in nature.

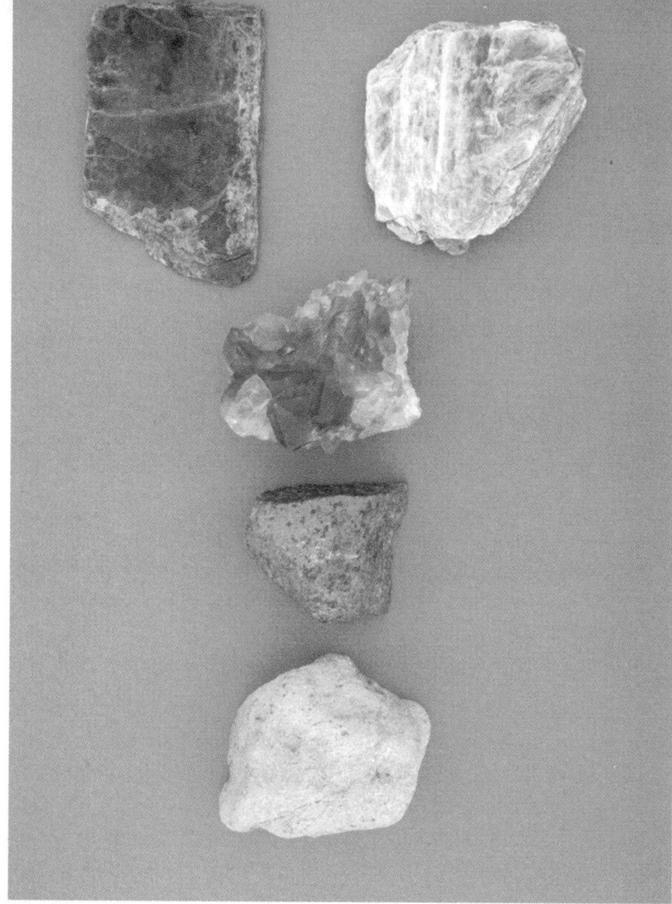

Figure 24.8 Each type of mineral in this photograph has a different internal structure. From top to bottom: biotite and muscovite micas, quartz, horneblend, and olivine.

The silicon-oxygen tetrahedrons exist as single units in some minerals and as multiple units in others. The difference in internal arrangements, such as in number and pattern, causes the minerals involved to have varying physical properties. Some examples of minerals with different internal structures appear in Figure 24.8.

The most common silicate minerals are the feldspars. This class of minerals is closely related to quartz in structure. However, aluminum atoms replace some of the silicon atoms in the tetrahedrons, and the elements potassium, sodium, and calcium are also present.

The identification of minerals and the rocks that they form is accomplished in several ways. X-ray studies are perhaps the most effective, but other methods are more commonly used. In one type of determination, known as petrographic analysis, sections about 0.3 millimeters thick are prepared and viewed under the microscope. Strangely enough, most specimens are transparent at this thickness, and a precise study of the crystals present can be accomplished.

Geologists have been able to identify minerals by means of correlated physical properties. In the following list, some of the more common properties used in mineral identification are described.

1. The internal structure of minerals affects the manner in which they tend to break. Some of the specimens have a tendency to break in definite directions along smooth planes of weakness in their internal structure. This property is known as *cleavage*. Figure 24.9 shows specimens of mica, feldspar, halite, and fluorite, which have one, two, three, and four cleavage planes, respectively.

Figure 24.9 The number of cleavage planes differs in each specimen shown in the photograph. Top, from left to right: mica (one cleavage plane); feldspar (two). Bottom, from left to right: halite (three); and fluorite (four).

2. A mineral that lacks cleavage usually breaks in some irregular manner. A common type of *fracture,* called *conchoidal,* is a hollow, arclike break. This type of fracture is found in quartz and in the rock obsidian. Specimens of other minerals, such as asbestos, separate (fracture) into fibers.
3. The *color* of freshly broken surfaces of some minerals may be an identifying property. Specimens such as galena and magnetite are reliably determined by color, whereas quartz, calcite, and many others usually contain impurities that produce a wider range of color in the specimen.
4. When crushed to powder form, many minerals appear to be white or a paler color than they would appear normally. Others in powdered form take on a color quite different from the color they appear to be when viewed as large fragments. The investigator carries a piece of rough, unglazed porcelain, which will powder most minerals as they are rubbed on its surface. The color of the powder that results is called the *streak*.
5. *Luster* is the manner in which a mineral reflects light. The two main types of luster are metallic and nonmetallic. Among the terms used to identify the luster of a specimen are the following: bright or dull metallic, nonmetallic, vitreous, earthy, silky, pearly, waxy, or resinous. Many other terms have been proposed and are used in determinative mineralogy.
6. Minerals resist scratching in varying degrees. This property of the mineral is called *hardness*. It is a difficult property to assess, since many variables are involved, including the amount of force used, the shape of the scratches produced, and the variations that might occur on different edges and faces. The mineralogist has established a relative scale of hardness, called *Mohs' scale* after the geologist who proposed the method of measure in the nineteenth century. The scale follows, and typical specimens are pictured in Figure 24.10.

Figure 24.10 Specimens that represent Mohs' scale of hardness. From top left to bottom right: talc, gypsum, calcite, fluorite, apatite, orthoclase feldspar, quartz, topaz, corundum, and diamond.

Figure 24.11 The rock-forming minerals found in Earth's crust. From top left to bottom right: orthoclase feldspar, plagioclase feldspar, quartz, pyroxene, muscovite mica, biotite mica, gypsum, halite, calcite, chlorite, serpentine, dolomite, and kaolinite.

1—Talc	6—Orthoclase
2—Gypsum	7—Quartz
3—Calcite	8—Topaz
4—Fluorite	9—Corundum
5—Apatite	10—Diamond

In order to determine the hardness of a mineral, the investigator uses the hardness scale to identify the softest mineral that will scratch the specimen. The geologist in the field may carry a kit that contains all the minerals listed on Mohs' scale of hardness, excluding the diamond, for obvious reasons. In addition, he or she generally makes use of other common materials with which to determine the relative hardness of a substance, such as a fingernail (2.5), a copper penny (3.5), and a knife blade (5.5).

7. Another reliable property of minerals used in identification is *specific gravity,* the mass of a given volume of a mineral compared to the mass of an equal volume of water. Quartz has a value of about 2.6–2.7; gypsum, 2.2–2.4; olivine, 3.2–3.6; and magnetite, 5.0–5.2. If metric (SI) units are used in the calculations, density and specific gravity would be numerically the same. Specific gravity has no units in any system of measurement since it is a ratio of two densities, and all units thus cancel.
8. Other useful properties that are commonly tested include electrical conductivity, magnetic tendencies, fusibility, solubility, reactivity, and the tendency of a substance to fluoresce under ultraviolet light.

All the minerals that have been identified are present among the rocks; however, only a few dominate and are the essential constituents. These very common minerals form a group called the *rock-forming minerals.* Some of the more abundant minerals that form the preponderance of the rocks are illustrated in Figure 24.11. The list includes feldspars, micas, gypsum, halite, and calcite, among others.

An *ore* is a mineral or a rock from which various metals and nonmetals can be profitably extracted. Water and volatiles, including molten quartz present in magma, may have caused formation of veins of ore minerals. The outer regions of batholiths are commonly examined for valuable ores since these areas cooled rapidly and solidified first. As the interior of the body cooled, many of the volatiles were separated from the magma, a condition that created a mechanism through which many of the elements could move and concentrate. This group of elements that do not fit the internal structure of the abundant rock-forming minerals includes gold, copper, lead, zinc, silver, and sulfur. They were carried as suspensions between the large crystals in the melt and were deposited in the lower extremities of the batholith by gravity. Other small particles of these elements were forced to concentrate along the edges of the great magmatic bodies as the fluids that contained them were squeezed outward by contraction and expansion of the larger minerals.

The miners of the ore minerals search for large igneous bodies and excavate them for the valuable veins they may contain. The veins take the form of thin, generally horizontal layers. They are often composed of the metallic elements that have reacted with sulfur and oxygen, both of which were contained in the gases that formed the volatiles. Iron atoms that reacted with oxygen formed magnetite, hematite, and limonite. Iron and sulfur also combined to form pyrite, a common crystal known as fool's gold because of its resemblance to the valuable element. Galena, the major ore of lead, is formed as lead sulfide and often contains silver as an impurity. Cassiterite (tin oxide), sphalerite (zinc sulfide), and bauxite (hydrous aluminum oxide) are the common ores of tin, zinc, and aluminum, respectively. The ores of uranium are quite valuable at the present time; pitchblende and carnotite are the most yielding and abundant ores of this fairly rare element. Pitchblende is basically related to magmas and is sought in regions where igneous rocks are abundant. Carnotite is found in sedimentary rock deposits, such as sandstone.

Rocks in General

The crust, the solid outer portion of Earth's lithosphere, is composed primarily of rocks. With few exceptions, **rocks** are *naturally formed aggregates of one or more minerals.* They ordinarily lack

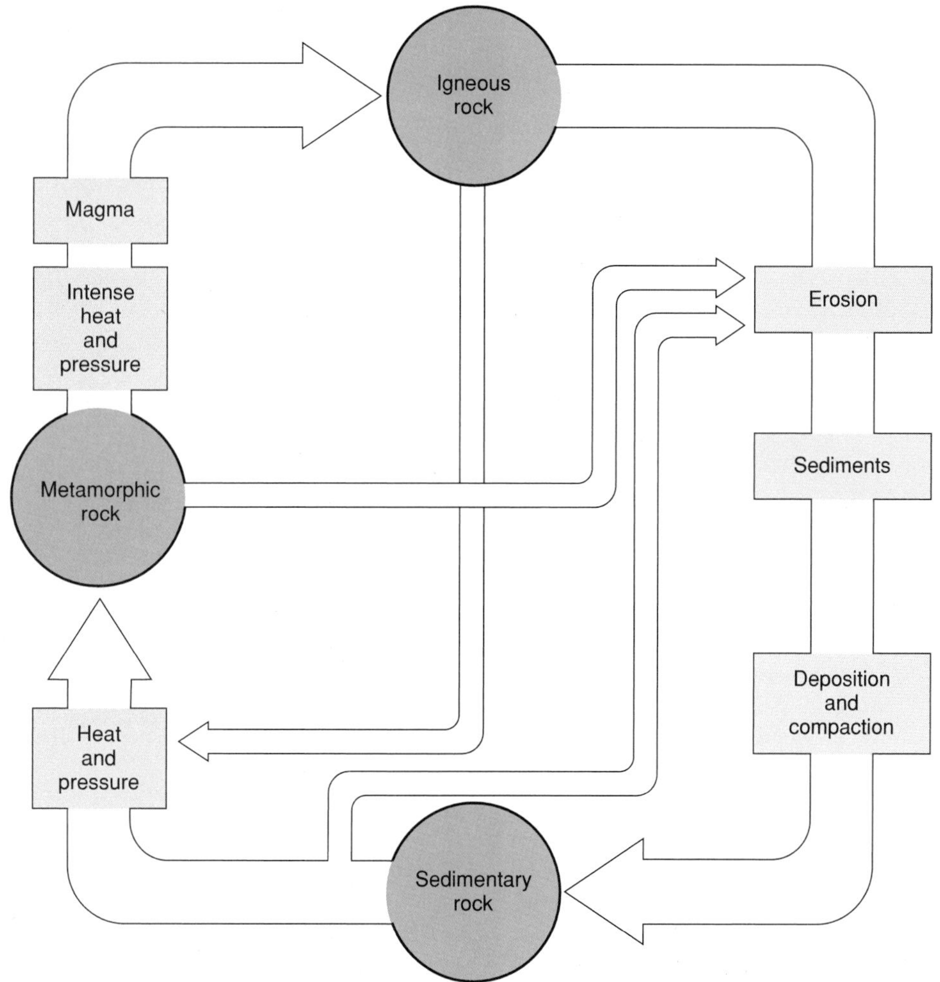

Figure 24.12 This schematic of the rock cycle shows the way in which different types of rocks are formed from other types of rocks. Igneous rocks and metamorphic rocks, through erosion, form sedimentary rocks. Metamorphic rocks are formed from igneous, sedimentary, or other metamorphic rocks. The melting of sedimentary or metamorphic rocks forms various igneous rocks to complete the cycle.

a definite chemical composition and given properties. The wide variations in physical properties and appearance that best describe each specimen depend on the amount and kind of minerals present, as well as the way in which the grains are held together.

Geologists are interested in rocks largely because of the information they provide about the physical conditions that existed on Earth when the rocks were formed and deposited. The deciphering of the rock record has yielded fascinating facts about Earth's history. The clues that the geologists have interpreted paint a picture of major calamities, such as devastating earthquakes, vast glaciation, great floods, and arid climates that occurred at various times. In Scotland, where the city of Glasgow now stands, was once a desert, and areas of Canada were warm enough in the past for tropical plants to flourish. However, the preponderance of rock layers, according to most geologists, was laid down by slow, uniform procedures and not by major catastrophies.

The rocks of which Earth's crust is composed are classified into three types in terms of their origin. Two of the types of rocks, igneous and metamorphic, were formed by geologic processes at various levels within Earth's crust. **Igneous rocks** are those that derived from a melt, and **metamorphic rocks** are those that underwent physical changes by virtue of high temperatures and great pressures to which they were exposed under Earth's surface. The third type, **sedimentary rocks,** were deposited on or near the surface of Earth by the action of water, wind, ice, and other mechanisms. Sedimentary rocks are mostly formed on Earth's surface from accumulations of gravel, mud, and sand, all of which are products of erosion of preexisting rocks. Still others of this type, such as limestone and gypsum, are composed mostly of material deposited from solution. Metamorphic rocks are, in effect, rocks of any type or composition that have been altered by great heat and pressure. The variety of metamorphic rocks that can exist, then, is unlimited. Possible interactions among the three types of rocks are illustrated in Figure 24.12.

Igneous Rocks

Igneous rocks result from the cooling of hot magma and lava. Most members of this category are formed from high-temperature melts at varying levels in Earth's crust. (Recall the previous discussion of extrusive and intrusive rocks.) The major classification of igneous rocks is done by the determination of their origins, textures, and mineral composition, the last of which reflects a specimen's color and density. *Texture,* the physical appearance of a rock that

Table 24.3

A classification scheme for selected igneous rocks. Volcanic rocks contain small but distinct variations in mineral composition, sometimes detectable only through chemical analysis.

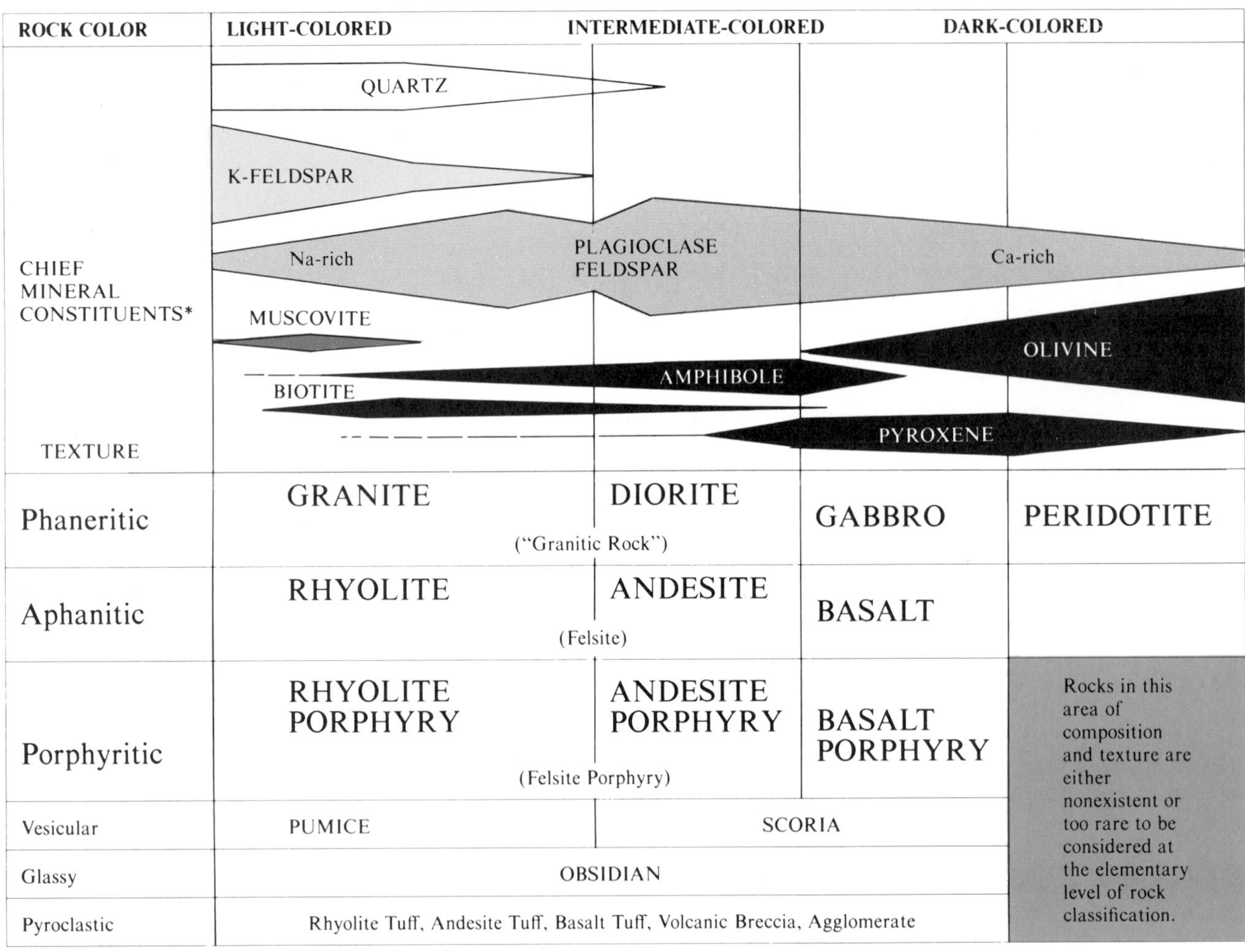

ROCK COLOR / TEXTURE	LIGHT-COLORED	INTERMEDIATE-COLORED	DARK-COLORED	
Phaneritic	GRANITE ("Granitic Rock")	DIORITE	GABBRO	PERIDOTITE
Aphanitic	RHYOLITE (Felsite)	ANDESITE	BASALT	
Porphyritic	RHYOLITE PORPHYRY (Felsite Porphyry)	ANDESITE PORPHYRY	BASALT PORPHYRY	Rocks in this area of composition and texture are either nonexistent or too rare to be considered at the elementary level of rock classification.
Vesicular	PUMICE	SCORIA		
Glassy	OBSIDIAN			
Pyroclastic	Rhyolite Tuff, Andesite Tuff, Basalt Tuff, Volcanic Breccia, Agglomerate			

involves the size, shape, and arrangement of its constituent grains, represents a most useful means of identification. Chemical composition is also used to classify the rocks whenever practicable. With the exception of volcanic glass, igneous rocks are crystalline aggregates that have formed by congealing rapidly or slowly from a melt. A combination of crystalline texture and silicate composition is typical of igneous rocks, since silicate minerals crystallize over a broad temperature range. The light-colored rocks of igneous origin tend to be composed of sialic (silicon-aluminum) minerals, including quartz, orthoclase feldspar, and muscovite mica. Those of darker colors contain simatic (silicon-magnesium) minerals of the ferromagnesian group, including augite, hornblende, olivine, and biotite. The variations in igneous rocks according to texture and mineral composition are shown in Table 24.3.

Norman L. Bowen (1887–1956) devised a scheme in the 1920s to explain how the many varieties of igneous rocks were formed from magma. He heated various minerals in the laboratory to create a silicate melt composed of elements in about the same proportions as might occur in typical magma. As the artificial magma cooled, he noted when various igneous rocks first appeared and concluded that the order of crystallization of igneous rocks is a function of decreasing temperature. The sequence is consequently known as the *Bowen reaction series*. A simplified schematic of the formation order is presented in Figure 24.13. According to the diagram, crystallization of the minerals takes place along two discrete paths. As liquid magma slowly cools, minerals at the top of the ferromagnesian branch on the left and the plagioclase branch on the right crystallize within the melt. The first ferromagnesian mineral to form as the temperature of the magma decreases is known as olivine. When the temperature of the remaining magma decreases further, the melting point of pyroxene is reached. At even lower temperatures, amphibole, then biotite, crystallize from the melt. After all of the biotite crystallizes, the remaining magma is void of the elements iron and magnesium. At approximately the

The Analysis of Rock Sample 12013

Lunar landings by U.S. astronauts have given the world's leading scientists a "hands-on" experience that will undoubtedly remain unprecedented for decades to come. Each mission brought back samples of the moon's crust, along with photographs and instrument data that have been used in an attempt to reveal many of the secrets about the lifeless but interesting satellite of Earth.

There is a high degree of similarity between lunar rocks and those of Earth's crust. Scientific investigations have confirmed that the lunar "seas" consist largely of basalt and fine-grained gabbro. The dominant material in the highlands is anorthosite, a lightweight igneous rock rich in plagioclase feldspar. Unlike most rocks found on Earth, lunar rocks have no water molecules chemically bound in them.

Geologists involved in NASA's Apollo program contend that, of all the known rock samples studied by scientists worldwide, a 4.1 billion-year-old lunar specimen brought back from the moon by an *Apollo* crew was examined with the greatest intensity. This lunar rock, known as "Sample 12013," was cut into small pieces and analyzed for age, trace elements, radioactive content, organic constituents, and other properties too numerous to mention. The results of the research performed on the samples were shared with scientists throughout the world at various international symposia. The widespread interest generated by this small sample of lunar breccia revealed the impact that planetary exploration has on all fields of scientific endeavors.

Even after several lunar expeditions, many controversies remain. Studies of all aspects of lunar geology completed at the present do not yield conclusive answers to such problems as the moon's overall age, its evolutionary pattern, and its origin. The exploration of the moon and other celestial bodies in our solar system presents a tremendous challenge to geologists. The laws, techniques, and conclusions held as valid by these scientists must prove to be applicable on a much larger scale than the scale on which they were based. Geophysical concepts about Earth, if correct, should apply to other celestial bodies as well ■

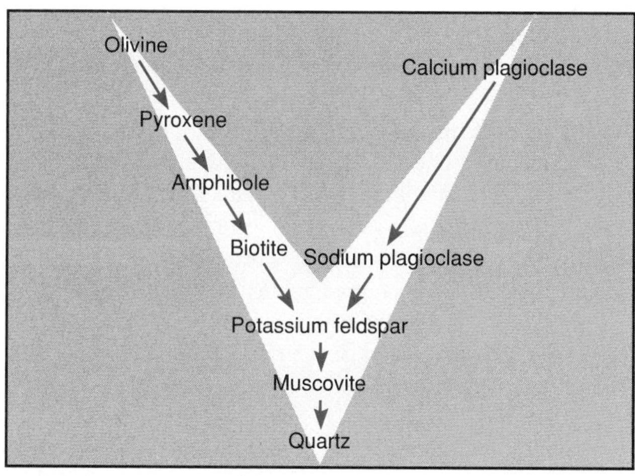

Figure 24.13 A schematic representation of Bowen's reaction series. If cooling is slow, the first-formed minerals do not separate from the magma but become chemically incorporated in the new minerals that form at cooler temperatures.

same time, the branch on the right of the diagram is undergoing a continuous crystallization process, with plagioclase feldspar representing the only mineral formed. The plagioclase with the highest calcium content is the first to crystallize as the magma cools. As the temperature of the remaining magma decreases, this plagioclase interacts with the remaining melt to form the variety of calcium and sodium feldspars associated with igneous rocks. The formation of plagioclase ceases once these two elements present in the magma are used up. The remaining melt crystallizes to form muscovite, orthoclase feldspar, and quartz. Typical specimens of the minerals included in the Bowen reaction series are shown in Figure 24.14. A brief discussion of various igneous rocks formed from these minerals follows.

1. Basalt is the most abundant of all extrusive igneous rocks. It has a fine-grained (*aphanitic*) texture and is formed as lava congeals after a volcanic eruption or as lava emerges from fissures in the volcano. Basalt is considered the subcrustal layer upon which the continents rest. The rock has a dark color and constitutes the upper part of Bowen's reaction series.
2. Gabbro is an igneous rock that is dark in color and has a coarse-grained (*phaneritic*) texture. It is composed primarily of plagioclase feldspar and pyroxene; therefore, it can form at high temperatures.
3. Rhyolite is the most abundant volcanic rock that has a composition high in potassium and sodium. It has an aphanitic texture and is light in color because of its low iron and magnesium content.
4. Granite is another igneous rock of phaneritic texture. It is the second most abundant igneous rock. Its two most common minerals are potassium feldspar and quartz. It is typically light in color and forms at relatively low temperatures. Granite grades into other igneous rocks, such as diorite and gabbro, as the content of feldspar increases and the amount of quartz decreases.
5. Obsidian is known as volcanic glass. It is one of the few rocks not composed of minerals, since none crystallize under the conditions in which it forms. It is characterized by its dark color and conchoidal fracture.

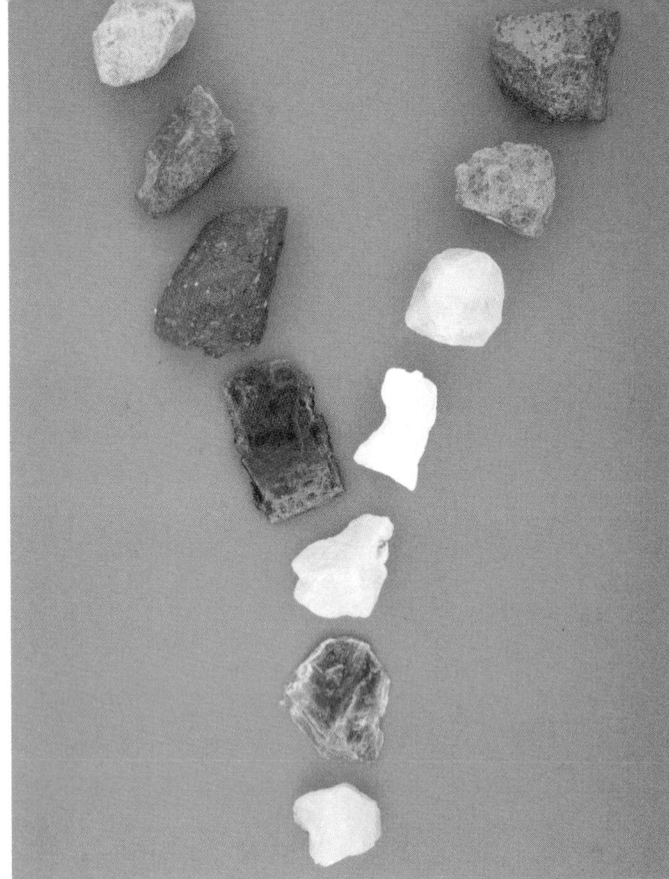

Figure 24.14 The igneous rocks shown comprise the Bowen reaction series. Crystallization of the upper members of the series occurs when the hot silicate magma cools to about 1400°C. The minerals in the lower part of the series solidify at about 570°C.

Sedimentary Rocks

As the name implies, a sedimentary rock is formed from sediment, an accumulation of mineral and/or organic matter that generally results from the degradation of rock materials. The small rock fragments that are produced are mechanically transported from the site of the original rock formation. Other types of sediment may have settled from a liquid as a chemical precipitate while actively being carried in solution, primarily by rivers and streams. Still other types of sediment were laid down biochemically, such as on ocean floors above which marine life once thrived.

The development of sedimentary rock from sediment occurs in discrete stages. First, the sediment is prepared for transport through chemical or physical weathering. Next, it is transported by agents of erosion such as running water, moving ice, wind, gravity, soil flow, or organisms to a site of *deposition,* the process by which the sediment is laid down to form sedimentary deposits. (Erosion and its agents are discussed in detail in the next chapter.) After being transported, the sediment is deposited through loss of velocity of its carrier, loss of volume, gravity, melting, or evaporation. Finally, the agents of *lithification,* which transform the sediment into rock, take over. Some of these agents are cementation, compaction, chemical changes, organic activity, and recrystallization.

Figure 24.15 Common sedimentary rocks. Top left to bottom right: shale, coquina, coal, conglomerate, fossiliferous limestone, rock gypsum, and sandstone.

The sites of deposition include the bases of various slopes, glaciated regions, and streams, as well as the mouths of rivers where large deltas form. Wind may deposit sediments in arid regions and along coastal areas.

Some of the most common sedimentary rocks are discussed briefly here and pictured in Figure 24.15.

1. Shale is the most common of all sedimentary rocks. It is formed from densely packed clay, mud, or silt and easily separates into layers. The composition of shale is essentially quartz, minerals of extremely fine grain, and micas. As one might suspect, most layers of shale contain an abundance of the remains of past life-forms.
2. Sandstone is composed of grains of sand that contain quartz as the common mineral. The color of a specimen is determined by the mineral that cements the grains together; it is generally red, brown, or green.
3. Conglomerates are cemented gravel-sized sediments or fragments that are often predominantly quartz. The fragments vary in size, with the space between larger particles filled with grains of sand.
4. Limestone is commonly formed by the chemical precipitation of calcium carbonate, a compound formed from the decomposition of shells and skeletons of marine organisms. Whole or partial shells may be preserved in the limestone.
5. Dolomite resembles limestone in many ways; however, magnesium has replaced the less active calcium present in limestone—a variation in composition that causes dolomite to be harder and heavier.
6. Gypsum is deposited in thick layers that alternate with other sedimentary rocks, which are also formed from the evaporation of salt water. Beds of rock salts, accompanied by gypsum, are found in Texas and New Mexico.
7. Chalk is the remains of small organisms. Large fragments of seashells are cemented together by other sediment to form *coquina.*
8. Coal is classified as a biogenic sedimentary rock—one formed by the accumulation and compaction of plant

matter. Coal deposits begin as thick, wet aggregates of slightly decomposed plants of all sorts. This spongy mass is called *peat* and is being formed today where proper conditions exist. As further decomposition and compaction continue, the plant matter undergoes chemical changes, giving off carbon dioxide and other gases to become *lignite,* a soft, brown coal. Additional compaction, brought about by the weight of other sedimentary layers, compresses the lignite to but a small fraction of the original thickness of the layer, forming the hard, black rock called *bituminous coal.* In eastern Pennsylvania and a few other regions, the coal layers were exposed to even higher temperatures and compressional forces. The deposits have been further compacted and more of the remaining gases removed, forming a valuable type of coal called *anthracite.* (This metamorphic rock is discussed in the next section.) Most of our coal deposits are located in Illinois, Kentucky, Ohio, Pennsylvania, and West Virginia. Over 300 million years ago, a major portion of these states was inundated by seawater, and between floodings primitive forests flourished on the swampy grounds. As millions of years passed, the geologic processes converted the decaying plant matter into deposits of this valuable natural resource.

Metamorphic Rocks

Metamorphic rocks may have any origin, but all have been visibly changed by great pressure and/or temperature while the rocks were in a solid state or by fluids that seeped through the solid rocks under these same conditions. Metamorphism involves any process that, as a result of geologic environment, produces changes in original texture or leads to new minerals, or both. The intensity of the metamorphism may be so great that the nature of the original rock is difficult to determine.

The changes in the rocks include partial or complete recrystallization, whereby the grain size of the minerals may become coarser and the crystalline alignment may differ from that of the parent rock. Banding or other parallel arrangements in structure that were not present in the original rock also develop. This property, referred to as *foliation,* causes the rocks to separate along their parallel sides; therefore, they possess rock cleavage.

Most metamorphic rocks are similar in composition to the rocks from which they were transformed. However, the process of *recombination* can create new minerals when rocks of different chemical composition interact.

Several factors determine the character of the rocks exposed to metamorphic processes. These include the composition and structure of the original rock, the processes involved, and the degree to which the processes exploit the original rocks. Some typical metamorphic rocks are described in the following list, and specimens of each are shown in Figure 24.16.

Figure 24.16 Common metamorphic rocks. Top left to bottom right: anthracite, gneiss, marble, quartzite, schist, and slate.

1. Anthracite, or hard coal, is the metamorphic form of coal. It is supplied by an area 200 kilometers eastward into the Appalachian Mountain region from large deposits of bituminous coal. Anthracite is difficult to ignite but is essentially dust-free and smokeless. It is of greater hardness than bituminous, lacks cleavage, and has a higher luster. Both types of coal occur in layers between shale and sandstone beds and were laid down in the same geologic period, the Pennsylvanian. The differences in the types of coal are due to the folding and resultant deformations that occur during mountain building, including the increase in temperature and pressure to which the original layer is exposed. Also, the gases have escaped from this layer more slowly, and the coal has undergone changes in chemical composition and physical properties. Anthracite, then, is a metamorphosed offspring of bituminous coal.
2. Gneiss (pronounced "nice") is the most coarsely banded metamorphic rock. Most gneiss contains bands of feldspar and quartz that are composed of unusually large granular particles. The specimen is light in color unless biotite mica, hornblende, or garnet is present in abundance.
3. Marble is formed from the metamorphism of limestone, dolomite, or both. The resultant specimen generally contains streaked light patches alternated with darker ones. It commonly contains pyroxene, amphibole, and serpentine in varying quantities.
4. Quartzite results from the alteration of quartz sandstone. Its appearance is glassy, and it has a texture resembling sugar. Unlike sandstone, it breaks across the grains rather than around them. Its great hardness and high resistance to chemical weathering make it ideal for use as a building material.
5. Slate is densely fine-grained and finely foliated; thus, it separates into thin sheets. The grains are visible only under a microscope, and the specimen shows some properties common to the sedimentary rocks in that it is layered and contains fossils. Most slate is the metamorphic form of shale. The black variety contains carbon, whereas the red and green types contain ferromagnesian minerals.
6. Schist is similar to gneiss except that the schist is foliated in thinner layers and may be dominated by micaceous

materials. Mica schist is the most common metamorphic rock. Most of this type of schist is the result of transformed shale and volcanic tuffs; therefore, it is similar to slate but coarser grained.

The Physical Laws of Geology

The early scholars who studied Earth had several purposes in mind: to determine the likely regions where the useful minerals might lie, to uncover the secrets of the history of Earth, and to learn about Earth's origin.

The bits of information gathered over the centuries led to the derivation of scientific principles, and ultimately, to the development of geology as a science. One of the earliest principles, uniformitarianism, previously discussed in this chapter, asserts that geological products such as minerals, rocks, and the visible physical environment are results of the same geological processes that take place on Earth today.

Nicolaus Steno (1631–1687), one of the earliest esteemed geologists, first stated the *law of original horizontality* in 1669. He declared that rocks formed from sediment and carried by water to depositional sites were horizontally laid down in strata, generally parallel to the surface on which the sediment accumulated. An outstanding illustration of this law is presented in Figure 24.17. Steno also stated the *law of superposition,* which contends that in a normal marine depositional sequence, the younger strata lie over the older, unless, of course, disturbances occurred at the site in the form of localized overturning after the sediments had accumulated. The law applies to both igneous rocks, particularly successive lava flows, and to sedimentary rocks (recall Fig. 23.4). In Figure 24.18, the previously horizontal layers have been exposed to tremendous internal forces, which created uplifting and resultant overturning. Such features as mud cracks and ripple marks ordinarily found on horizontal surfaces but now present on the sides of the vertical beds offer mute evidence of the stresses to which the layers were exposed.

Breaks in the geologic history of an area are often noted by scientists and are called **unconformities.** Layers of sedimentary rock may have been removed by erosional forces or deposition may have failed to occur. The time interval involved may range from thousands to millions of years.

Similar environmental conditions existed at the same geologic time over large areas of Earth's surface, just as is noted today. Geologists often apply the *law of continuity,* which indicates that a sedimentary rock unit was originally continuous but parts may be missing because of the action of the various agents of erosion. Some rock formations stretch over hundreds of square kilometers, others over even larger areas. The law of superposition aids in the correlation of identical depositional formations.

Another law, the *law of crosscutting relationships,* concludes that as one geologic body cuts another, the body that is cut is older than the body that does the cutting.

Several of these concepts are illustrated in Figure 24.19.

Figure 24.17 The variation in color and banding provide evidence of sedimentary deposition.

Our Mineral Resources

We have realized for quite some time that Earth's seemingly endless supply of minerals is limited. Earth is a closed system in that no new resources are being added—at least not at a rate comparable to that of consumption. The length of time that our resources last depends on how we conserve them. Technology must find ways to remove the minerals from neglected, low-yield sources and a way to reprocess industrial and municipal wastes (see Chapter 27). At the present rate of consumption, sometime during the next century our known available supply of the minerals rich in aluminum, copper, gold, lead, nickel, platinum, silver, tin, and uranium will practically be exhausted.

The minerals that constitute Earth's surface were first used by people of the Pleistocene epoch (perhaps 500,000 B.C.), as evidenced by paints that they used. Peoples of the Holocene epoch

a.

b.

Figure 24.18 (*a*) The folding and uplifting of these sedimentary rock layers reveal the orogenic forces in Earth's crust during various geologic times, if not presently. (*b*) The once-horizontal layers have now assumed a vertical position.

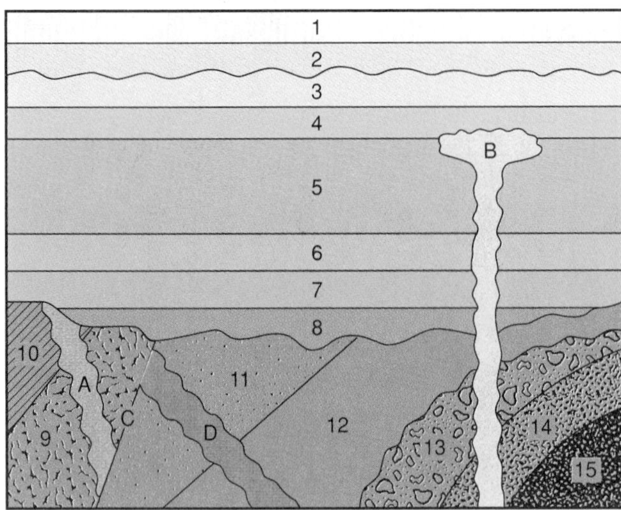

Figure 24.19 Intrusions supply evidence of crosscutting of older geologic bodies by younger ones. Layers 1 through 8 appear to be increasingly older (law of superposition). Intrusive body B is younger than layers 4 through 8 as well as layers 12 through 15 (law of crosscutting relationships). Intrusive bodies A and B, separated by fault C, may have been a single intrusive body.

(about 5000 B.C.) designed tools from chert, flint, obsidian, and quartz, as did their ancestors, but they also discovered the malleable property of copper and gold. With the Bronze Age (3000 B.C.) workers learned to cut stone for building blocks to build magnificent temples and pyramids. The practice of polishing gemstones to increase their value also was established during the Bronze Age.

Expeditions in quest of various minerals started about 2000 B.C., when prospectors ventured to the Red Sea, the Sinai Peninsula, and Africa from their homes in Egypt and Babylonia. Without the aid of machinery, they dug into the sides of the mountains to depths approaching 300 meters (about 1000 ft) in search of emeralds and turquoise.

The mining of gold and silver in Greece was a thriving industry at Cassandra from 2500 to 360 B.C. and was accomplished by tunneling through hills into the veins of the rare metals. Since the rock strata varied so greatly in different locations, there was undoubtedly an interest in creating a taxonomy to identify the differences. The available records represent the beginning of geology as a science and reveal the classification of minerals into sixteen categories. Aristotle, in his writings, discussed the oil fields and asphalt deposits of Albania and suggested how they might be used. *De Re Metallica,* a 12-volume treatise prepared by Georgius Agricola in 1556, included discussions of the probable origin of rocks and minerals, as well as guidelines on how to mine and smelt the metallic ores. The theories he proposed about erosion and other natural processes were vague, but the principles that they entailed are widely accepted by today's scientists.

Nicolaus Steno authored the first discussion of the history of Earth, based on fossils he observed embedded in the rocks of Earth's crust. William Smith (1769–1839) is noted for developing procedures to search for oil, gas, and water. Both investigators significantly contributed to current mineral exploration techniques.

In summary, the mineral resources of Earth have clearly been exploited for more than 4000 years. There is no second crop of minerals, yet more of these resources have been mined in the last 50 years than in all previous history. At the present rate of consumption, disregarding the increased need for the ores because of inevitable increases in population and additional uses, our next generation will find the supply essentially exhausted. We have no choice but to conserve the natural resources that still remain and to find better methods of recycling them.

Summary

The principle of uniformitarianism states that all changes of Earth's topography are the results of the same physical processes that are in operation today. This, in general terms, means that the present is the key to the past.

Earth's surface is constantly undergoing change as the result of numerous geologic processes. Gradation includes the various means by which the surface experiences degradation by erosion and is filled in by sedimentation, a process called aggradation. Gradation is generally attributed to the action of water, ice, and wind. Most of Earth's major mountain systems resulted from orogenesis. An episode of intense deformation of rock layers in a region is called an orogeny. Other mountain ranges result from igneous activity, the movement of molten rock and the formation of solid rock from the molten state in Earth's interior or on Earth's surface. Earthquakes are not the cause of geologic processes, but rather the result.

A mineral is a naturally occurring solid element or inorganic compound with an ordered internal structure and a chemical composition that can vary only within defined limits. A mineral, being homogeneous, exhibits uniformity and constancy in its chemical and physical properties. The most common physical properties of minerals are cleavage, fracture, color, streak, luster, hardness, and specific gravity. Most minerals are composed of two or more elements that have reacted to form a compound. Minerals are an integral part of the rocks; however, only a few of the many minerals are the typical constituents of the rocks. Some mineral deposits are mined for the valuable elements contained in them.

Rocks are generally aggregates of minerals. Their wide variation in physical properties is a result of the amount and kind of minerals they contain and the conditions under which they formed. Although geologists study the chemical composition and physical properties of rocks, their chief interest is the record that rocks reveal about the environment of Earth at the time the minerals aggregated into rock. The reading of the rock record by a trained geologist reveals many spectacular events in Earth's history. Earth's crust is composed of three main types of rocks, classified according to their modes of origin. Certain members of each type of rock grade into another member of the same origin. The three types of rocks are igneous, sedimentary, and metamorphic.

Igneous rocks are formed from the solidification of a melt. This type of rock is further classified according to texture and composition. The texture of intrusive and extrusive igneous rocks usually indicates the conditions under which they congealed. The size of the mineral grains in the rock generally reveals the pressure, temperature, rate of cooling, and volatile materials present during the period of solidification. Light-colored igneous rocks usually are composed of sialic minerals, while dark-colored specimens contain simatic minerals. The Bowen reaction series includes most of the main minerals of igneous rocks. Some common igneous rocks are basalt, gabbro, granite, obsidian, and rhyolite.

Sedimentary rocks are formed from material ultimately derived from weathering, chemical precipitation, or biochemical activity. The sediment is transformed into rock by such lithification processes as cementation, compaction, organic activity, and recrystallization. Some common sedimentary rocks are chalk, coal (other than anthracite), conglomerates, dolomite, gypsum, limestone, sandstone, and shale.

Metamorphic rocks are formed either from igneous or sedimentary rocks, by alteration of the rock, either in texture or in mineral composition while in the solid state, due to heat, pressure, or chemically active solutions. The changes may be in the form of recrystallization, foliation, or recombination. Some typical metamorphic rocks are anthracite coal, gneiss, marble, quartzite, schist, and slate.

The law of original horizontality states that sedimentary rocks were laid down in a horizontal fashion parallel to the surface on which the sediment accumulated. The law of superposition contends that the younger strata formed from sedimentary rocks or lava typically accumulate in such a manner that the younger strata overlie the older; therefore, their relative ages are known. The law of continuity points out that a sedimentary rock layer was continuous over an area of hundreds of square kilometers. A break in the rock layers in which various parts are missing constitutes an unconformity. The law of crosscutting relationships indicates that as one geologic body intersects another, the body that is cut is older than the one doing the cutting.

Earth is a closed system, meaning that no significant amount of new resources is being added to our supply of minerals that contain aluminum, copper, gold, lead, and other valuable metals. Our mineral resources have been exploited for centuries, but at the present rate of consumption, our supply will be essentially exhausted during the twenty-first century. Our remaining natural resources must be conserved.

Questions and Problems

Uniformitarianism

1. How might the concept of uniformitarianism be applied to determine how a given rock layer was formed?
2. Why do geologists conclude that most of the deep valleys observed today were formed by erosion rather than by earthquakes?

The Geologic Processes

3. Why doesn't Earth's surface become progressively smoother, due to the constant-action forces of gradation?
4. Describe three ways in which earthquakes may be scientifically predicted.
5. List four factors on which the extent of damage from an earthquake is dependent.
6. What is magma; how does its appearance differ from lava?

Minerals

7. The streak produced by a mineral is more reliable for purposes of identification than the color of the mineral. Why?
8. Table 24.2 lists the most abundant elements in Earth's crust. What minerals might you suspect to be the most abundant?
9. Why might some minerals be stable when located several kilometers deep in Earth's crust, but unstable on Earth's surface?

Rocks in General

10. How are rocks related to minerals?
11. What clues about Earth's past are revealed by the study of rock formations?

Igneous Rocks

12. How are igneous rocks formed?
13. What three factors are used to classify igneous rocks?

Sedimentary Rocks

14. A given layer of sandstone is chiefly composed of fine-grained fragments that are loosely adhered to each other. Another layer of this sedimentary rock is generally fine-grained also, but is well-compacted and does not fracture easily. State two reasons why the two specimens appear to be so unlike.
15. Why do shale deposits generally contain large amounts of past life-forms?
16. Why are some sandy beaches so dark-colored whereas others are sugary white?
17. Cite three ways in which sedimentary rocks are formed. Give an example of a rock formed by each process.

Metamorphic Rocks

18. According to Figure 24.12, what processes are necessary to transform igneous rocks into metamorphic rocks? What processes transform metamorphic rocks into sedimentary rocks?
19. In the process of metamorphism, what are the two major sources of heat?

The Physical Laws of Geology

20. What conclusions might be reached upon observing that two identical layers of rock are located 50 km apart, but that the layers are 150 m and 200 m, respectively, above sea level?
21. What laws of geology would lead one to conclude that a fault has occurred? What laws point out that folding has occurred?

Our Mineral Resources

22. Why is Earth considered a closed system?
23. How does the rate of exploitation of mineral resources over the past 50 years or so compare to the rate during the previous 4000 years?

25 Weathering and Erosion

Chapter outline

Key terms/concepts

A lake is formed in a depression left by glacial action in the Beartooth Mountains, Montana.

The surface of Earth has been boldly sculptured. Majestic mountain ranges thousands of kilometers long reach above the clouds like massive pillars supporting the sky; in striking contrast, immense chasms, such as the Grand Canyon, scar the surface of our planet. Not all the lands of Earth, however, have been carved into mountains and valleys, as is revealed by the vast plains that spread in all directions toward distant horizons. Could any artist have conceived on such a grand scale the varied and beautiful forms that nature has carved on the crust of planet Earth?

What causes this sculpturing of the land? Throughout Earth's history portions of the land have been uplifted above the surrounding landmasses. Whenever this transformation has occurred, water, wind, and ice have started immediately to erode or wear away the uplifted masses.

A continuing struggle exists between the forces within Earth's crust in which huge masses are uplifted to form mountains and plateaus, only to be worn down, slowly but relentlessly, by erosion. Fortunately for our aesthetic sensibilities, the struggle between land uplift and erosion gives our planet its multitude of surface features such as mountains, valleys, hills, lakes, plains, and plateaus. The process of erosion occurs very slowly, but Earth has plenty of time. Thus, a person who sees a few rocks fall down the slope of Mount Everest is witnessing the gradual destruction of the greatest mountain humankind has ever seen.

Weathering

The adage "Everything that goes up must come down" is beautifully illustrated by the process of **erosion.** The instant a mass is uplifted, gravity starts its relentless pull on the mass—a pull that never ceases until the mass is once again at the same level as its surroundings. As soon as a small piece of the mass breaks off, it inevitably finds its way downhill because of the pull of gravity, or it is dissolved in water and carried downhill by gravity. Because of the pull of gravity, the loosened material produced by weathering is transported to a lower level. What forces work to erode uplifted masses? Basically, there are three erosional agents: water, wind, and ice. Of the three, the most important is water.

The weathering of rocks is accomplished by chemical, mechanical, and biological processes. All three of these processes produce a breakdown of rocks. Climate is the major factor in weathering, and water, in the liquid or solid state, is the main instrument of weathering.

Much mechanical weathering is the result of running water or the freezing and thawing of water. Most chemical reactions producing weathering involve the work of water solutions. In general, the wetter the climate of a region, the greater the weathering due to chemical processes; and the higher the average temperature of the region, the faster chemical weathering occurs.

Biological weathering, which produces both chemical and physical weathering, involves the breakup of rocks by the mechanical action of the roots of plants and trees and the chemical action of microorganisms and various animal secretions. Further chemical weathering is caused by the different chemicals given off by flora of all kinds.

We have observed the effects of **chemical weathering** on many of the manufactured articles that we see every day. We have seen objects made of iron slowly rust by the action of oxygen and water, and we have observed the paint on houses slowly fade and peel. In a similar manner, the oxygen and carbon dioxide of the air attack rocks on the surface of Earth.

Carbonic acid, a compound formed by the reaction of carbon dioxide and water, is the chief agent in the chemical weathering of most rocks. In granite, carbonic acid attacks the feldspar minerals and reduces them to soluble carbonates and insoluble clay minerals and silicon dioxide (silica). Similarly, the iron content of various igneous rocks is oxidized in a process akin to that of the rusting of iron to produce iron oxide minerals such as hematite (Fe_2O_3) and limonite ($Fe_2O_3 \cdot H_2O$). Hematite and limonite are responsible for the red and yellow colors of much of our soil. Of all rocks, marble and limestone are most readily attacked by carbonic acid. The insoluble mineral calcite ($CaCO_3$), the main constituent of limestone and marble, readily reacts with carbonic acid to produce soluble calcium bicarbonate ($Ca(HCO_3)_2$), which is then transported in solution by water to different sites where $CaCO_3$ is deposited due to loss of CO_2 and H_2O. Some minerals, such as quartz, are almost completely immune to chemical weathering and, therefore, undergo erosional processes only by mechanical means.

Mechanical weathering results from the physical breakdown of rocks by such means as the expansion of freezing water in crevices or the expansion of plant roots in rock cracks. Just as an automobile engine block can be split by the force of expansion when water inside the block is changed into ice, the freezing of water that has collected in rock cracks and joints produces sufficient expansion pressure to widen existing cracks and sometimes to split off part of the rock. Likewise, roots of plants growing in rock cracks slowly grow larger and increase the size of the crack until a portion of the rock is eventually broken off. The process of weathering slowly disintegrates small rocks and mighty mountains alike. Although the destruction process is a very slow one, a person can see the result of this destruction in the buildup of loose debris at the foot of cliffs and mountains. These *deposits of loose material* are called **talus deposits** and are a direct measure of the disintegration that has already taken place (see Fig. 25.1).

As mentioned previously, there is a close relationship between the type of weathering that predominates in an area and the climate of that area. In general, chemical weathering predominates in humid regions, such as the eastern part of the United States. But mechanical weathering is more important in arid regions, such as the western part of our country, where large enough seasonal variations in temperature are experienced for water to freeze and thaw.

Figure 25.1 Talus deposits.

Figure 25.2 The Arches National Monument in Utah. Wind and rain have produced this example of differential weathering.

Some types of rocks are weathered more easily than others. When rocks that weather slowly are mixed with rocks that weather more readily, the resulting differential weathering can produce beautifully sculptured formations, such as that shown in Figure 25.2.

The process of weathering has been extremely important in the spread of life onto the lands of Earth. The result of weathering is still essential to our survival. The weathered products of Earth's rocks represent a large portion of the *soil* that is so necessary for the growth of agricultural products needed to sustain the enormous human population on our planet. Included in the soil is a significant amount of humus material, primarily the remains of partially decomposed plant matter.

The Action of Running Water

The most significant agent of erosion is running water. When water falls upon the land, a large percentage of the water finds its way back into the air by direct evaporation and by plant transpiration (water given off through a plant's leaves). About 20 percent of the water that falls upon the land returns to the oceans via streams and rivers. It is this water, called *runoff water,* that has such a great ability to erode the landscape. During the long journey that runoff water makes in its trip to the ocean, the running water carries with it dissolved chemicals from the chemical weathering of rocks as well as suspended material from the mechanical weathering of rocks. The suspended material carried by a river scours, abrades, and gouges the bed and banks of the river. It is the force of gravity that is responsible for the erosive cutting action of a river's suspended load. Consequently, the greater the difference in elevation of the head and the mouth of the river, the faster the water flows and the greater the erosional effect of the river's load.

The load of a river or stream may consist of anything from minute clay particles that have very little cutting power to boulders. The mighty Mississippi River alone carries over 1.3 billion kilograms (a weight of about 1.5 million T) of sediment each day to the Gulf of Mexico. Thus, the Mississippi River, along with the other rivers and streams that empty into it, removes a startling amount of fertile soil from the states through which it passes.

Why does a river flow in the lowest part of a valley? At first thought, we might be tempted to reason that since water seeks the lowest level, a river naturally flows in the lowest part of a valley. This explanation is only partly true, however. Actually a river flows at the bottom of a valley because the river has made that valley by erosion of what was probably once a level plain or plateau. At first, land had to be uplifted higher than the surrounding landmass to create a difference in gravitational potential. As rain fell upon this elevated plain, the water ran off in whichever direction would most quickly lead it to lower elevations. The runoff water would cut many gullies into the plain (see Fig. 25.3). Gullying on a small scale is often seen in fields along the highway where loose material has been quickly washed away by a heavy rain. The valleys formed from the initial gullying action on the plain appear in cross section to be V-shaped. V-shaped valleys are formed by the rapid downcutting of rivers, which is a direct consequence of the speed of flow of the rivers. V-shaped valleys, as shown in Figure 25.4, represent the youthful stage of valley formation and are characterized by rapids and waterfalls.

Figure 25.3 Gullies cut by running water.

Figure 25.4 A youthful valley. Note the V shape of the valley sides.

During the mature stage of valley formation, downcutting is appreciably reduced, and increased widening of the valley is produced by lateral (side) erosion of the river's banks. This increased lateral erosion in a mature valley causes the river's banks to be undercut. The undercut banks then slump into the river and are carried away. Whereas the rivers in youthful valleys flow in relatively straight lines, rivers in mature valleys follow curving paths (see Fig. 25.5).

In old-age valleys lateral erosion is still occurring, but downcutting has almost ceased. Rivers in old-age valleys meander along sinuous paths, as shown in Figure 25.6, and some meanders (looplike bends in a river's channel) are cut off to form oxbows (see Fig. 25.7).

Obviously, if the erosional effects of streams and rivers on the continents of Earth were the only geologic forces operating on the continents, all continents would eventually be eroded down to a **peneplain**—*land worn down to sea level.* At the present rate of erosion and transportation of eroded material, the continents would

Figure 25.5 A mature valley. Note the river's meandering path.

Figure 25.6 An old-age valley.

Figure 25.7 Oxbows in an old-age valley.

What's an Aquifer?

A body of saturated rock or sediment through which water can flow readily is known as an *aquifer.* The movement of groundwater represents a continuous and effective agent of erosion. The motion of groundwater continues mostly downward until the water reaches an impermeable layer, such as clay or shale. If the region is level, such as in the southeastern United States, the water may become trapped. The water table may rise to the surface and create swamps. The flow of groundwater in a lateral direction, particularly where there is a downward slope to the land, may produce springs. If the aquifer lies between two impermeable layers, the constant erosional action may cause a break and the underground pressure may create an artesian-type spring as the water is forced to the surface. Artesian wells are those drilled through one of the impermeable layers to take advantage of the pressure of the trapped water. Large aquifers may provide sufficient water to service many homes and irrigation projects. The numerous concerns for such aquifers include the possibility of lowering the water table below the capabilities of the installed pumping system. Another serious concern centers about the potential encroachment of dangerous pollutants, perhaps even some from those people who rely on the water to meet their needs ■

be reduced to a peneplain in less than 30 million years. Since Earth has been around for 4.6 billion years and the continents are not eroded to sea level, it is obvious that other geologic forces are working against the erosional effects of streams. As erosion is occurring, uplifting of the land in various regions is simultaneously occurring. There is an equilibrium between land uplift and land erosion. It is this equilibrium that is mainly responsible for our sculptured landscape.

Groundwater

As rain falls on the land, it is observed to run downhill into valleys and supply the water for streams and rivers. What is not so easily observed, however, is that much of the rainwater soaks into the soil and porous rocks of the ground to become **groundwater.** There is 35 times more water existing as groundwater in the United States than there is in all this country's streams, rivers, and lakes. In fact, enough water exists in the form of groundwater to cover an area equal to that of the United States to a depth of 30 meters.

Most of this groundwater eventually flows through layers of porous rock to emerge as river water, lake water, and natural springs. The weathered topsoil and porous rock layers of the ground act as giant sponges to soak up rainwater. The spaces between particles of soil and the pores and small cracks in layers of rock serve the same function. As the water flows down through the pores of Earth, the pores at a certain depth become completely filled with water. The topmost level at which the pores are saturated with water is called the **water table.** At the same time that groundwater moves downward, it also moves laterally through porous rock layers down mountain and hill slopes until eventually it merges with the level of water in lakes and rivers. The groundwater may also appear as springs if the rock layers carrying the groundwater outcrop on the surface.

Because the movement of groundwater is relatively slow, groundwater does not have the mechanical weathering ability that running water has in streams and rivers. On the other hand, since groundwater is in contact with rocks for such extended periods of time, the action of carbonic acid in the groundwater results in considerable erosional work in the form of chemical weathering. In some parts of the United States, such as Kentucky, where beds of limestone (which is easily attacked by carbonic acid) are quite extensive, the erosional work of groundwater can exceed the erosional work of streams and rivers. The chemical dissolution of limestone is responsible for the formation of caves, caverns, and sinkholes, which are the most apparent examples of the action of groundwater (see Fig. 25.8).

Figure 25.8 A sinkhole in Florida formed by dissolution of the limestone stratum beneath a house.

If water that contains dissolved calcium carbonate drips from the ceiling of a cave, the water partially evaporates to form icicle-shaped deposits of calcium carbonate, known as *stalactites,* which hang from the roof of the cave. Where the water droplets hit the floor of the cave, a deposit of calcium carbonate forms, resembling

Figure 25.9 This cavern and the calcium carbonate formations in it were produced by the action of groundwater.

an inverted stalactite. Such floor formations are called *stalagmites.* Both types of formations are shown in Figure 25.9.

Groundwater contains various dissolved chemicals that can be used to cement various rock particles together. Rock particles making up a loose sediment can be cemented together to form such hard sedimentary rocks as breccia, sandstone, and shale. Common cementing materials are calcite, silica, and iron oxides.

Wind Erosion

Southern France sometimes experiences red rain because the raindrops are colored with minute dust particles of red hematite. This red dust, which originates in the Sahara region of North Africa, is picked up by prevailing winds and carried across the Mediterranean Sea to become the red rains of southern France. Occasionally, individual dust storms pick up and transport hundreds of millions of tons of material over 1000 miles. It is quite likely that the dust in the area in which you live contains minute amounts of dust from almost every country in the world.

Winds carrying dust and sand particles can have an appreciable abrading and pitting effect on objects. In sand and dust storms the windshields of cars sometimes become so scarred that they are no longer transparent. Wind that is carrying sand has a natural sandblast action on objects. As a result of this sandblast action, natural features of rocks are sometimes carved into rather unusual shapes, as shown in Figure 25.10.

Figure 25.10 The result of wind erosion.

The material carried by the wind, besides being scattered all over the surface of Earth, is sometimes deposited as sand dunes (see Fig. 25.11) or sediments of extremely fine-grained material called **loess** deposits. Loess deposits are sometimes hundreds of meters thick.

The Work of Glaciers

On the upper slopes of very tall mountains that have a plentiful snowfall, not all the snow that falls during the winter melts or sublimes (a process by which a solid passes directly to the gaseous state without passing through the liquid state) during the summer.

The Sands of Time

There are relatively few notable regions on Earth where the surface is covered only by grains of loose material such as sand. Strong, sweeping winds can hold large amounts of the various small particles in suspension and transport them vast distances.

The tan, sometimes orange, skies occasionally visible over the western United States are due to the sand and other fine-grained sediment carried by the wind. The abrasive action of the fine particles is responsible for some of the unique rock formations in the region, such as smooth, intricately shaped pedestals, beautiful arches, and the unusual recessed features primarily located at the bases of cliffs.

The prevailing winds have been known to relocate large sand dunes without disturbing their shape. Moving dunes can creep across the desert region and completely inundate forests, buildings, and the like, all of which reappear later as the dunes move onward ■

Figure 25.11 Sand dunes—the result of wind deposition.

As a result, snow accumulates year after year. *The minimum altitude at which snow does not completely melt during the summer* is called the **snow line.** Whereas the snow line at the equator is about 6100 meters above sea level, the snow line at Antarctica and the north polar region is at sea level.

As snow slowly accumulates year after year in some mountain valleys above the snow line, the snow is compressed by its own weight. The resulting pressure slowly transforms the delicate flakes of snow into granules called *névé.* The névé is then changed into solid ice. Just as gravity is responsible for the movement of running water, gravity causes the huge mass of ice to move slowly downhill. Once the ice has started to move, it is known as a glacier (see Fig. 25.12).

There are two main classifications of glaciers: valley glaciers and ice sheets. The glacier that forms as described in the previous paragraph is a **valley glacier** in that it is *a stream of ice flowing down a mountain valley.* **Ice sheets,** on the other hand, are *huge, broad masses of ice that cover a very large land surface and move outward from the middle.* If an ice sheet is large enough to cover the major portion of a continent, it is called a *continental glacier.* Examples of continental glaciers are the Greenland and Antarctic ice sheets. The Antarctic continental glacier is by far the largest ice sheet on Earth. The Antarctic glacier covers an area approximately 1.5 times that of the United States and contains 90 percent of all the world's ice and 75 percent of all the world's fresh water.

Rocks firmly frozen into ice at the bottom of a glacier are slowly scraped along by the glacier's movement to gouge and polish the bedrock on which the glacier is moving. The glacier produces

Skating on Thin Ice

The water from melting glaciers and mountain snow deposits is critically important. In the state of Washington alone, almost 500 billion gallons of precious water is provided by over 750 glaciers of various sizes as melting occurs in the brief summer months, according to the U.S. Geological Survey. The meltwater is stored in lakes, reservoirs, and controlled-flow rivers or streams.

The winter rains permit the various streams to maintain a constant level. The hotter, drier climate of summer in the northwestern part of the United States forces the residents of the area to rely heavily on the meltwater for irrigation, livestock and human needs, and popular summer water sports.

The residents are constantly concerned that the rate of snowfall may deviate too far from the optimum. Too much snow would provide an excessive thermal barrier for the glaciers, and the rate of melting would decrease accordingly. On the other hand, with a relatively low snowfall, the rate of flow of the meltwater may be too great to store the meltwater adequately, and the overflow may bring about a water shortage for the following year ■

Figure 25.12 This Saskatchewan glacier is a tongue of the Columbia Ice Field in the province of Alberta, Canada.

more spectacular erosional work when it enters a youthful valley and widens the valley until its cross section is changed from a V shape to a U shape.

The general term for glacial deposits is *drift*. If the glacier melts, *meltwater deposits* result and are more or less sorted and layered. If material is deposited directly by the ice as it moves forward, the term *glacial till* is applied. **Moraines,** or till deposits, may be formed at the end (terminus), along the sides, or beneath the glacier (see Fig. 25.13).

The most spectacular example of glacial erosion and deposition in the United States is the formation of the Great Lakes. These lakes were formed by glacial action during the last ice age.

Figure 25.13 A terminal moraine at an edge of the glacier.

Summary

Weathering is the chemical and mechanical breakdown of rocks near Earth's surface. The chief agents in chemical weathering are carbon dioxide and water (carbonic acid). The main agents in mechanical weathering are water, wind, and ice.

The process of erosion is the transportation and deposition of material loosened or dissolved by weathering.

Running water erodes the land to produce valleys and hills. During this process of erosion, running water transports the eroded materials in suspension and in solution to lower levels. Groundwater erodes mainly by chemical solution of rock layers.

Wind transports small particles that erode rocks by abrading them with a natural sandblast action.

Glaciers erode by a scouring action resulting from rocks firmly frozen into the bottom of the glacier.

The ultimate cause of the transportation of material by erosional agents is the relentless force of gravity. Gravity causes eroded material to be pulled down from a higher level to a lower level, where the material is eventually deposited.

Questions and Problems

Weathering

1. How does chemical weathering differ from mechanical weathering?
2. In cold regions, does chemical weathering or mechanical weathering prevail? Why?
3. Describe the process by which limestone is attacked by carbon dioxide and water.
4. What is soil? What features determine its color, texture, and fertility?

The Action of Running Water

5. Why is a river not a reliable natural boundary, as it is sometimes considered; for example, between states or countries?
6. What aspects distinguish a youthful valley from an older one?
7. What is an oxbow? How is it formed?
8. How does a youthful valley compare with a mature valley?

Groundwater

9. What is the water table? How might it affect the type of homes constructed in a given area?
10. What is groundwater? How does it compare in abundance to all our other freshwater resources?
11. Caves and sinkholes are both produced by what process?
12. How do stalagmites differ from stalactites?

Wind Erosion

13. What are loess deposits? Where in the United States might you expect to find them?
14. How does wind erosion affect cars, trees, and crops?

The Work of Glaciers

15. How does the manner in which a glacier deposits sediment differ from the manner in which a stream ordinarily deposits its sediment?
16. What is névé? How is it formed?
17. How large is the Antarctic glacier? If it were to melt, what would be a major worldwide effect?
18. What is the snow line? Compare the snow line at the equator and at the north polar region.
19. A large boulder, foreign in composition to the region, was discovered near the site of a natural lake. How might the boulder have been transported?
20. What are talus deposits? Where are they generally located?

26 Continuing the Search for Energy

Chapter outline

Key terms/concepts

Windplants that furnish electric power are located in remote areas where winds are constant.

The decreased availability of some of our energy sources in the last several decades has made us realize that we must be prepared to confront impending energy shortages when our readily available fossil fuel reserve is essentially depleted. The Arab oil embargos in the 1970s and the 1991 conflict in the Persian Gulf have made us acutely aware of our established dependency on other nations for our crude oil supply. In fact, the United States has only 5 percent of the world's population, but we use 30 percent of all oil consumed. At times, the relatively short periods of oil glut have driven the price below the level at which searching for new oil reserves appears economically attractive. At other times, the shortage of oil, real or manipulated, has led to rapidly escalating prices and a great enthusiasm for oil exploration. The general curtailment of attempts to locate additional deposits of the other **fossil fuels;** namely, coal and natural gas, also points out the urgent need to conserve these vital natural resources. And, considering the predicted increase in use, how long will our shrinking supply of mineable coal and available natural gas last? Shortages of electrical power in metropolitan areas have already occurred and may become more frequent and severe. We even look to our domestic and industrial refuse, including raw sewage, as a source of heat energy to provide electrical power. Energy, so essential to every nation's welfare, may never again be as inexpensive or as available as before the 1970s. In fact, the next serious energy shortage is predicted to occur in the mid-1990s. We are facing constant energy crisis at the national level, and a well-informed public is necessary to convey the proper level of concern to our elected officials.

It is important to note that in the past several decades there has been a decided decrease in the reliance on oil whenever a viable alternative has become immediately available—an adaptation to the circumstances previously mentioned. The United States uses only about two-thirds as much residential and commercial heating oil as was annually consumed prior to 1979. A major portion of this decrease is attributed to better techniques of insulation and the substitution of natural gas and firewood for oil. In addition, major utilities have reduced their reliance on oil as a source of heat by about 50 percent during the same period, mostly by the conversion to coal. Presently, the consumption of gasoline in the United States is significantly lower than in the early 1980s, as a result of the changeover to smaller and more fuel-efficient automobiles and of the introduction of grain alcohol (ethanol) as an acceptable fuel alternative or additive. (It is noteworthy that larger, less economical automobiles have again increased in popularity.) However, crude oil still provides about 40 percent of the energy used in the United States and more than 50 percent of the world's energy.

Table 26.1
The services provided by approximately one kilowatt-hour of electricity.

Clothes dryer	24 minutes
Color television	5 hours
Dishwasher	18 minutes
Food processor	2 hours
Electric oven	48 minutes
Iron	21 minutes
100-W light bulb	10 hours
Microwave oven	1.5 hours
Radio	55 hours
Refrigerator	5 hours
Washing machine and required hot water	24 minutes
Videocassette player/recorder	6 hours

According to the current trend, by the turn of the century, half of the energy we use will be converted to electrical power. Electricity is used to furnish light and heat, as well as to run the electric motors in our many home appliances. It furnishes the power to operate our radios, telephones, and television sets, along with a host of other devices. We use it to provide the potential energy in our automobile batteries, to heat our water, and to cool our homes. This source of energy has become such an integral part of our lives that we would have little chance of survival in comfort without it. Some specific examples of the services we obtain from electricity are listed in Table 26.1.

Heat energy is also provided by the combustion of coal and natural gas, the other two major fossil fuels. The potential energy present in all three substances is necessary to heat other natural resources so that such elements as iron, copper, and other metals can be extracted from their ores. The elements and their alloys must be heated at least once again so that they can be pounded into sheets, stretched into wires, or molded into various shapes. The list of uses of heat energy, as one readily surmises, would appear to be endless.

Energy from Fossil Fuels

Energy can do work only as it is transformed from one type into another. As the second law of thermodynamics points out, in every conversion process some energy is unavailable to do work. In fact, the electrical energy provided the consumer is only a fraction of the energy that was needed to produce it. A measurable amount of energy is lost in electrical transmission, but most of the loss occurs at the generating plant. Most of our electrical energy is provided by steam-driven turbines at a power plant. The steam turbine generally consists of 24 paddle wheels on a common shaft. Each wheel, with its 5000 blades, is caused to rotate 3600 times per minute by the steam. The rotating turbine is coupled to a huge

A World without Gasoline

Even the racing vehicles that enter the annual Memorial Day Indianapolis "500" have been developed to perform more effectively without gasoline. *Methanol,* or wood alcohol, is often used at such spectacles, particularly since its combustion can produce more horsepower from the lighter engines than can gasoline.

The Environmental Protection Agency (EPA) has expressed considerable interest in methanol, since its use results in lower pollution levels than the use of gasoline or diesel fuel. Also, the EPA predicts that by the end of the twentieth century, methanol production costs will be price-competitive with production costs of the various fuels obtained from crude oil—perhaps even less expensive.

There are other reasons why methanol is considered superior to gasoline, such as cooler combustion (hence, more effective cooling systems); lower level of volatility (therefore, increased safety because it is harder to ignite); higher efficiency (because, upon ignition, it converts a greater amount of energy into useful work than does gasoline); and diversity of production (since it can be obtained from a wide variety of organic materials, including coal, natural gas, wood, peat, or even garden vegetables). However, automobiles that use methanol are difficult to start at temperatures below 10°C. This alcohol is quite toxic and it irritates the skin as well as dissolves paint. Also, methanol is environmentally suspect. It releases some potent carcinogens, a considerable amount of carbon dioxide, and contributes to high levels of smog formation. A mixture known as M85—85 percent methanol and 15 percent gasoline—overcomes some disadvantages. Engines using this combination start better at cold temperatures and the mixture burns with a yellow tint, as does gasoline, rather than with the transparent flame of methanol.

Ethyl alcohol, called *ethanol,* has been used as a mixture with gasoline for several years. In fact, vehicles have traveled an estimated billion miles on gasoline blends containing about 10 percent ethanol (gasohol) since it was introduced in 1978, according to the Iowa Corn Promotion Board. It is primarily made from corn or sugar cane, with a major by-product of the process used as a protein supplement for livestock feed. A bushel of corn produces about 2.5 gallons of ethanol, but the cost for obtaining it from either of these two sources is twice the cost of producing gasoline. However, the Department of Energy's Alcohol Fuels Program has reported great promise in obtaining ethanol from solid wastes, including paper and plastics. Economic incentives undoubtedly will bring about a significant increase in its use as a fuel by the turn of the century■

electric generator in which copper coils turn inside gigantic stationary electromagnets to produce alternating current.

As electricity is generated by means of the **steam turbine,** hot steam is released, hence energy is lost in the form of heat. Steam turbines are about 40 percent efficient, yet other types of generating plants are even less efficient and thus cause more pollution problems per energy unit available for our use.

By necessity, any given electrical power plant must operate constantly at no less than some minimum power output, resulting in a significant loss of generated electrical power if it is not used by a consumer. There is no way to store alternating current directly, so the *natural resources* consumed in the production of unused alternating current appear to be totally wasted. Various attempts are under way to conserve such electrical power that is produced in excess of the needs of consumers. For instance, Germany has developed a method of using the excess electrical power generated during off-peak periods to compress air in underground chambers. The compressed air is then used to turn air turbines and thus generate electrical power when greater consumption is required. In other countries the excess electrical power is used to pump water into reservoirs, where its potential energy can be used to generate electrical power when needed. Other indirect methods that appear to have merit are under investigation.

As previously discussed, most of our electrical power is generated by converting water to steam. The steam then is used to turn giant steam turbines that generate alternating current. Basically, steam is obtained by electrical power plants through the heat released during the burning of fossil fuels (see Fig. 26.1). According to the Department of Energy, the United States used over 2700 billion (2.7×10^{12}) kilowatt-hours of electricity in 1988. Slightly less than 72 percent of the electricity generated during that year was provided by the combustion of coal, crude oil, and natural gas. (Nuclear energy and hydroelectric sources provided approximately 20 percent and 8 percent, respectively. Renewable sources, including wind and solar energy, contributed but a small fraction of a percent of our electrical needs.) The heat produced by coal, crude oil, and natural gas, according to the typical units by which they are measured, is indicated in Table 26.2. Coal is relativey abundant; at the present rate of consumption, it should last several centuries. However, readily available oil and natural gas supplies are rapidly being depleted and may be practically exhausted before methods to extract them from shale deposits or other sources are effectively developed and put into operation. The overall use of crude oil has somewhat stabilized, yet our reliance on imported oil continues to grow. In 1985, the United States imported an average of 4 million barrels (168,000,000 gallons) daily.

a.

b.

Figure 26.1 (*a*) A power plant located along the Arkansas River burns fossil fuel to provide the electricity used by consumers in the surrounding area. (*b*) A closeup of smokestacks at a power plant operated by the Virginia Electric and Power Company.

Coal Facts

According to a presentation at a 1980 energy seminar sponsored by the National Society of Professional Engineers, the United States has 31 percent of Earth's coal reserves, about 250 billion recoverable tons. Coal represents over 90 percent of the United States' fossil fuel reserves, yet provides only 19 percent of our energy needs, whereas petroleum and natural gas provide 47 percent and 26 percent of our needs, respectively. Despite the immensity of our coal reserves and the centuries during which they could furnish a major amount of our needed energy, a seemingly more important aspect is that coal can provide the energy for our immediate needs. The availability of coal will enable scientists to continue research and to develop other alternatives that are economically competitive.

Coal and crude oil are the prime source for an almost endless number of organic compounds known as aromatics (see Chapter 14). Coal is converted into coal tar, a black, sticky fluid thicker than water, during the carbonizing process performed in industrial coke ovens. The amount of tar produced per ton varies from 40 to 160 pounds with an average yield of 88 pounds per ton. When crude coal tar is distilled, three principal products, or fractions, are obtained—light oil, 5 percent of the whole tar; middle oil, 17 percent; and heavy oil, 16 percent. The remaining 62 percent is a nondistillable residue called *pitch.* The light oil is primarily separated into benzene, toluene, xylene, and naptha, all of the same quality obtained from the distillation of petroleum. Benzene is used in making synthetic rubber, and styrene is used as a main ingredient in plastics. Toluene is used as a gasoline additive, in explosives, in saccharin, and in many plastics and paints. Other distillants provide the raw materials for dacron, plasticizers, antihistamines, vitamins, bactericides, and tranquilizers.

Napthalene is the major component in middle-oil fractions. It serves as an important raw material for insecticides, dyes, bakelite, nylon, and aspirin. The heavy-oil fraction known as creosote is used mainly in wood preservation, such as of railroad ties, utility poles, and construction materials. Heavy oil is very important in the manufacture of dyes. Pitch, the major part of coal tar, is very weather resistant. It is used to coat roofs and to make road-paving materials, such as blacktop.

The point of this narrative is to raise the following question: Can we afford to burn coal simply for its heat energy and lose forever the raw materials it contains? ■

Table 26.2
Energy provided by fossil-fuel combustion.

Fuel	Unit of measure	Calories produced	Btu equivalent
Coal	ton	5,650,000,000	22,400,000
Crude oil	barrel (42 gal)	1,460,000,000	5,800,000
Natural gas	1000 cubic ft	257,000,000	1,020,000

In 1987, this figure increased to 5 million barrels. By the mid-1990s, it is expected that we will be importing approximately 50 percent of the total oil we consume, as compared to 35 percent of the total oil we used in 1973.

Although much of our economy is based on this practice, the combustion of any of our fossil fuels to produce heat is an apparent waste of natural resources. These substances are composed of complex hydrocarbon compounds from which most of our synthetic materials are produced. Without them as building blocks, many of the products that we cherish—including medicines, cosmetics, building materials, and all plastic objects—will no longer be available. This reason in itself should bring about curtailment of the burning of fossil fuels.

Energy from Nuclear Fission

Nuclear fission provides a controlled release of energy to generate electricity according to specific need. In the United States, there are over 100 operating nuclear power plants that produce electricity, primarily located in industrial areas (see Fig. 26.2). (About 300 additional nuclear facilities provide electricity throughout other countries.) Also, 9 more nuclear power plants are under various stages of completion in the United States. The electrical energy made available by nuclear fission in the United States now furnishes about 20 percent of our total needs. It is encouraging to know that our reliance upon oil imported from OPEC nations specifically to produce electrical energy is reduced by about 180 million barrels annually through the availability of nuclear energy. Also comforting is the knowledge that the United States has about 25 percent of the world's available uranium supply for fueling our nuclear power plants.

Reactors are devices for starting and governing chain reactions that are self-sustaining. Reactors are currently used to furnish vast numbers of neutrons for scientific studies; to provide new elements through the process of neutron irradiation; and to provide heat for electrical-power generation, propulsion, and other industrial uses (see Figs. 26.3 and 26.4).

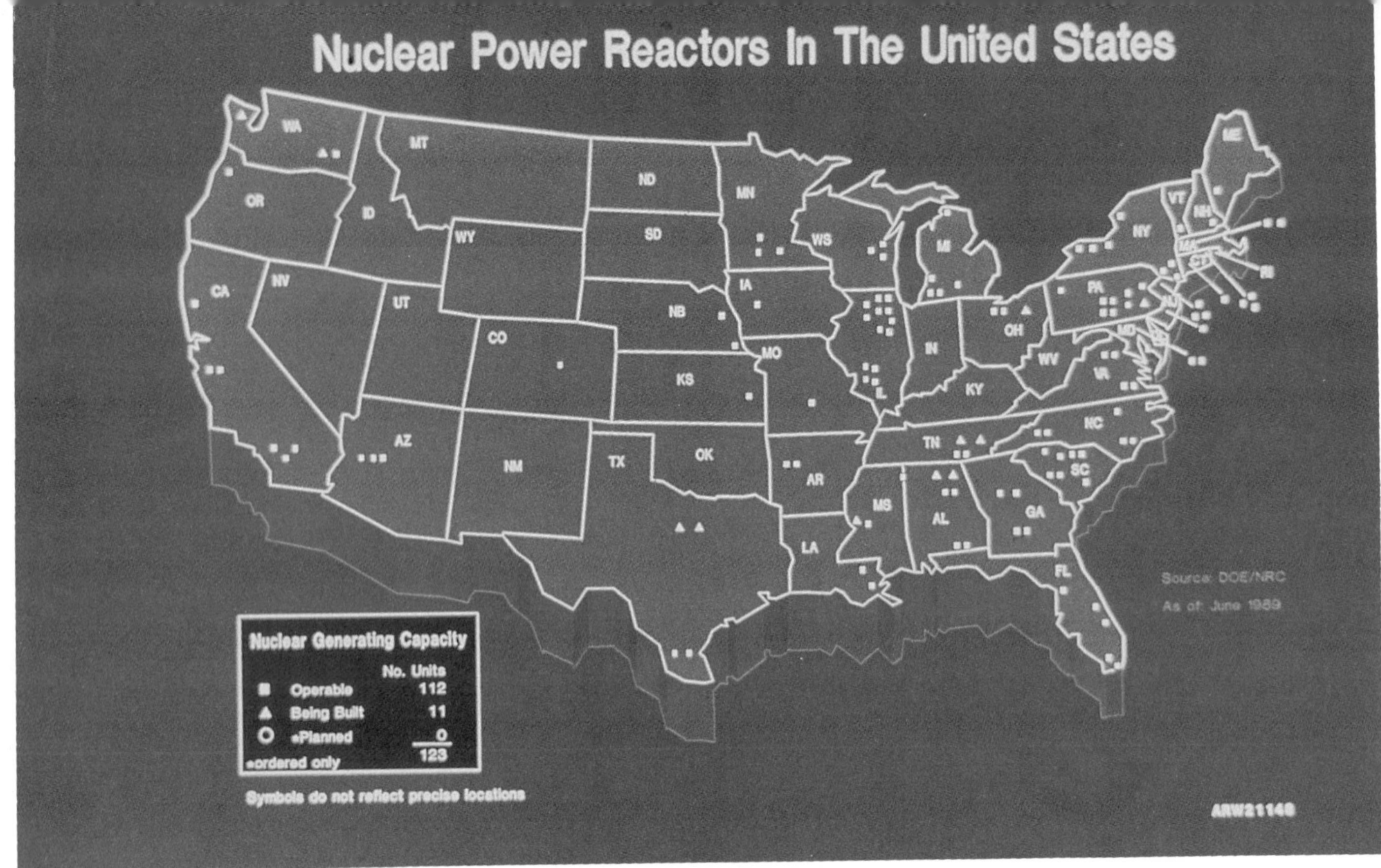

Figure 26.2 The approximate location of nuclear power plants within the United States in operation or in various stages of construction.

The typical **nuclear power reactor** is composed of five primary parts: (1) a core that furnishes the fuel; (2) a moderator that enhances the fission process by slowing the velocity of the neutrons; (3) a regulator that adjusts the number of free neutrons, thus directly affecting the fission rate; (4) a system that transfers the heat from the core; and (5) a type of shielding (and containment) to provide areas free from secondary radiation for operators and experimenters.

The material that fissions when it absorbs slow-moving neutrons provides the fuel for the reactor. Uranium-235 furnishes the fission energy for many reactors. These reactors make use of enriched uranium, which has a significantly higher percentage of uranium-235 than is ordinarily obtained from naturally occurring uranium ores. However, the percentage of enrichment is not sufficient to create the potential for a nuclear explosion, such as is available in a nuclear warhead. The fuel can be maintained in a liquid state, but more often is used as solid metallic uranium or uranium oxide, either of which may be in the form of pellets, plates, or cylindrical rods (see Fig. 26.5).

During the fission process, the neutrons that are released travel at high speeds. The probability of the neutrons colliding with other atoms (in order to continue the chain reaction) is greatly enhanced if the neutrons' speed is decreased. A substance called a moderator surrounds the fuel and permits free neutrons to collide with its atoms without the neutrons being absorbed; it thus slows the elementary particles by ensuring numerous collisions (see Fig. 26.6). Graphite, ordinary water, or heavy water often serves as the moderator.

The nuclear reactor is generally controlled by adjusting the number of free neutrons present in the core at a given time. Such substances as cadmium or boron provide control by absorbing a high percentage of those neutrons that collide with the cadmium or boron atoms. These two elements are manufactured and molded into the form of control rods. Varying the effective length of the rod introduced into the core adjusts the number of free neutrons left to interact with other matter.

The neutrons and the fission products that they produce collide with the moderating material or the highly absorbing control rods. The kinetic energy that the particles thus lose is converted to heat energy. In order to prevent the total amount of heat energy from becoming excessive, various coolants are circulated to carry the heat away from the central core. Such coolants as air, helium, carbon dioxide, water, heavy water, and liquid sodium or lithium are typically used. Research reactors operate at temperatures lower than 100°C; however, reactors designed to generate power operate at temperatures in excess of 300°C. In fact, a cubic meter of space within a reactor core can generate as much power as 1000 times this volume of raging flame inside a coal-fired boiler. The vast amount of heat in either case is converted into electrical energy or mechanical energy.

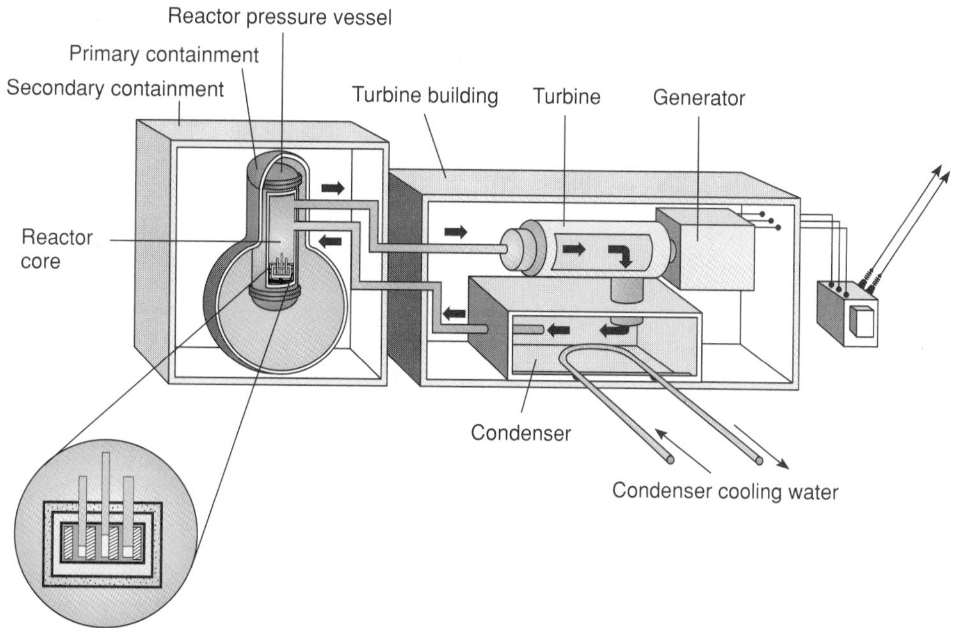

Figure 26.3 This schematic shows the production of electricity by nuclear power in a boiling water reactor (BWR). Coal- and gas-operated power plants basically differ only in the way that steam is produced. The reactor core is shown in more detail in Figure 26.6.

Figure 26.4 Browns Ferry Nuclear Plant, located near Athens, Alabama, uses three boiling water reactors, each of which is capable of providing 1100 MW of electrical power. The facility has recently undergone significant safety improvements and is the nation's second largest nuclear plant. Arizona's Palo Verde Nuclear Generating Station has a power capacity of 3810 MW.

Our Aging Nuclear Power Plants

A considerable number of our nuclear power plants will soon be reaching the end of the expected 40-year life span permitted by the Nuclear Regulatory Commission (NRC). By the year 2010 the operating licenses issued by the NRC to about 12 plants will have expired. If the increasing need for electrical power continues at the current rate, these facilities must be replaced or renovated without delay. However, because of safety and environmental controversies, the construction of new power plants to meet existing guidelines—fossil fuel as well as nuclear—has become progressively more difficult and almost cost-prohibitive. For these reasons, various methods of plant-life extension, or PLEX, are being explored.

Studies undertaken by the Electric Power Research Institute, along with the U.S. Department of Energy and Electricité de France, have determined that existing nuclear power plants could be made safe to operate for up to 70 years, if certain improvements are made. For instance, the utility would have to introduce an accelerated monitoring and maintenance program to ensure that the plant was in good condition and to minimize component wear. Also, steam generators would need to be replaced at certain intervals. In addition, PLEX programs involving the entire facility would have to be initiated and implemented. One of the largest age-related problems is metallic corrosion. This effect has caused utilities to replace at least 25 steam generators at U.S. plants and to install new recirculating water systems at about 23 plants worldwide. The effectiveness of the PLEX that is underway at the Yankee Rowe nuclear power plant in Massachusetts and at Minnesota's Monticello plant will determine how well the NRC accepts this means of increasing the operating life of other nuclear power reactors for perhaps an additional 30 years ■

Not all the energy produced by fission is released as heat energy. A significant amount of the total energy assumes the form of radiation. Therefore, the interior walls of the reactor are sealed with a heavy steel lining to prevent radiation damage. The outer regions of the reactor are constructed of concrete, water, or some other suitable protective barrier designed to protect the personnel in the immediate vicinity of the reactor. Further shielding may take the form of massive superstructures built to prevent extensive damage to surrounding areas in the event of an accident, such as that at the Chernobyl nuclear power plant in the Soviet Union in 1986.

Some reactors are designed to take advantage of a product of fission other than the heat energy provided by the process. The research reactor creates a beam of neutrons that is concentrated on a given area, or a region of high neutron density into which researchers may introduce various study materials through a number of ports. Physicists may use the reactor to study crystal structures and to investigate nuclear reactions. Chemists may use the reactor to determine the exact composition of various substances or the effects of radiation on chemical compounds. Biologists may use the reactor to study the genetic effects of radiation in plants and animals. Engineers, physicians, and other professionals have also found many uses for the reactor and have helped prove its worth to society.

Another type of reactor produces nuclear fuel to be used in other reactors. The breeder reactor converts various isotopes, such as uranium-238, into other fissionable material, including plutonium-239. This radionuclide is not present in natural ores, since its half-life is but 24,000 years. The breeder reactor has remained a controversial issue since the first successful experimental model began its operation. Since the early 1980s, various groups of citizens throughout the world have realized that plutonium-239 could be made readily available for use in the construction of atomic weapons and have voiced their opposition to constructon of breeder reactors. There are several commercial breeder reactors in operation in France and some in various other countries. However, only small experimental models have been constructed in the United States due to lack of adequate financial support and the various safety concerns that have been expressed.

The application of the nuclear reactor as a source of power has had a dramatic effect in the United States, as well as in remote regions that lack coal, oil, and water in sufficient quantity to furnish needed energy. In 1960 an electrical power unit was constructed only for demonstration of its potential; but by 1963, large units that could provide about 600,000 kilowatts of electrical power from heat energy were under construction. Five years later, the total capacity of nuclear-powered electrical energy under construction was in excess of 50,000,000 kilowatts—more than the total electrical energy required by the United States at the close of World War II. Today, nuclear power provides more electricity than was used in the United States in 1953, or about 20 percent of the 600,000,000 kilowatts currently required. However, as mentioned earlier in this chapter, several commercial reactors that were under construction have been converted to coal-burning operations or have been discontinued entirely. Undoubtedly influenced by various factions, members of Congress in the late 1980s offered resistance to the issuance of operating licenses for such completed nuclear power plants as the Shoreham facility in New York and the Seabrook plant in New Jersey. However, in 1990, the Seabrook nuclear facility was completed to the satisfaction of the Nuclear Regulatory Commission (NRC) and was granted an operating license to serve about 1 million homes and businesses. The Seabrook

a.

Figure 26.5 (*a*) The fuel area and control rods are constantly monitored by technicians for safety purposes. The reflections are caused by the water shielding. (*b*) The fuel area emits a blue glow caused by charged particles as they move through water at 2.25×10^8 m/s—faster than visible light travels in water. The spectacular glow is known as Cherenkov radiation. (*c*) A technician at Atomics International inspects a prototype fuel-rod assembly. The fuel rods of uranium monocarbide provide the temperature, heat-production, and burn-up characteristics necessary for a high-performance sodium-cooled power reactor.

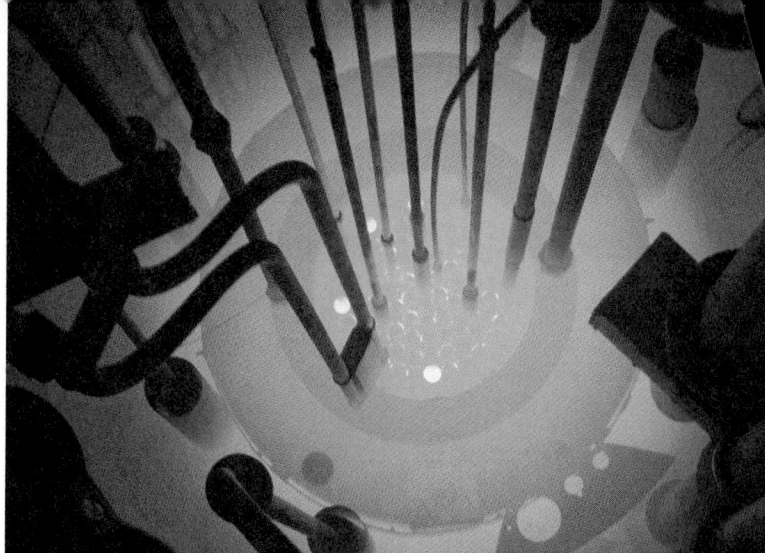

b.

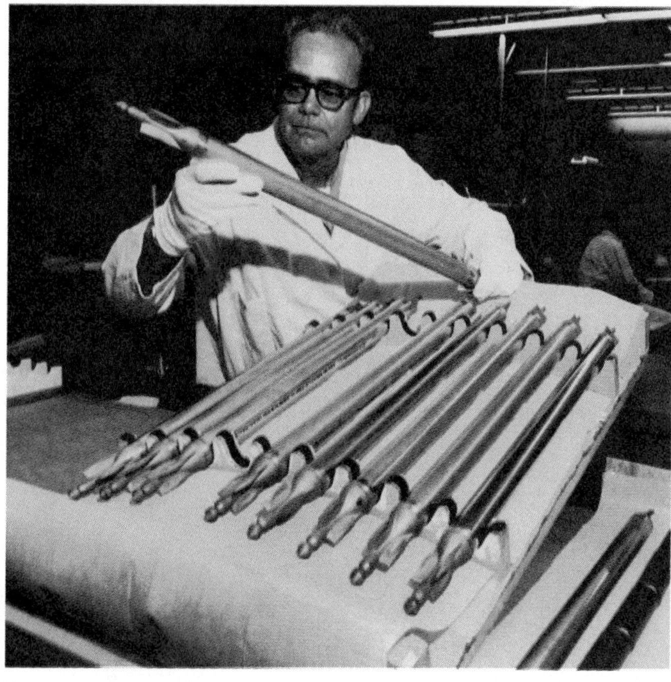

c.

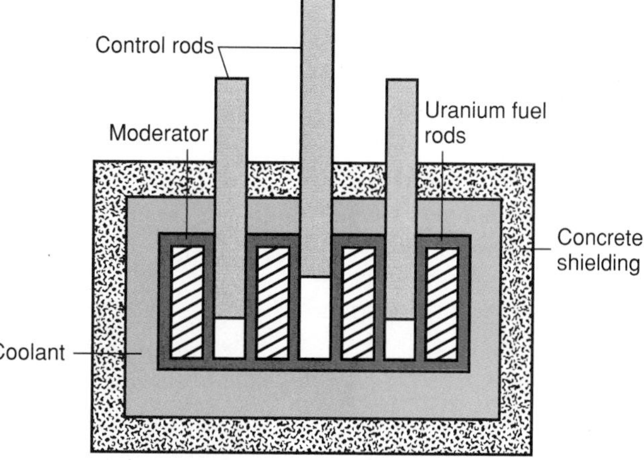

Figure 26.6 The moderator increases the efficiency of nuclear reactions by slowing the speed of the free neutrons so that fission reactions are enhanced. The cadmium control rods readily absorb the unused slow neutrons and further control nuclear reaction rates.

a.

b.

Figure 26.7 (*a*) New Jersey's Seabrook Nuclear Power Plant as it was undergoing construction. (*b*) The Diablo Canyon Nuclear Power Plant provides electricity from Santa Barbara to the Oregon coast. Compare the terrain surrounding the facility with that shown in (*a*).

facility is shown under construction in Figure 26.7a. In addition, the Comanche Peak 1 power plant in Texas was licensed for full power during the same year. It has become apparent that solutions to our needs for additional electrical energy must not only meet the standards of the technological realm, but of the economic and political, as well. Those states and organizations that furnish us with our coal, natural gas, and crude oil apply considerable political pressure in this regard.

In the 1980s, Cincinnati Gas and Electric converted its Zimmer nuclear power plant, the construction of which was nearly complete, to a gas-fired operation. The NRC found several problems with quality control during construction and insisted much of the plant be torn down and rebuilt before a license for operation would be issued. Public Service of Indiana was informed that severe

Table 26.3

A comparison of electric utility energy consumption, past and future.*

Energy source	(Quadrillion Btu's per year)		
	1982	*1990*	*2000*
Coal	12.7	17.3	23.0
Nuclear	3.0	6.5	7.9
Hydropower and geothermal sources	3.5	3.4	4.1
Natural gas	3.3	2.6	1.8
Oil	1.5	2.0	1.3
Wood	-	0.1	0.3
Solar and wind energy	-	-	0.8
Total energy used	24.0	31.9	39.2

*The U.S. Department of Energy predicts that by the year 2000, coal and nuclear energy will furnish 10 times more electricity than oil and natural gas.

flaws in the concrete work in its Marble Hill nuclear power facility could make the operation of the reactor unsafe; hence, construction was abandoned, although $2.5 billion had been invested. Also, in June 1988, the board of directors of the Long Island Lighting Company announced that they had reached an agreement with the state of New York to tear down the completed but unused Shoreham nuclear facility. Apparently, various experts determined it was not feasible to make adjustments in the $5.3 billion facility so that it would meet the necessary rigid specifications before it could be brought "on-line" to generate electricity. Although construction problems and economic conditions took their toll, several new reactors did manage to start operations during the 1980s.

The first nuclear generating plant to obtain an operating license after the Three Mile Island incident in May 1979 was the Sequoyah. This fission-type reactor provides electrical power for over 250,000 residents of Tennessee. Also, in the latter part of 1984, the NRC granted a full-power operating license to the controversial Diablo Canyon nuclear facility, removing the opposition it had faced since construction began about 16 years ago. The $5.1 billion plant is located in California, midway between Los Angeles and San Francisco. It suffered numerous construction delays, partially due to the discovery of a nearby earthquake fault located offshore (see Fig. 26.7b). It now provides electricity for residents from Santa Barbara to the Oregon coast.

Nuclear energy is aiding the economy by reducing electrical energy costs and conserving the diminishing supply of fossil fuels. For instance, only 1 gram of fissionable material provides 17,000 kilowatt-hours of electricity, assuming 75 percent efficiency for the conversion of heat energy to electrical energy. In comparison, 1 ton of completely fissionable material is equivalent in power release to 3,000,000 tons of coal or 12,000,000 barrels of oil. All things considered, electricity produced by nuclear power is cost-competitive with that provided by the combustion of coal. As shown in Table 26.3, the U.S. Department of Energy predicts that nuclear energy will provide twice the electricity by the turn of the century that it did in 1982.

Figure 26.8 A nuclear-powered steam turbine. Note the large pipes that carry steam from the turbine to a moisture-removal or separating system.

a.

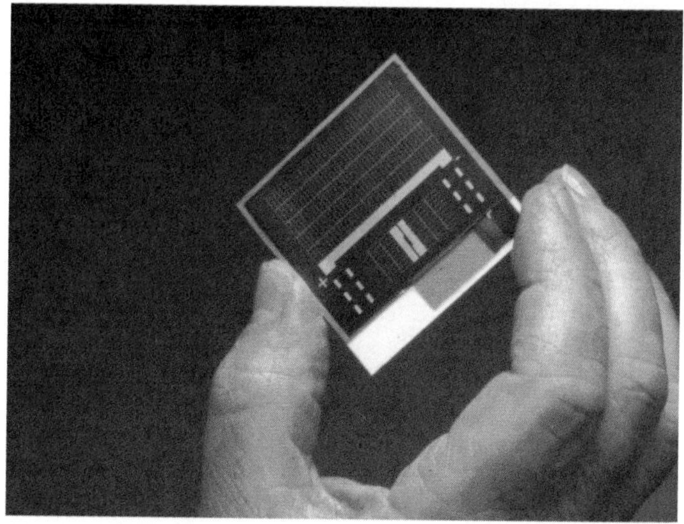

b.

Figure 26.9 (*a*) The roof of the Georgetown University's Intercultural Center supports the world's largest known roof-mounted photovoltaic system. This bank of 4000 modules provides 300 kW of electrical power for the university's use. (*b*) Photovoltaic cells convert sunlight directly into electricity. Thin-film solar cells, such as the one shown, provide electrical energy for small devices such as pocket calculators.

The steam turbines operated with the heat produced by nuclear reactors are physically much larger than those in which steam is derived from the burning of fossil fuels. The turbines at nuclear facilities turn at about 1800 revolutions per minute, half the rate of those at fossil-fuel sites. As a result, turbines at nuclear power plants generally can operate longer between major overhauls. Also, they operate effectively at lower temperatures and pressures than the steam turbines at fossil-fuel installations. The nuclear-powered steam turbine shown in Figure 26.8 is capable of producing about 1100 megawatts of electric power.

The additional heat (thermal) energy provided by the reactor may be used directly in various industrial processes. At some plant sites, for instance, the steam provided by the transfer of heat from the core to the water coolant is used to dry, evaporate, and distill industrial products at temperatures that approach 250°C. Future plans include the use of reactors to provide temperatures above 1650°C. Such a source of heat energy would replace the typical industrial furnace that is used to gasify coal and to furnish intense heat for other industrial processes.

The heat from nuclear reactors is also readily converted to mechanical energy and thus is immediately available as a source of power for locomotion. Such a power reactor was built into the hull of the submarine USS *Nautilus* and was tested in 1955. The United States now has well over 100 nuclear-powered ships that have traveled more than 30 million kilometers by means of pressurized water reactors. Advantages of nuclear-powered ships over conventionally powered ships include the elimination of the need to store fuel and the greater efficiency with which the engines operate. And because the routes do not include fueling stops, travel is more direct. The advantages, though numerous, are somewhat offset by the cost, which is substantially more than the cost of conventional engines.

Solar Energy

In addition to nuclear fissioning and the combustion of fossil fuels, there are many other viable means to obtain heat energy. The temperature and pressure needed to rotate the turbine components theoretically can be provided by the enormous amount of solar energy intercepted by Earth. The energy received from the sun is about 100,000 times greater than the entire world's present needs; however, this energy is exceedingly diffuse and difficult to concentrate. Presently, direct solar energy is practical on a small scale and assumes the form of heat in solar furnaces, solar cooking stoves and grills, solar water heaters, and solar-heated homes.

Several attempts at using **solar energy** to generate electrical power have met with varying degrees of success (see Fig. 26.9a). One technique uses an arrangement of lenses or mirrors to focus solar radiation onto a system of specially treated metal pipes. Various fluids with low coefficients of specific heat flow through the pipes and are heated to high temperatures. (Specific heat is discussed in Chapter 6.) The hot fluid is transported to insulated storage tanks and remains at temperatures in excess of 400°C. As additional heat is needed, the fluid circulates around coils that contain water. The hot water is used to provide heat or is converted to steam and directed to the steam turbine to produce electrical power.

Solar One, a 10-megawatt power plant near Barstow, California, in the Mojave Desert, was the first commercial solar-thermal facility constructed. It and Solar Plant 1, a 5-megawatt facility located near San Diego, represent successful attempts to make solar energy competitive with conventional methods of providing electricity. Both systems use solar energy to provide the heat required to convert water into the superheated steam that drives steam turbines. Solar One has a central tower that concentrates the sunlight received by a broad field of glass mirrors. Solar Plant 1 uses 24 plastic-coated mirrors that contain 700 concentrators. The mirrors reflect the sunlight that strikes them into individual receivers. The production of electricity by the conversion of solar energy into heat energy undoubtedly will increase in geographic regions where weather and climate make it feasible.

The *solar cell*, or *photovoltaic cell,* converts radiant energy directly into electrical energy. The energy conversion involves the photovoltaic effect; that is, solar energy is absorbed by semiconductors that contain added impurities, such as boron and phosphorus, and electrons are released within the cell. Where the impurity adds additional electrons to the semiconductor, a negatively charged area is created. When the impurity provides regions where electrons can settle, a positive semiconductor area is formed. A series of three alternating negative and positive areas can readily conduct an electric current when energy is added. The migration of the electrons in one direction and the positive ions in the other produces an electric potential, typically 0.5 volt. An array of these cells can produce a source of electrical power commensurate with the number of units in the series. Figure 26.9b shows a photovoltaic cell that converts sunlight into direct current electricity.

There are numerous improvements that must be made, such as significantly increasing the efficiency and substantially lowering the cost, if the solar cell is to become a viable source of electrical energy. Although many solar cells in use have an efficiency of 18 percent or less, scientists at the Electric Power Research Institute in Palo Alto, California, and other research centers have recently succeeded in converting more than 28 percent of the sunlight incident on a solar cell into electricity. Varian Associates, also of Palo Alto, reportedly has developed a layer of cells that have achieved an efficiency of over 30 percent. The cost to produce solar power has decreased from about $1.50 per kilowatt-hour in 1980 to the present rate of approximately $0.35 per kilowatt-hour, yet it still is several times more expensive than electricity generated by more conventional means. In general, energy experts maintain that efficiency still must double and costs drop in half before solar cells are inexpensive enough for utilities to consider using them. The Department of Energy, however, predicts that utilities will start building large solar plants by the mid-1990s. Their predictions are based on the increasing improvements in efficiency and lowering of cost brought about by the research conducted by industry and utilities. Noted also is the abundance of solar energy and its reliability. In addition, it is evident that there are no moving parts that will need replacement in solar cells. The process of directly obtaining electricity from solar energy releases no pollution into the environment, causes no acid-rain formation, and contributes nothing to the greenhouse effect that is of constant concern. Perhaps the most outstanding reason in opting for solar cells as a primary source of electricity is that the fuel, of course, is free and will last for billions of years.

According to a report released late in 1987, a team of researchers in Israel has developed a unique device that could result in increased application of solar cells. The team designed a photoelectrochemical cell, a unit immersed in an aqueous solution that converts sunlight directly into electrical or chemical energy. Thus, the portion of the device that furnishes the electrical energy, a crystal formed from a cadmium compound, can charge the other half of the cell. In turn, the electrochemical storage component can furnish electrical energy to the solar cell during darkness or in subdued light. The device produces about 1 volt of electrical potential at a solar conversion efficiency of approximately 15 percent. The device, then, is essentially a chemical system that stores energy and spontaneously releases it when needed. The overall value of a photoelectrochemical cell compared to the more familiar solid-state photovoltaic cell is yet to be determined, although sufficient benefits have surfaced to interest other investigating teams.

Other researchers have developed a device that converts solar energy into chemical energy, an integral part of a possible chemical system that could be used to economically transport energy over long distances. The Sandia National Laboratories in Albuquerque, New Mexico, have constructed a device that combines a solar collector with a chemical reactor. The collector uses sunlight to vaporize sodium metal. The relatively hot vapor is then introduced into the chemical reactor, where it condenses and releases its heat energy to create a chemical reaction between methane and carbon dioxide. The products of the reaction, hydrogen and carbon monoxide, are then transported at about room temperature through a closed system of pipes. Once the gases reach their destination, they are raised to a higher temperature by electric heaters, then used to heat various buildings or to generate electricity before returning to the collector to repeat the process. The potential value of this technique as a source of useful energy appears encouraging.

Solar energy is also being investigated as a major means to liberate the hydrogen bound in seawater. Researchers use a light-absorbing device made of a thin film of the semiconductor cadmium selenide deposited on nickel foil. When sufficiently excited by light, electrons in the semiconductor travel into the nickel and cause the surrounding seawater to decompose into hydroxyl ions and hydrogen. The hydrogen, then, is collected and serves as a vital fuel source, primarily in space exploration. Many scientists consider hydrogen as a viable replacement for our fossil fuels.

Given the depletion rate of our fossil fuels and the environmental problems associated with nuclear power, some scientists predict we will essentially have no choice but to rely heavily on solar energy in the future.

Geothermal Energy

In Earth's interior there is a ready source of steam. Deep in the crust, water contained in the rocks is converted to steam by the heat from Earth's mantle. (On the average, the temperature of Earth's crust increases 1 C° with each 18 meters of depth.) The

The Australian 3000

Automobiles that can effectively operate on solar power have been the dream of researchers for decades. In 1987, an experimental version, called the Sunraycer and developed by Hughes Electronics of Detroit, won the world's first international, transcontinental road race for solar-powered vehicles. The Sunraycer, described as "a streamlined cockroach," averaged almost 70 kilometers per hour over the established Australian route of 3000 kilometers and completed the race in 5.5 days.

The Sunraycer received its power from 7200 solar cells, each about twice the size of a postage stamp. The efficiency of the cells was about 16.5 percent, twice the efficiency of the cells currently used in solar-powered hand calculators.

In 1990, college engineering students from the United States and Canada conducted an event known as GM Sunrayce USA. Initially, 31 low-profile vehicles were entered in the race, which covered over 2575 kilometers (1600 mi) from Disney World in Florida to Detroit. One can safely assume that some of their innovations will have an impact on the design of such vehicles for the future.

During the same year, an ultralight aircraft known as the *Sun Seeker* made a transcontinental flight using solar cells to produce power for the electric motor, operated only during takeoffs and landings. After the craft was airborne, it rode thermals, columns of rising warm air, like a glider to make the 4000-kilometer journey. The pilot's flight helps substantiate the potential of solar power as an alternative energy source.

The major uses of solar cells by U.S. consumers today are to operate such devices as calculators, portable radios, tape recorders, watches, and television sets. In other countries or remote areas, solar cells provide the electrical energy for irrigation pumps, battery chargers, and emergency radio transmitters ■

a.

b.

Figure 26.10 (*a*) A geothermal plant provides electricity for residences and industries located in the area. (*b*) Other geysers close to the power plant could provide additional steam if needed.

steam, under great pressure, is trapped in the porous rocks and needs only to be located and brought to the surface to furnish heat energy for steam turbines. A great source of steam under Earth's surface was discovered in Italy and has been generating more than 350,000 kilowatts of electrical power since 1913. The geysers near San Francisco furnish the **geothermal energy,** as the heat is called, to produce 500,000 kilowatts of electrical power (see Fig. 26.10a, b).

There are at least 130 geothermal power plants in operation worldwide with a total capacity to generate 3200 megawatts of electrical power. Included in this number is the Salton Sea plant in southern California, which is among the most recent to begin operation. Other natural geothermal energy sources, located along crustal plate boundaries in Chile, Iceland, and Japan, are scheduled for further feasibility studies.

Further utilization of the heat energy present in Earth's interior shows much promise. Wells are drilled into the hot granitic regions of the crust. Water under high pressure is pumped into one well and rapidly converted into steam. The pressurized steam is allowed to escape through a second well and furnishes the driving force for a steam turbine. The steam cools and is returned as water into the injection well and is used again.

Drilling at such great depths has met with several problems, however, among which is the melting of the drill bits. A new technique using an incandescent device that melts through the rock layers rather than drills has proven very effective. With further development, the Subterrene, as the unit is called, may furnish the required steam with little or no resultant pollution.

Nuclear Fusion

The nuclear reactors currently in operation use the fission process to provide heat energy, but they also produce various long-lived radioactive by-products (see Chapter 15). Although the fusion of nuclei, a process that leaves little or no radioactive residue, is within the grasp of scientists, it is as yet beyond our capabilities to obtain the benefit this source of energy can provide. (The hydrogen bomb represents an uncontrollable fusion reaction in which temperatures reach millions of degrees; hence, it is known as a thermonuclear reaction.)

Theoretical physicists initially began a concentrated study of possible designs of fusion reactors during the middle of the century. These scientists calculated that three criteria must be met in order to obtain fusion, the source of energy present in the stars. The product of two of these requirements; namely, length of confinement time and density of a **plasma,** *a hot, ionized gas,* has come to be known as the *Lawson criterion.* In addition to meeting the Lawson criterion, the temperature of the plasma must be increased in order that the particles of ionized gas undergo a sufficient number of collisions to enhance the probability that they fuse into particles of greater mass. In the process of fusing, vast amounts of energy would be released. In order to attain the Lawson criterion at a temperature ranging from 1×10^8 to 5×10^8 degrees Celsius, the product of the ionic density and the confinement time must be on the order of 10^{14} seconds per cubic centimeter. Although the Lawson criterion has never been fulfilled, many scientists feel that it is within their capabilities, and continue efforts to achieve these conditions.

Investigators must continue to alter and to refine their basic approaches to the numerous problems that must be solved before fusion becomes a viable source of energy. For instance, to maintain the fusion process, the electrically charged plasma must be confined against its natural tendency to continuously expand. One possible method of accomplishing this feat is by **magnetic confinement.** In principle, a magnetic field should be able to contain and compress the plasma, but the proper shape of the magnetic field to accomplish this feat has yet to be determined. The magnetic field serves as an imaginary container; thus, it often is known as a "magnetic bottle." One possible solution uses a so-called plain magnetic mirror, a cylindrically shaped magnetic field that has constriction at the ends. The constrictions are designed to hold the charged particles of plasma by reflecting them toward the middle. Many particles, however, escape from the ends. Scientists at the Oak Ridge National Laboratory (ORNL) have developed a relatively effective process known as the *Elmo Bumpy Torus,* a device that uses a magnetic mirror that is toroidal, or doughnut-shaped. With this shape the ends are eliminated, but the curvature of the toroid induces particles to drift away from the center of the plasma, and thus they are effectively lost. Various solutions to this problem are under investigation, including the use of various techniques such as creating a "bumpiness" on the periphery of the area of confinement sufficient to balance toroidal drift. To provide the bumpiness, scientists use metal coils on the outside of the vacuum chamber that contains the plasma. Electrons are accelerated until their properties undergo significant relativistic changes (see Chapter 15). These relativistic electrons are used to form a ring around the interior of the chamber, and they create magnetic bumps that help control specific magnetic instabilities in the plasma.

The Princeton University Plasma Physics Laboratory (PPPL) has established as a major goal the development of nuclear fusion as a safe, economical means of generating electricity. This facility is deeply involved in the physics of magnetic confinement as it applies to fusion. The most advanced approach, funded by the U.S. Department of Energy, is the Tokamak Fusion Test Reactor (TFTR). The startup of this $314 million facility occurred in the latter part of 1982, after seven years of design, fabrications, and construction. It has already gained worldwide acclaim as an outstanding advancement in the field of nuclear fusion. PPPL's research goal is the magnetic confinement in a toroidal chamber of a plasma composed of the "heavy" isotopes of hydrogen (deuterium and tritium) at a density 1/100,000 that of air at sea level, a temperature of 100 million degrees Celsius, and an energy confinement time of one second—the physical conditions necessary for the production of fusion energy in a power reactor. This technique, called *neutral beam injection,* first proved successful in 1985. With this procedure, particles are made electrically neutral so as to penetrate the magnetic field and help raise the temperature and pressure of the plasma. In July 1986, PPPL physicists succeeded in producing plasma temperatures of 200 million degrees Celsius with the TFTR. This temperature is the highest ever recorded in a laboratory and is rated at 10 times that at the center of the sun. Another method involves *pellet injection,* the bombardment of a plasma with frozen hydrogen or deuterium pellets. The violent reaction that results causes a very brief but decided increase in density, a necessary condition to achieve the goal of developing a fusion reaction that will produce more energy than is required to initiate and maintain the process. A photograph of the Tokamak at the PPPL facility appears in Figure 26.11. Scientists there are encouraged with the progress being made. They feel that with newer techniques involving wall-conditioning, plasma-heating, and fueling strategies, the Tokamak will reach the breakeven point in the mid-1990s.

The second largest Tokamak in the United States is located at General Atomics. The interior of the DIII-D Tokamak vacuum vessel during a maintenance period is shown in Figure 26.12.

In 1990, a report from the National Academy of Sciences indicated that the U.S. program for the development of magnetic-confinement fusion had fallen considerably behind European programs. This situation is due in part to the success obtained with the Joint European Torus Tokamak. If additional funds are allocated to support the U.S. program, efforts will be made to construct and operate a facility known as a Compact Ignition Tokamak

Figure 26.11 The Tokamak Fusion Test Reactor at Princeton University's Plasma Physics Laboratory has made great strides toward making the fusion process available as a major source of energy.

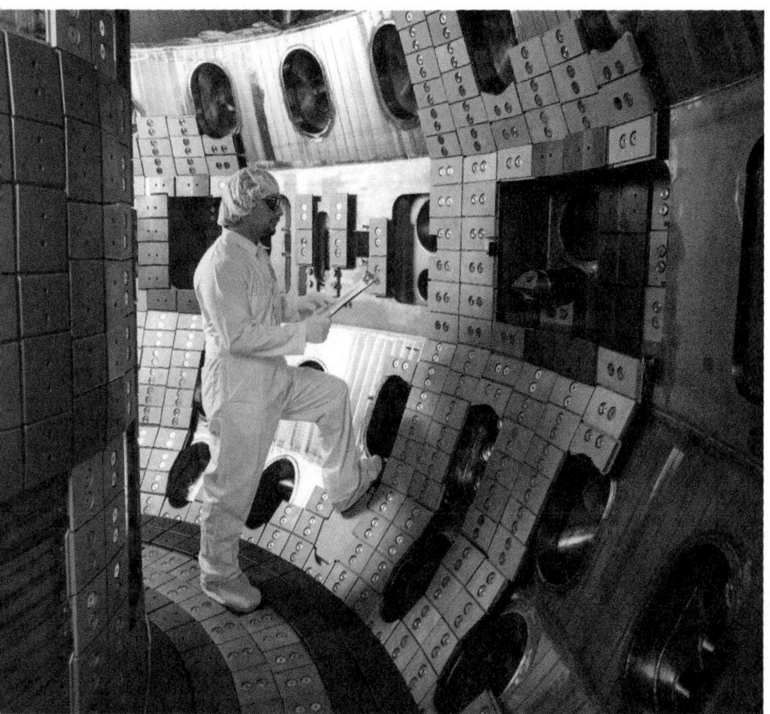

Figure 26.12 The interior of General Atomic's DIII-D Tokamak vacuum vessel. The many ports visible are used for plasma diagnostics, vacuum pumping, maintenance access, and auxiliary heating. DIII-D is presently generating physics data to be used in the design of fusion experiments and reactor prototypes of the future.

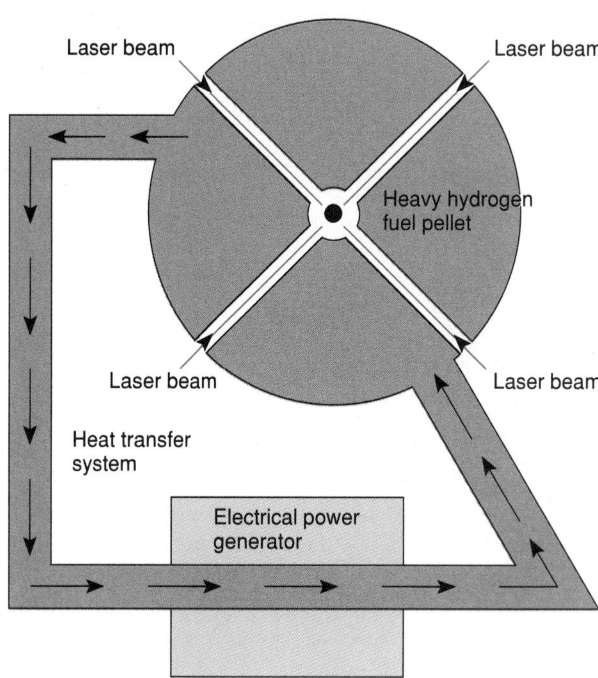

Figure 26.13 Laser beams from various angles strike the target to cause nuclear fusion. The heat generated by the fusion process is transferred by molten lithium or sodium to produce the steam needed to generate electricity.

by the mid-1990s. Despite years of effort, researchers have yet to develop such a system in which the energy output exceeds the energy required to obtain fusion.

The Lawrence Livermore Laboratory, near San Francisco, has in use one of the world's largest, most powerful lasers. This neodymium glass laser can provide 25 trillion watts of power on the target chamber from the 20 beams that originate from it. The unit, called Shiva, may eventually be extended to a power of 300 trillion watts. The great amount of energy that the unit can provide will make possible the high temperature and pressure necessary for fusion to be maintained. In addition, new discoveries in electron-beam technology should make fusion with the "e-beam machine" possible in the near future. This unit is, in principle, the same as the electron gun used to produce the image visible on an oscilloscope. The special type of oscilloscope that produces a picture on a TV set uses about 0.001 amperes and an electric potential of about 20,000 volts. The e-beam machine that causes fusion of heavy hydrogen atoms uses current of millions of amperes and an electric potential of millions of volts. Development of ion-beam generators, units similar to the e-beam device, has also been successful.

In addition, fusion energy specialists have employed a technique called **inertial confinement fusion,** a branch of research that replaces the "magnetic-bottle" method of confinement of target material. Inertial confinement fusion uses a small fuel-filled pellet that is bombarded from all sides by beams of high-energy electrons or ions, or by converging laser beams, to yield a tiny thermonuclear reaction. The small nuclear explosion that results can produce the necessary heat to maintain fusion of the heavy hydrogen atoms contained in the fuel pellet. A schematic of a laser-induced fusion reaction is illustrated in Figure 26.13.

This alternative to the major approach to fusion power appears very promising. Attempts are under way to use beams of light ions, such as from lithium, to implode deuterium and tritium fuel pellets, creating the fusion process within the target area. The

The Cold-Fusion Dilemma

Until recently, practically all efforts to gain the secrets of fusion power involved extremely high temperatures and intense pressures. However, in March 1989, researchers B. Stanley Pons and Martin Fleischmann announced their success in accomplishing a process known as *cold fusion* in their laboratories at the University of Utah. These scientists claimed to have consistently detected more energy released from their experiments than was furnished to conduct them at room temperatures; hence, the name "cold fusion." Their discovery could have provided the ultimate answer to most of our household and industrial energy needs. Unfortunately, the simple experiments that they had designed and that could be duplicated in other laboratories turned out to involve complex phenomena unaccounted for on the basis of classical nuclear physics and defying confirmation in many laboratories. Also, tritium, a potential fusion product, was detected in unused palladium electrodes that were furnished by the same metals processor that furnished those electrodes used in the experiments. Some scientists insisted the palladium contamination did not explain all the tritium measured in the confirmation attempts, since but a few tritium-positive experiments involved palladium from the original source. However, the case continues to be argued by cold-fusion advocates and skeptics alike ■

success of power-focusing experiments at Sandia National Laboratories, employing the light-ion method, may cause a change in major emphasis toward how the fusion process can be accomplished effectively. The Sandia experiments make use of a Particle Beam Fusion Accelerator (PBFA I). Electric charge is stored in giant capacitors called Marx banks along the outer circumference of the PBFA I. The electric discharge, created in pulses, is compressed in time and space as the pulses are accelerated toward a cylindrical diode in the center of the PBFA I. The system is capable of providing 1.5 trillion watts of electrical power per square centimeter on the target. A more sophisticated installation called PBFA II, when operational, may deliver over 100 trillion watts per square centimeter to the target, an amount that should exceed the required energy to bring about the first controlled thermonuclear fusion process in the laboratory.

If research results continue to improve at the present pace, controlled fusion reactors will be a reality by the end of the century. Power from thermonuclear fusion not only can provide vast amounts of electrical power, but also can be used in a multitude of ways. For instance, fusion can furnish the energy needed to remove salt and to obtain heavy hydrogen (both deuterium and tritium) from ocean waters. It can be used to concentrate and recycle domestic wastes, as well as to furnish the power for travel anywhere on Earth or in outer space. If fusion can be conquered, every known element can be produced from the ever-abundant hydrogen, as various numbers of atoms of the lightest element in existence react with each other in a nuclear fashion.

On the basis of our discussion about the use of fusion as a major source of energy in the future, the following advantages should be noted: (1) There is, in effect, a seemingly inexhaustible supply of deuterium fuel available for fusion reactors; (2) the rate of a fusion reaction can be more readily and precisely controlled than can be the rate of a fission reaction; (3) only small amounts of radioactive by-products are produced by the fusion reactions under study; and (4) the fusion reaction produces a portion of its own fuel, such as tritium, and this radioisotope of hydrogen has a relatively short half-life of 12.3 years.

Hydroelectric Power

All the previously discussed sources of energy provide the heat to create the steam used by the steam turbine to generate electricity. Other types of turbines do not use steam. For example, some turbines are placed at the bases of waterfalls or dams and caused to rotate by the force of the falling water. These installations provide *hydroelectric power* and produce about 15 percent of the electricity generated in the United States. Even though they create minimal pollution problems, these plants are quite limited in number and location because of the conditions that must exist for them to operate. **Hydropower,** as this source of energy is often called, makes use, indirectly, of the sun's energy. The heat from the sun forms the water vapor that rises into the atmosphere, destined to fall toward Earth in various forms of precipitation. Thus, some of the solar energy had been stored as potential energy in Earth's gravitational field. Hydropower plants are only feasible, then, where there is the constant attempt by water to seek a lower level, hence relinquishing even more potential energy.

Throughout the United States there are numerous ocean bays and inlets where volume substantially changes due to the tides. The potential energy of the falling and rising water, **tidal power,** can turn turbines that generate electricity. A tidal power plant located in France generates about 250,000 kilowatts of electrical power. A 20-megawatt single-turbine demonstration unit, built on the Annapolis River near the Bay of Fundy, represents the first North American tidal power generation plant. A 5000-megawatt plant is to be constructed nearby, if technological advances continue as planned and environmental concerns are resolved. However, if all available sites throughout the world were put to use, the amount of energy that could be provided by this method could generate only a very minor part of electrical energy requirements. Also, the nature and variations of tides during any given month make the use of tidal power impractical in an industrial society. On a cost-of-energy basis, it would take an estimated 10 years to gain back the energy used in the construction of such plants. Once this return had been achieved, however, the operational fuel cost would be negligible.

Windpower

Energy received from the sun causes air to move. The uneven heating of air masses in Earth's atmosphere creates the currents we know as wind. In many remote and rural areas of the Great Plains, windmills, vaguely resembling the classic models of the past, generate electricity as their blades rotate. As individual units, windmills can generate about 2000 watts of electrical power. A network of these devices, in sufficient numbers, could produce a large portion of the electrical needs of the western states. In fact, by the end of 1983, California had over 4500 wind generators of commercial grade in operation that could provide 300 megawatts of electrical power. The wind generators are so popular that many more thousands will be in operation by the turn of the century for use in that state alone. However, many possible technical difficulties might keep this generating system from being feasible on a large scale; for example, the effect of variable wind velocity and direction on the desired constant rate of electrical power generated. Also, a large unit built to provide electrical power to surrounding communities in western North Carolina was shut down and disassembled because of the many complaints about the noise that accompanied the electrical power generation.

Direct heat, however, can readily be obtained from windmills. A paddle wheel is caused to rotate inside an insulated chamber that contains only air. The mechanical energy of the rotating paddle is converted into heat, and the warmed air is forced into the home at temperatures slightly above the desired room temperature. Water can be heated to the boiling point quite easily by injecting it into the insulated chamber.

Windplants (wind farms), such as the facility in operation in Altamount Pass, east of Livermore, California, may be the forerunner of many such sites in the future. The farm is composed of 100 turbines, turned by three-bladed rotors, that collectively provide millions of kilowatt-hours of electricity (see Fig. 26.14). Other sites in various stages of construction are located throughout California and Hawaii, and areas in Montana, New Hampshire, and Wyoming are being evaluated as potential sites. New developments in aerodynamics, control systems, and materials have overcome many of the technical difficulties that have limited the feasibility of generating electricity from the energy provided by the wind. In many places, the wind is not capable of serving as a reliable source. Presently, wind turbines are not activated until the wind speed is at least 12 kilometers per hour and are automatically braked to a stop to prevent damage to the units at wind speeds of about 90 kilometers per hour.

Figure 26.14 Windplants (windfarms), such as the one shown, were developed by U.S. Windpower in the Altamont Pass for the Pacific Gas and Electric Corporation. These turbines can provide electricity for residents of areas where wind of sufficient speed is typically present, each providing 100 kW of electrical power.

The Fuel Cell

Several means have been developed to convert various types of energy directly into electrical energy. **Fuel cells** provide direct current by utilizing chemical energy. The hydrogen-oxygen fuel cell functions as a result of the reaction of hydrogen and oxygen to produce water. Hydrogen, the fuel, is bubbled through a porous electrode generally made of carbon. Oxygen, the oxidizer, is bubbled through a second carbon electrode. Very small quantities of platinum are added to each electrode to serve as a catalyst. The two electrodes are submerged in an electrolyte, such as a sodium hydroxide solution. As hydrogen diffuses through the carbon electrode, it concentrates on the surface of the electrolyte and reacts with the hydroxide (OH^{-1}) ions in the electrolyte to form water, releasing two electrons in the process. The equation is

$$H_2 + 2OH^{-1} \longrightarrow 2H_2O + 2e^-.$$

The electric charge that accumulates flows through an external circuit to the oxygen electrode. The oxygen diffuses through the carbon electrode and forms a layer of oxygen on the surface of the electrode. Here it combines with water and electrons to form hydroxide ions that migrate to the hydrogen electrode, where the cycle is completed. The equation for the reaction is

$$O_2 + 2H_2O + 4\,e^- \longrightarrow 4OH^{-1}.$$

The efficiency of fuel cells is about 60 percent, considerably more efficient than the use of hydrogen as a combustion fuel to produce the steam needed to operate the steam turbine. Techniques that use solar energy to decompose water are already available. The hydrogen and oxygen gases so produced are ready energy sources for fuel cells.

a.

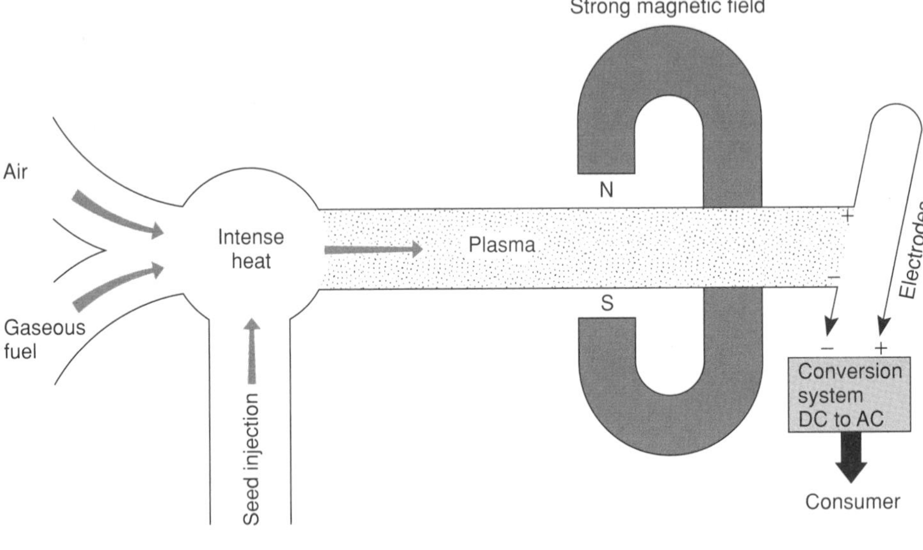

b.

Figure 26.15 (*a*) This technically sophisticated MHD generator, reportedly constructed with the precision of a fine watch, was designed by Modern Electric Power Products and Service Company and fabricated by Westinghouse Electric Company under the technical supervision of Argonne National Laboratory. (*b*) A schematic of how an MHD generator operates.

Fuel cells differ from wet-cell batteries in that the reactants, oxygen and hydrogen, are not incorporated in the plates, as are the reactants in automobile batteries. Instead, the two gases are continuously injected into the electrodes as the reaction proceeds. The water formed by the reaction, as mentioned, may simply evaporate, or it may be pumped out of the cell, perhaps to undergo ionization processes and thus be available to serve again as fuel for the cell.

The command module of the *Apollo* moon vehicle was powered by fuel cells. During the *Apollo 2* flight, about 400 kilowatt-hours of electrical energy was provided by fuel cells. Fuel cells are also being used aboard space shuttle missions and in underwater investigations. Currently under development are other types of fuel cells that can operate with hydrocarbon fuels, such as natural gas or fuel oil, with air serving as the oxidizer.

Magnetohydrodynamics

Magnetohydrodynamics provides a mechanism by which electrical power can be generated directly from heat energy. The first magnetohydrodynamic (MHD) generator, developed in the late 1950s, produced about 10 kilowatts of electrical power. The first large-scale United States MHD generator designed for electrical power production is shown in Figure 26.15a; part b of the figure is a schematic explaining its operation. Known as the *U.S. U-25 generator,* the unit weighs 20 tons and is 20 feet in length. Such units can generate 20,000 kilowatts or more of electrical power.

Gases are heated to very high temperatures by various means, one of which is by injecting them through the cores of nuclear reactors. At the proper temperature, the atoms within the gas are torn apart to form *plasma.* This excellent fluid conductor of electricity is "seeded" with such elements as potassium or cesium (or their compounds) that readily lose their electrons in a hot gaseous environment, enhancing the overall effect. The seeded plasma accelerates through the MHD generator in the presence of a magnetic field, and the intense electric field produced by the action concentrates at the electrodes of the generator, available to provide electrical power to the consumer. In the MHD generator, only the plasma is in motion. Hence, the absence of moving parts permits the efficiency of an MHD generator to be about 50 percent—significantly higher than the efficiency of conventional generating systems. Even though MHD generators are quite efficient and have very low pollution levels, further research on how to improve them has been considerably curtailed. In addition to severe cuts to budgets that sponsor such projects, some experts in the field have concluded that the problems encountered in the necessary refinement of the generator would likely require too great of a commitment.

Thermionic Conversion

Combustion temperatures of about 1650°C are possible in fossil-fuel power plants, yet the steam turbines used are typically designed to operate at under 500°C. Studies reveal that the greater the differential between the input and output temperatures, the more efficient the electrical generating system. The use of the available high temperatures can be accomplished with the process known as **thermionic conversion,** a principle that dates back to Edison's discovery in 1885 of how electric current can be caused to flow between electrodes of different temperatures in an evacuated tube. In thermionics, the vacuum is replaced by the gaseous state of an element that is a good electrical conductor, such as cesium. The gas neutralizes any buildup of electric charge that would disrupt current flow. As electrons "boil" off the heated emitter, they collect on the cooler electrode, called the collector, and return to the emitter through an electric circuit. During this latter part of the cycle, the electric energy that is produced is put to use.

Thermionic systems have already operated successfully for thousands of hours in solar applications, as well as with the heat generated by nuclear power plants. A major program is under way to furnish electrical power with thermionic converters, which have been used in certain Soviet spacecraft. In the United States, systems of prototype thermionic converters are being installed at the sites of coal-fired power plants and are in various testing stages.

Summary

A major portion of the energy we use is in the form of electricity. Electrical energy is provided largely through the combustion of our fossil fuels, such as coal, natural gas, and oil. The heat produced by the burning of these natural resources is used to convert water into steam, and electric power is generated through the utilization of the steam turbine. This device, a type of rotary engine, rotates at high speeds and drives a generator that produces our alternating current. Many predict a grim future for generations to come if we continue to consume the remaining supply of fossil fuels simply for the heat energy that they can provide.

Nuclear reactors also serve as a means to furnish heat energy for steam turbines. There are over 100 nuclear power plants in operation throughout the United States, and new ones are licensed annually to begin operation. The concentrated interest of various groups concerned with the safety of nuclear power generation has led to a careful scrutinizing of each new nuclear power plant. Several plants that have failed to meet inspection criteria have been converted to coal- or gas-burning facilities.

There are numerous alternative sources of energy that can provide electrical power. The energy provided from the sun, solar energy, appears to be a viable source where efficiency is basically unimportant. Heat provided from the depths of Earth's interior is readily available in specific geographic locations.

The fusion reactor is not yet a feasible source of heat energy to provide electricity. Although scientists have not yet been able to maintain a continuous fusion reaction for any appreciable length of time, the progress being made is encouraging, and the researchers involved predict that this long-sought process will eventually be within their grasp.

Waterfalls, rivers, and constant tidal action also are important, but quite limited, sources of electrical energy. In addition, construction costs currently are prohibitive.

Various regions of the continental United States are pursuing the feasibility of capturing windpower to produce electricity. Windplants, networks of turbines rotated by the wind, are in various stages of construction throughout California, Wyoming, and some regions along the East Coast.

Fuel cells are sources of direct current. Their operating fuel is primarily water; thus, they provide a source of electrical energy in situations where energy requirements are low but of strategic importance, such as in remote instrument stations.

Solar cells provide direct current also. Research continues to lower the costs of production and installation. These units offer electrical energy for seemingly endless periods of time, and they become cost-efficient when operating time is considered, since the energy they need to operate is provided by the sun.

Magnetohydrodynamics is a process by which electrical energy is obtained directly from heat energy. MHD units are under development and eventually may provide the electrical power necessary for small communities or, as a network, for metropolitan areas.

Thermionic conversion is a process that could add to the efficiency of various power plants, making use of excess heat energy that otherwise would escape into our environment. Thermionic converters have undergone tests in outer space and have proven worthy of further consideration.

Questions and Problems

Energy from Fossil Fuels

1. How do steam turbines generate electricity?
2. Coal is considered to be one of our most abundant natural resources. Why, then, do many conservation-minded people question the practice of burning coal in order to produce electrical energy?

Energy from Nuclear Fission

3. What are the major components of a nuclear power reactor?
4. In principle, how does obtaining electricity with nuclear fission differ from burning fossil fuels?

Solar Energy

5. How is solar energy directly converted into electricity?
6. How is thermal energy obtained from solar energy to heat our homes and commercial buildings?
7. The efficiency of the more popular solar cells is about 20 percent. At this low efficiency, why are these devices still considered a viable source of electrical energy?

Geothermal Energy

8. Describe how electricity is produced from geothermal energy.
9. What single factor apparently determines the availability of geothermal energy?

Nuclear Fusion

10. What major problems apparently must be solved as we attempt to accomplish controlled fusion?
11. What is the Lawson criterion? Briefly describe its role in the production of nuclear energy.
12. What role do lasers play in the attempt to maintain fusion reactions?

Hydroelectric Power

13. What environmental problems might result from the use of tidal power as a source of electrical energy?
14. What advantages does hydropower have over our major means of generating electricity?

Windpower

15. List the major factors you think could influence the feasibility of using windpower as a major source of electrical energy.
16. Why are many geographic regions not suitable for the installation of windplants?

The Fuel Cell

17. How do fuel cells provide electrical energy?
18. How do fuel cells differ from the wet cells in an automobile battery?

Magnetohydrodynamics

19. How does a MHD generator produce electricity?

Thermionic Conversion

20. Why are thermionic converters generally located near electrical power plants?

27 Concerns for Our Environment

Chapter outline

Key terms/concepts

The processing of crude oil can have detrimental effects on our air, land, and water, unless extensive precautions are taken.

From the very beginning of the relatively short time our species has inhabited Earth, humankind has been threatened by an increasing number of potential risks in a world that is constantly undergoing changes. Many of the changes that we face today are the result of our own past ecological practices. All over this planet, particularly in the more industrialized areas, society finds itself threatened with an environmental crisis that has resulted from steadily growing energy demands and pollution problems. **Pollution** is *the unnatural addition of contaminants to our air, land, or water.* It is the price paid by a world economy emphasizing ever-increasing growth as a primary goal. We have tainted the air we breathe, the water we drink, and the food we eat. The concentration of people in large cities, along with the factories that attract these people, results in the accumulation of wastes of all sorts—accumulations of unwanted substances that produce health hazards. During the late 1960s, environmental crises gained prominence. Industry became aware of the environmental problems that it had created, and factories are now attacking the pollution problems over which they have control. But industry cannot cope with the situation without the aid of each member of society. Each citizen must recognize the urgent need to do his or her part. Our only hope is to develop a concerned society before the problems become more severe.

In the past, the concern of farsighted individuals for our environment and the need to protect it was often met with indifference. The warnings of many great minds forecasting deterioration of our surroundings have gone unheeded from the days of Darwin and his contemporaries. But today most citizens, especially business people, politicians, scientists, teachers, and many young people, realize the importance of more rigid controls. The future holds promise for generations to come because of the earnest efforts of concerned individuals.

Energy and Environmental Pollution

The United States has less than 5 percent of the world's known crude oil reserves, and the other major nations, with the exception of Russia and Mexico, have even less. Over 50 percent of the world's known reserves of oil are located in the Middle East, although most U.S. imported oil does not come from the Persian Gulf. In order to gain greater self-reliance for crude oil (petroleum), we have had to take risks that constantly endanger various delicate environmental balances, such as in the construction and use of the Alaskan pipeline. Close monitoring is required to protect the environment and the wildlife it supports as new drilling projects are initiated throughout the Alaskan frontier. Even with the risks we have taken and the new sources that have been discovered, government geologists predict our reasonably priced domestic oil reserves will be depleted in less than four decades at the current rate of consumption. We have little recourse but to maintain at least the level at which we are currently importing crude oil. When millions of barrels of crude oil are handled daily, there is always the danger of accidental spills that could contaminate both land and water. Precautions must be taken as the well is being drilled, while the crude oil is being removed from the well, while it is being transported to the processing plant, while it is being refined, during the storage phase after processing, and finally while it is being transported to the bulk plant where the various products remain until they are transported by barge, pipe, rail, or truck to the retailer for further storage or direct sale to the consumer. The cost of the cleanup of an oil spill at any phase of the operation is tremendous, even assuming the technology exists to accomplish it. Unfortunately, there are various cases that exemplify the points under discussion. In 1968, the first of several major disasters that involved the supertankers designed to carry millions of barrels of crude oil occurred when the supertanker *Torrey Canyon* was dashed upon the rocks off the rugged coast of England. Over 30 million gallons of crude oil escaped from the gaping holes in its sides and washed ashore. This accident dramatically pointed out the potential danger from the mounting number of oil tankers transporting the valuable natural resource to all continents. The cost of the cleanup, primarily accomplished by spreading detergent over the oil slick, was an estimated $22 million. The attempt was largely successful, but more than 100,000 birds and thousands of tons of sea life perished.

An even larger and more damaging oil spill occurred in March 1978, when the supertanker *Amoco Cadiz* lost the use of its rudder and tore away from its tow by a German-owned tugboat. The vessels had entered the English Channel when the incident occurred, and the heavy seas caused the helpless tanker to break apart on the rocks off northwest France. About 68 million gallons of crude oil escaped from the 223,000-ton tanker and washed toward the beaches, representing the largest oil spill ever recorded in European waters. Many fish and birds perished, and the famous oyster beds along the Brittany coastline were heavily damaged. Although Amoco Corporation used practically all available means to contain the spill, over 100 French towns claimed that it adversely affected their commerce and well-being. As a result, the owners of the *Amoco Cadiz* were assessed damages by a U.S. district judge that totaled $155 million.

Our nation's worst oil spill occurred in March 1989 as the tanker *Exxon Valdez* ran aground on a reef and released almost 11 million gallons of crude oil into Alaska's Prince William Sound. Exxon will spend over $1 billion during the 1990s in an attempt to save the fragile fishing grounds and about 1800 kilometers of Alaskan coastline. Scenes from the cleanup effort are shown in Figure 27.1. The oil that was retrieved from the spill, minus some of the more volatile components that evaporated, was shipped to refineries in Seattle and Houston to be used primarily as a source of heat to dry cement and to cure lumber. About a year after the unfortunate accident, the 300-meter-long tanker was repaired and renamed the *Exxon Mediterranean.*

Also in 1989, an explosion aboard a tanker owned by Iran's national oil company spilled 19 million gallons of crude oil that seriously damaged the fisheries, resort beaches, and the breeding grounds of countless pink flamingoes along the coast of Morocco.

a.

b.

Figure 27.1 (*a*) Pressurized cold water was applied to the beaches to dislodge the crude oil from the *Valdez* accident. (*b*) Detergents were sprayed directly on the rocks in an attempt to dissolve the crude oil.

The extent of the damage as a result of a deliberate attack on the environment in the form of a gigantic oil spill into the Persian Gulf cannot yet be fully assessed. The spill in early 1991 released many times the 11 million gallons that poured from the *Exxon Valdez* into Prince William Sound. The oil reportedly spewed into the Persian Gulf from Kuwait's main supertanker loading station, as well as from five loaded tanker ships that held about 125 million gallons. Up to 250 million gallons may have flowed into the gulf, creating a record-sized slick that may have damaging effects on wildlife resources for years to come. The nature of the gulf itself exacerbates oil spills. It is comparatively shallow, with limited circulation; hence, natural flushing, which was a big help in the Valdez cleanup, cannot help very much in these cleanup efforts.

Threats of even greater environmental destruction loom in our future, for many larger tankers, capable of transporting considerably more than 400 million gallons of crude oil, are now in use and others are under construction. An accident that would release the entire cargo of such a vessel onto our shores could be devastating to all forms of life for decades.

The danger to the environment from accidents that involve smaller vessels along shipping routes and at docking sites is always present. As an example, a barge loaded with 318,000 gallons of high-density, processed oil sank in early 1988, well within sight of our west coast near Shannon Point, Washington, in about 40 meters of water. Presently, there is no evidence that the 9000 barrels of refined oil aboard are leaking, but the potential risk of a major problem will exist indefinitely. The possibility of recovery of the cargo or of returning the 60-meter vessel to the surface seems very slight at this time.

The blowout of a Mexican offshore oil well in the Gulf of Mexico in 1979 dramatized the vulnerability of our waters to pollution from sources beyond our borders, and thus our jurisdiction. The runaway oil well, called Ixtoc I and located in the Bay of Campeche, gushed up crude oil until about 140 million gallons had spilled into the waters of the gulf. This was more than twice the amount of crude oil spilled by the *Amoco Cadiz* incident. The damage to the Texas east coast was not as disastrous as was once feared because of the shielding effect of barrier islands and a seasonal reversal in gulf currents that directed the oil toward Mexico's Yucatan Peninsula. The detailed and prolonged media attention the accident received led to the enactment of more rigid policies governing oil and gas exploration. The incident also had a major impact on scientific spill research, international pollution damage compensation procedures, blowout prevention technology, and other related response actions. Improved safety procedures have been established on offshore drilling platforms and aboard oil tankers. Two production platforms that remove crude oil from under the ocean floor and two supertankers that transport it to processing plants are depicted in Figure 27.2.

One of the most promising changes in the construction of oil-bearing tankers scheduled to occur this decade is the addition of a double-walled hull that will lower the risk of the release of oil into our waters. Further action taken in the early 1990s to protect our environment was the significant increase in the pollution liability required of the owners of tanker ships that enter our ports.

The possibility of oil spills at sea will remain as long as oil is imported; hence, preparations to treat such spills continue. For example, increased numbers of professional crews are being trained to employ cleanup procedures most appropriate for particular situations, and industrial technicians are constantly striving to develop more efficient skimmers and booms to remove oil from the waters. The oil industry has developed chemicals that are effective in treating spills but that are not harsh on the environment. Several spills at the start of the decade, such as from the supertanker *Mega Borg* as it entered the harbor at Galveston, Texas, were treated with oil-eating bacteria. These microbes break up the crude oil and create tar balls that are not particularly damaging to the environment. This procedure was attempted only after numerous environment impact studies were completed.

Most major oil companies have had blemishes to their safety records during the storage phase of their operations. However, few, if any, accidental oil spills were as heavily publicized as the one that befell Ashland Oil Incorporated in January 1988. A 40-year-old cylindrical tank, over four stories tall, had been moved from

a.

b.

c.

d.

Figure 27.2 (*a*) Shell Oil's Beta Rig, an oil operation off the coast of Southern California. (*b*) The Seligi A production and compression platforms located off the east coast of Malaysia. (*c*) One of the large vessels that transports crude oil to processing plants. (*d*) The tanker *Esso Africa* as it moves through gentle seas.

Cleveland, Ohio, by the major petroleum firm and reassembled to become part of a storage terminal located at West Elizabeth, Pennsylvania. Workers had about completed the transfer of 4 million gallons of diesel oil to the tank when it ruptured. Almost instantly over 700,000 gallons of this fuel, a product obtained from crude oil, and one that is thicker than gasoline but thinner than motor oil, escaped the safety barriers surrounding the tanks and found its way into the Monongahela River near Pittsburgh, Pennsylvania, threatening the drinking water of residents in Pennsylvania, Ohio, and West Virginia. Ironically, even the residents in Kentucky near where the crude oil was initially refined were affected, but to a much lesser degree, as the 160-kilometer-long oil slick made its way into a second major midwestern waterway, the Ohio River. The spill took its predictable toll on the various forms of aquatic life and required many weeks until the flowing waters diluted it beyond detection. According to representatives from the Environmental Protection Agency who investigated the incident, the inland spill, costing Ashland Oil about $18 million in liability claims and cleanup, marked the ninth spill of about the same volume by various oil companies in the last two decades. However, since the rate of flow from the 1988 accident was almost instantaneous compared to other such spills, this situation was immediately beyond control. Although environmental groups had tried to alert decision-making agencies for years of the potential danger of permitting the installation of such holding tanks above ground, particularly near major rivers, no effective action had been taken for various reasons. The degree of safety of unloading the oil-bearing barges into storage tanks located near the riverfronts was considered a more important factor than the risk of the possible collapse of a storage tank. Laws to govern massive storage facilities, including the strict compliance as stated in construction permits and the attention to detail in following specific procedures for testing for evidence of metal fatigue or faulty welds, will surely be updated and more strictly enforced in the future. It is not unlikely that above-ground storage tanks will eventually be replaced with underground facilities that presumably would be safer.

The search for oil and natural gas has been extended to the floors of the oceans, under the arctic ice, and to the slopes of the mountainous terrain that mark our most remote of territories. The number of offshore drilling sites, such as are located in the Gulf of Mexico, continue to increase. These wells, drilled to varying depths into ancient marine deposits, can potentially contribute millions of barrels daily to our available supply. The risk from these coastal operations to our environment, however, is always present. The disruption of the delicate balance of certain environmental conditions so essential to the continued welfare of various forms of marine life, such as the dangerously overharvested oyster and shrimp beds, will continue to exist. The same statement is valid for the commercially sought, thus diminishing, schools of menhaden and other such "bait-fish" so abundant in the Gulf waters in the 1970s. The environmental impact from the presence of even minor amounts of oil of any sort has been witnessed by many of us. One such seemingly insignificant source can be traced to supertankers that unload their cargo, then thoughtlessly and illegally flush their storage tanks contaminated with the sludgelike remnants of their crude-oil shipments, even within sight of our shores, before taking on water for ballast in preparation for the return trip to their foreign bases. Residents and those who frequently vacation along our beautiful coastal waters have seen beaches strewn with literally truckloads of dead fish and other forms of marine life as a result of such inconsiderate (and illegal) practices. Unfortunately, the complaints registered by individuals generally fall on deaf ears, and many influential organizations have yet to take the necessary actions to ensure that our laws governing such practices are rigidly enforced. In addition, the penalties placed on those who are prosecuted often lack the degree of severity needed to discourage further abuse to our environment.

Another very delicate situation regarding our oil resources centers around the valuable Alaskan oil fields. The huge oil reserve, estimated in excess of 50 billion barrels, was first discovered in the state's harsh north-shore environment during the late 1960s. The crude oil obtained from the wells drilled in this region is transported by means of a pipeline constructed in 1977. In order to keep the oil from freezing, it must be heated at strategic points along its journey of approximately 1300 kilometers to maintain a minumum temperature of 60°C. The pipeline ends at Valdez, a seaport city located along the ice-free Prince William Sound in the northern Gulf of Alaska. Here the oil is loaded aboard waiting tankers that transport it to refineries located further south along our western coast.

The Alaskan pipeline was constructed without a positive recommendation from the teams of geologists who conducted the necessary environmental impact studies. Even geologists associated with the U.S. Geological Survey did not approve of the venture because of how readily the fragile environment of the arctic region could be seriously damaged by oil spills and melting of the permanent ice that covers the Alaskan tundra. However, strong lobbying by investors and the concern of Congress for the nation's economy brought about by the increasing trade deficit essentially caused the recommendations of the geologists and various environmental organizations to be cast aside.

Despite the constant vigilance connected with the project, there are always various dangers present. According to previous records, earthquakes that could seriously rupture the pipeline along its route pose a constant threat, although special construction techniques were used in regions where major tremors may be prevalent. However, during peak operation, each kilometer of the pipeline carries almost 7000 barrels. Emergency shut-off valves have been installed at intervals of 24 kilometers along the entire length of the pipeline; still, a serious rupture could release about 165,000 barrels (almost 7 million gallons) of crude oil onto the frozen tundra. One should note that a large portion of the pipeline is constructed above ground level, as illustrated in Figure 27.3. Up to this time, there have been only minor leaks in the pipeline. The most serious accident to date that involved the pipeline occurred in January 1981, when 5000 barrels escaped from a ruptured valve before workers could control the flow.

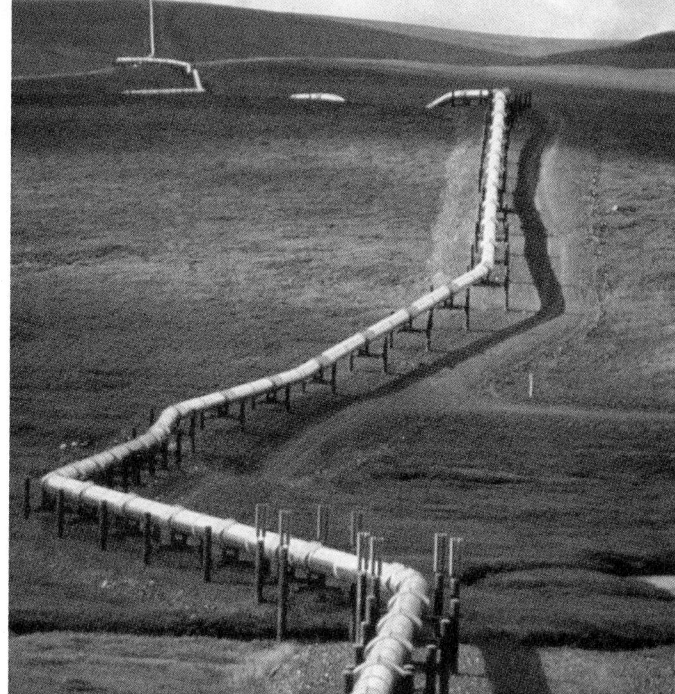

a.

b.

Figure 27.3 The trans-Alaska pipeline is constructed above the frozen tundra. This method permits constant surveillance along its route from the oil fields to the waiting vessels in Prince William Sound. Compare the green rolling landscape shown in (*a*) with the more rugged terrain in (*b*).

The arctic environment, as previously noted, is of surprisingly delicate balance. Even seemingly insignificant changes in its condition can have a decided effect on the climate of our entire continent. Oil spills over land cause minimal damage, and generally only near the site of the accident. Spills of equal amounts onto the arctic ice can cause it to darken, enabling it to absorb more solar energy, increase in temperature, and melt. The resulting blackened water would absorb more heat and bring about the melting of yet more ice. This effect could influence weather and wind patterns and create in the continental United States a drastic increase in rainfall for some regions and a serious drought for others. Before the United States, Canada, and other major nations pursue greater oil-exploitation operations in the arctic region, consideration of the possible detrimental effects must be seriously undertaken.

As a result of the oil embargoes of the 1970s, plans were accelerated to set into operation special plants that could process large quantities of oil-bearing rock located in various parts of the continental United States, primarily in the western states. This sedimentary rock, called **oil shale,** was known by the Native Americans as "the rock that burns." Oil shale contains the remains of marine material deposited about 50 million years ago. The decomposition and the compaction that followed have resulted in the formation of various limestone beds rich in oils, primarily *kerogen.* Some of these deposits were never buried deeply enough by further sedimentation for the weight of the layers above them to convert the kerogen into the more typical forms of hydrocarbons, as occurred with other deposits. This valuable source of oil can be mined from its site of deposition and heated to remove the oil and natural gas that it contains. Then, the two natural resources can be stored for future needs or immediately sent to a refinery for final processing. One barrel (42 gallons) of oil can be provided by the processing of about 1.9 tons of the higher grades of oil shale. Several corporations have conducted the necessary research and development that involve the mining of various shale deposits for the oil and gas they contain. Two experimental facilities originally set into operation during the 1970s are shown in Figure 27.4.

The largest known deposit of oil shale in the United States is the Green River Formation, located in Colorado, Utah, and Wyoming. The bed of shale extends over 4.5 million hectares (about 174,000 sq mi) and contains an estimated 2 trillion barrels of oil. About one-third of the oil present in this vast deposit is considered highly exploitable. The quality of the oil present in the shale deposit is such that it can be readily converted into gasoline or jet fuel.

The United States Synthetic Fuels Corporation (SFC), created by Congress in the 1980s, has as a major goal to end U.S. dependence on foreign oil. This agency has sponsored several synfuel projects, including the Union Oil Company's processing plant near Parachute, Colorado. This facility is capable of extracting 10,000 barrels of oil daily from the shale rock. Other such projects have been funded by the synfuels bill that financially sponsors the SFC.

Unfortunately, the disposal of the processed rock has been found to present a major problem. The oil shale must be crushed prior to the heating process, bringing about a significant increase in volume. Thus, it occupies more space after it has been processed than before it was mined. If the projects are put into a higher level of operation without this situation being resolved, excess piles of shale tailings could reach staggering proportions. In addition, many potentially dangerous chemicals could escape from the processed shale into streams, rivers, and groundwater supplies. Also, sulfur dioxide and other hazardous emissions may be released into the atmosphere during the heating process. The overall procedure has

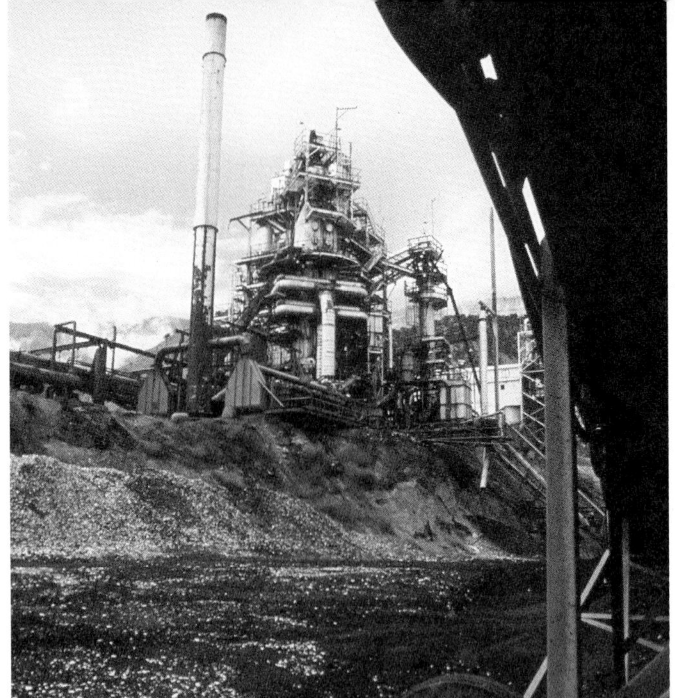

a.

b.

Figure 27.4 (*a*) The oil shale facility at Anvil Points, Colorado, was the site for process and environmental research conducted by the Paraho Development Corporation. Emissions and wastes from the operation were studied in detail. (*b*) The 100–m production headframe constructed by Occidental Oil was used to lower and raise workers and their equipment into the production shaft.

a.

b.

Figure 27.5 (*a*) The rotating drum removes the coal as it sweeps along the coal seam in a continuous mining process. (*b*) A specially designed loader removes the coal and residue from the underground mining operation.

also proved to be expensive; thus, major oil companies have demonstrated considerably less interest and have all but curtailed the research necessary to solve the problems that exist. They have concluded that the profit margin on oil and gas obtained in this manner could not compete with that of our other sources of domestic or imported supplies under existing circumstances. Eventually, the projects may be revived, but only after our less expensive sources are in greater danger of being depleted or at least are no longer so accessible. Still, it is somewhat comforting to know that such an alternative means of obtaining crude oil and natural gas exists and awaits future development.

The SFC and private speculation companies have also been involved in preliminary attempts to extract oil from deposits of *tar sand* located in Utah as well as in other states. Nature's creation of tar sands began 200 million years ago with the deposition and compaction of decaying organic matter. Tar-sand layers have been investigated by geologists for centuries, since they represent the outstanding deposits among the sedimentary rock layers that yield information about past environments and the types of life (including the dinosaurs) that thrived during the time that they were formed. This source of crude oil, as we find it today, assumes the form of beds of a black, sticky mixture of sand, water, and *bitumen,* a naturally occurring hydrocarbon. The processing of the tar sands involves the removal, primarily by heating, of the bitumen from the sand to which it adheres. The tar-sand deposit in Utah contains an estimated 28 billion barrels of oil, and even greater volumes have been located in the tar sands of Alberta in western Canada. However, the processing of tar sand to extract the valuable oil it contains creates numerous environmental problems, including the emission of sulfur dioxide and other obnoxious, if not hazardous, gases into the air. Also, the cost for restoration of the topsoil, along with its natural vegetation, once the various layers of sediment have been removed, has contributed to the lack of progress in obtaining oil from these deposits. It is likely that in the near future the United States and Canada will cooperate in a joint effort to solve the problems that exist so that the removal of oil from tar-sand deposits can become a viable operation.

Undoubtedly, coal is the most abundant and available source of energy in the United States. According to economists, coal is perceived to be as important to us as oil is to Saudi Arabia. At the current rate of consumption, about 700 million tons per year, the 215 billion tons considered to be readily accessible using current mining techniques should last for perhaps several centuries. However, the synthetic-fuels legislation put into action calls for an increase in the amount of coal to be extracted—an increase that could raise the annual consumption to as high as a billion tons. Our coal reserve, then, will disappear at an even more alarming rate.

A synfuel plant constructed at a cost of $2.5 billion and located near Bismark, North Dakota, converts coal into commercial-grade gas. **Coal gasification** has proved to be a very effective means to strengthen our natural gas reserve; and in case of a national energy crisis, the facility can serve as a model by which more such plants could be constructed near other coal reserves. In addition to serving as a source of oil and gas, hydrocarbons and methanol used by industry can be obtained from coal. The importance of these synthetic fuels to our nation's overall welfare can be gauged by the financial commitments made by Congress and others, as discussed earlier in this section.

The historical image of coal mining, with miners picking away at the walls of an underground mine, is not representative of mining today. Although half the coal is removed by the process called **deep mining,** very little is mined by hand. Two of the mechanical devices presently used are shown in Figure 27.5. The other half of the coal deposits is removed from coal veins near the surface by a technique known as **strip mining.** A large shovel, or "dragline," digs a long trench to expose the coal seam. One of the more modern shovels currently in use is shown in Figure 27.6. This large device can remove tons of soil from the underlying coal at one time. Smaller shovels remove the coal and deposit it into waiting trucks that transport it to the processing plant for further refining, and then to waiting railroad cars and barges or to a generating plant. The larger trucks and earth-movers can transport 100 tons of coal or soil with each load.

Unreclaimed land that has been strip-mined is not a pretty sight. Careless mining procedures from past removal operations have left hundreds of thousands of acres of land unproductive. The total effect depends on the topography of the area (whether it is flat or hilly). In mountainous regions the mined area reminds an observer of the lifeless Martian landscape depicted in the photographs taken from the unmanned landing craft. Rainfall causes major erosion, and nearby streams and lakes are filled with silt and deadly acidic materials that destroy all aquatic life. Often coal has a high sulfur content, and as water leaches through the residue, vast quantities of sulfuric acid (the acid used in automobile batteries) are produced. However, new state and federal laws have

a.

b.

Figure 27.6 (*a*) This large stripping shovel is used to scoop layers of soil and shale to expose the coal beneath. (*b*) The bucket on some stripping shovels can remove the tons of soil that cover a coal layer in a single scoop.

been enacted, and the enforcement of previously existing laws is bringing about considerable improvement. Reclamation efforts to correct the damage done over the years by strip-mining operations are under way.

At first glance, the environmental impact of underground mining appears to be minimal. The most obvious result is land subsidence, a depression caused by mine cave-ins. The problem is not severe, however, since mining laws require that roof supports remain in place after mining operations cease to lessen the possibility of roof collapse. However, mine acids from deep mines constantly build up and create a potential hazard. The greatest danger is water pollution from abandoned mines below groundwater levels. The leakage of contaminated groundwater into surrounding rivers has damaged aquatic life in thousands of miles of rivers and streams in coal-mining regions.

Much of the detrimental environmental effect results from the fires that burn uncontrollably in many underground mines. The noxious fumes of sulfur dioxide gas that are released, along with other air contaminants, befoul the areas of burning mines. A program instituted by the U.S. Bureau of Mines in 1949 has succeeded in bringing about 200 fires in abandoned mines under control.

Coal-mining operations rank high among the industrial producers of mineral wastes. The washing of coal before it is transported to consumers has accounted for the deposit of more than 2 billion tons of unsightly wastes on the banks of our rivers and streams, a portion of which is carried as dust throughout the communities near the processing plants. A large amount of the wastes is often carelessly ignited and adds to the pollution problems. Dams constructed of the coal wastes represent a grave danger to people living below them. The 1972 rupture of a makeshift dam of waste materials near Man, West Virginia, claimed the lives of 125 people who perished in the rushing water. Other such potentially dangerous structures exist throughout the country. Consider the accumulation of waste that results from the burning of coal. For instance, a coal-fired 1000-megawatt generating plant produces about 40,000 truckloads of ashes annually. These vast deposits constantly add to our solid-waste problems.

The most costly health problem attributed to the mining of coal is the dreaded black lung disease, a respiratory ailment brought about by continued inhalation of coal dust. Over 150,000 miners have been diagnosed as unfortunate victims. New state and federal regulations that govern maximum permissible levels of coal dust are designed to reduce the future incidences of this disorder.

Uranium mining is a much smaller venture than coal mining. In fact, there are only 300 uranium mines licensed by the Nuclear Regulatory Commission in current operation. The ore from which this valuable element is extracted is mined both deep in Earth's interior and near the surface. Early mining techniques included a number of careless procedures. For instance, wastewater that contained this radioactive element was dumped directly into rivers and streams, and homes were constructed directly on the mined residues, with the result that many people unknowingly received low doses of radiation exposure. Uranium miners face every hazard to which coal miners are exposed, but the exposure to the radioactive uranium ore adds another danger. Uranium disintegrates into *radon,* a heavy radioactive gas that concentrates in the mine shafts. Breathing this gas exposes sensitive lung tissue to alpha radiation, the charged particles of which increase the probability of developing lung cancer. Areas where uranium and its decay products might contaminate surfaces and the air must be carefully monitored (see Fig. 27.7). The rugged areas of the Midwest, including Colorado, appear to be the only regions of the continental United States that have a significant amount of uranium deposits.

The EPA monitors the level of naturally occurring radioactive gases and dust released into the air during the mining and processing of various minerals. For instance, phosphorus-processing plants, such as at Pocatello and Soda Springs, Idaho, are under close surveillance for any excessive amount of radioactive substances that might accidentally escape into the environment. It is comforting to know that the air quality in the surrounding area is consistently above the standards specified by the National Council on Radiation Protection and Measurement. There is also a continued effort by these industries to lower the level of radiation released into the atmosphere. At the present time, the potential presence of radon at numerous construction sites, including new and existing homes, is being closely monitored.

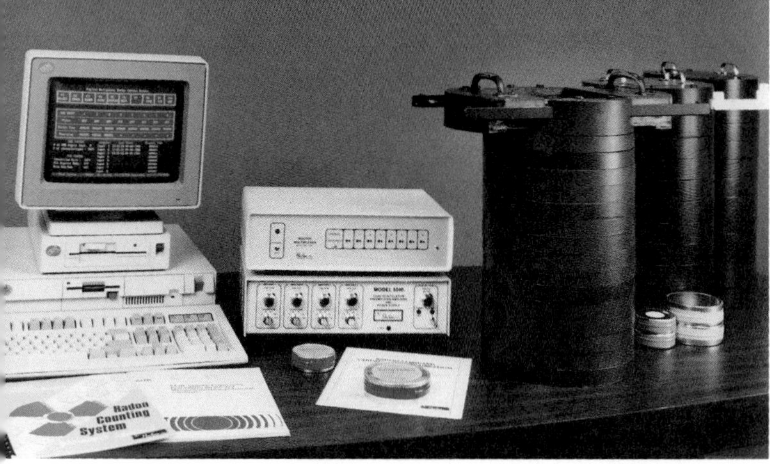

Figure 27.7 The radon detection system provides a constant vigil in areas where this radioactive gas could concentrate.

Figure 27.8 An aerial view of Baltimore Gas and Electric Company's Calvert Cliffs nuclear power plant on Chesapeake Bay. Compared with the effect of warm water on small rivers and lakes, the warm water discharged into large bodies of water has little detrimental effect on the environment.

In early 1988, the Nuclear Regulatory Commission directed 3M Corporation to suspend sales of a device that was found to release a radioactive material into the atmosphere. It had been installed at various factories both in the United States and abroad. The device, known as an air-gun ionizer, was found to unintentionally discharge polonium-210, a naturally occurring radionuclide. The radioactive substance was used in the air guns to eject a constant level of alpha particles that would neutralize static electrical charges present in the gases released by various industrial processes and to remove small amounts of ionized dust particles where a high degree of purity was required. As pointed out in an earlier chapter, alpha particles are not dangerous unless their source is inhaled or ingested. Therefore, a primary thrust of the NRC is to monitor those industries that manufacture containers for products that are ingested. Before sales were suspended, 3M had sold over 20,000 air-gun ionizers to several thousand companies. Presently, because of a design and wear problem that was not foreseen, any of the ionizers could potentially release the tiny ceramic spheres that contain the polonium, although they were assumed to have been chemically bonded within the device and in compliance with the applicable safety requirements.

The erosive actions of rain and melting snow add to the need for proper reclamation of surface-mined areas, such as at the sites of many uranium and phosphorus mining operations. Public concern has brought about greater efforts to correct the environmental damage due to the removal of Earth's natural resources. Most strip-mined areas can be reclaimed, if not to productivity, at least to a point of stability. These areas, by law, must be regraded to original contour and reseeded or leveled and covered by the topsoil set aside for that purpose during the initial part of the strip-mining operation. Suitable dams are being built in the eastern United States to contain the acid water until it can be neutralized. In western states the soil is quite basic, a condition brought about by the alkaline substances (primarily basic salts) that are present. The inhibiting effect of these substances must be neutralized if plant growth is to resume. All states require grading and seeding of mined areas. Some states refuse to refund posted bonds until inspection of the area after three growing seasons indicates the resumption of normal plant growth.

Thermal Pollution of Our Waters

Practically all the energy that we use eventually assumes the form of heat. Many of our home appliances—including toasters, clothes dryers, electric grills, and electric heating units—convert electrical energy into heat energy. Heat is required for many industrial processes. At other times, however, heat represents wasted energy and is considered a pollutant. *Thermal pollution,* as this undesired energy is called, is now being recognized as a serious hazard to our environment.

In an industrial process, some of the heat produced is used and the remainder is then released as additional refuse into the atmosphere or into the water that is used as a coolant. Heated water that is discharged back into the stream or river from which it was taken may be used by several industrial plants before it again reaches the *ambient temperature;* that is, the temperature of its immediate surroundings. The detrimental effect upon ecological relationships is thereby increased. An increase of normal water temperature of only 6 C° can be lethal to various species of plants and animals that serve as a food supply for other species. The result is death for many forms of animal and plant life. Also, warmer water can accommodate less dissolved oxygen than cooler water; therefore, oxygen-breathing fish and other varieties of aquatic animal life perish or move to regions containing more natural waters. The heat also increases the rates of chemical reactions and biological activity, both of which have a retarding effect on the ability of the water to purify itself. The addition of heat, then, magnifies other water pollution problems, as well as directly endangers our food supply.

Various industries have adopted numerous techniques to minimize their contributions to our thermal pollution problems. For instance, some power plants restrict the flow of discharged warm water, giving it time to cool before it returns to the large body of water that provided it, as shown in Figure 27.8. There, the remaining heat readily dissipates with no apparent damage to the environment. In other cases, the warm water is routed into a cooling lake where it cools by evaporation and by heat loss to the atmosphere. Another system makes use of a cooling tower where the heated water is sprayed into the air and cooled by the atmosphere or by giant fans. The last two methods, however, create fog under certain atmospheric conditions, and add to the expense of electrical power to the consumer. Another method utilizes the dry tower. In this type of cooling device, warm water is pumped through coils of tubing over which air is blown. The heat is dissipated in the same manner as the heat from the radiator of an automobile. The system is closed; thus, there are no evaporation problems. This process has gained considerable popularity among ecologists and other environmental scientists.

Not all studies are concerned with cooling the water from electrical power plants or other industries. Sometimes the warm water is put to beneficial use. For instance, when a carefully controlled amount of warm water is added to unusually cold water, the metabolic rate of fish increases, and production is improved in regions where low water temperatures had inhibited aquatic growth. Another innovative use of the warm water from generating plants is that of plant irrigation. Plants such as corn and beans grow much faster with warm water irrigation than with water of ambient temperatures. Also, warm water sprayed over strawberry plants and orange groves minimizes frost damage and produces a greenhouse effect about the plants. Other crops are being exposed to warm-water irrigation at the present time, and attempts to extend the growing seasons of many plants are also under consideration. There are many other proposed uses for the heated water, such as to furnish the heat energy for desalination plants and to provide the above-ambient temperatures necessary for sewage-treatment processes.

Although thermal pollution of our water represents no widespread hazard at the present, all branches of science must become involved in the location and design of future power plants so as to keep this type of pollution to a minimum.

Air Pollution

Of the 6 quadrillion tons of air that envelop Earth, 50 percent is within 5500 meters (18,000 ft) of the planet's surface. The cause for concern is that the lower level of our atmosphere is used, reused, and reused again, particularly the air closest to the surface. Disregarding water vapor and traces of various gases, the main elements that constitute our atmosphere are nitrogen (approximately 78 percent) and oxygen (approximately 21 percent).

Unwanted ingredients of the air are those particulate matter and gases that human activity has added to the lower levels. Soot and solid unburnable particles form the preponderance of the particulate matter. Also included are dust, paint pigments, and countless other contaminants that remain suspended in the air by virtue of the prevailing winds and other air currents. Air pollution is primarily retained, however, by an **inversion layer** that forms when the air is practically motionless. As a warm air mass slowly approaches cooler air, the warm air is turned upward and over the cooler air mass. The moderately warm air that surrounds Earth rises until it comes in contact with the progressively warmer air mass and becomes stationary. As this phenomenon occurs, wastes from home and industry add to the amount of pollution previously trapped in the lower levels, and the total amount of particulate matter and gases accumulate rapidly. Figure 27.9a represents a prime example of an inversion layer. Fortunately, weather conditions may prevent inversion layers from forming, even though high concentrations of air pollutants may be present, as illustrated in Figure 27.9(b).

An increased level of particulate matter may create a persistent pollution dome in the immediate atmosphere. This condition, often associated with a high-pressure cell, raises the possibility for an inversion layer to develop. The convection process generally in effect is halted by the formation of the dome, and the pollutants contained in rising air cannot be wafted away.

In a study conducted by the EPA in 1987, Denver, Colorado ranked first in terms of level of particulate matter present in the atmosphere among selected American metropolitan areas. At the time, the formation of inversion layers was quite common. Other regions rated in the same year according to level of particulate matter as well as to ozone level appear in Table 27.1.

If conditions are favorable, an inversion layer may form soon after sunset. The ground cools off quickly, as does the air immediately above it. Air farther from the surface of Earth retains its warmth; therefore, an inversion forms at the level of the contact region of the two air masses. The height of this inversion may range from 30 to 1000 meters.

Frequently, the inversion layer does not break apart readily, but remains intact for days. Then, areas of perhaps thousands of square kilometers experience an accumulation of polluted air, a condition that becomes hazardous to health. The layer subsides somewhat each day and constantly moves closer to Earth; thus, the situation becomes much more serious. Inversions are quite common over the Central Plains and are only slightly less common over the Atlantic Coast area.

Recent investigations indicate that air pollution may be increasing the average temperature of Earth. An overall temperature increase of 4 C° (7 F°) is thought to be enough to cause polar ice caps to melt appreciably, a condition that would flood many cities. Other studies suggest that the average temperature is decreasing. An overall decrease of 4 C° possibly could cause the climate of Miami to be similar to that of the city of Boston, and Seattle to be similar in climate to the coldest regions of Alaska. Either change could occur in less than four centuries, and the climate of the various continents would be affected accordingly. Interestingly enough, data collected by the TIROS-N series of weather satellites as they monitored Earth's entire lower atmosphere indicated no general warming or cooling trend over a 10-year period. The instruments aboard the satellites did detect both significant increases and decreases in global temperature, but each change lasted only several months. NASA scientists, however, point out that the temperature of Earth's surface and its lower atmosphere might not fully coincide over such a brief time span. Perhaps in decades to come, additional data gathered from weather satellites will ultimately settle the controversy over possible changes in global climate.

Various industries were directly affected by a court ruling in early 1988 that brought about more rigid smokestack emission controls. The federal appeals court ordered the EPA to rewrite,

a.

b.

c.

Figure 27.9 (*a*) Although air quality has significantly improved, inversion layers can still form above large cities as gaseous industrial wastes and auto exhausts concentrate. (*b*) Smog scenes still exist in various areas of the United States. (*c*) Prior to air-pollution reforms, industrial areas accounted for much of our air pollution.

Table 27.1

A comparison of levels of particulate matter and ozone present in the atmosphere over various metropolitan areas, as determined by the EPA in 1987.

Metropolitan area	Particulate ranking	Ozone ranking
Denver, Colorado	1	59
San Bernardino, California	2	1
El Paso, Texas	3	27
St. Louis, Missouri	4	7
Los Angeles, California	5	2
Phoenix, Arizona	6	39
Detroit, Michigan	8	59
Gary-Hammond, Indiana	8	46
Cleveland, Ohio	14	46
Chicago, Illinois	17	27
Birmingham, Alabama	17	46
Tacoma, Washington	19	59
Houston, Texas	21	4
Boston, Massachusetts	21	11
Charlotte, North Carolina	65	59
Toledo, Ohio	66	79
Buffalo, New York	73	47
Raleigh-Durham, North Carolina	80	46
Orlando, Florida	84	59
Monmouth-Ocean, New Jersey	88	20
West Palm Beach, Florida	89	85

for a third time, a portion of its smokestack rules. Previously, companies could rely on tall smokestacks to disperse particulate matter and various polluting gases, preventing excessively high concentrations of the various emissions from being pulled toward the ground. However, the winds that swirl around structures in the immediate vicinity create a "downwash" effect. The court decision was based on the premise that the EPA may have been too hasty in assuming that tall smokestacks satisfactorily reduced air pollution to acceptable levels. The ruling applied primarily to large coal- and oil-fired power plants, paper mills, industrial boilers, and certain smelters. The petrochemical processes of various oil-refining companies were also placed under closer observation to see if the contaminants they emitted, especially particulate matter such as calcium carbonate, as well as caustic mixtures of the oxides of aluminum, calcium, iron, and silicon, were creating health hazards and damaging the property of nearby residents. The resulting adjustments improved the air quality to the satisfaction of those who initially expressed the concerns.

Among the gases added to the atmosphere by industry are oxides of nitrogen and sulfur. Nitrogen oxides are unwanted by-products that accompany combustion. Their odors are obnoxious; they irritate eyes, noses, and throats; and they absorb sunlight. Sulfur oxides reduce ozone and accumulate in areas where fossil fuels and wood are burned, such as over power plants, coke-producing plants, and large apartment buildings. The black smoke from some past industrial processes contained various gases and particulate matter that often darkened the sky, as shown in Figure 27.9c.

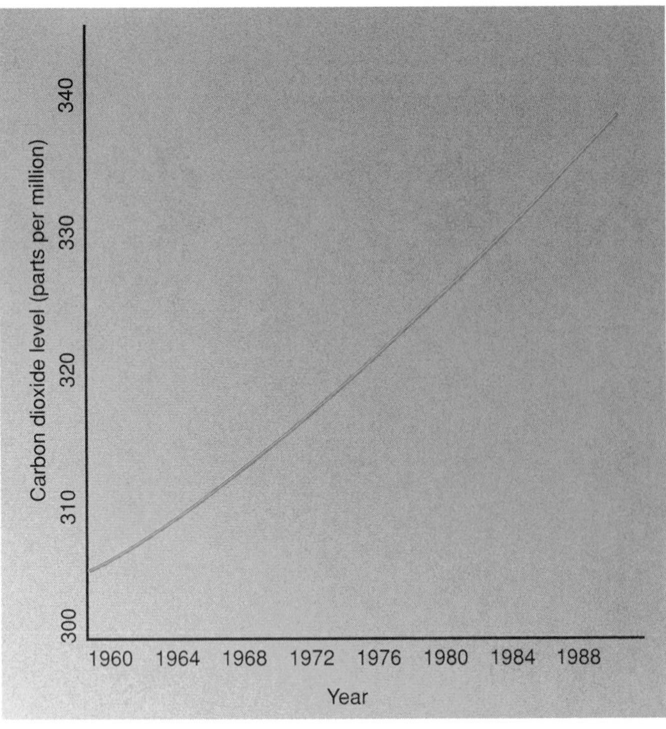

a.

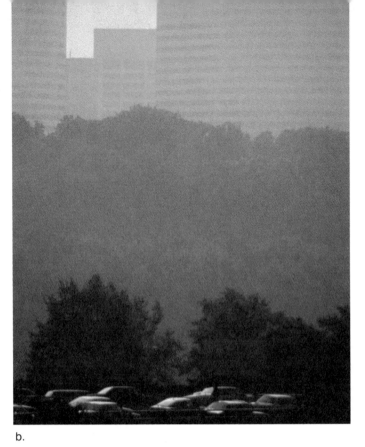

b.

Figure 27.10 (*a*) The approximate increase in atmospheric carbon dioxide, essentially brought about by human activities. (*b*) Although a hazy scene such as this one along a major freeway is becoming less frequent, the automobile remains the world's greatest source of air pollution.

A report released at the end of last decade by various investigating agencies indicated that for years, during the production of the electricity we required, our power plants had been emitting a combination of at least 60 million tons of sulfur dioxide, nitrogen oxides, and volatile organic compounds into the atmosphere annually. Most of these air pollutants were released by combustion of fossil fuels. Although this means of providing our electrical needs released a lesser amount of pollutants into our environment during the 1980s compared to the level emitted during the 1970s, our continued dependence on the combustion of fossil fuels to generate our electricity has been proven to be detrimental to our air quality. Even today, these electrical power plants, along with the major industrial operations they serve, produce almost half of the various pollutants added annually to our atmosphere. The Synthetic Fuels Bill recently passed by Congress also appropriated funds to study potential environmental effects caused by the combustion of fossil fuels. A portion of the report that resulted from this funding pointed out that there has been a significant increase in the carbon dioxide content of the air since 1960, as shown in the graph of Figure 27.10a. This environmental hazard, then, has been receiving attention for over 30 years, although efforts to lower the carbon dioxide concentration in our atmosphere seem to have had very little effect. (Incidently, many concerned citizens maintain that the stench released by paper mills located in such geographic regions as southern Alabama, western North Carolina, and central Ohio should also be investigated in similar studies, although in this case carbon dioxide is not the primary issue.)

During the early 1980s, some researchers did not consider the increased acidification of our surface waters to be directly related to the documented rise in air-pollution levels. Today, however, practically all scientists have discounted most other causes and have spoken out strongly for immediate reform before the environment is irreparably damaged. They conclude that sulfur- and nitrogen-based air pollutants emitted during the burning of fossil fuels account for most of the acidification plaguing various lakes and streams, particularly in the eastern United States. The precipitation over this area contains large amounts of these constituents and thus is known as *acid rain.* Precipitation in any form that has a high acid content is highly suspect in damaging most forms of aquatic life and woodlands. Canadian officials for decades have contended that our industrial emissions have detrimentally affected their country's forests. They, along with scientists from the United States, eventually were instrumental in the development of the National Acid Precipitation Assessment Program (NAPAP) of the 1980s. This $500 million federal program represented one of the largest research studies in our nation's history. Even before the NAPAP was initiated, sulfates, the leading component of acid rain and primarily formed from sulfur dioxides, were found to create a haze that decreases visibility by as much as 60 percent in some eastern regions (recall Fig. 27.9b). Scientists also concluded that organic particles added to the environment by industrial processes, along with the nitrogen oxides emitted by internal-combustion engines, created a similar haze and contributed to the acidity of the rain that falls on western states where much less coal is burned. This conclusion was upheld at an international NAPAP meeting conducted in 1990, attended by about 600 dedicated scientists.

Even though acid rain has received concentrated attention by a large number of NAPAP researchers, there remain important areas that apparently require further investigation. Among the topics that must be researched in greater detail is the determination of which chemical constituents present in rain, snow, or fog are most hazardous to humans and other forms of life. Also, researchers must seek the optimum pH range of precipitation for forests, lakes, and crops.

Another international group is continuing to study the fires deliberately set for agricultural practices and land-use conversion. Their task involves regions that contain the tall grasses of savannas and tropical rain forests, as well as the many farms in operation all over the world. To appreciate the magnitude of their endeavors, note Figure 27.16 (p. 588). The combustion of basically unwanted vegetation is called *biomass burning,* and it adds considerably more pollutants to the atmosphere than earlier researchers had assumed. High levels of nitrogen oxides and other gases released by such burning are converted by the atmosphere into nitric acid and organic acids, thus adding to the level of acid precipitation that falls on once-pristine areas, such as Africa and South America.

Although preliminary research projects indicated that crops were not particularly damaged by acid rain, later studies have revealed that at various stages of growth, corn plants bear a lesser number of kernels when exposed to acid rain than those plants that receive identical amounts of rainfall with significantly lower acid content. Further studies have also revealed leaf damage on such plants as tomatoes, soybeans, tobacco, and cotton that have been exposed to rain with relatively high acidity. A general survey has all but confirmed that there is evidence of serious forest damage throughout several continents, accompanied by a suspiciously high presence of such pollutants as acid-forming nitrates, sulfates, sulfur and nitrous dioxides, ozone, and the heavier metals.

Various scientific teams have dedicated a major portion of their time and efforts in an attempt to devise a cost-effective technique that would significantly lower damaging contaminants in environmentally sensitive areas. Scientists at the Argonne National Laboratory have developed several chemicals that can remove, through a filtration process and at a relatively low cost, more than 70 percent of the nitrogen oxide that accompanies other less troublesome gases released by the combustion of fossil fuels. In the atmosphere, nitrogen oxides can be converted into nitric acid, apparently the second-most damaging component of acid rain. Also, the wet-scrubber technique, used by many coal-burning power plants in the United States, is significantly improved when a fine liquid-chemical mist developed by the Argonne scientists is sprayed through combustion gases to remove the sulfur dioxide that forms sulfuric acid, considered the most devastating component of acid rain. The various techniques developed by Argonne scientists and others could save billions of dollars in our national efforts to preserve our forests and agricultural products.

Motor vehicles add more pollution to the air than any other source. A typical scene around certain metropolitan areas at various times is shown in Figure 27.10b. At the present time, there are 500 million registered automobiles on all continents that consume one-third of the world's production of oil. Over 100 million of these vehicles are located in the United States, and they add more than 200 million kilograms (a weight of 220,000 T) of gases and particulate matter to our immediate atmosphere each day. The auto industry continues to spend millions of dollars annually to develop methods to lower the levels of obnoxious gases and particulate matter emitted from automobile exhaust systems. In fact, all imported vehicles and those manufactured in the United States must be equipped with catalytic converters, efficient engines, and other devices that meet rigid pollution standards before they can be made available to the public.

Carbon monoxide, a by-product of incomplete combustion, is one of the most common harmful pollutants produced by the automobile. An estimated 66 million tons of the deadly gas are released each year from exhaust systems.

Other contaminants released by the internal-combustion engine are the nitrogen oxides. These compounds, formed by most combustion processes, require expensive and complicated devices to absorb them. Also released from automobile exhausts and those of other internal-combustion engines are some metallic toxins; however, the worst offender, lead, has all but been eliminated from exhaust emissions with the conversion to unleaded gasolines. Unfortunately, the lead and other metallic additives emitted into our environment by the combustion of almost countless gallons of leaded gasoline before the late 1980s will remain in our immediate environment for centuries to come.

Another type of atmospheric contaminant known as photochemical air pollution, or photochemical smog, has become a major concern for various geographic areas (again recall Fig. 27.9b). *Smog,* as the pollutant is better known, is a mixture of gaseous and particulate matter that results from reactions that occur in the atmosphere between substances placed there by incomplete combustion from motor vehicles and industry. Ultraviolet radiation from the sun initiates chemical reactions between nitrogen oxides and organic residues that include aromatic hydrocarbons and aldehydes (both are common categories of organic compounds). The accumulation rate is sometimes great enough to cause a haze to form, the density of which reduces visibility much the same as fog does.

In order to bring about a decrease in the health hazards from various air pollutants, numerous attempts have been made to replace the automobile and its internal-combustion engine, but none have made a contribution of any lasting significance. Mass transportation in metropolitan areas has been popular for decades, but irregular schedules, levels of passenger discomfort, and crowded conditions apparently have discouraged the expected increase in their sustained use by daily commuters. Trains that burn coal or diesel fuel and diesel-powered buses add their share of contaminants to the air, but by no means at the rate of the thousands of cars they have replaced. In many areas, they still remain a major means for people commuting to and from their jobs. Electric cars were introduced, since they did not contribute to the level of carbon monoxide, sulfur and nitrogen oxides, or hydrocarbon compounds

Hazards at the Office

The modern office is not without its dangers to its daily inhabitants. Various sources add to the contamination of the air where many people, more susceptible to the effects of poor air quality than most healthy factory workers, spend their working hours in windowless facilities. Such sources as ambient cigarette, cigar, and pipe smoke, odorless hydrocarbons from wet-process copiers, and the ozone released from dry-process copiers are all sources of office air pollution. Other sources of indoor pollution are found in plastic furniture, construction materials, and carpeting, along with the cleaning agents used on them. Air conditioners and heating systems that are not serviced regularly add dust, bacteria, and viruses to the polluted air. The obvious solution is sufficient ventilation to change the entrapped air at short, regular intervals, if not in a continuous fashion. The loss of heated or cooled air as it is exchanged creates an expense to the business that must be weighed against the health and comfort of the office employee. The dilemma seems unresolvable, so the unhealthy-environment syndrome remains ■

already present in the atmosphere above large cities. However, to produce the additional electrical energy to charge the batteries each night, an increase in level of contaminants from the power plants was soon noted. Electric-powered vehicles, like all other devices that use electric energy, release ozone, a molecule composed of three, rather than the more typical two, atoms of oxygen, into the air. Not only does ozone attack with zest such substances as rubber and asphalt, it is also harmful to most forms of life when present in concentrated amounts. Unfortunately, the ozone produced as batteries are charged or discharged is heavier than air; thus, it cannot move upward in our atmosphere to replace the thinning layer of ozone, particularly over the Antarctic region that protects Earth and its forms of life from the potentially dangerous ultraviolet radiation received from the sun.

A propane-powered vehicle may well be the least polluting type of engine available at the present time. The combustion of propane offers little waste since propane burns practically to completion and forms primarily carbon dioxide and water. More advantageous yet would be the development of an engine to use in our automobiles, airplanes, and other modes of travel that would obtain its energy from the reaction of hydrogen and oxygen to form water. Teams of scientists are currently investigating this possibility.

Most knowledge about the effects of air pollution on health has been accumulated during periods of acute inversion formations or after industrial accidents. Scientists have determined that air pollution is a complicated process, the severity of which depends on the climate, the density of traffic, home heating methods, and the topographic surroundings. A significant decrease in air quality would represent the most serious environmental deterrent to our health, and it must be monitored with extreme care. This attitude has gained support with regulations enacted by the EPA in 1989 and with the Clean Air Act introduced the same year. Provisions in this important legislation are designed to limit to some degree the amount of toxic organic compounds, including benzene, being released into our atmosphere. Industrial emissions have been absolved as the primary source of various toxic organics that have detrimentally affected our air quality. Instead, much of the undesirable nonoccupational emissions apparently can be attributed to tobacco smoke and auto exhaust systems, along with other consumer products such as household solvents and latex paints. The Clean Air Act, amended in November 1990 to include even more stringent controls, along with the constant involvement of the EPA, will have a great bearing on our welfare in the coming decades.

Noise Pollution

Another serious environmental problem is excessive noise. Within limits, noise is both unpreventable and necessary if our society is to function effectively. However, commercially produced noise is suspected of causing unfavorable conditions in our cities. The general public is aware of the annoyance created by power vacuum sweepers, pneumatic hammers, noon whistles and sirens, or noisy trucks and automobiles equipped with inadequate or altered muffler systems. Added to these sources of noise are the powerful sound systems privately installed in some vehicles from which emanate disturbingly loud levels of rap, rock, and country-western music.

Noise is purported to cause inefficiency, loss of sleep, nervous disorders, and heart disease, but reliable and sufficient proof of cause has yet to be established. However, science has undertaken the task of determining whether sounds of various frequencies and intensities are in any sense detrimental to good health. Undoubtedly, action to diminish these noises will be taken should research findings reveal them to be a definite health hazard.

In the past, various professions were linked to isolated cases of hearing loss, such as those found among blacksmiths, boilermakers, and other metal workers who extensively used noisy devices. During the last several decades, the loudness and frequency of sounds have been under study by the medical and other professions as a potential cause of hearing loss. More sophisticated testing devices to measure characteristics of sound and their effects on hearing have opened the way to concentrated research on the suspected correlation. A device to study sound intensity is shown in Figure 27.11. Comparisons of the intensities of various environmental sounds are presented in Figure 27.12.

Research reveals that sounds of equal frequencies can vary in annoyance according to loudness. In addition, sounds in the frequency range between 200 and 10,000 hertz can be heard at lower

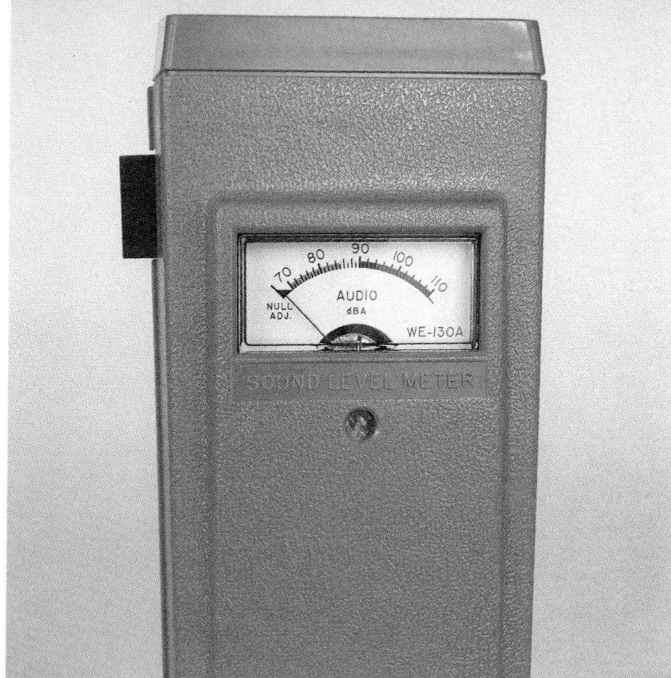

Figure 27.11 Instruments such as the one shown are used to monitor the loudness of sounds, measured in decibels.

Range	Decibels	Source
Dangerous to hearing	170	Rocket engine
	160	Jetliner, pre-takeoff
	150	Air raid / noon sirens, near source
	140	Amplified rock music, near speakers
	130	Thunder, immediate area
Extremely loud	120	Jackhammer, as heard by operator
	110	Jet plane at 300 meters
	100	Factory engines, as heard by operator
	90	Carpenter's skill saw
Loud	80	Busy street traffic
	70	TV commercials from set adjusted for normal listening
Quiet	60	Normal TV listening level
	50	Ordinary conversation
	40	Normal conversation, next room
Audible	30	City library
	20	Whisper
	10	Rustling of leaves
	0	Threshold of hearing

Figure 27.12 A chart of noise levels in decibels. Note the number of intense sounds produced in our homes. The United States is actively attacking this problem to lower the intensity of many sources of noise.

intensities than can sounds above 10,000 hertz or below 200 hertz. The range of greatest sensitivity lies between 500 and 4000 hertz.

The seriousness of the problem of hearing loss becomes evident when one realizes that about 10 million people in the United States wear hearing aids. Industry surveys have reported that 20 percent of the workers tested have impaired hearing. Those people who work in factories have twice the frequency of hearing loss as those in similar age groups who work in offices.

Often hearing loss is caused by a sudden loud noise, but there are many cases of severe pain leading to deafness among those people who are exposed to low-level noises over long periods of time. In fact, constant exposure to sounds of low intensity has been used as a torture device to gain information from prisoners of war.

Industrial and home appliances operate by the conversion of energy from one form to another. Sound, as well as heat and light, is often produced as wasted energy, the amount of which depends upon the efficiency of the device. Pneumatic hammers, for instance, are very inefficient. Each of the three forms of energy creates serious concern for industry, since too much of any one of them causes discomfort and inefficiency among workers.

Noise is considered *an unwanted sound;* hence, it is indeed subjective. Sounds that annoy some people cause no discomfort to others. Specifically, noise is a form of energy that is transmitted through some medium such as air, water, or solids and then is converted to sound by the mechanisms of the ear it reaches.

One of our most disturbing sources of noise is the jet engine. The engine, at advanced throttle, creates noise that exceeds 130 decibels, almost the limit of human endurance. Technical advances have found means to significantly muffle the sounds, but the devices required to do so are expensive and heavy. A plane equipped with the suppressive devices is forced to carry less cargo or else risk overburdening the engine. Therefore, costs to the consumer increase due to smaller payloads and to the costly installation of the muffler systems. Under present circumstances the roar of jets and the impending potential of the sonic booms they produce annoy many people who live on the routes over which the planes fly. Those people who live in the immediate regions of commercial airports are even more affected by the voluminous noises.

It was discovered in 1966 that the sonic boom is not the result of the plane's breaking the sound barrier (reaching velocities greater than the velocity of sound), a misconception held since World War II. In reality, the boom is produced by a tremendous pressure wave generated by the nose section and tail assembly of planes that exceed about 1060 kilometers (660 mi) per hour. The pressure wave eventually is transmitted through the air to Earth's surface and is interpreted as a thunderous boom.

Commercial and Domestic Wastes

Sewage is composed of the wastes from homes, industrial plants, hospitals, other public buildings, and similar sources. Not only human wastes but also industrial wastes are involved. Everyone is

Figure 27.13 Treatment plants, such as the one shown, are designed to process commercial and domestic wastes before they are released into the environment.

aware of the various types of refuse that accumulate at home, including waters from the bath and kitchen and the soap, oils, and organic materials that they contain. Added to this deluge of waste products is the industrial accumulation of undesirable materials from meat-packing plants, milk-processing facilities, canneries, breweries, and drug companies.

The addition of phosphates to laundry detergents has been essentially curtailed. This type of compound was used for a period of time to improve the effectiveness of water that had a high mineral content in removing dirt and stains from various materials. However, the buildup of phosphates as they made their way through the sewage systems into our rivers, lakes, and streams accelerated the unwanted growth of algae. As the algae died and decomposed, their remains reacted with the oxygen dissolved in the water and also produced hydrogen sulfide, an obnoxious, poisonous gas. Both actions are damaging to the more beneficial forms of aquatic life. As a result, scientific research led to the development of biodegradable additives, including various enzymes that appear to be harmless to our environment yet effectively improve the degree to which detergents can cleanse our clothes.

Many other problems in treating sewage yet remain. Some derive from the organic solids, including carbohydrates, fats, and proteins that are still present after the sewage is processed. Our larger metropolitan areas must deal with perhaps 500,000 kilograms of daily wastes that, regardless of the method of treatment, accumulate to staggering quantities (see Fig. 27.13). A major problem lies in disposition of the treated sewage, for as it is released into lakes and oceans, the various organic solids and other substances provide abundant food for algae and other microorganisms. This action also propagates viruses, protozoa, and pathogenic microbes that cause the spread of disease as they enter our streams, rivers, and groundwater. (Imagine how many times the water in our major waterways is used by consumers before it reaches the end of its journey into the seas!) Several dozen diseases can be communicated by solid wastes to unsuspecting consumers of contaminated water.

The Clean Water Act was finally passed by Congress in February 1987 amid considerable controversy over the cost associated with it. The law provides about $20 billion, primarily to construct new sewage-treatment plants. A smaller portion is to improve the water quality of Chesapeake Bay and to help states control chemical runoff from nonindustrial sites. Other funds conceivably will be provided by the act to sponsor research aimed at the development of more effective methods to treat sewage before it is released into our waters.

The impending danger from old methods of solid-waste disposal came to the forefront with the discovery in 1978 that toxic industrial wastes dumped at **Love Canal** in upstate New York in the 1940s and 1950s were responsible for many of the serious ailments that had befallen area residents. An abandoned canal, located near Niagara Falls, was the site of the dumping of many hundreds of waste-filled drums. In 1953, the canal was filled with dirt and the property was essentially donated for the purpose of constructing an elementary school and housing project on the site.

In 1978, blood tests conducted on various residents in the immediate vicinity of the Love Canal chemical disposal site revealed that they had experienced possible chromosomal damage. Further tests showed the presence of dioxin and other toxins that had seeped from the Love Canal disposal site into the basements of residents in nearby Niagara Falls, N.Y.

The Environmental Protection Agency, after much deliberation, finally determined that the best way to handle the cleanup of the Love Canal site was to incinerate contaminated sediments dredged from the Love Canal sewers and creeks, and to rebury the nontoxic residues in the old canal itself. A potential amount of 27,000 cubic meters of sediment was to undergo the treatment, marking the operation as the largest example of the "thermal-destruction" technique ever attempted.

The EPA projects that the program could take until about 1993 to complete, at a cost that ranges as high as $30 million. Federal and state programs have already contributed in excess of $200 million for the Love Canal cleanup and for the relocation of about 1000 families that had been living in the area. Even so, a carefully guarded area of about 30 acres close to the main disposal site will remain uninhabitable for years to come. It is noteworthy that in early 1991, despite strong opposition from environmentalists, the Love Canal Area Revitalization Agency was granted permission to sell renovated houses in the immediate vicinity—a neighborhood to be known as Black Creek Village.

Of the thousands of covered pits in the United States suspected of containing toxic wastes, about 900, according to the EPA, could possibly become serious health hazards. Another case similar to the Love Canal episode existed in Elkton, Maryland, where residents complained of sore throats, respiratory problems, and headaches, all reminiscent of the early symptoms experienced at Love Canal. In Rehoboth, Massachusetts, some 750 cubic meters of chemicals remaining from a solvent-distilling process were ordered removed from a dumpsite that extended almost under the owner's house. In Lowell, Massachusetts, some 15,000 drums of assorted toxic wastes were removed from a burial site within 200 meters of a housing development.

Some industrial companies in the past contracted private firms, called midnight haulers, to cart off wastes and dispose of them as they saw fit. The illegal dumping that followed was done quite often in swamps, sewers, pits, or abandoned wells to avoid paying for disposal at approved sites. In the early 1980s, one such hauler was convicted of dumping toxic liquid wastes along over 400 kilometers of North Carolina's highways. Other blatant violaters have been arrested for opening valves on their trucks to spill contaminants along streets, in fields and streams, as well as on interstate highways.

The need for full cooperation between the public, industry, and government regarding the disposal of wastes of all sorts has been recognized by many concerned citizens. The enforcement by organizations such as the EPA has done much to deter further damage to our environment. But chemical landfills apparently never lie dormant. When groundwater permeates the buried wastes, it constantly removes the soluble components of the wastes, producing a polluted mixture, called a **leachate,** that escapes from the site. The leaching spreads dangerous stable materials over a large area or pollutes other groundwater and our streams. An EPA estimate suggests that an average landfill site produces almost 5 million gallons of leachate annually in conjunction wth an average rainfall of 25 centimeters (10 in).

a.

b.

c.

Figure 27.14 Scenes from the past. (*a*) Although there is some controlled burning of solid wastes even today, the environmental impact from this procedure is minimal compared to its detrimental effects in the 1960s. (*b*) Much of the particulate matter that remained after the burning of solid wastes fell on surrounding areas. Note the light-colored ash deposited on rooftops, vehicles, and streets near the burning site. (*c*) Smoke from burning trash used to billow over the White House in a cloud visible for miles.

One manner of disposing of the solid wastes is to process it in an oxidation pond. The solids are suspended in water as they enter the pond. As the rate of flow of the water decreases, the heavy solids settle; then bacteria and microbes thrive on the organic matter that remains suspended in the water. Through this process the organic matter is consumed; then the remaining heavy solids are removed from the bottom of the pond and buried.

The long-accepted method of burning solid wastes in the open air, as shown in the various scenes of Figure 27.14 has been abandoned for the most part in practically all licensed dumping sites because of public reactions and the development of higher environmental standards. There are some violators, including sanitary landfills and lumber mills, that somehow have circumvented the various phases of the Clean Air Act of 1970, which expired in 1981. Clean Air Act amendments passed in 1990 gave the EPA authority to designate technologies that industries must use to limit pollution from almost 200 toxic substances. The agency also sought substantial voluntary cuts in emissions from various industries that released the 17 chemicals considered to pose the greatest threat to human health. These substances include benzene; carbon tetrachloride; various ketones; toluene; xylene; and such metals as cadmium, lead, mercury, and nickel, along with their compounds. Assuming industries will have no alternative but to dispose of these toxic substances through burial, their addition will significantly contribute to the mounting accumulation of other solid wastes. Furthermore, the number of suitable sites for new landfills is all but depleted in some states. In fact, the EPA has estimated that 25 states will have used all available landfill sites within their boundaries by the year 2010. Even now, some cities have had to refuse to accept the wastes from other cities within the same state—not to mention wastes from cities outside the state—for danger of overextending their existing facilities.

Fortunately, some areas have constructed incinerators that effectively filter the smoke they produce and thus can compact the volumes of those wastes that are not dangerous to our health or environment when burned. These sites undergo very rigid and regular inspections by federal and state health agencies.

In the summer of 1987, many of the national headlines chronicled the hapless journey of the barge loaded with refuse that was rejected from an overflowing landfill located at Islip, New York. The *Mobro 4000,* loaded to capacity with overripened trash, apparently exceeded the maximum amount that could be processed at other facilities within the immediate region. As a result, the barge moved from state to state along the eastern seaboard, in search of a site that would agree to process its contents. Finally, after sailing 6000 miles in a seemingly futile journey that lasted about six months, the boatload of refuse was reduced to about 400 tons of ashes in a New York City incinerator system. After this treatment, it was then considered acceptable for deposition by the same processing site at Islip, New York, that had initially rejected it, but only after the state had agreed to an expansion of the dumpsite.

At the present time, some states have started construction of incinerators and have initiated recycling programs, but others intend to continue their attempts to find sites that will process their trash. Certain metropolitan areas that have attempted this latter route will readily affirm that such a solution is not only difficult, but expensive, indeed.

Ever since the controversy and harsh public criticism that arose following Love Canal and similar incidents, safe methods of toxic-waste disposal have been a major objective of the various agencies in this and other countries. Recently, several companies have been formed for the sole purpose of disposing of toxic wastes. Two of these processors have proposed burning the wastes aboard specially designed ships miles out to sea. Numerous complaints about this suggested method have already been voiced, particularly from among those who live in coastal towns. The EPA must decide if this procedure is environmentally sound, and if such a disposal method is superior to land-based incineration or burial at sea or on land. There appear to be several advantages to the disposal of wastes by incineration at sea. First, this method may have the lowest cost of all those previously used. Also, the incineration of the toxic wastes at a slower rate than conventional methods can use may increase the percentage of the substances rendered harmless. As a result of slower incineration, the detrimental effects on the immediate environment may be decreased. There are several European countries that have used this disposal technique for a number of years with no obvious detrimental effects on the environment.

Another procedure for disposing of toxic wastes under study by the EPA involves the incineration of such substances in industrial boilers to make use of the valuable heat energy created by the process. For instance, the Pyropower Corporation of California developed a high-tech boiler that offered a new approach to combustion optimization and environmental impact. This corporation employs a technology called *circulating fluidized bed* (CFB) *combustion* to provide heat energy required by various utility stations. Through the control of temperature, turbulence, and time, the boiler system provides a high burning efficiency of such fuels as woodwaste products and other commercial wastes. Petroleum coke and highly contaminated fuels, including high-sulfur coal, have been used, with a release of gaseous and particulate matter reportedly well below acceptable EPA standards.

In both burning at sea and in the boiler system methods of treating wastes, the EPA found that about 99 percent of the waste was destroyed. With such high combustion efficiency and under strict monitoring, perhaps both methods can be used to handle solid wastes of all sorts.

The dangers from solid wastes are global concerns; they cannot be confined to a single country or region. In Bulgaria, for example, alarming levels of lead in the blood, nails, and teeth of children have been detected. Particulate matter, such as that released by a local battery factory, containing vast amounts of lead, as well as zinc and cadmium, continues to pollute the environment. Since 1959, flagrant neglect for the environment has caused the release of dangerous levels of toxic metals into the air, soil, and water to the point that it has detrimentally affected the health, intellectual development, disease immunity, and life span of milions of people as well as their livestock in Eastern Europe and the Soviet Union.

Some European countries are converting nontoxic solid wastes into cinders by burning the wastes under controlled conditions. The cinders are processed by crushing, then used to pave roads and to extend land areas into the seas. In the United States other projects are under way in which compressed solid wastes are treated with adhering plastics, molded into bricks, and used to construct buildings. Still others are investigating the feasibility of building islands in the calm, shallow inlets that encompass our heavily populated eastern seacoast.

Methods of reclaiming solid wastes are under serious study because of the immense storage problems and the rapid rate at which our natural resources are being consumed. At the present time, few, if any, efficient methods of reclamation of the constituents in solid wastes have been devised. However, a solution to the problem must be found before our resources are completely depleted. Future generations may consider our landfills, dumps, and auto graveyards as major sources of mineral deposits. In the future, our standard of living will be determined by our ability to salvage and recycle our waste products. The national cost of sewage collection, transportation, and treatment amounts to over $4 billion annually. Three-fourths of this amount is spent on collection and transportation alone; still, many cities have insufficient systems. More money must also be allotted for the treatment and possible recycling of the wastes to preserve our overall economy and our remaining resources.

The increase in the use of nonreturnable containers of all sorts continues to mount. Even though some states have discouraged the sale of soft drinks and other such commodities in disposable aluminum containers, the attempt at decreasing the amount of solid-waste accumulation in this category within the United States has been a dismal failure. Plastics, such as the synthetic polymer resin known as *styrofoam* used in the production of disposable cups, fast-food containers, and so forth, annually contribute over 20 million tons to our solid-waste problems. Such substances may take up to five centuries to degrade. However, some scientists contend that the materials used to replace syrofoam may pose an even greater threat to the environment. Add to the 20 million tons the volume from plastic garbage bags, frozen food containers, milk jugs, and the almost unimaginable number of disposable diapers used daily, and then consider the obvious consequences. Biodegradable plastics have been developed, but they are, as yet, too expensive to make a significant impact on the consumer market. According to scientists associated with various chemical corporations, the incineration of waste plastics would yield an energy that would rival an equal volume of oil, but hydrochloric acid and other such compounds would be released into the atmosphere by the burning of the plastic containers. Current research affiliated with this technique centers around the development of chemical scrubbers that would remove contaminants from the smoke produced by large-scale plastic incineration before release into the environment.

Although such waste-treatment techniques as electrodialysis, reverse osmosis, compaction, distillation, and dissolution are under various stages of development and consideration, the problem of treatment is severe and immediate. Our generation must confront the problem through combined efforts of science, technology, and citizen action if civilization as we know it is to continue.

Pollution from Hazardous Wastes

In the 1970s, the burning of fossil fuels as a source of heat energy was partially replaced by the energy from the atomic nucleus. Nuclear power, according to current plans, will provide a significant amount of our electrical energy in the future. This supplementary source of energy is relatively new, and there are great concerns and many controversies about its implementation. Some environmentalists realize that nuclear energy will solve some ecological problems but fear it will create others. The level of air pollution from nuclear reactors is much below that caused by coal-fired generating plants. In addition, a typical nuclear reactor uses 1.2×10^5 kilograms (a weight of 132 T) of enriched uranium annually and furnishes electrical energy equal to that provided by the combustion of 1.8×10^9 kilograms (about 2×10^6 T or 50,000 truck-loads) of coal. Additional implementation of nuclear energy to provide our electricity will permit us to conserve our fossil fuels and to remove from them the valuable natural resources they contain.

Nuclear generating plants are in operation through much of the world, and many more are in various stages of construction—but not without careful consideration of their environmental impact. There is no question that accidental exposure to certain levels of radiation is potentially harmful. At high levels of exposure, physical damage to our bodies occurs and serious illness or death may result. Even at lower levels of exposure, genetic effects detrimental to our offspring may occur. Moreover, radioactive by-products always accompany the fission process, and the disposal of nuclear wastes presents a serious environmental problem. Additional radioactive wastes are produced by various industrial processes, by medical applications, by nuclear fuel preparation, and by continued industrial and medical research.

Even though the nuclear industry exercises many precautions, there always remains the remote possibility that an accident will endanger public health and safety. Federal records initially released in part during 1986 from the files of the Atomic Energy Commission, an agency replaced by the U.S. Department of Energy (DOE), indicate that excessive levels of radioactive material escaped into the atmosphere from the Hanford nuclear weapons plant in south-central Washington. According to further data made public in 1990, significant quantities of iodine-131, a gamma-emitter, escaped from the facility as plutonium was being separated from its ore between 1945 and 1948. According to official estimates, about 5 percent of the people who lived in the area during this period absorbed a level of radiation over 1000 times the permissible level, as determined by the DOE. The radioactive iodine was primarily ingested through milk from cows that had fed on contaminated pasture grass. Hundreds of infants and children are estimated to have received even larger doses. However, far greater numbers of people received high levels of radiation from radioactive iodine and other radionuclides released during the Chernobyl accident. Over 5 million people live in the Ukraine and Byelorussia regions that were heavily contaminated. The cleanup of these areas will continue for many years at a cost that may exceed $400 billion.

Other cases of radiation exposure were made public at the end of the 1980s. Documents obtained from the Nuclear Regulatory Commission by a consumer advocacy group revealed that thousands of nuclear power plant employees received a measurable amount of radiation between 1986 and 1988. Presumably these individuals wore film badges and other detection devices that primarily measure accumulated doses of gamma radiation. The levels of exposure, however, were not released, but it can be assumed that some at least approached safety guidelines. The Council on Energy Awareness, an industry-sponsored group, responded to the report, but contended that the total number of individuals who receive measurable doses of radiation is constantly decreasing.

The Juárez Incident

The largest accidental release of radioactive materials into the North American environment occurred in early 1984. The incident, first detected by Los Alamos National Laboratory health physicists, triggered a search for the source in trucking companies, junkyards, and warehouses. The investigation next centered about the site of two Mexican foundries that had produced over 5000 tons of radioactively contaminated steel. The workers at these locations had been exposed to radiation much more intense than that released during the Three Mile Island reactor accident. The gamma-emitting radionuclide, cobalt-60, had unknowingly been mixed with the steel used in the production of construction rods, but the search was not over. Investigators traced the contaminated steel to a foundry in Chihuahua, Mexico. They located five more truckloads of the steel en route, and as a result, the cargoes were impounded before they were delivered for further distribution.

Ultimately, the bars of radioactive steel were found to be made from a shipment of scrap metal obtained from a junkyard located in Cuidad Juárez, a Mexican city across the Rio Grande River from El Paso, Texas. Agents from both nations investigated the site and found radioactive material scattered in the remaining junk, present in the dirt, spread throughout the office, and strewn along the roadside.

An audit of the records revealed that an employee of a local hospital had brought in a metal canister to sell as scrap. The steel cylinder was opened by parties unknown, and its 6000 tiny metallic pellets had been strewn throughout the scrap in the junkyard, perhaps by the electromagnet used to separate steel from other substances. The pellets of cobalt-60 had originally been used in a radiotherapy device for the treatment of cancer, but had decayed to a level of activity that was of limited benefit. The tiny pellets were picked up in the tire treads and shoe soles of the employees and customers, then carried many miles onto the highways and into homes.

Many of the pellets were located by radiation detection devices, but not before some of the workers had developed obvious signs of radiation damage, including bleeding noses, darkened fingernails, blisters and burns, as well as blood disorders. In addition to the rods, table legs were manufactured from the contaminated steel and designated for delivery to a Chicago distributor. Other steel products were also found to contain the radioactive cobalt, and they were returned to Mexico for proper disposal. The situation emphasizes the need for better international security to prevent further, perhaps more dangerous, occurrences ■

In June 1990, the International Commission on Radiological Protection, an independent organization in Didcot, England, recommended a significant reduction in some limits of exposure to ionizing radiation (see Chapter 15). The commission advised that the nuclear power plant employee's annual permissible dosage be decreased from 50 to 20 millisieverts. (A chest X ray exposes the patient to about 200 microsieverts.) No change in the annual exposure-level limit to the general public of 1 millisievert was proposed. However, the NRC tightened radiation exposure limits even further. As of January 1993, the general public is to receive no more than 0.1 millisievert (100 mrem) annually from nuclear facilities and the annual limit set for nuclear plant employees will be 5 millisieverts.

The most publicized nuclear accident that occurred in the United States was the serious crippling of the Three Mile Island (TMI) nuclear power plant near Harrisburg, Pennsylvania, in 1979. Large amounts of radioactive material threatened to escape into the environment. A faulty valve permitted water that served as a coolant to escape from the reactor after an unrelated malfunction had caused the reactor to become inoperable. The temperature of the reactor core reached dangerously near its melting point before the buildup of heat was brought under control. (Television cameras inserted into the core region of the crippled Unit-2 (TMI-2) reactor in 1986 reportedly discovered that a limited amount of the nuclear fuel did melt.) If a more serious meltdown of the core had occurred, however, the reactor building and its numerous safety features supposedly would have contained the materials and thus presented only a local safety problem. An investigating committee, in a report to the NRC, concluded that the radioactive materials that had escaped posed no apparent danger to public health, and that safety systems had functioned correctly to prevent more serious problems. The undamaged TMI-1 reactor has been serviced and refueled. It resumed operation in late 1986.

The worst commercial nuclear accident to date occurred in April 1986 at the *Chernobyl* nuclear power plant, near the Russian city of Kiev. As the facility's #4 reactor was undergoing a safety test, steam pressure in the reactor rapidly rose out of control until a non-nuclear explosion occurred that ripped apart portions of the reactor building. The core meltdown and extensive graphite moderator fire that followed contributed to the release of large amounts of radioactive material into the atmosphere. Although lethal exposure levels and serious injuries were confined to employees and nearby residents, low levels of fallout from the nuclear accident were detected over most of our planet. A considerable increase in cancer deaths inside the USSR, as well as in adjoining countries, is projected as a result of the fallout from the Ukranian reactor incident.

The Chernobyl reactor meltdown created several environmental conditions that apparently did not accompany any of the numerous nuclear detonations conducted by the various world powers. For instance, scientists have, for the first time, detected a well-defined layer of radioactive cesium (Cs-134 and Cs-137) in polar glaciers that originated from the atmospheric contaminants present in the fallout from the Cheronbyl reactor incident. As the particles settled from the atmosphere onto the surface of the ice, the radioactive layer continued to build. Also, various portions of Sweden that had received the higher levels of radioactive fallout experienced a significant increase in the frequency of lightning flashes during the 1986 thunderstorm season, compared to areas with lesser fallout levels. The increase in bolts of lightning presumably resulted from the greater amount of particles suspended in the atmosphere, causing increased electrical conductivity.

Although the Soviet government is fully aware of the seriousness of the Chernobyl plant accident, they still apparently have all intentions to proceed with further development and expanded reliance on nuclear-powered generating plants in the USSR. The designs of existing reactors, along with those to be constructed, however, will be considerably altered to meet the rigid safety specifications effectively adopted by the NRC in the United States. Also, Soviet officials intend to phase out all coal-generated electric power plants in the near future, after a detailed study revealed the degree to which the combustion of coal was detrimental to their environment.

In the immediate past, some U.S. reactor sites, particularly those involved in the production of nuclear warheads, have had to curtail operations because Department of Energy investigations into management and design safety came out with unfavorable results. These facilities include the uranium-processing plant located near Cincinnati at Fernald, Ohio; the nuclear reactor at Hanford, Washington; the Savannah River Plant at Aiken, South Carolina; and the Rocky Flats plant in Golden, Colorado. Several of these facilities might permanently shut down, since they are operating well past their projected 20-year life span, whereas others could resume operation as soon as present safety standards are met. Added to these concerns, the DOE has estimated that the cost of cleanup and storage of wastes from nuclear weapons production will exceed $20 billion over a 5-year period through fiscal 1995.

The safe disposal of radioactive wastes, such as from nuclear power plants and from the multitude of uses of radioactive materials described earlier, has always been of great concern. Radioactive wastes were for years dumped at sea or buried in a variety of methods at disposal sites scattered throughout the United States and other countries. In recent years, the radioactive wastes have been packaged in borosilicate glass and buried in geologically stable formations of salt and rock. However, the reinforced glass may not be able to withstand unforeseen distortions, movement, and high temperatures that might occur in Earth's crust. The released nuclear wastes may escape into our groundwater or migrate to the surface to be further scattered about by the agents of erosion. One method of disposal, called **synroc,** has been employed in limited cases. A geochemist at a university in Australia developed a procedure that involves the bonding of atoms in radioactive waste materials into the crystalline structure of a stable synthetic host rock. Nine parts of a melted constituent of selected minerals are mixed with one part of molten radioactive waste. The mixture is packed in heavy nickel canisters and compressed under alternating periods of hot and cold conditions until a predetermined density is reached. The canister is then buried deep in Earth's crust in granite with impermeable clay and mudstone packed carefully about it.

A process was developed at Batelle Laboratories that converts buried radioactive and other hazardous wastes, including the potentially contaminated surrounding soil, into an immobile mass. The procedure, called *in situ vitrification,* melts the soil that contains the wastes to form an impermeable glasslike rock, similar to the hard and strong obsidian, a naturally occurring volcanic rock. Electrodes are inserted into the adjacent soil and an intense electric current is applied to produce the chemically durable solid. The in situ process appears to be successful as a means to render old hazardous-waste burial sites safe.

Other suggested means of dealing with radioactive wastes, as well as nonradioactive hazardous wastes, include storing them until they are rendered harmless by time or are reclaimed for future use. Each system, however, appears to have at least one potential danger associated with it. For example, in 1988, the Department of Energy sought permission to develop a high-level nuclear waste burial site in Nevada's Yucca Mountain, where nuclear wastes could be interred well under the surface. At capacity, this site would have contained about 95 million gallons of highly radioactive wastes produced by the nation's nuclear arms facilities, industry, and commercial nuclear reactors. However, Nevada legislators ultimately contended that possible movement of contaminated groundwater posed too great an environmental threat. Also, Lathrop Wells, a potentially active volcano near the site, along with numerous unstable faults close by, were cited in the argument to withdraw the site from consideration. Furthermore, it was pointed out that, by common agreement, any region that was being considered should be void of any precious natural resources that might be worth mining. It is noteworthy that two of the nation's largest gold mines are located within 30 kilometers of the site proposed by the DOE. Disregarding the President's possible involvement through the Nuclear Waste Policy Act, it appears that the Energy Department must concentrate on locating an alternate site for the interim storage of high-level wastes. In fact, the DOE has moved toward opening another underground nuclear waste repository, the Waste Isolation Pilot Plant (WIPP) near Carlsbad, New Mexico. Obviously, the problem of dealing with all of our high-level nuclear wastes is years away from a solution.

Various degrees of consideration have been given to orbiting the most valuable of the hazardous wastes about the moon until techniques to extract the desired substances from the unwanted wastes have been developed. Hazardous wastes of no further value could be sent directly into the outer reaches of our solar system or beyond. However, there is always the remote possibility that the rocket systems could fail and the wastes thus return to contaminate Earth and its atmosphere. Others recommend that the wastes be buried below the sea floor in containers designed to resist chemical decomposition and tremendous pressure. Still additional suggestions include the burying of containers on some uninhabited island or in a remote region of Africa or Greenland. But, how can we accurately predict the many changes that will occur on Earth in the hundreds of centuries during which the hazardous wastes so processed will represent a danger to our planet's inhabitants?

The estimated 2.3×10^7 kilograms (a weight of 2.5×10^4 T) of radioactive wastes processed each year are but a portion of the hazardous wastes added to our environment. About 9 billion kilograms of nonradioactive wastes are generated by industrial processes each year; this amount represents about 10 percent of all industrial wastes. Approximately 90 percent of hazardous industrial wastes are in liquid or semiliquid form (sludge). The dangerous wastes that are created by various industrial processes are classified as inorganic toxic metals; inorganic compounds (such as acids, bases, and salts); synthetic organic compounds; flammable substances; and explosives.

Some of the toxic metal wastes result from mining and metallurgical processes, as well as from electroplating and metal-finishing techniques. Synthetic organic wastes include hydrocarbon derivatives, polychlorinated biphenyls, and phenols. The flammable wastes include contaminated organic solvents, oils, and plastics. The explosive wastes include obsolete munitions, such as bombs and artillery shells. Leading the list of nonflammable substances found to be hazardous to our health is the commonly used mineral *asbestos*. Research indicates that this construction material, used because of its fire-resistance, acoustical qualities, and beauty, is carcinogenic. Vast amounts of this mineral have been torn from the ceilings and walls of schools and office buildings, transported to a disposal site, and processed.

Biological wastes are also added to our environment from various sources. Pathological wastes from hospitals amount to over 1.6×10^8 kilograms annually. This waste includes living and dead tissue, outdated medicines, and used bandages. Chemical and germ warfare materials add to the hazardous wastes that must be processed for disposal. In addition, approximately 6.8×10^7 kilograms of residual salts from the neutralizing of dangerous chemicals, including nerve gases, require proper disposal. To add to our dilemma, containers used to store various hazardous materials have been found to be less inert to their contents and to the environment than was considered possible. For instance, nerve gases prepared for possible use in warfare have been stored for decades at various army depots and other facilities. The procedure for disposal of these substances has been complicated by the fear that the process of preparing them for transport to a disposal site may put undue strain on the cylinders, possibly causing them to rupture and spew their contents into the atmosphere. The detoxification and proper incineration or disposal procedures at the various storage sites could also create unexpected hazards to local residents. Moreover, a significant amount of hazardous wastes makes its way into our environment as the result of fires, accidents involving common carriers, and train derailments primarily attributed to our aging railroad system.

About 1986, the scope of the problems surrounding potential hazardous waste dumps became apparent. The original cleanup bill (commonly known as **"Superfund"**) was to be in effect from 1980 to 1985. Its primary intent was to dedicate $1.6 billion toward the cleanup of known hazardous waste dumps. However, the value of the Superfund was increased to over $9 billion as the full seriousness of the problems surfaced. These funds, obtained through various industrial taxes, were targeted to be spent between 1986 and 1991. Some of the 25,000 sites on record have been treated; however, the funds available are still not sufficient to pay for decontamination of all the targeted sites.

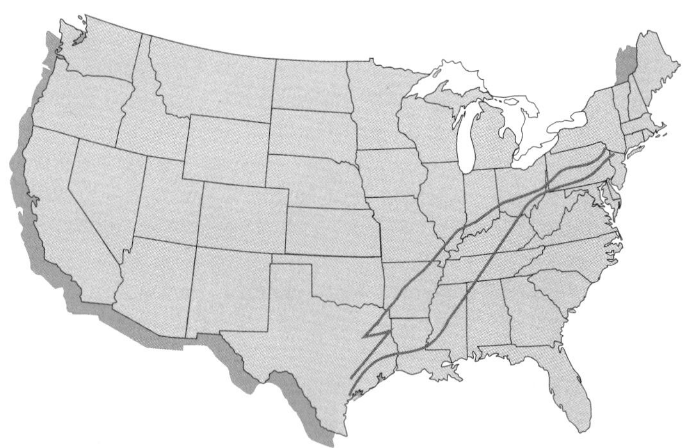

Figure 27.15 The approximate routes of two pipelines used by the Texas Eastern Gas Pipeline Corporation. Ninety or so sites along the routes are scheduled for cleanup from PCB contamination.

In 1984, Congress mandated an abrupt halt to the practice of land disposal of hazardous wastes, effective by the year 1990. As a result, Westinghouse Electric Corporation, in one of the larger toxic-waste legal settlements to date, agreed in the mid-1980s to spend about $100 million to clean up six dumping sites near Bloomington, Indiana, that were contaminated by large amounts of **polychlorinated biphenyls** (PCBs). Included in the project was a $25 million incinerator, designed to render the contaminated soil harmless, but requiring a decade or so to construct. In 1988, the Texas Eastern Gas Pipeline Corporation was assessed a record civil penalty for violations of federal environmental laws covering PCB disposal and agreed to spend about $400 million over the next 10 years to clean up approximately 90 sites along two pipelines located in 14 states, as illustrated in Figure 27.15. (In the past, PCBs were widely used in pipeline construction because of their excellent insulation capability.) Other PCB sites throughout the United States have been studied by the EPA, and the violators undoubtedly will be required to render the contaminated areas harmless to the public in the immediate future.

At the present time, technology is available to treat and safely dispose of most nonradioactive wastes that are considered hazardous. Physical, chemical, thermal, and biological processes are utilized to extract useful materials from hazardous wastes and put them to beneficial use. Other hazardous wastes are rendered harmless by chemical or physical means, and additional ones are destroyed by burning in special incinerators on land or aboard special ships at sea. The remaining dangerous residues are concentrated by various means, including evaporation, and further processed for final disposal.

The costs of environmental protection are generally assumed by the consumer, causing one to wonder how clean an environment we can afford. As more strict and complex regulations, some government-mandated and others self-imposed, are put into effect, we will become even more aware that the costs of environmental cleanliness must be assumed by all of us. We depend on an industrialized America to manufacture and transport the goods we wish to consume, although the consumption of some of these products puts the troubled environment further at risk. There are some who would argue that the materialistic values we have come to hold in such high esteem should be de-emphasized to avoid permanent damage to our environment.

Pesticides and Herbicides

A staggering amount of poisonous substances is added daily to our surroundings in the form of pesticides and herbicides. Pesticides, primarily insecticides, are applied to our soil, water, and atmosphere to destroy unwanted insects and other harmful animals. Herbicides are used to retard or destroy unwanted plant growth at home and on the farm. About 35 million acres of land are treated annually with pesticides and herbicides in the United States alone. Occasionally, a foreign insect will invade an area and endanger existing life of all sorts. The Mediterranean fruit fly threatened to ruin citrus and other crops in California and Florida in the 1980s. In an attempt to eradicate this new pest, insecticides have been sprayed over many square miles of fertile land, adding to the level of toxicity already used in pest control.

The application of pesticides was brought about because insects are a serious threat to our well-being. The insects have decided advantages over humans in that the insects have evolved through hostile environmental changes for 250 million years. The insect has adjusted to such a degree that some 40 species survive even in the antarctic region, and countless others thrive in desert areas. Nor does altitude destroy them, for some varieties have been found at 6000 meters above sea level. Others live in seawater, and some in hot springs that reach temperatures of steaming liquids. Still others have been found alive at $-40°C$. For all our attempts to annihilate various species of insects with pesticides, we have yet to meet with total success.

In 1977, approximately 40 million people in India contracted malaria, an alarming increase from the previous decade when this serious Third World health problem was nearly eradicated. This dreaded disease is caused by a microorganism carried by the Anopheles mosquito. In 1964, spraying of insecticides, such as DDT, to control the insect was halted due to economic problems and to the vast decline in the disease. The same situation occurred in Africa and Central America, since in these regions the annual incidence of malaria also plummeted. Farmers throughout the three continents continued to spray their crops frequently with the various insecticides and the Anopheles mosquito was continually exposed to the chemicals in varying amounts. Through genetic adjustments, about 40 species of this mosquito developed an immunity to a wide variety of insecticides. Cases of malaria in India increased to over 2.5 million in 1974, and the annual rate of occurrence in South America tripled from the 1960s to 1975. Various researchers suspect that ending the eradication programs, as well as changing agronomic practices, such as overspraying, brought about the return of this dreaded disease. New environmental manipulations, such as dredging the mosquitoes' breeding areas, are being attempted to prevent a serious recurrence of the disease. Medical researchers are also attempting to develop antimalarial drugs as they continue to search for the life-saving vaccine.

The storage of pesticides and herbicides is a necessary but potentially dangerous practice. In December 1984, a tremendous pressure buildup in an underground storage tank permitted the escape of a large, lethal cloud of methyl isocyanate into the atmosphere at a pesticide plant in central India partially owned by Union Carbide Corporation. The death toll from the accident was estimated at 2000, and many more people were seriously injured from inhalation and exposure to the pesticide. Less than a year later, a second Union Carbide plant, located at Institute, West Virginia, accidentally released a chemical substance derived from methyl isocyanate into the atmosphere, resulting in the hospitalization of 135 people. The growing concerns about the potential recurrence of such incidents has had decided effects on the intent of various corporations to produce, store, and apply such substances at additional locations in the United States and in other countries.

The indiscriminate applications of pesticides have also created reasons for serious concern. The chemical runoff from applications of pesticides onto farmlands with little regard for the possible consequences has destroyed or drastically affected many ecosystems. Animals suffering from grotesque tumors or adverse birth defects causing them to produce sterile offspring have been found to carry exceedingly high levels of certain pesticides. Also, in the mid-1980s, over 1300 residents along the Pacific coast of the United States became ill and several died as a result of eating California-grown watermelons that were laced with a pesticide marketed as *Temik*. This substance was licensed by the EPA as a pesticide for use on various fruits and such vegetables as potatoes, but was banned as suitable for use on melons. However, various growers disregarded the warning and applied the pesticide to their melon crops. Millions of watermelons were ordered destroyed due to the level of Temik they contained.

A similar case originated from the application of another popular pesticide by commercial growers to their apple orchards. In 1989, at the insistence of the EPA and other such supportive groups as the Natural Resources Defense Council (NRDC), Uniroyal Chemical Company discontinued U.S. sales of the pesticide known as daminozide and marketed under the copyrighted name of *Alar*. Data obtained from EPA studies in 1989 indicated that the pesticide used to treat apples, along with the by-product it formed when juices were extracted from the apples, caused cancer in laboratory animals when massive doses were ingested. The results suggested that this chemical represented a hazard to humans. However, a division of the Food and Drug Administration (FDA) contended that when fruits or vegetables were washed and/or peeled, the presence of pesticide residues was dramatically reduced.

Alar represents just one example of the continued concern by the NRDC and various factions for the use of pesticides on our foods. Others argue that the banning of agricultural chemicals may actually be dangerous to our health. And so the controversy goes on.

In contrast to pesticides, herbicides are chemical weed-killers, designed to control unwanted plant growth. At the close of World War II, the **chlorinated hydrocarbon** herbicide, 2, 4-D came into use to control broadleaf weeds. Later, 2, 4, 5-T, another chlorinated hydrocarbon, was developed to control woody species. Currently, about 40 weed-killers are available to be used in agriculture, forestry, wildlife, and right-of-way control. Large quantities of herbicides, including Agent Orange, were used as defoliants during the Vietnam War.

Another potent weed-killer, introduced in 1974 to destroy marijuana fields in Mexico and in Florida, is known as *paraquat.* It has also been used to clear new fields for cultivation. Paraquat is so effective in controlling weeds that it is considered one of the most versatile aids in agriculture, somewhat replacing the plow. The chemical also lowers soil erosion and retards evaporation of moisture, thus enhancing harvests. It is currently used on more than 10 million acres of crops in the United States, including soybeans, cotton, wheat, and corn. The use of this water-soluble herbicide is not without controversy, however. Paraquat may be dangerous to humans, whether swallowed, inhaled, or spilled on the skin. It may prove to be especially hazardous to the health of those who use it over lengthy periods of time, regardless of the precautions they take. In 1983, however, the EPA published a review of health-related research conducted on the use of the herbicide and ascertained that data were inadequate to warrant its abandonment from use in the United States.

Pesticides and herbicides can also have detrimental effects on the environment, according to most research reports available to the public. Both types of chemical compounds have migrated into our streams, wells, rivers, and lakes as the chemicals are absorbed by the soil, concentrated by natural processes, then carried off by the action of groundwater. The concentrated chemical runoff that enters slow-moving bodies of water has caused an increase in the growth of algae. This intensified growth has caused a corresponding decrease in oxygen content and, hence, the retardation of aquatic animal growth and survival.

Trees and plants are susceptible to about 1500 different diseases. Without herbicides and pesticides, such enemies as insects and blights would undoubtedly destroy many species. However, the direct dangers to human life from the chemical substances used are evident, as are the dangers of starvation and of the diseases carried by insects if they are not destroyed. Before insecticides, for example, millions of people succumbed to human diseases transmitted by ticks and mosquitoes. Later the pesticides were effective in minimizing the danger from typhus and various other plagues, since the pesticides annihilated the insects that carried the diseases. What is the solution to the problem? At present, the best solution seems to be rigid control of the quantities of pesticides used and the areas over which they are applied.

But what are pesticides? Chemists recognize five categories, classified according to active constituents:

1. Chlorinated hydrocarbons are the most commonly used class of pesticides. Among the most popular is DDT (dichloro-diphenyl-trichloro-ethane), which has been in use since 1937. The first major application of DDT, however, was to stop the spread of typhus among the military and civilians in occupied territories during World War II. Other members of this group are benzene hexachloride (BHC), chlordane, and heptachlor. Pesticides in this class have a low degree of solubility in water; therefore, they contaminate the area over which they are spread for years to come.
2. Another category of pesticides includes the inorganic pesticides that contain active metals, such as arsenic, lead, and mercury. These inorganic pesticides are effective in destroying insects only if used rarely, because insects and plant diseases, at the species level, develop an immunity to a given poison. For example, some strains of insects require doses of DDT 50 times the strength of those required to annihilate their immediate ancestors. If these poisons are used only rarely, some species are readily victimized by the inorganic compounds because they lack the natural resistance that is produced through overexposure to given poisons.
3. Organophosphate compounds are the by-products of war research. This group contains the most deadly insect poisons in use. Some examples are diazinon, malathion, and parathion. Like nerve gas, the action of these poisons is to paralyze nervous systems, hence muscular systems. The organophosphates are dangerous to a large spectrum of life, including many domestic animals. For instance, various flea collars available for dogs and cats contain organophosphates and, under certain conditions, are hazardous to animals and to humans.
4. Natural pesticides have been used for many centuries. For example, the powder produced from chrysanthemums has offered relief from the invasion of such insects as the ant and the housefly. In addition, rotenone (an extract from various plant roots) and nicotine have also been used as poisons in agricultural sprays with great success.
5. Another class of pesticides includes the petroleum distillates, which in the crudest form have destroyed the mosquito in the larval stage. The larvae are smothered by the oils, which are less dense than water, as they are spread over the swampy regions where mosquitoes breed. Other insects, including flies, have been controlled in the same manner.

At the present time, because of their limited applications, neither herbicides nor pesticides pose immediate danger to the general public. However, surveys conducted by the EPA in recent years have revealed detectable amounts of chlordane, dieldrin, heptachlor, and DDT in a high percentage of blood samples taken from people who have had no direct contact with the chemicals.

During the planting, continued maintenance, and harvesting of agricultural products, farmers have made it a practice to apply insecticides to the soil in order to ensure that their efforts are as productive as possible. The most popular and effective chemical presently used as a soil fumigant is known as ethylene dibromide (EDB). The chemical is also reportedly under close scrutiny as a potential hazardous substance by the FDA. Substitute chemicals are being sought that would be safer to the farmer and consumer alike.

The FDA is also investigating the feasibility of using gamma radiation to control insects in harvested crops. Wheat and flour are irradiated to destroy insects and larvae; also, potatoes are exposed to gamma radiation to inhibit sprouting, prolonging the vegetable's storage life. The FDA is contemplating awarding permission to irradiate other produce, such as spices, dried onions, and garlic, in order to prolong shelf life. Although the military has used the technique for some time, the FDA is presently reviewing the possibility of using gamma rays to treat meat, poultry, and seafood to destroy the bacteria that cause food poisoning. There is no radioactive residue produced by the process; however, changes in taste and texture in various products may limit the acceptability of this means of decreasing the amount of chemical additives in our food chain.

Genetically engineered farm products are destined to have a decided effect on our everyday lives as we move closer to the turn of the century. Consider the benefits we will reap when various organisms such as genetically engineered bacteria, viruses, fungi, and insects will be available that can effectively control unwanted insects, undesirable weeds, and damaging plant diseases. Presently, about one-third of the crops grown in the United States, valued at about $50 billion, are damaged or destroyed by various pests. Not only would we experience a considerable decrease in the cost of pesticides and the environmental impact they have, but the increase in crop yield would decrease the amount of land required to meet our agricultural needs. Geneticists presently are developing crops that are not as susceptible to freezing temperatures and that will be able to "fix" their own nitrogen; thus, these plants will be able to transform the molecules of this gaseous element already present in the atmosphere into a form suitable for their immediate use. The annual savings on nitrogen fertilizer alone would be about $4 billion.

Herbicide-resistant crops represent a genetically engineered agricultural product that could have a complicated effect on both our economics and our environment. Crops could possibly be developed that are not affected by various herbicides, so farmers could grow crops in areas presently too inhabited by troublesome weeds to cultivate. However, weeds that could resist various herbicides would multiply rapidly. More powerful herbicides would be required to remove the weeds that survive, creating a treadmill effect. Greater use of specific types of herbicides would prevent crop rotation of those plants sensitive to the specific herbicide. Soil erosion and problems associated with the control of various pests where the same crops are constantly grown would develop. Even now, soybeans cannot be planted after a corn crop is harvested if residues of certain herbicides are present in the soil. There remains serious doubt that certain crops resistant to some proven herbicides will ever be acceptable.

As our interests in the biotechnology industry continued to increase, the federal government, in 1986, set forth very rigid rules and guidelines for regulating the adjustments in agricultural practices that are certain to occur. The enforcement of specific regulations was assigned to five federal agencies:

1. The Environmental Protection Agency (EPA) was assigned to determine what genetically engineered organisms may be released into the environment.
2. The United States Department of Agriculture (USDA) was assigned to decide what genetically engineered organisms may be used in conjunction with agricultural products of all sorts.
3. The National Institutes of Health (NIH) were assigned the task of ensuring that the various genetically engineered organisms introduced into our environment were not detrimental to public health.
4. The Occupational Safety and Health Administration (OSHA) was assigned the task of determining that genetically engineered organisms would bring about no detrimental effects in the workplace.
5. The Food and Drug Administration (FDA) was assigned responsibility for evaluating the safety of genetically engineered organisms used in our medicines and food products.

We must heed the warnings of nature and exercise care in the use of chemical compounds and the advances made in the biotechnical fields. Herbicides and pesticides have served valuable purposes—both types of substances have helped us maintain a high level of food production as well as an environment in which the presence of unwanted plants and insects can be controlled. The threat to the general public from both types of chemicals lies in the continuing buildup of unwanted and uncontrollable concentrations that may be harmful to all life-forms.

People and the Environment

Biomedical researchers and others are making rapid progress in seeking better means to treat diseases and health problems of all sorts. They are also concentrating on gerontology, the branch of knowledge that involves aging processes. As a result of their efforts, life spans have been significantly increased; thus, a new realistic concern for our environment has come into view. A decided increase in population over the past two centuries is attributed to a decline in death rate, not to the more controllable birthrate. Studies indicate that our population will grow moderately for the next 60 to 70 years, despite a projected decrease in birthrate. Even at a modest 1 percent increase in rate of growth, the increased population will require additional food, shelter, energy, employment opportunities, and the like that must be furnished if we are to exist as a society in the manner to which we are accustomed.

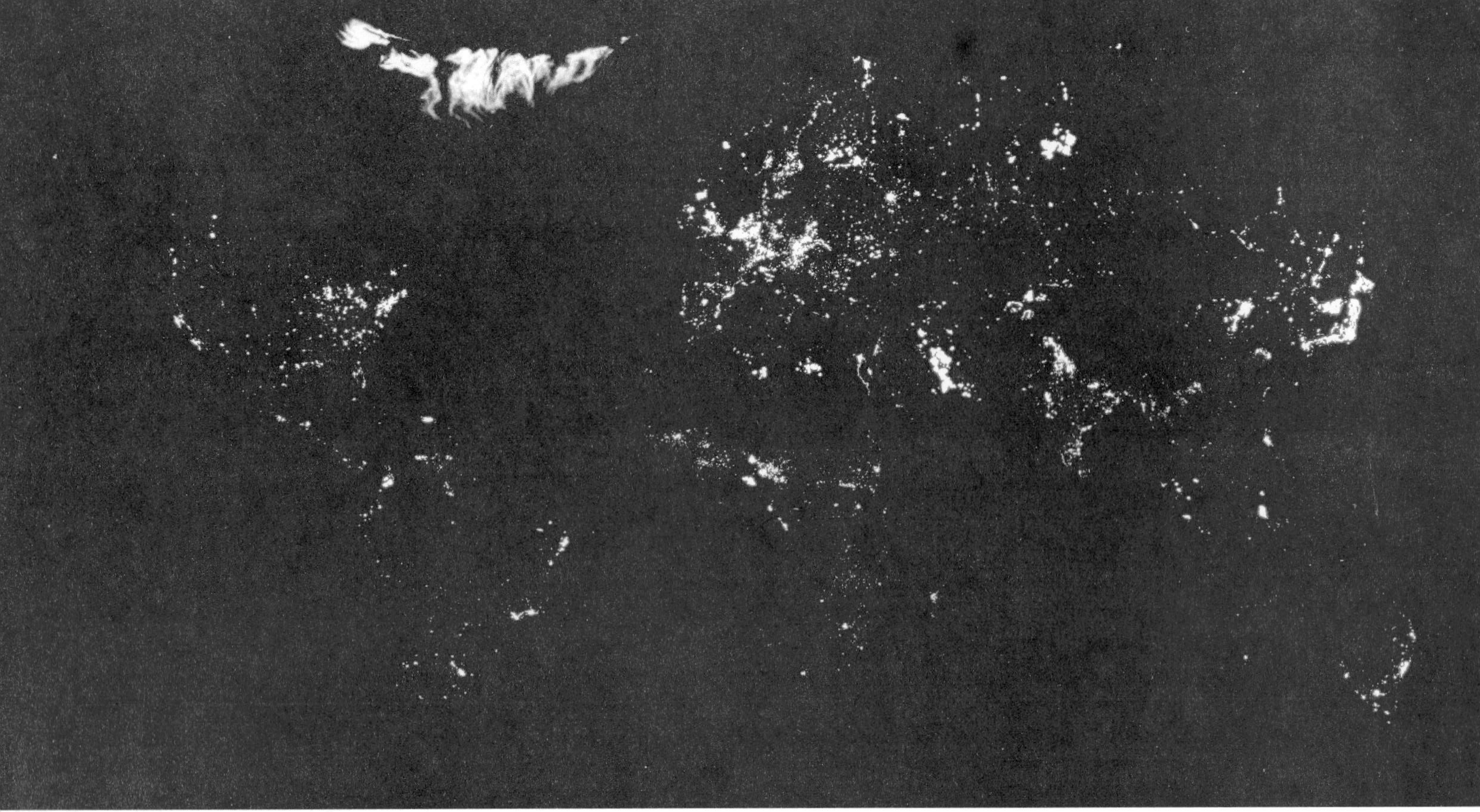

Figure 27.16 Most of the continents are outlined by artificial light along with the many fires that burn over portions of our planet. Approximately 5 percent of Earth's land masses are set ablaze each year, but only 1/100 of this portion is due to wildfires. A magnificent display of Aurora Borealis (the northern lights) is shown above North America.

We need only to consider the tragic effects caused by overpopulation in various Third World countries to realize the looming social and economic problems we must resolve.

Science can contribute to solutions, but it alone doesn't solve specific problems. For instance, the search for an inexpensive, environmentally safe source of energy is primarily limited by social beliefs rather than by scientific progress. Science, along with technology, has provided solutions to many problems that involve overpopulation, yet the solutions are not always acceptable to all segments of society. An adjustment may have to occur in our cultural habits and attitudes if we are to avail ourselves of the new strategies science and technology can provide.

A Nightly View of Earth

Our astronauts are offered a beautiful view of the portion of our planet illuminated by the sun. Perhaps to our surprise, they can also distinguish at least the outline of land masses that should be in total darkness. This situation exists because of the level of artificial lighting that floods various areas of each continent. Also, the nighttime view of some continents is considerably enhanced by perhaps a million fires—most purposely set to destroy unwanted vegetation on grasslands and savanna, making way for new tender grasses and controlling insects and pests. Consider the amount of carbon dioxide, noxious fumes, and particulate matter these fires add to the atmosphere each day. If the view of each portion of Earth were available during a moonless night and we were able to see all continents collectively, a space traveler could experience the outstanding scene designated as Figure 27.16. In this photograph created by the computer enhancement of satellite transmissions, the continents are practically outlined by artificial lighting and the fires. Necessary overexposure has caused some bright light sources, such as gas flares above fires, to appear much larger than they really are. Also visible above the North American continent is a display of the northern auroral light. Light from these sources is known as *light pollution,* and it effectively disturbs our view of the night skies. Many of the world's astronomical observatories were constructed in remote areas where there was once minimal light interference; but, with few, if any, exceptions, the surrounding areas have become inhabited. Our view with the naked eye of the dark sky is considerably inferior to that of our ancestors. Because of artificial light and air pollution, we can see but a fraction of the stars and other celestial bodies observed by them. No doubt the nightly view of future generations will be even less spectacular.

The Challenge Is Ours

Each individual should be assigned the task of protecting our environment now and in the future. The nature of the task would appear to be simple; however, following through successfully presents a formidable challenge. First, we must control and lower pollution levels in all of our environment. Second, we must preserve our natural resources as much as possible and maintain the beauty of our environment that remains. Finally, we must use our natural resources wisely and learn to reuse those that can be separated from unrecoverable waste.

Pollution results mainly from the use of natural resources in such a way that they are not recoverable. The total waste products remain in our environment for long periods of time once they are placed there. For civilization to continue and at least to maintain its present standard of living, the industrial and domestic wastes must be handled properly. The solution seems straightforward:

1. Care must be exercised to see that such solid wastes as garbage, junk, and trash are disposed of in a manner that will not contaminate our environment—land or water.
2. Smoke and dust particles must be removed at the source, prior to their release into the atmosphere. Hydrocarbon types of aerosol propellants have already been discontinued; other gaseous contaminants must be closely monitored.
3. Commercial and domestic wastes that serve as a haven for bacteria must be converted into substances that inhibit unwanted bacterial growth. Hazardous wastes must be rendered harmless before they are released into our environment.
4. Proper control must be maintained over the chemicals that are used as insecticides, herbicides, cleaning agents, and food preservatives to provide protection for both domestic animals and plants as well as for all species of wildlife.

All of these solutions can be implemented only through extensive research and financial investment. As new techniques for waste-product treatment are discovered, new regulations must be provided by society. Necessary educational programs must be initiated to make all citizens aware of the seriousness of environmental pollution. Each individual must realize his or her need to share the expense and the burden of treating wastes properly. Each of us must also develop an awareness of the demands of all society; we must have as an ultimate aim the rebeautification of cities, parks, and rural areas. All pollution problems are serious, even potential ones. We have taken great strides recently, but only when we fully appreciate our environment as the greatest resource of all will the desired progress be made.

Summary

The acquisition and transportation of our fossil fuels and other natural resources have had detrimental effects on our environment. Still other environmental hazards are created as these valuable materials from Earth are used to produce electrical energy.

Industrial areas, with their large populations, require that sources of electrical energy be developed nearby. Any system that generates electrical energy for large communities releases energy in the form of heat directly into the atmosphere and into the water. Heated water retards and destroys most ordinary aquatic life, yet it is conducive to various unwanted plant and animal growth, including algae. Industry is overcoming this problem by cooling the heated water before its release into streams and lakes, as well as by using it in the irrigation of various crops and in other beneficial ways.

Solid wastes are often combustible, yet the burning of solid wastes adds unwanted gases and particulate matter to our atmosphere. Inversion layers, suspended layers of air caused by unusual temperature variations, concentrate the gases and particulate matter and hold them over many densely populated areas. Industry adds many other contaminants, including nitrogen and sulfur compounds, to our atmosphere. However, most of our air pollution is attributable to the automobile and its internal-combustion engine. Automobiles with electric motors or with engines that consume propane are more than just speculative replacements for the internal-combustion engine. Air pollution is not only a health hazard, but it also reportedly affects our climate in adverse ways.

Various studies indicate that noise, unwanted sounds of all sorts, affects our health and comfort in numerous ways. Sound, heat, and light energies are often produced during the conversion of one form of energy to another. Public reactions to noise have not gone unheeded. The thunderous sounds from jet engines, faulty mufflers, pneumatic hammers, and excessively loud sound systems are among those sources of greatest concern.

Wastes from homes and industry pose serious problems to our environment. The most critical problems from domestic wastes are the discharges of soaps and organic solids into our waters and the accumulation of nonreturnable containers in our solid-waste disposal sites.

Many hazardous materials, once released into the air, dumped into our waters, and heedlessly buried under our soil, have been recognized as serious pollutants of our environment. Various countermeasures are under way to see that these hazardous substances are properly processed before disposal is completed.

Pesticides are chemical compounds that are applied to our immediate environment to destroy unwanted insects and other harmful animals. Chemical compounds designed to control or to destroy undesirable plant growth are called herbicides. Herbicides and pesticides that attack only specific forms of life have been developed. However, some such chemical compounds disintegrate into other compounds that are harmful to humans and to various species of animals, insects, and plants that appear to be beneficial to our society. A consideration of all effects must be maintained to guide users of these chemicals in employing them to our advantage.

Light pollution considerably hinders our view of the night skies. It is created by artificial lights and the vast number of fires set primarily to destroy unwanted vegetation, various insects and other pests, as well as to prepare fields for planting.

We have entered a period when decisions about biotechnical advances must be considered. We must carefully consider the effects of genetic engineering upon our environment. It is our challenge to be certain that the possible eradication of various forms of animal, insect, and plant life will not adversely affect the portions of our environment already endangered.

To prevent further damage to our environment, we must employ measures to lower pollution levels and exercise extreme care in the consumption of our remaining natural resources. We must also resort to better means for solid, liquid, and gaseous waste disposal if future generations are to survive and to prosper.

Questions and Problems

Energy and Environmental Pollution

1. Why are oil spills in our waters of greater concern than similar spills over land?
2. List and briefly discuss two major problems that accompany the procedures used in strip mining.
3. What factors have somewhat curtailed the attempt of industries to offset our need for imported oil through processing of our available supply of oil shale?

Thermal Pollution of Our Waters

4. How does an increased water temperature in the vicinity of an electrical power plant adversely affect aquatic life?
5. List and briefly discuss three means of processing the warm water from power plants before it is returned to our environment.

Air Pollution

6. List at least three major air pollutants that ordinarily accompany the production of our electricity.
7. What is photochemical smog? Why does it form in some industrial areas and not in others?
8. Identify three major sources of air pollution in metropolitan areas, other than electrical power plants.
9. What is acid rain? Describe its composition, sources, and effects on our environment.
10. Why are temperature inversions potentially dangerous to our health and to our environment?

Noise Pollution

11. In what ways does noise pollution affect our health?
12. What are sonic booms and what causes them?

Commercial and Domestic Wastes

13. How do commercial and domestic wastes affect our environment if proper disposal methods are not applied?
14. How will increased recycling of our nontoxic wastes help alleviate our environmental problems?

Pollution from Hazardous Wastes

15. What are hazardous wastes? List four distinctly different categories of these wastes.
16. Briefly discuss the rationale for processing radioactive wastes to separate them according to half-lives before they undergo further handling and storage.
17. List five national or international waste-processing policies that have been initiated within the past 10 years to protect our environment.
18. Mercury, asbestos, lead, and many other such substances have been used in various beneficial ways for many generations. Why, then, are these substances now classified as materials hazardous to our health?

Pesticides and Herbicides

19. Cite three major problems that might develop if there were no rigid controls over the applications of certain insecticides and pesticides.
20. Do you think that scientists will eventually develop the means to genetically control any form of animal, insect, or plant life? Why is this such a controversial issue?
21. Cite the major advantages and disadvantages that result from the application of a pesticide or herbicide that is soluble in water.
22. How do insects and other pests become immune to insecticides that ordinarily have been proven capable of destroying them on contact?

People and the Environment

23. How does even a modest increase in population add to our environmental concerns?

A Nightly View of Earth

24. With reference to Figure 27.16, what geographic areas on each continent have the highest light density? How do these areas correlate with the location of the more industrialized nations?

The Challenge Is Ours

25. Briefly discuss three ways in which we, as individuals, can help protect our environment.

APPENDIXES

Appendix 1

Systems of Measurement

n order to gain a better understanding of our natural world, we must be aware of the units used in making quantitative measurements. The fundamental concepts of length, mass, and time serve to define most other mechanical quantities.

The Sumerian civilization, which flourished between 5000 and 6000 years ago, reportedly was the first organized community to develop standardization of measurement. Certain stones designated by the high priests became the standards of weight throughout the kingdom. The standards were later adopted by the Babylonians, who in turn passed along a modification of these standards to the Greek civilization.

Our English System of Linear Measurement

Many of the units of measurement in use today were first standardized by ancient civilizations. A selected few of these units are described in this appendix. The standards were determined from a variety of sources.

The *inch* was standardized from three grains of barleycorn, round and dry, taken from the middle of the plant and placed end to end. Each grain, then, averaged one-third of an inch along its longer axis. Today this value still remains as the difference between respective shoe sizes. For instance, the difference in length between a size 7 and a size 8 shoe is the length of a grain of barleycorn. The difference between sizes 6 and 6½, then, is one-half the length of a grain of barleycorn. At the time the length of an inch was set, the measurement called the foot already had been established as 39 grains of barleycorn in length. Interestingly enough, our shoe sizes grade into categories from infants to youths to adults by thirteens (one-third of 39).

The actual foot from which the standard *foot* was derived belonged to the emperor Charlemagne. A member of Charlemagne's court noted that the great emperor's footprint corresponded closely to the distance from the servant's elbow to his wrist and decided that this measurement would make an excellent standard, since he could duplicate it with reasonable accuracy.

The *cubit,* used by various tradespeople even today, was standardized as a measure of the distance from the elbow to the tip of the longest finger, although the individual on whom the standard was based is unknown. However, the pyramids, as well as the tombs of the ancient Egyptians appear to have been constructed in even cubits.

The distance from the royal nose of King Henry I to the end of the next-to-last joint of his longest finger became the standard *yard.* The *fathom* is equal to twice this distance; therefore, it is derived from the combined span of both arms.

The *rod,* a unit standardized long ago, was established from the total length of the left feet of 16 men. These men were selected from a church congregation on a given Sunday and were asked to stand heel to toe until the measurement could be made. The rood, as the unit was originally known, is equal to the *pole* and is 16½ feet long.

The length of the *furlong* is 40 rods (660 feet, or one-eighth of a mile). This distance was derived from the average length of a furrow on a typical field to be plowed with a team of oxen. (A field 1 furlong square encompasses 10 *acres.*) The furlong is still a popular unit in horse racing. A typical horse race is officially listed as being 6 furlongs (¾ mi).

Several classics, including the Kentucky Derby, are 10 furlongs (1¼ mi) in length.

The *mile* was standardized from the stride of the marching Roman soldier. The typical stride of 5 feet multiplied by 1000 became the mile. The original mile, then, was 5000 feet long. Britain and the United States refused to accept the Roman mile since the furlong, a well-established unit, did not relate well to this value. For that reason the British and American mile became 5280 feet long, eight times the length of the furlong. The U.S. mile of 1760 yards was not universally accepted—presently the mile in Sweden is 1700 yards; in Italy, 1614 yards; in Russia, 1155 yards; and in Iceland, 2240 yards.

The *thumb,* the *finger,* the *hand,* the *link,* the *nautical mile,* the *league,* and numerous other linear units add to the complexity of our English system.

Our English System of Weight

Early attempts to standardize units of weight were left to individual communities. Generally, stones and boulders were selected as benchmarks against which to compare items to be bartered by weight. The *libra,* a unit developed by the ancient Romans, is the forerunner of our pound, as indicated by the abbreviation for the pound, *lb.* The English system of measurement is highly complex. Consider, for instance, which is heavier—a pound of feathers or a pound of gold. And again, which is heavier—an ounce of feathers or an ounce of gold? The (perhaps surprising) answers to these questions can be obtained from the information in the following equivalencies. All three systems of weight measurement are based on the weight of a specific number of grains of wheat. (Consider the degree of variation that potentially exists as one strives to duplicate such a standard.)

Avoirdupois weight—This system is used to measure the weight of ordinary commodities.

16 drams = 1 ounce = 437.5 grains (of wheat)
16 ounces = 1 pound = 7000 grains
14 pounds = 1 stone
100 pounds = 1 hundredweight
2000 pounds = 20 hundredweight = 1 ton
2240 pounds = 1 long ton

Apothecaries' weight—This system is used in the pharmaceutical industry.

20 grains = 1 scruple
3 scruples = 1 dram
8 drams = 1 ounce
12 ounces = 1 pound = 5760 grains

Troy weight—This system is used to weigh precious stones and valuable metals.

1 carat = 3.086 grains
24 grains = 1 pennyweight
20 pennyweight = 480 grains = 1 ounce
12 ounces = 1 pound = 5760 grains

All of us realize the need to use a common language in order to be understood by others. The same need prevails for understanding the units of measurements, regardless of occupation and interests. Obviously, standards that are based on seeds, stones, and parts of the human body, are no longer satisfactory; our need for precision is too great.

The Decimal System and the Meter

In 1670, the Frenchman Gabriel Mouton proposed a decimal system of measurement based on an arc of 1 minute of a great circle of Earth. This system received limited support. Late in the eighteenth century, international scientists developed a completely new decimal system based on multiples of ten. A standard unit of length was proposed by various members of the organization—a measure that was based on the length of a pendulum having a period of 1 second. Other scientists rejected this proposal, pointing out that a pendulum swings faster near the North and South poles than it does at the equator. This standardizing technique also presupposed a universally defined second—at the time a questionable measure. Finally, in 1792 a portion of a meridian (it was erroneously assumed at that time that all meridians were of equal length) was agreed upon from which the new standard could be determined. This unit initially was based upon a ten-millionth part of the distance from the North Pole to the equator along the meridian that passes through Paris. This standard of linear measurement was called the *meter* (in French, *metre*) from the Greek *metron,* meaning "measure." It became widely accepted by those nations that were involved with France in foreign and domestic trade.

The unit, as standardized, was not acceptable to Great Britain or the British colonies, primarily because of their conflicts with France during the Napoleonic wars. If a well-defined alternate standard had been proposed and accepted by most other countries, undoubtedly Great Britain and the United States would have adopted the standard also, since both countries were actively involved in trade with other nations.

The United States strongly considered adopting a decimal system of measurement in 1786, the year that Congress decided to eliminate use of the English pound-shilling-pence money system and to set up a decimal system of money. Later, President Thomas Jefferson favored the metric system of measurement, but he, like Congress, refused to accept a standard based on a meridian that passed through France, and which was thus not readily accessible to all nations.

Another setback to the early adoption of the metric system occurred when the United States was not invited to the World Conference on Weights and Measures held in France in 1790. Still another event prompted Congress to decide against the changeover to the metric system. The creators of the metric system proposed a new calendar starting with the year 1. They designed a new week of 10 days, neglecting, however, to stipulate one day as the Sabbath. As a result of their actions, the entire metric system became known as an atheistic system of measurement. But for all these circumstances, the United States doubtlessly would have adopted the metric version of measurement in the late nineteenth century. Since 1893 the meter has, in fact, been the basis for the standard of length in the United States. The yard, in that year, was officially defined as 3600/3936 meter. In 1959 the yard was redefined as 0.9144 meter.

The metric system gained popularity in most countries soon after its adoption by the French Academy of Science for several reasons:

1. The metric system offered simplicity of measurement.
2. The units for the measurements of length, mass, and volume are related.
3. The metric system, being a decimal system, permits manipulations of the powers of ten by simple relocation of the decimal point. The need to deal with common fractional measurements does not exist in the metric system.

The basic units of measurement in the metric system are the meter (m), the gram (g), and the liter (L), in addition to the second (s). Through the addition of the prefixes in Table A to any of the base units, quantities can be measured in a practical manner regardless of magnitude.

Most measurements made in metric units involve only the following five prefixes: *mega-, kilo-, centi-, milli-,* and *micro-*. For instance, measurements of length other than those in meters are most often expressed in kilometers (km), centimeters, (cm), millimeters (mm), and micrometers (μm). Measurements of mass are commonly expressed in kilograms (kg), grams (g), milligrams (mg), and so forth.

Table A
Prefixes used in the metric system.

Prefix	Symbol	Signifies	Powers of ten
Tera-	T	1,000,000,000,000	10^{12}
Giga-	G	1,000,000,000	10^{9}
Mega-	M	1,000,000	10^{6}
Kilo-	k	1000	10^{3}
Hecto-	h	100	10^{2}
Deka-	da	10	10^{1}
Deci-	d	0.1	10^{-1}
Centi-	c	0.01	10^{-2}
Milli-	m	0.001	10^{-3}
Micro-	μ	0.000,001	10^{-6}
Nano-	n	0.000,000,001	10^{-9}
Pico-	p	0.000,000,000,001	10^{-12}

SI Units

In 1960, the General Conference of Weights and Measures, held in Paris, defined the International System of Units (SI) now in use practically throughout the world. In fact, only three countries do not primarily make use of the metric system and the SI units. These countries are Myanmar (formerly, Burma); Brunei; and of course the United States. Even in our country, various government agencies and industries have adopted the metric system because of trade agreements with other countries and the need for better precision of measurement.

Let us now consider the base units used in the metric system, along with their means of standardization.

The **meter (m)**—Originally the meter was standardized as 0.000,000,1 (one ten-millionth) the distance from the equator to the North Pole of Earth, as measured along the meridian that passes through Paris. Later, to make more precise measurements possible, the standard was set as 1,650,763.73 times the wavelength of the orange light emitted by the atoms of the gas krypton-86. A better standard, 10 times as precise as its predecessor, has recently been adopted by scientists. The meter is presently defined as the distance light travels through a vacuum in 1/299,792,458 of a second. In effect, the meter, a unit of distance, is now defined in terms of time.

The **kilogram (kg)**—The mass of 1 cubic decimeter of water at 4°C (the temperature at which water is at its maximum density) was the original standard for the kilogram. Presently, the mass of a platinum-iridium cylinder preserved at Sevres, France, serves as the standard for the base unit of mass. The kilogram is the only base unit still defined by a metallic object.

The **second (s)**—This time unit was first standardized as 1/86,400 of the mean solar day, and later as 1/31,556,925.47 of the year 1900. Today the standard second is derived from the duration of 9,192,631,770 cycles of simulated radiation from the cesium-133 atom. The great precision with which time can be measured by use of an "atomic clock" was the basis for redefining the meter according to time.

The **ampere (A)**—The ampere is defined as that constant current, within controlled conditions, that is produced between two electrical conductors that produce an electrical force of 2.00×10^{-7} newton per meter between them. In a practical sense, it is the rate of flow of 1 coulomb of electric charge per second.

The **kelvin (K)**—The kelvin is a unit of thermodynamic temperature equal to 1/273.16 the thermodynamic temperature of the triple point of water; that is, of the fixed point at which ice, liquid water, and water vapor coexist in equilibrium. At atmospheric pressure ice melts at 273.16 K, and water boils at 373.16 K. The Thirteenth General Conference of Weights and Measures (1967–1968) accepted the name *kelvin* (K) to replace the names *degrees Kelvin* (°K) and *degrees* (deg), contending that the unit of thermodynamic temperature and the unit of temperature interval are one and the same. Therefore, the unit should be denoted by a single name (*kelvin*) and a single symbol (*K*). Some scientists commonly speak of 50 K as "50 kelvins," 200 K as "200 kelvins," and so on.

The **mole (mol)**— One mole is the amount of a substance that contains the same number of elementary entities as there are atoms in 0.012 kilogram of carbon-12.

The **candela (cd)**—The candela is the standard unit for measuring the intensity of visible light. It is based on a standard that takes into account the eye's response to light. The candela is the luminous intensity, in a given direction, of a source that emits monochromatic (one-color) radiation of 540×10^{12} hertz (green light) and that has a radient intensity in that direction of 1/683 watt per steradian. (A steradian is a measure of an angle in three dimensions, with one steradian being the angle of a cone that would have a spherical "cap" of an area r^2, where r is the distance from the apex of the cone to the "cap.") An easier unit to use is the *lumen,* which measures the total luminous energy radiated from an object. An object radiating uniformly in all directions with an intensity of 1 candela radiates 12.57 lumens (lm).

Other units are derived from the base units, such as the newton and the joule, discussed below. Some additional derived units, along with their expressed equivalences, appear in Table B.

The **newton** (N)—The newton is the official unit by which forces of all sorts are measured. This derived unit is the force that will impart an acceleration of 1 meter per second per second to a standard 1-kilogram mass.

The **joule** (J)—One joule is a measure of work done by a force of 1 newton acting over a distance of 1 meter. The joule also serves as a unit of heat and of other types of energy. One calorie, the thermochemical unit, is currently defined as equal to 4.184 joules. The international steam table calorie is defined as equal to 4.1868 joules.

Measurements of Area and Volume

The unit of *area* is a square, the sides of which are a standard unit of length. In the English system it is a square 1 inch, 1 foot, or 1 yard on a side. The areas of the respective squares are 1 square inch (1 in^2), 1 square foot (1 ft^2), and 1 square yard (1 yd^2). In the metric system the square would be, for example, 1 centimeter or 1 meter on a side, and the corresponding units of area would be 1 square centimeter (1 cm^2) and 1 square meter (1 m^2). The area of a surface is the number of squares of any of these units that would fit on the designated surface. To determine the area of a rectangular surface, multiply the dimensions of the sides; for a triangle, multiply one-half the base by the height; and, for a circle, multiply the radius by itself, then that product by the constant π (3.14).

The *volume* of an object is a measure of the space it occupies. The dimensions of the object are expressed in standard units. In the English system a cube, the dimensions of which are 1 foot on an edge, has a volume of 1 cubic foot (1 ft^3). In the metric system the edges of the cube are measured in centimeters or meters. The volume of the space that the object occupies is expressed in cubic centimeters (cm^3) or cubic meters (m^3). The volume of liquids in the English system is expressed in quarts, gallons, and other such units. The units of volume of liquids and gases in the metric system are expressed in milliliters or liters. A liter (L) is the volume of a liquid or gas contained in a cube having an inner volume of 1000 cubic centimeters. A milliliter (mL) would be contained in a cube having inner dimensions all of 1 centimeter. The volume of a cube with inner sides all of 1 centimeter would be 1 cubic centimeter; thus, 1 mL = 1 cm^3. (Some professions still use the abbreviation cc—for cubic centimeters—as the unit of liquid volume.)

Table B
Derived SI units.

Quantity	Unit	Symbol	Expressed equivalence
Space and time			
Area	square meter	m^2	
Volume	cubic meter	m^3	
Velocity	meters per second	m/s	
Acceleration	meters per second squared	m/s^2	
Angular velocity	radians per second	rad/ s	
Angular acceleration	radians per second squared	rad/s^2	
Frequency	hertz	Hz	
Mechanics			
Density	kilograms per cubic meter	kg/m^3	
Momentum	kilogram-meters per second	kg·m/s	
Force	newton	N	$kg \cdot m/s^2$
Torque	newton-meter	N·m	
Work, energy	joule	J	N·m
Power, radiant flux	watt	W	J/s
Pressure and stress	pascal	Pa	N/m^2
Heat			
Celsius temperature (By definition, O°C = 273.16 K)	degree Celsius	°C	1 C° = 1 K
Quantity of heat	joule	J	N·m
Electricity and magnetism			
Electric charge	coulomb	C	A·s
Electric potential	volt	V	W/A
Electric resistance	ohm	Ω	V/A
Electric capacitance	farad	F	C/V
Magnetic flux	weber	Wb	V·s
Light			
Luminous flux	lumen	lm	cd·steradian (sr)
Illumination	lux	lx	$lumen/m^2$

Selected Regulations from the General Conference of Weights and Measures

The International System of Units, also known as Le Système International d'Unités (SI), established by the General Conference of Weights and Measures provides very useful direction in the way in which metric measurements are to be expressed. Following is a general summary of some of the rules:

1. Symbols for units are written in a single standard form regardless of whether the number of units is less than one, one, or more than one; that is, there is no plural form for symbols: 0.05 mL, 1.00 kg, 10.0 cm.
2. All symbols are printed in roman (upright) rather than italic (*slanted*) type. (However, because the symbol for liter, l, might be misread as "one," a capital rather than a lowercase L is often used.)
3. A space is provided between quantity and symbol: 100 kg, 50.0 cm, 274 K. Several exceptions are noted: 10°C, 70°5′10″, and 100′ (ft).
4. Only symbols derived from proper names are capitalized: m for meter and cd for candela; but K for kelvin, N for newton, and J for joule.

5. In most cases, symbols for units are preferable to writing the names of units in full: cm^3 rather than cubic centimeters, m^3 rather than cubic meters, mL rather than milliliters.
6. For derived units other than density, no prefix is attached to units in the denominators of concepts expressed as fractions: N/m^2, W/A, C/V, $kg \cdot m/s^2$. Density measures in g/cm^3 and g/mL are acceptable as indicated.

The Systems of Units

There essentially are three systems of units in common use today:

1. The English system of units used in the United States is slowly being replaced in many cases. In this system the foot is generally considered the fundamental unit of length; the inch, yard, and mile are commonly used. For relatively short measures, the inch is employed; for very long distances the measure is made in miles. The English system ordinarily expresses the weight of an object rather than its mass. The common unit of weight is the pound; however, light objects ordinarily are weighed in ounces and very heavy objects in tons. The base time unit is the second, with other time units of minutes, hours, and days used where applicable. This foot-pound-second system of units is conventionally referred to as *fps*.
2. The System Internationale (SI) system of units is the major of the two systems of metric measurements. If the dimensions of an object are determined in meters, by convention the mass of the object is measured in kilograms. Its time unit, the second, is universally used in all systems of measurement. The SI system is stressed in this textbook to simplify a formal introduction of the metric system and its various units to many students who have not studied it in detail in previous science courses.
3. For small objects measured in the metric system, the dimensions are determined in centimeters and the mass is measured in grams. The time unit is the second. The centimeter-gram-second (*cgs*) system is used in the laboratory and the pharmacy, as well as in the canning and bottling industries.

In problem solving, all units of measurements should be expressed within the same system. The proper units of mechanics within the three systems of measurement are expressed in Table C.

Table C
Units of mechanics.

Unit	Length	Mass	Time	Velocity	Acceleration	Force	Work	Power
Symbol/ Formula	*l*	*m*	*t*	*v*	***a***	***F***= *m**a***	*W*= ***F**s*	*p*= *W*/*t*
SI system	meter (m)	kilogram (kg)	second (s)	m/s	m/s^2	$kg \cdot m/s^2$ or newton (N)	N·m or joule (J)	joule/s or watt (W)
cgs system	centimeter (cm)	gram (g)	second (s)	cm/s	cm/s^2	$g \cdot cm/s^2$ or dyne (d)	dyne·cm or erg	erg/s or 10^{-7} watt (W)
fps system	foot (ft)	slug	second (s)	ft/s	ft/s^2	pound (lb)	ft·lb	$\frac{ft \cdot lb/s}{550} = hp$

Table D

Conversions, Physical Constants, and Units

Length

1 inch = 2.54 centimeters
1 foot = 0.305 meter
1 yard = 0.914 meter
1 mile = 5280 feet = 1.61 kilometers
1 centimeter = 0.394 inch
1 meter = 100 centimeters = 39.4 inches = 3.28 feet = 1.09 yards
1 kilometer = 1000 meters = 0.621 mile
1 astronomical unit (AU) = 1.50×10^8 kilometers = 9.29×10^7 miles
1 light year (LY) = 9.46×10^{12} kilometers = 5.88×10^{12} miles = 0.307 parsec

Area

1 square inch = 0.00694 square foot = 6.45 square centimeters
1 square foot = 144 square inches = 0.0929 square meter
1 square yard = 1296 square inches = 9 square feet = 0.836 square meter
1 square mile = 640 acres = 2.59 square kilometers
1 square centimeter = 10^{-4} square meter = 0.155 square inch
1 square meter = 10^4 square centimeters = 10.8 square feet

Volume

1 cubic inch = 0.000579 cubic foot = 16.4 cubic centimeters
1 cubic foot = 1728 cubic inches = 0.0283 cubic meter
1 cubic yard = 27 cubic feet = 4.66×10^4 cubic inches = 0.765 cubic meter
1 cubic meter = 10^6 cubic centimeters = 1000 liters = 35.3 cubic feet
1 quart = 2 pints = 946 milliliters = 0.946 liter
1 gallon = 4 quarts = 231 cubic inches = 3.79 liters
1 liter = 1000 cubic centimeters = 1.06 quarts = 0.265 gallon

Mass

1 slug = 14.6 kilograms
1 kilogram = 1000 grams = 0.0685 slug
1 atomic mass unit (u) = 1.66×10^{-27} kilogram = 1.66×10^{-24} gram
1 electron mass = 9.11×10^{-31} kilogram = 9.11×10^{-28} gram = 5.46×10^{-4} u

Conversion between Weight and Mass on Earth

A mass of 1 slug weighs 32.2 pounds.
A mass of 1 kilogram weighs 9.80 newtons or 2.21 pounds.
A mass of 1 gram weighs 980 dynes or 0.0353 ounce.
A 1-ounce weight has a mass of 28.4 grams or 0.0284 kilogram.
A 1-pound weight has a mass of 454 grams or 0.454 kilogram.
A 1-ton (short) weight has a mass of 907 kilogram.

Velocity and Speed

1 ft/s = 0.305 m/s
1 mile/hour = 1.47 ft/s = 1.61 kilometers/hour = 0.447 m/s
1 m/s = 1000 cm/s = 3.28 ft/s
1 kilometer/hour = 0.278 m/s = 0.621 mile/hour = 0.912 ft/s
Speed of light = 3.00×10^8 m/s (more precisely: 299,792,458 m/s in a vacuum)
= 186,000 miles/second
Speed of sound in air at 0°C (32°F) = 331 m/s = 1090 ft/s = 741 miles/hour

Table D

Conversions, Physical Constants, and Units *continued*

Acceleration

1 ft/s^2 = 0.305 m/s^2 = 30.5 cm/s^2
1 mile/hour/second = 1.47 ft/s^2 = 1.61 kilometers/hour/second = 0.447 m/s^2
1 m/s^2 = 1000 cm/s^2 = 3.28 ft/s^2
Acceleration due to gravity on Earth = 9.80 m/s^2 = 980 cm/s^2 = 32 ft/s^2

Force

1 pound = 16 ounces = 4.45 newtons
1 newton = 0.223 pound

Temperature

32° Fahrenheit = 0° Celsius (centigrade) = 273 kelvins
0 kelvin = −273.16° Celsius = −459.72° Fahrenheit

Pressure

Air pressure at sea level = 1.01 × 10^5 newtons/square meter
= 14.7 pounds/square inch
Water pressure increases 9810 newtons/square meter for each meter of depth or 0.433 pound/square foot for each foot of depth.

Heat and Work or Energy

1 calorie = 4.18 joules = 0.00442 Btu
1 foot-pound = 1.36 joules = 1.36 newton-meters

Power

1 horsepower = 550 foot-pounds/second = 746 watts
1 watt = 1 joule/second = 0.239 calorie/second = 3.80 Btu/hour
1 kilowatt = 1.34 horsepower

Activity

1 curie = 3.70 × 10^{10} disintegrations/second = 3.70 × 10^{10} becquerels

Metric Prefixes

g = giga = ×10^9	c = centi = ×10^{-2}
M = mega = ×10^6	m = milli = ×10^{-3}
k = kilo = ×10^3	μ = micro = ×10^{-6}
d = deci = ×10^{-1}	n = nano = ×10^{-9}

Densities of Various Common Substances

Material	**g/cm^3*	*lb/ft^3*	*Material*	**g/cm^3*	*lb/ft^3*
alcohol	0.79	49.3	gold	19.3	1200
blood	1.04	65.0	ice	0.917	57.5
carbon tetrachloride	1.588	99.09	iron	7.85	490
gasoline	0.69	41.2	lead	11.3	708
kerosene	0.8	50	silver	10.5	655
mercury	13.6	849	wood (maple)	0.55	34.1
milk	1.03	64.3	dry air at sea level	0.001,29	0.081
sea water	1.025	63.96	humid air at sea level	0.001,19	0.074
water	1.00	62.4	carbon dioxide	0.001,98	0.123
aluminum	2.70	168	carbon monoxide	0.001,25	0.077
copper	8.89	555	hydrogen	0.000,09	0.0056
cork	0.25	15.6	oxygen	0.001,43	0.088

*(To express density in the *mks* system (*kg/m^3*), multiply density in the *cgs* system by 1000. Note also that 1 g/cm^3 = 1 g/mL.)

Exercises

In order to check your comprehension of the metric system as discussed in this appendix, complete the following. Then, compare your answers with those selected and listed in Appendix 4.

1. Change the following measurements to the submultiple, base unit, or multiple indicated:

(a) 486.4 cm to m
(b) 348 m to cm
(c) 86.4 cm to mm
(d) 49.6 mm to cm
(e) 1944 mm to m
(f) 1864 mL to L
(g) 448 mL to L
(h) 0.65 L to mL
(i) 13.6 kg to g
(j) 1959 g to kg

2. Although there are no references to conversions from one system of measurement to another in this text, some instructors opt to have their students perform such exercises for various reasons. The necessary conversion factors appear on the inside back cover.

(a) 23.0 ft to m
(b) 80.0 km to mi
(c) 440 lb to kg
(d) 320 g to lb
(e) 45.0 ft^3 to m^3
(f) 14.0 qt to L
(g) 100 mi to km
(h) 100 m to ft
(i) 5.00 lb to g
(j) 12.0 oz to g

3. Express each of the following measurements in terms of cm^3:
(a) 54.0 mL (b) 6.00 L (c) 54.0 g of water, 4° C

4. With reference to Table C in this appendix, answer the following questons:
(a) What is the accepted unit of length in the SI system?
(b) What is the accepted unit of mass in the SI system?
(c) What is the accepted unit of velocity (and speed) in the SI system?
(d) What is the accepted unit of acceleration in the SI system?
(e) What is the accepted unit of time in the SI system?

5. According to Table C in this appendix, what two units in the appropriate system of measurement are multiplied or divided by each other to obtain each equivalency that follows?
(a) one watt (W)
(b) one newton (N)
(c) one joule (J)
(d) one unit of work in the *fps* system
(e) one horsepower (hp)

Appendix 2

Mathematics Refresher

The principles that underlie physical science are generalizations based upon observations. The genius of such scientists as Newton, Dalton, the Curies, and Einstein is that they were able to recognize "patterns" in their observations. Although there are exceptions, most observations on which scientific principles are based are quantitative in nature; that is, the observations involve measurements made to some degree of precision. Moreover, the measurements require some degree of mathematical manipulation. The following examples are typical of those encountered in this text. It is recommended that the student use a hand-held calculator when performing numerical calculations.

Addition

Numbers and quantities can be added in any order as long as the units of measurement are identical. For example, the sum of 3 m, 4 m, and 150 cm = 3 m + 4 m + 1.5 m = 8.5 m. Also, $X + Y = Y + X$, and $3 + 5 = 5 + 3$.

Calculator Tips

On most calculators, the numbers and addition signs may be keyed in just as if read aloud. Thus, key in 4 + 5 + 1.5 and press the = key to get the sum. When adding a set of numbers, it is not necessary to press the = key until the final answer is desired. The calculator operator must keep up with what types of units are being added.

Subtraction

The operation of deducting one number or quantity from another is called subtraction. If the positive number 8 is deducted from the positive number 12, the difference is the positive number 4. That is, $12 - 8 = 4$. To subtract a negative number from a positive number, change the sign of the negative number and add. For instance, $13 - (-6) = 13 + 6 = 19$. Algebraically, $X - (-Y) = X + Y$. Also, $-X - (+Y) = -X - Y$.

Calculator Tips

As in the previous example, the numbers and signs may be keyed in just as if read aloud. Thus, key in 12 − 8 and press the = sign to get the difference. If several numbers are to be added and subtracted, the = sign need be pressed only after all numbers and signs are entered. When *adding or subtracting a negative number,* the [+/−] key may be used to change the sign of the number and the calculator will give the proper answer if the + and − signs are entered in the ordinary way. For example, key in 12 − 8 and press = to subtract 8 from 12, and key in 13 − [+/−] 6 and press = to subtract −6 from 13.

Multiplication

Multiplication is a mathematical operation by which a number is added to itself a specified number of times. Multiplication can be indicated either by a multiplication sign, as in 5×7, or by parentheses, as in (5)(7) or 1.3(4.1). The following examples are used to point out manipulations involved in multiplication of numbers and of quantities. Note the manner in which positive numbers and negative numbers are treated.

$(5) \times (4) = (4) \times (5) = 20$
$(X) \times (Y) = (Y) \times (X) = XY = YX$
$(-5)(-4) = (-4)(-5) = +20$
$(-X)(-Y) = (-Y)(-X) = XY = YX$
$(-5)(+4) = (+4)(-5) = -20$
$(-X)(+Y) = (+Y)(-X) = -XY = -YX$

Also,

$$\begin{aligned}(4\text{ m}) \times (3\text{ m}) &= (4) \times (3) \times (\text{m}) \times (\text{m}) \\ &= 12\text{ m}^2 \\ (3\text{ cm})(2\text{ cm})(4\text{ cm}) &= (3)(2)(4)(\text{cm})(\text{cm})(\text{cm}) \\ &= 24\text{ cm}^3 \\ (5\text{ N})(2\text{ m}) &= (5)(2)(\text{N})(\text{m}) \\ &= 10\text{ N}\cdot\text{m}\end{aligned}$$

The units to be multiplied do not have to be the same, but they may be. For example, meters times meters gives meters squared, a measure of area; but meters may be multiplied by force to give torque (Chapter 4) or work (Chapter 5).

Division

The process by which we can determine how many times one number or quantity is contained in another is called division. Division is typically indicated by either a horizontal or a diagonal line between numerator and denominator.

For example, 3 divided by 4 is $\frac{3}{4}$ or 3/4. (The number 3 is the numerator and the number 4 is the denominator.) F divided by M is $\frac{F}{M}$ or F/M. If the numerator, denominator, or both are fractions, division is accomplished by inverting the denominator and multiplying. For instance, 4/5 divided by 2/3 is written $\frac{\frac{4}{5}}{\frac{2}{3}}$ and simplified to $\frac{4}{5} \times \frac{3}{2} = \frac{12}{10} = 1\frac{2}{10} = 1\frac{1}{5}$. Further, 10 m divided by 2 s $= \frac{10\text{ m}}{2\text{ s}} = \frac{5\text{ m}}{1\text{ s}} = 5$ m/s. Also, 6 m divided by 2 m $= \frac{6\text{ m}}{2\text{ m}} = 3$. (Since m divided by m = 1, there is no unit assigned to the answer.)

The units to be divided do not have to be the same, but they may be. For example, if we wish to find out how many fence posts are needed for a fence 100 meters long if there is to be a post every 4 meters, we would divide 100 meters by 4 meters to obtain 25 and add 1 for a post at the start for a total of 26. The number 26 in this case has no dimensions, but tells us how many posts are needed. Determining the speed of an object provides an example of different units being divided. Speed is the distance the object has traveled in meters divided by the time it has taken to travel that distance in seconds, so that the units come out in meters per second.

Calculator Tips

When multiplying or dividing, all numbers in the numerator (the top of a fraction) are multiplied together and all numbers in the denominator (the bottom of the fraction) are divided into the numerator. For example, to find

$$\frac{(25.3)(30.1)}{(19.7)(2.33)},$$

key in 25.3 × 30.1 ÷ 19.7 ÷ 2.33 = to get 16.6 (rounded off). Multiplication and division on a calculator are done *before* addition and subtraction. To compute $(2 + 3)/4$, either enter the parentheses as shown (if the calculator has that capability) or enter 2 + 3 and press = to complete the addition before entering the division by 4 (÷ 4); otherwise, the calculator will divide 3 by 4 first and the incorrect answer of 2.75 will appear instead of 1.25, which is correct.

When multiplying or dividing by a negative number, just key in the usual way, except press the [+/−] key after entering the negative number. Thus, $(5)(-8)$ is entered 5 × 8 [+/−] = and the result is −40.

Constants and Variables

A constant is a quantity, the value of which does not change in a given problem. A variable is a quantity that may assume any one of a set of values in a given problem. For example, if you travel on a train that runs at a uniform speed of 60 kilometers per hour, the speed of the train is constant; that is, it does not change. The distance that you go will depend on how long you ride the train; thus, distance is variable. The following table points out how the distance (d) varies with time (t); speed $v = 60$ km/h, a constant:

t (hours)	½	1	2	3	4	5	10
d (kilometers)	30	60	120	180	240	300	600

Notice that any two values of t with the corresponding values of d form a proportion. For example, 3:4 = 180:240, is read as "3 is to 4 as 180 is to 240."

Direct Proportion

One quantity, say x, is said to be *directly proportional* to another quantity, say y, if increasing y by multiplying it by some number makes x the same number of times as large. For example, doubling y would cause x to double, tripling y would cause x to triple, and so forth. Mathematically, this is expressed

$$x \propto y.$$

If x is known for at least one y, the proportional sign can be replaced with an equal sign and a constant called the *constant of proportionality.* In the previously cited example, it is known that after 1 hour the train has traveled 60 kilometers, so the distance d can be computed from the equation

$$d = 60\,t,$$

where d is in kilometers and t is in hours. The distance d is said to be *directly proportional* to the time t, and the *constant of proportionality* is 60. In more general terms, if the speed of any object is constant, then $d = vt$, where v is the speed of the object, and d is proportional to t.

Inverse Proportion

The quantity x is said to be *inversely proportional* to y if, when y is multiplied by some number, x is found by dividing by that number. For example, when y is doubled, x is one-half its original value; or when y is made three times as large, x is one-third as large, and so forth. Mathematically, this is expressed as

$$x \propto 1/y.$$

(Notice that, stating this equation in words, x is proportional to the *inverse of y,* the inverse being $1/y$.) Again, if x is known for some value of y the proportionality sign can be replaced with an equal sign and a constant of proportionality, say C; thus,

$$x = C/y$$

Returning to the idea of the previous example associating distance, speed, and time, suppose that the speed of several objects is to be computed by measuring the time it takes an object to move 10 meters. Then

$$v = 10/t.$$

and v is said to be inversely proportional to t. An object that takes 2 seconds to move the 10 meters has a speed of 5 meters per second. Another object that takes 4 seconds, twice as long, has a speed of 2.5 meters per second; that is, a speed that is half as great.

Equations

The solution of problems in science involves the collection of relevant data and an algebraic interpretation of the way in which the data are related. Generally, if one quantity cannot be measured directly, logic is used to determine its value from known or measurable quantities. Many general rules aid in determining the unknown quantities. Algebraic expressions of scientific principles are called equations. In equations (or formulas) letters are used to represent words or numbers. Several rules for solving equations are illustrated here.

1. Any quantity can be added to or subtracted from both sides of an equation.

If $A+10=40$
$A+10-10=40-10$
$A=30$

Likewise, if
$5B=20+4B$
$5B-4B=20+4B-4B$
$B=20$

and

$A-10+15=30+15$
$A-10+15-15=30+15-15$
$A-10=30$
$A-10+10=30+10$
$A=40$

2. Both sides of an equation may be multiplied or divided by any quantity. Three examples are

$\frac{1}{5}A=6$
$(\frac{1}{5}A)(5)=(6)(5)$
$A=30$

$8A=56$
$\frac{8A}{8}=\frac{56}{8}$
$A=7$

$5ab=15a$
$\frac{5ab}{5a}=\frac{15a}{5a}$
$b=3$

3. A fractional equation that states that two ratios are equal is called a proportion. Equations expressed as proportions are simplified by cross-multiplying and canceling.

$$\frac{a}{b}=\frac{c}{d}$$

$$\frac{a}{b}\times bd=\frac{c}{d}\times bd$$

$$a\times d=b\times c$$

$$ad=bc$$

So, if

$$\frac{a}{4}=\frac{2}{8}, \text{ then}$$

$$a\times 8=4\times 2$$

$$8a=8$$

$$\frac{8a}{\cancel{8}}=\frac{\cancel{8}}{8}$$

$$8a=8$$

$$a=1$$

4. Both sides of an equation may be raised to the same power or have the same root taken.

$$\sqrt{X}=10$$

$$(\sqrt{X})^2=(10)^2$$

$$X=100$$

and

$$X^2=25\text{ m}^2$$

$$\sqrt{X^2}=\sqrt{25\text{ m}^2}$$

$$X=5\text{ m}$$

Examples of Symbol Manipulation in Equations

When we want to find the numerical value of a term in an equation, it is often convenient to solve the equation by manipulating the symbols before inserting the numbers each symbol represents. Some examples from straight-line motion follow.

Average velocity of an object:
The average velocity for an object moving in a straight line having a constant acceleration (Chapter 3) is

$$\bar{v} = 1/2\ (\bar{v}_i + \bar{v}_f),$$

where $\bar{v}_i$ is the initial velocity and $\bar{v}_f$ is the final velocity. We know (from Chapter 3) that $\bar{v}_f = \bar{v}_i + \boldsymbol{at}$ where $\boldsymbol{a}$ is the acceleration and t is the time. Thus, if we substitute this expression for v_f in the previous equation, we obtain

$$\bar{v} = 1/2\ (\bar{v}_i + \bar{v}_i + \boldsymbol{at})$$
$$= 1/2(2\bar{v}_i + \boldsymbol{at}).$$

So

$$\bar{v} = \bar{v}_i + 1/2\ \boldsymbol{at}, \text{ giving the average}$$

velocity at any time if the initial velocity and the acceleration are known.

Displacement of an object:
The final displacement $\boldsymbol{d}_f$ of an object depends on its initial displacement $\boldsymbol{d}_i$, the average velocity $\bar{v}$, and the time t according to (Chapter 3):

$$\boldsymbol{d}_f = \boldsymbol{d}_i + \bar{v}t.$$

Substituting the result just obtained for the average velocity $\bar{v}$ into this equation gives

$$\boldsymbol{d}_f = \boldsymbol{d}_i + (v_i + 1/2\boldsymbol{a}t)\ (t).$$

So

$$\boldsymbol{d}_f = \boldsymbol{d}_i + v_i\, t + 1/2\boldsymbol{a}t^2,$$

giving the final displacement in terms of the initial displacment, the initial velocity, the acceleration, and the time.

Eliminating time from velocity and accleration equations:
Consider the displacement equation in the case where the initial displacement is zero; that is,

$$\mathbf{d}_f = \bar{\mathbf{v}}t,$$

or

$$\mathbf{d}_f = 1/2(\mathbf{v}_i + \mathbf{v}_f)t.$$

Now, if an object is moving in a straight line, the boldface symbols are not needed if we remember to include signs for direction. Hence, the equation can be written

$$d = 1/2\ (v_f + v_i)(t).$$

(This is necessary in the discussion that follows because vector multiplication is beyond the scope of this text.) Notice that the subscript on d has been dropped. In this equation, d is the total displacement of the object during the motion. Now consider the definition of acceleration, once again letting the signs represent directions and dropping the boldface notation:

$$a = \frac{v_f - v_i}{t}.$$

Multiplying both sides of this equation by t and dividing both sides by a gives

$$t = \frac{v_f - v_i}{a}.$$

Substituting t from this equation into the previous equation for displacement gives

$$d = 1/2\ (v_f + v_i)\,\frac{(v_f - v_i)}{a}.$$

Multiplying both sides by $2a$ gives

$$2ad = (v_f + v_i)\ (v_f - v_i)$$
$$= v_f^2 - v_i^2 .$$

Adding v_i^2 to both sides gives

$$v_i^2 + 2ad = v_f^2 .$$

This is usually written with the sides interchanged so that the final velocity is on the left; thus,

$$v_f^2 = v_i^2 + 2ad .$$

Scientific Notation (Powers of Ten)

Very large numbers and very small numbers are more readily manipulated when they are expressed in "scientific notation," sometimes called powers of ten. Any number may be written as a number between 1 and 10 (the coefficient) multiplied by a number written as a power of ten (the exponent). For example, 2800 may be written as 2.8×1000 or as 2.8×10^3, since $1000 = 10^3$. Also, 5,410,000 may be written as $5.41 \times 1{,}000{,}000$ or as 5.41×10^6. Numbers less than 1 may be written using negative powers of 10. A decimal fraction such as 0.000,045,3 may be written as $\frac{453}{10{,}000{,}000}$ or as $\frac{453}{10^7}$. The latter expression, upon bringing the power of ten up to the numerator, becomes 453×10^{-7}, or $4.53 \times 100 \times 10^{-7} = 4.53 \times 10^2 \times 10^{-7}$, finally becoming 4.53×10^{-5}. Lest the student conclude that the conversion of numbers into scientific notation is overly complicated, the following simple rule is stated: *To convert a number to scientific notation, simply count the number of digits the decimal is moved, letting the power of ten equal* **that number if the decimal is moved to the left** *and letting the power of ten equal* **the negative of that number if the decimal is moved to the right.** Thus in the first example given, the decimal was moved three places to the left and the power of ten was 3. In the third example, the decimal was moved five places to the right and the power of ten was -5.

The rules for combining numbers written in scientific notation also are easy to comprehend. In addition or subtraction, the exponents of all numbers written as powers of ten to be added or subtracted must be the same. For example,

$$(5.6 \times 10^3) + (4.2 \times 10^3) = 9.8 \times 10^3 = (5.6 + 4.2)(10^3) = 9.8 \times 10^3$$

and

$$(3.65 \times 10^4) + (4.3 \times 10^3) = (3.65 \times 10^4) + (0.43 \times 10^4) = 4.1 \times 10^4.$$

Also,

$$(4.64 \times 10^2) - (2.42 \times 10^{-2}) = (4.64 \times 10^2) - (0.0242 \times 10^2) = 4.62 \times 10^2.$$

To multiply or to divide numbers written as powers of ten, the coefficients are multiplied or divided. In multiplication the exponents are combined (added or subtracted) according to sign. In division, the sign of the exponent in the denominator (that is, the number being divided into the other number) is changed and then combined accordingly:

$$(5.2 \times 10^3)(1.5 \times 10^4) = 7.8 \times 10^7$$
$$(2 \times 10^{-3})(6 \times 10^5) = 12 \times 10^2 = 1.2 \times 10^3$$
$$(6 \times 10^5) \div (2 \times 10^2) = 3 \times 10^3$$
$$\frac{(8 \times 10^6)}{(4 \times 10^{-3})} = 2 \times 10^9.$$

Calculator Tips

Numbers multiplied by powers of ten may be entered into most calculators by entering the number that multiplies the power of ten and then pressing EXP to enter the power of ten. Thus, 4.32×10^8 is entered by keying in 4.32 EXP 8. Some calculators display this entry by leaving a space between the coefficient number and the power of ten; for example, 4.32 08. Others use smaller numbers to indicate the power of ten, so $4.32\ ^{08}$ is displayed. The sign of the exponent is changed by keying +/− after the exponent is entered and before any calculations are made. Thus, 1.60×10^{-19} is entered 1.60 EXP 19 [+/−]. When entering a power of ten by itself, say 10^{20}, enter 1 EXP 20, since $10^{20} = 1 \times 10^{20}$. (Some calculators will give the wrong power of ten unless the 1 is entered.)

When performing any calculations using powers of ten, the calculator will automatically keep up with the proper power of ten. Thus, adding $2.36 \times 10^4 + 1.32 \times 10^6$ will give 1.3436×10^6. You probably need to round this off to 1.34×10^6 (see the following section on significant figures).

Finally, most calculators allow you to display very large or very small numbers in power-of-ten notation. Usually this is accomplished by pressing the key labeled ENG. For example, if the number 3426711. is displayed on the screen, pressing ENG (or the equivalent on your calculator) displays 3.426711 06 or $3.426711\ ^{06}$. Thus, it is not necessary to count the decimal places, but you probably need to round off the coefficient of the power of ten as discussed in the following section.

Significant Figures (Digits)

Much of a scientist's time is spent in making measurements. The accuracy of the measurements is indicated by the number of digits recorded. Quite often a doubtful figure is retained; namely, the last digit. A linear measure of 56 centimeters indicates that the distance was measured to the nearest centimeter. A recording of 56.3 centimeters indicates greater accuracy than does a recording of 56 centimeters. A recording of 56.0 centimeters would indicate the same degree of accuracy as 56.3. If the number were recorded as 56.00 centimeters, an even greater degree of accuracy would be indicated, and so on. No calculation should be made that is more accurate than the data recorded. The solutions to calculations, therefore, should never be expressed more accurately than the least reliable recorded data. For example, if a series of measurements were to be simply added, we could get the following answer: 13.232 cm + 11.26 cm + 18.2943 cm = 42.7863 cm. However, in the second measurement in the series, the 6 is a doubtful figure; therefore, the second digit to the right of the decimal point in the answer is also a doubtful figure. The data, then, with respect to the measurements obtained, should be combined 13.23 cm + 11.26 cm + 18.29 cm = 42.78 cm. Also, disregarding the number of significant figures (or digits) in each measurement, the density of a substance with a measured mass of 18.6427 grams and with a volume of 5.62 cubic centimeters may be calculated to be 3.3172 g/cm^3. However, the 2 in the divisor is a doubtful figure. In order to adjust for significance, the mass would be rounded off to as many significant figures as the volume; namely, three. The density of the substance, to an accuracy of three significant figures, then, should be calculated as 18.6 g/5.62 cm^3 = 3.31 g/cm^3. (Refer to Chapter 3 for additional discussion of significant figures.)

The Use of Units in Problem Solving

In the solution of a problem in science, we almost always find units of measure. These units can be treated as algebraic expressions; that is, they may be added, subtracted, multiplied, or divided as are integers. Many common units, such as newton-meters or grams per cubic centimeter, indicate the manipulations performed on the data. That is, newton-meters, the unit of work in the SI system of measurement (abbreviated N·m), is determined from the product of the force exerted on an object and the distance the object is lifted (or the force moves). The unit grams per cubic centimeter (g/cm^3), is a measurement of the density of an object in the *cgs* system. It is obtained by dividing the mass of an object by its volume. The word "per" points out that the first measurement indicated has been divided by the second. If the algebraic manipulation of the units produces the correct expression for a given type of data manipulation, generally the problem has been approached properly.

Graphs in Problem Solving

Measurements obtained from an experiment can be made more meaningful and the relationships that exist between measurements can be more readily understood if the data are expressed in the form of a graph. A graph presents a pictorial view of the results of an experiment and allows for further interpretation and projections of the initial data obtained.

The quantity the investigator can control is known as the *independent variable* and should be plotted along the horizontal or x axis. The *dependent variable,* the quantity that changes as the independent variable is altered, is plotted along the vertical or y axis.

The spaces on the graph paper should be numbered so as to spread the data across the width and along the length of the graph paper to the greatest extent possible, allowing for easier interpretation of the graph.

A quick glance at a graph reveals information that may not be so obvious from simple tabulated data. As an example, the following data were taken as a spring was caused to stretch when additional 50-gram masses were attached to it:

Total Mass, in Grams	*Increased Length of Spring, in Centimeters*
50	10
100	20
150	30
200	40
250	50
300	60
350	76
400	92

After scales for the dependent variable and independent .variable are chosen, the data are plotted with various symbols, large enough that they can be distinguished when they are connected by a line of best fit. The graph of the information appears in Graph A. The spring, according to an interpretation of the graph, may have been stretched beyond its limit of resilience.

Consider also the following data collected as a ball rolls down an inclined plane. A graph of this data is presented as Graph B.

Time, in Seconds	*Total Distance the Ball Rolls, in Centimeters*
1	14
2	50
3	120
4	200
5	320

A smooth curve has been drawn to include as many symbols as possible. If all the plotted points cannot be connected, the curve is continued so as to average the distance between them. Because the line that connects the points or averages them is not a straight one, the ball must have undergone a change in its rate of motion, or accelerated.

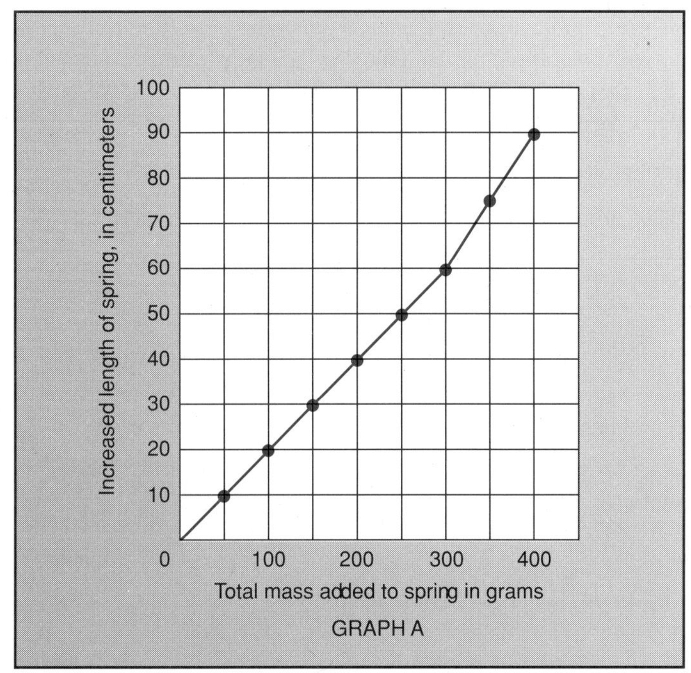

GRAPH A

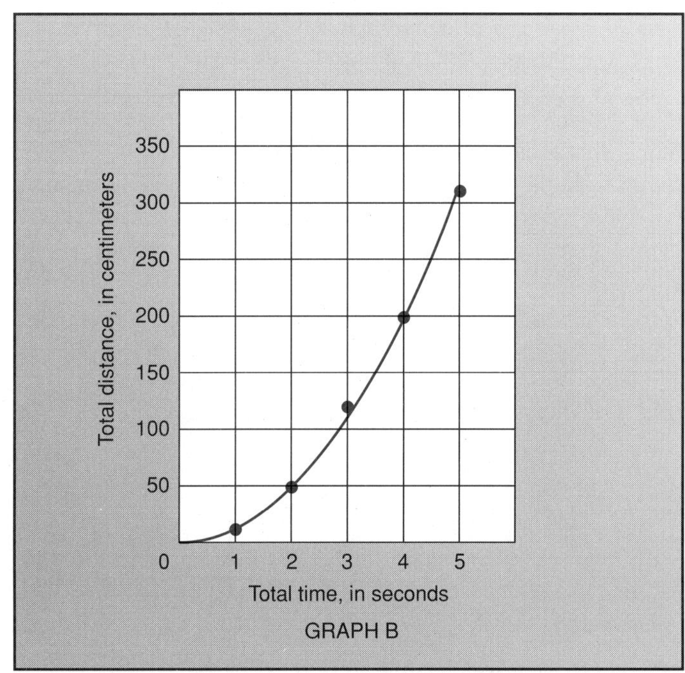

GRAPH B

Exercises—Appendix 2

The following exercises use the mathematical principles explained in this appendix. Necessary conversion factors are found either in Appendix 1 or inside the back cover of the text. Answers to selected problems are given in Appendix 4.

1. Complete the following manipulations using scientific notation:

(a) $(9.5 \times 10^8) \times (2.0 \times 10^5) =$
(b) $(6.6 \times 10^9)(3.1 \times 10^{-6}) =$
(c) $(8.8 \times 10^{10}) \div (2.2 \times 10^6) =$
(d) $(5.5 \times 10^{11}) \div (1.1 \times 10^{-9}) =$
(e) $63{,}000{,}000{,}000 \times 0.000{,}000{,}021 =$

2. With regard to the rules dealing with significant figures, express the answers to the following problems in appropriate form:

(a) $136.31 + 56.24 + 89.56 =$
(b) $113.66 - 107.84 =$
(c) $64.624 \times 36.17 =$
(d) $14.44 + 21.238 + 8.642 =$
(e) How many significant figures are there in
1000 = ; 0.0001 = ; 245.647 = ; 10,800 = ;
543 = ?

3. In an experiment designed to determine the relationship between the bending of a rigid bar and the load placed on it, the following loads were successively applied: 25 kg, 50 kg, 75 kg, 100 kg, 125 kg, 150 kg, 175 kg, and 200 kg. At each load, the bending in of the bar was noted to be 0.14 cm, 0.28 cm, 0.42 cm, 0.56 cm, 0.70 cm, 0.86 cm, 1.04 cm, and 1.24 cm, respectively.

Noting graphs A and B, plot the preceding data on linear graph paper. What value appears to be the elastic limit of the rigid bar; that is, the applied load that probably would bend the bar beyond its ability to return to its original shape?

4. Solve the following equations for the unknown value indicated:

(a) $16 + 32 + X = 98$, $X =$
(b) $F = \dfrac{120}{44.0}$, $F =$
(c) $10\ X = 800\ \text{cm}/200\ \text{s}$, $X =$
(d) $13.0\ \text{kg/m}^3 = \dfrac{56.0\ \text{kg}}{v}$, $v =$
(d) $2400\ \text{W} = (120\ \text{V})\ (\text{I})$, $\text{I} =$

5. With regard to graph B, how much farther in the second second did the ball roll than in the first second; how much farther did the ball roll in the fourth second than in the second second?

6. How do direct proportions differ from inverse proportions? Cite three examples of each type.

Appendix 3

Abbreviations and Symbols

a

acceleration = $\boldsymbol{a}$
acceleration due to gravity = $\boldsymbol{g}$
alpha particle = α
alternating current = AC
ampere = A
amplitude modulation = AM
area = A
astronomical unit = AU
atomic mass = A
atomic mass unit = u
atomic number = Z

b

becquerel = Bq
bel = B
beta particle = β
British thermal unit = Btu

c

calorie = cal
candela = cd
Celsius = C
centi- = c
centripetal acceleration = $\boldsymbol{a}$
centripetal force = $\boldsymbol{F}$
charge = Q
coefficient of friction = μ
coefficient of linear expansion = α
compressed area = C
coulomb = C
cubic centimeter = cm^3
cubic foot = ft^3
curie = Ci
current = I

d

deci- = d
decibel = dB
delta = Δ
density = ρ
direct current = DC
distance = s
dyne = d

e

effort = E
electromotive force (emf) = ε
electron = e^-
electron volt = eV
energy = E

f

Fahrenheit = F
feet per second per second = ft/s^2
focal length = F
foot = ft
force = $\boldsymbol{F}$
frequency modulation = FM
fulcrum = f

g

gallon = gal
gamma ray = γ
gram = g
gray = Gy

h

half-life = $t_{1/2}$
height = h
hertz = Hz
horsepower = hp
hour = h

i

index of refraction = n
integrated circuit = IC
intensity = I

j

joule = J

k

kelvin = K
kilo- = k
kilogram = kg
kilowatt-hour = kWh
kinetic energy = $K.E.$

l

length = l
light-emitting diode = LED
liquid crystal display = LCD
liter = l or L

m

mass = m
mass number = A
mechanical advantage = $m.a.$
mega- = M
meter = m
micro- = μ
mile = mi
milli- = m
million electron volts = MeV
milliroentgens = mR
minute = min

n

nano- = n
neutrino = ν
neutron = n
neutron number = N
newton = N
number of whole-number multiples = n

o

ohm = Ω
ounce = oz

p

pascal = Pa
period = T
Planck's constant = h
positron = e^+
potential energy = $P.E.$
pound = lb
pounds per square inch = lb/in^2
power = P
pressure = P
proton = p^+

q

quart = qt

r

radiation absorbed dose = rad
radius = r or R
rarefied area = R
resistance = R
roentgen = R

s

second = s
sievert = Sv
specific heat = c
speed = v

t

temperature = T or t
tension = T
thermal energy = Q
time = t
ton = T
torque = T

v

velocity = $\mathbf{v}$
velocity of light = c
volt = V
voltage = V
volume = v or V

w

watt = W
wavelength = λ
weight = w
work = W

y

yard = yd
year = y

Appendix 4

Answers to Selected Problems

Chapter 3

7. 15 mi/h, eastward; 5 mi/h, eastward

9. 23.8 m/s; no; no

11. (a) 0–1 s, 2.00 m eastward;
1–2 s, 3.00 m eastward;
2–3 s, 4.00 m eastward;
3–4 s, 5.00 m eastward;
4–5 s, 5.00 m eastward.
(b) 0–1 s, 2.00 m/s eastward;
1–2 s, 3.00 m/s eastward;
2–3 s, 4.00 m/s eastward;
3–4 s, 5.00 m/s eastward;
4–5 s, 5.00 m/s eastward.
(c) 3.80 m/s eastward
(d) 1.00 m/s^2 eastward
(e) 0

13. (a) 59.2 m/s^2 upward
(b) 80.6 m upward

15. (a) 4.55 m/s^2 eastward
(b) 146 m eastward

17. (a) 17,700 N westward
(b) The weight of the truck and the upward force of the road. They are equal and opposite, so their sum is zero.

19. (a) 74.0 kg (b) 74.0 kg (c) 123 N

Chapter 4

9. (a) 19.6 m/s^2
(b) 4.90 m/s^2
(c) 3.00 kg
(d) 29.4 N on Earth, 58.8 N on a planet with twice the mass and of the same size, 14.7 N on a planet with twice the mass and twice the radius.

11. (a) 14.2 m
(b) If the floor level with the ground is known as the first floor, the student lives on the fifth floor.

17. (a) 8.33 m/s^2
(b) 15,000 N
(c) 3350 lb

25. 1.03×10^9 h; 111,700 y

27. 150 N toward the center of Earth

29. 1.54 km/s

Chapter 5

1. 223 N

2. 13,400 N

5. 3500 lb

8. 98,000 J

13. (a) 6950 g

Chapter 6

6. 39,500 cal

9. 12,000 cal

15. (b) 0.0124 m^3

23. 46,000 cal

Chapter 7

12. 9.00×10^{-8} W/m^2

15. 0.639 m

19. 1020 m

Chapter 8

17. Upward, 50,000 N/C. It is twice as much; that is, 100,000 N/C.

9. 2.25 A

11. 8.33 V

13. 78.0 W

19. 0.833 A

25. 134 V

Chapter 9

3. 40.0 W

5. 1200 W; 17,200 cal

13. 12.0 V

Chapter 10

5. 6.32×10^{14} Hz

9. 71.8 km

19. 1.50×10^8 m/s

25. It would be about 1/1000 as long.

Chapter 11

3. 13.3 lm/m^2

5. 20.0 cm

6. concave, 0.600 m

10. 10 cm; −12 cm (a virtual image)

12. He should buy the $f/1.4$ lens. About 1/200 s.

15. Maximum "useful" magnification is 144.

25. 122 mm

Chapter 13

26. 6

35. A solution with a pH of 4 has 100 times more hydrogen ions than a solution with a pH of 6.

36. 10

Chapter 15

4. 1802 MeV

12. (a) 28,650 y

25. 173 kg

Chapter 17

8. 8 y

Chapter 22

30. (a) $-20°$ F (b) $-60°$ F
(c) $-40°$ F

Chapter 23

19. 0.036 percent

20. About 2 billion years

Appendix 1

1. (a) 4.864 m
(e) 1.944 m
(h) 650 mL

2. (a) 7.01 m
(e) 1.27 m^3
(i) 2270 g

3. (b) 6000 cm^3

5. (c) $\boldsymbol{F} \times s$

Appendix 2

1. (a) 1.9×10^{14}
(c) 4.0×10^4
(e) 1.32×10^3

2. (c) 2337
(e) 4; 1; 6; 5; 3

3. 150 kg

4. (b) 2.73
(d) 4.31 m^3

Appendix 5

Atomic Numbers, Names, Symbols, and Atomic Weights of the Elements

Table of Atomic Weights to Four Significant Figures (Scaled to the relative atomic mass of ^{12}C = 12 exactly)

Values quoted in this table are reliable to ±1 or better in the fourth significant figure except for the five elements for which the larger indicated uncertainties apply. Each element that has neither a stable isotope nor a characteristic natural isotopic composition is represented in this table by one of that element's commonly known radioisotopes identified by its mass number (in superscript preceding the chemical symbol) and its relative atomic mass, in the Atomic Weight column.

Atomic Number	Name	Symbol	Atomic Weight	Atomic Number	Name	Symbol	Atomic Weight
1	Hydrogen	H	1.008	27	Cobalt	Co	58.93
2	Helium	He	4.003	28	Nickel	Ni	58.69
3	Lithium	Li	6.941±2	29	Copper	Cu	63.55
4	Beryllium	Be	9.012	30	Zinc	Zn	65.39±2
5	Boron	B	10.81	31	Gallium	Ga	69.72
6	Carbon	C	12.01	32	Germanium	Ge	72.61±2
7	Nitrogen	N	14.01	33	Arsenic	As	74.92
8	Oxygen	O	16.00	34	Selenium	Se	78.96±3
9	Fluorine	F	19.00	35	Bromine	Br	79.90
10	Neon	Ne	20.18	36	Krypton	Kr	83.80
11	Sodium	Na	22.99	37	Rubidium	Rb	85.47
12	Magnesium	Mg	24.30	38	Strontium	Sr	87.62
13	Aluminum	Al	26.98	39	Yttrium	Y	88.91
14	Silicon	Si	28.09	40	Zirconium	Zr	91.22
15	Phosphorus	P	30.97	41	Niobium	Nb	92.91
16	Sulfur	S	32.07	42	Molybdenum	Mo	95.94
17	Chlorine	Cl	35.45	43	Technetium	^{99}Tc	98.91
18	Argon	Ar	39.95	44	Ruthenium	Ru	101.1
19	Potassium	K	39.10	45	Rhodium	Rh	102.9
20	Calcium	Ca	40.08	46	Palladium	Pd	106.4
21	Scandium	Sc	44.96	47	Silver	Ag	107.9
22	Titanium	Ti	47.88±3	48	Cadmium	Cd	112.4
23	Vanadium	V	50.94	49	Indium	In	114.8
24	Chromium	Cr	52.00	50	Tin	Sn	118.7
25	Manganese	Mn	54.94	51	Antimony	Sb	121.8
26	Iron	Fe	55.85	52	Tellurium	Te	127.6

Table of Atomic Weights *continued*

Atomic Number	Name	Symbol	Atomic Weight	Atomic Number	Name	Symbol	Atomic Weight
53	Iodine	I	126.9	79	Gold	Au	197.0
54	Xenon	Xe	131.3	80	Mercury	Hg	200.6
55	Cesium	Cs	132.9	81	Thallium	Tl	204.4
56	Barium	Ba	137.3	82	Lead	Pb	207.2
57	Lanthanum	La	138.9	83	Bismuth	Bi	209.0
58	Cerium	Ce	140.1	84	Polonium	^{210}Po	210.0
59	Praseodymium	Pr	140.9	85	Astatine	^{210}At	210.0
60	Neodymium	Nd	144.2	86	Radon	^{222}Rn	222.0
61	Promethium	^{145}Pm	144.9	87	Francium	^{223}Fr	223.0
62	Samarium	Sm	150.4	88	Radium	^{226}Ra	226.0
63	Europium	Eu	152.0	89	Actinium	^{227}Ac	227.0
64	Gadolinium	Gd	157.2	90	Thorium	TH	232.0
65	Terbium	Tb	158.9	91	Protactinium	Pa	231.0
66	Dysprosium	Dy	162.5	92	Uranium	U	238.0
67	Holmium	Ho	164.9	93	Neptunium	^{237}Np	237.0
68	Erbium	Er	167.3	94	Plutonium	^{239}Pu	239.1
69	Thulium	Tm	168.9	95	Americium	^{243}Am	243.1
70	Ytterbium	Yb	173.0	96	Curium	^{247}Cm	247.1
71	Lutetium	Lu	175.0	97	Berkelium	^{247}Bk	247.1
72	Hafnium	Hf	178.5	98	Californium	^{252}Cf	252.1
73	Tantalum	Ta	180.9	99	Einsteinium	^{252}Es	252.1
74	Wolfram (Tungsten)	W	183.8	100	Fermium	^{257}Fm	257.1
75	Rhenium	Re	186.2	101	Mendelevium	^{256}Md	256.1
76	Osmium	Os	190.2	102	Nobelium	^{259}No	259.1
77	Iridium	Ir	192.2	103	Lawrencium	^{260}Lr	260.1
78	Platinum	Pt	195.1				

Prepared by N. N. Greenwood and H. S. Peiser, on behalf of the Committee on Teaching of Chemistry of the International Union of Pure and Applied Chemistry (IUPAC) in consultation with the IUPAC Commission on Atomic Weights and Isotopic Abundances. (Note that information about elements 104 to 109 has not been included by the IUPAC at this time.)

Glossary

aberration The apparent change in position of a star caused by Earth's motion about the sun.

absolute zero The temperature at which atoms or molecules have minimum kinetic energy ($-459.69°F$, or $-273.16°C$, or 0 K).

acceleration The rate at which velocity changes with time. Acceleration may involve either a change in magnitude, direction, or both.

acceleration due to gravity The rate of change of velocity of a free-falling object in the absence of air resistance or other source of friction; about 9.8 m/s^2 downward on Earth.

acid A substance that produces hydrogen ions in water solution and turns blue litmus pink. (Contrast **base.**)

acid rain Precipitation that contains potentially dangerous amounts of sulfuric and nitric acids, formed from the oxides emitted by burning fossil fuels.

actinide series The sequence of increasing atomic numbers in period seven that appear at the bottom of a periodic chart. The atomic number of the series begins after actinium, $Z = 89$.

action The bringing about of the motion of a body by an applied force.

activation The process of producing artificial radioisotopes by bombarding stable elements with charged particles or neutrons.

activation analysis A procedure that permits the detection and measurement of artificially produced radioisotopes for the purpose of identification of trace elements present.

activation energy The minimum amount of energy needed to overcome repulsive forces between molecules in a reaction. Also, the quantity of energy needed to initiate a chemical reaction.

activity In radioactivity, the rate of disintegration of atomic nuclei.

activity series A list of metals indicating their relative ability to displace hydrogen from water or acids. Any metal in the series can replace any metal below it from a solution of a salt of the metal.

addition polymerization A reaction during which polymers are formed by the addition of one unsaturated monomer molecule at a time to the end(s) of a growing polymer chain.

addition reaction A reaction in which two or more molecules react to form a single molecule of product.

aerodynamics The study of forces that act upon a solid or liquid body as it moves relative to a gas, such as air.

aggradation The modification of Earth's surface toward uniformity by deposition.

airfoil A body designed to provide a specific reaction force when in motion; for example, an airplane wing or propeller blade.

air mass A large body of air that stays over a certain area long enough to acquire the humidity and temperature of that region.

air pollution The contamination of the atmosphere by the addition of harmful gases, smoke, and/or particulate matter.

alar The commercial name for the pesticide daminozide, used primarily on apples.

albedo The fraction of incident light reflected from the surface of a celestial object.

alchemy Medieval chemistry chiefly aiming to achieve the transmutation of base metals into gold and to find a means of indefinitely prolonging life.

alcohol Any one of a number of organic compounds having an —OH group attached to a carbon atom.

aldehyde Any one of a number of organic compounds containing the $-\overset{\overset{\displaystyle O}{\|}}{C}-H$ group.

alkali metals A family of soft, light, highly active metals with similar chemical properties. The alkali metals are located in group IA in the periodic chart of the elements.

alkaline cell A source of direct current that uses an alkaline electrolyte, such as potassium hydroxide.

alkaline earths Any element in group IIA of the periodic chart.

alkane A hydrocarbon that contains only single bonds.

alkene A hydrocarbon that contains two bonds between two adjacent carbon atoms.

alkyl halide An organic compound that contains one or more halogen atoms.

alkyne A hydrocarbon that contains three bonds between two adjacent carbon atoms.

alloy A mixture of two or more metals, such as tin and copper, or of a metal or metals with a nonmetal, such as iron and carbon.

alnico An alloy containing aluminum, nickel, cobalt, and iron; used in making permanent magnets.

alpha particle The nucleus of a helium atom, a particle ejected by many heavy radioisotopes. The alpha particle is composed of two protons and two neutrons.

alpha radiation A stream of alpha particles ejected from a radionuclide.

alternating current An electric current that periodically reverses in direction. A complete reversal is generally known as a cycle. Typically, alternating current is produced at 50 or 60 cycles per second (hertz).

amber A brown translucent resin that readily accumulates an electric charge.

ambient temperature The temperature of the surrounding environment.

amides Organic compounds in which the —OH of the carboxylic acid has been replaced with an $—NH_2$.

amine An organic compound containing a carbon-to-nitrogen single bond. This definition does not include amide.

amino acid An organic compound containing an amino group and a carboxylic acid group.

amorphous solid A solid that is noncrystalline; therefore, it has neither definite form nor structure.

ampere The rate of flow of electric charge equal to 1 coulomb per second.

amphibian A cold-blooded, smooth-skinned vertebrate of the class Amphibia. As a larva it breathes with gills but metamorphoses into an adult form with air-breathing lungs.

amplifier A device capable of increasing the magnitude or power level of a physical quantity, such as electric current.

amplitude The maximum distance through which vibrating particles in a wave disturbance move from a rest position.

amplitude modulation (AM) The changing of the amplitude of a transmitting radio wave in accordance with the strength of the audio or other signal.

analysis The resolution of a whole into parts; the separation of compounds and mixtures into their constituent substances to determine the nature and/or proportion of the constituents.

Andromeda galaxy (M31) A galaxy considered quite similar to the Milky Way. It has a brilliant central nucleus and conspicuous spiral arms.

angle of incidence The angle formed by a ray of light falling on a surface and the perpendicular or normal to that surface at the point of incidence.

angle of reflection The angle between the direction of propagation of a wave reflected by a surface and the line perpendicular to the reflecting surface at the point of incidence.

angle of refraction The angle between the direction of propagation of a wave that is refracted by a substance and the line perpendicular to the surface at the point the ray exits.

angular diameter The angle formed when an imaginary line is drawn from one edge of a celestial body to the observer and back to the opposite edge of the celestial body.

angular momentum The momentum of a mass moving in a circular path.

animal starch A natural polymer, called glycogen, that is made up of glucose units. The glycogen molecule is smaller and more branched than the starch molecule.

anode A positively charged terminal in an electrical device.

Antarctic Circle The parallel of 66.5 degrees south latitude.

anthracite coal A form of hard coal, difficult to ignite, but affording much heat with little smoke.

aperture The diameter of the primary lens or mirror of an optical instrument.

aphanitic Of fine-grained texture, such as the igneous rock basalt.

aphelion In the orbit of an object revolving around the sun, that point at which the object is farthest from the sun.

Apollo space program A NASA program, dedicated primarily to landing astronauts on the moon.

aquifer A permeable body of rock or sediment through which groundwater can readily move.

Arctic Circle The parallel of 66.5 degrees north latitude.

armature The rotating coils of wire in a motor or a generator.

aromatic compound A compound similar to benzene characterized by a cyclic structure containing alternating double and single bonds. Aromatic compounds do not normally undergo addition reactions.

aromatic hydrocarbon A compound that incorporates a ring nucleus, such as benzene, in its structure.

artificial transmutation The change in identity of an element, brought about by neutron or charged particle bombardment.

asbestos A nonburning grayish mineral that occurs in fibrous form. Asbestos was widely used in insulation; however, it is suspected of being either a true carcinogen or a carrier of carcinogenic trace elements. It also has been found to create serious respiratory ailments.

asteroid One of ten of thousands of planetoids that orbit the sun between the orbits of Mars and Jupiter.

asthenosphere The layer or shell of Earth below the lithosphere. The asthenosphere is the uppermost portion of the mantle.

astrology The study of the positions and geometric relationships of the sun, moon, and planets, with respect to the zodiac. Astrologers believe the positions and aspects of stars and planets influence human affairs and terrestrial events.

astronomical unit (AU) A unit of length equal to the mean distance of Earth from the sun; approximately 1.5×10^8 kilometers.

astronomy The science of celestial bodies, including their size, motion, structure, emissions, origin, fate, relationship, and composition.

atmosphere The mass of gases and particulate matter that surround an object such as Earth.

atom The smallest part of an element that can exist either alone or in combination.

atomic concept The well-established and proven idea that matter is composed of discrete particles called atoms.

atomic crystal lattice An orderly, three-dimensional arrangement of atoms in a crystal.

atomic energy Energy released by nuclear fission or nuclear fusion. Also known as nuclear energy.

atomic mass The mass of a neutral atom, generally based on the average abundance of nuclides for a given element.

atomic mass unit (u) The standard unit of atomic mass, established as equal to one-twelfth the mass of carbon-12.

atomic number The identifying number of an element that corresponds to the number of protons present in the nucleus of the element. The symbol for atomic number is *Z*.

atomic pile A structure made of graphite, uranium, and uranium oxide blocks that furnish the neutrons capable of producing controlled fission in a nuclear reactor.

atomic theory The theory that matter is composed of particles called atoms and that these particles represent the limit to which matter can be subdivided.

atomic weight The average relative weight of the isotopes of a given element compared with $^{12}_{6}C$ atoms, the accepted standard.

attractive The term used with reference to a force that tends to pull two physical entities together. For example, the force between a positive and a negative charge is attractive. (Contrast **repulsive.**)

audiosonics Sound waves, the wavelengths of which are between 0.0017 and 20.96 meters, and thus are in our range of hearing.

aurora australis The glow about Earth produced in the Southern Hemisphere by the solar wind.

aurora borealis The glow about Earth produced in the Northern Hemisphere by the solar wind.

Australopithecus africanus A hominid that appeared during the Pliocene. *Australopithecus africanus* is not considered to be part of the *Homo* line.

autotroph An organism that can synthesize its food from inorganic compounds.

autumnal equinox The time during autumn (Northern Hemisphere) when the sun crosses the equator.

average speed The numerical result obtained by dividing the distance traveled by the elapsed time.

average velocity The change in displacement divided by the elapsed time, including both magnitude and direction.

Avogadro's number The number of units (6.02×10^{23}) in 1 gram atomic weight or 1 gram molecular weight of a substance.

axis A line about which a body rotates.

background radiation In the absence of a radioactive sample, that detected activity attributed to cosmic radiation, electronic noise, and naturally occurring radioactivity present in building materials.

balanced equation A chemical equation that has the same number of atoms or ions of each element on each side.

barred spiral galaxy A spiral galaxy in which curved spiral arms begin from the ends of a protrusion that passes through the nucleus rather than originating themselves from the nucleus.

Barringer Meteorite Crater A famous crater near Winslow, Arizona, created by the impact of a meteorite.

barycenter The center of mass of two bodies about which they mutually revolve due to gravitational attraction.

baryon Any one of a number of particles found in the atom that are massive when compared to other particles.

basalt A fine-grained igneous rock consisting of feldspar and ferromagnesian materials.

base A substance that produces hydroxide ions in water solution and turns pink litmus blue. (Contrast **acid.**)

base (semiconductor) A region in a semiconductor triode that, by means of applied voltages, controls the current that passes from the emitter to the collector.

batholith A great mass of intruded igneous rock, the exposed area of which is at least 100 square kilometers.

battery A source of electric potential composed of two or more cells.

beat A series of outbursts of sound followed by intervals of comparative silence.

becquerel The unit of disintegration rate of a radionuclide equal to 3.7×10^{10} disintegrations per second.

bel A unit for expressing in logarithms the ratios of sound levels; after Alexander Graham Bell.

Bernouilli effect The tendency for the pressure of a stream of fluid (gas or liquid) to be reduced as its velocity of flow is increased.

beta particle An electron (or positron) emitted during the disintegration of a radioactive atom.

beta radiation A stream of beta particles, positrons or negatrons, ejected from a radionuclide.

big-bang theory The theory of cosmogony that the universe started with a great explosion and that the universe has been expanding ever since.

bimetallic strip A device composed of two dissimilar metals that have significantly different coefficients of thermal expansion, bonded together.

binary system Two stars mutually revolving about their barycenter.

binding energy The energy that holds the particles in an atom's nucleus.

biodegradable A term used to describe a substance capable of being converted to water-soluble substances by biological means.

biological science The study of living things, including their forms, growth processes, evolutionary development, and death. Also, the study of living things from the immediate past, such as those forms discovered suspended in amber and other such preservatives.

biomass burning The intentional burning of basically unwanted vegetation.

bitumen A naturally occurring hydrocarbon associated with old deposits of sand and water.

bituminous coal A soft form of coal that yields pitch or tar when burned.

blackbody An ideal body or surface that completely absorbs all incident radiant energy.

black dwarf The final stage in the evolution of a star. The star no longer emits light.

black hole The result of uninhibited gravitational collapse of a star. Since the escape velocity of a black hole exceeds the speed of light, neither matter nor electromagnetic radiation can escape from a black hole.

blue giant A large star with a surface temperature so high that it appears bluish in color.

Bode's law An empirical scheme by which a sequence of numbers can be obtained indicating the approximate distances of many of the planets from the sun.

boiling point That point (temperature and pressure) at which the vapor pressure of a liquid slightly exceeds the pressure of the atmosphere upon the liquid.

Bowen reaction series The sequence in which minerals solidify from a cooling basaltic magma.

Boyle's law The law that the product of the volume of a gas times its pressure is a constant at a fixed temperature.

British gravitational system A measurement system based on the foot, the second, and the pound as a unit of force.

British thermal unit (Btu) The amount of heat required to raise or lower the temperature of 1 pound of water 1 Fahrenheit degree.

bronze An alloy made of copper and a small amount of tin.

Brownian motion The random motion of small particles caused by collisions with molecules.

bubble chamber A device in which the movements and interactions of charged particles can be observed as visible tracks in a superheated liquid. The tracks appear as gas bubbles that form along the paths of the particles.

bubble structure of the universe The apparent existence of large spherical regions (bubbles) within the universe that are devoid of galaxies.

caldera A volcanic depression that is larger than the original crater.

caloric A substance thought to exist to which the phenomena of burning and oxidation were attributed.

Calorie The amount of heat required to raise the temperature of a kilogram of water 1 Celsius degree; a kilocalorie.

calorie The amount of heat required to raise the temperature of 1 gram of water 1 Celsius degree.

canali Surface features supposedly observed on the planet Mars thought to be channels, but misinterpreted in language translation to be canals, suggesting intelligent life on the planet.

candela The standard unit of luminous intensity, based on the intensity of a standard source and the response of the eye to light. The luminous intensity of a source in a given direction in candelas can be found by dividing the number of lumens emitted by the source, if it is uniformly radiating in all directions, by 4π.

capacitor A device used to store electric charge. In an electronic circuit, capacitors may be used to regulate charge flow or to store information.

carbohydrate An organic compound containing many alcohol (—OH) groups and usually an aldehyde ($-\overset{O}{\overset{\|}{C}}-H$) or a ketone ($-\overset{O}{\overset{\|}{C}}-$) group. Sugars and starches are carbohydrates.

carbon cycle The process whereby carbon nuclei fuse with hydrogen nuclei to eventually produce helium nuclei and regenerated carbon nuclei. Much energy is released in the process.

carbonated beverage A beverage that contains carbon dioxide under pressure.

Carboniferous period The combined periods during the Paleozoic era known as the Mississippian and Pennsylvanian, and during which most of Earth's coal was formed.

carbonyl group The ($-\overset{O}{\overset{\|}{C}}-$) group present in aldehydes and ketones.

carboxylic acid An organic compound that has a $-\overset{O}{\overset{\|}{C}}-OH$ group.

Cassini division A gap, such as between the B ring and the A ring of Saturn.

catalyst A substance that speeds up or slows down the rate of a chemical reaction, yet undergoes no permanent chemical change itself.

cathode A negatively charged terminal in an electrical device.

cathode-ray tube (CRT) An electronic tube using electrons (cathode rays) emitted from a cathode that strike a phosphorescent surface inside the tube to produce an image on the face of the tube. The picture tube in an ordinary television set and the display terminal of most computers are CRTs.

celestial sphere The apparent sphere of the sky, half visible at any time to an observer on Earth.

cell A source of electric potential, generally 1.5 or 2 volts.

cellulose A natural polymer of glucose; the chief substance composing the cell walls or woody part of plants.

Celsius degree The unit of temperature interval equal to the kelvin.

Celsius temperature scale A temperature scale on which the freezing and boiling points of pure water under standard conditions are defined to be 0°C and 100°C, respectively; formerly called the Centigrade scale.

Cenozoic era That period of geologic time from 65 million years ago to the present.

center of curvature At a given point on a curve, the center of the circle the curve represents at that point.

center of gravity The single point at which the force of gravity appears to act on a body.

center of mass The single point on a body at which all of its mass appears to be concentrated.

centrifugal force A fictitious force that seems to pull bodies outward when they are moving in a circular path. The centripetal force seems real to an observer moving in a noninertial reference frame.

centripetal acceleration The acceleration produced by a center-seeking force impressed on a body. The resultant change in velocity may be in direction only.

centripetal force The center-seeking force that causes a body to assume a circular path.

Cerenkov radiation The blue glow produced by charged particles as they travel through water faster than does visible light.

Ceres The largest of the known asteroids or planetoids that orbit the sun between Mars and Jupiter.

chain reaction A sequence of nuclear events brought about by the inelastic collisions of particles with nuclei. It is the process for nuclear fissioning.

charge To give an electric charge or a quantity of electrical energy to a device, such as a battery, or some other object.

charge-coupled device (CCD) A semiconductor device containing many individual cells, each of which acquires a charge proportional to the intensity of the light that strikes it. CCDs are used to transform images into electrical impulses.

Charles' law The law that the volume of a fixed mass of gas at constant pressure varies directly with the absolute temperature.

chemical change A change, in the chemical composition of a substance.

chemical property One of the characteristics of a substance that describes the manner in which the substance reacts chemically. For example, a chemical property of water is that water can be decomposed to form two new substances, hydrogen and oxygen.

chemical weathering The breakdown of substances such as rocks by various chemicals usually in water solution.

chemistry The science that deals with the composition, structure, properties, energy, and physical transformations of matter in all its various forms.

chemosynthetic autotroph An organism that can synthesize its food from carbon dioxide and certain inorganic compounds, such as hydrogen sulfide, in the absence of light.

Chernobyl The site of a nuclear power plant near Kiev, Russia, where the chemical explosion of a nuclear reactor in 1986 released radioactive materials into the atmosphere.

chill factor An expression of the effect of wind speed on body tissue at a given air temperature and expressed as a temperature equivalent.

chlorinated hydrocarbon A compound, such as DDT, composed of an organic compound that contains chlorine.

chromatic (color) aberration Distortion of an image by a lens because the edge of the lens acts as a weak prism. Chromatic aberration causes the edges of images to appear colored.

chromatic scale A musical scale composed of notes arranged in half steps.

chromosphere The part of the sun's atmosphere directly above the photosphere.

circulating fluidized bed (CFB) A term used to describe a technology for the combustion of various woodwaste products and other commercial wastes with a minimum release of unwanted gaseous and particulate matter.

circumference The distance around a circular path.

cirrus cloud A wispy, fibrous form of cloud.

cleavage The tendency of a mineral to break along definite planes.

closed electric circuit A closed path around which electric charges may flow.

closed pipe A musical organ component that, when a stream of air is directed across it, produces a sound, the wavelength of which is four times the length of the pipe.

cloud A mass of very small droplets of water or ice crystals condensed from the air. Only slight upward air currents are necessary to support the buoyant mass.

cloud chamber A device used to detect the presence of radioactivity.

cluster A group of galaxies in the same region of space gravitationally bound to each other.

coal gasification The process by which gas is extracted from coal.

coefficient A number, constant for a given substance, used to determine the change brought about in the substance as a condition is varied.

coefficient of friction A ratio of the amount of force required to roll or slide an object and the amount of force necessary to lift it.

coefficient of linear expansion The fractional change in length of a solid brought about by a change in temperature of 1 degree.

coefficient of volumetric expansion The fractional change in volume of a liquid or solid brought about by a change in temperature of 1 degree.

coherent light Light in which all photons are in phase with each other.

cold front The contact zone between a moving cold air mass and a warm air mass in which the cold air horizontally replaces the warm air.

cold fusion The process of obtaining fusion in which extremely high temperatures are not required.

collector The region of a transistor through which a primary flow of charge leaves the base of the transistor.

collider A facility designed to accelerate charged particles, such as electrons, positrons, and protons, and then cause them to interact through head-on collision.

collimated light Light that has been brought to a focus in such a way that it forms parallel rays.

combined gas law See **general gas law.**

combustion The rapid oxidation of a substance in which heat and light energy are released.

comet A swarm of solid material and gases revolving about the sun in a highly elliptical orbit.

complementary colors Two specific colors of light that can combine to produce white light, or two specific pigments that can combine to produce dark brown or black.

completed circuit A system of electrical conductors that provides an uninterrupted path from the cathode to the anode of the source of electricity.

compound A chemically combined unit of two or more different atoms.

compression In sound, that part of a wave form in which the density of the vibrating medium is greater than normal.

concave Hollow or rounded inward; used to describe a lens thinner in the center than at the edges.

concave lens A lens with a negative focal length. Incident parallel rays of light diverge (spread out) after passing through it.

concave mirror A hollow and curved reflecting surface shaped like a section of the inside of a sphere. Incident parallel rays of light converge (come together) after striking it.

concentrated solution A homogeneous combination of compounds or elements in which there is a large quantity of the dissolved material present.

conchoidal A surface shaped like the inside of a bivalve shell.

condensation The change in a form of matter's state from a vapor to a liquid.

conduction The transmission of thermal energy from molecule to molecule.

conductor Any substance through which electric charge (or heat) readily flows. Conductors are generally metallic in nature.

conservation of angular momentum The principle that the angular momentum of any group of objects must remain constant if no outside torques act on the group.

conservation of charge The principle that charge is neither created nor destroyed. In ordinary processes this principle is upheld, but in some exotic experiments it can be shown that charge is related to other quantities which, taken as a group, are conserved.

conservation of energy The principle that energy cannot be created or destroyed, but can be changed from one form to another.

conservation of mass The concept that mass can neither be created nor destroyed; it is not valid for numerous microscopic phenomena.

conservation of momentum The principle that, providing no external force or unbalanced external torque acts on a system, the momentum of the system must remain constant.

constellation A configuration of stars named after some object, animal, ordinary mortal, or mythological being. Currently there are 88 constellations defined by the International Astronomical Union.

constructive interference The increase in the amplitude of a wave when two or more waves pass through the same region at the same time and are in phase with each other.

contact force The force one body exerts on another when the two bodies are in contact with one another.

contact method The transfer of static electric charge from one surface to another by direct contact.

continental glacier An ice sheet large enough to cover the major portion of a continent.

continental tropical air masses Air masses that originate over northern Mexico and the southwestern United States.

control rod The component used to adjust the rate of fissioning of a nuclear reactor.

convection The transfer of thermal energy in a gas or liquid by moving currents in the fluid.

convection currents Air movements created by cool air as it warms and moves upward and warm air as it cools and moves downward toward Earth.

converging lens or mirror A lens or mirror that causes parallel rays of light to converge (come together).

convex Curved or rounded outward; used to describe a lens thicker in the center than at the edges.

convex lens A lens with a positive focal length. Incident parallel rays converge (come together) after passing through it.

convex mirror A hollow and curved reflecting surface shaped like a section of the outside of a sphere. Incident rays of light diverge (spread out) after striking it.

core (of Earth) The iron and nickel-rich innermost zone of Earth. The outer core is molten, and the inner core is solid.

Coriolis effect The effect of Earth's rotation that deflects moving objects, such as projectiles or air currents, to the right in the Northern Hemisphere and to the left in the Southern Hemisphere.

corona The outermost portion of the sun's atmosphere. The ivory-white corona is seen as a halo around the totally eclipsed sun.

corpuscular theory (light) The theory that light is composed of a stream of particles.

cosmogony Theories of the origin of the universe.

cosmology Study of the general structure and evolution of the universe.

coulomb A measure of electric charge that equals the total electric charge carried by 6.25×10^{18} electrons.

Coulomb's law An equation relating the force between two like or unlike charges. The force between two charges is directly proportional to the product of the charges and inversely proportional to the square of the distance between them.

covalent bond A chemical bond formed by two atoms that share a pair of electrons.

crater A basinlike depression over a vent at the summit of the cone of a volcano. Also, a structure formed by the force of a meteorite striking a celestial object.

critical angle The minimum angle of incidence for which total internal reflection occurs.

critical mass The minimum amount of a fissionable material necessary to sustain a chain reaction.

Cro-Magnon Inhabitants of Earth about 35,000 years ago and classified as the same species (*Homo sapiens*) as recent humans.

crust (of Earth) The outer layer of Earth. The crust varies in thickness from about 5 to 50 kilometers.

crustal plates Large portions of Earth's crust that are in relative motion with respect to each other. Plates may be as large as a single continent.

crystal A homogeneous solid with an orderly internal structure.

crystal lattice An orderly, three-dimensional arrangement of atoms, ions, or molecules in a crystal.

crystalline solid A solid composed of interlocking crystals that are arranged in an orderly fashion.

cumulus cloud A billowy, round form of cloud.

cuneiform writing Wedge-shaped characters pressed or carved on clay tablets.

curie The total number of disintegrations per second from approximately 1 gram of radium. The accepted value is 3.7×10^{10} disintegrations per second. The curie also represents that quantity of any radioisotope that is decaying at the rate of 3.7×10^{10} disintegrations per second.

Curie temperature The temperature above which a magnet appears to lose its magnetic properties. For iron, the Curie temperature is 770°C; for nickel, it is 358°C.

current The rate of flow of electric charge through a conductor.

cycloalkane A hydrocarbon in which the carbon atoms are arranged to form rings. Cycloalkanes can have noncyclic carbon chains attached to their ring structures.

daughter The immediate product of radioactive decay of an element.

decibel The practical unit of measurement of sound intensity equal to one-tenth bel.

deep mining The removal of various ores, such as coal, from underground mines by way of tunnels, sloping shafts, or vertical shafts.

degradation The wearing away of Earth's surface by the action of erosional agents.

delocalized electrons Electrons that occupy a volume of space encompassing the area surrounding three or more nuclei.

density The mass of a substance per unit volume. Commonly, density is a measure of weight per unit volume. The density of water, for example, is 1 gram per milliliter, or 62.4 pounds per cubic foot.

deposition The process by which sediments are laid down to form sedimentary rock layers.

De Re Metallica The publication prepared by Agricola in 1556 that included a discussion of various geologic principles in use today.

destructive interference The decrease in the amplitude of a wave when two or more waves pass through the same region at the same time and are out of phase with each other.

deuterium The isotope of hydrogen that has a neutron and a proton in its nucleus, designated D or $^{2}_{1}H$.

dew Moisture condensed from the atmosphere onto cool surfaces, usually at night.

dew point The temperature at which air becomes saturated with water.

diamagnetic A term used in reference to the property of certain metals and their alloys whereby, under various conditions, they are repelled by both poles of a magnet.

diameter The straight-line distance through the center of a circular path.

diastolic pressure The reduced pressure the heart exerts on the blood during the relaxation phase. (Contrast **systolic pressure.**)

diatonic scale A musical scale with eight notes to the octave.

dielectric A poor conductor of electric current, such as mica and most plastics.

differential rotation The type of rotation of a fluid body, such as the sun or Jupiter, in which the higher latitudes have a lower rotational speed than those near the object's equator.

diffraction The spreading or bending of a light wave around an obstacle or through a narrow opening in such a manner that parallel fringes of light and dark bands are produced.

diffraction grating An optical device composed of numerous narrow grooves or slits that interfere with light to the extent that a spectrum of colors is produced from an incident beam.

diffusion The reflection of light from an irregular reflecting surface.

dike A tabular body of igneous rock injected, while molten, into fissures.

dilatancy The formation of microscopic cracks in rock layers under study in earthquake zones.

dilute solution A homogeneous combination of compounds or elements in which there is only a small quantity of the dissolved material present.

diode A semiconducting crystal having two regions with different electrical properties, or a two-electrode electron tube.

direct current An electric current, pulsating or steady, that flows in one constant direction.

direct proportion The comparative relationship between two quantities in which one increases as the other increases.

dispersion The separation of a beam of light into its component colors (or frequencies). The separation is usually accomplished by passing the beam through a prism or diffraction grating.

displacement The position of a point relative to some reference point. It is specified by giving the straight-line distance from the reference point and the direction..

displacement reaction A chemical reaction in which atoms or ions of one substance take the place of other atoms or ions in a compound.

distance The spatial separation of two points, measured as the length of a hypothetical line that joins them.

diverging lens or mirror A lens or mirror that causes parallel rays of light to diverge (spread out).

DNA (deoxyribonucleic acid) A large molecule present in cells that contains the genetic code and transmits the hereditary pattern.

doldrums The area along the equator in which there is little or no wind.

Doppler effect The variation in frequency or pitch of sound waves created by the relative motion of the listener and the source. The principle also applies to electromagnetic waves emitted from a source moving relative to an observer.

Doppler radar A microwave transmission technique by which shifts in frequency of the waves reflected from moving objects can be detected.

dormant volcano An inactive volcano; one that potentially could become active at some time in the future.

dosimeter A pocket-sized ion chamber used to determine the amount of radiation exposure.

drag Resistance caused by friction between a body, such as an airplane, moving through a fluid, such as air, parallel and opposite to the direction of motion.

drift A general term applied to all material, such as clay, sand, gravel, and boulders, transported by a glacier and deposited directly by or from the ice or by running water from the glacier.

duration A characteristic of sound related to the length of time a sound lasts.

dynamic equilibrium (chemical) A state in which no net change occurs in a system but in which reactants and products are constantly being formed.

dynamics The branch of mechanics dealing with the motion of a body as influenced by forces that act on it.

dynasties Families and relatives that banded together under a common ruler.

dyne The force that causes a mass of 1 gram to accelerate 1 centimeter per second per second.

eardrum The thin sheet of tissue at the end of the auditory canal that is caused to vibrate by sound waves.

earthlight Earth's equivalent of moonlight. Earthlight shines upon the nightside of the moon in the same way that moonlight shines upon the nightside of Earth. Also called *earthshine.*

earthquake A violent disturbance in Earth's crust caused by the slippage of one block of crustal material relative to another. Also associated with volcanic eruptions.

Earth-type planets The relatively small inner planets that resemble Earth in density and size.

echo A reflected sound wave.

eclipsing binary Two stars mutually revolving about their barycenter in such a way that one star fully or partially eclipses the other during each full revolution.

ecliptic The apparent annual route of the sun against the background of stars.

Ediacara Formation An area in the Ediacara Hills of South Australia, in which fossils at least 700 million years old have been found.

Edison effect The phenomenon produced by an electron tube in which the tube will conduct negative but not positive electric charge.

effective value The current or voltage produced by an alternating-current source equivalent to that produced by a direct-current source. It is equal to 0.707 times the maximum value.

efficiency The ratio of useful work to total work.

effort An applied force.

ejecta Fragments expelled from an erupting volcano.

electric charge A basic characteristic of some elementary particles that causes them to attract or repel each other. There are two types of electric charge, usually called positive and negative.

electric current The flow of electric charge in which energy is transported from one point in an electric circuit to another. The unit by which current is measured is the ampere.

electric field The influence of electric charges on a region of space. A charge placed in an electric field has a force acting on it.

electricity The flow of electric charge through a conductor. Static electricity is an accumulation of ions (including electrons) on the surface of an insulating substance.

electric potential The potential energy per unit electric charge. Electric potential is measured in volts and is often referred to as voltage.

electric potential energy The energy gained by an electric charge as work is done against an electric force in moving the charge from one point to another.

electric power The rate at which electric energy is converted into other forms of energy; equal to the product of current and voltage.

electrode A positively or negatively charged wire or other terminal in an electrical device, such as a vacuum tube or battery.

electrolyte A chemical compound that will conduct electric current when molten or dissolved in water or other solvent.

electromagnet An electrically produced magnet generally made of a soft-iron core and a coil of insulated wire through which a current passes.

electromagnetic energy The energy associated with electromagnetic radiation. It is directly proportional to the frequency of the photon released.

electromagnetic waves A disturbance that moves outward from any electric charge that is accelerated or that oscillates.

electromotive force A force acting on electric charges causing them to move against the electrostatic force, thus increasing their electric potential.

electron The negatively charged elementary particle that orbits the nucleus of an atom.

electron capture The process in which a nucleus absorbs one of its orbiting electrons.

electron current Current produced by an excess of electrons, as in a crystal.

electronegativity The ability of an atom or a group of atoms to attract electrons.

electronic Of or relating to electrons or the utilization of them in communication and other fields.

electron volt The unit used to measure the energy released during radioactive decay and atomic processes.

electroscope A device used to detect the presence of electric charge.

element A substance that has the same number of protons in all its atoms.

element (electrical) A single portion of an electrical circuit. For example, the wire that is heated in an electric iron or in an electric oven is called the element.

elementary particle A particle assumed to be indivisible. Elementary particles are the basic constituents of all matter.

elliptical galaxy A galaxy with a shape that varies from spherical to near lens-shaped.

emitter A transistor region that introduces electric charge into the base.

endothermic reaction A chemical reaction in which energy is absorbed.

energy The ability or capacity to do work. The unit of energy is identical to the unit of work within a system of measurement.

energy level The discrete energy that an electron (or other particle) can assume in a system, such as an atomic system.

energy of activation See **activation energy.**

enriched uranium A quantity of the element uranium to which has been added a concentrated amount of the isotope U-235.

entropy A measure of the disorder of a system; an indication of the unavailable energy in a closed thermodynamic system.

enzyme Any one of a number of proteinlike substances that act as organic catalysts.

epicenter The point on Earth's surface immediately above the focus of an earthquake.

epoch A subdivision of a geological period.

equilibrium The condition in which no apparent change occurs in the state of a system until conditions are altered. The system is said to be in balance.

equinox See **autumnal equinox** and **vernal equinox.**

era A major subdivision of the geologic time scale. An era is composed of periods that are subdivided into epochs.

erg The work done by a force of 1 dyne acting through a distance of 1 centimeter.

erosion The breakdown of rock and the transportation of the resulting loosened material from a higher to a lower level, mainly by the action of water, wind, and ice.

escape velocity The minimum velocity a space vehicle must attain to overcome the effect of the gravitational attraction of Earth or other celestial body.

ester Any one of a number of organic compounds that has a $—CO_2R$ group.

ether Any of various organic compounds that have a C—O—C bond.

excited state The quantum state to which a nucleus, electron, atom, ion, or molecule is raised by acquiring energy from electromagnetic radiation or collisions.

exosphere The outermost layer of the atmosphere, which reaches thousands of kilometers into space.

exothermic reaction A chemical reaction in which energy is released.

extinct volcano The site of a volcano that has been inactive for extremely long periods of time, perhaps centuries.

extrusive igneous rock Any igneous rock that solidifies at Earth's surface.

Fahrenheit A temperature scale on which the boiling point of pure water under standard atmospheric pressure is 212 degrees and the freezing point is 32 degrees.

families Groups of elements with similar chemical properties. Families appear in a vertical fashion on a periodic chart.

fat An ester formed from the reaction of carboxylic acid molecules with glycerol. Fats are solid at room temperature.

fault A fracture in the rock layer along which lateral (horizontal) or vertical displacement has taken place.

fauna Animal life in general.

fermentation The process used by some organisms to obtain energy through the breakdown of organic compounds in the absence of oxygen.

fiber optics The use of thin, transparent fibers of glass or plastic to transmit light without appreciable loss. Fiber optics relies on total internal reflection of light within the fibers and allows light to be "piped" to and from otherwise inaccessible locations.

filament The fine metal wire in a light bulb or vacuum tube that becomes incandescent when heated by an electric current.

film badge A sensitive film used to determine radiation exposure levels.

final velocity The velocity of an accelerated object at the end of a specified time.

firedamp Methane gas often present in coal mines.

First law of thermodynamics The principle that the total amount of energy always remains the same in the transfer of heat from body to body; that is, energy can neither be created nor destroyed. (See **law of conservation of energy,** an equivalent statement.)

fission The splitting of a nucleus into fragments of approximately equal mass. As a result, several lighter atoms are formed along with the release of energy and neutrons. (See also **nuclear power.**)

flora Plant life in general.

fluid friction The friction present in liquids and gases as an object moves through the fluid.

fluorescence The property of some materials enabling them to absorb invisible radiation, such as ultraviolet light, and emit light in the visible portion of the spectrum.

fluorescent lamp A lamp that utilizes a fluorescent coating on the inside of an evacuated glass envelope exposed to ultraviolet radiation as a source of visible light.

focal length The distance from a lens or mirror to its focal point.

focal point The point to which parallel rays of light incident on a lens or mirror converge or from which they appear to diverge.

focal ratio The ratio of the focal length to the aperture (diameter) of a lens. The intensity of an image is proportional to the inverse of the square of the focal ratio.

focus The point at which light rays converge or from which they appear to converge.

fog A concentration of minute, visible water droplets suspended in the air near Earth's surface.

folding The bending, flexing, or wrinkling of rock layers while the layers are in a plastic state.

foliation Parallel alignment of textural and structural characteristics of metamorphic rocks.

force A push, pull, or other influence on a body that tends to cause it to be accelerated.

formula unit The simplest unit indicated by the formula of an ionic compound.

formula weight The sum of weights of the ions indicated by the simplest formula of an ionic compound.

fossil Any remains, traces, or impressions of prehistoric life.

fossil fuels Sources of heat energy formed from the remains of life-forms, such as coal, natural gas, and crude oil.

fractional distillation The process of separating a mixture of chemical substances based on the differences in their boiling points.

fracture The lack of cleavage in a mineral; the general appearance of a freshly broken surface of a mineral.

free-fall The ideal falling motion of a body acted upon only by Earth's gravitational attraction.

freezing The solidifying of a liquid by the removal of thermal energy.

frequency The number of waves, complete or fragmented cycles, that arrive at a given point per unit of time, usually the second.

frequency modulation (FM) Modulation of a radio wave by varying the carrier wave's frequency rather than its amplitude.

friction The force that resists the rolling or sliding of one object on another. Air friction retards the acceleration of a freely falling body.

fringe (optics) Alternating regions of light and dark formed by constructive and destructive interference of waves.

front The boundary between two air masses.

***f*-stop** The focal ratio of a lens as adjusted by a variable diaphragm.

fuel cell A source of electric potential in which hydrogen generally serves as the fuel.

fulcrum The point or object about which a lever rotates.

functional group The atom or group of atoms that determines the major chemical and physical properties of a molecule.

fundamental tone The harmonic component of a tone that has the lowest frequency and usually the greatest amplitude.

fusion The combining of light nuclei to form a heavier nucleus. Energy is released during fusion, a process that takes place only at extremely high temperatures. (See also **nuclear power.**)

galaxy A large collection (millions to hundreds of billions) of stars bound together by gravitation.

gamma radiation A stream of high-frequency photons emitted by a radionuclide.

gamma ray Electromagnetic radiation of high frequencies emitted by some nuclei of radionuclides.

gamma ray spectrometer The instrument most commonly used to determine the energy of gamma radiation emitted from a specific radionuclide.

gas The fluid form of a substance in which it can indefinitely expand.

Geiger-Müller counter The most common electronic device to detect beta and gamma radiation.

general gas law The combination of Charles' law and Boyle's law, as they apply to confined gases. (Ideal gases obey these laws exactly; they are obeyed by real gases only to a limited extent.)

general relativity An extension of the theory of special relativity to include gravitation and related acceleration phenomena.

geocentric concept The idea that Earth is the center of the solar system.

geology The scientific study of Earth, the processes that shape the planet, and the life recorded in its rocks. Geologists also study the moon as well as other planets along with their satellites.

geosyncline According to the recently discounted geosynclinal theory, elongated regions of shallow ocean basins that appeared to be constantly sinking, then continually replenished with sediment.

geothermal energy A source of energy provided by heat from Earth's interior.

glacial till Sediment that has been transported and deposited by melting glacial ice.

glacier A large mass of ice, consisting of recrystallized snow, resting on land. Gravity causes a glacier to flow under its own weight.

globular cluster A spherical collection of stars. Globular clusters are associated with our Milky Way galaxy.

Glomar Challenger A research vessel used to study the ocean floor.

gluon One of several particles that appear to have no mass and to be involved in the interactions between quarks.

glycogen A natural polymer of glucose that has about 10 times the number of glucose units per molecule that starch has. Sometimes called *animal starch,* it serves a function for animals similar to that of starch for plants.

gradation The lowering or smoothing of Earth's surface by processes that produce sediment. The wearing down of rock layers is called *degradation* and the building up of low areas is called *aggradation.*

gram atomic weight The atomic weight of an element expressed in grams.

gram formula weight The formula weight of an ionic compound expressed in grams.

gram molecular weight The molecular weight of a compound expressed in grams; also called a *mole.*

granite A coarse-grained igneous rock consisting of the minerals feldspar, quartz, and mica.

gravitation The force that every object in the universe exerts on every other object because of its mass.

gravitational field The influence of a mass (or masses) on a region of space. A mass placed in a gravitational field has a force acting upon it. For example, there is a gravitational field in the space around Earth.

gravitational mass The mass of a particle as it determines the force it experiences in a gravitational field. (Current experimental evidence indicates that gravitational mass and inertial mass are equal.)

gravity The force of attraction that causes all freely falling bodies at Earth's surface to accelerate about 9.80 m (32 feet) per second per second. Gravity varies among celestial bodies.

gray The amount of radiation that will deposit 1 joule of energy in 1 kilogram of material.

The Great Pyramid The largest of three pyramids located at Giza. It contained the original burial site of Khufu (Cheops); his body, however, was removed many years ago.

greenhouse effect Atmospheric heating resulting from the selective absorption of long-wavelength radiation by carbon dioxide, water vapor, and other gases in the atmosphere. Also called *hothouse effect.*

ground (electricity) A conduction path, intentional or accidental, between an electric circuit or equipment and Earth, or some conducting body serving in place of Earth.

ground state The stationary or lowest energy of a particle or a system of particles.

groundwater Water present in porous rock strata and soils.

groups The vertical columns of the periodic chart.

gyroscope A wheel or disk mounted so as to spin about an axis that is free to turn in any direction.

half-life The time required for one-half of a given number of atoms of a radioisotope to undergo decay.

halogen Any element in column groups VIIB of the periodic chart. The halogens are fluorine, chlorine, bromine, iodine, and astatine.

hardness The property of a mineral that indicates its ability to resist being scratched.

the harmonic law A summary of Kepler's third law of planetary motion.

hazardous wastes Wastes of all sorts that appear to be dangerous to our health or the environment.

health physics The science concerned with the protection of all people from the harmful effects of ionizing radiation by means of area and personnel monitoring, protective equipment, and safety procedures.

heat The energy that flows when a body of higher temperature interacts with a body of lower temperature until thermal equilibrium is reached.

heat index A measure of a person's comfort at a given temperature and different relative humidities.

heat of fusion The amount of heat necessary to bring about a change in state of a substance from a solid to a liquid with no change in temperature.

heat of vaporization The amount of heat necessary to bring about a change in state of a substance from a liquid to a gas with no change in temperature.

heliocentric concept The idea that Earth and all the other planets revolve around the sun.

helix A coil formed by wrapping insulated wire around a uniform tube or cylinder.

hemisphere One of the halves of Earth or other celestial objects as divided by an equator into northern and southern portions.

herbicide A chemical applied to retard or destroy plant growth.

hertz The unit of frequency equivalent to cycles per second and abbreviated as Hz.

Hertzsprung-Russell (H-R) diagram A plot of luminosity (more accurately, absolute magnitude) against temperature for a group of stars.

heterotroph An organism that cannot synthesize organic compounds from inorganic sources. It must acquire food from external sources.

hole A site within a semiconductor where an electron would normally be bonded between two atoms, but is missing because of thermal vibrations or a deficiency in the number of electrons. A hole acts as a positive charge carrier.

hole current The flow of electric charge that behaves as if it were composed of positively charged particles.

hologram A pattern of interference fringes recorded on a photographic plate from a scene illuminated with coherent light (as from a laser). Passage of the original beam of light through the hologram reproduces the waves as if they had come from the original scene; hence, a true three-dimensional image results.

Homo erectus The humanlike inhabitant of Earth after *Homo habilis.*

homogenous mixture A mixture that is of uniform composition throughout.

Homo habilis The humanlike inhabitant of Earth about 2 million years ago.

homologous series A series of compounds in which each member differs from the next member by a constant amount. The constant difference between successive members of an alkane homologous series is a CH_2 unit.

Homo neanderthalensis The Neanderthal, the most famous of all extinct *Homo* species.

Homo sapiens The scientific name for human beings.

horizontal velocity The rate of change of position along a path parallel to some surface such as that of Earth.

horse latitudes The areas around 30 degrees north and south latitude.

horsepower The rate of doing work equal to 550 foot-pounds per second.

hot fusion The term applied to the fusion process under study in today's research fusion reactors.

humidity The amount of water vapor present in the atmosphere. The percentage of the water vapor present compared with the amount the air could hold at that given temperature is called **relative humidity.**

hurricane A tropical cyclone with winds in excess of 120 kilometers per hour; called a typhoon in the western Pacific and Indian oceans.

hydrocarbon An organic compound containing only carbon and hydrogen.

hydrodynamics A branch of science dealing with the motion of fluids and the forces that act on bodies immersed in fluids.

hydroelectric power (hydropower) A source of energy provided by falling water or by the friction of water or steam to generate electricity.

hydrologic cycle The constant circulation of water through evaporation from the oceans, rivers, lakes, and streams, through the atmosphere, to the land by rain and snow, and back by run-off to the ocean in a never-ending cycle.

hydronium ion An ion formed by the reaction of a hydrogen ion and a neutral water molecule. The formula for the hydronium ion is H_3O^+.

hyperopia (hypermetropia) The condition of farsightedness. Light rays, particularly from close objects, are focused behind the retina.

hypothesis An assumption made to test a set of logical or empirical consequences.

ice age A cold period characterized by extensive glaciation.

ice sheet A glacier forming a continuous cover over a land surface with the ice moving outwards in many directions. (See also **continental glacier.**)

igneous activity All movements of molten rock and the formation of solid rock layers from the molten state.

igneous rock An aggregate of rock, generally silicate in nature, formed by the cooling and solidification (congealment) of magma.

illumination The number of lumens per square meter illuminating a scene.

image An optical reproduction of an object (or another image) by a lens or mirror.

incandescence The emission of visible radiation from a hot body, as in a light bulb.

incident light Direct light that strikes a surface.

index of refraction A comparison of the velocity of light in a vacuum (or, approximately, air) with its velocity in another medium.

induction The process by which an electrical conductor becomes charged by an electrically charged body or a magnetizable body becomes magnetized by a magnetized body.

inertia The property of a body that enables it to resist changes in its state of motion; that is, changes in its velocity.

inertial confinement fusion The process by which fusion is attempted by the exposure of a small fuel-filled pellet to laser or charged-particle beams.

inertial mass The mass of an object as determined by Newton's second law, in contrast to the mass as determined by comparison to the gravitational force.

inertial reference frame A reference used for making measurements of position, velocity, and acceleration in which Newton's first law of motion is valid.

inferior planet A planet having an orbit closer to the sun than Earth's orbit is.

infrared A term used to describe electromagnetic radiation, the wavelength of which is slightly longer than the visible color red, from 700 to 1200 nanometers.

infrasonic Sound frequencies below the range of normal hearing.

inner planets Mercury, Venus, Earth, and Mars.

inorganic With few exceptions, compounds that do not contain the element carbon.

inquiry A questioning or search for truth, order, and knowledge about the natural world.

in situ vitrification A process that uses electric current to melt hazardous wastes and soil to form an impermeable glasslike mass.

instantaneous speed The speed at some particular instant of time.

instantaneous velocity The velocity at some particular instant of time, including both magnitude and direction.

insulator Any substance through which electric charge (or heat) resists flow.

integrated circuit A tiny complex of electronic components, such as diodes and transistors, placed on a single semiconductor.

intensity The brightness of light or the loudness of sound.

interference (wave) The increase or decrease in the amplitude of a wave caused by the combined effect of two or more waves passing through the same region at the same time. (See **constructive interference; destructive interference.**)

intrusive igneous rock An igneous rock that appears to have congealed from magma forced into surrounding rock.

inverse proportion The comparative relationship between two quantities in which one decreases as the other increases.

inversion layer A stationary layer of air formed by increasing air temperature with altitude. The warm rising air is held suspended by the lighter warmer air above it.

ion An atom or group of atoms that has lost or gained one or more electrons so that the original atom or group of atoms is electrically charged.

ionic bond The electrostatic force of attraction that holds two or more ions together.

ionic crystal A homogeneous solid with an orderly internal structure. Ionic crystals are composed of ions, such as Na^+ and Cl^-.

ionic layer The sublayer of the thermosphere that reflects radio waves.

ionization The process whereby ions are formed from atoms or a group of atoms.

ionizing radiation Particles or energy emitted during radioactive decay. Ionizing radiation affects the electrical balance of atoms with which it interacts.

ionosphere The portion of Earth's upper atmosphere in which many atoms are ionized.

irregular galaxy A galaxy with no apparent symmetry.

isobar A line on a map or chart connecting points of equal pressure.

isomer One of two or more compounds having the same empirical formula but different structures. For example, methyl ether (CH_3OCH_3) and ethyl alcohol (CH_3CH_2OH) both have the empirical formula C_2H_6O.

isomerism The existence of different compounds (*isomers*) that have identical molecular formulas but different molecular structures or different arrangements of atoms in space.

isostasy The equilibrium between adjacent blocks of crust that rest on the plastic mantle. The state of balance supports landmasses.

isotope One of two or more atoms of a single element that contains different numbers of neutrons. For example, the isotopes protium (with one proton and no neutrons) and deuterium (with one proton and one neutron) are both atoms of the element hydrogen since each has one proton in its nucleus.

jet stream Narrow, meandering streams of high-speed winds located in the upper portion of the troposphere. The winds usually blow from west to east and on occasion exceed 400 kilometers per hour.

joule The force of 1 newton acting through a distance of 1 meter. Also, the work done by moving 1 coulomb of electrical charge through an electric potential of 1 volt. One joule of work (1 watt-second) is equal to 10^7 ergs.

Joule's law The expression in units of thermal energy for the power expended when a current passes through a conductor. The energy is equal to the product of the resistance of the conductor and the magnitude of the current squared.

Jovian planet Jupiter or any of its accompanying three giant gaseous planets within our solar system.

Juno The third largest of the asteroids or planetoids that orbit the sun between Mars and Jupiter.

kelvin A unit of thermodynamic temperature equal to 1/273.16 of the absolute temperature of the triple point of water.

Kelvin temperature scale The "absolute" temperature scale, with 0 kelvin being the lowest possible temperature.

Kepler's laws of planetary motion The three laws stated by Kepler during the early seventeenth century to describe the motion of planets and other celestial objects.

kerogen A variety of oil associated with the decomposition and compaction of limestone.

ketone An organic compound containing the

$$R-\overset{\displaystyle O}{\overset{\|}{C}}-R \text{ group.}$$

kilowatt-hour A unit of energy or work equal to 1000 watt-hours. Abbreviated kWh.

kinematics The study or description of motion.

kinetic energy The energy an object has as a result of its motion.

Kuiper belt A proposed belt of comets orbiting the sun at a distance of about 40 AU. (The suspected Oort Cloud of comets is thought to extend as far as 100,000 AU from the sun.)

laccolith A mass of igneous rock intruded between sedimentary beds to produce a bulging effect on layers above it.

lanthanide series The sequence of increasing atomic numbers in period six that appear near the bottom of a periodic chart. The atomic number of the series begins after lanthanium, $Z = 57$.

lapse rate The decrease in temperature of the air with altitude.

Large Magellanic Cloud One of the neighboring irregular galaxies visible with the unaided eye from southern latitudes.

laser A device that creates a narrow, intense beam of coherent, monochromatic light.

latitude A north-south coordinate on Earth's surface.

lattice The structure or framework of a crystalline substance.

lava An extrusion of magma from Earth's crust that has been deposited on Earth's surface.

law A statement describing how nature behaves under specific conditions.

law of areas A summary of Kepler's second law of planetary motion.

law of conservation of charge The law that charge is neither created nor destroyed in ordinary processes. Exceptions to this law occur in some nuclear decay processes and in some experiments using high-energy particle accelerators.

law of conservation of energy The law that the total energy of a body or system is neither increased nor decreased by any process. Energy can be transformed from one form to another and transferred from one body to another, but the total

law of conservation of momentum The law that states that, in an isolated system (a group of bodies that have no outside forces acting on them), the total momentum cannot change.

law of continuity The conclusion that a sedimentary rock unit was originally continuous but that parts may have been removed by erosion and faulting.

law of crosscutting relationships The law that points out that, as one geologic body cuts (or intersects) another, the body that is intruded is older than the body that does the cutting.

law of definite proportions The law that the elements that make up a chemical compound are always combined in the same proportions by weight.

law of ellipses A summary of Kepler's first law of planetary motion.

law of faunal succession The conclusion that fossils present in different rock layers present a record of succession in time according to the law of superposition.

law of original horizontality The conclusion that beds of sediment deposited in water were formed as horizontal or nearly horizontal layers.

law of redshifts The law that the distance to a galaxy is proportional to its velocity away from Earth.

law of reflection The law that the angle of a light ray reflected from a smooth surface is equal to the angle of the incident ray, when both angles are measured with the normal.

law of superposition The law that within a sequence of undisturbed rock layers, the layers become progressively older from top to bottom.

law of universal gravitation The law that each particle of matter in the universe attracts every other particle of matter.

Lawson criterion The product of the length of confinement time and the density of a plasma in reference to fusion reactions.

leachate A solution or soluble product that mixes with water and percolates out of the ground or into the groundwater.

left-hand rule The rule that if the left hand is wrapped around a conductor with the thumb pointing in the direction of electron flow, the curled fingers will point in the direction of the magnetic field lines around the conductor.

lepton An elementary particle that is not composed of quarks and that has a mass less than that of a proton.

lever A rigid object that transmits or modifies force or motion; one of the simple machines.

lift The component of total aerodynamic forces acting on a body perpendicular to the undisturbed fluid flow relative to the body.

light The visible part of the electromagnetic spectrum.

light-emitting diode (LED) A semiconductor diode that converts electric energy into visible light.

light pollution Light from fires and artificial sources, diminishing our view of the night sky.

light year (LY) The distance traveled by light in one year, about 9.3×10^{12} kilometers (5.7×10^{12} mi).

lignite A soft, brownish-black coal in which the texture of the original wood is often visible.

linear collider A facility designed to accelerate beams of charged particles such as electrons, positrons, and protons in a straight path, then cause them to strike various charged particles accelerated in the opposite direction.

linear expansion The change in length, width, or height of a solid generally associated with a temperature change.

linear momentum The momentum of a mass moving in a straight line.

linear motion Motion along a straight-line path.

lines of forces Imaginary lines in a field of force, such as in an electric, magnetic, or gravitational field.

liquid A substance that, unlike a solid, generally flows readily. A liquid has no definite shape but, unlike a gas, does not expand indefinitely.

liquid crystal A liquid that has some degree of molecular orderliness.

liquid crystal display (LCD) A display, often found on timepieces and calculators, in which the polarizing properties of liquid crystals are used to control transmission or reflection of light.

lithification The processes by which unconsolidated rock, generally sedimentary in origin, is consolidated.

lithosphere The strong, solid outermost shell of Earth. The lithosphere includes the crust and the upper part of the mantle.

local group The cluster of about two dozen galaxies of which our galaxy, the Milky Way, is a member.

lodestone A magnetic variety of the mineral magnetite.

loess A fine-grained, silt-sized material deposited by wind action.

longitude An east-west coordinate on Earth's surface.

longitudinal wave A wave motion in which the disturbance is in the same direction in which the wave travels. Sound is generally considered a longitudinal wave.

long wave (earthquake) A disturbance created by earthquakes and confined to Earth's crust.

loudness A subjective measure of sound; the attribute of sound that determines the effect of its magnitude on the auditory system.

Love Canal The abandoned canal near Niagara Falls, New York that served as a commercial dumping site and subsequently was found to contain excessive amounts of toxic industrial wastes.

lumen The SI unit of luminous flux, equal to 12.57 candelas.

luminosity A measure of the total electromagnetic energy output of a star. Often the luminosity of the sun is taken to be 1 for comparison.

luminous flux The total amount of energy emitted by a source in the visible region of the spectrum per unit of time.

lunar eclipse The blocking out of our view of the moon when the moon, Earth, and the sun are in a straight line in that order and the shadow of Earth falls on the moon. This phenomenon can occur only during the full-moon phase. (See also **solar eclipse.**)

luster The appearance of the surface of a mineral with regard to its light-reflecting qualities.

L wave A type of earthquake wave confined to Earth's crust. (See also **P wave; S wave.**)

machine A device designed to transmit or modify the application of power, force, or motion.

Magellan The spacecraft launched from the space shuttle *Atlantis* to orbit about the planet Venus.

magma Molten rock cooled deep in Earth's crust that has formed crystals and gases.

magmatic differentiation A layering in some igneous rocks, created as the heaviest of the first congealed minerals concentrate in the lower portion of a rock mass.

magnetic confinement The containment of a plasma within a given volume of space by magnetic forces.

magnetic declination A measure of the difference between true north and the compass reading at a given geographic position.

magnetic domains Small regions in a material such as iron that form a magnet when aligned with other regions so that their magnetic fields are reinforced.

magnetic field The region that surrounds a magnet in which magnetism can be detected.

magnetic induction The process by which a piece of steel is magnetized by stroking it with a permanent magnet.

magnetic north pole The point on Earth's surface in northern Canada to which the north-seeking end of a compass will always point.

magnetic reversal A condition in which Earth's magnetic poles exchange polarity.

magnetic south pole The point on Earth's surface south of Australia that attracts the south-seeking end of a compass.

magnetism The force by which a magnet attracts or repels other magnetic materials.

magnetohydrodynamics The means by which electrical power is produced directly from the motion of electrically conducting fluids in the presence of electric or magnetic fields.

magnetron A pulsed microwave radiation source for radar, and a continuous source of electromagnetic waves for microwave cooking. In a magnetron, electrons, generated from a heated cathode, are subjected to the combined force of an electric field and a magnetic field, causing them to move in such a manner as to produce electromagnetic waves in the microwave region.

magnification (optical) The ratio of the size of an image to the size of the object that the image represents.

magnitude The number representing the size of a physical quantity (without regard to direction, if the quantity has direction).

main sequence A diagonal band representing most of the stars that appear on an H-R diagram.

mantle The layer of Earth between the crust and the core. The mantle is about 2900 kilometers thick.

maritime polar air masses Air masses that originate over the northern Atlantic Ocean or over the North Pacific Ocean.

maritime tropical air masses Air masses that originate in the Caribbean Sea, Gulf of Mexico, or subtropical Pacific Ocean.

marsh gas Methane gas formed on the bottoms of swamps.

maser A device that generates electromagnetic waves in the microwave region; used in astronomy and communications.

mass The quantity of matter that a body contains. Also, the property of a body that causes it to resist changes in velocity (see **inertial mass**), and the property that causes it to attract other bodies (see **gravitational mass**).

mass defect The difference between the atomic mass of an atom and the total mass of its individual particles.

mass-energy equivalence The relationship between mass and energy, expressed by $E = mc^2$.

mass number The total number of nucleons in a nucleus, a number represented by the symbol *A*.

matter Anything that occupies space and is perceptible to the senses in some manner.

mechanical advantage The ratio of the force that performs useful work to the force applied to the machine.

mechanical energy In mechanics, the sum of the kinetic energy and the potential energy of a body.

mechanical weathering The physical disintegration of rocks without a change in chemical composition.

mechanics The branch of physics that investigates the behavior of physical systems influenced by interactions with the environment.

melt A mixture of molten rocks and/or minerals.

meltwater deposits The deposits of layered materials formed as glaciers melt.

meson Any one of various particles, the masses of which are greater than that of leptons but less than that of protons.

mesosphere The layer of the atmosphere immediately above the stratosphere, extending from 55 to 80 kilometers above Earth.

Mesozoic era The period of geologic time from 225 million years ago to 65 million years ago.

Messier catalog The first published catalog listing nonstellar objects in the night sky (nebulas, star clusters, etc.), compiled by Charles Messier in the late eighteenth century. Objects listed in the catalog are referred to by their "M" number, M1 being the first object in the catalog, and so forth.

metal Any of a class of chemical elements that are all solids at room temperature (except mercury), typically lustrous and malleable, and generally good conductors of electricity and heat.

metallic bond A type of bonding that results from the nuclei of metal atoms being submerged in a "sea" of electrons. The electrons in a metallic bond are highly delocalized.

metallurgy The science of separating metals from their ores and of making alloys.

metalsmith A craftsperson engaged in designing and making such products as helmets, shields, swords, and vases from various metals.

metamorphic rock Any rock that has been changed in texture or composition by pressure and/or heat.

meteor A piece of naturally occurring material that enters Earth's atmosphere and produces a luminous streak in the sky.

meteorite A meteoroid that strikes Earth before vaporizing.

meteoroids Pieces of matter that travel through outer space; they become meteors when they interact with Earth's atmosphere.

meteorology The science of the atmosphere and its phenomena, including weather and climate.

meteor shower The phenomenon observed when members of a group of meteors encounter Earth's atmosphere and their luminous paths appear to diverge from a common point in the sky. The particles causing a display are most likely dust particles associated with dead comets.

metric absolute systems The systems of measurement composed of the *mks* (multiples) or *cgs* (submultiples) of base and derived metric units.

microphone An instrument used to intensify sounds and to transmit them by transforming them into variations of electric current.

microwave An electromagnetic wave with a wavelength between about 0.3 and 30 centimeters. Microwaves may overlap wavelengths generally considered to be infrared and radio waves.

microwave guide A metallic device that confines and directs electromagnetic waves in a specific direction.

mid-ocean ridge A continuous, seismic, median mountain range extending through the North and South Atlantic Oceans, the Indian Ocean, and the South Pacific Ocean. According to the theory of seafloor spreading, the mid-ocean ridge is the source of new crustal material.

milk sugar The common name for the sugar found in the milk of mammals, called lactose.

Milky Way A band of diffused light across the night sky containing many of the stars that make up the Milky Way galaxy.

mineral A naturally occurring inorganic substance with a definite chemical composition and an ordered internal structure.

mirage An optical effect caused by light as it enters various layers of air of different optical densities, often appearing as a pool of water.

mixture A substance containing two or more elements or compounds. A mixture may be homogeneous, such as a solution of salt and water, or it may be heterogeneous, such as a mixture of sugar and flour.

moderator The material used in a nuclear reactor to decrease the speed of neutrons so that they can be more readily absorbed by nuclei.

modulation The process of regulating the amplitude, frequency, or phase of a carrier wave in radio or television.

Mohs' scale The relative scale of hardness commonly used by geologists in the study of minerals.

mole The amount of a substance that contains the same number of units as there are atoms of carbon in 0.012 kilogram of the isotope carbon-12. One mole of a substance contains 6.02×10^{23} atoms or molecules. (See **Avogadro's number.**)

molecular crystal A homogeneous solid composed of molecules, such as ice, sugars, and aspirin (acetylsalicylic acid), with an orderly internal structure.

molecular weight The relative average weight of a molecule of a substance. The molecular weight of a substance is also the sum of the atomic weights of all the atoms in one molecule of the substance.

molecule Two or more atoms chemically bound together.

momentum The product of the mass of a body and its velocity.

monomer A small molecule from which polymers are made.

moraine A mound of unsorted and unstratified glacial till deposited by the action of glacier ice.

music The blending of sounds that is pleasing to a listener.

myopia The condition of the eye in which light rays from distant objects are focused too far in front of the retina for clear vision.

nanometer A unit of length equal to 10^{-9}, or one-billionth, meter.

natural pesticide A class of naturally occurring substances that effectively controls insects and pests of other types.

natural resources Industrial materials, such as crude oil, natural gas, coal, wood, metallic ores, supplied by nature.

natural science The combined area of knowledge that encompasses biological science and physical science.

neap tide The lowest tide of a given month, where the moon is either in its first or third quarter.

nebula A cloud of interstellar gas and/or dust.

negative catalyst A substance that reduces the normal rate of a chemical reaction without undergoing any permanent change.

negative ion An atom or group of atoms that has gained one or more electrons and thus has acquired a negative electric charge.

neutral-beam injection A procedure in which particles are made electrically neutral so they can penetrate the magnetic fields and help raise the temperature and pressure of plasma.

neutralization The process by which an acid and a base react to produce a salt and water. The reaction should produce a neutral solution (pH of 7).

neutrino An uncharged particle of negligible mass that accompanies a positron during beta decay. The antineutrino accompanies the electron during beta decay.

neutron The uncharged elementary particle found in the nucleus of an atom.

neutron activation The process of causing an atom to become radioactive through the absorption of neutrons.

neutron bombardment The technique of projecting neutrons at a target nucleus to produce an unstable nucleus, permitting the identification of the original nucleus by analysis of the particle and/or energy it has released.

neutron number The number of neutrons in a nucleus, represented by the symbol *N*. $N = A - Z$.

neutron star A very small star (only a few miles in diameter) that has used up its nuclear fuel and has gravitationally collapsed. The star consists entirely of neutrons.

névé The ice produced by the constant force of pressure from overlying snow in a glacier.

newton The force that causes a mass of 1 kilogram to accelerate 1 meter per second per second.

newton-meter The SI unit of work equivalent to the joule.

Newton's first law of motion The law that a body remains in its same state of motion, either at rest or moving at a steady speed in a straight line, if the (vector) sum of the forces acting on it is zero. This is also called the *law of inertia.*

Newton's second law of motion The law that the acceleration of a body is proportional to and in the same direction as the (vector) sum of the forces acting on it and inversely proportional to its mass.

Newton's third law of motion The law that, if one body exerts a force on another body, the second body exerts an exactly equal and opposite force on the first body.

nichrome A conducting wire with high electrical resistance, composed of nickel and chromium.

nimbus cloud A cloud from which precipitation is occurring or is threatening to occur.

noble gas Any of the elements of group VIII (column 18) of the periodic chart, from helium down to radon. Noble gas molecules have filled outer electron energy levels. Also called *inert gases.*

noise An unwanted sound, displeasing to the listener; hence, subjective in nature. Noise generally is produced by an object that is vibrating in an irregular fashion.

normal An imaginary line drawn perpendicular to the surface of a medium at the point of incidence of a ray striking the medium.

north pole (magnet) The region of a magnet from which the lines of its magnetic field diverge into the space outside the magnet. The northern end of a bar magnet free to rotate in Earth's magnetic field is its north pole.

nova A star that suddenly increases its luminosity by hundreds or thousands of times and then fades back to near its original luminosity.

nuclear force The strong attractive force between nucleons, considered the effective force that binds the nucleus together.

nuclear power Power generated by the controlled nuclear processes of fission or fusion. Generally, the useful heat produced by the nuclear power plant is used to generate electrical energy.

nuclear reactor A device that supports a self-sustaining chain of nuclear reactions under controlled conditions.

nucleon A particle, such as a neutron or a proton, found in an atomic nucleus.

nucleosynthesis The reaction by which light elements are converted to heavier elements among the stars.

nucleus The positively charged core of an atom. Although the nucleus is extremely small compared to the atom, the nucleus contains most of the atom's mass.

nuclide A species of atom characterized by its number of protons (Z), its number of neutrons (N), and the energy content in its nucleus.

occluded front A mingled cold front and warm front. A cold occlusion exists when a cold front undercuts a warm front. A warm occlusion results when a cold front is forced aloft by a warm front.

occultation An eclipse of a planet or star by the moon or another planet.

octane number A rating based on an arbitrary scale in which isooctane is rated 100 and n-heptane, is rated zero. A mixture of 90 percent isooctane and 10 percent n-heptane has an octane rating of 90.

octave The interval in pitch between two tones, the second of which is twice the frequency of the first.

octet rule The rule that points out that there is unusual stability of all elements that have eight electrons in their outermost shell.

ohm (Ω) The unit of electrical resistance. It is the resistance that permits a current of 1 ampere to flow if an electrical potential of 1 volt is applied to the circuit.

Ohm's law The law that points out the relationship of electric current to electric potential and electric resistance; that is, $I = V/R$.

oil An ester formed from the reaction of carboxylic acid molecules with glycerol. Oils are liquids at room temperature. Petroleum oil is a mixture of long-chain hydrocarbons.

oil shale A shale with a high content of organic matter from which oil may be extracted by various processes.

old-age valley A surface feature of Earth characterized by continual lateral erosion and meandering rivers or streams that sometimes create oxbows.

Oort cloud A suspected spherical mass of cometary nuclei that orbit the sun at a distance of perhaps 100,000 AU.

opaque Not transparent to visible light.

open cluster A relatively loose cluster of stars gravitationally bound together. Open clusters generally lie within the plane of the Milky Way galaxy.

open pipe A musical organ component that, when a stream of air is directed across it, produces a sound, the wavelength of which is twice the length of the pipe.

opposition A direction directly opposite to that of the sun as viewed from Earth.

optical fiber A long, thin thread of a transparent substance used to transmit light.

optics The study of visible light. (Sometimes radiation in the spectral regions near the visible spectrum is included as a part of optics.)

orbital velocity The instantaneous velocity at which a satellite or other orbiting body travels around another body or the system's barycenter.

ore A mineral or rock from which metals and nonmetals can be profitably extracted.

organic A term used in reference to most compounds of carbon.

organic acid An organic compound that contains the carboxyl group $-\overset{\overset{\displaystyle O}{\|}}{C}-O-H$.

organic chemistry The chemistry of carbon and most of its compounds.

organic compound Any compound containing carbon, with the exception of a few such simple compounds as carbon dioxide and mineral carbonates.

orogenesis The mountain-building processes that result in displacement or deformation of Earth's crust.

orogeny A sequence of events occurring within a specific time frame and generally leading to intense deformation of rock layers.

oscillating-universe theory The theory that the expansion of the universe will ultimately be reversed by gravitational attraction, causing the universe to collapse and begin anew with another "big bang," endlessly repeating the cycle.

oscillator A device that generates high-frequency alternating current in electronic circuits.

oscilloscope An instrument that uses a cathode-ray tube to make visible the instantaneous values and waveforms that rapidly vary with time.

outer planets Jupiter, Saturn, Uranus, Pluto, and Neptune.

overtone Any one of a number of vibrations of higher frequency than the fundamental tone produced by a vibrating source of sound. Each overtone is a whole-number multiple of the fundamental.

oxidation The combining of oxygen with another element or compound, or the loss of electrons by an element or compound.

ozone layer A sublayer of the stratosphere composed of molecules of ozone, the triatomic form of oxygen (O_3).

pair production The conversion of energy into mass through the interaction of a particle and its antiparticle.

Paleozoic era That period of geologic time from about 600 million years ago to 230 million years ago.

Pallas The second largest of the asteroids or planetoids that orbit the sun between Mars and Jupiter.

Pangaea Former supercontinent composed of all the continental crust of Earth, and later (about 200 million years ago) fragmented to form the present continents.

parabolic mirror A concave mirror that produces parallel rays of light from a source placed at its focal point and, conversely, focuses parallel rays of light at its focal point.

parallax The apparent displacement of the position of an object with respect to a reference point, caused by an actual shift in the point of observation.

parallel Extending in the same direction and at a constant distance apart.

parallel circuit The arrangement of electrical devices within a circuit so that there are alternate paths through which current may pass from one point to another.

parametric amplifier (paramp) A device in which an optical or infrared beam draws power from a laser beam and amplifies it.

paraquat A potent herbicide frequently used to clear new fields for cultivation.

parsec A unit of distance equal to 3.27 light years. An object at this distance would have a stellar parallax of 1 second of arc.

peat Partially decayed, moisture-absorbing plant matter found in ancient swamps and used as a fuel or plant cover.

pellet injection The bombardment of a plasma with frozen hydrogen or deuterium pellets.

peneplain Land worn down almost to sea level by geologic processes.

pentatonic scale A musical scale composed of notes that correspond to those produced by the black keys on a piano.

penumbra The partially lighted area surrounding the completely dark shadow of a body.

peridotite An igneous rock consisting of feldspar and a high concentration of ferromagnesian minerals.

perihelion In the orbit of an object revolving around the sun, that point at which the object is nearest the sun.

period The time during which a vibrating object such as a pendulum, completes one cycle or vibration; a subdivision of a geological era.

period (chemical) A horizontal row of the periodic chart. Each period ends with a noble gas, such as neon or argon.

period (of a satellite) The length of time it takes a satellite to make one complete revolution.

periodic chart A table that presents the chemical elements in order of their atomic numbers and chemical properties. (Also called *periodic table.*)

periodic law Stated in a modern fashion, the law that, if the elements are listed in order of increasing atomic numbers, elements with similar properties recur at definite intervals.

permeability The ability of iron or other magnetic material to concentrate magnetic lines of force.

perpetual motion Motion that continues indefinitely after a force has overcome the inertia of the object and has produced a minimum acceleration of the object on which it acts.

Perseid shower The annual meteor shower that occurs on or about August 11, as Earth's orbit intersects the ring of debris left from a relatively large comet that disintegrated after an encounter with the sun.

perturbation The deviation of a celestial object from its calculated path.

pesticide A chemical applied to control unwanted animals and insects.

pH The negative power to which the number 10 must be raised in order to equal the effective concentration of hydrogen ions in a solution.

phaneritic Of coarse-grained texture, such as the igneous rock granite. (Compare **aphanitic.**)

philosopher A person engaged in logical and critical study of the source and nature of human knowledge.

philosopher's stone A magical implement supposedly capable of transmuting various metals into gold.

philosophy Theory or investigation of the principles or laws that regulate the universe and underlie all knowledge and reality.

phosphor A substance, such as a metallic halide, that emits primarily visible light when subjected to various frequencies of electromagnetic radiation.

photochemical smog Smoke and particulate matter that contaminate the atmosphere, formed by light-induced chemical reactions.

photochromic materials Substances capable of changing color when exposed to light or other electromagnetic radiation.

photoelectric effect The emission of electrons from a metal as a result of light striking the metal's surface.

photoelectrochemical cell A device immersed in an aqueous solution that converts sunlight directly into electrical or chemical energy.

photon A massless particle; a quantum of electromagnetic energy.

photosphere The visible surface of the sun.

photosynthesis The formation of carbohydrates in plants from water and carbon dioxide by the action of sunlight on chlorophyll.

photovoltaic cell A device that detects or measures electromagnetic radiation by generating an electric potential.

photovoltaic effect The production of an electric potential in a semiconductor, such as silicon, or at the junction between two different materials, upon absorption of radiant energy.

pH scale A scale used to indicate the concentration of hydrogen ions in a solution. (See **pH.**)

physical change A process, such as boiling or freezing, that does not alter the chemical composition of a substance.

physical property One of the identifying characteristics of a substance, such as color, odor, boiling point, texture, and density.

physical science The study of inanimate matter and of energy and its interactions with material things.

physics The science dealing with the properties, changes, and interactions of matter and energy.

pi The constant value that relates a circle's circumference to its diameter, equal to approximately 3.14.

pile A battery composed of similarly constructed cells; an assembly of fuel rods and moderators used in nuclear reactors.

pitch The position of a musical note on a musical scale. It is determined by the frequency of the sound impulses produced by the source.

Planck's constant A fundamental physical constant, the ratio of the energy of a photon and its frequency. It is equal to 6.63×10^{-34} joule-second.

plane mirror A mirror with a flat reflecting surface.

plane polarized wave An electromagnetic wave, the electric field vectors of which are aligned with accompanying waves.

planetary nebula A shell of ejected expanding gas surrounding some stars.

plasma Extremely hot gases composed of electrically charged particles, in reality fragments of atoms and nuclei. Plasma is sometimes considered the fourth state in which matter can exist.

plate The positively charged terminal (anode) in a vacuum tube; a large, solid portion of Earth's crust.

plate tectonics The theory that the lithosphere is made up of rigid plates. Some of the plates move relative to each other over the surface of Earth; the mechanisms by which large parts of Earth's crust are formed, move, and are destroyed.

plutonism The process by which igneous rocks crystallize or solidify deep in Earth's crust.

polar easterlies The prevailing winds in the latitudes from 90 degrees to 60 degrees. These winds are caused by cold polar air masses that move toward the equator.

polarization The alignment of electromagnetic waves in such a manner that all light waves are parallel to each other.

polarization (electrical) A separation of positive and negative charges on a body having a net charge of zero.

polarizing (electricity) Causing the positive and negative electric charges of a body to move and concentrate with other like charges.

polar molecule A molecule with a covalent bond in which one atom of the bond has a relative excess of negative charge and the other atom has a relative excess of positive charge.

pole (magnetic) The end of a magnet where a magnetic field appears to concentrate.

pollution The unnatural addition of contaminants to our air, land, and water.

polyatomic ions Ions containing more than one atom.

polychlorinated biphenyls (PCBs) Organic compounds primarily used by utilities in electrical transformers and as insulation materials, but which have been found to be hazardous to human health.

polymer A large molecule produced by the chemical combination of many small molecules. Thus, the polymer polyethylene results from the reaction of thousands of individual ethylene ($CH_2 = CH_2$) molecules.

polymerization The process by which smaller molecules are made to join together into more complex compounds by means of heat and use of catalysts.

polysaccharide Carbohydrate molecules containing many monosaccharide units.

population I stars Stars associated primarily with the spiral arms of galaxies. The stars in population I have a wide range of ages but are generally young and bright, having low concentrations of elements heavier than helium compared to those in population II.

population II stars Stars associated primarily with the nuclei and coronas of galaxies. The stars in population II are old stars with higher concentrations of elements heavier than helium compared to those in population I.

positive catalyst A substance that speeds up the rate of a chemical reaction without undergoing any permanent change.

positive ion An atom or group of atoms that has lost one or more electrons and thus has assumed a positive electric charge.

positron A positively charged electron (beta) emitted during the disintegration of a proton.

potential energy The energy stored in a body or system by virtue of its position in a field of force or due to the arrangement of its parts.

potter's wheel A device used to shape soft clay into vases and the like before the clay is hardened.

pound A unit of force or weight in the British gravitational system.

power The rate of doing work or the rate of energy usage. Power is commonly measured in watts and horsepower.

Precambrian era That period of geologic time from the time of formation of Earth to about 600 million years ago.

precession (of Earth) The conical motion of Earth's spin axis. It takes Earth about 26,000 years to precess once.

preferential absorption The manner in which a material absorbs only specific wavelengths of light and reemits others.

presbyopia A form of farsightedness caused by a gradual decrease in elasticity of the lens of the eye.

pressure The magnitude of the force exerted by a gas or liquid on a given area of its container.

prevailing westerlies Winds of the horse latitudes (from 30 degrees to 60 degrees) that blow diagonally eastward away from the equator.

primary colors The three basic colors of light that, combined, produce white light or, when combined in various proportions, any color in the spectrum; the three basic colors of pigments that produce dark brown or black.

principle A law of nature that explains a natural action; a fundamental truth upon which others are based.

principle of equivalence A statement of fact that the amount of inertial mass a body contains is exactly equal to the body's gravitational mass.

projectile A body given an arbitrary initial velocity and then released. Near the surface of Earth the primary forces acting on the projectile are its weight and frictional forces as it moves through the air. (The frictional forces are usually neglected in elementary calculations.)

prominences Gigantic arching columns of extremely hot gases ejected outward from the sun.

proportionality constant A number having a fixed numerical value and specific units that correctly relates certain quantities to each other.

protein An organic polymer consisting of many amino acids.

protocells Synthetic microspheres of protein material that exhibit some enzymatic activity.

proton The positively charged elementary particle found in the nucleus of all atoms.

proton-proton chain The fusion process in which four hydrogen nuclei are fused at very high temperatures into a single helium nucleus, with the liberation of considerable energy; a primary source of energy production in stars.

protoplanet A planet during the early stages of its formation.

pulley A simple machine composed of a small wheel with a grooved rim in which a rope is guided to lift objects by an applied force.

pulsar A rapidly rotating and highly magnetic neutron star that gives off light and radio waves of very rapid pulses.

pulsating variable A star that continually expands and contracts. This pulsation of size produces a pulsation of the star's luminosity.

pumice A spongy, light, porous volcanic rock of low specific gravity.

P wave A longitudinal or compression-type earthquake body wave. P waves can travel through solid or liquid material (See also **L wave; S wave.**)

pyroclastics Fragments of rocks formed during explosive eruptions of volcanoes.

Pythagorean theorem A theorem in geometry that the square of the length of the hypotenuse of a right triangle equals the sum of the squares of the lengths of the other two sides.

quality A property of sound waves that depends on the number and prominence of overtones present.

quantum A definite unit of electromagnetic energy, such as a photon.

quantum theory The theory that energy changes in bound systems can only occur in discrete steps called quanta. The size of these discrete steps is so small that they are most easily observed in systems of atomic and molecular scale or smaller.

quark One of the presumed basic subnuclear particles of which elementary particles are composed.

quasar A starlike object that exhibits a very large redshift and is amazingly luminous. *Quasar* is a contraction of *quasi-stellar radio source.*

rad The amount of radiation necessary to deposit energy equal to 1.00×10^{-5} joule in 0.001 kilogram of material.

radar A device that emits and detects ultrahigh-frequency electromagnetic waves.

radiation The transmission of energy by various particles and/or electromagnetic waves.

radiation belts See **Van Allen radiation belts.**

radio A device for receiving and transmitting electromagnetic waves through space.

radioactive The property of various nuclei that indicates they are unstable.

radioactive decay The spontaneous emission of particles and/or energy from the nucleus of an atom.

radioactive isotope (radioisotope or radionuclide) An unstable isotope of an element that gains stability through the release of particles and/or energy.

radioactivity A particular type of radiation, such as alpha, beta, or gamma emission, by a radioactive substance.

radioisotope A radioactive isotope of an element. All elements have at least one unstable isotope that undergoes radioactive decay.

radionuclide A radioactive atom characterized by its number of protons and neutrons; also generally known as a radioisotope.

radio waves Electromagnetic waves with wavelengths ranging from the order of meters to hundreds of meters. They are used in transmitting information either by amplitude modulation or frequency modulation.

radon A radioactive element produced in the decay series of uranium. Radon exists as a gas at ambient temperatures and is a noble gas.

rarefaction In sound, that part of a wave form in which the density of the vibrating medium is reduced at some instant.

rate of reaction The rate (speed) at which a chemical reaction occurs, normally expressed as the change in the concentration of one component of the reaction.

ray The path followed by light (or material particles that are moving in a "beam.")

reaction The equal and opposite force that results when a force is exerted on an object, according to Newton's third law. Also, a chemical change.

reaction rate See **rate of reaction.**

real image An image that is actually formed by rays of light. If a screen is placed in the plane of a real image, it will generally become visible.

recombination The process by which new minerals are created when rocks of different chemical composition interact.

rectifier A device that permits electric current to flow in only one direction. Alternating current is essentially converted by the device to direct current; thus, the current is said to be rectified.

red dwarf A small, cool star on the main sequence in an H-R diagram.

red giant A large, relatively cool star of high luminosity.

redshift The shift of the wavelength of light from stars and galaxies toward longer wavelengths. The redshift is produced by the Doppler effect.

red supergiant A huge red star of very high luminosity.

reduction The removal of oxygen from an element or a group of atoms or the gain of electrons by an element or a group of atoms.

reflection The rebounding of energy, such as heat, light, or sound, from a surface.

refraction The bending or abrupt change in direction of energy, such as light, as it passes from one medium into another of different optical density.

refractive index The ratio of the velocity of light in a vacuum to that in a specified medium.

relative humidity The amount of water vapor in the air compared to the amount that fully saturates the air at a given temperature.

relativistic mass The mass of a particle moving at a specific velocity. The increase above rest mass is detectable at one-tenth the speed of light or faster.

relativity A theory that asserts the equivalence of mass and energy.

reluctance A measure of magnetic resistance. The reluctance of an air core is greater than that of a core of iron.

rem The unit of effective radiation dose; the **r**oentgen **e**quivalent **m**an.

repulsive The term used to describe a force that tends to push two physical entities apart. For example, the force between two positive charges is called repulsive.(Contrast **attractive.**)

resistance The opposition offered by a substance or object to the passage of electric current through it; the opposition, such as a weight or force, to the applied force.

resistor An electronic component designed to have a definite resistance and used to limit current flow to various parts of a circuit.

resonance The intensification and enriching of a musical tone by supplementary vibration from a vibrating source that has the same or a simple multiple frequency. Also applies to physical systems, such as bridges or electrical wires.

resultant The geometric sum of two or more vectors.

retrograde rotation The rotation of a planet on its axis that is the reverse of that of the other planets. Only Venus and Uranus exhibits retrograde (clockwise) rotation.

reverberation Multiple echoes created in a confined space.

reversible reaction A chemical reaction in which the products formed also react to produce the original reactants.

reversing layer The lower portion of the sun's chromosphere.

revolution The motion of a body about a closed orbit.

Richter scale A numerical scale of earthquake magnitudes.

right-hand rule (magnetism) For a conducting wire, the rule that if the fingers of the right hand are placed around a wire so that the thumb points in the direction of conventional current flow, the curled fingers will indicate the direction of the magnetic field produced in the region around the wire.

Roche's limit The minimum distance from a primary body that a satellite can approach the body without being disintegrated by tidal forces caused by the primary body.

rock A naturally formed aggregate of one or more minerals. A rock ordinarily lacks a definite chemical composition and given properties.

rock-forming minerals The relatively few common minerals of which most rocks are composed. All identified minerals are found among the rocks, but only a few are dominant.

roentgen A measurement of exposure dose of X- or gamma radiation that will produce 2.082×10^9 ion pairs in one cubic centimeter of dry air.

rolling friction A force that opposes the motion of any object that is rolling over the surface of another.

Rosetta stone A slab of basaltic rock with inscriptions in Greek, Egyptian hieroglyphics, and Demotic, discovered in 1799 near Rosetta, Egypt. It provided the key to deciphering many ancient writings.

rotary motion Motion about a central axis, such as the motion of a rotating wheel.

rotation The turning of a body about a fixed axis, such as a wheel about its center (hub).

rotational inertia The property of a body that resists any change in its rotation.

rotational motion The change in orientation of a body as opposed to the change in location of the body.

runoff water The water that returns to the oceans via streams and rivers after falling upon the land (approximately 20 percent).

R-value A rating that indicates the resistance to heat flow in a specific material.

saccharides A general term for carbohydrates. From the Latin *saccharum,* meaning "sugar."

salt One of the two products of the neutralization reaction between an acid and a base. The other product is water.

San Andreas fault A fracture and associated movement of the rock layer that stretches along about 1000 kilometers of the western portion of California.

satellite A celestial body, artificial or natural, that orbits about a larger body.

saturated air Air that contains all the water vapor it can hold at a given temperature.

saturated hydrocarbon A hydrocarbon that has no more than a single covalent bond between carbon atoms.

saturated solution A homogeneous combination of compounds or elements in which there is the maximum amount of solute dissolved in the solvent possible at a given temperature.

scalar quantity A physical quantity that does not have a direction associated with it. Examples of scalar quantities are candlepower, speed, mass, and volume.

scattering A process in which a molecule absorbs electromagnetic waves and releases them in a different direction.

science The systematized knowledge derived from observation, study, and experimentation undertaken to determine the nature or principles of what is being studied.

scientific method A system presumably used by a researcher to conduct a scientific investigation.

scintillation A burst of electromagnetic energy that occurs in some crystals and liquids upon their interaction with gamma radiation or X-radiation.

seafloor spreading A theory that there is a slow migration of the ocean floor away from a midocean ridge. The mechanism is a significant factor in continental drift.

seat of electromotive force A device, such as a battery or generator, containing an electromotive force.

Second law of thermodynamics The law that only possible processes that can occur in an isolated system are those during which the net entropy either increases or remains constant.

sedimentary rock Rock formed from the accumulation of sediment that has undergone cementation, compaction, chemical change, or other processes of lithification.

seismic waves Waves of energy released during an earthquake.

semiconductor A solid material, the electrical conductivity of which is between that of a conductor and that of an insulator. Silicon and germanium, generally insulating materials, become conductors when impurities are present in the crystal lattice of the substance.

series circuit The arrangement of electrical devices in which all the current in the circuit must flow through every component.

sial Crustal material primarily composed of silica and aluminum.

sidereal day The length of time it takes Earth to make one rotation on its axis. The sidereal day is 23 hours and 56 minutes.

sievert The SI unit of radiation dose equal to the amount of radiation exposure received from 1 milligram of encapsulated radium.

significant digits (figures) The number of digits in a measurement or in a computation that reflects the probable accuracy.

sill A tabular body of igneous rock injected while molten between sedimentary or igneous beds or along foliation planes of metamorphic rocks.

sima Crustal material primarily composed of silica and magnesium.

simple machine The most basic of devices designed to transmit or modify the application of power, force, or motion.

SI system of measurements The internationally accepted system of measurement of physical quantities.

sinkhole A depression caused by ground collapse above a hollow or hole formed by solution.

sliding friction A force that opposes the motion of any object that is sliding over the surface of another.

slug The mass of an object, measured in the English system. A mass of 1 slug is accelerated 1 foot per second per second (1 ft/s^2) when a force of 1 pound acts on it.

Small Magellanic Cloud One of the neighboring irregular galaxies visible with the unaided eye from southern latitudes.

smog A mixture of gaseous and particulate matter that forms from the reactions that occur in the atmosphere, particularly between substances released by internal-combustion engines and industrial processes.

snow line At any given location, the altitude above which snow does not completely melt during the summer.

soil The loose top layer of Earth's surface in which plant life can grow.

solar cell An electrical device that converts radiant energy into electrical energy.

solar cycle The variations in sunspot activity that appear to reach a maximum of activity every 11 years.

solar day The apparent period of rotation of Earth as measured in relation to the sun. The solar day is 24 hours.

solar eclipse The blocking out of the sunlight that occurs when Earth, the moon and the sun are in straight line and the shadow of the moon falls on Earth. This phenomenon can occur only during the new-moon phase. (See also **lunar eclipse.**)

solar energy The energy derived from the sun.

solar flare A temporary outburst of light from a region of the sun's surface.

solar wind A radial flow of charged particles, including electrons and protons, ejected from the sun.

solenoid An electrically energized coil of insulated wire that produces a magnetic field within the coil.

solid A substance, neither liquid nor gaseous, having a definite shape and a specific volume.

solstice See **summer solstice; winter solstice.**

solute The part of a solution that is considered dissolved in the other.

solution A mixture at the molecular level.

solvation The surrounding of solute molecules or ions by solvent molecules.

solvent That part of a solution in which the solute, such as a salt, is dissolved.

sonic boom A noise caused by a shock wave produced by an aircraft traveling at or above sonic speed.

Sothic The star Sirius, called *Sothic* by the ancient Egyptians, and used by them to determine the length of time known as a year.

sound Successive compressions and rarefactions representing longitudinal waves that travel through a medium such as air. Sound often represents wasted energy.

source region The surface from which an air mass derives its characteristics.

south pole (magnet) The region of a magnet toward which the lines of magnetic field converge from the space outside the magnet. The southern end of a bar magnet free to rotate in Earth's magnetic field is its south pole.

space velocity The velocity of a star relative to the sun.

special relativity The theory that space and time are not independent of each other, based on two postulates: (1) that the speed of light is constant and is independent of the source or the observer, and (2) that the laws of physics, in a mathematical form, are invariant in all inertial systems.

specific gravity The ratio of the density of a substance to the density of some standard material. The standard for liquids is water and the standard for gases is air.

specific heat The amount of heat required to change the temperature of 1 gram of a substance 1 Celsius degree.

speckle interferometry A technique that improves astronomical views by adjusting for turbulence in Earth's atmosphere.

spectrum A beam of electromagnetic waves that has been separated into components according to wavelengths. Often, a continuous sequence or range of energy is referred to as a spectrum because the array can be readily separated according to some varying characteristic.

spherical aberration Distortion in an image arising from light rays arriving in a lens with spherical surfaces at different distances from the axis.

spherical mirror A curved mirror that is part of a spherical surface. The inside of the spherical surface reflects as a converging mirror, while the outside reflects as a diverging mirror.

spiral galaxy A galaxy flattened by rotation to produce pinwheel-like arms.

spontaneous emission Emission of a photon by an atom when it spontaneously makes a transition to a lower energy level.

spring tide The highest tide of the month, occurring when the moon is either new or full.

stadia Units of distance based on the longest dimension of the famous Greek sports stadium and used by Eratosthenes in determining Earth's polar circumference with surprising accuracy.

stalactites Stony projections of calcium carbonate from the roof of a limestone cave.

stalagmites Stony projections of calcium carbonate from the floor of a limestone cave.

starch A natural polymer of glucose. A reserve food supply that plants store in various parts of their structures such as fruit, seeds, and roots.

static electricity The accumulation of electric charges, positive or negative, produced by friction between insulating materials.

statics The science of mechanics dealing with forces that act on a body in equilibrium.

stationary front A contact zone between two air masses where neither of them is moving.

steady-state theory A theory of cosmogony that presumes the universe to be infinite and unchanging. In this theory matter is constantly being created. During the 1960s, evidence began to accumulate in favor of the "big-bang" theory.

steam turbine A rotary device that is driven by the pressure of steam discharged at high speed against vanes attached to the rotor. This action causes the rotor to turn, producing useful mechanical energy.

stellar parallax The apparent displacement of a star when viewed from different positions of Earth's orbit.

stereophonic sound Sound presented to the listener from two sources so as to generate the sensation of spatial location for the source of the sound. Stereophonic sound requires two microphones and two separate channels to convey the spatial information to the listener.

stimulated emission Emission of a photon by an atom when it is stimulated (by another photon) to make a transition to a lower energy level.

stirrer A slowly rotating fan from which microwaves reflect toward the cooking area of a microwave oven.

stock A mass of intruded igneous rock, the exposed area of which is less than 100 square kilometers (40 square mi).

stratosphere The layer of Earth's atmosphere immediately above the troposphere.

stratus cloud A stratified or layered cloud formation.

streak The appearance of a mineral in powdered form.

strip mining A mining operation in which the surface layers of rock and soil are removed before the layer of ore is excavated.

strong nuclear force A force of attraction other than the gravitational force between nucleons, regardless of electrical charge.

subduction zone A region where two plates converge and one plate is deflected downward into the asthenosphere.

sublimation The process by which a solid passes directly into the vapor state without going through the liquid state. The process is reversible.

substitution reaction A reaction in which one or more atoms of a molecule are replaced by a different atom.

summer solstice The time during the summer (Northern Hemisphere) when the sun is farthest north of the equator.

sunspots Dark spots on the photosphere of the sun. The spots appear dark because they are cooler than the surrounding photosphere.

supercluster A collection of clusters of galaxies that are, themselves, clustered.

superconductor Materials that lose all measurable electrical resistivity near absolute zero.

Superfund The original appropriation set aside by the U.S. Congress in the mid-1980s to clean up hazardous waste sites.

superior planet A planet with an orbit farther from the sun than Earth's orbit is.

supernova A star that explodes and increases its luminosity hundreds of thousands to millions of times.

supersaturated solution An unstable homogeneous combination of compounds or elements in which there is more solute present than the amount of solvent can normally dissolve at a given temperature.

supertanker A sea vessel capable of transporting perhaps 100 million gallons or more of crude oil from its source to the processing plant.

S wave A transverse or shear-type earthquake body wave. S waves can travel only through solid material. (See also **L wave; P wave.**)

symbol (chemical) A "shorthand" designation of the name of an element, consisting of one or two letters.

synchronized orbit The direction and rate of revolution a satellite must assume in order to remain directly over a fixed point, such as on Earth.

synchrotron radiation Radiation emitted by charged particles when they are accelerated near the speed of light in a strong magnetic field.

synroc A solid composed of a stable synthetic crystalline substance into which radioactive wastes are incorporated for storage.

synthesis The combination of parts into a whole; the formation of a complex chemical compound by the combining of two or more simple compounds, elements, or ions.

systolic pressure The maximum pressure the heart exerts on the blood during the contraction phase. (Contrast **diastolic pressure.**)

talus deposit A deposit of loose rocks at the base of cliffs or mountains.

tar sands Asphalt-cemented sand deposits from which petroleum can be obtained.

technology Scientific findings put to use to achieve a practical purpose, such as to provide objects for human sustenance and comfort.

television The technique of transmitting scenes by light waves that have been converted into electronic impulses, then reconverted into electronic beams that are projected against a luminescent screen, reproducing the original image.

Temik A pesticide licensed by the EPA for use on various fruits and vegetables.

temperature That property determining whether or not thermal energy is transferred between two objects in thermal contact with each other.

tension The force exerted by a stretched object on its support.

terminal velocity The maximum velocity that a moving object attains because of the combined forces in action on it. For a falling object, terminal velocity is reached when the downward force of gravity is precisely balanced by the counteracting upward forces exerted on the object by the gas or liquid through which it is falling.

terrestrial planets The four inner planets, including Earth.

tetrahedron A solid figure with four triangular faces.

texture The physical appearance of a rock that involves its size, shape, and arrangement of constituent grains.

theory A tentative explanation for the order and behavior of phenomena observed in nature.

theory of relativity The theory of science that recognizes the universal character of the speed of light and the consequent dependence of space, time, and other mechanical measurements on the motion of the observer performing the measurements. (See also **general relativity** and **special relativity,** its two main divisions.)

thermal energy The energy a body contains as a result of the kinetic energy of its molecules.

thermal neutrons Neutrons of low kinetic energy that more readily undergo inelastic collisions than do those of higher kinetic energy.

thermal pollution A change in the quality of the environment brought about by the release of heat from commercial, domestic, and industrial processes.

thermionic conversion A principle that denotes how electric current is caused to flow between electrodes of different temperatures in an electron tube.

thermionics The science dealing with the emission of electrons or other charged particles from surfaces that have been heated.

thermodynamics The branch of physics that deals with heat and mechanical energy and the conversion of one into the other.

thermonuclear reaction The process through which stars obtain their energy.

thermosphere The layer of the atmosphere directly above the mesosphere. It extends from 80 to 400 kilometers above Earth and contains a large quantity of ionized oxygen and nitrogen.

thermostat A temperature-operated, heat-regulating device used to control mechanical or electrical equipment automatically.

Three Mile Island The site of a nuclear power plant accident that threatened the safety of residents near Harrisburg, Pennsylvania, in 1979.

thrust The forward force produced as a reaction to the escaping gases in a jet or a rotating propeller.

tidal power The potential energy of the falling and rising waters in ocean bays or inlets to turn turbines that generate electricity.

till Unsorted and unstratified drift deposited directly by and under a glacier without subsequent reworking by water from the glacier.

timbre The distinctive quality given a sound by the addition of various overtones to the fundamental tone.

Tokamak A fusion reactor that confines hot hydrogen gas by means of a magnetic field in an attempt to reach the proper temperature, density, and confinement time for fusion to occur.

tornado A violent, destructive whirling wind accompanied by a funnel-shaped cloud.

toroidal chamber A doughnut-shaped magnetic mirror used to confine the plasma in a nuclear fusion reactor.

torque A twisting or wrenching effect on a body produced by a force that tends to rotate the body about some axis.

total internal reflection Reflection without loss that occurs within a medium having a higher index of refraction than its surroundings, if the angle of incidence exceeds the critical angle.

trade winds Winds that blow from the northeast in the Northern Hemisphere and the southeast in the Southern Hemisphere between 5 degrees and 30 degrees north and south latitudes.

transformer A device for increasing or decreasing the voltage of alternating current.

transistor A semiconducting crystal having three or more regions with differing electrical characteristics. Transistors are used to control currents in electronic circuits.

transition elements The elements in columns IB and IIIA through VIIIA of the periodic chart.

translatory motion Motion along a straight line.

translucent A term used to describe a medium that lets light pass through but diffuses it so that objects on the other side cannot be seen clearly. (Compare **transparent.**)

transmutation The process by which one element is changed into another. The emission of any charged particle by an atom's nucleus will produce instant transmutation.

transparent A term used to describe a medium that lets light pass through without appreciable scattering so that objects on the other side are clearly visible. (Compare **translucent**.)

transpiration The passage of water vapor from a living body through membranes or pores, such as those of the leaves of plants.

transverse wave A wave motion in which the disturbance is at right angles to the direction in which the wave travels. Electromagnetic waves are considered transverse waves.

triode A semiconducting crystal having three regions with differing electrical characteristics, or a three-element electron tube. Triodes are used to control currents in electronic circuits.

triple-alpha process The fusion of three helium nuclei at extremely high temperatures into one carbon nucleus, with the liberation of considerable energy.

Tropic of Cancer Parallel of 23.5 degrees north latitude.

Tropic of Capricorn Parallel of 23.5 degrees south latitude.

troposphere The layer of Earth's atmosphere near Earth's surface, where most weather occurs.

tsunami Seismic sea waves produced by volcanic action or other disturbances in Earth's crust.

tunnel diode A semiconducting crystal with a narrow channel of given electrical properties surrounded by material having different electrical properties. Charge flow through the channel can be controlled by "electrically" varying the width of the channel.

typhoon A violent cyclonic storm, such as those that occur in the China Sea and the western Pacific and Indian oceans. The storm is also known as a hurricane along the Atlantic and Gulf coasts of the American continents.

ultrahigh frequencies (UHF) Electromagnetic waves used in television, the frequencies of which are from 470 to 890 megahertz (channels 14 to 83).

ultrasonic A term used in reference to sound frequencies above the range of normal hearing.

ultraviolet radiation Electromagnetic waves in the range of 200–430 nanometers.

umbra The completely dark shadow of a body.

unconformity The conspicuous absence of layers of sedimentary deposition that have been removed by erosional processes.

uniformitarianism A principle that asserts that the present is a key to the past.

unsaturated air Air in which the water-vapor content is less than the air can hold at a given temperature.

unsaturated hydrocarbon Hydrocarbons that have more than one bond between adjacent carbon atoms.

unsaturated solution A homogeneous combination of compounds or elements in which there is less solute dissolved in the solvent than is possible at a given temperature.

vacuum A space completely void of matter.

valley glacier One of two basic categories of glaciers. A valley glacier is a large body of ice that flows down a mountain valley. (See also **ice sheet,** representing the other basic classification.)

Van Allen radiation belts Doughnut-shaped areas surrounding Earth that contain charged particles trapped from the solar wind by Earth's magnetic field.

variable star A star having a light output that varies with time.

vector An arrow drawn to scale used to represent a vector quantity.

vector quantity A physical quantity that possesses both magnitude (size) and direction. Examples of vector quantities are force, velocity, and weight.

velocity A vector quantity denoting the time rate of change of position of a body in reference to a specified direction.

vernal equinox The time during spring (Northern Hemisphere) when the sun crosses the equator.

very high frequencies (VHF) Electromagnetic waves used in television, the frequencies of which are from 54 to 216 megahertz (channels 2 to 13).

vibration The back-and-forth motion of a body. Vibrations are generally measured in cycles.

vibratory motion A rhythmic motion back and forth, such as that of a swinging pendulum.

Viking The spacecraft that landed on the surface of the planet Mars.

virtual focus The point from which parallel rays seem to diverge (spread out) after passing through a diverging lens or striking a diverging mirror.

virtual image An image that appears only when the eye follows diverging rays back to their apparent source. The illusionary image cannot be projected on a screen.

viscosity A measure of resistance to fluid flow. Fluids of high viscosity flow more slowly than fluids with low viscosity.

volcanic neck An intrusive rock formation composed of magma that solidified in such a manner as to form a pipelike body.

volcanism Volcanic action on the surface leading to the formation of rock layers.

volcano A landform that develops during the accumulation of molten rock (lava) around a central vent.

volt The unit of measure of electric potential.

voltaic cell A primary cell composed of two dissimilar metal electrodes in a solution that acts chemically on one or both of them to produce a voltage.

volume A measurement of the size of a body; the amount of a substance a body will hold.

volumetric expansion The change in volume of a gas, liquid, or solid, ordinarily brought about by a temperature change.

Voyager The spacecraft that approached the planets Jupiter, Saturn, and Uranus before continuing into outer space.

warm front The contact zone between a moving warm air mass and a stationary cold air mass.

warping The slight bending of a rock layer over a long area.

water cycle See **hydrologic cycle.**

water table The level below which the ground is saturated with water.

watt The unit of power in the *mks* system; equal to 1 joule per second.

wave equation In mechanics, an equation that specifies how the velocity of a wave equals the product of its frequency and wavelength.

wavelength The distance from any given point on a wave to the corresponding point on the successive wave.

weather The condition of the troposphere over a short period of time and for a designated geographic area.

weathering The physical, chemical, and biological disintegration of rocks near Earth's surface.

weather map A map compiled from information of various frontal systems, low- and high-pressure areas, and other data such as temperature, barometric pressure, precipitation, cloud cover, and wind speed and direction.

weight A measure of the gravitational attractive force that Earth or other celestial body exerts on an object.

white dwarf A small star (about the size of Earth) that has exhausted most of its nuclear fuel and has collapsed to a small size.

wind Air in motion, generally in a horizontal direction.

windplants (windfarms) A series of wind-powered turbines that provide electrical power for domestic and industrial use.

winter solstice The time during the winter (Northern Hemisphere) when the sun is farthest south of the equator.

work The product of the magnitude of an applied force and the parallel displacement (distance) through which the force acts. The SI unit of work is the joule.

X-ray diffraction A technique that determines the manner in which X rays are spread or bent by a crystal lattice and indicates the orderliness of the spacing between atoms or ions in a crystal lattice.

Z particle One of three carriers of the weak nuclear force that explains certain kinds of radioactive decay. It is produced during the collision of a highly accelerated electron and positron.

zenith The point in the sky directly overhead.

zero-point energy The kinetic energy retained by the molecules of a body at a temperature of absolute zero.

ziggurat A general term applied to ancient Assyrian and Babylonian temple towers. These lofty pyramid structures were constructed in step-type stages with an outside staircase that led to a shrine at the top.

zodiac An imaginary region in the heavens encompassing the paths of the various planets and dividing the heavens into 12 constellations or signs, each taken for astrological purposes to extend 30 degrees of longitude.

Credits

Photo Credits

Chapter 1

Opener: Courtesy of Brookhaven National Laboratory

Chapter 2

Opener: Courtesy Dr. Lamar B. Payne; **2.1:** Charles A. Payne, William R. Falls and Charles J. Whidden; **2.3:** Professor Robert T. Lierman; **2.4:** Courtesy of Dr. Lamar B. Payne; **2.6:** Courtesy Professor Robert T. Lierman; **2.7:** Courtesy of American Iron and Steel; **2.8:** Charles A. Payne, William R. Falls, Charles J. Whidden; **2.9A:** Courtesy: Dr. Bill Booth; **2.9B:** Courtesy of Dr. Bill Booth; **2.10:** Courtesy of Dr. Lamar B. Payne

Chapter 3

Opener: 3.1A: Charles A. Payne, William R. Falls, and Charles J. Whidden; **3.1B:** Courtesy of Delta Air Lines, Inc.; **3.7A:** Charles A. Payne, William R. Falls, and Charles J. Whidden; **3.8A:** Courtesy of Delta Air Lines, Inc.; **3.10, 3.12, 3.14, 3.17, 3.18, 3.20, 3.21, 3.22, 3.23:** Charles A. Payne, William R. Falls, and Charles J. Whidden

Chapter 4

Opener, 4.3, 4.13: Charles A. Payne, William R. Falls, and Charles J. Whidden; **4.17, 4.19, 4.20, 4.21:** NASA

Chapter 5

Opener: © Thomas Kitchen/Tom Stack and Associates; **5.1, 5.2, 5.3A–C:** Charles A. Payne, William R. Falls, and Charles J. Whidden; **5.4A:** Courtesy of Central Scientific Company; **5.4B:** Courtesy of Central Scientific Company; **5.4C, 5.5, 5.6A, 5.6B, 5.7A, 5.8, 5.9:** Charles A. Payne, William R. Falls, and Charles J. Whidden

Chapter 6

Opener: American Iron and Steel Institute; **6.5, 6.10A, 6.10B, 6.10C:** Charles A. Payne, William R. Falls, and Charles J. Whidden

Chapter 7

Opener: © H. Armstrong Roberts; **7.1 A&B, 7.9, 7.10 A,B,C, 7.11A–C, 7.12A–D:** Courtesy of Sargent-Welch Scientific Company, Skokie, IL.; **7.13:** © Wide World Photos, Inc.; **7.15, 7.17, 7.18:** Courtesy of Sargent-Welch Scientific Company, Skokie, IL.

Chapter 8

Opener: Charles A. Payne, William R. Falls, and Charles J. Whidden; **8.3A:** Courtesy, U.S. Department of Agriculture; **8.6A, 8.8, 8.10, 8.20, 8.21, 8.22, 8.23 A,B,C, 8.24B:** Charles A. Payne, William R. Falls and Charles J. Whidden; **8.29A:** Courtesy of Eveready-Union Carbide; **8.29C:** Charles A. Payne, William. R. Falls, and Charles J. Whidden

Chapter 9

Opener: Charles A. Payne, William R. Falls and Charles J. Whidden; **9.2:** Courtesy of General Electric Company; **9.4A, 9.5:** Charles A. Payne, William R. Falls and Charles J. Whidden; **9.6:** Courtesy of Consumer Products Division, Union Carbide Corp.; **9.7 A&B, 9.8, 9.9A, 9.11:** Charles A. Payne, William R. Falls and Charles J. Whidden; **9.12B:** Courtesy of General Electric Company; **9.13B:** Courtesy of GTE Lighting Products; **9.14:** Courtesy of Westinghouse Electric Corporation; **9.15 A&B:** Courtesy of GTE Lighting Products; **9.15 C&D:** Courtesy of General Telephone and Electronics Corporation; **9.16 A,B,C:** Charles A. Payne, William R. Falls, and Charles J. Whidden; **9.17 A&B:** Courtesy of General Electric Company; **9.19 A&B:** Charles A. Payne, William R. Falls, and Charles J. Whidden; **9.20:** Courtesy of Sharp Electronics Corporation; **9.23A, 9.24, 9.25B:** Charles A. Payne, William R. Falls, and Charles J. Whidden; **9.26B–D:** Courtesy of General Telephone and Electronics Corporation; **9.27A:** Charles A. Payne, William R. Falls, and Charles J. Whidden; **9.28A:** Courtesy of Metrologic Instruments; **9.32:** Courtesy of Phillips Industries

Chapter 10

Opener, 10.9: Charles A. Payne, William R. Falls and Charles J. Whidden; **10.10:** Courtesy of Kodak; **10.12, 10.13, 10.14, 10.15, 10.17, 10.18, 10.20A:** Charles A. Payne, William R. Falls and Charles J. Whidden; **10.22:** Courtesy of Metrologic Instruments; **10.24 A&B, 10.25 A&B 10.26:** Charles A. Payne, William R. Falls and Charles J. Whidden

Chapter 11

Opener: © Bruce Iverson; **11.1, 11.5A, 11.9A, 11.12, 11.13A, 11.17A, 11.20, 11.21, 11.24, 11.26A, 11.29, 11.31B, 11.32C, 11.36B–D, 11.37 C&D, 11.39, 11.43:** Charles A. Payne, William R. Falls and Charles J. Whidden; **11.44:** Courtesy of Bell Laboratories; **11.46A:** Charles A. Payne, William R. Falls and Charles J. Whidden; **11.46B:** AT & T Bell Laboratories Inc.; **11.48:** Charles A. Payne, William R. Falls and Charles J. Whidden; **11.49:** Observatory/National Astronomy and Ionosphere; **11.50A:** Charles A. Payne, William R. Falls and Charles J. Whidden

Chapter 12

Opener: © Paul Silverman/Fundamental Photographs; **12.1:** Professor Albert Crewe, University of Chicago

Chapter 13

Opener, 13.4, 13.5: Charles A. Payne, William R. Falls and Charles J. Whidden

Chapter 14

Opener, 14.1, 14.2: Charles A. Payne, William R. Falls and Charles J. Whidden

Chapter 15

15.6, 15.7: Brookhaven National Laboratory; **15.8A, 15.9, 15.10:** Courtesy of the Nucleus, Inc., Oak Ridge, Tennessee; **15.11:** Courtesy of Thermilab; **15.15:** Courtesy of Westinghouse Electric Corporation; **5.16:** Courtesy of Slac/ Stanford University

Chapter 16

Opener: Courtesy Dr. Carl V. Ramey; **16.1A, 16.1B:** Robert C. Mitchell; **16.2:** Charles A. Payne, William R. Falls and Charles J. Whidden

Chapter 17

Opener: NASA; **17.2, 17.3, 17.4A–C:** Jet Propulsion Lab/NASA; **17.5:** JPL/NASA Hale Observatory; **17.6:** JPL/NASA; **17.11:** NASA; **17.13:** NASA/Jet Propulsion Laboratory; **17.14:** NASA; **17.15B:** Soviet Academy of Sciences/ Brown University; **17.16:** Both: Jet Propulsion Laboratory/CA. Institute of Technology; **17.17, 17.18A, 17.18B, 17.19:** NASA; **17.20:** U.S. Geological Survey, Flagstaff, AZ; **17.21A–D, 17.22, 17.23, 17.24, 17.29A–F, 17.30A–C, 17.31A–F:** NASA; **17.32:** Official Naval Observatory; **17.33:** Courtesy of the Hale Observatories; **17.34:** William Liller; **17.35:** Courtesy of Hale Observatories; **17.39:** Johnson Space Center—NASA; **17.40 A&B:** USGS

Chapter 18

Opener: Tersch, Inc.; **18.1A:** NASA; **18.1B:** Tersch, Inc.; **18.5A:** Charles J. Whidden; **18.6A:** Mike Burchett; **18.6B:** Charles J. Whidden; **18.7 A&B:** Courtesy Lowell Observatories; **18.9:** Tersch, Inc.; **18.10:** Courtesy of Lowell Observatories; **18.12:** Courtesy of Palomar Observatories; **18.13:** Courtesy of Hale Observatories; **18.14:** Charles A. Payne, William R. Falls, and Charles J. Whidden; **18.15:** Hale Observatories, CA. Institute of Technology; **18.16A:** The Kitt Peak National Observatory/S. Maran; **18.17A:** Tersch, Inc.; **18.17B:** Charles A. Payne, William R. Falls, and Charles J. Whidden; **18.19:** Charles J. Whidden; **18.20 A&B:** Dennis Dicicco/Sky and Telescope

Chapter 19

Opener: Courtesy of the Hale Observatories; **19.1:** Dennis Dicicco/Sky and Telescope; **19.2, 19.3, 19.4 A&B:** Courtesy of the Hale Observatories; **19.6:** Tersch, Inc.; **19.7:** Courtesy of the Hale Observatories; **19.8:** Charles A. Payne, William R. Falls and Charles J. Whidden; **19.9:** Courtesy of Mount Wilsona and Law Components Observations; **19.11, 19.13:** Charles J. Whidden

Chapter 20

Opener: Courtesy NOAA/Inter Network, Inc.; **20.17:** Charles A. Payne

Chapter 21

Opener: NASA; **21.2:** William R. Falls; **21.15:** Both: Courtesy of NOVA Scotia Tourism; **21.20, 21.21, 21.22:** NASA; **21.23:** Courtesy of the Author/Randy Falls

Chapter 22

Opener, 22.2: Charles A. Payne; **22.11:** Hubbard Scientific Co., Northbrook, IL #1/J. A. Day; **22.12:** Courtesy: NOAA; **22.13:** #31/J. A. Day, **22.14:** #5/J. A. Day, **22.15:** #6/J. A. Day, **22.16:** #11/J. A. Day, **22.17:** #12/J. A. Day, **22.18:** #15/J. A. Day, **22.19:** #7/J. A. Day, **22.20:** #9/J. A. Day, **22.21:** #2/J. A. Day/ Hubbard Scientific Co., Northbrook, IL; **22.26:** NASA; **22.27:** NOAA; **22.28:** NASA; **22.29:** NOAA

Chapter 23

Opener: Charles A. Payne; **23.1:** Courtesy of Dr. Sidney Fox, Institute of Molecular and Cellular Evolution, University of Miami; **23.5, 23.6:** Courtesy of Professor E. S. Barghoornn, Harvard University; **23.8:** Courtesy of Professor M. S. Glpessner, University of Adelaide, South Australia; **23.9, 23.10:** Smithsonian Institute; **23.11:** Courtesy: Field Museum of National History/Chicago; **23.12, 23.13, 23.15:** Smithsonian Institute; **23.16:** Field Museum of Natural History, Chicago, IL; **23.17:** American Museum of Natural History/Charles R. Knight Artist; **23.18:** Field Museum of Natural History, Chicago, IL; **23.19:** Peabody Museum of Natural History, Yale University. Painted by Rudolph F. Zallinger; **23.20:** Painting by Rudolf Freund, Courtesy of the Carnegie Museum of Natural History; **23.21:** Field Museum of Natural History (Neg. No. CK24T) Chicago, IL; **23.22:** Field Museum of Natural History; **23.23:** Neg. #1017, Courtesy Dept of Library Services, American Museum of Natural History/NY; **23.24:** Charles A. Payne, William R. Falls and Charles J. Whidden; **23.26:** Woods/Hole Oceanographic Ens., Robert Bullard; **23.28:** American Museum of Natural History, Charles R. Knight

Chapter 24

24.3 A–D: US Forest Service; **24.4:** Michael Buck, Photographer. State of HI Dept of Land and Natural Resources; **24.5 A&B:** G. K. Gilbert, U.S. Geological Survey; **24.6A:** U.S. Department of Interior; **24.6B:** Photo Courtesy of National Geophysical Data Center; **24.8, 24.9, 24.10, 24.11, 24.14, 24.15, 24.16:** Charles A. Payne, William R. Falls and Charles J. Whidden; **24.17, 24.18 A&B:** Randy Falls

Chapter 25

Opener: Carla Montgomery; **25.1:** U.S. Department of Agriculture; **25.2:** National Park Service; **25.3:** E. A. Hudson U.S. Department of Agriculture; **25.4, 25.5:** W. T. Lee, U.S. Geological Survey; **25.6:** G. W. Stase, U.S. Geological Survey; **25.7:** J. R. Fakley, U.S. Geological Survey; **25.8:** Courtesy of U.S. Geological Survey; **25.9:** Courtesy of Luray Caverns, Luray, Virginia; **25.10:** H. E. Gregory, U.S. Geological Survey; **25.11:** W. C. Mendenhall, U.S. Geological Survey; **25.12:** H. E. Malde, U.S. Geological survey; **25.13:** I. C. Russell, U.S. Geological Survey

Chapter 26

Opener: Pacific Gas and Electric Co.; **26.1 A&B:** Virginia Electric and Power Company/USDA; **26.4:** Tennessee Valley Authority; **26.5A–C:** Courtesy of General Atomics; **26.7A:** Pacific Gas and Electric Co., San Francisco; **26.7B:** Nuclear Regulatory Commission; **26.8:** Courtesy: Westinghouse Electric Co.; **26.9 A&B:** Dept. of Energy; **26.10 A&B:** Pacific Gas & Electric Co., San Francisco; **26.11, 26.12:** Princeton Plasma Physics Laboratory; **26.14:** Pacific Gas and Electric Co., San Francisco; **26.15A:** Westinghouse Electric Co.

Chapter 27

Opener: Courtesy of Randy Falls; **27.1A:** U.S. Department of Agriculture/Forest Service Incident Team; **27.1B:** Robert Schaefer, State of Alaska, Office of the Governor; **27.2A:** Shell Oil Co.; **27.2B:** Exxon Corp; **27.2C:** Conoco/A Subsidiary of Dupont; **27.2D:** Exxon Corp; **27.3A:** Alaska Dept. of Natural Resources; **27.3B:** DNR/ Dept. of Interior; **27.4 A&B:** Dept. of Energy; **27.5A:** Consolidated Coal Company, A Dupont Subsidiary; **27.5B:** U.S. Bureau Of Mines; **27.6A:** Consolidated Coal Co., A Subsidiary Of Dupont; **27.6B:** U.S. Bureau Of Mines; **27.7:** Tennelec; **27.8:** Baltimore Gas and Electric Company; **27.9A–C, 27.10B:** EPA; **27.11:** Courtesy of Randy Falls; **27.13:** S. C. Delaney, EPA; **27.14A:** R. W. Reed, U.S. Department of Agriculture; **27.14B:** P. E. Farnes, U.S. Department of Agriculture; **27.14C:** Jack Hayes, U.S. Department of Agriculture; **27.16:** Poster Copyright 1986 Hansen Planetarium Text & Image Copyright 1985, W. T. Sullivary III

Illustration Credits

Norm Frisch: figures 2.2, 8.25a,b, 11.33a,b, 17.30, 18.4, 18.7a–e, 18.10, 19.6, 20.7, 20.13, 20.14, 20.15, 21.2, 21.13, 21.15a,b, 22.7a,b, 22.8, 23.4, 23.7, 24.1, 27.15.

Hans & Cassady, Inc.: figures 1.2, 1.3, 1.4, 1.5, 1.6, 3.1a,b, 3.2a,b, 3.3a,b, 3.4, 3.7a,b, 3.9a,b, 3.10, 3.11, TA4.1, figures 4.1, 4.2, 4.3a,b, 4.5, 4.8, 4.10, 4.12, 5.1a–c, 6.2, 6.3, 6.4, 6.6, 7.2, 7.6, 7.20, 8.1, 8.2a–d, 8.3b,c, 8.4, 8.5a,b, 8.6b, 8.7, 8.9a,b, 8.11, 8.12, 8.13a–d, 8.14, 8.15, 8.18, 8.24a–c, 8.26a–d, 8.27, 8.28, 8.30a–e, 9.1, 9.3a,b, 9.20, 9.21, 10.1a,b, 10.2a–d, 10.3, 10.4, 10.5, 10.6, 10.7, 10.16a,b, 10.18, 10.23a–d, 11.2, 11.3a,b, 11.4, 11.5b, 11.6, 11.7, 11.8, 11.9b, 11.10, 11.11a,b, 11.13b, 11.14, 11.15, 11.16, 11.17b,c, 11.18, 11.20a–c, 11.21, 11.22a,b, 11.23, 11.25, 11.26, 11.27a,b, 11.28a,b, 11.30a,b, 11.32a,b, 11.34a,b, 11.35, 11.36a, 11.38, 11.40a,b, 11.41a,b, 11.42a–c, 11.47a,b, 11.51, TA12.1, TA12.2, TA12.3, TA 12.4, TA 12.5, figure 12.3, TA 12.6, TA12.7, TA12.8, TA12.13, TA12.14, figure 13.1, TA13.2, TA13.3, TA13.4, TA13.5, figure 13.3, TA14.3, TA14.4, figures 15.4, 15.15a,b, 15.8b, 17.7, 17.8, 17.9, 17.28, 17.28, 17.29, 18.2a–c, 18.3a,b, 18.5, 21.9, 21.10, 22.1, 22.6, TA22.1, figures 22.22b, 23.25, 27.10.

Index

A

B

C

D

E

F

G

H

I

J

K

L

M

N

O

P

Q

R

S

T

U

V

W

X

Y

Z